I'VE BEEN USED
PRICE 5GS2810
Recycle your money
Buy used books
A.S. BOOKSTORE
CSU - CHICO

BOTANY
AN INTRODUCTION TO PLANT BIOLOGY

NEW YORK
CHICHESTER
BRISBANE
TORONTO
SINGAPORE
JOHN WILEY & SONS

SIXTH EDITION

BOTANY

AN INTRODUCTION TO PLANT BIOLOGY

T. ELLIOT WEIER

C. RALPH STOCKING

MICHAEL G. BARBOUR

THOMAS L. ROST

University of California
Davis, California

This book was set in Helvetica. The designer was Rafael Hernandez. The drawings were designed and executed by John Balbalis with the assistance of the Wiley Illustration Dept. Lilly Kaufman supervised production.

Library of Congress Cataloging in Publication Data:

Main entry under title:

Botany: an introduction to plant biology.
Fifth ed.: Botany / T. Elliot Weier, C. Ralph Stocking, Michael G. Barbour.
Includes indexes.
1. Botany. I. Weier, T. Elliot (Thomas Elliot), 1903-
QK47 B776 1982 580 81-10304
ISBN 0-471-01561-X AACR2

Printed in the United States of America

10 9 8 7 6 5 4

This edition is dedicated to
Katrina Weier
companion and critic

PREFACE

In this edition as in previous editions, we present the reader with a dynamic plant—an organism full of color, intricate pattern, and beauty. New insights continue to be uncovered into the relationships among plants and into the interrelationships between structure and function of plant organelles, cells, tissues, and organs.

For this sixth edition, we have revised the entire text. Some chapters have been completely rewritten, while in others we have made additions and revisions to incorporate some of the new findings and interpretations that have occurred in the ever-changing field of botany. Examples of some of the major revisions included in this edition are listed below. The five kingdom system of classification of organisms has been introduced to illustrate our changing concepts of the evolutionary relationships among organisms. However, in this book we continue to study organisms that traditionally have been considered plants. The chapter on cell division has been rewritten to include a comprehensive treatment of the phases that each cell must undergo in preparation for cell division and DNA replication, *i.e.,* the cell cycle.

The variations and significance of differential responses of plant genotypes to several nutrient stresses are included in the discussion of mineral nutrition of plants. The rapidly developing understanding of the process of photosynthesis has necessitated revision of our treatment of photosynthesis, particularly with respect to electron transport in the thylakoid membranes and the pH gradient and formation of ATP.

The material on bacteria has been completely rewritten, with a new emphasis on diversity within the group: diversity in terms of ultrastructure, metabolism, habitat, and morphology. Both the bacteria and the blue-green algae are now presented in the same chapter to stress the unifying prokaryotic character that holds these two groups together.

The survey of the algae has been expanded with material on economical importance of various divisions and by the inclusion of many new illustrations. The result is two algal chapters, rather than one as in the last edition. The two fungal chapters have also been revised, largely from a taxonomic viewpoint. In line with generally accepted ideas of mycologists, we recognize three divisions. Additional attention is paid to certain pathogenic fungi, mycorrhizal fungi, the Fungi Imperfecti, and lichens.

New material has also been introduced in the chapters dealing with the other groups of plants. For example, leafy liverworts, the economic importance of *Sphagnum,* phylogenetic relationships among bryophyte classes and between bryophytes and other groups are all discussed.

Particularly significant is the use of over 250 four-color illustrations in nearly 50 color plates. Each illustration has been evaluated for its teaching effectiveness, and we have attempted to balance the use of color throughout the book. Thus, where stained sections of plant material especially reveal differences in tissue characteristics, we have included numerous colored illustrations that correspond to the microscope slide material most commonly used by students in the laboratory. The physiology sections contain illustrations of color differences associated with mineral deficiencies, etiolation, and dye reduction in the Hill reaction. Floral characteristics and differences in pigmentation among diverse groups of plants are illustrated in sections dealing with ecology and taxonomy.

To insure rapid and easy reference, figure numbers of the color illustrations are printed in boldface in the text and are accompanied by the page number of the appropriate plate.

Again it is a pleasure to acknowledge the very significant contribution of Alice Baldwin Addicott whose skill and artistry are evident in the many new drawings and diagrams. Special thanks go to our colleagues who generously gave of their time and help in making certain sections of this revision possible: Bruce Bonner, growth and development; Emanuel Epstein, mineral nutrition; Jim Doyle, evolution; Norma Lang, algae, including blue-green algae; Karol Paterson, leaf regeneration; John Tucker, angiosperms; Mark Wheelis, bacteria; Kenneth Wells, fungi.

We owe a special debt to our reviewers: Neal Barnett, University of Maryland; Brian Capon, California State University, Los Angeles; Brian Davis, Virginia Polytechnic Institute; Patricia Gensell, University of North Carolina, Chapel Hill; Frank Gleason, Santa Rosa Junior College; Margaret Goodhue, University of Wisconsin, Stevens Point; Richard Jensen, St. Mary's College; Sue Madison, Louisiana State University; Richard Pippen, Western Michigan University.

In addition, we thank the many other colleagues who willingly assisted us by contributing stimulating ideas and excellent illustrations in the preparation of this edition.

Davis, California

T. Elliot Weier
C. Ralph Stocking
Michael G. Barbour
Thomas L. Rost

CONTENTS

BOTANY
AN INTRODUCTION TO PLANT BIOLOGY

CHAPTER 1

INTRODUCTION

An educated person is one who has a fair knowledge and an appreciative understanding of the environment and of the part human beings play in modifying it. Our environment is most complex. It embraces everything that surrounds us, both living and nonliving. It includes the air we breathe, the food we eat, the water we drink; it includes temperature, rainfall, light, humidity of the atmosphere, wind, soil, and atmospheric pressure; it involves the varied plant and animal life, both wild and domesticated; and it includes, as well as plants and animals, the human beings with whom we are associated including their social and economic systems.

This environment, in its totality, is called the **biosphere.** This book concerns one component of the biosphere: plants. This first chapter will briefly discuss some of the roles that plants play in helping to stabilize the biosphere. It will point out ways in which botanists study plants and their interrelationships with other components of the biosphere. Chapter 2 sketches some of the similarities and diversities among plants, as well as briefly relating the botanists' efforts to arrange plants into a logical system. This is plant taxonomy and it draws upon knowledge from many fields of botany: anatomy, plant biochemistry, heredity, and morphology. This effort, which is the central theme of Chapter 2, results in a more or less satisfying separation of plants into groups that appear to be genetically related through the sharing of a common ancestor. All details of anatomy, cytology, and physiology used here will be discussed in more depth in subsequent chapters and are of importance here only insofar as they serve to clarify the groupings of plants.

THE BIOSPHERE

Our environment constitutes the biosphere of the planet Earth. It is a thin layer of water, soil, and atmosphere that supports life as we know it. The soil and water teem with microorganisms, but relatively few large organisms inhabit the soil. On the other hand, many species, including some of the largest forms, inhabit the water. But it is the narrow band of atmosphere about the land that supports all plant life.

The temperature range within our solar system is from nearly absolute zero to 10^7°C, and the wavelengths of radiant energy range from cosmic rays of 10^{-13} to 10^{-9} cm to long radio waves of 10^4 to 10^5 cm. Life requires a temperature between 0 and 50°C, with the optimum around 30°C. The human eye is sensitive to radiant energy from the sun between 400 and 700 nm in wavelength (4 to 7×10^{-5} cm). Plants, the sole converters of radiant energy into chemical energy, use mainly blue and red light for this purpose. However, plant growth responds to blue light and other wavelengths in a variety of ways (Chapter 20). Thus, life is tolerant to a very narrow temperature band and responds to an equally narrow band of radiant energy.

The solar radiation reaching the earth's surface is modified by the composition of the atmosphere, which is a mixture of 21% oxygen, 78% nitrogen, 0.03% carbon dioxide, and less than 0.006% of three other gases. At sea level, the column of atmosphere above 1 cm^2 weighs 1.03 kg and will support a 1 cm^2 column of water 9.76 meters high (see Fig. 5.16). There is abundant life in the biosphere from sea level to an elevation varying from about 2.5 miles (4.02 km) at the equator to about 1 mile (1.61 km) in the northern Canadian Rockies. In the oceans, abundant life is found from sea level to a depth of 100 ft (30 m).

But the biosphere has not always been constituted as described. Life probably originated in an atmosphere devoid of molecular oxygen. Thus, the first living organisms were anaerobic forms and probably never advanced much beyond simple filaments or masses of cells. Since an atmosphere lacking molecular oxygen and hence ozone (O_3) would be very inefficient in screening out ultraviolet radiation from the sun, these primitive organisms may have been subjected to large amounts of ultraviolet radiation, which has a damaging effect on life. In time, there arose organisms capable of carrying out photosynthesis in which molecular oxygen is produced. The oxygen escaped to the atmosphere, and ozone formed in the atmosphere from oxygen filtered out much of the harmful ultraviolet radiation. Early primitive plants played a dominant part in establishing oxygen in the atmosphere, and present-day plants are essential for its maintenance there.

However, there is growing concern that several of our activities, (1) the use of aerosols, which may bring about a depletion of the protective ozone layer in the upper atmosphere, (2) the utilization of enormous amounts of fossil fuels, and (3) the utilization of nuclear energy, are bringing about changes in the biosphere. We already know that these changes are presently harmful to some forms of life, including man himself. We do not know the permanent long-term effects of such changes in the biosphere, but they pose a very important problem for contemporary biologists, planners, and legislators.

CERTAIN BIOLOGICAL UNIVERSALS

If we assume that the original anaerobic organisms were similar in appearance and function to those of today, and if the first aerobic and photosynthetic forms are also similar to the simplest photosynthetic forms of today, then certain general principles about life emerge. These principles may be called **biological universals.** First, *all life is cellular* (Fig. 2.1). Since viruses cannot live without host cells, we shall not consider them true living organisms. Second, the use and transfer of energy (photosynthesis and respiration) within all cells involves the same compound, known as **adenosine triphosphate,** or **ATP** for short (Chapter 5). Third, some steps in respiration are identical in all aerobic organisms and in some anaerobic forms (Chapter 14). Fourth, some steps in energy-transforming reactions in all organisms and in all cells are associated with membranes (Chapter 4). Fifth, the green pigment, **chlorophyll,** is required in the conversion of the radiant energy of sunlight to energy in food in all photosynthetic plants, and the light-trapping steps in photosynthesis are always associated with membranes (Chapter 4). And sixth, genetic information that carries inheritance from one generation to the next is stored in the same compound in all organisms. This is **deoxyribose nucleic acid,** or simply **DNA** (Chapter 5).

In other words, if the present-day single-celled algae, which we consider to be primitive organisms, are similar to their ancestors of 1 to 3 billion years ago, certain important basic processes of life become apparent: Respiration, photosynthesis, and inheritance are basically the same in all organisms and have been so from the beginning. If this assumption is false, we have no starting point, and an understanding of past life will be much more difficult to attain.

Today, and through much of the past, there are two fundamentally different patterns of cell organization. In most organisms, the processes of respiration, photosynthesis, and inheritance are confined to their own special compartments within the cell: mitochondria, chloroplasts, and nuclei, respectively (see Figs. 2.1*B* and 4.3). But in the bacteria and blue-green algae these three activities, as well as others, occupy communal space in a single noncompartmented cell (Fig. 2.1*A*). Since these more primitive forms lack a true nucleus, they are known as the **prokaryotes.** The forms with a true nucleus are the **eukaryotes.**

PLANTS AND PLANT SCIENTISTS

In the plant kingdom only the blue-green algae and the bacteria are prokaryotes and only the fungi, and most bacteria together with a few isolated species of parasitic or saprophytic higher plants lack chlorophyll. The metabolic processes of all green plants are similar; they are all producers of chemical energy from radiant energy.

In all of life on earth, plants are the only producers. All consumers, particularly people, are dependent upon plants for food, fiber, wood, energy, and oxygen. A knowledge of plants, their habitats, structure, metabolism, and inheritance is thus the basic foundation for human survival.

What are the problems that concern the plant scientists in their efforts to accumulate significant knowledge of the plant life of the biosphere? Plants must first be named. Early botanists, such as the Greek, Theophrastus (300 BC), were largely concerned with naming plants. In fact, many current scientific names for plants are the same as, or similar to, those given to the same plants by Greek or Roman botanists. Concomitant with naming plants came an increase in the numbers of known plants. Plant explorers accompanied the expansion of empires. As a result, botanic gardens and herbaria, storehouses of botanic information, were established in several countries. In order to retrieve information there must exist some system of classification. Modern methods of classifying plants date from the Swedish botanist, Linnaeus (1707–1778). New species of plants are still being described.

Some other important problems claiming the attention of plant scientists are as follows: (*a*) What are the principles underlying the great diversity of plants? (*b*) What are the mechanisms that control the precise patterns of growth? (*c*) Since plants are the only producers, what kinds of plants, growing under what conditions of temperature, soil, moisture, and nutrients will produce most abundantly? (*d*) How are plants related to their environment, and how can the harmful changes in the biosphere brought about by man be reversed? These problems can only be studied with experiments.

It is through experimentation that we hope to derive answers to these questions. In 1770, the Dutch physician van Ingenhousz asked the question, "Will a glowing sliver burst into flame in an atmosphere in which plants have been confined in the dark?" The answer was, "No." Will it burst into flame in an atmosphere in which plants have been confined in the light? Now the answer was, "Yes." What is the difference between the two atmospheres?

Asking the right question is frequently the most difficult step. Generally, it should be a simple question, hopefully with not more than two possible answers. The procedure must be within the realm of available experimental technique. The experimental system set up for study of the question must be under reasonably good control. The fewer the variables there are in the system, the better. And, of considerable significance, the answers to the right questions lead to a greater understanding of the whole problem and, in addition, to other questions.

Plant Taxonomists

Taxonomists wish to complete an inventory of the earth's plant resources; they want to categorize, for easy reference, the sum total of existing plant variation. As the inventory becomes complete (and it is far from complete today), they hope to organize this diversity into some evolutionary scheme. What sort of patterns exist and do these patterns mean genetic relationship and evolution from common ancestors? Which plants or plant characteristics are the most primitive, that is, imitating early, now-extinct species; and which characteristics are advanced, that is, indicating a recent time of evolution? Taxonomists use morphology, anatomy, biochemistry, hybridization, and chromosome studies to arrive at tentative conclusions about the **phylogenetic** (evolutionary) relationships between species.

Taxonomy is basic to botany, for all botanists use taxonomic information in their research. The physiologist, for example, must know the particular species he or she works with. Since species differ in their physiology, the results of the physiologist's work would be meaningless if the names of the study organisms were not known; other workers could neither corroborate nor extend the findings. The anatomist, developmental morphologist, and cytologist all face the same necessity of knowing the species with which they work.

Plant Ecologists

Plant ecologists want to understand how plants are adapted to their environment. What are the environmental forces that regulate plant growth and what kinds of responses can plants make to them? Living organisms and their environment are interdependent in complicated, often unexpected ways. Can human beings learn to imitate this natural balance or to manipulate it for their own ends? Is it possible, with the vegetation we have, to meet several pressing needs at once: those of agriculture, industry, recreation, conservation? What is the wisest, most efficient way to use our plant resources? Have we changed our environment dangerously by pollution? What level of pollution can be safely tolerated? Are plants adapting to pollution?

The ecologist also wishes to inventory the earth's plant resources. Unlike the taxonomist, he is interested in categorizing groupings of species (i.e., communities) rather than individual species. How is it that certain species repeatedly group together to the exclusion of all other species? What is the geologic history of these communities, and what are their paths of migration through time?

Ecologists are often generalists, utilizing information from taxonomy, physiology, anatomy, geology, and meteorology to search for answers to questions like those posed above.

Plant Morphologists

The great diversity of plant forms is readily apparent. There are seaweeds, mosses, ferns, pine trees, corn plants, palms, and dandelions. Obviously, corn plants bear seeds (Chapter 16), but do ferns, mosses, and seaweeds as well? If they do not, how do they propagate (Chapters 24, 27, 28)? Both pines and palms have single straight trunks, but the palm has large divided leaves, found at the top only, and the pine has many lateral branches clad with needles. Why? (Chapter 17.) Botanists interested in plant form are known as morphologists. Their interests lie in the diversity of plant form and the manner in which cells are arranged to give this form. All plants require water and food; are water and food conducted and transported in cells of similar form in plants of the ocean and plants of the desert? How do the reproductive structures of plants of the ocean, marsh, and desert differ (Chapters 24, 28)? Are the cellular details of the stems of mosses, ferns, pine trees, and palms alike (Chapters 7, 29)?

One branch of morphology, morphogenesis, is concerned with the mechanisms involved in how a particular precise form develops. When a fertilized egg cell in the flower of a pea plant starts to grow, it follows a pathway similar to that followed by fertilized eggs of all other seed plants. Eventually a pea seed develops similar to all other pea seeds but different from every other kind of seed. When the seed is planted, it germinates and grows just as do the seeds of all other species of plants, but the pea seed always produces pea plants, and only pea plants. Cells divide in precise planes and enlarge in precise locations within the developing seed at regular rates. Differentiation of cells to form tissues also occurs in a regular fashion. The morphogeneticist has learned how to modify some of these steps and would like to know more of the manner in which they are controlled and directed within the plant itself. A better knowledge of these things will give us a deeper insight into the developmental process of all cells.

Plant Physiologists

While it is true that the basic processes, respiration and photosynthesis, are similar in all plants, the end products of the synthetic processes in plants are as variable as there are different species of plants. Why? There are hard wheats and soft wheats, and the flour made from them is different. Some wheat seeds need to be held at a low temperature before they will germinate. Some lettuce seeds will not germinate until they have been briefly exposed to red light. Again, why? The answer to these and similar questions lie in the realm of the plant physiology.

Morphologists and physiologists frequently unite in searching for aspects of form that have a chemical basis. A vegetative shoot of *Chrysanthemum* will not flower as long as it receives more than 10 hr of daylight. Soon after the vegetative shoot receives a day-night cycle of only 9 hr of daylight and 15 hr of darkness, flower buds form. What are the morphological changes that take place in the *Chrysanthemum* shoot as it changes from the vegetative to the flowering condition? What chemical substance, that is, hormone, is induced by the change in light period and how does it bring about the change in the growth pattern?

One of the major goals of the plant physiologist is to understand all the biochemical and physical processes that are the basis of life, and that are responsible for the great diversity of life form and activities. The more the physiologist knows about life reactions, the more he or she is able to regulate and control these activities and so shape plant growth and development. Is it possible to predict or even regulate when a desirable plant will flower? Can fruit without seeds be produced? Why are some plants bushy-shaped dwarfs and others of the same species tall and vinelike?

Cytologists

Cytologists, or cell biologists, as they are now more popularly known, are concerned with problems at the cellular level. It might seem that since the basic phenomena of inheritance, photosynthesis, and respiration are essentially identical in all organisms and since these three processes are confined to specific organelles within the cell, understanding them would be easy. We do know that the genetic code present in the nucleus directs the synthesis of specific proteins. But we do not know how protein is actually synthesized in the cell. We are not sure of the physical arrangement of DNA in a chromosome. We would like to know the precise relationship between the steps in photosynthesis and respiration and the molecules of the membranes associated with these processes. Respiration and photosynthesis supply energy for cellular synthesis. Protein synthesis probably takes place in all parts of the cell, even in the mitochondria and chloroplasts that supply energy for synthesis elsewhere in the cell.

Since protein synthesis is coded by DNA (Chapter 6), and mitochondria and chloroplasts have their own DNA, what is the relationship between the nuclear or chromosomal DNA and mitochondrial and chloroplastic DNA in the synthesis of the important array of enzymes characteristic of these organelles? What other organelles may be involved? How? Questions are being answered slowly. The cell physiologist and the biochemist find themselves seeking solutions to similar questions but seeking answers by different approaches.

CONCLUSIONS

Since plants (except fungi and some bacteria) are producers and animals (including man) are consumers, we must know about plants. Universities and colleges must continue to educate people knowledgeable in the ways of plants, just as they train specialists in space, medicine, law, and social theory. For without plants there would be no life.

CHAPTER 2

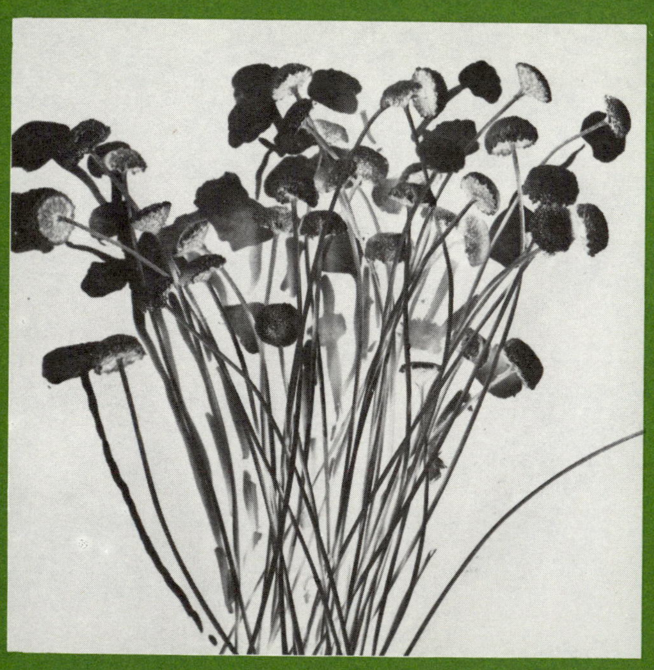

CLASSIFICATION— THE RELATIONSHIP OF PLANTS

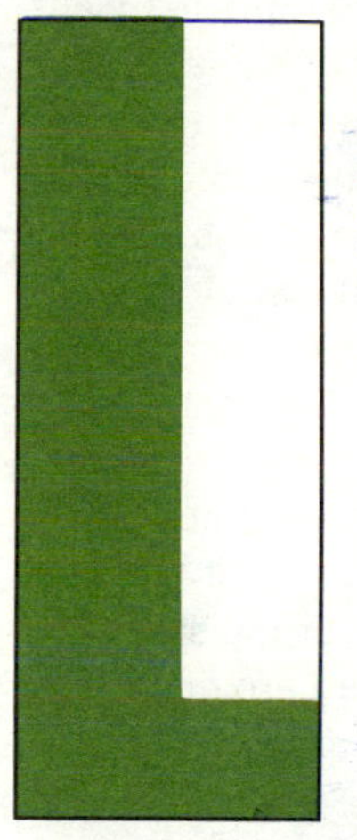

ife, since its origin over one to three billions of years ago, has evolved a fantastic variety of forms as organisms have developed innumerable adaptations to cope with the wide range of changing environmental conditions to which they have been subjected. Classification is both a practical necessity and an important intellectual goal of biologists. They attempt to place organisms into groups that depict genetic relationships. Such a classification is said to be a **natural** or **phylogenetic classification.** Grouping plants by habitats (such as desert or marsh plants), or by the size of their leaves, or color of their flowers, has limited usefulness and is said to be an **artificial classification.** Since we do not know the exact genetic, and physiological information) and to construct relationship among plants, an accurate natural system of classification is impossible to construct, however there have been several attempts to do so. As we shall see, these are useful systems but it should be remembered that each of them is simply the investigator's best attempt to use all of the information about organisms that is available (fossil records, morphological, biochemical, genetic, and physiological information) and to construct a mentally satisfying system of classification. There is no one correct system.

In the most accepted system of biological classification, the largest unit is the **kingdom.** However, there is no agreement among biologists on how many kingdoms should be used to classify the biological world. Several alternatives have been proposed. However, in this book we shall consider only the traditional two kingdom system and a recently proposed five kingdom system.

KINGDOMS IN THE LIVING WORLD

In the two kingdom system, all life is considered either as belonging to the **plant kingdom** or the **animal kingdom.** But what is a plant? Certainly early in an introductory botany text there should be a clear, precise definition of a plant. Plants and animals appear to have so many different and distinguishing characteristics that one would at first assume that there should be no difficulty in defining a plant. A working definition might be that plants are "those organisms which have chlorophyll and are capable of trapping the radiant energy of sunlight, transforming it into stored energy in food. Also included as plants would be those organisms lacking chlorophyll but which are morphologically very similar to certain groups of the least complex plants containing chlorophyll." Plants, then, make their own food while animals ingest organic matter as food. Another difference between plants and animals is that plants generally are nonmotile whereas animals usually are capable of moving from place to place. Furthermore, most plant cells have rigid cell walls while animals cells do not. Such easily observable and contrasting characteristics led early naturalists to divide living things into two kingdoms, plants and animals. This concept of two kingdoms was convenient, useful, and was generally accepted. Most people today would accept the idea that the biological world consists of animals and plants. Even childhood games such as "Animal, Mineral, or Vegetable" help to impress this idea on our thinking. However, as more and more has been learned about the characteristics of the multitude of life forms, it has become apparent that the grouping of all organisms into two kingdoms is highly artificial, and there are many difficulties in classifying some organisms as either plants or animals.

Imagine that you arrived on earth from outer space. Your task is to classify all life forms. One useful method of classification is to group similar things together in the hope that you will be placing closely related organisms into the same group. You are not hampered by a preconceived idea that there are only two major groups (kingdoms), plants and animals. Your problem is to identify similarities and differences in the organisms and to place similar organisms into groups. You study the larger, more obvious forms of life first and note that the nonmotile green forms make their own food and the motile nongreen forms engulf food. But you also study some nonmotile forms such as molds, mushrooms, and other fungi that do not contain chlorophyll and they neither engulf food nor make their own but instead absorb food from dead organic matter or else are parasites and absorb their food from other organisms. You might conclude that, on earth, life forms should be classified into three large groups or kingdoms on the basis of how they obtain food. In fact, it has been proposed that living organisms should be classified into three different kingdoms; plants, animals, and fungi.

However, if you continued your study of life on earth, you would find that there are some unicellular microscopic forms of life that are similar to each other and that differ from all other organisms in that their cells lack nuclei. Such organisms, those whose cells lack nuclei, are called **prokaryotes** in contrast to those whose cells have nuclei, **eukaryotes** (Fig. 2.1). Might it not be logical to place these relatively simple organisms with similar, apparently primitive, structural characteristics into a separate kingdom even though one group, the bacteria, lack chlorophyll and another group, the blue-green algae contain chlorophyll? Finally, in your study you discover other organisms that are primarily unicellular but do have true nuclei. Many of these forms are capable of movement by means of complex threadlike protoplasmic structures, **flagella**. These organisms do not appear to fit readily into any of your other four kingdoms—a fifth kingdom?

Early in the nineteenth century, Haeckel proposed that biologists should place simple forms of life into a separate kingdom, Monera, and by 1900 biologists were grouping organisms into four kingdoms, **Monera, Protista, Plantae,** and **Animalia.** Recently some modern biologists with greatly increased knowledge of the similarities and differences in life forms and with a greater knowledge of evolutionary relationships among organisms than early biologists had also have suggested a revision of our traditional two kingdom concept. They have proposed that a classification having five kingdoms would more logically serve to show evolutionary relationships among organisms. Although none of these systems is wholly satisfactory, that proposed by the late R. H. Whittaker has considerable merit in expressing broad relationships in the biological world. Almost as soon as this proposal was made, suggestions for its modification were proposed by other biologists! Table 2.1 summarizes the five kingdoms, **Monera, Protista, Fungi, Plantae,** and **Animalia** proposed by Whittaker.

However useful as such groupings are in helping us to recognize and study possible evolutionary relationships among organisms, it is important as botanists that we not only study the higher plants but that we have a knowledge and understanding of the nature of organisms that are most closely related to them. Consequently, in this text, we shall consider those organisms that traditionally have been considered as plants in the two kingdom system. For convenience we also include fungi.

LEVELS OF CLASSIFICATION

Over 550,000 species of plants inhabit the earth and most of the waters to a depth of approximately 30 meters. What is a **species?** How are species arranged or classified, to bring order to this vast assemblage of plants? See **Fig. 2.2** (page 9).

The Swedish botanist Linnaeus divided all flowering

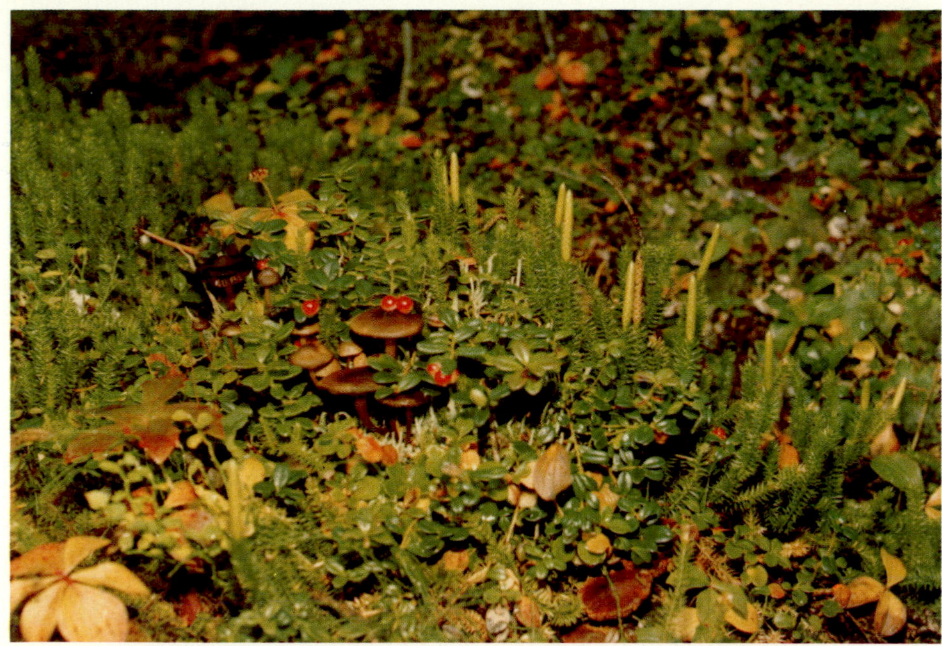

Figure 2.2
Cranberry, club moss, and a mushroom living in close association. They have similar environmental requirements. Are they similar?

Figure 2.5
A and *B*, two clones of irises *(Iris germanica)*. These two clones freely interbreed and thus constitute a species. Notice, however, that the flower colors of the two clones are different. *C*, a single population of Dutch iris *(Iris xiphium)*. These plants interbreed freely, but only very rarely do they cross with members of the populations shown in *A* and *B*. *C* is a second species in the genus *Iris*. *D*, a population of *Freesia*. These plants have some characteristics similar to those of *Iris*. Both *Iris* and *Freesia* are genera in the family Iridaceae.

Figure 2.8
Pigments, hence color, are of importance in distinguishing divisions in the thallophytes. *A, Amanita muscaria,* or the fly mushroom, is in the Eumycota; it lacks chlorophyll but does have a bright red pigment in the cap. *B, Ulva,* or sea lettuce, of the division Chlorophyta has both chlorophylls *a* and *b*. Because its thallus is only two cells in thickness, the green color shows best through overlapping folds of the blade. *C,* the Phaeophyta, represented by *Postelsia,* or the sea palm, have chlorophylls *a* and *c* and a brown pigment. *D,* the Rhodophyta have only chlorophyll *a* and a red pigment that shows well in the blades of *Iridophycus*. *E,* the blade of *Porphyra* are two cells thick; because of the thinness of the blades and a frequently poor development of the red pigment, this species of Rhodophyta is not as red as other members of the division. *F,* a narrow strip of intertidal zone shows the green alga *Ulva,* a brown alga resembling *Fucus,* and the red alga *Gigartina*. (All with the assistance of David Brown.)

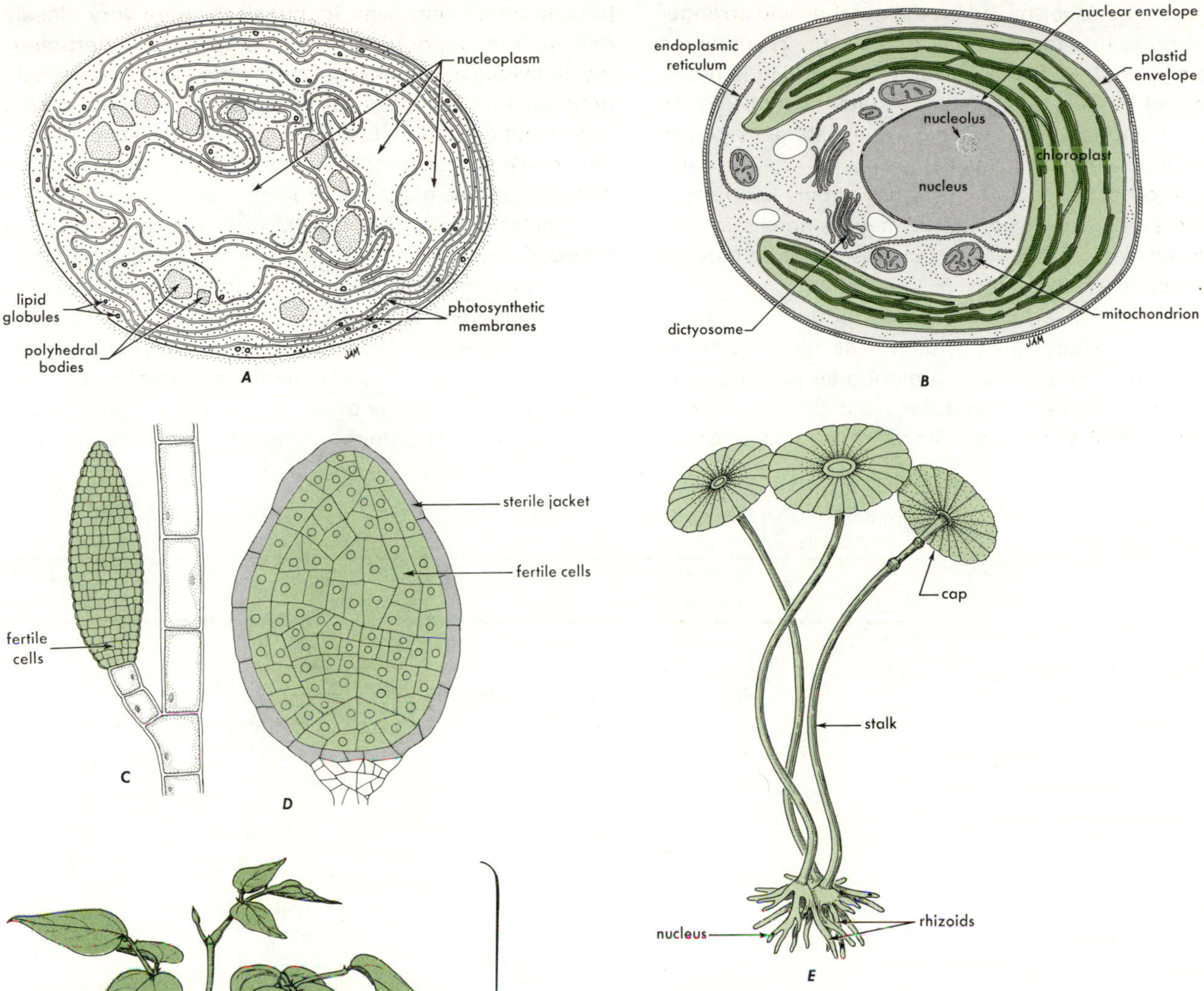

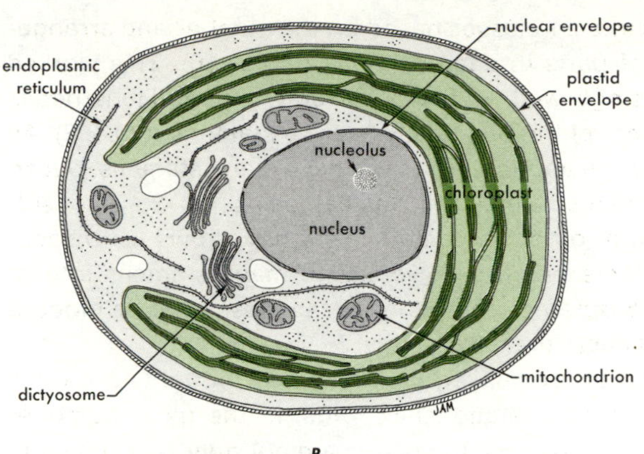

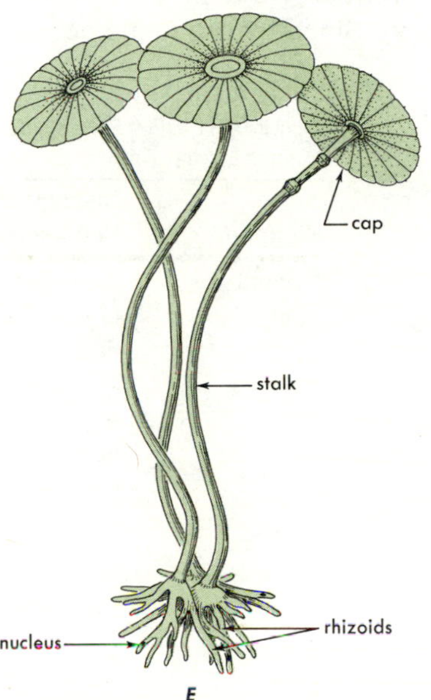

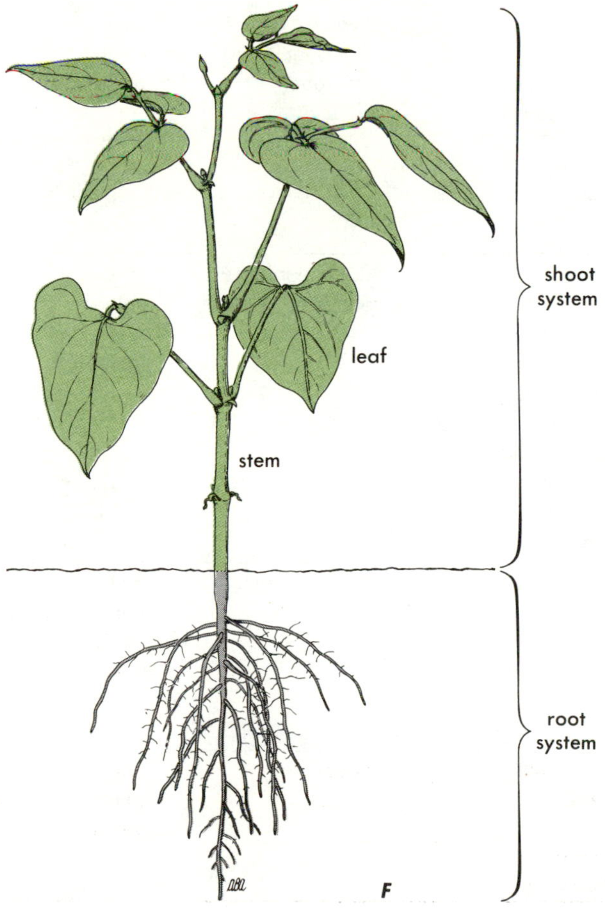

Figure 2.1
Diagrams to show differences between groups of plants. *A*, a prokaryotic cell (blue-green alga). *B*, a eukaryotic cell (green alga, *Chlorella*). *C*, a group of reproductive cells lacking a protective jacket (the brown alga, *Ectocarpus*), *D*, a group of reproductive cells provided with a protective jacket (the liverwort, *Riccia*), *E*, a thallophyte (the single-celled green alga. *Acetabullaria*), *F*, a seed bearing vascular plant (bean, *Phaseolus vulgaris*).

plants into 24 classes based on the number and arrangement of parts in each flower (Chapter 15). This worked surprisingly well. All plants currently grouped together as members of the pea family have 10 stamens (Fig. 2.3). All buttercups and their relatives have many stamens, but so do some roses, mallows, and camellias (Fig. 2.4). Consideration of other characteristics today allows us to separate plants bearing flowers with many stamens into at least 35 different categories or families. What do modern taxonomists mean by similar?

Plants are placed in categories thought to have a close genetic relationship. Thus, in the most homogeneous populations, the plants are not only very similar in many aspects of their morphology but they *interbreed freely* (**Figs. 2.5*A*,*B*,** page 9). Since their genetic constitutions are similar, even identical, they are very closely related. Such a population of closely related interbreeding individuals constitutes a **species.** The most basic requirement of a biological classification is that it show a true genetic relationship. Several different species may have some characteristics in common and an occasional mating may take place between members of two different but similar species **(Fig. 2.5*C*).** Such similar species are grouped together in **genera** (s. genus), which are thus groups of genetically related species. A comparison of some genera will show that they have traits in common **(Fig. 2.5*D*).** Genera with similar traits are grouped in **families.** They are therefore also genetically related, but less closely than are plants of the same genus. Plants in the same family have common characteristics that indicate

Table 2.1

SUMMARY OF A FIVE KINGDOM SYSTEM OF CLASSIFICATION OF ORGANISMS (ADAPTED FROM WHITTAKER, 1969)

Kingdom	Distinguishing Characters and Representative Organisms	Nutritional Mode	Estimated Time of Origin
Monera	Prokaryotic cells, unicellular, or colonial unicellular. Plastids, mitochondria, and lacking advanced (9+2 strand) flagella. Asexual reproduction by fission or budding; protosexual phenomena also occurs. Bacteria, blue-green algae.	Predominantly absorption, some groups photosynthetic or chemosynthetic	3 to 4 billion years ago
Protista	Unicellular, multicellular, eukaryotic organisms that constitute a heterogenous group. Many forms motile by means of advanced (9+2 strand) flagella. Euglenophyta, Chrysophyta, Pyrrhophyta, Protozoans, flagellated fungi.	Diverse—photosynthesis, absorption, ingestion	1000 million years ago
Fungi	Uni- or multinucleate with eukaryotic nuclei dispersed in a walled, often septate mycelium. Primarily nonmotile, living in or on food supply. Both sexual and asexual reproduction. Chloroplasts and photosynthetic pigments lacking. Includes chytrids, water molds, bread molds, sac fungi, club fungi.	Absorption of food	1000 million years ago
Plantae	Multicellular eukaryotic organisms containing plastids. Generally nonmotile, living anchored in a substrate. Multicellular algae, liverworts, hornworts, mosses, vascular plants.	Photosynthesis	500 million years ago
Animalia	Multicellular organisms with eukaryotic cells lacking a rigid cell wall, plastids, and photosynthetic pigments. Reproduction chiefly sexual.	Ingestion	700 million years ago

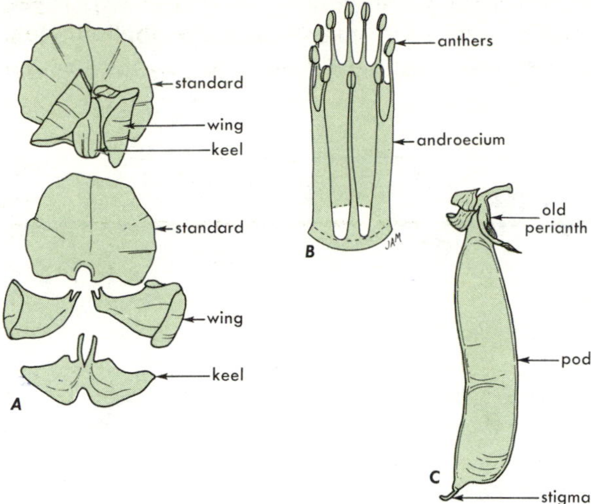

Figure 2.3
Diagram showing traits that are important in the classification of the sweet pea (*Lathyrus odoratus L.*). *A*, flower, exploded view. *B*, 10 stamens, 9 attached together, 1 free. *C*, pod.

they may all have evolved from the same ancestor. For instance, members of the pea family have a characteristic flower shape (Fig. 2.3), flower parts in fives, pods containing seeds, and dissected leaves.

Families are grouped together in **orders,** and orders may be grouped in **classes.** At the top level it is customary to divide all plants into **divisions,** or **phyla.** Differences in photosynthetic pigments (Chapter 23), manner of leaf development (Chapter 31), structure of the conducting, or vascular, tissues (Chapter 28), and methods of reproduction (Chapter 24) are the most important elements (Table 2.2) in separating plants into divisions.

The word similar, in a natural classification, means organisms having genetic constitutions indicating relationship. This system is illustrated in the complete classification of the sweet pea given below. Class and subdivision endings are not standardized.

SWEET PEA

	Name	Ending
Species designation	*odoratus*	—
Genus	*Lathyrus*	—
Family	Fabaceae[a]	aceae
Order	Rosales	ales
Subdivision	Dicotyldeonae	ae
Division	Anthophyta	phyta

[a] Leguminosae is a less accepted synonym.

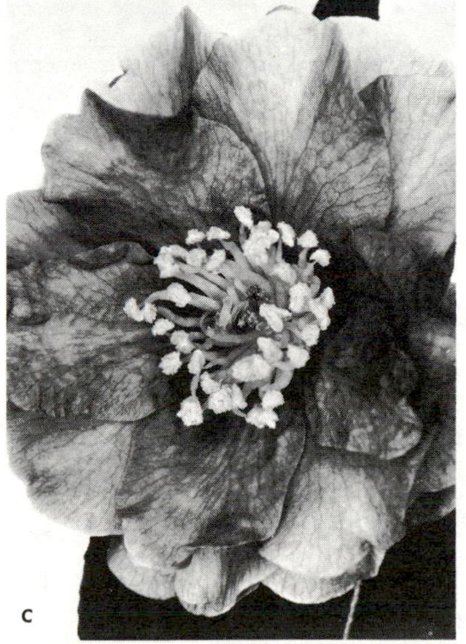

Figure 2.4
There are 35 different families having flowers with many stamens. Three flowers are shown, *A Rosa*, Rose family; *B*, *Hibicus*, Mallow family; *C*, *Camellia*, tea family.

SOME CONVENIENT GROUPS OF PLANTS

Broad groups of plants may include several divisions. While these categories are convenient, they do not have taxonomic importance. A large group of primitive plants, mainly aquatic and filamentous, may be set apart from other plants by a very distinctive characteristic: the sexual cells (eggs, sperms, or gametes) are not protected by a jacket of vegetative or sterile cells (Fig. 2.1*C*). These plants may be referred to collectively as the **thallophytes.** The thallophytes constitute 9 algal, 3 fungal, and a single

Table 2.2
SYNOPSIS OF PLANT DIVISIONS COVERED IN THIS BOOK

	Divisions	Common Names or Examples
A. Plants without roots, stems, or leaves, and no protective jacket of vegative cells around the developing reproductive cells		thallophytes
B. Plants lacking chlorophyll[a]		
C. Prokaryotes	Bacteria	bacteria
C. Eukaryotes		
D. Vegetative stage of naked cytoplasm	Myxomycota	slime molds
D. Vegetative stage with a cellulose cell wall, mostly filamentous.	Oomycota	fungi
D. Vegative stage with a chitin cell wall, mostly filamentous	Eumycota	fungi
B. Plants with chlorophyll		algae
C. Chlorophyll *a* alone (c in some Xanthophyta)		
D. Prokaryotes	Cyanobacteria	blue-green algae
D. Eukaryotes		
E. With water-soluble blue and red pigments	Rhodophyta	red algae
E. With plastid pigment xanthophyll (chl.c in some)	Xanthophyta	
C. Chlorophylls *a* and *c*		
D. Large seaweeds with a cell wall of cellulose	Phaeophyta	brown algae
D. Small, mostly unicells, cell wall of cellulose, two anterior unequal flagella	Chrysophyta	golden algae
D. Small, mostly unicells, cell wall of silica	Bacillariophyta	diatoms
D. Small unicells lacking a cell wall or with cellulose plates; two lateral unequal flagella (includes Cryptophyta)	Pyrrhophyta	dinoflagellates and Cryptophyta
C. Chlorophylls *a* and *b*		
D. Unicells without a cell wall	Euglenophyta	*Euglena*
D. Greatly diversified forms with cell walls (includes Charophyta)	Chlorophyta	green algae and Charophyta
A. Protective jacket of vegetative cells present; roots, stems, and leaves may be present		
B. Plants lacking vascular tissue	Bryophyta	liverworts, mosses
B. Plants with vascular tissue		vascular plants
C. Plants without seeds		lower vascular plants
D. Roots absent	Psilophyta	*Psilotum*
D. Roots present		
E. Small leaves, no gap at union with stem, leaves do not unroll as they open		
F. Stems not jointed	Lycophyta	club mosses
F. Stems jointed	Sphenophyta	horsetails
E. Leaves with well-developed gap at union with stem, leaves unroll as they open	Pterophyta	ferns
C. Plants with seeds		seed plants
D. Seeds not covered		gymnosperms
E. Seeds in cones		conifers
F. Trees palmlike	Cycadophyta	cycads
F. Trees conical	Coniferophyta	conifers
E. Seeds not in cones like those above		
F. Leaves fan-shaped and deciduous; trees	Ginkgophyta	*Ginkgo*
F. Leaves not fan-shaped; shrubs, vines, or prostate perennials	Gnetophyta	*Ephedra, Gnetum, Welwitschia*
D. Seeds covered	Anthophyta	angiosperms
E. Two-seed leaves	Dicotyledonae[b]	dicotyledons
E. One-seed leaf	Monocotyledonae[b]	monocotyledons

[a] A few bacterial species do have chlorophyll.
[b] Subdivisions

prokaryotic group. The only unifying characteristics of all the remaining plants are that their reproductive cells are protected by a jacket of sterile cells (Fig. 2.1*D*) and they possess an embryo stage in their life cycle. A small group of these plants lack vascular conducting tissue. They are the **liverworts** and **mosses** (bryophytes) (see Fig. 2.6). All other plants have developed a specialized conducting tissue, and are known as **vascular plants.** The more advanced vascular plants bear seeds and are frequently referred to as the **seed plants** **(Fig. 2.2*F*).** These are rather precisely separated into **gymnosperms** (see Fig. 2.13), whose seeds are borne naked on a cone scale, and **angiosperms** (see Fig. 2.14), whose seeds are protected within an ovary.

Figure 2.6
The hair cap moss, *Polytrichum*, is a representative of the Bryophyta.

PLANT DIVISIONS

We have seen that plants constituting a species are very closely related. And we have said that all members of a family arose, in the primitive past, from a common ancestor. Do all members of a division have a common ancestor? Are divisions related? Do all plants have a common ancestor? A partial answer comes from the fossil record.

Plants of past eons have left their remains, or impressions, on geologic strata. These **fossils** (Fig. 2.7) are found in sedimentary rocks, in volcanic ash, and in quantity in coal. By comparing the morphology of present-day plants with the morphology of fossils from past geologic periods, it is possible to establish broad lines of relationship between the divisions. Unfortunately, fossils are few in number, do not reveal the entire plant, and are preserved only in certain habitats. Therefore, the fossil record must be used with caution. This is why agreement among scientists on the evolutionary relationships (phylogeny) of plant divisions is difficult to achieve.

Divisions of Thallophytes

Thallophytes, a very large group of primitive plants, do not have a jacket of sterile protecting cells around their

Figure 2.7
An imprint of a leaf on *Pecopteris*, a common plant growing during the Pennsylvanian geologic period about 300 million years ago. (Courtesy of Field Museum of Natural History.)

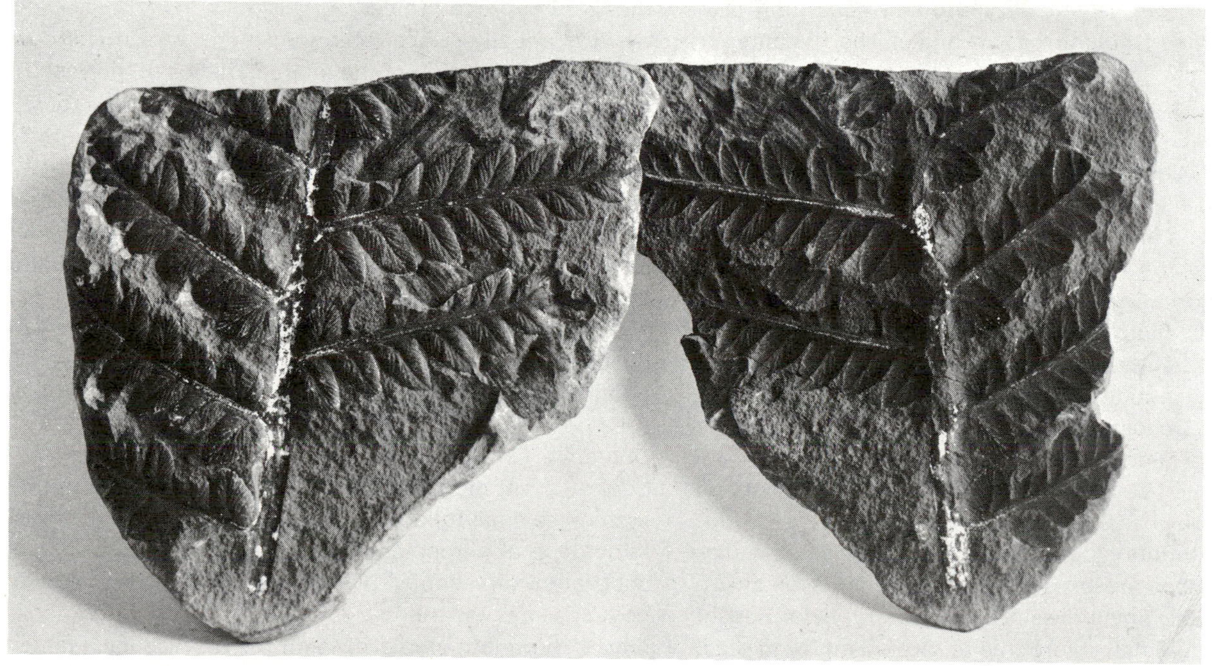

reproductive cells (Fig. 2.1*C*). Those without chlorophyll are the **bacteria** and **fungi** **(Fig. 2.8*A* page 10)**; those with chlorophyll are algae **(Figs. 2.8*B,C,D,E,F*)**.

Algal Divisions

The algae may be separated further into groups on the basis of photosynthetic pigments. Some of these groups have division rank. A large group of algae contain chlorophylls *a* and *b*. These algae are easily divided into two divisions. One of these, the **Euglenophyta,** consists of single-celled forms **(unicells)** that lack a cell wall. The other division is a large one, the green algae (**Chlorophyta,** **Fig. 2.8*B***). Because higher plants also contain chlorophylls *a* and *b*, it is thought that primitive Chlorophyta are ancestors of the higher plants (Chapter 23). Chlorophyll *b* is lacking in all other algae except the Euglenophyta.

A second large algal group is known to contain only chlorophyll *a* together with accessory water-soluble pigments, one blue and the other red (Chapter 21). This assemblage of forms is divided into two divisions. One division, the blue-green algae (**Cyanobacteria,** Fig. 2.1*A*), are prokaryotes and hence lack true nuclei. The other division, the red algae (**Rhodophyta,** **Fig. 2.8*D***), is composed of eukaryotes that inhabit the coastal waters of the oceans in the tropic and temperate zones. While both divisions contain only chlorophyll *a* and similar accessory pigments, their structure is so different that any relationship between the two divisions must be very distant (Chapters 21, 23).

A third group definitely contains chlorophylls *a* and *c* plus certain other distinguishing plastid pigments that separate them into four divisions. One division comprises the seaweeds—kelps—of ocean coasts. These are brown algae, or **Phaeophyta** **(Fig. 2.8C)**. A second division in this group is the **Chrysophyta,** consisting of a small number of golden brown algae. The diatoms, or **Bacillariophyta,** comprise a third large division in this group. The diatoms are quite small, but anyone who has looked at a microscope slide of water under magnification has probably seen them, for they are extremely common unicells and filaments. Their cell walls are impregnated with silica, and are so constructed that the two halves fit together like the top and bottom of a petri dish. The fourth division having chlorophylls *a* and *c* is comprised of small cells, many of which lack cell walls. It is known as the **Pyrrhophyta.** These algae are not common, but occasionally a sudden increase in their population to vast numbers causes "red tides," poisoning many miles of ocean coastlines.

The algae in a fifth group were thought to contain only chlorophyll *a*, and it was separated from the Cyanophyta and Rhodophyta because the water-soluble pigments were not present. This group consists of only a single relatively small division, the **Xanthophyta.** Recent work indicates that chlorophyll *c* is present, in at least many species, and that the Xanthophyta may best be characterized by certain xanthophyll plastid pigments.

Fungal Divisions

Three thallophyte divisions lack chlorophyll; thus, the plants cannot make their own food and are therefore called **heterotrophs.** A primitive subdivision of fungi, the **Myxomycota,** are the slime molds that, in their vegetative stage, are nothing but naked masses of protoplasm. More advanced fungal divisions are the **Oomycota** and the **Eumycota** **(Fig. 2.8*A*)**. Some of them very closely resemble the algae. The thallophyte divisions are summarized in Table 2.3.

Classification of Land Plants

There is general agreement that the land plants take their origins from the ancient green algae. One group of land plants lacking a vascular system forms a natural division of liverworts and mosses, the **Bryophyta.** All other land plants have conducting, or vascular, tissue. General agreement is lacking as to their natural relationship.

All vascular plants have a system for conducting water and nutrients throughout the plant. They all have roots, stems, and leaves (Fig. 2.1*F*), except for some primitive living genera which lack roots (see Fig. 2.9). Also, they have all developed a cuticle on their aerial portions that serves as protection against desiccation. The classification problem facing us is: Did all these characteristics, which distinguish vascular plants from all other plants, arise just once in the long evolutionary history of plants? If so, all plants living today have evolved from the same vascular ancestral form. This theory of plant evolution is known as a **monophyletic origin** of vascular plants; all such plants would then belong to one division: the **Tracheophyta** (Table 2.4).

It is equally conceivable that there were numerous steps in the evolution of the characteristics of the vascular plants, and that the cuticle, vascular tissue, roots, stems, and leaves may have arisen in several different periods and localities. The plants living today, then, would not have a common vascular ancestral form. There would be several common ancestral forms, at least one for each suggested division. This is known as a **polyphyletic origin.**

It therefore becomes difficult to present a system for the classification of the vascular plants in an introductory botany course. Two such systems are compared in Table 2.4. The first column summarizes a monophyletic system. Column 2 shows a polyphyletic system used in a textbook of plant morphology, which is more similar to the system we shall follow in this book (column 3). There is fairly general agreement among botanists on the main subgroups of vascular plants; the differences in these columns lie in the number of divisions. According to column 1 there is a single division for all vascular plants,

Table 2.3

SUMMARY OF THALLOPHYTE DIVISIONS

Number	Division	Common Name	Chlorophylls	Description
1	Cyanobacteria	Blue-green algae	*a*	Prokaryotic cells
2	Pyrrhophyta		*a,c*	Mostly unicells
3	Chrysophyta	Golden algae	*a,c*	Some unicells
4	Bacillariophyta	Diatoms		Many unicells, silica in walls
5	Phaeophyta	Brown algae	*a,c*	Marine forms, some very large
6	Rhodophyta	Red algae	*a*	Mostly marine forms, many with a feathery plant body
7	Xanthophyta	Golden-green algae	*a,c*	Many unicells, some simple filaments
8	Euglenophyta	Euglena	*a,b*	Motile unicells without cell walls
9	Chlorophyta	Green algae	*a,b*	Mostly rather small forms of great diversity
10	Charophyta	Stoneworts	*a,b*	Differentiated into nodes and internodes, sometimes put into Chlorophyta
11	Bacteria	Bacteria	none or bacteriochlorophyll	Prokaryotic cells, a few with bacterial chlorophyll
12	Myxomycota	Slime molds	–	Amorphous masses of protoplasm in vegetative stage
13	Oomycota	Egg fungi	–	Fungi with hyphae and cellulose walls
14	Eumycota	True fungi	–	Fungi with hyphae and chitin walls

Table 2.4

COMPARISON OF SYSTEMS OF CLASSIFYING THE MAJOR DIVISIONS OF THE LAND PLANTS

Monophyletic (H. P. Banks, 1968)	Polyphyletic (Scagel et al., 1967)	Used in This Text	Common Names
Bryophyta	Bryophyta	Bryophyta	Bryophytes; mosses and liverworts
Tracheophyta			Vascular plants
	Psilophyta	Psilophyta	Psilophytes
Rhyniophytina	*Rhynia*	*Rhynia*	*Rhynia*
Zosterophyllophytina			
Psilophytina	*Psilotum*	*Psilotum*	*Psilotum*
Lycophytina	Lycophyta	Lycophyta	Lycopods, club mosses
Sphenophytina	Arthrophyta	Sphenophyta[a]	*Equisetum*, horsetails
Trimerophytina			
Pterophytina			
Cladoxylopsida			
Filicopsida	Pterophyta	Pterophyta	Ferns
Coenopteropsida			
Progymnospermopsida	Pteridospermophyta	Pteridospermophyta	Seed ferns
Cycadopsida	Cycadophyta	Cycadophyta	Cycads
	Ginkgophyta	Ginkgophyta	*Ginkgo*
Coniferopsida	Coniferophyta	Coniferophyta	Conifers
Gnetopsida	Gnetophyta	Gnetophyta	Gnetum
Angiospermopsida	Anthophyta	Anthophyta	Angiosperms

[a] Sphenophyta is an older name for this division; it is retained here because of confusion arising from similarity of the new proposed name, Arthrophyta, with Anthophyta and the animal phylum Arthropoda.

Figure 2.9
A potted plant of *Psilotum nudum*, a living species closely resembling a fossil form found in sedimentary rocks laid down 400 million years ago.

thus implying a single ancestor. According to column 2, there are ten vascular plant divisions, implying at least ten ancestors.

The fourth column gives the common names for these groups of plants. While these names refer as precisely to a group of plants as do the scientific names, the common names do not suggest relationships between the plant groups they designate.

Divisions of Land Plants

The **mosses and liverworts** (**Bryophyta,** Fig. 2.6, Chapter 27) are easily separated from all other land plants. Instead of true roots, they have rhizoids that act as absorbing organs. They also have small leafy stalks that produce sperm and egg cells. Frequently, in mosses, a brown upright stalk bearing a capsule is seen growing from the low leafy stalk; this is the offspring of sperm and egg fusion. The mosses, which typically lack vascular tissue, are thought to have arisen either from the Chlorophyta or by reduction from primitive vascular plants. They have not given rise to any other land plants.

The most primitive living vascular plants are known as the **psilophytes (Psilophyta),** and there are only two living genera, *Psilotum* and *Tmesipteris. Psilotum* is easily cultivated and is fairly common as a potted plant in botany greenhouses (Fig. 2.9). The name *Psilotum* means naked, or smooth, and refers to the simple dichotomously branching stem, with small scale-like leaves and with rhizoids rather than roots (Chapter 28). *Psilotum* appears very similar in structure to a fossil (*Rhynia*) from Scotland found in sedimentary rocks that were deposited 400 million years ago. Two other very ancient groups of plants, the **club mosses** or **lycopods (Lycophyta),** and the **horsetails** or *Equisetum* **(Sphenophyta),** once dominated the earth's vegetation with many tree forms. We shall consider three of the five genera of club mosses, and a single genus of horsetails, *Equisetum.* Both are ancient plants, being easily related to fossils found in rocks nearly 400 million years old.

Equisetum has hollow, jointed stems that have an encircling sheath of small leaves at each joint. Spores, or reproductive cells, are borne in specialized cones at the summit of some of the upright stalks (Fig. 2.10).

The **club mosses (Lycophyta)** more closely resem-

Figure 2.10
Horsetails, *Equisetum telemateia L.;* this is the only genus in the division Sphenophyta.

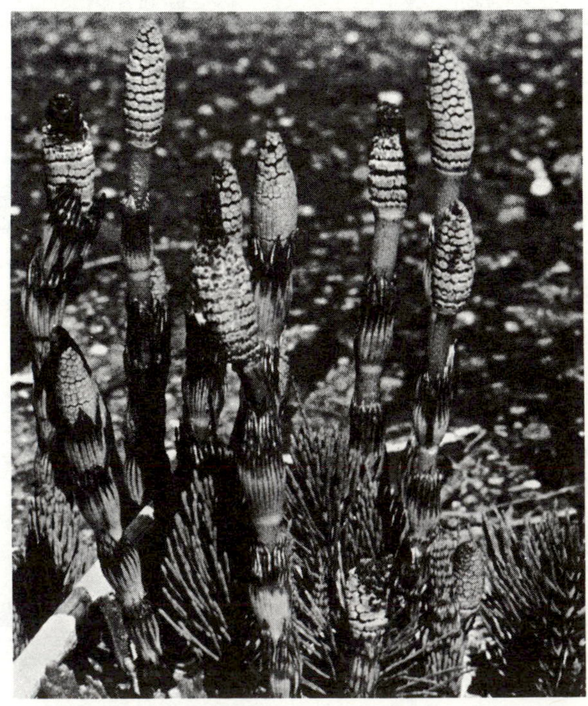

ble the higher plants (Fig. 2.11). One genus is known as ground pine and has been used extensively for Christmas wreaths. The club mosses are separated from the other groups of vascular plants by their distinctive reproductive structures; their vascular system; the single veins in their small leaves; and the union of stem and leaf without any break in the vascular tissue of the stem (Chapter 28).

The living **ferns (Pterophyta)** are generally small plants and are characterized by their highly dissected, characteristic leaves, which unroll at the tip as they open. Seeds are not formed by any living fern (Fig. 2.12) but fossil forms did produce seeds.

The remaining six divisions of plants (Table 2.2) are seed plants. The first, the seed ferns, are now extinct. Four of these divisions have seeds borne upon scales in a cone (Fig. 2.13) or as naked seeds on short stems; these are known collectively as the **gymnosperms.** The remaining division of seed plants has seeds enclosed in a seed case or ovary (Fig. 2.14); these are the **angiosperms (Anthophyta).**

Two of the divisions of gymnosperms, *Ginkgo* (**Ginkgophyta,** see Fig. 29.2) and the **cycads** (**Cycadophyta,** see Fig. 29.18) are the most primitive of the living seed plants. *Ginkgo biloba* is the only species in its division. It grows to be a large tree, has been cultivated for many centuries in Chinese temple gardens, and is currently finding much favor in western countries. It is extinct in the wild. *Ginkgo* has fan-shaped leaves with many veins radiating from the petiole and a break or gap in the vascular cylinder where the vascular tissue from the leaf joins the stem. Naked seeds are borne on the end of short stems.

The **cycads (Cycadophyta)** comprise 11 to 12 genera, two of which are extinct. Living cycads range in size from small forms to large trees resembling palms.

The most conspicuous division of gymnosperms are

Figure 2.11
There are five genera in the Lycophyta. This is *Lycopodium obscurum L.,* or running pine.

Figure 2.12
The fern, *Polypodium vulgare,* representing the Pterophyta.

Figure 2.13
An open cone of the pinon pine, *Pinus monophylla,* showing two seeds in place upon a scale. A gymnosperm.

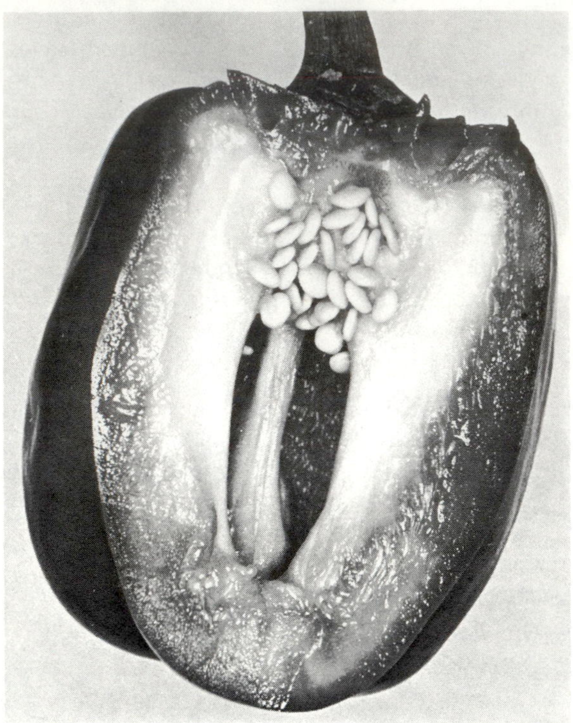

Figure 2.14
A fully matured ovary of a bell pepper, *Capsicum frutescens var grossum,* showing seeds within the matured ovary. An angiosperm.

the **conifers (Coniferophyta).** Together with the angiosperms (Anthophyta), they dominate modern vegetation. The needle or scale leaves of the conifers have but one or two veins, and in some forms there does not appear to be a break in the vascular cylinder where the vascular tissue from the leaf joins with the vascular cylinder of the stem. In contrast, leaves of the **angiosperms (Anthophyta)** have many veins and there is always a gap at the junction of leaf vascular tissue with the stem vascular tissue. We may divide the angiosperms into two subdivisions. The seeds of one have two seed-leaves, or **cotyledons,** as seen in the peanut or bean. These are the **Dicotyledonae.** The seeds of the other group (onions, lilies) have a single seed-leaf. These are the **Monocytoledonae.**

The association between the divisions is shown by the key on page 14.

THE BINOMIAL

The species designation of sweet pea is *odoratus* and it belongs to the genus *Lathyrus* in the Leguminosae family. To call it just *odoratus* would not be very meaningful, for it does not show relationship with anything else. To call it simply *Lathyrus* would not separate it from other species in the genus *Lathyrus.* The scientific name of every species is a binomial consisting of two words: the genus plus the species (*Lathyrus odoratus*).

Frequently, the genus name refers to the ancient name of the same or a similar plant. *Lathyrus* is the Greek word for pea. An effort is usually made to make the species name descriptive. *Odoratus* is the Latin word designating that something is sweet-smelling. So, altogether the Greek genus and Latin species translate to sweet-smelling pea. There are also black peas, *L. niger;* cultivated peas, *L. sativus;* beach peas, *L. maritimus;* splendid peas, *L. splendens;* and so on. Since an effort is made to keep the scientific names of plants and animals in the correct Latin or Greek form, they are really foreign terms and as such are always written in *italics.* As the species designations are generally descriptive, they may be used in conjunction with other genera. We have, for instance, black pepper, *Piper nigrum;* black mustard, *Brassica nigra;* and black walnut; *Juglans nigra.* There is sweet alyssum, *Alyssum odoratum,* and cultivated oats, *Avena sativa.* A genus name, however, is used for only one group of species and cannot be used for any other group, plant or animal.

The Swedish botanist, Linnaeus (Fig. 2.15) was largely responsible for establishing the binomial system, not only for plants but for all living organisms. In fact, the full name of the sweet pea is *Lathyrus odoratus* L. The cultivated pea should likewise be *Lathryus sativus* L. The L. after these two names indicates that they were first described by Linnaeus. The large numbers of plant names followed by the letter L. are a living tribute to the greatness of Linnaeus' taxonomic contributions. And today when the sweet pea binomial appears in the literature, every plant scientist, no matter what language he speaks, knows that reference is made to a specific pea with certain characteristics.*

It is not uncommon, particularly in cultivated plants, to have races develop within a species. Indeed, in the many years devoted to breeding sweet peas, such races have arisen. One has a condensed, bushy type of growth. Bailey gives this the name *Lathyrus odoratus nanellus* Bailey or dwarf sweet pea. This is a trinomial.

Several thousand new species of plants are discovered every year. The plant specimens may have been collected in parts of the world previously unexplored by botanists, although new species are being discovered every year in areas traversed many times by botanists. These specimens ultimately fall into the hands of systematic botanists who are interested in and have given special attention to the group (family or genus) to which

* Realizing that many species of plants have several common names, thus giving rise to confusion, the American Joint Committee on Horticultural Nomenclature has prepared a book entitled *Standardized Plant Names,* which in the second edition is described as "a revised and enlarged listing of approved scientific and common names of plants and plant products in American commerce or use." In the preface to the second edition the statement is made that "the purpose of *Standardized Plant Names* is to bring intelligent order out of the chaos in names of plants and plant products existing the world over. Such standardization, supported by adequate authority, will not only promote satisfactory understanding between those who sell and those who buy, but will also improve the multifarious relations, scientific, educational and social, into which the advancing plant consciousness of America has grown."

Figure 2.15
Linnaeus (1707–1778), the Swedish botanist who initiated the binomial system of nomenclature. (Courtesy of the Library of The New York Botanical Garden.)

the plant in question belongs. The botanists set out to identify the species. In so doing, they compare it with allied forms, of which they have herbarium specimens or published descriptions. They may find that the plant has characters similar to those of a species heretofore described; or they may find that the plant has characters so different from those of any known species that they conclude that it is a new species (**species novum,** sp. nov.). In other words, it is a species that has never been described. Accordingly, the botanist describes this new species and gives it a name. The description is published in one of the many recognized botanical journals, and the specimen or specimens that were used in making the description are properly labeled and placed in one or more of the herbaria of the world. This specimen is known as the **type specimen.**

HERBARIA AND FLORAS

There are about 550,000 species of plants, and they may be divided as we have just indicated. No attempt has been made to describe all of these plants in one treatise. It requires, for instance, ten large volumes to describe the vascular plants in Engler and Prantl's current *Genera Plantarum.* For general convenience, books describing regional floras are in common use. We have, for instance, for the United States:

Leroy Abrams (1940–1960). *Illustrated Flora of the Pacific States.* 4 vols. Stanford University Press. Stanford, Ca.

L. H. Bailey (1949). *Manual of Cultivated Plants.* Macmillan Co., New York, 1116 pp.

H. A. Gleason (1952). *Illustrated Flora of the Northeastern United States and Adjacent Canada.* 3 vols. Lancaster Press, Lancaster, Pa.

A. S. Hitchcock (1950). *Manual of the Grasses of the United States.* 2nd ed. U.S. Government Printing Office, Washington, D.C., 1051 pp.

J. K. Small (1933). *Manual of the Southeastern Flora.* University of North Carolina Press, Chapel Hill, N.C., 1554 pp.

We have noted that through the use of such manuals one could, with sufficient knowledge and skill, identify any unknown plant. This is sometimes difficult, even for an expert. For the agriculturist, cattle rancher, conservationist, tomato grower, or orchidist struggling with forage plants, poisonous plants, or weeds, the task of identifying plants is likely to be difficult and time-consuming.

Specimen plants, stored in herbaria, make identification of unknown plants easier and more definite. Such specimens have been identified by specialists; sometimes they must be sent long distances to these storehouses of information about plant classification. Here, unknown plants may be accurately identified. Herbaria are generally associated with universities, botanic gardens, and certain government agencies. Several of the larger herbaria are listed below.

Herbarium	Number of Specimens
Royal Botanic Gardens, Kew, England	6,500,000
Komarov Botanic Institute, Leningrad, USSR	4,500,000
Conservatoire et Jardin Botaniques de Geneve, Switzerland	4,000,000
U.S. National Herbarium, Washington, D.C.	3,000,000
New York Botanical Garden, New York	3,000,000
Missouri Botanical Garden, St. Louis	2,200,000
National Herbarium, Melbourne, Australia	1,500,000

BOTANIC GARDENS

Botanic gardens are of particular interest because they store the plants themselves (Fig. 2.16). Furthermore, these gardens are not confined to the dissemination of knowledge about plants, but are also concerned with dissemination of the plants themselves. Some of the great botanic gardens of today, like the Kew Gardens in En-

Figure 2.16
Botanic gardens are storehouses and disseminators of information. *A*, the extensive collection of monocotyledons at the Brooklyn Botanic Garden is displayed in a large and attractive border planting. *B*, The Childrens' Garden at the Brooklyn Botanic Garden gives practical information about plants to the children of the neighborhood. (*A*, courtesy of Brooklyn Botanic Garden; *B*, courtesy of Mr. Bertram Wiener.)

gland, have a long history of distributing valuable crop and ornamental species to areas far from their native habitats.

Collecting and distributing valuable plants has sometimes involved questionable activities such as smuggling. For example, by the mid-nineteenth century, rubber had become a valuable, desired plant product. The best rubber trees (*Hevea brasiliensis*) were native to Brazil, and the Brazilian government definitely wished to maintain its monopoly. Export of rubber tree cuttings or seeds was strictly forbidden.

The India Office of England, however, was determined to initiate rubber plantations in the English colony of Ceylon. In 1876 it secretly commissioned an English rubber worker in Brazil named H. A. Wickham to smuggle out rubber tree seeds, agreeing to pay him $50 per thousand. Wickham proved to be very efficient and soon collected some 70,000 seeds, packing them in covered baskets which he labeled simply, "Botanical specimens for Her Majesty's Gardens at Kew." He eventually found a willing ship captain with available space and slipped the seeds past the customs office at Belem without incident.

Once in England, he turned the seed over to botanists at Kew Gardens, who carefully germinated and nurtured 2400 seedlings from the 70,000 seeds. Most of these plants were shipped within a year to plantations that had been made ready in Ceylon. Wickham had his picture taken in a Ceylon plantation in 1912, when it was producing a princely $18,000 worth of rubber per acre per year! Not surprisingly, he was knighted in 1920. Today, 95% of the world's rubber comes from the Far East. Other plantations have been started in Africa.

Botanic gardens serve as a definite link in the distribution of knowledge about plants to the general public. In addition to actual displays of growing plants, informative lectures and booklets dealing with plants are available to the interested and advanced amateur.

Among the oldest and best-known botanic gardens in the world (with the dates of establishment in parentheses) are the following:

Royal Botanic Garden of Padua, Padua, Italy (1545)
National Museum of Natural History, Paris, France (1635)
Royal Botanic Garden, Edinburgh, Scotland (1670)
University Botanic Gardens, Cambridge, England (1762)
Botanic Garden of Harvard University, Cambridge, Mass. (1907)
Botanical Garden of Rio de Janeiro, Brazil (1808)
Botanic Gardens of New South Wales, Sydney, Australia (1816)
Government Botanic Gardens, Buitenzorg, Java (1817)
U.S. Botanic Garden, Washington, D.C. (1820)
Royal Botanic Gardens, Kew, England (1841)
Missouri Botanical Garden, St. Louis, Mo. (1859)
The New York Botanical Garden, New York City (1895)
Brooklyn Botanic Garden, Brooklyn, N.Y. (1910) (Fig. 2.16)

These gardens have plantings of species from all parts of the world, arranged by plant families or according to habitats. They contain outdoor plantings and often greenhouses controlled to grow tropical plants. Their reference libraries may contain thousands of volumes of books, pamphlets, and photographs, and the herbaria may be extensive.

BOTANICAL JOURNALS

Today, the greatest increase in our knowledge of plants and animals comes from the biology, botany, and zoology departments of many universities. Since most of this work receives support from public funds, it must be available to all who desire to know about it. As a consequence, this new information is published in some 2000 scientific journals. A number of examples of such journals is given in Table 2.5.

Table 2.5
REPRESENTATIVE JOURNALS PUBLISHING THE RESULTS OF BOTANICAL RESEARCH

Library of Congress Call Number	Journal	Published by	Country of Origin	Date Established
I. General				
QK 1 A55	*American Journal of Botany*	Botanical Society of America	U.S.	1914
QK 1 A65	*Annals of Botany*	Oxford University Press	England	1887
QK 1 C35	*Canadian Journal of Botany*	National Research Council, Canada	Canada	1922
QK 1 D4	*Deutsche botanische Gesellschaft, Berichte*	Deutsche botanische Gesellschaft	West Germany	1882
QK 1 F5	*Flora*	Gustav Fischer Verlag	East Germany	1817
QK 1 J3	*Japanese Journal of Botany*	National Research Council of Japan	Japan	1923
QK 1 L45	*Linnean Society of London, Journal of Botany*	Linnean Society of London	England	1856
QK 1 T6	*Bulletin Torrey Botanical Club*	Torrey Botanical Club	U.S.	1870
QK 1 E3	*Economic Botany*	Society for Economic Botany	U.S.	1947
QK 1 R8	*Doklady Botanical Science*	Academy of Sciences of the USSR	Russia	1921
II. Cells and Cellular Biology				
QK 1 R38	*Revue de Cytologie et de Biologie Végétales*	Centre National de la Recherche Scientific	France	1937
QH 301 C5	*Chromosoma*	Springer-Verlag	Germany	1939
QH 581 A1J6	*Journal of Cell Biology*	Rockefeller University Press	U.S.	1955
QH 591 P6	*Protoplasma*	Springer-Verlag	Austria	1927
QH 301 C9	*Cytologia*	Wada-Kunkokae Foundation	Japan	1929
III. Taxonomy				
QK 1 R5	*Rhodora*	New England Botanical Club	U.S.	1899

Table 2.5
REPRESENTATIVE JOURNALS PUBLISHING THE RESULTS OF BOTANICAL RESEARCH *(continued)*

Library of Congress Call Number	Journal	Published by	Country of Origin	Date Established
QK 1 B675	*Brittonia*	American Society of Plant Taxonomists	U.S.	1931
QK 1 P465	*Phyton*	Fundacion Romulo Raggio	Argentina	1948
IV. Ecology				
QH 301 J6	*Journal of Ecology*	British Ecological Society and Blackwell Scientific Publications	England	1913
QH 301 E34	*Ecology*	Ecological Society of America	U.S.	1920
QK 901 A1V4	*Vegetatio*	Association of International de Phytosociologie	Netherlands	1948
V. Inheritance and Evolution				
QH 301 G4	*Genetics*	Genetics Society of America	U.S.	1916
QH 301 H4	*Hereditas*	Nordick Genetikertörening	Sweden	1920
QH 301 E9	*Evolution*	Society for the Study of Evolution	U.S.	1946
VI. Plant Physiology				
QK 1 P48	*Plant and Cell Physiology*	Japanese Society of Plant Physiologists	Japan	1959
QK 1 P5	*Plant Physiology*	American Society of Plant Physiologists	U.S.	1926
QK 1 P452	*Physiologie Végétale*	Gauthier-Villars	France	1963
VII. Morphology, Including Algae and Fungi				
QK 600 M9	*Mycologia*	Mycological Society of America	U.S.	1909
QK 564 J6	*Journal of Phycology*	Phycological Society of America	U.S.	1965
QK 564 P46	*Phycologia*	International Phycological Society	U.S.	1961
QK 534 B7	*Bryologist*	American Bryological Society	U.S.	1898

CHAPTER 3

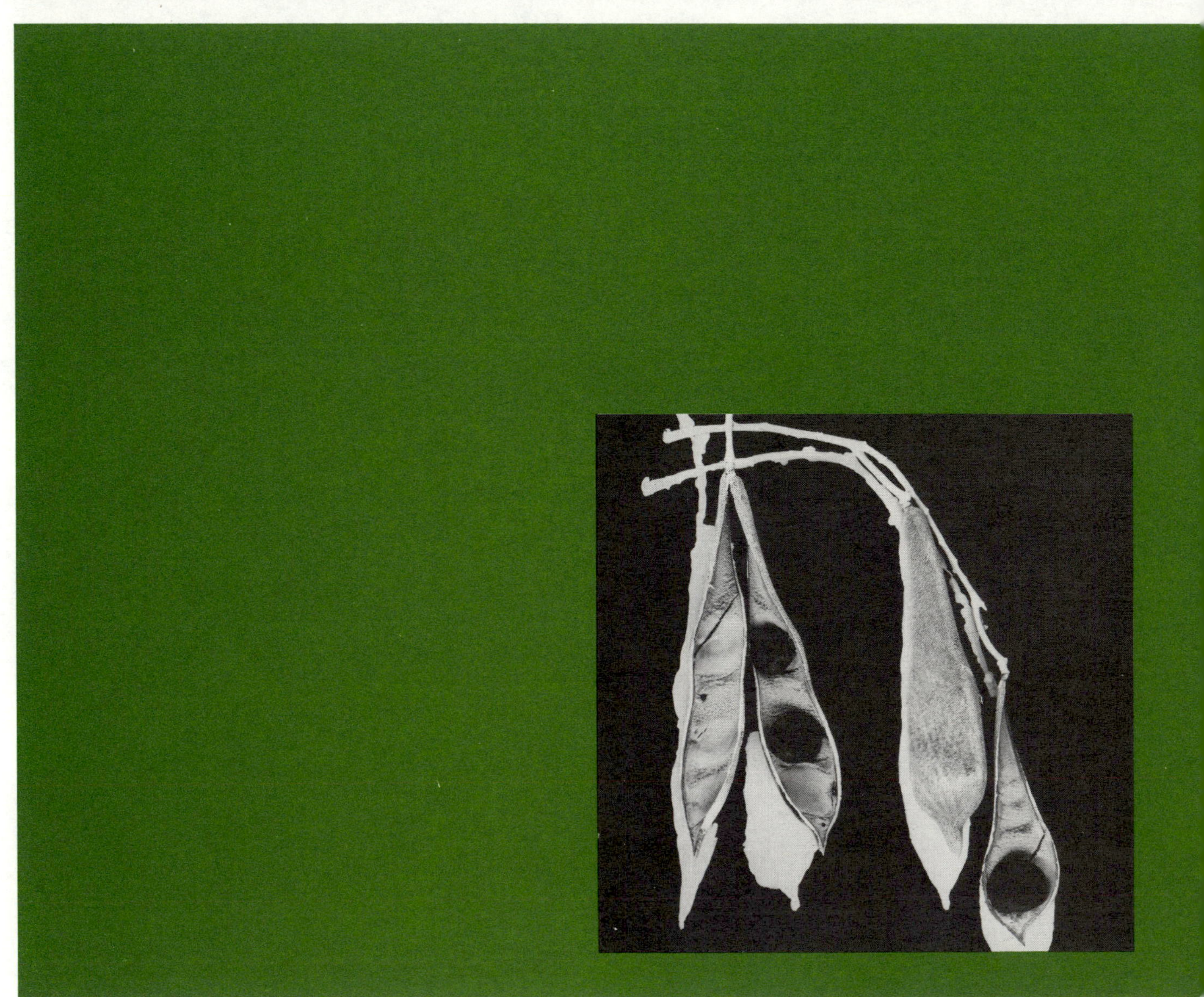

THE PLANT BODY OF SEED PLANTS

In this chapter we shall give attention to the general structures and functions of a familiar seed plant, the common garden bean. Any one of many thousands of seed plants would serve our purposes as well. It is recognized that seed plants are very complex organisms, as compared with unicellular, filamentous, and platelike plant bodies or even multicellular plants, like the mushrooms. Other types of plants have different types of plant bodies (Figs. 2.1, **2.8,** 2.9). They will be discussed in some detail after the structure and function of the plant body of seed plants is understood.

The bean seed is borne in a pod on a parent plant. When the bean seed is removed from the pod, under suitable growing conditions, it germinates and a young plant (seedling) emerges (Fig. 3.1*D*) having roots, stems, and leaves. The plant *grows,* extends its roots into the soil and its stems with their leaves into the atmosphere, and after a number of weeks attains adult size. Flowers are formed, then pods "set," and in the pods are seeds ("beans"). The bean plant has then completed a **life cycle.** Many of the physiological processes involved are complex and require a knowledge of chemistry and physics to be understood. A bean plant is an individual—it has a birth, a young life, an adult life, a period of reproduction, and then it dies. It has **organs** that perform the various functions necessary to maintain its life. These organs—**roots, stems, leaves,** and **the flower parts**—are composed of **tissues,** groups of cells that carry on different activities (Figs. 3.1*E, F,* 3.2). The cells are the small compartments that make up tissues; they possess living material **(protoplasm)** and hence are the *smallest living units of plant structure and function* (see Fig. 4.2).

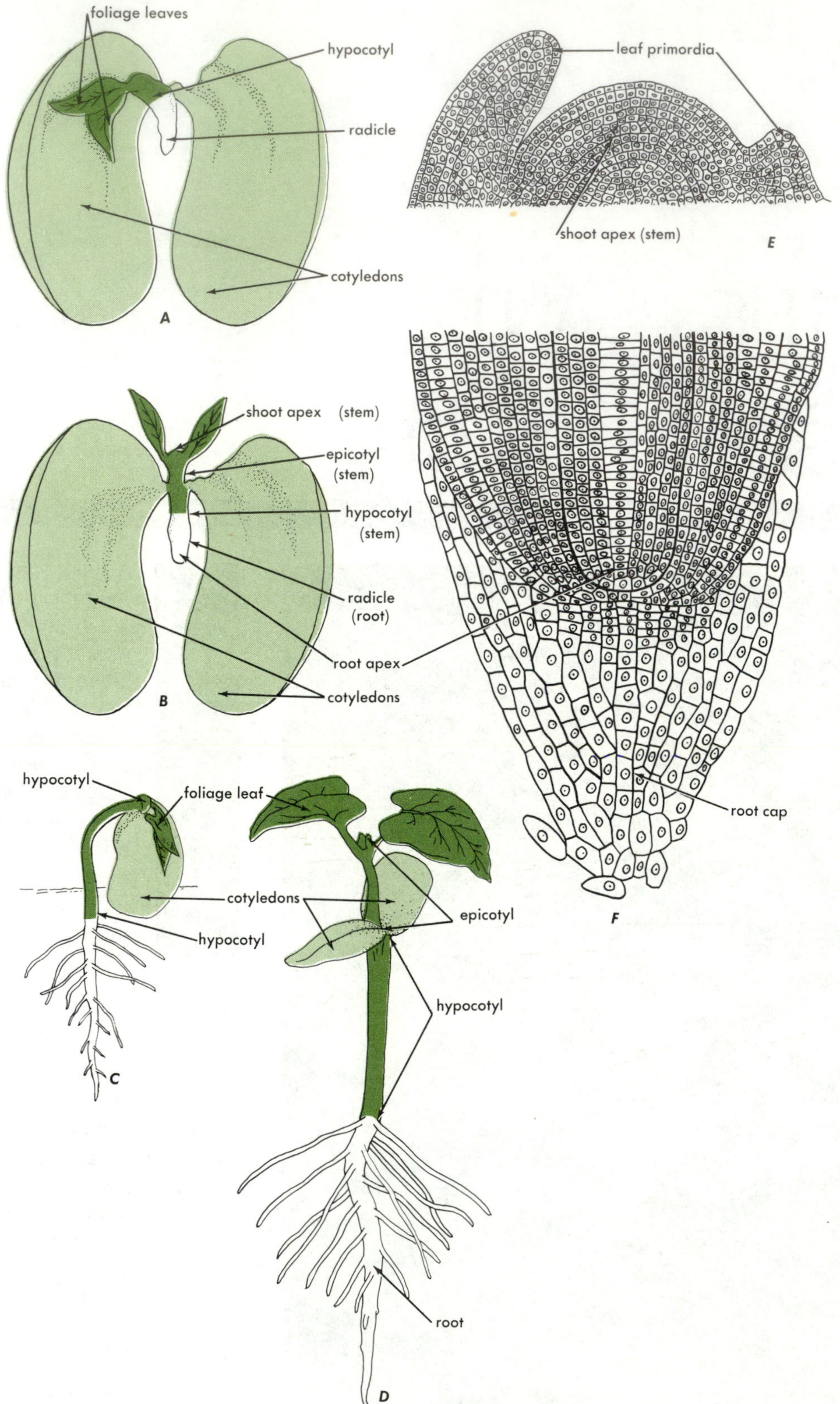

Figure 3.1
The body of a seed plant (common bean). *A*, dormant embryo showing natural position of the organs, ×4. *B*, embryo straightened to show the axis of the plant body and the relative position of the organs, ×4. *C* and *D*, stages in the development of the plant body, ×1. *E*, shoot apex, ×100. *F*, root apex, ×100. (*A, C, D* redrawn from R. M. Holman and W. W. Robbins, *A Textbook of General Botany*, John Wiley & Sons, New York.)

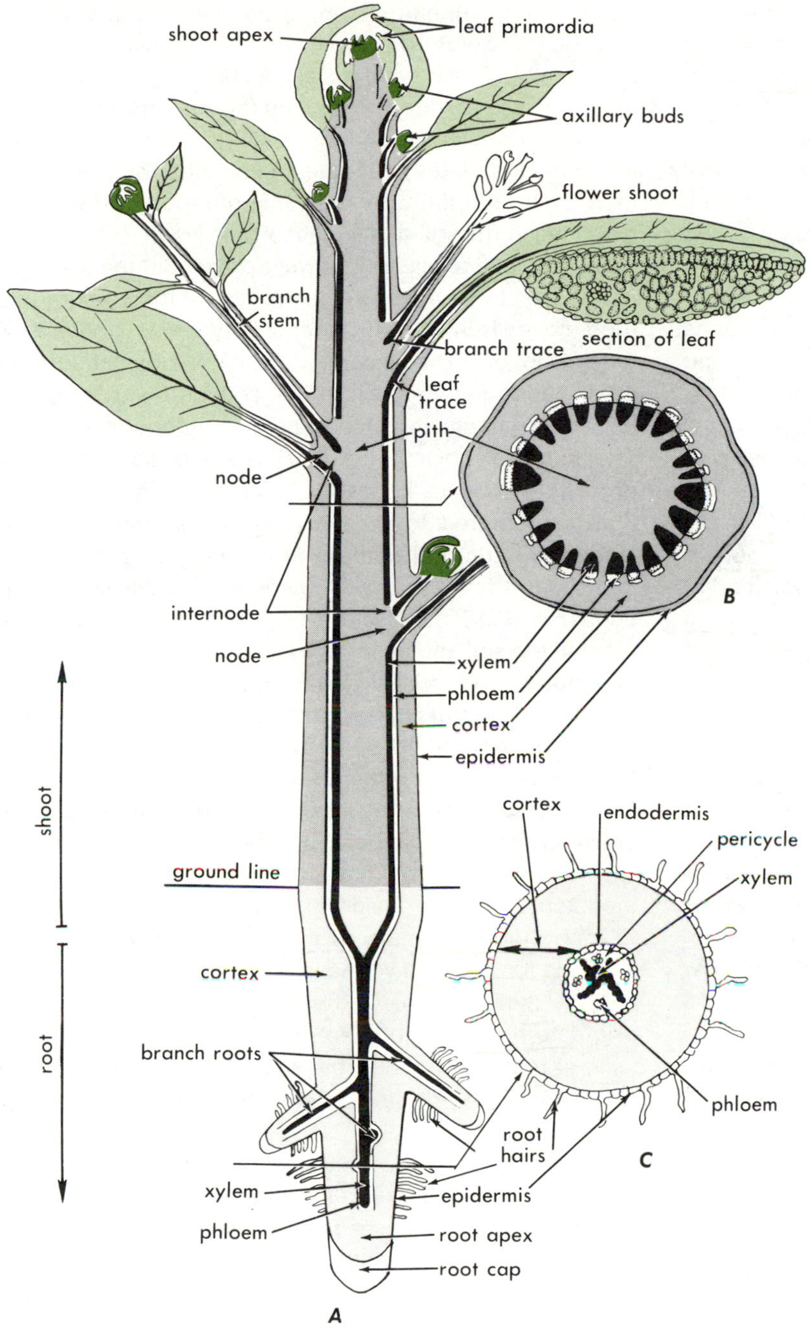

Figure 3.2
A, the principal organs and tissues of the body of a seed plant. *B*, cross section of stem. *C*, cross a section of root. (*A*, redrawn from R. M. Holman and W. W. Robbins, *A Textbook of General Botany*. John Wiley & Sons, New York.)

A plant body like that of the bean (Fig. 3.1*D*) possesses all the organs and tissues necessary to maintain life. The stem (*a*) supports the foliage leaves and flowers, (*b*) conducts water and mineral salts from the roots, (*c*) conducts plant foods from tissues where they are manufactured to tissues where they are needed for growth or where they are stored for future use, and (*d*) may store foods. The roots (*a*) anchor the plant in the soil, (*b*) absorb water and mineral salts from the soil, and (*c*) may store food. The leaves (*a*) manufacture foods and (*b*) give off water in the form of water vapor. These activities involve both chemical and physical processes, which will be discussed in later chapters.

One cannot have a clear understanding of the structures and functions of a plant without a knowledge of its anatomy, such as is revealed by the use of a compound microscope. All-important at the very beginning is an understanding of cells—the units of plant structure and function. Chapters 4 to 6 deal with the cell.

The gross development of the seed plant body is

shown in Figs. 3.1*A* to *D*. After a bean seed is soaked in water for several hours, the outer skin or seed coat is easily removed. The body exposed is the **embryo** (Figs. 3.1*A*,*B*). The term "embryo" as used in biology refers to a living organism in the early stages of development. In most plants and the higher animals, the embryo is more or less dependent upon the maternal parent for its food. The bean embryo is a very young plant. There are two large **cotyledons** or seed leaves, several young **foliage leaves,** a **stem** (composed of **epicotyl** and **hypocotyl**), and a **rudimentary root (radicle).** The two cotyledons contain stored food consisting of protein, lipid, and starch. Although the young foliage leaves are small, their veins can be easily seen with a lens. Figures 3.1*A*,*B* show the bean embryo spread open so that the relationship of the different parts is more easily observed.

The embryo plant body has two regions of importance to future growth—the **shoot tip** and the **root tip** (Figs. 3.1*E*,*F*). The shoot tip includes (*a*) at the very end of the young plant, between the rudimentary leaves, a dome-shaped mass of undifferentiated cells constituting the shoot apex and (*b*) cells directly below the apex that are enlarging and in early stages of elongation and differentiation. The root tip, at the lower end of the young plant, is composed of similar tissues: (*a*) the root apex of undifferentiated cells, (*b*) the elongating and differentiating cells above the root apex, and (*c*) a root cap protecting the root apex. The undifferentiated cells that make up these two tips are capable of **division, differentiation,** and **growth. Meristematic cells** are cells capable of active cell division with a resulting increase in the number of cells. Until the death of the plant, there are always meristematic tissues at the tips of stems and roots.

As long as they are alive, all seed plants contain localized regions of meristematic tissue, enabling them to grow and differentiate new stems and roots. Indeed, we shall see that a continual growth of new roots is essential for continued absorption of water and nutrients. Plants are thus said to have open growth, and it has been demonstrated that many fully differentiated individual plant cells contain a full genetic complement and are able to give rise, through mitosis, to new individuals. Animals, on the other hand, do not have meristematic tissue; only the undifferentiated cells of embryos are able to divide and to differentiate into specialized cells. With but few exceptions, mature animal tissue can only produce new cells similar to those of their own tissue type. Moreover, while well-fed animals may increase somewhat in girth, they do not *continually grow* in girth or in length as does a tree or vine.

During the early stages in the development of the seedling bean plant, the cells are nourished by food stored in the cotyledons. Before this food moves from the cotyledons to other parts of the plant, it is changed to sugar. Sugar, unlike starch, is soluble in the water of cells, and in solution it moves from the cotyledons to all other cells of the embryo, especially to those actively growing cells of the shoot tip and the root tip. As soon as the cotyledons are raised into the light by the lengthening of the shoot, they become green. Then they begin the manufacture of food. By this time, the roots are well-established in the soil and are absorbing water and mineral salts. The foliage leaves that were present in the seed expand; they, too, engage in the process of foodmaking. As food moves from the cotyledons, they shrivel and after a time drop off. The seedling, with its roots absorbing water and mineral salts from the soil, and with its green leaves manufacturing food, is now an *independent plant.* The plant, as shown in Fig. 3.1*D*, is equipped with all the organs and tissues necessary to carry on the processes that are required to keep it alive and growing. At this stage the structural framework of the adult plant is established. It is an axis with meristematic tissue at both ends (Fig. 3.2). The lower portion of the axis anchors the plant in the soil and absorbs water and mineral nutrients. The upper end of the axis produces leaves for photosynthesis, and flowers and fruits for reproductive purposes. The central portion of the axis supports the photosynthetic and reproductive parts in the light and air of the biosphere and serves as a communication pathway between the absorbing and synthetic ends of the axis.

The axis may branch. Branch roots arise from certain meristematic cells inside the main root (Fig. 3.2*A*). Branch shoots originate in undifferentiated cells found in an angle formed by a young leaf at the point of its attachment to the stem. This is the leaf **axil** (Fig. 3.2*A*). Leaves originate from the undifferentiated cell or meristematic cells at the upper end of the shoot or shoot apex (Figs. 3.1, 3.2*A*). These undeveloped leaves are known as **leaf primordia.**

A diagram of a seed plant (Fig. 3.2) will serve to summarize the foregoing discussion and illustrate the structural organization of a seed plant. The plant body is made up of two parts: the shoot and the root. The shoot is composed of two kinds of organs—stems and leaves. In the seedling there are two kinds of leaves, the cotyledons or seed leaves, which are temporary in most species, and the foliage leaves, which are relatively longer-lasting. The stem, in contrast to the root, is divided into **nodes** and **internodes,** regions that alternate throughout the length of the stem. Nodes are the slightly enlarged portions where leaves and buds arise and where branches originate from the buds. An internode is the region between two successive nodes. As we shall see later, the anatomy of nodes and internodes differs. In Fig. 3.2, structures and tissues are shown that will be discussed in later chapters.

CHAPTER 4

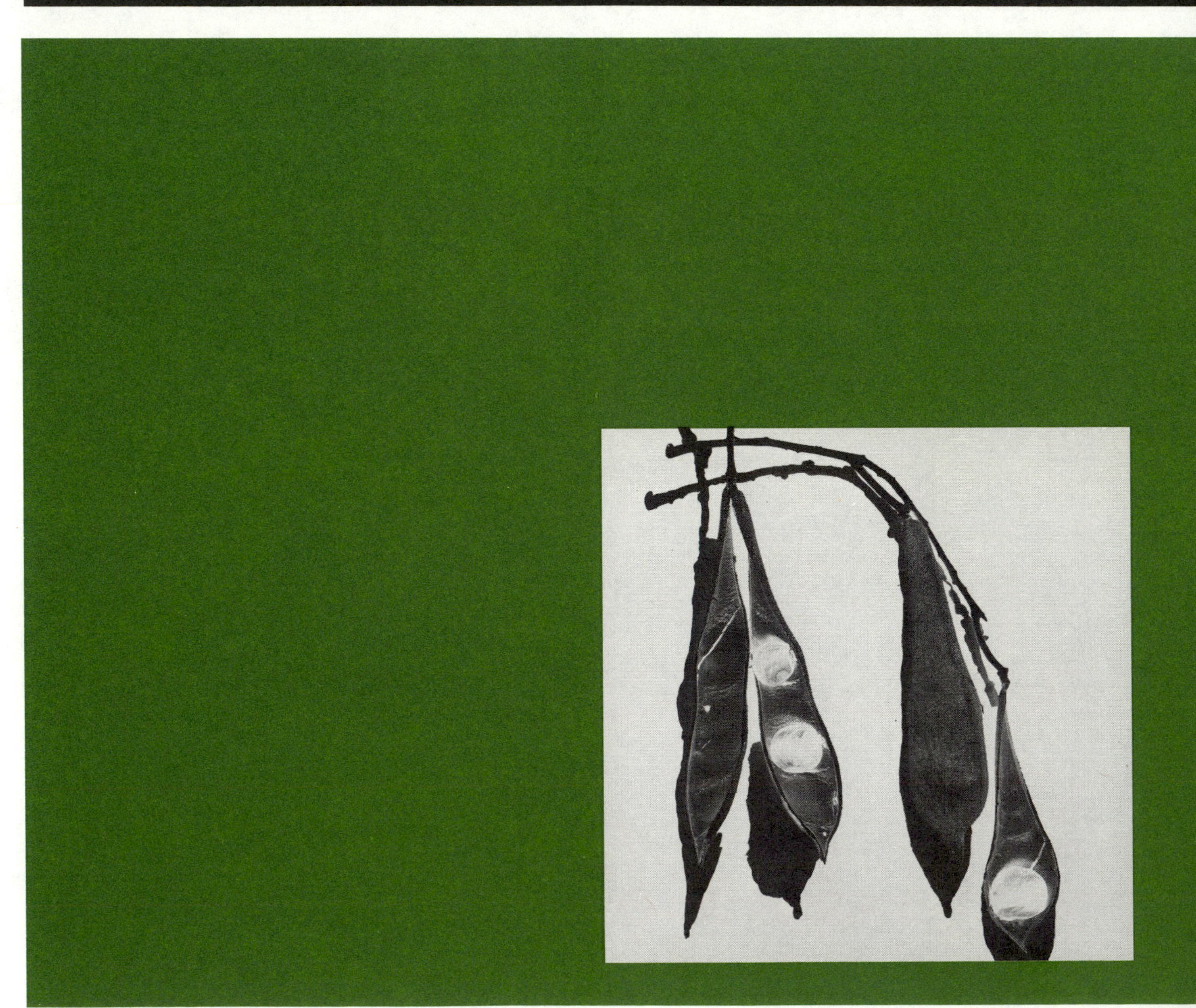

THE METABOLIC PLANT CELL

Galen, the last of the great Greek doctors, who lived in Asia Minor during the second century AD, thought that all tissues, such as spleen, brain, and kidney of animals, leaves, stems, and roots of plants, were a "sensible element, of similar parts all through, simple and uncompounded." Others before him had thought that animal tissues were simply coagulated "juices" seeping through the walls of the intestine. No one dreamt that these tissues and their plant counterparts had a fabulous and complicated structure, a structure that could not be seen until the invention of a microscope, which Zacharias Jansen, a spectacle maker of Holland, accomplished in 1590. This instrument was later improved by Robert Hooke, an Englishman, who was not only interested in optics but was also an architect and an experimenter with flying machines. Hooke, who lived from 1636 to 1703, examined all sorts of natural objects with his improved microscope. Among these were thin slices of cork (the dead outer bark of an oak). Figure 4.1 is an illustration of the cork tissue as Hooke saw it under his microscope. This figure was published in 1664 in an article entitled *Micrographia, or Some Physiological Descriptions of Minute Bodies Made by Magnifying Glasses.* The term **cell** was first used by Hooke to denote in cork "little boxes or cells distinct from one another . . . that perfectly enclosed air." He estimated that a cubic inch of cork would contain about 1259 million such cells.

Because of the prominence of cell walls in plant tissue, the cell early became the unit of structure and of life for botanists. However, during these early years, zoo-

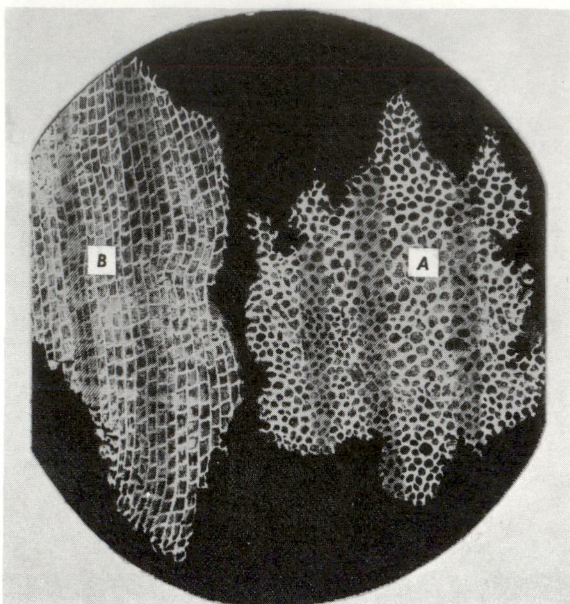

Figure 4.1
Cork tissue as Robert Hooke observed it under his microscope. (From Hooke, *Micrographia* (1664), The Council of the Royal Society of London for Improving Natural Knowledge.)

logists considered the tissues as the true centers of life of the body. There were supposed to be 21 different tissues, able to change within themselves, depending upon the organ in which they were located. Life was thought to reside in these tissues. However, numerous observations of protozoa and animal tissues led to a gradual accumulation of evidence that cells also existed in animals. In 1838 the German zoologist Theodor Schwann and the Belgian botanist Matthias Jacob Schleiden collaborated in a paper entitled, *Microscope Investigations on the Similarity of the Structure and Growth in Animals and Plants.*

This paper established on a firm basis *the theory* that the cell is a basic unit of structure *in both plants and animals.* However, another 30 years of research were necessary for the general acceptance of such a new idea and for the establishment of *the fact* that *all organisms are composed of cells,* that cells are indeed the basic units or building blocks of life. Cells are maintained by the division of cells. Thus, meristematic cells of shoot apices (see page 116) have a direct line of descent going back to the first cellular units that evolved billions of years ago.

CELL CONCEPT

Cells may be considered from the viewpoint of their (*a*) structure (Chap. 4), (*b*) activity (Chap. 6), or (*c*) chemical organization (Chap. 5). Knowledge of the interrelationships of all three approaches is necessary for an understanding of the cell as a whole.

Cells are dynamic living units maintained in equilibrium with their surroundings only through the expenditure of energy. Cells are continually renewing their substance. Cells divide. To keep pace with the increasing number of cells, the mass of protoplasm must also increase. Cells also accumulate materials to be used in their own metabolism.

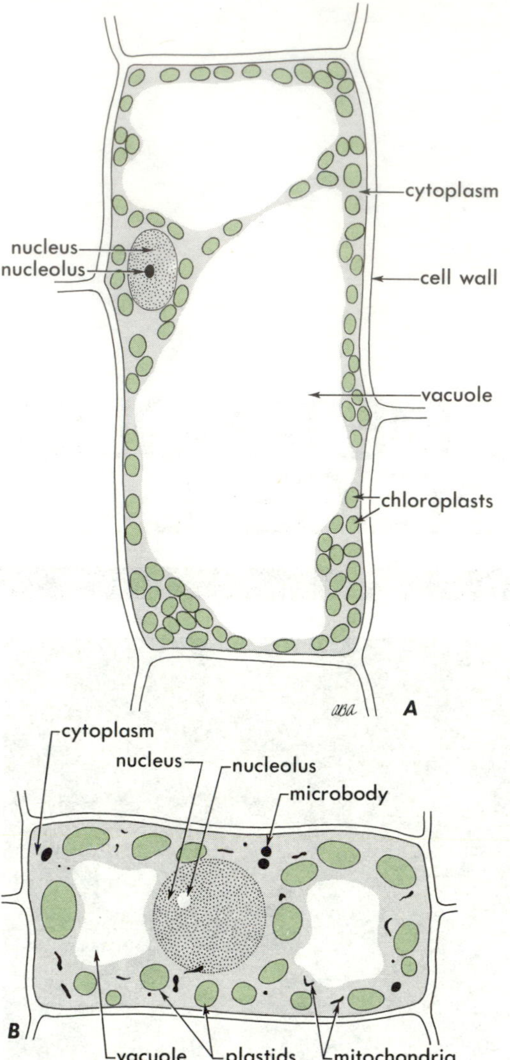

Figure 4.2
Plant cells. *A*, diagram of a living cell from an *Elodea* leaf. When viewed on the light microscope, organelles in living cells, other than chloroplasts, lack contrast. The nucleus and nucleolus are generally visible. Sometimes one of the categories of microbodies may be seen. Mitochondria are seen only with perfect lighting conditions, × 700. *B*, a cell from a young leaf of *Elodea*, fixed to show plastids and mitochondria, × 500.

The products, or accumulations, of protoplasmic activity (water, sugar, cellulose, hormones, and similar items) are found in the dynamic cooperating unit of protoplasm, called the **protoplast.** The protoplast of a plant cell secretes about itself a **cell wall** of cellulose, which frequently later receives a deposition of secondary products (Fig. 4.2). All of the protoplast, plus the cell wall, is the **plant cell.**

The activities of whole cells may be studied. Their

ability, through the expenditure of energy, to accumulate materials in greater concentration than occurs in the surrounding media is a subject of considerable interest. Subcellular structures may be isolated from cells, and their activities or chemical characteristics may be studied in test tubes.

Whatever the approach to the study of cells, it is important to remember that energy is required to maintain a living cell as an individual entity. The maintenance of a steady living state, or equilibrium, through the expenditure of energy is known as **homeostasis.** Disruption of the source of the necessary energy must result in the death of the cell.

We have already noted the existence of two fundamentally distinct types of cells, the prokaryotic cells that characterize more primitive organisms, and the eukaryotic cells of all other animals and plants. A detailed consideration of the prokaryotic cell will be reserved for a later chapter. The following discussion is concerned mainly with the structure of the eukaryotic plant cell (Fig. 4.2).

In eukaryotic cells, many of the metabolic processes of the cell are separated and occur within discrete membrane-bound, organized, protoplasmic structures, called **organelles.** Each kind of organelle has a characteristic structure and performs the same general functions wherever it occurs, whether in plant or animal cells. *This is one of the important unifying facts of biology.* The chart below lists the major structural components of the plant cell and shows the relationships among them. A more detailed list of the parts of the cell is given in Table 4.1. Also included are the dates when the parts were first observed by the light microscopist (**boldface** type) and by the electron microscopist, the approximate dimensions of the parts, and a notation as to their function.

Table 4.1
THE PARTS OF THE CELL[a]

Cell Part	Dimension	Function
Plant cell, **1880**		
Primary cell wall, **1665**	2–5 μm	Protection, strength
Cellulose microfibrils, 1948	10–25 nm (indefinite length)	Mechanical strength
Cellulose molecules, 1922 (x ray), 1949	0.834 × 0.8 nm	Strength
Amorphous matrix		
Middle lamella	2 μm	Adhesion between cell walls
Protoplast, **1846**	0.025–2 mm	Homeostatic unit
Protoplasm, **1840**		Synthetic and developmental reactions under direction of nucleus
Cytoplasm, **1882,** 1957		
Chloroplast, **1702,** 1953	2–20 μm	Photosynthesis
Chromoplast, **1900,** 1958	2–10 μm	Accumulation of carotenes and similar pigments
Amyloplast, **1884,** 1955	2–25 μm	Starch storage
Leucoplast, **1883,** 1957	2–10 μm	
Mitochondrion, **1897,** 1947	2 × 2–10 μm	Respiration
Dictyosomes, **1927,** 1956	3 μm	Enzyme synthesis
Endoplasmic reticulum, 1957	17 nm (indefinite length)	Protein synthesis
Ribosomes, 1955	20 nm	Protein synthesis
Spherosomes, **1919,** 1967	2 μm	Lipid synthesis
Microbodies, 1965	0.1–2.0 μm	Compartmentalization of various enzymes
Microfibrils, 1959	28 nm (indefinite length)	Various functions
Tonoplast, **1877,** 1960	8 nm	Regulation of exchange between vacuole and cytoplasm
Plasmalemma, **1880**(?), 1954	8 nm	Regulation of exchange between cytoplasm and external solution
Crystals, 1963	10 μm	Unclear, possibly storage
Plasmodesmata, **1879,** 1957	2 μm	Protoplasmic bridge between cells
Nucleus, **1831**	5–30 μm	Contains genetic information necessary for normal cell development and activity

Table 4.1
THE PARTS OF THE CELL (continued)

Cell Part	Dimension	Function
Nuclear envelope, **1907,** 1955	25 nm	Separates nucleoplasm from cytoplasm
Nucleoplasm, **1879**		
Chromatin, **1880**		
Nucleoproteins, **1869**		
Nucleic acids, **1889**		
DNA, **1924**		
DNA helix, 1953	1.8 nm (indefinite length)	Carries genetic code
Unit fibers, 1963	12.5 nm (indefinite length)	Encompasses DNA helix and nucleoproteins
Nucleolus, **1882,** 1958 (animal, 1952)	1–5 μm	RNA synthesis
RNA		Transferal of information from DNA to cytoplasm
Ribosomes, 1956	15 nm	Protein synthesis
Chromosome, **1888,** 1955	2–200 μm	Vehicle carrying DNA helix in replication and distribution
Kinetochore		Region of chromosome to which spindle fibers are attached
Centromere, **1925**		
Chromatid, **1900**	1–10 μm	One-half of a chromosome
		Cytoplasmic structure involved in moving chromosomes
Fibers, **1881,** 1960	Various indefinite lengths	During mitosis
Microtubules, 1960	28 nm (indefinite length)	
Equatorial plate, **1875,** 1958	Various	First stage in separation of derivative cells
Vacuole, **1835,** 1957		Various functions important in water economy of cell
Water		
Inorganic salts		
Various organic solutes		
Crystals		

[a] Date in **boldface type** refers to the year first reported with the light microscope; date in lightface type indicates the year first seen with the electron microscope.

Although eukaryotic plant and animal cells are remarkably similar in having, in general, similar organelles, they differ in several basic ways:

(a) Plant cells specialized to carry out photosynthesis have one to several **chloroplasts** (see Fig. 4.14), organelles that contain enzymes, pigments, and intermediate substances required for photosynthesis. Chloroplasts are found in green plant cells but not in animal cells.

(b) With few exceptions, plant cells have a cell wall composed of **cellulose,** which frequently may contain secondarily deposited materials. Cell walls give greater rigidity and lessened mobility to plants.

(c) Most plant cells develop a large central aqueous **vacuole** (see Fig. 4.22).

(d) Synthesis of materials may occur at very high rates in some animal cells, for example in gland cells, where the organelles involved in synthesis (endoplasmic reticulum and dictyosomes) are especially numerous.

TECHNIQUES OF STUDYING CELL STRUCTURE

Units of Measurement

At this point a consideration of a few measurements is in order (Table 4.1). We have already stated that Robert Hooke calculated that there would be 1259 million cells in a cubic inch of cork. This figure is based on his estimate of about 1100 cork cells along a line 1 inch long. One thousand cells along 1 inch would give us about 40 cells for the length of 1 mm, so a single cork cell is $\frac{1}{40}$ mm in diameter. It now becomes convenient to use a smaller unit of measurement, the micrometer (μm) (Table 4.2). There are 1000 μm in 1 mm, and a cell $\frac{1}{40}$ mm in

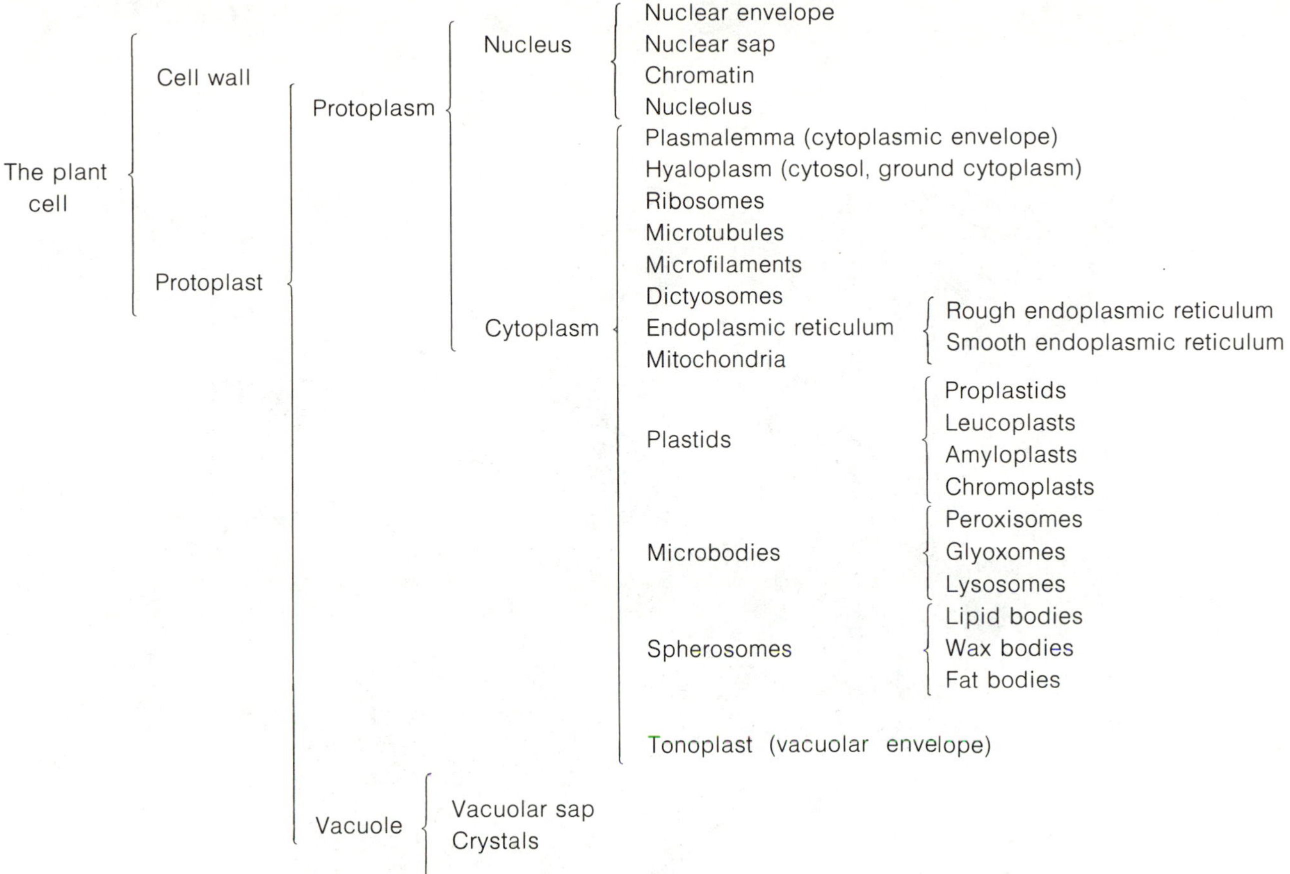

diameter would be 25 μm in diameter. This is about correct for a cork cell, although many plant cells are larger. With modern techniques of electron microscopy it is now possible to study the detailed ultrastructure of cells. To do this electron beams are used. Such an electron beam may have a wave motion with a distance of 0.0000005 μm from crest to crest. Such small distances call for a still shorter unit of measurement. This is the nanometer, abbreviated nm.* There are 1000 nm in 1 μm.

Light Microscopy

Living cells maintained in the proper medium may be observed for many hours with the light microscope. In a living leaf of *Elodea* or a young bean plant (*Phaseolus vulgaris*), the chloroplasts are the most prominent organelles. They are embedded in the clear hyaline cytoplasm that is confined to the periphery of the cell, or to thin strands crossing the clear central vacuole (Fig. 4.2*A*). A single nucleus is embedded in the cytoplasm and may be seen flattened against one of the walls. In the living cell it generally appears structureless, although with careful focusing, one to four small denser bodies, the **nucleoli** (nucleolus s.), can be seen (Fig. 4.2*A*). In addition, small particles about 0.54 μm in diameter may be seen in the living cytoplasm. These may be microbodies or mitochondria (Fig. 4.13).

The cytoplasm in a leaf cell, especially when warmed by the light of a microscope, will show active streaming, carrying both the chloroplasts and the microbodies with it. With time, the velocity increases up to 5 to 10 mm per min. This rapid flowing of the cytoplasm is a dramatic indication of its plasticity, and should be recalled when studying microscope slides and electron micrographs, for it is sometimes difficult to reconcile the activity of flowing cytoplasm with the static and frequently complicated images seen in electron micrographs.

Because the protoplasm of the cell has an index of refraction close to that of water (the protoplasm bends light rays to the same degree that water does), many details of cell structure are not apparent in living cells. Enhancement of detail may be attained by studying living cells with phase contrast optics, Nomarski interference optics (Fig. 6.12), polarized light, or fluorescent light.

Still more detail is obtained by killing (fixing) cells in various mixtures of alcohol, acetic acid, chromic acid, formaldehyde, or osmium tetroxide. The latter two compounds are of particular interest, because in killing cells they do not coagulate the proteins. The killed tissue is dehydrated, embedded in paraffin, and cut into sections

* An angstrom (Å), the unit formerly used, was 1/10,000 μm. There are 10 Å in 1 nm.

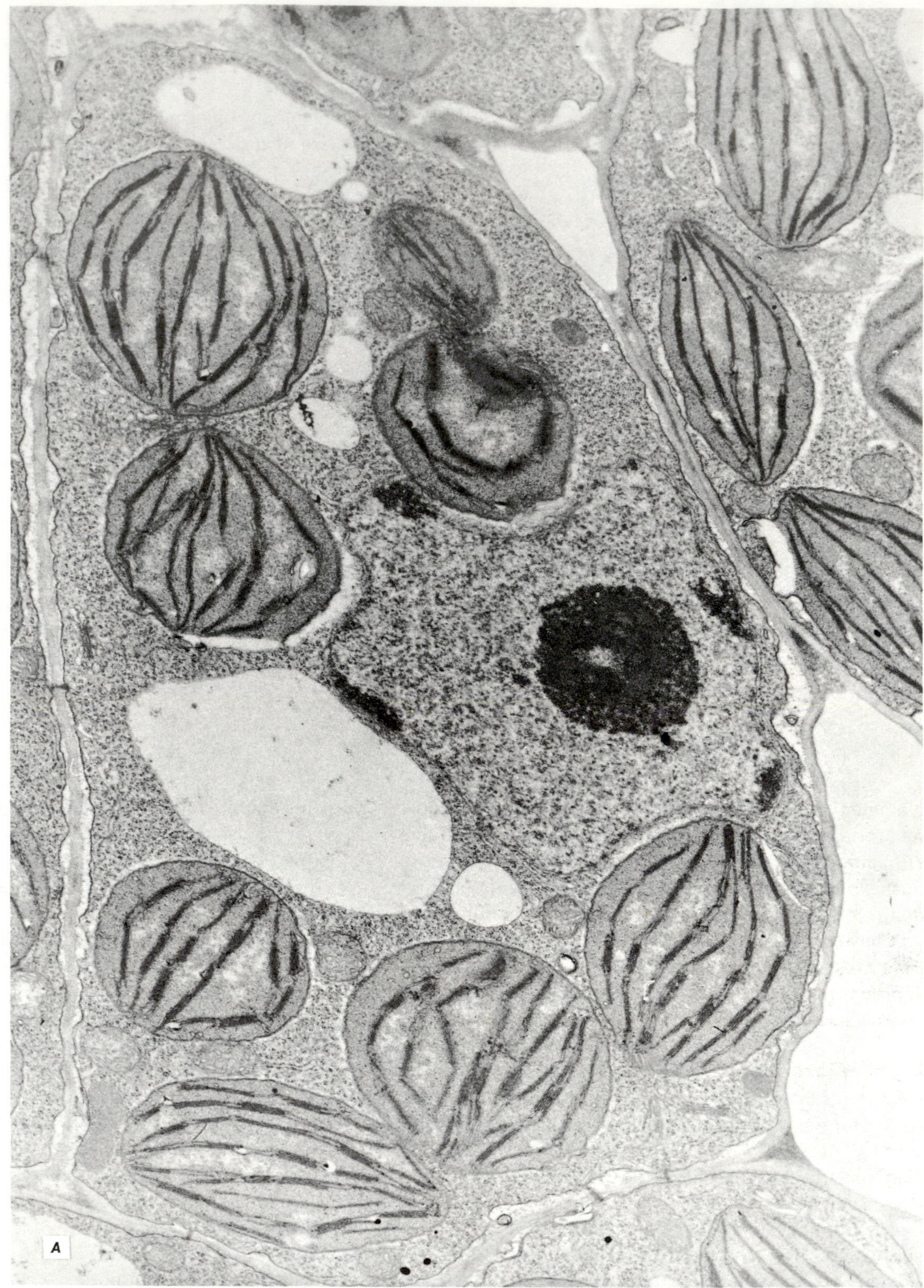

Figure 4.3

A cell from a leaf of a bean seedling (*Phaseolus vulgaris*). *A* is an electron micrograph; *B* is a diagram of *A*. Chloroplasts, nucleus, and vacuoles are same size as in *A*; other organelles are slightly enlarged or otherwise accentuated. Ribosomes not drawn. Original magnification of *A* is ×10,000.

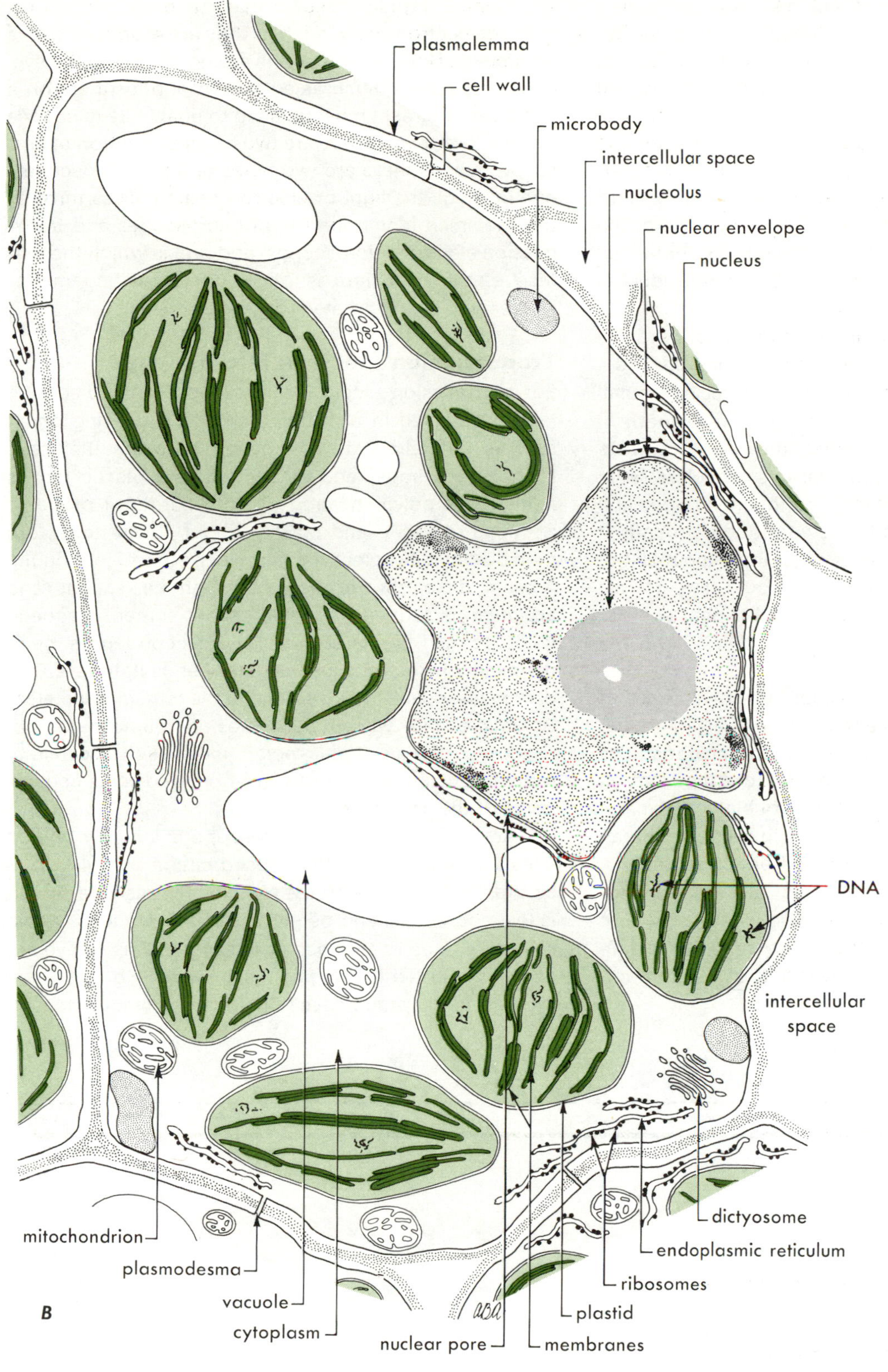
plasmalemma
cell wall
microbody
intercellular space
nucleolus
nuclear envelope
nucleus
DNA
intercellular space
dictyosome
endoplasmic reticulum
ribosomes
plastid
membranes
nuclear pore
cytoplasm
vacuole
plasmodesma
mitochondrion
B

from 5 to 20μm in thickness. When this tissue is mounted on a glass slide and properly stained, much detail becomes apparent. A variety of chemicals, including stains, are available that react with specific compounds that are present in the cell walls or in the internal contents of cells. These reaction products may give characteristic color reactions. Thus, the investigator has a tool that enables him not only to see the cell in considerably more detail, but also to determine the location of some of the chemical components of the cell (Chapter 5). Polarized light and fluorescent light may also be used to identify some cellular substances (Chapter 7).

Under the best conditions one can detect particles 0.25 μm in diameter with the light microscope. Just as a large ocean wave is not modified as it passes by a small post, particles smaller than one-half the wavelength of light that strikes them do not deflect the advancing wave front of light and, therefore, the particle cannot be seen. The shortest wavelengths of visible light are between 400 and 500 nm long (violet and blue light); consequently if one wished to detect particles smaller than about 200 to 250 nm (0.25 μm) in diameter one would have to use wavelengths shorter than those in visible light.

Electron Microscopy

With the short wavelengths (0.0005 nm) of a beam of electrons, resolution of distances as small as that separating atoms is theoretically possible. But the difficulties in making perfect magnetic lenses to focus the electron beam and the difficulties in the nature of biological material itself means that lines or particles in cells closer than 2 nm can not be distinctly separated from each other regardless of the magnification obtained. Because the best figure obtainable with the light microscope is 0.25 μm, or 250 nm, the electron microscopes allow more than a 100-fold increase in magnification over that of the light microscope.

Plants, or plant tissues, may be examined with a beam of electrons very much as they are examined with a light beam. However, electrons are not visible to the eye, and some device, such as activation of phosphors on a fluorescent screen, must be used to enable us to see the beam of electrons. There are two types of electron microscopes just as there are two types of light microscopes: one in which the light or electron beam passes through the specimen (compound light microscopes and transmission electron microscopes) and one in which the surface of the specimen is observed (dissecting microscopes and scanning electron microscopes).

Transmission Electron Microscopy

For transmission electron microscopy, small pieces of tissue are killed in a chemical killing and fixing agent such as glutaraldehyde and osmium tetroxide; the tissue is then dehydrated and embedded in a plastic that is heated in a polymerization oven to harden it properly. Sections of tissue and plastic, about 50 nm thick, are then placed in the electron microscope in a high vacuum. A beam of electrons is passed through the sections, and an image is formed on a fluorescent screen. Magnetic fields, formed by the electron microscope lenses, bend the beam of electrons just as glass bends light rays. The dark masses seen on the screen of the transmission electron microscope will not be images of organic matter but shadows of a high-atomic-weight metal such as osmium, which has become preferentially bound to some components of the protoplasm.

Actually, an electron microscope is a highly refined television set tuned onto a closed circuit (Fig. 4.4). The contrast in the image on the electron microscope screen is low and details can be studied best on photographic prints prepared to enhance contrast. A cell from a young bean leaf (*Phaseolus vulgaris*) is shown in Fig. 4.3. Since the cells at this stage have not elongated and contain all

Table 4.2
SOME DIMENSIONS[a]

	Micrometers	Nanometers
1 inch	25,000	2,500,000
1 mm	1,000	1,000,000
1 μm		1,000
Resolution of unaided eye	100–300	
Resolution of light microscope	0.25	250
Practical resolution of electron microscope for biological material		2
Wavelength of blue light	0.50	500
Wavelength of red light	0.70	700
Wavelength of electron beam 75 KV (KV is 1000 volts)		0.05
Diameter of cork cell	25	
Volume of cork cell	15,625 μm^3	
Diameter of leaf cell	200–400	
Diameter of vessel in xylem	500	
Diameter of *Scenedesmus*	10	
Length of *Scenedesmus*	30	

[a] One foot magnified 100,000× is nearly 19 miles; 1 inch magnified so each angstrom in it measures 1 inch in length appears about 46,000 miles long.

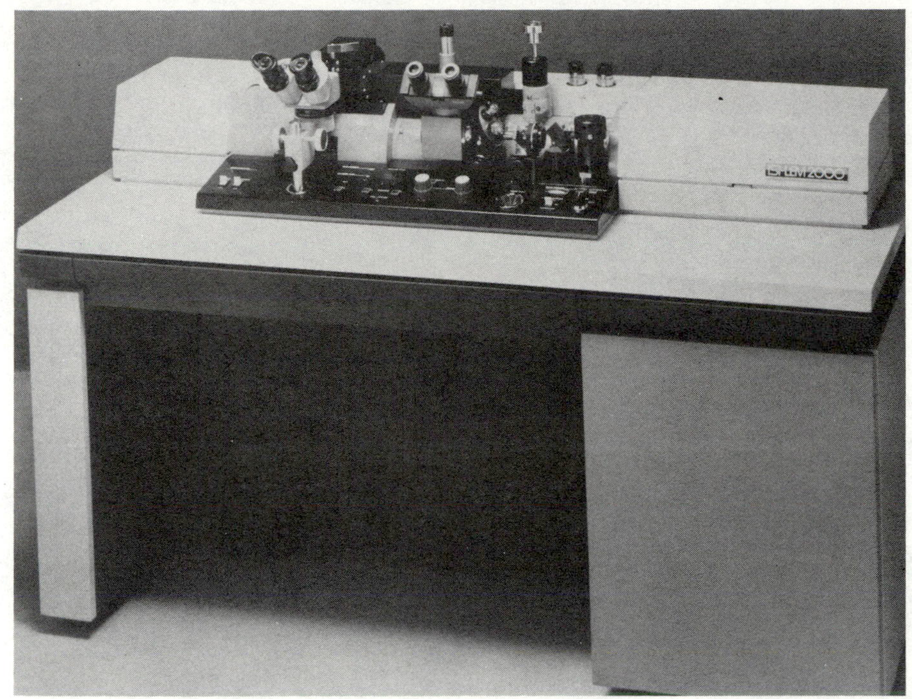

Figure 4.4
An electron microscope. In this instrument a scanning electron microscope (SEM) and a transmission electron microscope have been combined on one console. (Courtesy of International Scientific Instruments Inc.)

the organelles, they are satisfactory examples of plant cell structure as seen by the electron microscope. Figure 4.3*A* illustrates the relative size and abundance of the organelles. Figure 4.3*B* is a diagram of this cell that should be used to definitely locate the organelles in the micrograph. In the diagram, mitochondria, dictyosomes, and the endoplasmic reticulum are enlarged. The free ribosomes are not indicated. The tonoplast is accentuated.

Freeze-Etch Technique

A recent development in technique makes it possible to examine the ultrastructure of frozen cells without chemical fixation. Cells or isolated organelles are rapidly frozen at the temperature of liquid nitrogen. They are placed under a vacuum and the frozen suspension is kept at −147°C. The top of the frozen drop is chipped away and some of the top ice is allowed to evaporate. This leaves $\frac{1}{3}$ nm of frozen tissue extending up out of the frozen block. A very thin layer of carbon and platinum is now deposited over the tissue and ice surface. This layer of carbon and platinum, actually the impression of the exposed tissue, may now be placed on a grid and observed with the electron microscope (Fig. 4.5). The technique is referred to as **freeze-etch.** Bacteria may be quickly frozen to −147°C and upon thawing still survive, so bacterial cells examined by the freeze-etch technique still may be active.

Scanning Electron Microscopy

For viewing with the scanning electron microscope, an object is mounted on a small steel plate and placed in a vacuum in the path of a beam of electrons. Plant parts to be studied are usually killed, dehydrated carefully, with a final drying in liquid CO_2, and then shadowed or coated with a thin layer of gold. Thus prepared, the tissue has normal appearance, does not further deteriorate, and has a better surface from which the electrons of the beam are scattered. These electrons together with secondary electrons given off by the specimen itself are collected to form the image on the viewing screen.

STRUCTURAL DETAILS OF PLANT CELLS

Perhaps the most notable thing about cells is their compartmentalization into membrane-bound organelles. In considering the various parts of the cell in more detail, we should always keep in mind that the complex activities of the living cell result from the coordinated activities of the cell parts whose structural details we are now considering.

Cytoplasm

The term cytoplasm was used by early cell biologists to designate all of the cell's protoplasm except the nucleus. However, it was soon apparent that instead of being relatively homogeneous, the cytoplasm included a variety of membrane-bound organelles bathed by a more fluid cytoplasmic matrix. This cytoplasmic matrix has been called **hyaloplasm,** cytosol, or groundplasm. Though not very precise, these terms generally are used to designate the cytoplasm in which the organelles are bathed. Used in this way, the hyaloplasm would include some membranes, ribosomes, microfilaments, and microtubules as well as the generally fluid mixture of a wide range of water soluble organic and inorganic solutes. The physical properties of cytoplasm may vary widely depending upon the physiological state of the cell and

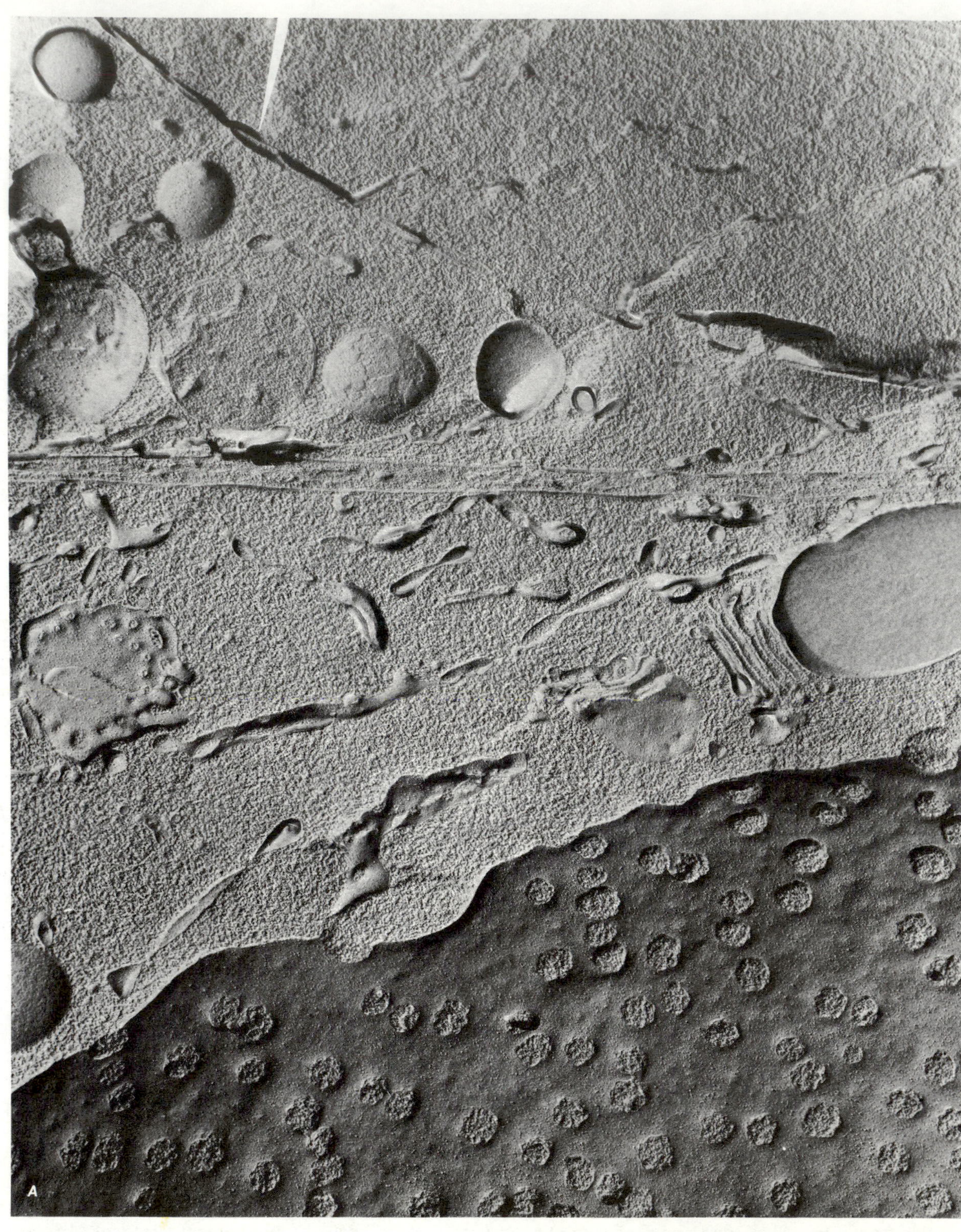

the environmental conditions. Cytoplasm may be relatively fluid (sol-like) and exhibit a movement or streaming, or it may be more viscous, and even gel-like.

Membranes

In addition to the membranes bounding the organelles, the interfaces of the cytoplasm–vacuole and the cytoplasm–cell wall are characterized by special cytoplasmic membranes that exercise some control over the passage of substances into and out of the cytoplasm. These membranes are not visible with the light microscope, but, because of their activity, their existence has long been recognized. The membrane separating cytoplasm and vacuole is known as the **tonoplast;** that bounding the

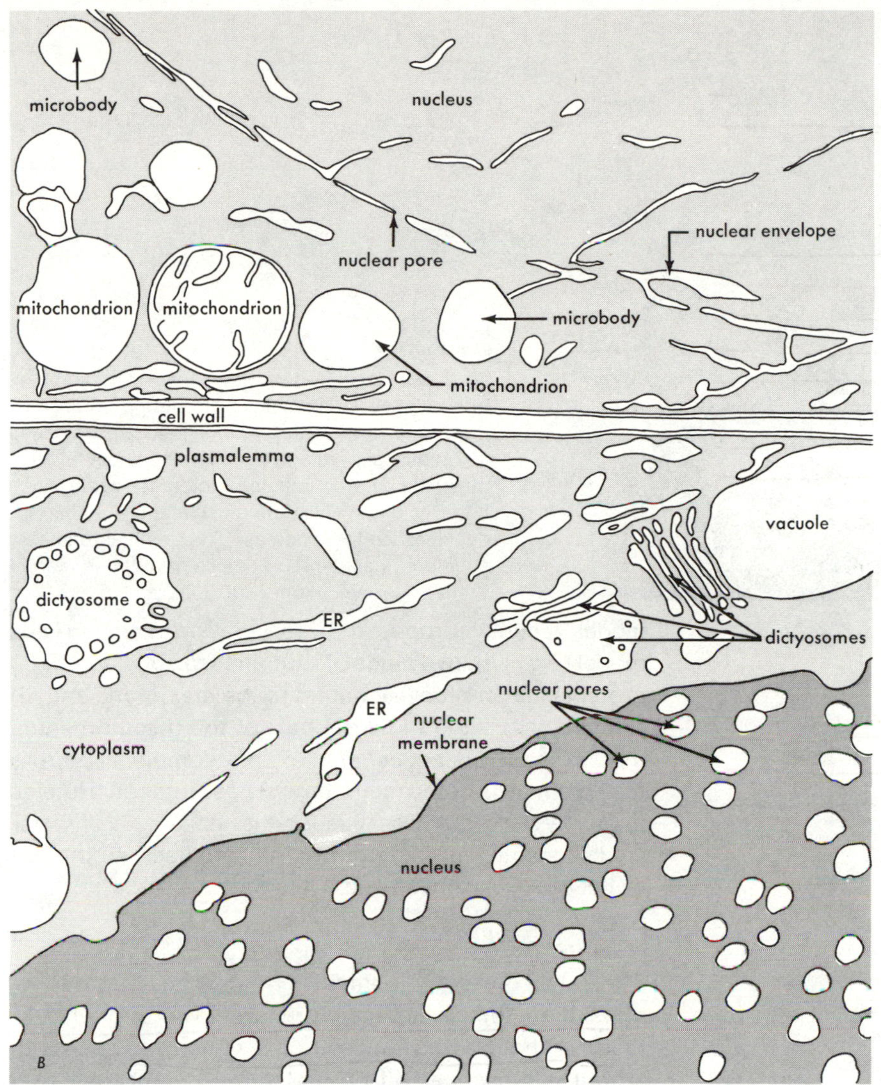

Figure 4.5
Freeze-etch preparation of an onion root tip cell, showing the organelles as they appear when the cells are examined after freezing, rather than after chemical fixation. Bacteria and other cells, after a similar cold treatment, are viable. The similarity between the chemical fixation image and the freezing (living) image is striking. *A*, freeze-etch preparation, ×10,000. *B*, diagram of *A*, with labels (ER represents endoplasmic reticulum), ×7400. (*A*, courtesy of D. Branton.)

outer surface of the cytoplasm is the **plasmalemma** (Fig. 4.6).

Note that in the electron micrographs, each of these membranes consists of a light line sandwiched between two dark lines. The approximate widths of the three membrane components are 2, 3.5, 2 nm, respectively. Note that the plasmalemma is more heavily stained and the two darker components may not have identical widths. These membranes are known as **unit membranes.** The major chemical components of these membranes are lipids (fatty substances) and proteins. Early experiments demonstrated that a membrane consisting of a bimolecular layer of lipid sandwiched between monomolecular layers of protein could account for many of the properties of cellular membranes. Davson and Danielli proposed a model of a cellular membrane as diagramed in Fig. 4.7*A*. These workers proposed that the protein layers were in the more natural form of globular proteins. The idea that the protein is fully extended (Fig. 4.7*B*) was proposed by Robertson who, in studying the membranes of the rather inert myelin sheath protecting nerve fibers, noticed that there was not space enough for the globular proteins. The difference in the chemical activity of globular proteins and fully stretched proteins is significant and would greatly influence the activity of the membranes. The problem of membrane structure has great significance because many important cellular reactions are membrane-bound. Electron transport activity in

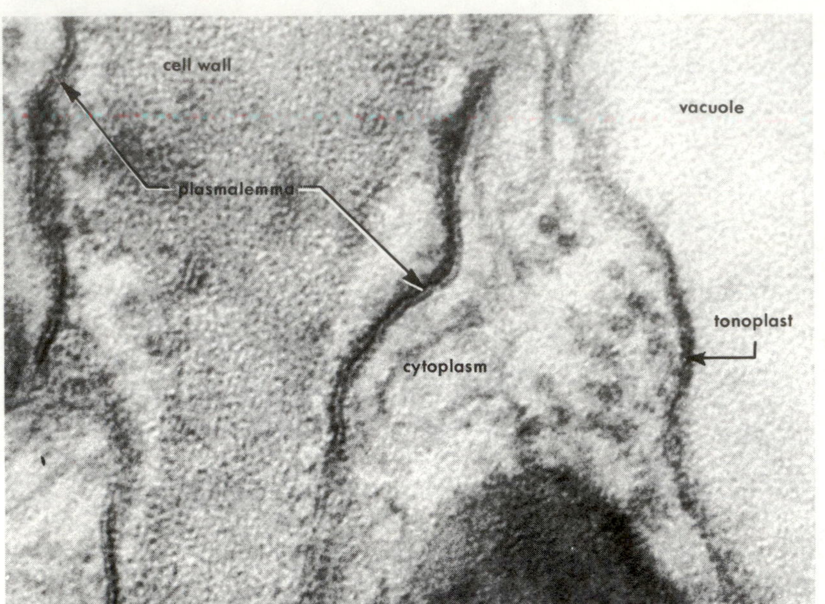

Figure 4.6
A cell wall bounded on each side by a thin peripheral layer of cytoplasm. Bordering portions of the vacuoles of each protoplast are present. The tonoplasts separate cytoplasm and vacuole. The plasmalemma bounds the protoplast, separating it from the cell wall. The plasmalemma stains heavier than the tonoplast. The unit membrane structure is evident in portions of all three membranes. Magnification ×100,000.

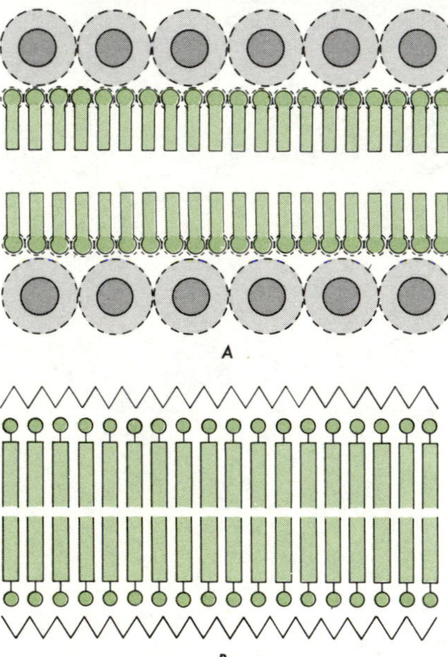

Figure 4.7
The unit membrane. *A*, the model proposed by Davson and Danielli in 1935. It shows a bimolecular layer of lipid bounded by a monomolecular layer of globular proteins. *B*, the model proposed by Robertson in 1959. Here the bounding monomolecular layer is fully stretched protein.

both photosynthesis and respiration is bound to membranes. The binding of enzymes to membranes relegates the enzymes to a particular spot in a team of cooperating enzymes.

More recently it has become increasingly clear that many of the activities of membranes in cells, such as functional diversity and regulation of solute transfer and electron transport, cannot be explained by the Davson–Danielli or the Robertson models of membrane structure. A model of biological membranes that seems to account for many of their properties is the lipid-globular mosaic model recently proposed by several workers. In this model (Fig. 4.8), two kinds of proteins are considered: (1) proteins that are loosely bound to the membrane and (2) proteins that are an integral part of the membrane and that, in many instances, may extend completely across the lipid matrix of the membrane. These integral proteins thus may serve as passages for the movement of water soluble solutes into and out of the cytoplasm of the cell or between organelles and hyaloplasm.

Ribosomes

The hyaloplasm contains numerous particles, **ribosomes,** that are intimately involved with protein synthesis. Although ribosomes are too small to be seen with a light microscope, with the electron microscope they appear as densely stained granules, roughly angular, and from 17 to 20 nm in diameter (Fig. 4.9). There may be as many as 500,000 ribosomes in a single cell. The enzyme, **RNAase** specifically degrades **RNA** (ribonucleic acid, see Chapter 5). Treating tissue with RNAase results in the disappearance of these granules, with no change in the other organelles; thus there is good evidence that RNA is present in ribosomes. Ribosomes are present in plastids and in mitochondria where protein synthesis also occurs. They appear to be lacking in nuclei (Figs. 4.3*A*, 4.20). Some micrographs show them attached to the cytoplasmic side of the nuclear envelope. If they do occur within the nucleus, they are probably confined to the nucleolus (Fig. 4.19*B*). Ribosomes are frequently associated with the cytoplasmic membrane system, the endoplasmic reticulum (Figs. 4.3, 4.9*A*). Under certain circumstances they appear in groups, frequently as helical aggregations. These arrays are designated as **polyribosomes** or simply **polysomes** (Fig. 4.9*B*).

Endoplasmic Reticulum

In electron micrographs, the cross section of the en-

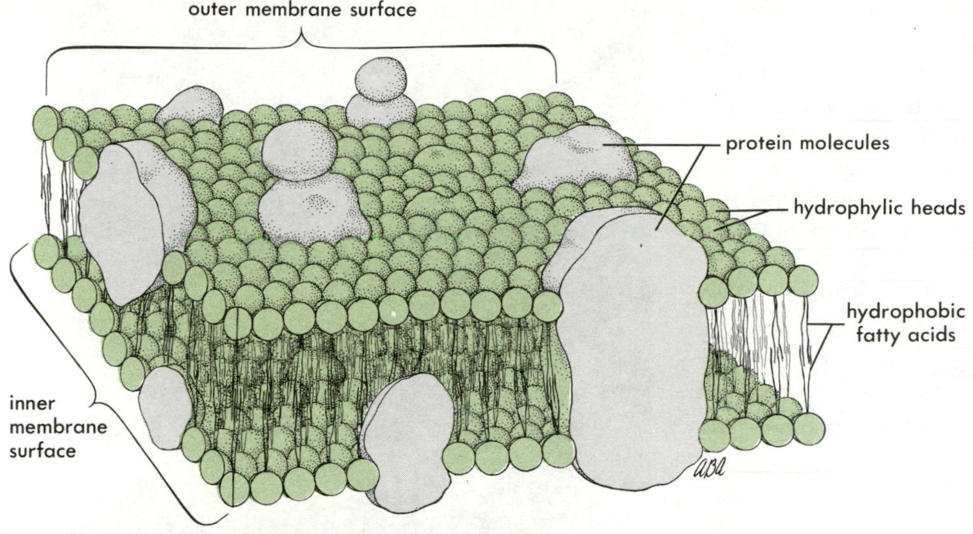

Figure 4.8
The lipid-globular mosaic model of a cell membrane.

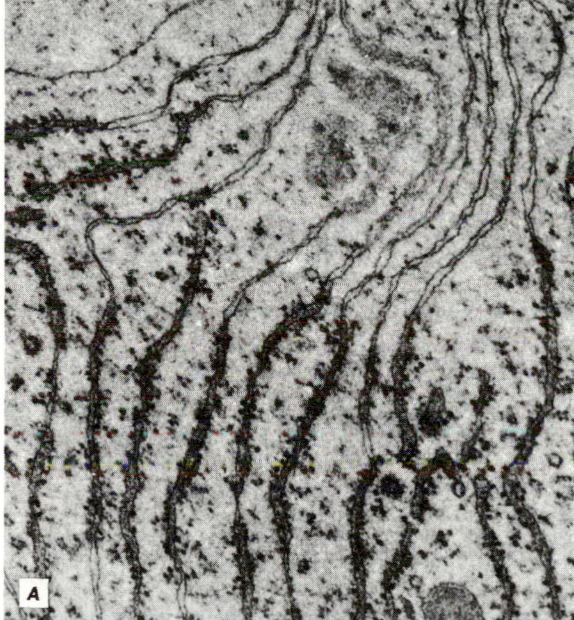

Figure 4.9
Endoplasmic reticulum. *A*, endoplasmic reticulum in an apical cell of the alga *Chara*. Both rough and smooth endoplasmic reticulum are present and are continuous with each other. A microbody is also visible, ×15,000. *B*, endoplasmic reticulum in a meristematic cell from a root tip of wheat. Both rough and smooth endoplasmic reticulum may be seen. There are also free ribosomes and some in helical aggregations that form polyribosomes, ×30,000. *C*, endoplasmic reticulum in a sieve element and companion cell of curcurbit phloem. The endoplasmic reticulum may be seen in the plasmodesmata that pass through the cell walls connecting the two cells, ×60,000. (*A*, courtesy of J. B. Pantastico; *B*, courtesy of P. Bartels; *C*, courtesy of K. Esau and J. Cronshaw, *J. Ultrastruct. Res.* **23,** 1. © 1968 by Academic Press. Reprinted with permission of the publisher.)

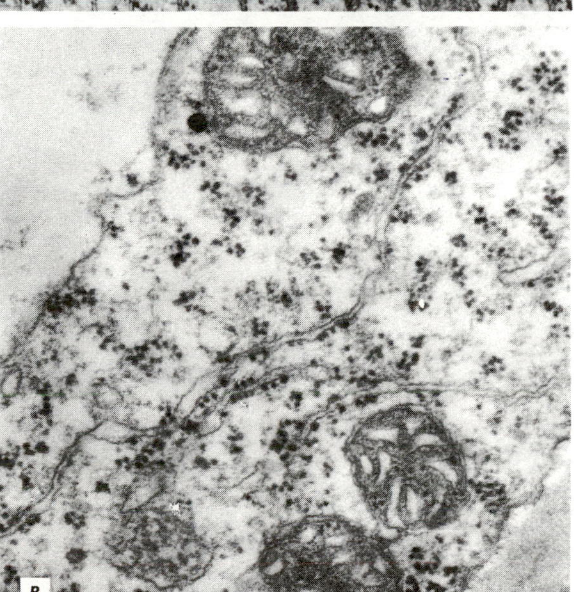

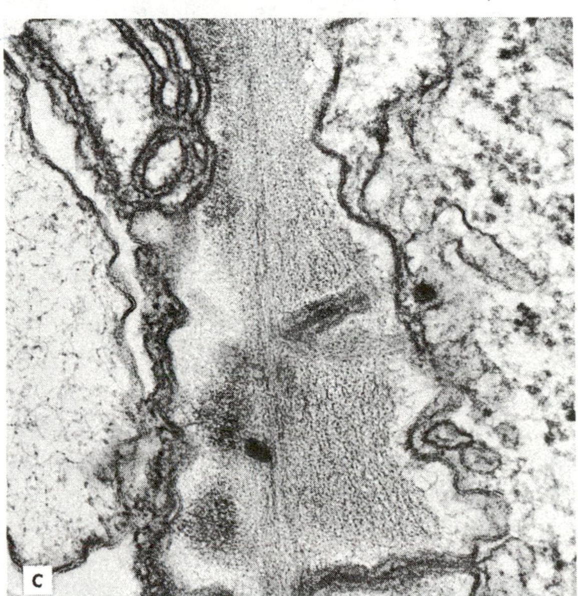

doplasmic reticulum appears as two parallel membranes separated by a narrow light space about 4 nm wide (Figs. 4.5, 4.9). These profiles are actually sections through extensive flattened vesicles that form branched, interconnected, closed systems. Their ends are never open to the cytoplasm. However, in some instances they may be connected to the outer membrane of the nuclear envelope. The space between the membranes of the endoplasmic reticulum may vary considerably under different conditions and in different cell types, appearing tubular, vesicular, or as flattened as sheetlike sacs, **cisternae.** The endoplasmic reticulum frequently has ribosomes attached to its outer, or cytoplasmic surface. This is known as **rough endoplasmic reticulum. Smooth endoplasmic reticulum** lacks ribosomes and is not as common as rough endoplasmic reticulum (Fig. 4.9*A*).

Protein synthesis, particularly that concerned with membrane formation, is associated with the rough endoplasmic reticulum. For example, endoplasmic reticulum is the most highly developed in cells active in protein synthesis, such as the pancreas of animals and the glandular hairs of *Drosera* leaves. It is poorly developed in mature leaf cells (Fig. 4.3*A*). The endoplasmic reticulum is associated with **plasmodesmata,** or the cytoplasmic connections from cell to cell across cell walls (Fig. 4.9*C*). It has been suggested that the endoplasmic reticulum may function as a communication system through the cell.

Dictyosomes

Dictyosomes, or **Golgi bodies,** are about 2μm in diameter and 0.5 μm in width. They are thus large enough to be seen with the light microscope. There are as many as 400 dictyosomes in a plant cell. Collectively they may be referred to as the Golgi apparatus, or Golgi zone. This name is derived from the Italian neurologist, C. Golgi, who first described them in nerve cells in 1898. Dictyosomes are composed of from 5 to 15 circular flattened vesicles, or **cisternae,** aligned in stacks (Figs. 4.5, 4.10). In cross sections, many small vesicles appear at the margins of the cisternae. However, face views reveal that the margins of each cisterna form a coarse net with true vesicles only at the extreme outer regions (Fig. 4.10). The peripheral vesicles, in most cases, contain granules of secretory material elaborated in the dictyosome cisternae, possibly in collaboration with the endoplasmic reticulum.

The Endomembrane System

The membranes of the rough and smooth endoplasmic reticulum, the dictyosomes, the outer membrane of the nuclear envelope, and various cytoplasmic vesicles form a structurally and developmentally continuous system in the cell, the endomembrane system. This system of membranes appears to be involved in the synthesis of plasmalemma and tonoplast membranes, cell wall formation,

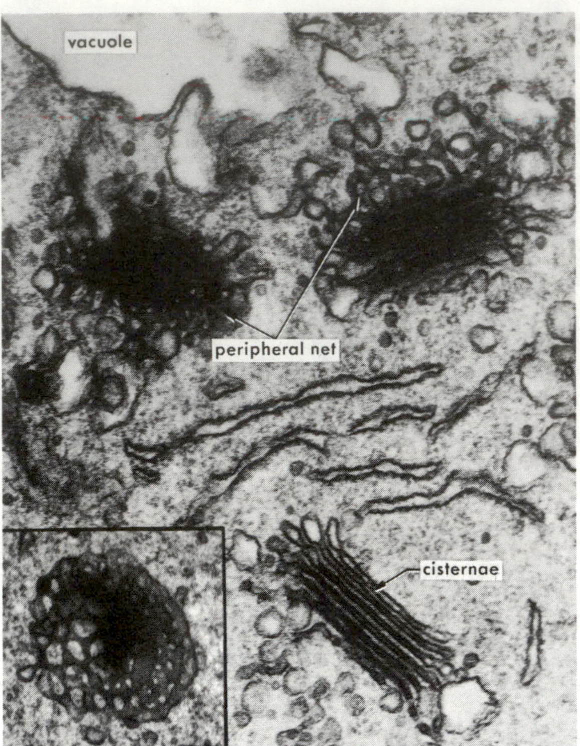

Figure 4.10
Four dictyosomes in a cell from the root tip of a water plant *(Hydrocharis)*. One dictyosome has been sectioned at right angles to its cisternae, and another one almost parallel to its cisternae. Note the peripheral net around this second dictyosome. The two other dictyosomes have been sectioned obliquely to the cisternae and some material appears to be passing from them to the vacuole, ×15,000. (Courtesy of E. G. Cutter.)

Figure 4.11
Diagram to show membrane flow from ER through a dictyosome to the plasmalemma.

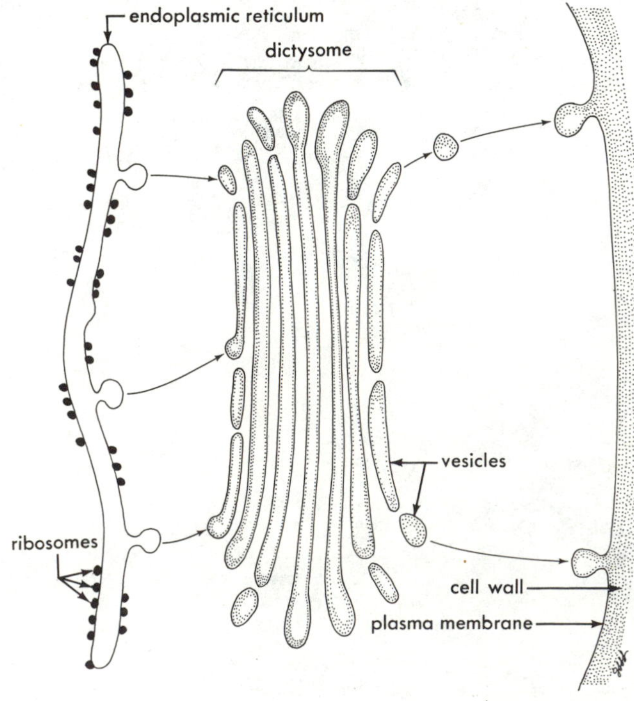

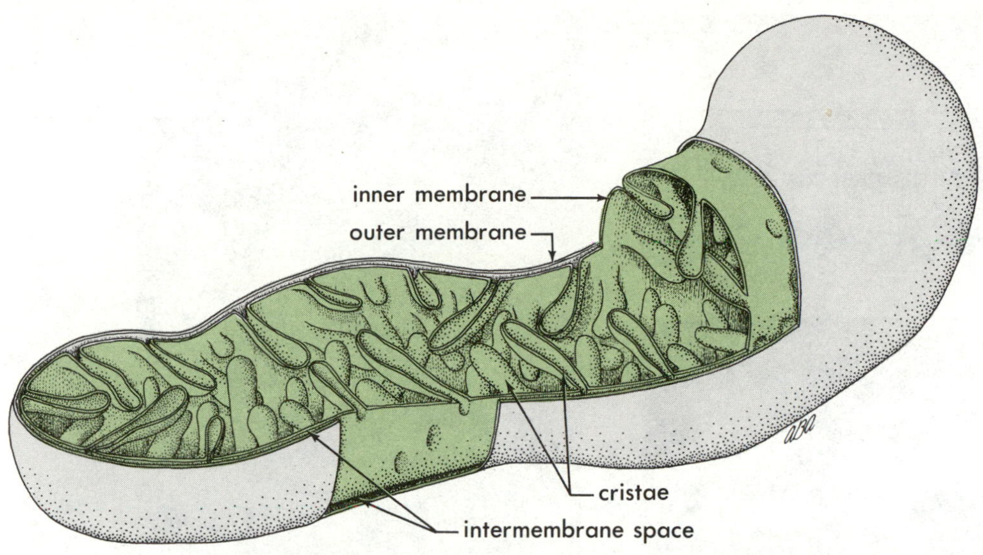

Figure 4.12
A plant cell mitochondrion.

and the synthesis of some proteins, and various plant secretions, for example, slimes, mucilages, and oils. It appears that there can be membrane flow (exchange of membrane material) involving the various parts of the endomembrane system (Fig. 4.11).

Mitochondria

Although some cells of less complex plants may have few **mitochondria,** an active cell of a higher plant may have as many as 100 to 1000 mitochondria completely surrounded by the hyaloplasm. The average diameter of a mitochondrion is between 0.5 to 1.5 μm, and they range in length from 3 to 10 μm. Since the cytoplasm in which the mitochondria are embedded may be rapidly streaming, the mitochondria are constantly being moved from one area of the cell to another. At any one instance, they may be close to the nucleus, or the plastids, or the microbodies, or any of the other organelles in the cell. This dynamic state results in a distinct advantage to the cell since the mitochondria, which are involved in the liberation of useful energy in the cell, are thus brought close to all parts of the cell where the energy may then be more efficiently used to drive cellular reactions.

The mitochondrial envelope that separates the internal mitochondrial substance from the hyaloplasm is double and each component is a distinct membrane. The outer membrane forms a continuous barrier around the mitochondrion. The inner membrane contains a series of tubular extensions or folds, the **cristae** (Figs. 4.12, 4.13). The narrow space between the two membranes of the envelope is thus separated from the **mitochondrial matrix.** The form and amount of the cristae varies in mitochondria from different sources. They are numerous in mitochondria from animal cells and generally are parallel to each other and at right angles to the long axis of the mitochondrion. Cristae in plant mitochondria frequently are sparse and irregularly arranged (Figs. 4.12, 4.13). The inner membrane of the envelope including the cristae contain the biochemical machinery that functions in aerobic respiration. The internal aqueous matrix of the mitochondria contains various solutes as well as ribosomes and DNA fibrils.

Plastids

Chloroplasts

The most easily observed organelles in green plant cells are **chloroplasts,** plastids containing the green pigment chlorophyll. Chloroplasts vary greatly in number and shape from one kind of plant to another. In algae particularly, a wide range in number and size of chlorplasts as well as chloroplasts with bizarre shapes may be observed. Generally most of the chloroplasts of higher plants are found in leaves where from fewer than 20 to more than 50 plastids may occur in a single cell. Chloroplasts, like mitochondria, are bounded by an envelope composed of two closely associated membranes. The plastids have an aqueous matrix, the **stroma,** in which many of the enzymes and other substances necessary for photosynthesis are dissolved. In addition, the stroma contains numerous ribosomes as well as DNA fibrils about 5 nm in diameter. Throughout the stroma is a complex system of membranes in which are located chlorophyll, accessory pigments. and other chemicals involved in trapping light energy and converting it into chemical energy during the process of photosynthesis. These membranes, sometimes called **lamellae** (singular lamella) may occur in cylindrical stacks, **grana** (singular granum). Electron microscopic studies show that the lamellae make up a series of flattened sacs, **thylakoids,** that are stacked in the grana one on top of another (Fig. 4.14). At irregular intervals, thylakoids may extend from

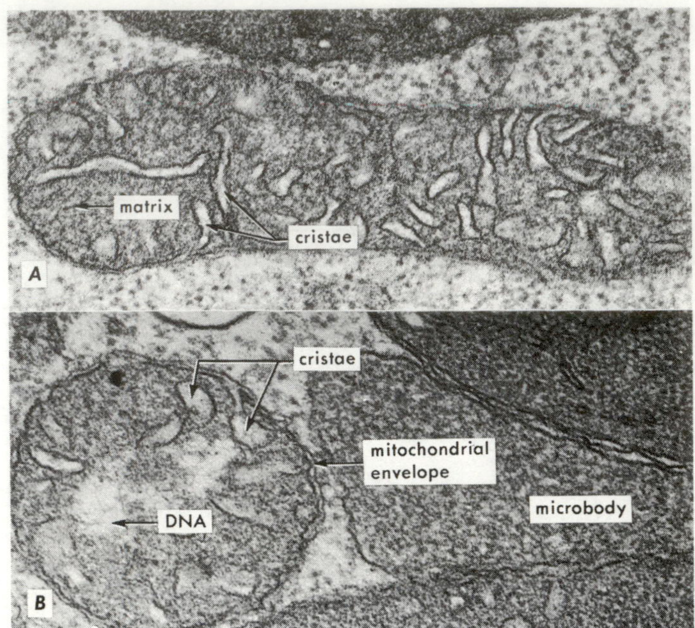

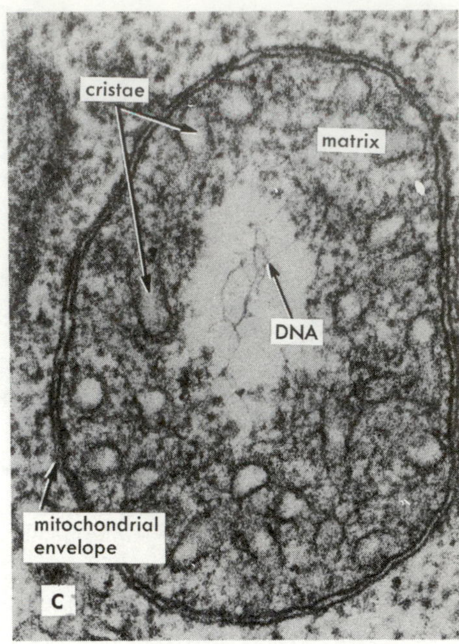

Figure 4.13
Mitochondria. *A*, mitochondrion from a young leaf of a safflower *(Carthamus tintorius)* seedling, ×60,000. *B*, mitochondrion and a microbody from a 7-day-old seedling of bean *(Phaseolus vulgaris)*. The mitochondrion is spherical; the association between a crista and the inner component of the envelope is distinctly seen. The light areas have fine fibrils' in them, probably of DNA. The double envelope of the mitochondrion may be compared with the single membrane bounding the microbody, ×70,000. *C*, mitochondrion in a cortical cell of the brown alga *Egregia menziesii;* the double envelope, cristae, and DNA strands are apparent, ×75,000. *(A,* courtesy of A. H. P. Engelbrecht; *C,* courtesy of T. and A. A. Bisalputra, *J. Cell Biol.* **33,** 511 © 1967 by The Rockefeller University Press. Reprinted with permission of the publisher.)

one granum to another through the stroma. These interconnections are called **stroma lamellae** or **frets.**

Plastid Development

A beautiful picture of the way in which plastids are formed, enlarge, differentiate, and mature is seen if one studies a young grass leaf such as corn from the base to the tip of the leaf blade. The oldest cells of the leaf are at the tip and the youngest, dividing cells, are near the base of a grass leaf blade. Consequently, along an entire leaf blade, one can observe the structural changes that take place as cells mature. Near the base of the leaf, the only plastids, **proplastids,** found in the cells are small, about 1 to 3 μm in diameter. They have only a few short internal lamellae some of which appear to be attached to the inner membrane of the plastid envelope. In addition, small crystallinelike associations of membrane material, **prolamellar bodies,** may be observed in these young plastids. Proplastids are difficult to identify with a light microscope but are readily seen by electron microscopy (Fig. 4.15). Farther up the leaf blade, cells may be found in which the proplastids have begun to differentiate into chloroplasts. The prolamellar bodies disappear, and the internal membrane system continues to develop into grana and stromal lamellae. Here also the plastids may be seen to be dividing. Chlorophyll is present in the internal membranes of the chloroplasts and the leaf in this region is light green. As the cells mature, these young chloroplasts develop an extensive system of granal and stromal lamellae. The leaf becomes deep green. The proplastids have developed into mature chloroplasts. A similar series of developmental changes occur in the development of chloroplasts from proplastids in other types of leaves although not all proplastids contain prolamellar bodies.

In contrast to the orderly development of a proplastid into a chloroplast in the light, if a leaf develops in the dark, it is yellowish-white instead of green and instead of chloroplasts, plastids lacking chlorophyll are formed. However, they may contain small amounts of a green pigment, protochlorophyll, which is transformed into chlorophyll in the light. These plastids are called **etioplasts** (Fig. 4.16*B*). Instead of grana and stroma lamellae, etioplasts have large elaborate prolamellar bodies. However, when the leaves of a dark-grown plant are exposed to light, etioplasts are transformed into chloroplasts. Together with newly synthesized chlorophyll, protein, and lipids, the material composing the prolamellar bodies is used in the formation of the internal lamellar systems of the chloroplasts.

Various Kinds of Plastids

Colorless plastids, **leucoplasts,** are found in many parts

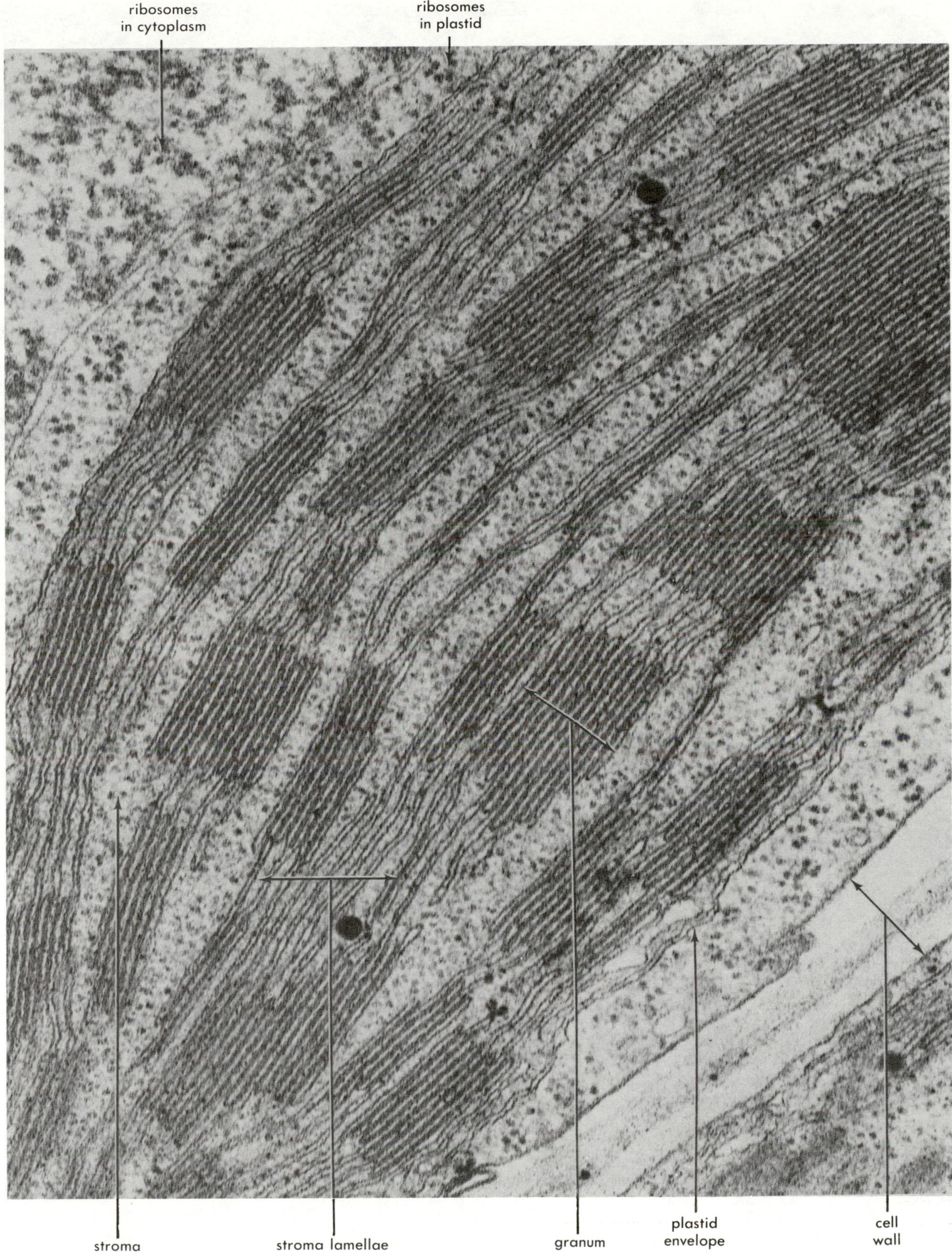

Figure 4.14
Detail of chloroplast structure from a corn (*Zea mays*) leaf, ×75,000. (Courtesy of R. Leech.)

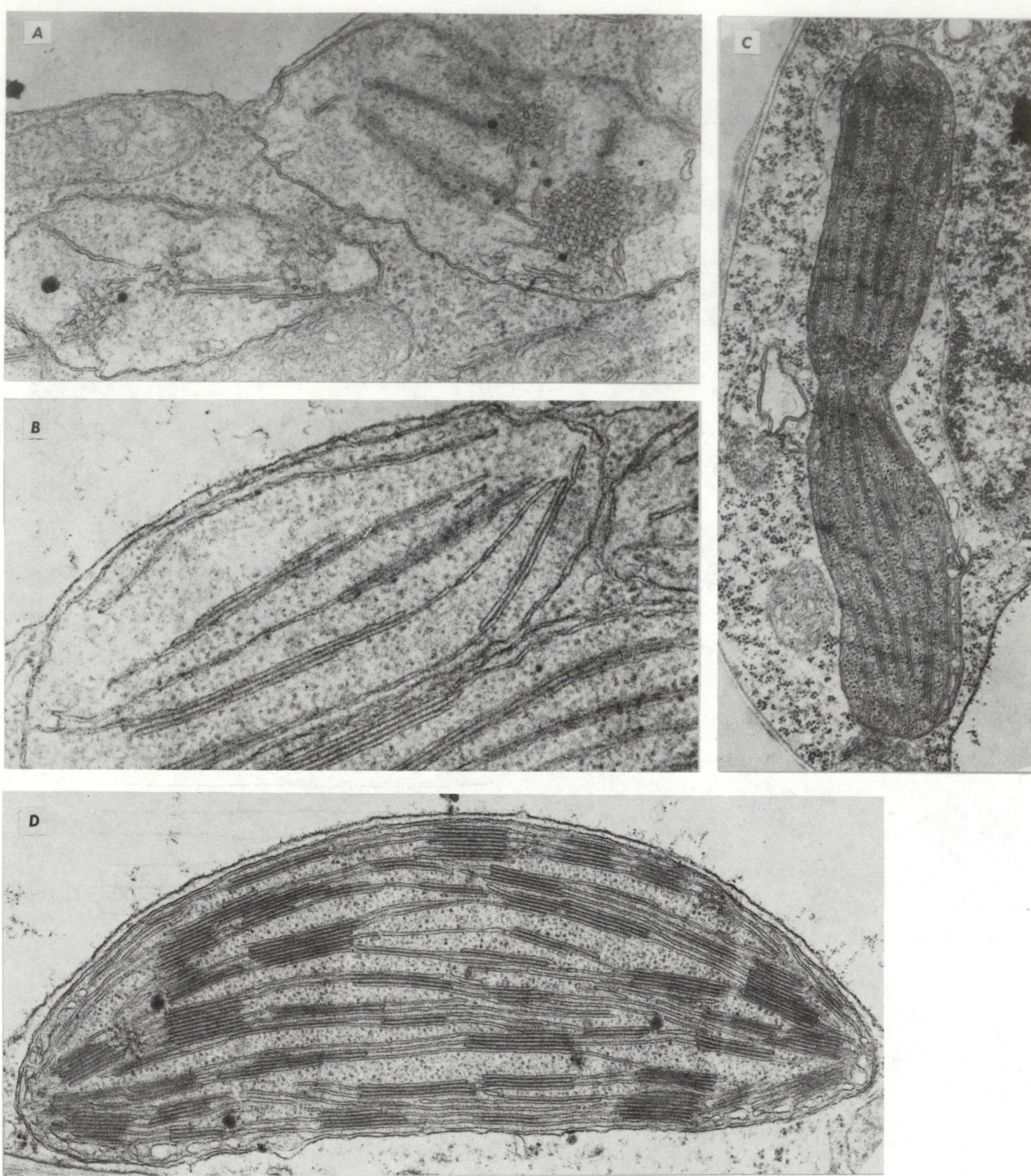

Figure 4.15
Electron micrographs showing plastid development in a green corn (*Zea mays*) leaf. *A,* proplastids with prolamellar bodies, ×36,000. *B,* young plastid lacking grana, ×22,500. *C,* dividing plastid, ×22,500. *D,* mature mesophyll plastid, ×27,000. (Courtesy of R. M. Leech. M. G. Rumsby, and W. W. Thompson, *Plant Physiol.* **52,** 240. © 1973 by the American Society of Plant Physiology.)

of the plant body, for example, epidermal cells of leaves, onion bulbs, storage tissue of apples, and other nonpigmented tissues. Leucoplasts are about 3 to 5 μm in diameter. They may be involved in the synthesis and storage of a variety of substances such as starch, protein, and lipids. Leucoplasts that store large amounts of starch

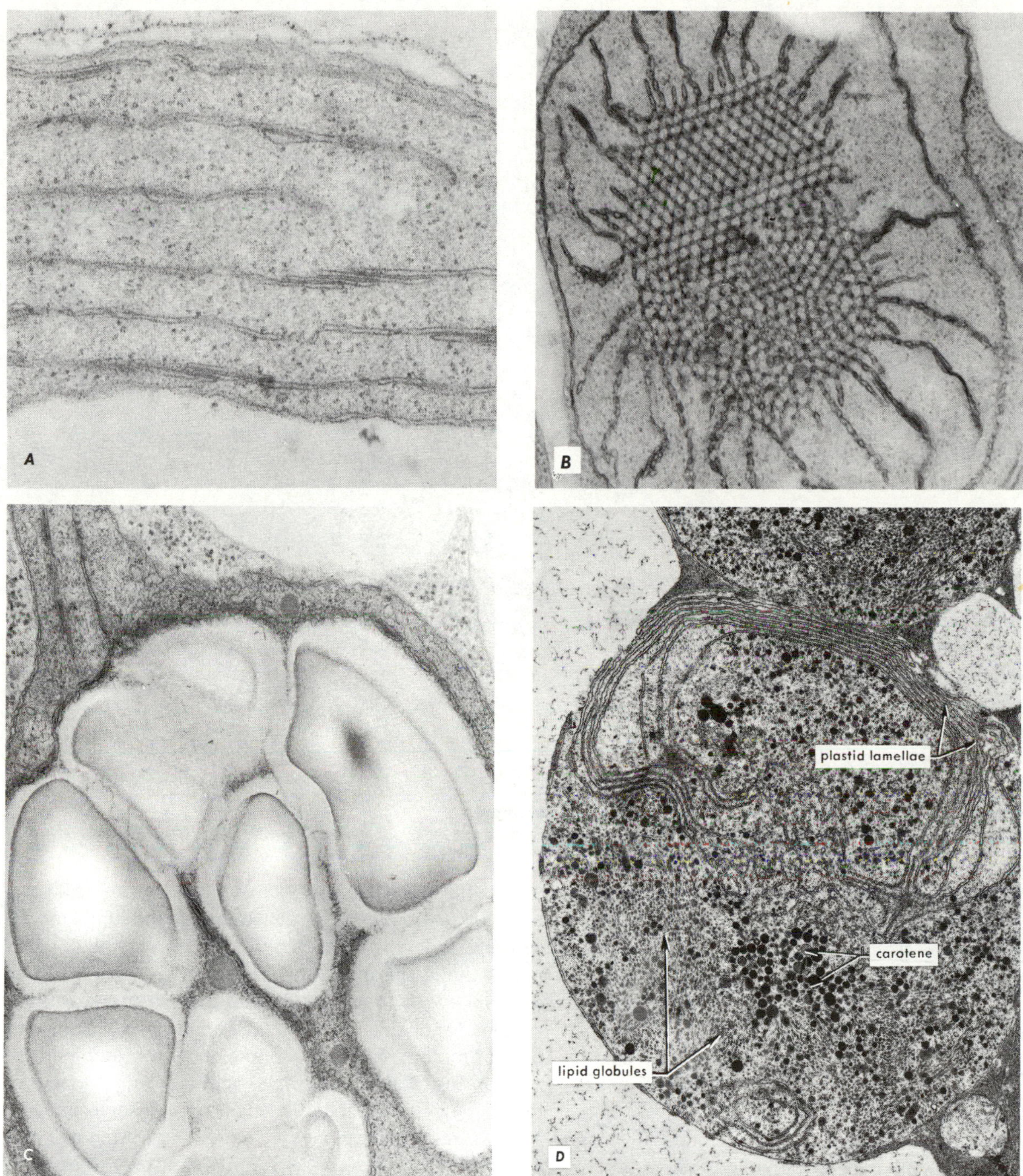

Figure 4.16
Four types of plastids. *A,* a small leucoplast from an inner white leaf of endive *(Cichorium endiva),* ×20,000. *B,* an etioplast from a yellow leaf of a dark-grown bean seedling *(Phaseolus vulgaris),* ×20,000. *C,* an amyloplast from a bean seedling *(Phaseolus vulgaris,* ×20,000. *D,* a chromoplast from a mature red pepper *(Capsicum),* ×21,000. (*D,* courtesy of A. R. Spurr and W. H. Harris, *Am. J. Bot.* **55,** 1210. © 1968 by Botanical Society of America.)

may be called **amyloplasts** (Fig. 4.16*C*) and frequently occur in roots and other nonphotosynthesizing cells.

As fruits ripen on trees and leaves prepare to fall in the end of summer, they change from green to red, orange, or yellow. This occurs because there is a destruction of chlorophyll in the chloroplasts, accompanied by an accumulation of yellow or red pigments known as the **carotenoids.** Plastids with a dominance of the red and yellow pigments are called **chromoplasts** (Fig. 4.15*D*).

Microbodies

Preparations of a number of plant and animal tissues show a variety of spherical organelles that vary from about mitochondrial size to very much smaller. These organelles can be distinguished morphologically from mitochondria, since they have a single outer membrane rather than a double membrane envelope that is characteristic of the mitochondria. Cisternae and cristae are absent from microbodies, and the central area frequently appears rather dense under the electron microscope and may contain a variety of crystals (Figs. 4.5, 4.13).

Microbodies not only are morphologically different from other cellular structures, but also can be separated from them by centrifuging a mixture of ground tissue containing all organelles on a density gradient of sucrose solution.

Considerable research is being carried out at the present time to determine the exact chemical and physiological significance of these organelles. Evidence is accumulating that many different enzymes may be found in single membrane-bound spherical organelles in both plant and animal cells. To some extent, the types of enzymes found depend upon the tissue and kind of cells being studied. Thus, it appears that there is not just one kind of microbody but that the term "microbody" as defined above includes a fairly large number of different kinds of organelles that contain different enzymes and perform different functions in the cell.

The beginning student can appreciate some of the complexities in terminology when learning that microbodies isolated from leaves and containing a complex of oxidative enzymes have been called **peroxisomes;** those isolated from castor bean seeds and containing other oxidative enzymes have been called **glyoxysomes;** liver and various other animal tissues have yielded microbodies called **lysosomes** (high in hydrolytic enzymes).

Spherosomes

These spherical organelles are approximately the same size as microbodies (about 2 μm), have a single limiting membrane, but are specialized for the storage of lipid materials. They are sometimes known as lipid bodies, fat bodies, or wax bodies depending upon their storage component. Spherosomes are found in many different living plant cells, but are particularly abundant in cells of cotyledons of many seeds (Fig. 4.17). In some instances spherosomes appear to originate from the endoplasmic reticulum.

Microtubules and Microfilaments

The hyaloplasm of almost all kinds of cells contains three types of submicroscopic fibrils: **microtubules,** between 18 and 25 mμ in diameter; **microfilaments,** 4 to 7 mμ in diameter; and other filaments, 8 to 10 mμ in diameter. Some of the physical properties of cytoplasm, such as elasticity, and many of its structural characteristics appear to be due to the presence of these structures. Microtubules are tubular fibrillar structures of indefinite length, with cylindrical walls of highly organized globular protein (tubulin) subunits surrounding an electron transparent core about 10 nm in diameter (Fig. 4-18). A variety of cellular functions frequently associated with some aspect of motility is now generally accepted as being dependent on microtubule action. Microtubules occur in the cilia or flagella of motile cells of all eukaryotes. They are involved in chromosomal movement during cell division (Fig. 6.13). They and microfilaments play roles in the movement of vesicles through the cytoplasm. Their close association with cell walls, just internal to the plasmalemma, and their orientation parallel with each other and at right angles to the long dimension of the cell wall of many elongating cells have led to the suggestion that they may be associated in some way with the direction of deposition of cellulose microfibrils during wall synthesis.

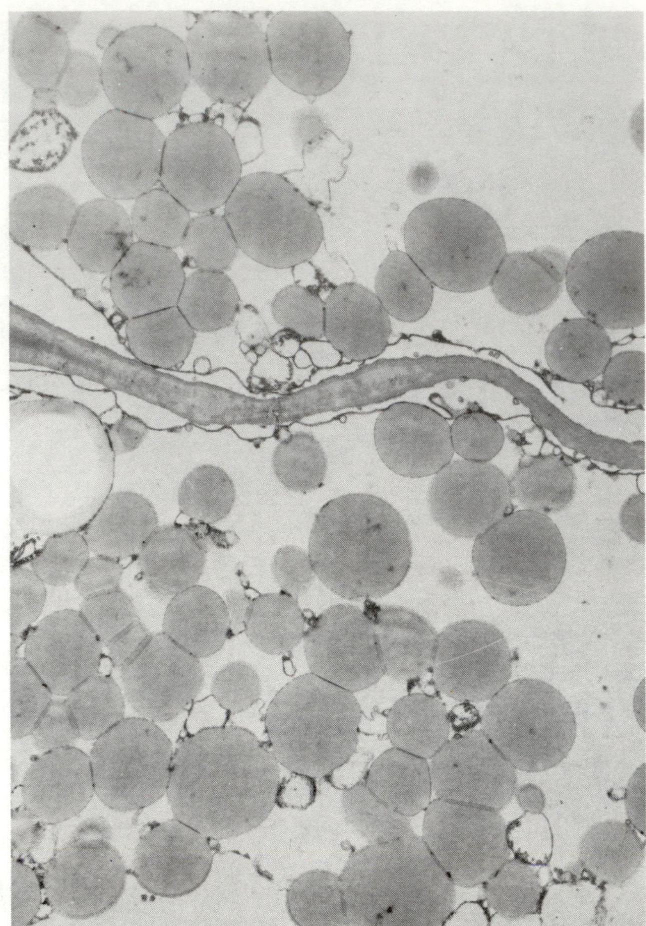

Figure 4.17
Spherosomes. Section of Jojoba (*Simmondsia chinensis*) cotyledon showing wax containing spherosomes, ×4200.

Nucleus

Structure

Cells of all eukaryotic plants characteristically have a

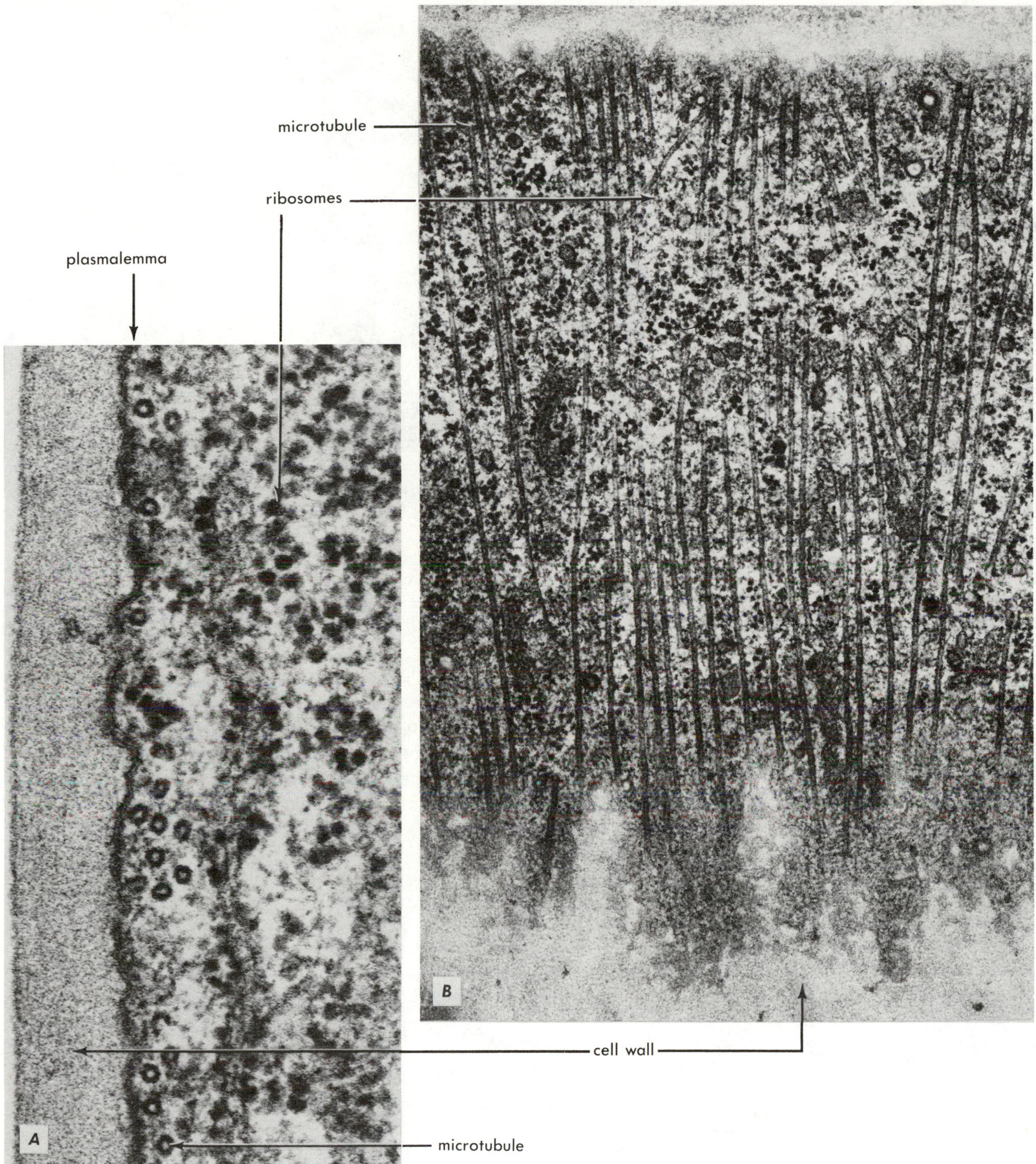

Figure 4.18
Electron micrographs of *Phleum pratense* root tip. *A*, section showing cell wall and microtubules in cross section. *B*, section cut parallel to microtubules. (Micrographs by B. E. S. Gunning.)

single spherical organelle, the **nucleus.** In meristematic cells of higher plants, the nucleus usually is 7 to 10 μm in diameter while in more mature, larger cells, the nucleus may be 35 to 50 μm in diameter. Very large nuclei, up to 1 mm in diameter, are found in some specialized cells of a few plants while very small nuclei, 1 μm in diameter, are found in the cells of some fungi. The nucleus is bounded by an envelope made up of two membranes. A fluidlike substance, the **nucleoplasm,** fills the nucleus and in it is found a mass of thin fibrils about 10 nm in diameter, **chromatin** (Fig. 4.19). The fibrils contain DNA and protein (particularly a protein called histone) and become tightly coiled to form the familiar chromosomes when the cell divides. One or more roughly spherical structures of ir-

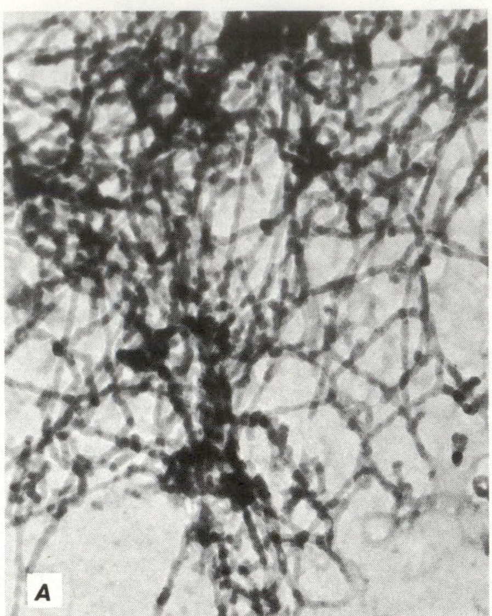

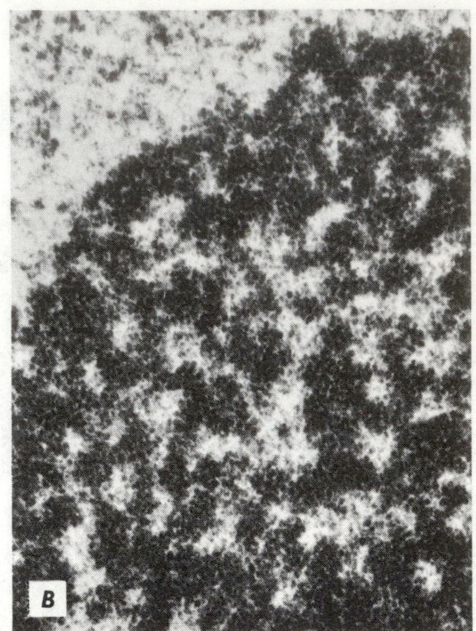

Figure 4.19
The nucleoplasm. *A,* fibers, about 17.5 nm in diameter, are a constant feature of the nucleoplasm obtained from isolated nuclei of barley *(Hordeum vulgare)* endosperm processed by the critical freezing point method. *B,* a view of a section of a nucleolus from the root tip of the water plant, *Hydrocharis.* Note the aggregation of polysomelike particles and the absense of a limiting membrane between the nucleolus and the nucleoplasm. All magnifications ×30,000. (*A,* courtesy of S. L. Wolfe, *J. Cell Biol,* 37, 610. © by The Rockefeller University Press, Reprinted with the permission of the publisher. *B,* courtesy of E. G. Cutter.)

regular outline and of variable size, the nucleoli (nucleolus singular), are also embedded in the nucleoplasm. These nucleoli contain a dense mass of granular material. They contain RNA and protein and are associated with synthesis of cytoplasmic ribosomes.

We have already mentioned that the outer membrane of the nuclear envelope may be continuous with other components of the endomembrane system of the cell. Thus, some parts of the endoplasmic reticulum may be connected with the outer nuclear membrane in such a way that the space, **lumen,** between the membranes of the nuclear envelope may be in direct continuity with the

Figure 4.20
The nuclear envelope. *A,* a section of a nucleus from a meristematic cell of the shoot apex of *Chenopodium album.* The section is at right angles to the envelope, which is distinctly double and shows definite interruptions. *B,* a section parallel to the nuclear envelope from a cell in a squash cotyledon. The pores are very distinct, ×20,000. *(A,* courtesy of E. M. Gifford, Jr., and K. Stewart. *B,* courtesy of J. Lott.)

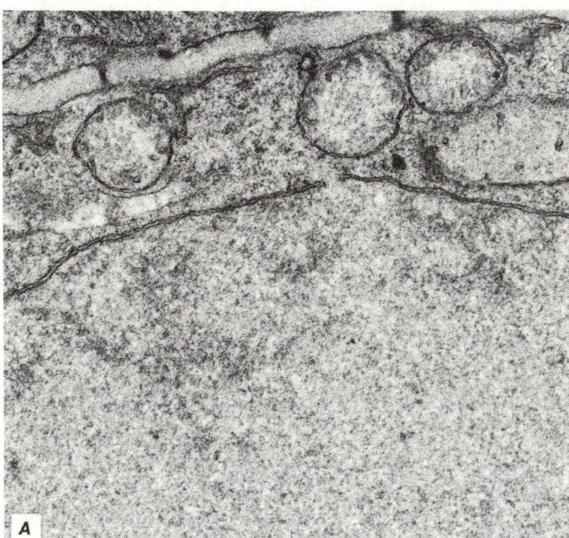

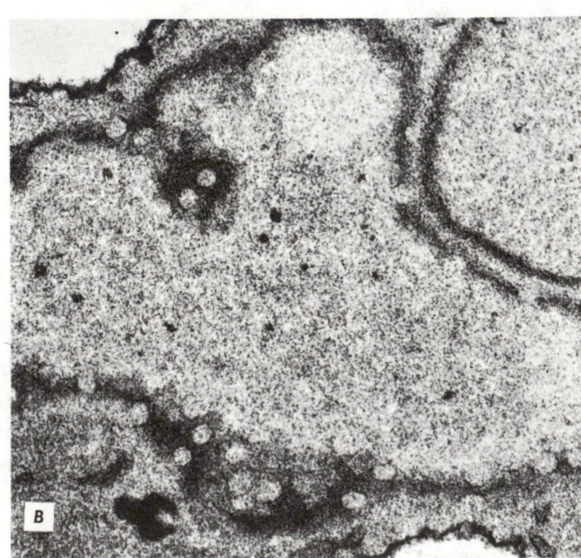

lumen of the endoplasmic reticulum. However, because the inner membrane of the envelope is continuous, it constitutes a barrier between the lumen of the endoplasmic reticulum and the nucleoplasm. A further important feature of the nuclear envelope is the presence of pores (Fig. 4.5) about 65 to 70 nm in diameter. The two membranes of the envelope are joined together at the sites of the pores, while the pores themselves are filled with a dense material that, according to some investigators, effectively blocks the free diffusion of large molecules into or out of the nucleus. Figure 4.20*B* shows a tangential cut through a small portion of the envelope of a nucleus in a cell of a squash cotyledon. The pores, in face view, are very evident. They are also evident in the surface of the nuclei when the freeze-etch technique is used for preparation (Fig. 4.5*A*).

Function

New information about the nucleus and nuclear activity has come mainly from two sources: (*a*) studies of the growth of organisms in which nuclear control of differentiation has been subjected to experimental techniques and (*b*) biochemical studies of the nucleus and of inheritance. These investigations have furnished us with an insight into nuclear activity and structure that is far beyond the details of the cytoplasmic organization we have just discussed.

While it is well-established that genetic information is stored within, and distributed from, the nucleus, the details of how this occurs are not all clear. Cells deprived of nuclei may continue normal activities for as long as 3 months. The green alga, *Acetabularia* (Figs. 2.1*E*, 4.21) is a single-celled plant 2–4 cm. tall. It shows considerable differentiation for a single cell. It has rhizoids for attachment, an upright stalk, and an expanded cap, which differs from species to species. The nucleus is situated at the bottom of one of the rhizoidal branches (Fig. 4.21*A*). It may easily be removed by cutting off the tip of the rhizoid containing it (Fig. 4.21*B*). Such an enucleated *Acetabularia* plant has been kept alive in the light for 7 months, although *normal synthetic activities* ceased after 2 weeks. When the cap and a portion of the stalk are removed from an enucleated cell (Fig. 4.21*B*), a new cap will be regenerated. The shorter the piece of stalk that is removed, the better the regeneration. It appears that materials capable of inducing the formation of a new cap are released by the nucleus and accumulate in the upper portion of the stalk. Here they are readily available for the renewal of a cap, should the original cap be damaged or lost. These experiments would seem to indicate that the presence of a nucleus is not immediately necessary for the normal day-to-day metabolic activities of *Acetabularia.* There must have been sufficient genetic information present in one form or another, in the cytoplasm, to provide for the immediate regeneration of the new cap. It can be shown in *Acetabularia* that this material is used up. There must be a continual renewal of the cytoplasmic informational materials. The nucleus is also necessary for cell division. We shall see later on (Chapter 6) that growth and differentiation do not occur without cell division. In the periods between cell divisions, the nucleus is actively preparing for a subsequent division.

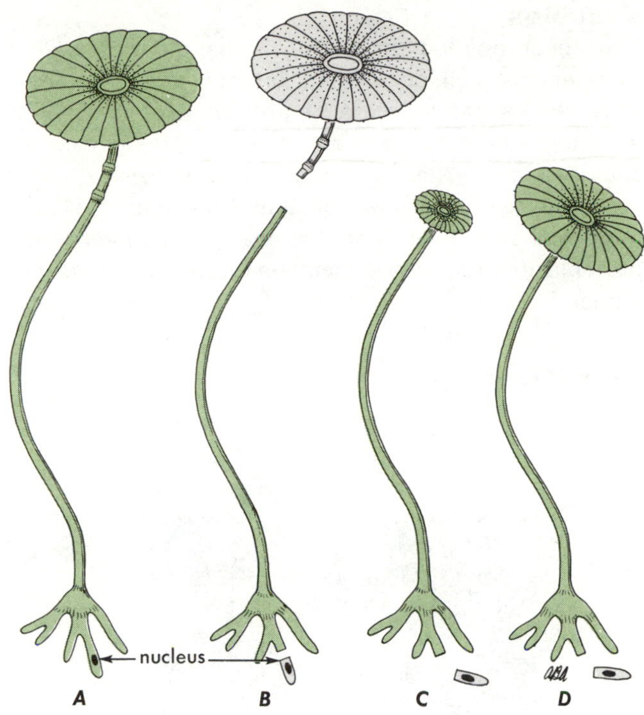

Figure 4.21
Regeneration of a new cap by an enucleated cell of the green alga *Acetabularia*. *A*, a cell of *Acetabularia;* note nucleus at tip of rhizoid. *B*, nucleus and cap have been cut off. *C*, new cap grows. *D*, growth of a new cap completed; it may be smaller than the original cap.

This line of reasoning leads to an enigma—the lack of a nucleus in the food-conducting cells of the angiosperms, the sieve tube members of the phloem (p. 000). Mature and actively conducting angiosperm sieve tubes lack a nucleus. Such cells normally function for periods not longer than 3 years, but in palm trees they must function for 200 years! The only other enucleated functioning cell is the mammalian erthrocyte, and its life span is about 3 months.

Nonprotoplasmic Portions of the Cell

All the organelles and particulates so far discussed have been an integral part of the protoplasm. Energy has been expended within the protoplasm to maintain its state of homeostasis. The cell also contains regions outside the protoplasm which are maintained in a state of equilibrium just as are organelles. The vacuoles and crystals within the protoplast, as well as the cell walls bounding the exterior of the protoplast, are such extraprotoplasmic regions.

Vacuoles

It has been pointed out previously that plant cells differ from animal cells in having one or more regions—**vacuoles**—separated from the protoplasm of the cell by a limiting single membrane, the **tonoplast** (Fig. 4.6). We shall see later (Chapter 5) that vacuoles have a higher energy potential (solute concentration) than the solutions outside the cells and that energy expended by the protoplasm is required to maintain this higher energy potential.

Figure 4.22
Stages in the growth of a cell. Progressively older cells shown from *A* through D, ×2000.

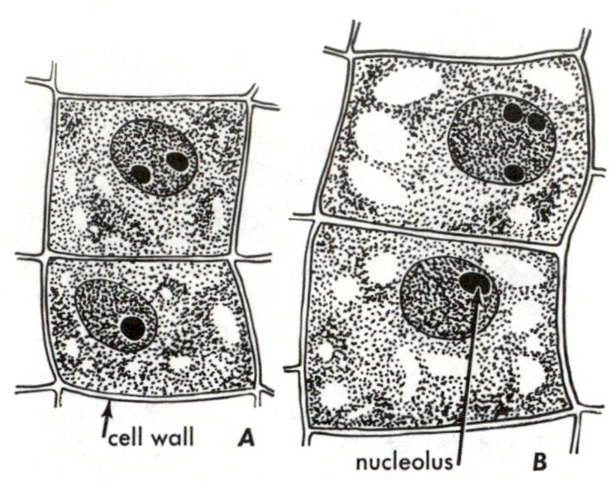

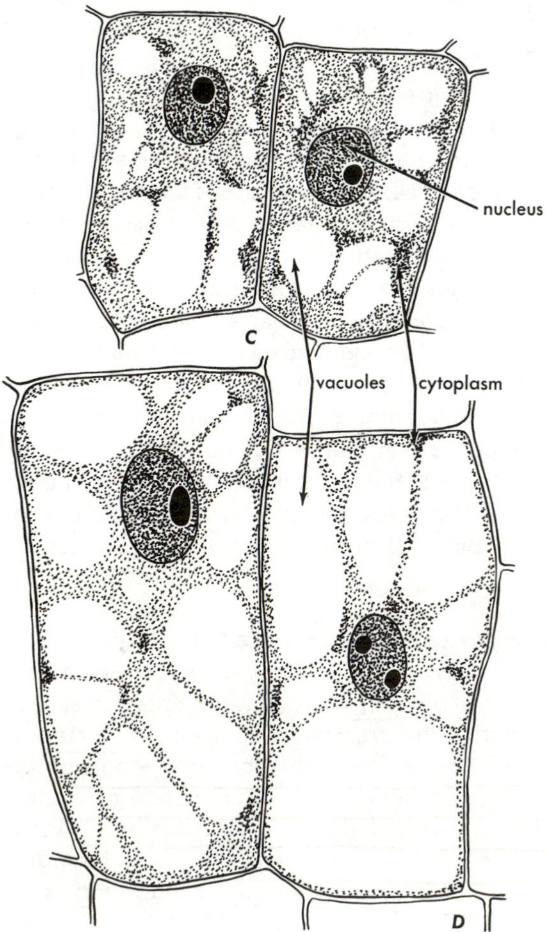

In many young cells, the cytoplasm occupies much of the space in the cell. Small vacuoles are, however, present (Fig. 4.22*A*). As the cell grows larger, the small vacuoles within the cytoplasm increase in size, coalesce, and become fewer in number (Fig. 4.22*B*, *C*). Finally, when the cell has attained its mature size, only a few large vacuoles, or even only one, may remain (Fig. 4.22*D*). The protoplasmic contents (nucleus and cytoplasm) of the cell lie compressed against the cell wall. The nucleus may occupy a position near the center of the cell, in which event it is embedded in cytoplasm and connected with the cytoplasm around the cell wall by strands of cytoplasm (Fig. 4.22*D*)

Vacuoles do not contain pure water, but rather a highly dilute solution of many substances. This aqueous solution in the cell is termed **cell sap.** Among the substances dissolved in the water, and thus constituents of cell sap, are varying amounts of the following: (*a*) atmospheric gases, (*b*) inorganic salts, (*c*) organic acids, (*d*) salts of organic acids, (*e*) sugars, (*f*) water-soluble proteins, alkaloids, and certain pigments. Generally the cell sap is slightly acid.

The most common pigments of the vacuolar sap are the **anthocyanins.** These pigments are responsible for the red, purple, or blue of the petals of many flowers or of other parts of plants. The yellow coloration of poppy flowers is also due to yellow vacuolar anthocyanin-like pigments called anthoxanthins. The red color of roots and leaves of garden beets is due to another vacuolar pigment called betacyanin.

Crystals

Crystals were among the first objects seen in plant cells. Only 11 years after Hooke discovered cork cells, Leeuwenhoek, in 1675, wrote a letter describing needlelike calcium oxylate crystals that he observed in *Arum.* Cells with crystals can be found in almost all plants and in many different plant tissues. Crystals frequently develop inside of vacuoles and vary in chemical composition and in form (Fig. 4.23). The most common crystals are of calcium oxalate; it is generally held that they are an excretory product of the protoplast formed by the union of calcium and oxalic acid. This acid is soluble in cell sap and is toxic to the protoplasm if it attains a high concentration in the cell. By its union with calcium, the soluble oxalic acid is converted into the highly insoluble calcium oxalate, which will not injure the protoplasm. However, there have been many alternative speculations as to the function of crystals. In addition to calcium oxalate, crystals of other calcium salts frequenty occur, for example, carbonate, sulfate, phosphate, citrate, tartrate, malate. Fine crystals of silica are sometimes found in some species of plants particularly grasses.

Crystalline deposits of protein are often observed in the nucleus, cytoplasm mitochondria, chloroplasts, or microbodies. In isolated chloroplasts (Fig. 4.24), they are

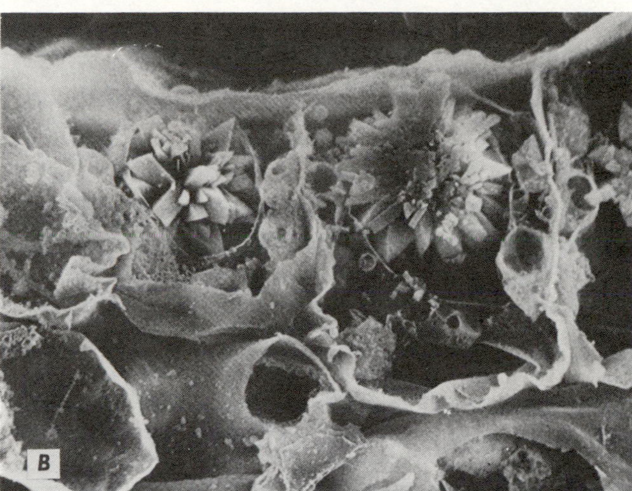

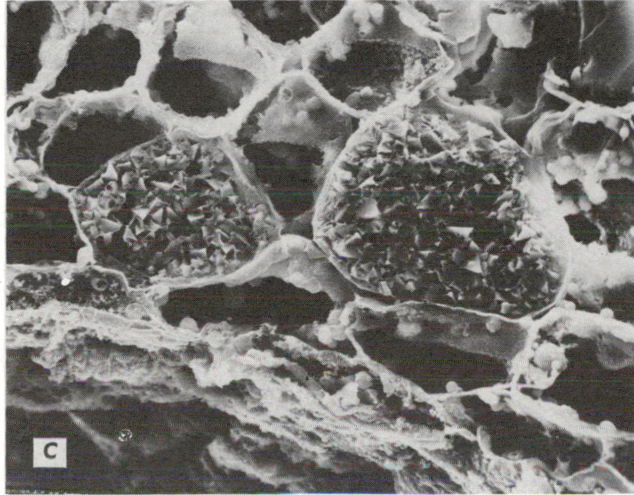

Figure 4.23
Types of crystals found in cells. *A,* raphide crystals in *Psychotria punctata* callus, ×2000. *B,* druse crystals in *Peperomia* sp. leaf cells, ×2700. *C,* crystal sand in *Capsicum annuum* anther, ×1150. *(A,* courtesy V. R. Franceschi and H. T. Horner, Jr., *Zeit. Pflanzenphysiol.* **92,** 61. © 1979 by Gustave Fisher Verlag. *B,* courtesy H. T. Horner Jr.; *C,* courtesy of H. T. Horner, Jr., and B. L. Wagner.)

generally noted when the isolation medium has a high molar concentration.

The Cell Wall

Outside the plasmalemma, and surrounding the entire protoplast, is a rigid wall. When, as is usual, protoplasts are separated from each other by walls, the walls are cemented together by an intercellular substance, the **middle lamella** (**Fig. 4.25,** page 67), which is characterized by **pectins** and certain other substances. The first wall formed by the protoplast is the **primary wall,** and it is composed mainly of cellulose. A **secondary wall** is formed sometimes after living cells have stopped enlarging, and the complete mature cell wall may finally come to have a thickness many times as great as the primary wall. In some tissues, the secondary wall is stratified and composed of several layers. In others, the cells do not lay down secondary wall material, in which event the common wall between two adjacent protoplasts is composed of the middle lamella, with primary wall material on each side. The secondary wall may be of cellulose or of cellulose impregnated with other substances. Some of these substances, notably lignin, add to their hardness and lend decay resistance to wood; others, like suberin and cutin, are fatty or waxy substances and protect leaves and stems against water loss. In addition, certain other materials may enter into the composition of the cell wall —gums, tannins, minerals, pigments, proteins, fats, and oils. It should be emphasized that in mature, hard tissues, such as wood, lignin may be deposited not only in the secondary wall but also in the primary wall and middle lamella.

Although the walls of cells vary considerably in composition in different species, and from one part to an-

Figure 4.24
Crystals within the cytoplasm. Crystals, probably of protein, in a chloroplast isolated from a leaf of bean *(Vicia faba).* Similar ones are frequently seen, not only in chloroplasts, but also in mitochondria and in the cytoplasm, ×60,000. (Courtesy of L. K. Shumway.)

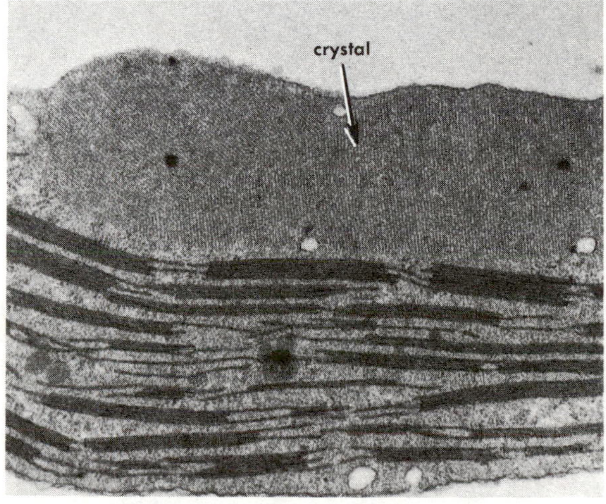

other in the same individual plant (**Fig. 4.25,** page 67), **cellulose** constitutes the greatest percentage of the material of which most cell walls are made.

SUMMARY

1. The cell may be considered the unit of life because it is the smallest unit known to be able to replicate itself and to develop in an orderly fashion from a simple beginning to a very specialized end.
2. Cells may be divided into protoplast and cell wall.
3. Protoplasm is divided into many compartments by an array of particles, membranes, and organelles.
4. The plasmalemma is the membrane separating the protoplast from the cell wall, and the tonoplast is the membrane separating the vacuole from the cytoplasm.
5. Ribosomes are small particulates having a high RNA content.
6. The endoplasmic reticulum consists of an array of extensive flattened vesicles. It may or may not have ribosomes associated with the outer surface of the vesicles. It appears to be involved in protein synthesis.
7. Dictyosomes are stacks of from 3 to 10 flattened cisternae. Each cisterna is surrounded by a peripheral net. Dictyosomes appear to be involved in the synthesis of various cellular products.
8. Mitochondria are small organelles about 0.5 μm in width and from 1 to 3 μm in length. They are bounded by a double envelope. The inner component invaginates to form cristae that are surrounded by a homogeneous matrix. Respiration is localized in the mitochondria.
9. Chloroplasts are approximately 5-10 μm in size. They are bounded by a double envelope. The internal chloroplast lamellae aggregate to form cylindrical grana that are connected by intergranal lamellae. The photochemical reactions of photosynthesis are membrane-bound; the enzymatic reactions of photosynthesis are located in the stroma.
10. In addition to chloroplasts, cells may contain leucoplasts, amyloplasts, proplastids, and chromoplasts.
11. Microbodies are all bounded by a single membrane. They are variable in size and in morphology. They are closely associated with various types of intracellular enzyme activities.
12. Spherosomes are single membrane bounded vesicles involved in the synthesis or storage of fats, waxes, or oils.
13. Microtubules and microfilaments are submicroscopic fibers, ranging in diameter from 4 to 26 nm, found in the cytoplasm of virtually all cells. Many functions particularly those involving the movement of cytoplasmic components (e.g., chromosomes and vesicles) have been ascribed to them.
14. The nucleus is bounded by a double-membraned envelope provided with pores. The nucleoplasm appears to be characterized by a closely packed array of unit fibers about 22.5 nm in diameter and of indefinite length. While cells may live and even differentiate for a short time without a nucleus, a nucleus is required for the continued life of a cell and for cell division. Vacuoles contain aqueous solutions within the protoplast, separated from the cytoplasm by the tonoplast.
15. Crystals, frequently of calcium oxalate or protein, may be found in all parts of the cell.
16. The cell wall, which bounds the protoplast, is formed of cellulose fibrils embedded in an amorphous matrix. Other compounds (suberin, pectin, cutin, and lignin) may become a part of it.

The type of cell just summarized is highly compartmentalized. It is known as a eukaryotic cell. More primitive cells are not compartmentalized; DNA, photosynthetic processes, and respiratory activity all share a common cytoplasm. These are prokaryotic cells.

CHAPTER 5

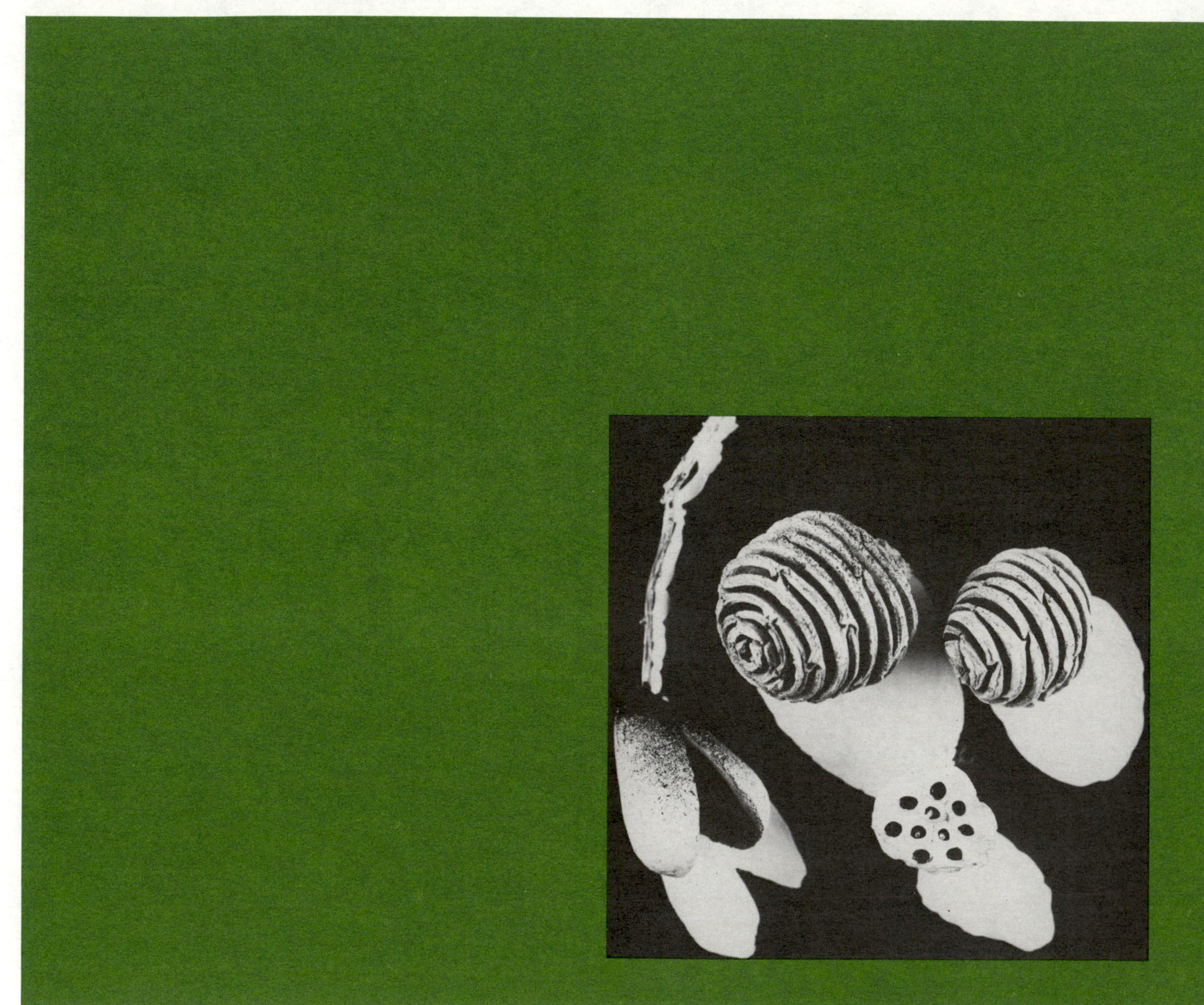

CHEMISTRY AND PHYSIOLOGY OF THE CELL

Plants receive radiant energy from the sun. They convert it into chemical energy in the form of food, from which it is released to be used in the multiple activites of the living cell. These cellular activities—growth, cell division, and differentiation—all occur in the protoplasm of the cell. The astonishing properties of the protoplasm in which all this occurs are inherent in its structure. The structure of protoplasm may be studied morphologically under microscopes or chemically in test tubes. Significantly, since the advent of the electron microscope, the distinction between chemistry and morphology is becoming less and less clear. In other words, the chemical elements (Table 5.1) group together in a little-understood manner to give the visible structure of protoplasm and the intricate transformations of energy that occur within it. Therefore, in discussing the activities of protoplasm, we must first take into consideration the arrangements of atoms and of molecules that are at the bases of its chemistry and morphology.

We have within the living cell a wide variety of inorganic molecules, as well as carbon-containing molecules. Although a discussion of all these compounds would involve a whole course in plant chemistry, we should have a general understanding of the diversity of these chemical compounds in the cell and of the major roles that the more important compounds play in the life of the cell. In general, we may divide the chemicals in a plant into two groups: *(a)* inorganic (substances taken into the cell from the outside) and *(b)* organic (carbon-containing compounds synthesized by the living cell from inorganic raw materials).

Inorganic Ions and Molecules in the Cell

Water is the solvent in which ions and molecules entering and leaving the cell are dissolved, and in which foods, nutrients, and organic compounds move through the cells and are transported throughout the entire plant. Water is the most abundant inorganic compound in the cell. It permeates the cell wall, makes up more than 90% of the fresh weight of protoplasm, and is the chief constituent of the cell vacuole. However, for such a small, simple molecule, water has unusual properties. These properties, which are very important to the life of the cell, are related to the fact that water molecules are **polar,** that is, have positive and negative regions. Although the water molecule has no net charge, the electrons of the two hydrogen atoms are strongly attracted to the oxygen atom in the molecule, hence two positive charges, associated with the two hydrogen atoms, are separated from two negative charges, associated with the oxygen atom (Fig. 5.1). The positive pole of a polar molecule will attract the negative poles of other polar molecules. Consequently, water molecules bind to each other as well as to other polar molecules in the cell. Water molecules may then be bound to four adjacent water molecules through the positive charges on the hydrogen and the negative charges on the oxygen (Fig. 5.1). These bonds, **hydrogen bonds,** although not as strong as covalent bonds between the hydrogen and oxygen atoms in the molecule, are responsible for the binding of water to many polar molecules in the cell. Only the most electron-negative atoms, such as oxygen, nitrogen, chlorine, and fluorine, are capable of forming hydrogen bonds. Oxygen and nitrogen are the most important hydrogen bonding atoms in organic molecules in the cell. Even seeds dried in air contain some water much of which is bound to oxygen- and nitrogen-containing groups in such cellular organic molecules as cellulose, sugar, starch, protein, and nucleic acids.

Many inorganic mineral salts are essential for the growth of plant cells (Chapter 11) and are obtained by the plant from its external environment, particularly from the soil or water in which it is growing. These inorganic salts are found in the cell generally in low amounts. They serve as raw materials that, together with carbon dioxide and water, are used to synthesize the many complex organic materials in the cell. Some inorganic materials in the cell, such as potassium, may accumulate to relatively high levels as inorganic **ions** (charged atoms or groups of atoms).

Organic Compounds in the Cell

Most of the dry weight of a plant cell is organic matter synthesized from inorganic raw materials by a cell itself

Table 5.1
SOME OF THE COMMON CHEMICAL ELEMENTS FOUND IN PLANTS

Element	Symbol	Approximate Atomic Weight	Common Valence Numbers	Common Ions in Which Elements Are Found
Elements Essential for Plant Growth				
Carbon	C	12	−4, +4	CO_3^{2-}, HCO_3^{2-}
Hydrogen	H	1	+1	H^+, H_3O^+, OH^-
Oxygen	O	16	−2	OH^-, NO_3^-
Phosphorus	P	31	−5	$H_2PO_4^-$, HPO_4^{2-}, PO_4^{3-}
Potassium	K	39.1	+1	K^+
Nitrogen	N	14	−3, +5	NH_4^+, NO_3^-
Sulfur	S	32.1	+6, −2	SO_4^{2-}
Calcium	Ca	40.1	+2	Ca^{2+}
Iron	Fe	55.9	+2, +3	Fe^{2+}, Fe^{3+}
Magnesium	Mg	24.3	+2	Mg^{2+}
Boron	B	10.8	+3	BO_3^{3-}
Zinc	Zn	65.4	+2	Zn^{2+}
Manganese	Mn	54.9	+2, +3	Mn^{2+}
Molybdenum	Mo	96	+3, +5	Mo^{3+}, MoO_2^+
Chlorine	Cl	35.5	−1	Cl^-
Copper	Cu	63.6	+1, +2	Cu^+, Cu^{2+}
Some Other Elements				
Sodium	Na	23	+1	Na^+
Silicon	Si	28.1	+4	SiO_3^{2-}
Aluminum	Al	27	+3	Al^{3+}

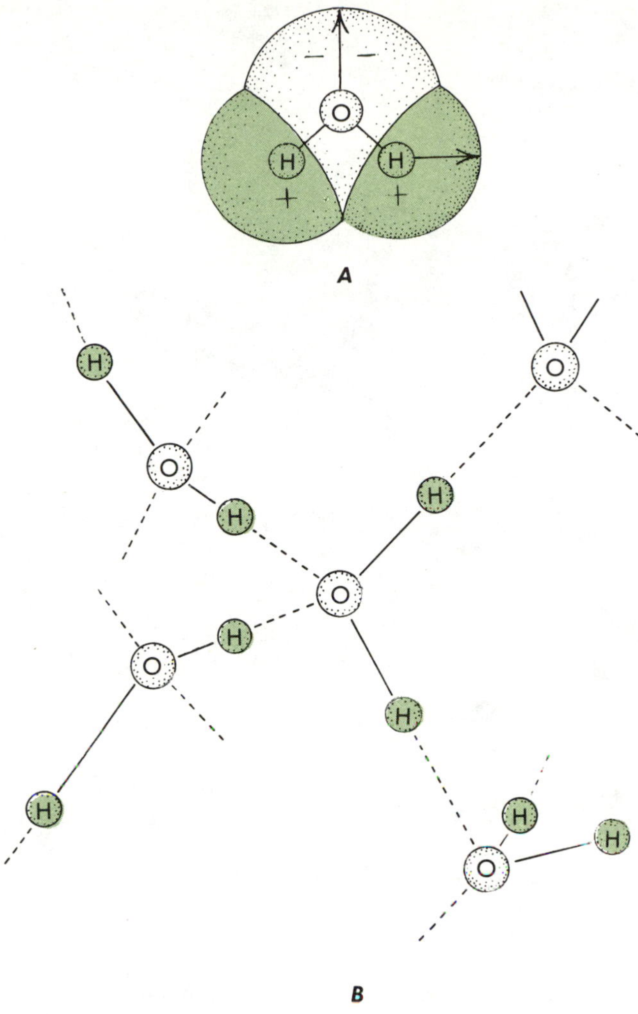

Figure 5.1
A, the polar nature of water molecule. *B*, hydrogen bonding of several water molecules.

or by other cells in the plant and moved to that cell. A wide variety of organic compounds from relatively simple sugars to extremely complex proteins occurs in the cell. Some of these molecules are made up of hundreds of atoms. Organic compounds make up the food for the cell, the structural components of the wall and the protoplasm, and many of the more special compounds such as hormones, pigments, and enzymes.

Carbohydrates

Carbohydrates are organic compounds containing carbon, hydrogen, and oxygen atoms with the general formula $(CH_2O)_n$, where *n* may be any number. They make up the major part of the cell wall and are, in addition, one of the three general types of food. The major structural carbohydrate in the wall is cellulose, whereas the principal carbohydrate foods are sugars and starches. A **food** is any organic material that serves the plant as a source of energy and as a source of carbon from which new compounds may be made. A food also supplies animals with the same source of energy and carbon.

Sugars. Let us consider briefly the structure of a sugar molecule. In one group of simple sugars, the hexoses, each molecule contains six carbon atoms and has the general formula $C_6H_{12}O_6$. For example, glucose has the straight-chain structure shown in Fig. 5.2. The $-CH_2OH$ end is an **alcohol** group; the $-CHO$ end is an **aldehyde** group. The sugar molecule may occur in two forms; when not in solution the carbon atoms form a straight chain (Fig. 5.2). When the sugar is in solution, four or five of the carbon atoms (depending on the kind of sugar) and an oxygen atom form a closed ring (Fig. 5.3). In the straight-chain form, note the difference in the end carbon atom of glucose and fructose. As noted, the $-\mathrm{C}\begin{array}{l}\diagup \mathrm{H}\\ \diagdown\!\!\diagdown \mathrm{O}\end{array}$ of glucose is an aldehyde. The $-\overset{|}{\mathrm{C}}=\mathrm{O}$ of fructose characterizes a **ketone.** Aldehydes are of considerable importance in plant metabolism, and we shall meet with them in other topics; that is, nucleic acid and lignin.

It is obvious that the -OH and -H groups of the sugar molecule may be arranged in different positions in the ring without changing the relative numbers of carbon, hydrogen, and oxygen in the formula. In fact, a shifting of these atoms, as in the examples of glucose and fructose, results in sugars of different properties. Actually, sixteen different hexoses are possible and all sixteen are known, although only a few occur naturally in plants. Molecules such as these sugars, with the same chemical composition but with different chemical properties, are called **isomers.** The most common hexoses in plants are glucose and fructose. Sugars with three carbon atoms are called trioses; with four carbon atoms, tetroses; with five and seven carbon atoms, pentoses and heptuloses. **Ribose,** a constituent of nucleic acids and other complex protoplasmic molecules, is a pentose sugar (Fig. 5.2).

Two molecules of simple sugars, **monosaccharides,** may combine with each other. Thus, if two molecules of glucose are united, with the loss of a water molecule, a **disaccharide** is formed. One such disaccharide is known as **maltose.** The union of the monosaccharides, fructose and glucose, results in the commonest of all sugars, **sucrose** (Fig. 5.4). Sucrose can easily be split into fructose and glucose. One water molecule is required to split a molecule of sucrose and the process is called **hydrolysis.** While many different kinds of disaccharides are possible, sucrose, our common table sugar, is by far the most abundant sugar in plants, where it serves as a soluble stored or movable food. Maltose occurs in a free state only to a limited extent, but may be readily obtained by the hydrolysis of starch.

Polysaccharides. Three or more monosaccharide molecules may join to form tri-, tetra-, or **polysaccharides.** The latter are composed of the union of many simple

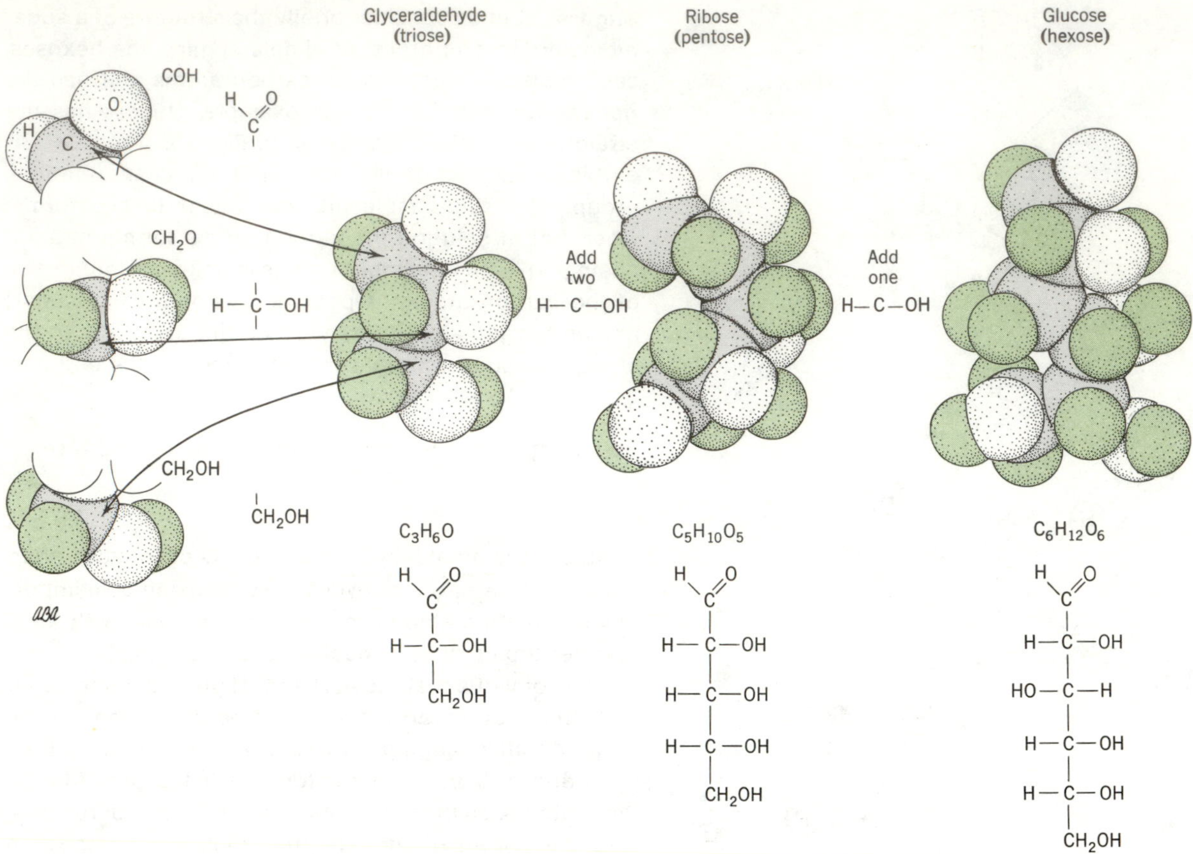

Figure 5.2
Structural formula and three-dimensional representations of 3-,5-, and 6-carbon sugars.

sugar molecules with the loss of a water molecule for each pair of simple sugar molecules united. Polysaccharides are not generally soluble in water, nor are they sweet. **Starch** and **cellulose** are the two most abundant polysaccharides in the plant. Each is composed of a long chain of many glucose molecules. In starch, the chain may be coiled because of the way the glucose units are linked together and some chains are branched, while in cellulose the chains are unbranched and more or less straight. Cellulose is a major structural material in the wall, while starch is a reserve water-insoluble food that is stored in many cells. Some cells, instead of storing starch, store a polysaccharide called **inulin** that is composed of fructose units instead of glucose units.

The union of relatively simple molecules, like sugar, into long-chain gigantic molecules composed of the repetition of simple units is a common chemical process known as **polymerization.** We shall meet it again in our discussion of proteins and nucleic acids.

The formula of cellulose is given in Fig. 5.5.

Lignin. The secondary walls of various wood cells are mostly cellulose but may be permeated by a second substance: lignin (**Fig. 4.25,** page 67). Lignin reduces infection, rot, and decay. It also is responsible for some of the strength of wood. Between 20 and 30% of the cell walls in wood is lignin. The unit structure in lignin is not a sugar but a complex alcohol. It does not polymerize in straight chains as glucose does in cellulose, but, instead forms a firm net. Lignin is thus a complex polymer with many slight variations in its molecular structure. Some plant taxa may contain only a specific type or variant of lignin. For example, oxidation of lignin from monocotyledonous plants yields, among other things, three different types of aldehydes; lignin from dicotyledonous wood yields two of these types; and lignin from gymnosperm wood yields but one of these. The lignin in these three plant groups must therefore have different structures. Primitive vascular plants contain relatively little lignin, and oxidative treatment suggests that it has a simple chemical structure. Lignin is among the most chemically inert of plant substances and it remains in fossils of woody stems (see Chap. 31).

Lipids

All living plant cells contain **lipids** (fats and fatlike substances) in their cytoplasm, and in some cells, fatty compounds are present in the cell walls. These fats serve as food. Phospholipids (lipids containing phosphorus) and other complex lipids are found in protoplasmic membranes. **Cutin** and **suberin** are lipid materials that func-

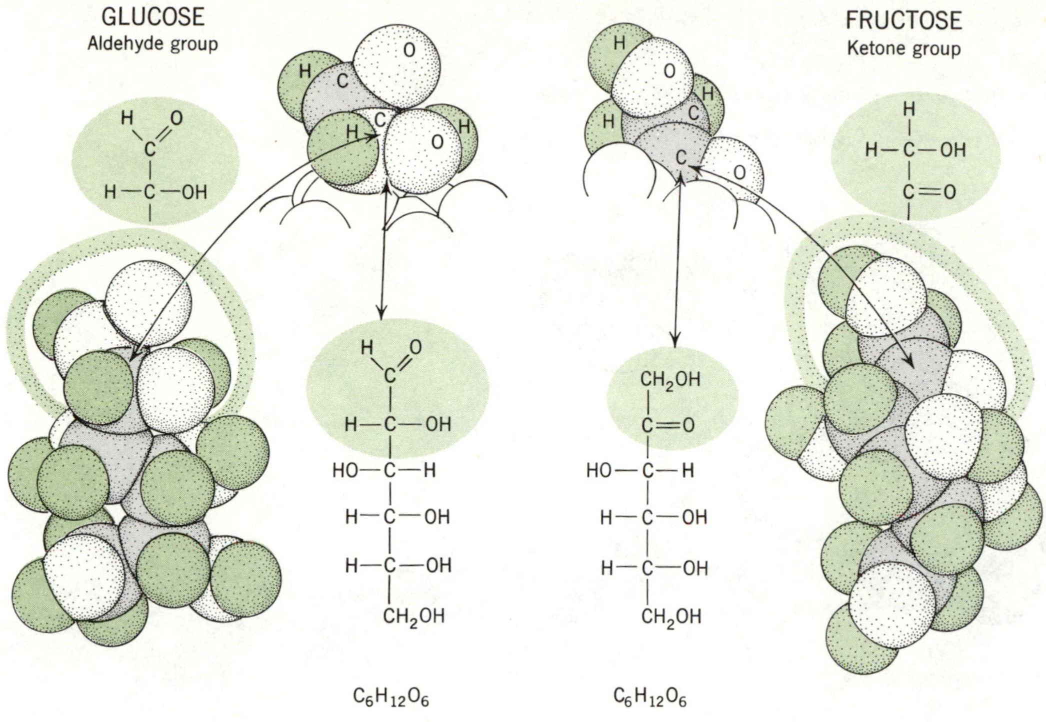

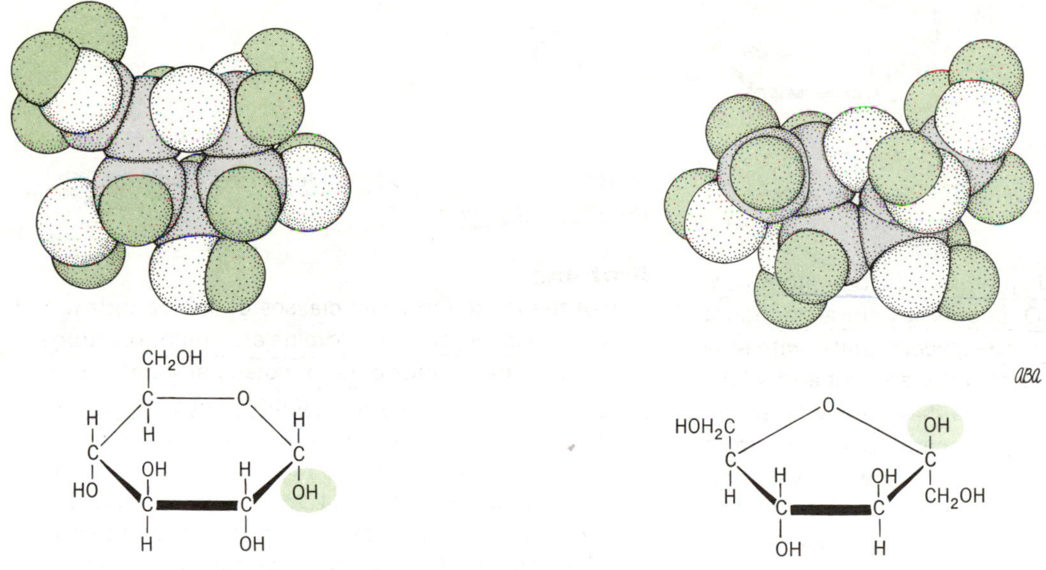

Figure 5.3
Structural formulas and three-dimensional representations of glucose and fructose.

tion in protecting the plant against excessive water loss. One common property of lipids is their relatively poor solubility in water.

Fats, the most abundant and simplest of plant lipids, are composed of fatty acids united with the three-carbon-atom alcohol, **glycerol.** Fatty acids have very little oxygen and are generally relatively long chains of carbon with only hydrogen atoms as side attachments. In lauric acid, the fatty acid shown in Figure 5.6, the carbon chain is composed of 12 carbon atoms.

SUCROSE

Glucose + Fructose ⟶ Sucrose + Water

$$C_6H_{12}O_6 + C_6H_{12}O_6 \longrightarrow C_{12}H_{22}O_{11} + H_2O$$

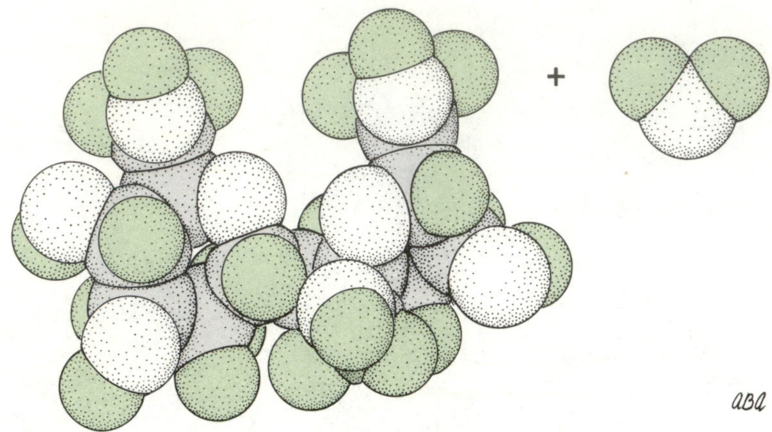

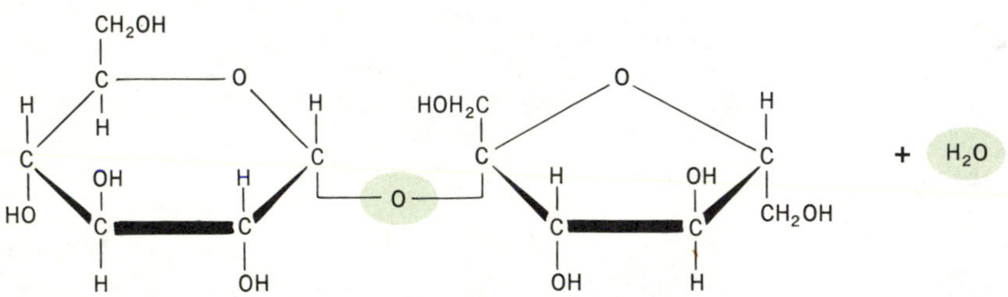

Figure 5.4
Formation, structural formula, and three-dimensional representation of sucrose.

Glycerol is a rather simple three-carbon-atom compound. Its formula is $C_3H_8O_3$ and its structural configuration is shown in Figure 5.6. When glycerol unites with three fatty acids, three molecules of water are lost and a fat is produced. Notice that in the fats there are only six oxygen atoms, while there may be many hydrogen atoms. This arrangement leaves many spots where oxygen may be introduced when fats are broken down in the cell. Fats are easily oxidized and as foods are very good sources of energy.

Waxes, which are fatty acids esterified to fatty alcohols, are sometimes found in plant cuticles and as a stored food in the seeds of at least one plant, the johoba (*Simmondsia chinensis*).

Figure 5.5
Structural formula of glucose units in cellulose.

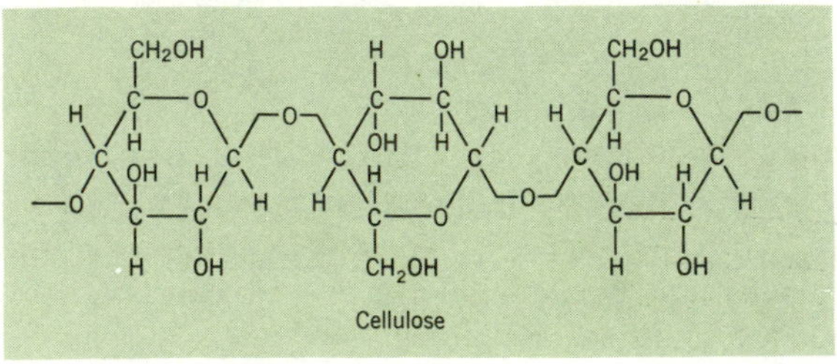

Cellulose

Proteins

One of the most important classes of compounds found in protoplasm is **protein.** Proteins are complex nitrogen-containing compounds of high molecular weight. Egg albumin and gelatin are good examples of proteins. While there are many kinds of plant proteins, they do not occur in such large masses in plants as they do in animals; hence, plant proteins are not so commonly known.

When proteins are taken apart, it is found that they are all formed of 20 different units, the **amino acids.**

Each amino acid has at least two linked carbon atoms: —C—C—. To the terminal carbon atom there are always attached (*a*) an oxygen atom, =O, and (*b*) an hydroxyl group, —OH. This terminal group of atoms, $-\mathrm{C}\begin{smallmatrix}/\!\!/\mathrm{O}\\ \backslash\mathrm{OH}\end{smallmatrix}$, is known as a carboxyl group. Its charge is negative, or acidic. To the second carbon there is always attached, at least one hydrogen atom, —H, and an amino group, $-NH_2$. The charge on the amino group is positive, or basic. The terminal structure of an amino ($-NH_2$) acid (COOH) thus is

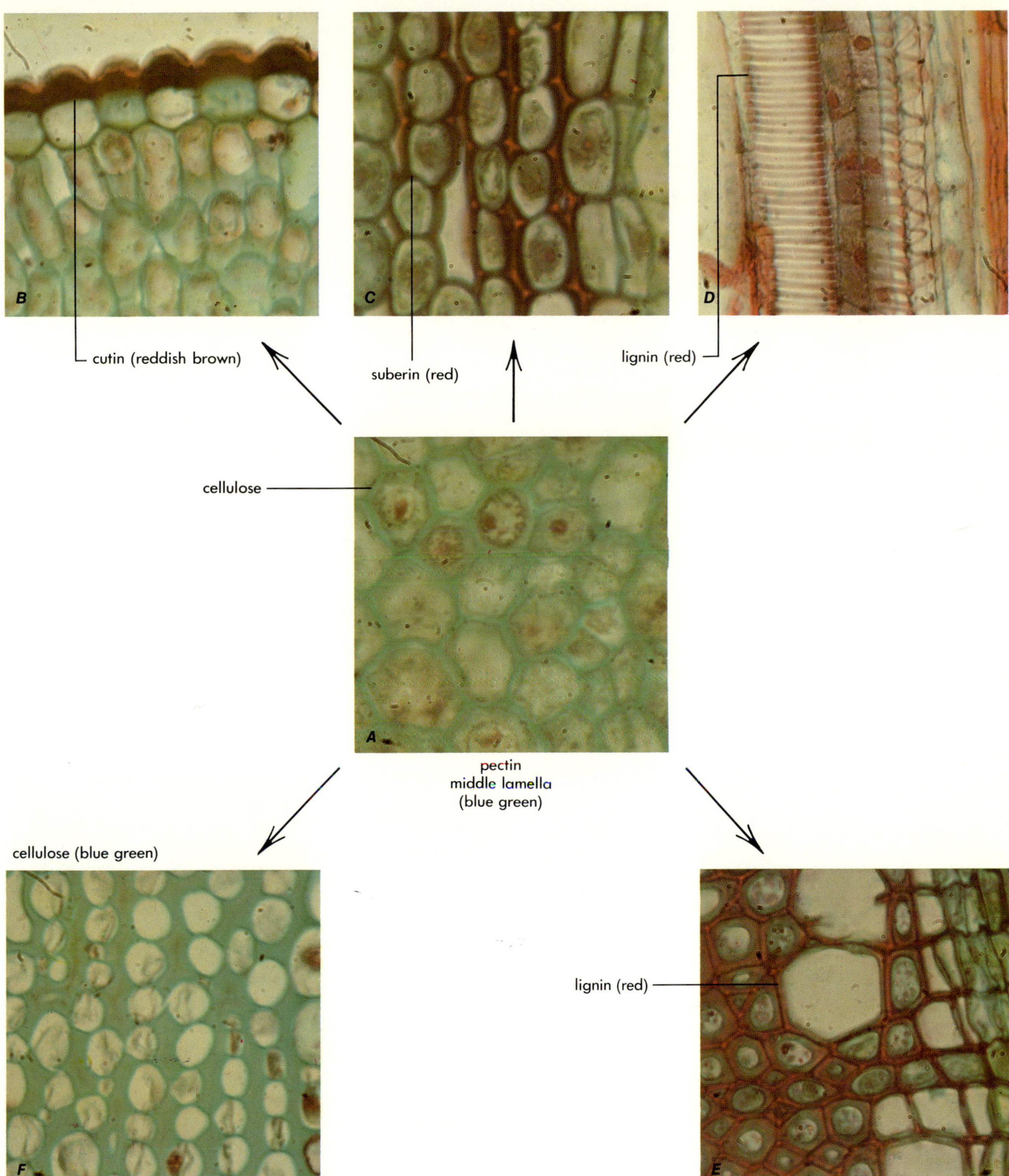

Figure 4.25
Development of cell walls. *A*, cells with cellulose (blue-green) walls. Cells are held to each other by pectin in the middle lamella. *B*, cutin (red) may be laid down on the outside of an epidermal cell. *C*, in another location, suberin (red) may be deposited within the cellulose primary cell wall to form a cork cell. *D*, During rapid elongation, lignified cellulose (red) is laid down as rings or as a spiral band. *E*, when elongation stops, the lignified cellulose forms a complete box—the secondary wall—about the cell (except for pits). *F*, some cells may form thick cellulose blue-green walls. Each specialized cell is derived from its own antecedent meristematic cell and never gives rise to another specialized cell. Note characteristic staining of chemicals in the walls.

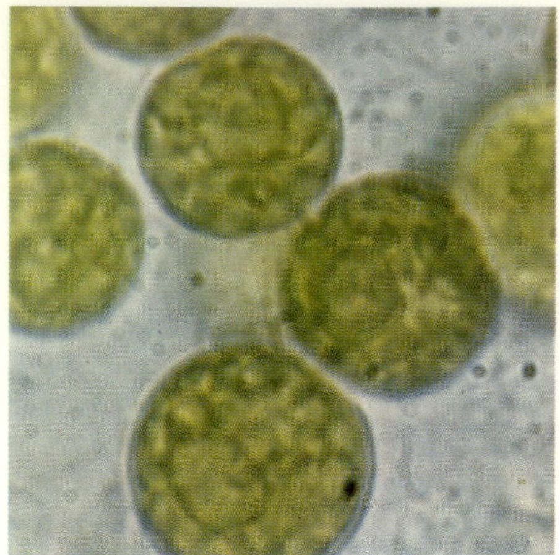

Figure 5.13
A free-floating algal cell.

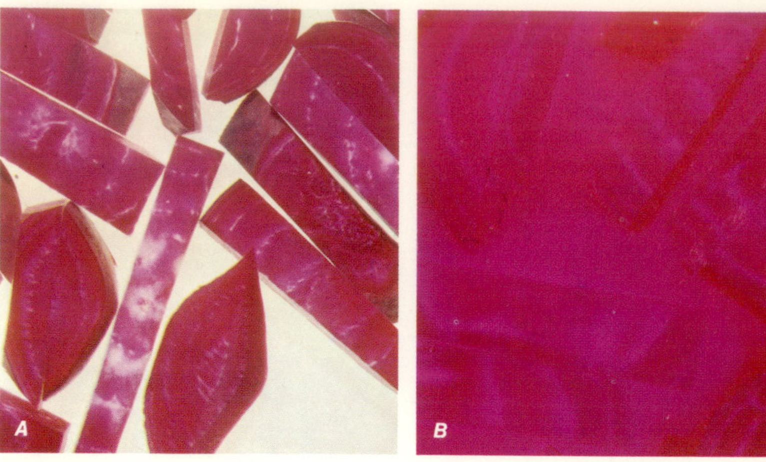

Figure 5.18
Osmosis in living and dead beet strips in water. *A,* in living cells, the membrane retains natural pigments within the cells. *B,* the pigments rapidly diffused from slices of beets placed in boiling water for 30 seconds, killing the cells.

Figure 5.17
Plasmolysis in the aquatic plant *Elodea*. *A,* living untreated cells; note the distribution of chloroplasts around sides, top, and bottom of the cell, with a vacuole occupying the rest of the cell. *B,* cells of *Elodea* are able to concentrate substances, such as the dye neutral red, in the vacuole. *C,* cells containing neutral red plasmolyzed with 10% Ca $(NO_3)^2$, reduction in size of vacuole indicates outward diffusion of water, but the dye is retained in the vacuole. *D,* in cells recently killed with formaldehyde, the protoplast may withdraw slightly from the cell wall, chloroplasts may form irregular clumps, and the nucleus frequently becomes visible. *E,* neutral red dye does not accumulate in dead cells, nor *(F)* do dead cells become plasmolyzed when placed in $Ca(NO_3)_2$.

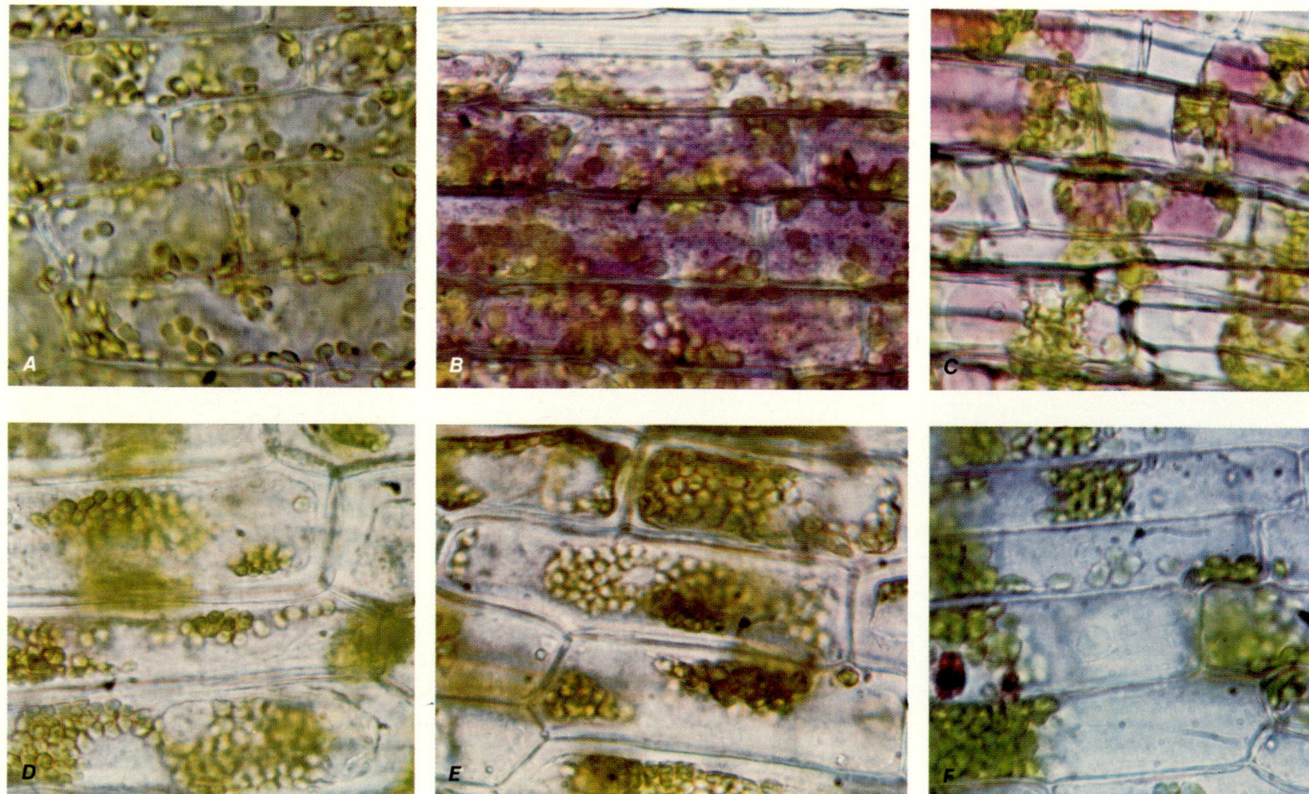

A Glycerol + 3 fatty acids ⇌ 3 water + fat

B

```
   H
   |
H—C—OH  +  HO—C(=O)—CH2—CH2—CH2—CH2—CH2—CH2—CH2—CH2—CH2—CH2—CH3  ⇌
   |
H—C—OH  +  R′
   |
H—C—OH  +  R″
   |
   H

                          H
                          |
        3 H2O   +     H—C— lauric acid
                          |
                      H—C— fatty acid R′
                          |
                      H—C— fatty acid R″
                          |
                          H
```

Figure 5.6
Structural formula for a fatty acid (lauric acid), glycerol, and a fat. *A,* in words; *B,* structural formula. R′ and R″ represents fatty acids.

```
  H     O
  |    //
 —C—C
  |    \
 NH2    OH
```

There remains one unsatisfied charge on the amino carbon. This is satisfied by a specific side chain which we may designate as R. There are 20 different side chains that occur in common amino acids; thus each amino acid is characterized by its own side chain. The simplest of these amino acids is glycine, whose side chain is a single hydrogen atom (Fig. 5.7). The structural formulas of three other amino acids tryptophane, arginine, and cystine, are also shown in Fig. 5.8.

Polymerization of these amino acids takes place in the formation of protein, and a long chain is formed in which the side chains of the amino acids extend outwardly from the backbone of the molecule (Fig. 5.9).

Consider that each of the amino acids is a symbol so that we can form an alphabet with them of more than 20 characters. From the 26 letters of our alphabet, all of the countless words of all of the known languages can be formed. Furthermore, the great majority of these words have no more than five letters and very few of them have more than 20 letters. Even the simplest proteins have 1000 or more amino acids, and the common number of amino acids per protein is of the order of 100,000. Proteins are **macromolecules.** The possible number of different kinds of proteins formed from these 20 different amino acids, taken in any order, is enormous. One of the most exciting fields of modern biology is in the determination of the exact sequence of amino acids in protein molecules and in determining how living cells regulate and control the specific kinds of protein that they produce.

Proteins are the basic building block of protoplasm. They account for its great activity and are the reason for our previous statement that the protoplasms of similar cells are always somewhat different.

Long protein molecules can fold, bend, and coil in many ways (Fig. 5.11). Particularly important is a folded or helical structure that many chains assume. In the cytoplasm of the cell, the long protein molecules tend to react with one another to form weak cross-linkages, and this increases the viscosity of the solution. The streaming of protoplasm in the cell must be viewed as involving the breaking and reforming of many weak cross-linking bonds among the protein molecules. The primary valence bonds in the side chains or backbones of protein molecules remain intact under these conditions of streaming.

Histones. DNA in eucaryotic cells (see Chapter 6) is complexed closely with both acidic proteins and **histones (basic proteins)** to form **chromatin.** The histones complexed with DNA are low molecular weight proteins (8000 to 26,000 m.w.) that contain large amounts of the amino acids lysine and arginine in regularly spaced positions. These amino acids are positively charged at pH 7.0 and are able to form ionic bonds with the negatively charged groups on the periphery of the DNA molecule. The precise role of the histone-DNA interaction is not completely understood, but certain aspects of the protein regulation of gene (DNA) expression are being investigated. One interesting idea stems from the observation that isolated chromatin in certain animal cells exists in a curious "string-of-beads" fashion (Fig. 5.10). It has been suggested that these beadlike structures consist of an aggregate of histone surrounded by coils of DNA. These "beads" are called **nucleosomes,** and occur in regularly spaced 20 to 30 nm sequences. The spaces between nucleosomes are associated with a specific histone called **H1.**

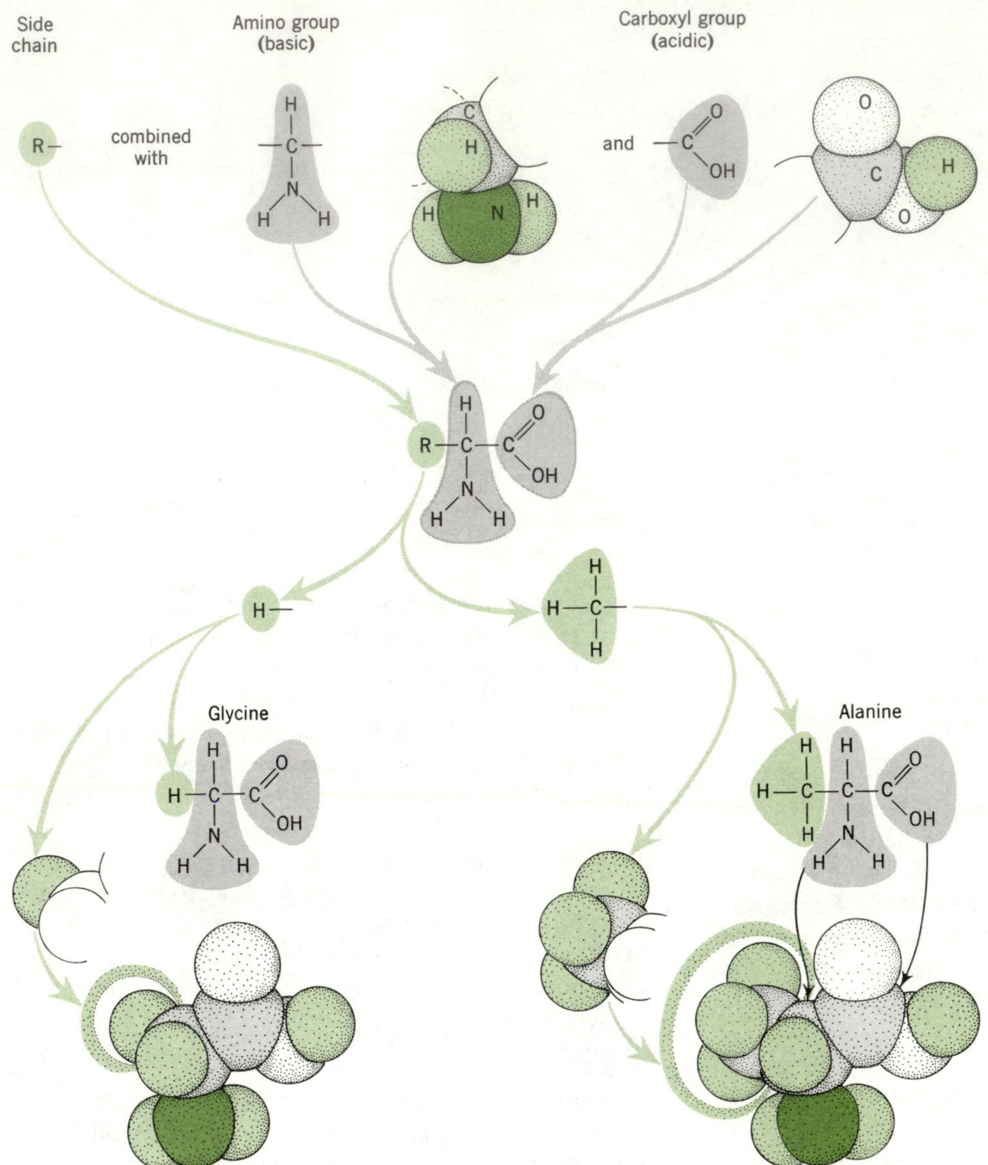

Figure 5.7
Structural formulas and three-dimensional representations of several amino acids. R represents a side chain.

The nucleosome organization probably serves two useful functions. One function is to provide a means to systematically coil the DNA fiber so that the long DNA strand can be packed in an organized way into the cell nucleus. The second function concerns the regulation of DNA expression. We shall discuss histones again in the next chapter on cell division.

Enzymes. While there are many ways in which proteins play a part in the economy of the cell, let us consider for the moment only the important part they take in controlling the chemical reactions of the cell. For example, we may look at one step in the energy flow in the cells of plants and animals, in which the energy stored in foods is released to perform useful work. Burning gasoline in a flask releases uncontrolled and therefore wasted energy. Burning gasoline in a car also releases energy, but the design of the motor controls the release so that useful work is performed. Still, much of the energy is lost as heat, as the temperature of the motor illustrates. Sugar is burned in the cells of your body and in the cells of a plant, again with the release of energy and the performance of useful work, although some energy is lost as heat. A primary function of certain specific proteins is to control and to direct chemical reactions in the cell, limiting them to very specific pathways and bringing them about at relatively low temperatures.

Substances that increase the rate of chemical

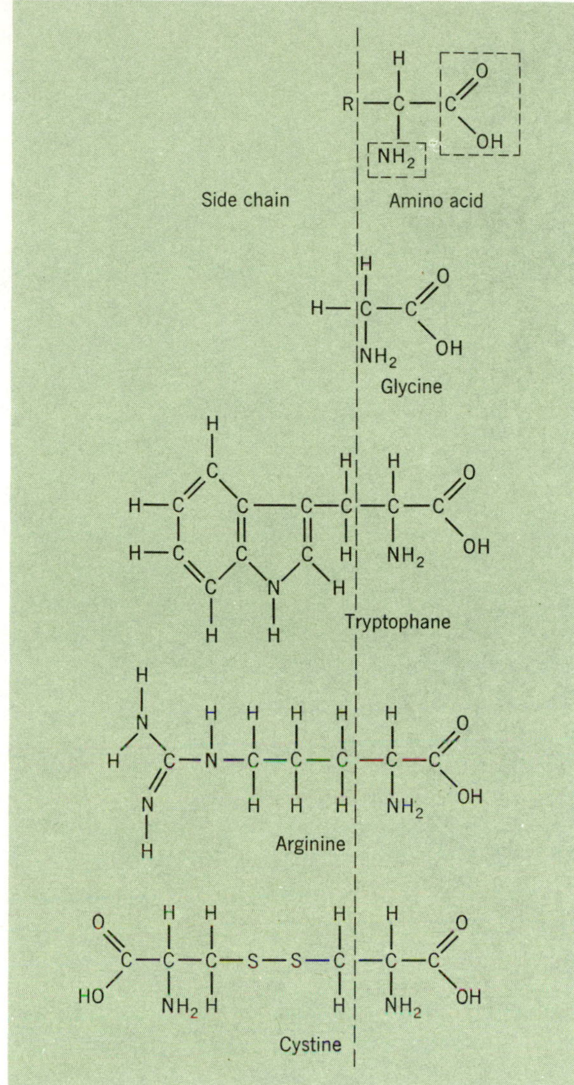

Figure 5.8
Some other amino acids.

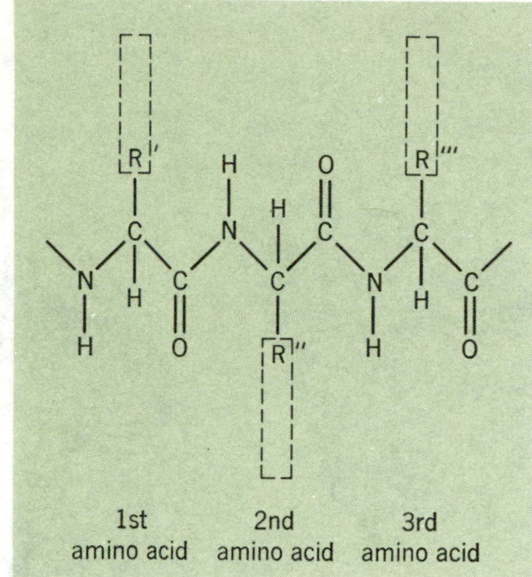

Figure 5.9
A polypeptide chain is a part of a protein molecule. The backbone of the protein molecule is formed by many amino acids joined by the union of the amino group (NH_2) of one amino acid to the acid group (COOH) of another amino acid by the removal of a water molecule. The R groups represent side chains of the different amino acids.

change without themselves appearing to change in the reaction are **catalysts.** There are both *inorganic catalysts* and *organic catalysts.* Organic catalysts produced by organisms are called **enzymes.** Enzymes are proteins. They are frequently very specific in their activity, catalyzing only one, or a few, chemical changes. The shape of the enzyme molecule plays an important role in both enzyme specificity and activity. Before an enzyme can affect a chemical reaction, it must unite with the reacting molecule or molecules. Consequently, some part of the three dimensional structure of the enzyme molecule matches the structure of the reacting molecules in a way analogous to the matching of a key to a lock.

Figure 5.11*A* is a pair of stereo drawings of a space-filled model of an enzyme from yeast, yeast hexokinase. This model illustrates how an enzyme has a specific three-dimensional shape into which the substrate fits. Thus, in this enzyme, the cleft that divides the molecule into two lobes is the binding site for the substrate, glucose. Figure 5.11*B* shows that when the substrate, in this case glucose, complexes with the enzyme, it causes a conformational change in the enzyme molecule.

The structure of the enzyme specifically limits the type of combining material to only those molecules having a shape that complements that of the enzyme mole-

Figure 5.10
The possible relation between histones and DNA. (Redrawn from D. E. Olins and A. L. Olins *Amer. Sci.* **66,** 704. © 1978 by Sigma Xi. Reproduced by permission of the publisher.)

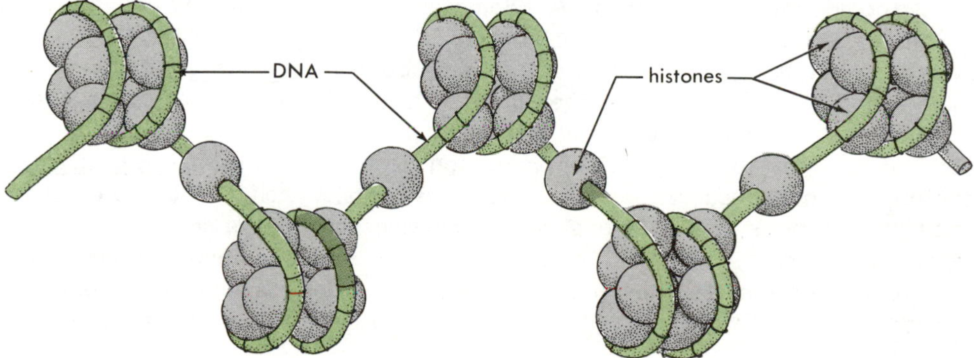

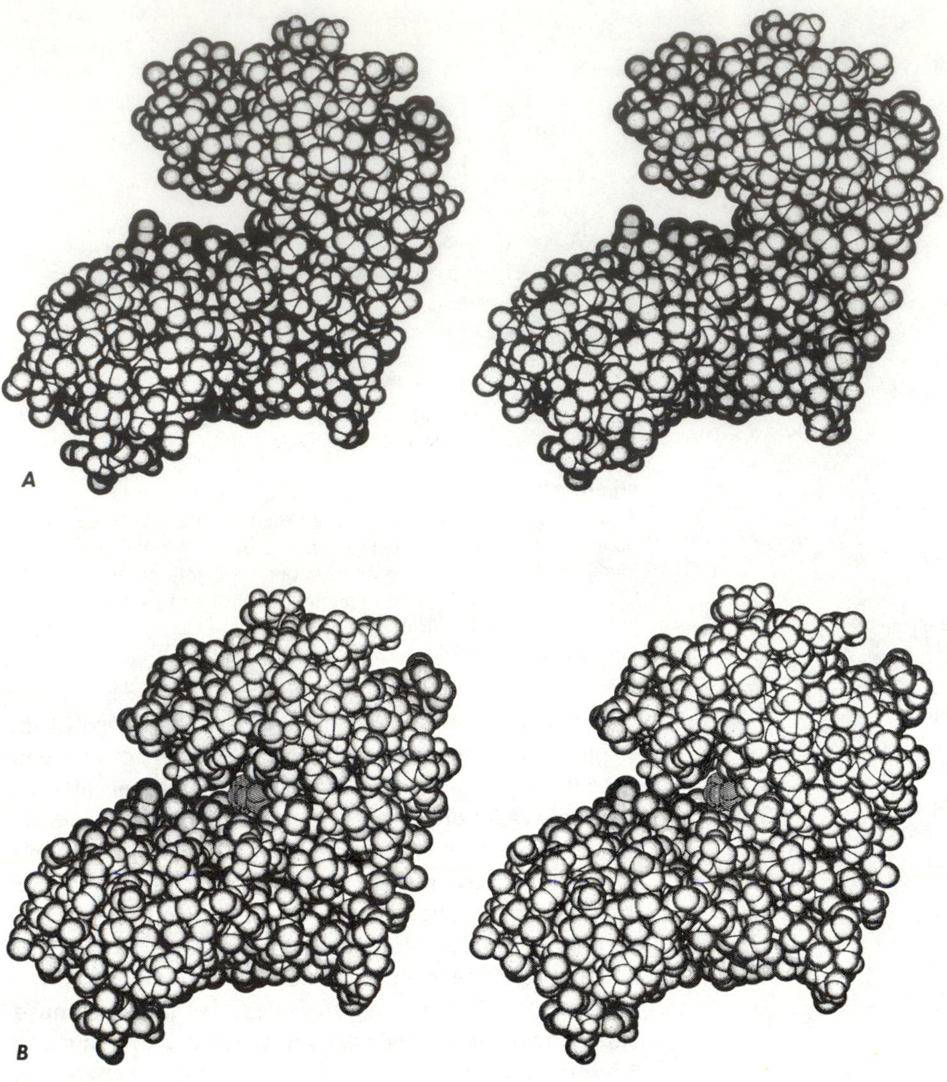

Figure 5.11
A stereo drawing of model of yeast hexokinase. *A*, in its native state. *B*, a stereo drawing of yeast hexokinase complexed with glucose. Note active site of enzyme is closed when enzyme is complexed with glucose, and the glucose molecule is almost completely buried in the cleft of the enzyme molecule. (Courtesy of C. M. Anderson, F. H. Zucker, and T. A. Steitz, *Science,* **204,** 375. © 1979 by Amer. Assoc. Adv. Sci.)

cule. Thus, only highly selected molecules are brought into close proximity by combining with the enzyme molecule. Once they are closely associated, the combining molecules may react with each other, or with other molecules. At the completion of the reaction, the newly formed compound or compounds, the products of the reaction, are released from the enzyme molecule. The enzyme molecule itself, being little if any changed during the reaction, is then ready for further activity. Thus, an enzyme greatly modifies the rate of a reaction without being much changed itself. Since enzymes regulate and control the biochemical reactions that are fundamental to all life, the importance and significance of enzymes is readily appreciated.

Enzymes have certain optimum conditions for their greatest activity. They may be destroyed by heat and by some metals and certain other substances. Many of the numerous steps involved in photosynthesis, respiration, and digestion are catalyzed by enzymes. Enzymes are active in practically all cellular processes.

Coenzyme Activators. Some enzymes depend only on their protein structure for their activity, while others are unable to function in the absence of certain nonprotein substances called **cofactors.** In some instances, metal ions may serve as cofactors and either bind the substrate to the enzyme or actually serve as the catalytic group itself. One group of enzymes that have metal-containing active sites are the cytochromes. We will see that cytochromes are very important in the oxidation–reduction reactions of photosynthesis and respiration. In some instances the nonprotein cofactors may be organic molecules called **coenzymes.** Coenzymes sometimes are bound tightly to specific enzymes, but in other cases the coenzyme may be very loosely bound to the enzyme. If the coenzyme is only loosely associated with the enzyme, it may actually diffuse away and take part in other enzyme reactions; that is, the same coenzyme may act in two different enzyme reactions. Thus, if the coenzyme is capable of accepting and then giving off hydrogen atoms (or electrons), it might serve as a carrier of hydrogen from one reaction to another.

Many **vitamins** serve as active groups of coenzymes. Thus, members of the vitamin B complex, riboflavine, thiamine, nicotinamide, and pantothenic acid all are found as active groups of certain coenzymes and in this way serve essential roles in the life of plants and animals. Four coenzymes that will be particularly important in our study of plant metabolism are listed in Table 5.2. You will note that the first three of these, NAD, NADP, and FMN, serve as carriers of hydrogen (electrons) between two different reactions in the cell. The fourth, CoA, acts as a carrier of an organic group containing two carbon atoms. It receives the group in one enzymatic reaction and may act as a coenzyme in another enzyme reaction where the two-carbon-atom group is used.

Pigments in the Cell

Pigments are colored compounds, and since they are colored they must somehow absorb certain wavelengths of light. Plant pigments may be divided for convenience into water-soluble pigments that are usually found in the vacuoles of the protoplasts, and lipid-soluble pigments that occur in the plastids. Vacuolar pigments (anthocyanins and anthoxanthins) have already been discussed on p. 56. Several kinds of other pigments occur in the plastids. The principal ones in the chloroplasts of higher plants are the green chlorophylls, chlorophyll *a* (Fig. 5.12) and chlorophyll *b*, and the yellow and red carotenoids, the carotenes and the xanthophylls.

We have noted that carotene apparently serves to trap the energy of certain wavelengths of light and to pass it on to the chlorophyll molecule. It also is an effective agent against photoxidation in the cell and a precursor of vitamin A. The carotenes and xanthophylls are

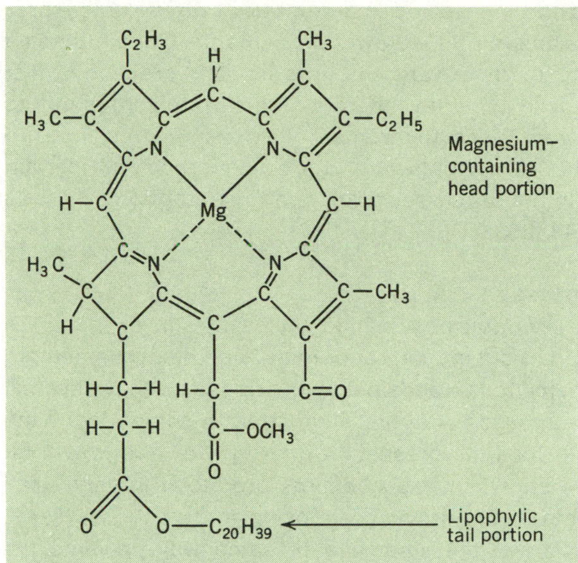

Figure 5.12
Chlorophyll *a* molecule.

responsible for much of the fall leaf coloring and the color of ripe fruits.

There is an interesting group of water-soluble pigments important in the trapping of light in the blue-green and red algae. These are known as the phycobilins. They are able to trap light that is only poorly absorbed by chlorophyll and pass the trapped energy on to chlorophyll for the normal steps of photosynthesis. The phycobilins are responsible for some of the color of the blue-green and red algae.

Chlorophyll has a molecular structure of moderate complexity. It consists of a head and a tail region. At the center of the head is a single atom of magnesium that is surrounded by four complex rings of carbon and nitrogen. The tail is a fat-soluble alcohol composed of a long carbon chain flanked by atoms of hydrogen. The structural formula of chlorophyll *a* is shown in Fig. 5.12, other chlorophylls—*b*, *c*, and *d*—differ slightly.

Other very important pigments occur in plant cells but are inconspicuous. One such pigment is phytochrome. It is found in extremely small amounts in

Table 5.2
SOME COENZYMES IMPORTANT IN METABOLISM AND HAVING VITAMINS AS REACTIVE GROUPS

Coenzyme	Abbreviation	Vitamin Associated With Activity	Reaction in Which Coenzyme Functions
Nicotinamide adenine dinucleotide	NAD	Nicotinamide	Transfer of hydrogen (electrons)
Nicotinamide adenine dinucleotide phosphate	NADP	Nicotinamide	Transfer of hydrogen (electrons)
Flavine mononucleotide	FMN	Riboflavine	Transfer of hydrogen (electrons)
Coenzyme A	CoA	Pantothenic acid	Transfer of two-carbon-atom groups (acetyl groups)

leaves and is involved in a number of physiological processes including the flowering of plants. Other pigments that occur in minute amounts in cells are riboflavine, which is important in the bending response of plants to light, and the cytochromes, yellow pigments that are found in the cytoplasm and are concerned with certain steps in the processes of respiration and photosynthesis (pp. 243, 263).

Hormones

It is a common observation that if the tip of a plant is pinched or cut off, new shoots develop below the injured area. Why do the buds below the tips grow only after the tip is removed? Are they stimulated in some way by the injury? Does the presence of the tip prevent growth? It can be shown that a chemical produced in very small amounts in the growing shoot tip and youngest leaves moves down the stem and in some way prevents the growth of the lateral buds. When the tip is removed, this chemical decreases in concentration in the stem and the lateral buds begin to grow.

Thus, plants, like animals, produce chemical messengers that move in very low concentrations from one part of the plant to another and bring about many important changes in growth and development. Such naturally produced growth regulators are called **hormones.**

We shall consider the different plant hormones in detail when we study plant growth and development (Chapter 20).

In addition to the compounds that we have discussed, many other important substances are found in the cytoplasm, nucleus, and vacuoles of the cell. The chemistry of the nucleus in particular will be considered in more detail in the following chapter.

PHYSIOLOGY OF THE CELL

We have studied the structure and some of the chemistry of the individual plant cell and we have seen that most plants are composed of countless numbers of these cells. A knowledge of the structure of a cell, a tissue, or an organ is essential to an understanding of their activities or functions. An automobile is composed of several thousand parts, each of which has a particular composition, shape, and location upon which depends its special role in the proper operation of the machine. The human eye is a complex organ in which the many individual parts are "put together" in such a way that the function of sight is possible. Likewise, the cell is a most complex structure—a *living structure.* It is the structural and functional unit of the plant. It has work to do; that is, it has functions or activities. **Physiology** is the study of the functions or activities of cells or organisms.

Our study of the physiology of cells will be simplified if we consider a chlorophyll-containing plant such as a unicellular alga **(Fig. 5.13,** page 68) floating freely in water. This one-celled plant carries on all the functions that are necessary to maintain its life. Within this microscopic bit of living substance a number of physical and chemical processes occur simultaneously. Each cell is (*a*) absorbing materials such as water, mineral salts, and gases from outside its own body, and simultaneously losing materials to the external environment; (*b*) building foods from the materials and light energy that it absorbs; (*c*) digesting foods (changing complex foods to a simple form); (*d*) respiring, thus releasing energy for various activities; (*e*) building protoplasm from the foods; (*f*) growing; and (*g*) producing new cells.

We shall briefly discuss each of these physiological processes. However, it is not possible to elaborate on several of them without a knowledge of the structure of the plant as a whole. Therefore, the following discussion merely defines some of these processes, reserving a more complete discussion for later chapters.

Since the life activities of all cells are carried out in a water environment and often involve the movement of solutes throughout the individual cell, between cells, and between the cell and its external environment, it is appropriate that we first consider how a cell obtains water and solutes. Why does water move into some cells and out of others? Can the cell regulate the flow of water through it? Does the cell select the solutes that it absorbs? How does the structure of the cell influence water or salt absorption? Is energy necessary for either process? The answers to these and similar questions lie in the ultimate relation between cellular structure and function.

Permeability of Plant Membranes

Diffusion

All molecules are in motion and may move from place to place. In liquids and gases, molecules move freely; they will move from a region of higher concentration to a region of lower concentration. This is **diffusion.** It is a very important basis for physiological activities in plants.

Permeable Membranes

Let us now consider the green plant floating in water. Bathing the cell on all sides is a dilute but exceedingly complex solution of molecules and ions that are constantly bombarding the outer surface of the cell, the cell wall. This wall, though apparently continuous when viewed through a microscope, is in reality quite porous, with microcapillary spaces existing between the interwoven cellulose **microfibrils** (Fig. 5.14). Many of these spaces are filled with substances such as pectates, but many are filled instead with liquid. Cellulose and pectic substances in the cell wall are polar. Consequently, water is found throughout the wall, forming a continuous pathway through which solute particles can freely diffuse. The process in which water is attracted to **(adsorbed on)** the surfaces of the cellulose microfibrils, causing them to move apart and the wall to swell, is called **imbibition.**

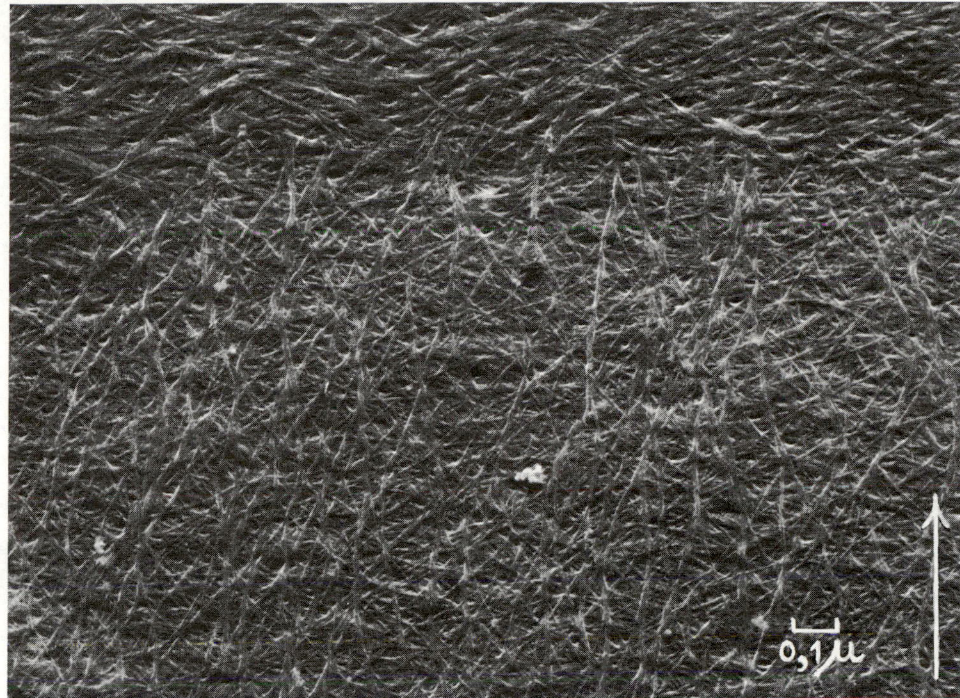

Figure 5.14
Electron micrograph of the cell wall of a cotton fiber, ×60,000. (Courtesy of P. A. Roelofsen.)

Cellulose imbibes water and the cell wall thus allows the relatively free passage of water and dissolved materials. The cell wall is **permeable.** Cell walls found in the outer surface of higher plants may be impregnated with fatty materials such as cutin and suberin. These walls do not imbibe water, and are **impermeable** to water and dissolved substances, though they may be permeable to substances soluble in fat.

Differentially Permeable Membranes

Lying in intimate contact with the wall of a living cell is the outer cytoplasmic or plasma membrane, the **plasmalemma,** which is the first important barrier to the free passage of the molecules and ions that bombard the cell from the outside. This limiting surface of the protoplast is characterized by a different physical and chemical composition from the rest of the protoplasm. It is not a rigid, unchanging surface, but an ever-changing dynamic "guardian of life," for if it is destroyed, the cell dies. Recall that this membrane is believed to be a lipid-protein mosaic. Water can pass readily through the plasmalemma, but the passage of many dissolved materials is restricted or prevented. It is **differentially permeable,** allowing some substances to pass freely, others to pass slowly, and still others to pass hardly at all.

The inner cytoplasmic membrane, the **tonoplast,** surrounding the vacuole, is also differentially permeable. It is more fatty in nature than the plasmalemma and consequently, certain substances diffuse through the cell wall and the plasmalemma into the cytoplasm but not into the vacuole. As we have already seen, the vacuolar sap frequently contains high concentrations of plant products (such as anthocyanins) that are not found in the cytoplasm. Destruction of the cytoplasmic membranes—by immersing the cell in alcohol, for instance—renders the membranes permeable, allowing the outward diffusion of vacuolar materials, and results in the death of the cell.

Within the cell itself, solutes are not free to diffuse at random throughout the protoplast. In addition to the tonoplast that restricts the movement of materials between the cytoplasm and the vacuole, diffusion of materials into and out of membrane-bound organelles is restricted. Moreover, the colloidal nature of the cytoplasm itself presents a delicate but ever-changing architecture within which the foods, vitamins, hormones, enzymes, and inorganic ions are so distributed that the intricate yet precise processes of life proceed.

Movement of Substances through Differentially Permeable Membranes

Membranes in Nonliving Systems

The differentially permeable membranes of the cell play very important roles in the maintenance of life activities. Certain features of their behavior can best be explained by reference to the simple appartus called an **osmometer** (Fig. 5.15). The cylindrical sac of an osmometer is made of parchment paper, plastic, or other material that is differentially permeable. This sac is filled with a concentrated sugar solution and stoppered with a tight-fitting

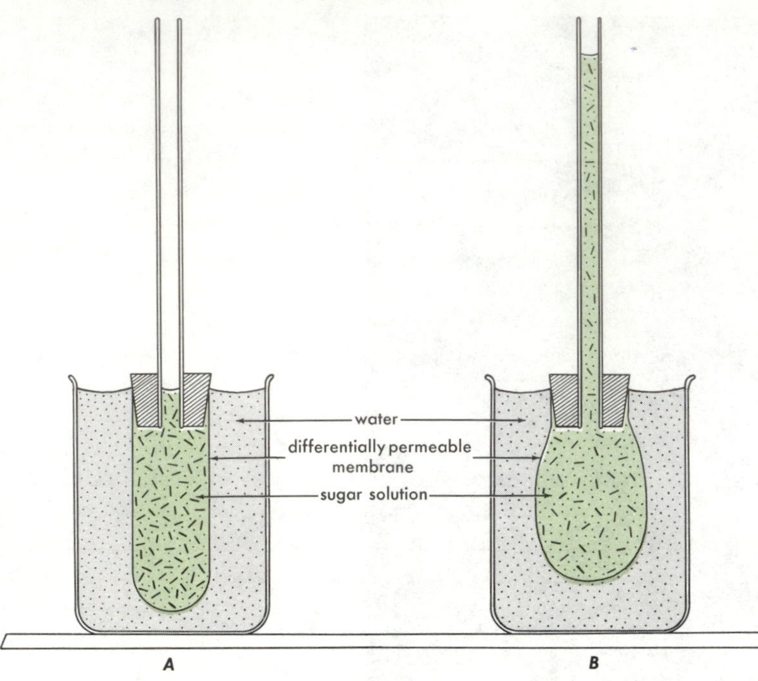

Figure 5.15
Diagram of osmometer and diffusion through a differentially permeable membrane. *A*, before osmosis has occurred; *B*, after osmosis.

rubber stopper through which a glass tube is fitted; the sac is then immersed in distilled water (pure solvent). In following this discussion we should keep in mind that the sap of a living algal cell is usually more concentrated in total solutes (salts, sugars, and the like) than is the pond water in which it may be floating. Movement of water into the plant cell obeys the same laws as those governing movement of water into an osmometer.

Turgor Pressure

Outside the osmometer are water molecules, and inside are water molecules and sugar molecules. The membrane, being differentially permeable, permits the free movement of water molecules inward or outward and retards or prevents the free movement of sugar molecules outward. After a short time the liquid rises in the tube, and the sac becomes distended, or **turgid.** The pressure of the solution against the sac, the **turgor pressure,** is a measurable pressure opposed by an equal and opposite pressure of the wall of the sac against its contents (the **wall pressure**). Evidently, water molecules are moving inward more rapidly than outward. Under these conditions the **activity (chemical potential)** or **"free energy"** of water molecules outside the sac is greater than the activity or "free energy" of water molecules inside the sac; thus, water is driven into the sac. As water continues to diffuse into the sac, turgor pressure is built up inside it. This pressure increases the activity of water molecules in the sac, and the tendency for water to diffuse back out of the sac gradually increases.

This outward diffusion of water finally equals the inward diffusion. At this point, there is no further net movement of water into the sac, and the system is at equilibrium. Water molecules are passing through the membrane in both directions at the same rate. An important conclusion to be drawn from this demonstration is that (*a*) the *activity of water is reduced by the presence of solute particles* that reduce the concentration of the water, and (*b*) the *activity of water is increased by the development of a physical turgor pressure in the solution.* These two factors, solutes and turgor pressure, thus have opposite effects on the activity of water. In our example, equilibrium was attained when these two factors balanced each other and water within the sac had the same activity as the pure water outside.

The absolute value for the activity or chemical potential of water in a cell or in a solution is not easily measured. *Differences* in activity, however, can be determined, and pure water is the standard reference by which these differences can be conveniently measured. This difference between the activity of water molecules in pure water at atmospheric pressure (i.e., in an open container) and the activity of water molecules in any other system (i.e., water in a cell, in a solution, or in the sac) may be called the **water potential** of the cell or solution. Pure water has a water potential of zero under standard conditions of pressure and temperature. Since pure water is used as a standard and the activity of water in a cell or a solution is usually less than that of pure water, the water potential of a cell or a solution is usually a negative number.

The activity of water molecules, and thus the water potential in a system, is affected by several factors. The activity is decreased by the addition of solutes; it is increased by the physical pressure of cell walls on the aqueous cell contents and also by an increase in temperature. Water will diffuse from regions of high water potential to regions of lower water potential (i.e., from a dilute solution toward a more concentrated solution, when solute concentration is the only factor affecting the water potential). The water potential may be regarded as the driving force that causes a net movement of water in any system.

Osmosis

The movement of water (a solvent) through a differentially permeable membrane is called **osmosis.** Osmosis is not mysterious, but is simply a special example in which the differentially permeable membrane of a sac, or of a cell, separates the internal water that has a given water potential from the water of the surrounding environment that has a different water potential. Osmosis occurs (i.e., water moves across the membrane) along the water potential gradient from a region of higher to a region of lower potential of water. The cell gains or loses water until the potentials of water on both sides of the membrane are equal.

Although we may consider osmosis to be a special case of diffusion, the exact mechanism involved in the passage of water through the micropores of the membrane is still in doubt. The rates of water movement, when there are differences in water potential across membranes, have been calculated and found to be too high to be accounted for by simple diffusion. It has been suggested that perhaps some water molecules are moved in bulk along the water potential gradient across the membrane, possibly in response to the hydrogen bonds linking the water molecules together.

As we have seen, the presence of solute particles (i.e., molecules or ions) lowers the activity of the water in which they are dissolved. This effect on the water potential is proportional to the total number of dissolved particles (molecules and ions) in a given volume of water. Ten grams of sugar dissolved in 50 ml of water will lower the water potential approximately twice as much as will 5 grams of sugar dissolved in 50 ml of water.

This reduction of the activity of water caused by the presence of dissolved particles in it can be measured. If this solution is placed in an osmometer with pure water outside, osmosis will occur and turgor pressure will develop as long as there is any difference in the potential of water across the membrane. The theoretical maximum turgor pressure that could develop in this system equals the amount of pressure that the water potential in the solution is reduced by the solute particles in it.

Osmotic Pressure

The term **osmotic pressure,** sometimes called osmotic concentration or osmotic potential, is used to express these two properties of solutions, namely (*a*) the amount of reduction of the water potential in the solution caused by the presence of dissolved particles, and (*b*) the potential maximum turgor pressure that could develop as a result of osmosis. The osmotic pressure of a solution is usually expressed in units of atmospheric pressure. One atmosphere of pressure is approximately 15 lb per sq in. or in the metric system one atmosphere pressure approximately equals one bar (1 **bar** = 0.987 atm) or 1.03 kilograms per sq cm (Fig. 15.16). The osmotic pressure of a solution is not an actual physical pressure that exists in the solution. We might say that it is a power rating of the solution, just as an automobile engine has a rating of 150 horsepower. We do not mean that when the motor is not operating it is doing the work of 150 horses. It is only when the proper conditions exist (i.e., when the motor is operating), that the ability to do work—the power—actually results in the work being done. A solution having an osmotic pressure of 5 bars and standing in a beaker has the capacity to develop a turgor pressure of 5 bars only when placed in an ideal osmometer with pure water outside. This turgor pressure will develop because the water potential of pure water is 5 bars higher than the water potential in the solution. If, however, the solution were placed in an osmometer and immersed in another solution having an osmotic pressure of 3 bars, osmosis would occur only until a turgor pressure of 2 bars had developed in the osmometer, because in this case the water potential in the solution outside the sac would only be 2 bars higher than the water potential inside the sac.

Membranes in Living Cells

Water moves along a gradient from the external environment into the plant cell—whether it is the single cell of an alga, a root cell of a flowering plant, or a seed—along a gradient of water potential. Usually, the concentration of

Figure 5.16
Diagram showing the relationship of one atmosphere pressure to the height of mercury and water columns that it will balance.

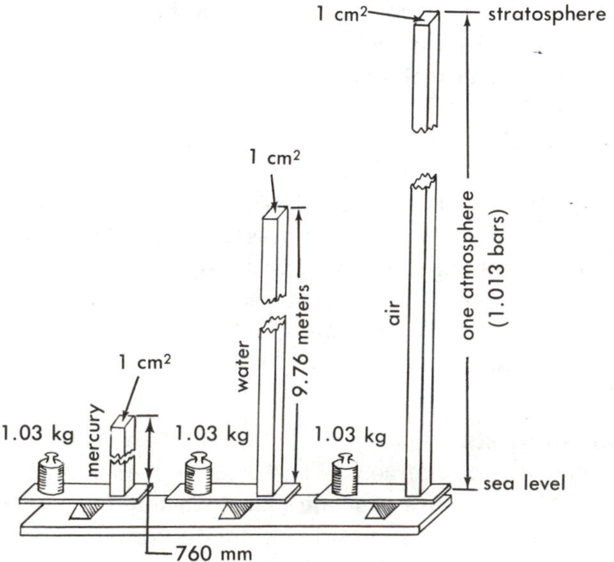

the osmotically active solutes (the osmotic pressure of the cell sap) and the turgor pressure in the cell largely determine the water potential in the cell. In certain instances, such as air-dried seeds, there is present a high percentage of colloidal material such as starch, protein, and cellulose, all of which have a great affinity for water and hence a large capacity for imbibing water. When these seeds are planted in moist soil, the water in the seeds has a very low potential compared with the water in the soil. Consequently, water diffuses into these seeds, causing them to swell. In this case, the colloidal materials largely contribute to the low activity of water in the seed.

The activity of water in a pond is primarily determined by the concentration of dissolved material, but in the case of water in the soil an additional factor is involved, namely, the presence of colloidal soil particles that imbibe water and thus lower its activity. Nevertheless, the direction of water movement will be from a region where the water potential is high (near zero) to a region where the water potential is lower (more negative).

We have seen that turgor pressures result because of differences in solute concentrations. Although the pressure outward is often considerable, the protoplast is prevented from bursting by the elastic cell wall, which resists stretching and exerts a pressure inward. The cell wall serves the same purpose that the casing of an automobile tire does in preventing the bursting of the inner tube.

It is obvious that in a cell the *turgor pressure outward* against the wall is equal to the *wall pressure inward* against the cell contents. A cell may have varying degrees of turgidity. The maximum turgor pressure that can be developed in a cell is equal to the osmotic pressure of the cell contents. That is, when a cell has absorbed all the water it can, osmotic pressure, turgor pressure, and wall pressure are equal. Normally, most living cells are in a condition in which turgor pressure is somewhat less than the maximum pressure possible. This means they are capable of taking in more water by osmosis. However, if they lose water, cells may become *flaccid,* and have a very low turgor pressure.

The crispness of the leaves and the rigidity of the young parts of plants are due to the turgid condition of the individual cells. A young bean seedling stands erect chiefly because of the turgor pressure in all the cells of the stem. In such a young plant, strengthening tissue is not plentiful. If the cells of the seedling lose water rapidly, the whole plant becomes flaccid and droops, but it may recover if water is resupplied.

Plasmolysis

If a cell or group of cells is immersed in a solution that has a higher solute concentration than that of the cell sap, water diffuses outward, and the turgor pressure in the cell is reduced. The volume of the cell decreases somewhat, but more striking is the withdrawal of the protoplast from the cell wall and the decrease in the size of the vacuole. This phenomenon is called **plasmolysis.** As shown in **Fig. 5.17,** page 68, the space between the cytoplasm and the cell wall in a plasmolyzed cell is filled with the plasmolyzing solution in which the cells are immersed. If plasmolyzed cells are immersed in water or in a solution whose concentration is less than that of the cell sap, the cells regain their turgor; water molecules diffuse inward. If cells remain long in a state of pronounced plasmolysis, however, death ensues. A normal, healthy, and functioning cell is one in a turgid condition.

It is instructive to plasmolyze living cells that contain a pigment dissolved in the vacuolar sap. For this purpose, cells of *Elodea* with neutral red in the vacuole may be used. When they are immersed in a strong solution, such as one of table salt, plasmolysis soon follows. It will be seen, however, that the red pigment is retained in the vacuole (**Fig. 5.18,** page 68). Obviously, the tonoplast is impermeable to the outward movement of this dye. If the cells are heated or treated with various chemicals, such as chloroform, alcohol, or ether, the red pigment readily diffuses from the protoplast. Cytoplasmic membranes manifest differential permeability *only when the cell is alive;* when the cell is killed, cytoplasmic membranes lose their differential permeability.

A cell may die if plasmolysis is pronounced and prolonged. For example, if heavy application of ordinary salt is placed on the soil where weeds are growing, so that the root cells are surrounded by a solution of high concentration, water diffuses from the cells, and they become severely plasmolyzed. If this state is prolonged, the roots die. The salt killed, not because it was toxic to root cells, but rather because severe plasmolysis was brought about by the high concentration of the soil solution. In parts of the western United States where rainfall is low and the evaporation rate is high, salts of the soil may accumulate on the surface and form what are known as "alkali flats." In such soils, the concentration of the soil solution may be so high that ordinary crop plants cannot grow; only species that are especially adapted to tolerate high salt concentration are able to survive. Plants of this type are known as **halophytes** (see Chapter 18).

Absorption of Dissolved Substances

In addition to water, the cell absorbs various inorganic ions such as potassium, calcium, nitrates, phosphates, and sulfates; and the gases, oxygen and carbon dioxide, both of which are soluble in water.

If water molecules are diffusing through the cell wall and the cytoplasmic membranes and are entering the vacuole at a certain rate, it does not follow that any particular solute particle is entering the vacuole at the same rate. Moreover, the individual ions in a solution bathing a cell do not necessarily enter the cell at the same rate. Although loss of water in vapor form from a plant does affect the proportions of water and solutes absorbed, it appears that the kind and quantity of ions absorbed are chiefly determined not by the volume of water absorbed but by the nature of the solutes and by certain chemical and physical properties of root cells.

It has been found experimentally that the concentration of a solute particle (ion or molecule) may be greater in the vacuole than in the solution outside the cell (Table 5.3). How, then, do we harmonize *simple diffusion* with that in which solute particles move from a place where they are in low concentration to a place where they are in high concentration? It has been demonstrated that the **accumulation** of solute particles by plant cells is usually attended by high respiration rates. It may be assumed that the energy released by respiration is utilized by the cell to perform the labor of *forcing* the solute particles to move *against a concentration gradient,* and this same energy maintains the concentration difference. Thus, the living cell performs work, the energy for which is derived from respiration. Ions that are accumulated by root cells are apparently "pumped" into the vacuole of the cell from the soil solution (page 226).

SUMMARY OF ABSORPTION

1. All substances that enter the cell, gases included, are in solution in water.
2. The cell wall is normally permeable to all small molecules and ions in solution.
3. The cytoplasmic membranes are differentially permeable.
4. Cell colloids imbibe water and swell.
5. Osmosis is the movement of water through a differentially permeable membrane. Water diffuses along a gradient of water potential.
6. Any particular solute may accumulate in the cell and thus maintain in the vacuole a higher concentration of that substance than exists outside the cell. To accomplish and maintain this higher concentration, the cell must perform work. The energy for this work is furnished by respiration.
7. The diffusion of water and different solutes is independent.
8. Solute particles decrease the activity of water molecules; the solution then has a low water potential.
9. When water diffuses into a cell, the internal turgor pressure increases.
10. Turgor pressure increases the potential of water.
11. Water enters the cell when the water potential in the cell sap is less than that of the water in the solution outside the cell. Water leaves the cell when the water potential inside the cell exceeds that in the solution outside the cell.
12. Cells immersed in a solution of high concentration become plasmolyzed. Water diffuses out of the protoplast, and the protoplast withdraws from the wall, being replaced by the plasmolyzing solution.
13. Respiratory energy may be utilized by a cell to accumulate ions.

METABOLISM

The sum total of all the chemical reactions that go on in the plant body is called **metabolism.** Food manufacture, formation of cell walls, and synthesis of protein all involve many chemical reactions and result in the production of new materials. Such synthetic metabolic reactions, **anabolic reactions,** contrast with the breakdown reactions such as are involved in respiration or the oxidation of foods. The phase of metabolism that involves breakdown or destructive reactions is known as **catabolism.** Both aspects of metabolism are essential for normal cell activities, since anabolic reactions are necessary for the continued production of new materials, and catabolic reactions release, in an available form, energy stored in foods.

Photosynthesis

Every day the combined oil wells of the world deliver several billion barrels of oil for running the cars, boats, trains, and factories of our highly industrialized civilization. The energy this oil so conveniently stores came

Table 5.3

THE CONCENTRATION OF VARIOUS IONS IN THE VACUOLAR SAP OF *NITELLA CLAVATA* AND IN THE POND WATER IN WHICH IT WAS GROWING[a]

Ion	Cell Sap Concentration, Millequivalents per Liter	Pond Water Concentration, Millequivalents per Liter	Accumulation Ratio, Sap Concentration/ External Concentration
Ca^{2+}	13.0	1.3	10.0
Mg^{2+}	10.3	3.0	3.6
Na^{+}	49.9	1.2	41.6
K^{+}	49.3	0.51	96.7
Cl^{-}	101.1	1.0	101.0
SO_4^{2-}	13.0	0.67	19.0
$H_2PO_4^{2-}$	1.7	0.008	212.5

[a] Data from Hoagland and Davis (1929), *Protoplasma.*

originally from the sun. It was "captured" by plants, and because of certain geological conditions was trapped and stored in the earth's crust for many millions of years. The process by which plants in the past captured, and today continue to capture, the radiant energy of the sun is known as **photosynthesis,** a term that literally means putting together (*synthesis*) by means of light (*photo*).

The principal features of this physiological process will be mentioned here, but a more complete discussion will be given in Chapter 13. Photosynthesis is a process that goes on only in cells that have chlorophyll and only when these cells are **illuminated.** Certain blue-green algae and purple bacteria are exceptional in carrying on photosynthesis in the absence of chloroplasts, but chlorophyll bound to membranes is present.

In photosynthesis, the simple compounds **water** and **carbon dioxide** are united to form **sugars** and **oxygen.** Glucose is one of the principal sugars produced during photosynthesis.

The oxygen may go into solution in the cell sap or diffuse out of the cell or be used in another cellular process, respiration. Carbon dioxide is dissolved in the water in which aquatic plants live. As carbon dioxide is taken out of solution and used in photosynthesis, its concentration in the cell sap becomes less than that in the water outside the cell. As a result, diffusion of that gas into the cell goes on as long as active photosynthesis continues. In land plants, green cells obtain carbon dioxide from the atmosphere, but this gas must go into solution in the imbibed water of the cell wall and cytoplasm before it can diffuse to the chloroplasts.

Photosynthesis is an Energy-Storing Process

Energy is required to bring about the synthesis of glucose from carbon dioxide and water. Light is the source of energy for photosynthesis. If the glucose molecule is burned, oxygen is consumed and carbon dioxide and water are formed. The heat energy evolved in burning represents the release of the glucose molecule's chemical potential energy, which is equal to the light energy transformed and stored as chemical energy in the photosynthetic process.

Photosynthesis may be simply represented as follows:

$$\underset{\text{carbon dioxide}}{6\,CO_2} + \underset{\text{water}}{6\,H_2O} + \underset{\text{kcal}}{686} \rightarrow \underset{\text{sugar}}{C_6H_{12}O_6} + \underset{\text{oxygen}}{6\,O_2}$$

A kilocalorie (kcal) is the amount of heat required to raise the temperature of one liter of water 1°C.

Glucose may be utilized in a number of different ways: (*a*) broken down in the process of respiration, yielding energy; (*b*) converted into some closely related carbohydrate, such as sucrose (cane sugar, $C_{12}H_{22}O_{11}$), or cellulose (cell-wall building material), or starch (a reserve food supply); (*c*) converted into fatty substances; (*d*) united with the nitrogen, sulfur and phosphorus derived from various inorganic salts absorbed from the water of the soil or the lake or stream in which the plant lives, thus forming proteins; and (*e*) employed as the chemical basis for a number of other substances that may be found in the cell.

Digestion

Enzymes are involved in the processes of building and degrading complex organic molecules in the cell. The breaking apart of complex carbohydrates, fats, and proteins into simpler compounds (sugars, fatty acids, glycerine, and amino acids) occurs easily in the presence of specific enzymes and without the release of large amounts of energy. This is known as **digestion.** Since water molecules are generally used in the digestion process, the process is chemically one of hydrolysis.

Plants as well as animals **digest** *foods.* However, plants have no special organs for digestion; it is carried on *in any cell that may store food,* even temporarily. In most green cells, photosynthetic activity may proceed at such a rate during the day that the food (glucose) accumulates faster than it is used in those cells. When this occurs, the glucose may be changed temporarily to starch, which appears as granules in the chloroplasts. Glucose is soluble in the water of the cell sap, but starch is insoluble. When the time comes for this temporary starch reserve to be used by the cell or transported out of the cell, it must be changed back into sugar. This chemical transformation of the insoluble starch into soluble glucose is one example of digestion. The equation for starch digestion is as follows:

$$\underset{\text{starch}}{(C_6H_{10}O_5)_n} + \underset{\text{water}}{n\,H_2O} \rightarrow \underset{\text{glucose}}{n\,C_6H_{12}O_6}$$

Any other foods that are insoluble in the cell sap, such as proteins and fats, must be digested (rendered soluble) before they can diffuse and nourish the cell. Enzymes *facilitate digestion.*

Respiration

Many processes taking place in plants require energy. These include: the absorption of mineral salts against an osmotic gradient, the synthesis of complex compounds such as proteins, the maintenance of the protoplasm in a living state, cell division and cell growth, movement, photosynthesis, and translocation. Energy for photosynthesis is supplied by sunlight; energy for the translocation of water comes from the sun, which evaporates water from leaf surfaces. All other processes obtain energy from respiration.

Respiration may be defined as the **oxidation** *of organic substances, with the release of energy, within cells.* In its simplest form, oxidation involves chemical union of some substance with oxygen. However, in its broadest sense, oxidation covers the energy-releasing reactions in which molecular oxygen itself may not be involved. In the burning of a match, oxygen of the air unites with wood, heat energy is released, and the wood is broken down into carbon dioxide and water. We have seen that photosynthesis is the reverse: carbon dioxide and water unite, oxygen is released, and energy is stored. Photosynthesis represents a type of reaction termed **reduction.** Actually, the two reaction types are much more complicated than expressed here and will be considered in more detail in later chapters.

Sugar is the usual organic substance oxidized in all plant cells. The energy stored in the sugar is, as we have learned, derived from sunlight. Sugar is a **food,** and all foods are energy-rich compounds whose molecules are so constructed that the energy they store may be released with comparative ease.

Respiration is a cellular process. All living cells respire. Although respiration is a complicated chemical process, the overall reaction occurring in cells may be described as the reverse of photosynthesis.

This statement may be simplified as follows:

$$C_6H_{12}O_6 + 6\,O_2 \rightarrow 6\,CO_2 + 6\,H_2O + 686$$

$$\text{sugar} + \text{oxygen} \rightarrow \text{carbon dioxide} + \text{water} + \text{kcal}$$

Energy Flow

When sugar is burned, carbon bonds are broken and heat is generated. The heat may be used to do work. If it were possible to use the rearrangement of the atoms of the sugar molecules directly to do work, little heat would be generated, and our machine would be highly efficient. This is exactly what happens in living cells during the process of respiration; reactions that release energy are **coupled** with reactions that absorb energy. Molecules are rearranged, new materials are formed, and work is done. These **coupled reactions** are key reactions in biosynthetic processes.

When carbon bonds are broken in the cell during the oxidative reactions of respiration, small amounts of energy may be made available for work. Some phosphate bonds, on the other hand, are able to store large amounts of energy, and when broken release it easily to drive other processes along in **coupled reactions.** Two most important compounds associated with energy transfer in cells are **adenosine diphosphate (ADP)** and **adenosine triphosphate (ATP).** There is also an **adenosine monophosphate (AMP)** with only one phosphate group on the molecule. When a second phosphate is added to the single phosphate of AMP in the formation of ADP, a bond is formed that concentrates much more energy than is present in the phosphate bond of AMP. When a third phosphate is added to form ATP, an additional large amount of energy is concentrated in this bond. These two bonds that hold the two terminal phosphate groups of the ATP molecule are known as **high-energy bonds** and are designated with the symbol $\sim$P. These bonds may be easily broken by hydrolysis, and each then releases about 8000 cal of energy per mole of ATP hydrolyzed. When the carbon-oxygen bonds in a mole disaccharide or in a fat molecule are hydrolyzed, only about 1000–2000 are released.

Adenosine triphosphate is the principal compound that appears to be universally formed during respiration in the cells of microorganisms, plants, and animals, and in which the energy from the oxidized food is temporarily stored. ATP has been called the energy currency of the cell. It may be formed from ADP and inorganic phosphate in the mitochondria during respiration, or it may be formed in the chloroplasts of green cells exposed to light during photosynthesis. From its site of synthesis, ATP may move to other locations where it gives up its energy in the performance of cellular work and it is converted to ADP and inorganic phosphate. ADP may now migrate back to the mitochondria where, in cooperation with respiration, it is again combined with inorganic phosphate to form ATP, and the process is repeated.

Assimilation

The cell takes the food (carbohydrate, fats, and protein) and from them builds protoplasm (living substance). *The conversion of foods into the living material (protoplasm) of the cell is called* **assimilation.** The chemical changes involved in this conversion are not understood, but energy is used in the process. Except in a very general way, the chemical nature of protoplasm itself in unknown.

Animal cells and nongreen plant cells can also convert foods into protoplasm. Also, any plant cell, with or without chlorophyll, can synthesize fats and proteins and can change glucose into starch and other carbohydrates. But only chlorophyll-bearing cells can manufacture glucose from carbon dioxide and water.

Control of Metabolic Reactions

Although protoplasm is complex, the wide variety of chemical reactions that occur in it simultaneously during the life of the cell are not haphazard. There are intricate spatial and time interrelationships among them. The regulated sequential nature of cellular reactions is one of the attributes of life. How are these reactions so precisely regulated and controlled in such a minute space as the volume of a cell? From the point of view of the size of an enzyme molecule, this volume is enormous; but our efforts toward miniaturization of electrical circuits in sophisticated electronic equipment can hardly be compared to the ultimate in miniaturization, the living cell.

Control by Compartmentation

Part of the regulation and order of metabolic reactions is dependent on cell structure. We have already seen that there are many organelles and membranes within a cell and that certain enzymes are confined, during the life of the cell, to very specific locations. This compartmentalization of enzyme activities can be readily demonstrated by separating the cell into its various parts. Then it is found that specific metabolic reactions occur only in certain parts of the cell. For example, a sugar molecule synthesized in the chloroplast by photosynthesis may move into the cytoplasm, where it is changed and partially oxidized by the action of enzymes in the cytoplasm during a series of reactions called **glycolysis.** The oxidized product of glycolysis then moves into a mitochondrion where it is further oxidized. In this example, the sugar molecule has passed through three compartments (chloroplast, cytoplasm, mitochondrion) and has been treated uniquely in each. This beautifully precise compartmentalization in the cell is an important part of the mechanisms that control metabolic reactions. Membranes themselves often contain specific enzymes and may play a role in the regulation of enzymes.

Control by Enzyme Synthesis and Breakdown

Another aspect of metabolic control depends on genetic information in the DNA molecule that ultimately regulates the timing and the amount of synthesis of enzymes. This precise regulation of enzyme synthesis is of paramount importance in determining the growth and development of the plant from seed to maturity. In some instances, hormones may affect the rate of production of specific enzymes. In other instances, the presence of a substrate in a cell may cause the increased production of an enzyme. For example, if a plant is growing on an inorganic medium containing ammonium ions as a source of nitrogen, the plant may be unable to utilize nitrate ions rapidly when they are placed in the solution. However, in a short time, a rapid increase in the plant's ability to use nitrate is often observed. It can be demonstrated that the absorption of nitrate from the culture medium has stimulated the synthesis of an enzyme, nitrate reductase. This enzyme is said to be **"induced"** in the cell by the **substrate** (nitrate). In a somewhat similar way, the accumulation of the final products of a series of reactions may repress the production of an enzyme that is involved in the formation of those products.

In each of these three cases, hormone action, induction, and repression, the regulation may be brought about by effects on the DNA, RNA, protein synthesis mechanism. The amount of an enzyme in a cell is continually changing not only because of the synthesis of new enzyme molecules, but because of the continual breakdown of the enzyme. The rate of degradation is characteristic of each enzyme and is dependent on the presence in the cell of protein hydrolyzing enzymes, **proteases.** The rate at which an enzyme is broken down may be an important regulatory mechanism.

Control by Enzyme Regulation

A more rapid mechanism of affecting enzyme reactions is achieved by certain **regulatory enzymes** whose activities may be modified because of their sensitivity to certain ions or molecules, **modulators,** produced or utilized during metabolic reactions. Regulatory enzymes frequently react at branch points in a series of interrelated metabolic reactions. Presumably the modulator combines reversibly with a specific site on the enzyme causing a change, an **allosteric effect,** in the physical shape of the enzyme molecule, thus affecting the activity of the enzyme. Often the end product of a series of metabolic reactions can act as a modulator of an enzyme catalyzing one of the early steps in the reaction sequence. This regulation of the course of a reaction through the action of one of the products of the reaction is known as **feedback.** It is a basic type of control in electronics. In cells, the concentrations of certain key metabolities play similar regulating roles in many phases of metabolism.

As an example of feedback control, we can look at some effects of ATP in the cell. During respiration, ATP is formed. The energy in ATP may be used to do useful work in the cell. The more work that is done, the more rapidly ATP is used. When little work is done, ATP tends to accumulate in the cell. It is known that a high level of ATP in the cell decreases the rate of a least one enzyme reaction in respiration. Consequently, an increase in ATP in the cell causes a decrease in the rate of respiration. Conversely, when a cell does more work, the level of ATP decreases, its inhibitory effect on the enzyme reaction involved in respiration decreases, and respiration increases. More ATP is generated. Thus, the rate of respiration is in part controlled by the level of cellular ATP, which in turn is influenced by the rate at which work is being done.

Figure 5.19 illustrates some of the ways in which control of metabolic reactions is achieved in the cell.

SUMMARY OF CELLULAR PHYSIOLOGY

1. Metabolism is the sum total of all catabolic or breakdown reactions and all anabolic or synthetic reactions in the cell.
2. Cells absorb water and mineral salts from the external environment.
3. The numerous chemical reactions that go on in the cell are controlled by organic catalysts called enzymes.
4. Green plant cells are able to synthesize carbohydrates from water, carbon dioxide, and light energy, through the process of photosynthesis.
5. Carbohydrate foods can be transformed into fats and, with the addition of certain chemicals, into proteins.
6. Complex foods are rendered simple through the process of digestion.

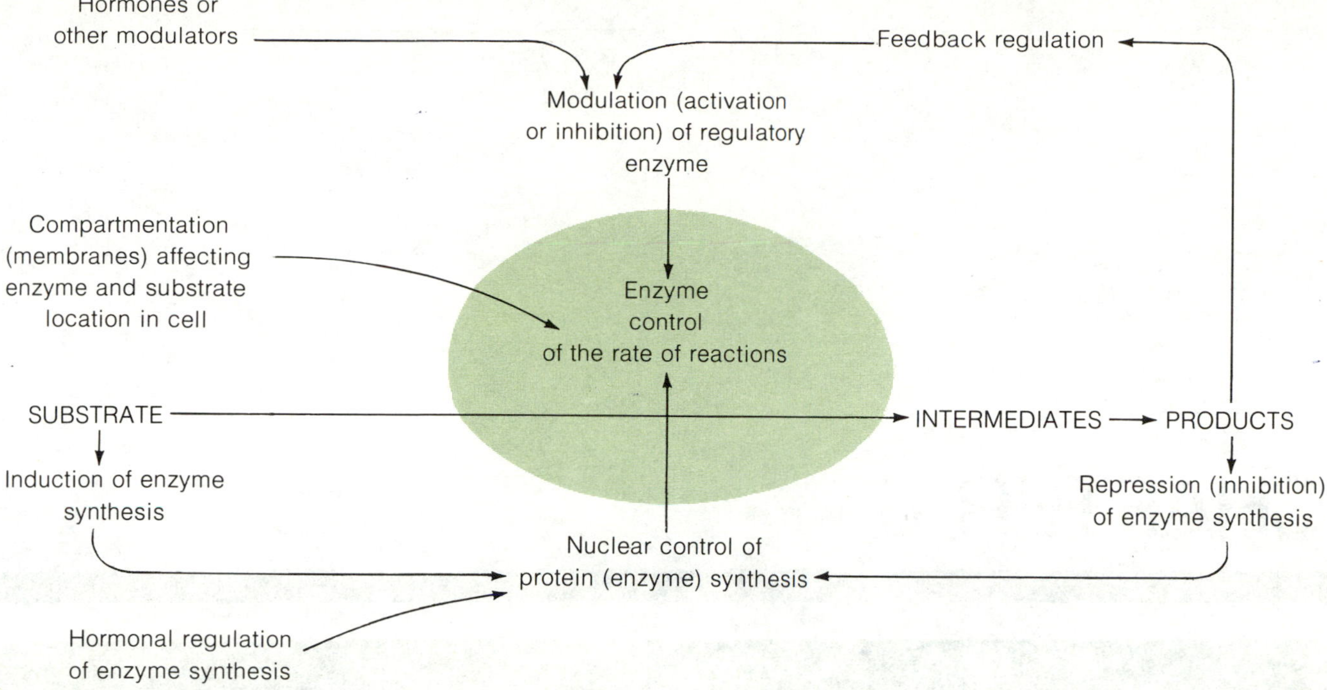

Figure 5.19
Some mechanisms regulating metabolic reactions in the cell.

7. Respiration is the oxidation of organic substances within living plant cells.
8. Protoplasm is built up through the process of assimilation of foods.
9. Energy flows from sunlight through chlorophyll into sugar molecules where it is stored as chemical energy during the process of photosynthesis.
10. Some chemical energy in sugar is transferred to ATP during respiration.
11. ATP is used by the cell to drive the energy-consuming reactions of metabolism.
12. The control of metabolism may depend in part on compartmentalization, enzyme synthesis, enzyme sensitivity to ions and molecules, and may involve a feedback mechanism.

CHAPTER 6

THE CELL CYCLE

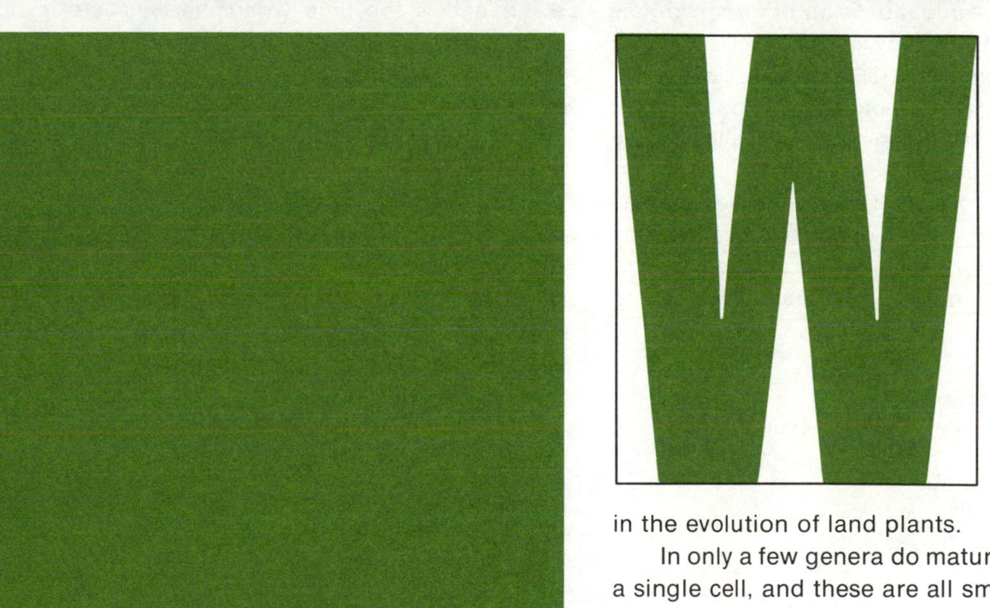

Why must cells divide? No successful invasion of the land was ever made by autotrophic plants whose cells do not divide. It would appear that the mechanical strength of many cells and the increased flexibility of specialization made possible by a cellular pattern are vital survival factors in the evolution of land plants.

In only a few genera do mature plants occur with but a single cell, and these are all small plants. Perhaps the largest of them, *Acetabularia* (Fig. 4.21) attains a height of about 2 cm. Even in *Acetabularia,* cell division is a prerequisite for sexual reproduction. Cell division enables an increase in size and complexity of the plant body.

Cell division is apparently required because a single nucleus with a full complement of genes can react with only a small amount of cytoplasm. Increase in size and complexity of the plant body requires cell division. The resulting increase in number of cells makes possible tissue specialization for the different structural and physiological activities including absorption, conduction, reproduction, photosynthesis, and support. Cell division is always correlated with active growth. Growth is the direct result of (*a*) cell division, and (*b*) cell elongation.

All diversity in cellular and biological form results from the sequence of four relatively simple organic bases found in the DNA of the nucleus and chromosomes. The DNA molecule is a double-stranded helix present in the cell nucleus. The process of DNA replication is essentially the passing on of identical DNA helices to derivative cells.

In this chapter we shall consider the mechanisms of cell division and DNA replication. In addition, we shall outline the phases that each cell must undergo in preparation for these events.

THE CELL CYCLE

Description

Organs of the plant body are composed of cells and aggregates of cells, called tissues (see Chapter 7). **Meristems,** which are regions that initiate these tissues, wherever they are located, consist of cells that are actively dividing and beginning to enlarge and differentiate. Those cells that are dividing or are preparing to divide are said to be progressing through the **cell cycle.** One cell cycle would equal the time from the formation of an initiating cell to its subsequent division to form two new derivative cells (Fig. 6.1). The four parts of the cell cycle were discovered and named in 1953, by two scientists, A. Howard and C. Pelc. Their terminology for the stages of the cell cycle **(G_1-S-G_2-*M*)** have been universally adopted for both animal and plant cells. Traditionally, the first three phases of the cycle are referred to collectively as **interphase.** It is appropriate to consider interphase as the preparation period for division. **Cell division** is composed of two portions—**mitosis (nuclear division)** and **cytokinesis (cytoplasmic division)**. We shall now consider the entire cell cycle in detail.

G_1-S-G_2 (Interphase)

The stages are as follows (Fig. 6.1): The **G_1 phase** or **pre-DNA synthetic phase** represents the time during the cell cycle in which the cell prepares itself metabolically to go through DNA synthesis. This preparation will involve such things as the accumulation and synthesis of specific enzymes needed to control DNA synthesis and the production of the DNA base units so that a supply or pool is on hand when synthesis begins.

The second phase of the cycle is the **S phase** or **DNA synthesis phase.** It is during this portion of the cycle that the DNA molecules are actually duplicated. If the preparatory events in G_1 are not completed, the *S* phase cannot occur. Specific events of *S* will be discussed later in this chapter.

The **G_2 Phase** or **Premitosis phase** follows DNA synthesis and is the time when the cells are involved in metabolically preparing for mitosis. For example, energy storage for chromosome movement, the production of mitosis-specific proteins and RNA, and the formation of microtubule subunits needed to make spindle fibers must be completed before mitosis can begin.

The **M phase, mitosis** is the last cell cycle phase and involves the actual separation of the DNA strands, which were duplicated, into homologous chromosome sets, and their equal division into derivative cells. The stages of *M*

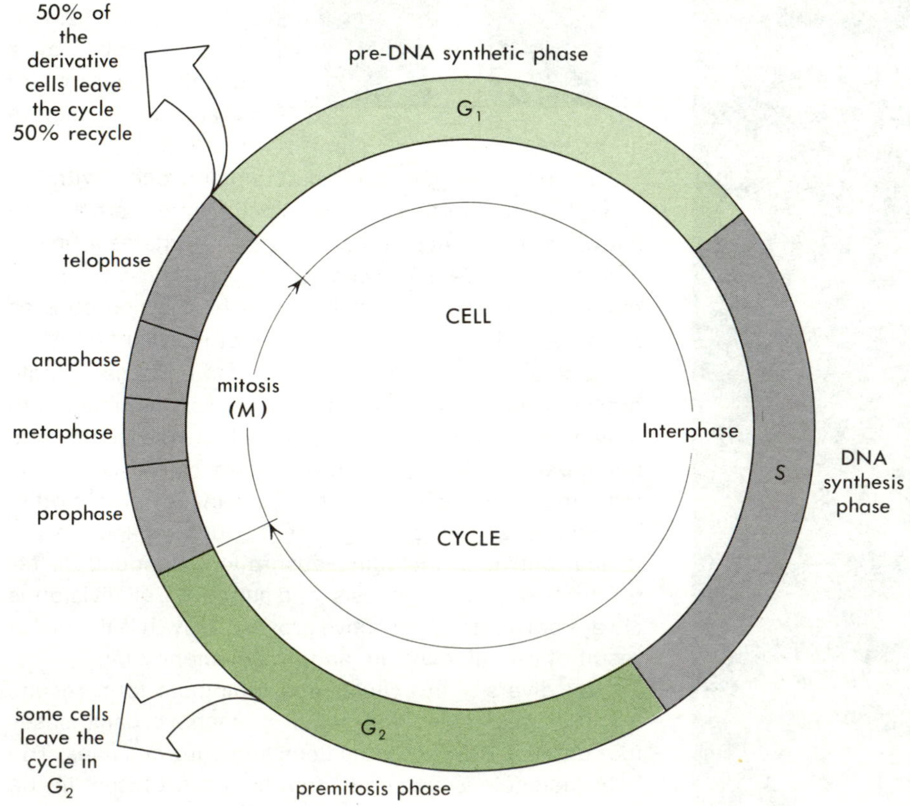

Figure 6.1
The cell cycle.

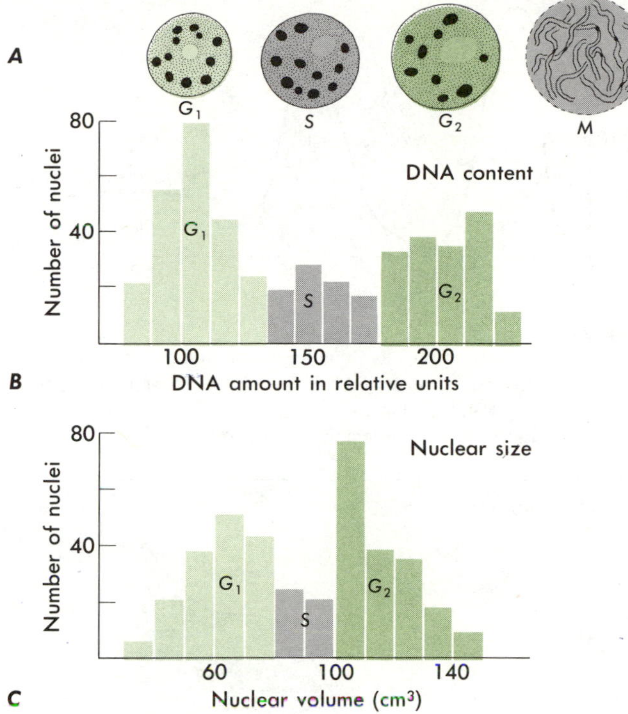

Figure 6.2
Difference between onion *(Allium flavum)* nuclei in each cell cycle phase based upon *(A)* structure, *(B)* DNA content, and *(C)* nuclear volume. (*Redrawn from W. Nagl, Cytobiologie* **4,** 398. © 1970. Reprinted with permission of the publisher.)

and the behavior of chromosomes will be discussed in the next section.

The cells that progress through the cell cycle are usually located in meristems. A meristem is a localized region of the plant body that is highly specialized for this function. The tips of roots and shoot tips, especially at the sites of new leaf formation, are examples (Fig. 3.1).

Cycling cells in each phase of the cycle are unique, and in some cases can be distinguished. A look at onion root tip cells (*Alium*) is useful in pointing out these differences (Fig. 6.2). First, at least in species of onion, there are structural differences between the cells in each cycle stage. The size and configuration of the nucleus and the distribution of DNA changes, because during *S* the DNA is exactly doubled. A cell in G_1 will have half the amount of DNA as a cell in G_2. Figure 6.2*B* shows this in a histogram where the number of nuclei containing different DNA contents are plotted and compared. Cells in G_1 have the least DNA, those in *S* have an intermediate concentration, and G_2 cells have the most. A third way to distinguish cells in different cell cycle positions is simply by nuclear size. Since a nucleus in G_2 has twice the DNA of one in G_1. It must also be larger (Fig. 6.2*C*).

All of the cell cycle events take time. The time involved for this progression is dependent upon the amount of DNA contained in the nuclei of any particular plant. Plants with higher DNA amounts in their nuclei have longer cell cycle times than plants with lower DNA amounts. Three representative plants, their nuclear content, and cell cycle times are shown in Table 6.1. Note that the time in mitosis (*M*) is always shortest, proportionally, in comparison to other cycle phases.

Metabolic Requirements for Cell Cycle Progression

In order for cells to progress through the cycle certain specific metabolic requirements must be met. Protein synthesis, for example, is continuous from telophase through the cycle to prophase of the next cycle. Certain proteins, especially enzymes, must be synthesized at very specific times in order to catalize necessary reactions at the required times. RNA synthesis is likewise continuously required except during the actual division of chromosomes. In the following discussion we explore these requirements in greater detail.

Histones

DNA in eukaryotic cells is complexed closely with both acidic proteins and histones (basic proteins) to form chromatin. Chromatin is coiled tightly together during prophase of mitosis to form the chromosomes. The histones complexed with DNA are in the form of beadlike structures called **nucleosomes** (see Chapter 5). The nucleosome organization serves the function of coiling the DNA molecule and also possibly to control the regulation of gene expression. A suggested mechanism for this regulatory system was discussed in Chapter 5.

Acidic proteins apparently function by uncoiling the nucleosome, detaching the histone, thereby opening a site for RNA synthesis. The systematic control of this process will trigger protein synthetic events associated not only with cell cycle behavior, but with all cell differentiation and other metabolic events as well. Another bit of evidence, although circumstantial, correlating histone with cellular control functions is the curious fact that histone synthesis always parallels DNA synthesis exactly (Fig. 6.3).

Table 6.1
CELL CYCLE DURATIONS

	DNA Amount (picograms)	*CT* (hours)	G_1	*S*	G_2	*M*
Helianthus annuus (sunflower)	6.3	7.8	1.2	4.5	1.5	0.6
Pisum sativum (pea)	7.9	14	5	4.5	3	1.2
Vicia taba (longpod bean)	24.3	18	4	9	3.5	1.9

Data from J. Van't Hof. 1974. The duration of chromosomal DNA synthesis, the mitotic cycle and meiosis in higher plants. In *Handbook of Genetics*, Vol. II, R. C. King (Ed.). Plenum, New York.

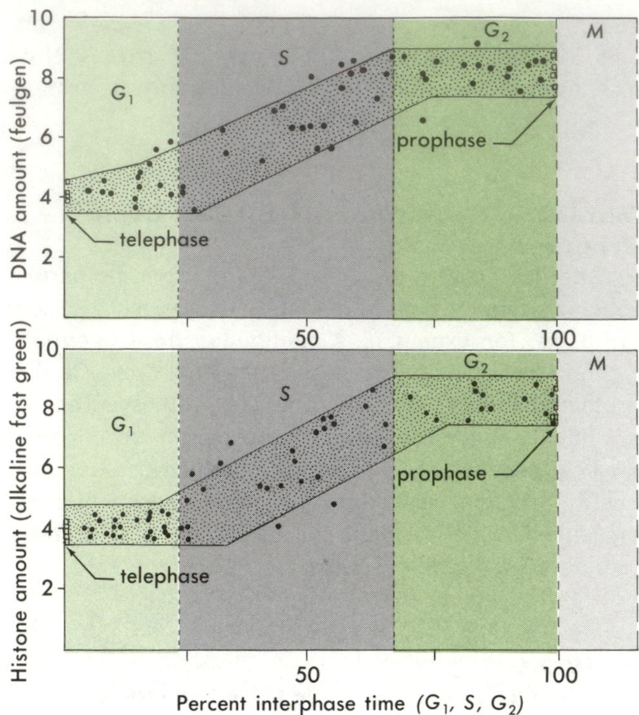

Figure 6.3
Graphs to show that nuclear DNA synthesis and nuclear histone synthesis occurs at the same time in the cell cycle. (Data from J. Woodward, E. Rasch, and H. Swift, *J. Biophy. Biochem, Cytol.* **9,** 445. © 1961 by The Rockefeller University Press. Reprinted with permission of the publisher.)

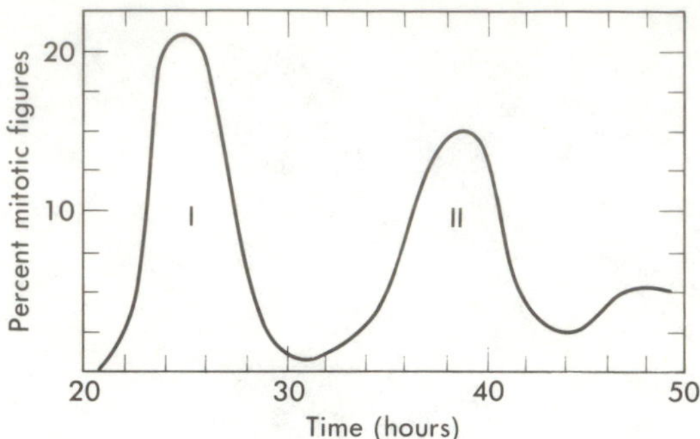

Figure 6.4
Synchronous divisions for two cycles in Jerusalem artichoke tuber sections. (Redrawn from H. E. Street, Ed. *Plant Tissue and Cell Culture.* © 1973 by University California Press, Berkeley, p. 246. Reprinted with permission of the publisher.)

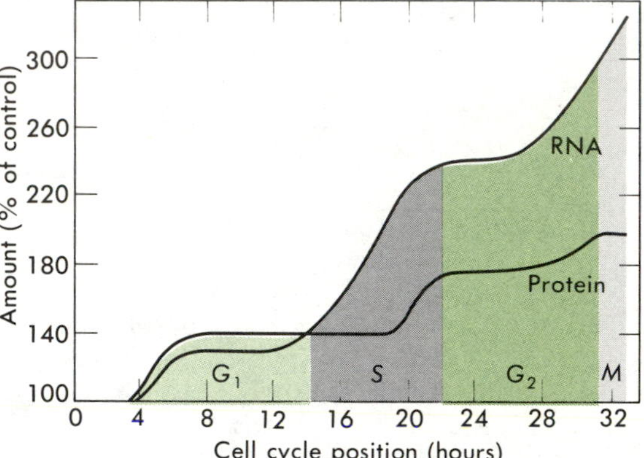

Figure 6.5
Protein and RNA accumulation in dividing Jerusalem artichoke cells. (Data from J. P. Mitchell, *N. S. Ann. Bot.* **33,** 25. © 1969 by Oxford University Press. Reprinted with permission of the publisher.)

Other Proteins and Cell Cycle Control

Some proteins are needed by cells for various structural purposes, such as those found in membranes (see Chapter 4). Others have enzymatic roles in metabolic pathways, such as respiration, that must function during the entire life of the cell. Still other proteins are associated specifically with cell cycle events. The problem with investigating cell cycle protein requirements is that in any random population of cells, say in a root tip, meristem cells are present in each stage of the cell cycle (G_1-S-G_2 and M). How can one reach into a population and figuratively pull out a cell in a specific stage in order to examine its protein composition? Many scientists have attempted to answer this question experimentally. By far the most successful and useful method is to use plant **organs** (leaves or roots, for example) which, for a time, contain naturally synchronous cell populations (that is, populations made up of cells, all in the same phase). One elegant example is the tuber (underground storage stems) of the Jerusalem artichoke (*Helianthus tuberosum*). When thin slices of these tubers are placed onto tissue culture medium, the cells around the periphery of the slice all progress through the cell cycle at the same time (synchronously) for two cell cycles (Fig. 6.4). This is extremely useful because it allows the investigator to correlate protein synthetic activity precisely with cell cycle position. One scientist who has used this system has data that clearly show three periods in the cell cycle when protein synthesis activity increases—early G_1, late S, and late G_2 (Fig. 6.5). Exactly what proteins are synthesized at these three times in the plant cell cycle are not completely known, but some information is available from animal cells. In order to synthesize DNA, specific enzymes must be available. In sea urchin eggs, scientists have demonstrated the correlation of the activity of the enzyme DNA polymerase with DNA synthesis (Fig. 6.6). The peak occurrence of the enzyme precedes DNA synthesis and drops to zero immediately after DNA synthesis is accomplished. Other protein such as that required for spindle fibers in mitosis can be correlated with specific cell cycle position. Spindle protein is present in plant cells at all times, but it is particulary abundant just preceding mitosis.

RNA Synthesis

RNA synthesis takes place continuously throughout the cell cycle except during mitosis. It is necessary in order

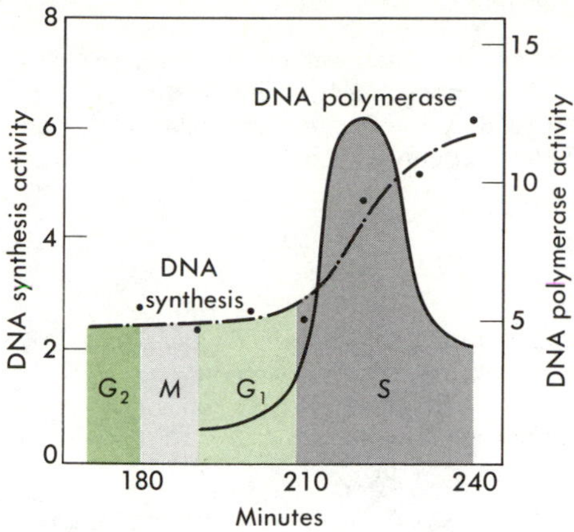

Figure 6.6
Relationship of DNA synthesis activity to DNA polymerase activity. Note that DNA polymerase activity starts in G_1 and peaks in mid-*S*. (After D. M. Prescott, *Reproduction in Eukaryotic Cells*. © 1976 by Academic Press, N.Y. Reprinted with permission of the publisher.)

to code for proteins needed for progression through all cell cycle phases. The chromatin during mitosis is tightly coiled and consequently sites for RNA transcription are not available. Note in Fig. 6.5 that just as with protein synthesis, RNA content increased in three steps corresponding to G_1, late *S*, and late G_2. Presumably certain specific RNAs are synthesized at these times. Studies done on cell cycle progression in cells treated with RNA synthesis inhibitors show a curious phenomenon. In pea (*Pisum satium*) root tips, for example, cells will continue to cycle (replicate DNA) for approximately 10 hours in the presence of the RNA synthesis inhibitor, Actinomycin D. Sunflower (*Helianthus annuus*) root tips will continue to cycle for more than 24 hours after the RNA inhibitor treatment. This apparently means that, at least in root tip cells, a supply of already synthesized RNAs for cycle activities is always available.

Energy Requirements

Cycling cells require a constant supply of carbohydrates, from photosynthesis in intact plants and from added sucrose in cultured plant cells or organs, in order to respire and synthesize ATP. Without this carbohydrate, cells can no longer cycle and become arrested. The phenomenon of cell cycle arrest in starved cells is not a random process and, in fact, is a species-specific regulatory process in both plant and animal cells. Researchers at Brookhaven National Laboratory made the original observation that if plant root tips are grown in sucrose-deficient medium, their cells stop progressing through DNA synthesis (*S*) and mitosis (*M*) and become arrested in G_1 and G_2. The ratio of G_1 and G_2 cells is species-specific—for example, in peas 50% stop in G_2 and 50% in G_1, while in sunflower 90% stop in G_1 and 10% stop in G_2 (Fig. 6.7). Progression of these arrested cells can be resumed by adding sucrose back to the medium.

Figure 6.7
The cell cycle distribution of pea root tip meristems after normal culture and starvation treatments. Sucrose starved root tips become arrested in G_1 and G_2 of the cell cycle.

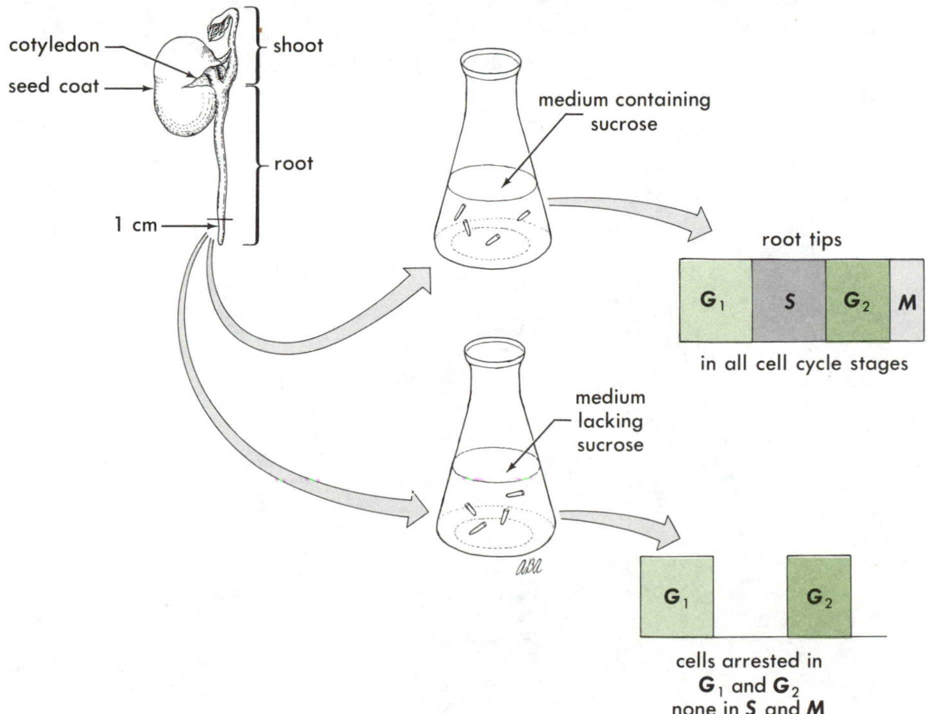

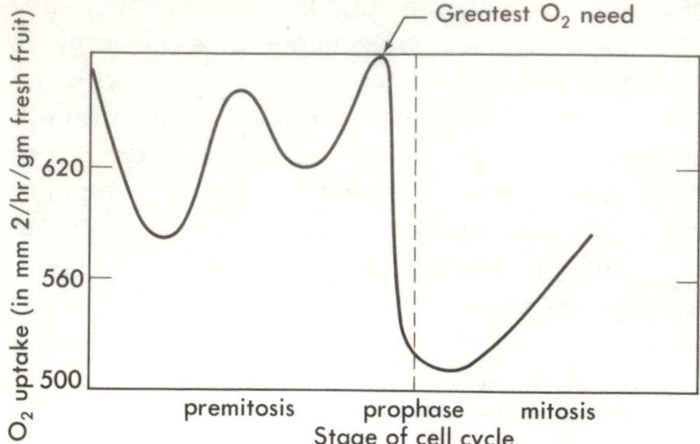

Figure 6.8

The relative oxygen requirement correlated with cell cycle phase in *Trillium* microspores. (Redrawn from H. Stern and P. L. Kirk. *J. Gen. Physiol.* **31,** 243. © 1948 by The Rockefeller Institute of Medical Research. Reprinted with permission of publisher.)

Figure 6.9

Meristematic cells in the shoot apex of *Chenopodium album*. An electron micrograph is shown with an accompanying drawing (Fig. 6.10). Note large nucleus, mitochondria, dictyosomes, proplastids, microbodies, and absence of large vacuoles, ×10,000. (Courtesy of E. M. Gifford and K. Stewart.)

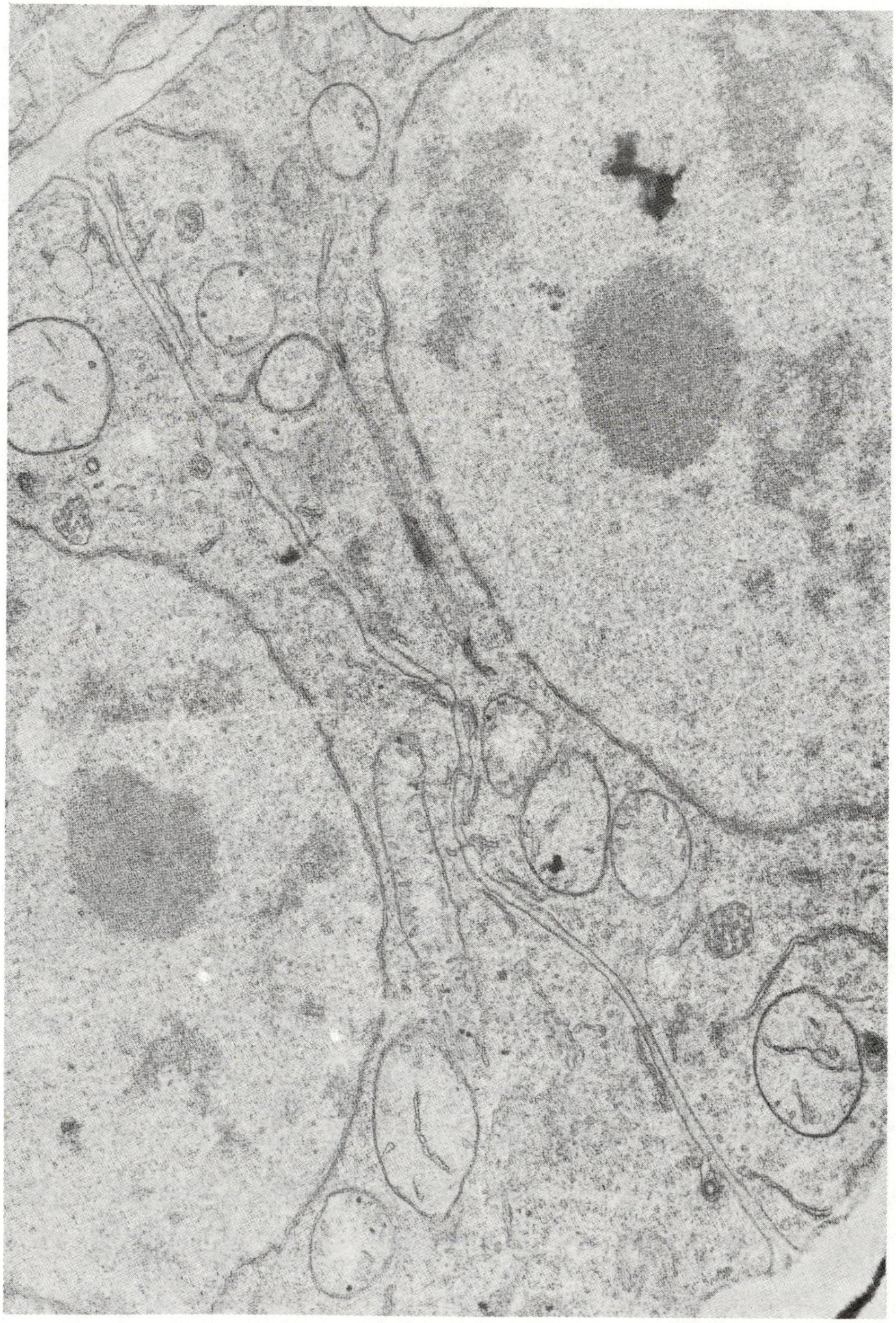

Other Environmental Effects

If roots are exposed to anaerobic conditions, growth soon stops and cycle progression ceases with most cells accumulating in G_1 and G_2. Lack of oxygen will also stop DNA synthesis, but it may not affect mitosis. The graph for *Trillium* microspores shows that O_2 is needed to some extent all during the cell cycle (Fig. 6.8). During mitosis the O_2 requirement is very low while during G_1 the O_2 requirement is quite high. The peak in late G_2 is very important. This rather drastic increase in O_2 requirement just before mitosis begins means that at the end of G_2 the cell is preparing itself for mitosis by storing up metabolic energy.

Cell cycle progression is also temperature dependent. As temperature increases the cell cycle duration tends to become shorter (Table 6.2). The phase of mitosis likewise shortens with increasing temperature.

Figure 6.10
Diagram to accompany Fig. 6.9.

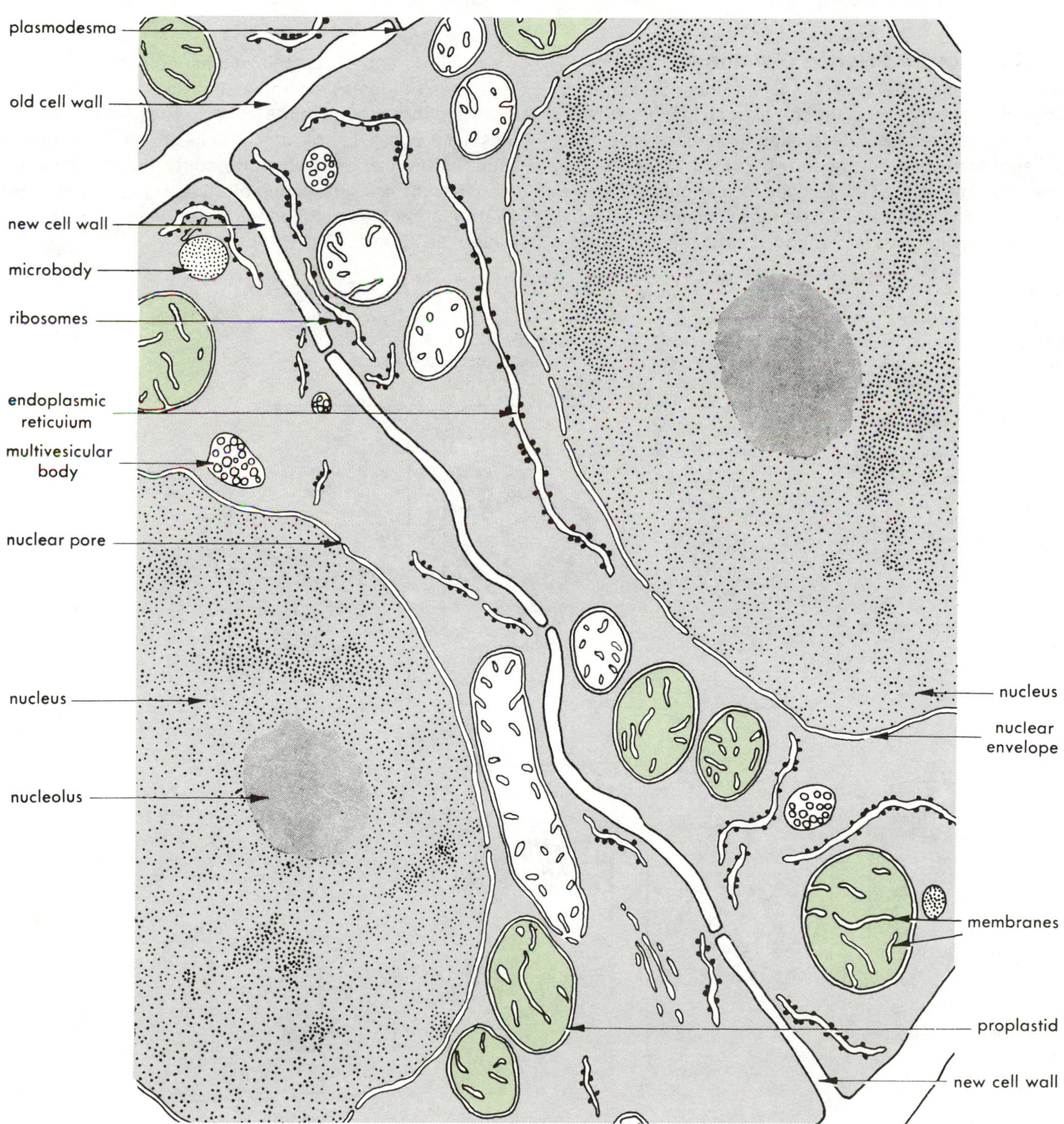

Table 6.2
DURATION (HOURS) OF THE CELL CYCLE AND MITOTIC PHASES IN *ALLIUM CEPA* ROOT TIP CELLS AT DIFFERENT TEMPERATURES

Temperature	10°C	15°C	20°C	25°C
Cell cycle time (G_1-S-G_2-M)	54.6	29.8	18.8	13.5
Prophase	3.1	1.7	1.0	0.7
Metaphase	0.8	0.5	0.3	0.2
Anaphase	0.6	0.3	0.2	0.1
Telophase	1.9	1.1	0.7	0.5
Total mitotic time	6.4	3.6	2.2	1.5

Data adapted from Gimenez-Martin et al. (1977) Cell Division in Higher Plants. In *Mechanisms and Control of Cell Division.* T. L. Rost and E. M. Gifford, Jr., (Eds.), Dowden, Hutchinson and Ross, Inc., Stroudsburg, Pa.

Cell Structure Prior to Division

An interphase cell in meristematic tissue has characteristic differences from an interphase cell in mature tissue. Portions of interphase cells taken from the shoot apex of *Chenopodium album* are shown in Figs. 6.9, 6.10. They may be compared with the young leaf cell of Fig. 4.3. Meristematic cells of apical regions are almost spherical. There is no large vacuole, and the nucleus occupies a major proportion of the cell volume. Mitochondria, dictyosomes, endoplasmic reticulum, and ribosomes are similar to those found in the leaf cell. The plastids, which lack chlorophyll in the apical cells, differ from those found in the leaf cell, which do contain chlorophyll. In meristematic cells, the young plastids, or proplastids, are slightly larger than mitochondria and possess only a few paired membranes (Fig. 6.10).

Figure 6.11
Mitosis, diagrammatic representation; *A* to *D,* stages in prophase. Two pairs of chromosomes are represented, the green pair have a median kinetochore, while in the gray pair, the kinetochore is close to one end. The chromosomes shorten and thicken, and chromatids become apparent. *E,* metaphase. *F,* early anaphase. *G,* late anaphase. *H,* telophase.

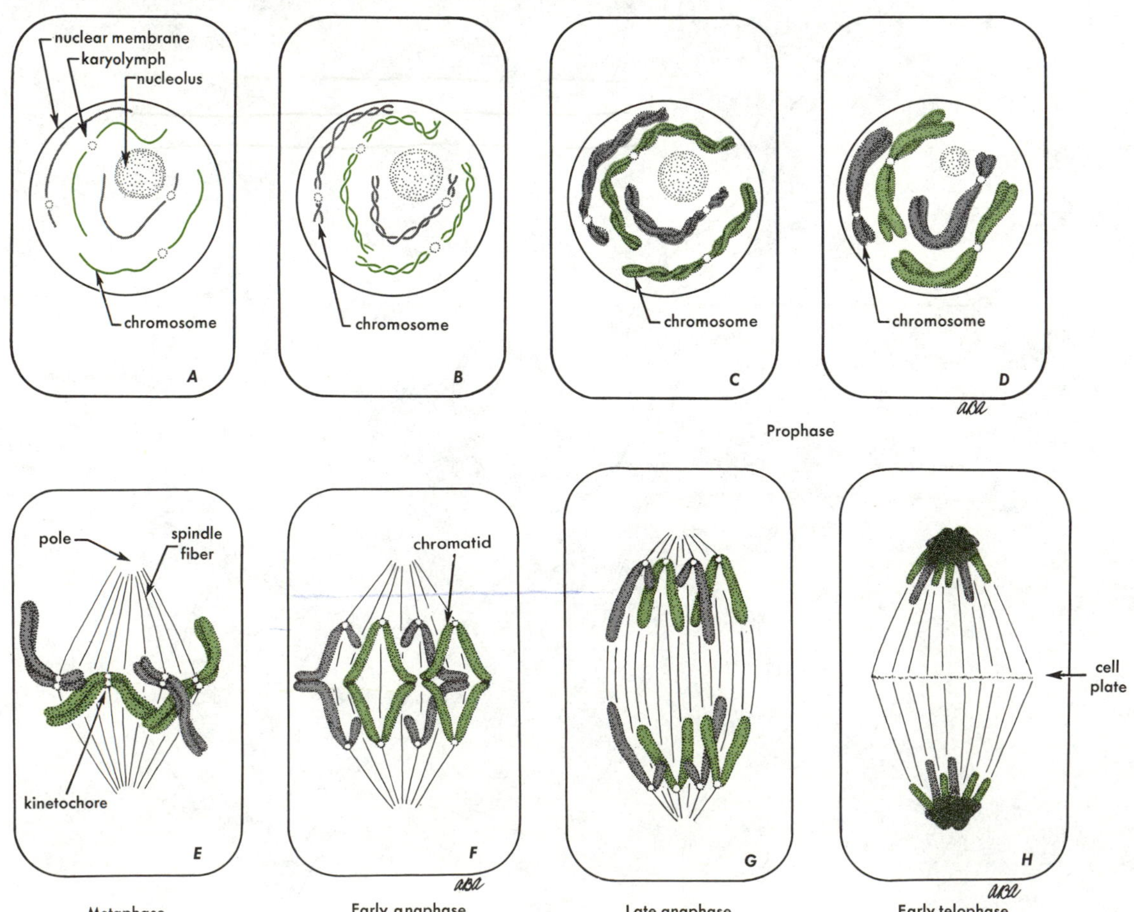

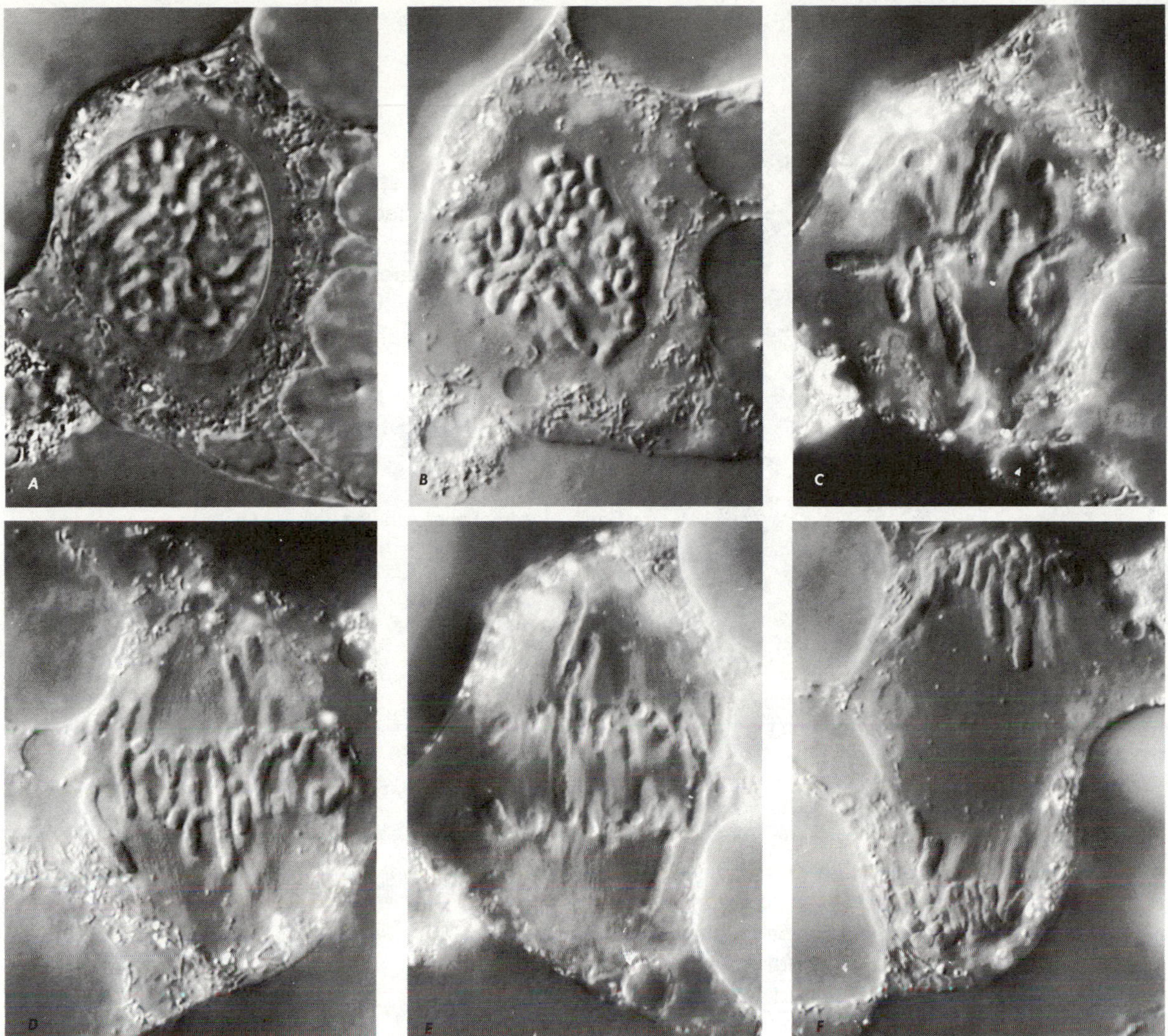

Figure 6.12
Mitosis as followed in living unflattened cells of the endosperm of seeds of the blood lily *(Haemanthus katerinae).* Photographs in Nomarski system optics. *A,* early prophase; the nuclear membrane is still present. Note the clear zone of cytoplasm surrounding the nucleus. *B,* the nuclear envelope has disappeared, and the clear zone still surrounds the chromosomes. A few short spindle fibers are already present. *C,* full metaphase; the coiled chromatids are distinct. The two chromatids moving to the upper pole have separated, and a few spindle fibers may be seen. *D,* early anaphase; the chromatids are moving to opposite poles of the cell. *E,* mid anaphase; spindle fibers are in evidence between the chromosomes and the poles of the cell. *F,* telophase; the chromatids have aggregated at the opposite poles of the cell. A nuclear envelope has not yet formed. Time intervals after *A, B,* 14 min; *C,* 64 min; *D,* 74 min; *E,* 93 min; *F,* 107 min, ×2000. (Courtesy of A. Bajer, *Chromosoma* **25,** 249. © 1968 by Springer-Verlag. Reprinted with permission of the publisher.)

Thin sections of interphase nuclei frequently show the nucleoplasm to be evenly and finely granular or coarsely granular in some cases (Fig. 6.10). Ribosomelike particles in spiral aggregations appear to form the basic nucleolar structure. Note the absence of a barrier between the nucleolus and nucleoplasm (Fig. 6.9).

Nuclei isolated from young barley root tips and spread on distilled water will burst. The resulting film may be picked up, carefully dried, and observed on the electron microscope. The nuclei appear (Fig. 4.19) to consist of loosely coiled or folded fibers about 17.5 nm in diameter. Both thin-sectioned and isolated nuclei indicate that the basic structure in interphase nuclei is an array of fibers called **unit fibers.**

Mitosis (Nuclear Division—The M Phase)

Prophase

The unit fibers in the interphase nucleus are long, slender, and seem tangled. We do not know their exact disposition in the interphase nucleus nor their precise relationship with chromosomes. But the onset of mitosis, as seen in the light microscope, is heralded by the presence of definite chromatin threads (Figs. 6.11*A*, 6.12*A*).

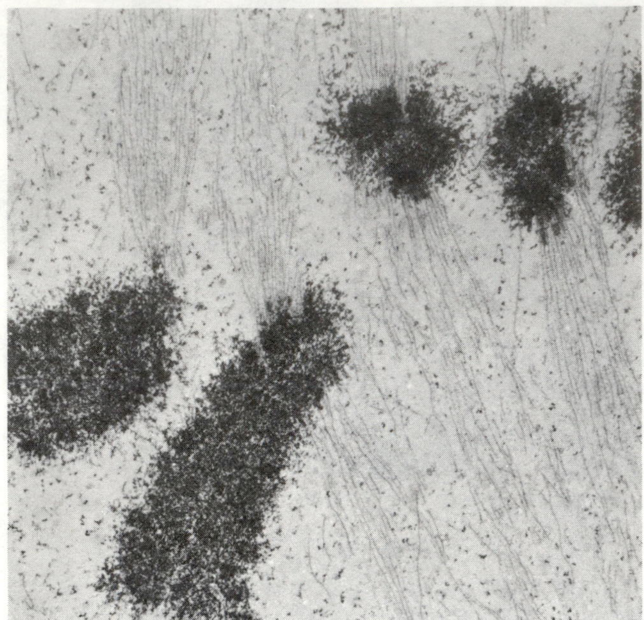

Figure 6.13
Electron micrograph of thin section of metaphase chromosomes in the endosperm of *Haemanthus katherinae,* showing the aligned chromosomes with microtubules (spindle fibers) attached to the kinetochore, ×40,000. (Courtesy of A. Bajer.)

These threads gradually shorten and thicken and become easier to see (Fig. 6.12*B*). They stain more heavily with certain dyes. For this reason they are called colored (*chromo*) bodies (*soma*) or **chromosomes.** Each chromosome is derived from an individual thread. The nucleolus slowly decreases in size and finally disappears during this stage (Figs. 6.11*A* to *D*).

It becomes apparent that at late prophase each chromosome is composed not of one but of two threads coiled about each other (Figs. 6.11*B, C, D*). The nuclear membrane disappears toward the end of the prophase, but there is reason to believe that the nucleoplasm may not mix with the general cytoplasm of the cell and that the chromosomes may remain embedded within it. In onion root cells (*Allium cepa*) the duration of prophase is 1 hour at room temperature (20°C, Table 6.2).

Metaphase

Forces active within the cell now arrange the chromosomes, or at least a specialized portion of each chromosome (the **kinetochore**), in the equatorial plane of the cell (Figs. 6.11*E,* 6.12*D*). The nucleoplasm appears to elongate slightly, and **spindle fibers** appear attached to the kinetochores. As the nucleoplasm elongates, the chromosomes, or at least the kinetochore, is moved to the equatorial plane of the cell (Figs. 6.11*E,* 6.12*D,* 6.13). Spindle fibers extend from the chromosomes to the opposite poles of the cells. Other fibers apparently reach to the poles but are not attached to the chromosomes. The electron microscope demonstrates that these spindle fibers have the characteristics of microtubles (Fig. 6.13). This structure—chromosomes, microtubules, or spindle fibers—plus any adherent nucleoplasm is called the **spindle apparatus.** Note also that the cytoplasm within the spindle apparatus is completely devoid of organelles. A **clear zone** is necessary here to allow the free movement of the chromosomes. This phase takes about 0.3 hour to complete in onion root cells.

The chromosomes are now distinct bodies of two closely associated halves, each half known as a **chromatid** (Fig. 6.11*F*). In plants, such as corn, which have been intensively studied, each chromosome can be recognized and numbered. There are 20 chromosomes in corn, but only 10 different types that can be distinguished by

Figure 6.14
Cross section of microtubules from root cells of the water fern, *Azolla pinnata*. Scale marker = 1 μm. (From B. E. S. Gunning, A. R. Hardham, and J. E. Hughes, *Protoplasma* **143** and 145. © 1978 by Springer-Verlag. Reprinted with permission of the publisher.)

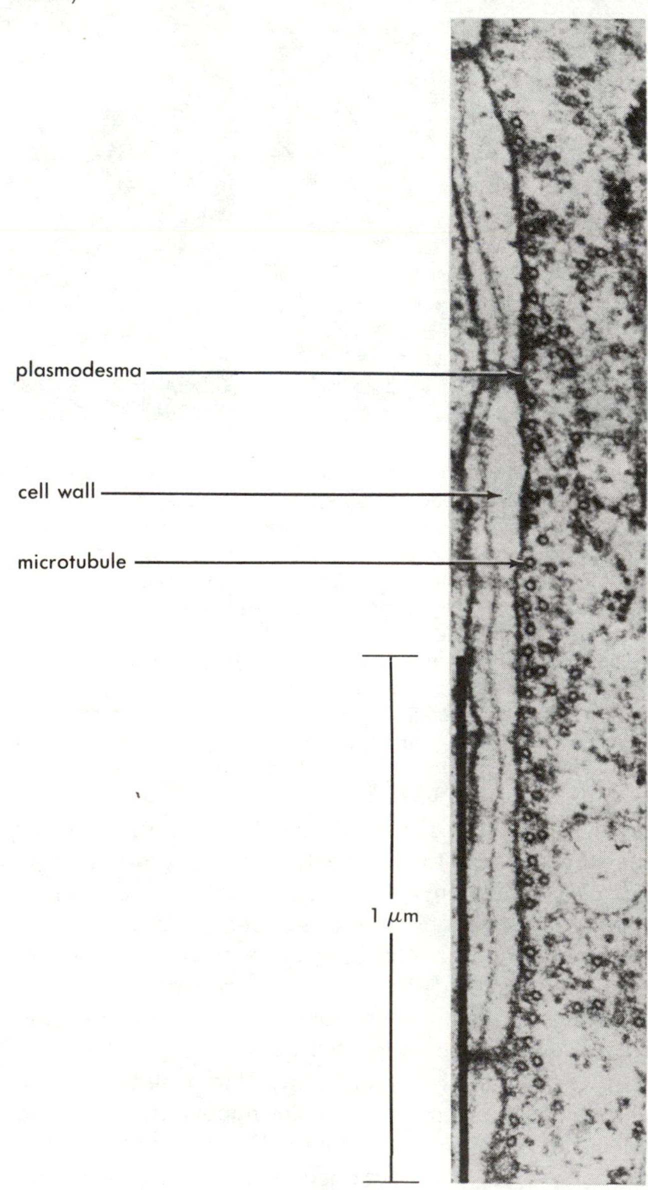

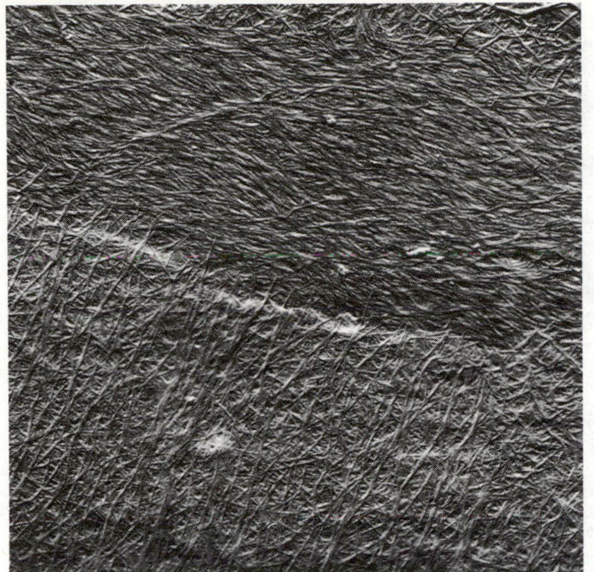

Figure 6.15
The wall of a hair cell from the base of a sedge (*Juncus*) leaf. Note that the innermost fibrils are parallel with the girth of the cell and that the outermost cellulose fibrils are parallel with the long axis of the cell, ×53,000. (Courtesy of A. L. Houwink and P. A. Roelofsen, *Act. Bot. Neerland,* **3,** 385. Reprinted with permission of the publisher.)

their size and form. There are, thus, two chromosomes of each type. The 20 chromosomes of corn may be arranged in 10 pairs. Maps have been prepared of corn chromosomes showing the relative positions of the genes along them. Since the chromosomes split longitudinally, each gene is replicated and each chromatid contains a full set of genes.

Microtubules and Cell Division Polarity. Microtubules as you remember from Chapter 4 play an important role in orienting cell wall microfibrils (Figs. 6.14 and 6.15). Electron micrographs of cells taken immediately adjacent to the cell wall show the close juxtaposition of both components (Fig. 6.14). Microtubules also play an important role in cell division. One of these roles concerns their regulation of the plane of cell division; another concerns their role in chromosome movement.

Several cell biologists, examining many different types of cells, have reported a curious redistribution of microtubules into a narrow (1.5–3.0 μm) band around the inside periphery of the cell prior to its division (Fig. 6.16, 6.17). This **pre-prophase band** of microtubules consists of many (150 or more) short segments that may be stacked two or three deep or may be in a monolayer. These microtubules encircle the cell in a position that marks or predicts the exact plane of division when that cell divides. Figure 6.18 shows a hypothetical example of a pre-prophase band position in a series of cells in different developmental stages. Pre-prophase bands are apparently established by some kind of internal signaling mechanism to insure the correct polar orientation of new cells. If the spindles in a meristematic tissue were oriented at random, an irregular mass of tissue would result. This does occur when a single cell or small group of cells from a carrot root in tissue culture produces an irregular mass of cells called a **callus** (Fig. 6.19*A*). For orderly growth there must be a precise orientation of the mitotic spindles. A polarity is established so that, in gen-

Figure 6.16
Meristematic cells in a young leaf of tobacco (*Nicotiana tabacum*) showing the pre-prophase band of microtubules. *A,* section at right angles to the plane of the future cell plate shows cross section of microtubules, ×10,000. *B,* section in plane of future cell plate shows microtubules encircling nucleus, ×10,000. *C,* three-dimensional interpretation of *A* and *B.* (Redrawn after K. Esau and J. Cronshaw, *Protoplasma,* **65,** 1. © 1966 by Springer-Verlag. Reprinted with permission of the publisher.)

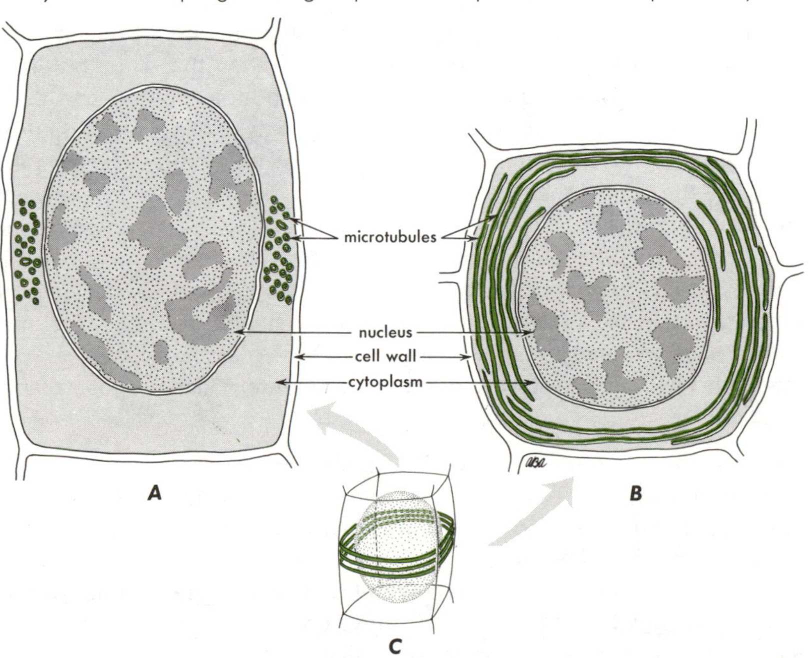

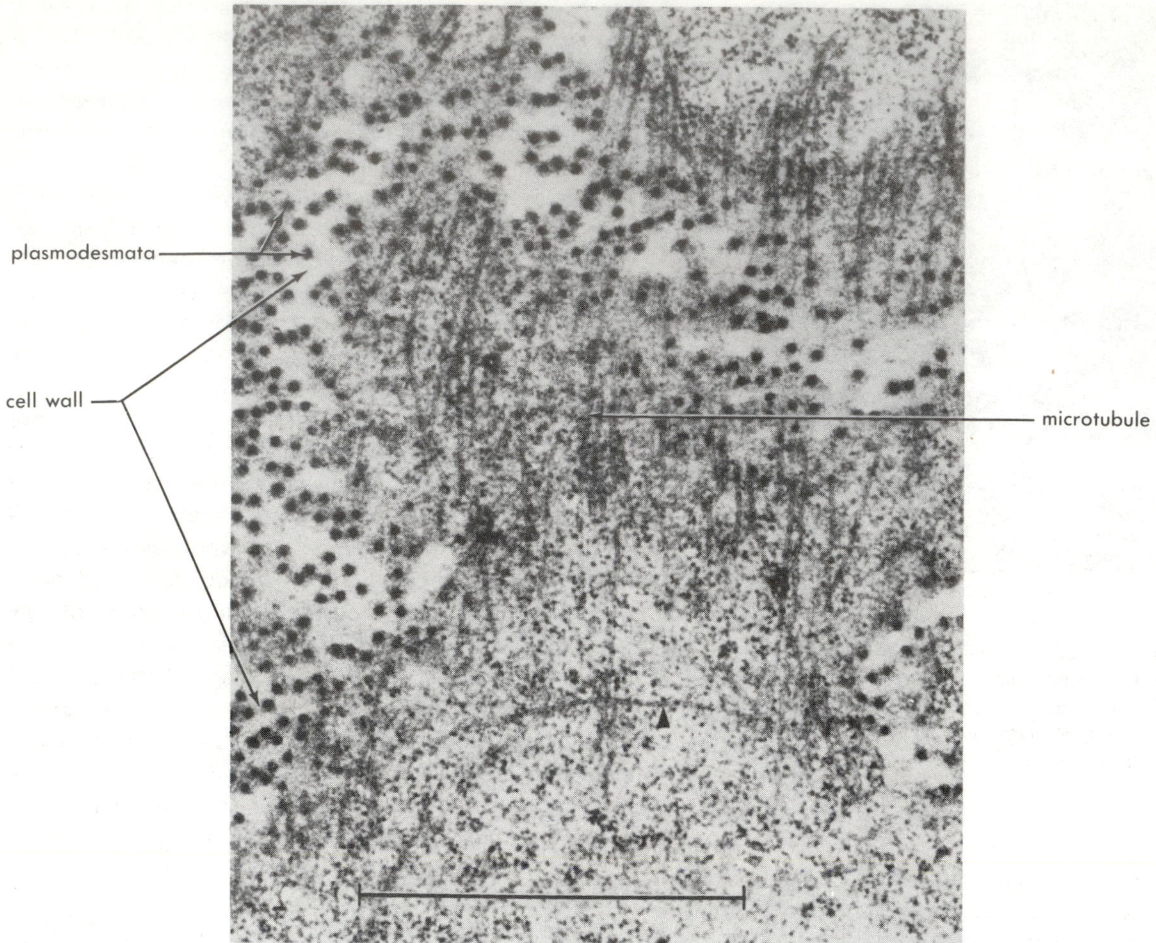

Figure 6.17
Glancing section through pre-prophase band of microtubules in *Azolla* root cell. Scale marker = 1.0 μm. From B. E. S. Gunning, F. L. Hardham, and J. E. Hughes. *Protoplasma,* **143,** 145. © 1978 by Springer-Verlag. Reprinted with permission of publisher.)

eral, the axes of spindles are parallel with each other and with the axis of the shoot or root (Fig. 6.19*B*) When the spindle axes are oriented parallel to the root-shoot axis, the divisions are called **anticlinal.** Less commonly, when the axes are perpendicular to the root-shoot axis, increasing the girth, the divisions are called **periclinal** divisions (Fig. 6.19*B*)

What is responsible for polarity? We do not know. Polarity seems to be inherent in cells and in the tissue of which the cells are a part. Changes in polarity may be induced by hormones. In some instances, the cellular organelles will pass largely to one end of the interphase cell.

Chromosome Movements. The mechanism of chromosome movement and the functional role of the microtubules composing the spindle fibers are not understood. At present, there are three major theories that are used to explain this phenomenon. The **assembly-disassembly** theory contends that chromosomes move because force is generated by the depolymerization of microtubule protein at one end of the spindle. The removal of subunits at both poles of the spindle would shorten the spindle and cause the chromosomes to separate. The microtubule **sliding theory** contends that the microtubules slide over each other, thereby facilitating chromosome movement by a physical mechanism. And the third theory, the **zipper hypothesis,** explains chromosome movement by an actual zipping and unzipping motion between microtubules attached to the chromosomes and the continuous but parallel microtubules that are not attached to chromosomes. Whether any of these theories, or possibly a composite theory, proves to be correct remains to be seen.

Anaphase

The chromosomes do not remain long in the equatorial plane. The individual chromatids of each pair soon separate from each other and move to opposite poles of the cell. This period is of shortest duration (0.2 hour in onion root cells) and is called **anaphase** (Figs. 6.11*F*, *G*, 6.12*D*,

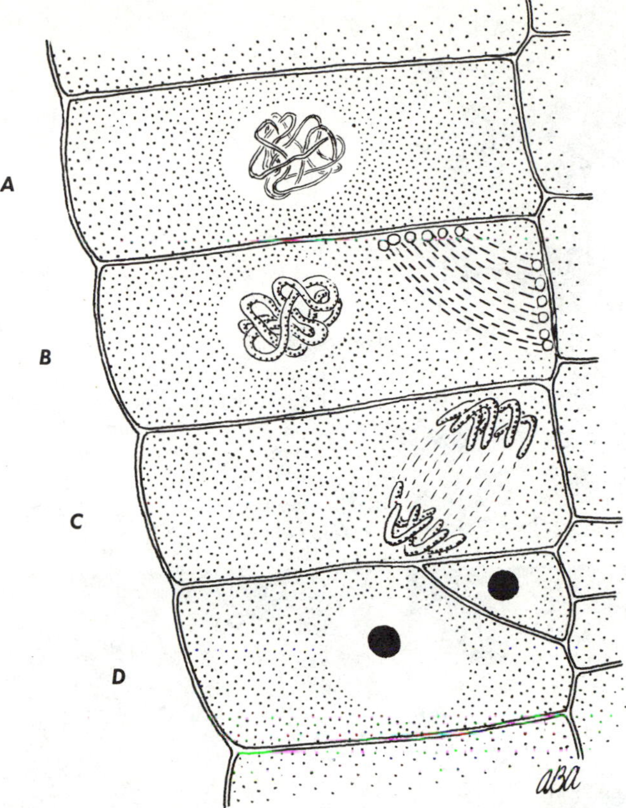

Figure 6.18
A, cell in interphase—G_1, *S*, or G_2 of cell cycle. *B*, cell possibly in G_2—pre-prophase band of microtubules in position to "mark" the plane of mitosis. *C*, anaphase of mitosis—note that the plane of division is at right angles to the pre-prophase band position. *D*, mitosis is completed and the new cell wall separating derivative cells forms in the same position previously occupied by the pre-prophase band.

E.) Figure 6.20 is a stereo pair of photomicrographs of an anaphasie mitotic figure from a kangaroo rat.

Telophase

When the divided chromosomes have reached the opposite poles of the cell, they group together and spin out into long, thin threads of chromatin. The nuclear membrane and the nucleolus again become apparent. This period of transition from chromosomes to thin threads in the interphase nucleus takes about 0.7 hour in onion root cells and is known as the telophase. The chromatids that have gone into the constitution of the new derivative nuclei grow, forming full-sized chromosomes once again (Figs. 6.11*H*, 6.12*F*, 6.21*A*, *B*).

Interphase

Even in the most rapidly dividing cells, the telophase nucleus gradually returns to the G_1 phase. Either a new division must be prepared for or the genetic information must be transmitted to the cytoplasm for the synthesis of new proteins, wall materials, and structure that accomplish differentiation in the new cell.

Cytokinesis

In the great majority of cases, the division of the nucleus is followed by the division of the cytoplasm. Light microscopy demonstrated that a cell plate forms at the equatorial plane of the cell at right angles to the spindle fibers (Figs. 6.21*A*, *B*). At the equatorial region, the microtubules are gathered together in bundles and each bundle is surrounded by an amorphous material (Fig. 6.21*C*). Vesicles appear in this region and eventually fuse to form the first cell wall component dividing the derivative protoplasts (Fig. 6.21*D*). This region, formed of pectin, is known as the **middle lamella**. The remaining new cell wall is formed by the deposition of cellulose by each protoplast on its side of the middle lamella. The resulting structure is the **primary cell wall.**

DNA Synthesis (The S Phase)

The DNA Molecule

The molecular structure of the DNA double helix is shown in Fig. 6.22. Each strand forming part of the backbone of the helix is made up of alternating sugar, S (deoxyribose), and phosphate, P, groups. These back-

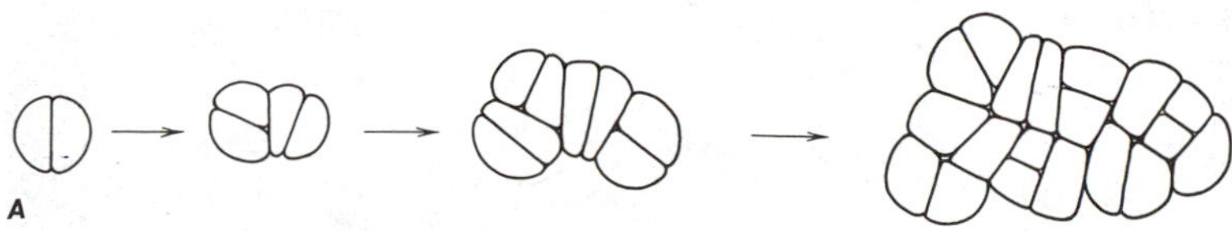

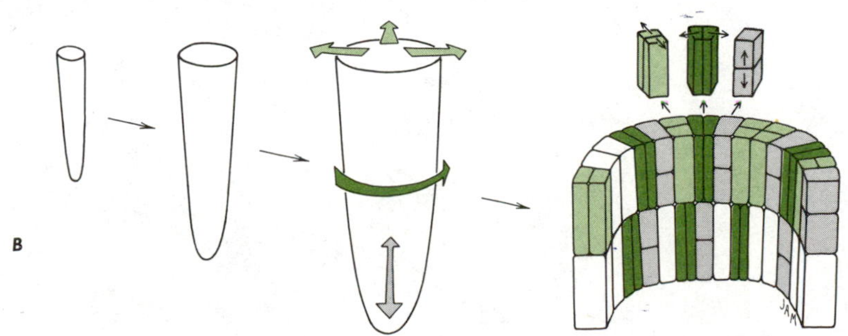

Figure 6.19
Influence of polarity of cell divisions on the development of form. *A*, nonpolarized cell division results in an irregular mass of cells. *B*, cell division parallel with the axis of the root (gray arrow and cells) results in growth in length; cell division parallel with the circumference of the root (dark green arrow and cells) increases the circumference of the root; cell division at right angles to the circumference of the root (light green arrows and cells) increases the diameter of the root.

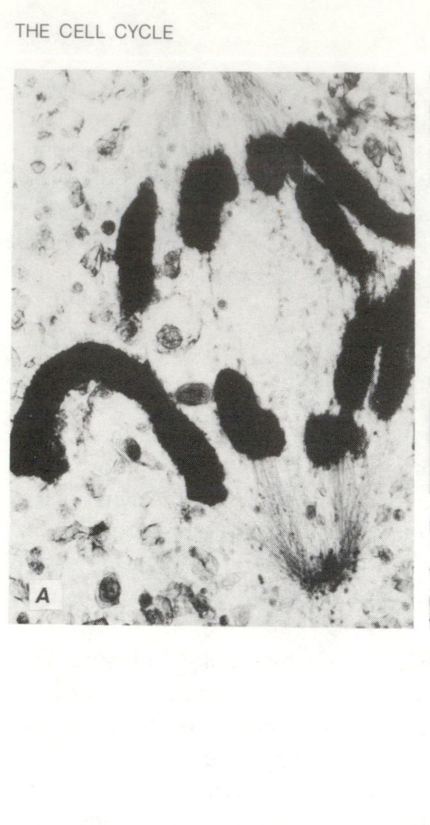

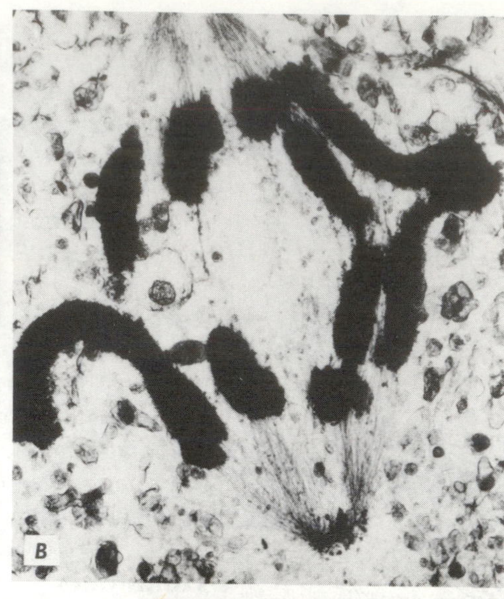

Figure 6.20

Stereo pair of anaphase mitotic figure from a Kangaroo Rat to show the three-dimensional array of microtubules. To see stereo you must view these photos together with stereo viewing glasses, ×6,000. (Courtesy of J. Richard McIntosh, University of Colorado.)

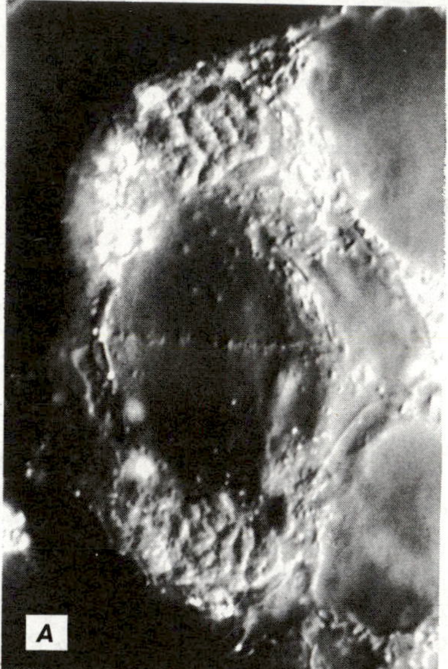

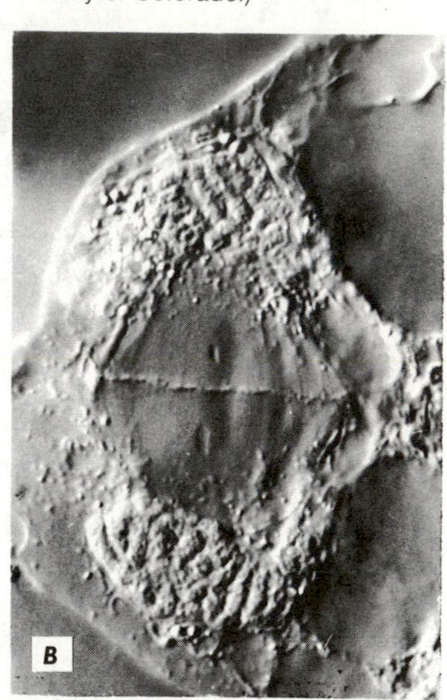

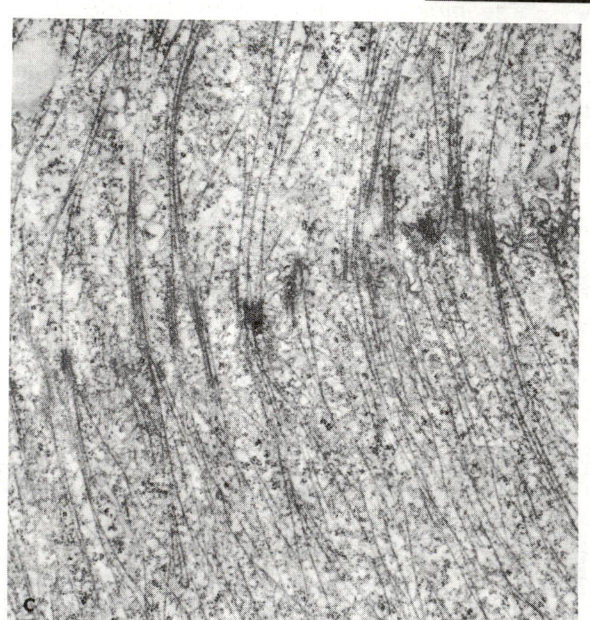

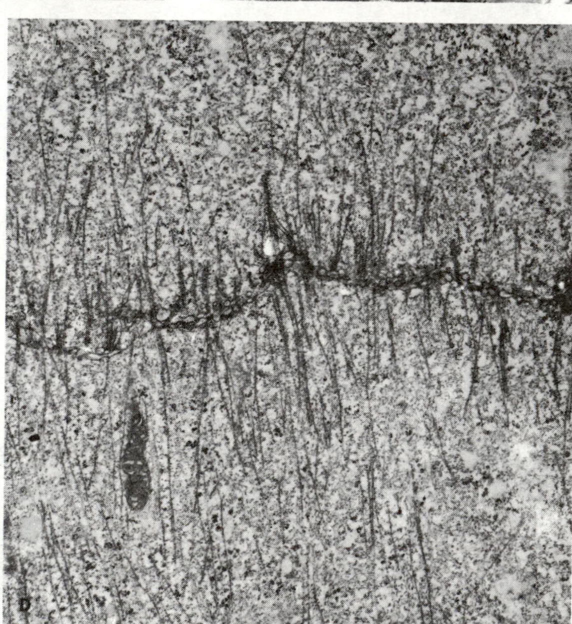

bone strands are held together by hydrogen bonds between the nitrogen-containing organic bases (**purine** and **pyrimidine**). The bases of one strand are bonded to complementary bases of the other strands (Figs. 6.23*A*, *B*). When a purine or a pyrimidine base is joined with a ribose sugar and a phosphate, the resulting compound is a **nucleotide.** Although millions of nucleotide units may polymerize to form the gigantic nucleic acid molecule (Fig. 6.22), only four different nucleotides are found in a nucleic acid molecule. The two pyrimidine bases occurring in DNA are **cytosine** and **thymine;** the two purine bases are **adenine** and **guanine** (Fig. 6.23*C*). Notice in Fig. 6.22 that the bases have a linear arrangement along the background strands of the helix and the double helix is stabilized by hydrogen bonding between complementary base pairs. The sequence of purines or pyrimidines in one of the strands of the spiral determines the sequence of the bases in the other strand, since adenine of one strand always pairs with thymine of the other strand, and guanine of one strand always pairs with cytosine of the other. To illustrate this arrangement and the great variety within the nucleic acids that may arise from this simple set of four compounds, let us use the letters A, G, T, and C. We may first write them down in a line using each as many times as we wish and in any order we wish. Now, below these write again A, G, T, and C, but be careful that A always matches T and G always matches C.

Thus, if the first strand is represented by

A G C A A C A T C

the second strand would be

T C G T T G T A G

The two complementary bases (AT or GC) joined by hydrogen bonds between the complementary strands in the DNA molecule are frequently called **base pairs.**

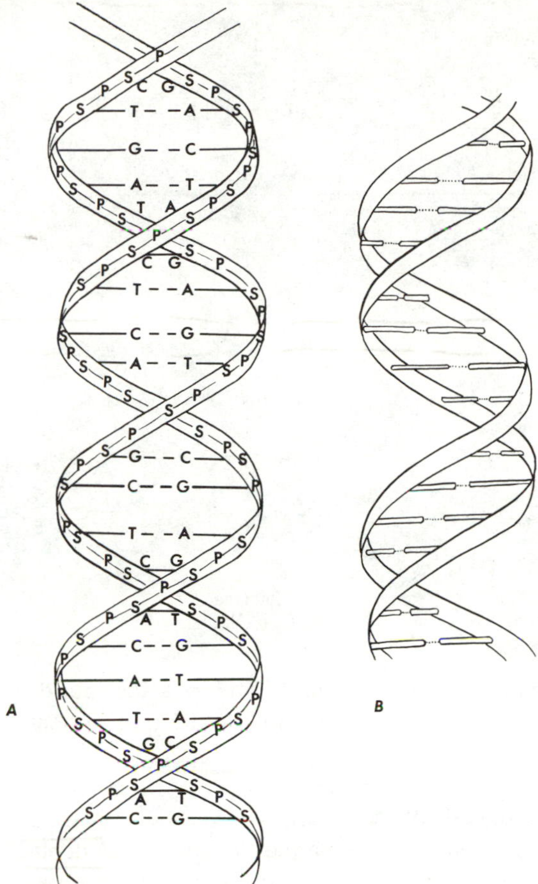

Figure 6.22
Schematic diagram of the double helix in a portion of a DNA molecule. Alternating sugar, *S* (deoxyribose); and phosphate, P, groups make up the backbones of the strands. Attached to each sugar unit is one of the purine or pyrimidine bases, adenine, A; thymine, T; guanine, G; cytosine, C. The strands of the helix are held together through hydrogen bonds between the base pairs, adenine to thymine and guanine to cytosine. *A*, one strand is displaced along the axis of the helix for convenience of diagramming; *B*, diagram of the parallel strands in the helix.

Figure 6.21 *(opposite)*
The formation of the cell plate in the endosperm of *Haemanthus katherine. A* and *B*, living unflattened cells; photography by Nomarski system optics. *A*, telophase; the chromosomes are tightly clustered at opposite poles of the cell, and the cell plate has started to form; compare with *C*, ×4000. *B*, late telophase; the cell plate is almost continuous, and fibrils may be seen extending poleward a short distance from the plate; compare with *D*, ×4000. *C* and *D*, comparable stages of cell plate formation as seen with the electron microscope. *C*, early stage in plate formation; the microtubules are clustered into groups that are associated with the denser material, apparently in the plane of the future cell plate, ×26,000. *D*, late stage; the microtubules are fewer in number and are not clustered; vesicles are present and apparently fusing to form a continuous separation phase between the sister cells, ×26,000. *(A* and *B*, courtesy of A. Bajer, *Chromosoma*, **25,** 249. © 1968 by Springer-Verlag. *C* and *D*, courtesy of P. K. Hepler and W. T. Jackson, *J. Cell Biol.* **38,** 437. © 1968 by the The Rockefeller University Press. Reprinted with the permission of the publishers.)

Enzymes regulate metabolic reactions in the cell. Each different kind of enzyme is a different protein and regulates a different reaction in the metabolism, growth, and development of the cell. Since each kind of protein has its own exact sequence of amino acids, the way in which the cell regulates the sequence of amino acids in proteins is very important.

The exact sequence of the amino acids in a newly formed protein molecule is specifically determined by the genetic information contained in the cell's nucleic acid molecules. We now know that the basis of this genetic information is the sequence of bases that each part of the DNA molecule contains. One of the most exciting accomplishments of modern biology has been the cracking of this genetic code.

Each amino acid is coded for by a short nucleotide sequence. In fact the four bases, in combinations of

Pyrimidine	Purine
Cytosine	Adenine
Thymine	Guanine
Uracil	

A

Adenine
($C_5H_5N_5$)

Guanine
($C_5H_5N_5O$)

C

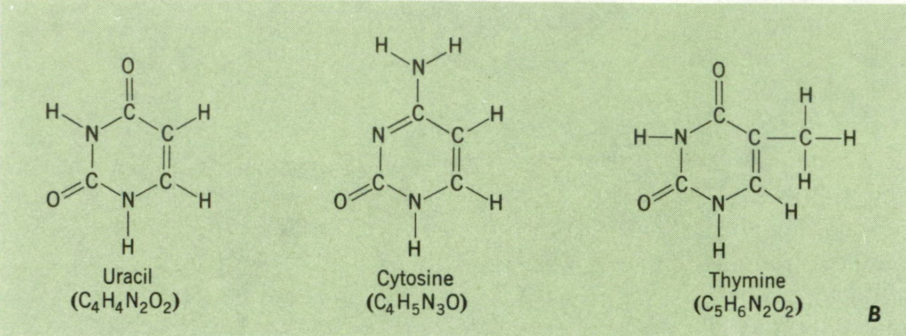

Figure 6.23

General formulas for purine and pyrimidine bases. Each particular base has characteristic groups attached at the unfilled valence bonds in the diagrams. These groups are shown in the complete formulas of *B* and *C*.

three, specify for each of the 20 amino acids found in proteins. The genetic code is made up of three "letters." It is a triplet code.

Replication of DNA

We shall now consider the manner in which DNA itself is replicated. From each double helix two new double helices must form, without at the same time, disturbing the precise ordering of the base pairs (Fig. 6.22). The dividing double helix may be several million nanometers long and, since a single base pair occupies the space of only a few nanometers, there are many base pairs along a double helix.

Precise information telling us the exact time and location of the synthesis of new DNA may be obtained by using radioactive precursors of nucleic acid. A commonly used substance of this type is radioactive thymidine. It seems to enter cells and to be directly incorporated into nuclei of dividing cells of both plants and animals. Root tips of *Tradescantia* (a common house plant) grown in a culture solution containing radioactive thymidine (a form of the DNA base thymine) show that interphase may be divided into three periods (the cell cycle). During the first stages the nucleus increases in volume, but there seems to be no incorporation of radioactive thymidine; this is G_1. Later, radioactive material is incorporated into newly synthesized DNA (*S* phase). A period follows in which the nucleus maintains a constant size without any further incorporation of labeled thymidine (G_2). Refer back to Table 6.1 to reexamine the time it takes for a cell to transverse each stage of the cell cycle.

But these observations do not tell us the manner of replication of the double-stranded DNA helix. It is possible that a whole new double-stranded helix might be replicated, thus totally conserving the old double helix intact. Or the coiled double strands could somehow separate and a new strand could be formed by each old strand. This is semiconservative; half of the new helix contains old DNA. Or the replication could take place at random throughout the DNA molecule.

Careful experiments conducted by Dr. J. Herbert Taylor when he was at Columbia University, demonstrated that DNA replication was, in fact, semiconservative. The mechanism for this replication can be explained in the following way.

Assume that a double helix is present in early interphase at the onset of replication. This helix may untwist, separating the two strands (Fig. 6.24). Each strand now serves as a template for the formation of a new strand. Thus, following replication with radioactive thymidine available, each chromatid will contain a strand bearing the radioactive material. In a subsequent division in the absence of radioactive thymidine, the radioactive strand will serve as a template for a new nonradioactive strand. The old nonradioactive strand will also serve as a template for a new nonradioactive strand.

SUMMARY OF THE CELL CYCLE

1. Cell division increases the numbers of cells.
2. All cells have the potential to divide but are normally blocked from doing so.
3. Preparation for division takes place during interphase (G_1–S–G_2).
4. Nuclear division is called mitosis; cytoplasmic division is called cytokinesis.
5. The nuclei of derivative cells are genetically identical.
6. The cytoplasm of derivative cells may be unlike each other.
7. In general, mitotic spindles are oriented parallel with each other and with the axis of the shoot. The

pre-prophase band of microtubules apparently regulates this orientation.

8. The stages of mitosis are prophase, metaphase, anaphase, and telophase.
9. The spindle, which is the mechanism for chromosome movement, is constructed from many microtubules and is involved in chromosome movement by some mechanism.
10. The middle lamella is the first cell wall component to form between the two new derivative cell protoplasts.
11. Vesicles collect around the microtubules in the equatorial plane of the cell. They fuse to form the middle lamella.
12. Cellulose formed by the two derivative protoplasts and deposited on the middle lamella forms the primary wall.
13. The DNA double helix is composed of deoxyribose sugars, phosphate groups, and purine (adenine and quanine) and pyrimidine (cytosine and thymine) organic bases.
14. Two complementary base pairs—one purine to one pyrimidine—join adjacent DNA strands by hydrogen bonds.
15. DNA replication is semiconservative (one strand is retained and one strand is newly synthesized) and occurs by a controlled mechanism.

PROTEIN SYNTHESIS

RNA

So far we have followed the phases of the cell cycle, including G_1, DNA synthesis (S), G_2, and mitosis (M). We have also discussed the DNA molecule, and the manner in which it is enabled to serve as a storehouse of genetic information. Now we shall consider the manner in which this information is transmitted to the sites of protein synthesis in the cytoplasm. RNA is involved in this transfer of information. There are three major types of RNA: **messenger RNA (*m*RNA), transfer RNA (*t*RNA),** and **ribosomal RNA (*r*RNA)** (Fig. 6.25). They are all synthesized in the nucleus under the direct influence of nuclear DNA. In the cases of *m*RNA and *t*RNA, the important DNA is located in the chromosomes. On the other hand, a portion of the DNA helix residing in the nucleolus is implicated in the synthesis of *r*RNA.

The molecule structure of RNA is similar to that of DNA in some ways and very different in others. Both have a similar backbone chain of alternating phosphate and sugar groups. Both have sequences of two pyrimidine and two purine bases linked to the sugar so that, in the resulting nucleotide, the bases have a position lateral to the main length of the chain. However, in RNA, the pentose sugar **ribose** is used rather than deoxyribose. And the pyrimidine base **uracil** bonds with adenine and re-

SUMMARY OF METABOLIC EVENTS REQUIRED FOR CELL CYCLE PROGRESSION

$G_1 \longrightarrow S \longrightarrow G_2 \longrightarrow M$

Event	Times of maximum activity	Duration
Protein synthesis	X X X	Continuous
Thymidine kinase		G_1
DNA Polymerase		G_1
Spindle protein	X	Continuous (maximum G_2)
Histones (nucleo)		Only S
RNA	X X X	Continuous (not in M)
Oxygen	X	Continuous
Energy source		Continuous

X = time of maximum activity

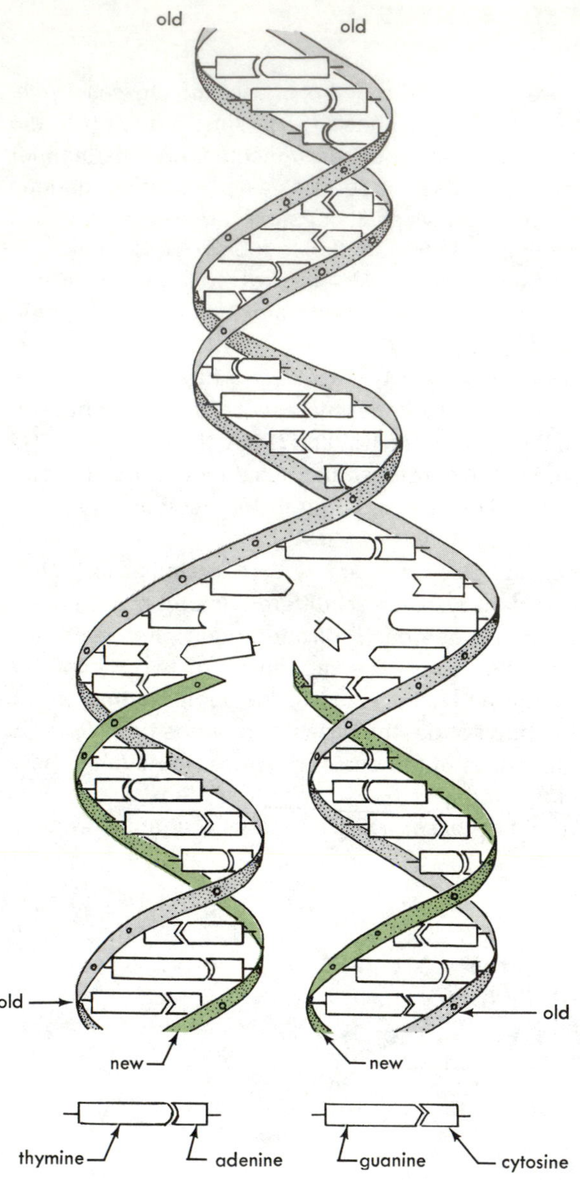

Schematic diagram showing how one "old" double-stranded molecule of DNA could be replicated into two new double-stranded molecules of DNA. Each new molecule has one old and one newly synthesized strand (semiconservative replication).

places the thymine that occurs in DNA. Perhaps the most striking difference between the two molecules is that a double-stranded helix is characteristic of DNA, while the RNA molecule exists as a single long strand, sections of which fold back to form a modified double strand.

The synthesis of RNA is directed by the DNA molecule. In the process, the coded information carried in the sequential arrangement of the nucleotides in the DNA molecule is transcribed in new RNA molecules. RNA synthesis is initiated and terminated by an enzyme called RNA polymerase.

The synthesis of RNA molecules resembles the replication of DNA (*a*) in that the initial purine bases are joined with complementary pyrimide bases or vice versa and (*b*) in the linking of new sequences of bases by ribose and phosphate groups to form a single continuous strand of RNA. This results, as in DNA synthesis, in a new strand bearing the complementary bases of the old strand in exactly the same sequence. A significant difference in the process of replication of DNA and synthesis of RNA lies in the disjoining of the newly formed sequences of RNA nucleotides from the DNA template to give a molecule of single-stranded RNA.

Codons

Proteins are composed of amino acids linked together by peptide linkages (Fig. 5.9) to form complex molecules. The problem before us is to understand how a series of nucleotide bases in the DNA molecule can be responsible for the specific and accurate sequence of amino acids in a protein. A sequence of three nucleotides on a DNA molecule, known as a **codon** (Fig. 6.26), codes for one amino acid. Sixty-four different codons may be formed from different combinations of four nucleotides. Three codons are punctuation codons marking the starting and end points of a given protein. All remaining codons code for the 20 or so amino acids; several different codons may code for the same amino acid. The new molecules of RNA contain fewer nucleotides and are much shorter than the DNA molecules that serve as templates. Indeed, we may think of the DNA molecule as a paragraph. A sequence of codons coding for a protein would be a long sentence. Each codon itself represents a word (Fig. 6.27).

*t*RNA

There are three different types of RNA cooperating to synthesize a protein. *m*RNA carries the code for a specific codon sequence from the DNA to the site of protein synthesis. *t*RNA transfers amino acids from the cytoplasmic amino acid pool to the site of synthesis in the ribosome. *r*RNA somehow fashions the peptide linkage between two adjacent amino acids.

*t*RNA (Fig. 6.26) is easily isolated and is frequently also called soluble RNA. There are about 60 different *t*RNAs, each composed of about 80 nucleotides. A specific enzyme and ATP are involved in loading each *t*RNA molecule with its specific amino acid. We may diagram these steps by assuming that the molecules involved are shaped, or keyed, to match each other as are the pieces of a puzzle (Fig. 6.28). The first step, catalyzed by an enzyme, is the coupling of ATP with the amino acid (Fig. 6.28). The combination enzyme–amino acid is now keyed to a matching molecule of *t*RNA. The *t*RNA thus brought into contact with the amino acid becomes attached to it and the enzyme is released to catalyze other amino acid-*t*RNA combinations.

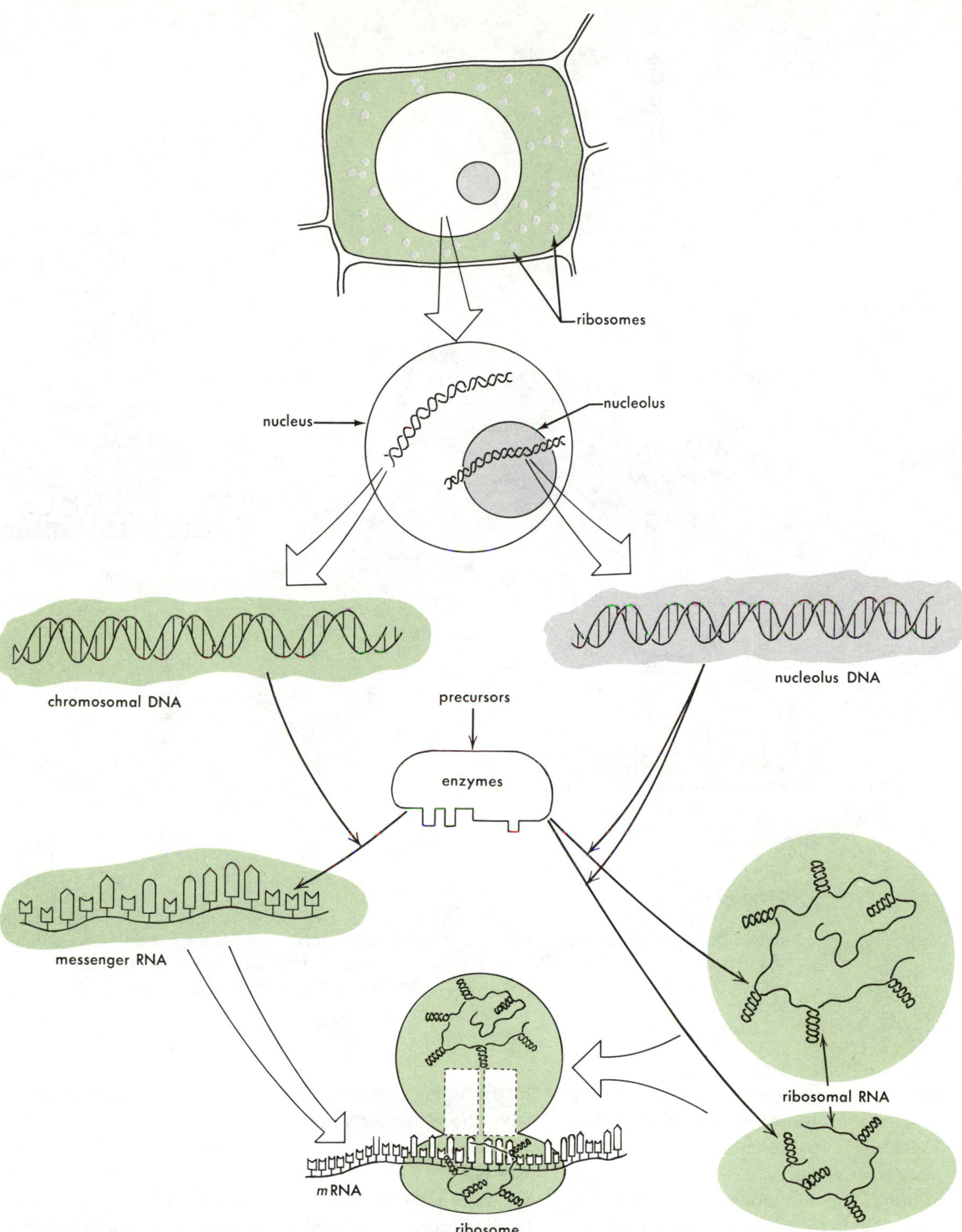

Figure 6.25
The synthesis of *m*RNA and *r*RNA and their relationship. *m*RNA and *r*RNA are synthesized from precursors in reactions catalyzed by enzymes. Chromosomal DNA codes the *m*RNA. DNA in the nucleolus codes the *r*RNA. *m*RNA is single-stranded. *r*RNA is also single-stranded, but it doubles back upon itself to form regions of a double helix. *r*RNA strands are of two sizes and they become associated with proteins to form ribosomes that appear to have two components of different sizes. The *m*RNA becomes associated with the smaller component. The larger component has sites that function in the formation of the polypeptide chain (see Fig. 5.9).

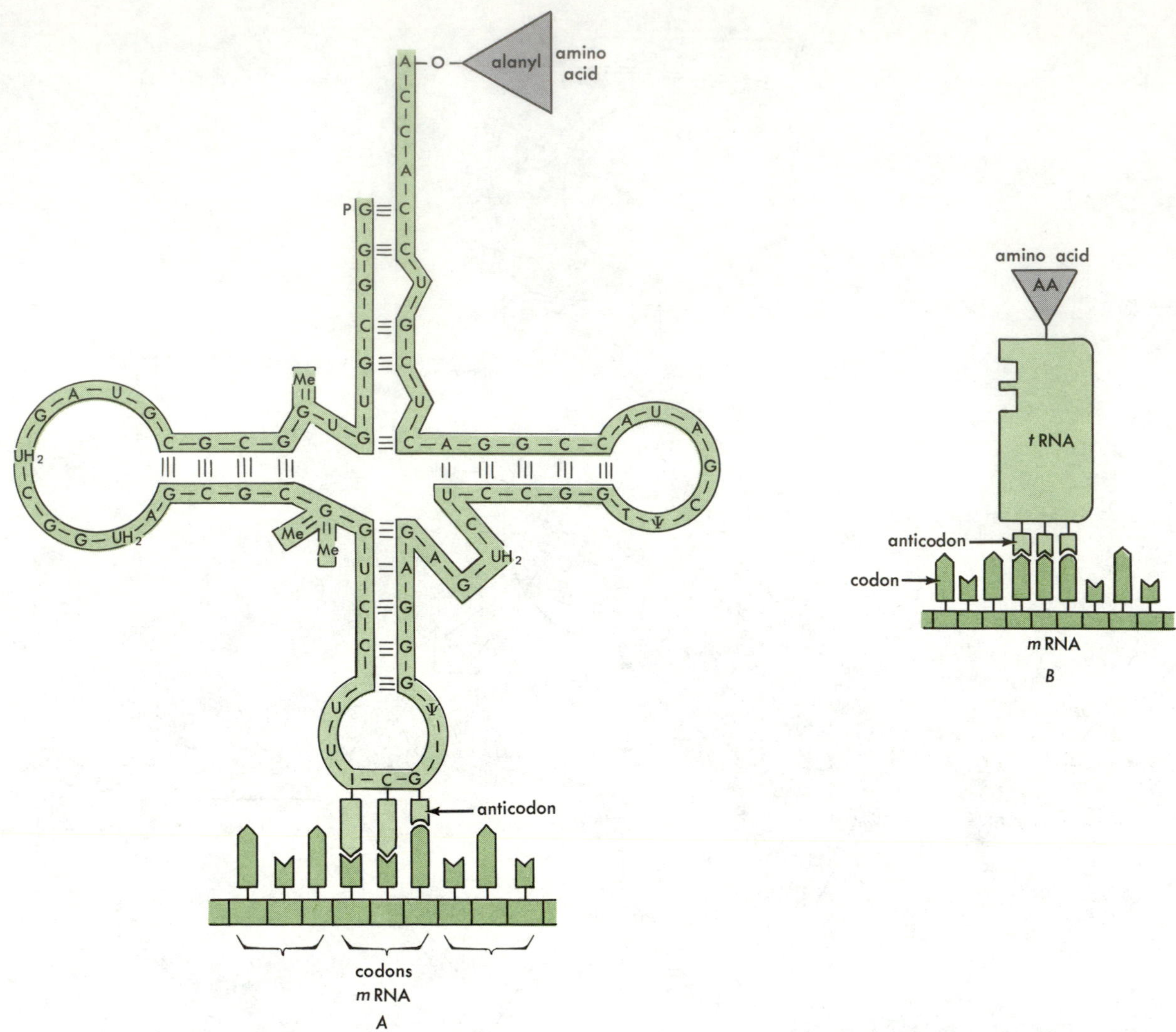

Figure 6.26
Transfer RNA. *A*, two-dimensional representation of the molecule showing a single strand of *t*RNA folded back upon itself with specific cross linkages (≡) between some nucleotides. One end of the molecule is keyed to bind with a specific amino acid. The opposite end (an anticodon) is coded to match a codon on *m*RNA. *B*, conventional shape frequently used to denote *t*RNA.

Three nucleotides of the *t*RNA molecule are free to combine with their complementary nucleotides on the *m*RNA molecule. Recall that in DNA and RNA synthesis purine combines with pyrimidine or vice versa. A similar relationship holds here. The nucleotides on the *t*RNA molecule are complementary to those on the *m*RNA molecule. Since the latter are known as a codon, the complementary nucleotides on the *t*RNA molecule are known as an anticodon (Fig. 6.26).

A conjectural structure of a *t*RNA molecule is shown in Fig. 6.26. Note that it is composed of a single strand, which doubles back upon itself forming definite regions of double strands. The anticodon site is indicated as well as the attachment point for the amino acid.

*t*RNA carries the amino acid to the ribosome. *m*RNA becomes associated with a ribosome. Subsequent steps in protein synthesis between *t*RNA and *m*RNA occur when both are associated with a ribosome.

*m*RNA

Molecules of *m*RNA vary greatly in size depending largely upon the length of the code message they carry. They may code for one, or several, proteins. The smallest may have a molecular weight of around 500,000 and carry about 900 nucleotides; others may have 12,000 nucleotides with a molecular weight of 4,000,000. There are many types of *m*RNAs; they are difficult to isolate, and only a few are known. The best examples come from bacteria with a virus infection. Some other well-known *m*RNAs have been isolated from red blood and from muscle cells. *m*RNA may be thought of as a long backbone helix of phosphate-ribose linkages to which are at-

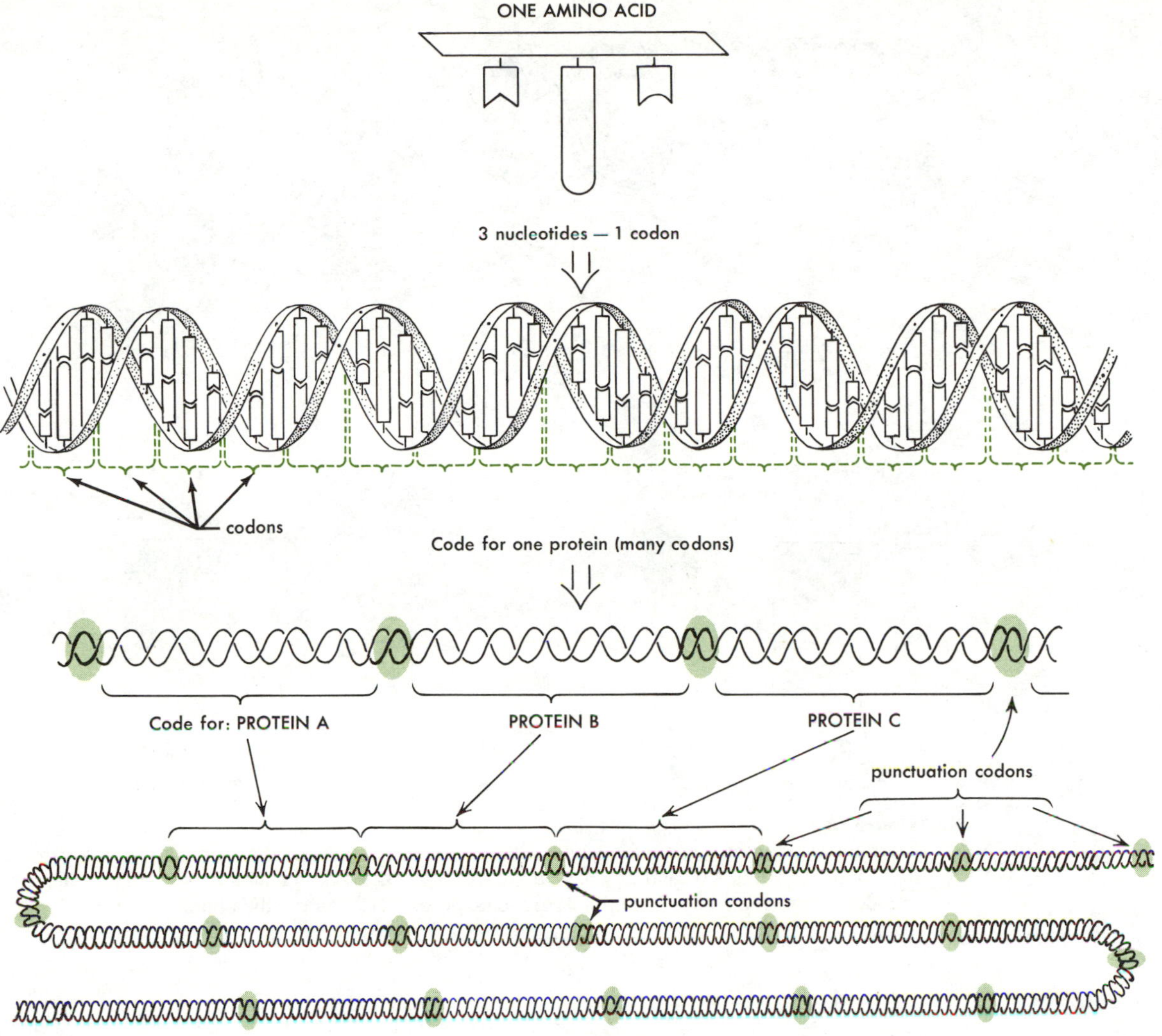

Figure 6.27
A sequence of three nucleotides is a codon. One codon codes for one amino acid: many codons code for a single protein. A single DNA molecule carries the code for many proteins. The protein codes are separated by special "punctuation" codons. The codon may be likened to a word, the protein to a sentence, and the DNA molecule to a paragraph.

tached the nucleotides, each three forming a codon (Figs. 6.25, 6.29).

*r*RNA

As the name suggests, *r*RNA is associated with ribosomes (Fig. 6.25). It is formed on a DNA template just as are the two other types of RNA. There are two subunits in the *r*RNA molecule, both of which are required for the activity of *r*RNA. The nucleolus appears to play some part in bringing the two subunits together before they leave the nucleus (Fig. 6.25). The relative size of the *r*RNA subunits seems to depend upon their species origin. The *r*RNA of *Escherichia coli* exists in three forms ranging in molecular weight from about ½ million to more than one million. All three complex with proteins, about 50 in all. The molecule of *r*RNA is composed of sections of both double and single helices (Fig. 6.25). X-ray pictures of the molecule appear distorted, and a complete molecular model for *r*RNA has not been entirely proposed. Ribosomal RNA seems to act mainly in a structural manner by bringing *m*RNA, with its loaded *t*RNA, into position for establishing the peptide linkage between two amino acids. This step occurs in the ribosome.

Peptide Linkage.

The peptide linkage (Fig. 5.9) forms when the strand of *m*RNA with its loaded *t*RNA associates with a specific synthesizing site in the ribosome. This association may start at one end of the *m*RNA strand or at specific sticky points along the strand (Fig. 6.29). Only short portions of the *m*RNA strand become associated with a ribosome, so several ribosomes may be attached to the same strand of

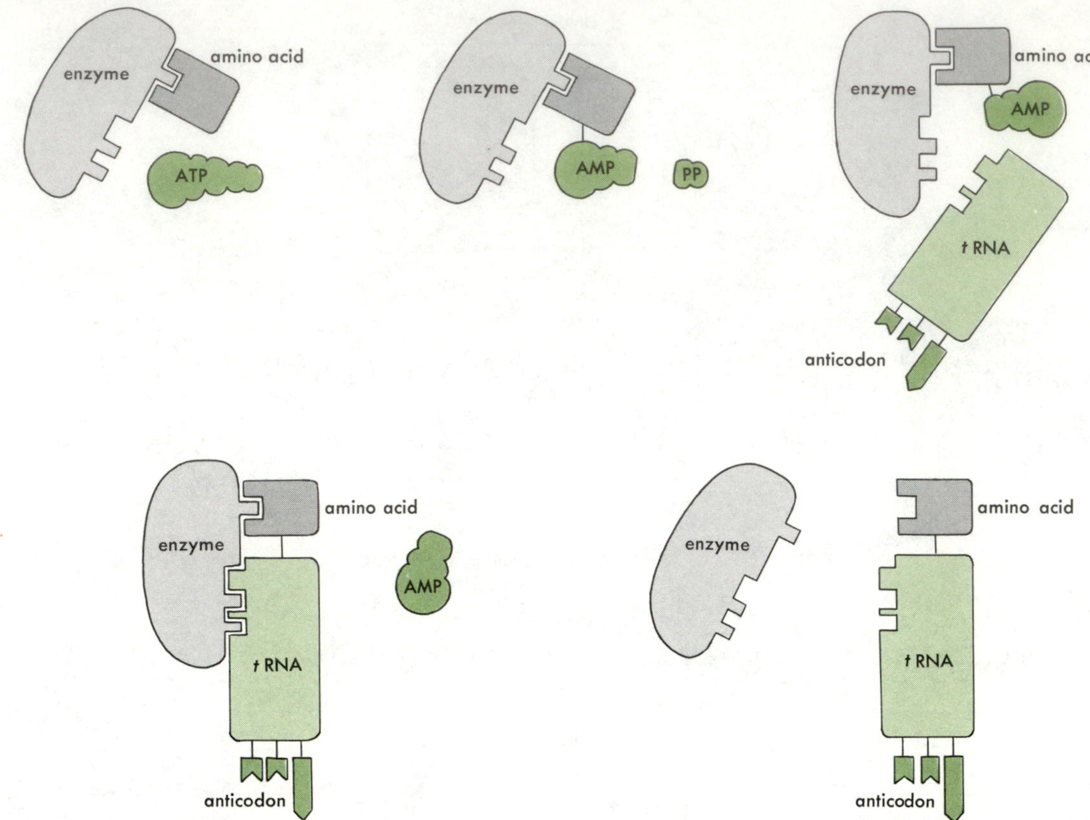

Figure 6.28

Diagram showing one manner in which *t*RNA could select a specific amino acid from an amino acid pool. Note that ATP and an enzyme are involved. The binding sites on the enzyme are specific for a certain amino acid and for the molecule of *t*RNA bearing the anticodon for that amino acid.

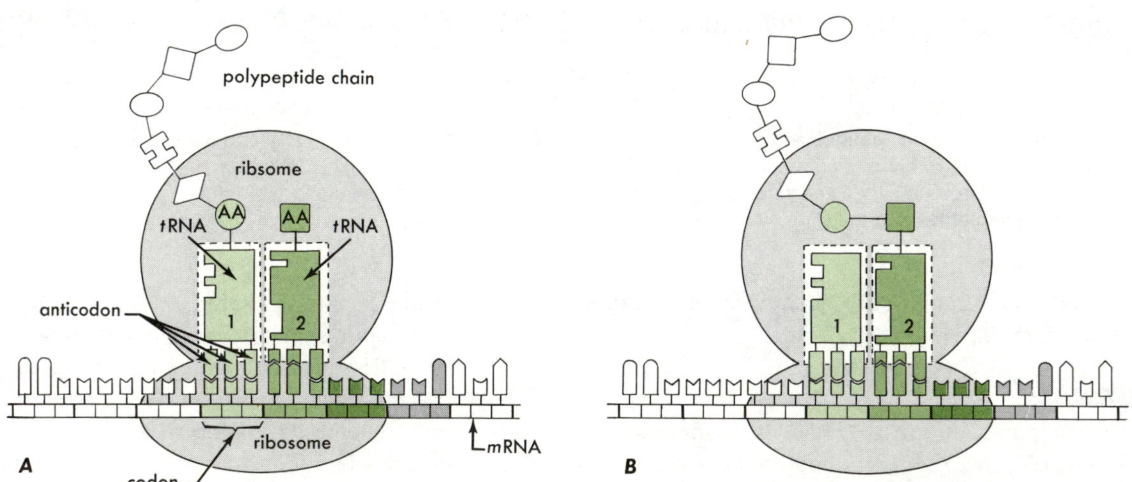

Figure 6.29

Protein synthesis.

A, ribosome with *m*RNA strand in place in lower component, *t*RNA molecules in place in active sites, short polypeptide chain. The amino acid is bound to *t*RNA molecule 1. The polypeptide linkage with the amino acid on *t*RNA molecule 2 has not been formed. Note anticodon of the *t*RNA molecules and matching codons of the *m*RNA molecule.

B, peptide linkage has formed between amino acids in the active sites, and bond linking amino acid with *t*RNA molecule 1 has broken.

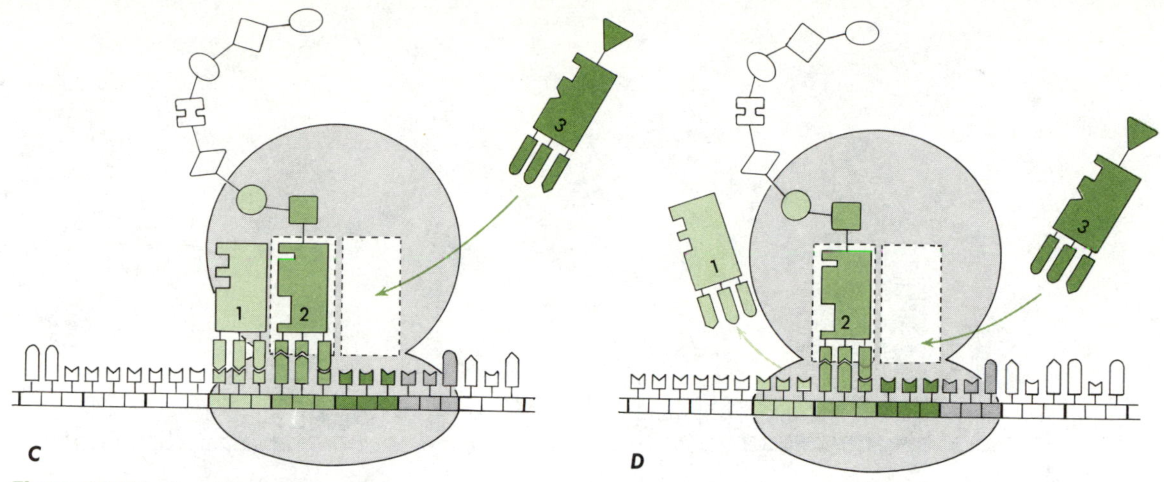

Figure 6.29 *(continued)*
*C, t*RNA molecule 1 has left the reactive site, the ribosome has moved along the *m*RNA strand a codon to the right. *t*RNA molecule 2 is now in the left-hand reactive site and the right-hand reactive site is empty. *t*RNA molecule 3 is approaching the ribosome.
*D, t*RNA molecule 1 leaves the ribosome and *t*RNA molecule 3 prepares to enter it.

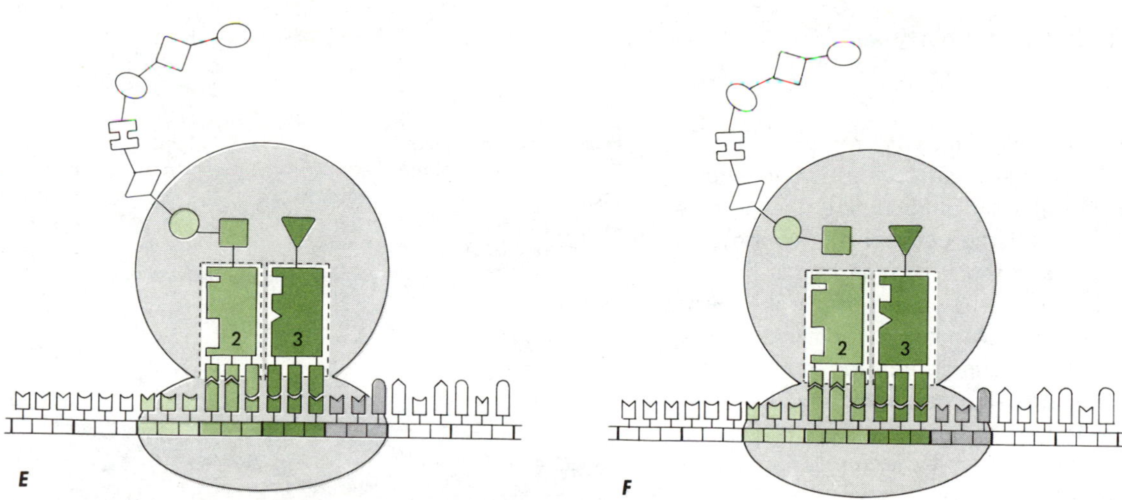

Figure 6.29 *(continued)*
*E, t*RNA molecules 2 and 3 now occupy the reactive sites. Bond between *t*RNA molecule 2 and associated amino acid is intact.
F, peptide linkage forms and bond between *t*RNA molecule 2 and its amino acid is broken.

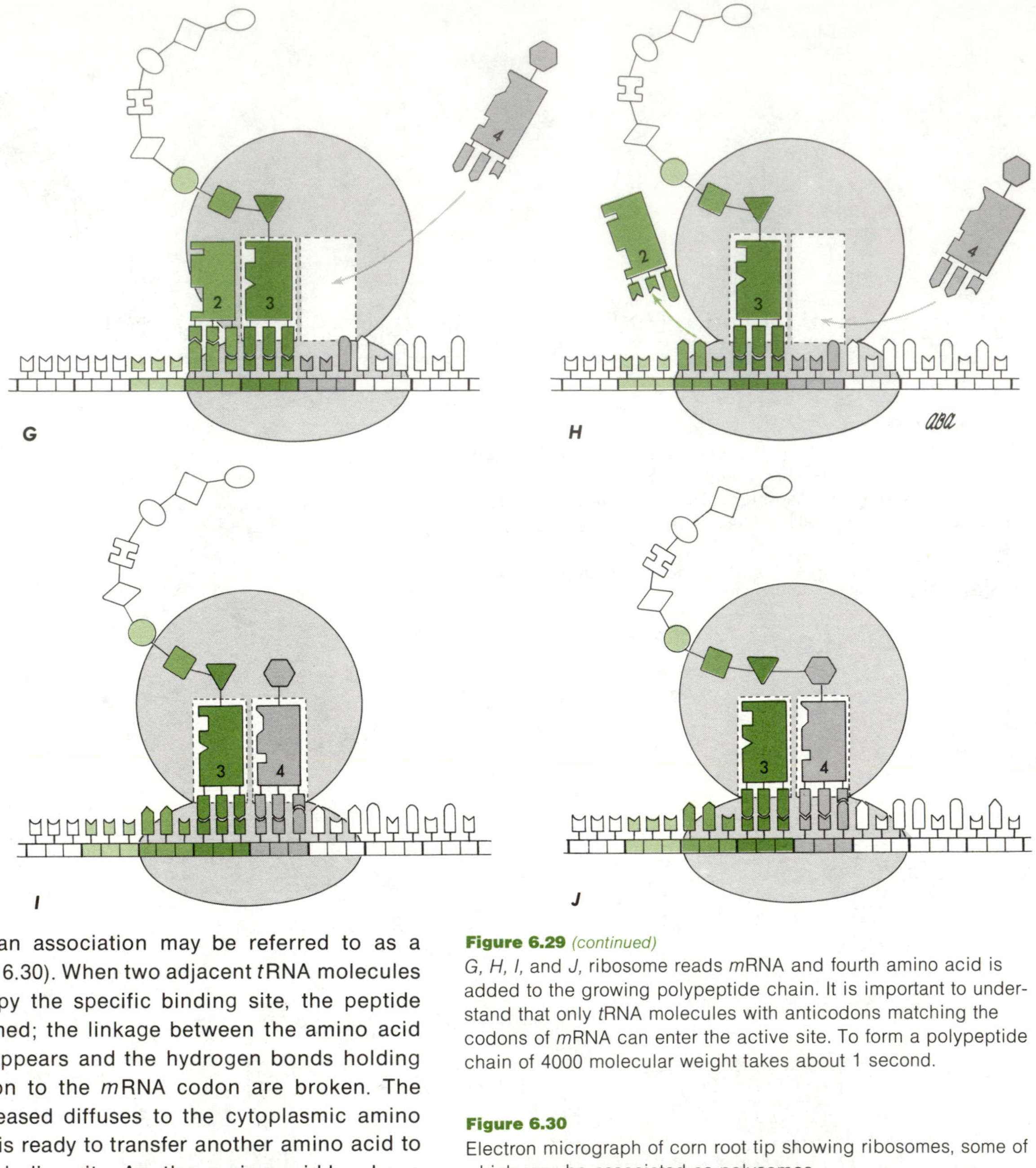

*m*RNA. Such an association may be referred to as a polysome (Fig. 6.30). When two adjacent *t*RNA molecules come to occupy the specific binding site, the peptide linkage is formed; the linkage between the amino acid and *t*RNA disappears and the hydrogen bonds holding *t*RNA anticodon to the *m*RNA codon are broken. The *t*RNA thus released diffuses to the cytoplasmic amino acid pool and is ready to transfer another amino acid to the ribosome binding site. Another amino acid has been added to the growing protein molecule. The process continues and as the protein molecule lengthens, it becomes free from the ribosome.

Under optimal conditions it takes about 10 sec for a polypeptide with a molecular weight of 40,000 to be synthesized in the bacterium *Escherichia coli.*

SUMMARY OF PROTEIN SYNTHESIS

1. RNA is a single-stranded molecule, composed of a backbone chain of phosphate and ribose units to which are attached the nucleotides uracil, guanine, pyrimidine, and cytosine.
2. There are fewer nucleotides in a molecule of RNA than there are in a molecule of DNA.

Figure 6.29 *(continued)*
G, H, I, and *J,* ribosome reads *m*RNA and fourth amino acid is added to the growing polypeptide chain. It is important to understand that only *t*RNA molecules with anticodons matching the codons of *m*RNA can enter the active site. To form a polypeptide chain of 4000 molecular weight takes about 1 second.

Figure 6.30
Electron micrograph of corn root tip showing ribosomes, some of which may be associated as polysomes.

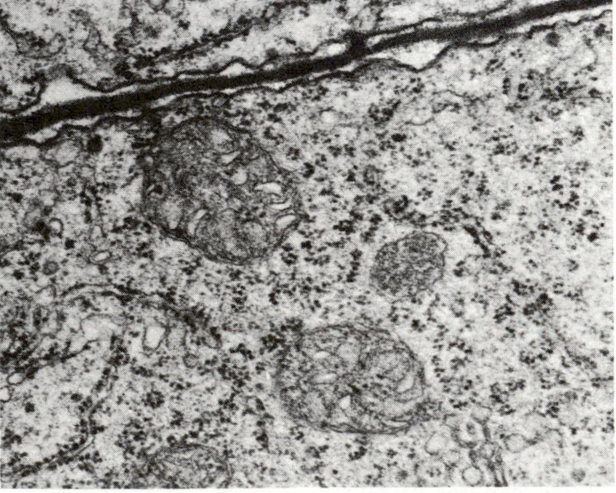

3. There are three types of RNA, all coded by the nucleotides on the DNA molecule.
4. The three types of RNA are transfer RNA (*t*RNA), messenger RNA (*m*RNA), and ribosomal RNA (*r*RNA).
5. There are 60 types of *t*RNA, each carrying about 80 nucleotides.
6. *t*RNA transfers an amino acid from a cytoplasmic amino acid pool to a specific codon on a molecule of *m*RNA associated with a ribosome.
7. There are many types of *m*RNA, each molecule being coded for a specific sequence of amino acids as determined by the nucleotide sequence of the DNA strand involved in the synthesis of *m*RNA.
8. The molecule of *r*RNA is composed of two subunits, both complexing with protein. The molecular structure is not known.
9. A strand of *m*RNA and the amino acid-loaded molecules of *t*RNA are associated with a ribosome. At a specific combining site the codon and anticodon unit and the peptide linkage form. The protein molecule grows and the molecule of *t*RNA returns to the cytoplasmic amino acid pool.
10. *In vitro* studies on protein synthesis indicate that it takes about 10 sec to form a polypeptide with a molecular weight of 40,000

CHAPTER 7

CELL TYPES, TISSUES, AND PRIMARY GROWTH IN STEMS

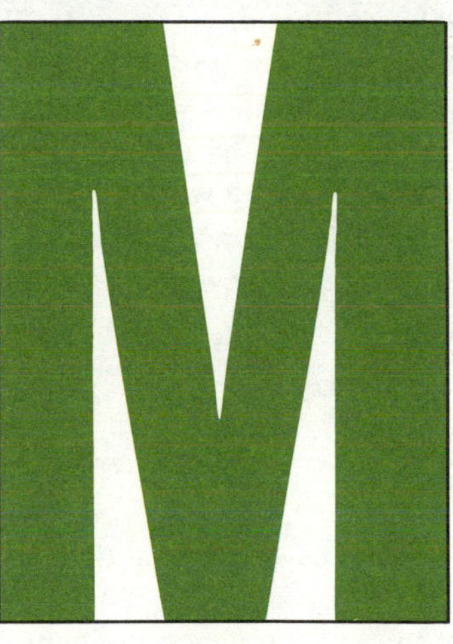

echanical strength of cellular structure obviously is a prerequisite for growth of plants on land. Only when stems developed supporting tissue could they rise into the atmosphere and display leaves, facilitating the trapping of radiant energy of sunlight. Increase in plant body size required anchorage in the soil and a conducting system.

Also required were means of protecting structures from dessication and for assuring survival over periods unfavorable for growth. With the successful development of specialized structures to meet these requirements, there evolved the dominant present-day land flora.

The vegetative bodies of dominant land plants consist of three types of organs—stems, leaves, and roots. In this and the following three chapters we shall examine their external form **(morphology)** and internal structure **(anatomy).** One serious problem in presenting this material is that photographs and drawings are two-dimensional, and words are often inadequate to explain the pattern and symmetry of plant parts. So, while reading these sections, you should visualize the plant's internal structure in three-dimensions. Try to use your "mind's eye" as a kind of X-ray vision to see the length, width, and breadth of cells and tissues and their interconnections. It will become apparent, as you study plant structure, that all organs are integrated functionally and structurally. This means that the entire plant acts as one unit, and is not simply an aggregate of isolated parts.

STEM FUNCTIONS

Stems provide mechanical support for leaves and other appendages. They raise leaves toward the sun, thus facilitating photosynthesis. Flowers and fruit are held in positions that allow for efficient pollination and seed dispersal. **Stems support.** The manner by which stems function as supporting organs is diverse and varied (**Fig. 7.1**, page 117).

Stems provide a pathway for movement of water and mineral nutrients from roots to leaves, and for transfer of foods, hormones, and other metabolites from one part of the stem to another. **Stems conduct.**

The normal life-span of plant cells is 1 to 3 years. Water and mineral salts in dilute solution move in dead cells, but this movement depends upon the activity of living cells in leaves and roots generally less than 3 years old. Stems in herbaceous perennials and in 2000-year-old redwoods, or bristle-cone pines (**Fig. 7.1*A***), annually provide new living tissue for normal metabolism of the plant. **Stems produce new living tissue.**

All stems contain cells capable of storing food, or other plant products, or water, but some stems are particularly adapted for storage and may be variously enlarged (**Fig. 7.1*E***). **Stems store food and water.**

Stems thus have four major functions: (*a*) support; (*b*) conduction; (*c*) the production of new living tissue; and (*d*) storage.

STEM MORPHOLOGY

The organs of plants can usually be distinguished from one another by their morphology or anatomy. Stems are elongated organs that form the axis on which the lateral appendages, leaves, and buds are attached. The place on a stem where leaves and buds are attached is called a **node,** the portion between any two adjacent nodes is an **internode.** Some plants have erect stems (**Fig. 7.1*A***), others have horizontal, creeping stems (**Fig. 7.1*C***), some are vines (**Fig. 7.1*B***), and some, rosette plants (**Fig. 7.1*F***), have stems that do not elongate except at flowering time. All stems whether short or long, horizontal or erect, are distinguishable as stems by the presence of nodes and internodes (Fig. 7.3).

Buds

The basic structure of vascular plants is an axis with potentially long-lived apical meristems at opposite ends. Specialized protection for the root apex involves a root cap, which serves to protect the root as it grows through the soil. More elaborate protection is needed for stem apical meristems, particularly in perennial stems that overwinter. In such cases, the shoot meristem is dormant during the unfavorable season. It may be protected from desiccation by small modified leaves known as **bud scales.** Such an unelongated stem consisting of short internodes and immature leaves and protected by bud scales is a **bud** (Fig. 7.2). In a few instances, protective scales are lacking and the unprotected dormant apex is a **naked bud.** Actively growing shoot apices are also supplied with a protective covering of young leaves. Such active shoot apices are not generally called buds.

Arrangement of Buds

A 3-year-old twig of walnut in winter condition is shown in Fig. 7.3. The tip of the twig generally bears a large **terminal** leaf bud. At regular intervals along the stem, other buds may be seen; they are called lateral buds. Note that below the base of each **lateral** bud there is a scar that was made when a leaf fell from the twig; this is a **leaf scar.** Vascular bundle scars may be seen within each leaf scar; strands of food and water-conducting tissues passing from the stem into the leaf stalk were broken when the leaf fell, leaving these scars. Buds and leaves are usually borne in this relationship to each other; buds form in the angle made by the stem and the leaf stalk. This angle is termed the **leaf axil,** and consequently these buds may also be called axillary buds.

The passage of strands of conducting tissue into leaves and buds must mean that the internal structure of the stem is so arranged as to make possible a continuous connection of leaves and buds with all other parts of the stem. The node is thus a specialized stem region—a region to which are attached buds and leaves and within which conducting strands unite with other conducting strands of the stem.

Protecting the young immature cells within the bud is a series of overlapping scales; **bud scales.** They are usually shed when the bud develops into a new shoot, and they also leave scars, bud-scale scars. The part of a stem or twig between sets of terminal-bud-scale scars is generally formed during one growing season. For instance, growth made by the twig this year is set off from growth made last year by means of a ring or girdle or terminal-bud-scale scars (Fig. 7.3). When scales of a terminal bud fall off in spring, they leave a number of closely crowded scars that form a distinct ring. Examination of several-year-old twigs shows that growth in length may vary from year to year. This is shown by the different spacings between the terminal-bud-scale scars. It is sig-

Figure 7.2 (*opposite*)
Buds. *A*, flower and leaf buds of *Camellia japonica*, ×1. *B*, mixed buds of pear *(Pyrus)* on short fruit spur, ×2. *C*, flower and leaf buds of almond *(Prunus amygdalus)*, ×1. *D*, a single leaf bud of apricot *(Prunus armeniaca)* with two accessory flower buds at a node, ×2. *E*, the naked staminate bud of walnut *(Juglans regia)*, ×2. *F*, a flower bud of lilac *(Syringa)* with two opposite leaf buds, ×2. *G*, terminal mixed bud of buckeye *(Aesculus californica)*, ×1½. *H*, with bud scales opening, ×½. *I* and J, stages in expansion over a 24-hour period, ×½.

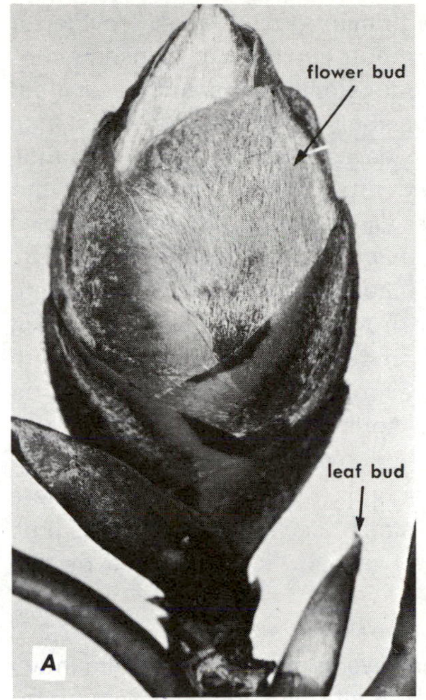
flower bud
leaf bud
A

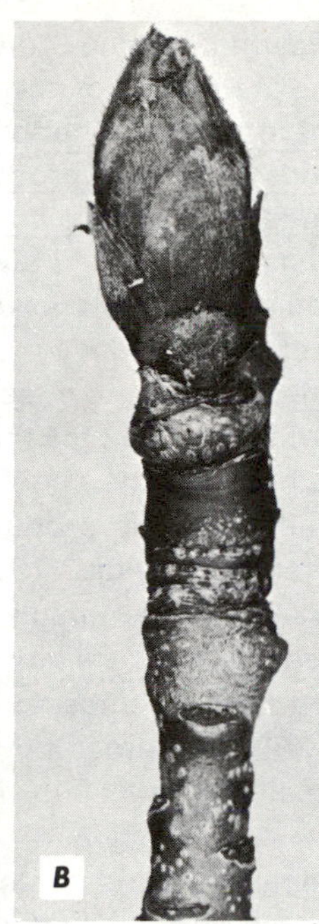
B

flower bud
leaf bud
C

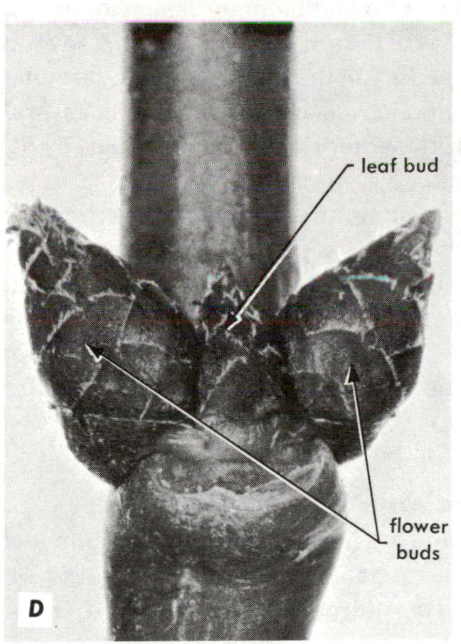
leaf bud
flower buds
D

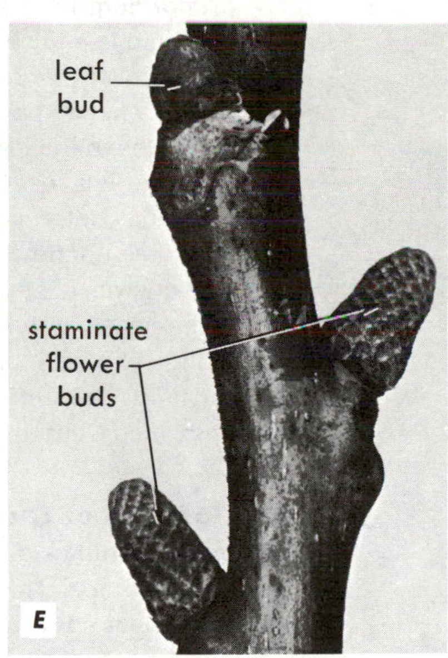
leaf bud
staminate flower buds
E

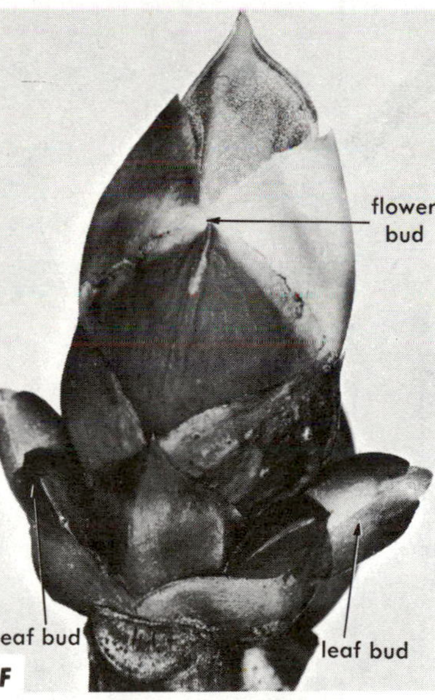
flower bud
leaf bud
leaf bud
F

G

H
I

J

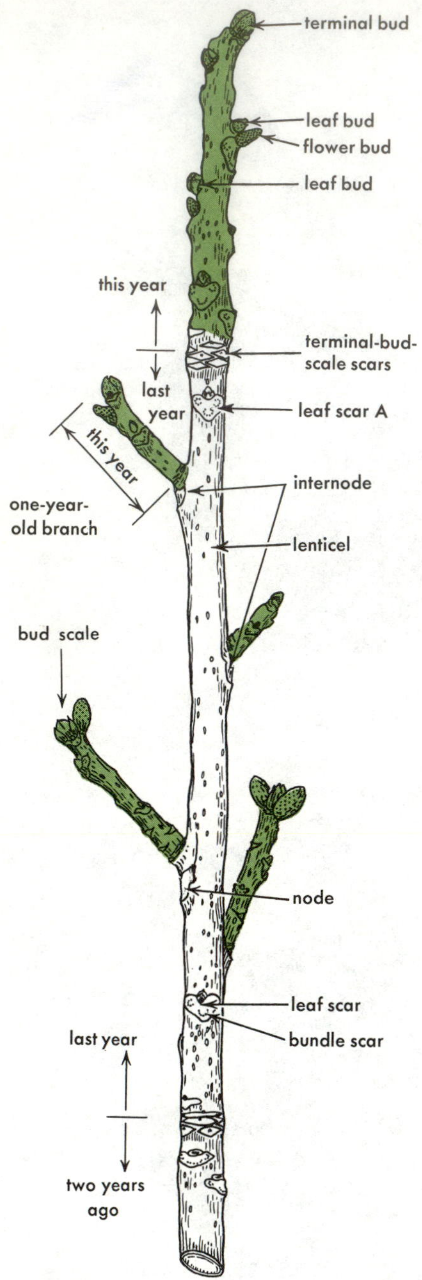

Figure 7.3
Three-year-old twig of walnut *(Juglans regia),* ×1.

nificant that there is no increase or decrease in the length of any portion of a stem after that portion is 1 year old.

The slightly raised areas on the bark are **lenticels** (Fig. 7.3). They are composed of cells that fit loosely together, with air spaces between, which permit passage of gases inward and outward.

Kinds of Buds

We shall see (Chapter 15) that growth terminates in shoot apices that produce flowers. Yet the production of flowers early in the season is frequently advantageous for woody plants; therefore, at the time of resumption of growth in spring, new primary growth must provide for the production of new shoots and flowers. Consequently, there are buds that produce only stems and leaves **(leaf buds),** buds that produce only flowers **(flower buds),** and buds that produce both leaves and flowers **(mixed buds).** *Camellia japonica* always produces, frequently at the same node, both flower and leaf buds (Fig. 7.2*A*). Apples and pears produce mixed buds on short spurs (Fig. 7.2*B*). These spurs grow less than 1 inch a year, but a severe pruning will induce the growth of a long vegetative shoot rather than the usual production of a short shoot with fruit. Almonds produce both flower and leaf buds on short spurs (Fig. 7.2*C*). Apricots often exhibit three buds at a node; the middle bud is a leaf bud and the two **accessory** buds are flower buds (Fig. 7.2*D*). In walnut, there are both leaf buds and staminate catkin flower buds (Fig. 7.2*E*). The larger flower bud of lilac is flanked by two small leaf buds (Fig. 7.2*F*)

A terminal bud of buckeye is shown in Fig. 7.2*G*. Separating bud scales reveals the largest rudimentary leaves (Fig. 7.2*H*). If when all rudimentary leaves are removed, there remains a small floral apex and leaf bud primordium in an axil of a rudimentary leaf (Fig. 7.4*B*). This bud is a mixed bud. Figure 7.2*G* to *J* shows several stages, over a 24 hr period, in the expansion of a terminal bud. If when expanding leaves are removed, only several small unexpanded rudimentary leaves remain (Fig. 7.4*A*). This bud is a leaf bud.

What determines whether a shoot tip will produce leaves alone, flowers alone, or both leaves and flowers? The answer is still very elusive. But it is known that, in some cases, induction of a floral apex is related to day length, to a hormone, and to two specific wavelengths of red light. The stimulus is transmitted from leaves to shoot apex (see Chapter 20).

Position of Buds on a Woody Twig

In the walnut twig there are just *one leaf bud* and *one leaf at each node.* This arrangement of buds and leaves on the stem is spoken of as **alternate** (Figs. 7.5*A*, 7.6*A*, *C*, *E*). It is the most common type of bud and leaf arrangement. Ash, maple, lilac, and many other plants have two leaves opposite each other at each node, and a bud in the axil of each leaf (Fig. 7.5*B*, 7.6*D*, *F*). This arrangement of leaves and buds is spoken of as **opposite.** When three or more leaves and buds occur at each node, as in *Catalpa*, leaf arrangement and bud arrangement are said to be **whorled** (Fig. 7.5*C*, 7.6*B*).

Some plants have several buds in or near the leaf axil. For example, apricot often has a group of three buds in the leaf axil: a central bud, which develops into a side branch, and two lateral ones, floral buds (Fig. 7.2*D*). All but the central one are called **accessory buds.** Walnut may also have more than one bud in the leaf axil.

Not infrequently, buds may arise on the plant at places other than leaf axils. They may appear on stems,

Figure 7.4
Stereoscan micrographs of shoot apices of buckeye *(Aesculus californica)*. *A,* shoot apex, enveloped by opposing rudimentary leaves, ×250. *B,* floral apex, with adjacent small shoot apec, ×40. (Courtesy of D. Hess.)

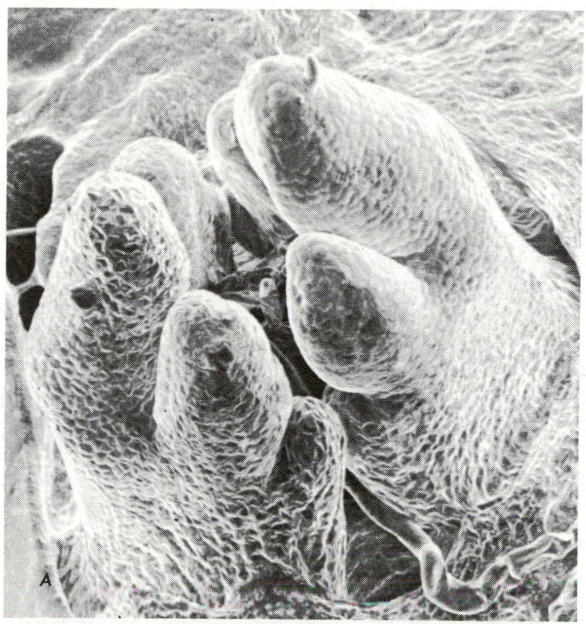

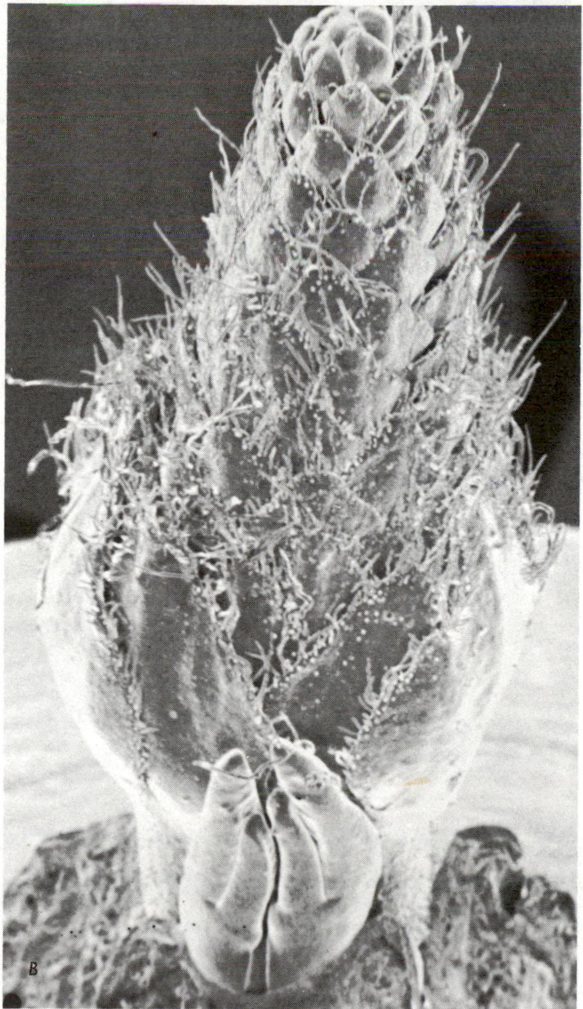

roots, or even leaves, and give rise to new shoots. Such buds are called **adventitious buds.** Their formation may be stimulated by injury, such as occurs in pruning.

From the above discussion it is seen that buds may be classified by their **arrangement on the stem,** which may be (*a*) **alternate,** (*b*) **opposite,** or (*c*) **whorled;** by their **position on the stem,** which may be (*a*) **terminal,** (*b*) **lateral (axillary),** (*c*) **accessory,** or (*d*) **adventitious;** and by the **nature of the organs into which they develop,** which may be (*a*) **leaf,** (*b*) **flower,** or (*c*) **mixed.**

Dormant or **latent buds** arise in a regular fashion in the leaf axil, but their development is usually inhibited by the dominance of the terminal bud. As a rule, the terminal bud of a stem is the most active and grows more vigorously than any of the axillary buds. Usually, the lowest lateral buds on a year's growth of the shoot

Figure 7.5
Twigs showing three methods of bud and leaf arrangement. The position of leaves is shown by the leaf bases and scars. *A,* alternate, walnut *(Juglans regia)*. *B,* opposite, lilac *(Syringa vulgaris)*. *C,* whorled, catalpa *(Catalpa bignonioides)*, ×½.

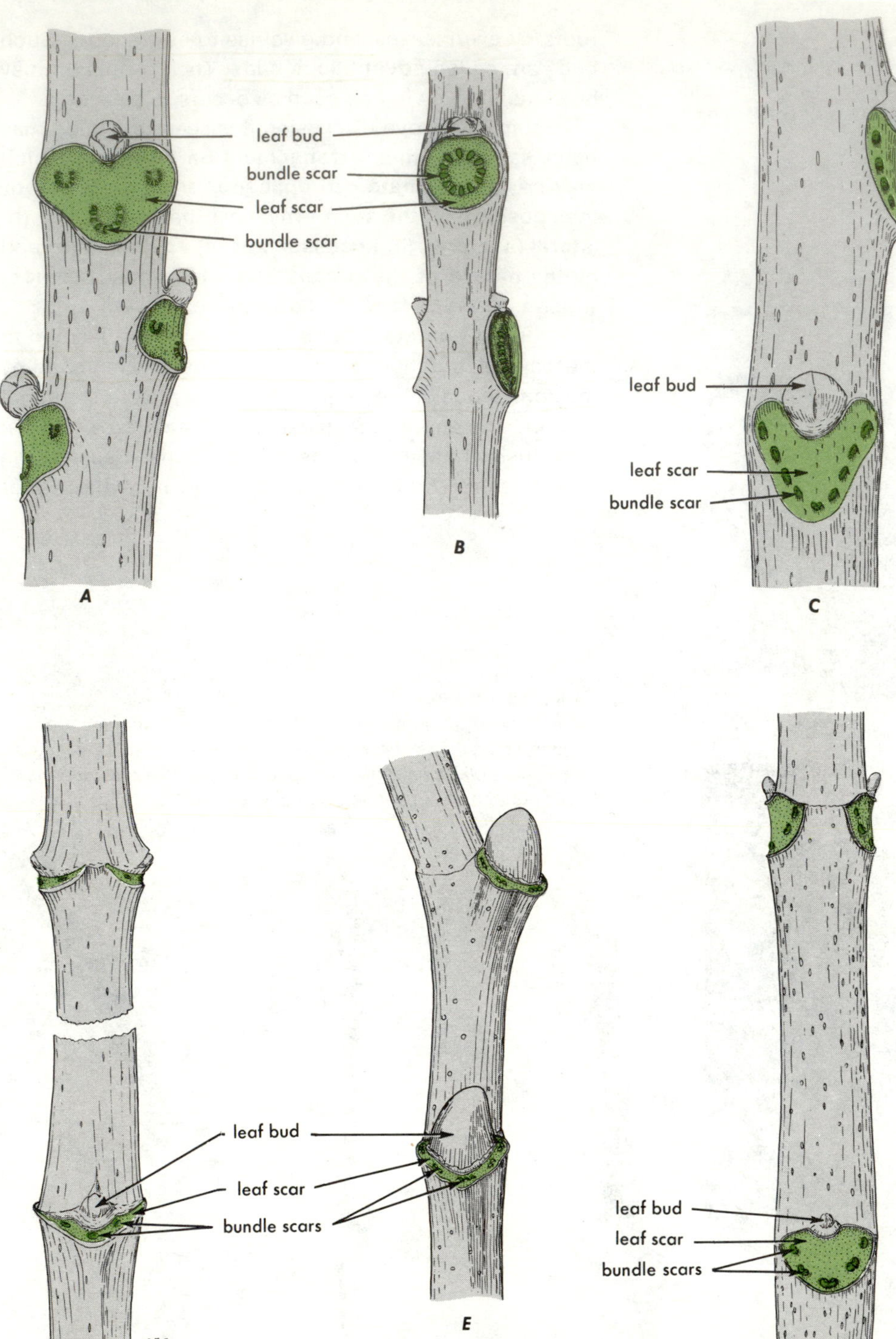

Figure 7.6
Leaf scars and bud arrangement of different species of woody plants. *A*, walnut *(Juglans regia)*, ×2. *B*, catalpa *(Catalpa bignonioides)*, ×2½. *C*, tree of heaven *(Ailanthus altissima)*, ×2. *D*, box elder *(Acer negundo)*, ×2. *E*, European plane *(Platnus acerifolia)*, ×3. *F*, buckeye *(Aesculus californica)*, ×2.

remain dormant and do not develop into branches. If the terminal bud is removed, however, as may be done in pruning, lateral buds, otherwise dormant, may become active. This phenomenon, apical dominance, is discussed in Chapter 20.

DEVELOPMENT OF TISSUES OF THE PRIMARY PLANT BODY

The Primary Meristems

Plant organs (leaves, stems, roots, and flowers) are obviously different from each other morphologically. But, if

Figure 7.1
Stems vary in structure and function. Some are annuals, others are perennials of great age. *A,* redwood *(Sequoia sempervirens);* some are vines, *B,* fig *(Ficus pumila);* some are prostrate, *C,* crab grass *(Digitaria sanguinalis);* some carry out photosynthesis. *D,* horsetail tree *(Casuarina equisetifolia);* others may store large quantities of food or water, *E,* prickly pear *(Opuntia occidentalis);* some, rosette plants, may elongate only during flowering, *F,* dandelion *(Taraxacum vulgare).*

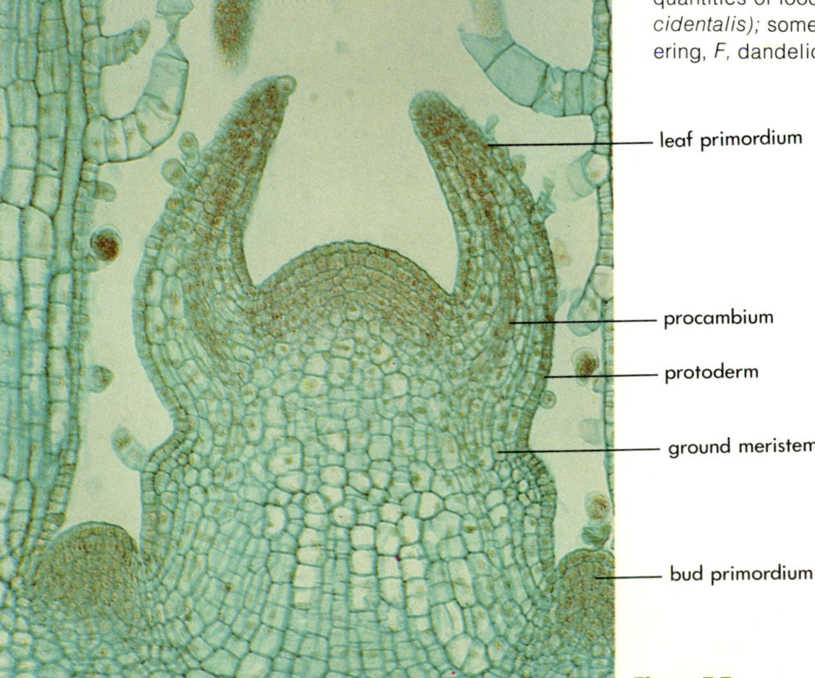

Figure 7.7
Longitudinal section of the shoot apex of bean *(Phaseolus vulgaris),* ×90.

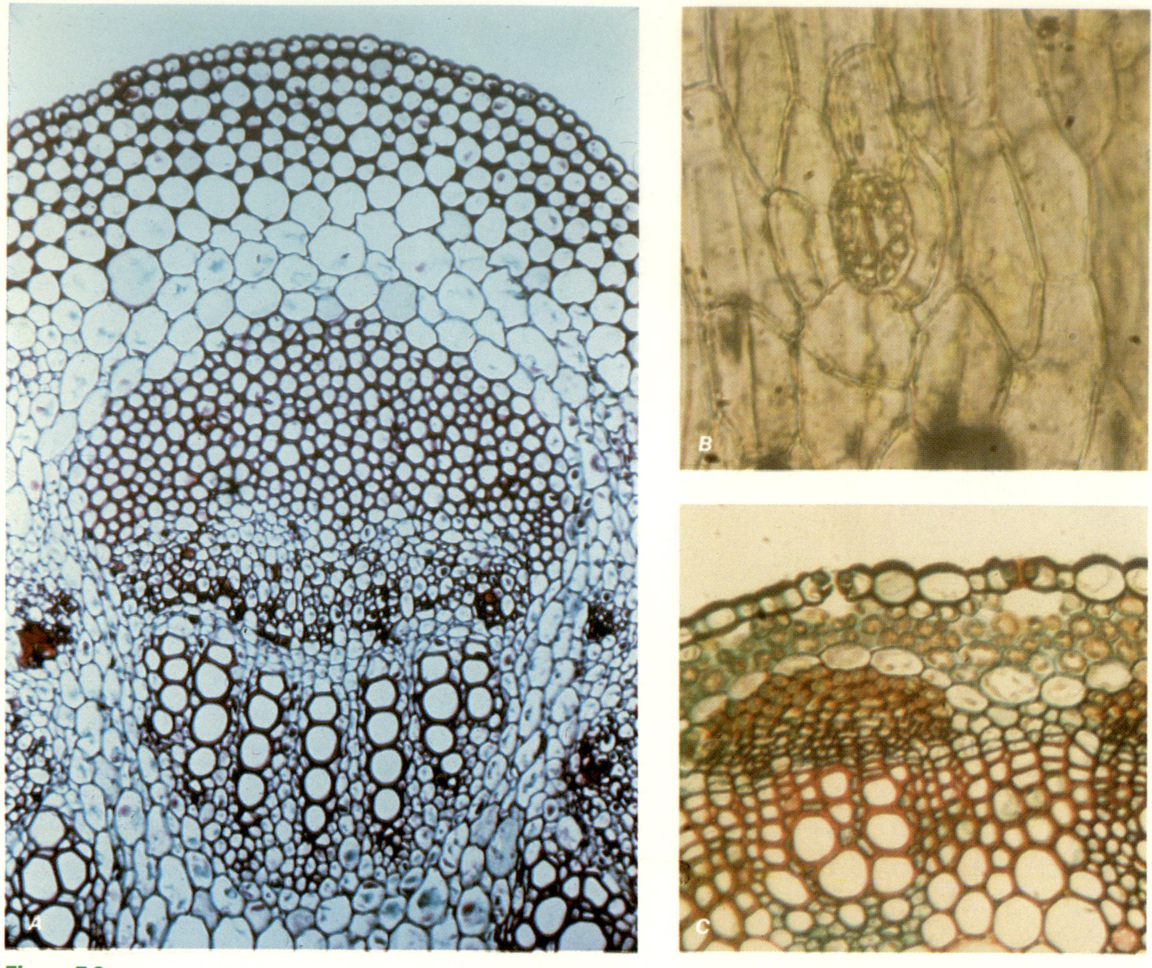

Figure 7.9

A, cross section of stem of sunflower *(Helianthus annuus)* showing a vascular bundle. *B,* surface view of epidermis from bean *(Phaseolus vulgaris)* stem showing epidermal cells, guard cells, and stoma. *C,* cross section of alfalfa *(Medicago sativa)* stem showing epidermal cells, guard cells, stoma, and cortex with substomatal chamber, ×400.

Figure 7.10

A, glandular hair from *Geranium* sp. *B,* stereoscan view of epidermal hairs of tomato stem *(Lycopersicon esculentum),* ×130. (*B,* courtesy of D. Hess.)

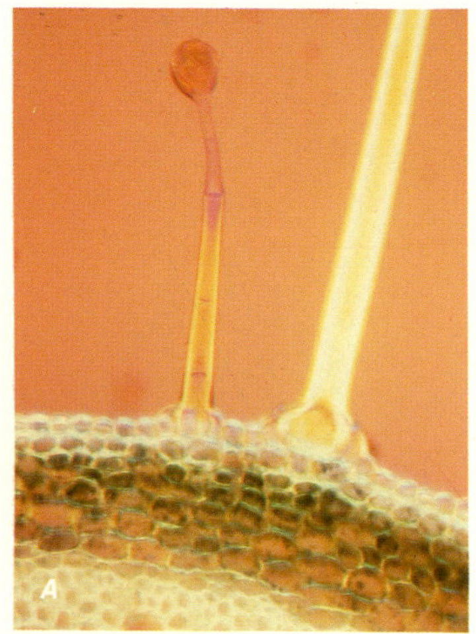

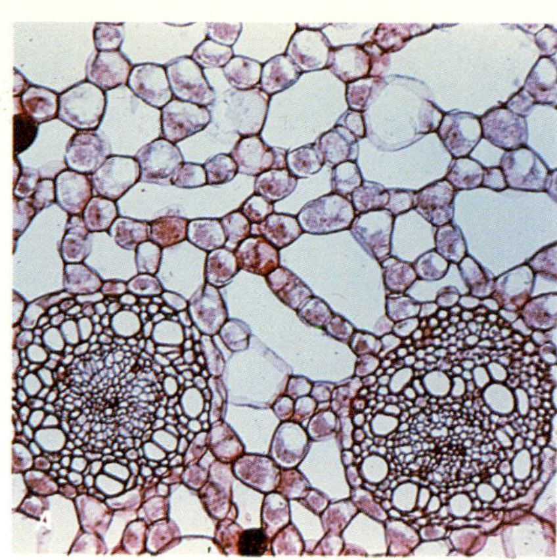

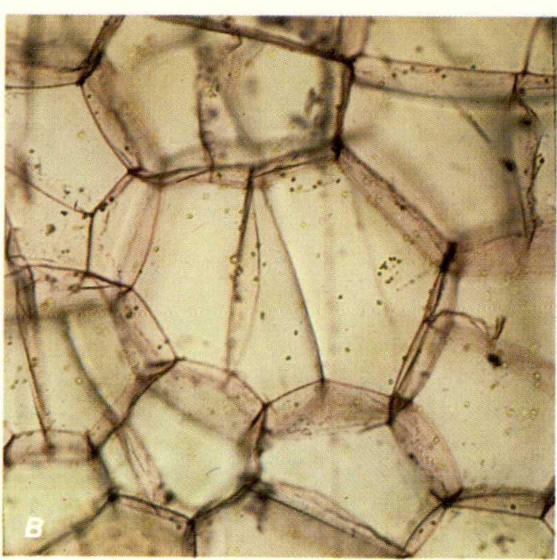

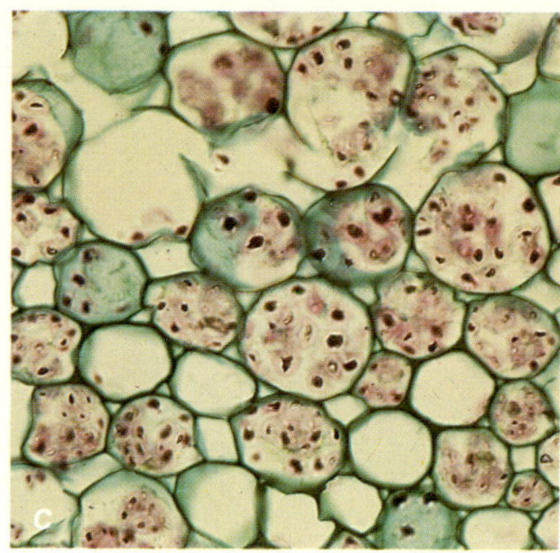

Figure 7.11
A, aerenchyma tissue from rhizome (underground stem) of *Acorus calamus*. *B*, parenchyma cells in stem of *Impatiens balsamina*. *C*, storage parenchyma (starch grains are stained red) in cortex of *Ranunculus* sp. root.

Figure 7.14
A, section of Marigold (*Calendula* sp.) stem showing collenchyma cells with thick pink-stained walls. *B*, red-stained sclereids (stone cells) from the fruit of pear (*Pyrus* sp.). *C*, sclereids (star-shaped) from phloem of Douglas fir stem (*Pseudotsuga menziesii*). *D*, vascular bundle of stem of *Sanseveria cylindrica* showing blue-stained fibers.

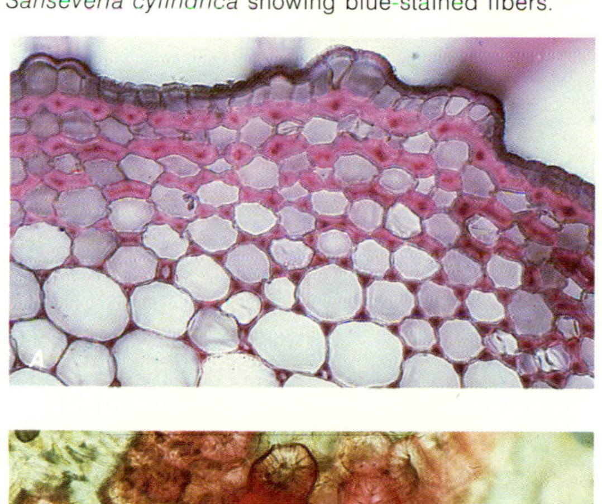

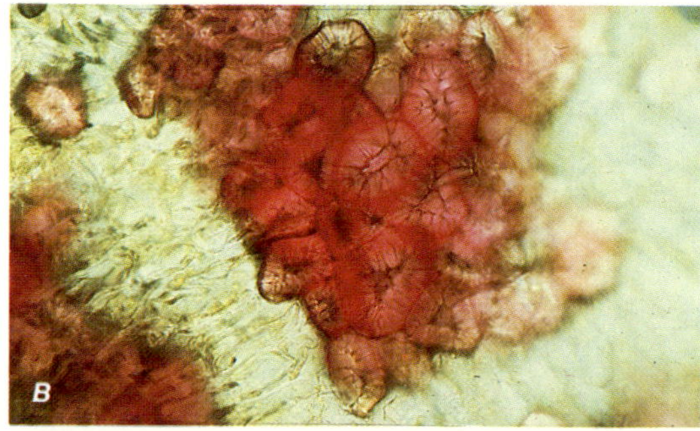

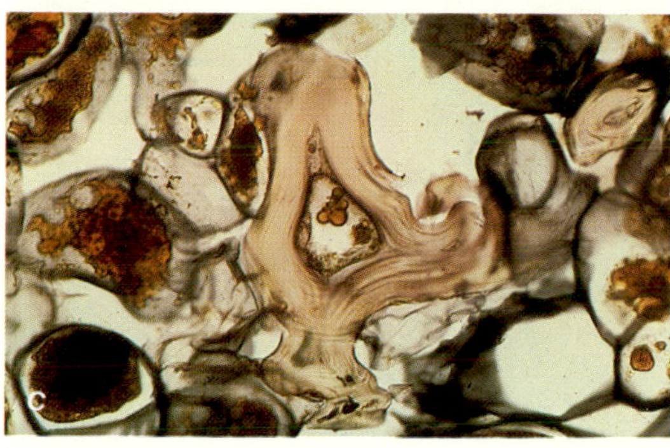

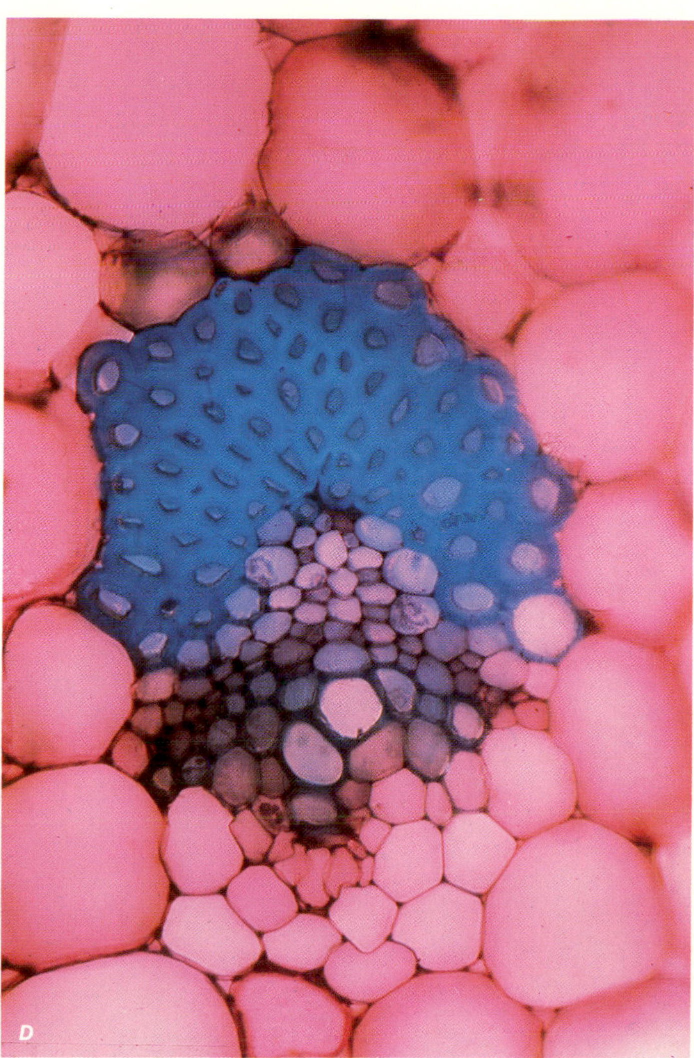

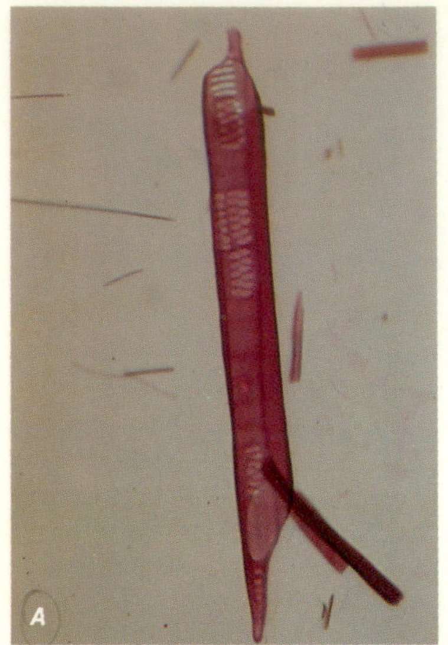

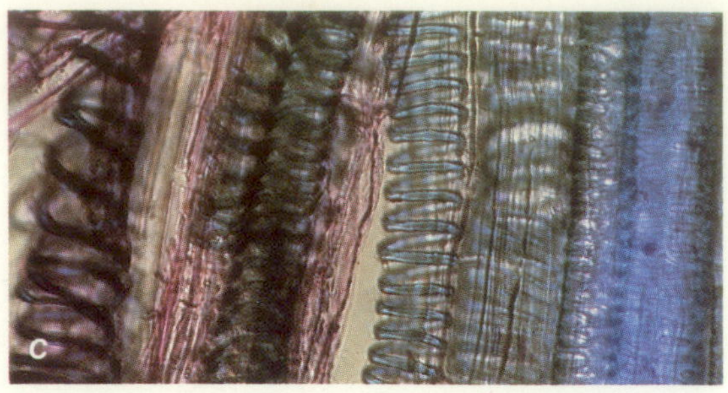

Figure 7.20
A, cells from macerated xylem of basswood tree *(Tilia americana)*, note the red-stained vessel element with open end walls. *B*, section of black walnut wood *(Juglans nigra)* through polarized light showing a portion of a vessel consisting of several connected vessel elements. *C*, squashed petiole of celery *(Apium graveolens)* to show the variable patterns of vessel element secondary cell walls. Compare with Figure 7.21*A*.

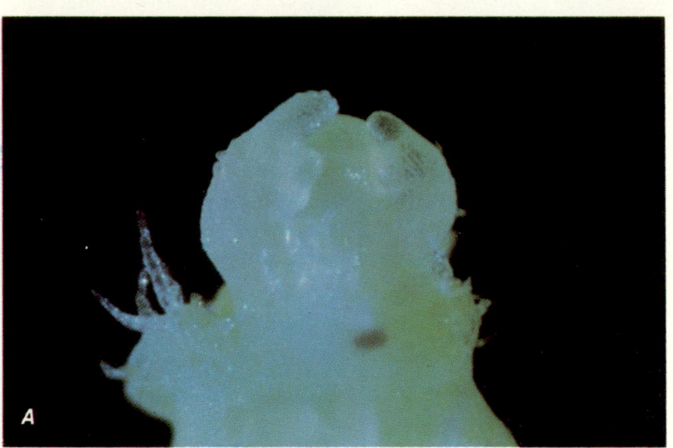

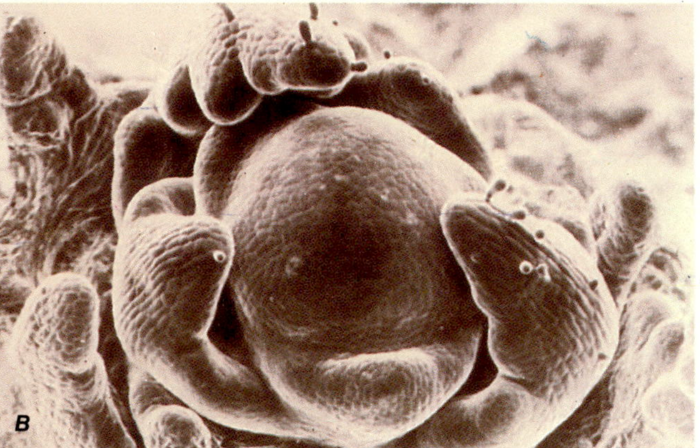

Figure 7.24
A, exposed shoot tip of nasturtium *(Tropaeolum* sp.) showing the round apical dome and two overlapping leaf primordia. *B*, stereoscan view of a shoot tip of *Adonis aestivales*. *(B*, Courtesy of Dr. J. Lin.)

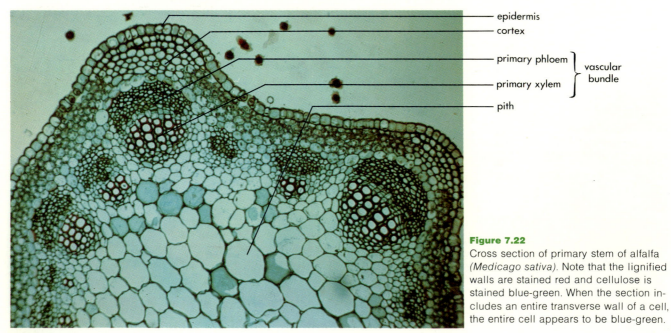

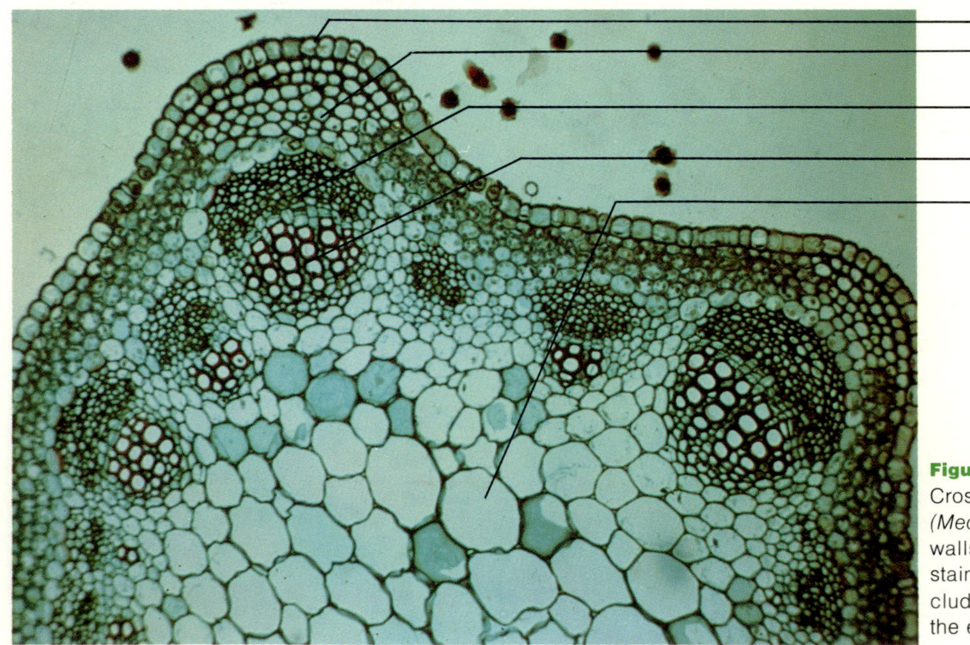

Figure 7.22
Cross section of primary stem of alfalfa *(Medicago sativa)*. Note that the lignified walls are stained red and cellulose is stained blue-green. When the section includes an entire transverse wall of a cell, the entire cell appears to be blue-green.

we examine their internal anatomy it is apparent that all organs are composed of similar structural units—**cells** and **tissues.** Each cell type is modified to make it ideally suited to perform one or more specific functions. Tissues are organizations of cells of one or more types that have a common origin and a common collective function. If a tissue is composed of only one cell type performing but a single function, it is a **simple tissue.** Frequently, however, several cell types will be organized into a tissue performing more than one function, such as support and conduction; in this case, we may speak of a **complex tissue.**

Cell types and tissues develop by a process called **differentiation.** A differentiating cell progresses through a series of steps that result in the cell becoming mature and functional. Each different cell type "experiences" slightly different differentiation steps. The formation of new cells, and the initiation of differentiation in plants takes place in specific regions called **meristems.** A longitudinal section of a terminal bud is shown in **Fig. 7.7,** page 117. Note the nodes with **leaf primordia,** the very short internodes, and the **bud primordia** that occur in the axils of the more advanced leaf primordia.

Let us examine the cells and tissues of this bud. At the apex of all shoots and roots, whether terminal or lateral, there are dome-shaped masses of cells called **apical meristems.** Cells of this apical meristem have large nuclei, a compact cytoplasm, and small vacuoles. In **Fig. 7.7,** they have been stained a brownish red. While the bud is dormant these cells are relatively quiescent, but when conditions are favorable some cells divide rapidly and the daughter cells increase in size, becoming in some instances larger than the cell from which they originated. This increase in number and size of cells results in elongation of the young shoot. But in conjunction with increase in number and size of cells, differentiation occurs; that is, cells change morphologically and physiologically from the meristematic cells from which they arose. Thus, a short distance (usualy a few millimeters) below an apical meristem we recognize three fairly distinct **primary meristematic tissues,** namely **protoderm, ground meristem,** and **procambium** (also stained brownish-red in the leaf primordia in **Fig. 7.7**). These three tissues are derived directly from the apical meristem. They in turn will differentiate into the three primary tissues: **epidermis, ground tissues** (pith and cortex), and **vascular tissues** (xylem and phloem).

The term **shoot tip** is applied to the youngest part of the shoot and includes apical meristematic cells associated with leaf primordia. Some of these meristematic cells are elongating and others are in early stages of differentiation. Rapid elongation of stems results from the formation and growth of thousands of new cells in the shoot tip. It is in the shoot tip that we see the origin of leaf primordia, bud primordia, and primary meristematic tissues.

Now let us consider in detail each of the primary meristems and tissues derived from them in the plant body.

Protoderm

This is the outermost layer of cells derived from the apical meristem (**Figs. 7.7,** 7.8). It develops into **epidermis**—the special primary tissue that covers and protects all underlying primary tissues. The epidermis prevents excessive water loss and yet allows for the exchange of gases necessary for respiration and photosynthesis.

Ground Meristem

The ground meristem comprises the greater portion of meristematic tissue of the shoot tip **(Fig. 7.7).** Its cells are

Figure 7.8
A to *F*, the pattern of vascular development and the positions of the three primary meristems.

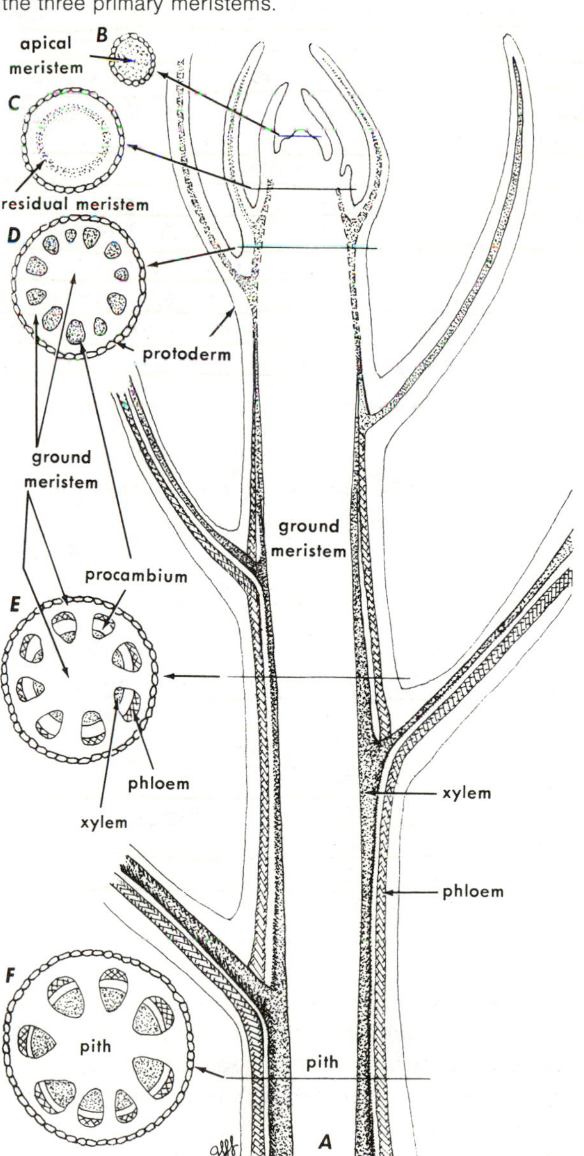

relatively large, thin-walled, and isodiametric. The regions forming from the ground meristem are (*a*) **pith,** in the very center of the stem, and (*b*) **cortex,** in a cylinder just beneath the epidermis and surrounding the vascular tissue. Sometimes pith and cortex are connected by pith rays, also formed from ground meristem (Fig. 7.8*E*). A note of clarification must be added here. Traditionally speaking, the pith and cortex are not tissues but, instead, are regions that are composed of the ground tissues, parenchyma, sclerenchyma, and collenchyma (see below).

Procambium

Procambium cells usually appear first as strands among ground meristem cells (**Figs. 7.7,** 7.8). In cross section, strands appear as isolated groups of cells arranged in a circle. Sometimes, as in most dicotyledonous roots, a continuous **procambium cylinder** is formed. As seen in a transverse section of a single procambium strand, procambium cells are smaller than those of the surrounding ground meristem; in longitudinal section, they are much longer. Procambium cells give rise to **primary vascular tissues** (Figs. 7.8, **7.9,** page 118). These primary tissues carry out several functions and are divided rather rigidly into two groups according to these functions. Food is conducted in the outer group of primary vascular cells, which is **primary phloem.** Water and mineral salts are conducted in the inner group of primary vascular cells, which, together with strengthening cells, constitute **primary xylem.** In many stems, a meristematic region remains between primary xylem and primary phloem to become **vascular cambium.**

Primary Tissues

Primary tissues of the stem are differentiated from the three primary meristematic tissues—protoderm, ground meristem, and procambium—and these three are derived from the apical meristem of the shoot tip. In woody plants, we must look for primary tissues of the stem a very short distance behind the stem tip. Even before the end of the first season's growth, differentiation of these primary tissues from primary meristematic tissues is completed, and **secondary tissues** may be formed in abundance. Whereas primary tissues are derived from primary meristematic tissues of the shoot, secondary tissues are the result of production of new cells by the **vascular cambium** and by the **cork cambium.** The origins and nature of these types of cambia are discussed later.

The Epidermis

The epidermis is usually a single superficial layer of cells that covers all other primary tissues, protecting them from drying out and, to some extent, from mechanical injury. It is the limiting layer of cells between the plant and its environment. In surface views **(Fig. 7.9*B*),** epidermal cells are elongated in the direction of the stem's length; in transverse section, they are usually isodiametric **(Fig. 7.9*B,C*).** Its protoplasm forms a thin layer that lines cell cavities and normally remains living for a long period.

The outer tangential wall of cells exposed to air is usually thicker than the other walls, and its surface layer is usually coated with a waxy substance called **cutin** (see Chapter 10). This superficial layer is the **cuticle,** which is quite impermeable to water and gases. There may be cracks or other imperfections in the cuticle, however, and water vapor may pass out of the plant at those points. The inner walls, parallel to the stem surface, are thinnest, and radial walls, at right angles to the surface, often taper in thickness toward the inner wall.

Young stems usually possess specialized epidermal cells. By means of certain stains and chemical reactions cutin can be distinguished from the cellulose in the cell walls **(Fig. 7.9*C*)** called **guard cells (Fig. 7.9*B,C*).** Between each pair of guard cells is a small opening, or **pore,** through which gases enter and leave the underlying stem tissues. Two guard cells plus the pore make one **stoma (pl. stomata)** (see Chapter 10). Guard cells differ from other epidermal cells by their crescent shape and the fact that they contain chloroplasts. Stomata are common in the epidermis of leaves, floral structures, and fruits as well as of stems.

Epidermal appendages, such as hairs, may occur on young stems (**Fig. 7.10*A,B*,** page 118). These structures will also be discussed in the chapter on leaves.

The Cortex

This complex region, derived from ground meristem, forms beneath the epidermis as a cylindrical zone that extends inward to the primary phloem (Fig. 7.8). The following tissues or cell types may be found within it: **parenchyma, collenchyma, sclerenchyma,** and **secretory tissue.**

Parenchyma. The principal tissue of the cortex is parenchyma (Figs. **7.9*A*,** 7.13*G*). It consists of isodiamtric parenchyma cells with thin walls, made mostly of cellulose, and with protoplasts that remain alive for a long time. We may speak either of a parenchyma tissue, where many parenchyma cells are found together, or of individual parenchyma cells.

Parenchyma tissue is characterized by the presence of intercellular air spaces, which vary greatly in size; in some parenchyma tissues they are difficult to find, while in others they are very apparent (**Fig. 7.9*A*,** page 118). Parenchyma tissue with large air spaces, such as found in stems of aquatic plants is called **aerenchyma** (**Fig. 7.11*A*,** page 119). Because parenchyma cells retain active protoplasts, they function in the storage of water and food, in photosynthesis, and sometimes in secretion (**Figs. 7.11*B,C*,** 7.13*G*). When the parenchyma cells contain chloroplasts, they are collectively referred to as

chlorenchyma. Parenchyma cells can be reprogrammed to differentiate into different cell types. One example would be the changes parenchyma cells undergo in response to being wounded. In this case, the wounded cells on the surface simply die, and the inner cells divide and form a layer that is quite similar to the bark on the outside of stems.

Parenchyma cells are not confined only to the cortex and pith of the stem, but also occur in practically all other types of tissue and in other organs of the plant.

Crystals of calcium salts, generally calcium oxalate (see Chapter 4), are commonly found in different forms in parenchyma cells. Figure 7.12 shows some of these interesting structures.

Collenchyma. The outermost cells of the cortex of young stems, lying just beneath the epidermis, often constitute a tissue known as **collenchyma.** This tissue may form a complete cylinder, or it may occur in separate strands. Collenchyma cells are elongated, often contain chloroplasts, and are living at maturity. The walls of collenchyma cells are composed of alternating layers of pectin and cellulose. In the most common type of collenchyma, the cell walls are thickened at the corners (Figs. **7.9*A*,** 7.13*D,E,* **7.14*A*,** page 119). These thickenings are quite flexible and will stretch without resistance, in much the same way as heated plastic. Collenchyma is, therefore, an ideal strengthening tissue because it strengthens and at the same time allows normal tissue growth. Collenchyma serves as a strengthening tissue in young expanding stems and also in the petioles of leaves.

Sclerenchyma. The main functions of **sclerenchyma** cells are support and, in many cases, protection. Their shape and the thickness and toughness of their walls contribute to the ability of these cells to support and protect the stem. Thickness and toughness of walls are increased by deposition, within the original cellulose wall, of a substance known as **lignin.** Lignin is made by the protoplast and deposited in the wall. When this secondary wall is completely formed, the protoplast usually dies. There are two types of sclerenchyma cells: (*a*) **sclereids** and (*b*) **fibers** (Figs. 7.13, **7.14**).

Of various types of sclereids, the most common are stone cells. Stone cells are more or less isodiametric and have minute canals, **pits,** extending through the thickened walls (Fig. 7.13*H*). These pits may be separated only by primary walls from ends of similar pits in adjacent cells. Other types of sclereids are branched, resembling very irregular stars (Fig. 7.13*F*). Some sclereids are derived from parenchyma cells by a pronounced thicken-

Figure 7.12
Stereoscan of *Psychotria punctata* showing three parenchyma cells containing bundles of raphide crystals, ×2000. (Courtesy V. R. Franceschi and H. T. Horner, Jr., *Zeit Pflanzenphysiol.* **92,** 61. © by Gustav Fisher Verlag. Reprinted by permission of the publisher.)

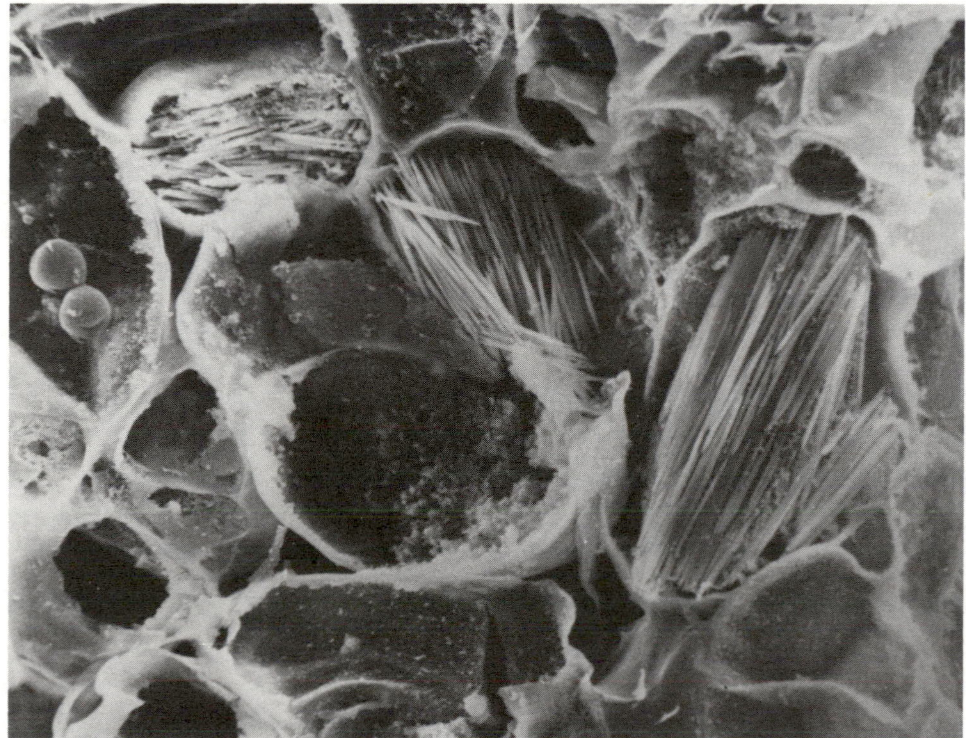

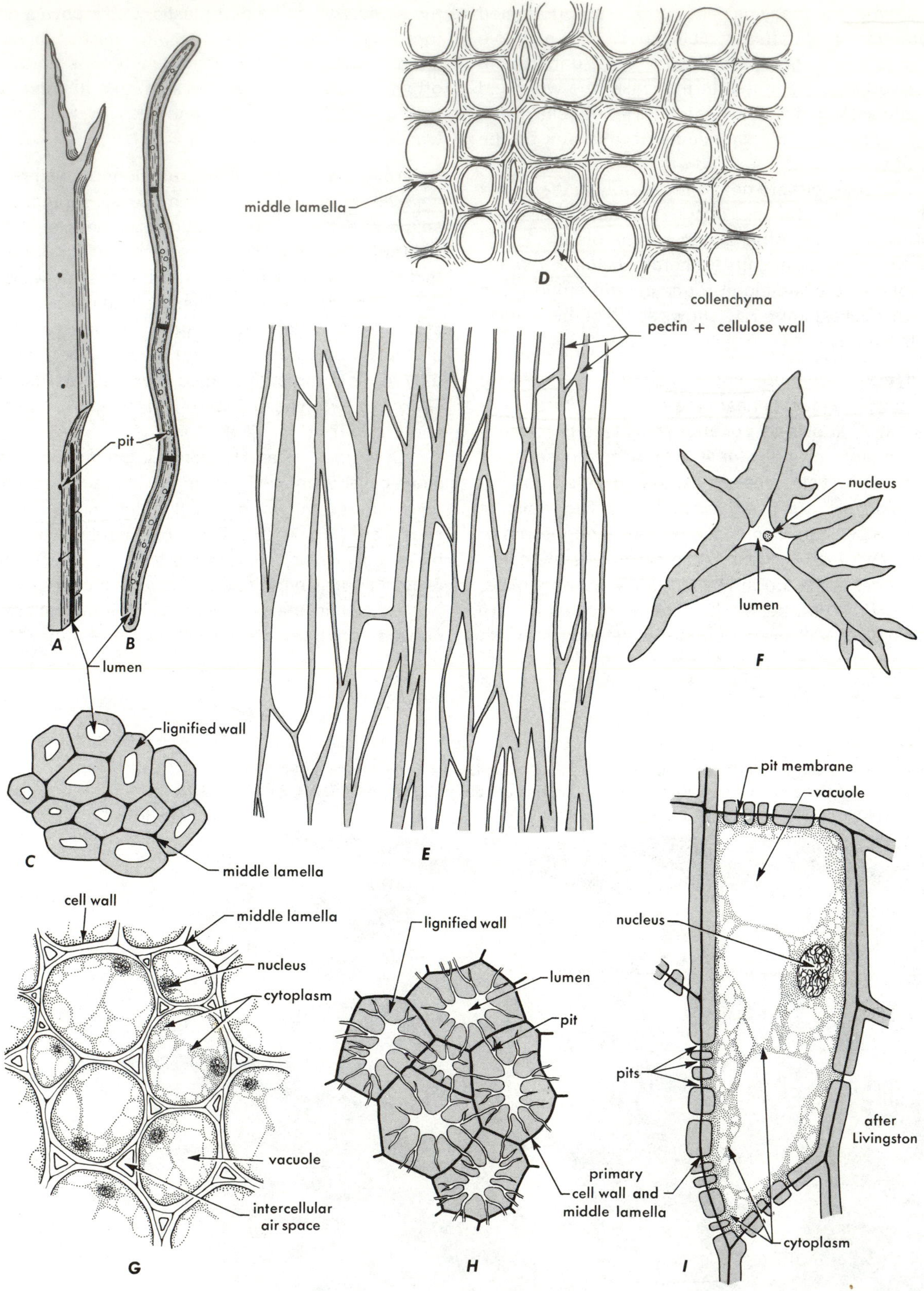

Figure 7.13
Cell types and tissues. *A* and *B*, fibers in longitudinal view. *C*, fibers in cross section. *D*, collenchyma in cross section. *E*, collenchyma in longitudinal view. *F*, sclereid. *G*, parenchyma. *H*, stone cells. *I*, woody parenchyma.

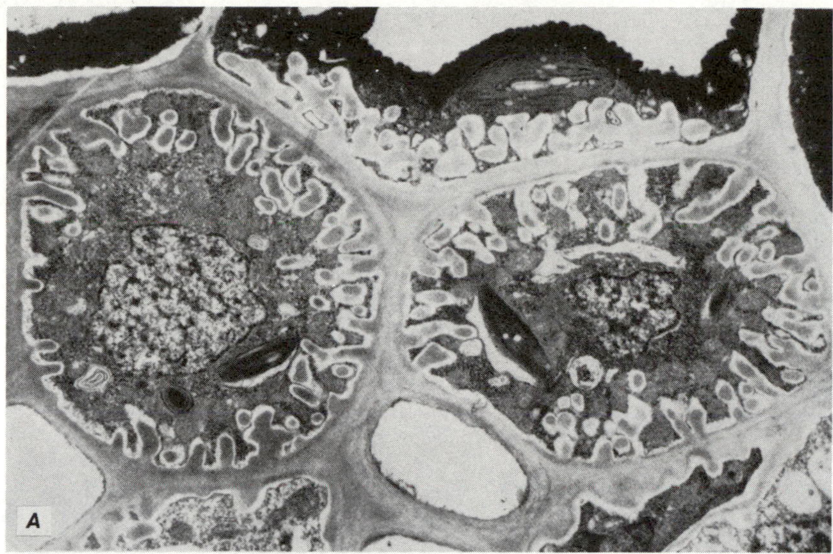

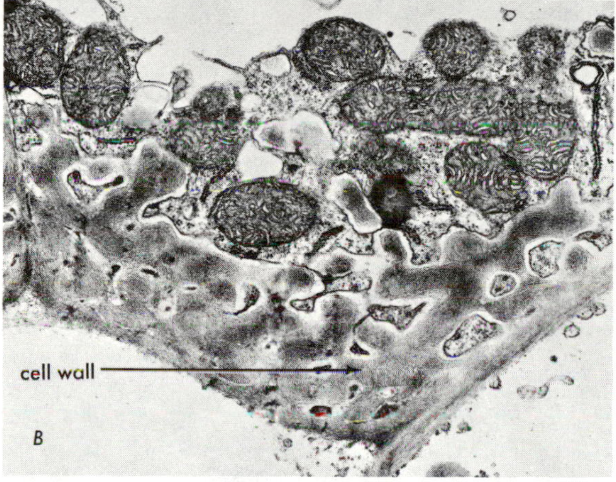

Figure 7.15
Transfer cells. *A*, around vascular bundle in leaf of *America corsica*, ×1900. *B*, xylem parenchyma transfer cells from the cotyledonary node of a seedling of *Lactuca*, ×275. (*A*, From J. S. Pate and B. E. S. Gunning, *Protoplasma* **68,** *140*. *B* Courtesy of J. C. Pate and B. Gunning, *Ann. Rev. Plant Physiol.* **23,** 173, by Annual Reviews, Inc. Reprinted by permission of the publishers.)

ing of cell walls; others arise from separate meristematic cells.

Sclereids occur not only in the cortex of stems but also in the hard shells of fruits, seed coats, and bark, in pith of some stems, and in certain leaves (Fig. 7.13*B,C*).

Fibers are elongated, strengthening cells that are thick-walled, and usually pointed at the ends (Figs. 7.13, **7.14*D***). Their walls may or may not be lignified. The walls may become so thick that the cell cavity, the **lumen,** almost disappears. Simple pits form in their thick walls (Fig. 7.13*A*). Fibers are sometimes very elastic and can be stretched to a great degree without losing their ability to return to their original length. Protoplasts of fibers often disappear as the cells attain maturity.

Secretory Cells. Secretory cells are parenchyma-like and contain dense protoplasm. They secrete various substances, such as resinous materials (Fig. 8.12) and nectar in many flowers. Many epidermal hairs are secretory cells **(Fig. 7.10*A*).**

Transfer Cells. Frequently, cells located in positions of active solute transfer will show an irregular extension of the cell wall into the protoplast. Since the plasmalemma follows the contour of the wall, its surface is greatly extended. Mitochondria appear to aggregate adjacent to these areas of increased membrane surface. The morphological picture thus presented (Fig. 7.15) suggests an adaptation that facilitates transport from one cell to another, or from the interior to the exterior of the plant. These cells have been called **transfer cells.**

The Primary Vascular Tissues

The term **"vascular"** pertains to tissues that conduct various substances in liquid form.

In vascular plants, water and different water-soluble inorganic salts from soil, as well as food substances, are conducted throughout the plant in well-defined vascular tissues.

In a young stem, very near the tip (**Figs. 7.7,** page 117, 7.8), vascular tissues occur as separate bundles, **primary vascular bundles.** Each primary vascular bundle is differentiated from a procambium strand and consists of primary xylem and primary phloem. If secondary growth is to occur, a thin band of meristematic tissue destined to become vascular cambium remains between primary xylem and primary phloem (discussed in Chapter 8).

The Primary Phloem

Phloem in angiosperms may possess several types of cells: **sieve-tube members, companion cells, fibers, sclereids,** and **parenchyma.** A **sieve tube** is a vertical row of elongated cells; each cell is known as a **sieve-tube member.** These are the conducting elements of phloem (Fig. 7.16) that translocate sugars and other

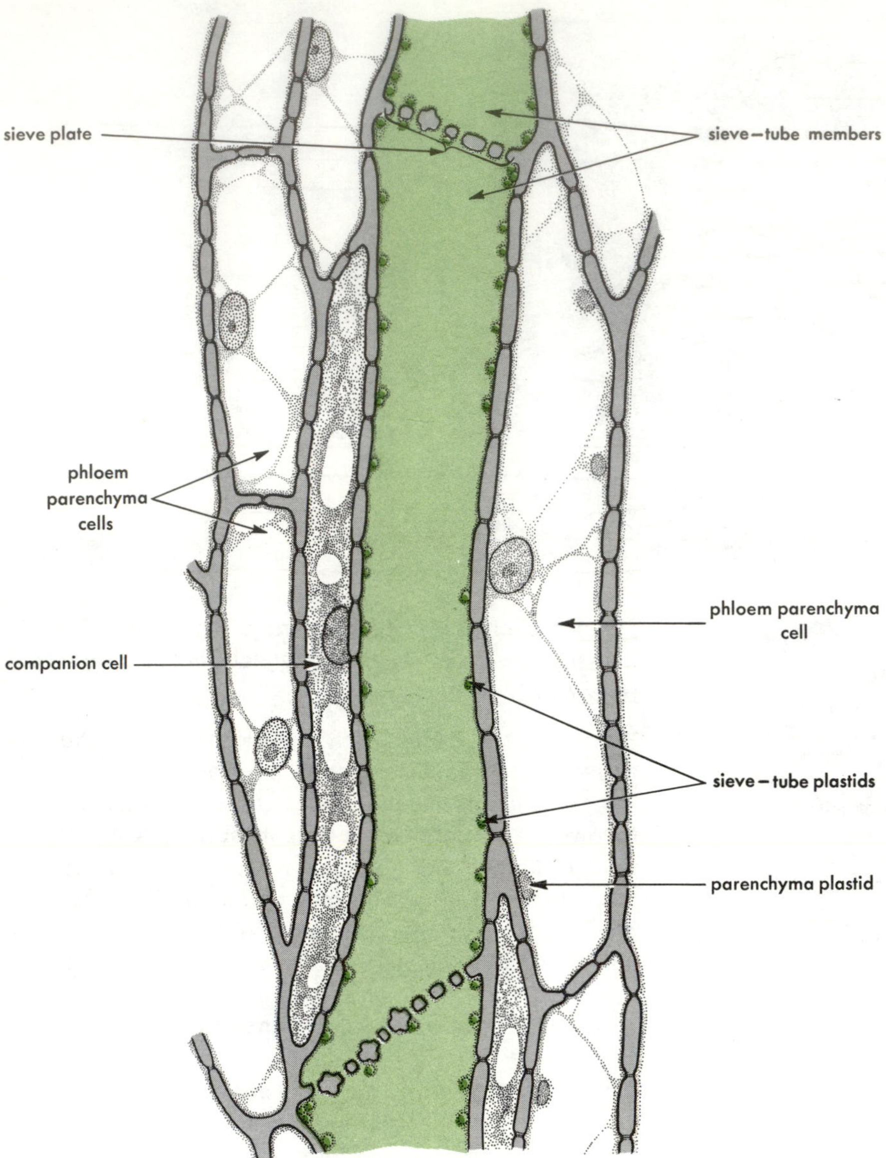

Figure 7.16
Phloem tissue from the stem of tobacco *(Nicotiana)*, ×400.

organic solutes throughout the plant body (see Chapter 12). Among angiosperms, a sieve-tube member and a companion cell originate by division from the same procambial cell. Young sieve-tube members have the usual complement of organelles: nucleus, plastids, mitochondria, and dictyosomes (Fig. 7.17). As a sieve-tube member matures, its protoplast becomes greatly modified. Its nucleus usually disintegrates. Plastids lose most of their internal membranes, but usually retain starch. The plasmalemma remains intact. The mitochondria become small. The cytoplasm, much reduced in amount, exists chiefly as a thin peripheral layer. The vacuole membrane, tonoplast, also disintegrates, and the endoplasmic reticulum (ER) sometimes aggregates in parallel arrays along the cell wall.

The central part of the cell is occupied by a mass of strands or tubules. This mass may be seen with the light microscope, and has been called **slime.** Since it is now known to be a protein, it is more correctly referred to as **P-protein** (Fig. 7.17). At maturity one or more companion cells lie adjacent to each sieve-tube member. Since companion cells have a normal protoplast with a full complement of organelles (Fig. 7.16), these cells possibly regulate the metabolic activity of the sieve-tube members that have no nucleus. Plasmodesmata connect the protoplasts of companion cells and sieve-tube members.

A characteristic structural feature of mature sieve-tube members is the **sieve plate.** It may occur in the end or side walls (Fig. 7.17*B*). The end wall between two adjacent sieve-tube members is thickened and strands of cytoplasm pass through pores adjoining them. With the exception of a few trees such as palms, where they live longer, sieve-tube members live and function about 1 to 3 years. In perennial plants, new ones are formed annually.

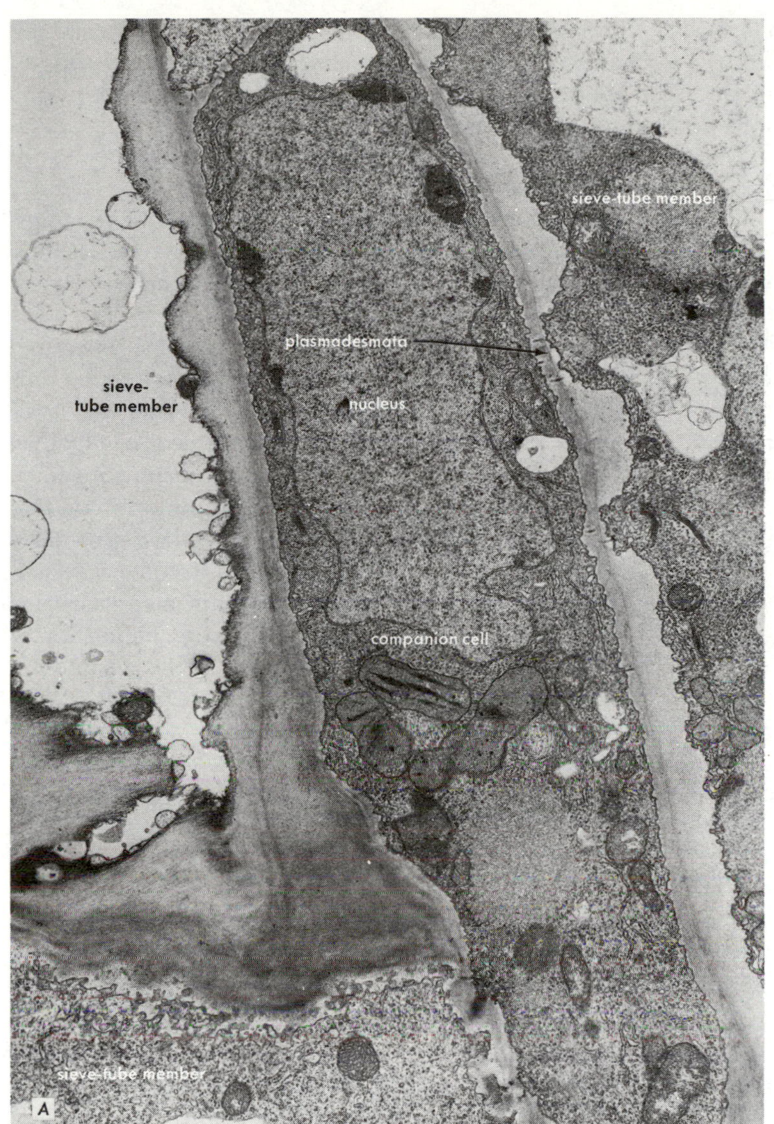

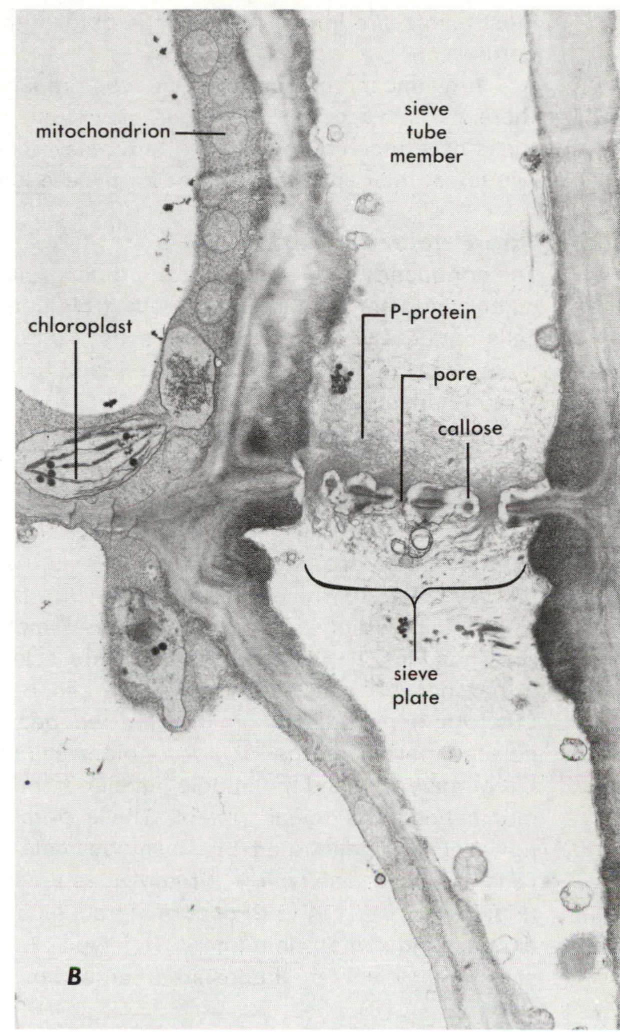

Figure 7.17
A, longitudinal section of three sieve-tube members and one companion cell. The nucleus and various organelles are present in the central companion cell. Most organelles are missing in the left sieve-tube member, but are present in the one on the right, *Cucurbita maxima,* ×8300. *B, Beta vulgaris* petiole phloem to show P-protein and callose around pores of the sieve plate, ×10,600. *C, Beta vulgaris* petiole 1 hour after cold treatment stress; note the massive amount of callose plugging the sieve plate, ×10,200. (*B* and *C,* courtesy of V. Franceschi.)

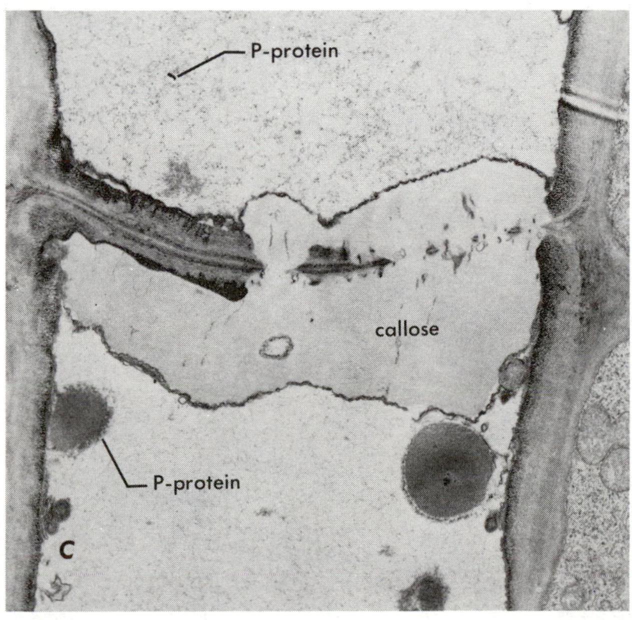

In many studies on the structure of mature sieve-tube members, a carbohydrate known as **callose** is seen around the margins of pores in the sieve plate (Fig. 7.17*B*). In some instances, protein may also collect at the sieve plate. Obviously, this would block the pores of the sieve plate and obstruct movement of food materials. It has been demonstrated in other tissues that callose forms very rapidly in response to wounding and other environmental stresses (Fig. 7.17*C*).

Fibers of phloem have the same general characteristics as those described as occurring in the cortex. In most plants, are longer fibers in the phloem than in the cortex.

In gynmosperms, companion cells are lacking, and there are **sieve cells** rather than sieve-tube members; these have tapered end walls without sieve plates; they overlap at their ends, but do not form sieve tubes.

The Primary Xylem

The conducting cells that occur in primary xylem of vascular plants are **tracheids** and **vessel elements.** These cells conduct water and mineral salts. Associated with them may be **fibers** (xylem fibers) and **parenchyma** (xylem parenchyma).

A tracheid is a single elongated cell more or less pointed at its ends (Fig. 7.18). Functioning tracheids are not alive. The tracheid wall may not be the same thickness throughout. All the wall may be thickened except for numerous small, circular, or oval areas called **pits** (Fig. 7.18*A*). There are two types of pits in xylem cells: **simple pits** and **bordered pits.** Simple pits as shown in Figs. 7.13*A* and *I* occur in fibers, sclereids, and in parenchyma cells when they have secondary walls. Pits form opposite each other in the secondary walls of adjacent cells. A pit-pair is not a hole in the wall, since the primary wall and the middle lamella of the two communicating cells remain intact. These primary layers, however, are penetrated by plasmodesmata while the cells are living. The type of pit known as a **bordered pit** (Figs. 7.18*A,E;* 7.19*A* to *C*) occurs in tracheids, vessel elements, and some xylem fibers. This type of pit is more structurally complex. It consists of an expanded border of the secondary cell wall that extends over a small pit chamber. Within the chamber is a diaphragm-like primary cell wall. In gymnosperms, the center portion, the **torus,** is thickened and impregnated with a waxy material (Fig. 7.19*C*) and may regulate water and water vapor flow (Chapter 12).

A **vessel element** is a *single unit* with oblique, pointed, or transverse ends. A **vessel** is a series of vessel elements differentiating end to end, with perforated end walls (**Fig. 7.20*A,B,*** page 120, 7.21*A* to *C*). Vessels are often several centimeters long, and in some vines and trees they may be many meters in length. Before the protoplasts disappear, vessel walls become thickened, forming a secondary wall; the thickening material is laid down on primary walls in various patterns so that in some places the secondary walls are thick and in others thin (**Figs. 7.20*C,*** page 120; 7.21*A*). The material deposited is cellulose; later, the layers of cellulose become lignified. The end walls of the vessel elements also dissolves before the protoplasts disappear. Thus, the deposition of thickening material forming the secondary walls and the dissolution of end walls are functions of living cells. After these events have taken place, the protoplast dies. Consequently, the vessels that are functioning in the rapid transport of water and mineral solutes have no living contents.

The secondary walls of vessels in angiosperm stems are desposited in several different patterns (Figs. **7.20*C,*** 7.21*A*): **annular, scalariform, reticulate** and **pitted.** The ends of the vessel elements (Figs. **7.20*A,B;*** 7.21*A,B*), are generally on a slant and, although open, they may have

Figure 7.18
Tracheids and vessel elements from secondary xylem. *A,* tracheid from spring wood of white pine *(Pinus). B,* tip of tracheid from wood of oak *(Quercus). C,* tip of vessel element from wood of *Magnolia. D,* tip of vessel element from wood of basswood *(Tilia). E,* diagram of bordered pit.

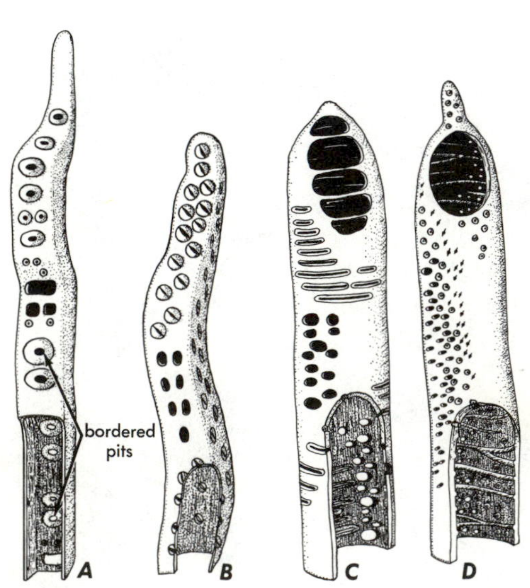

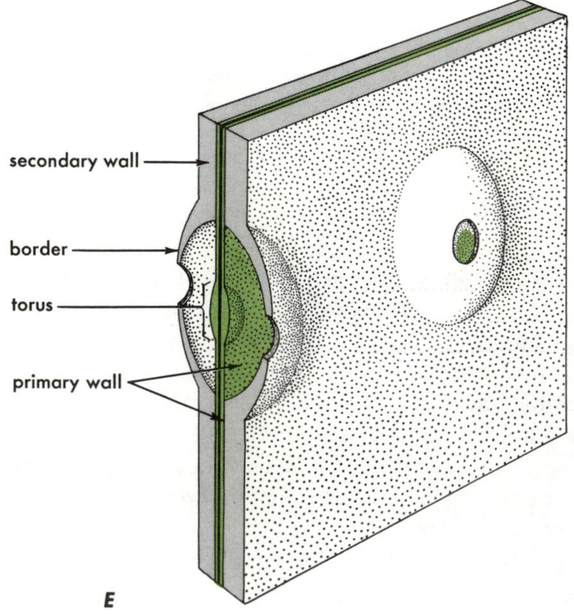

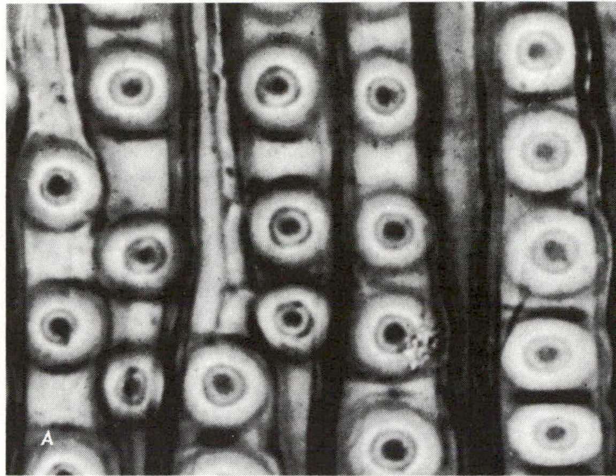

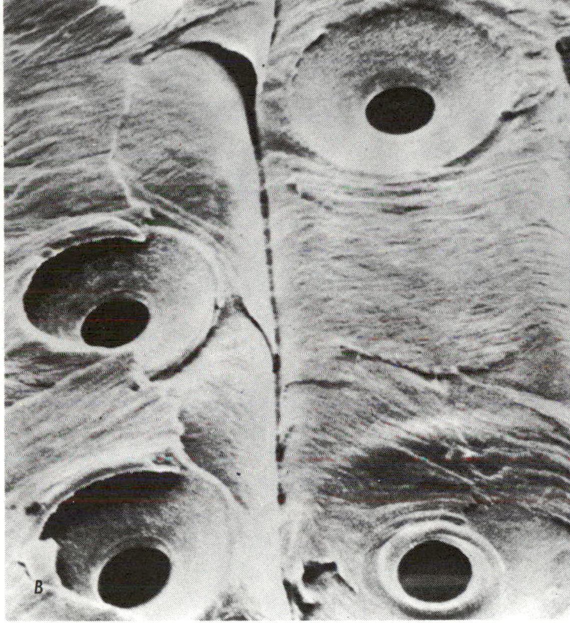

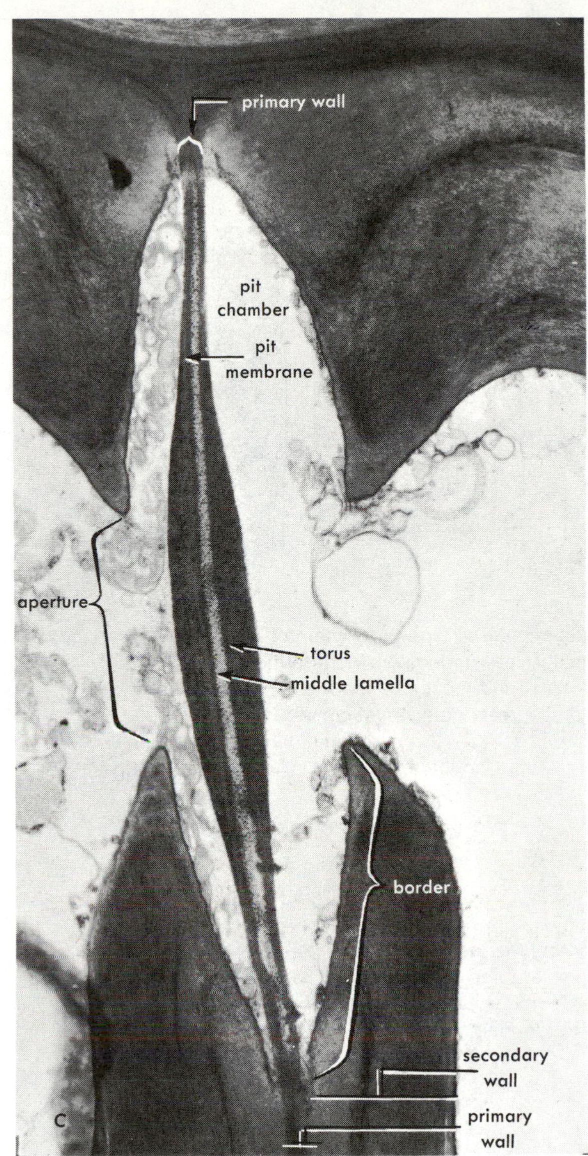

Figure 7.19
Bordered pits. *A* and *B,* in tracheids of pine wood. *A,* light micrograph, ×1800. *B,* stereoscan micrograph, ×2400. *C,* electron micrograph of bordered pit in ground hemlock *(Taxus canadensis),* ×4300. (*A,* courtesy of Artschwager; *B,* courtesy of S. Cook and D. Hess; *C,* courtesy of M. Ledbetter.)

bars of wall material across them. The shape of vessel elements may indicate evolutionary relationships among plants. Vessel elements do not occur in small veins in leaves, and are lacking in most gymnosperms and in the lower vascular plants. In these forms, tracheids occur in elongating regions, and they may have annular and spiral types of secondary walls.

Xylem parenchyma cells outlive vessel elements, tracheids, and most xylem fibers. They function in the storage of water and foods, which, as we have learned, is one of the principal functions of parenchyma wherever it occurs in the plant. Parenchyma may also conduct materials for short distances.

Xylem fibers are similar to the fibers described elsewhere.

The Pith and Pith Rays

Pith is composed of large-celled parenchyma with numerous intercellular spaces **(Figs. 7.22,** page 120, 7.23). Storage of food is its principal function. In some stems, primary vascular bundles are separated by wide strips of parenchyma that extend from the pith to the cortex. Such strips of parenchyma are sometimes called **pith rays** and may be considered radial extensions of the pith (Fig. **7.22**). Like pith, one of their functions is food storage; they also conduct materials short distances radially. They merge with parenchyma of the cortex so that no line of separation is visible. In many woody species primary vascular bundles are separated by very narrow rays of perhaps two to four cells in width (Fig. 8.10).

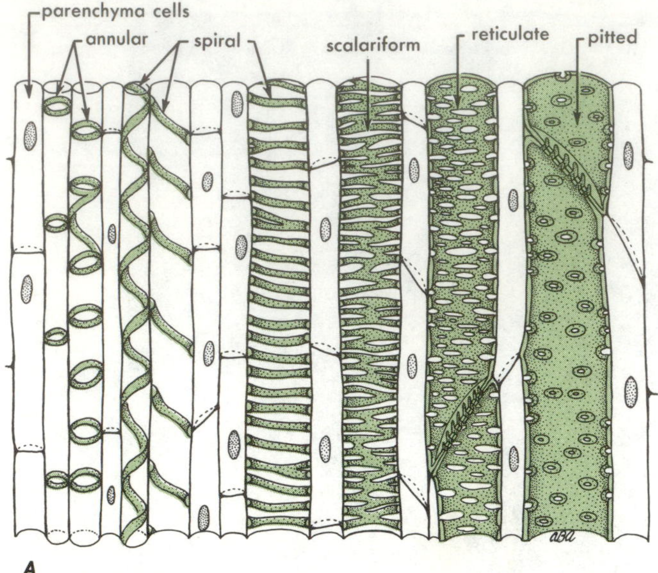

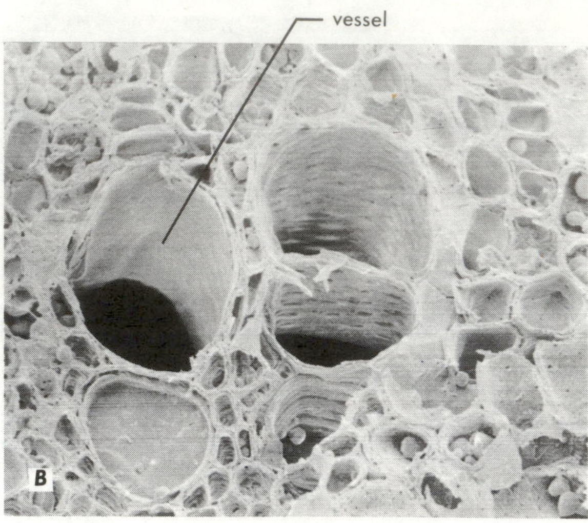

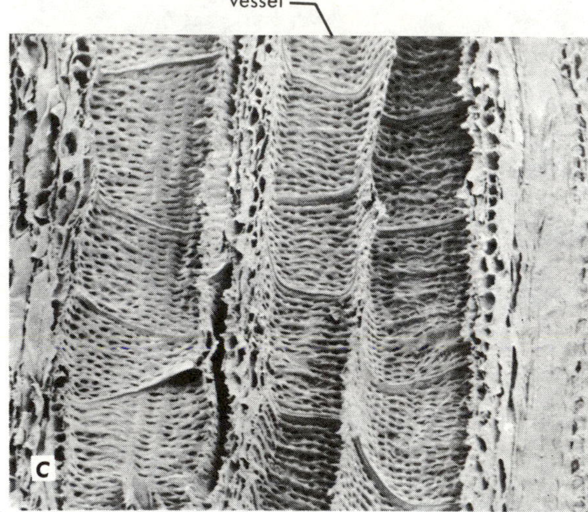

Figure 7.21
Types of vessel elements in primary xylem of an elongating branch. *A*, annular vessel elements were formed first and are therefore the oldest and most stretched. When elongation has stopped, the pitted vessel elements, formed last, will not be stretched. *B*, cross section. *C*, longitudinal section. (Note the rims of old cross walls now no longer present.) (Courtesy of D. Hess.)

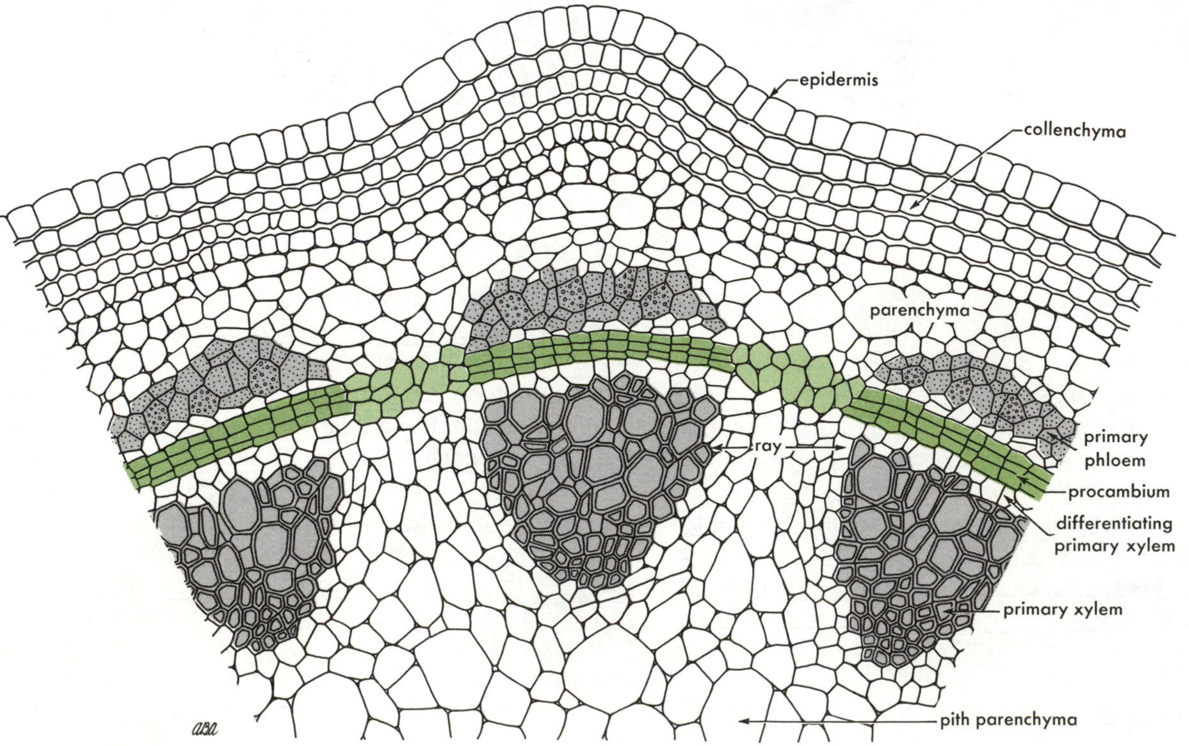

Figure 7.23
A stem at the close of primary growth. Some procambium is still present and the last primary xylem and primary phloem cells are still undergoing differentiation. Ray parenchyma separates vascular bundles. Cortex consists of collenchyma and parenchyma. Epidermis is intact.

A Summary of Primary Development

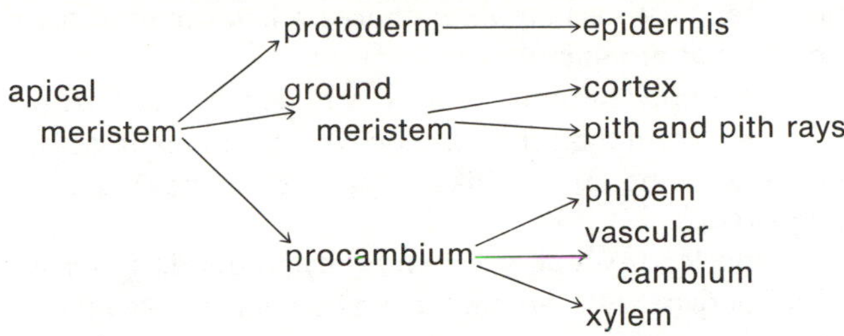

The term **stele** is applied to the part of the stem that includes primary vascular tissues, pith, and pith rays. The **primary plant body** is composed of the above primary tissues.

The main functions of these primary tissues may be summarized as shown below.

Epidermis: Protects underlying tissues
Vascular tissues
Phloem: Conducts food
Vascular cambium: Produces secondary phloem and secondary xylem
Xylem: Conducts water and mineral salts, and gives strength to stem
Cortex: Stores food and, in young stems, manufactures food, strengthens, and protects
Pith: Stores food
Pith rays: Store food, and conduct water, mineral salts, and food radially

THE DICOTYLEDONOUS STEM

Primary Growth

The tip of the stem consists of small immature leaves enclosing a dome-shaped apical meristem (**Figs. 7.24*A,B*,** page 120). The apical meristem is composed of dividing cells, arranged in various ways, that give rise to the leaves, buds, and the primary meristematic tissues (**Figs. 7.7,** page 117, 7.8).

The apical meristem is potentially immortal. In long-lived tropical vines and trees, it may continue actively producing new primary tissues for hundreds of years. In temperate climates, it has rhythmic activity, with growth and differentiation into primary meristems alternating with rest periods during which metabolism is reduced and growth and differentiation are retarded. In redwoods (*Sequoia sempervirens*), some junipers (*Juniperus occidentalis*), and bristle-cone pine (*Pinus aristata*), the apical meristem persists for several thousand years. Most plants, however, do not live for such long periods. Many are annuals, the primary meristem differentiates in one year into floral apices, flowers are produced, and the plants die. Many others, due to inherent genetic factors, susceptibility to wood decay, or environmental factors have more limited life spans.

Beneath the apical meristem and derived from it are the three primary meristematic tissues—protoderm, ground meristem, and procambium. The ground meristem starts to differentiate first. These tissues form a cylinder near the outside of the stem and a core in the inside (Fig. 7.8*C*), that will continue to differentiate into the cortex and pith, respectively.

Between these two tissues, near the apex, is a ring of **residual meristem** cells that retain the cellular characteristics of the apical meristem (Fig. 7.8*C*). This ring will become a cylinder of discrete procambium strands that later differentiate into primary xylem and phloem (vascular bundles). Each bundle is separated from others by regions of parenchyma cells (Figs. 7.8*D,* **7.22,** page 120, 7.23). The formation of vascular bundles from the residual meristem is apparently in response to leaf development. The mechanism is not exactly understood, but it may involve the production of some substance by the leaf primordia (immature leaves on the shoot tip) and its transport downward to the residual meristem ring. When the substance reaches a certain critical concentration, it apparently induces the residual meristem cells to form bundles of procambium and then xylem and phloem. Each bundle is called a **leaf trace** and, as it matures, will lead from the stem into a leaf, connecting it to the axis of the stem (see Chapter 10, Leaf Development). The vascular system of the stem actually consists, for the most part, of interconnected leaf traces. Other traces, however, connect to buds (bud traces), and some traces end at the apical meristem without connecting to either a leaf or a bud.

The role that leaves play in vascular differentiation and in the pattern of vascular bundles has been determined by elegant experiments. In one of these experiments, the leaves and leaf primordia were removed around the apex (Fig. 7.25*A*). As the stem elongated the new leaf primordia that formed were destroyed. Anatomical examination of the stem that differentiated during the weeks of the experiment revealed that the "vascular tissue" remained as an unbroken cylinder (Figs. 7.25*B,C*). The cells making up the cylinder did not differentiate completely, and vascular bundles did not form. If leaves were later permitted to develop, the newly formed vascular tissue did have bundles and did differentiate into xylem and phloem. This experiment demonstrated that leaves are needed for vascular differentiation and for the formation of procambial strands and vascular bundles (see also Chapter 20).

Tissues of both herbaceous and woody stems originate in the same fashion. Primary tissues are arranged in bundles in alfalfa **(Fig. 7.22)** as they are in *Sambucus* (Fig. 8.14). Slightly different arrangements occur in other species.

These primary tissues are, from the epidermis inward: epidermis; cortex, composed of collenchyma, groups of fibers, and parenchyma; phloem with sieve tubes, companion cells, fibers, and parenchyma; xylem with vessels, tracheids, fibers, and parenchyma; pith and

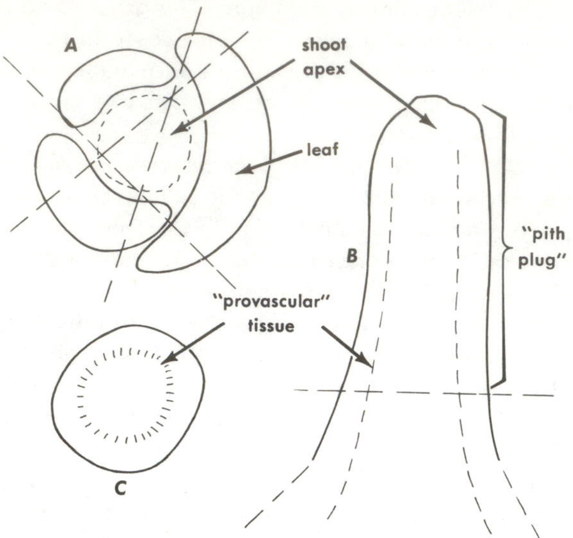

Figure 7.25
An experiment to show the effect of removing leaves on vascular development in the stem of *Geum* sp. In *A,* top view of shoot tip, dotted lines indicate cuts to remove young leaves and leaf primordia that are not yet visible. *B,* pith plug that forms several days after removing leaf primordia. *C,* cross section of pith plug to show the provascular tissue that forms in a complete circle or cylinder when leaves are removed.

pith rays composed of parenchyma cells. The outermost cells of the primary phloem in alfalfa develop into fibers.

THE MONOCOTYLEDONOUS STEM

Primary Growth

The angiosperms can be conveniently divided into two major groups—dicotyledonous and monocotyledonous plants (Chap. 30). The primary difference between these two groups is that seeds of monocotyledonous plants, for example, grasses, contain only one cotyledon (seed leaf) while dicotyledonous seeds contain two. In addition, they have certain other characteristics in their plant bodies that are sometimes different.

Monocotyledonous stems are usually uniform in thickness from the top to the bottom of the plant, as opposed to being tapered like most dicot stems (**Fig. 8.22,** page 148).

With few exceptions, monocotyledons do not have secondary growth. Some, such as palms and *Pandanus,* that do achieve considerable size, do not produce a woody stem suitable for lumber. There are many perennial monocotyledons, and mechanisms have been devised to circumvent the generally short life span of primary tissues. Because of this, the external appearances and cellular anatomy of the monocotyledons show considerable variation. At this point, we shall consider only monocotyledonous stems as they appear in the grasses, particularly corn (*Zea mays*), an annual plant.

The apical meristem of a typical monocotyledon looks like a small dome in a broad depression. Young leaves grow out of the depression and extend over the apical meristem. Beneath it is the inverted saucer-shaped primary thickening meristem (Figs. 7.26*A,B*). This meristem may extend down the stem a short way, so that it forms new cells upward to allow for an increase in length and outward to allow for an increase in girth. Leaf traces traverse the primary thickening meristem to connect it to already existing bundles.

With but few exceptions, procambium strands, and hence vascular bundles, in stems of monocotyledonous plants are *scattered* throughout the ground meristem or at least through its outer region (Figs. **7.27,** page 145, 7.28). However, in some grasses like wheat and barley, the stems are hollow, and vascular bundles are in a definite ring. In most monocotyledonous stems, *all* procambium cells differentiate into primary xylem and

Figure 7.26
A, longitudinal section of *Iris* shoot apex to show the primary thickening meristem, ×88. *B,* longitudinal diagram of the shoot apex of corn *(Zea mays.).* (Redrawn from K. Esau, *Plant Anatomy,* 2nd ed. © 1965 John Wiley & Sons, Inc. N.Y.)

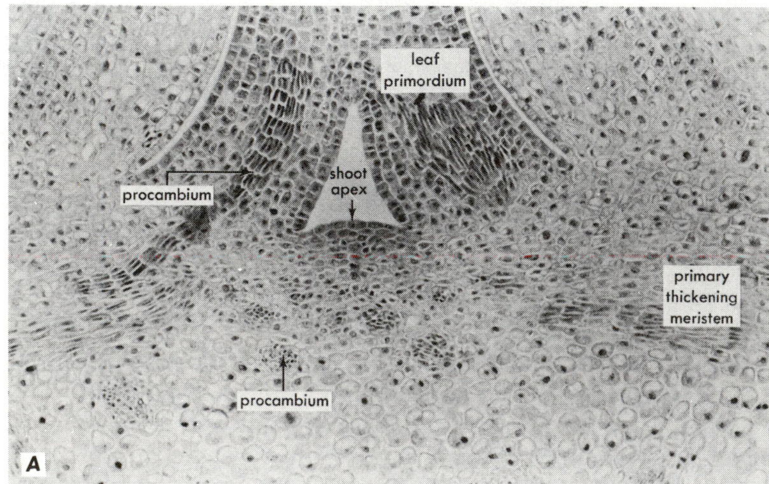

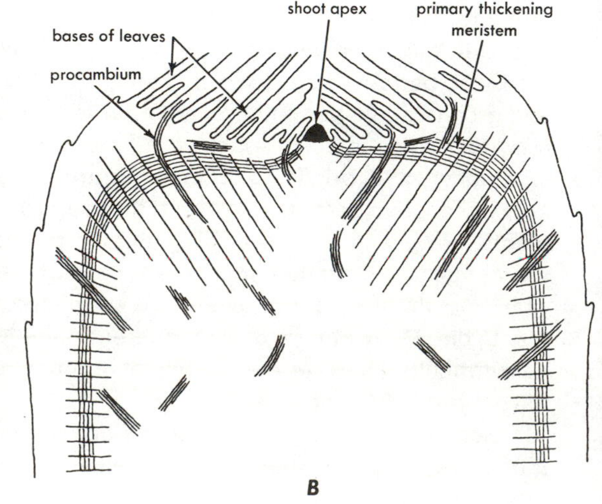

primary phloem elements; there is no vascular cambium and, as a consequence, no production of secondary tissues. Bundles of this sort that are "closed" to further growth are called **closed bundles** (Fig. 7.28). As has been observed, bundles of other seed-bearing plants that possess a vascular cambium are "open" to further growth; they are known as **open bundles**.

The terms pith and cortex are often not applicable in monocot stems. The tangential vascular bundle pattern in the nodes of monocot stems is complex and is called a nodal plate **(Fig. 7.27*C*)**. In transverse sections of internodes of most grasses (barley, rye, wheat), the following tissues are evident **(Fig. 7.27):** (*a*) a single layer of epidermal cells, (*b*) strengthening tissue variously arranged beneath or near the epidermis, (*c*) ground parenchyma, and (*d*) vascular bundles. Stomata occur in the epidermis. Strengthening tissue consists of elongated fibers with thick lignified walls. In some grasses, vascular bundles are arranged in two rings near the periphery of the stem. In many grass stems, there is usually a continuous ring of fibers some distance from the epidermis, with the small bundles of the outer row embedded in it. On the outer sides of the small bundles of the outer ring occur strands of fibers that reach to the epidermis. Bands of chlorophyll-bearing ground parenchyma are enclosed between these strands. Ground parenchyma may extend to the center of the stem, as in corn (Fig. 7.28), sorghum, and sugar cane. Or its central portion may become destroyed during the growth of the stem, leaving a hollow pith cavity, as usually happens in wheat, oats, barley, and rye.

Figure 7.28
Cross section of corn *(Zea mays)* stem vascular bundle, ×140. (Steroscan micrograph courtesy of S. Cook.)

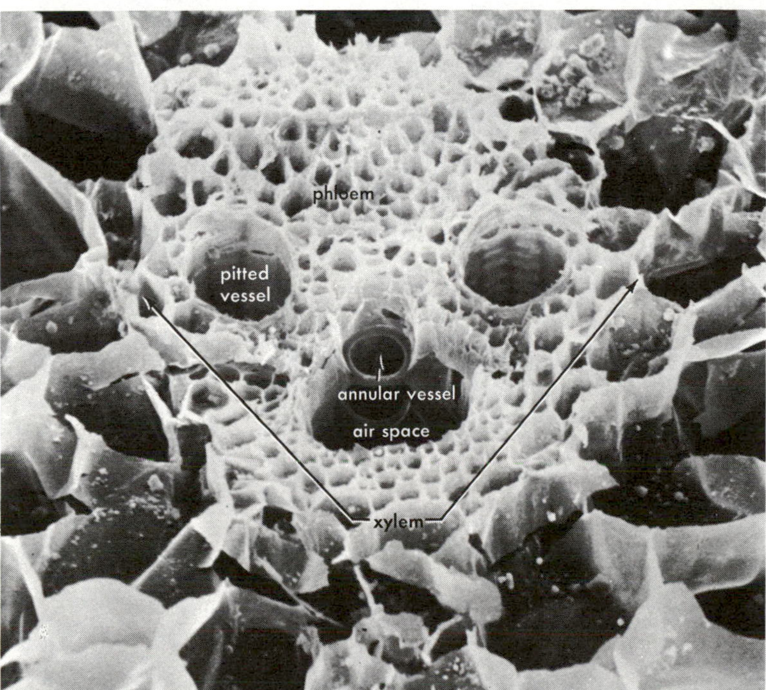

The vascular bundle of corn stems is an example of a common type of bundle occurring in monocotyledonous stems (Fig. 7.28). Large pitted vessels of the xylem are prominent features. There are usually two of these, and one or two smaller annular or spiral vessels may also be present. In addition, older bundles invariably contain a large air space or intercellular passage. Close examination of this space is likely to reveal within it a lignified ring or portion of a ring. This is evidence that the air space was originally an annular vessel. Elongation of the young stem, which has stretched and broken the vessel, gives rise to the relatively large air space. Vessels, together with the intervening fibers, tracheids, and parenchyma cells, comprise the xylem tissue, which is always located on the side of the bundle toward the center of the stem. Phloem forms a regular pattern of thin-walled cells exterior to xylem. Sieve plates may usually be seen. Companion cells are small, generally square, or rectangular in cross-section, and, because of their cell contents, stain more heavily than the sieve tubes. The bundle is generally surrounded by lignified fibers forming a tissue called the **bundle sheath**.

In grasses, a tissue at the base of each internode usually remains meristematic long after the tissues in the rest of the internode are fully differentiated. Thus, such plants contain a meristem at the apex of the shoot and also a meristem at the base of each internode. Each internode has its own growing zone. These internodal meristems are called **intercalary meristems**, and growth of the cells derived from them is termed **intercalary growth**. The flowering stems of such cereals as wheat, oats, barley, and rye shoot up very quickly. This rapid elongation results not only from the growth of cells derived from the apical meristem, but also from the growth of cells derived from intercalary meristems at the base of each internode.

SUMMARY OF STEMS AND PRIMARY TISSUES

1. Buds are characteristic of woody stems. Bud scales enclose and protect rudimentary leaves surrounding an apex. The apex may be either a vegetative shoot or a floral apex. A floral apex terminates growth. Woody plants each year require new vegetative growth, which is normally accompanied by flowers.
2. Primary growth brings about the elongation of stems and establishes the basic pattern of cells and primary tissues characteristic of the particular stem and upon which the functioning and future growth of the stem depend.
3. Primary meristematic tissues from the apical meristem are the protoderm, ground meristem, and procambium.
4. The primary plant body is composed of (*a*) the epidermis, which may be differentiated into epidermal cells, guard cells, and epidermal hairs; (*b*) cortex, which is composed of collenchyma, sclerenchyma (fibers and sclereids), and parenchyma; (*c*) vascular tissue, composed of xylem (fibers, tracheids, vessel elements, and parenchyma) and phloem (fibers, sieve-tube members, companion cells, and parenchyma); (*d*) pith, composed largely of parenchyma and sometimes accompanied by sclereids; (*e*) pith rays, composed of parenchyma cells.
5. The functions of these cells and tissues are as follows. Parenchyma is used for the storage of water and food and the conduction of materials for short distances. Collenchyma, sclereids, fibers, and tracheids are the strengthening or mechanical tissue elements. Tracheids and vessels (series of vessel elements) conduct water and mineral salts. Sieve tubes (series of sieve-tube members) conduct foods produced by photosynthesis.
6. Collenchyma and sclerenchyma may occur in patches or completely surround the stem just underneath the epidermis. Sclerenchyma is frequently associated with vascular bundles.
7. Primary vascular tissues in dicotyledonous stems usually consist of a ring of vascular bundles. These bundles form an interconnected system that connects leaves and buds to the plant body.
8. Monocotyledonous stems consist of ground tissues containing randomly distributed vascular bundles. Some monocot stems have a hollow pith region with vascular bundles in a ring.
9. In grasses, the tissue at the base of internodes that remains meristematic is called intercalary meristem.

CHAPTER 8

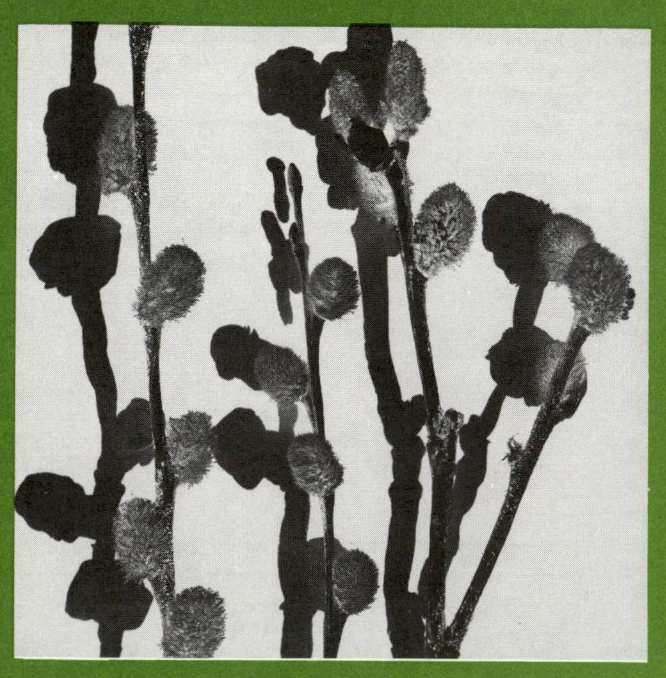

SECONDARY GROWTH IN STEMS AND STEM MODIFICATIONS

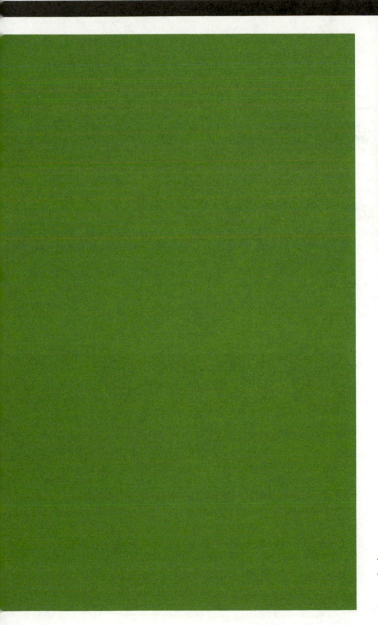

Primary growth is responsible for increases in stem length and laying down the basic tissue pattern. Secondary growth is responsible for development of new secondary vascular tissue, increasing the girth of the stem, and providing a continuous connection of living cells between newly formed primary tissues of the developing shoots and roots. This device makes possible attainment of great age by individual plants, although living tissue may be only 3 years old or less.

SECONDARY GROWTH

Formation of Vascular Cambium

This discussion is concerned with perennial woody stems like those of common shrubs and trees, including pines and their allies; but even some herbs, such as alfalfa, may develop some secondary tissues before the end of their relatively short lives.

Indeed, formation of a complete circle of cambium in alfalfa makes this plant a convenient one for an explanation of the manner in which cambium develops (Figs. 8.1, **8.2**, page 146). This occurs in the following manner: Not all procambium between primary xylem and primary phloem is used up; some remains (residual procambium) and proceeds to divide to form phloem toward the exterior of the stem and xylem toward the interior. It becomes a vascular cambium and forms **secondary xylem** and **secondary phloem**. This causes the stem to increase in girth (Figs. 8.1, **8.2**). Since this first cambium forms within vascular bundles, it is called a **fascicular cam-**

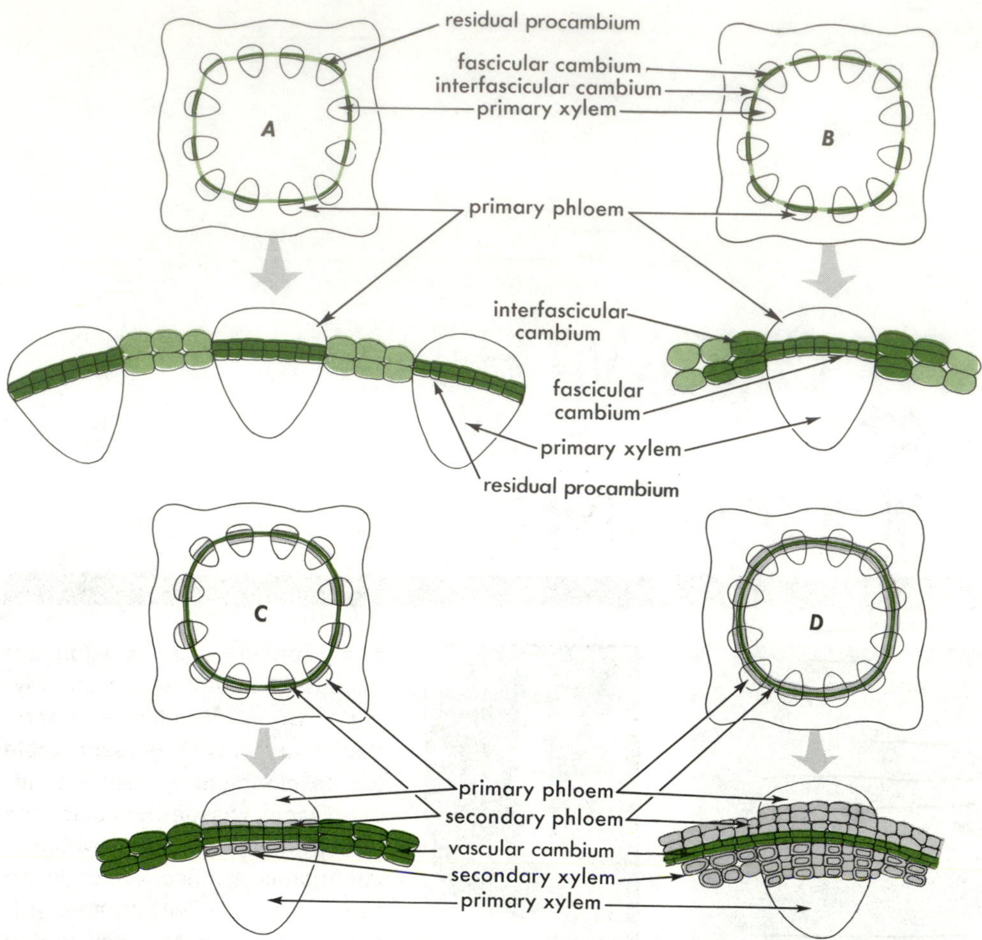

Figure 8.1
Formation of a complete cylinder of vascular cambium. *A,* at completion of primary growth, some meristematic cells remain between primary xylem and primary phloem (residual procambium shown in dark green). Parenchyma cells appear in pith rays between vascular bundles (light green). *B,* some of the parenchyma cells of the pith ray become meristematic (interfascicular cambium; dark green) those dividing cells between the primary xylem and phloem are now referred to as fascicular cambium. *C,* parenchyma cells of pith between the meristematic cells of the bundles have returned to a meristematic state, forming a cylinder of vascular cambium (dark green). *D,* complete cylinder of secondary xylem and secondary phloem (light gray) is formed by the vascular cambium. Parenchyma cells are shown in light green, vascular cambium in dark green, and secondary tissues in light gray.

bium. What happens between the vascular bundles? In stems such as alfalfa, in which vascular bundles are separated by wedges of parenchyma, certain of these cells may become meristematic (Figs. 8.1*A*, **8.2*A***). Thus, cambium within bundles may be joined by cambium between bundles, forming a complete cambium ring (Figs. 8.1*B,C*). Cambium formed between bundles is called **interfascicular cambium.** The circle of **vascular cambium** thus formed in alfalfa and other herbaceous stems is not active for long. In woody stems it remains active as long as the tree lives, over 3000 years in redwoods. Interfascicular cambium thus formed joins fascicular cambium to form a complete ring of vascular cambium (Fig. 8.1*C*). It should be noted that fasciular cambium cells are derived from residual procambium cells that have never ceased to be residual meristematic. Interfascicular cambium originates from cells that have differentiated from residual meristem and perhaps ground meristem into a parenchyma tissue (see Fig. 7.8).

This development results in a cylinder of vascular cambium completely surrounding xylem and surrounded on the outside by phloem (Fig. 8.3*A*). In cross section, these cambial cells are uniformly bricklike with thin walls (Fig. 8.3*C*). Close examination of a woody stem reveals that the cells that make up wood are actually oriented in two ways (Figs. 8.3*A,B*). Some cells are elongated parallel with the axis of the stem, making the **axial system.** The axial system is composed of vessel members, tracheids, fibers, and parenchyma. In the phloem, sieve-tube members, fibers, companion cells, and parenchyma make up the axial system. The source of these cells in the vascular cambium are **fusiform initials.** A sec-

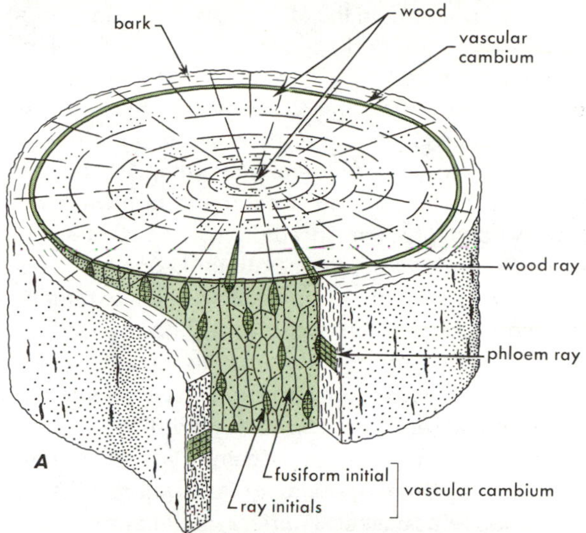

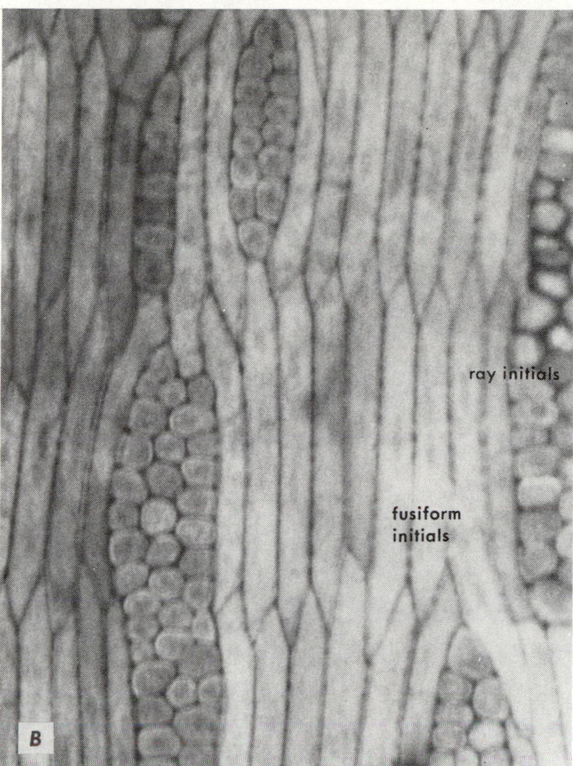

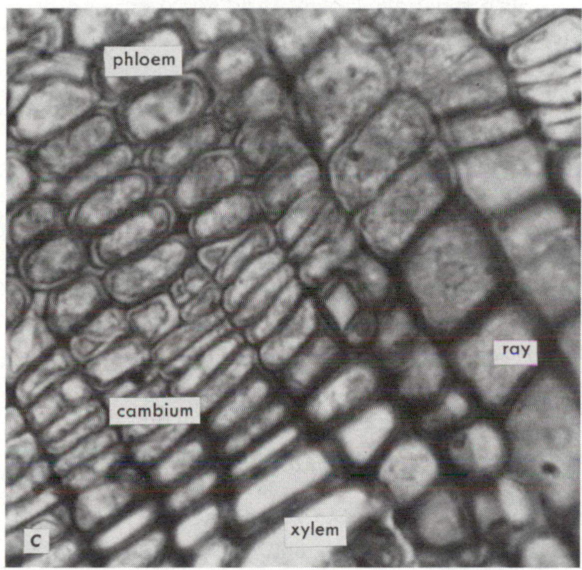

Figure 8.3
Above and left: Vascular cambium. *A*, showing relationship of cambium and cambial intitials to the stem. *B*, tangential section. *C*, cross section of quiescent cambium of locust *(Robinia)*, ×3-3¾.

ond system, the **ray system**, is composed of cells oriented at right angles to the axis of the stem (Figs. 8.3*A*,*B*). The **rays** that make up this system develop from **ray initials** in the vascular cambium. Rays are living channels through which nutrients and water move laterally in stems with secondary growth. Xylem rays are made up of ray tracheids, and ray parenchyma and phloem rays are made of phloem parenchyma. Ray cells may remain alive for several years, perhaps 10 or more.

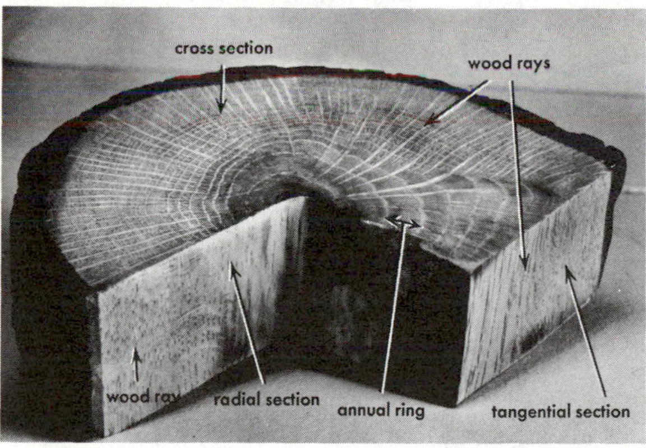

Figure 8.4
Portion of stem of oak *(Quercus)*, showing cross (transverse), radial, and tangential longitudinal sections and their gross characteristics, × ½.

Sections of Woody Stems

A woody stem may be viewed in three different planes (Fig. 8.4). A cross-wise view is logically termed a cross or transverse section. There are two types of longitudinal sections. A section cut in a plane passing through both the center of the stem and the circumference is cut along a radius and is termed a radial section. The other longitudinal section is cut at right angles to a radius, some distance from the center, and is therefore called a tangential section.

Differentiation of Phloem and Xylem from Vascular Cambium

Figure 8.5 shows stages in differentiation of vascular cambial cells to form secondary phloem and secondary xylem. Cambial cells divide most frequently by a tangential wall, forming two derivative cells. As a rule, the inner cell, next to the xylem, develops into a secondary xylem cell. The outer cell, next to the phloem, remains meristematic and divides again. When new phloem elements are produced, an inner cell retains the power of division,

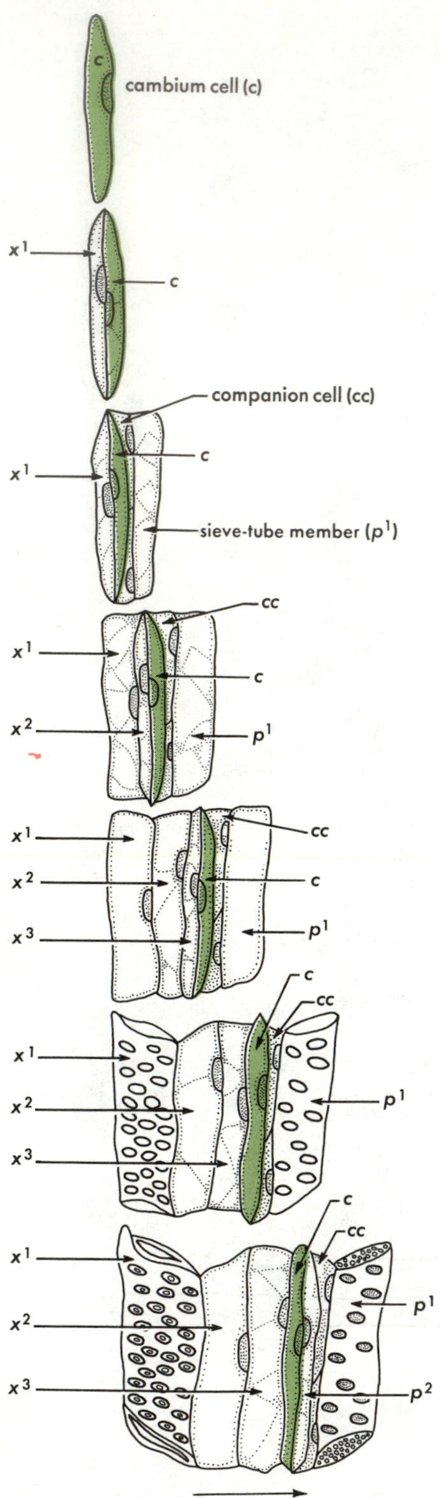

Figure 8.5
Diagram as seen in radial section, showing stages in differentiation of vascular cambium cells (*c*, cambium; *cc*, companion cell; p^1, p^2, phloem; x^1, x^2, x^3, xylem).

and an outer cell gives rise to one or more secondary phloem cells. Generally, more secondary xylem than secondary phloem is produced during the season. Cambium cells continue thus to divide throughout a growing season, adding secondary xylem on the *outside* of old xylem, and secondary phloem on the *inside* of old phloem. Thus, secondary xylem is superimposed upon primary xylem, and secondary phloem tends to exert pressure on primary phloem and cortex and to push them outward. Although cambial cells, by repeated divisions, are differentiating into xylem and phloem elements, some cells remain meristematic, and so a cambium is always between xylem and phloem. In addition, as the stem increases in circumference, the cambium keeps step with this increase by radial divisions of cambial cells, thus increasing the number of cells and enlarging the circumference of the cambium cylinder. During the winter season, cambium is inactive, only to begin divisions again in spring after bud break and new leaf growth.

The resulting cellular composition of secondary xylem and phloem is, in general, quite similar to that of the primary vascular tissues. It differs in that annular and spiral vessels are usually lacking, and secondary phloem contains more fibers.

Tissues of the Woody Stem

Anyone who has examined the stump of a cut tree has observed **annual rings** (Figs. 8.6, **8.7,** page 147). One annual ring represents the amount of secondary xylem growth for one season. Thick rings mean maximal growth has occurred during a season, and thin rings mean minimal growth. The density (width) of these rings gives an indication of past climate conditions.

Microscopic examination of an individual growth ring shows that during the early part of the growing season, cells are large and have relatively thin walls. This part of the growth ring is called the **early wood (spring wood)**. Later in the season the cells become smaller in diameter and have thicker walls; this is the **late wood (summer wood)** (**Figs. 8.7*C*,** 8.9, 8.10). In certain trees, large vessel members form only in the early wood and only small vessel members occur in the late wood; this organization is called **ring porous** (**Fig. 8.7*C***). In other trees the occurrence of vessel members is uniform throughout the growing season; this is **diffuse porous**

Figure 8.6
Cross section of a branch of mulberry *(Morus)*, ×½.

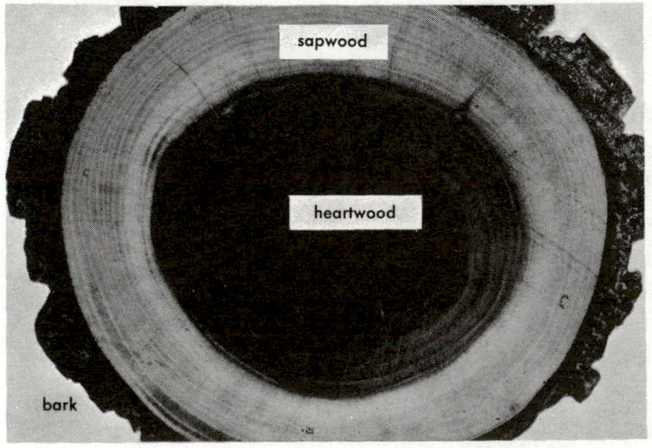

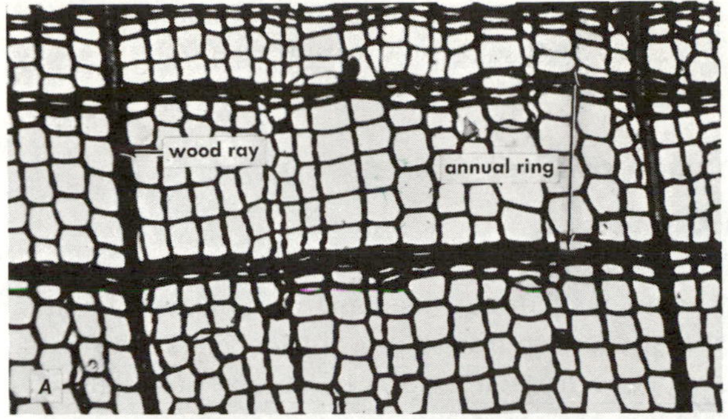

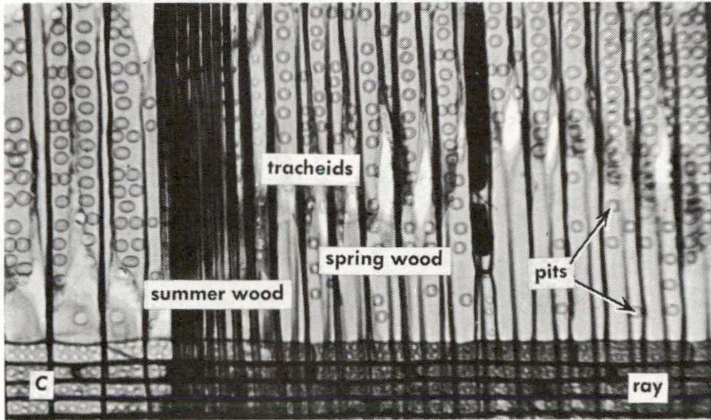

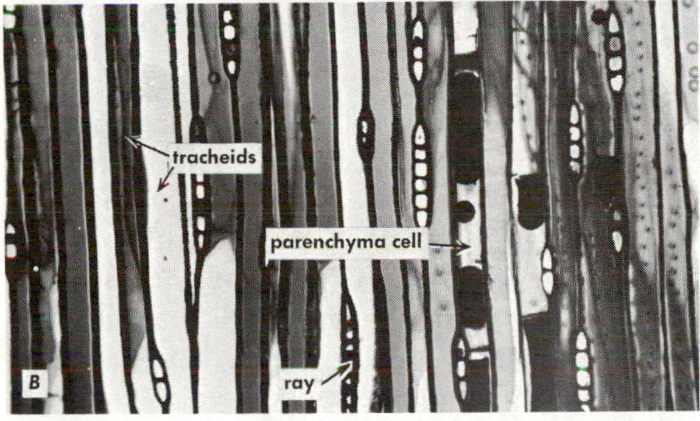

Figure 8.9
Sections of wood or redwood *(Sequoia sempervirens). A,* cross section. *B,* tangential section. *C,* radial section. ×100.

wood (Fig. **8.8,** page 147). Many trees in tropical areas, where growing conditions are uniform throughout the year, show no annual rings in their wood.

After a tree or branch is several years old, the inner part of the stem usually becomes inactive and filled with resins. This dense, central core is called the **heartwood.** Good barrels and wooden tubs are made from heartwood, because the resin makes this wood impermeable to water. Outside the heartwood is a light-colored, less dense region of active wood called **sapwood** (Fig. 8.6).

Two things contribute to the formation of heartwood. One is the curious formation of structures called **tyloses.** Tyloses are ingrowths of the primary wall of parenchyma cells that grow through the pits of vessel members. The walls of tyloses eventually become enlarged and may form secondary walls that completely plug up the xylem. Species of trees that have abundant parenchyma cells adjacent to vessel members readily form tyloses; however other species with more scattered parenchyma do not (Fig. 8.11). The second contributing factor to forming heartwood is the activity of rays. Rays apparently play an important role in transporting resins from the active sapwood into the central heartwood. Heartwood is a repository for metabolic by-products.

Gymnosperm wood is anatomically simpler than angiosperm wood. It is composed almost entirely of tracheids in the spring wood and of fiber-tracheids (cells that are intermediate between fibers and tracheids) in the summer wood. Generally there are no vessel elements in Gymnosperms.

A three-dimensional reconstruction of a block redwood (*Sequoia*) xylem is shown in Fig. 8.12*A*. This reconstruction was made by studying three stem sections similar to those shown in Fig. 8.9. A cross section from a small piece of *Sequoia* is shown in Fig. 8.9*A*. It is extremely regular, almost like a wire netting. Tracheids ap-

Figure 8.10
Sections of wood of oak *(Quercus borealis). A,* cross section. *B,* tangential section. *C,* radial section, ×40.

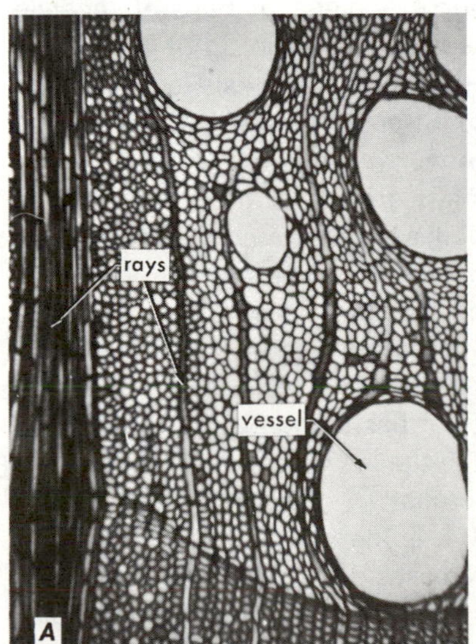

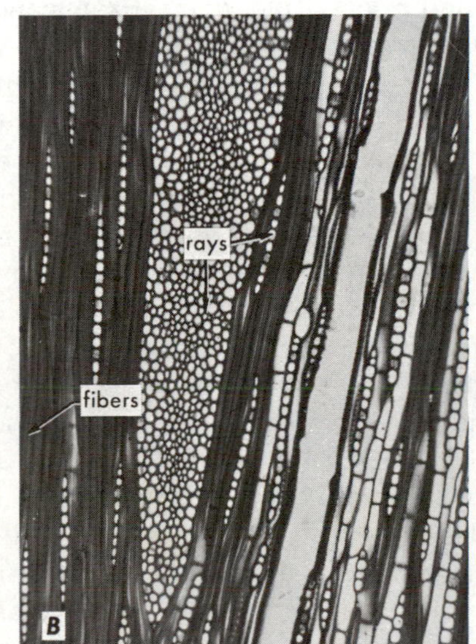

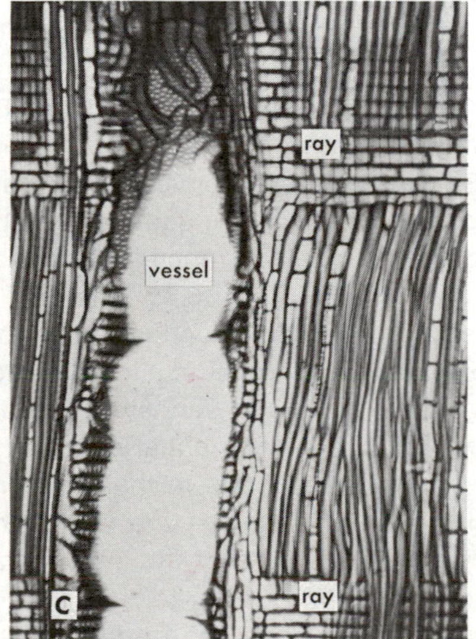

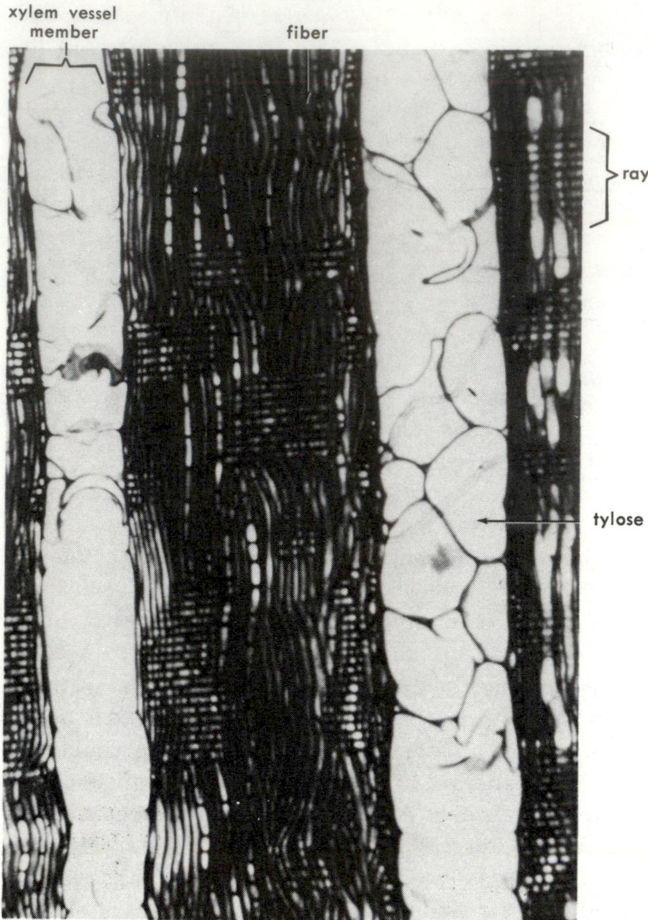

Figure 8.11
Radial section of wood from white oak *(Quercus alba)* that shows tyloses. Tyloses are ingrowths of parenchyma cells that form secondary cell walls and may plug up xylem vessel members, ×35.

pear in cross section as square hollow cells. Those formed in the spring are largest in diameter; as the season progresses, newly formed tracheids are smaller in diameter with thicker walls. The heavy-walled cell is called a **fiber-tracheid.** Note **xylem rays** or **wood rays** composed of parenchyma cells, and occasional small square cells with dark-staining contents. The latter are parenchyma cells. The radial view (Figs. 8.9, 8.12*A*) shows mainly tracheids, with a small strand of fiber-tracheids. Pits are present in radial walls of tracheids. An elongated parenchyma cell is made apparent by its dark-staining contents. Wood parenchyma cells are represented by dark green in Fig. 8.12. They are shown in tangential section and cross section. A horizontal ray in the radial section lies perpendicular to a tracheid. Pits occur between ray cells and tracheids. The ends of the rays are apparent in the tangential section (Fig. 8.9*B*). Here, fiber-tracheids have pits in their tangential walls. Pits in radial walls of tracheids are just visible. In this view, several parenchyma cells with heavily stained contents are present.

Although *Sequoia* does not produce resin, in many other gymnosperms **resin ducts** may be abundant. These secretory ducts are long hollow tubes containing an internal layer of cells that produce resin (Fig. 8.12*B*).

Formation of Cork (Periderm)

Cork Cambium

Figure 8.13 shows the origin of cork cambium and the development of secondary tissues from it. Cork cambium in most plants arises from outer cortical cells. The outer derivative cells from the cork cambium generally differentiate into cork cells, thus forming a layer of **cork** beneath the epidermis. The inner cells are known as **phelloderm,** and they are parenchyma-like cells. Cork cambium may originate also from epidermal or phloem cells.

Cork tissue (Figs. 8.13, 8.14), is composed of flattened, thin-walled cells with no, or small, intercellular spaces. A fatty substance called **suberin** is deposited in the walls, rendering the cells almost impermeable to water and gases. Hence, this tissue provides protection for the stem against excessive loss of water and also against mechanical injury. The protoplasts of cork cells are short-lived.

Bark

Bark (Fig. 8.16) includes all tissues outside the vascular cambium, and its exact cellular composition will depend upon the age and species of the twig or tree trunk being examined. In a young stem, the bark is made up of the following tissues, in this order, from the *outside* to the *inside:* **cork, cork cambium, phelloderm** (if present), **cortex,** and **phloem** (Figs. 8.13, 8.14). Microscopic examination may reveal the presence of epidermal cells still clinging to the cork. In old stems, the epidermis, cortex, and primary phloem become separated from the adjacent inner tissues by successively deeper layers of cork formation. When bark is peeled from wood, some cells of the cambial zone adhere to the bark, and some adhere to the wood. Examination of a piece of bark of incense cedar (Fig. 8.15) shows that it is made up of a series of layers of two different-appearing tissues. Recalling that the vascular cylinder of a tree is increasing in diameter, let us examine the manner of formation of this bark.

At the close of primary growth, primary vascular tissues, in strands, formed a ring around the peripheral area of the stem (Figs. 7.23, 8.1). The central portion of the stem was occupied by pith. Cortex and epidermis surrounded the ring of vascular strands. Fascicular cambium joined with interfascicular cambium to produce a ring of secondary vascular tissue increasing the girth of the stem (Fig. 8.16). As this takes place, what will happen to the cortex and epidermis?

Cells composing the epidermis are fully differentiated, and except in rare cases do not divide. As the stem increases in diameter through the activity of vascular cambium, the epidermis must be stretched and torn.

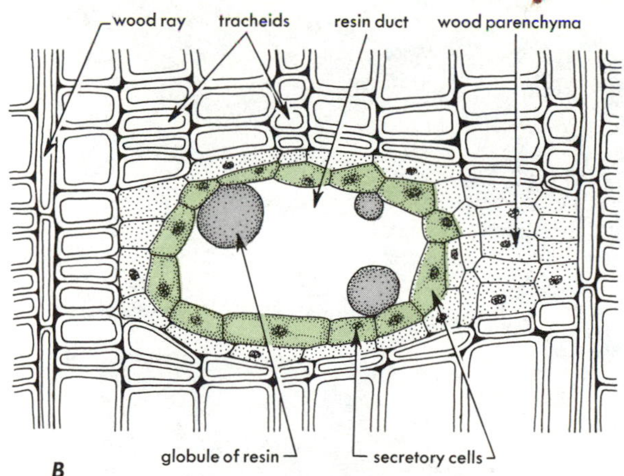

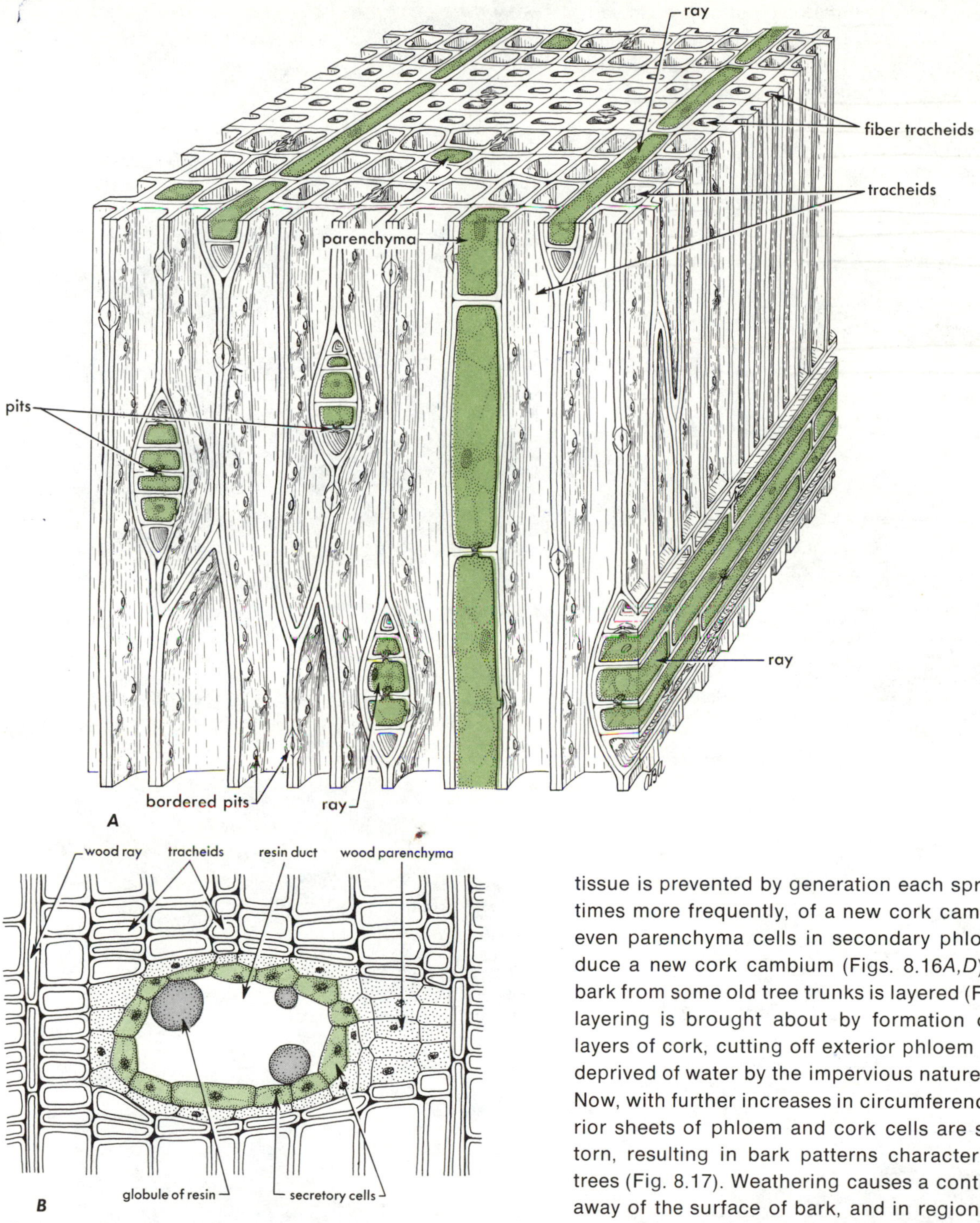

Figure 8.12
A, block diagram of secondary xylem of redwood *(Sequoia sempervirens). B,* resin duct in pine wood *(Pinus)* as seen in cross section.

When this happens, delicate underlying cells would be exposed to drying out and weathering were it not for development of a cork cambium and a layer of cork as previously described (Figs. 8.13, 8.16*A,C*).

As the stem continues to grow in girth, the cylinders of cork, in turn, will be ruptured. Exposure of underlying tissue is prevented by generation each spring, or sometimes more frequently, of a new cork cambium. Finally, even parenchyma cells in secondary phloem may produce a new cork cambium (Figs. 8.16*A,D*). This is why bark from some old tree trunks is layered (Fig. 8.15). This layering is brought about by formation of successive layers of cork, cutting off exterior phloem which is thus deprived of water by the impervious nature of cork cells. Now, with further increases in circumference, dead exterior sheets of phloem and cork cells are stretched and torn, resulting in bark patterns characteristic of many trees (Fig. 8.17). Weathering causes a continual wearing away of the surface of bark, and in regions where sand particles are carried by wind, trees have a beautiful polished appearance. Commercial cork is bark of an oak (*Quercus suber,* Fig. 8.17*B*). It is not layered, because it develops in a somewhat different fashion; here cork cambium never develops within phloem tissue. Thus, outer bark of the cork oak may be comprised almost entirely of cork cells.

Lenticels

An impervious layer of cork would effectively cut off the oxygen supply of the living tissues beneath it if groups of parenchyma cells called **lenticels** did not develop in various places (Fig. 8.14). They frequently originate beneath the epidermal stomata of young stems. When a cork cambium is formed, it produces ordinary parenchyma cells below these stomata in an outward direction. The resulting loose aggregation of parenchyma tissue bursts through the epidermis to form the lenticel. The air spaces between the parenchyma cells permit gaseous interchange.

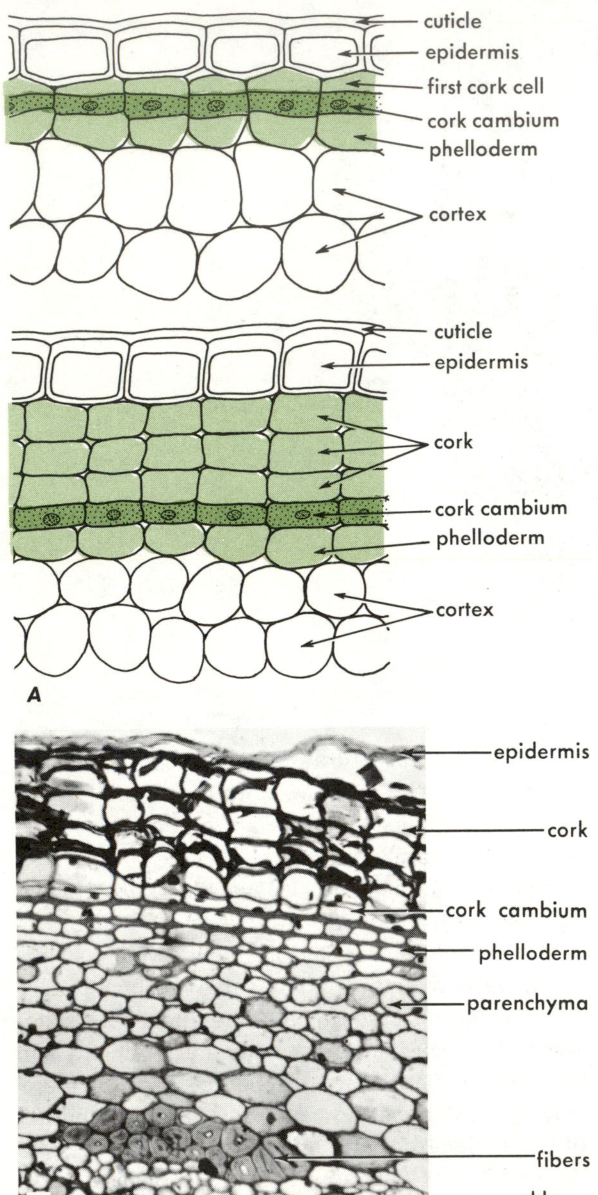

Figure 8.13
A, diagram showing origin of the first cork cambium and the first layer of cork. New cork cambia and new layers of cork form in a similar manner each spring. *B,* light micrograph showing cork cambium and newly formed cork in an elderberry stem *(Sambucus).*

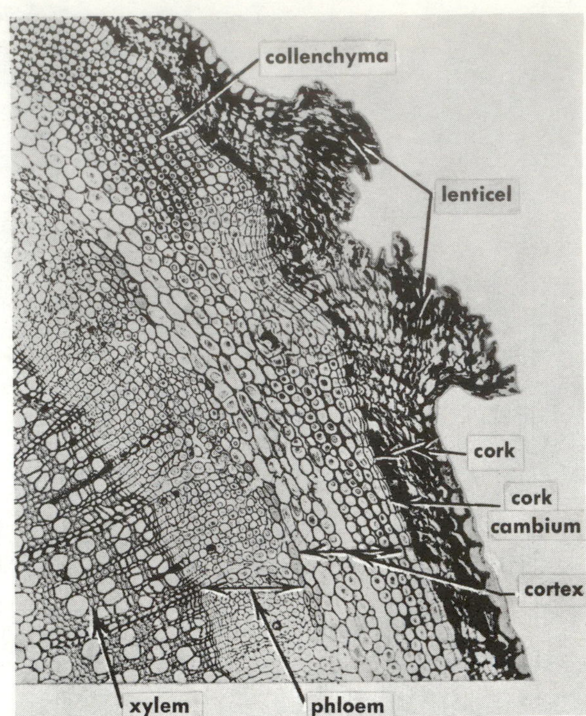

Figure 8.14
Cross section of a portion of young stem of elderberry *(Sambucus).* From the outside to the inside of the stem, the following tissues may be seen: cork, cork cambium, cortex, secondary phloem, and xylem. The vascular cambium makes up the two layers of thin-walled cells on the phloem side of the xylem. Note the lenticel in the cork, ×100.

Figure 8.15
Bark of incense cedar *(Libocedrus decurrens). A,* section of bark. *B,* three-dimensional view, ×½.

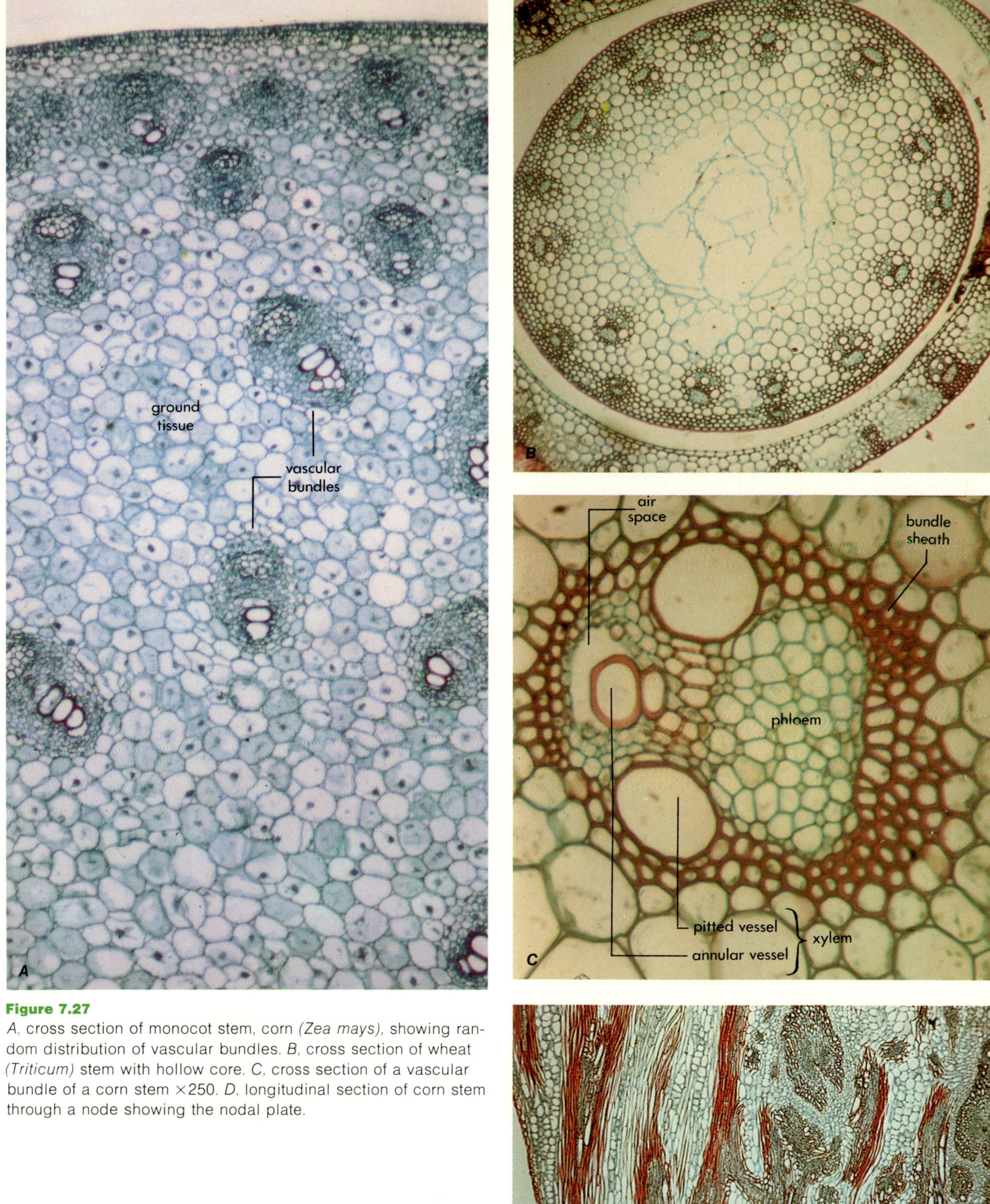

Figure 7.27
A, cross section of monocot stem, corn *(Zea mays)*, showing random distribution of vascular bundles. *B*, cross section of wheat *(Triticum)* stem with hollow core. *C*, cross section of a vascular bundle of a corn stem ×250. *D*, longitudinal section of corn stem through a node showing the nodal plate.

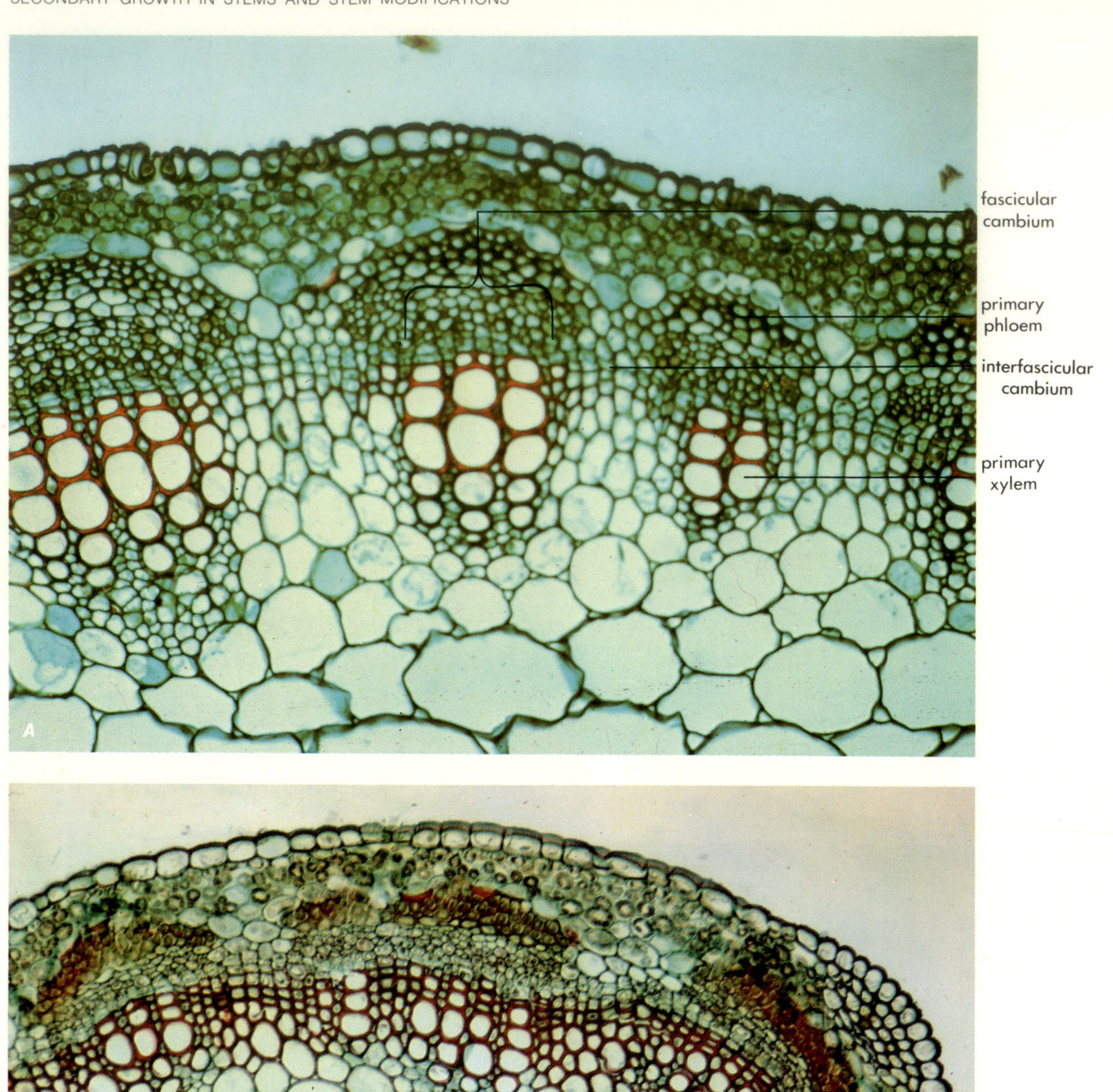

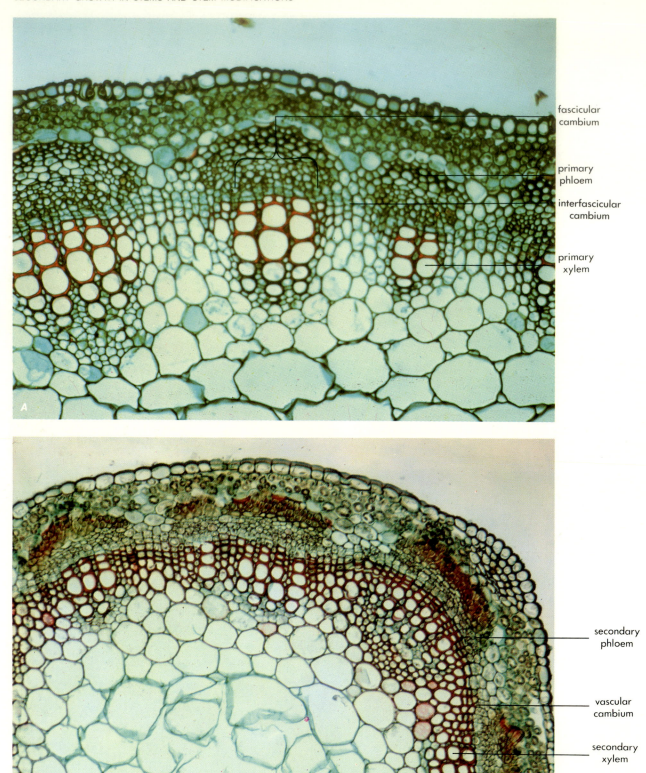

Figure 8.2
Cross sections of a stem of alfalfa *(Medicago sativa)*. *A*, showing the fascicular cambium within the bundle and interfascicular cambium between the bundles. *B*, some secondary vascular tissue has been produced, ×300.

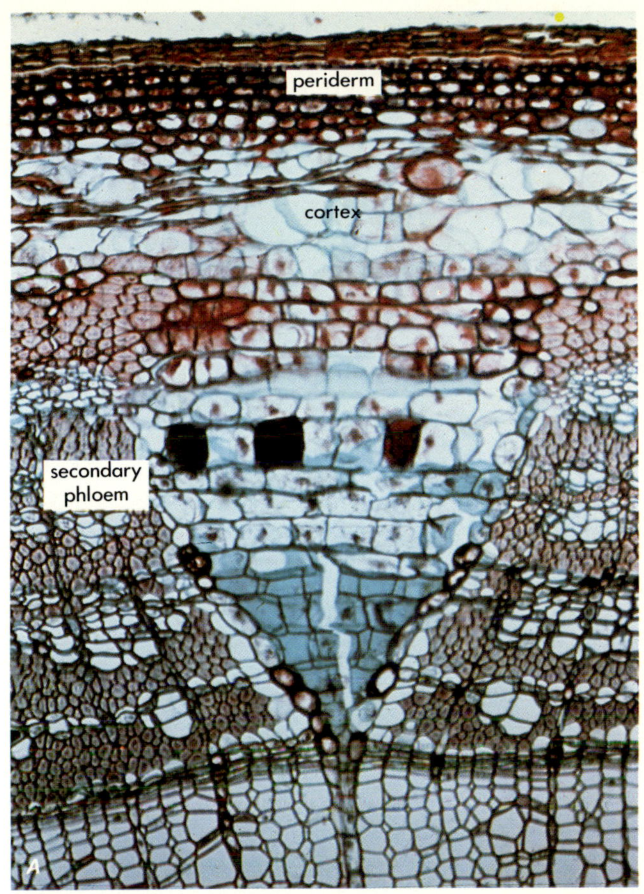

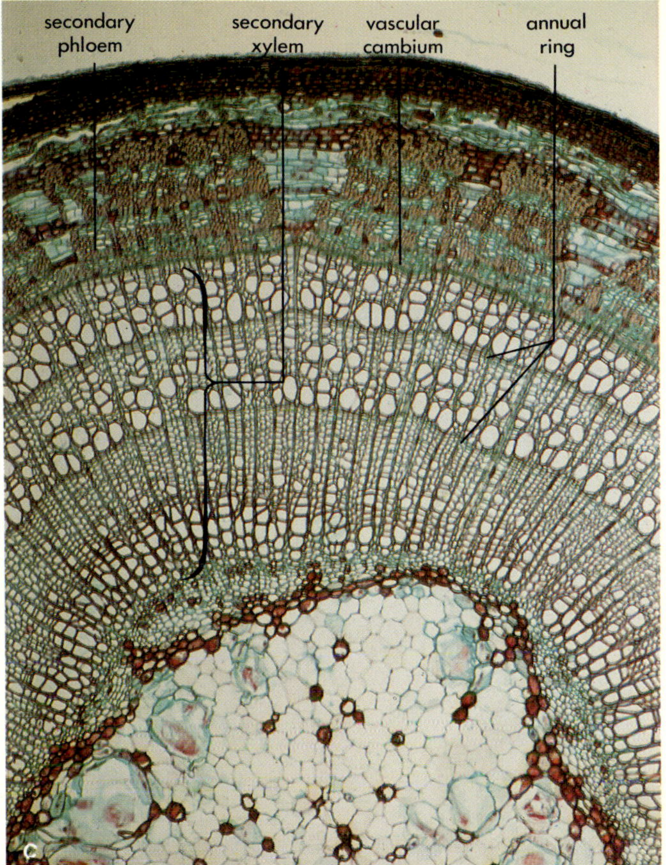

Figure 8.7
A, cross section of basswood stem *(Tilia americana)* through the secondary phloem. From the outside are layers of cork cells, followed by cortical parenchyma. *A*, wedge of parenchyma cells (a dilated ray) lies between two regions of active phloem cells. These cells consist of alternating layers of red thick-walled fibers, thin-walled sieve-tube members, and companion cells. *B*, cross section of basswood through secondary xylem showing an annual ring and portions of two others. *C*, lower magnification of a three-year-old basswood stem to show the annual rings and ring porous wood.

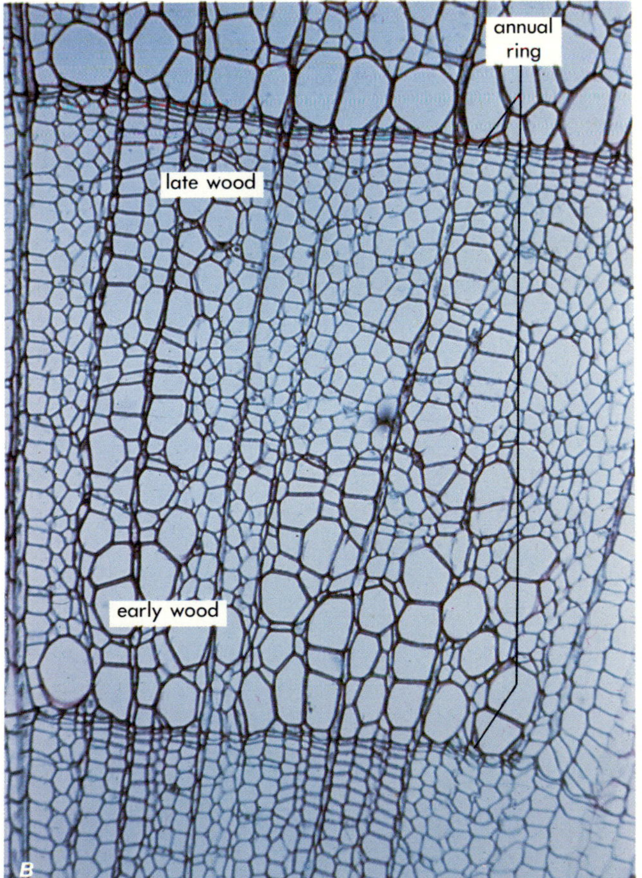

Figure 8.8
Maple *(Acer saccharum)* cross section of diffuse porous wood.

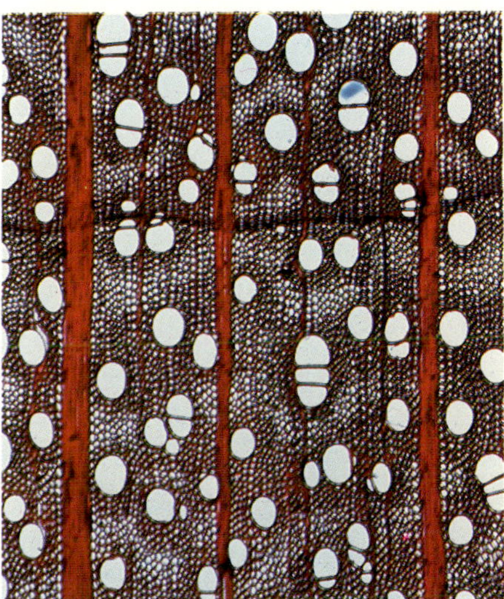

Figure 8.22
Arborescent monocotyledons. *A*, California fan palm *(Washington filifera)* has long-lived phloem but no secondary growth. *B*, the Joshua tree *(Yucca brevifolia)* has secondary growth but produces closed vascular bundles from meristematic tissue resembling procambium strands.

Figure 9.5
Specialized roots. *A*, banyan *(Ficus bengalensis)* tree showing an extensive adventitious root system. (The tree is growing in Honolulu). *B*, extensive adventitious root system of mangrove *(Rizophora mangle)*. Note the many air roots sticking up from the mud. (The tree is growing in the tidal zone of Australian tropical coast.) *C*, adventitious roots of *Ficus pumila* forming adhesive pad.

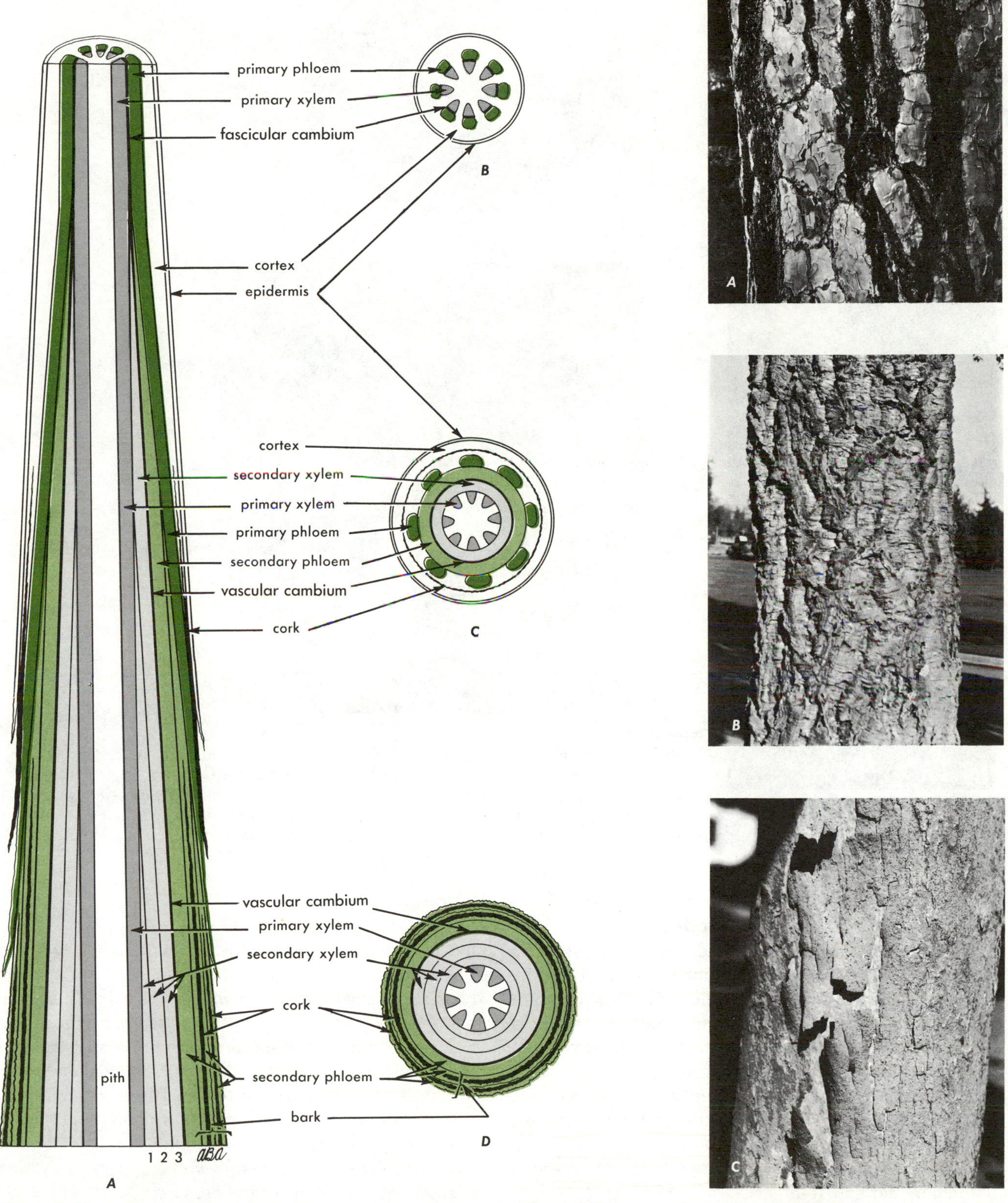

Figure 8.16
Changes in a stem as it increases in age and in girth. *A*, longitudinal section through a three-year-old stem. *B*, *C*, and *D*, cross section at indicated levels.

Figure 8.17
Bark of three species of trees. *A*, western yellow pine *(Pinus ponderosa)*. *B*, cork oak *(Quercus suber)*. *C*, plane tree *(Platanus orientalis)*.

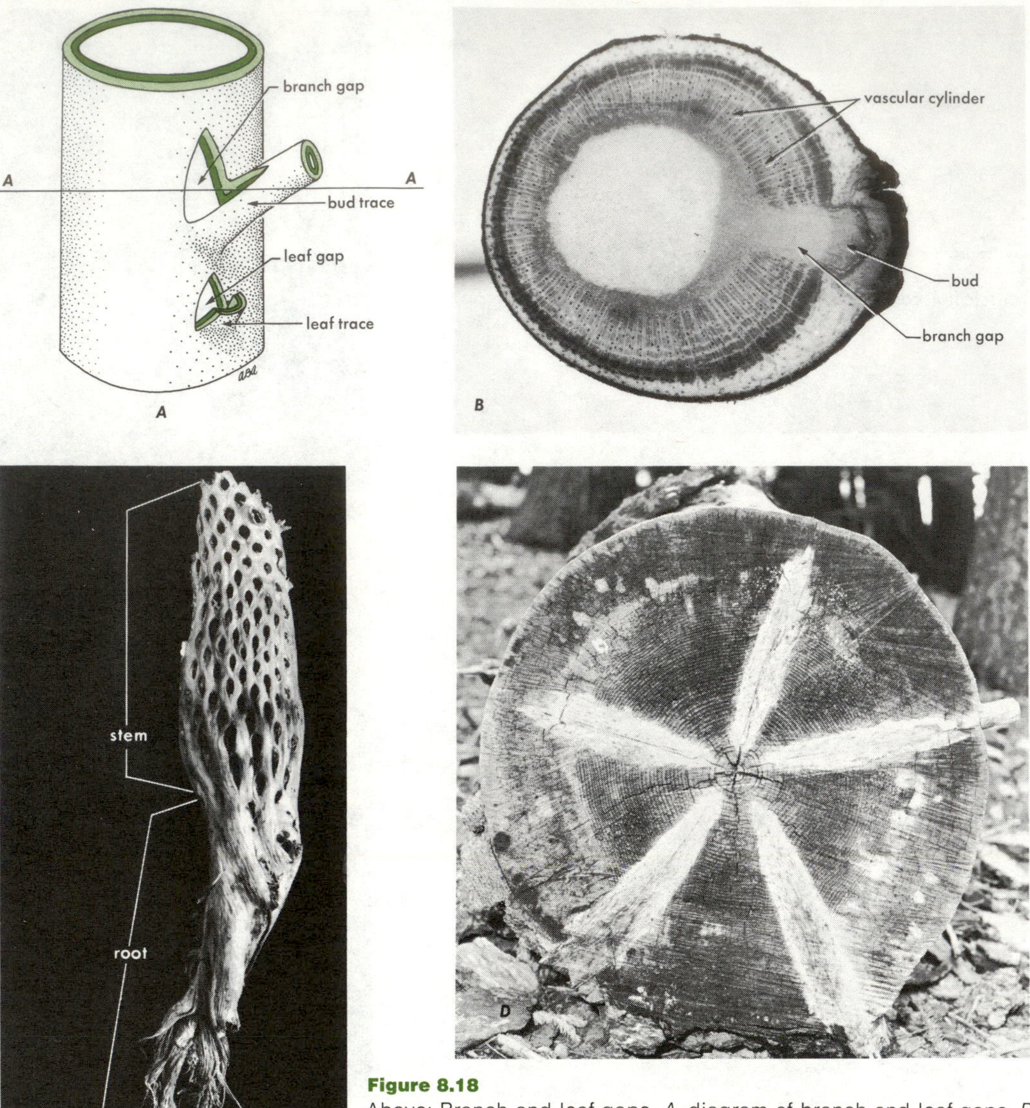

Figure 8.18
Above: Branch and leaf gaps. *A,* diagram of branch and leaf gaps. *B,* free-hand cross section of a twig of fig *(Ficus carica)* through bud and accompanying node. Gap in vascular cylinder composed of parenchyma cells is apparent, ×15. *C,* rotted cabbage *(Brassica oleraceae)* with only secondary xylem remaining, note the leaf and branch gaps in the stem, ×½. *D,* branches form a continuous pathway from the center of the trunk to the exterior, thus constituting a gap in the xylem cylinder of the trunk *(Pinus jeffreyi),* ×1/10.

Anatomy of the Stem at Nodes

The point of attachment of leaves on a stem is called the node. Since both buds and leaves are joined by vascular tissue to conducting tissue of the stem, anatomically the node must be characterized as a region where vascular bundles are connected to bring about a continuity of vascular tissue in leaves, buds, and stem. It is often a rather complex region. Bundles leading to leaves and buds are referred to as **leaf traces** and **bud traces,** respectively. Xylem of traces is continuous with xylem in leaf and stem, and phloem of the trace is continuous with phloem in leaf and stem.

In stems with secondary growth, the sites where leaf and branch traces diverge from the vascular cylinder of the stem, a gap exists immediately above the point where the trace departs. This is called a **leaf gap** or a **bud gap.** These are shown diagrammatically in Fig. 8.18*A.* Branch gaps are already present in twigs in the winter condition and they may be seen by cutting a free-hand cross section through a bud (the future branch) and its accompanying node. Figure 8.18*B* shows a free-hand cross section through a bud and node of a young branch of fig (*Ficus carica*). The section makes a nearly longitudinal cut through the bud (the future branch), and the continuity of vascular tissue of the bud with the vascular cylinder is apparent. The plane of the cross section of Fig. 8.18*B* is roughly indicated by the line *A-A* on Fig. 8.18*A.* Gaps are also prominent features of herbaceous stems after undergoing secondary growth. This condition is seen in

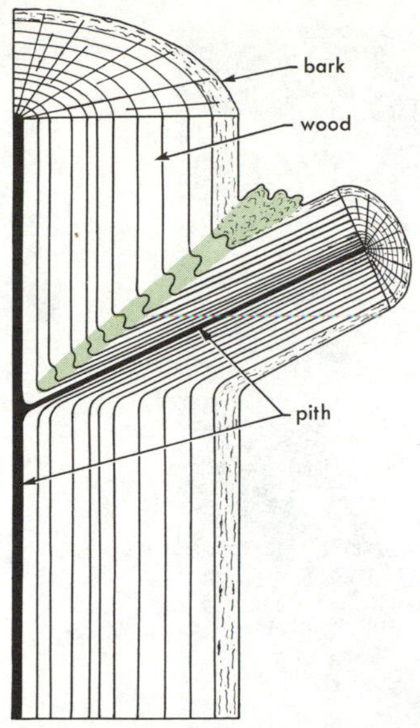

Figure 8.19
Relationship of wood and bark in main stem to wood and bark in a branch. (Redrawn from A. J. Eames and L. MacDaniels, *Introduction to Plant Anatomy*. © 1947 by McGraw-Hill Book Company. Reprinted with permission of the publisher.)

the picture of a cabbage stem (Fig. 8.18*C*) from which pith, phloem, and cortex have rotted away. The gaps originally composed of thin walled parenchyma cells are now holes through the xylem cylinder.

As secondary development of branch and trunk proceed, both will increase in diameter; new vascular tissue of the stem must be laid down in such a manner as to bury the younger portions of the branch in the trunk (Fig. 8.18*D*). Furthermore, there must be a crowding of new tissue in the upper acute angle formed by the branch and the stem. This is shown diagrammatically in Fig. 8.19. This entire process results in a knot. Can you explain why it is that in some gymnosperm lumber knots are held firmly in place, while in others they fall out?

SECONDARY GROWTH IN MONOCOTYLEDONOUS STEMS

Most monocot stems lack secondary growth. Palm trees increase their thickness in the apex by the activity of the primary thickening meristem (Fig. 7.26, **8.22*A*,** page 148). Further down the stem, parenchyma cells continue to divide and enlarge, allowing for continued lateral stem enlargement. This process is called **diffuse secondary growth** since it does not involve an actual lateral meristem. In some palms, the stem looks much thicker at the base than at the top of the stem, but this thickening is caused partly by the overlapping leaf bases of old leaves, and partly by the presence, in some palms, of great masses of adventitious aerial roots.

Other monocots like *Yucca, Agave, Cordyline, Sansevieria,* and *Dracaena* have true secondary growth that involves a secondary meristem **(Fig. 8.22*B*).** This secondary meristem forms below the primary thickening meristem and extends to the base of the plant. It divides and forms only parenchyma cells to the outside (secondary cortex). To the inside, it forms parenchyma cells and secondary vascular bundles (Fig. 8.23). The secondary vascular bundles consist of a ring of xylem surrounding the phloem, whereas the primary bundles consist of xylem and phloem side by side.

Secondary growth occurs in the roots of only one genus of monocot—*Dracaena.* A barklike layer is present in some monocots. It is derived from cortical parenchyma cells that divide to form files of suberin-filled parenchyma cells.

INDUSTRIAL PRODUCTS FROM TREES

The most obvious use of wood is as lumber. Trees, however, are grown in many countries for other purposes, some of which do not involve the destruction of the trees. For instance, turpentine, natural rubber, and chewing gum are derived, respectively, from the sap exudates of southern pine, particularly the long leaf pine (*Pinus palustris*) and the slash pine (*P. cubensis*); the rubber tree (*Hevea brasiliensis*); and the sapodilla tree (*Achras sapota*). In all three cases, gashes are cut in trunks of trees and sap is collected in buckets or on sticks, after which it is processed in factories into the well-known articles of commerce.

When wood itself is utilized, the trunk may be sawed into lumber or completely disintegrated into the cells of which it is composed. In manufacture of ordinary lumber, the log moves against the saw so that tangential cuts rip each board from the log. Large rays of oak wood give boards cut from these trees a characteristic pleasing appearance if they are cut in a radial plane (Fig. 8.20*B*).

Sheets for plywood manufacture are produced in a different manner. Logs are steamed and then placed on a large lathe and made to turn against a large and sharp knife which literally peels off thin sheets of wood from the slowly revolving log (Fig. 8.21). Note (Fig. 8.20) the different appearances that these processing procedures give to wood. Can you account for the "grain" in Fig. 8.21, in terms of spring and summer wood?

For other industrial uses, wood is not sawed into boards but placed into machines that tear it apart, breaking it down in some instances into fibers and tracheids to manufacture paper. Dacron shirts may have started as trees, although other cellulose products may have been the original raw material. In papermaking, lignin is undesirable and must be removed from the fibers and tracheids. It is a waste product, and huge piles of it have accumulated outside paper mills. Finding a use for it has been a major research project of a number of large paper companies. It is now being marketed for use as, among other things, a soil conditioner, an aid in drilling oil wells, and an adhesive in manufacture of plywood and linoleum.

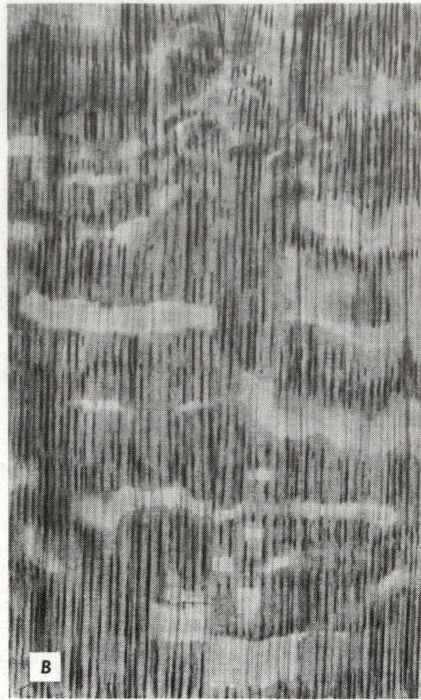

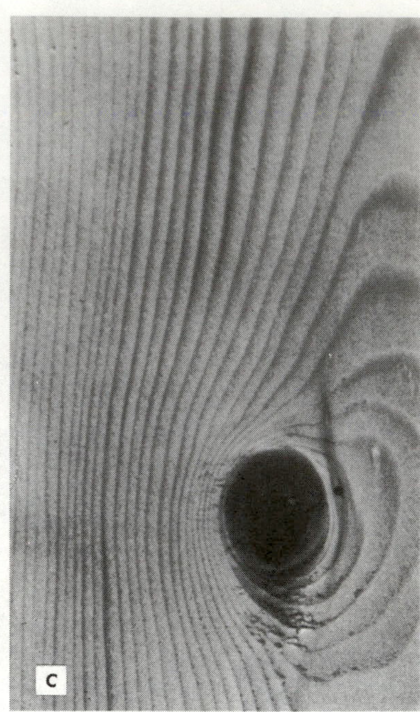

Figure 8.20
Characteristics of finished woods. *A,* plywood, a perfect tangential section. *B,* quarter-sawed oak, a radial section. *C,* knotty pine, ×¼.

Recycling newspring, magazines, and other paper products offers an excellent means for us to preserve our natural resource of trees. Unfortunately, the processes involved in recycling often result in the release of toxic inks and chemicals, originally used in paper manufacture, into the environment. Hence we have a paradox where a conservation measure, recycling paper, can result in a perhaps more serious problem, environmental pollution, unless careful handling and manufacturing practices are observed.

PRODUCTION OF OTHER MATERIALS BY STEMS

A great range of substances are produced in stems. A few of these may be mentioned.

Mucilaginous substances (complex carbohydrates very similar to gums) accumulate in stems of some plants, such as many ferns, cacti, and other succulents.

Tannins, substances that impart an astringent, bitter taste to tissues in which they are present, occur in bark and wood of many plants. They are derived commercially chiefly from bark of hemlock, tan oaks, mangrove, and wattle (certain *Acacia* species), and from wood of chestnut.

Latex is a milky secretion that occurs in many different families. It is a very complex mixture containing such materials as water, resin, oil, proteins, gums, tannins, sugars, alkaloids, and salts of calcium and magnesium. Latex may occur in special cells or in tubes **(latex tubes** or laticifers), which are distinct from the vascular system. Common latex-producing plants are the fig, dandelion, spurge, and milkweeds. Of particular economic importance is latex of those plants from which crude natural rubber is obtained. The most important rubber-yielding species are as follows: (*a*) *Para rubber tree* (*Hevea brasiliensis*), a native of the tropical forests of the Amazon and Orinoco river valleys in South America, but since 1929 grown in large plantations, chiefly in Ceylon, Malaya, Java, and Sumatra; (*b*) *Panama rubber tree* (*Castilloa elastica*), a native of Mexico and Central America; (*c*) *Manihot glaziovii,* a native tree of Brazil, yielding "ceara rubber"; (*d*) *guayule* (*Parthenium argentatum*), a native American shrub which, during World War II, was grown in large plantations in the western United States; and (*e*) *kok-saghyz,* or Russian dandelion (*Taraxacum kok-saghyz*), a plant grown quite extensively in Russia, and to some extent, experimentally, in the United States during the period of rubber shortage.

Resins are complex substances, secreted by glands into special ducts known as **resin ducts** (Fig. 8.12*B*). Crude resins contain, in addition to resinous materials, a considerable amount of essential oils. The principal product derived from resin is turpentine, obtained exclusively from coniferous trees.

STEM MODIFICATIONS

Stems may become adapted for functions other than support, conduction, and production of new growth. They may, for instance, become attachment organs for vines; they may carry on photosynthesis, store food or water, and develop protecting devices.

Rhizomes

A rhizome is a horizontal, underground stem. In most *Iris* species, leaves and flowering stalks are produced at the growing rhizome tip. The leaves die a relatively short distance back from the growing tip, so that many iris plants bear senescent leaves. Roots are also formed at nodes, and they may remain for the life of the rhizome. Other rhizomes, like *Canna* (Fig. 8.24*A*), produce upright leafy stems with terminal flowers at every third node. The intervening nodes are marked by only small sheath leaves.

Corms

A shortened vertical, thickened underground stem is a corm (Fig. 8.24*C*). In *Gladiolus,* it consists of a short stem with much stored food. Nodes are, as usual, indicated by leaves; some bases are shown in Fig. 8.24*B*. Small buds occur in axils of some of these leaves. In a median section of a corm (Fig. 8.24*C*) one can distinguish between stored food and the central portion containing a single bud that will produce a single leafy, flowering shoot.

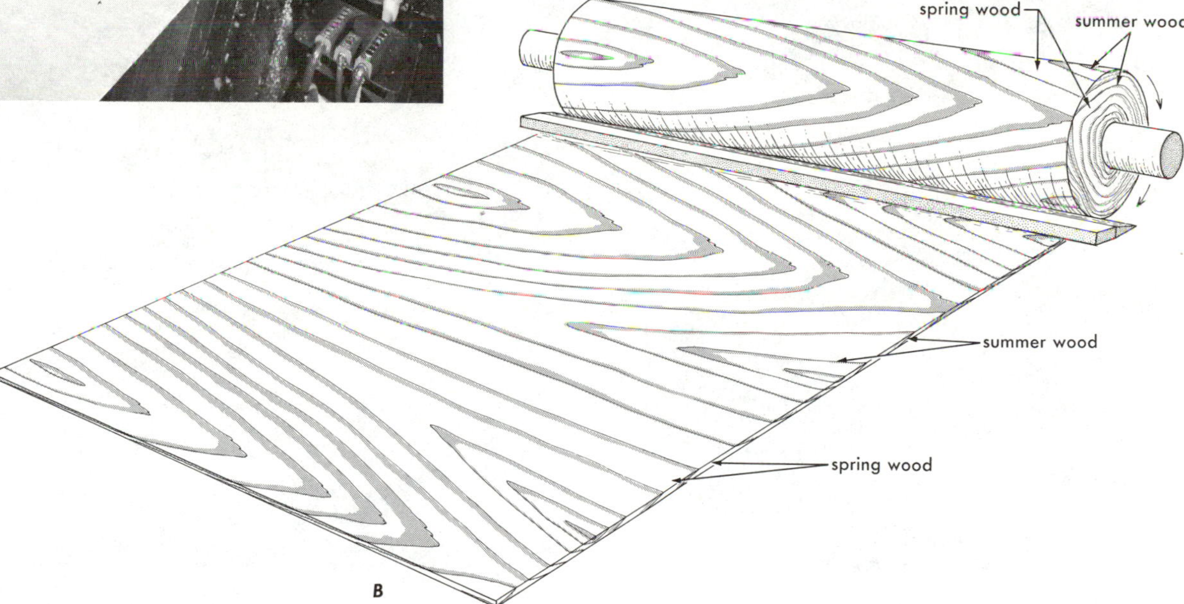

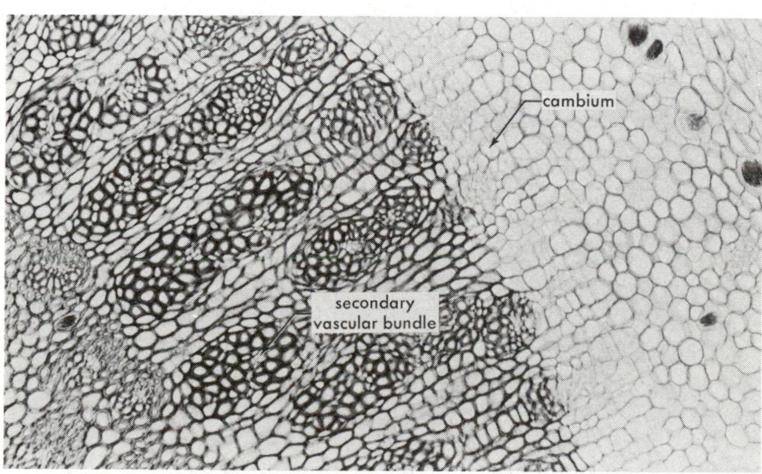

Figure 8.21
Manner of preparing plywood. *A,* sheet of veneer being peeled from a slowly revolving log. *B,* diagram showing relation of grain in plywood to annual growth rings of the log. (*A,* courtesy of American Forest Institute.)

Figure 8.23
Cross section of a portion of monocotyledonous stem, *Dracaena* sp., that has true secondary growth. Note that the cambium forms discrete secondary vascular bundles and parenchyma cells to the inside of the stem, and parenchyma cells only toward the outside, ×13.

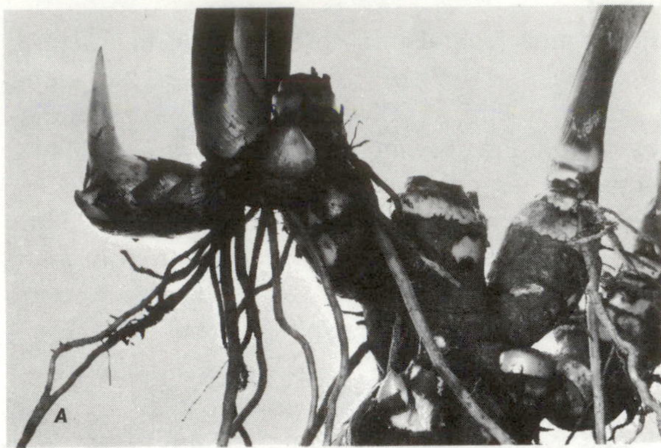

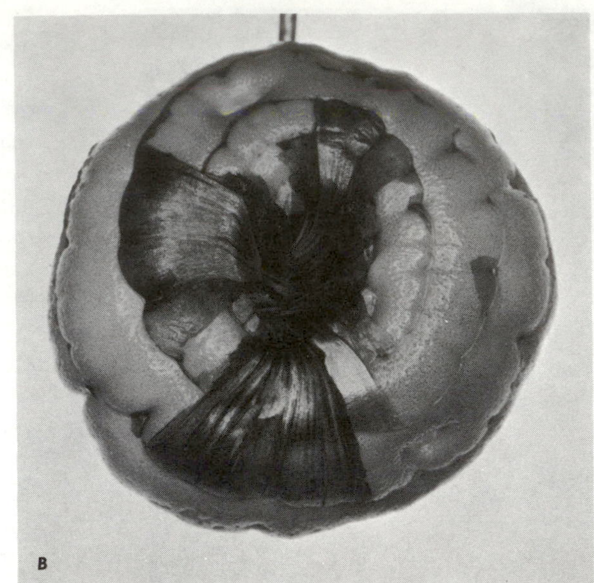

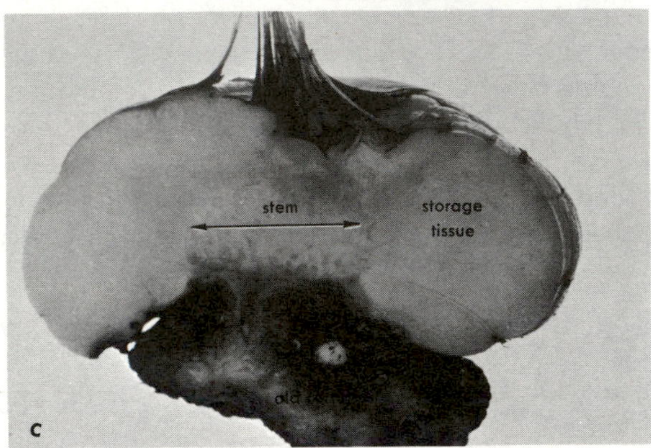

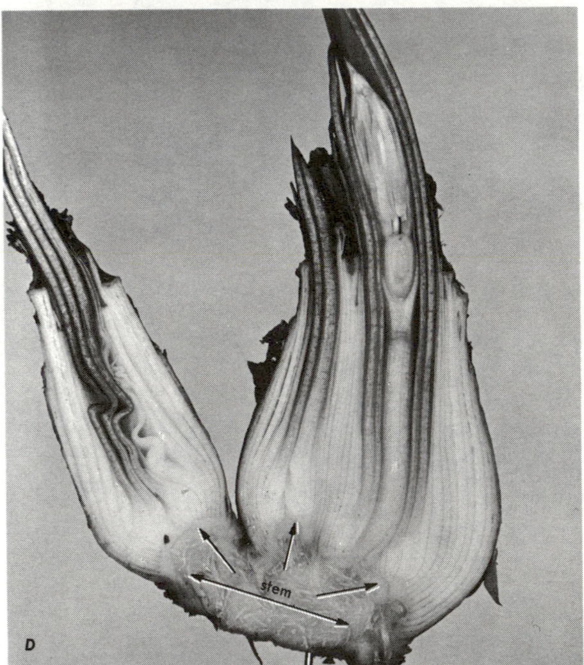

Figure 8.24
Plants illustrating continued development with only primary growth. *A,* roots and shoots are produced at every third node by the rhizome of *Canna,* ×½. *B,* corm of *Gladiolus,* ×1. *C,* corm sectioned to show short stem with storage tissue of current corm and disintegration of corm of preceding year, ×1. *D,* longitudinal section of young daffodil *(Narcissus)* plant, showing two bulbs, one with two shoots, united by the short stem, ×1. *E,* potato tuber *(Solanum),* note spiral arrangement of the eyes and the short sprout, ×1.

Food stored in a dormant corm is used in the production of the leafy shoot. New corms will develop from axillary buds. In addition, short underground stems may form, each giving rise, at its tip, to a single small corm.

Bulbs

A bulb differs from a corm in that food is stored in leafy scales. The stem portion is small and has at least one central terminal bud that will produce a single upright leafy stem. In addition, there is at least one axillary bud that will produce a bulb for the subsequent year. In the longitudinal section of the sprouting daffodil bulb shown in Fig. 8.24*D,* the stem is producing three leafy stalks, one of which is forming a new bulb.

Food stored in the leafy scales of a bulb is used up by the initial growth of a leafy shoot. Food to be stored for a new bulb is supplied from a leafy shoot. The table onion is a good example of a commercially valuable bulb.

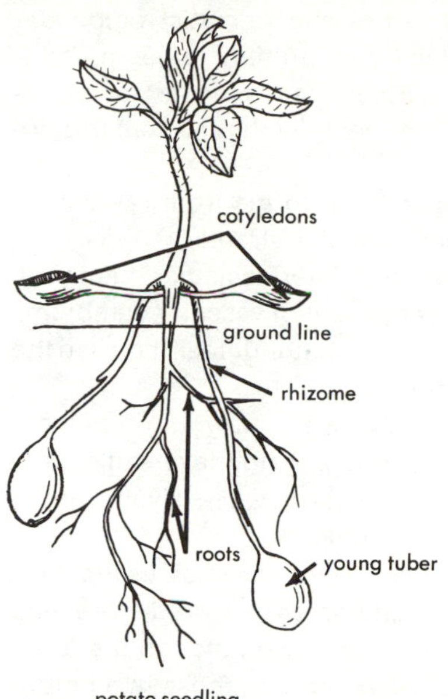

Figure 8.25
A potato seedling showing development of young tubers at the end of slender rhizomes.

Tubers

Tubers are enlarged terminal portions of slender rhizomes (Figs. 8.24*E*, 8.25). The potato, (*Solanum tuberosum*), is a good example. The potato plant possesses three types of stems: (*a*) ordinary aerial stems, (*b*) slender underground rhizomes, and (*c*) their enlarged tips, tubers. In the mature potato, the scar left where the tuber was broken from the rhizome is clearly visible. On the potato tuber there are nodes and internodes, lateral buds, and a terminal bud. Buds develop into shoots (Fig. 8.24*E*). The "eyes" of the tuber are groups of buds; each group along the sides represents a lateral branch with undeveloped internodes. At the unattached "seed end" of the tuber, the "eye" is in reality a terminal branch on which only one bud is strictly terminal. In an elongated potato it is possible to make out the spiral arrangement of the eyes, because there is only one eye at a node.

Stem Tendrils

Tendrils are slender, coiling structures that are sensitive to contact stimuli and attach the plant to a support. Tendrils are of two morphological sorts: leaf and stem. In the trumpet flower (*Bignonia*), for example, several uppermost pairs of leaflets have no blades, but instead form very slender leaf tendrils. In grapes (*Vitis*) and Virginia creepers (*Parthenocissus quinquefolia*), tendrils are modified stems (Fig. 8.26*B*), as evidenced by their presence at nodes in leaf axils.

In the Virginia creeper, each tendril ends in a knob that flattens out when it comes in contact with a surface to which it adheres.

Cladodes

These are stems that are leaflike in form, are green, and perform the functions of leaves. They may bear flowers, fruit, and temporary leaves. Examples of plants with cladodes are *Ruscus* (Fig. 8.26*D*), *Asparagus* (Fig. 8.26*E*), *Smilax*, various species of cacti, and some orchids (e.g., *Epidendrum*).

Spines and Thorns

Most spines and thorns of plants are modified stems or outgrowths of stems. Leaf spines, however, occur in certain plants, such as barberry (*Berberis*) and black locust (*Robinia pseudoacacia*), and in a few cases even roots become modified as spines. Good examples of stem thorns are those of fire-thorn (*Pyracantha*) (Fig. 8.26*F*) and honey locust (*Gleditsia*) (Fig. 8.26*G*). They are borne in the axils of leaves as ordinary branches are. Sometimes thorns bear leaves (Fig. 8.26*F*), which is further evidence that they are stems. Prickles, on the other hand, such as

Figure 8.26
Types of stem modifications. *A*, twining vine. *B*, stem tendril. *C*, rose stem prickles. *F*, and *G*, thorns. *D* and E, leaflike stems called cladodes (cladophylls).

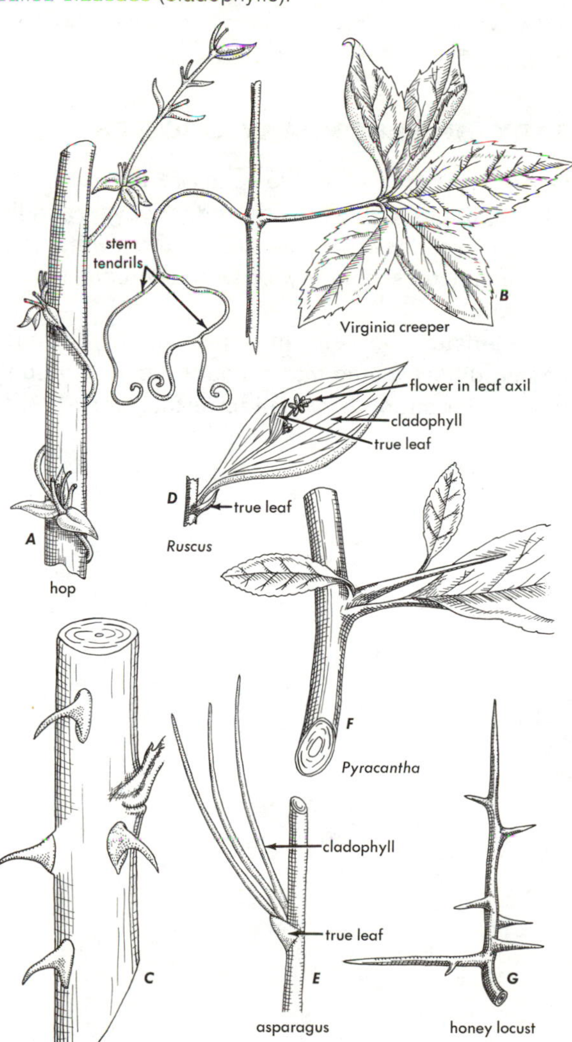

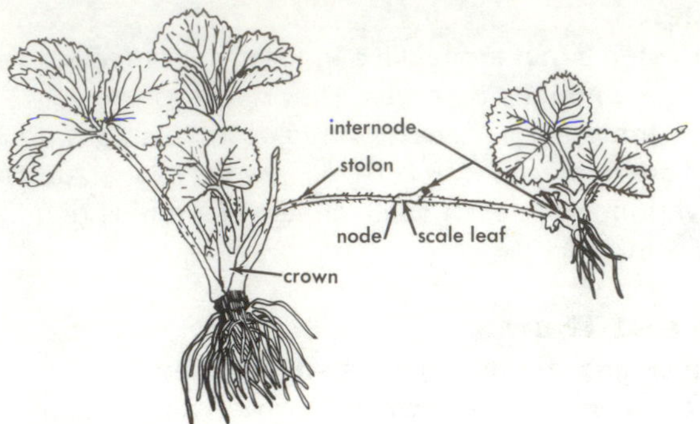

Figure 8.27
Runner of strawberry *(Fragaria)*, ×1/8; roots and shoots are produced at every other node.

those on rose stems, are merely epidermal outgrowths, somewhat like hairs.

Stolons

Bermuda grass (*Cynodon dactylon*) has above-ground horizontal stems called stolons. These stems creep along the ground, and at each node, shoots and roots arise. In strawberry (*Fragaria*) (Fig. 8.27) stolons, roots and leaves arise at every other node.

SUMMARY OF SECONDARY GROWTH

1. The predominantly short life span of 3 to 5 years for cells, in general, limits the size and longevity of individual plants having only primary growth. The mechanism of secondary growth overcomes this limitation.
2. If all meristematic cells in a procambium strand become differentiated into primary vascular tissue, the vascular strand is closed to further growth.
3. If there remains an active meristematic region between primary xylem and primary phloem, continued growth is possible. These meristematic cells (residual procambium) become the fascicular cambium.
4. An interfascicular cambium forms from ray parenchyma cells between vascular strands.
5. Union of fascicular and interfascicular cambium produces a complete cylinder of vascular cambium.
6. Divisions in cambium are longitudinal, so that the stem now increases only in girth.
7. Cork cambium is short-lived; new cork cambia may arise each year, producing new layers of cork.
8. Cork cambium may originate in successively deeper tissues from secondary phloem.
9. The production of secondary vascular tissue thus makes possible the attainment of great size and great age by individual trees, even though the functioning cells may be no older than those in a perennial herbaceous plant such as an *Iris*.
10. Most monocot stems have only primary growth that arises from a special primary thickening meristem. This meristem causes increase in both height and girth.
11. Palms supplement their lateral growth by the division of parenchyma cells throughout the stem.
12. Other monocots, such as *Agave* and *Sansevieria*, have true secondary growth, with a type of cambium that produces secondary vascular bundles and parenchyma cells.
13. In woody and herbaceous stems, vascular tissues are arranged in bundles, generally forming a circle. In monocotyledonous plants, bundles are irregularly distributed throughout the stem.
14. Rhizomes, bulbs, corms, stolons, and tubers are all modified stems.

CHAPTER 9

ROOTS

FUNCTIONS OF ROOTS

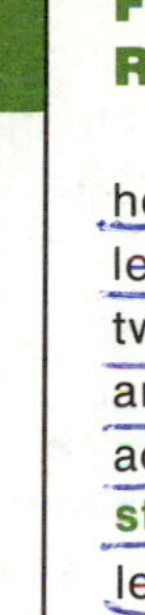

The roots of a plant, considered collectively, form the **root system.** The two principal functions of the roots are **anchorage** and **absorption.** In addition, roots of all plants usually **store** a certain amount of food, at least for a short time. Roots of such plants as sugar beet, carrot, sweet potato, and others are specialized food storage organs. Roots also perform the function of **conduction.** Water and mineral salts absorbed from soil, and foods that may be stored in roots, are conducted to stems and then to leaves and other organs above ground. Foods manufactured in leaves are conducted by stems to main roots, and then by the latter to branch roots, so that these foods are carried to the extremities (growing tissues) of all the smallest roots.

You need only walk through a stream bed and observe the exposed root systems of large trees, or attempt to pull weeds, to get a first-hand idea of the anchorage function of roots (Fig. 9.1). Plants with unusual roots provide anchorage in atypical ways. In ivy (*Hedera helix*) and in some species of fig (*Ficus pumila*) roots develop on stems to form an unusual adhesive pad allowing these vines to cling to vertical surfaces. Parasitic plants, like dodder, anchor themselves by sinking haustorial roots into their host's vascular tissue and thus tap its water and nutrient supply.

In a normal, healthy plant there is a balance between shoot system and root system. Of particular importance is the relation of total leaf surface to total root surface (Fig. 9.4). There is a balance between total surface ex-

Figure 9.1
Exposed roots of white elm.

posed to the sun, from which energy is absorbed and used in the manufacture of carbohydrates, and total root surface in contact with the soil solution, from which the plant absorbs water and mineral nutrients. The root system must be able to supply the shoot with sufficient water and mineral nutrients, and the shoot system must manufacture enough food for maintenance of the root system.

All roots, even slender ones whose primary function is absorption, may have a small amount of food stored temporarily in them. For example, when sugar moves into roots more rapidly than it can be utilized by growing cells, it may be converted to starch and as such be stored for a time, particularly in cortical cells. During the dormant season, rather large quantities of starch are stored in woody roots of orchard trees. This food constitutes a reserve that is called upon when flowering and active growth is resumed in spring. The roots of the wild morning glory (bindweed) and other perennials store large quantities of food. This stored supply enables the plant to send up new shoots when the "tops" of the plants are destroyed. The seasonal trend in storage of foods in roots of wild morning glory is seen in Fig. 9.2. It will be noted that, in undisturbed plants, readily available carbohydrates rapidly build up during summer, whereas, in plants cultivated at 2-week intervals, food storage is greatly hindered.

Food storage in roots may occur in cortex, phloem, and xylem. Among our native plants, the most striking examples of fleshy-rooted plants occur in arid regions. Such roots generally contain a large quantity of stored water that can be used by leafy shoots during periods of drought.

In a herbacious plant, water is responsible for about 95% of the plant's weight. Water is needed for all root processes and for every metabolic reaction. Soil water contains dissolved salts and minerals, such as potassium, sulfur, phosphorus, calcium, and magnesium,

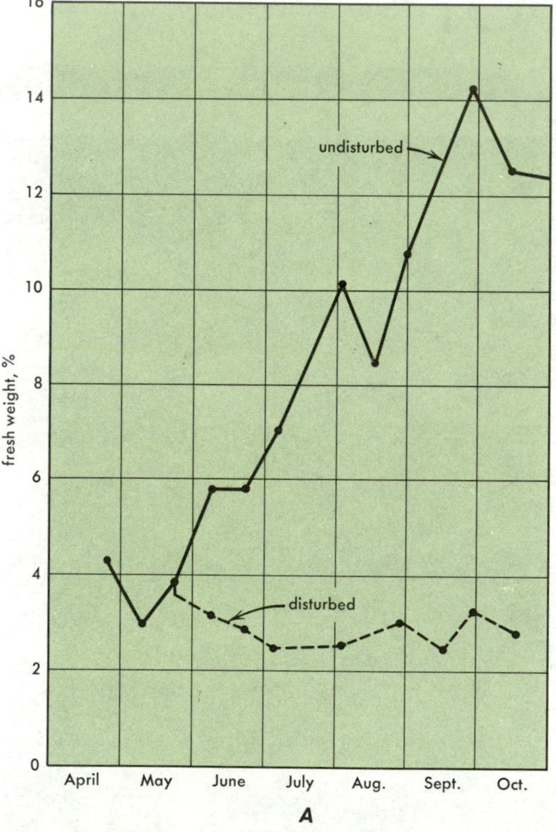

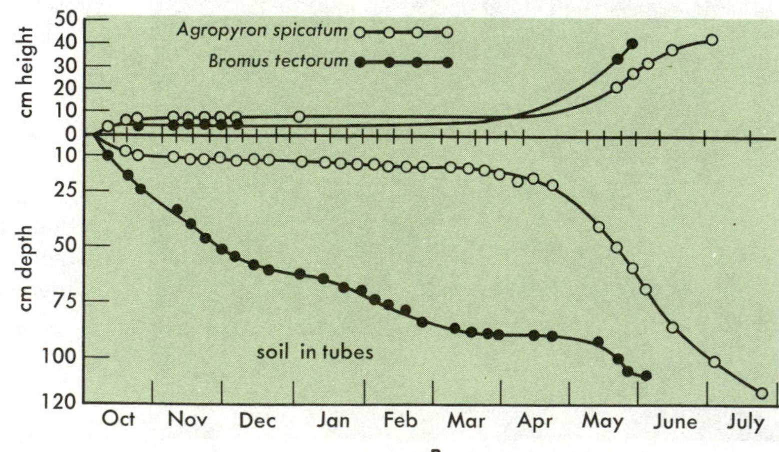

Figure 9.2
A, seasonal trend in the storage of reserve carbohydrates in morning glory *(Convolvulus arvensis)* (from Barr). *B,* seasonal leaf and root growth of *Agropyron spicatum* and *Bromus tectorum* in glass tubes in the field. (Redrawn from G. A. Harris, *Ecological Monographs,* **37,** 89. © 1967 by Ecological Society of America.)

which are needed by the plant (Chapter 11). For these reasons, plants have developed an elaborate system for the absorption and conduction of water (Chapter 12). Root hairs penetrate between small soil particles and are in intimate contact with them and the water film that surrounds them (Fig. 9-7). The cell walls of the root hairs and epidermal cells are made of cellulose and pectin. The pectic coat on the outside of the cell wall makes the root hairs gummy, facilitating their adherence to soil particles. Water is absorbed by the root hairs and epidermal cells and is then passed on by diffusion and active selective processes to the conducting elements.

TYPES OF ROOT SYSTEMS

When pulled up, many grasses and small garden plants take with them a massive clump of soil. This happens because the **fibrous root system** of these plants consists of several main roots that branch to form a dense mass of intermeshed lateral roots (Fig. 9.3*A*,*B*).

Vegetables with a large storage root, like the carrot, have a **tap root system** that consists of one main root from which lateral roots radiate (Fig. 9.3*C*, *E*). Some desert plants have a rapidly growing tap root system that enables them to penetrate the soil quickly to reach deep sources of water. The depth of root penetration of various plants is quite remarkable. A University of Nebraska botanist, John Weaver, exhumed the entire root systems of 45 species of native plants from the Midwest prairie. Only four species maintained root systems primarily in topsoil; most of them had roots that extended to 4 m deep. Even in very hard soils, the depth of root penetration well exceeded the height of the above-ground plant. A typical annual plant, like corn or rye, will build an immense fibrous root system in one growing season. In a

Table 9.1
RYE ROOT SYSTEM MEASUREMENTS

Root	Number	Length (m)
Main roots	143	65
Secondary roots	35,600	5,181
Tertiary roots	2,300,000	174,947
Quarternary roots	11,500,000	441,938
Total	14,000,000	609,570

study on rye, a single plant 50 cm tall, with 80 tillers (shoot branches), had a root system amounting to 210 m^2 of surface area compared to only 4 to 5 m^2 of the above-ground portion (Table 9.1).

In trees and shrubs, most functioning roots are localized in the upper 1 m of soil, with the majority of the feeder roots in the upper 15 cm (Fig. 9.4). However, these plants may extend lateral roots well beyond the expansion of overhead branches. In areas of closely packed trees, or on sandy soils, competition and low surface moisture may reduce the amount of area covered and encourage deeper root penetration.

ROOT SYSTEMS AND PLANT COMPETITION

Where rainfall and temperatures permit, plants of the same and different species grow close together and

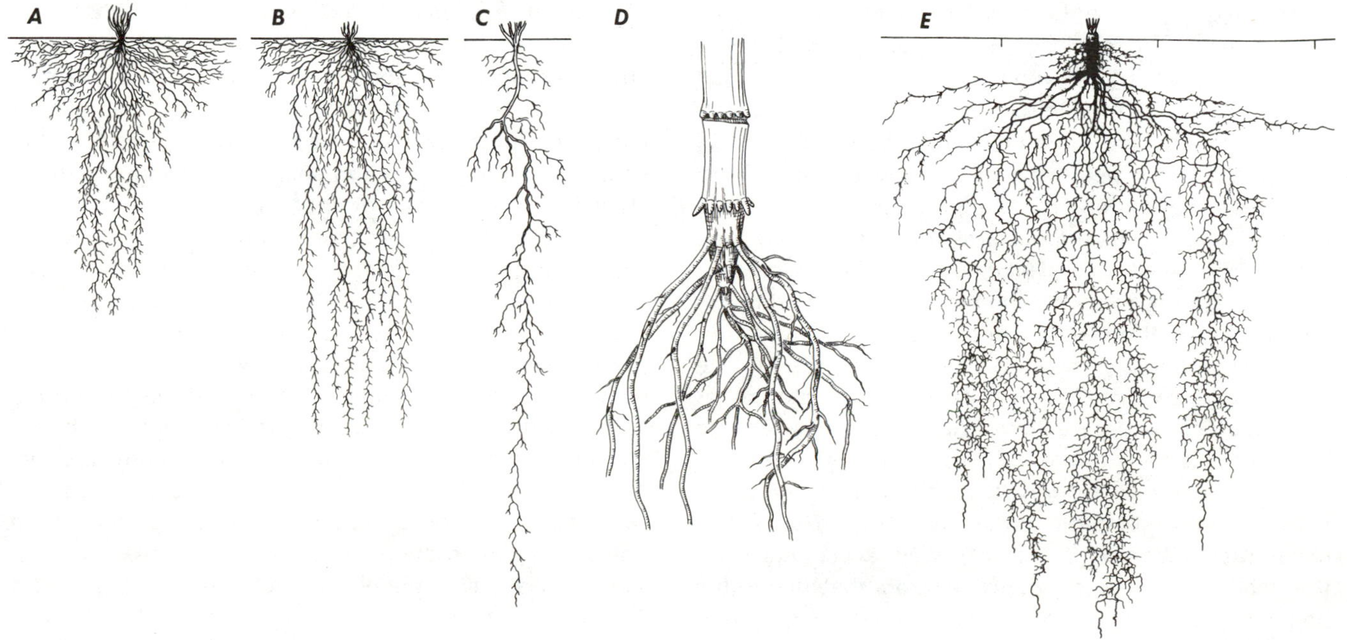

Figure 9.3
Different types of root systems. *A*, shallow, spreading, fibrous root system. *B*, fibrous root system penetrating the soil evenly from 1 to 1.5 m. *C*, tap root system, in which main primary root penetrates soil 2.5 m or more. *D*, fibrous root system developed from adventitious roots growing at lower nodes of stem. *E*, tap root system in carrot *(Daucus carota)*.

Figure 9.4
Root system of Cox's Orange Pippin on M. II, excavated when 16 years old from brick-earth soil. Note main scaffolding of roots about 25 cm deep and vertical "sinkers." (From Rogers and Head in W. S. Whittington, *Root Growth*. Plenum Press, N.Y. also courtesy of East Malling Research Station).

compete for available soil water, mineral nutrients, and light energy. As will be discussed in Chapter 18, plants reduce competition by utilizing different parts of the environment: Some utilize full sunlight, while others grow normally in shade; some are ephemeral and complete their life cycle only during periods of abundant moisture, while others are perennial and grow slowly even during periods of water shortage. In addition, as documented by the American ecologist J. E. Weaver, root systems of different species occupy and utilize different parts of the soil profile.

Weaver and his students carefully excavated root systems of hundreds of prairie plants and concluded that there were three general categories of rooting depths. As shown in Fig. 9.3, some grassland species, like blue grama (*Bouteloua gracilis*), possess a very shallow root system. Most of the roots are within the top 15 cm of soil. Another group of species is like buffalo grass (*Buchloe dactyloides*), having an even distribution of roots to a soil depth of about 1.5 m. A third group is like loco weed (*Oxytropis lamberti*), having an even distribution of roots to soil depths much below 1.5 m. Loco weed happens to have a tap root system, but there are some grasses with fibrous root systems that also fall into the last category. These three categories of plants, then, reduce competition for moisture when growing together by utilizing different regions of the soil profile.

Competition for moisture may be the reason one species is replacing another in the northern intermountain region of the United States. About 125 years ago, the dominant plant of this grassland area was bluebunch wheatgrass (*Agropyron spicatum*), a perennial. About the middle of the nineteenth century, an annual grass called cheatgrass (*Bromus tectorum*) was accidentally introduced to the area from Europe. From that time to the present, graziers have noticed an enormous increase in abundance of cheatgrass and an equally impressive decrease in the abundance of bluebunch wheatgrass. What caused the shift?

Both species have a similar life cycle. They germinate (or break dormancy if perennial) in fall, grow slowly during winter, grow rapidly in spring, form flowers in early summer, and die in June (or begin dormancy in mid-July if perennial). G. Harris of Washington State University studied growth and survival of these two species during a year, starting from seed for both. He found that the presence of cheatgrass greatly reduced growth and survival of bluebunch wheatgrass. In one field trial, bluebunch wheatgrass and cheatgrass were sown together in two plots; in one plot the amount of cheatgrass sown was much higher than in the other. After a normal growth period (October–June), the number and weight of wheatgrass plants in each plot were tabulated. Table 9.2 shows that survival and plant weight was much lower when cheatgrass was more abundant than when it was scarce.

Harris then planted seeds of each species in long glass tubes filled with soil. The tubes were inserted in the field, level with surrounding soil. Every month the tubes were lifted and depth of rooting was measured. The graph in Fig. 9.2*B* shows that root growth during winter was much greater for cheatgrass than for wheatgrass. At the start of rapid spring growth in April, cheatgrass roots had penetrated 90 cm, in contrast to 20 cm for wheatgrass. This difference in depth allowed cheatgrass to absorb water from a much greater part of the soil profile than wheatgrass could. Consequently, when the upper soil became dry in early summer (with both species, drawing water from it), only wheatgrass suffered, for water still remained available at greater depths (where only cheatgrass roots penetrated). Harris measured soil water availability in the two soil regions at this time and found water potential was below −15 atmospheres in the upper region (near the wilting point of many species), but was only −1 atmosphere in the lower region (essentially still saturated).

The recent success of cheatgrass, then, seems due to its winter root growth, which gives it a spring and early summer advantage over wheatgrass that started from seed at the same time. Cheatgrass each year increases in density, and each year this results in greater competition for moisture in the upper soil and greater stress on wheatgrass.

DEVELOPMENT OF ROOT SYSTEMS

The seeds of higher plants contain a small undeveloped plant, the embryo. When the seed germinates, the embryonic root or **radicle,** extends by the division and elongations of its cell to form the **primary root** (Fig. 9.7). Tap root systems develop from only one enlarged primary root, which then forms **lateral** or **secondary** roots. Further branching of secondary roots give rise to succeeding orders of roots, for example, tertiary and quarternary. Fibrous root systems develop in a different way. The embryo of some grasses, like barley, consists of not only one radicle, but also of several additional embryonic roots called **seminal (seed) roots.** The seminal roots emerge soon after the radicle during germination. They elongate rapidly, and soon it is not possible to distinguish the primary root. Usually the seminal roots do not persist, but in some grasses they may function throughout the plant's life.

Roots that develop from organs other than roots are called **adventitious roots.** In many instances, adventitious roots develop at the nodes of the stem. There are several common examples of adventitious roots. In a young corn plant, soon after the emergence and development of a rudimentary root system, **prop roots** develop from the shoot node nearest the soil level (Fig. 9.3*D*). These prop roots function as roots, but also assist in the support of the plant in the soil. Banyan (*Ficus bengalensis*) trees grow in saline mud in tropical lagoons and tidal marshes. Branches of these trees form adventitious prop roots that extend down into the soil where they enlarge and actually prop up the large branches (**Fig. 9.5*A*,** page 148). The banyan is considered a sacred tree in some parts of India. Indian merchants have in the past held open-air bazaars among the prop roots and expansive branches of the banyan.

Table 9.2
SURVIVAL AND PLANT WEIGHT OF BLUEBUNCH WHEATGRASS AFTER A SEASON'S GROWTH IN TWO LEVELS OF CHEATGRASS COMPETITION

Relative Cheatgrass Competition	Survival of Wheatgrass, %	Average Dry Weight of Wheatgrass, g
Low	86	10.4
High	39	1.8

Aerial prop roots are produced by many tropical trees. One of these, the common mangrove (*Rizophora mangle*) (Fig. **9.5*B***), is native to low tidal shores and marshes in tropical and subtropical regions. Aerial roots arising from trunks and lateral branches may form impenetrable thickets. Note in **Figure 9.5*B*** the small air roots extending up from the mud. These air roots (*pneumatophores*), like the "knees" of bald cypress (*Taxodium distichum*) (Fig. 11.6) increase oxygen availability to the roots submerged in the swamp.

In some instances, adventitious roots do not form at nodes. In *Ficus pumila,* a climbing fig, roots develop in clusters on stem internodes. The root cluster becomes flattened against a surface forming a flat adhesive pad **(Fig. 9.5C).** The root hairs that grow on them may be involved in secreting a sticky substance that causes the root cluster to cling.

The split leaf philodendron (*Monstera deliciosa*) is another example of a plant with adventitious aerial roots. In *Monstera* the aerial roots emerge from stems and grow straight out from it. After a time, the aerial roots bend downward and eventually touch the soil. *Monstera* aerial roots never branch unless they have been seriously wounded, or when they touch and penetrate the soil.

PROPAGATION OF ROOTS AND BY ROOTS

Pieces of stem, like a cane from a blackberry plant, or a branch of willow, can be made to root from their cut ends simply by placing them in moist soil. Leaves from *Begonia* and several other species also can be rooted simply by soaking them in water. Many commercially important ornamentals, in fact, are reproduced by root propagation from the leaves or stems (Chapter 15).

Ornamental crab apple, cherry, plum, quince, hawthorn, and a number of other plants are usually propagated by root cuttings. Shoots that arise from cuttings are from adventitious buds (Fig. 9.6). Injury to roots may induce development of such buds. The tap root of a dandelion may be cut into many small pieces, and each section capable of producing new shoots.

ANATOMY OF ROOTS

Accompanying similarities and dissimilarities in the functions of roots and stems are corresponding variations in structure. We shall find, in the root, meristematic and vascular tissues similar to those found in the stem. Both primary and secondary growth occur. There are three tissues in the root (root cap, endodermis, and pericycle) that are not present in the stem, and the arrangement of vascular and meristematic tissues is slightly different. A cuticle often does not develop on root epidermal cells. Instead, some root epidermal cells become hair cells and facilitate absorption (Fig. 9.7). The general

Figure 9.6
Although roots do not ordinarily produce buds, adventitious stems do form on some roots as in *Rumex acetosella*.

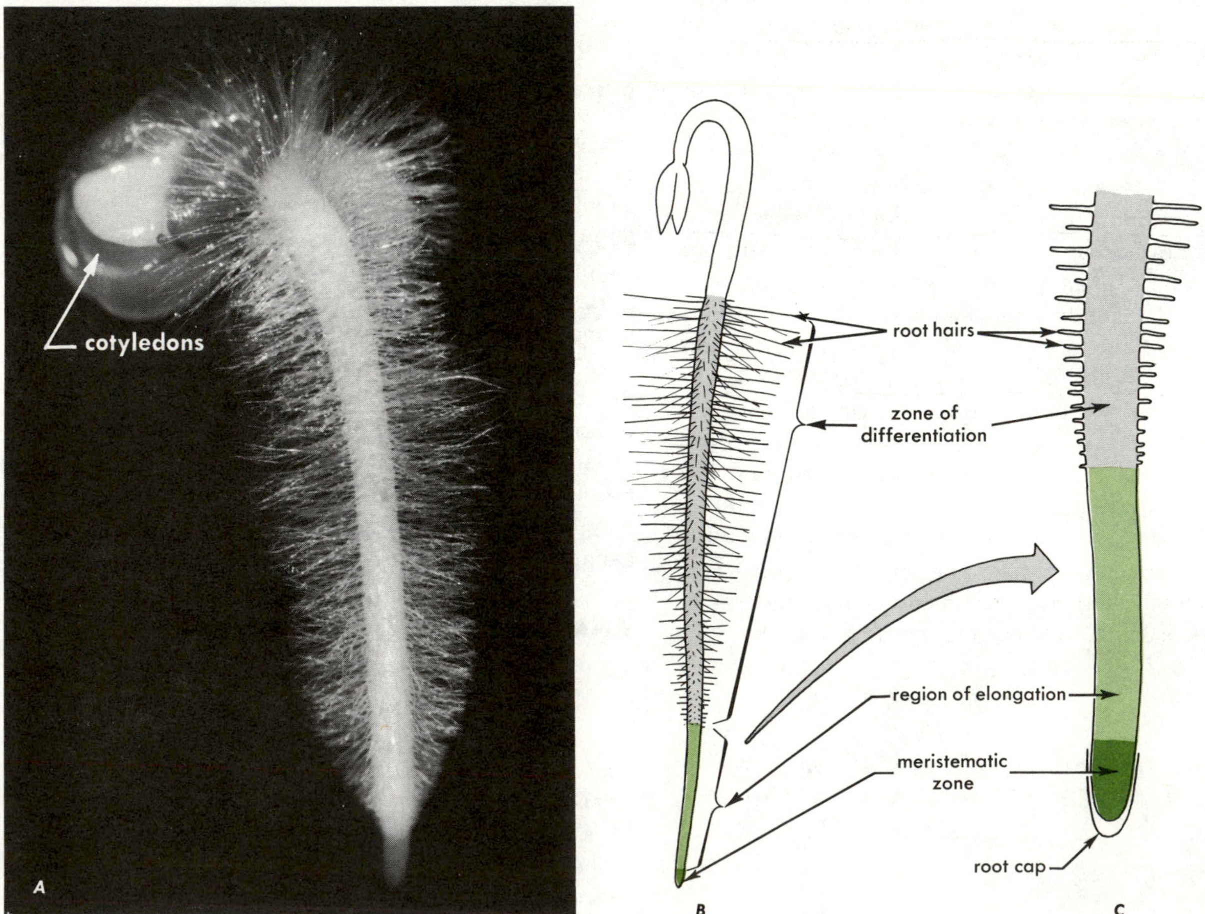

Figure 9.7
Radish seedling *(Raphanus sativus)*. *A*, photograph showing root hair growth, ×2. *B* and *C*, diagrams showing growth regions.

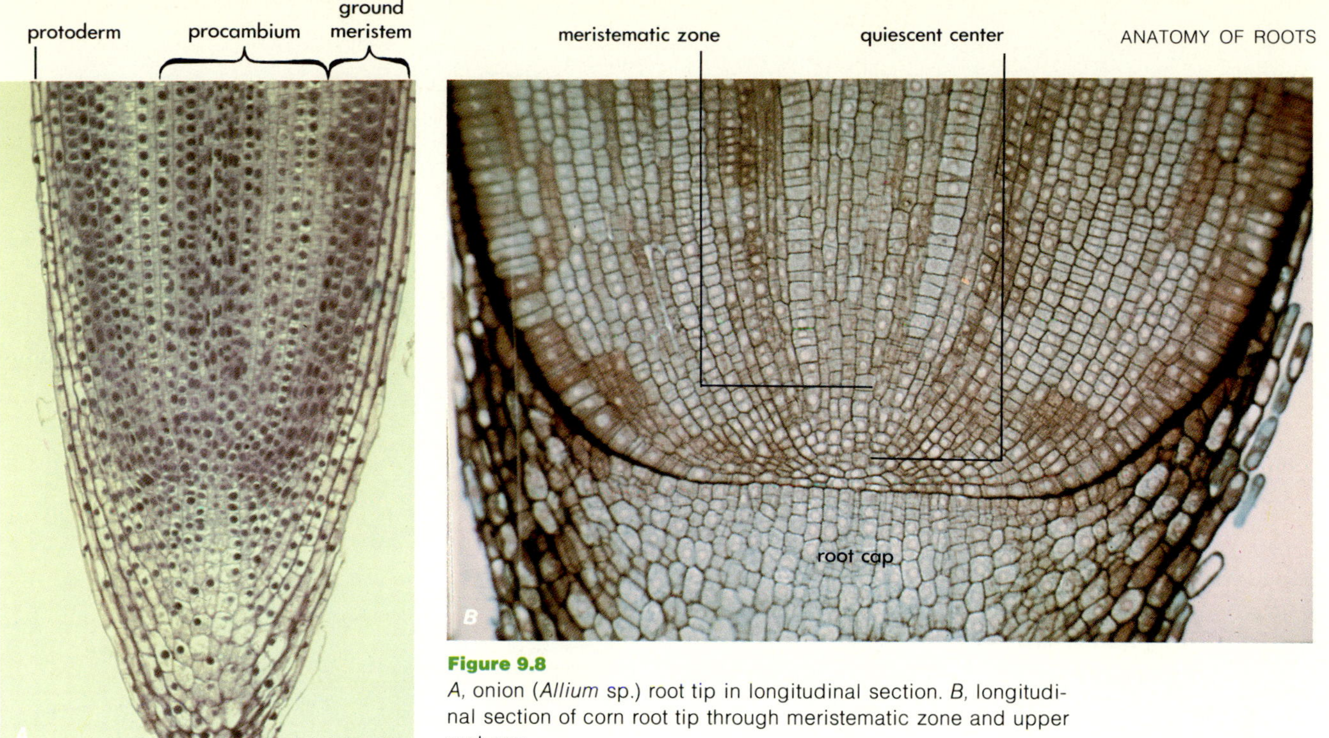

Figure 9.8
A, onion (*Allium* sp.) root tip in longitudinal section. *B*, longitudinal section of corn root tip through meristematic zone and upper root cap.

Figure 9.9
Diagram of a median longitudinal section through the meristem of a hypothetical root. The quiescent center cells divide very slowly in the center and more rapidly near the periphery. All three primary meristems (procambium, ground meristem, and protoderm) plus the root cap are derived from the apical initial cell region.

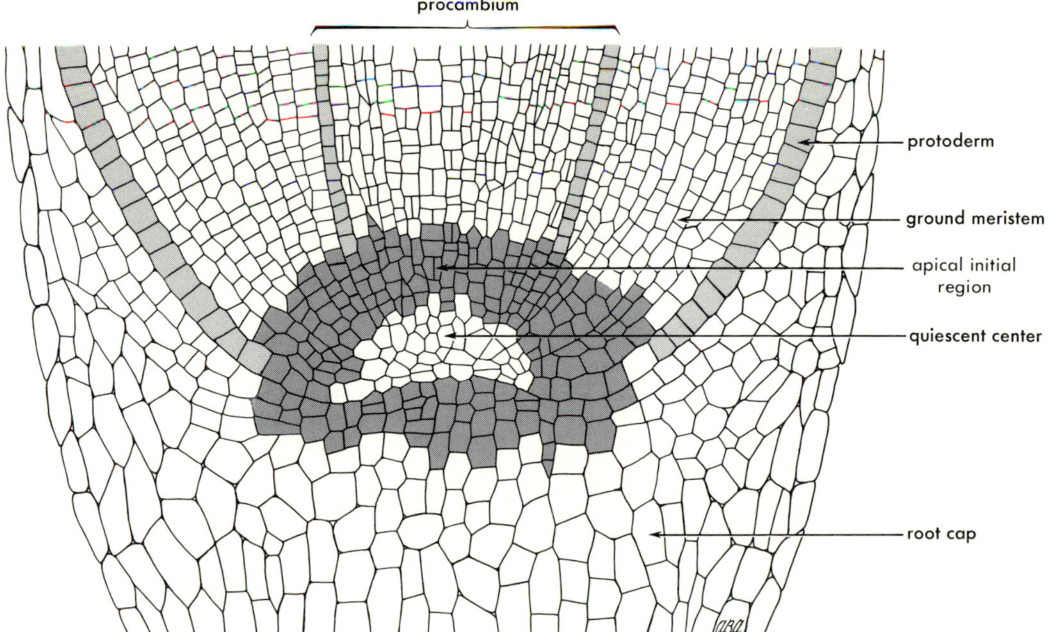

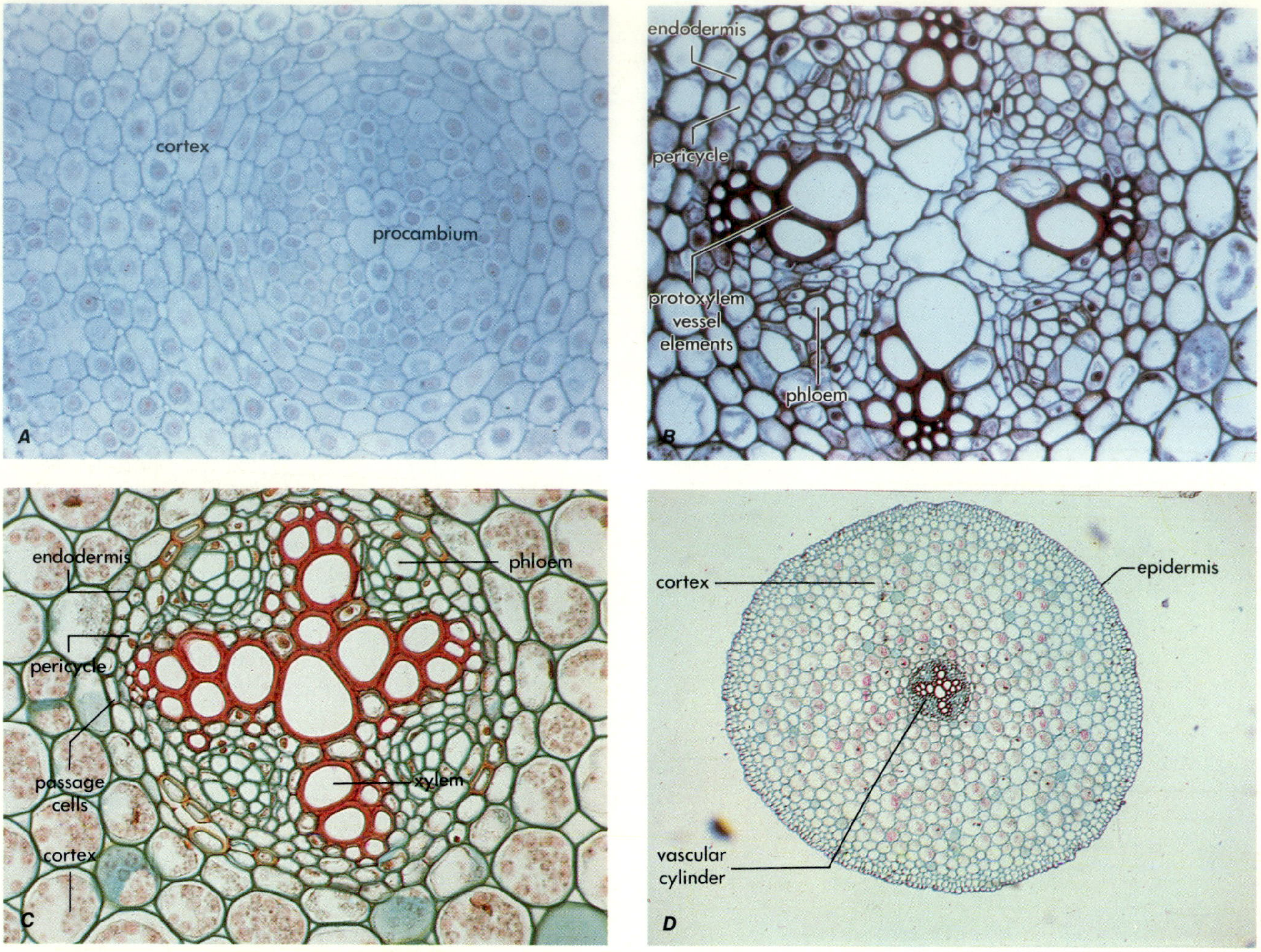

Figure 9.16

Buttercup *(Ranunculus* sp.) roots showing the stages of differentiation in primary growth. *A*, immature section taken from near the meristematic region. The layers of tissues and regions can already be seen. The metaxylem and protoxylem cells in the central stele are clearly formed. *B*, the cells of the protoxylem have developed secondary walls (shown here stained red with safranin). This particular root is tetrarch with four protoxylem points with phloem between. *C*, fully mature root with all primary tissues differentiated. Note that the endodermis cells adjacent to the protoxylem poles are lacking secondary walls (passage cells). *D*, this is the same root section as *C*, but shows all tissues.

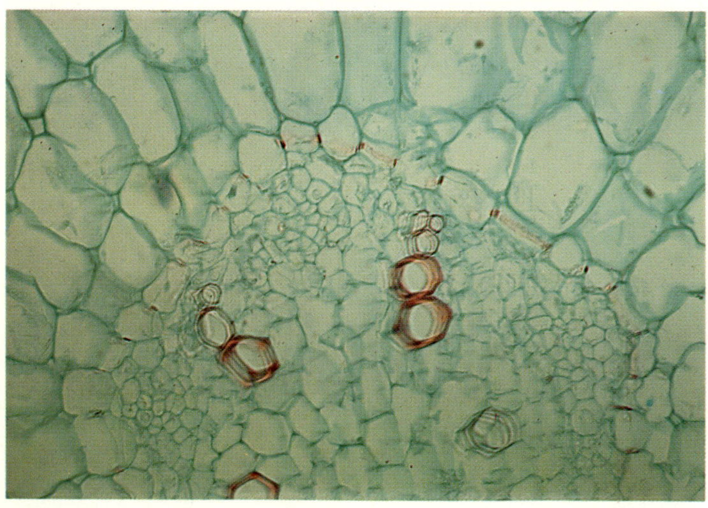

Figure 9.14

Transverse section of a *Nymphoides* sp. root to show the Casparian strip within endodermal cell walls. Note the red-stained Casparian strip.

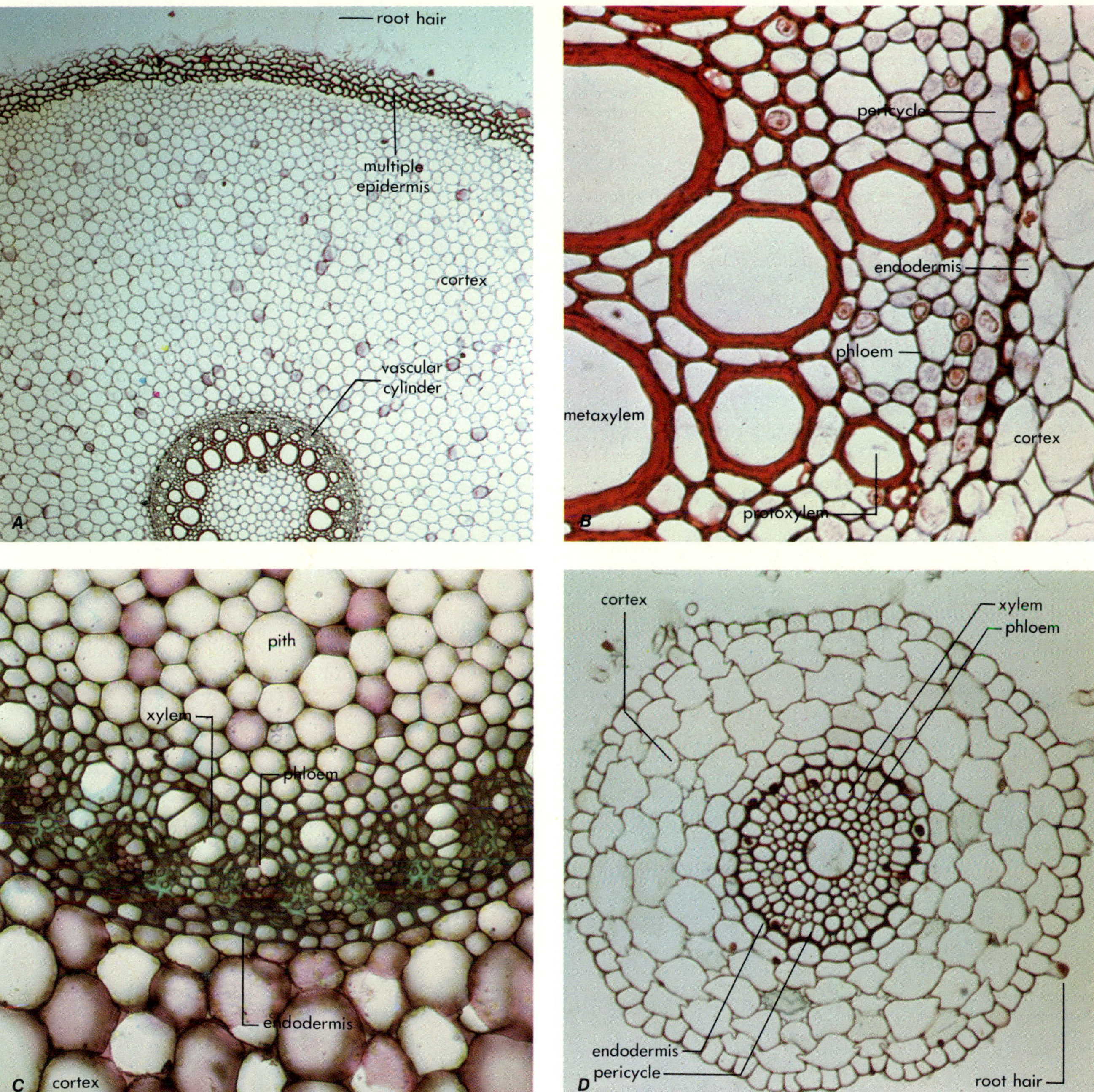

Figure 9.17
A, Asparagus officinale root in transverse section showing all tissues and root regions. The vascular cylinder (stele) has a pith of all parenchyma cells. *B, Asparagus* root at higher magnification to view the cell layers in greater detail: the endodermis, pericycle, xylem, and phloem of the polyarch stele are clearly defined. *C, Epidendron* sp. (orchid) root in transverse section to show details of the stele. *D, Triticum* sp. (wheat) root; this is an example of a monocotyledonous root lacking a parenchymatous pith.

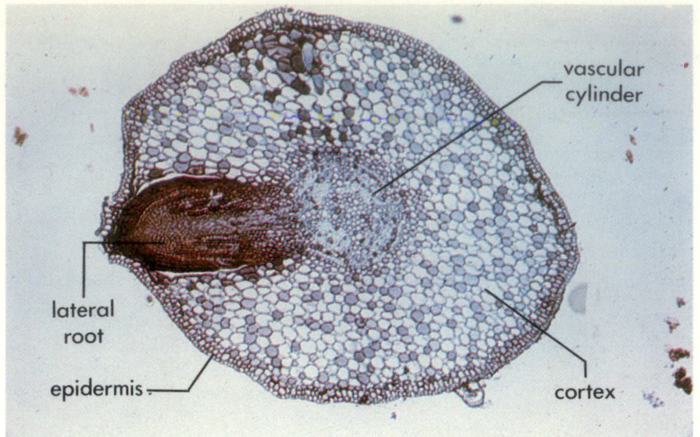

Figure 9.19
Willow *(Salix* sp.) root transverse section showing a large lateral root soon after its emergence.

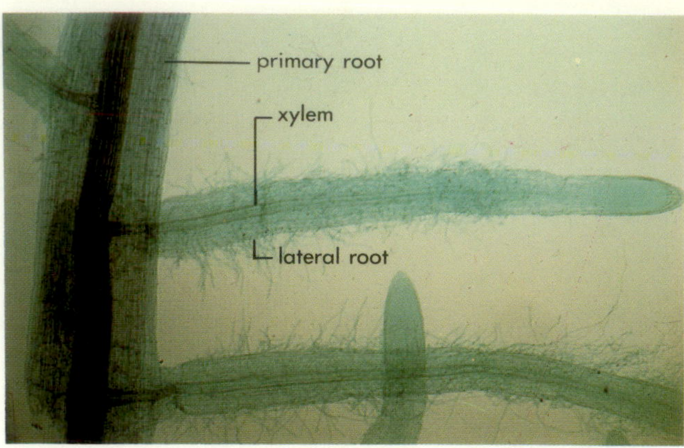

Figure 9.20
Radish (*Raphanus* sp.) root cleared to show the continuous vascular connection between primary and lateral roots.

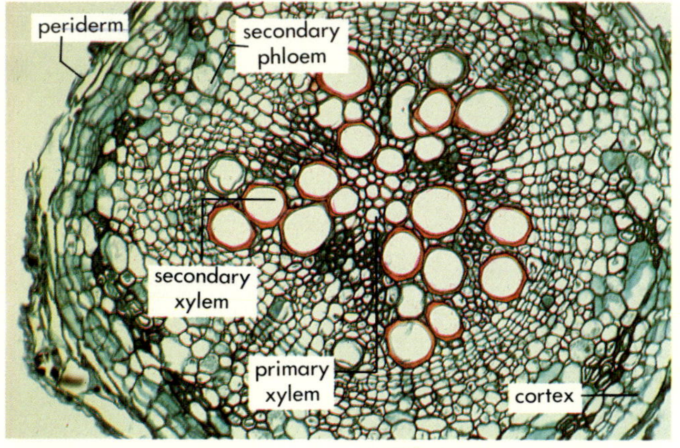

Figure 9.22
Alfalfa *(Medicago sativa)* root transverse section with some secondary tissues and periderm (cork) on its periphery.

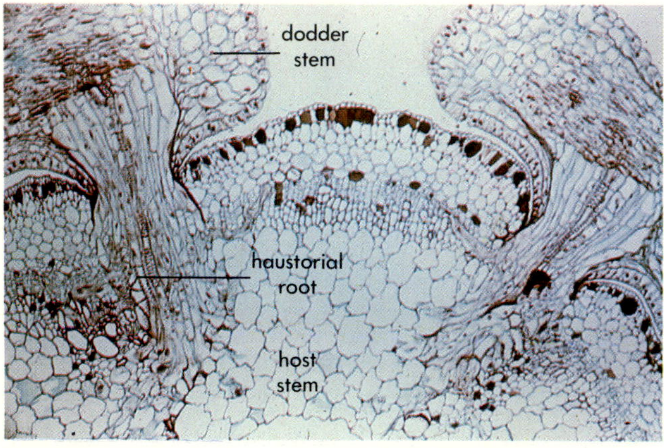

Figure 9.24
Transverse section through a stem showing penetrations by haustorial roots of the plant parasite dodder *(Cuscuta* sp.)

structural features of roots can best be studied by means of seedlings that have been grown either in sand or on filter paper in a moist atmosphere. Figure 9.7 shows such a seedling. Observe here the portion of the root that is clothed with root hairs; this region is known as the **root-hair zone.** Note that root hairs do not extend to the very tip of the root. Root-hair zone and hairless tip constitute a region of especial interest. It is in this part of the root that (*a*) growth in length, (*b*) most absorption of materials from soil, and (*c*) development of primary tissues take place. Below the root hairs there occur, in descending order, the **region of elongation,** a **meristematic zone,** and the **root cap. Differentiation** starts in the upper cells of the meristematic zone and extends upward through the region of elongation into the lower zone of root hairs.

Root Apex

In our discussion of root structure let us begin with the embryonic root, the radicle (Fig. 9.7). The **root apical meristem** (meristematic zone) at the tip of the radicle is composed of small, regularly shaped cells that are capable of dividing. The emergence of the radicle during germination is dependent on the initiation of cell division in this region (**Figs. 9.8*A*, *B*,** page 165). After the radicle emerges, the cells of the root apical meristem continue to divide; on the average, one derivative cell of each division remains as part of the meristem and divides again, while the other differentiates into a mature, specialized cell.

The central portion of the apical meristem is composed of a pellet-shaped region of cells known as **quiescent center (Figs. 9.8*B*, 9.9,** page 165**).** These cells divide at an extremely slow rate. They act as a regulatory center by releasing dividing cells, **initials,** just fast enough to continuously maintain the root's shape and to keep pace with its growth. The function of the quiescent center is not fully understood. One hypothesis, however, is that it may be the location where growth regulators are synthesized and released to control development of the root as new cells are made.

Some initials divide and produce cells to form the **root cap** (**Fig. 9.9**). The root cap is a thimble-shaped region of short-lived parenchymalike cells that surrounds the apical meristem and precedes it as the root elongates and forces its way through the soil. The root cap not only protects the apical meristem and lubricates its passage through the soil, but also is the site that perceives gravity and controls the direction of root growth. Root cap cells are constantly being sloughed off at the very tip, but at the same time new cells are being added to it by apical meristem cells. In the **apical meristem,** cells are actively dividing, adding new cells to the root cap and others to the region of elongation. Rapid growth in root length, however, is largely the result of elongation of cells behind the apical meristem. This is the **region of elongation** (Fig. 9.7). Thus, it is seen that a very short portion of the root at the tip, usually 2 to 5 mm long, is constantly being forced through the soil. The method of growth in length of the root just described explains the protective function of the root cap; the delicate cells of the apical meristem are protected from mechanical injury as they are pushed through the soil. The region of elongation grades into the region of differentiation of primary vascular tissue internally and the formation of root hairs in the epidermis. Thus, the mechanism for absorption (root hairs) differentiates as conduction tissue is forming from procambium (Figs. 9.10, 9.11).

It is significant that elongation of cells does not occur in the root-hair zone. If elongation did occur in this zone, root hairs that wrap around and adhere to soil particles would be torn loose.

Initial cells behind the quiescent center divide to form the primary meristems: **protoderm, ground meristem,** and **procambium (Figs. 9.9,** 9.10**).** These cells continue to divide, so that the region of cell divisions may ex-

Figure 9.10
Median longitudinal view of a root tip showing the primary meristems and primary tissues and regions that develop from them.

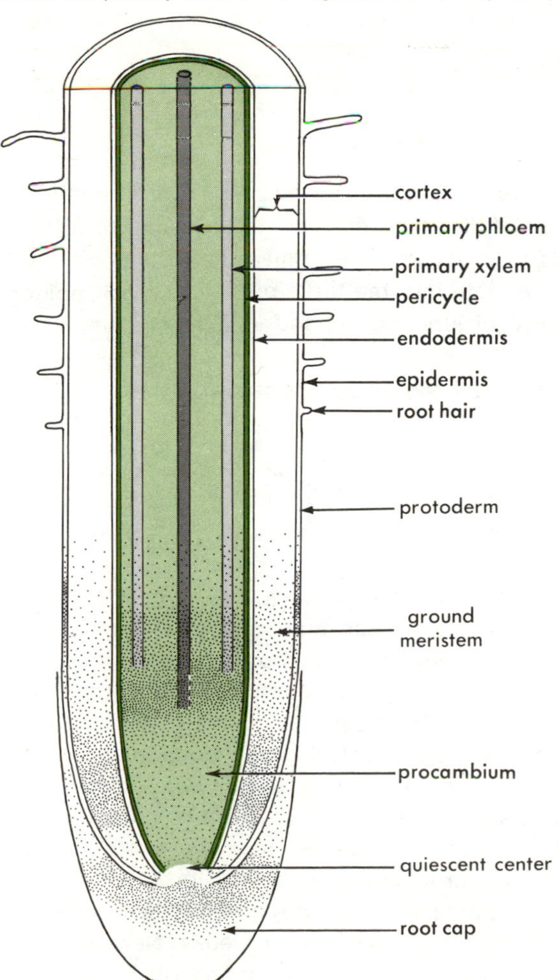

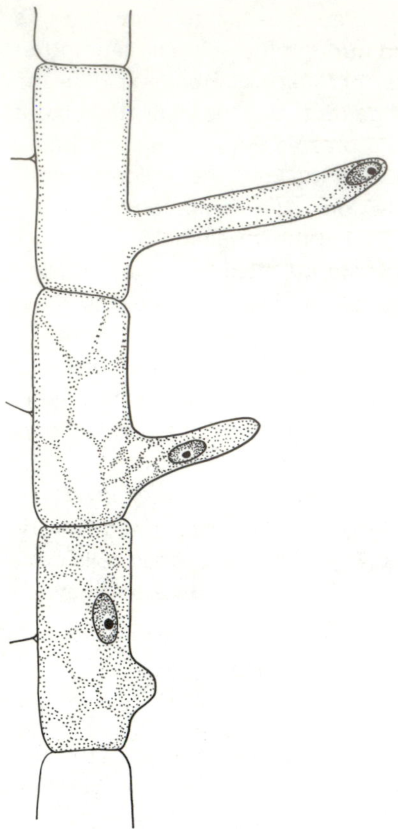

Figure 9.11
The development of root hairs. Note that the external epidermal cell wall protrudes and that the cell cytoplasm and nucleus move into the root hair near the tip.

tend as far as 1.5 mm from the tip of the root. The primary meristems of the shoot and the root are similar. Derivative cells from these three primary meristems elongate and differentiate into the three primary tissues: **epidermis, cortex** (ground tissue), and **vascular tissue.**

Formation of Primary Tissues

Epidermis

The protoderm consists of a single layer of uniformly shaped meristematic cells that differentiate into epidermal cells. The epidermis of roots is devoid of stomata and a significant cuticle. Above the root apex some cells of the protoderm develop into **root hairs** by the extension of their cell wall into the surrounding soil. When grown in moist air, each root hair has the form of a slender tube, but in soil, it may be greatly contorted in its growth between and around soil particles. The root hair and the epidermal cell from which it grows constitute a single cell. The walls are thin, composed principally of cellulose and pectic substances. The protoplast has a high water content. The nucleus is usually near the end of the hair and there is a large central vacuole (Fig. 9.11).

In most plants, the life of any one root hair is short; it functions only for a few days or weeks. New hairs are constantly forming at the anterior end of the root-hair zone (Fig. 9.1), while those at the posterior end are dying. Thus, as the root advances through the soil, fresh, actively growing root hairs are constantly coming into contact with new soil particles. In the rye plant, it is estimated that new root hairs develop at an average rate in excess of 100 million per day. Root hairs of such plants as redbud, honey locust, and a number of others may persist for several years.

Although nearly all ordinary land plants possess root hairs, a few, such as the firs, redwoods, and Scotch pine, are devoid of them. Also, many aquatic plants have no root hairs. Moreover, land plants (corn, for example) that normally develop root hairs when the root system grows in the soil or in moist air, develop no root hairs when the roots grow in water. In plants devoid of root hairs, absorption is accomplished entirely through typical thin-walled epidermal cells. Root-hair development is often inhibited by a concentrated soil solution and by high or low soil temperatures. Root hairs develop in light and dark about equally well if moisture and oxygen are adequate.

The epidermis in roots is usually one cell layer thick, but in the aerial roots of certain plants such as orchids, a multiple epidermis called a velamen develops.

Cortex

This region, relatively thicker than that in stems, is derived from ground meristem. The cortex is composed chiefly of storage parenchyma with large intercellular spaces (Fig. 9.12) and occasionally schlerenchyma. In many species secretory cells and resin ducts are present. The innermost layer of the cortex is a single row of cells, the **endodermis** (Fig. 9.13), which is usually a conspicuous feature of roots. As a rule, in the primary state, endodermal cell walls are thin except for a bandlike thickening running around the cell on radial and transverse walls. This thickened strip, the **Casparian strip,** is suberized (Figs. 9.13, **9.14,** page 166).

Cellulose walls are permeable to water. Cell walls and intercellular spaces of the cortex may freely imbibe water and dissolved nutrients from the surrounding soil solution. The suberized strip that forms a complete ring around the endodermal cell in radial and transverse walls prevents water from reaching the pericycle except by passing through the protoplast of the endodermal cell (Fig. 9.15). Electron micrographs show that the plasmalemma of the endodermal cell is fused to the Casparian strip.

Endodermal cells in mature regions of the root often develop partial secondary cell walls. Passage cells adjacent to xylem points sometimes do not form secondary walls until late in the root's development.

Vascular Cylinder or Stele

The **vascular** tissue develops from the procambium and forms the central core of the root. At an early stage of

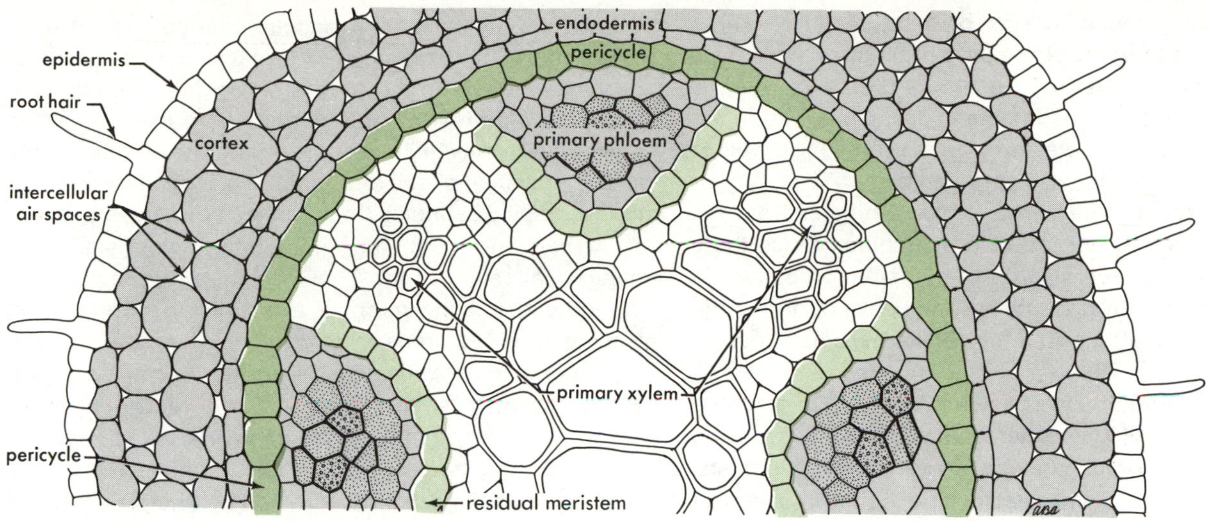

Figure 9.12
Cross section of the root showing the primary tissues.

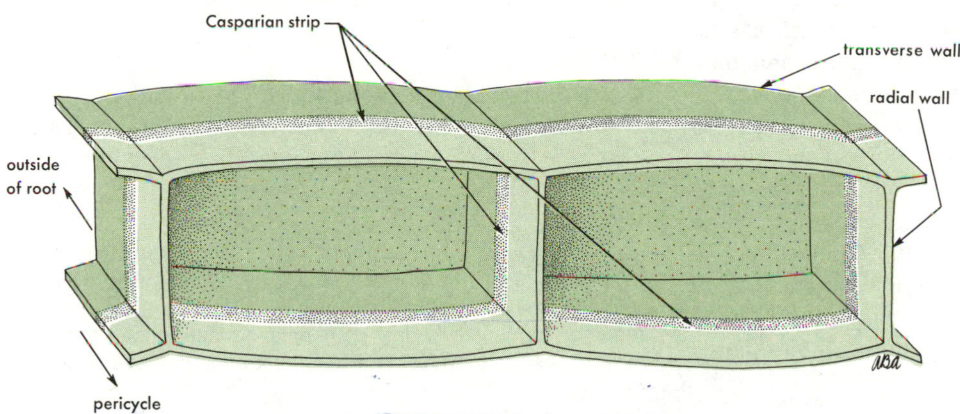

Figure 9.13
Three-dimensional view of two endodermal cells. The suberized Casparian strip is shown in gray. Water from solution outside the root wets the cellulose walls until it reaches the Casparian strip. Since water can not cross the Casparian strip, it must pass into the cytoplasm of endodermal cells.

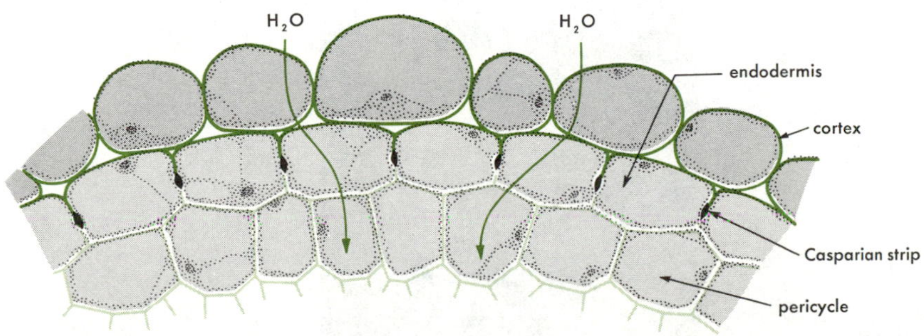

Figure 9.15
The relationship between cortex, endodermis, and pericycle. Water present in the walls of cortical cells cannot diffuse across the suberized Casparian strip. Water inside the endodermal layer has been filtered by the protoplasts of the endodermal cells.

development, a special layer of cells is differentiated from the outer region of the procambium cylinder. This is the **pericycle** (**Figs.** 9.12, **9.16*B*, 9.16*C*,** page 166). It persists as a type of meristematic tissue involved in the formation of lateral roots and in some roots the development of both vascular and cork cambia.

The **vascular cylinder** or **stele** originates from the remaining internal portion of procambium. Since there is generally no internal ground meristem as occurs in tips of shoots, pith does not usually develop in dicotyledonous roots, although it may occur in roots of monocotyledons.

In roots primary xylem usually consists of a central mass or "core" of xylem elements with several radiating arms, between which are groups of phloem elements. Between these two tissues are one or more layers of residual procambial cells, represented by the light green line on Fig. 9.21*A*. In roots with secondary growth, these cells give rise to vascular cambium (which is shown in Fig. 9.21*B* as a light green line). Many roots have no pith, but in most monocotyledons (**Fig. 9.17,** page 167) and some herbaceous diocotyledons, the central core of the stele is parenchyma that resembles pith of stems. Whereas pith of stems is derived from ground meristem, that of roots is perhaps derived from procambium.

The first matured xylem-conducting elements, the **protoxylem,** develop at the points of the radiating xylem (**Figs. 9.16*A,B*,** page 166). Dicot roots with two protoxylem points are called diarch roots, those with three points are triarch, and so on (**Figs. 9.16*B*,*C*)**; most monocot roots have more than five protoxylem points and are therefore called polyarch (**Fig. 9.17,** page 167). **Metaxylem,** the last primary xylem to mature, differentiates toward the center of the vascular cylinder (**Fig. 9.16*C***). The protoxylem is modified so that it is capable of transporting water while the root is elongating; that is, protoxylem cells must be able to withstand the forces that move water, and still be flexible enough to elongate. This is accomplished by the formation of a secondary cell wall in the shape of annular rings or as a spiral (see Chapter 7). When the root has completely elongated, the metaxylem cells mature. They no longer are required to elongate and, consequently, form thick secondary cell walls with pitted areas through which lateral exchange may take place. Protoxylem cells often become crushed while the metaxylem develops. Xylem of roots consists of vessel members, tracheids, parenchyma, and fibers.

Phloem cells form in the spaces between the protoxylem arms (**Figs.** 9.12, **9.16*B*, 9.17B).** The protophloem is the first part of the vascular system to become functional. These cells form at the periphery of the phloem and function primarily during root elongation. Metaphloem tissue develops toward the inside and functions during the plant's adult life. Phloem of roots consists of parenchyma, fibers, sieve-tube members, and companion cells.

Origin of Lateral Roots

It will be recalled (p. 115) that lateral or branch stems originate from cell layers at or near the shoot tip. In con-

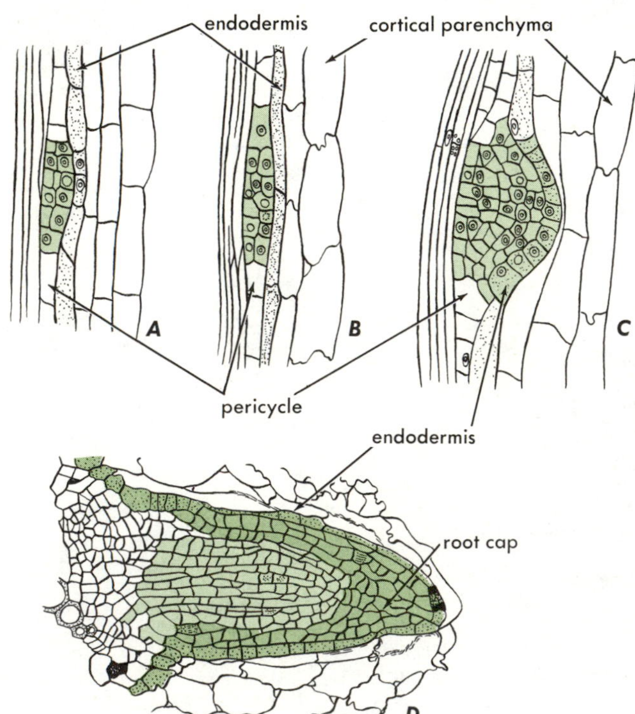

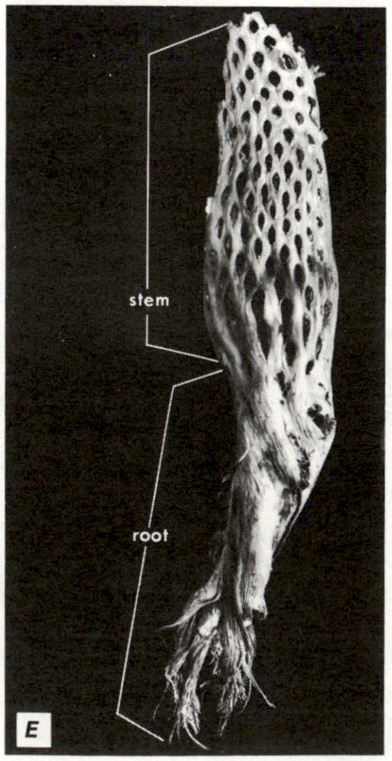

Figure 9.18
Lateral roots. *A*, initiation of lateral root of carrot *(Daucus carota)* by division of cells in the pericycle. *B* and *C*, formation of lateral root primordium. *D*, young root pushing through cortex, ×50. *E*, roots forming in this manner do not leave a gap in the vascular cylinder, as is shown by the xylem framework of a rotted cabbage *(Brassica oleraceae)*, ×½.

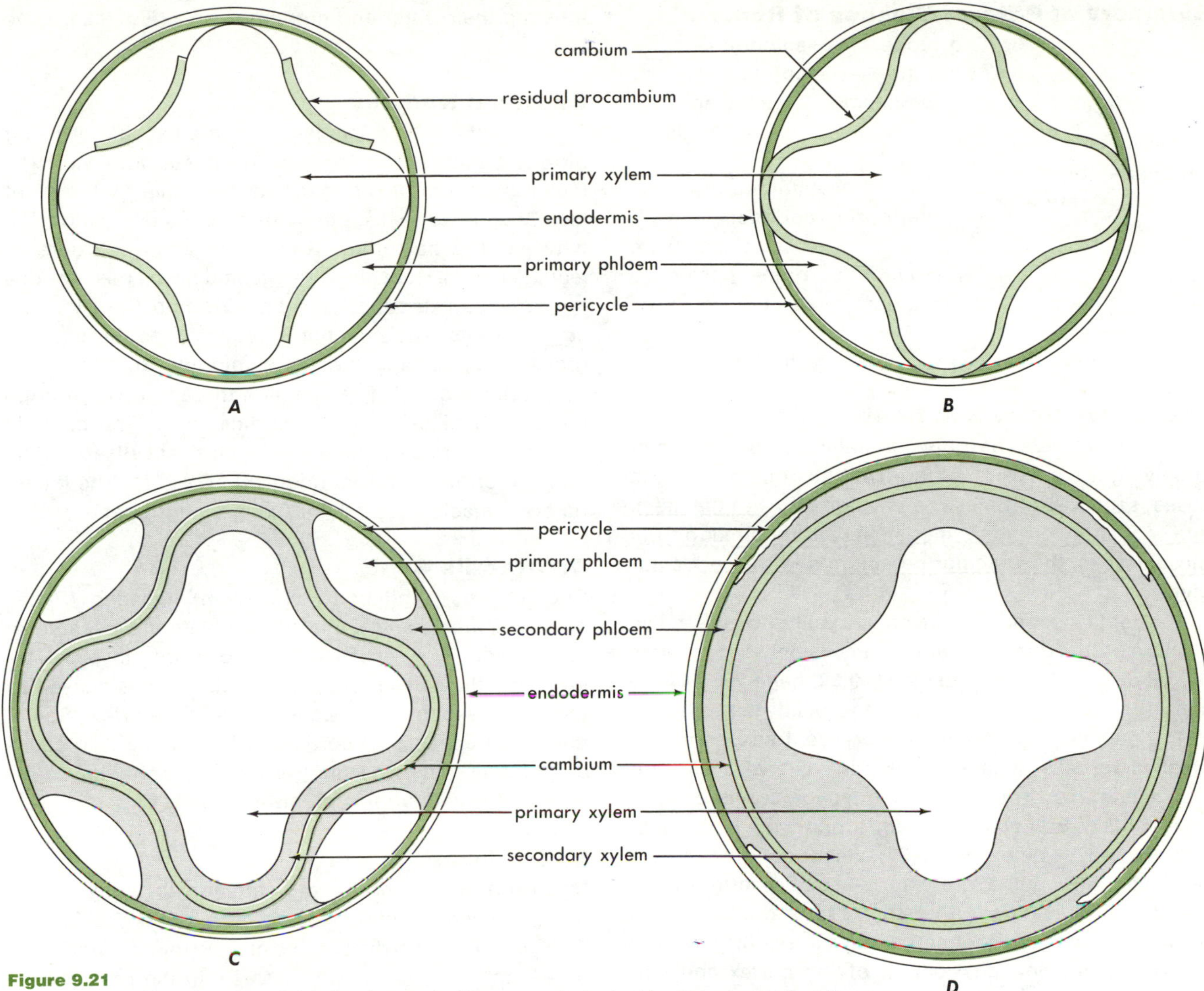

Figure 9.21
Diagrammatic representation of the development of the secondary plant body in a root. *A*, at the completion of primary growth an arc of residual procambium cells remains (light green) and a complete circle of pericycle is present (dark green). *B*, the residual procambium (light green) joins with the pericycle cells outside the xylem arms to form a continuous cylinder of vascular cambium (light green). *C*, the vascular cambium forms secondary xylem internally and secondary phloem externally (light gray). The primary phloem is being pushed outward and a still complete ring of pericycle (dark green) is still associated with it. *D*, a smooth circle of vascular cambium forms, producing secondary xylem and phloem. The primary xylem remains in the center of the stem, the primary phloem has been crushed, and the pericycle that remains will form the cork cambium.

trast, lateral or branch roots in gymnosperms and angiosperms originate from cells of the pericycle. **Fig. 9.18** shows the manner of origin of branch roots. Often, the point of origin of a branch root is opposite a primary xylem strand. For example, in the beet root, there are two primary xylem strands and two vertical rows of branch roots.

Initiation of lateral roots at particular locations is controlled by chemical growth regulators that cause pericycle cells to begin to divide (Figs. 9.18 and 9.19). The **lateral root primordia,** which result, continue to form new cells that in turn elongate. Endodermal cells subtending this region often also divide for a short time, contributing cells to the tip of the new lateral root. As it expands, the lateral root pushes its way through and destroys the cortical cells and the outer epidermis. As it emerges, the new lateral root becomes organized into an apical meristem with primary meristems identical to its parent root (**Fig. 9.20,** page 167).

This method of root formation leaves the xylem portion of the vascular cylinder intact, in contrast with the stem where leaf and bud gaps results in openings in the xylem cylinder. This difference is shown in Fig. 9.18*E*, which illustrates the portion of cabbage xylem where root and shoot merge. The lateral roots have a vascular system that is continuous with that in the main root (**Fig. 9.20**).

Summary of Primary Tissues of Roots

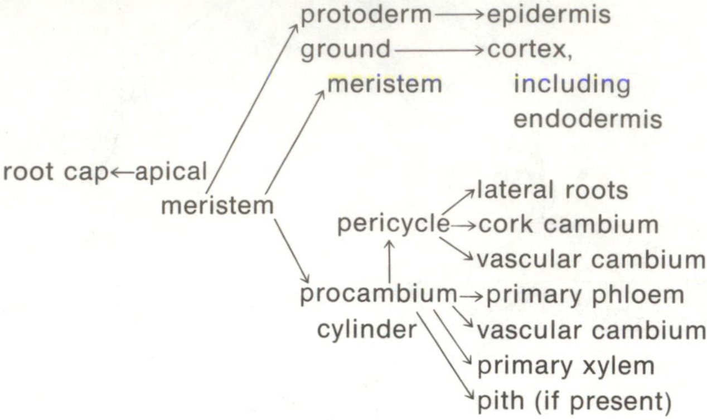

Secondary Growth in Roots

The primary tissues of roots generally are not capable of supplying the needs of a long-lived plant body. In these cases, secondary tissues are developed. As in the stem, two secondary or lateral meristems are required to form these tissues, the **vascular cambium** and the **cork cambium.**

In roots, secondary growth is initiated by the active division of (residual) procambium cells between the arcs of xylem and phloem (Figs. 9.21, **9.22,** page 168). These cells form secondary xylem to the inside and phloem to the outside. After a time, the crescent-shaped region of dividing cells joins with the pericycle, which in turn also begins to divide. This connection forms a complete ring of vascular cambium around the entire vascular cylinder (stele) (Fig. 9.21). Secondary tissues are produced during the growing season to form annual growth rings, just as in the stem (Fig. 9.23). Continued growth expands the root and finally causes the splitting, sloughing off, and destruction of the cortex and epidermis. However, early stresses of expansion, plus the activity of growth hormones, stimulates the still intact pericycle to resume division and form the cork cambium. This meristem forms cork cells to the outside, and phelloderm cells to the inside, just as in the stem (Figs. **9.22,** 9.23). Monocotyledonous roots generally do not have any secondary growth, but at least one monocot, the Dragon's blood tree (*Dracena draco*), does produce secondary tissues.

SPECIALIZED ROOTS

Roots of a great many plants do not have the general characteristics common to most roots. We have already discussed examples of adventitious roots that arise from nonroot origins, the roots of parasitic plants, and the uniquely shaped roots of clinging ivy vines. Other plants, especially those forming partnership with microorganisms, have specialized root structures.

Haustorial Roots

Parasitic plants, like dodder, anchor themselves by sinking haustorial roots into their hosts' vascular tissue and thus tap their water and nutrient supply (**Fig. 9.24,** page 168).

Bacterial Nodules

Some plants, like certain legumes, are capable of fixing nitrogen, that is, changing N_2 in the soil into the NH_4^+ (ammonia) form of nitrogen that is usable by the plant (see Chapter 10). This is done by an unusual relationship between the bacterium *Rhizobium* and the roots of legume plants. Root cells are infected by the passage of a thin infection thread of the bacteria into the root hair cells of roots. The infection thread passes through the root hair by causing the cell membrane to enfold as it penetrates the cell. The infection thread grows through the epidermal cell into the cortical cells. The bacteria then divide and also stimulate the cortical cells to divide, thereby forming the root nodule (Fig. 9.25*F*). The bacteria are the actual agents for fixing the nitrogen.

Contractile Roots

Roots of the dandelion (*Taraxacum*), ginseng (*Panax quinquefolium*), and some other plants are capable of contracting. The result is that above-ground parts are kept near the soil surface. This contraction is caused by the radial expansion of cells in the root cortex. Sometimes, the collapse of cortical cells is thought to cause contraction. The vascular tissue in contracted roots forms a twisted, undulated mass (Fig. 9.25*A*).

Mycorrhizae

Mycorrhizae are short roots common to many plants; they represent an association with a soil-borne fungus. Two types of mycorrhizal roots may be found, depending on whether the fungus penetrates into the root cells or not. **Ectotrophic** types are found in roots of such trees as

Figure 9.23
Cross section of an old root of cherry (*Prunus avium*), ×⅞.

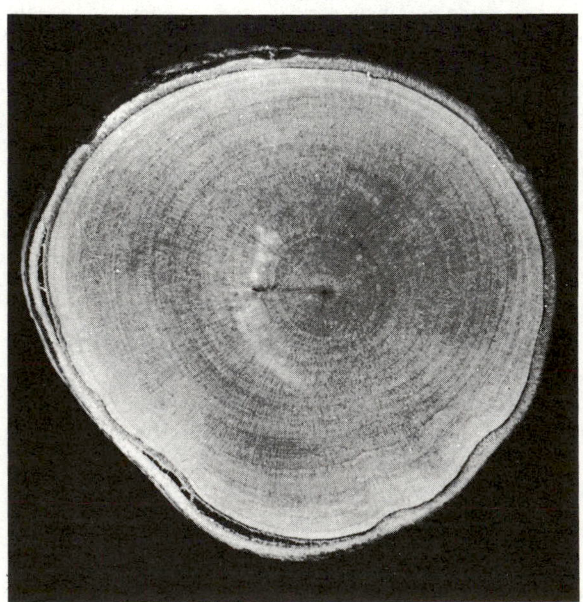

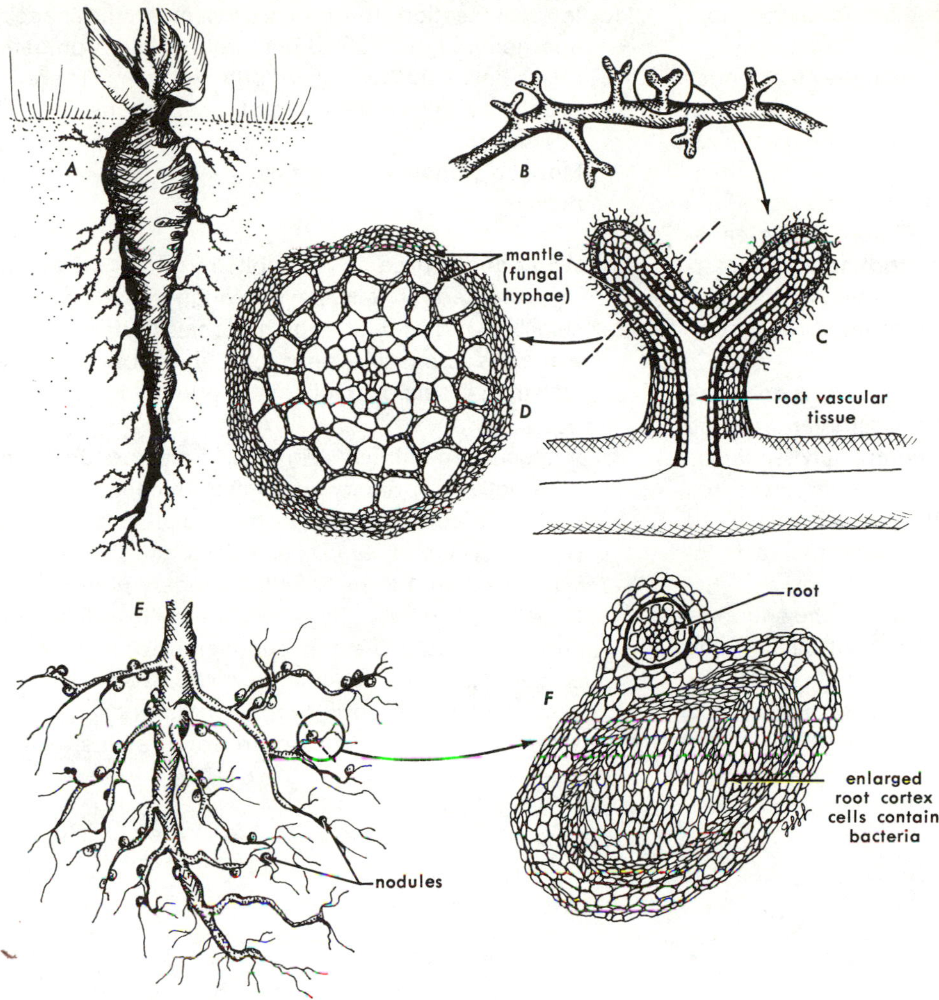

Figure 9.25
Views of root modifications. *A*, contractile roots. *B*, branched habit of mycorrhizal root. *C*, longitudinal section through one of the Y-branched mycorrhizal roots. *D*, cross section through one root showing the fungal mantle. *E*, soybean *(Glycine max)* root with bacterial nodules. *F*, cross section of root and nodule.

pines (*Pinus*), birches (*Betula*), willows (*Salix*), and oaks (*Quercus*). This type causes a drastic change in the root shape (Fig. 9.25*B* to *D*). These mycorrhizal roots are about 0.5 cm long, have no root cap, and exhibit a simple monarch stele. In addition, the fungus penetrates between the cell walls of the cortex and forms a covering sheath, or mantle, of fungal hyphae (Chapter 23) around the entire root. Mycorrhizal roots are short and forked and sometimes are borne as tight clusters. **Endotrophic** mycorrhizae do not form a mantle over the root, and the fungus actually enters the cortex cells. Mycorrhizal roots are more efficient in mineral absorption, but they are apparently not absolutely essential for the growth of the usual host plants. This is known, because plants that are artificially fed adequate nutrients can grow without mycorrhizae. Mycorrhizae also may be beneficial to their host plants by secreting hormones or antibiotic agents that reduce the potential of plant disease.

SUMMARY

1. The principal functions of roots are absorption of water and nutrients, conduction of absorbed materials and food, and anchorage of the plant in the soil.
2. The primary root develops from the radicle in the embryo. It generally penetrates the soil to some depth and may either form a tap root system or develop an extensive series of lateral roots to form a fibrous root system.
3. More frequently, the fibrous root system is formed by adventitious roots arising from nodes of the stem. This is the general situation in monocotyledons and occurs in many dicotyledons; it is most generally associated with horizontal stems.
4. Adventitious roots may arise in the internodal regions of the stem.

5. Shoots may arise from adventitious buds developing on roots.
6. Roots differ from all stems because they lack nodes and internodes.
7. The root tip is divided into four zones of specialization: (*a*) root cap, which protects the (*b*) meristematic region as it moves through the soil; (*c*) a region of elongation; and (*d*) a region of differentiation characterized externally by root hairs and internally by the formation of primary vascular tissues.
8. Soil water and nutrients move easily through epidermal and cortical cells.
9. The endodermis is the innermost cell layer of the cortex. A suberized band, the Casparian strip, in radial and transverse walls, completely encircles endodermal cells. Water cannot move across the Casparian strip. Therefore, all water with dissolved nutrients must pass through the protoplasts of endodermal cells.
10. The pericycle, a row of cells internal to the endodermis, represents the outermost row of cells of the vascular cylinder. Cells of pericycle may eventually form part of vascular cambium or cork cambium and give rise to lateral roots.
11. In cross section, the primary xylem is star-shaped and generally triarch to pentarch. Primary phloem arises between the arms of primary xylem. There is usually no pith in the roots of dicotyledons.
12. Monocotyledon roots usually have a pith and are polyarch.
13. Water and inorganic nutrients enter root hair and other epidermal cells, pass through the cortex, are selected and regulated by protoplasts of endodermal cells, cross the pericycle, and move directly into the vessels or tracheids of primary xylem.
14. A vascular cambium originates from procambium cells between primary xylem and phloem and from pericyclic cells exterior to the radiating points of primary xylem. This vascular cambium forms secondary xylem internally and secondary phloem externally. The resulting increase in diameter stretches and tears the endodermis, cortex, and epidermis. A cork cambium develops from the pericycle and forms cork. Woody roots consequently are very similar in structure to woody stems.

CHAPTER 10

LEAVES

reen plants and a few species of bacteria are the only producers that use sunlight as an energy source to produce food; all other inhabitants of the earth consume what the green plants produce. In seed plants, leaves are the principal organs of food production. **Chloroplasts** within leaf cells are the organelles that trap light energy and convert it into chemical energy by the process of **photosynthesis.**

The water economy of plants calls for the absorption of much more water than can be metabolized. This excess water is returned to the atmosphere by leaves. A second role of leaves is **transpiration.** The structure of leaves is uniquely adapted to carry out these two primary roles.

Leaves are of many shapes and sizes (Fig. 10.1, 10.2). Practically all leaves have veins (vascular bundles) for support and conduction, and a ground tissue with chloroplasts. Shape and size are such constant traits for many categories of plants that one may frequently identify an unknown plant simply by its particular type of leaf (Fig. 10.2). Leaf shape, however, is not always a good diagnostic trait for use in plant identification, because it may vary within a species, and because leaf shape may sometimes be altered by its environment.

Leaf blades provide large surfaces for the absorption of light energy and carbon dioxide, both of which are required for photosynthesis. They are thin and hence no cells lie far from the surface. This form facilitates the absorption of light energy and the exchange of carbon dioxide, oxygen, and water vapor between the intercellular spaces of the leaf and the atmosphere. The water

Figure 10.1
Different kinds of leaves. *A*, poplar *(Populus deltoides)*. *B*, castor bean *(Ricinus communis)*. *C*, oak *(Quercus lobata)*. *D*, rose *(Rosa odorata)*. *E*, Virginia creeper *(Parthenocissus quinquefolia)*. *F*, faba bean *(Vicia faba)*, ×½.

that leaves utilize and transpire comes from roots and enters the leaf in the xylem tissue of veins. Food manufactured in leaves is transported out of the leaf in the phloem tissue of the same veins. In addition, veins may add support to the leaf.

LEAF COMPONENTS

The leaves of dicotyledonous plants are generally different from those of monocotyledonous plants. A typical foliage leaf of a plant belonging to the dicotyledons is composed of two principal parts: (*a*) **blade** and (*b*) **petiole** (Fig. 10.1). The blade is thin and expanded, the petiole slender. The thin blade is supported by a distinct network of **veins** that are composed of vascular tissues and fibers. In addition to forming a supporting framework for softer tissues of the blade, veins carry water, mineral salts, and food to and from the leaf.

The Blade

Variable external features of the leaf blade are its overall shape: **apex, margin,** and **base.** This range of variation is great, and many terms are employed by taxonomists to describe leaf shape accurately.

Figure 10.2
Leaves of different species of *Populus*. Leaf characters alone often enable one to identify a species.

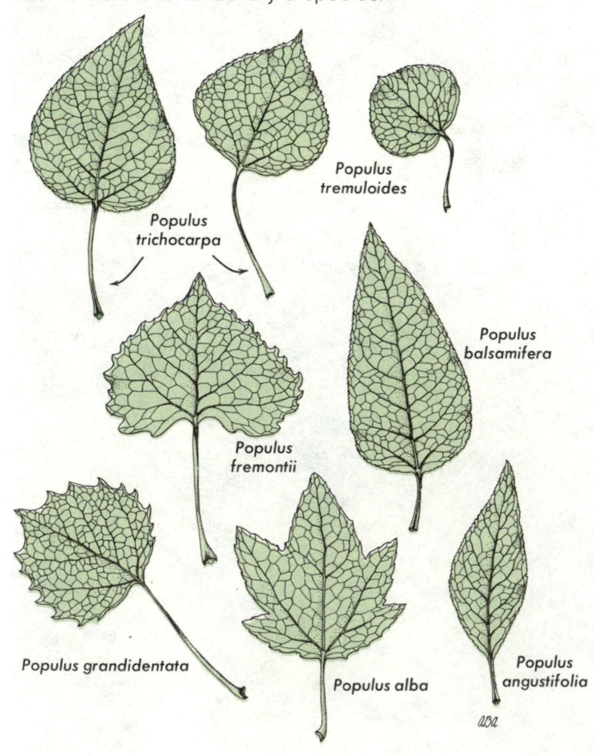

Shape

The leaf may be long and slender (linear), or oval, heart-shaped (cordate), or triangular (deltoid) (Fig. 10.2).

Apex

The apex of a leaf may be pointed, rounded, or flattened.

Margin

The margin may be entire that is, with no indentations whatsoever; it may be toothed (dentate) (Fig. 10.1*A*), scalloped, wavy, or cut into a number of lobes. Marginal indentations may be shallow, or they may be deep clefts extending almost to the midrib. **Lobed** leaves that are pinnately veined are said to be pinnately lobed (Fig. 10.1*C*); those that are palmately veined are palmately lobed (Fig. 10.1*B*).

Base

The base of the leaf may be rounded, heart-shaped (cordate), or flattened (truncate).

The Petiole

Some leaves, such as those of peas (*Pisum*), beans (*Phaseolus*), roses (*Rosa*), tulip trees (*Liriodendron*), and many others, have two small, leaflike outgrowths at the base of the petiole, known as **stipules** (Figs. 10.1*D*,*F*). The petiole itself may be long or short, rounded or occasionally flat. It is usually attached to the base of the leaf blade, but in some plants, such as *Tropaeolum* and castor bean (*Ricinus*), it is attached at the middle of the blade on the underside. This leaf is called a **peltate** leaf (Fig. 10.1*B*). The petiole is sometimes absent, as in *Zinnia,* the blade being mounted directly on the stem. Such a leaf is said to be **sessile** (Fig. 10.21*B*, juvenile *Eucalyptus*).

Monocotyledonous leaves such as grasses do not have distinct petioles. Instead, the leaf is divided into two parts, **sheath** and **blade.** The blade is the typical thin, expanded portion. The sheath is green, perhaps nearly as large as the blade, and it completely sheaths the stem (Fig. 10.3). In many species such as corn the sheath extends over at least one complete internode. If the region of union between the blade and the sheath is examined carefully, a small flap of delicate tissue extending upward from the sheath may be seen closely enveloping the stem. This is called the **ligule** (Fig. 10.3*B*). It may, in some cases, serve to keep water and dirt from sifting down between stem and sheath. In many species, of which barley (*Hordeum*) (Fig. 10.3*C*) is a good example, the base of the blade, at its union with the sheath, is carried around the stem in two earlike points, the **auricles**. Ligule and auricles may both be present, or one or the other may be absent.

Simple and Compound Leaves

There are two kinds of leaf blade configuration: (*a*) **simple** and (*b*) **compound.** In a simple leaf the blade is all in one unit (Figs. 10.1*A* to *C*). In a compound leaf the blade is composed of a number of separate leaflike parts, the **leaflets** (Figs. 10.1*D* to *F*). Inasmuch as leaflets have the characteristics of a simple leaf, it may sometimes be doubtful whether the structure is a simple leaf or a leaflet, especially if the leaflets are large or very numerous. A primary distinction is that buds occur in the axils of leaves, but not in the axils of leaflets.

When the leaflets of a compound leaf arise from the **rachis** (continuation of the petiole), as do the pinnae of a feather, the leaf is said to be **pinnately** compound, like the rose and bean (Fig. 10.1*D*); when the leaflets diverge

Figure 10.3
Leaves of members of the grass family. *A*, crabgrass *(Digitaria sanguinatis)*, showing sheath and blade. *B*, corn *(Zea mays)*, showing ligule (sheath pulled away from stem). *C*, ligule and auricles of barley *(Hordeum vulgare)*, about ×¾.

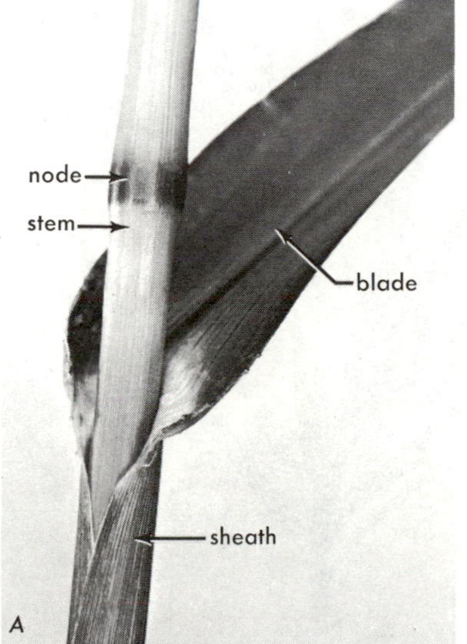

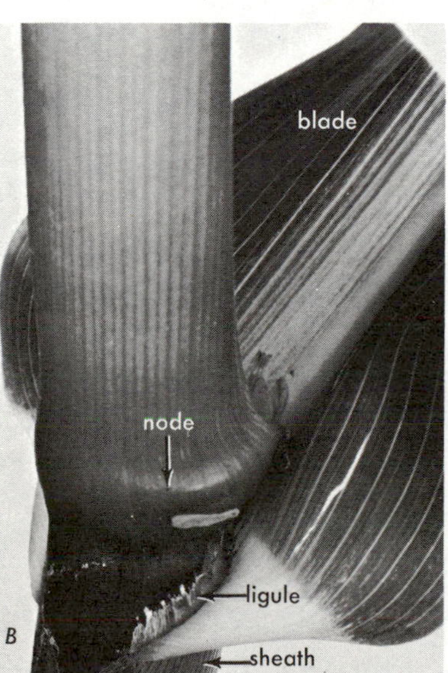

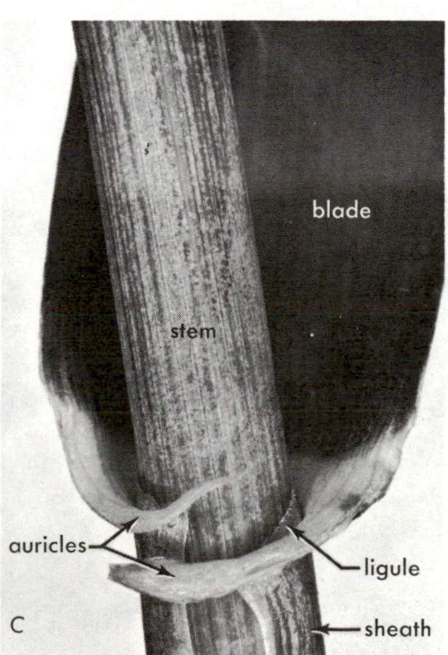

from a common point at the tip of the petiole, the leaf is said to be **palmately** compound, like the horse chestnut (*Aesculus*) and Virginia creeper (Fig. 10.1*E*). Compound leaves may be once, twice, or thrice compound.

Venation

The arrangement of veins of a leaf is called venation. There are two principal types of venation: (*a*) **parallel venation,** usually characteristic of monocotyledonous leaves (Figs. 10.4*A*, **10.6*A*,** page 185), and (*b*) **netted venation,** usually characteristic of dicotyledonous leaves (Figs. 10.4*B*, **10.6*B* to *D***). In netted venation there are one or more prominent veins from which smaller veins branch off to join with other small veins, thus forming a conspicuous net (**Figs. 10.6*B* to *D***). In parallel veined leaves, there usually is one or a few large veins with many veins branching from them. These veins run parallel to each other and do not branch further (**Fig. 10.6*A***). It is important to note that there are some dicotyledonous plants with parallel veined leaves and some monocotyledonous leaves with netted veins.

GYMNOSPERM LEAVES

The leaves of all gymnosperms native to the United States and Europe are needlelike (Fig. 10.5), or scalelike. They are discussed in some detail in Chapter 29, on the conifers. The needles of the dawn redwood (*Metasequoia glyptostroboides*) and the fan-shaped leaves of *Ginkgo biloba* are shown in Figs, 10.5*A*,*B*. Both of these trees are native to China and represent very primitive seed plants. *Ginkgo* is known only in cultivation, and the dawn redwood, like the California redwood, has a very restricted distribution. The leaves of many gymnosperms of the southern hemisphere are linear (Fig. 10.5*D*) or expanded.

Figure 10.5
Gymnosperm leaves. *A*, dawn redwood (*Metasequoia glyptostroboides*). *B*, *Ginkgo biloba*. *C*, pine (*Pinus*). *D*, *Podocarpus*, about ×3/4.

Figure 10.4
Leaves, showing parallel and netted venation. *A*, *Canna* leaf, about ×1/5. *B*, maple (*Acer*) leaf, about ×1/2.

ANATOMY OF THE FOLIAGE LEAF

The anatomy of a leaf blade is best shown in section. In Fig. 10.7 we observe three principal tissues: (*a*) **epidermis,** (*b*) **mesophyll** (ground tissue), and (*c*) **veins** or **vascular bundles.** The epidermis usually consists of a single layer of cells that covers the entire leaf surface. It protects the tissues within the leaf from drying out and from mechanical injury. The mesophyll is composed of parenchyma cells, most or all of which contain chloroplasts and thus are able to carry on photosynthesis. Veins possess xylem and phloem elements, which conduct water, inorganic salts, and foods. Fibers and sometimes collenchyma may be associated with conducting elements of midrib and larger lateral veins.

The petiole has its own specialized structures that enable it to support the leaf blade, conduct food, water, and inorganic salts, and disconnect itself from the stem at the close of the growing season without exposing living stem tissue to drying out or to infection.

Epidermis

The epidermis covers the entire leaf surface and is continuous with the surface of the stem to which the leaf is attached. In most leaves the epidermis is a single layer of cells. It may consist of several kinds of cells: (*a*) ordinary **epidermal cells,** (*b*) **hair cells,** (*c*) **guard cells,** (**Fig. 10.8,** page 185). Ordinary epidermal cells show a variety of shapes, depending on the species (**Fig. 10.8,** 10.12). Epidermal cells are generally covered on their outer surfaces by a waxy cuticle (**Fig. 10.9*A,B*,** page 186) secreted by their protoplasts. The deposition of the cuticle frequently forms a pattern, quite specific for the leaves on which it is found (Fig. 10.10). The cuticle is thicker on the upperside of the leaf than on the underside.

Figure 10.7
Three-dimensional diagram of a section of a foliage leaf, ×20.

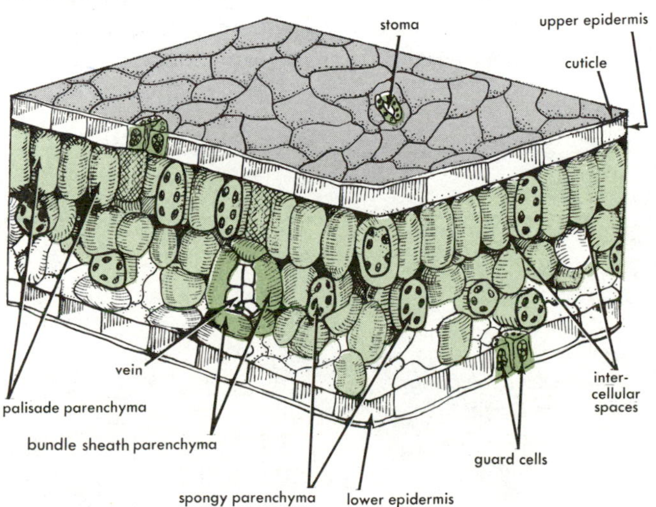

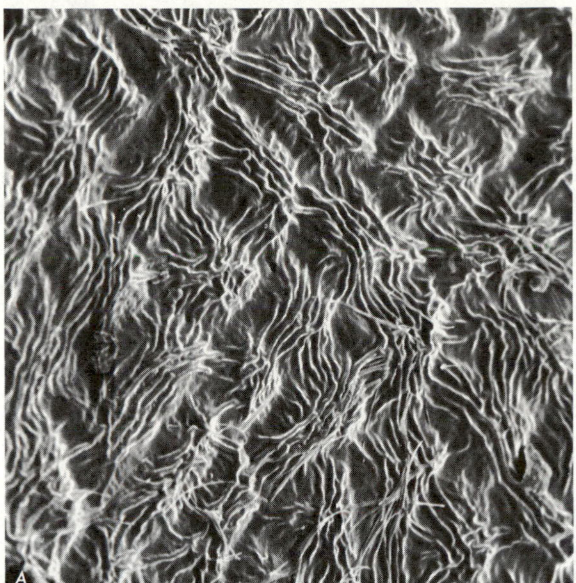

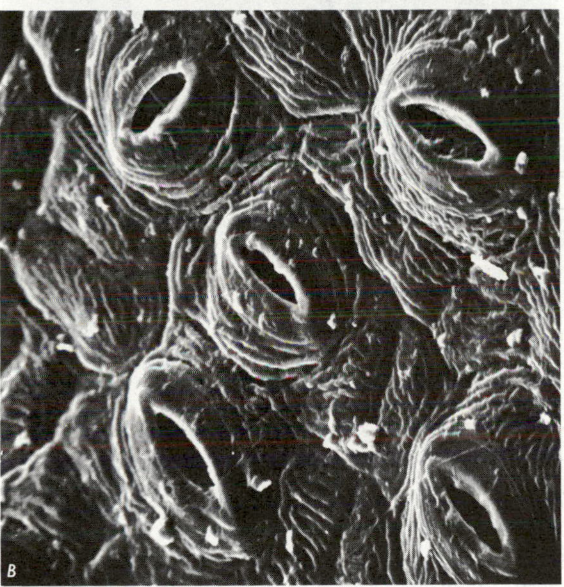

Figure 10.10
Cuticle on the epidermis of *Cnidoscolus* as seen with the scanning electron microscope. *A,* upper surface, showing cuticular ridges, ×650. *B,* lower surface at higher magnification, showing cuticular ridges between stomata, ×1100.

Several different types of **hairs** grow out from the epidermis of leaves and resemble hairs from the epidermis of stems. They may be unicellular or multicellular, simple or branched, scalelike or glandular (Figs. 10.12, 10.13). The unicellular hair, the simplest kind, may be branched or unbranched. Multicellular hairs may consist of a single row of cells, but may also be branched (Fig. 10.13*A*). At the upper end glandular hairs bear a single large cell or a group of cells, some of these cells excrete ethereal oils, which often impart a stickiness to leaves (Figs. 10.12*A*, 10.13*C*)

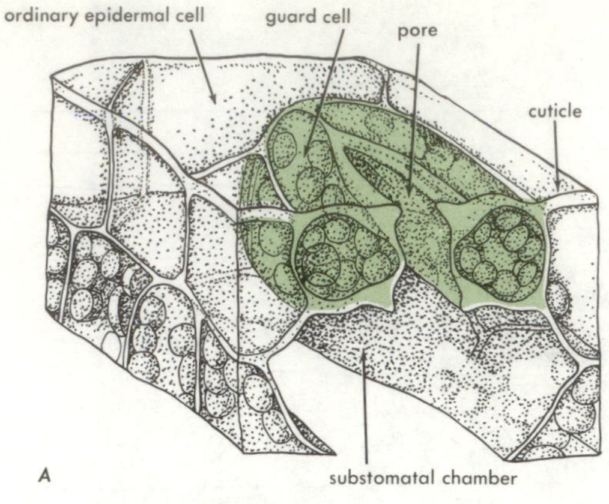

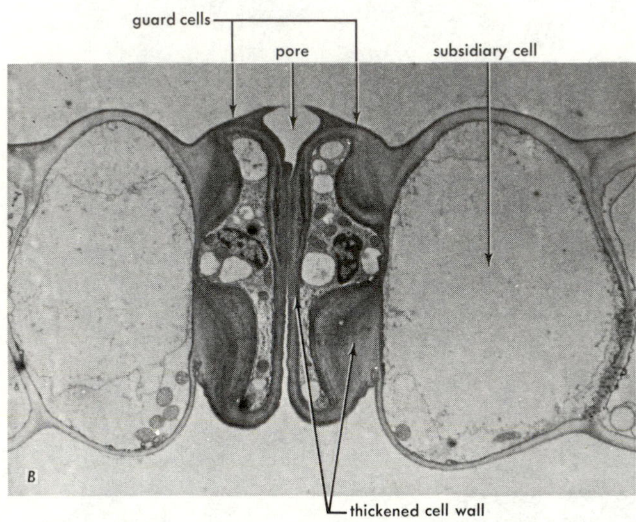

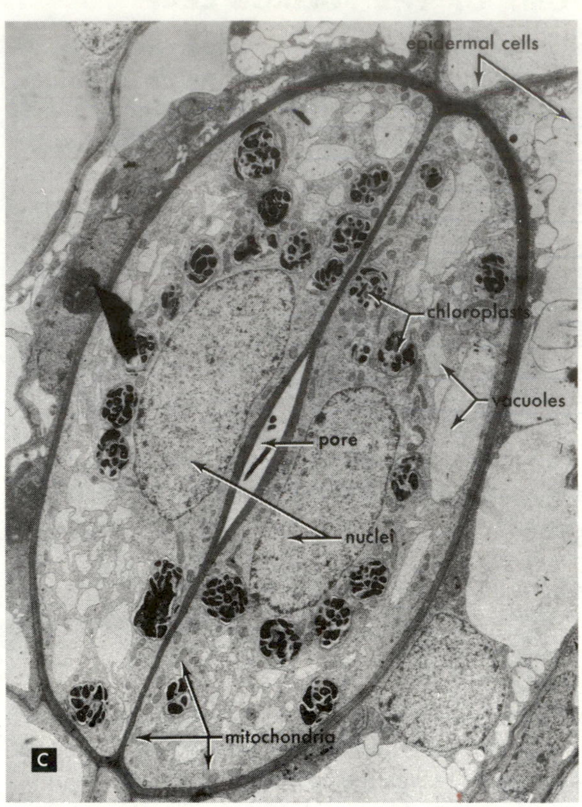

Figure 10.11
Views of stomata. *A*, diagram. *B*, corn *(Zea mays)*, *about* ×1000. *C*, rice *(Oryza sativa)*, ×900.

A **guard cell** is a special type of epidermal cell. Guard cells occur in pairs, separated by an opening or pore (Fig. 10.11). Two guard cells plus a pore are called a **stoma** (plural = **stomata**). The size of stomata varies considerably according to species and even in any one plant. Here are some representative measurements in micrometers (length × breadth): bean, 7 × 3; geranium *(Pelargonium)*, 19 × 12; corn *Zea)*, 19 × 5. The number of stomata per unit area varies widely depending on the species of plant and the environmental conditions. Usually more stomata are found on the lower (abaxial) than on the upper (adaxial) surface. Table 10.1 gives the average number of stomata per square centimeter on the upper and lower surfaces of some common plants.

Stomata are widespread in the plant kingdom. With the exception of a few submerged aquatic plants, stomata are present in all angiosperms and gymnosperms. Functional stomata have been found in liverworts, mosses, horsetails, club mosses, ferns, and cycads. In the angiosperms, they can occur on stems, petals, stamens, and pistils as well as on leaves. The major group of green plants that lack stomata is the algae. Stomata are structurally similar in the many different groups in which they are found.

Guard cells occur in pairs; each is crescent-shaped or semicircular in form, as seen in a surface view (Fig. 10.11*C*). Chloroplasts occur in guard cells but are lacking in ordinary epidermal cells. Both kinds of epidermal

Table 10.1
AVERAGE NUMBER OF STOMATA PER SQUARE CENTIMETER

Plant	Upper Epidermis	Lower Epidermis
Alfalfa (*Medicago*)	16,900	13,800
Apple (*Malus*)	0	29,400
Bean (*Phaseolus*)	4,000	28,100
Cabbage (*Brassica*)	14,100	22,600
Corn (*Zea*)	5,200	6,800
English Oak (*Quercus*)	0	45,000
Nasturtium (*Tropaeolum*)	0	13,000
Oat (*Avena*)	2,500	2,300
Potato (*Solanum*)	5,100	16,100
Tomato (*Lycopersicon*)	1,200	13,000

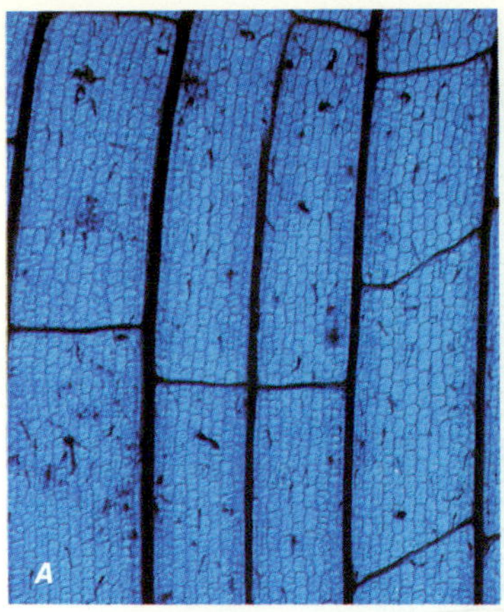

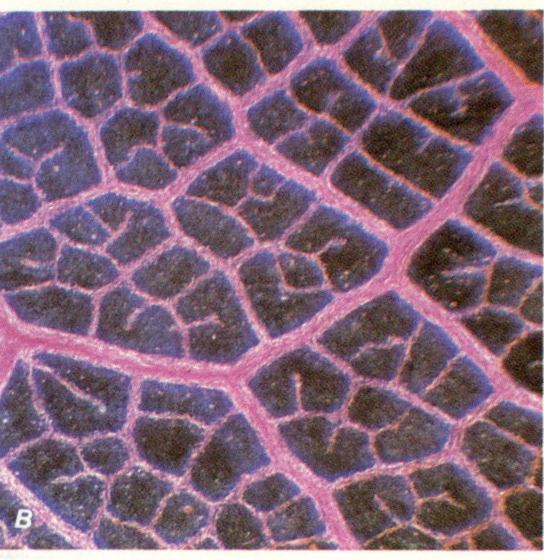

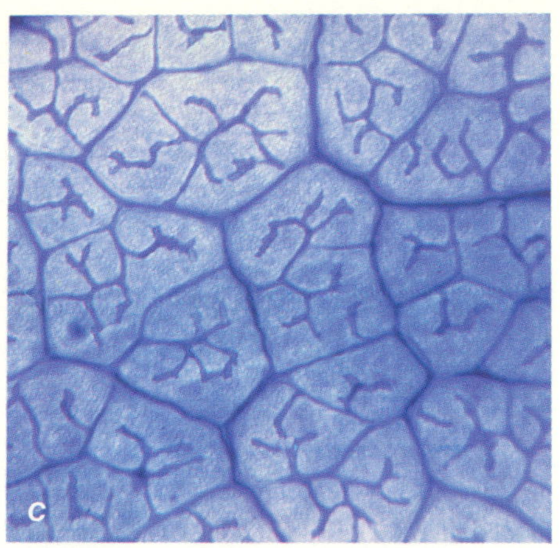

Figure 10.6
Venation patterns in leaves. *A*, parallel venation in the monocotyledonous plant, *Orthoclada* sp. *B*, netted venation, *Acer* sp. *C*, netted venation showing different branching pattern, elm (*Ulmus*, sp.). *D*, netted venation (*Oxalis* sp.) to show the vein ending.

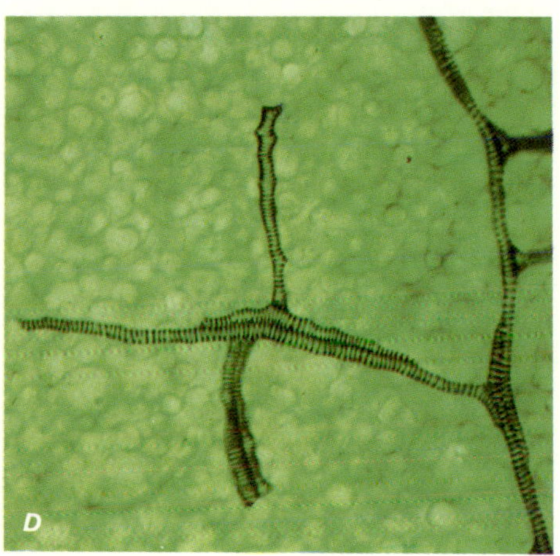

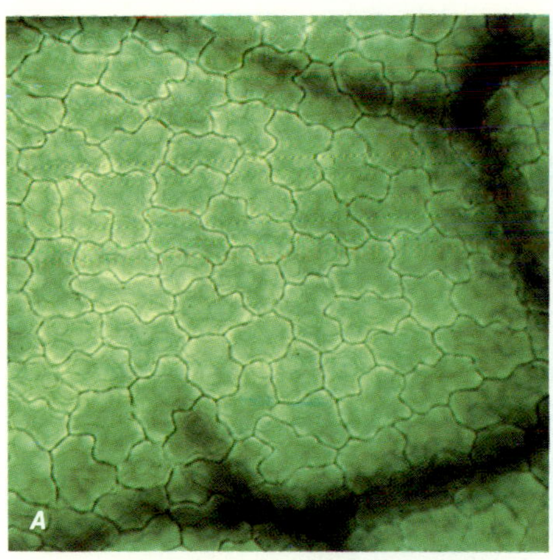

Figure 10.8
Angiosperm leaf epidermal structures. *A*, cleared leaf of *Acacia* sp. seen from the upper (adaxial) surface. *B*, cleared leaf of *Acacia* sp. but from the lower (abaxial) surface that contains stomata. *C*, stoma from leaf of *Crassula argentea*. *D*, section through the surface of a corn *(Zea mays)* leaf showing one stoma plus its subsidiary cells.

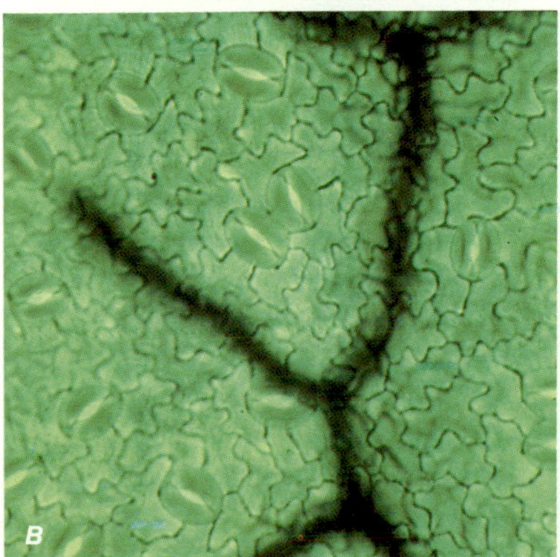

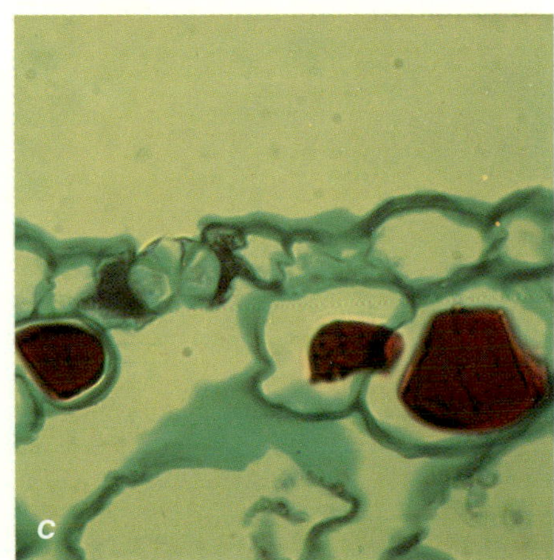

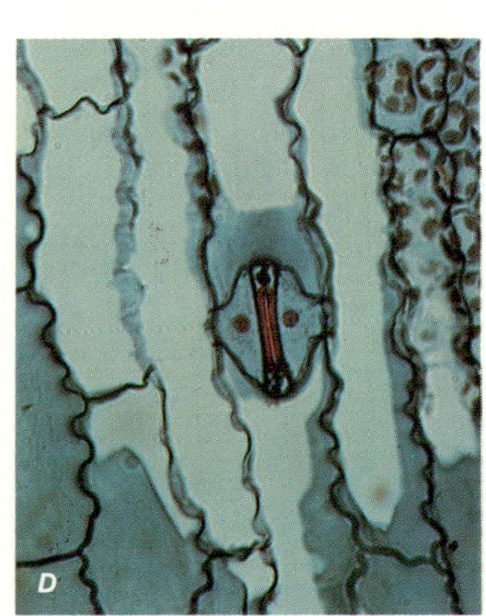

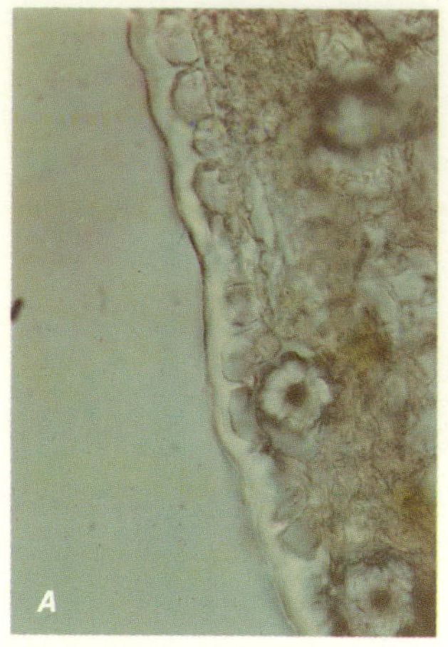

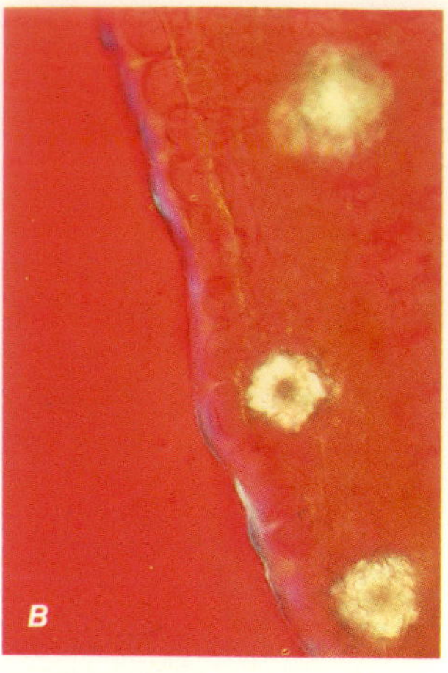

Figure 10.9
A, fresh unstained section of a jojoba leaf *(Simmondsia chinensis)*; note the thick outside epidermal wall plus cuticle. *B*, same section photographed through polarizing filters; note the different birefringent pattern of the waxy cuticle nearest the outside. Three bright structures in mesophyll cells are crystals.

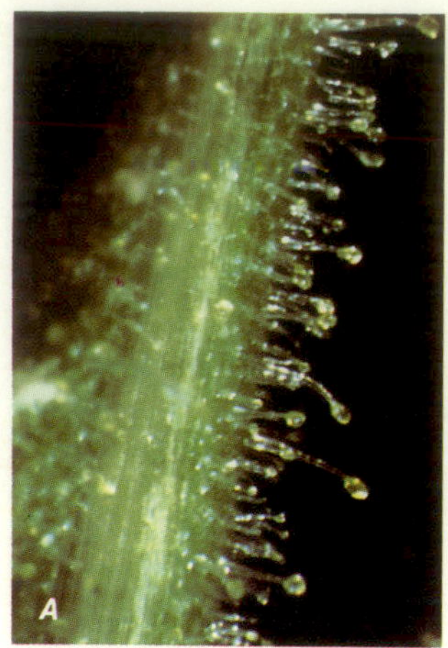

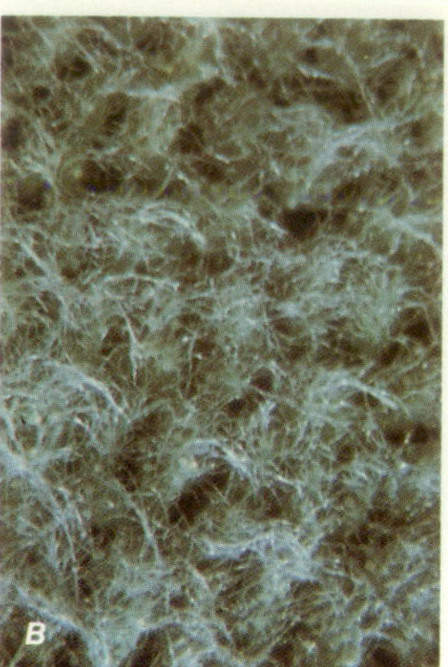

Figure 10.12
A, glandular hairs on the leaf surface of *Phaseolus* (also see Fig. 10.13*C*). *B*, mass of trichomes on surface of sage *(Salvia officinalis)*.

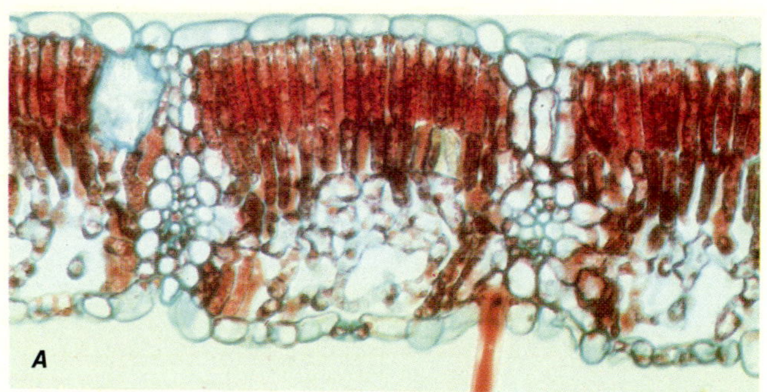

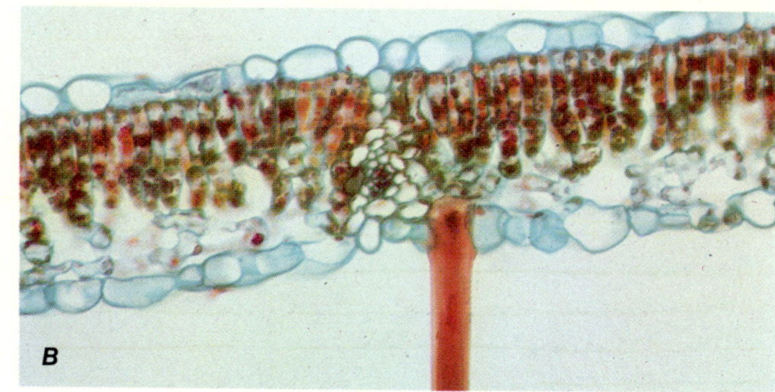

Figure 10.22
A, transverse section of sun leaf (*Acer sp.*). *B*, a shade leaf.

Figure 10.24
Leaf environmental types, xerophytic adaptations. *A*, *Nerium oleander* with stomatal crypts and sunken stomata. *B*, *Crassula argentea* leaves store water in mesophyll cells. *C*, *Ficus* sp. leaf with multiple epidermis that stores water and insulates against solar radiation.

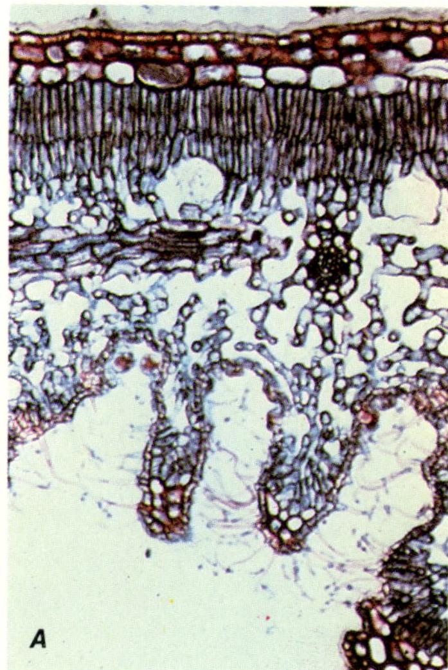

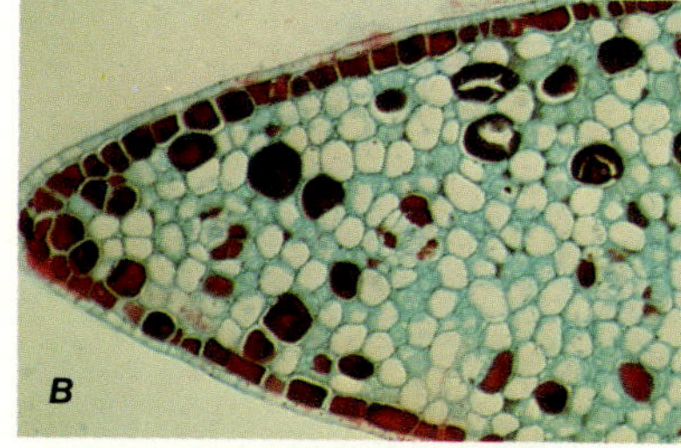

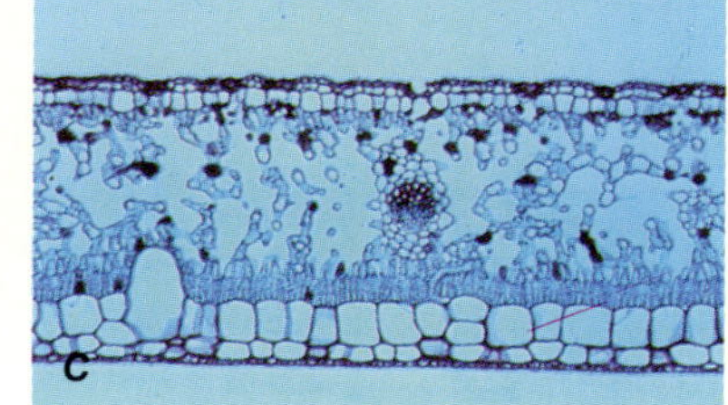

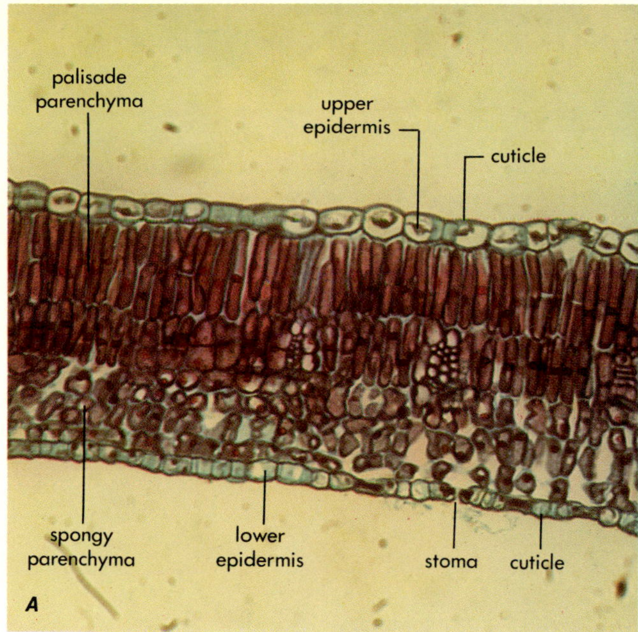

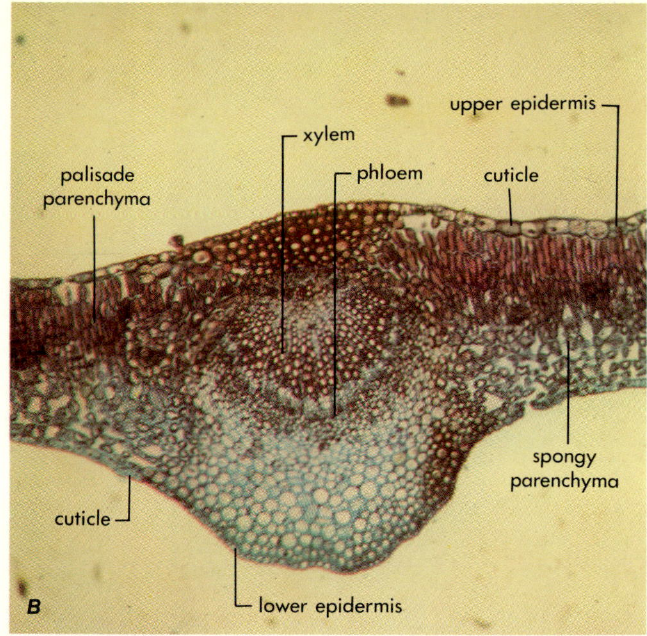

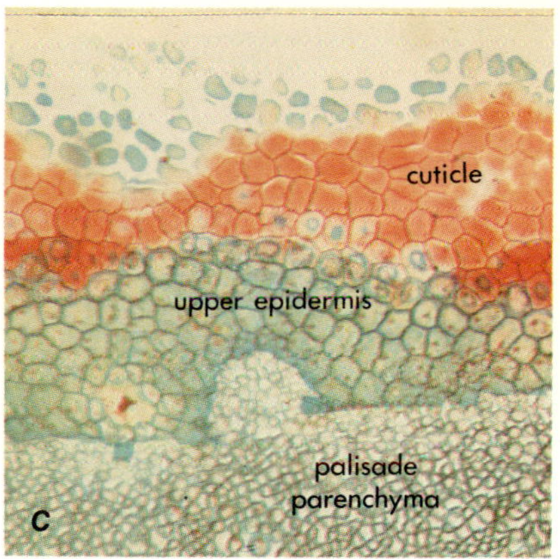

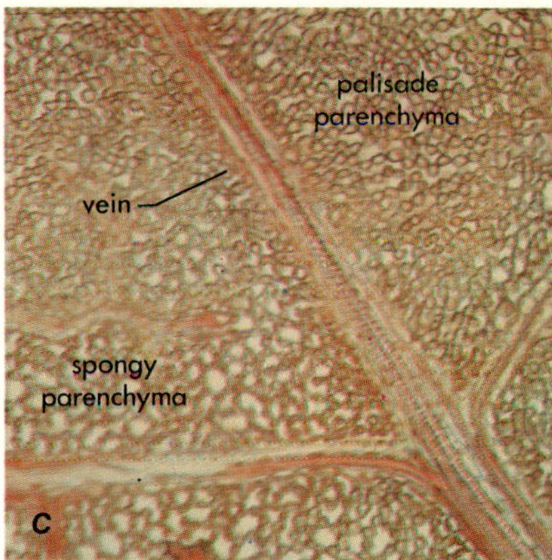

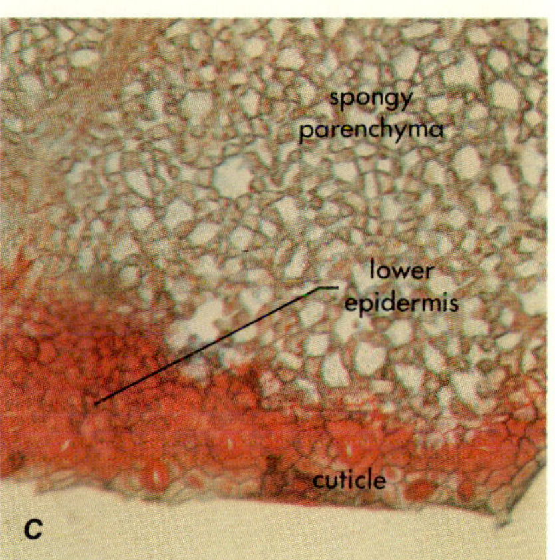

Figure 10.14
Photomicrographs of sections of a lilac *(Syringa vulgaris)* leaf. *A*, cross section of leaf, ×50. *B*, photomicrograph of a cross section of a midrib of leaf, ×40. *C*, three sections cut parallel to surface of leaf at different levels, ×75. (Slides courtesy of Triarch Products.)

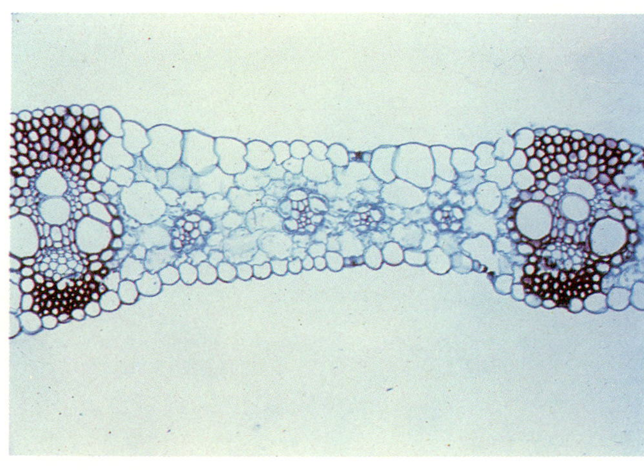

Figure 10.17
Transverse section of a corn leaf showing large bundle sheath parenchyma cells surrounding the small veins.

Figure 10.27
Insectivorous plant leaves. *A*, Venus fly trap *(Dionaea muscipula)*. *B*, the trap, a modified leaf, will close on an insect that trips the trigger hairs on the leaf surface. Hairs secrete enzymes that digest the insect. *C*, *D*, sundews. *C*, *Drosera capillaris*. *D*, *D. rotundifolia* growing in a sphagnum bog. Glandular hairs on leaf surfaces secrete a sticky substance attractive to insects. Insects become stuck; secretory enzymes produced by hairs digest insects.

Figure 10.28
Insects fall into the attractive pitchers that contain water and digestive juices. *A*, *Sarracenia* sp. *B*, *Sarracenia purpurea*. *C*, *D*, *Darlingtonia californica*. *D*, spiny hairs inside pitcher on walls inhibit insects from escaping.

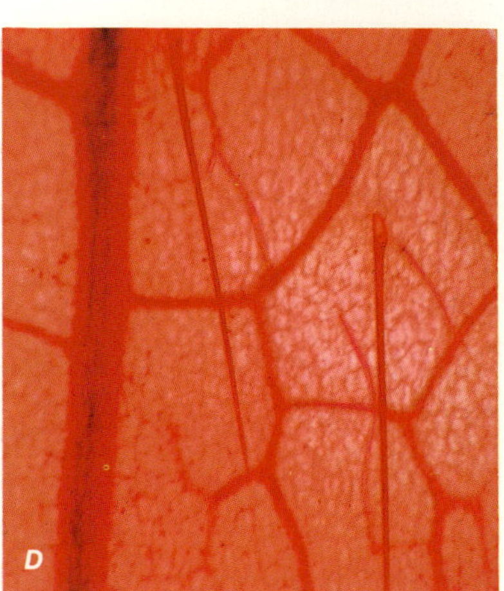

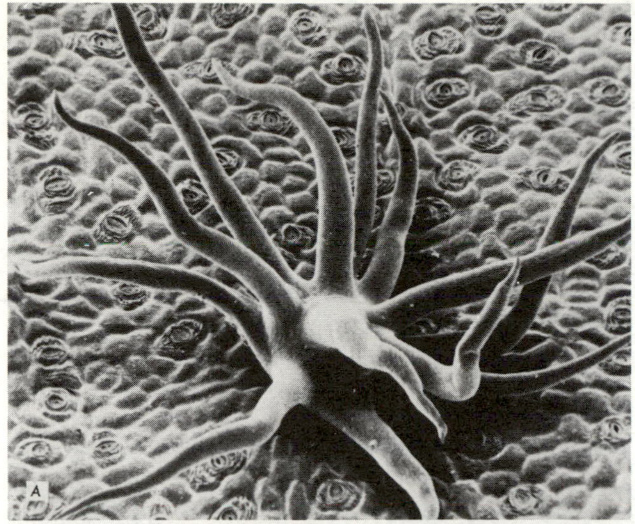

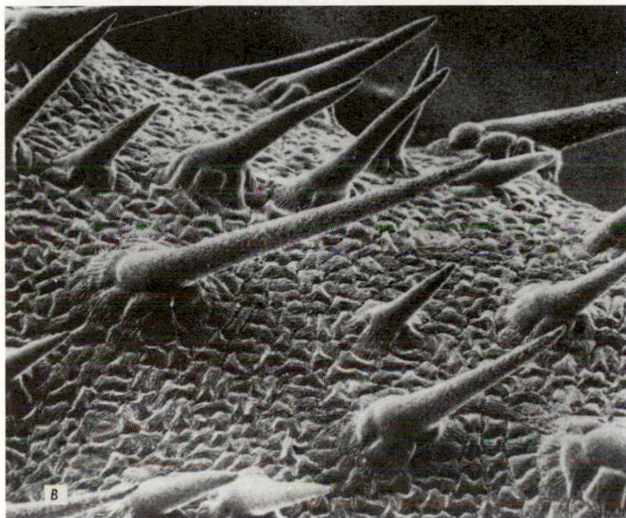

cells have long-lived protoplasts, and are covered with cuticle on their outer surfaces.

A cross-sectional view of the guard cells shows that the walls are typically unevenly thickened, with the thicker, less elastic walls, adjacent to the pore (Fig. 10.11). The stomata are the only openings in the leaf epidermis, and it is chiefly through them that gases pass into or out of the leaf. Although the cuticle is nearly impermeable to gases, small amounts of gas pass directly through the outer wall and the cuticle of epidermal cells.

Opening and Closing of Stomata

The physical opening and closing of stomata is a result of changes in guard cell turgor, particularly with reference to turgor in adjacent epidermal cells. Water movement from these adjacent epidermal cells into guard cells results in an increased turgor in the guard cells and an elastic stretching of the guard cell walls. In two associated guard cells, thin walls adjacent to epidermal cells stretch more than the thicker, inelastic wall bordering the pore. In addition, the cellulose microfibrils making up the guard cell walls tend to radiate out around the circumference of the elongated guard cells. This unique arrangement of these inelastic cellulose microfibrils restricts the radial expansion of the guard cells but allows an increase in guard cell length. These two structural features, the thickened guard cell wall adjacent to the pore and the radiating arrangement of the microfibrils, results in the formation of an elliptical aperture during the turgor expansion of the guard cells.

The change in shape of guard cells can easily be visualized by imagining that you are blowing up a long balloon with one side slightly thinner than the other. As air pressure increases in the balloon, the thin side will stretch more rapidly than the thicker side and the balloon will assume a kidney shape. The central regions of two similar balloons, attached only at their ends and with their thicker walls parallel and touching, would separate when blown up. Thus, if the structural features of guard cells are understood, it is easy to see how changes in turgor can cause either an opening or closing of stomata. But how are changes in turgor brought about? If one observes the influence of the external environment on stomatal opening, it soon becomes apparent that stomatal opening and closing can be triggered by a number of factors: Light generally causes opening and darkness, closing; wilting, or a water deficit in leaves, or an increase of CO_2 around leaves may induce closure; increase in temperature frequently results in stomatal opening.

Since such widely different environmental changes may result in opening or closing of stomata, it is evident that the osmotic regulatory mechanism of this stomatal

Figure 10.13
Trichomes: *A,* branched trichome on *Aleurites* leaf, ×350. *B,* simple trichomes on *Croton* leaf, ×350. *C,* glandular trichome on *Phaseolus* leaf, ×565.

action is complex—in fact, it is still not clearly understood. One of the earliest suggestions made to explain changes in turgor associated with light-induced stomatal opening was that sugar, an osmotically active solute, would be produced by photosynthesis in the light in guard cells but not in other epidermal cells (which lack chloroplasts). This would result in a decrease in water potential in the guard cells. Thus, water would diffuse into the cells, increasing turgor and causing stomatal opening. It was soon discovered, however, that the effects of light were very rapid and, in some instances, light intensity that induced opening was too low to produce significant photosynthesis.

An extension of this photosynthesis theory was based on observations that in light the sap of guard cells became less acid and the total amount of starch decreased. It was thought that in the presence of phosphate in the cell, a decrease in acidity induced an enzymatic breakdown of starch to a sugar phosphate which in turn formed sugar and liberated the phosphate. Water would diffuse in and stomata would open.

Although this theory is based on a number of valid observations, it cannot be used to explain all the reactions of the stomatal apparatus to environmental changes. It has been observed that both the opening and closing reactions may require cellular energy that may be supplied by the generation of ATP in the normal processes of photosynthesis and respiration. Stomatal opening in the light has been found to be correlated with an active accumulation of potassium ions in the guard cells. A decrease in the level of CO_2 in the intercellular spaces (and guard cells) induces K^+ absorption by the cells, water absorption, and stomatal opening. Such a condition would exist when stomata are closed or partly closed and light strikes the leaf. Energy for this ion accumulation comes from ATP formed during photosynthesis.

In some succulent plants the response to light signals seems to be reversed. The stomata open at night when the tendency for water loss would be reduced and close during the day when water stress would be greatest. In these plants carbon dioxide absorbed at night and trapped in organic acids is later utilized in photosynthesis during the day when the stomata are closed. This is an effective adaptation for survival under the stress of desert life (see Chapter 13). In some instances, organic acids (e.g., malic acid) are synthesized in guard cells. This results in an influx of K^+ ions, a lowering of the osmotic potential, an influx of water, an increase in turgor pressure, and stomatal opening.

Stomata frequently close under conditions of water stress. Under these conditions, abscisic acid (ABA) increases in the water moving to guard cells. Abscisic acid is a powerful inducer of stomatal closure. Thus, it appears that stomatal closure during water stress results from a signal by a chemical messenger.

Mesophyll

Mesophyll is the photosynthetic tissue between the upper and lower epidermis. It is parenchyma tissue, and is traversed by veins. Chloroplasts are present in the mesophyll cells, which, in some species, may be differentiated into two distinct layers: **palisade parenchyma** and **spongy parenchyma** (Figs. 10.7, **10.14,** page 187). The palisade parenchyma is just below the upper epidermis and usually consists of from one to several layers of narrow cells with their long axes at right angles to the leaf surface. The spongy parenchyma extends from the palisade parenchyma to the lower epidermis. Cells of the spongy parenchyma are irregular in shape and loosely arranged.

Intercellular spaces are much larger in spongy parenchyma than in palisade parenchyma. Since the air spaces between mesophyll cells are interconnecting, many cells are in contact with an intercellular space. Thus, most food-making cells have free access to carbon dioxide and oxygen. Gaseous exchange is facilitated by large air spaces, called substomatal chambers, which are generally present beneath each stoma (Fig. 10.11*A*)

A section cut obliquely through the leaf shows, essentially, cross sections of these leaf tissues. Such a view is given in **Fig. 10.14*C*.** The cuticle occupies the upper portion of the figure, with sections of cells of the upper epidermis just beneath it. Below them, palisade parenchyma cells have a circular outline and fit together very loosely. One obtains the impression of a greater amount of air space in the palisade tissue than is apparent in the cross section of leaf. Anastomosing veins, with their occasional endings, are very apparent.

Palisade parenchyma is, in general, found above veins, with spongy parenchyma below them. Note also that in this plane, cells of spongy parenchyma fit together to form an open network. The lower epidermis is at the bottom of the section (**Fig. 10.14*C***) and guard cells may be seen within it.

Intercellular spaces are prominently developed in the mesophyll but are much larger in spongy parenchyma than in palisade parenchyma. The air spaces between mesophyll cells are interconnecting, and many cells are in contact with an intercellular space. Thus, most food-making cells have free access to carbon dioxide and oxygen.

Vascular System

The veins, or vascular bundles, form a network that extends throughout the leaf. The conducting elements are xylem and phloem. Veins conduct water, mineral salts, and foods, and also mechanically support the mesophyll tissue. In addition to the midrib and larger lateral veins that are visible to the naked eye, innumerable minute branch veins can be seen with the aid of a microscope (**Fig. 10.6,** page 185). The large veins contain vessels, tracheids, sieve tubes, companion cells, and also some supportive tissue (**Fig. 10.14*B***). Veins in most leaves have only primary vascular tissue, but in some secondary growth may also occur.

In larger veins that have both xylem and phloem elements, the xylem is toward the upper surface of the leaf, and the phloem is toward the lower surface (**Fig. 10.14*B***). Smaller veins have few vascular elements and few or no supportive cells. The very end of a vein is usually a single tracheid with a spiral wall **(Fig. 10.6*D*)**. Veinlets are usually surrounded by one or more layers of parenchyma cells, the **bundle sheath,** which may or may not possess chloroplasts (Figs. 10.7, 10.15*A*). Water and solutes must pass from conducting elements of the veinlet through the bundle sheath in order to reach the mesophyll. The smallest veinlet has an unbroken connection with vascular elements of the midrib, petiole, and stem to which the leaf is connected. Plants such as corn and other tropical grasses use a photosynthetic pathway called C_4 photosynthesis (Chapter 13). Leaves of these plants usually are characterized by the presence of an enlarged bundle sheath around small minor bundles (Figs. 10.15*B,* **10.17,** page 187) and lack mesophyll differentiation into palisade and spongy types. In addition the bundle sheath cell chloroplasts contain few or no grana.

The Petiole

In the petiole, phloem and xylem maintain their relative positions; as in the stem the phloem is on the underside of the petiole (and leaf blade) and xylem is on the upperside. One or more vascular bundles are embedded in parenchyma. Fibers may be associated with vascular tissues of bundles and, not infrequently, groups of collenchyma cells occur beneath the epidermis (Fig. 10.16).

LEAF DEVELOPMENT

Leaf development is intimately associated with the differentiation of the young shoot apex.

The leaf is initiated on the flanks of the apex, as a slight bulge resulting from the enlargement and division of several cells in the outer layers of the apical meristem

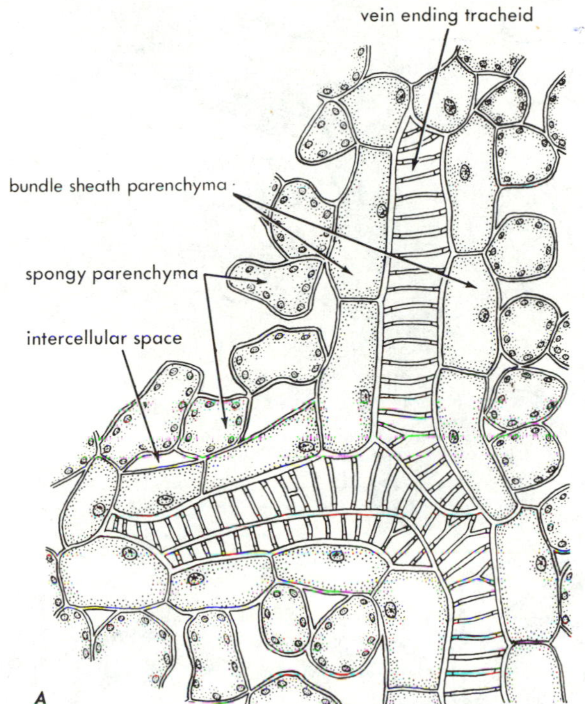

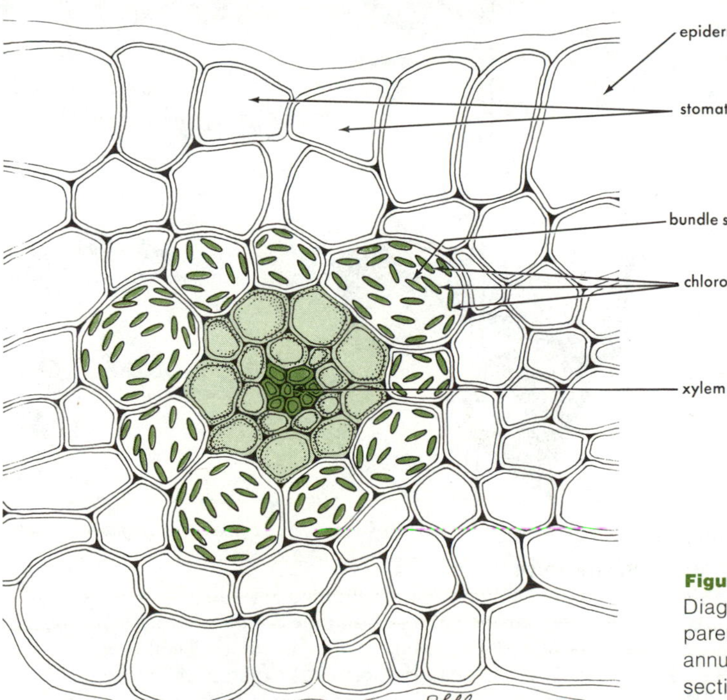

Figure 10.15
Diagrams of leaf vascular bundles. *A,* parallel section, note the parenchyma tissue bordering the vein and the presence of only annular wall thickenings in the tracheids at the vein tip. *B,* cross section of a leaf with C_4 photosynthesis; note the prominent bundle sheath cells.

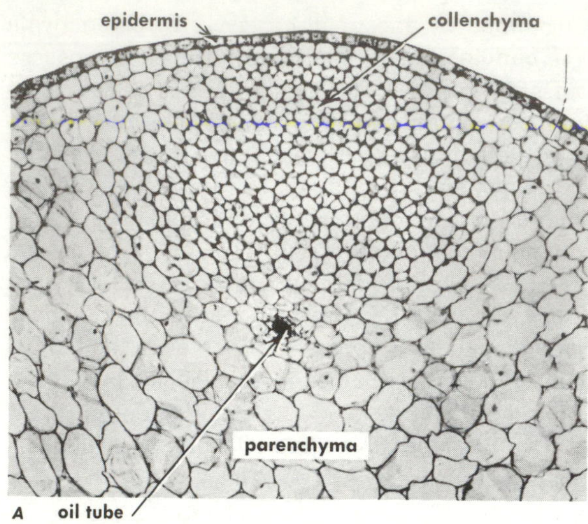

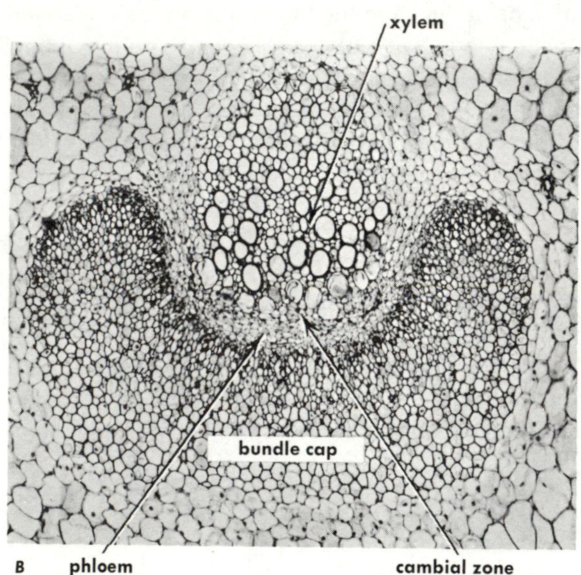

Figure 10.16
Celery *(Apium graveolens)* petiole. *A,* outer tissues. *B,* vascular bundle (vein), ×50. (Courtesy of K. Esau.)

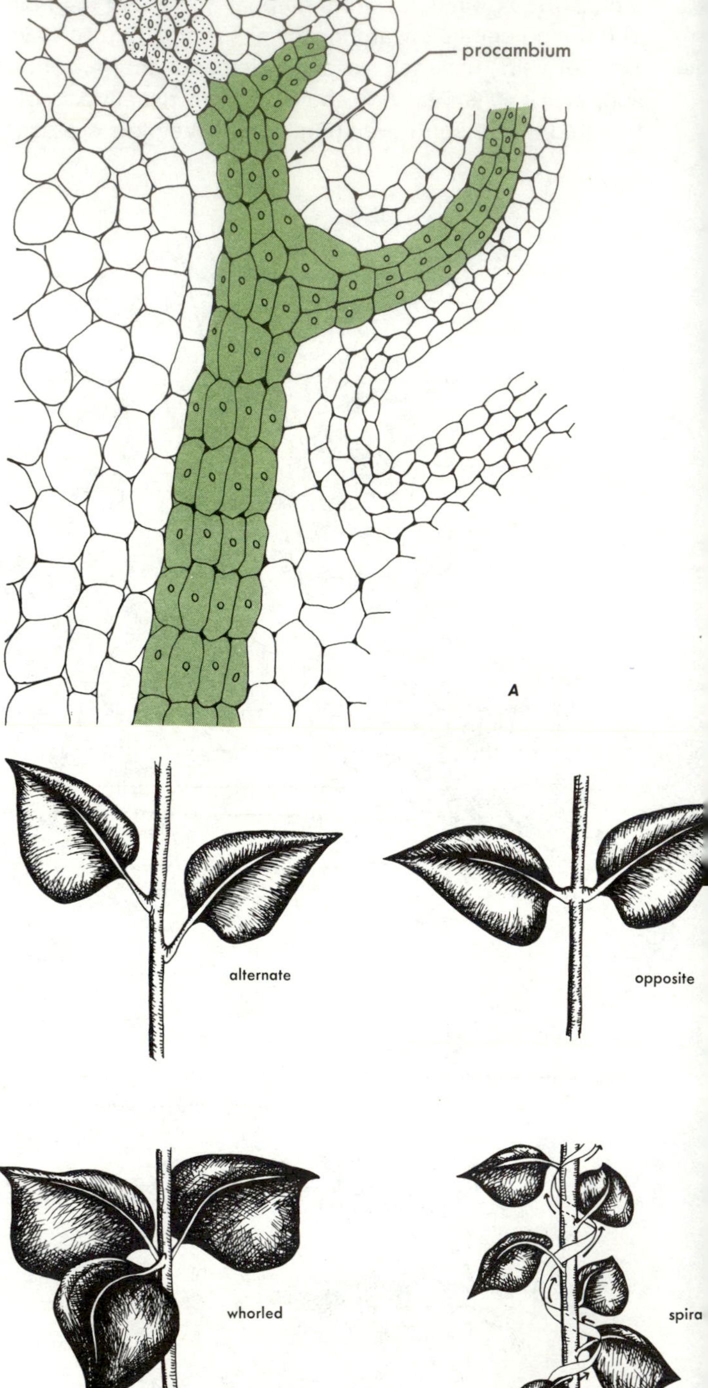

Figure 10.18
A, Diagram of a shoot apex showing the direct relationship between the initiation of a leaf and procambium strand. The second leaf shows a definite primordium and the third leaf is well-formed. The procambium strands associated with these leaves are shown. *B.* Leaf arrangement on stem axis alternate, opposite, whorled, spiral.

(Fig. 10.18*A*). This bulge continues to enlarge until it forms a thin, elongated **leaf primordium** (Figs. **7.24*A,B*,** 10.19*A*). The flanks of the primordium (the marginal meristem) will initiate the formation of the leaf blade (Figs. 10.19*B,C*). This region of growth remains active until the leaf has fully formed. Most cell divisions are completed early in leaf development; cell enlargement accounts for most of the increase in leaf size.

During the early stages of leaf formation, the cells immediately beneath each primordium also become meristematic. These cells, the procambium, will form the vascular tissue that connects the leaves to the stem's vascular system (Fig. 10.18*A*).

One problem that has continually intrigued plant biologists is how the plant regulates the site of new leaf

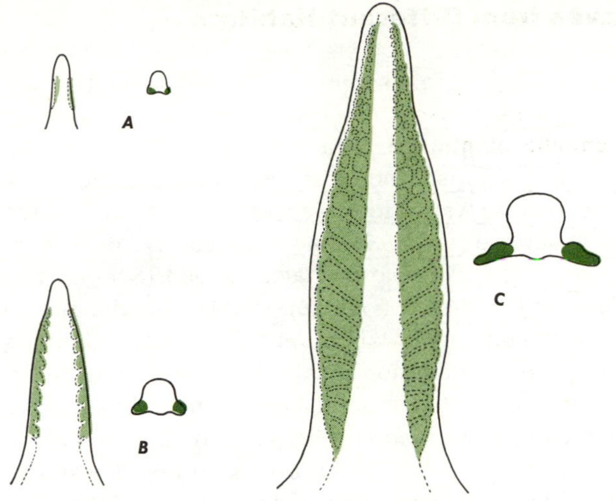

Figure 10.19
Growth of a young leaf. *A*, pencil stage, formation of marginal meristems. *B*, activity of marginal meristems initiates lateral expansion of a young leaf. *C*, continued lateral growth through the activity of marginal meristems.

Figure 10.20
Different leaf shapes induced in leaves of tomato *(Lycopersicon esculentum)* by single gene mutations. (Courtesy of Tomato Genetics Cooperative.)

initiation. The leaves of vascular plants are distributed on the stem in four base patterns **(phyllotaxy)–opposite, alternate, whorled,** and **spiral** (Fig. 10.18*D*). The position of a new leaf primordium is closely correlated with the position of the next older primordium. This has been proven by the experimental removal or isolation of the youngest visible primordium on a shoot apex, causing the next formed primordium to develop closer to it. It is also known, at least in ferns, that if the procambium beneath a suspected leaf primordium is severed, the primordium will not develop.

LEAF SHAPE

Internal Control

Leaf shape is under direct genetic control. Several single gene mutations of tomato (*Lycopersicon esculentum*) result in striking changes in leaf shape (Fig. 10.20). Leaves are also affected by internal factors, for example growth regulators, or by environmental conditions. Any situation where more than one leaf form exists on a plant is known as **heterophylly.** When treated with the hormone gibberellic acid, thistle plants (*Centaurea*), for example, formed abnormal, entire leaves instead of normal, lobed leaves.

Leaf shape may also vary considerably with the physiological age of plants. Thus, some plants (ivy–*Hedera helix,* and some species of *Eucalyptus* are examples) have distinctive juvenile leaves for the first few years of growth. Subsequent adult leaves of a different form are characteristic of the older or adult plant (Fig. 10.21). The age-related response is usually associated with flowering, that is, only adult plants are capable of producing flowers and subsequently fruits and seeds.

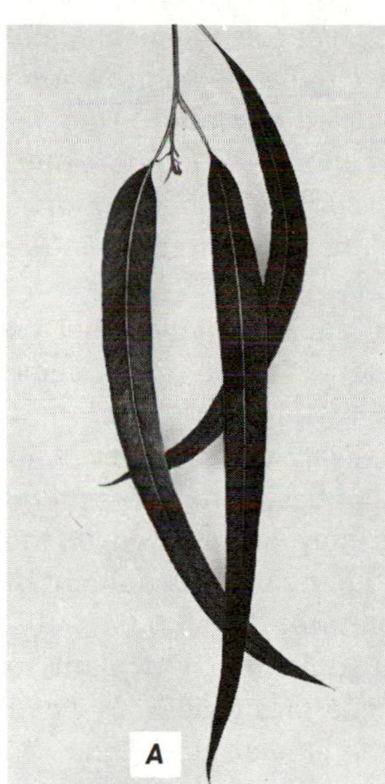

Figure 10.21
A, adult form of *Eucalyptus* sp. leaves ×¼. *B*, juvenile form ×½.

Environmental Factors

Leaf shape is also influenced by environmental factors such as light, moisture, and temperature. The total intensity, wavelength, and daily duration of light have separate but interrelated effects. Plants growing in intense sunlight **(sun leaves)** usually have thick leaves with a thick palisade tissue and a dense, spongy parenchyma (**Fig. 10.22*A*,** page 186). Their intercellular spaces are small, and the epidermis is heavily cutinized and generally glossy, with the stomata confined to the lower epidermis. These leaves may also have woolly epidermal hairs. Leaves of the same species growing in shade **(shade leaves)** have contrasting traits (**Fig. 10.22*B***). They are thin, possessing a single palisade layer and a spongy parenchyma with many intercellular spaces. The epidermis has a thin cuticle that is usually dull, and stomata may be present on both upper and lower epidermal surfaces.

Light is required for the normal development of leaves, including the differentiation of chloroplasts to a state where they are capable of carrying on photosynthesis. In darkness, the leaves of a typical dicot remain small and pale yellow while the stems grow long and slender. This condition is called **etiolation.** Regular daily illumination that is intense enough to support photosynthesis is required for the continued healthy existence of leaves. Excessive shading usually results in the death of a leaf.

Submerged leaves of semi-aquatic plants may be vastly different in shape from aerial leaves of the same plants (Fig. 10.23). Water leaves tend to be deeply lobed, very thin, and tend to have relatively undeveloped palisade mesophyll, and little vascular tissue. Air leaves tend to be thicker. Similar changes may be induced in the aerial portion of the shoot by reducing the CO_2 content of the air and lowering the temperature.

Nitrogen starvation may also induce the plant to acquire leaf traits similar to those of sun leaves; water stress may change the plant in a similar way. Thus, we see that environmental stress is an important, but complex, factor.

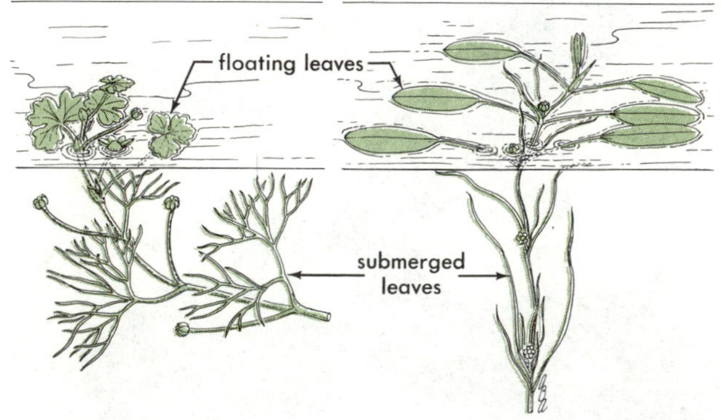

Figure 10.23
Two examples of aquatic plants with both air and water leaves.

Leaves from Different Habitats

Leaves of angiosperm plants have evolved at least in part on the basis of their survival strategies in certain climates. Plants capable of growing well in dry habitats (**xerophytic plants**), for example, maintain leaves that have a characteristic structure usually including some means to store water, to inhibit its evaporation through the leaf surface, or to withstand dryness by some manner. One good example is that of jojoba (*Simmondsia chinensis*) (**Fig. 10.9,** page 186). In this case the cuticle and outer epidermal wall is very thick, thus inhibiting evaporation. A reduction in air spaces between mesophyll cells is another modification to inhibit water loss. Leaves of *Nerium oleander* (Fig. 10.24*A*) have adapted to a dry climate by evolving leaf crypts on their lower surface. These crypts contain the stomata and are interlaced with long trichomes that trap moisture and create an artificial moist microclimate, thereby inhibiting stomata opening. Olive (*Olea* sp.) leaves inhibit water loss by the presence of a thick mat of surface trichomes while *Crassula argentea* (**Fig. 10.24*B*,** page 186) has solved the water problem by evolving a means to store water in mesophyll cells. Fig leaves (*Ficus* sp.) have evolved a capacity to store water, but in a multiple epidermis of large, thin-walled cells (**Fig. 10.24C**). Large calcium carbonate crystals called cystoliths are found in some of the cells of this tissue.

Leaves that grow submerged in, floating on, or living in water or wet soil (**hydrophytic plants** have also evolved characteristic leaf structures. In this case since water loss isn't a problem, mechanisms to conserve water have no value. Instead leaves with few or no stomata have evolved, and many plants have leaves with limited vascular tissue and mesophyll development. Some hydrophytic leaves are also very thin and sometimes tend to be deeply lobed or narrow.

Other types of leaves, on plants living in moderate climates (**mesophytic plants**), have anatomy of a more moderate form such as in lilac (**Fig. 10.14,** page 187).

LEAF ABSCISSION

The separation of plant parts from the parent plant is a normal, continually occurring phenomenon. Leaves fall, fruits drop, flower parts wither and fall away, and even branch tips or whole branches may be separated normally from some plants. The fall of leaves after the stress of drought and in the autumn are common examples.

Separation or **abscission** is the result of activity in a special region known as the **abscission zone** that typically is found at the base of the petiole (Figs. 10.25 *A* to *D*). The cells of the abscission zone are smaller and more densely filled with cytoplasm than cells in adjacent regions. Vascular elements are shorter, fibers are typically lacking, and in general there is little lignification of cell walls in the abscission zone. These anatomical features make the zone a region of weakness.

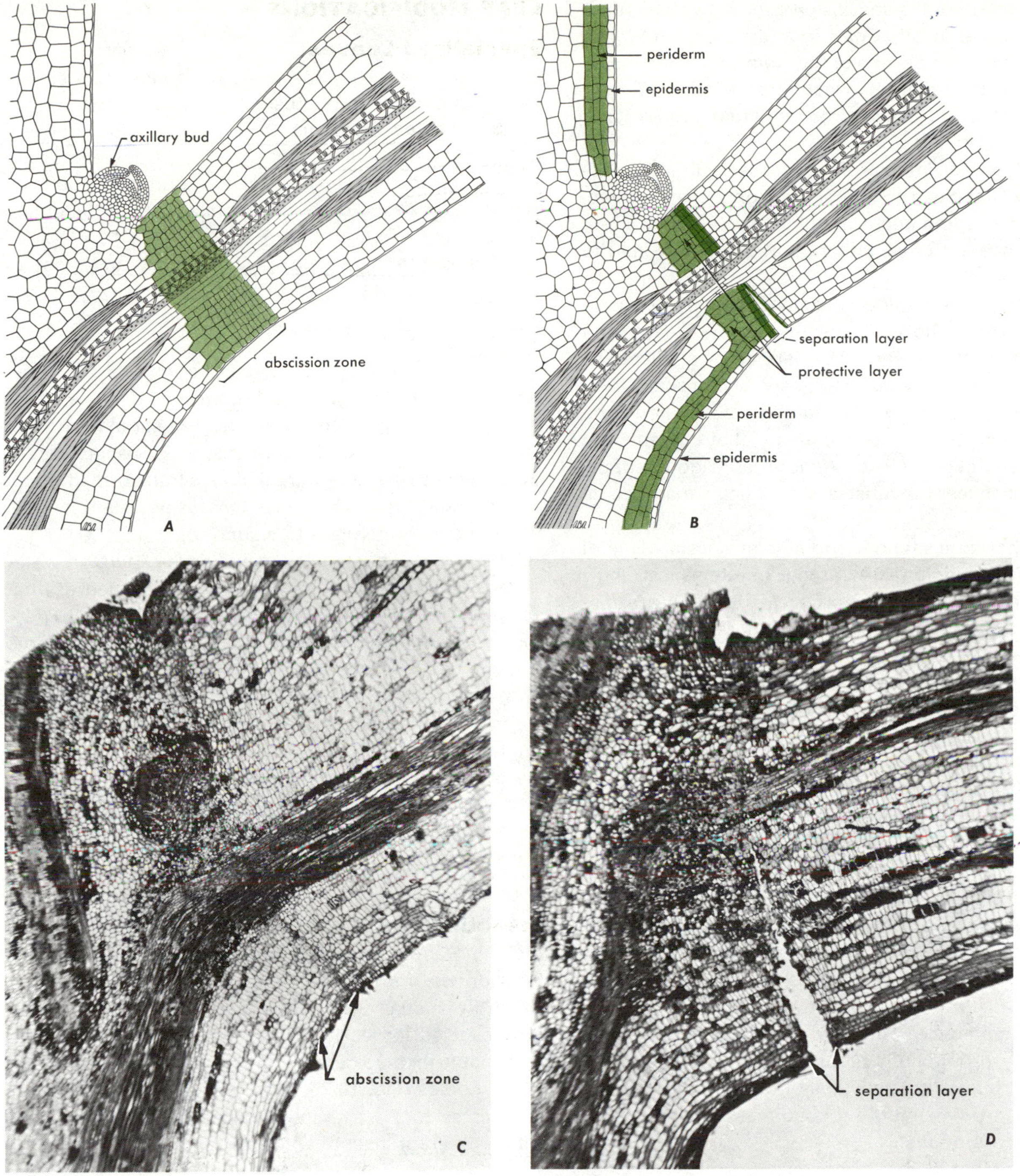

Figure 10.25
Leaf abscission. *A* and *C*, formation of the abscission zone. *B* and *D*, separation within the abscission zone and initiation of periderm. (*A* and *B*, courtesy of F. T. and A. B. Addicott; *C* and *D*, courtesy of Hall.)

As leaf abscission approaches, a layer of cells becomes physiologically active across the abscission zone. This is called the **separation layer.** The activity of the cells results in the secretion of pectinases and cellulases that digest the middle lamella and at least parts of the primary cell wall, permitting separation of the cells and, in consequence, the abscission of the entire leaf. At the same time, the active cells closest to the stem usually divide and form a **protective layer** very similar to the cork that develops in the cortex of a stem. Typically the protective layer joins the cork of the stem and becomes continuous with it (Fig. 10.25*B*).

The function of the abscission zone is twofold: (*a*) to separate the leaf or other part from the plant and (*b*) to protect the exposed portion of the stem from infection by bacteria or fungi and from insect penetration. Leaf scars covered by their protective layers are often quite distinctive in their form (Fig. 7.6).

A healthy leaf inhibits its own abscission. It is abscised only after it can no longer function, because of factors such as shading and drought stress. Injury, disease, or senescence are other factors that commonly lead to abscission.

It has been known since 1933 that **auxin (IAA)** can inhibit abscission. In horticulture, abscission of leaves and fruits can be easily prevented or delayed by applications of synthetic auxins such as naphalene acetic acid (NAA). Auxin coming from a healthy leaf to its abscission zone is now recognized as the messenger that prevents the onset of separation activity. As long as a good supply of auxin reaches the abscission zone, separation is prevented.

Other hormones tend to promote abscission. One of these is **abscisic acid (ABA).** In many systems it strongly promotes abscission (other properties of ABA are discussed in Chapter 20). However, ABA can only act on abscission after the retarding effects of IAA have been lowered. Thus, there is a balance between the two hormones. Abscission is favored when the relative amounts of ABA rise; abscission is prevented when the relative amounts of auxin are high. **Ethylene (ETH)** also promotes leaf abscission. In the leaf blade it lowers the levels of IAA by increasing the natural inactivation of IAA and by interfering with the transport of IAA from the leaf blade to the abscission zone. However, ethylene is not produced in significant amounts in senescent leaves. So, this action may be of only minor importance in nature. Ethylene does have a second more significant function in the abscission zone. Here it is produced when the hormone message from the leaf blade indicates that abscission must commence. Its function is to promote respiration (energy production) required for the synthesis of enzymes and the other activities of the separation layer. Thus, ethylene is a kind of "second messenger," speeding the process of abscission.

The two other major hormones, gibberellin and cytokinin, also influence abscission, but in less definite ways. In general, abscission is controlled by an interaction of all the hormones. The hormones that promote growth in general inhibit abscission, and the hormones that retard growth tend to promote abscission. The balance between these opposing influences determines whether or not any particular leaf will be abscised.

Environmental cues are important in synchronizing abscission with the seasons. These cues are most commonly cold temperature or short days that in turn induce hormonal changes that then affect the formation of the abscission zone (Chapter 20).

LEAF MODIFICATIONS

Specialized Leaves

Leaf shape may be considerably modified to perform functions other than photosynthesis (Fig. 10.26).

Bud scales are short, thick, sessile, often covered with dense hairs on the outer surface, and sometimes are waxy or resinous. When present, they protect the delicate meristematic tissue of the shoot tip and the rudimentary leaves from drying out (Fig. 7.2).

The **spines** of various species of cacti and those of *Fouquieria* represent entire transformed leaves. In the black locust (Fig. 10.26*A*) the spines are stipules, rather than leaves.

In some species of *Lathyrus,* the *tendrils* are transformed leaflets; in other species, the whole leaf is transformed into a single tendril, and leaflike stipules perform the normal functions of the leaves. In *Bignonia capreolata* (trumpet vine), the third leaflet is transformed into a tendril (Fig. 10.26*D*). Both leaf tendrils and stem tendrils serve to attach the plant to a support.

Leaves are sometimes modified as food or water storage organs. The thick, fleshy bases of leaves that comprise much of the daffodil (*Narcissus*) bulb (Fig. 8.24*D*) accumulate large quantities of food. Succulents found in deserts and saline soils have thick, fleshy leaves with special water-storage tissue.

In certain species of *Acacia* two leaf types form on a single branch. One type is a pinnately compound leaf, and the other is a modified flattened leaf called a phyllode. This leaf modification, previously considered an enlarged petiole, is actually a leaf that has thickened dramatically in the vertical plane (Fig. 10.26*C*).

Insectivorous Plants

A striking and bizarre adaptation of leaves for a special function occurs in insectivorous plants. These plants, which live usually in mineral poor bogs or wet areas, have evolved leaves with forms and structural features that enable them to capture, digest, and obtain food from insect bodies **(Fig. 10.27,** page 188, **10.28,** page 188. Let us discuss four well-known insectivorous plants, each with a different insect-capturing strategy.

The leaves of the Venus fly trap (*Dionaea muscipula*) are modified at their tips into a specialized hinged trap (**Fig. 10.27*A***). Flies, ants, and other insects are attracted to these leaves. Inside the hinged portion of each leaf are several long trigger hairs. As the unsuspecting insect walks along the leaf surface, it strikes or pushes these hairs inadvertently. After at least three hairs have been touched, a hydraulic response is stimulated and the trap rapidly closes (**Fig. 10.27*B***). Cells in the hinge region of the leaf apparently can lose water very rapidly allowing the leaf to close. Long projections along the leaf margin cage the insect in place and many short glandular digestive hairs secrete digestive enzymes and slowly digest the

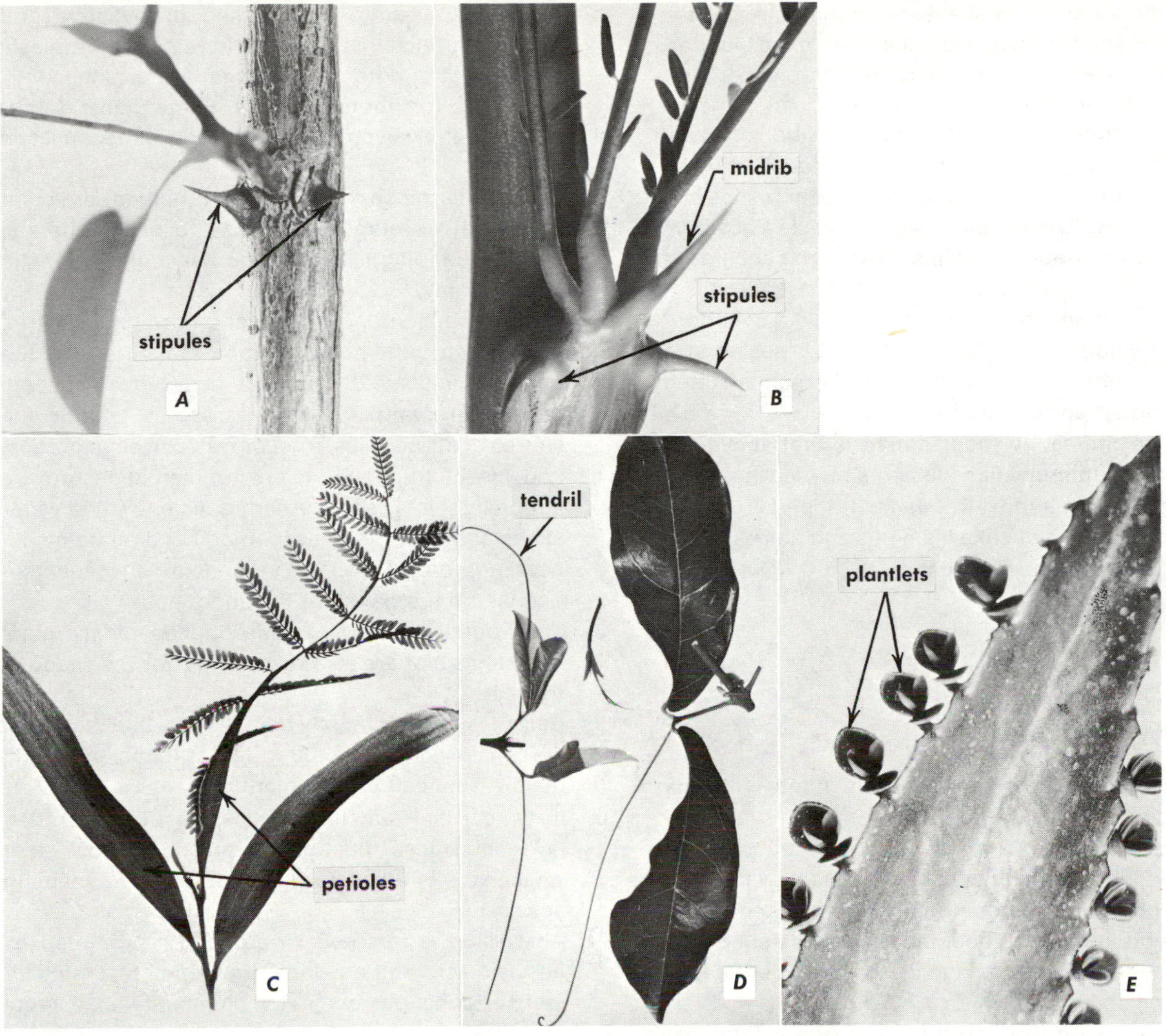

Figure 10.26
Leaf modifications. *A*, stipular spines of the black locust *(Robinia pseudoacacia)*. *B*, spines from stipules and midrib in *Parkinsonia*. *C*, phyllodes of *Acacia*. *D*, tendrils from leaflets of *Bignonia*. *E*, plantlets on leaf of *Bryophyllum*, ×¼ to ×1.

trapped insect. After several days, the trap opens once again for new business.

The sundew (*Drosera* sp.) has a different, though equally nefarious scheme to capture insects. These plants have leaves sometimes 60 cm. in length that are covered on their upper surface with multicellular tentacles **(Fig. 10.27*C*)**. A fly landing on a sundew leaf immediately becomes stuck to some of the tentacle tips **(Fig. 10.27*D*)**. As the insect struggles to escape, it stimulates nearby tentacles to bend toward the captive and consequently the insect becomes hopelessly entangled. As the insect is gradually digested, its minerals and nutrients are absorbed by the tentacle cells and transported back to the sundew plant body.

Pitcher plants (*Sarracenia* and *Darlingtonia*) have amazing leaf adaptations consisting of an elaborate pitcher-shaped leaf, which contains a pool of water and digestive juices. Insects are attracted to the pitcher by its often fetid smell or by secretion of nectar from glands located on the inner rim of the pitcher; they land on the slippery pitcher lip and fall into the pool within **(Fig. 10.28)**. *Darlingtonia* **(Fig. 10.28*C*)** has an additional anti-escape mechanism in the form of short, spiny trichomes pointing downward on the inside pitcher wall **(Fig. 10.28*D*)**. These trichomes provide a barrier to keep any insects from crawling out. When they tire, they fall back into the pool and are digested!

The bladderwort, *Utricularia* is a fascinating insectivorous plant with a unique trapping mechanism, the bladder. The aquatic species are mostly free floating plants. The bladder, a lens-shaped or flattened pear-shaped vesicle, is equivalent to a segment of a leaf. At the anterior end of the bladder, there is a semicircular opening that is closed by a valve seated on a lower threshold and attached to the sides and the upper margin. In *Utricularia vulgaris*, a very common species, four to six stiff tapering bristlelike hairs that make up the tripping mechanism project from the door of the bladder.

When the bladder is set, the contents are under a reduced pressure, and the sides of the bladder are concave and distorted from their normal convex position. If any of the hairs of the tripping mechanism are touched when the trap is set, the door is distorted, and water rushes into the trap as the walls snap back to their relaxed position. Any small insect or aquatic animal next to the opening is sucked into the bladder; the door is then closed. This entire process may take place within less than a second.

When the door of the trap is seated, it is sealed by a glandular secretion of mucilage. Water is then secreted from the bladder and the trap is reset.

A further interesting modification is that above the door, branched antennaelike filaments resembling algae are attached. Animals most frequently trapped are those that feed among the algaelike filaments and come in contact with the trigger mechanism.

SUMMARY—LEAVES

1. The chief functions of leaves are photosynthesis and transpiration.
2. Angiosperm leaves are generally supplied with flat, thin blades that are attached to the stem by petioles or sheaths. Veins strengthen the blades and transport food and water. Leaf blades may be simple or compound; leaf margins may be entire, dentate, or lobed.
3. A cross section through the blade of a typical leaf shows the following tissues: upper epidermis, mesophyll, differentiated into palisade parenchyma and spongy parenchyma, and lower epidermis. Generally, a waxy cuticle coats the epidermis. Guard cells of the epidermis form stomata that control gas exchange.
4. Mesophyll tissue is comprised of palisade and spongy parenchyma tissues and veins. Chloroplasts are present in both parenchyma tissues. Mesophyll is adapted for photosynthesis. Some plants do not differentiate mesophyll into spongy and palisade regions.
5. Xerophytes, mesophytes, and hydrophytes are plant types that develop in response to dry, moderate, and wet conditions. Each type has a characteristic leaf anatomy.
6. Leaf shape is under genetic and hormonal control, and is also influenced by light, moisture, and the physiological age of the plant. Leaves may become modified, serving as bud scales, spines, tendrils, the source of propagules, food or water storage organs, and insect traps. Such environmental factors as strong light intensity, nutrient deficiency, and water stress can affect leaf morphology and anatomy.
7. Leaf primordia develop in a definite spatial sequence on the flanks of the shoot apex.
8. The young leaf lamina is produced by marginal meristems that are also responsible for the shape of the leaf.
9. Cell divisions are mostly completed while the leaves are enclosed in the bud; leaf expansion results mainly from cell enlargement.
10. In typical leaf and fruit abscission the separation layer produces pectinases and cellulases that weaken the middle lamellae and cell walls, enabling abscission.
11. Abscission is retarded by auxin and promoted by abscisic acid and ethylene. Environmental cues affect hormonal balance in the plant and hence affect abscission.
12. Several plants have developed highly specialized leaves, for example, bud scales, spines, and phyllodes adapted to perform specialized functions.
13. Insectivorous plants have evolved very specialized leaf forms adapted to capturing insects and digesting their bodies.

CHAPTER 11

SOIL AND MINERAL NUTRITION

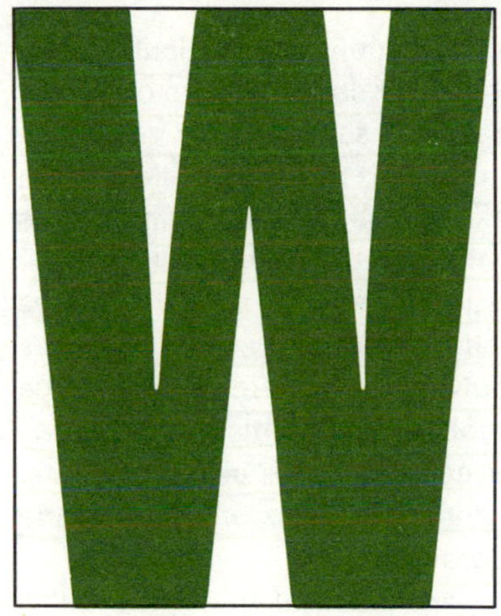

We have been studying the morphology of seed plants—the structure of their organs, tissues, and cells. With this knowledge to help us, we are now ready to consider in more detail the physiology of the plant—the ways in which the various parts function together in the processes of absorption, conduction, transpiration, photosynthesis, and respiration.

SOIL—THE ENVIRONMENT OF ROOTS

Soil may be defined as the weathered superficial layer of the earth's crust that typically is made up of decomposed and partly decomposed parent rock material with associated organic matter in various stages of decomposition.

Soil is the natural medium in which the roots of most plants grow. From soil the plant absorbs water and solutes necessary for its development. If a soil is fertile, it contains in a readily available form all the chemical elements essential for plant growth. It is through soil that the agriculturist can effectively alter the environment of roots and thus control plant growth, at least partially. The time and method of fertilizer used, as well as cultivation and irrigation practices, are all directed toward increasing production of plant products through effects that these practices have on soil and root relationships and ultimately on the growth and development of the plant.

There are many kinds of soils and many different soil conditions. The character of natural plant covering and behavior of crops depend upon soil conditions as well as upon climatic conditions. Environmental factors that operate through soil are called **edaphic factors;** those that act upon the plant through the atmosphere are called **climatic factors.**

Soil Formation

Each environment creates a soil type unique to it. These soil types have their own history of development, morphology, and chemical attributes. Soil is a complex system that includes mineral (inorganic) matter, organic matter, water, air, and organisms.

Most soils consist largely of mineral particles formed by a slow continual process of weathering of the parent rock. Mineral matter is derived from fragmented rock. The kind of parent rock (granite, lava, sandstone, limestone, shale, etc.) and the degree of weathering determines the nature of the mineral or inorganic components of the soil.

Weathering may involve simply **mechanical breaking** of the parent rock. For instance, the action of strong winds or wave action may hurl sand against rock outcroppings and wear them down, or water collecting in pores of the rocks may expand at freezing temperatures to create fissures that ultimately fragment the rock. **Chemical weathering** results in more profound changes in the mineral matter itself. Atmospheric gases, such as CO_2 or SO_2, become dissolved in rainwater and produce acidic solutions that dissolve the parent rock material. Plant roots secrete weak acids. Certain algae, bacteria, and lichens hasten the decomposition of some of the least resistant of the mineral matter.

Weathering of the parent material under the influence of climate and organisms proceeds along a definite series of stages from a young soil in which the processes of soil development are continuing to a mature soil that has approached a state of equilibrium.

Soil formation may proceed relatively rapidly if the type of parent material and type of climate are favorable. Conditions that hasten the rate of soil development are a warm, humid climate, forest vegetation, flat topography, and a parent material that is easily broken down. Fort Kamenetz, built of limestone slabs in the Ukrainian part of Russia, was abandoned in 1699. Today, a mature soil 10 to 40 centimeters thick has developed from the limestone. Conditions which retard soil development are a cold, dry climate, grass vegetation, steeply sloping topography, and parent rock material that is not easily broken down. It has taken from 1000 to 10,000 years for mature soils to develop in northern areas scoured by the Wisconsin glaciation.

It may take hundreds of years for soils to develop, but under adverse conditions, soil erosion can rapidly result in the complete loss of the soil resource. The erosion of the soil is a natural process; however, where a stable plant cover is present or under careful soil management, erosion is a slow, gradual process more or less in equilibrium with the slow processes of soil formation. Under intensive agriculture and poor soil management, the eroding influences of water and wind may rapidly remove the 15 to 20 cm of fertile topsoil. Yearly, thousands of acres of good agricultural land are lost in this country by this process.

In a wet temperate region, soil development is principally caused by percolating rainwater that continuously dissolves nutrient salts in the upper portion of the soil and carries them, along with particles of finely divided, chemically weathered mineral matter and bits of organic matter, downward in a process called **leaching.** In time, the soil consists of a series of superimposed layers or **horizons** that differ in color, texture, and chemical attributes (Fig. 11.1).

The upper, or *A*, horizon is rich in organic matter only in the few uppermost inches, where freshly deposited organic litter is decomposing. The lower portion of

Figure 11.1
Profile of a typical soil of a wet temperate region. The *A* horizon, except for a thin region just beneath the litter, is leached of organic matter, clay, and salts. These leached substances accumulate in the *B* horizon. The *C* horizon, removed from weathering processes, represents parent material from which the *A* and *B* horizons have been formed.

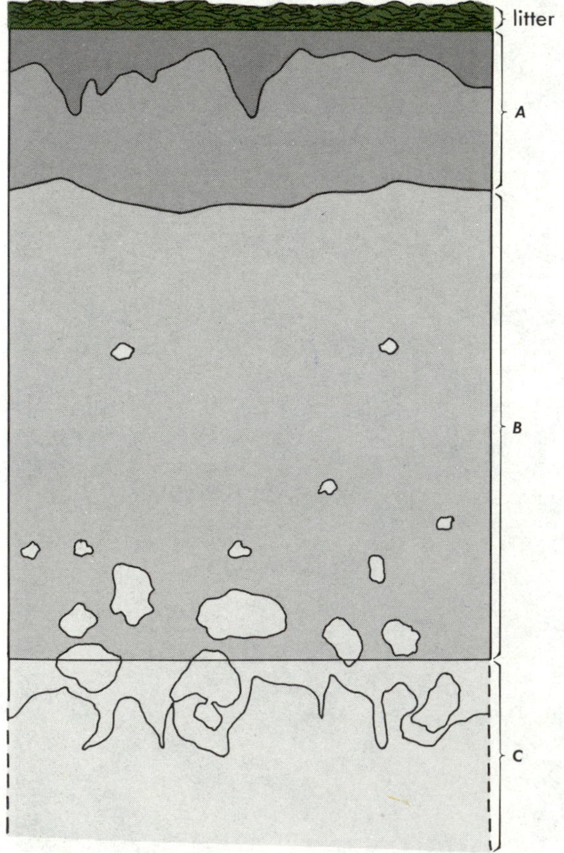

the *A* horizon is sandy and light-colored, and may be deficient in nutrients, organic matter, and clay. Gradually, 30 to 60 cm below the surface, the *B* horizon begins. The *B* horizon is often reddish-brown from oxidized iron that has accumulated in it; it is relatively rich in mineral nutrients and clay and usually has a more neutral pH than the *A* horizon. The *B* horizon, then, is a layer of accumulation. Finally, 90 to 120 cm below the surface, the *C* horizon, made up of slightly weathered parent material, appears. The relative thickness of the *A* and *B* horizons depends upon the climate and amount of plant cover.

Mineral Matter of the Soil

The mineral fractions in most soils are primarily responsible for the texture of the soils.

In referring to the size of the mineral fractions of a soil, it is customary to speak of coarse sand, fine sand, silt, and clay. The size of soil particles decreases in the order given (Table 11.1). Clay is not only the smallest mineral soil particle, but it is also the most changed from the parent rock by chemical weathering. The clay fraction, together with some of the organic material, is of colloidal dimensions and, as we will see, gives to soils some of their most important properties, such as the ability to hold water and nutrients.

Organic Matter of the Soil

The organic matter of the soil is derived from plants and animals. Throughout the centuries of soil formation, the plants and animals that lived in and on the soil have left their residues. Annual herbs die each year; the whole plant, including the roots and tops, contributes to the organic matter of the soil. Also, trees and shrubs shed their leaves, twigs, bark, and fruits, and their roots die. All this plant material decomposes, and thus, with unimportant exceptions, soils are a mixture of mineral matter and organic matter, the latter in various stages of decomposition.

We might be led to believe that during the centuries the organic material in the soil would gradually increase. This material, however, is being continually decomposed by bacteria and fungi. Cultivation hastens the loss of organic matter from a soil by increasing aeration.

Generally, for good crop growth, most soils should have considerable organic matter. Soils naturally low in organic residues may be improved by adding barnyard manure or other fertilizers, or by plowing under green manure crops. Since plants are unable to absorb complex organic molecules, of what value is the added organic matter? Organic matter improves the physical condition of most soils, especially those with much clay. There is less tendency for the soil to pack. It is more readily penetrated by air, water, and roots, and it is easier to cultivate. Added organic matter may increase slightly a soil's ability to hold water, but added organic matter may not increase the amount of water the plant is able to extract from the soil. The gradual decomposition of organic matter in the soil continually liberates essential elements in forms that are available to plants. Although all of the elements utilized by plants may eventually cycle back into inorganic compounds in the soil and perhaps be reabsorbed by other plants, sulfur, phosphorus, and nitrogen are among the quantitatively most important. The exact series of changes undergone by each element as it is cycled in this manner are specific to the individual element. However, the general details can be simply represented as shown in Fig. 11.2.

The amorphous, dark-colored, partially decomposed organic matter in soils is called **humus.** The chemical nature of humus is very complex; in part, humus is a polymer of rings and straight chains of C, H, O, and N.

Mineral and Organic Soils

Although most soils contain both mineral and organic matter, in the majority of soils, organic matter constitutes only 2 to 5% of the soil by weight. These predominantly **mineral soils** may be classified according to the relative amounts of the sand, silt, and clay they contain. Soils that contain roughly equal amounts of sand, silt, and clay are called **loams.** Depending upon the relative amounts of each component in the soil, we speak of clay, clay loam, silt loam, loam, sandy loam, loamy sand, and sand (Table 11.2). Soil texture and the suitability of a soil for plant growth are greatly influenced by its sand, silt, and clay content.

Organic soils are defined as soils having a layer, 30 cm or more thick, that contains more than 30% organic matter. Such soils occur in wet, cool depressions that become acidic and anaerobic as the rate of litter accumulation becomes greater than the rate of decomposition.

Table 11.1
CLASSIFICATION OF SOIL MINERAL MATTER ACCORDING TO SIZE OF PARTICLES

Type of Particles	Range in Diameter of Soil Particles, mm
Coarse sand	2.0–0.2
Fine sand	0.2–0.02
Silt	0.02–0.002
Clay	0.002 and smaller

Table 11.2
COMPOSITION OF THREE SOILS ACCORDING TO SIZE OF MINERAL PARTICLES

Soil Type	Coarse Sand, %	Fine Sand, %	Silt, %	Clay, %
Sandy loam	67	18	6	9
Loam	27	32	21	20
Clay	1	9	22	68

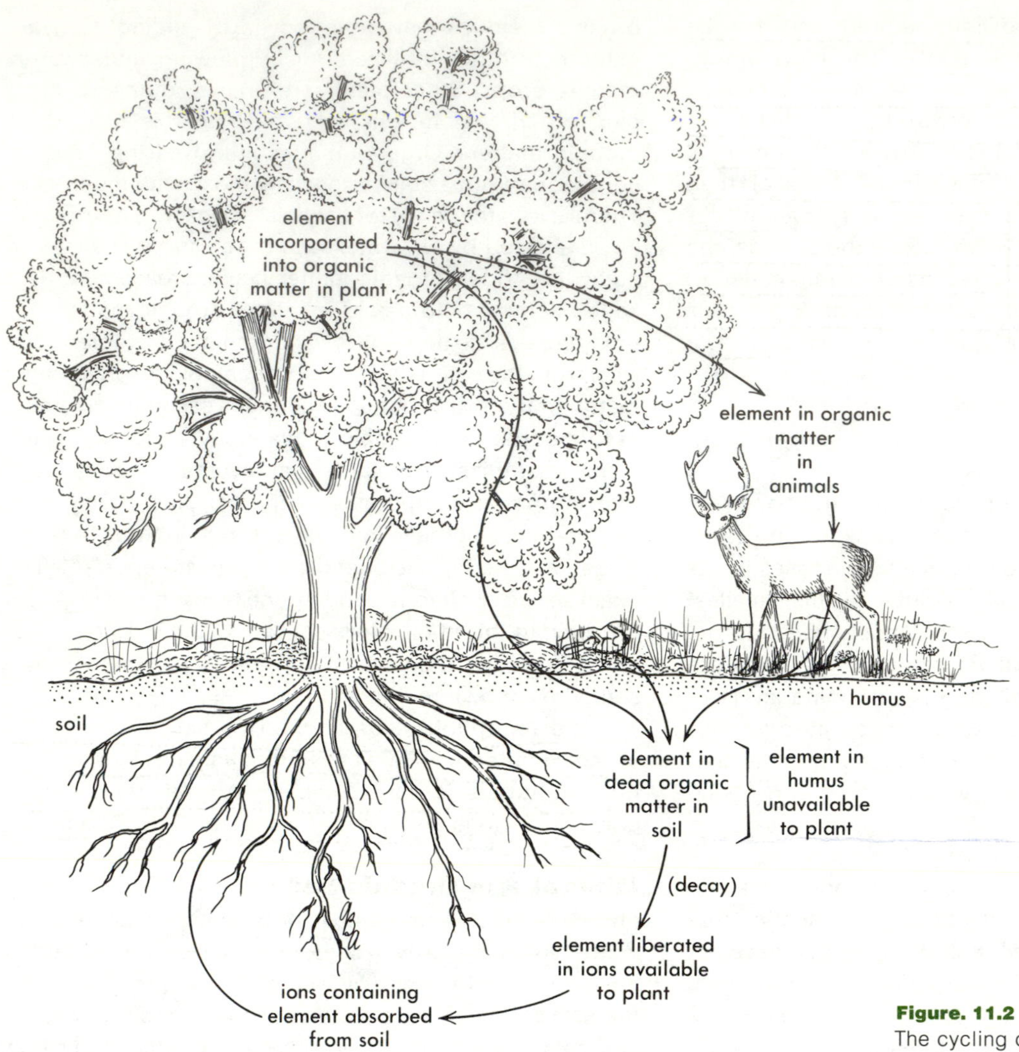

Figure. 11.2
The cycling of essential elements in nature.

Organic soils are divided into two categories, **peat** and **muck,** depending on the degree of decomposition. In peat, most of the original plant material can be identified as to species, while in muck the original plant material has been decomposed beyond recognition. Peat accumulates to such an extent that, in treeless areas, it is cut out in blocks and used for fuel.

Soil Water and Its Dissolved Substances

An important difference among various soils is their ability to hold water. It is greatest in clay and organic soils, and least in coarse sand.

If water is applied to a soil in the field, the spaces between the soil particles **(pore spaces)** become filled only for a short time to the depth wetted. With drainage, the water begins to move downward under the influence of gravity. After a while, this movement downward stops; the soil particles hold a certain amount of water against the pull of gravity. The amount of water held by the soil after drainage is called the **field capacity** of that soil. When a soil is at field capacity, or even at a moisture content well below this amount, a film of water completely surrounds each soil particle, and water also exists in soil capillaries in the form of wedges between soil particles. Clay particles and organic matter in the soil constitute **colloidal systems,** and such particles hold water by **imbibition.** Water imbibed by soil particles is much more difficult to remove from the soil than water that exists as a film or as wedges.

The soil, then, is a reservoir of water, but in addition it is the reservoir of mineral nutrients required for plant growth. The water holds in solution various inorganic salts (nitrates, phosphates, sulfates, etc.) and other water-soluble substances. The composition and concentration of this solution is ever-changing, depending upon the higher plants and microorganisms growing on and in the soil. All substances that enter the plant must be in solution. This makes possible their passage through the cell wall and cytoplasmic membranes of root hairs.

Because the living cells of the root have the ability to absorb and accumulate solutes, the soil solution, which is very dilute in most agricultural soils, is usually much less concentrated than the cell sap. The water potential in the soil is usually higher than that in the plant. As long as this relationship is maintained, water will diffuse inward. Absorption of ions by roots from the very dilute soil solution would soon exhaust all the available ions were it not for the continual release of nutrients from the solid phase of the soil. In the soil, nutrient elements may exist in a relatively unavailable form. For example, most organic nitrogen compounds are not directly available to plants; that is, nitrogen in the form of either plant or animal proteins cannot be absorbed by green plants. However, these organic compounds are acted upon by soil organisms, and some of the products of decomposition do become directly available. Organic sulfur and phosphorus compounds are also broken down by living organisms to inorganic compounds. In general, it can be said of soils that at all times chemical changes are taking place that set free inorganic nutrient substances that will dissolve in soil water and thus assume a form that can be taken up by the roots.

Cation Exchange

Soil particles themselves, particularly finely divided clay and organic matter, act as giant negatively charged ions which attract a cloud largely of positively charged ions such as Ca^{2+} and K^+. Thus, colloidal soil particles, while not in solution themselves, serve as reservoirs to which many of the various cations essential for plant growth are loosely attached and from which these ions may be released into solution by a process known as **cation exchange** (Fig. 11.3).

Roots in the soil are covered by a thin film of the dilute soil solution. As respiration occurs in the living root cells, CO_2 is produced. Carbon dioxide combines with water, releasing hydrogen ions (H^+), bicarbonate ions (HCO_3^-), and carbonate ions (CO_3^{2-}):

$$CO_2 + H_2O \rightleftarrows H_2CO_3 \rightleftarrows H^+ + HCO_3^- \rightleftarrows H^+ + H^+ + CO_3^{2-}$$

These ions may pass out of the root cell into the surrounding soil solution. Here, the hydrogen ions may exchange with the cations associated with the negatively charged surfaces of soil colloids.

The cations now in solution may diffuse to the root surface and be absorbed. Actually, the cell wall itself is negatively charged and may bind cations in the same way that the soil colloids do. By the process of cation exchange, the soil continually supplies cations to the root in exchange for hydrogen ions. The soil itself becomes more acid in the process.

Negatively charged ions such as nitrate, NO_3^- and sulfate, SO_4^{2-}, are not held by the soil particles, but are in solution. Phosphate, PO_4^{3-}, presents a different picture; these ions enter into the composition of certain clays or are easily precipitated and do not usually exist in high concentrations in soil solution.

Mineral elements essential for plant growth cannot move rapidly in the soil for any appreciable distance, but plant roots have evolved the capacity to develop a large surface to contact the soil. Root hairs are particularly important in this regard. In addition, roots have the capacity to grow in and penetrate soil. In this way, the root system of an individual plant can tap the water and mineral reserves in a large mass of soil.

Roots of many plants develop a symbiotic association with soil fungi that may serve this need for an extensive soil surface contact. In this association **(mycorrhiza)** a fungus invades the tissues of the root. Its filamentous body may extend out into the soil, and through its own absorptive capabilities it extracts minerals from the soil and brings them into the host plant. In return, the fungus obtains various organic materials from the host. In this association, the fungus takes over the role of root hairs, and the formation of root hairs is strongly suppressed. Although mycorrhizae are especially common on forest tree roots in nutritionally poor soil, they are widespread throughout the plant kingdom. Illustrations and further details of mycorrhizal assoaciations are found in Chapters 9 and 26.

Acidity of the Soil

The acidity of a soil influences the physical properties of the soil, the availability of certain minerals to plants, and the biological activity in the soil. It consequently strongly influences plant growth. Certain plants such as azaleas, camellias, and cranberries grow best in acid soils; most other plants do best in soils near neutrality, while a few plants grow satisfactorily in slightly alkaline soils (Table 11.3). Soil acidity is a measure of the concentration of hydrogen ions in solution and is usually expressed in pH units. The pH scale is a logarithmic scale in which each unit represents a hydrogen ion concentration ten times more, or less, than the next unit. Thus, a solution having a pH of 7, which is the neutral pH, has ten times the hydrogen ion concentration of a solution of pH 8, a solution that is slightly alkaline. The same solution of pH 7 has one-tenth of the hydrogen ion concentration of a solution of pH 6, a slightly acid solution.

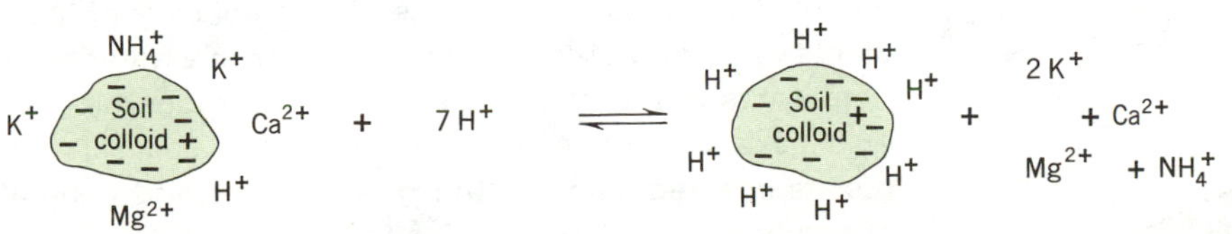

Figure 11.3
Cation exchange reactions in soil.

Table 11.3
REPRESENTATIVE VALUES FOR THE ACIDITY (pH) OF SOILS IN WHICH CERTAIN PLANTS GROW WELL

Relative Acidity	pH Range	Plants That Grow Well in Soils of This Acidity
High acidity	3.2–4.6	*Sphagnum* moss in peat bogs
Moderate acidity	3.5–5.0	Blueberry
Slight to moderate acidity	4.5–6.0	Azalea, camellia
Nearly neutral	6.0–7.5	Many forest and field plants
Alkaline	7.5–9.1	Greasewood, rabbit brush, stonewort, various plants growing around alkali flats

Acid soils, such as those beneath northern coniferous forest with acidic foliage, range from pH 3.5 to 5.0; agricultural soils in the humid areas range from pH 5.0 to 7.0; and soils of arid or saline regions may range as high as pH 11.0, but 8.0 to 9.0 is more common. Acidity *per se* —that is, H^+ or OH^- ions—is not directly responsible for limiting plant growth. Rather, soil pH affects the availability of plant nutrients in two ways. In an acid soil, hydrogen ions (H^+) replace other **cations** (like K^+, Ca^{2+}, and Mg^{2+}) on the negatively charged clay particles, allowing the nutrient ions to be easily leached from the soil. Soil pH also affects the solubility of plant nutrients. As shown in Fig. 11.4, some nutrients, like iron, manganese, and aluminum, increase in solubility as pH decreases. Aluminum may reach toxic levels in some acid soils. Other nutrients, like calcium and magnesium, increase in solubility as pH increases. Soil pH may also affect plant growth indirectly by suppressing bacterial growth at pH extremes.

Figure 11.4
Relationship of relative nutrient availability to soil pH.

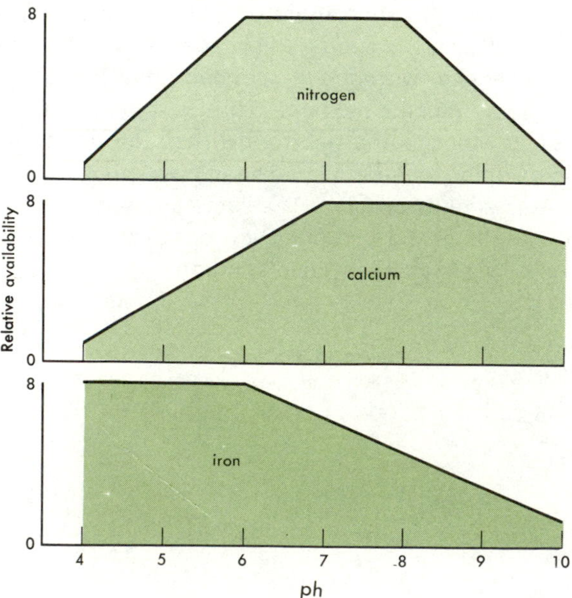

Acid soils may be improved by the application of lime (calcium hydroxide or calcium carbonate), which tends to neutralize the acid. Alkaline soils may be improved by the addition of sulfur or ammonium sulfate. If elemental sulfur is applied, it may become oxidized to sulfate and make the soil more acid. The ammonium of ammonium sulfate is either converted into nitrate by bacterial action or is absorbed more rapidly than sulfate by the plants, resulting in a net increase in sulfate and acidity in the soil.

Air of the Soil

We have spoken of the pore space in soils. The air of the soil contains the same gases as the atmosphere above the soil. However, the air of the soil is considerably richer in carbon dioxide and poorer in oxygen than that of the atmosphere. In the latter, the average percentage by volume of carbon dioxide is 0.03%; in soil air, the percentage may go as high as 5%. This high percentage of carbon dioxide in soil air is due to the respiration of soil microbes and, to a lesser degree, the respiration of roots themselves. The soil air usually has a relative humidity near 100%, a condition very favorable to the growth of soil organisms.

The living cells of roots must have oxygen to support respiration. Normally, the oxygen necessary for respiration reaches the roots by absorption through the root hairs and other epidermal cells. This inward diffusion of oxygen from the soil air ordinarily goes on readily because of the thinness of the walls of these cells and the absence of a cuticle. Roots at all levels in the soil secure oxygen directly from the air in the pore spaces and, to a limited extent, from the oxygen dissolved in water; in ordinary land plants, there is no system of air-conducting tubes that conveys oxygen from the atmosphere through the plants to the extremities of the roots.

The amount of air in the soil depends not only upon the pore space but also upon the water content of the soil. If water occupies the pore space, air is forced out. A water-soaked (saturated) soil contains practically no air save that which is dissolved in water. If the soil about the roots is continuously water-soaked, roots die because of insufficient oxygen and possibly as a result of the accumulation of carbon dioxide. Most species of land plants will grow normally with their roots in a water solution, if it is well-aerated by bubbling air through it (Fig. 11.5). Evidence indicates that inadequate soil aeration results in a reduction in the rate of water and mineral absorption.

Whereas most land plants, including agricultural plants, will not long survive with the root system submerged in unaerated water or surrounded by a soil that is

Figure 11.5
Effect of aeration on root growth in tomato *(Lycopersicon esculentum)*. Left: plants growing in complete nutrient solution through which air was bubbled; right: plants growing in same solution without aeration. (Courtesy of D. R. Hoagland.)

water-soaked, some plants flourish under such conditions. Among them may be cited rice, various swamp and marsh plants, and the bald cypress (*Taxodium distichum*, Fig. 11.6). Almost all such plants contain in the stem and roots large communicating air spaces. Thus, air absorbed into the leaves and stems may reach the living cells of the root in sufficient quantity. Bald cypress and certain species in tropical mangrove swamps develop special root branches, **pneumatophores,** that grow upward until their ends are above the water level. These special root branches have a central core of loose tissue through which air moves downward to the submerged organs.

Organisms of the Soil

Ordinary soils teem with living organisms, both plants and animals. These organisms are an important part of the environment of roots. The soil flora includes bacteria, fungi, and algae. Living roots of higher plants also are an important part of the soil. When the roots die, they leave a considerable quantity of organic matter. The soil fauna includes protozoa, nematodes, earthworms, various insects, and burrowing animals. Most of these are beneficial to man in some way or another; others may be harmful. Certain bacteria are absolutely essential in the maintenance of soil fertility. This point will be pursued further in Chapter 21, but it should be indicated here that certain bacteria and fungi that live in the soil are responsible for the decomposition of organic matter (plant residues, manure, and other organic fertilizers), and for nitrogen fixation. As a result of their activity, they keep up the supply of soil nitrogen, one of the chief factors in soil fertility. Earthworms affect soil structure by moving and mixing the soil; they pass large quantities of soil through their bodies. Charles Darwin, in 1885, published *The Formation of Vegetable Mold,* in which the importance of earthworms in soils is stressed. He points out that earth worms in an acre of soil may pass through the bodies annually as much as 15 tons of dry earth.

Temperature of the Soil

The rate of absorption of both water and solutes by roots may be reduced by extremes of temperature. Plants na-

Figure 11.6
Aerial pneumatophores of bald cypress *(Taxodium distichum)*. (Courtesy of U.S. Forest Service.)

tive to cool climates absorb these substances more freely at low temperatures than do plants of warm climates. A plant may wilt in a soil containing ample water if the soil temperature sinks below or rises above, a certain degree. In cold, dry climates, winter-killing may be the result of a cold soil that slows up absorption, accompanied by a high transpiration rate. It is believed that in winter-killing, the plant is as frequently killed by direct drying as by actual freezing. Plants short enough to be buried beneath snow are protected from desiccation and from extreme cold.

All chemical and biological activities are influenced by soil temperature. The soil temperature is not always the same as the temperature of the air above it. It may be lower or higher than the air temperature (see Chapter 18).

MINERAL ELEMENTS ESSENTIAL FOR PLANT GROWTH

For over 100 years physiologists have been studying the mineral nutrition of plants. Extensive investigations have been carried out to determine which elements are essential, how the plant absorbs them, how they are utilized, why they are essential, and what effects are produced in the plant when a particular essential element is lacking (see **Fig. 11.7,** page 221).

Plants utilize elements in four basic ways: (*a*) *The elements may form part of structural units,* such as carbon in cellulose or nitrogen in protein. (*b*) *Elements may be incorporated into organic molecules important in metabolism,* like magnesium in chlorophyll or phosphorus in ATP. (*c*) *Elements may function as enzyme activators,* necessary as catalysts in certain enzymatic reactions. Magnesium is used as an enzyme activator in several of the enzymatic steps of glucose degradation in the process of respiration. (*d*) *Elements in ions help to maintain the osmotic balance,* for example, potassium in guard cells.

What methods have been employed to determine the chemical elements that are essential to plant growth? Of course, accumulated experiences of agriculturists have shown that applications of various nutrients to soil result in healthier plants, increased yields, and better quality. Plant physiologists have used more critical methods—careful weighings and measurements of quantities of elements used and results obtained. The solution-culture method has long been employed to determine what elements are indispensable in plant growth. Most plants that normally grow in soil can be grown to maturity (fruit and seed production) in aerated water to which soluble essential nutrient salts have been added **(Fig. 11.7).** By growing plants in a nutrient solution containing all elements believed to be essential, except the particular element being investigated, the response of plants when this element is absent can be ascertained. In this case, all plants grown in deficient solutions were first grown in a complete nutrient solution for 2 weeks before being transferred into the deficient solution.

From such experiments a list of elements that are known to be essential for the growth of plants has been compiled. Long recognized as essential to the continued growth and development of green plants are the following ten elements, with their appropriate chemical symbols: carbon (C), hydrogen (H), oxygen (O), phosphorus (P), potassium (K), nitrogen (N), sulfur (S), calcium (Ca), iron (Fe), magnesium (Mg). If the symbols are arranged in a line, they can be used to remember the elements thus:

C HOPK'NS CaFe Mg

"See Hopk'ns Cafe, mighty good." Of these ten elements, the last seven are in the nutrient medium; that is, their various salts are dissolved in the soil water. Relatively large quantities of nitrogen, sulfur, phosphorus, potassium, magnesium, and calcium are required; very small quantities of iron meet all plant needs for this element. These elements, with the exception of iron, have been called the **macronutrient elements,** because they are needed in relatively large amounts.

More recently, it has been discovered that at least six additional elements are needed by higher plants but in much smaller amounts. The **micronutrient elements** (trace elements or minor elements) include, in addition to iron (Fe), the elements boron (B), manganese (Mn), copper (Cu), zinc (Zn), chlorine (Cl), and molybdenum (Mo). Table 11.4 shows the relative amounts of macro- and micronutrients removed in kilograms per hectare by wheat during a growing season. Such elements are required in such minute amounts that their importance remained long undetected because the air about the plants, the containers holding the growth medium, or impurities in the chemicals used supplied enough of them to support plant growth. There are other elements that are required by some plants in very small amounts. Sodium, for example, is required in very small amounts by such C_4 plants as the desert shrub *Atriplex,* but is not required by most other species. It is not considered as

Table 11.4
AMOUNTS (IN KILOGRAMS PER HECTARE) OF MACRO- AND MICRONUTRIENTS REMOVED FROM THE SOIL IN ONE GROWING SEASON BY A WHEAT CROP

	Macronutrients						Micronutrients				
Element	N	K	P	Ca	S	Mg	Fe	Mn	B	Zn	Cu
Amount	85	47	17	13	12	9	0.8	0.6	0.3	0.2	0.03

generally essential for most higher green plants. However, it is still likely that not all of the elements essential for plant growth, have been identified. For example, there is recent evidence that silicon may be essential for the growth of at least some plants.

A plant growing in soil or in solution cannot distinguish between elements that are essential to it and those that are not essential or that might be harmful. If the element or ion containing it is in solution, it will probably be absorbed by the plant. Thus, we find in the plant almost all the elements present in the soil. Gold has been isolated from plants growing in gold-bearing soils. Selenium, an element poisonous to livestock, is absorbed by certain plants in sufficient amounts to be harmful to animals grazing on them.

Macronutrient Elements

Carbon, Hydrogen, and Oxygen

The absorption of water, carbon dioxide, and oxygen brings into the plant large quantities of the elements carbon, hydrogen, and oxygen, from which is formed the major part of each of the hundreds of organic compounds found in the plant. We have seen how in the process of photosynthesis these elements are combined into carbohydrates. Simple fats also are composed entirely of these three elements, whereas proteins, alone of the foods, contain appreciable amounts of other elements, particularly nitrogen and, to a lesser extent, sulfur and phosphorus.

Nitrogen

This element is a component of proteins, which form an essential part of protoplasm and also occur as stored foods in plant cells. Nitrogen is also a part of other organic compounds in plants, such as chlorophyll, nucleic acids, amino acids, alkaloids, and at least some plant hormones.

Ordinary green plants cannot utilize elemental nitrogen, which represents about 78% of the air. It has been estimated that above every hectare of land surface there are about 130,000 to 135,000 metric tons of this gas. Chief sources of nitrogen for green plants are the nitrates, such as sodium nitrate (Chilean saltpeter), potassium nitrate, ammonium nitrate, and calcium nitrate. Green plants can also utilize ammonium salts, and evidence indicates that some plants can derive nitrogen from certain low-molecular-weight nitrogenous compounds.

Nitrogenous fertilizers, both natural and commercial, are usually the most important fertilizers applied to growing plants. The chief commercial nitrogenous fertilizers are: (*a*) those with nitrogen in the nitrate form; (*b*) those with nitrogen in ammonia or its compounds; (*c*) those with nitrogen in organic compounds, such as tankage and cottonseed meal; and (*d*) those with nitrogen in the amide form, such as urea and calcium cyanamide. Organic nitrogen in complex molecules cannot be used directly by plants but must first be converted into available forms through the action of organisms in the soil.

Since nitrogen fertilizers are so important in modern agriculture and the amount of energy used in their production, distribution, and use is so large, modern geneticists and cell physiologists are intensively investigating the feasibility of developing crop plants that will be able to utilize atmospheric nitrogen. One approach is to alter the plant's genetics, through DNA manipulation, so that the plant will grow in association with strains of bacteria that are capable of utilizing atmospheric nitrogen. Further details of this exciting area of molecular biology are discussed in Chapter 21.

The rate of growth of plants is influenced to a large degree by the available nitrogen. In contrast to some other nutrients (e.g., calcium), nitrogen is very mobile in the plant and can be translocated from mature to immature regions of a plant. An early symptom of nitrogen deficiency is a yellowing of leaves, particularly the older leaves; then follows a stunting in the growth of all parts of the plant **(Fig. 11.7*E*)**. An excess of available nitrogen results in vigorous vegetative growth and a suppression of food storage and of fruit and seed development. The nitrogen cycle in nature is discussed more fully in Chapter 21.

Sulfur

Sulfur forms a part of the molecules of proteins. Plant proteins containing sulfur may have from 0.5 to 1.5% of this element. The sulfhydryl, -SH, group is a very important group essential for the action of certain enzymes and coenzymes. In addition, sulfur is a constituent of ferredoxin (an important compound in photosynthesis) and of some lipids found in chloroplasts.

Ordinary green plants cannot utilize elemental sulfur or sulfur containing organic compounds; they must first be oxidized to sulfates, which are normally absorbed by the roots.

Certain agricultural soils may become deficient in sulfur. The most noticeable responses to applications of sulfur fertilizers are an increased root development and a deeper green color of the foliage. Sulfur deficiency symptoms in tomato are shown in **Fig. 11.7*F*.**

Phosphorus

This element is also a component of some plant proteins, phospholipids, sugar phosphates, nucleic acids, ATP, and NADP. The highest percentages of phosphorus occur in parts of the plant that are growing rapidly, such as meristematic regions and maturing fruits and seeds. Some of the very important roles that phosphorus plays in plant metabolism in relation to energy transformation and biosynthetic reactions are discussed in Chapter 20.

Methods have been developed by which elements

are "tagged." For example, radioactive phosphorus in the form of phosphate has been introduced into the culture solution in which tomatoes were growing. By means of the Geiger-Müller counter, or by radiograms, the path of radioactive phosphorus through the plant has been followed. After a time, the greatest accumulation of phosphorus was found in the young developing fruit; very little occurred in the ripe fruit. Within the fruit itself the greatest accumulation of phosphorus was in the developing seed.

Applications of phosphorus to soils deficient in this element promote root growth and hasten maturity, particularly of cereals. Phosphates are the principal source of phosphorus for plants. A tomato plant grown in culture solution without phosphorus is shown in **Figs. 11.7*B,C*.**

Potassium

Potassium is the only monovalent cation that is essential for plant growth. This element, indispensable to plants, accumulates in tissues that are growing rapidly. It will migrate from older tissues to meristematic regions; for example, during the maturing of a fruit crop there is a movement of potassium from the leaves into the fruit. In plants, potassium occurs chiefly in inorganic form.

The primary role of potassium in the cell is as an enzyme activator. Over 40 different enzymes have been found to require monovalent cations for maximal activity. In plant cells, potassium serves this role. However, in contrast to most enzyme activators that are micronutrients, potassium is effective only at the relatively high concentrations that it occurs in plant cells. Processes in the plant that appear to require an adequate supply of potassium are: (*a*) normal cell division, (*b*) synthesis and translocation of carbohydrates, (*c*) synthesis of proteins in meristematic cells, (*d*) reduction of nitrates, (*e*) development of chlorophyll, and (*f*) stomatal opening and closing. In addition to affecting changes in sugar concentration in guard cells, illumination has been shown to stimulate the accumulation of potassium ions in guard cells. This light-stimulated ion accumulation is correlated with stomatal opening. When the leaf is darkened, the guard cells lose potassium and stomata close. This accumulation of potassium by illuminated guard cells can be dramatically observed by means of an electron probe analyzer (see page 190).

Any water-soluble inorganic compound of potassium, such as potassium sulfate, potassium phosphate, or potassium nitrate, can be utilized by plants as a source of potassium. Potassium deficiency symptoms in tomato are shown in **Fig. 11.7*D*.**

Calcium

All ordinary green plants require calcium. It is one of the constituents of the middle lamella of the cell wall, where it occurs in the form of calcium pectate. Calcium affects the permeability of cytoplasmic membranes and the hydration of colloids. Calcium may be found in combination with organic acids in the plant in the form of crystals (see page 56). There is also evidence that calcium favors the translocation of carbohydrates and amino acids and encourages root development. Calcium deficiency is frequently characterized by a death of the growing points **(Fig. 11.7G)** because it is not readily translocated in the plant from mature to immature regions. General disorganization of cells and tissues also results from calcium deficiency. This effect is consistent with one of the key roles of calcium—maintaining the normal structure or cell membranes.

Magnesium

Magnesium is a constituent of chlorophyll, where it occupies a central position in the molecule (Fig. 5.12). Chlorophylls are the only major compounds of plants that contain magnesium as a stable component. Many enzyme reactions, particularly those involving a transfer of phosphate (energy metabolism), are activated by magnesium ions. Consequently, magnesium deficiency rapidly affects many aspects of metabolism.

A deficiency of magnesium results in the development of pale, sickly foliage, an unhealthy condition known as **chlorosis** **(Fig. 11.7*I*).** This disease is a common one of economic plants, and in many instances, applications of magnesium have effected a cure. However, chlorosis can be caused by deficiencies in other essential elements such as iron.

Micronutrient Elements

A tomato plant grown in a nutrient medium lacking all the micronutrient elements except iron is shown in **Fig. 11.7*J*.**

Iron

A number of essential compounds in plants, such as cytochromes and ferredoxin, contain iron in a form that is bound rather firmly into the molecule. Iron plays a role in being the site on some electron carriers where electrons are absorbed and then given off during electron transport. The iron atom is alternately reduced and then oxidized. Several cytochromes, both in mitochondria and in chloroplasts, contain iron. Other iron-containing proteins include ferredoxin in the chloroplasts, and certain enzymes such as catalase and peroxidases, Thus, iron plays very important roles in energy conversion reactions both of photosynthesis and respiration.

Although iron is not a component of the chlorophyll molecule, it is essential for the synthesis of chlorophyll. Chlorosis may be caused by iron deficiency as well as by a deficiency of magnesium **(Figs. 11.7*B,H*).** The quantity of iron required is very small. As an example, chlorosis of pineapples in Hawaii, due to the unavailability of iron from the soil, is remedied by spraying the plants with iron salt solutions. Orchard trees suffering from iron chlorosis may be cured by injecting iron compounds into the trunk, or by applying various kinds of iron salts to the soil.

Boron

That plants require boron is well-established. Symptoms of boron deficiency include inhibition of root and shoot elongation, inhibition of flowering, darkening of tissues, and various growth abnormalities and disturbances. Some of the physiological diseases of plants due to boron deficiency are internal cork of apples, top rot of tobacco, cracked stem of celery, browning of cauliflower, and heart rot of sugar beets. Very small applications (20 to 60 kilograms) of sodium tetraborate (borax) per hectare may be sufficient to cure these diseases. In contrast, fertilizers containing the macronutrient elements are frequently applied to fields at rates of several hundred kilograms per hectare.

Although plants require boron for normal development, the quantity in the soil or culture solution must be very small or injury will result. In fact, in certain agricultural sections, severe injury to crops occurs because of excessive amounts of boron in the soil or in the irrigation waters. Plants vary in their tolerance of boron. Borates are used as weed-killers.

Although the exact function of boron in plant metabolism is still obscure, boron does play a regulatory role in carbohydrate breakdown. Boron deficiency tends to shift metabolism from glycolysis to the pentose phosphate pathway (pp. 258, 263).

Zinc

Zinc is essential to the normal development of a variety of plants; probably it is required by all plants. As with other micronutrients, proof has come from careful water-culture experiments; also, certain diseases of plants have been cured by zinc applications. Two well-known diseases caused by zinc deficiency are "little leaf" of deciduous fruit trees (Fig. 11.8) and "mottle leaf" of citrus trees. These abnormal conditions are corrected by spraying the trees with zinc salts, by injecting dilute solutions of zinc salts into the trunks, or by driving zinc brads into the trunks. These corrective measures show how minute are the quantities of zinc required by plants for normal development.

The very pronounced effect of zinc deficiency on growth, especially internode elongation, is a consequence of the importance of zinc in auxin synthesis. Indole acetic acid, auxin (see Chap. 20), is necessary for cell elongation. It is found to be in low concentration in zinc-deficient plants even before visible symptoms appear. If zinc is added to the culture solution, the auxin level increases and this is followed by a resumption of growth. Zinc is also an activator or a part of several enzymes.

As is the case of a number of the micromineral nutrients, large quantities of zinc are toxic to plants.

Figure 11.8
Disease of peach *(Prunus persica)* known as "little leaf," caused by a deficiency of zinc. Branch at left untreated; branch at right cured by driving in zinc-coated nails. (Courtesy of E. L. Proebsting.)

Manganese

The most striking symptom of plants with a deficiency of manganese is chlorosis (Fig. 11.9), but it is a somewhat different type from that caused by iron deficiency. Whereas in iron chlorosis the young leaves may become yellow or white with prominent green veins, manganese chlorosis results in the leaf taking on a mottled appearance. Spraying or dusting crops suffering from manganese deficiency with as little as 24 kilograms of manganese sulfate per hectare frequently effects a cure.

Figure 11.9
Chlorosis of tomato *(Lycopersicon esculentum)* leaf caused by a deficiency of manganese in the nutrient solution. (Courtesy of D. R. Hoagland.)

Manganese's importance as an activator of several enzymes of the Krebs' cycle of aerobic respiration (Chapter 14) explains some of the disruptive effects of manganese deficiency on metabolism. A further function of manganese in green cells is concerned with the oxygen-liberating steps of photosynthesis (Chapter 13).

Copper

Abnormalities in the growth of many plants, especially those in marsh and peat soils, have been corrected by the application of copper compounds (Fig. 11.10). Copper is also a constituent of certain enzyme systems, such as ascorbic acid oxidase and cytochrome oxidase. In addition, copper is found in plastocyanin, part of the electron-transport chain in photosynthesis.

Molybdenum

Molybdenum is now recognized as an essential micronutrient element. For example, it was found that molybdenum is required for normal growth of tomato seedlings, that only 0.01 parts per million (ppm) of the element in the nutrient solution is needed, and that concentrations exceeding 10 ppm cause injury. For some Australian soils, only 140 grams of MoO_3 per hectare, applied once every 10 years, produces optimum forage improvement.

Molybdenum is important in enzyme systems involved in nitrogen fixation and nitrate reduction. Plants suffering molybdenum deficiency can absorb nitrate ions, but are unable to utilize this form of nitrogen. If nitrogen in the form of ammonium is given to these plants, many of the deficiency conditions are less severe.

Chlorine

Chlorine is present in the soil solution as the very soluble, negatively charged chloride ion, Cl^-. Because of the very small quantities required by plants and its almost universal occurrence in soils, chlorine is never added intentionally as a fertilizer, although it occurs in small amounts in all fertilizers. Rainfall may contain enough chlorine to supply plant requirements, and some plants may obtain enough from the air to prevent deficiency symptoms. It is very difficult to prepare a nutrient solution completely free of chlorine. Consequently, it is difficult to demonstrate chlorine deficiency symptoms in plants. Although the functions that chlorine plays in plant metabolism are obscure, it is known that chlorine takes part in the reactions of photosynthesis leading to oxygen evolution.

Figure 11.10
Tomatoes *(Lycopersicon esculentum)* growing in solution with all essential chemical elements except copper. Leaves at left were sprayed with solution containing copper. (Courtesy of D. R. Hoagland.)

VARIATIONS IN RESPONSE TO MINERAL STRESS

Although most plants require the same essential elements for their normal growth and development, plants differ in their ability to absorb nutrients from relatively unavailable sources and in their tolerance and efficiency of utilization of absorbed elements. This genetic variability in plants is very important in providing a basis for adaptability of wild plants to survive in a specific environment. Figure 11.11 shows some differential responses of plant genotypes to several nutrient stresses. In most instances little information is available about the mechanisms responsible for these differences. However, in the case of iron stress it is known that plants that are efficient in obtaining iron from alkali soils, where it is relatively unavailable to other plants, respond to iron deficiency by biochemical or physiological changes in their roots. These reactions may include the release of hydrogen ions or reducing compounds from the roots, a reduction of iron (Fe^{3+} to Fe^{2+}) at the root surface, or an increase in organic acids in the roots.

High levels of salt in the soil, **soil salinity,** frequently coupled with a low level of availability of water, are responsible for restricting plant growth in many regions of the world. Soil salinity becomes an increasingly important problem in otherwise productive land when the continual accumulation of salt from irrigation water may eventually make the land unproductive. The two major factors usually affecting plant growth in these areas are (1) high osmotic concentrations resulting in low water potentials and (2) the presence of potentially toxic levels of sodium and other ions. Sodium chloride or sodium carbonate are often found associated with soil salinity. In addition to the problems of high concentration of salts and toxic substances, plants growing in saline soils often suffer from low availability of potassium or oxygen. As we have seen (p. 78) some plants, halophytes, are able to survive even under high salt levels. Some halophytes are salt-secreters, with salt glands and visible salt crystals on the leaves, others are accumulators (some succulent plants are included in this group), with large vacuoles and large cells. Although adult halophytes show resistance or tolerance to salt spray or soil salinity, very few show optimal growth at even moderate salinity. Typically, germination is best in low salinity and may occur when the soil salinity is decreased following rain.

Figure 11.11
Differential response of plant genotypes to several nutrient stresses. *A*, zinc-deficiency symptoms in Sanilac but not Saginaw navy beans when grown in Shano soil (Zn-deficient soil). *B*, iron-deficiency symptoms in ys/ys but not W79 corn when grown in nutrient solution with 0.6 ppm iron. *C*, manganese-toxicity symptoms in Forrest and Bragg but not in T203 and Lee soybeans when grown in Zanesville soil (pH 5.3). (Courtesy of J. C. Brown.)

Even mesophytic plants exhibit wide ranges in their ability to grow in saline conditions. Recently extensive studies have been conducted aimed at selecting highly salt tolerant strains of agriculturally important plants, for example, barley and tomato. **Figure 11.12,** page 222, shows some of the variability in strains of barley to high levels of salt. Such variability in response to nutrient availability, to toxic compounds, or to high salt concentrations is a very important but still relatively little explored genetic resource for the plant scientist interested in selecting or breeding for more efficient plants.

SUMMARY

1. Soil is a complex system composed of rock particles and plant and animal remains in various stages of decay, and in which an abundance of soil microorganisms is growing. Throughout the soil structure there is an interconnecting system of large and small pores filled with air or with water and dissolved materials.
2. All the essential elements except carbon, hydrogen, and oxygen are absorbed through the roots, generally in the form of ions from the soil. Hydrogen and oxygen are absorbed by roots as water, and carbon enters the leaves as carbon dioxide.
3. Biologically produced hydrogen ions may replace cations associated with the negatively charged soil colloids: clay and organic matter.
4. The macronutrient essential elements are carbon, hydrogen, oxygen, nitrogen, sulfur, phosphorus, potassium, calcium, and magnesium.
5. The micronutrient essential elements are iron, zinc, copper, boron, manganese, chlorine, and molybdenum.
6. Essential elements fill four roles: part of structural molecules, part of "working" molecules in metabolism, as enzyme activators, and important to osmotic balance.
7. Plants show genetic variability in response to nutrient availability, toxic levels of compounds, and high salt concentration.

CHAPTER 12

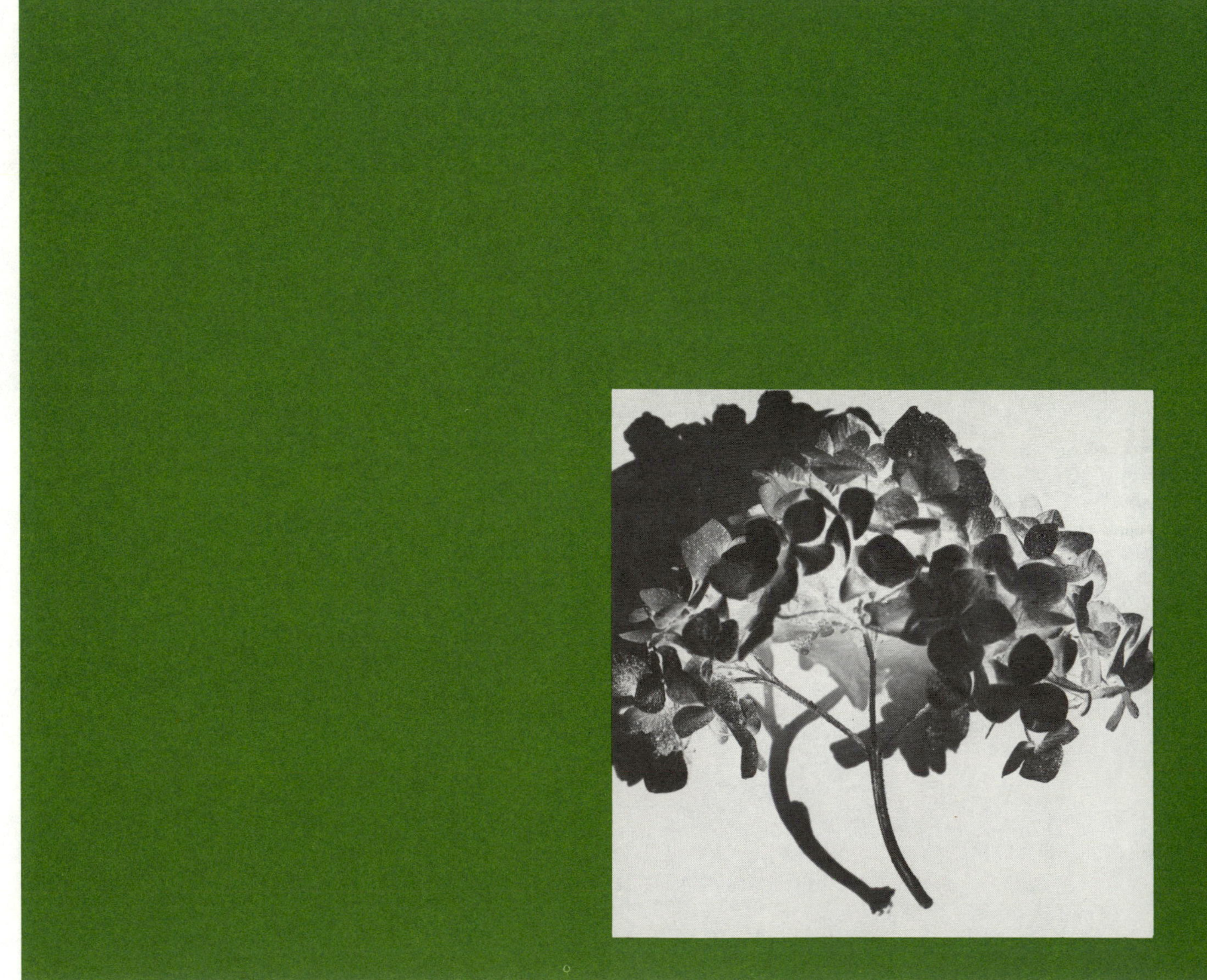

TRANSPIRATION, CONDUCTION, AND ABSORPTION

Plants, with their root systems in moist soil and their shoots in the atmosphere, require for their life only relatively few simple raw materials (water and inorganic ions containing the essential elements from the soil and carbon dioxide and oxygen from the atmosphere). How are these materials absorbed by the plant and subsequently moved throughout its body to sites of utilization? How are the products of photosynthesis and other organic compounds moved from sites of synthesis to regions of utilization and storage? These are some of the questions that we shall consider in this chapter. Since all of these materials are absorbed and moved in solution, we shall first consider how the plant maintains the flow of water throughout its body.

WATER IN THE PHYSIOLOGY OF THE PLANT

The Roles of Water

The most abundant compound in an active cell is water. It plays a varied role in the life of a plant, for it serves as (*a*) a raw material in the synthesis of organic compounds, (*b*) the solvent in which vital reactions take place, (*c*) the medium through which solutes move from cell to cell, and (*d*) the source of turgor in plant cells. Plant tissues vary in water content; those actively growing have a water content equal to as much as 85 to 95% of their fresh weight, while a dormant structure such as a seed may have a water content as low as 5 to 10%. The entire shoot system of an herbaceous plant such as barley may be 82% water (**Fig. 12.1,** page 222).

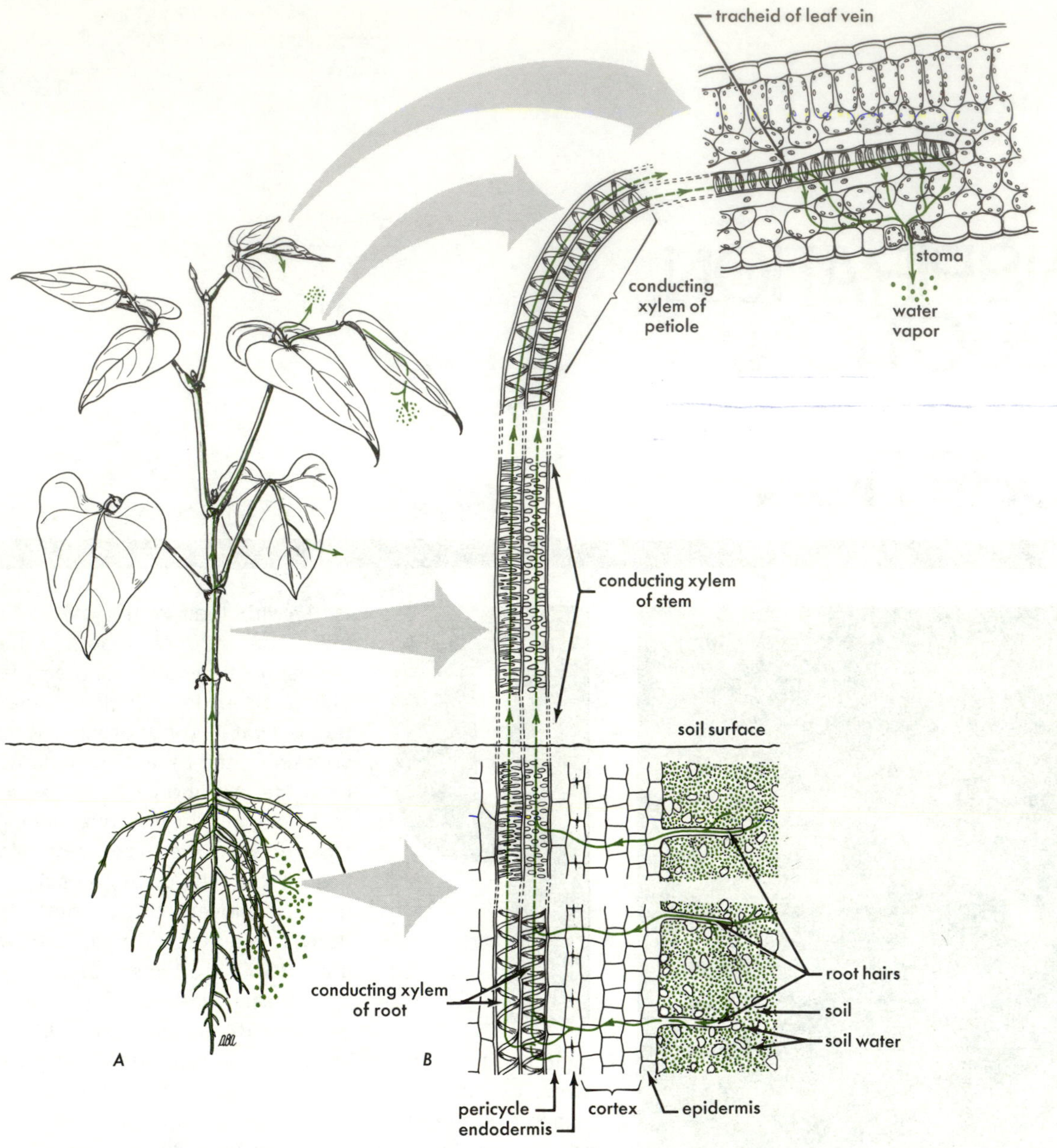

Figure 12.2
A, diagram showing water movement in a whole bean plant. *B,* details of xylem pathway.

A plant from which the water has been removed is made up chiefly of organic material synthesized by the plant primarily from carbon dioxide, water, and inorganic nitrogen. The small quantity of mineral matter contained in a plant can be readily demonstrated by burning the dry tissue. The carbon, hydrogen, and oxygen, as well as the nitrogen and some of the phosphorus and sulfur, pass off in the combustion process, while most of the minerals remain in the ash, combined with a small amount of oxygen (**Fig. 12.1**).

While minerals that are once absorbed by a plant are not generally lost again in large quantities, except during leaf fall, water is constantly being lost from the aboveground portions of the plant. Because this water loss profoundly influences the rates of water absorption and movement, it will be studied first.

The Continuity of Water in the Plant

Try to visualize all of the water in an entire plant (Fig. 12.2). You should see an interconnected network of water in the form of the plant. The lower part of this network extends into the soil and is in intimate connection with the soil water. It extends throughout the entire root system connecting every cell and permeating, with a few exceptions, every cell wall. The water system extends upward, not only as thin threads in the vascular system, but laterally through the permeable walls into every cell. It is only separated from the surrounding atmosphere by the thin cutinized or suberized walls of the outer surface of the stem. Even here, there are lenticles, stomata, and cracks in the cuticle through which water evaporating from the walls of underlying cells might pass. This interconnected system of liquid water extends into every apical bud cell and into every leaf cell and out into the walls of the palisade and spongy parenchyma in contact with the intercellular spaces of the leaves. One might visualize all of the liquid water in the plant as one giant molecule with the individual water molecules held together by hydrogen bonds (see page 62). Thus, the absorption or loss of a water molecule at any point in the system would affect the entire system.

Transpiration and Guttation

By far the greater part of the water lost from the plant passes into the atmosphere as invisible water vapor. This loss of water in vapor form from a living plant is called **transpiration.** Although a small amount of water may be transpired directly through the cuticle, **cuticular transpiration,** most of the water lost during the day diffuses out through the stomata, **stomatal transpiration.** Leaves are the principal transpiring organs.

A knowledge of leaf structure is essential to an understanding of the mechanism of transpiration. It will be recalled that the mesophyll of most leaves is a very loose tissue with large intercellular air spaces. Even within the palisade parenchyma, a large portion of the walls of most cells is exposed to intercellular air spaces (**Fig. 12.3,** page 222). These are connected with similar air spaces throughout the leaf and, through the stomata, with the outside air. The extensive intercellular surface, loose internal structure, and numerous stomata of the leaf allow for rapid gaseous exchange between the internal leaf cells and the outside air. An appreciation of this structure is gained when we consider a large leaf such as a squash leaf that may be 70% internal air space, and may have 20 times as much total internal cell surface bordering this air space as leaf surface exposed to the outside air. In addition, a squash leaf may have as many as 60 million stomata connecting this internal air with the outside environment and may have 6000 vein endings per square centimeter of leaf surface.

Process of Transpiration

Water permeates the living turgid leaf cells, filling the vacuoles, making up a large part of the cytoplasm, and penetrating the walls. Thus, cell walls bordering on intercellular air spaces are moist, and because of this, air in these spaces is almost saturated with water vapor. The water potential or vapor pressure of water in these spaces is then generally higher than in the outside atmosphere. When the air around a leaf is not saturated but relatively dry, water vapor molecules tend to diffuse from the saturated air in the leaf through the stomata into the less saturated outer air where the vapor pressure, and hence the water potential, is lower. This loss in water results in a slight drying of the air in the intercellular spaces (**Fig. 12.3**). Water molecules then evaporate from the wet walls and diffuse into this drier air in the leaf. As water leaves the walls, the walls in turn become drier and imbibe water from the enclosed cytoplasm and vacuole.

Thus, a diffusion gradient for water exists from the vacuole through the cytoplasm, the wall, and the intercellular air spaces to the drier outer air surrounding the leaf. Water moves along this gradient (see Chapter 5), in liquid form within the cell, and in vapor form once it evaporates from the cell walls. Each cell, however, is in contact with several other cells and is only a few cells distant from the water-filled xylem elements of a vein.

As water is lost from a cell by outward diffusion, the cell loses some of its turgor and the concentration of solutes in it increases; hence the water potential within the cell sap decreases. The cell is then able to gain water by osmosis from adjacent more-saturated cells. These in turn gain water from adjacent cells and eventually from the water-filled tracheid of a vein ending. A water potential gradient soon becomes established, along which water moves from the vein through one or several cells to a cell wall bordering an intercellular space, into and through the intercellular spaces, and through the stomata to the outside air.

Quantity of Water Transpired

The amount of water transpired by plants is very great. A single corn plant in Kansas, between May 5 and September 8, transpired 460 liters of water. An hectare of such plants (14,800 plants) would transpire, during that season, over 3 million liters of water, which is equivalent to a sheet of water 28 cm deep over the entire hectare. It has been estimated that an hectare of red maple trees, growing in a soil with ample moisture, may lose in a growing season an amount of water sufficient to cover each hectare with 28 cm of water. A soil clothed with plants is depleted of its moisture at a much more rapid rate than one that is bare. Nearly all the water loss from a soil below the first 15 to 20 cm results from absorption and transpiration by plants.

Of the total quantity of water absorbed by the roots of plants, as much as 98% of it escapes from the plant by transpiration. The small quantity of the water that is retained by the plant includes the water of vacuoles and protoplasm, that in cell walls and in the conducting elements, and that entering into chemical combination.

Loss of water through the stomata is an unavoidable consequence of photosynthesis, for the stomata must be open to allow CO_2 to enter. Through various modifications, however, some plants lose less water per gram CO_2 fixed than others. This ratio of grams H_2O lost per gram CO_2 fixed is called the **water use efficiency.** Most plants have a water use efficiency of about 600, but plants with the C_4 photosynthesis path are more efficient with an efficiency near 300, and succulents such as cacti very efficient with a value of 50.

Evapotranspiration

In evaluating an area of land in terms of the amount of water needed to produce maximum plant growth, one must take into consideration both evaporation of water

Figure 12.4
A lysimeter is a device for measuring the evaporation and transpiration from a given area of turf. The area being measured is delineated by the circle, and the size of the area may be judged by the individual present. A laboratory beneath the turf is crowded with electronic instruments. (Courtesy of W. O. Pruitt.)

from the soil surface and transpiration of water. The total amount of water that a unit of land area loses to the atmosphere through evaporation and by transpiration from plants growing on the land is called **evapotranspiration**.

A number of rather complex and expensive instruments have been designed to determine evapotranspiration. One of these, the **lysimeter,** is shown in Fig. 12.4. The lysimeter, essentially, is a large metal tank containing a soil mass that may weigh (together with the soil water) as much as 45,000 kilograms. It is mounted in a pit so that the soil surface in the tank is level with the surface of the soil in the field. The tank is so arranged that a sensitive and automatic balance will continuously record any changes in weight of the container and its contents. The sensitivity of the balance may be to the nearest kilogram, which is a sensitivity equal to 0.002 cm water applied to the soil surface. All water draining through the soil in the lysimeter is collected and measured. Hence, if plants are grown in the lysimeter, a continuous record of weight changes and of the volume of runoff or water percolating through the soil is recorded, and the evapotranspiration can be calculated. The gain in weight due to the growth of the plants must be taken into consideration, but this weight change is usually relatively small compared to the changes caused by moisture gain or loss.

Under conditions of adequate water supply to an area covered with vegetation, the plants will act as channels through which water passes from the soil to the atmosphere, and the amount of water lost by evaporation and transpiration will be primarily a function of the climate. Many attempts have been made to use weather data to calculate the maximum or potential evapotranspiration for a given area. These calculations have been only partly successful.

Factors Affecting Transpiration Rate

The factors influencing the rate of transpiration include two principal groups: **environmental factors** and **morphological factors.** The former include conditions external to the plant, whereas morphological factors include structural features and habits of growth of the plant, sometimes referred to as internal factors.

Environmental Factors. The important environmental factors that influence transpiration rate are (*a*) relative humidity of the atmosphere, (*b*) air movements, (*c*) air temperature, (*d*) light intensity, and (*e*) soil conditions. These factors affect transpiration through their effect on the vapor pressure of water in the intercellular space or of water in the air.

Atmospheric humidity. The humidity of the air surrounding the plant is an important external factor affecting the rate of transpiration. The drier the air above the plant, the greater the transpiration rate. For example, air with a relative humidity of 20% is considered dry because it has the capacity for holding a great deal more water; air with a relative humidity of 80% is regarded as moist because its capacity for holding more water is not great.

The air in intercellular spaces of a turgid leaf is very moist—in fact, usually almost saturated. A diffusion of water vapor from intercellular spaces through stomata to the air surrounding the leaf takes place as long as the vapor pressure of water in the intercellular spaces exceeds that of the air outside. Diffusion becomes more rapid as the difference between these two pressures increases. When the difference is great, we say the **diffusion gradient** is steep. If the relative humidity of air surrounding the plant is low, the diffusion gradient between moist intercellular spaces and outside air is steep, and hence transpiration is rapid.

Air movements. As transpiration occurs, there is a tendency for a moist **boundary layer** of air to form next to the leaf surface, particularly in still air (Fig. 12.5). This will decrease the diffusion gradient between the leaf and the atmosphere, and transpiration will consequently decrease. Air movement carries away much of this layer of humid air, replacing it with drier air, resulting in an increase in transpiration. The more rapid the air movement, or the smaller the leaf blade, the faster the moist air will be carried away and the faster the rate of transpiration. If the wind is quite strong, stomata may close, probably as a result of excessive water loss, and transpiration is then reduced.

It is a common observation that on a warm, bright, windy day, plants may lose water to the point of wilting even though their roots are supplied with adequate moisture. On a cool, cloudy day with little wind movement, however, seldom is the water loss sufficient to result in

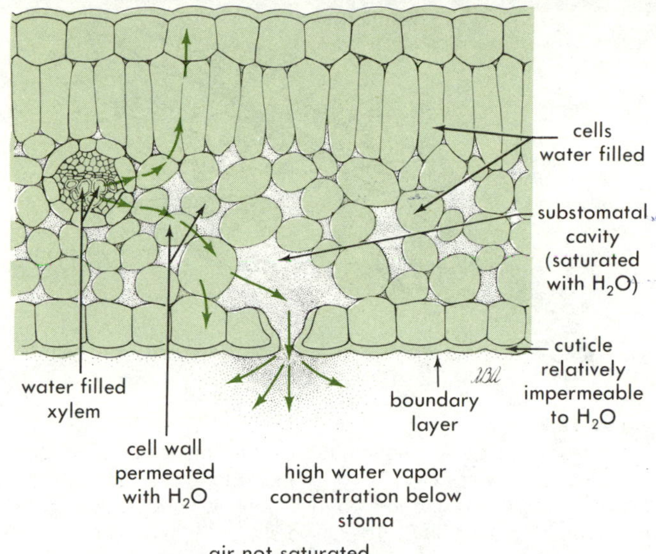

Figure 12.5
A cross section of a leaf showing the path of water loss due to transpiration. Note the presence of a boundary layer.

wilting. Even though plants may wilt during the hottest part of the day, this may be but a temporary condition, as is shown by the fact that the next morning the leaves are again fresh and turgid. Return to the turgid condition indicates that available water in the soil is not exhausted.

Air temperature. In direct sunlight, the temperature of a leaf is usually higher than that of the air about it; the difference may be as much as 10°C. About 80% of the radiant energy of sunlight that falls upon green leaves may be absorbed by them. Part of this absorbed energy is changed to heat and raises the temperature of the leaf; part is utilized in vaporizing water; and another small part is used in photosynthesis. When the leaf temperature rises, the vapor pressure in intercellular spaces becomes greater than that in the air surrounding the leaf; as a result, the diffusion gradient becomes steeper, and the rate of transpiration increases.

Light intensity. As light intensity increases, the internal temperature of the leaf is raised, with the result that water loss is accelerated. Another effect of illumination upon transpiration is that it stimulates, at least in many species of plants, the opening of stomata (p. 189). When the stomata are closed, transpiration virtually ceases. Thus light, through its effect upon leaf temperature and upon the opening of stomata, influences the rate of transpiration.

Soil conditions. Any soil condition that influences absorption of water by the roots affects the transpiration rate. When soil becomes very cold, absorption of water is retarded, even though soil water may be available. An increase in the concentration of the soil solution, as often occurs in alkali soils, reduces the rate of water intake. Poor aeration of the soil may result in diminished absorption. The availability of soil water also determines the rate of intake. The rate at which water is absorbed by the roots greatly influences the rate of transpiration. If water loss exceeds water absorption, wilting will occur. Under these conditions, the mesophyll cells do not give up water vapor as freely as when they are well-supplied with water. Consequently, the rate of transpiration is reduced.

Structural Factors. On days when water loss is high, various kinds of plants growing side by side, exposed to the same environmental conditions, respond differently with respect to wilting. Some may be severely wilted; others may show no signs of distress. It is apparent that some plants have better provisions for transpiration regulation than do others. Thus, plants have certain structural features and habits of growth that influence the rate of transpiration.

Certain plants are able to survive in extremely dry habitats, whereas certain other plants succumb. Plants of dry habitats often possess roots that penetrate deeply into the soil; or, like many desert annuals, they have an extensive surface root system that can rapidly absorb limited rainfall from spring or summer showers. Many desert plants possess special water storage tissue, as is well-shown in so-called succulents, that is, plants with fleshy stems of leaves (see **Fig. 12.6,** page 222).

Anatomical features that are advantageous from the point of view of prevention of water loss are (*a*) the cuticle of leaves, young stems, and fruits, (*b*) sunken stomata, (*c*) distribution of stomata, and (*d*) reduction of the transpiring surface. In addition, stomatal behavior is an important factor in controlling water loss.

Cuticle. Most of the water lost from a plant passes out through the stomata. Some water, perhaps 10% at most, is lost through the cuticle. Various modifications in leaf structure reduce cuticular transpiration; for example, thickening of the outer wall of the epidermal cells and the presence of a waxlike material, cutin, in this wall. Most plants of arid and semi-arid climates have a thicker cuticle than do those of humid climates (**Fig. 12.7,** page 222).

Stomatal behavior. We have seen in Chapter 10 that a stoma may open and close as a result of changes in the turgor of guard cells. When guard cells become turgid, the stomatal aperture widens; when they become flaccid, the aperture narrows or completely closes. Often after a period of rapid transpiration, there is a deficit of water in all leaf cells, and stomata close, thus reducing the water loss, due to transpiration. This is an advantageous behavior. But, from the standpoint of photosynthesis, the stomata must serve also as the site of entrance for carbon dioxide. If they close during the daytime because of excessive transpiration, the diffusion inward of carbon dioxide, as well as the diffusion out-

Figure 12.8
Guttation from tips of barley *(Hordium vulgare)* leaves.

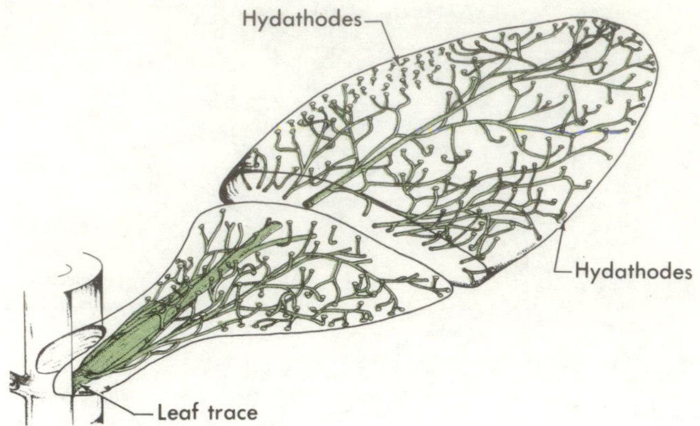

Figure 12.9
Sterodiagram of the venation pattern and hyathodes of *Crassula argentea*. Some veins have been excluded for clarity, and vein thickness has been exaggerated for emphasis, ×½. (T. Rost, *Bot. Gas.* **130,** 267. © 1969 by The University of Chicago Press. Reprinted by permission of publisher).

ward of water vapor, is restricted. Stomatal closure, in this event, works to the disadvantage of the plant by decreasing photosynthesis, although it is a favorable behavior in that it limits water loss. However, since the photosynthetic CO_2-trapping mechanisms in C-4 plants is much more efficient than that of the C-3 plants, partial stomatal closure does not reduce photosynthesis as much in a C-4 plant. As shown in Chapter 13, one characteristic of C-4 plants is their adaptation to arid conditions. This is an example of a biochemical rather than a morphological adaptation.

Certain succulent species open their stomata at night, when air humidity is high, and close them during the day. These species have a very favorable water use efficiency. Light is a factor that stimulates stomata to open in most other types of plants, but not for these succulents.

Sunken stomata. In some plants **(Fig. 12.7),** the stomata are below the general level of the leaf surface. When this condition occurs, the water vapor must diffuse through a relatively long and sometimes tortuous passageway, with the result that the diffusion rate is lessended. Note also in **Fig. 12.7** that crypts in which stomata occur are lined with an abundance of epidermal hairs that would reduce air movement into or out of the crypts.

Distribution of stomata. The leaves of many plants have stomata only on the undersurface, or, if on both surfaces, mostly on the lower. This distribution is shown in Table 10.1. Hence, loss of water from leaves that have stomata only on the lower surface, or fewer on the upper than the lower, is usually less than loss from leaves that have an equal number on both the upper and lower sides.

Reduction of transpiring surface. In most plants, the principal organs of transpiration are leaves. Therefore, any decrease in leaf surface will reduce transpiration and conserve water absorbed by roots. In corn and other grasses, and also in oleander, the leaves roll up during drought, thus exposing less surface to the air than do full expanded leaves. Cacti and some euphorbias have no foliage leaves; in these plants the transpiring surface is restricted to the stem surface.

Epidermal hairs. The leaves of many plants are clothed with epidermal hairs. In some plants, such as common mullein and dusty miller (*Centaurea cineraria*), the hairs may form a dense, cottony covering. There is no clear evidence that surface hairs reduce transpiration; in fact, experiments with some plants have shown that water loss from the leaves is greater when the hairs are present than when they are lacking. The hairs may serve mainly to reflect light, keeping leaf temperature low.

Guttation

The loss of liquid, as contrasted with the loss of vapor, from leaves of *intact plants* is termed **guttation.** This will occur especially in herbaceous plants when conditions favor rapid absorption of water and low transpiration. For instance, when a well-watered, vigorously growing tomato plant is placed under a bell jar, transpiration ceases as the atmosphere in the jar becomes saturated. Continued water absorption then results in a slow exudation of water from the tips of the leaves (Fig. 12.8). Many plants have specialized structures called **hydathodes** at the tips, margins, or surfaces of their leaves, through which the liquid passes outward (Fig. 12.9). Experiments with barley plants show that, when the roots are immersed in distilled water, guttation is very slight or

Figure 11.7
Tomato plants grown in nutrient culture solutions to show visual symptoms of mineral deficiencies. *A*, complete solution. *B*, deficiency symptoms in leaves. *C*, minus phosphorus. *D*, minus potassium. *E*, minus nitrogen. *F*, minus sulfur. *G*, minus calcium. *H*, minus iron. *I*, minus magnesium. *J*, minus all micronutrients. All plants subjected to deficient solutions were first grown in a complete nutrient solution for two weeks.

Figure 11.12
Differential response of barley *(Hordeum vulgare)* genotypes to salt stress 20 days after germination in a nutrient solution containing 400 mM sodium chloride. (The concentration of sodium chloride in sea water is roughly 500 mM.) The lines H (U. C. Signal), CMS (a selection from CM), N (Numar), and CM (California Mariont) germinated and established seedlings. B (Briggs) and A (Arivat) failed at this salt stress. F_2 is progeny of a cross between A and CM; segregation is evident: note the uneven performance of the plants. (Courtesy of E. Epstein and J. Norlyn.)

Figure 12.11
The presence of Ca^{2+} may modify the toxic effects of Na^+ by reducing the absorption of Na^+ by the roots. Plants were grown in the presence of 50 milliequivalents of sodium ion and in varying amounts of calcium. From left to right, 0.0, 0.1, 0.4, 1.0, 4.0, and 10.0 mM $CaSO_4$. (Courtesy of E. Epstein).

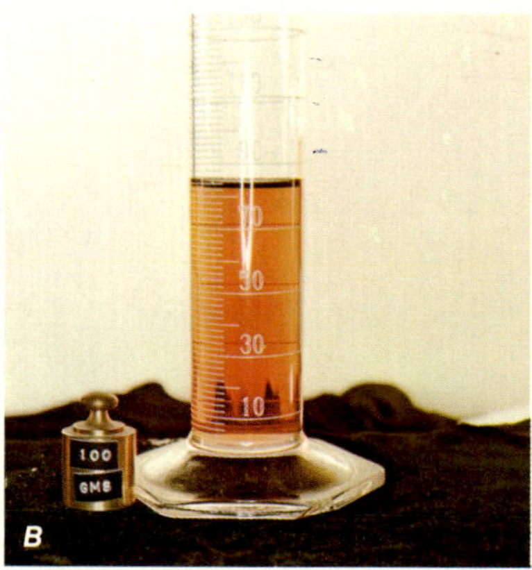

Figure 12.1
Moisture content, dry weight, and ash content of fresh grass leaves. *A,* 100 g fresh leaves yield *B,* 82 g water and *C,* 18 g dry leaves. *D,* the dry leaves yield, on burning, 1.5 g ash.

Figure 12.3
Cross section of a pear *(Pyrus)* leaf showing extensive internal surface and intercellular air spaces and thin cuticle.

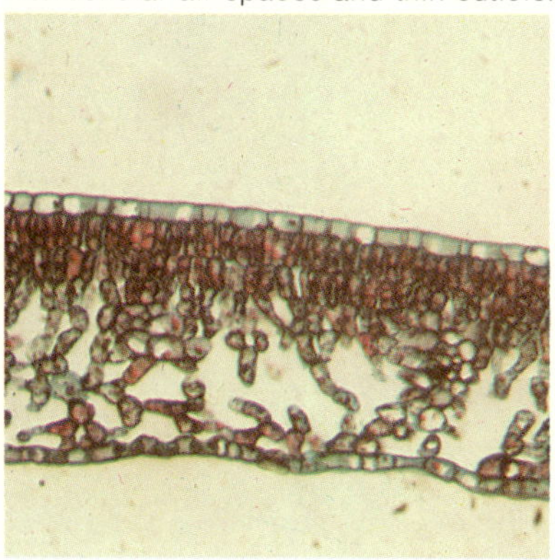

Figure 12.7
Cross section of Oleander *(Nerium oleander)* leaf. Note thick cuticle and absence of stomata except in crypts that also have an abundance of epidermal hairs. Compare with thin cuticle of pear (Fig. 12.3).

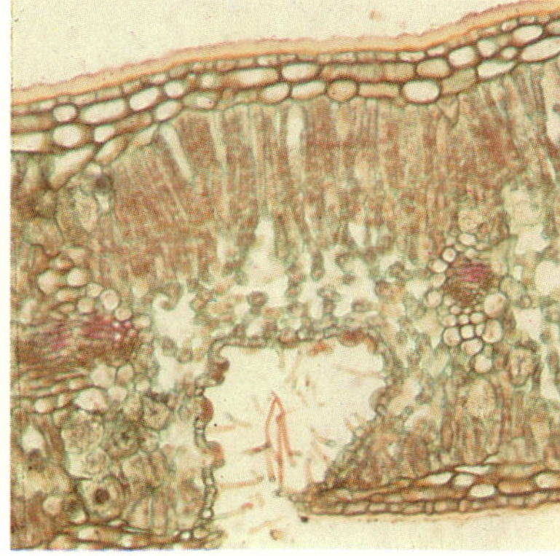

Figure 12.6
Cross section of *Yucca* leaf showing tissue adapted for water storage.

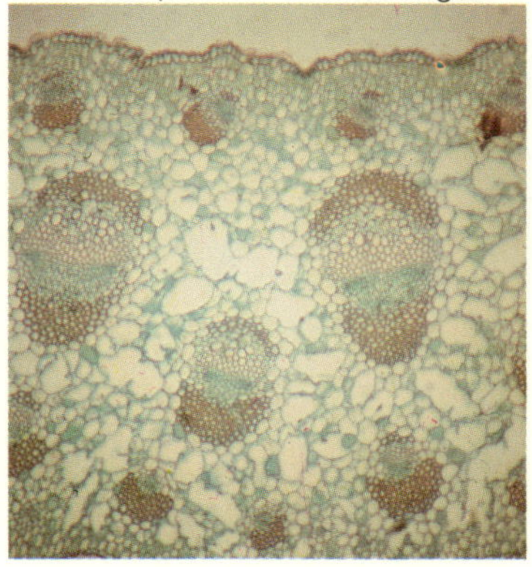

Figure 12.10
Deposit on blades of grass following the evaporation of guttation fluid. Deposit composed largely of glutamine with a small amount of potassium chloride and undetermined organic matter. (Courtesy of Curtis.)

ceases altogether. This is true even though the water is aerated amply. In dilute salt solution, but without aeration, guttation is also slight. On the other hand, if the roots are immersed in a dilute salt solution, with good aeration and favorable temperature, guttation is rapid and continues for a long period in a humid atmosphere. Guttation is associated with salt absorption and salt movement into the xylem. The liquid of guttation is not pure water but a dilute salt solution (Fig. 12.10).

Summary of Water Loss

1. Transpiration is the loss of water vapor from plant surfaces, largely those of leaves.

2. Evaporation removes the water vapor from the external surface of the leaf.

3. The amount of water transpired during the life of a plant is large, from 200 to 1000 times its dry weight.

4. The rate of transpiration is affected by the relative humidity of the air, air movements, air temperature, light intensity, and soil conditions.

5. Morphological details such as the cuticle, epidermal hairs, and location and distribution of the stomata also influence the rate of transpiration.

6. Stomatal opening and closing may regulate transpiration. Some succulents open their stoma at night and close them by day, in contrast to most other plants.

7. Guttation is the escape of water as liquid from leaves; it may occur when the relative humidity is high. Guttation fluid contains solutes.

CLASSIFICATION OF PLANTS ACCORDING TO THEIR WATER NEEDS

In many agricultural sections of the country, and particularly in semi-arid and arid regions, water is the principal limiting factor in crop production. Water is a most important factor in determining both the distribution of plants over the earth's surface and the character of the individual plant. Probably no single factor is so largely responsible for the diversity of plants in various habitats as is the difference in supply of water. So important are the water relationships of plants that various attempts have been made to classify plants on the basis of these relationships. One such classification divides plants into (*a*) **xerophytes,** which are able to live in very dry places, (*b*) **hydrophytes,** which live in water or in very wet soil, (*c*) **mesophytes,** which thrive best with a moderate water supply, and (*d*) **halophytes,** which have modifications that enable them to grow in very salty soils.

Plants with xerophytic characteristics, which limit transpiration or in other ways balance water outgo and water intake, occur in different climatic zones, but those of deserts are most typical. Xerophytic species do not necessarily have a lower transpiration rate than do mesophytes when water is ample, but they do possess one or more characteristics that enable them to survive periods of drought. The most effective of these are thick cuticle, stomatal closure, reduction of the transpiring surface, and water storage tissue. Mesophytes may not be able to long survive low soil moisture, whereas plants with xerophytic characteristics will.

PROCESSES CONCERNED IN ASCENT OF WATER

Water in the plant moves upward from the roots to the stems and thence into leaves, from which it escapes into the atmosphere. That the path of rapid movement is through the xylem is shown by "ringing" or "girdling" the stem, that is, removing a complete ring of bark (thus all of the active phloem). This procedure has little immediate effect on the movement of water; the plant does not wilt. The path of movement of water may also be demonstrated by placing the base of a leafy branch that has been cut underwater into an aqueous solution of a dye, such as eosin. If after a short period the stem is split, it can be observed that the walls of vessels and tracheids are stained by the dye, whereas other tissues are not stained.

Transpiration Pull

It is estimated that a force of 20 bars (Fig. 5.16), over 20 kilograms per cm², is necessary to lift water to a height of over 100 meters through the xylem of a tree. This estimate takes into consideration the resistance to water movement offered by the xylem, as well as the weight of the water column. How does water get to such a height in a tree? What are the forces involved? Although plant physiologists are not agreed on the relative importance of the various processes, the explanation that is now most generally accepted as most plausible will be described here. It is spoken of as the **transpiration-pull or water-cohesion theory.**

Evaporation

When evaporation of water occurs from the cell walls in a leaf during transpiration, forces are brought into play that result eventually in the movement of water through the plant. The diffusion of water molecules from the wet cell wall surfaces into the intercellular spaces of the leaf mesophyll results in partial drying of the walls.

Imbibition

The walls of a leaf cell, losing water as a result of evaporation, imbibe water from the enclosed cell contents or from adjacent cell walls with a higher water content. Since the pectic and cellulosic components of the walls are polar, they bind water molecules very firmly. These **adsorptive** or **imbibitional forces** are very great. Although the first water that is lost from a saturated mesophyll wall is held loosely, the forces holding the remaining water in the wall are stronger. Thus, during transpiration, the water potential in the wall gradually decreases because some of the imbibitional forces are not satisfied.

Osmotic Forces

This decrease in the water potential in the wall results in the movement of water from the protoplast into the wall. As water moves from the protoplast of a mesophyll cell into the walls, the turgor of the cell decreases. This causes a decrease in the water potential of the cell, and water will move into the cell from adjacent cells down a water potential gradient. Because of the relatively high osmotic concentration in mesophyll cells, these cells can have a lowered water potential an still retain considerable turgor pressure. The disturbance in water equilibrium is transmitted from cell to cell through both cellular contents and walls. Water moves into these cells from the tracheids in the veinlets of the leaves. Thus, a force equivalent to the imbibitional force created in the leaf cell walls, as a result of the loss of water from them during transpiration, is pulling on the water in the tracheids. Because of the relatively low osmotic concentration in the tracheids, as the water potential in the tracheids gradually decreases, the contents may be subjected to reduced pressures and, because water molecules are linked by hydrogen bonds, they may actually be pulled along.

Water Continuity and Cohesion

Water in the tracheids of leaf veinlets is continuous from the leaves to the roots. This continuity exists throughout the life of the plant. In a sense, the water columns "grow" with the plant. However, living cells of the root, xylem elements themselves, and living leaf cells offer resistance to the free flow of water. This resistance, and the gravitational pull on the water itself, are overcome by the forces developed in the leaf when water is moving up in the plant. The question now arises: Do these unbroken columns of water have sufficient tensile strength to prevent them from being broken by the forces acting on them? Experiments using both water and plant sap indicate that this requirement is met. As a matter of fact, plant sap in vessels and tracheids may withstand tensions greater than 300 bars without being broken. Thus, it is evident that the attractive forces of the water molecules for each other, **cohesion,** and of cell walls for the water molecules, **adhesion,** are very great. It is because of these attractive forces that the liquid columns possess **tensile strength** (resistance to breaking when placed under a strain).

The columns of water are raised by the force of evaporation of water from the leaf cells walls and the imbibitional forces that develop there.

Summary of the Transpiration-Pull Theory

1. The evaporation of water from the walls of mesophyll cells of leaves results in a decrease in the water potential in these walls.
2. This disturbance in the water balance of the cells causes a series of changes that set in motion the entire train of water through the plant.
3. The living leaf cells next to the tracheids in veinlets eventually lose water to neighboring cells.
4. Water moves from the veinlets into the adjacent cells. This results in the upward movement of water in continuous liquid columns in the xylem.
5. Continuity of water is maintained even under conditions of considerable strain because of the strong cohesive forces between water molecules.

ABSORPTION

Absorption of Water

The large quantities of water moved through the plant body and lost by transpiration, then, must be replaced continually if the plant is to live.

In the absorbing region of the root (Figs. 9.7, 9.9), the phloem does not lie outside the xylem but occurs in groups of cells that alternate radially with xylem masses or with the radiating arms of a single central mass of xylem. Hence, it is possible for the water and mineral salts to be absorbed by epidermal cells or root hairs; pass through the cortex, endodermis, and pericycle into the primary xylem; and move upward without traversing the phloem (Fig. 12.2).

Mechanisms of Water Absorption

The uptake of water by roots involves two mechanisms, which may or may not act simultaneously:

1. When transpiration is slow and water is available in the soil, absorption of water may exceed transpiration. This condition results in a pressure being developed in the xylem, and may lead to guttation. If the shoot of a plant is cut off a few inches above the soil level, absorption of water by the roots will often force the sap out of the cut ends of vessels and tracheids, a phenomenon known as **bleeding.** It appears that this type of absorption depends upon the *activity of living root cells.* The pressure developed in the xylem is called **root pressure.**

This mechanism of absorption of water by roots is generally explained as resulting from the osmotic movement of water into the root. As a result of the activities of the living root cells, soil solutes are absorbed and accumulate in the cell in high concentrations (pp. 226). The total concentration of the cell sap of the root hairs, as well as of the other root cells, is thus normally greater than that of the soil solution. The total amount of solutes present in the xylem elements of the root is also generally higher than that of the soil solution. The greater amount of solutes in the cell sap reduces the potential of water in the cell sap below that of the water in the soil solution. Under this condition, movement of water is inward. The rate of movement inward depends upon the difference in the water potential of the two solutions.

2. When transpiration exceeds absorption, no root pressure can be demonstrated. In the xylem elements of plants in this condition, the water is under **tension,** and the pressure on the water in the vessels is lower than atmospheric pressure. Under these conditions, a high rate of transpiration results in a water deficit in the plant tissues. Water then moves into the roots passively. In fact, under such conditions, water may even be absorbed by dead roots. Water absorption may then result from osmotic forces originating in roots or from forces that originate in the leaves—that is, forces set in motion by the loss of water in transpiration.

Factors Affecting Water Absorption

We have seen that roots grow in a very complex environment—the soil. The activities of roots are influenced to a marked degree by this environment. Many environmental factors determine the nature and extent of the root system, the rate of growth of roots, the rate at which they absorb water and mineral nutrients—in fact, all their activities.

Unavailable Soil Moisture. Plants cannot absorb all the water in a soil. There is always some water in the soil that is not available to the plant; it is held so tightly by the soil particles that the roots cannot absorb it rapidly enough to prevent wilting. If such a wilted plant is placed in a darkened moist chamber and will not recover, it is said to be **permanently wilted.** The only way to bring about its recovery is to add water to the soil. At the time that the plant has permanently wilted, the percentage of water left in the soil may be determined. This percentage (based upon the dry weight of soil) is called the **permanent wilting percentage of the soil.** This water represents that which the plant cannot absorb readily from the soil. Only that part of the total soil moisture above the permanent wilting percentage is considered as **available water** for plant growth.

Different kinds of mesophytic plants growing in the *same soil* reduce it to approximately the same permanent wilting percentage when the soil water potential reaches approximately −15 bars. A few plants, for example, creosote bush (*Larrea divaricata*) can absorb water to a water potential of −60 bars. However, there is very little water in the soil at such negative soil water potentials. Consequently, the actual difference between the amount of water held in a soil at −15 bars and −60 bars may be almost negligible.

Influence of Soil Type. *Different kinds of soils,* however, differ in their permanent wilting percentage. Let us perform an experiment to illustrate this point. Fill one container with a clay soil, and another with a sandy soil. Then wet each soil to field capacity. We know the clay soil will hold more water than the same volume of sandy soil. In each grow the *same kind of plant.* Allow each to reach permanent wilting, and then determine the permanent wilting percentage of the two types of soil. It will be found that the permanent wilting percentage of clay soil is much higher than that of sandy soil.

Moisture Content of Soil. From the standpoint of plant growth, the readily available water of a soil is the important consideration. This is really of more significance than the water-holding capacity of a soil.

Under field conditions, soon after a rain or irrigation, followed by drainage, the soil to a certain depth (depending upon the amount of rain or irrigation water applied) is up to its field capacity of moisture. The roots of plants begin to absorb the water, and soon the soil is below its field capacity. The soil moisture constantly decreases in the zones occupied by roots. Moreover, the movement of water by capillarity, from moist soil to drier soil immediately surrounding the roots, is so slow that it may be regarded as negligible in influencing water available to the roots.

Root Growth and Soil Moisture. Since the capillary movement of water in soils is too slow to meet the requirements of growing plants, roots must continually penetrate into new soil if they are to utilize the water it holds. The total root elongation of a plant may be very great. Root extensions and the formation of new root hairs are constantly bringing about new contacts with moist soil.

The erroneous impression is sometimes held that roots "go in search of water." A more accurate statement is that they grow only in moist soil (soil above the permanent wilting percentage). In practice, this would mean that if the upper foot of soil is moist, and dry soil lies below it, the roots of plants will be confined to the upper, moist soil. In irrigation practice, this relationship of soil moisture and root growth is important. In irrigating a garden or lawn with a hose, one is easily misled into believing, because the soil surface is wet, that sufficient water has been applied. Examination may reveal that at depths of 12 cm or more the soil is very dry—too dry for the growth of roots. In growing plants, it is nearly always desirable to stimulate maximum root growth and a root system that attains its normal depth and spread. This may be accomplished in part by keeping the soil moist to the proper depth.

The loss of water from soils at levels below the first few centimeters is essentially due to transpiration from plants. Soils are dried out because of the presence of absorbing roots. One of the principal reasons for removing weeds from a growing crop is that they deplete the soil of water.

Solute Concentration of Soil Solution. As the concentration of the soil solution approaches that of the cell sap, the rate of water absorption declines. When the concentrations of the two solutions are the same, water intake ceases. And, if the soil solution becomes more highly concentrated than the cell sap, owing to excessive applications of fertilizers or of saline water, water will be withdrawn from the root cells. Moreover, root growth is inhibited at high salt concentrations, and the roots are not able to extend into new soil areas.

Absorption of Solutes

Solute particles diffuse relatively independently of water. Solutes move by diffusion from the soil solution into the wet cell walls of root cells. Here, some of the cations may exchange for cations associated with the negatively charged sites on the wall. Although solutes may more relatively freely through the wall, ions meet a barrier to free diffusion, the plasmalemma, when they reach the protoplasmic surface. Cell membranes are not readily permeated by charged ions. Nevertheless, cells do absorb ions and actually accumulate them to concentrations many times higher than their concentration in the external medium. How do solutes enter cells? What is the method used by the cell to accumulate solutes?

Active Solute Absorption

It has been known for a long time that some ions may be taken up and stored in living cells so that their concentration is 1000 or more times that in the external medium. Such an accumulation against a concentration or diffusion gradient cannot be accomplished by simple diffusion. The ions are not precipitated or bound in the cell, but are still free ions. Diffusion would result in equal concentrations of a particular ion throughout the whole system. Accumulation of ions requires energy. It is an active process. What is the source of the energy used?

The major source of energy to nongreen cells such as root cells is food that is respired. Several experimental observations indicate that solute absorption is linked to respiration.

1. The absorption of many ions is markedly decreased under low oxygen conditions and essentially ceases when oxygen is completely removed from the cell's environment. This parallels the effect of oxygen an aerobic respiration.

2. Increasing the temperature around roots from 10 to 30°C increases the rate of respiration and the rate of solute absorption similarly. Simple diffusion is relatively less sensitive to temperature change.

3. It is possible to use certain chemicals to inhibit respiration. Such studies show a close correlation between the production of ATP during respiration and ion absorption.

4. Further evidence of the role of ATP in the absorption of ions by cells is found in the observation that the plasmalemma contains enzymes, ATPases, that hydrolyze ATP molecules and release the terminal phosphate groups. In the process, energy is also liberated. The activity of these enzymes is stimulated by some of the cations, for example, Ca^{2+}, Mg^{2+}, K^{+} that are absorbed by the cell. For example, the stimulation of ATPase by K^{+} is correlated with the rate of K^{+} uptake by roots supplied with the same concentration of K^{+}. Many experimental observations such as these point to the fact that inorganic ions are accumulated in plant cells by the expenditure of energy released through aerobic respiration.

The Carrier Hypothesis

But how do ions pass the plasmalemma that represents a barrier to their movement? One hypothesis that has been advanced to explain this problem suggests that there are definite chemical compounds, proteins called **carriers** (transporters, translocases, permeases) in the membrane. According to this idea, an ion that comes in contact with the plasmalemma may combine with a carrier that has a specific attraction for that ion. The ion associated with the carrier then moves across the membrane and is released into the cytoplasm. The carrier is then free to combine with another ion of the same kind. The results of many experiments on the relation between the concentration of an ion and its absorption are in agreement with this carrier hypothesis.

Once an ion reaches the cytoplasm, it is free to diffuse or to be carried by protoplasmic streaming to other parts of the cytoplasm. It may diffuse from one cell to another through plasmodesmata. However, since cell organelles are separated from the rest of the cytoplasm by membranes, one might postulate that these membranes too may have specific carriers for specific solutes.

Selectivity of Ion Absorption

Under many conditions the absorption of ions by roots is highly selective. For example, in the presence of low concentrations of Ca^{2+}, the absorption of K^+ may be unaffected by the presence of a similar ion such as Na^+. Under these conditions, the roots act as if there are carriers in the plasmalemma that specifically bind K^+ but not Na^+. Under other conditions however, different ions appear to compete for the same carrier. Thus, rubidium competes with potassium for binding and transport, and Ca^{2+} absorption may be strongly influenced by the presence of Sr^{2+}. An example of the influence of one ion on the effects caused by another ion is seen in the ameliorating effect of calcium ions on the injurious effects of sodium ions on plant growth (**Fig. 12.11,** page 222).

Summary of Absorption

1. Absorption of water by roots when transpiration is low or absent occurs only in living root systems.
2. This absorption of water results when the solute concentration of the xylem sap is higher than that of the external solution.
3. Root pressure, guttation, and bleeding are expressions of active absorption.
4. Passive absorption of water may result from the transmission to the roots of forces that originate in the leaves; it may even take place if the roots have been killed.
5. A plant is said to be permanently wilted when it will not recover in a saturated atmosphere unless water is added to the soil.
6. The percentage of moisture remaining in soil in which a plant is permanently wilted is called the permanent wilting percentage of that soil.
7. Only that part of the total soil moisture that is above the permanent wilting percentage is available for plant growth.
8. The rate of water absorption is influenced by the available water in the soil, the air in the soil, the soil temperature, the concentration of the soil solution, and the rate of transpiration.
9. The absorption of ions by root cells may occur against a concentration gradient. This process is called accumulation and is dependent upon energy released by the living cells during respiration.
10. ATP produced during aerobic respiration is the important energy source for accumulation.
11. The carrier hypothesis states that solute pass through cell membranes by combining with specific carriers in the membranes.

CONDUCTION

Within a living plant, substances are constantly moving from one place to another. Although they are interrelated, we can detect four types of movement based on the rates at which movement occurs: (*a*) slow *diffusion* of molecules and ions; (*b*) moderate movement of materials carried by *cytoplasmic streaming* in living cells: (*c*) more rapid *flow* of material *in the sieve tubes;* and (*d*) very *rapid conduction* of water and mineral solutes *in the xylem.*

Diffusion and Protoplasmic Streaming

The movement of water or of solute molecules and ions into a cell across the cell wall is a relatively slow process. Diffusion occurs along activity gradients and may take place through plasmodesmatal connections between cells as well as directly through the permeable cellulose walls of most cells. Once a solute molecule or ion diffuses into a cell, its rate of movement may be increased manyfold as it is picked up by the protoplasmic steam. Cytoplasm may stream at rates of a few centimeters per hour. The highest rate of protoplasmic streaming that has been measured was 48.6 cm per hr, observed in the protoplasm of a slime mold.

In terms of the dimensions of a single cell, cytoplasmic streaming is frequently very rapid, since a substance may be transported across the length of a cell in a matter of seconds. In terms of the whole plant, however, this is a very slow process. It would take days, for instance, for a molecule in the leaf to be carried upward to the growing shoot tip or downward to the roots. Figure 12.12 diagrams the way in which one solute may be moving by diffusion and protoplasmic streaming in one direction through a series of cells and another solute may be moving in another direction. Diffusion and protoplasmic streaming are the chief methods by which solutes move through living plant tissues, except in the sieve tubes.

Conduction in the Phloem

The rapid translocation of foods throughout the plant takes place chiefly in the phloem and more specifically in sieve tubes. That the phloem is the major path of food transport is borne out by the results of ringing experiments followed by chemical analyses of tissues and cells above and below the ring. When a ring of bark is removed from a stem down to the cambium (thus removing the phloem), the carbohydrate content and especially the sugar concentration of the sap in the leaf, bark, and wood *above* the ring increases after a few hours. *Below* the ring, the total carbohydrate content and the sugar concentration in the sap of the bark and the wood decreases. These conditions would not prevail if foods were moving downward in the xylem tissue.

This principle has been used to increase the size and sugar content of certain fruit. For example, if a narrow (2 mm) ring of bark is removed all of the way around a grape stem a few centimeters below a developing cluster of grapes, all of the sugars produced by photosynthesis in leaves above the girdle will be unable to leave that part of

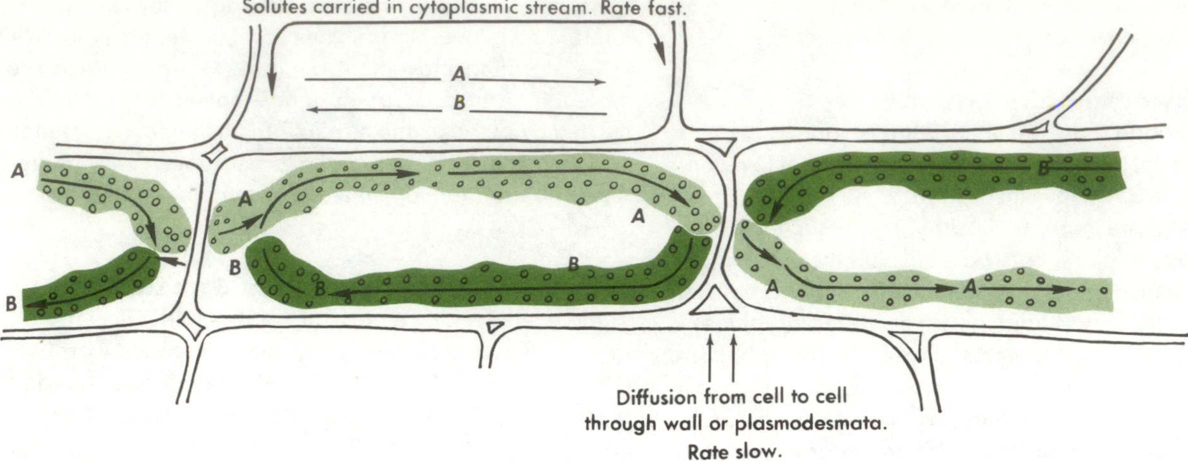

Figure 12.12
Two solutes may move in opposite directions by diffusion and protoplasmic streaming in the same tissue at the same time.

the branch. The local blocking of phloem transport would result in the production of larger grapes with a higher sugar content above the girdle. Such a practice would be detrimental to the roots since it would reduce the supply of food moving into them. However, if the ring of bark removed is not too wide, new cells would be produced bridging the gap and reestablishing the continuity of the translocation pathway.

Both carbohydrates and nitrogenous substances move to tissues that are utilizing them. This means that foods may move upward or downward in the phloem. The direction of food movement is usually from a region of the shoot where excessive food is present to a region where a lesser amount is present. Food may move from the place of manufacture (leaves) to a storage place (roots, fruits, seeds) or to a region of growth (buds, cambium). It may also move from storage tissue to a region of growth.

Translocation of material in the phloem may reach a rate of 100 cm per hr. When this is compared to the much lower rates of protoplasmic streaming, it is evident that normal protoplasmic streaming alone is insufficient to cause this rapid transport. In fact, evidence indicates that the mature functional sieve-tube members have highly modified cellular contents (absence of nuclei and definite vacuoles, Chapter 7) and that protoplasmic streaming does not occur in them.

The contents of the sieve-tube members of one sieve tube make up a continuous liquid system. The pores in the sieve plates provide relatively large connections between adjacent sieve-tube members. It has been shown that at maturity the sieve tube is highly permeable in the longitudinal direction.

The concentration of sugars (sucrose is the most common) in the sieve tubes is usually quite high and may reach 25%. In addition, phloem sap contains numerous nitrogenous compounds especially amino acids, hormones, and various inorganic ions. The process that results in this accumulation of solutes in the sieve tubes to concentrations 1.5 to 2 times that in the surrounding cells is called vein or phloem loading. It has been shown that sugars produced by photosynthesis in leaf mesophyll cells may be secreted into the surrounding cell walls and then may move into companion cells, or in some instances possibly into transfer cells, from which phloem loading probably occurs. Phloem loading is an energy (ATP) dependent process since sucrose and other compounds found in the sieve tubes cross the sieve-tube elements' membranes against concentration gradients. The solution in the sieve tube is under positive pressure. This hydrostatic pressure is built up by the diffusion of water into sieve tubes in regions of the plant where food is produced or where insoluble food is solubilized. According to one of the most generally accepted theories, the **mass flow theory,** water and solutes move together under pressure (as one mass) in the sieve tubes in a longitudinal direction. This movement occurs from food production or solubilization regions to sites of food storage or utilization where solutes are removed from the sieve tubes. When solutes are removed, the water potential increases, and water diffuses out of the sieve tubes into adjacent cells and into the xylem. This results in a net movement of solution along the sieve tube (Fig. 12.13).

A physical model demonstrating this theory may be easily set up in the laboratory (Fig. 12.14).

Lateral Transport

There is lateral (radial) movement of foods in stems along the vascular rays, bringing the foods into close contact with conducting elements of both phloem and xylem. Thus, the rays carry foods from sieve tubes radially into the cambium and xylem.

In some plants, such as the sugar maple, the xylem parenchyma cells serve as places where large quantities of carbohydrates are stored. In the spring, the presence

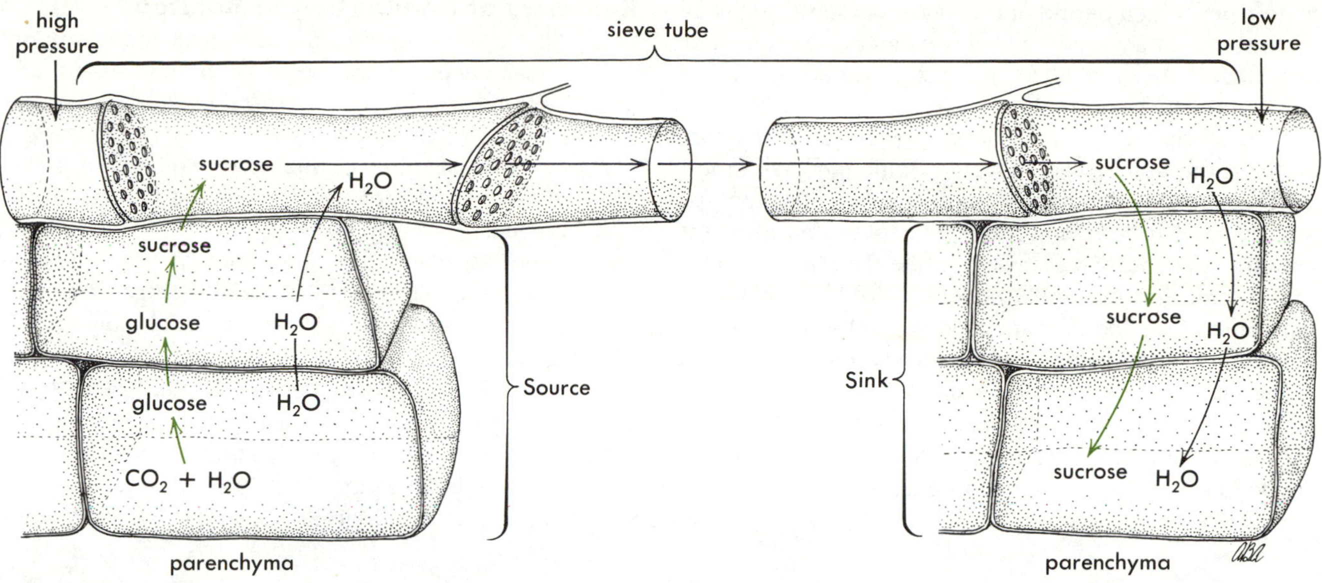

Figure 12.13
The mass flow theory. Sucrose is actively transported into the sieve tubes at the food source region in the plant (e.g., leaves or storage organs) and removed at the sink region (e.g., regions of food utilization or storage). Water follows by osmosis and raises the pressure in the sieve tubes at the source region and lowers the pressure at the sink region. The sieve tube contents flow *en masse* from high to low pressure regions.

in the xylem of soluble carbohydrates results in a high osmotic concentration. This, in association with cold nights and warm days, causes sap to bleed from cuts made in the trunk in the spring of the year.

Conduction in the Xylem

If a stem is girdled by removing a layer of bark around the circumference of the stem down to the cambium, the mineral (inorganic) ions that are absorbed from the soil and conducted in the stems and roots move upward past this girdle. Girdling may or may not decrease this upward movement, depending upon the species and other factors. This fact would serve to indicate that normally the greater amount of ion translocation in the stem is in xylem elements. An especially convenient method of studying the path of movement of ions in stems is now being used. Certain ions, such as potassium, phosphate, and bromide, are rendered radioactive and added to the solution in which the plants are growing. By appropriate means, the path of movement of these radioactive **tracers** has been followed accurately, even for short periods of time, both in girdled and ungirdled stems. The results of these experiments seem to show that inorganic solutes move upward chiefly in xylem tissues. As the ions move upward, however, they may transfer radially from xylem through vascular rays into phloem tissues; and they may also accumulate in living cells along their path of movement. When radioactive phosphorus was supplied to the nutrient solution in which tomato plants 2 meters high were growing, it could be detected throughout the plants after 40 min. There is also evidence that certain minerals carried upward in the xylem to the leaves may be moved from the leaves and conducted downward via the phloem.

Figure 12.14
Physical demonstration of the mass flow theory.

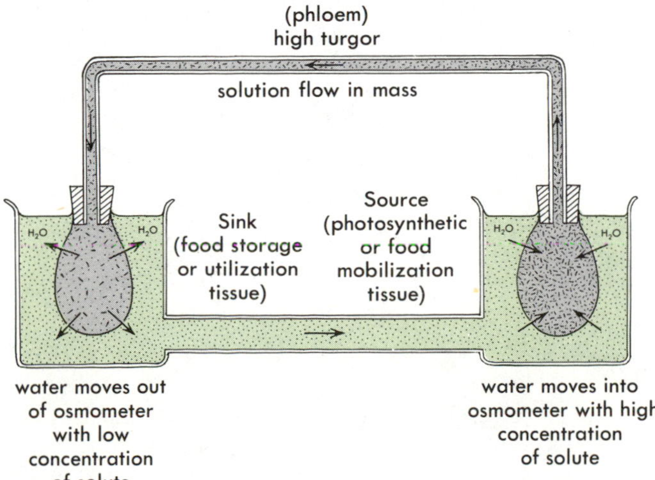

Frequently, not all of the xylem vessels or tracheids function in the conduction of water and mineral ions at any one time. In trees particularly, the heartwood ceases to play a role in conduction (p. 141). Sometimes, only the outermost layers of sapwood contain xylem elements filled with water.

The rate of movement of water and solutes in the xylem depends in large measure upon the rate of transpiration. If plants are growing in a very humid atmosphere, the movement of water through the xylem is very slow; if they are in a dry, warm atmosphere, which heightens the rate of transpiration, the movement may be as rapid as 40 meters per hr. In the larch tree, the rate of water movement was found to be approximately 1.8

meters per hr at a period of the day when water loss was highest; but in the early morning, when the rate of water loss was at the minimum, the rate of water movement was but 0.5 to 1 cm per hr.

Absorption of inorganic ions by roots is *not proportional* to the absorption of water. Some increase in ion absorption accompanies an increase in water absorption, but not in the same ratio. Moreover, the different ions are not absorbed at the same rate. Thus, the amount of water moving through the plant is not the only factor that determines the amount of ions or of any particular ion which is absorbed and conducted through the plants to the leaves. Other factors, such as the photosynthetic activity of leaves, respiration, and other chemical processes in the plant, determine to a degree the quantity and kind of ions absorbed and transported.

Summary of Conduction of Solutes

1. Normally, the greater amount of mineral salt (ion) translocation in the stem is in the vessels and tracheids of the xylem. The movement is usually upward.
2. Ions may also be conducted in the phloem.
3. The major path of food movement is through the phloem.
4. Food movement is from a region of high food content or place of manufacture or storage to a place of food utilization, or from a place of manufacture to storage tissues.
5. Rapid lateral movement of solutes takes place along the vascular rays.
6. The mass flow theory states that solutes and water move together in the sieve tubes as a result of the development of internal pressure in the sieve tube.

CHAPTER 13

PHOTOSYNTHESIS

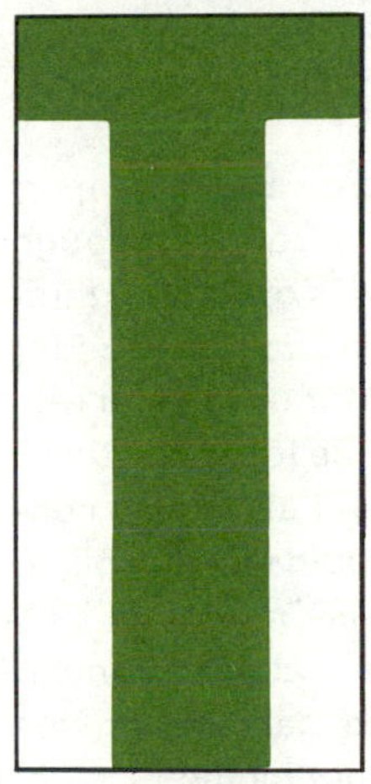

The maintenance of life in a plant or animal cell requires the continual use of energy. This energy directly or indirectly comes from the sun. Two major processes that are carried out by terrestrial green plants and that use energy directly from the sun are transpiration and photosynthesis. Both of these processes use large amounts of light energy, but only in photosynthesis is a significant amount of energy from light actually stored for future use. Light influences other processes such as flowering, seed germination, certain growth curvatures, stomatal movements, and pigment production, but in these cases only very small amounts of light energy are involved.

During transpiration, energy from the sun evaporates water from cell walls. Although this results in a movement of water in the xylem, this energy is neither stored nor used to bring about the vital reactions involved in the synthesis of foods, in assimilation, growth, and reproduction. On the other hand, in photosynthesis, energy from the sun is transformed into chemical energy. In most photosynthesizing plants, photosynthesis results in the utilization of carbon dioxide and water and the production of complex sugar molecules and oxygen gas. During this process, light energy from the sun is changed into energy holding atoms together in the sugar molecule. Neither carbon dioxide nor water will burn, nor would either make a very good meal. Carbohydrates (sugars, starches, cellulose), on the other hand, burn easily and release considerable energy in the process. Furthermore, many carbohydrates are excellent foods.

The great importance of photosynthesis lies in the transformation of low-energy compounds, carbon dioxide and water, into compounds (sugars) containing large amounts of energy. Energy stored as chemical energy in foods (carbohydrates, fats, and proteins) is continually released in living cells during the process of respiration and is lost as heat or converted into energy-rich bonds (for example, ATP). Many energy-requiring reactions carried out by living cells occur at the expense of energy temporarily stored in ATP. *Photosynthesis stores energy. Respiration releases it,* enabling cells to perform the work of living.

Perhaps some day man will use other sources of energy to drive the energy-requiring steps in the production of food and fiber, but today photosynthesis is man's only source of food. Will the capacity of plants to carry out photosynthesis finally determine the number of people who can live on the planet Earth? There are now about 3.5 billion people living on earth; in 1900 there were about 2 billion, and in 1800 a little over 1 billion (Fig. 13.1). If the population increases at its present rate, it will double by the year 2000, reaching between 6 and 7 billion.

If all the land surface of the earth could support plants, it would potentially be able to produce enough food for 1000 billion people. This figure is of course unrealistic, since only a small part of the land can be used for agricultural purposes. Land is needed for urban and recreational use, and some land is unsuitabe for any of these uses. Even if only 7% of the earth's land surface is considered agriculturally productive, it should still be able to produce enough food to support 79 billion people.

Such calculations lead to two far-reaching conclusions: (*a*) the number of people who can live on the planet Earth ultimately will be limited not by the production of food but by the amount of space that a man needs to live in and to work in with reasonable comfort, and perhaps by his contamination of his environment; and (*b*) starvation of thousands of people today is occurring in a world in which technical means of preventing this are known and available, given sufficient sources of energy.

Figure 13.1
Graph showing the world population projected to the year 2000.

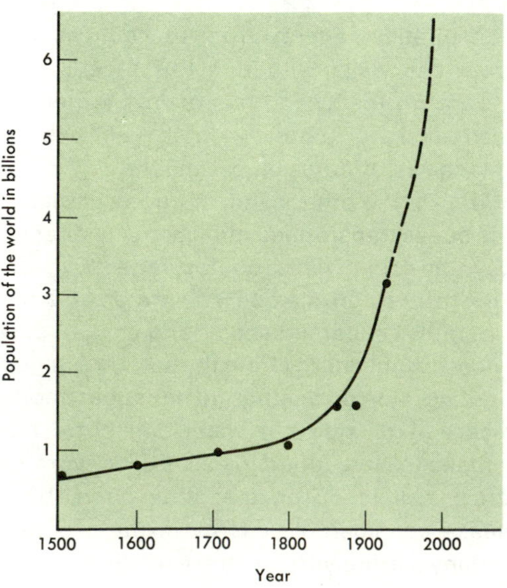

THE DISCOVERY OF PHOTOSYNTHESIS

Our present understanding of photosynthesis is closely linked with the development of modern scientific thought and modern chemistry. Prior to the early seventeenth century, the general belief was that plants derived the bulk of their substance from soil humus. This idea was overthrown by a simple experiment performed by Flemish physician and chemist Jan van Helmont, who planted a 5 lb willow branch in 200 lb of carefully dried soil. He supplied rain water to the plant as needed and in 5 years it grew to a weight of 169 lb. The soil had lost only 2 oz according to his measurements; consequently, van Helmont reasoned that the plant substance must have come from water. This was a logical deduction, though we now know it was not entirely correct.

Our knowledge of photosynthesis begins with the observations of a religious reformer, philosopher, and spare-time naturalist, Joseph Priestley. In 1772, Priestley reported that he had found almost accidentally that a sprig of mint could restore confined air that had been made impure by burning a candle. The plant changed the air so that a mouse was able to live in it. The experiment was not always successful, probably because Priestley, who did not know the role of light, did not always have adequate light. Seven years later, Jan Ingen-Housz noticed that air was revitalized only when green parts of plants were in the light.

It was 3 years later, in 1782, that a Geneva pastor, Jean Senebier, discovered another important part of the process—that "fixed air," carbon dioxide, was required. Thus, in the new chemical language of Antoine Lavoisier, it could be said at this time that green plants in the light use carbon dioxide and produce oxygen.

But what was the fate of the carbon in the carbon dioxide? Ingen-Housz answered this question in 1796 when he said that the carbon went into the nutrition of the plant. In 1804, 32 years after Priestley's early observations, the final part of the overall reaction of photosynthesis was explained by the Swiss botanist and physicist, Nicolas Th. deSaussure, when he observed that water was involved in the process. Now the experiment that van Helmont made almost 200 years earlier could be explained:

$$\text{carbon dioxide} + \text{water} \xrightarrow[\text{green plants}]{\text{light}} \text{oxygen} + \text{organic matter}$$

Almost 50 years elapsed before the carbohydrate nature of the organic matter formed during photosynthesis was recognized.

THE PRODUCTS OF THE REACTION

Oxygen

The evolution of oxygen during photosynthesis in higher plants can be readily demonstrated. If an inverted glass funnel is placed in water over a mass of green water plants and the funnel is completely filled with water and then closed, gas bubbles given off by the plants may be collected in the funnel tube. If a glowing splinter is inserted into the tube, it will glow more brightly, showing that the gas contains oxygen.

The volume of oxygen liberated in photosynthesis is approximately equal to the volume of carbon dioxide absorbed. This ratio, $O_2/CO_2 = 1$, is known as the **photosynthetic quotient.** Under like conditions of temperature and pressure, equal volumes of different gases have the same number of molecules; hence for every molecule of carbon dioxide absorbed during photosynthesis, one molecule of oxygen is liberated. Thus, the equation

$$6\,CO_2 + 6\,H_2O \xrightarrow[\text{green plant cell}]{\text{light energy}} C_6H_{12}O_6 + 6\,O_2$$

is supported by experimental data.

Sugar

The demonstrations that carbon dioxide and water are the raw materials and that oxygen is one of the end products of photosynthesis are easy to make and may be carried out in any botany laboratory. Proof that sugar is the other major product of photosynthesis was more difficult to obtain before $^{14}CO_2$ was available.

It has been pointed out that the photosynthetic quotient O_2/CO_2, is equal to 1; in other words, the amount of carbon dioxide taken up by the leaf is equal to the amount of oxygen given off. This condition is met by the elaboration of a carbohydrate as is seen from the overall equation.

Starch grains occur in the chloroplasts of the higher plants, and if leaves containing starch are kept in darkness for some time, the starch grains will disappear **(Fig. 13.2*A,B*, page 239).** If these leaves are exposed to light, starch reappears in the chloroplasts. Starch, a carbohydrate, is the *first visible product* of photosynthesis.

Some plants, such as certain varieties of *Geranium* and of *Coleus,* have variegated leaves, that is, leaves with white bands or blotches. Microscopic examination of mesophyll cells of the white areas shows them to be devoid of chloroplasts. These cells, however, are living and functioning. If a variegated leaf is given the starch test after several hours exposure to light, starch will be found only in the cells that contain chloroplast **(Figs. 13.2*C,D*).** This test demonstrates that photosynthesis and starch storage have occurred in association with the chloroplasts.

A GENERAL EQUATION FOR PHOTOSYNTHESIS

If we look at the overall equation for photosynthesis,

$$CO_2 + H_2O \xrightarrow{\text{light}} \underset{\text{carbohydrate}}{(CH_2O)} + O_2$$

one might conclude that the oxygen liberated is released from the carbon dioxide molecule. However, van Neil, working at Stanford University, made a comparative study of photosynthesis which occurred in a number of organisms belonging to different groups of plants. The green and purple sulfur bacteria are able to use hydrogen sulfide instead of water to reduce carbon dioxide to form sugar. Sulfur instead of oxygen is liberated.

$$CO_2 + 2\,H_2S \xrightarrow{\text{light}} (CH_2O) + H_2O + 2\,S$$

Some other bacteria and some algae can use hydrogen instead of water to reduce carbon dioxide:

$$CO_2 + 2\,H_2 \xrightarrow{\text{light}} (CH_2O) + H_2O$$

By comparing similar reactions that occur in a variety of forms in different organisms, one is often able to clarify some of the puzzling parts of complex reactions. In the two examples above, CO_2 was still used up, but O_2 was not released. Van Neil concluded that a general equation for photosynthesis should be written as follows:

$$\underset{\text{carbon dioxide}}{CO_2} + \underset{\text{hydrogen donator}}{H_2A} \xrightarrow{\text{light}} \underset{\text{carbohydrate}}{(CH_2O)} + \underset{\text{water}}{H_2O} + A$$

The hydrogen donator, H_2A, can be H_2O, H_2S, H_2, or any other substance capable of donating hydrogen to CO_2 in the process of photosynthesis. The donation of hydrogen is a reduction reaction and requires an input of energy. When H_2A gives up its hydrogen, it becomes oxidized to A.

The validity of van Neil's hypothesis was demonstrated when Ruben, Hassid, and Kamen used the heavy isotope of oxygen (^{18}O) and determined that when water molecules contained ^{18}O, then the oxygen liberated by photosynthesis contained ^{18}O. In contrast, if carbon dioxide contained ^{18}O, then the oxygen liberated did not contain the heavy oxygen isotope. The oxygen liberated in photosynthesis comes from *water.* Although there are some theoretical objections to this experiment, other experiments have confirmed that *the oxygen liberated during photosynthesis comes from water.* A more complete equation for photosynthesis that indicates this fact is:

$$6\ C^{16}O_2 + 12\ H_2{}^{18}O \xrightarrow[\text{green plant cell}]{\text{light energy}} C_6H_{12}{}^{16}O_6 + 6\ {}^{18}O_2 + 6\ H_2{}^{16}O$$

LIGHT AND DARK REACTIONS

Between 1883 and 1885 a German physiologist, Engelmann, in a remarkably simple experiment, demonstrated which *quality* (color) of light is effective in photosynthesis. He placed together on a microscope slide an algal filament and some bacteria that would migrate toward high concentrations of dissolved oxygen. He then placed the alga in different wavelengths of light. The bacteria were most active near the algal filament in red and in blue light. So red and blue light were trapped and supplied energy to drive photosynthesis and liberate oxygen (**Fig. 13.3,** page 239).

At approximately the same time (1883), J. Reinke, another German scientist, was studying the effect of changing the *intensity* of light on photosynthesis. Reinke observed that the rate of photosynthesis increased proportionally to the increase in the intensity of light only at low-to-moderate light intensities. At higher light intensities, the rate of photosynthesis was unaffected by changing the light intensity; the reaction was then proceeding at a maximum rate and was unaffected by light—it had become light-saturated.

A further study and a more comprehensive interpretation of this phenomenon was carried out in 1905 by F. F. Blackman, a British plant physiologist. He reasoned that photosynthesis actually is made up of light-sensitive and light-insensitive reactions (Fig. 13.4). Thus, photosynthesis may be divided into two general parts: (*a*) photochemical reactions, and (*b*) "dark reactions," or enzyme reactions. These enzyme reactions are sensitive to temperature changes, but are independent of light and can occur either in light or dark. It is apparent that photosynthesis has many hidden complexities.

Figure 13.4
Interaction of light and temperature on the rate of photosynthesis. When light is limiting, the rate of photosynthesis is independent of temperature. When temperature is limiting, the rate of photosynthesis is independent of light intensity.

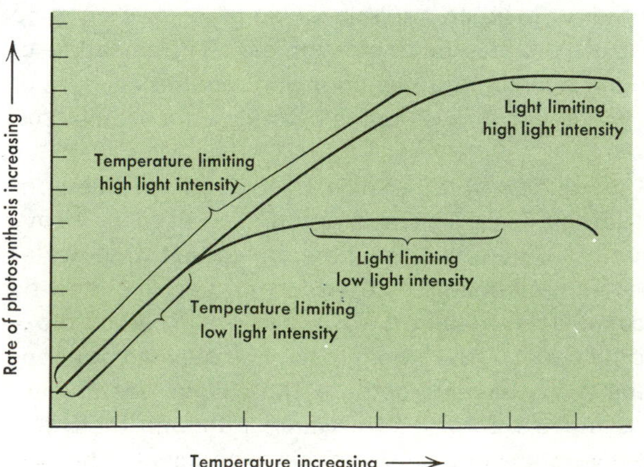

TAKING PHOTOSYNTHESIS APART

In order to understand a complex series of reactions such as those that make up photosynthesis, it is necessary to take the process apart as much as possible and to study each reaction separately. As we have seen, Blackman's study on the effects of light and temperature on the rate of photosynthesis was an example of one method of demonstrating the involvement of two types of reactions. Later, the use of the heavy isotope of oxygen to trace the oxygen in the reaction added more knowledge to our understanding of the process. Another isotope, radioactive carbon, ^{14}C, was used by Melvin Calvin to work out the sequence of enzymatic reactions concerned with the carbon cycle of photosynthesis.

A further method of studying the individual steps in photosynthesis is to free chloroplasts or parts of chloroplast from the cell and to study the role they may play in the complex process of photosynthesis. One of the earliest successful attempts to do this was made by Robin Hill in Cambridge, England. He demonstrated, in 1932, that chloroplasts isolated from the cell could still trap light energy and liberate oxygen. Later, Arnon, at the University of California, and others, demonstrated that isolated chloroplasts could convert light energy to chemical energy and use this energy to reduce CO_2.

CHLOROPLASTS—THE SITE OF PHOTOSYNTHESIS

Since photosynthesis is intimately dependent upon the structure of the chloroplasts of eukaryotic cells, it will be helpful to review the structure of the chloroplast. Although chloroplasts in many plants have a characteristic globular shape, in some plants, especially the algae, they may have a wide variety of forms (see Chapter 23). Chloroplasts seen with the light microscope in cells in a perfectly natural habit appear a homogeneous green. Such chloroplasts, in a filamentous stage in the life history of a moss, are shown in **Fig. 13.5,** page 239.

A section of a plastid in a mesophyll cell of corn leaf is shown in Fig. 13.6. The double nature of the envelope is apparent. The stroma is granular; some of the darker, larger granules are chloroplastic ribosomes. There are two types of membranes, those forming the grana, and those interconnecting the grana, the stroma lamellae (frets).

We know that all biological membranes are high in lipids and proteins. In the case of the chloroplasts, it is

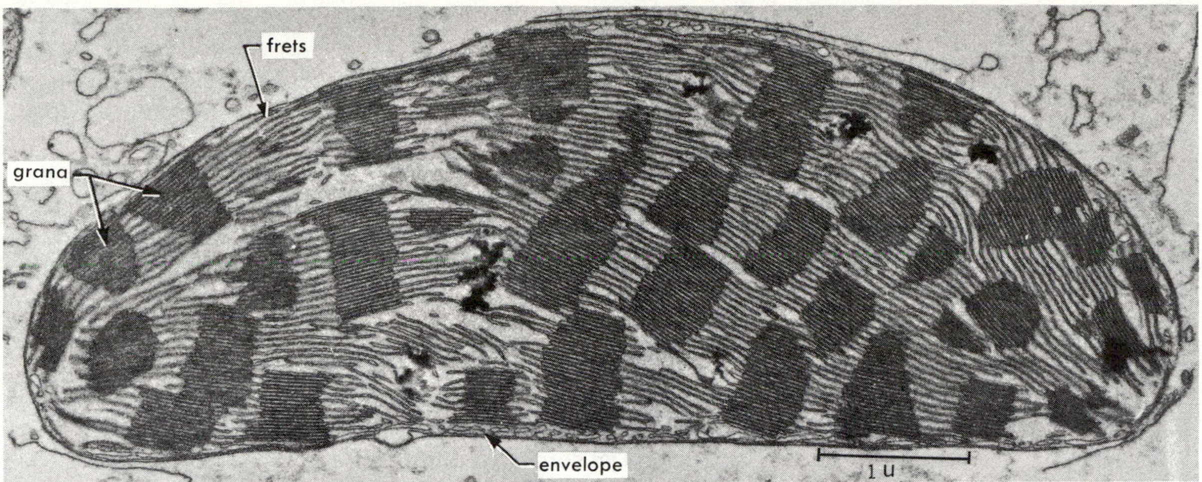

Figure 13.6
Electron micrograph of chloroplast from corn (*Zea mays*) mesophyll.

believed that the chlorophylls and accessory pigments are in close contact with lipid material and with some of the photosynthetic enzyme systems in the grana compartments in a definite protein, lipid, pigment pattern. Such a structural arrangement would facilitate the trapping of light energy and its transfer into chemical energy during photosynthesis. As we will discuss later, light energy, when trapped by chlorophyll, initiates a series of reactions in which electrons from chlorophyll may flow to other compounds (electron acceptors) found in the internal membranes, thylakoids, of chloroplasts.

Chloroplasts, like other cell organelles, are very fragile and are difficult to isolate from the cell in an intact and functional state. However, special methods have been worked out for their isolation. When cells are cut or ground carefully in a dilute, buffered sugar solution, many chloroplasts remain intact and can be separated from the rest of the cell contents by placing the ground material in a centrifuge and spinning it at high speed. If the correct speed is chosen, large and heavy particles, nuclei, starch grains, and cell wall fragments will collect at the bottom of the tube (Fig. 13.7). If the liquid containing the chloroplasts and other smaller cellular components is now poured into another tube and centrifuged again at a slightly higher speed, the choroplasts, free of most of the other cellular organelles, will separate out at the bottom of the tube. When chloroplasts are isolated, two different kinds may generally be seen on a microscope slide **(Fig. 13.8*A*,** page 240**)**. They may be homogeneous or show the presence of darker spots or grana. Electron micrographs of these plastids show some to be complete, with internal membranes embedded in stroma and surrounded by the outer envelope. The others have lost the envelope and stroma and consist only of a system of membranes **(Fig. 13.8*B*)**. Under certain conditions, these latter membranes will carry out part of the reactions of photosynthesis and will liberate oxygen as Hill demonstrated. **Figure 13.8*C*** illustrates the Hill reaction with isolated chloroplasts. Chloroplasts in the light transfer hydrogen from water to a hydrogen acceptor. In this case, the hydrogen acceptor is the blue dye dichlorophenol-indophenol (DCIP) that is reduced to the colorless form. At the same time, oxygen is liberated.

$$\text{DCIP (blue)} + H_2O \xrightarrow[\text{chloroplasts}]{\text{light}} \text{DCIP-}H_2\text{ (colorless)} + \tfrac{1}{2}O_2$$

Only the complete chloroplasts will both liberate oxygen and carry out the enzymatic steps that convert CO_2 into a carbohydrate. Although carbon dioxide fixation by isolated chloroplasts occurs at a slow rate compared to the rate in an intact cell, it demonstrates that all the reactions of photosynthesis go on in the chloroplast. Chloroplasts isolated in this way have been used to study many of the reaction steps in the complex process of photosynthesis.

LIGHT—THE ENERGY USED IN PHOTOSYNTHESIS

The Nature of Light

Before we discuss the way in which light is used in the process of photosynthesis, we should briefly consider the nature and properties of light.

White light, as it comes to us from the sun, is composed of waves of different lengths, ranging from relatively long waves of red light, through successively shorter waves to violet light. When passed through a glass prism, white light is resolved into these colors. The band of colors is the **visible spectrum.** The complete visible spectrum is composed of the following colors (starting with the longest rays): red, orange, yellow, green, blue, indigo, and violet. Wavelengths exist that we are unable to perceive with our eyes. Beyond the red are still *longer*, invisible rays, the infrared; and beyond the violet are

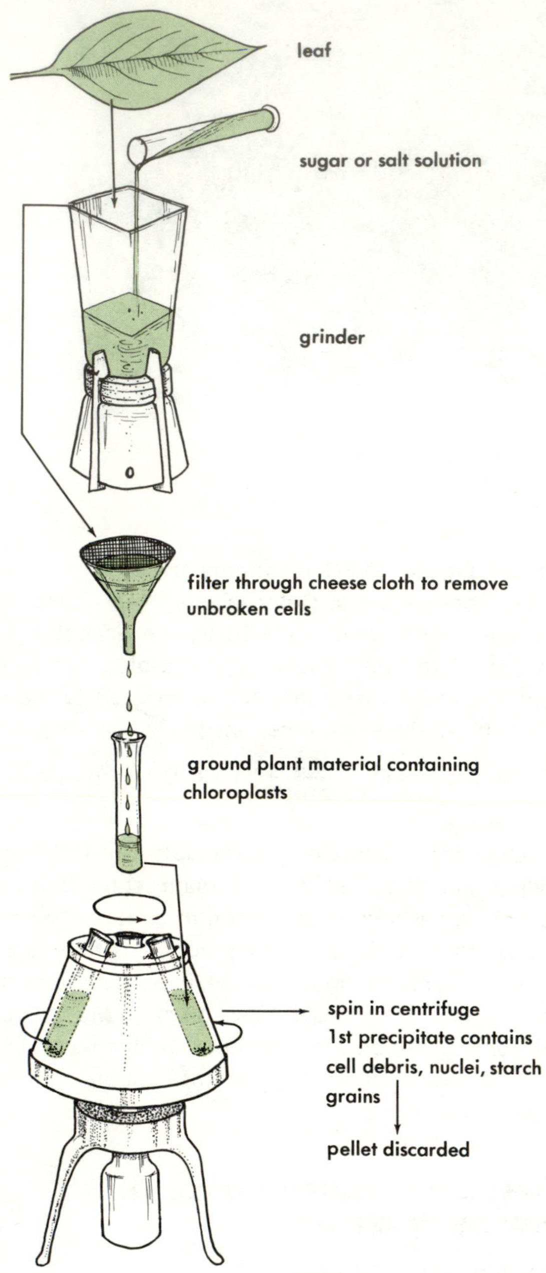

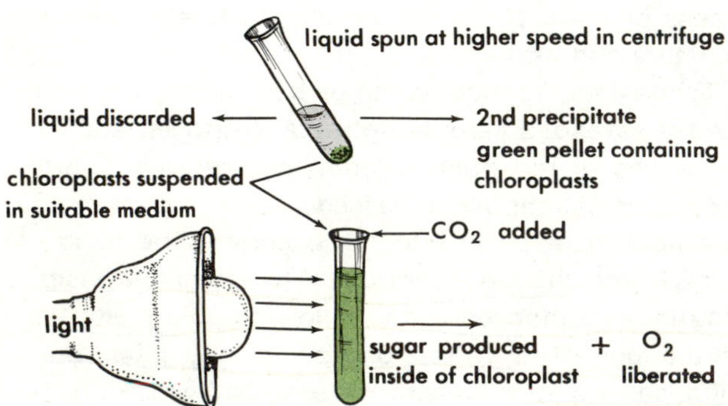

Figure 13.7
A method of isolating chloroplasts.

shorter, invisible rays, the ultraviolet. Thus, the visible spectrum represents only a part of the radiant energy that comes to the earth from the sun. But only a part of the visible spectrum is effective in photosynthesis **(Fig. 13.9,** page 240).

Numerous experiments have shown that light not only acts as if it were traveling in waves but also resembles particles or packets of energy called **photons** or **light quanta.** The energy of photons determines the color of light.

Absorption of Light Energy

The fact that chlorophyll is green to the eye is evidence that some of the blue and red wavelengths of white light are absorbed, leaving proportionally more green to be transmitted or reflected and seen. *It is the absorbed light that is used in photosynthesis.* The part of the white light that is absorbed by chlorophyll can be determined if the light is passed through a solution of chlorophyll and then caused to fall upon a glass prism so that it is broken up into its component colors. The spectrum thus formed, the **absorption spectrum** of chlorophyll **(Fig. 13.9),** is different from the spectrum of white light that has not passed through a chlorophyll solution. Thus, the positions of dark bands, **absorption bands,** in the chlorophyll spectrum indicate which wavelengths are absorbed most strongly by the chlorophyll. They show that much of the red, blue, indigo, and violet are absorbed; these are the wavelengths that are utilized most in photosynthesis. Part of the red and most of the yellow, orange, and green are scarcely absorbed at all unless the chlorophyll solution is very concentrated. If a green leaf, instead of a chlorophyll solution, is placed between a light source and a prism, the absorption bands are quite similar to, though not identical with, those from the chlorophyll solution **(Fig. 13.9).** The leaf contains the yellow carotenoid pigments that absorb blue light; also, chlorophyll in the leaf is probably in association with protein and fatlike materials, while chlorophyll in the solution is not. These factors contribute to the difference between the absorption spectra of a green leaf and of a chlorophyll solution.

THE CONVERSION OF LIGHT ENERGY INTO CHEMICAL ENERGY

What happens when light is absorbed by a chlorophyll molecule? How is it used in the process of photosynthesis? We know that molecules are composed of atoms having positively charged nuclei and one or more electrons spinning in certain definite positions, **orbitals,** around the nuclei. Energy is required to move an electron away from the positively charged nucleus. When a unit of light energy, a **quantum,** is absorbed by a molecule, the energy causes an electron of an atom in the molecule to be moved into an orbital farther from the

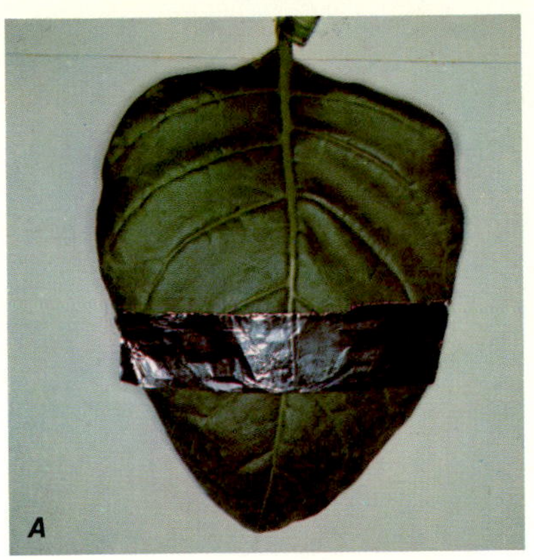

Figure 13.2
Light is necessary for photosynthesis. *A*, leaf from a bean plant that was kept first in darkness for 24 hours; a portion of the leaf was then covered by foil as shown, and the plant was kept in the light for 8 hours. *B*, the iodine test shows that starch was formed only in the portion of the leaf receiving 8 hours of light. Chlorophyll is required for photosynthesis. Variegated leaf from *Coleus*. *C*, living leaf. *D*, same leaf after extraction of chlorophyll and iodine staining to show starch. Compare the blue areas indicating the presence of starch with the distribution of chlorophyll (green and brown parts of the living leaf).

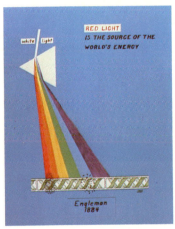

Figure 13.3
A schematic drawing showing Engelmann's experiment demonstrating the kinds of light effective in photosynthesis.

Figure 13.19
Corn *(Zea mays)*, a C_4 plant (right), with its low CO_2 compensation point is able to survive at a lower CO_2 concentration than bean *(Phaseolus vulgaris)*, a C_3 plant (left), when they are grown together in a closed chamber in light for 10 days.

Figure 13.5
Surface view of leaf of moss *(Bryum)* leaves showing numerous chloroplasts, ×400.

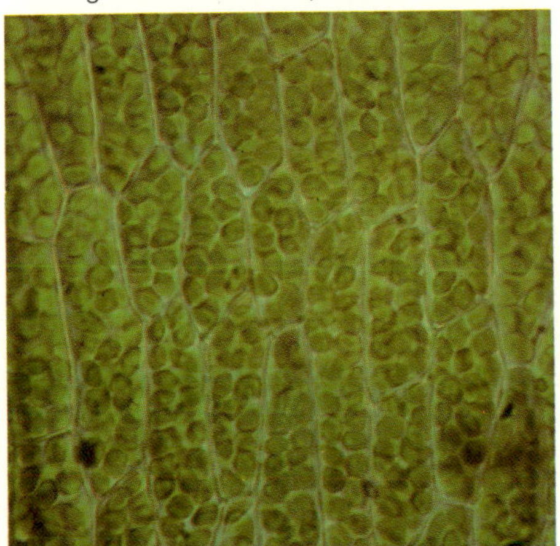

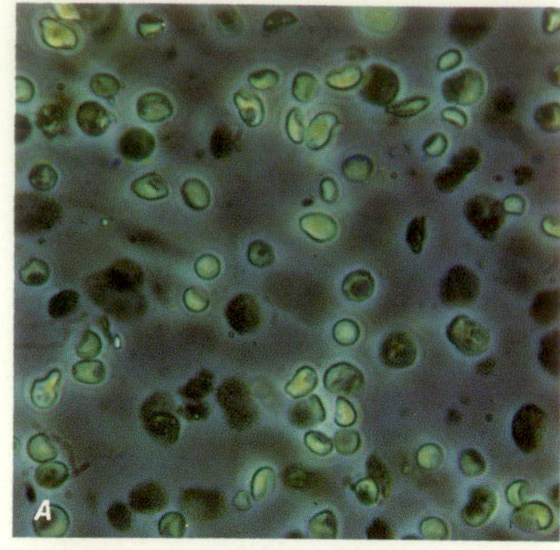

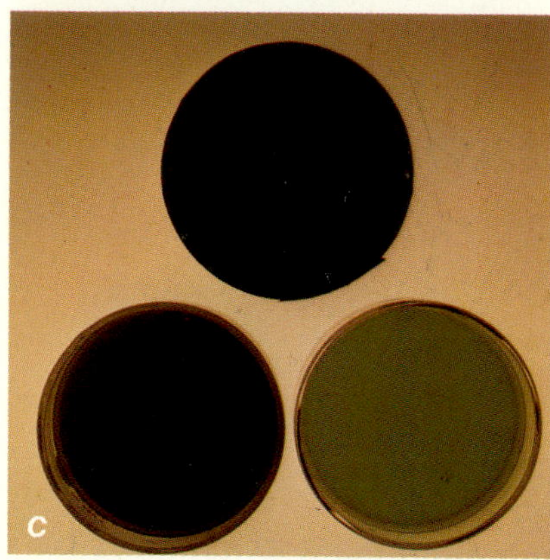

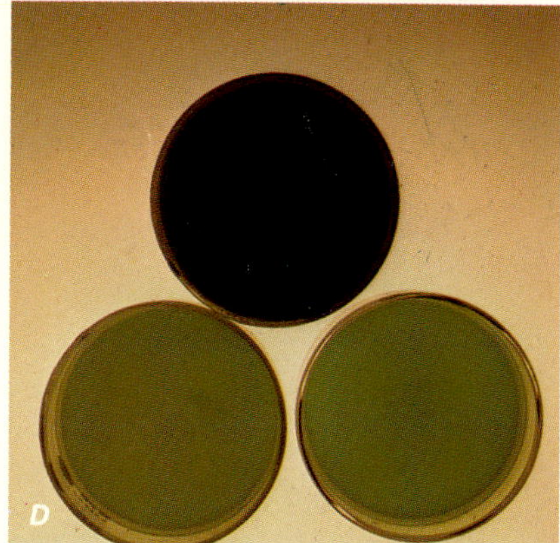

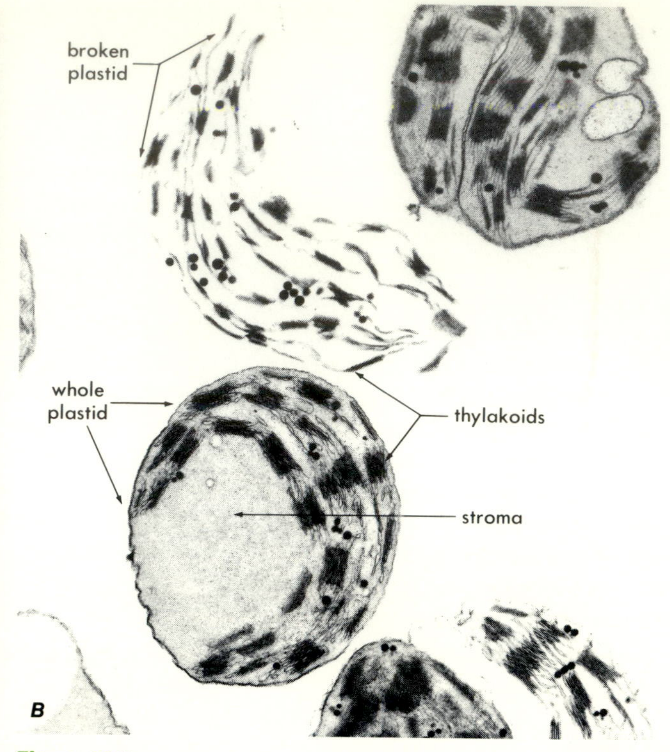

Figure 13.8
A, a preparation of chloroplasts isolated from spinach, *(Spinacia oleracea)* leaves, ×1,000. Both intact and broken chloroplasts are present. *B*, electron micrograph of isolated chloroplasts of *Vicia faba*, showing intact plastids with stroma and grana and plastids that have lost their outer envelope and stroma, ×2500. *C*, *D*, Hill reaction. All dishes contained a suspension of chloroplasts. Dishes at top and left contain the blue dye, dichlorophenol indophenol. *C*, At start, top dish was covered with black paper. *D*, in the light, chloroplasts transfer hydrogen from water to the dye reducing it to a colorless form (dish at left). In the dark (top dish), the dye is unchanged.

Figure 13.9
Left: visible white light passing through a prism is broken up into a range of wavelengths from the 390 nm of violet to the 780 nm of red. Right: the sunlight passing through a leaf placed in front of the prism reveals dark bands in the spectrum between 390 and 500 nm and between 650 and 740 nm, indicating that chlorophyll has absorbed light of these wavelengths.

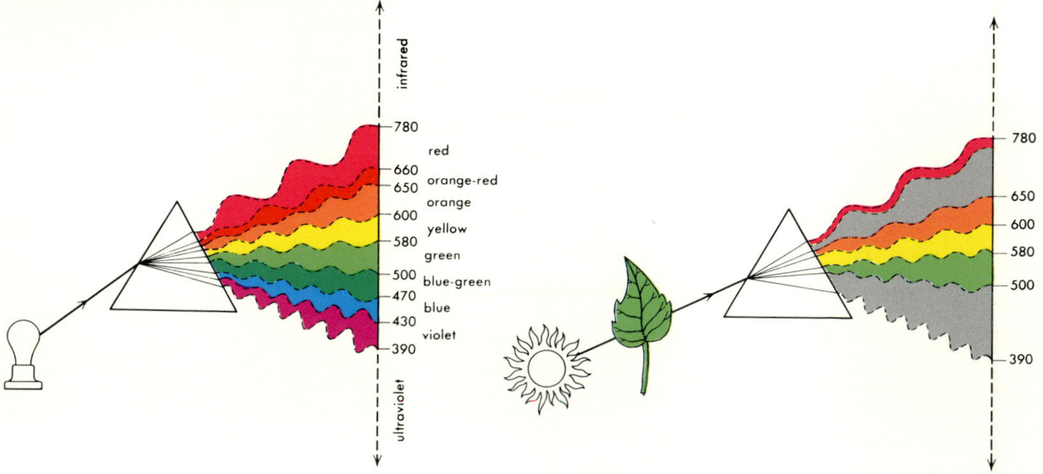

atomic nucleus. The molecule now is said to be in an **excited state** (Fig. 13.10). This is an unstable condition, and the electron tends to move rapidly, usually within a billionth (10^{-9}) of a second, back into its original orbital. As it does this, the trapped energy is released. It may be given off as heat, or as light (fluorescence). It is evident that in the process of photosynthesis, energy that has excited the chlorophyll molecule is in some way used to bring about a chemical reaction rather than being lost simpy as light or heat.

Although a knowledge of the minute details of the steps involved in the conversion of light energy to chemical energy during photosynthesis is not necessary for an appreciation of the significance of the process, a consideration of the major steps in the reaction as we now know them will help us to understand how the details of such a complex reaction are being unravelled. Eventually, when more is known about the process, man should be able to take this knowledge and to use light energy to produce food directly from the raw materials, carbon dioxide and water.

The Two Photosystems

Over 99% of the chlorophyll *a* molecules in the thylakoids and all of the functional chlorophyll *b*, carotenoids and other accessory pigments (phycobilins in various blue green and red algae) act as light harvesting pigments. They absorb light energy and transfer the resulting excitation energy to other pigment molecules and eventually to a special form of chlorophyll *a* located in a **reaction center.** This process of transferring the energy absorbed by many pigment molecules to a reaction center is a fundamental aspect of energy conversion in photosynthesis (Fig. 13.11).

Several lines of evidence indicate that there are two light reactions involved in photosynthesis in plants that evolve oxygen. For example, it was found by Emerson that the red light alone (light of 680 nm wavelength and longer) is very inefficient in photosynthesis, but the addition of light of shorter wavelength increases the efficiency of the long wavelength of light. Thus, it was postulated that there were two light reactions, one sensitive to longer-wavelength light, and the other to shorter-wavelength light.

It is now known that there are indeed two kinds of reactions centers located in thylakoid membranes, and there may be several hundred of each kind in each thylakoid (Fig. 13.12). These reaction centers are called P_{680} and P_{700} because they absorb light with wavelengths of about 682 and 703 nm, respectively. They each contain chlorophyll *a* that has special absorption properties because of its association in the membranes with specific proteins. Each reaction center is associated with a group of functional light-harvesting pigments. Hence, there are two kinds of pigment systems in thylakoid membranes, **photosystem I** and **photosystem II.** In addition to having P_{700} as its reaction center, photosystem I has a high proportion of chlorophyll *a* relative to chlorophyll *b* in its light-harvesting pigment system and is sensitive to longer wavelength light. In contrast, photosystem II, sensitive to shorter wavelength light, contains P_{680} and has a high proportion of chlorophyll *b* in its light-harvesting system.

Figure 13.10
The fate of the energy in a quantum of light absorbed by a pigment molecule.

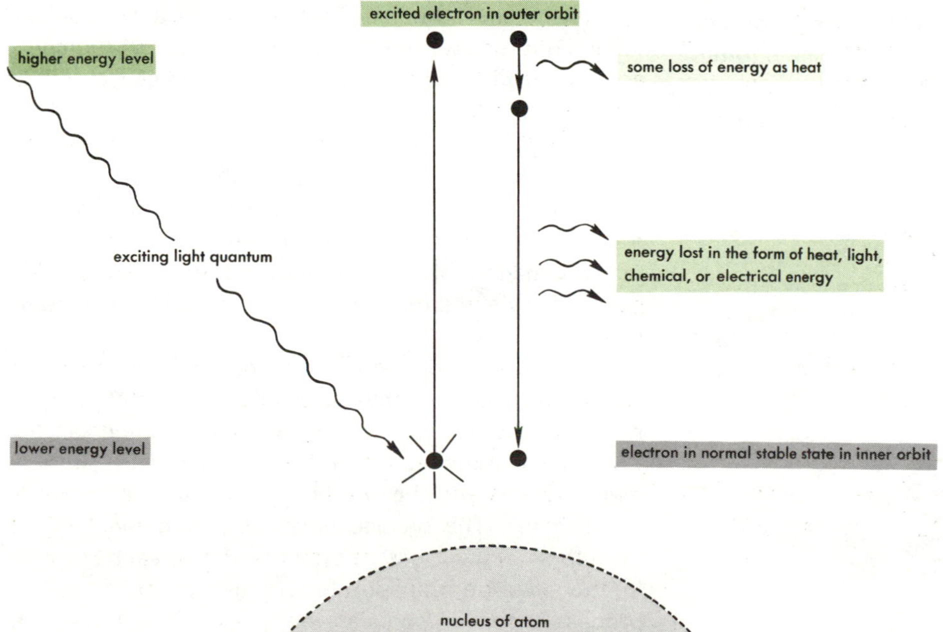

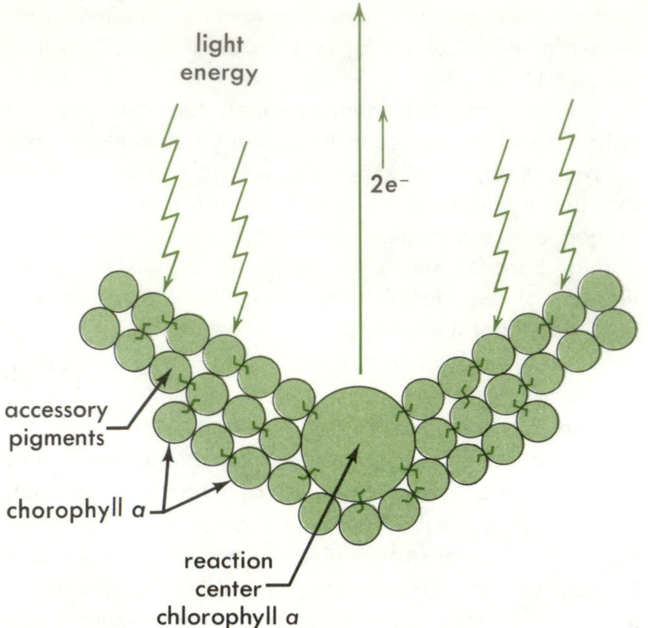

Figure 13.11
Accessory pigments absorb light energy and transfer it ro reaction centers where a chlorophyll *a* molecule is excited and loses an electron. (Highly diagramatic.)

Separating the Pigment Systems

In attempts to determine the relation of the photosystems to the ultrastructure of the thylakoid membranes in the chloroplasts in which they occur, investigators have used various techniques to break the membranes up into smaller particles and to test these particles for the presence of one or both of the pigment systems. One of the most successful attempts has been accomplished by breaking up isolated chloroplasts with high-frequency sound and isolating the different-sized particles by density centrifugation. It was found that a preparation of small particles could be obtained that had the properties of photosystem I but not photosystem II, while another preparation of particles still contained both photosystems. Some research workers believe that the photosystem I particles were located in the stroma lamellae, while the other particles containing both photosystem I and photosystem II were located in the granal partitions. This is one of many aspects of photosynthesis about which there is still much controversy, and active research is being conducted.

Figure 13.12

Proposed moael for the arrangement of photosystems in the thylakoid membrane. Each complex is a group of proteins to which chlorophyll molecules are bound. The Photosystem I and Photosystem II complexes include reaction centers. Adapted from P. A. Armond, L. A. Staehelin, and C. J. Arntzen, *J. Cell Biol.* **73,** 400. © 1977 by The Rockefeller University Press. Reprinted with permission of the publisher.)

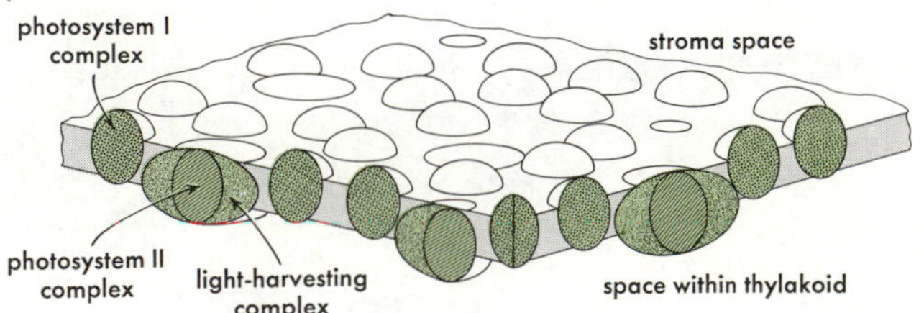

Electron Transport in the Thylakoid Membranes

Although photosynthesis occurs in plants that vary greatly in structural characteristics, there is a remarkable similarity in the chemical reactions of photosynthesis in all species. Photosynthesis is an oxidation-reduction process involving the transfer of large amounts of energy in relatively small energy steps. These energy transfers are brought about by the flow of electrons from reducing agents, electron donors, to oxidizing agents, electron acceptors. Unless outside energy is put into the system, the normal tendency is for electrons to move from a reduced form of an **electronegative** compound to an oxidized form of a more **electropositive** compound.

In addition to the two photosystems, thylakoid membranes contain a series of chemicals that are capable of receiving and giving off electrons. The quantatative term for the comparison of their tendency to accept or release electrons is the electrode potential, or oxidation-reduction potential. This is shown by the scale on the left in Fig. 13.13. These compounds have two very important characteristics: (1) They vary from each other in their ability to gain and release electrons. (2) They are associated in a very important and precise way in the membrane so that they form a chain of molecules along which electrons flow. Just as electrons moving along a copper wire in an electrical circuit can cause work to be done, electrons moving along a chain of electron carriers in a membrane can drive chemical reactions.

Let us see how this might occur in a chloroplast. When light is absorbed by the light-harvesting pigments associated with photosystem I, P_{700} becomes excited. In the excited state, the special chlorophyll *a* molecule (P_{700}) has a tendency to lose an electron to a suitable electron acceptor near it in the thylakoid membrane. Once the electron acceptor has gained an electron, it is able to transfer the electron to another carrier in the chain. Note that the names of a number of electron carriers that occur in thylakoid membranes are given in the scheme shown in Fig. 13.13.

It should be pointed out that this scheme shows the sequence of carriers in the chain but says nothing about how the electron carriers are actually arranged in the thylakoid membranes. You will also note that the two pigment systems and the two light reactions are shown in this scheme. The precise physical arrangement in the thylakoids of the pigment systems and of each member in the electron transport chain are essential in the process of converting solar energy into chemical energy

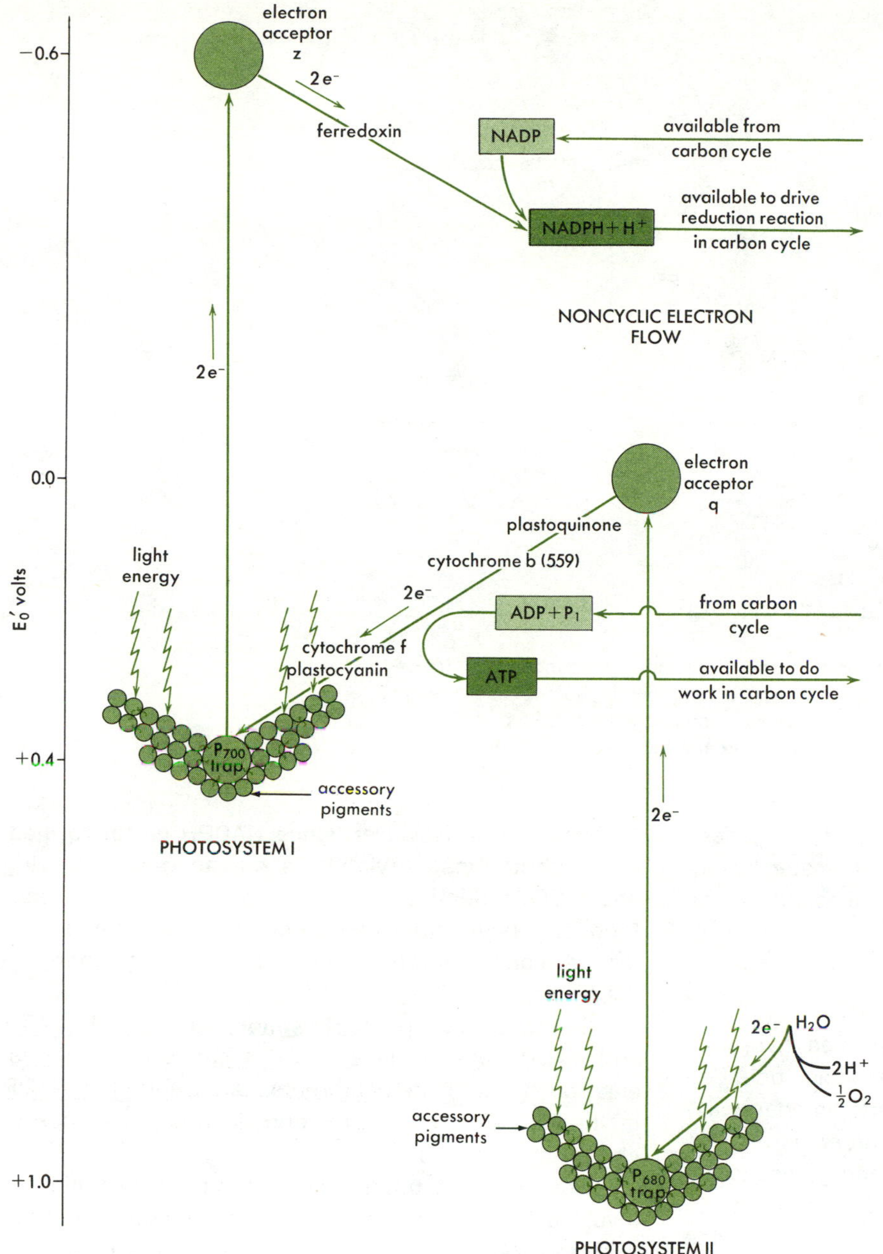

Figure 13.13
Diagramatic arrangement of photosystems I and II. The two systems are connected by an electron-transport chain involving plastiquinone, cytochromes, and plastocyanin. Cyclic electron flow involves only photosystem I. ATP may be generated, but oxygen is not evolved nor is NADP reduced. Noncylic electron flow involves both systems. Electrons from water move through the entire system of electron carriers. NADP is reduced, oxygen is liberated, and ATP may be formed.

in the chloroplasts. One scheme that has been suggested is shown in Fig. 13.14.

The Formation of NADPH

In this scheme, the electron that has been lost by P_{700}, after its excitation, moves through electron carriers toward the stromal side of the thylakoid membrane where it is finally transferred to NADP. The NADP also picks up H^+ from the stroma and the energy-rich NADPH molecule is formed. P_{700}, which is now ionized and in an oxidized state, has a strong affinity for electrons. Energy, from additional light absorbed by the light-harvesting pigments of photosystem II, excites the special form of chlorophyll *a* (P_{680}) in this photosystem. P_{680} in this excited state becomes ionized by losing an electron to the first molecule in the chain of electron carriers that transfers the electron to P_{700}. In its turn, the ionized chlorophyll *a* in photosystem II (P_{680}) is now in a highly oxidized state and gains an electron indirectly from water with the accompanying formation of oxygen. The exact steps in the ox-

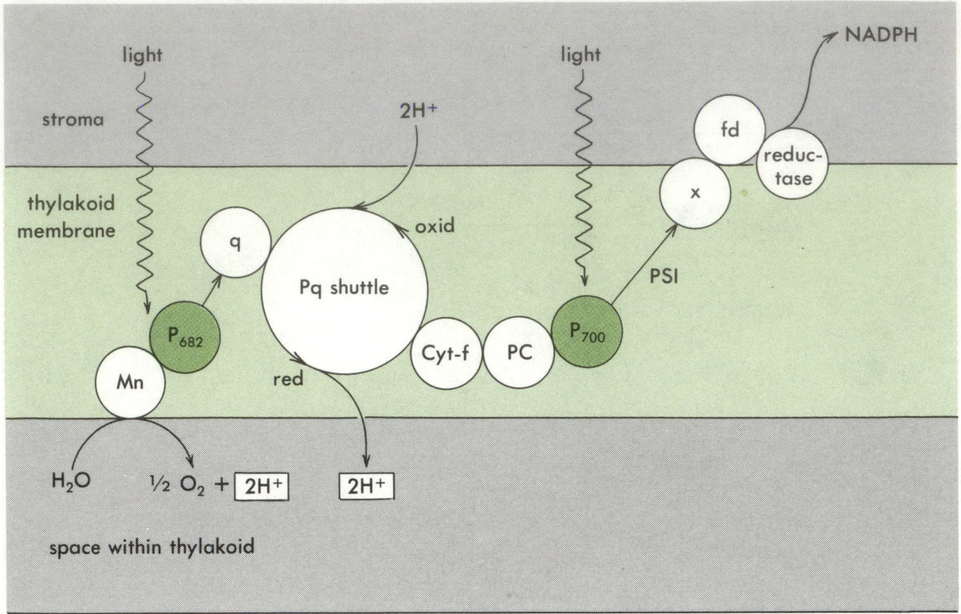

Figure 13.14
Proposed flow of electrons and H^+ in the thylakoid membrane. Overall, light is thought to drive a flow of H^+ across the membrane in one direction and a flow of electrons in the opposite direction. RC I and RC II are reaction centers; circles and PQ represent compounds that take part in electron transport chains. Sizes of the parts are not drawn to scale and locations are schematic.

ygen liberation process are not known, but manganese and at least one enzyme are involved. As indicated by Fig. 13.14, this reaction is thought to take place near the inner surface of the thylakoid membrane.

The H^+ Gradient and ATP Formation

During the series of reactions just described, hydrogen ions are moved from the stroma into the space bounded by thylakoid membranes. This represents an energy accumulation, and the resulting **H^+ gradient** across the membrane is believed to provide the energy necessary for the formation of the energy-rich ATP. According to this idea, an enzyme, ATPase, bound to the stromal side of the thylakoid membrane, in the presence of the H^+ gradient forms ATP from inorganic phosphate and ADP. In the process, H^+ is released to the stroma and OH^- to the space within the thylakoid where water is formed. Consequently the H^+ gradient across the membrane is decreased.

The production of ATP as a result of photochemical reactions like those just described is called **photophosphorylation.** Two types of photophosphorylation are recognized (Fig. 13.13). In **cyclic photophosphorylation,** the flow of electrons from light-excited chlorophyll molecules to electron acceptors proceeds in a cyclic fashion through the acceptors and back to chlorophyll. Only photosystem I is involved in this process. No oxygen is liberated and no NADP is reduced, since it does not receive electrons. ATP is formed during the electron flow, and light energy is converted into chemical energy in the ATP molecules. However, since NADPH is not formed, cyclic photophosphorylation is not adequate to bring about CO_2 reduction and sugar formation. Cyclic photophosphorylation has been demonstrated, but its significance in photosynthesis is limited to supplying energy in the form of ATP.

In **noncyclic photophosphorylation,** both ATP and NADPH are produced. In this series of reactions, electrons from excited chlorophyll are trapped by NADP in the formation of NADPH and do not cycle back to chlorophyll (Fig. 13.13).

Note that both photosystem I and photosystem II are involved in noncyclic photophosphorylation, and both ATP and NADPH are formed. These are the chemicals that are used to drive the CO_2 reduction reactions of photosynthesis.

Summary of Reactions in Thylakoid Membranes

1. Light energy driving the two photoreactions provides all of the energy.
2. All other reactions occur spontaneously since in them electrons are moving toward electron acceptors having a greater affinity for electrons.
3. Energy released in these reactions is used in the formation of NADPH and to produce a H^+ gradient across the membrane. This gradient in turn drives the ATP synthesizing reaction.
4. The two compounds, NADPH and ATP, are the chemicals in which the energy absorbed as light by

chloroplasts is first stored as chemical energy.

5. Energy stored in these two chemicals drives the energy-consuming reactions that are involved in the reduction of CO_2 during the formation of sugar.
6. ATP and NADPH are not accumulated in the cell in any appreciable amount, but are continually being formed and broken down as energy is transferred through them from chlorophyll to sugar molecules.

THE NECESSITY OF ENZYMES FOR PHOTOSYNTHESIS

The complete process of photosynthesis will not take place in the dark, yet as we have seen, many of the steps in the process are controlled by enzymes and are therefore not sensitive to light but are particularly sensitive to changes in temperature. Even in the production of NADPH, enzyme reactions as well as light reactions are involved. The process of carbon dioxide reduction to carbohydrates involves many enzyme reactions. All the enzymes directly concerned with photosynthesis occur in the chloroplast. Many of them, especially those linked to the carbon cycle, are water-soluble and are found in the stroma of the chloroplast.

One of the most extensively studied enzymes of photosynthesis (and the enzyme that probably occurs in higher concentration than any other enzyme in many leaf cells) is **ribulose bisphosphate carboxylase.** This is one of the most important enzymes in plants, since it catalyzes the first step in the entrance of CO_2 into organic compounds:

$$\text{carbon dioxide} + \text{ribulose bisphosphate} \xrightarrow[\text{carboxylase}]{\text{ribulose bisphosphate}} 2\ \text{phosphoglyceric acid}$$

In this reaction, CO_2 combines with the five-carbon atom sugar phosphate in the plastid to produce two molecules of the three-carbon atom acid, phosphoglyceric acid (PGA). This is a spontaneous reaction with little change in energy being involved.

CARBON DIOXIDE REDUCTION AND THE FORMATION OF SUGAR

Some Techniques

Physiologists have long been interested in determining the sequence of carbon compounds that are formed in photosynthesis. Although early physiologists analyzed the carbohydrate content of photosynthesizing leaves and could easily recognize that starch as well as sugars usually accumulate during photosynthesis, it was not until two sensitive analytical techniques were developed that the actual sequence of reactions was understood. These new methods were the use of carbon dioxide in which the carbon was radioactive and the use of a special process, **chromatography,** with which the investigator could easily and accurately separate minute amounts of different organic compounds from one another. These techniques permitted the investigator to follow carbon in the series of reactions taking place in the plant (Fig. 13.15).

The problem confronting physiologists was to determine the sequence of compounds (*A, B, C,* etc., Fig. 13.16) through which carbon atoms taken into the leaf pass during the synthesis of sugar. When the radioactive isotope of carbon (^{14}C) became available, scientists treated leaves with carbon dioxide containing radioactive carbon ($^{14}CO_2$) and determined the carbon compounds formed. This may easily be done by the method outlined in Fig. 13.15. It is first necessary to prepare "maps" of the rates of movement of known compounds across the large sheets of filter paper. Once positions (on the filter paper) of known compounds, under controlled experiments, have been located, a drop of an unknown extract may be placed on the corner of the filter paper. After the usual washing and rotating, the presence of a spot at a definite position on the filter paper is taken to mean that a certain specific compound was present in the unknown sample. A group of scientists at the University of California under the direction of Melvin Calvin were particularly successful in using this technique for working out the early carbon pathways of photosynthesis. Calvin eventually received the Nobel Prize for this work.

The method finally adopted by this group as most successful was to grow algae in an inorganic nutrient medium, place such an algal culture in a glass flask in the light, expose the cells to radioactive $^{14}CO_2$, and kill the cells immediately in boiling alcohol. A similar technique of treating leaves with $^{14}CO_2$ could also be used. The alcohol extract contained essentially all of the radioactivity incorporated into the cells after short time intervals. By varying the time to which the plant was exposed to $^{14}CO_2$, the first compounds synthesized during exposure to ^{14}C could be determined. Not only is photosynthesis very rapid, but also the first products are quickly changed into other substances. Thus, very short exposures to $^{14}CO_2$, in the range of 5–10 sec, were used. Figure 13.17 shows an autoradiogram of the products formed when a tobacco leaf is treated with $^{14}CO_2$ for 10 sec.

If photosynthesis proceeds for an hour or so in an atmosphere containing radioactive carbon dioxide, most of the labeled carbon will be found in carbohydrates (sugar or starch). If photosynthesis is stopped after only a few seconds, most of the labeled carbon is found in a three-carbon-atom acid containing phosphorus, **phosphoglyceric acid** (PGA).

Phosphoglyceric acid, formed during photosynthesis, is thus an intermediate between carbon

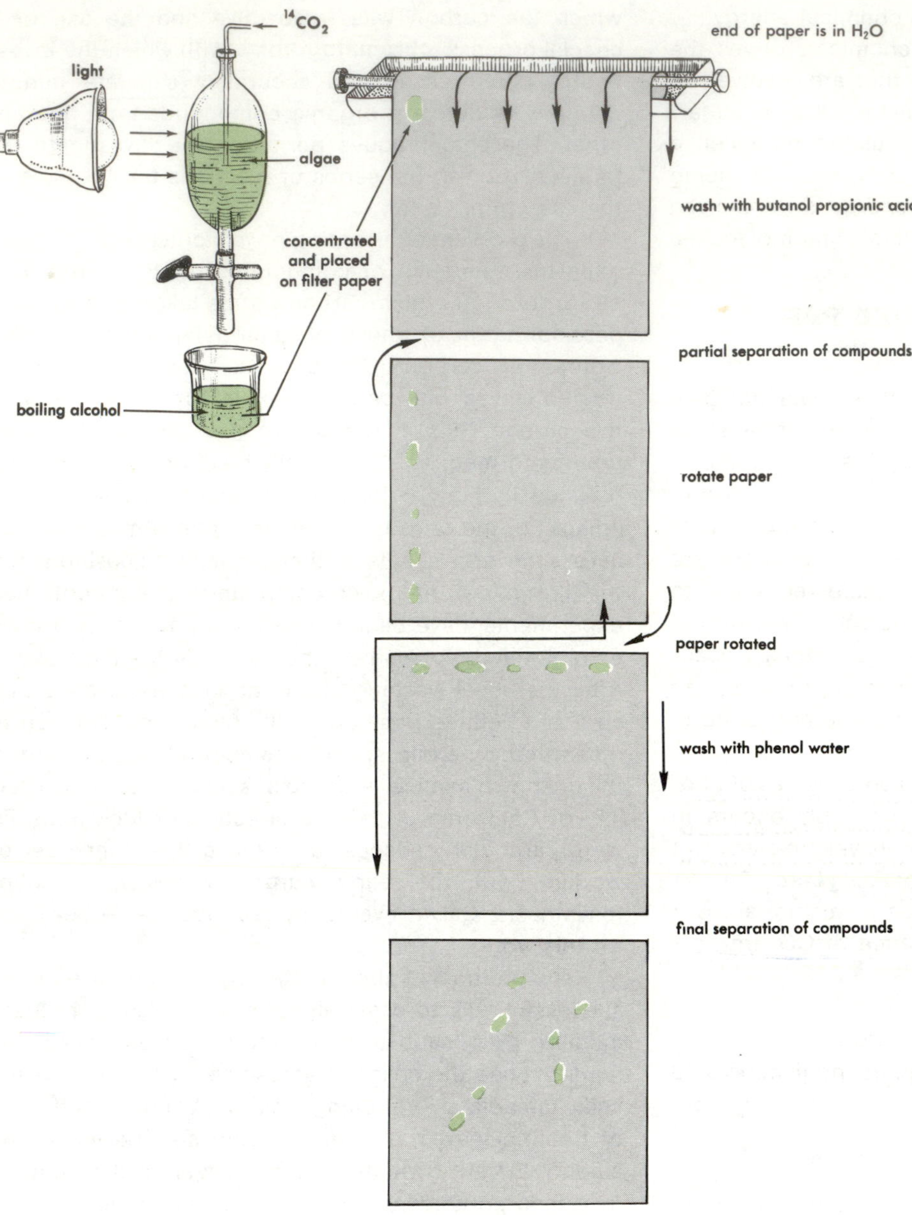

Place on x-ray film in dark. Develop film after two weeks.
Radioactive compounds on the paper are indicated by
darkened areas on the film.

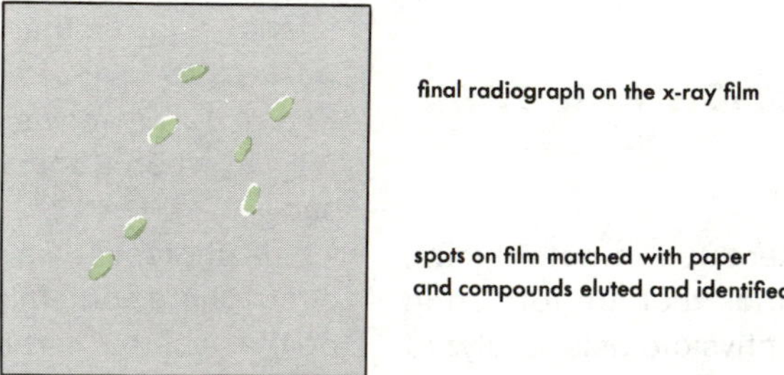

Figure 13.15
Steps in the development of an autoradiogram as used in the study of photosynthesis. Radioactive carbon is supplied in the leaf, and metabolites are separated by paper chromatography. ^{14}C produces a latent image on X-ray film which, when developed, shows the presence of the initial ^{14}C in the metabolites formed by photosynthesis in the chloroplasts after exposure to $^{14}CO_2$.

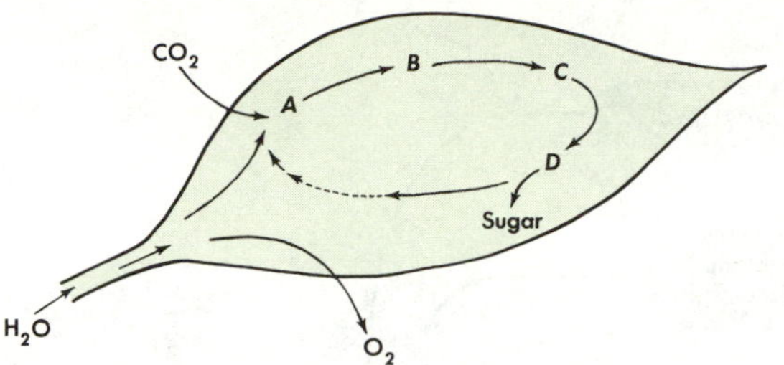

Figure 13.16
The carbon from carbon dioxide passes into several intermediate compounds (*A, B, C,* etc.) before sugar is formed in photosynthesis.

dioxide and sugar. By stopping photosynthesis at increasingly longer intervals (from a few seconds to several minutes), investigators have found that many carbon compounds become labeled with radioactive carbon.

The C_3 Pathway

It has been possible by means of such experiments to draw up an outline representing the various reactions that occur (Fig. 13.18). Here again it should be emphasized that a rote memorization of the sequence of all these reactions and of the intermediate compounds is not particularly useful, but an understanding of the roles that this complex series of reactions play in the metabolism of the plant is important. The key points to note in this carbon cycle are the following:

1. Catalyzed by the enzyme ribulose biphosphate carboxylase, CO_2 combines with **ribulose bisphosphate,** a five-carbon-atom sugar phosphate continually being produced in the cell. Two molecules of phosphoglyceric acid (PGA), a three-carbon-atom compound, are produced.
2. The carbon cycle is associated at two points with high-energy products of photosynthesis, NADPH and ATP.
3. Two molecules of PGA are reduced to form two molecules of a three-carbon-atom sugar phosphate, phosphoglyceraldehyde. The energy that drives this reaction comes from NADPH and ATP, and energy from these molecules is stored in the newly formed triose sugar. ADP and NADP are regenerated.
4. Once the PGA is reduced, it may form another triose phosphate, dihydroxyacetone phosphate. The two triose phosphates may be combined to form a six-carbon-atom sugar phosphate, fructose diphosphate. The plant by this process has essentially added one CO_2 molecule to a five-carbon-atom sugar to produce one molecule of a six-carbon-atom sugar.
5. As this process continues, some of the fructose phosphate may be transformed through other reactions into other carbohydrates, including sucrose and starch.
6. Some of the fructose phosphate molecules are used to form more of the five-carbon-atom sugar phosphate, ribulose bisphosphate. This in turn accepts more CO_2. Thus, a cycle of carbon compounds exist, with CO_2 from the air and hydrogen from water entering the cycle, and various sugars being produced.

Figure 13.17
Autoradiogram of the products of photosynthesis after treating a tobacco leaf with $^{14}CO_2$ for 10 sec.

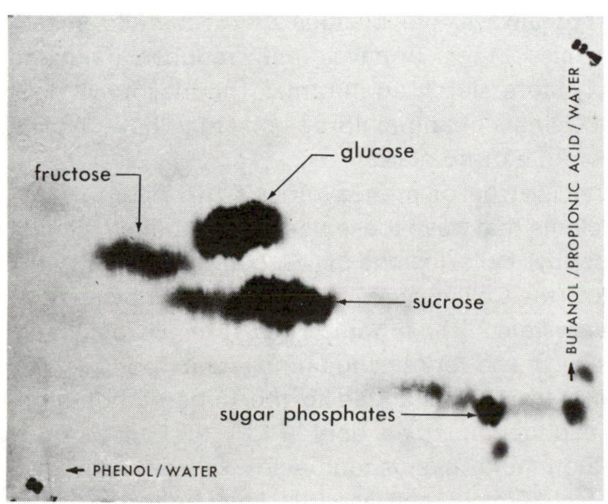

This carbon cycle of photosynthesis is often called the Calvin cycle or the C_3 path of photosynthesis and appears to be of almost universal occurrence in plants that photosynthesize. However, it is reasonable to expect that during the evolution of plants in a wide variety of ecological situations, adaptations and modifications in metabolic processes would have occurred. Indeed, we do find that variations on the basic theme of photosynthesis have evolved. Two examples are found in succulent plants and plants of tropical origin.

ENVIRONMENTAL STRESS AND PHOTOSYNTHESIS

Succulents and the Trapping of CO_2 at Night

Plants have adapted in a variety of ways to extreme environmental conditions existing in desert areas. In order to survive in these xeric conditions, plants must either endure recurrent drought or avoid drought by such means as carrying out the active part of their life cycle rapidly during the brief rainy periods. One group of plants, the succulents, has developed methods of storing and conserving water. The parenchyma tissue in succulent plants is highly developed, vacuoles are large, and intercellular spaces are reduced. When moisture is available, succulents absorb and store large quantities of water, and during periods of drought, they resist the loss of water to the environment.

In contrast to most mesophytic plants, many succulents have closed stomata during the day and open stomata at night. This adaptation is advantageous in reducing water loss during the day, when water stress is high, but it could be very disadvantageous for pho-

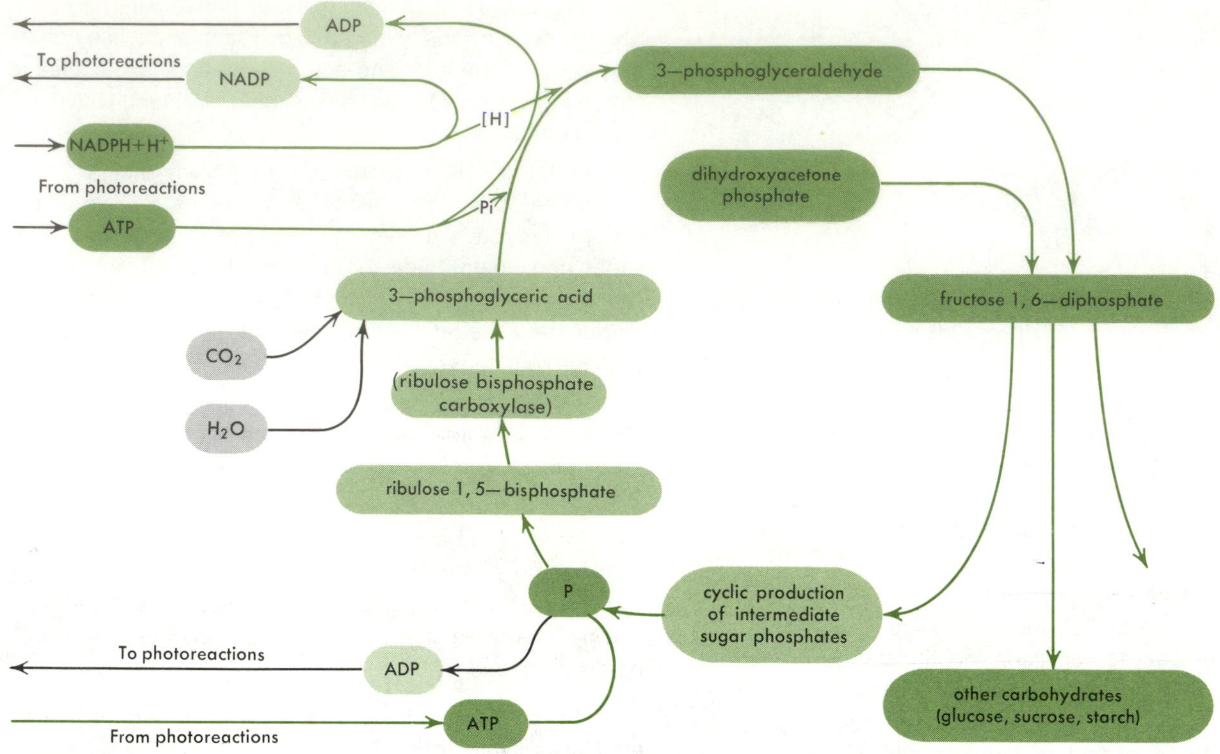

Figure 13.18
Some steps in the carbon cycle of photosynthesis, the Calvin, or C_3, cycle.

tosynthesis by reducing CO_2 uptake. However, these plants show a particular type of carbon metabolism. This is called **Crassulacean acid metabolism** (CAM), since plants belonging to the Crassulaceae were first seen to possess this process.

In the photosynthetic organs of at least some succulent plants, the total amount of organic acids in the vacuoles increases rapidly at night when stomata are open, and decreases during the day when stomata are closed. One of the major acids accumulated is malic acid, which is formed when CO_2 is absorbed in the dark. During the day, acid rapidly disappears, and carbon from it is incorporated into carbohydrate, probably through the addition of CO_2 to ribulose bisphosphate to yield 3-phosphoglyceric acid in the usual C_3 cycle of photosynthesis. These plants thus have an effective mechanism of trapping CO_2 at night and releasing it and using it for photosynthesis inside the leaf in the day when the stomata are closed.

The C_4 Pathway and Increased Efficiency of CO_2 Absorption

A second variation in the carbon cycle of photosynthesis is shown in some plants of tropical origin. If we place a plant in the light in a closed chamber, we can determine how efficiently the plant can lower the CO_2 content of the air in the chamber. At some point, the CO_2 produced by respiration will just balance or compensate for the CO_2 absorbed during photosynthesis. The percentage of CO_2 remaining in the chamber under these conditions is known as the **CO_2 compensation point.** Thus, if a bean and a corn plant are placed together in a chamber in the light, the corn plant will successfully compete with the bean for the limited CO_2 (**Fig. 13.19,** page 239). Both will eventually die of starvation, but the bean plant will die before the corn plant does. Plants such as corn have very low CO_2 compensation points. In general, they seem to be adapted to grow in habitats with high light intensities and high temperatures, and they are more efficient in the use of water. They also have certain structural features in common such as the arrangement of large parenchyma cells in a sheath around the veins. Often, though not always, the plastids of the bundle sheath parenchyma cells lack or have greatly reduced grana and frequently store starch. In contrast, the mesophyll cells between veins contain chloroplasts that have typical grana but little or no starch.

A special type of photosynthesis has been discovered in plants that have these specialized bundle sheath and mesophyll cells. Instead of CO_2 being taken up by the reaction of the Calvin cycle, CO_2 is first fixed by another enzyme system, **phosphoenolpyruvate carboxylase.** This enzyme has a very strong affinity for carbon dioxide. In this system, CO_2 is added to the three-carbon-atom acid phosphoenolpyruvic acid (PEP) to form a four-carbon-atom acid, oxaloacetic acid. Subsequently, two other C_4 acids (malic and aspartic) may be formed from oxaloacetic acid. It is believed by many plant physiologists that the C_4 acids act as effective shuttles and move from mesophyll plastids to bundle sheath plastids where they release CO_2. They then return as a C_3 acid to the mesophyll where another CO_2 can be picked up. The CO_2 released in the bundle sheath plastid can then enter the normal Calvin, or C_3, carbon cycle to form sugars (Fig. 13.20).

An interesting aspect of this idea is that there is a division of function between the two kinds of parenchyma cells. Mesophyll chloroplasts act as traps of CO_2 which is then (in the form of C_4 acids) shuttled to bundle sheath plastids, where it is reduced to sugars. This hypothesis requires a very extensive and rapid movement of solutes over considerable cellular distance and across a number of membrane barriers. This system was first extensively studied by Hatch and Slack, working in Australia, and it is known as the **Hatch-Slack, or C_4, pathway** of photosynthesis. It differs from the C_3, or Calvin, system in that it ensures a very efficient absorption of CO_2 and gives rise to a very low carbon dioxide compensation point.

EFFICIENCY OF PHOTOSYNTHESIS

Although the green leaf is the major organ in which light energy is used in the production of food used by man, it is not particularly efficient in utilizing the sun's energy. We speak of the efficiency of a machine, such as a diesel engine, a gasoline motor, or an electric motor. We calculate the energy value of fuel used and compare it with energy output of the machine. We know that much of the energy is lost. The ratio of energy outgo to energy intake represents efficiency. Of the total radiant energy that falls upon green leaves, about 80% is absorbed (Fig. 13.21). Of the remaining 20%, a part is reflected from the leaf surface and a part passes through the leaf. Part of the radiant energy absorbed is changed to heat and raises the temperature of the leaf; a large part of that absorbed is used up in transpiration; the remainder is utilized in photosynthesis and stored in carbohydrate molecules. Thus, only about 0.5 to 3.5% of all the light energy that falls on a leaf is used in the process of photosynthesis. By standards of machine efficiency, this is a very low percentage. But the supply of solar energy is continuous and abundant.

However, there is reason to believe that the leaves of higher plants, under the most favorable conditions, should be about ten times as efficient as this in coverting solar energy into chemical energy by photosynthesis. Intensive research is needed to raise the low efficiency of photosynthesis of plants to the efficiency that is theoretically possible.

PHOTOSYNTHESIS AND FOOD PRODUCTION

The large amount of research effort that has been devoted to solving the problems of photosynthesis has as its final goal the accumulation of adequate basic knowl-

Figure 13.20
Distribution of C_4 and C_3 pathways in mesophyll and bundle sheath plastids of corn (*Zea mays*).

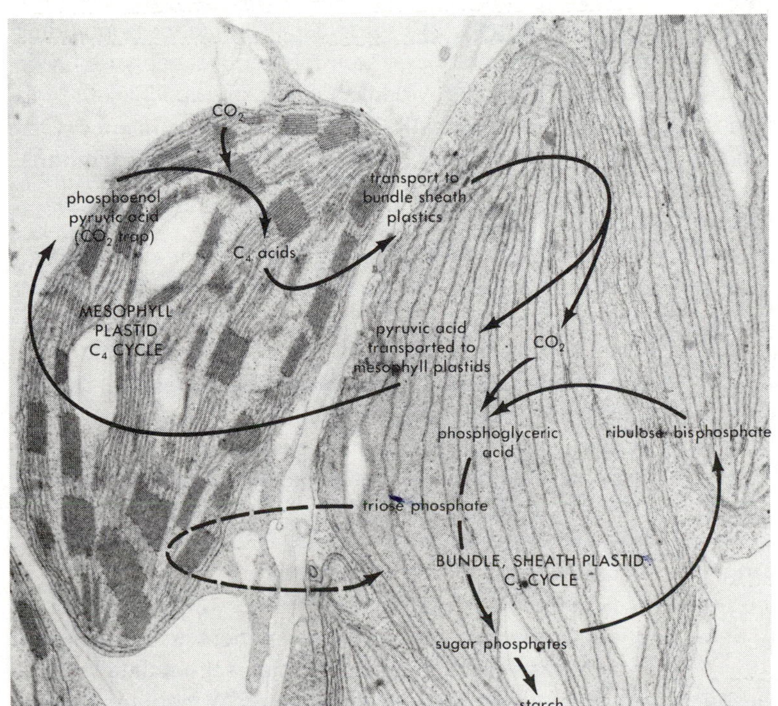

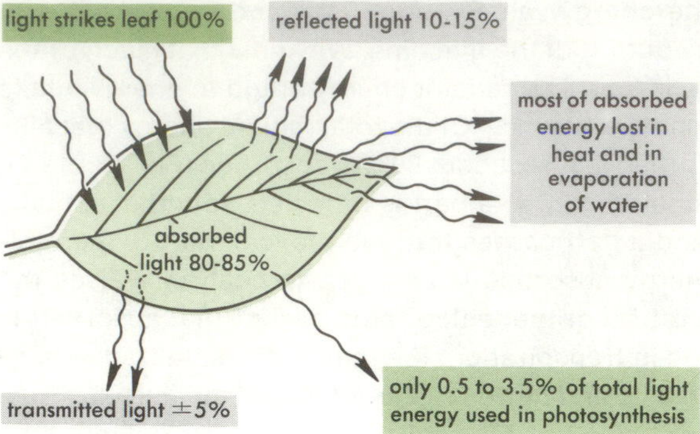

Figure 13.21
Diagram showing what happens to light that strikes a leaf.

edge on the subject so that we will understand how nature uses energy to produce food from the abundant raw materials carbon dioxide and water. Once we understand this, a direct consequence will be our ability to artificially duplicate photosynthesis and hopefully to solve some of the problems of food production in the world. Also some of the lessons that we learn may be useful in helping our search for alternatives to nuclear and fossil fuel.

Another approach to the solution of the problem of producing enough food for the world population is to develop, through plant breeding, varieties of plants that are very highly productive. This approach already has been so successful in some areas (e.g., the development of highly productive types of rice and wheat) that we speak of the "green revolution." The importance of this approach was recognized by the award of the Nobel Prize to Gustav Borlaug, who succeeded in breeding new strains of high-producing grains. Unfortunately, these strains require high levels of fertilizer application – an expensive and pollution-creating practice. There are also potential genetic problems associated with these grains (see Chapter 17).

A third approach to increasing the productivity of plants is to apply the principles of plant physiology to the cultural practices used in growing plants. This is the conscious or unconscious desire of everyone who grows plants whether he is a manager of an extensive farm or a home gardener.

PHOTORESPIRATION

Scientists have wondered for a long time whether respiration proceeds in green cells at the same rate in the light as it does in the dark. Recently, it has been shown that light actually stimulates respiration in some plants. However, in these cases, respiration is different from the usual aerobic respiration carried out in these cells in the dark. Instead of mitochondria, specific microbodies called **peroxisomes** (Chapter 4) are involved in photorespiration.

Evidence indicates that glycolic acid (a two-carbon-atom acid) is an early product of photosynthesis in many plants. Glycolic acid production is stimulated under conditions of high O_2 and low CO_2 levels. Glycolic acid is able to diffuse from chloroplasts to peroxisomes, where it is rapidly oxidized. Since this oxidation is not coupled to the production of reduced nucleotide or to ATP synthesis, all the energy released is lost. Thus, it appears that photorespiration is a wasteful process. It has been shown that under some conditions over 30% of the carbon reduced during photosynthesis may be reoxidized to CO_2 during photorespiration. This is a very severe loss.

Some plants that have the C_4 acid cycle of photosynthesis show essentially no photorespiration under natural conditions. These plants, such as sugar cane and corn (*Zea mays*), are among the most efficient plants in trapping light energy. Investigations are currently underway to determine whether new strains of crops or new cultural practices can be found so that crop plants will have lower rates of photorespiration and consequently higher yields.

CONDITIONS AFFECTING THE RATE OF PHOTOSYNTHESIS

The rate of photosynthesis is of great importance in food and fiber production and will affect the yield of any given crop.

Internal Factors

The rate of photosynthesis is influenced by a number of **internal factors**, that is, conditions inherent in the plant itself. Chief of these are (*a*) the structure of the leaf and its chlorophyll content; (*b*) accumulation within the chlorophyll-bearing cells of the products of the photosynthesis; and (*c*) protoplasmic influences, including enzymes. Considering the interplay of these various internal factors, it is evident that different species of plants vary considerably in their photosynthetic efficiency, even when growing under the same environmental conditions.

Leaf Structure

The structural features of the leaf influence the amount of carbon dioxide that reaches the chloroplasts. These features include size, position, and behavior of the stomata, and the amount of intercellular space. Also, the intensity and quality of the light that reaches the chloroplasts are influenced by thickness of the cuticle and epidermis, the presence of epidermal hairs, the arrangement of mesophyll cells, the position of chloroplasts in the cells, and so forth.

Products of Photosynthesis

With an increase in the concentration of the products of photosynthesis in mesophyll cells, there is a decrease in the photosynthetic rate. Starch may accumulate in chloroplasts during the day, when sugar is manufactured at a more rapid rate than it is transferred from the cell; a result of this accumulation may be a retardation in the rate of photosynthesis.

Protoplasm

The rate of photosynthesis is affected by conditions associated with the protoplasm itself. If the cells have a water deficit and the protoplasm is dehydrated, photosynthesis slows down. Moreover, a disturbance of certain enzyme activities influences the photosynthetic rate. We have already discussed the effects of photorespiration on the rate of photosynthesis.

External Factors

These factors as they influence the rate of photosynthesis are, to a degree, under the control of the plant grower. Crop yields can be influenced by modifying these factors. The principal external conditions that affect the rate of photosynthesis are (*a*) temperature; (*b*) light intensity, quality, and duration; (*c*) carbon dioxide content of the air; (*d*) water supply; and (*e*) mineral elements in the soil.

Temperature

Plants of cold climates carry on photosynthesis at much lower temperatures than do those of warm climates. The process is known to occur in certain evergreen species of cold regions, even at temperatures below 0°C. Algae in the water of hot springs may carry on photosynthesis at a temperature as high as 75°C. Most plants however, function best between temperature of 10 and 35°C. If there is adequate light intensity and a normal supply of carbon dioxide, the rate of photosynthesis of most ordinary land plants increases with an increase in temperature up to about 25°C; above this range there is a continuous fall in the rate as the temperature is raised. At these higher temperatures, the time of exposure is of importance. At a given constant high temperature (for example, 40°C), the rate of photosynthesis decreases with time.

Under conditions of low light intensity, an increase in temperature beyond a certain minimum will not produce an increase in photosynthesis (Fig. 13.4). These conditions may occur in the winter in greenhouses. If the temperature is raised too high, the plants will suffer because the rate of photosynthesis has not been changed but respiration has been increased by the higher temperature.

Light

In discussing the effect of light upon the rate of photosynthesis, three elements must be considered: (*a*) intensity, (*b*) quality (wavelengths), and (*c*) duration. With proper temperature and sufficient carbon dioxide, carbohydrates produced by a given area of leaf surface increase with increasing light intensity up to a certain point (optimum light intensity), after which their production decreases. It is not the intensity of light that falls upon the leaf surface that is of importance, as much as it is the intensity to which the chloroplasts are exposed. The light intensity diminishes from the leaf surface to the chloroplasts, owing to surface hairs, thick cuticle, thick epidermis, and other structural features.

Intense light appears to retard the rate of photosynthesis. Many plants that live in deserts and other places where the light is very bright often have structural adaptations which tend to diminish the intensity of light that reaches the chloroplasts. The usual light intensity in arid and semi-arid regions is well above the optimum for photosynthesis in many plants, especially introduced crop plants. In these regions on days when the sky is overcast, the light intensity is probably nearer the optimum for photosynthesis than on clear, sunny days. Leaves on the surface of plants receive light of greater intensity than those beneath that are shaded. Therefore, some of the leaves receive light of optimum intensity, whereas others may receive light either above or below the optimum.

Under many natural conditions the quality of light striking green plants may vary widely. Such variations are particularly important for seedling survival and growth on forest floors. In these situations, the canopy of leaves of the forest absorb predominantly red and blue light (see Figure 18.3, Chapter 18). Even lower leaves of taller plants, screened from direct sunlight by upper leaves, receive light rich in green wavelengths. Plants, such as marine algae, growing in deep water are subjected to light rich in the shorter (blue-green) wavelengths. Many such plants have developed accessory pigment systems adapted to absorb blue-green light and utilize it in photosynthesis (see Figure 23.7, Chapter 23).

In each of these instances the ability of the plant to adapt to the particular quality as well as quantity of light it receives is important for its survival.

Carbon Dioxide

The carbon dioxide utilized by land plants is absorbed by the leaves from the atmosphere. About 78% of the atmosphere is nitrogen and about 21% is oxygen, the remaining small percentage being composed of carbon dioxide, argon, and traces of hydrogen, neon, helium, and other gases. Surprisingly, only about 0.03% of the atmosphere is carbon dioxide.

Carbon dioxide in the air surrounding the leaves ultimately reaches the chloroplast. Its inward diffusion path is through the stomata to the intercellular spaces, through walls of palisade and spongy parenchyma cells, to the cytoplasm, and thence into the chloroplasts. The walls of palisade and spongy parenchyma cells contain water. Carbon dioxide is readily soluble in water; hence, carbon dioxide passes through the cell walls in aqueous solution.

In the process of photosynthesis, the cells remove

carbon dioxide from solution in the cell sap. As a result, there is diffusion of that gas inward from the wet cell walls. This loss of carbon dioxide from the cell wall allows the water in the wall to dissolve more carbon dioxide from the air in the intercellular spaces. The carbon dioxide content of the intercellular spaces is lowered below that of the outside atmosphere. Diffusion of carbon dioxide inward through stomata tends to make up this deficiency. Thus, during active photosynthesis in the chloroplasts, a diffusion gradient for carbon dioxide is set up between the outside atmosphere and the chloroplasts. At the same time, oxygen liberated in photosynthesis is used in respiration, or it diffuses outward in aqueous solution through the cell walls to intercellular spaces and thence through stomata to the atmosphere surrounding the leaf.

Let us consider the problem this way: If all the chlorophyll-bearing cells of the plants of the world are constantly taking carbon dioxide from the atmosphere during daylight, the quantities of this gas used must be enormous, and there must necessarily be processes in nature that are continually replenishing this supply. It is known that the amount of carbon dioxide in the air is low (0.03%) and that it remains fairly constant. It has been estimated that an hectare of corn (24,700 plants) during a growing season of 100 days will accumulate 6265 kilograms of carbon; all this carbon is derived from the carbon dioxide of the atmosphere. It would require 22,974 kilograms of carbon dioxide to furnish this quantity of carbon. These estimates serve to emphasize the fact that enormous quantities of carbon dioxide are used in the photosynthetic process of green plants.

Obviously, the amount of carbon dioxide is limited. The present atmospheric supply would be used up in about 22 years were it not constantly being renewed. Several natural processes are continually releasing carbon dioxide to the atmosphere:

1. The living cells of all plants (both green and nongreen) and of all animals release carbon dioxide in the respiratory process (Chapter 14).

2. The dead bodies of plants and animals, and the excretions of animals, contain large quantities of carbon and other elements in the form of organic compounds; in the decay of these compounds, resulting from the activities of bacteria and fungi, large quantities of carbon dioxide are released to the atmosphere.

3. Carbon dioxide is also added to the atmosphere when wood, coal, oil, gas, or any other carbon compound burns.

4. Carbon dioxide is released to the atmosphere from mineral springs and volcanoes.

5. The oceans are important reservoirs of carbon dioxide, and carbon dioxide probably escapes from the oceans whenever its concentration in the atmosphere decreases.

If light intensity and temperature are favorable, the carbon dioxide of the atmosphere frequently limits the rate of photosynthesis. This may be particularly true in greenhouses kept closed in the winter. Under these conditions, the carbon dioxide in the air may be reduced much below the 0.03% average.

It has been determined experimentally that at usual temperatures and light intensities an artificial increase of carbon dioxide up to a concentration of 0.5% may give an increased rate of photosynthesis, but only for a limited period. It appears that this high level of carbon dioxide is injurious to plants; after 10–15 days' exposure, the plants, show injury.

Water

Although the water content of an actively photosynthesizing cell is high and large amounts of water are lost from the cell by transpiration, only about 1% or less of the water absorbed by the roots is actually used in photosynthesis. However, the rate of this process may be changed by small differences in water content of the chlorophyll-bearing cells. In some instances, the rate of photosynthesis is increased by mild dehydration (15% water loss) and retarded by vigorous drying (45% water loss). Since stomata tend to close when the plant is deprived of water, conditions of drought tend to reduce the rate of photosynthesis. Thus, though water is one of the raw materials in the process, it rarely, if ever, is directly a limiting factor in photosynthesis.

Minerals

The chemical formula of chlorophyll *a* is $C_{55}H_{72}O_5N_4Mg$ and, of chlorophyll *b* is $C_{55}H_{70}O_6N_4Mg$. Thus, it is seen that the synthesis of chlorophyll depends upon a supply of nitrogen and magnesium, both derived from salts in the soil. Moreover, chlorophyll is not formed unless iron is available, although this element is not a component of the chlorophyll molecule. Leaves of plants deficient in nitrogen, magnesium, or iron are pale and yellow, a condition termed **chlorosis.** This abnormal condition may also be caused by other factors, but when it occurs, the rate of photosynthesis is lowered.

SUMMARY OF PHOTOSYNTHESIS

Some of the important points that have been mentioned concerning photosynthesis are summarized in the schematic diagrams shown (Figs. 13.13, 13.14, 13.18).

1. Photosynthesis is the primary energy-storing process of life in which light energy is stored as chemical energy in organic compounds.
2. Carbon dioxide and water are the raw materials.
3. The products of photosynthesis are sugar and oxygen.
4. Light energy is absorbed by chlorophyll in the chloroplasts and drives the reaction of photosynthesis.

5. Two light reactions are involved. Absorbed light energy is transferred to a reaction center causing a chlorophyll molecule there to lose an electron.
6. Electrons move along chains of electron carriers.
7. Some of the energy is used to produce NADPH and ATP.
8. Electrons move from water to replace those lost by chlorophyll in the reaction center P_{680}. H^+ and O_2 are formed.
9. Light reactions and electron transport occur in thylakoid membranes.
10. Enzymes involved in the carbon cycle of photosynthesis are found in the stroma.
11. CO_2 is incorporated into organic compounds and, in a series of reactions, sugar molecules are formed. The hydrogen and electrons from NADPH and the energy from ATP are used in these reactions.
12. The two major pathways of carbon fixation are the C_3 and C_4 pathways.
13. Plants with low photorespiration rates are more efficient in fixing CO_2.
14. Photosynthesis may be limited by CO_2 supply, light, temperature, minerals, and the plant's hereditary efficiency.

CHAPTER 14

RESPIRATION

Every living cell in order to stay alive must break down complex organic molecules (foods) and obtain energy from them. This energy is supplied by cellular respiration. Differential permeability cannot be maintained without a supply of energy. Roots will not accumulate solutes, protoplasm will not move without energy from respiration. The synthesis of new cellular material such as amino acids, proteins, fats—growth itself—all require energy from respiration. The term respiration was first used to indicate the exchange of gases between an organism and its environment. Even today, respiration and breathing of animals are often popularly considered synonymous. Nevertheless, breathing is only an outward indication of the fundamental chemical reactions going on in the animal cells and characteristic of all life, that is, the breakdown of food. This breakdown process, as we have seen in Chapter 5, is through oxidation reactions and often, though not always, yields carbon dioxide and water as end products. Today, the term respiration has come to have a wider and more fundamental meaning than mere gaseous exchange or breathing. In its broadest sense, **respiration** is defined as the oxidation of organic substances within cells and is accompanied by the release of energy. Some biochemists reserve the term respiration for the final sequence of reactions that terminate in the use of molecular oxygen. However, we will use respiration in its broader chemical sense as defined above.

The energy stored in food (largely sugars) is energy from sunlight that has been converted to chemical energy during photosynthesis. We shall see that some of the steps of respiration are the reverse of reactions occurring in photosynthesis. We will again meet sugar phosphates, three-carbon-atom compounds, ATP, and NADPH. But the energy pathways are very different. In photosynthesis, energy is stored during food synthesis. In respiration, energy is released during food breakdown.

COUPLED REACTIONS

In order for a cell to be able to use the energy stored in food, a mechanism must exist in the cell whereby some of the energy is captured in a form that can be delivered to the energy-consuming reactions of the cell. In other words, energy-yielding reactions are coupled or linked to energy-requiring reactions. Such mechanisms frequently result in the transfer of energy from food to the energy carriers ATP and NADH or NADPH (Fig. 14.1). To simplify Fig. 14.1, only ATP and NADPH have been shown. Whether NAD or NADP is reduced depends on the enzyme system.

Thus, cellular respiration is not a simple oxidation; it is a chain of chemical reactions marvelously linked to all cellular processes. When 1 gram molecular weight of glucose (180 grams) is burned to CO_2 and H_2O, 686 kcal of energy are released. Obviously, this amount of energy released at once would be destructive to the cell. Instead, energy transformations in cells occur slowly and in small steps through a sequence of many reactions. Each of these small reactions is catalyzed by its own specific enzyme. During several of these reactions, some of the energy in the food is used to combine a phosphate group with ADP, and ATP is formed. This phosphorylation reaction results in the storage of about 7 kcal of energy in the terminal phosphate bond in each mole of ATP formed. Since ATP is in solution in the cell, it may diffuse from one location to another, carrying its stored energy with it.

A second important link in the transport of energy between energy-yielding reactions of respiration and energy-requiring reactions of synthesis involves the reduction of NADP or NAD to NADPH or NADH and then their oxidation back to NADP or NAD. NADP, an electron-acceptor, can accept two electrons (and energy) given up by food during a few specific steps in respiration. The resulting NADPH molecule contains about 53 kcal (per mole) more energy than does NADP. This compound, like ATP, can diffuse from one point to another in the cell and carry energy from a source, oxidation of food, to a site where it is used to drive synthetic reactions. For example, the synthesis of hydrogen-rich compounds such as fatty acids uses energy obtained when NADPH is oxidized to NADP.

Thus, energy-yielding reactions are coupled to energy-requiring reactions by energy-transporting substances such as ATP and NADPH. Both ATP and NADPH yield energy that is used in the performance of cellular work. They become ADP and NADP. In these forms, they

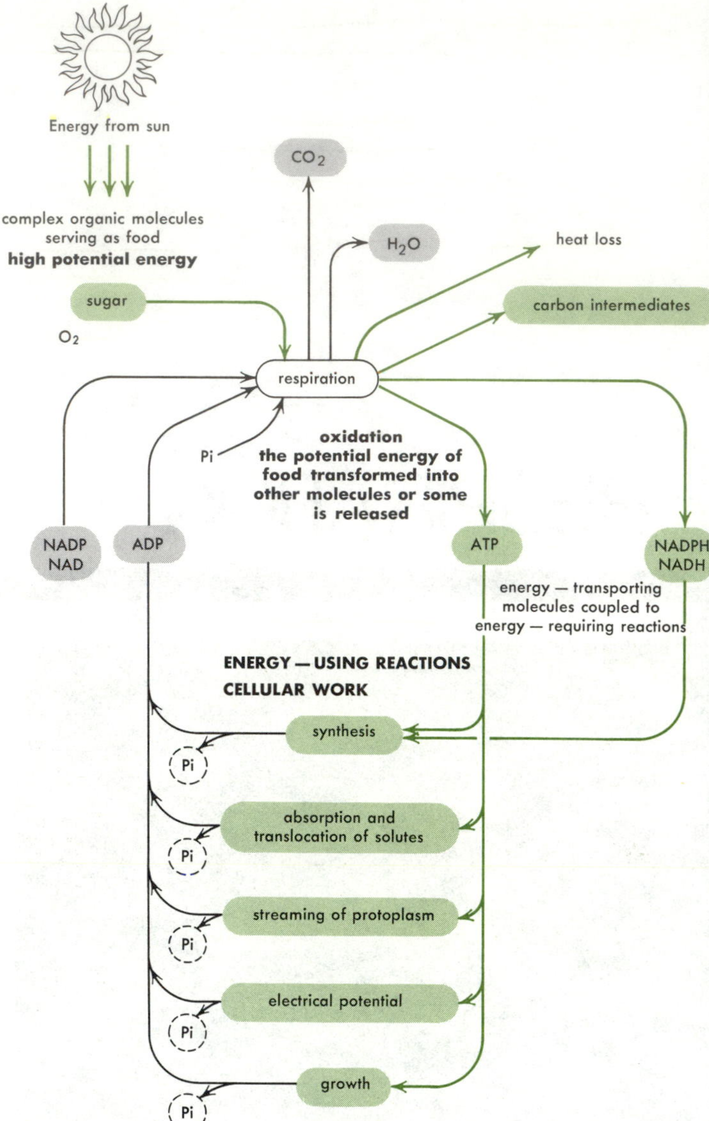

Figure 14.1
Energy from sunlight is trapped in food during photosynthesis. Some energy in the food is transferred during respiration to molecules (NADPH, ATP) that carry energy to sites of work in the cell.

may move back to the site of energy-yielding reactions and capture energy again to become ATP and NADPH (Fig. 14.2).

Large amounts of ATP and NADPH are never formed in the cell at any one time, since the total amounts of ADP and NADP from which they are formed are small. The cell stores its energy reserves in the form of food.

If a cell is at rest and is not using energy rapidly, most of its ADP may be converted to ATP. Respiration may be very slow under these conditions. If such a cell is stimulated to do work (for example, to synthesize new materials, or to increase the rate of protoplasmic streaming, or the amount of salt uptake), then ATP is broken down. Some of the energy it contained is used to do work in the cell. ADP and inorganic phosphate are released.

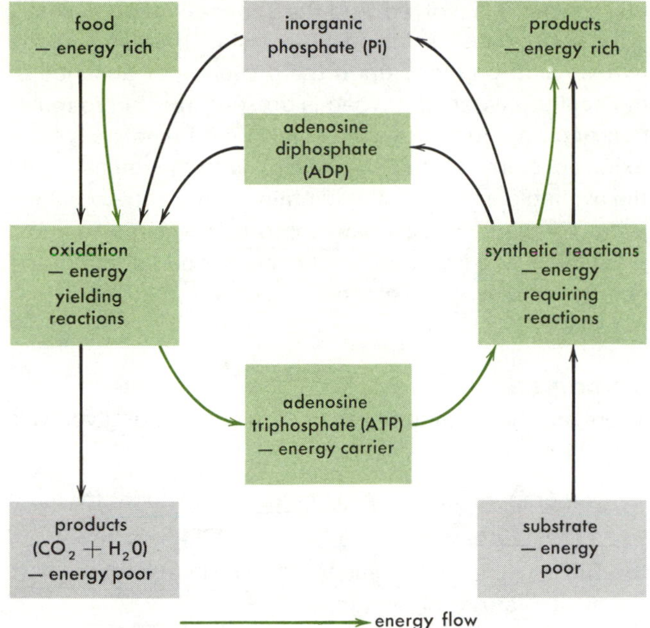

Figure 14.2
The coupling of energy-yielding reactions to energy-requiring reactions through ATP production and utilization.

They become available to the respiratory machinery, where they may be synthesized again into ATP, trapping more energy. Under these conditions, the rate of respiration may be increased. This is one example of many control mechanisms that regulate the rates of various metabolic reactions in the cell (see page 82).

When we consider the total energy cycle in the living cell, we see that stored chemical energy is moved from one part of the plant to another in the form of potential energy in the complex molecules that serve as food. At the cellular level, some of the potential energy in the food is trapped during respiration in high-energy, reactive molecules.

Respiration plays another very essential role in the life of the cell, for it is during the many steps in the respiratory breakdown of food that essential intermediate compounds are formed. Many of these intermediate compounds are used by living cells as sources of carbon building blocks from which new compounds such as proteins, fats, nucleic acids, and vitamins are synthesized (Fig. 14.3). *Respiration through the oxidation of food* (*a*) *supplies energy in an available form* and (*b*) *produces usable intermediate carbon compounds essential to the continued growth and metabolism of cells.*

THE OVERALL PROCESS OF RESPIRATION

Let us now consider in more detail the mechanism by which the living cell breaks down food and traps energy in the form of the energy-transport molecules ATP and

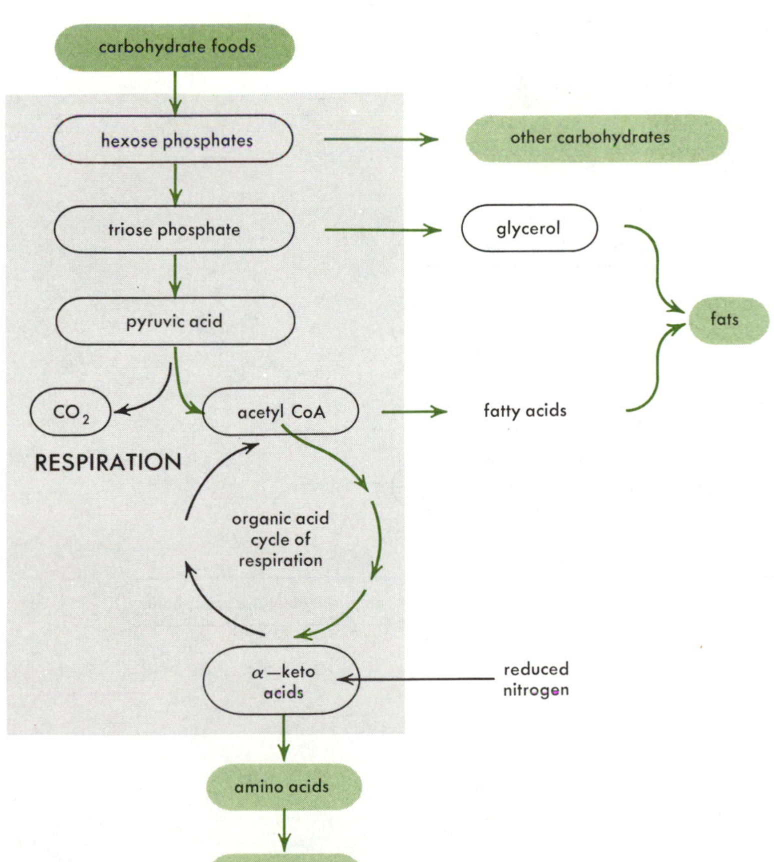

Figure 14.3
Some interrelationships between respiration and the synthesis of carbohydrates, fats, amino acids, and protein.

NADP. If one looks at the complete oxidation of a simple hexose (six-carbon-atom) sugar in the presence of molecular oxygen, only CO_2, H_2O, and energy (heat) are found as final products. This overall process may be written:

$C_6H_{12}O_6$	+	6 O_2	⟶	$6CO_2$	+	6 H_2O	+	686 kcal
sugar	+	oxygen	⟶	carbon dioxide	+	water	+	energy

This equation for respiration indicates the two reacting substances, sugar and oxygen, and the two usual end products, water and carbon dioxide; it tells nothing of the intermediate steps or of other possible end products. Biochemists and physiologists have been able to take the process of respiration apart and thus have determined the individual steps in the reaction. They have characterized the enzymes and cofactors involved. However, it is more important for us to understand how the essential steps in the process fit together and result in the slow release and trapping of energy than it is to try to remember all the steps of the process.

Respiration may be separated into distinct phases, each of which involves many reactions. In the first phase, for every molecule of the six-carbon-atom sugar glucose, which is oxidized, two molecules of **pyruvic acid** (a three-carbon-atom organic acid) are formed. No molecular oxygen is involved, and the energy change is small. This first phase is called **glycolysis.** The fate of the pyruvic acid depends upon the presence or absence of molecular oxygen. If oxygen is present, **aerobic respiration** occurs and the pyruvic acid formed by glycolysis is oxidized, stepwise, to CO_2 and H_2O, with the release of all the available energy that it contains. If molecular oxygen is not present, **anaerobic respiration** ensues in most plant cells, and CO_2 and alcohol are formed. Most of the energy remains in the alcohol and only a small amount is released.

Glycolysis

There are three distinct steps in the process of glycolysis (Fig. 14.4).

1. The preparation of the six-carbon-atom sugar for reaction by the addition of phosphorus—**phosphorylation.**
2. The splitting of the sugar into two three-carbon-atom fragments—**sugar cleavage.**
3. The oxidation of the fragments to form an intermediate product of respiration—**pyruvic acid formation.**

Preparation of Sugar for Reaction—Phosphorylation

The six-carbon-atom sugar glucose is the most common sugar respired in the cell. Although glucose is rich in

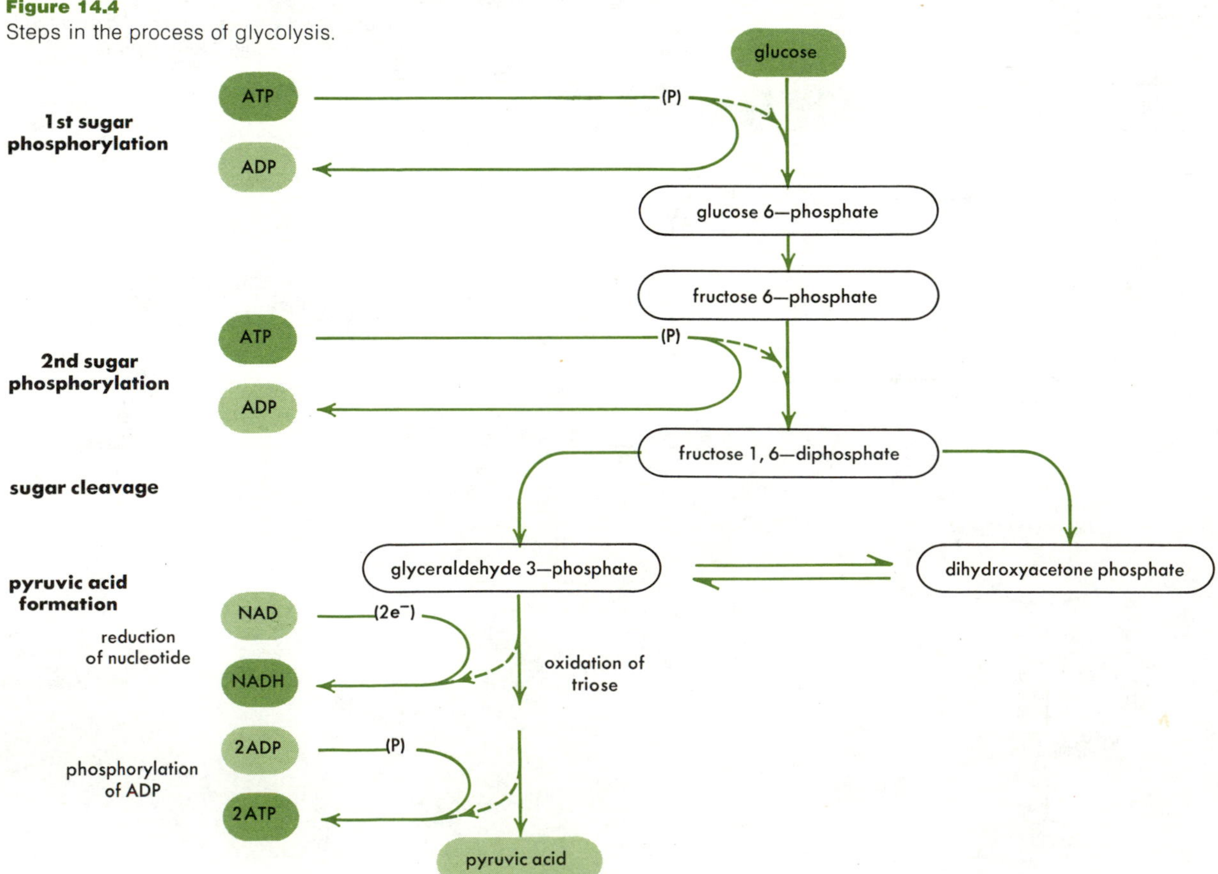

Figure 14.4
Steps in the process of glycolysis.

energy, it is stable and does not react readily with oxygen at temperatures at which life reactions occur, nor is it readily broken down into intermediate products. Before glucose can be broken down and its stored energy released, it reacts enzymatically with ATP and acquires a reactive phosphate group. This occurs at two points in glycolysis so that two ATP molecules are involved in providing energy and phosphate in the **phosphorylation** of the sugar molecule. The phosphorylated sugar also undergoes a series of internal rearrangements that results in the formation of another six-carbon-atom sugar with a phosphate group on either end, fructose 1,6-diphosphate.

Sugar Cleavage

An enzyme present in the cytoplasm of the cell is capable of catalyzing the splitting of fructose 1,6-diphosphate into two different three-carbon-atom sugars (trioses), dihydroxyacetone phosphate and D-glyceraldehyde 3-phosphate.

Pyruvic Acid Formation

These trioses are in equilibrium and may be converted one into the other. One of them, glyceraldehyde 3-phosphate, is oxidized to **pyruvic acid** and as it is broken down, the equilibrium shifts so that more glyceraldehyde 3-phosphate is formed. Thus, both triose phosphates are available for the reaction.

Several enzymatic steps are involved in pyruvic acid formation. Although molecular oxygen does not take part, the triose phosphate is oxidized by the transfer of two of its electrons and hydrogens to a hydrogen acceptor. In this case, NAD and not NADP is the usual hydrogen acceptor. Some of the energy originally in the triose is now in the newly formed NADH. Also, during the oxidation of each triose phosphate molecule to pyruvic acid, the phosphate group is lost and energy is used to form two molecules of ATP from two molecules of ADP and two phosphates. A total of four ATP and two NADH molecules are formed during the oxidation of two trioses.

The importance of glycolysis for the cell lies in the

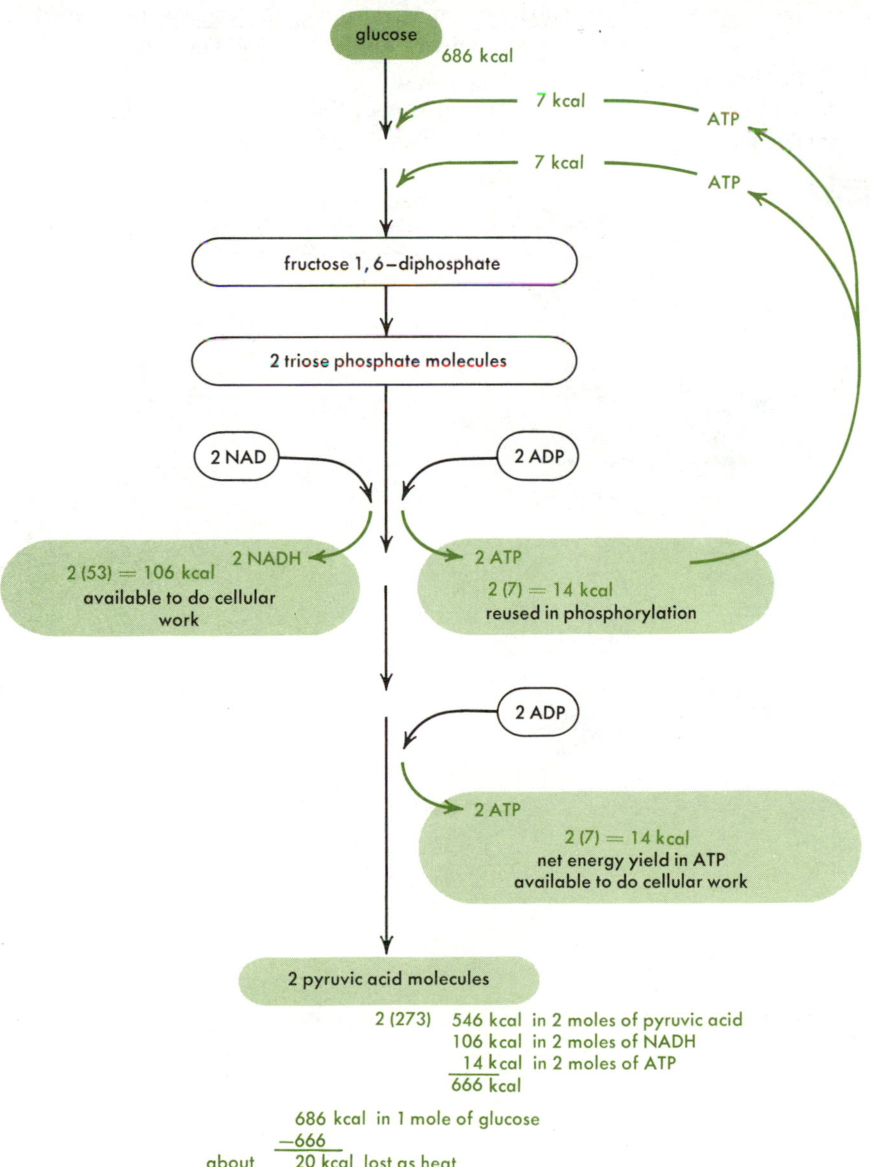

Figure 14.5
Energy summary of glycolysis of 1 mole of glucose. All energies are approximate kilocalories per mole.

fact that during the stepwise phosphorylation, cleavage, and oxidation of sugar to pyruvic acid, *usable intermediate compounds are formed,* and although only a small amount (approximately 17%) of the *energy has been trapped in energy-carrier molecules* (ATP and NADH), only about 3% has been lost as heat (Fig. 14.5). The major part, almost 80%, of the energy still lies trapped in the final product, pyruvic acid.

The fate of the pyruvic acid depends on whether molecular oxygen is or is not available to the cell.

Anaerobic Respiration

Normally, higher plants are not able to live long in the absence of oxygen. Under this condition, the food is not sufficiently oxidized to yield enough energy for life processes, and certain products formed may be poisonous. In contrast, some fruits, notably apples, may be held for long periods in an atmosphere containing very small amounts of oxygen and continue to give off carbon dioxide. Yeast may live actively in an atmosphere with very small amounts of oxygen and produce relatively large amounts of carbon dioxide and alcohol.

But, yeasts have a limited tolerance for alcohol. When the alcohol concentration in the medium in which they are living reaches about 12%, the yeast cells are killed. This is why wines and other naturally fermented alcoholic beverages do not have an alcohol content above about 12%. Yeast grows much more luxuriantly under aerobic conditions than in the absence of oxygen. Indeed, there are only a relatively few organisms that grow only under strictly anaerobic conditions.

Some fungi, bacteria, and many animal cells may live and grow slowly under anaerobic conditions, where they can produce many products in addition to alcohol. When oxygen is not supplied rapidly enough to vigorously exercising muscles in your body, the muscle cells may produce, not alcohol, but lactic acid. It is the presence of lactic acid that makes you stiff after hard muscular work.

Under anaerobic conditions in higher plants, the pyruvic acid formed during glycolysis is usually converted into carbon dioxide and ethyl alcohol. This process, called **alcohol fermentation** (see also Chapter 21), takes place in two steps. First, one carbon dioxide molecule is enzymatically split off from each pyruvic acid. This leaves a two-carbon-atom compound, acetaldehyde. Next, NADH, formed during glycolysis, is now used to reduce acetaldehyde to alcohol. NAD is released to take part in glycolysis again.

Thus, during anaerobic respiration, the energy trapped in NADH is used to form alcohol. It is significant

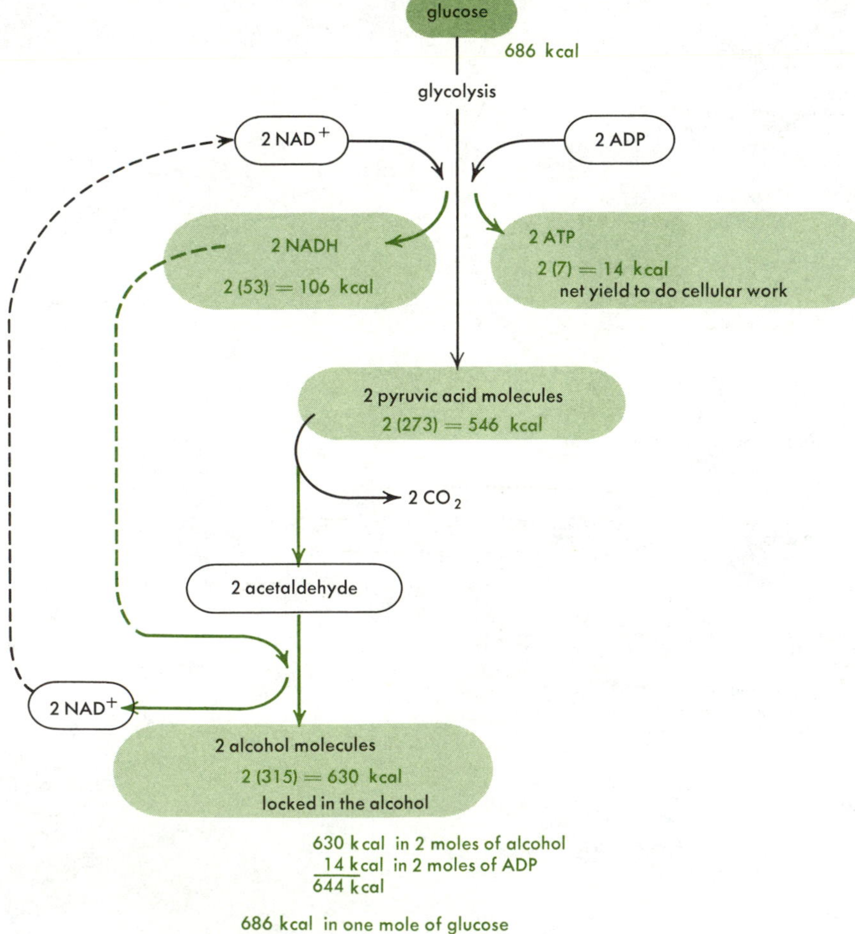

Figure 14.6
Energy summary of the anaerobic oxidation of 1 mole of glucose.

in the life of the cell that for each glucose molecule fermented to alcohol a net of only two ATP molecules are available to do work in the cell. This is about 14 kcal, or less than 3% of the energy available in 1 mole of glucose. About 6% is lost as heat, while almost 84% is still locked in the alcohol and is unavailable to the plant (Fig. 14.6). In addition, the alcohol itself may be toxic to the plant cell. It is obvious that *anaerobic respiration is not a very efficient way of utilizing energy in the food that is respired.*

Some plants that grow with roots in wet, anaerobic soil, accumulate malate instead of ethanol, an adaptation that avoids the toxic accumulation of ethanol.

Aerobic Respiration

Since glycolysis results in the release of only about 20% of the energy available in a glucose molecule, 80% still is locked in the two pyruvic acid molecules formed. In contrast to anaerobic respiration, aerobic respiration in a plant cell yields adequate amounts of energy so that the cell can carry out the energy requirements of life. During aerobic oxidation of pyruvic acid to CO_2 and H_2O, between 40 and 50% of the energy is trapped in a potentially useful form in ATP or reduced nucleotide. How is this accomplished?

We may conveniently divide aerobic respiration into three main steps.

1. *Entrance of carbon into the organic cycle of respiration.* CO_2 is released from pyruvic acid, the remaining two-carbon-atom "fragments" enter into the organic acid cycle, and NADH is formed.
2. *The organic acid cycle of respiration.* The two-carbon-atom "fragment" is oxidized, CO_2 is released, and ATP, NADH are formed.
3. *Electron transport and terminal oxidation.* Reduced coenzymes are oxidized, ATP and water are formed, and molecular oxygen is used.

Entrance of Carbon into the Organic Acid Cycle

One of the most complex series of reactions of respiration is involved in the oxidation of pyruvic acid. In these reactions, several vitamins, particulary those of the vitamin B complex (niacin, thiamine, pantothenic acid), serve as coenzymes or parts of coenzymes (page 72). In the initial stages, for each molecule of pyruvic acid oxidized a molecule of CO_2 is formed, and two electrons, two hydrogen atoms, and about 53 kcal of energy are transferred to NAD, forming NADH. Water molecules enter into the reactions. The remaining part of the pyruvic acid is now a two-carbon-atom "acetate fragment" ($\mathrm{H{-}\overset{H}{\underset{H}{C}}{-}C{\overset{O}{\diagup\!\!\diagup}}{-}}$ or acetyl group). It becomes associated with another coenzyme, coenzyme A, forming the reactive complex acetyl-CoA. In the presence of a specific enzyme, coenzyme A is capable of combining with (accepting) such acetyl groups and transferring (donating) them to other acceptor molecules.

In this way, the two-carbon-atom acetyl group (still containing about 20% of the energy present in the original glucose molecule) may be transferred from one series of reactions to another in the cell (Fig. 14.7). This is analogous to the transfer of electrons and hydrogen (energy) by the coenzymes NADH and NADPH.

The actual entry of the reduced carbon into the organic acid cycle, where its energy can be utilized to form ATP, is accomplished by the donation of the acetyl group from acetyl-CoA to an acceptor molecule, **oxaloacetic acid,** found in mitochondria. Oxaloacetic acid is a four-carbon-atom acid that combines with the acetyl group and forms the six-carbon-atom acid, citric acid. The acetyl group has now entered the organic acid cycle of respiration where the mitochondrial machinery can act on it and release its potential energy.

Although pyruvic acid is the usual donor of acetyl groups to coenzyme A, it is not the only donor. The breakdown of fats and proteins (amino acids) results in the formation of acetyl groups that also may be donated to coenzyme A and then enter the acid cycle. Thus, fats and proteins may also serve as foods and supply energy for the energy-requiring processes of life.

The Organic Acid Cycle of Respiration

The new acid, citric acid, formed when the acetyl group of acetyl-CoA combines with the oxaloacetic acid, is gradually broken down. This process involves a cyclic series of reactions in which seven other organic acids including the acetyl acceptor, oxaloacetic acid, are formed (Fig. 14.8). During these reactions, pairs of electrons (and hydrogen atoms) are transferred to electron carriers, the coenzymes NAD (NADP) and FAD (flavine adenine dinucleotide); see page 73. All of the atoms brought in with the acetyl group are gradually removed. The carbon and oxygen are released as the molecules of CO_2 are produced. A new oxaloacetic acid capable of accepting another acetyl group is formed.

Energetically, in one turn of the cycle, the potential energy in one acetyl group is released or trapped in other molecules. Part of it is lost as heat, but the rest is transferred to one molecule of ATP, three molecules of NADH (NADPH), and one molecule of $FADH_2$. The ATP is free to carry its energy to other parts of the cell where work is being done, but the energy in NADH and $FADH_2$ generally is used in the synthesis of more ATP in the final stages of respiration. Of the total available energy in the original acetyl group, about 66% is trapped in ATP and reduced nucleotides that are produced when the acetyl group is oxidized.

Because this is a repeating process and because organic acids are involved, it has been called the organic

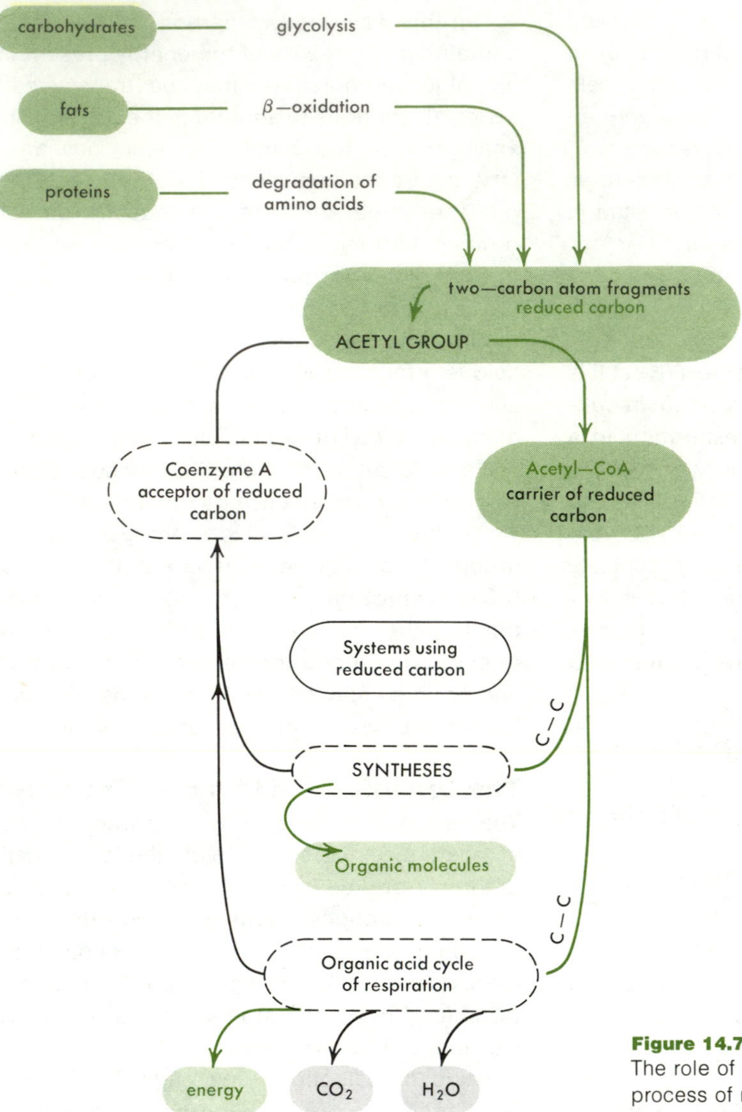

Figure 14.7
The role of coenzyme A as a carrier of acetyl groups in the process of respiration.

acid cycle of respiration. Sometimes it is called the **citric acid cycle,** the TCA cycle, or more frequently the **Krebs cycle,** after the physiologist whose research contributed a great deal to our knowledge of respiration. It is believed that in many plants, animals, and microorganisms, such a cycle occupies a central position in aerobic respiration.

The important results of the organic acid cycle to the cell are listed below.

1. Intermediate compounds in the cycle may act as starting materials for pathways of synthesis such as protein and lipid synthesis.
2. Hydrogen is transferred to energy carriers (NADH, FAD) forming energy-rich reactive compounds.
3. Energy-rich ATP is formed.

Although all the energy originally associated with the reduced carbon atoms in the sugar molecule (food) is released or transferred when the Krebs cycle is completed, only a relatively small amount, less than 10%, is directly trapped in the ATP formed during glycolysis and during the Krebs cycle. About 56% is first trapped in reduced nucleotides. How does the cell utilize the energy trapped in the reduced nucleotides?

The Final Stages of Oxidation—Electron Transport

The energy present in the reduced nucleotides usually is not utilized directly by the cell. In this case, the nucleotides are oxidized stepwise, and part of their energy is temporarily trapped in ATP. In this final series of reactions, pairs of electrons are transferred from the reduced coenzymes through an electron transport system of car-

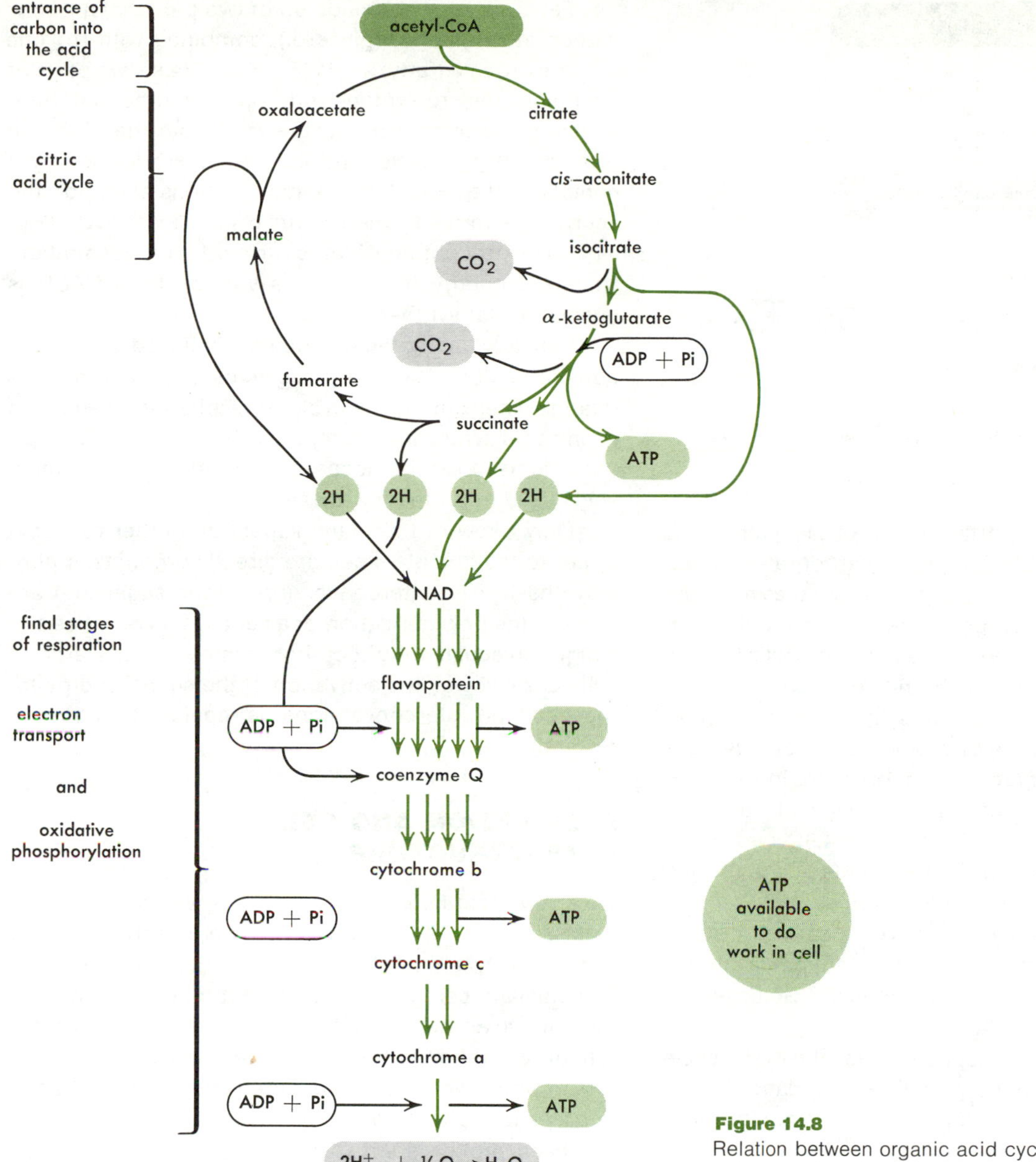

Figure 14.8
Relation between organic acid cycle, electron transport, and oxidative phosphorylation.

riers that alternately are reduced and then oxidized. Several of these electron carriers are **cytochromes** and contain iron atoms that are alternately reduced to the ferrous (Fe^{2+}) form and then give up their electrons, becoming oxidized to the ferric (Fe^{3+}) form. The last step involves the transfer of a pair of electrons to an oxygen atom. Two hydrogen ions from the cellular environment then combine with the oxygen. Water is formed. If free oxygen is not present in the cell, this last transfer of electrons could not take place. The flow of electrons in the chain would cease and the entire aerobic respiratory sequence would stop.

In the electron transport process, three molecules of ATP may be synthesized for each NADH or NADPH that is oxidized, and two molecules of ATP may be synthesized for each $FADH_2$ oxidized. This ATP synthesis that is linked to oxidation reactions of respiration is known as **oxidative phosphorylation.**

Complete aerobic respiration of a glucose molecule thus involves a stepwise breakdown of the sugar molecules. Six CO_2 molecules and six H_2O molecules are formed. Six O_2 molecules are absorbed and about 40% of the potential energy in the glucose is transferred to some 36 molecules of ATP, which can be utilized to do work in the cell. About 60% of the energy is lost as heat during the various steps of respiration.

Alternate Pathways of Respiration

Although aerobic respiration is the most common method by which plant cells oxidize foods, alternate pathways do occur. One such pathway is known as the **pentose phosphate pathway** (PPP), or the hexose mono-

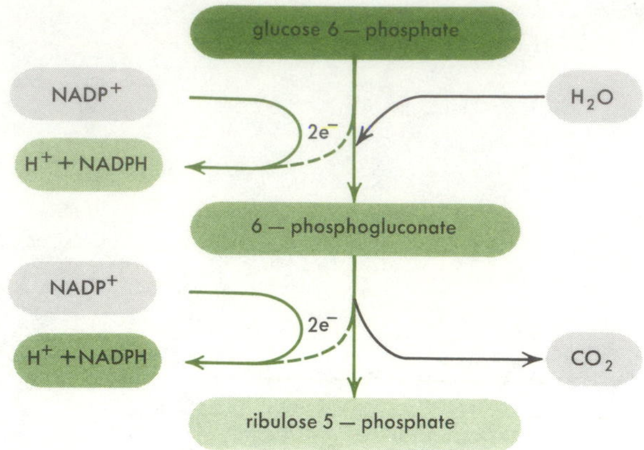

Figure 14.9
The first steps in the pentose physphate pathway.

phosphate shunt. An important result of this pathway is the production of NADPH and a five-carbon-atom sugar phosphate, ribose phosphate, in the cytoplasm of the cell. While the ribose phosphate may be broken down in later reactions, the NADPH is available for utilization in important synthetic reactions in the cell.

Biochemically, the PPP starts in the same way as glycolysis, that is, with the phosphorylation of a glucose molecule to produce glucose 6-phosphate. In the PPP, however, in the presence of NADP and a specific enzyme, glucose 6-phosphate is oxidized (Fig. 14.9). Two molecules of NADPH are formed. One molecule of CO_2 is liberated, and a molecule of ribulose 5-phosphate is produced. The rest of the cycle involves a series of sugar transformations that is essentially identical to those found in the Calvin cycle of photosynthesis. Glucose may be regenerated in the cycle.

Young growing tissues appear to use the Krebs cycle as the predominant pathway of glucose oxidation, while aerial parts of the plant and older tissues seem to utilize the PPP as well as the Krebs cycle.

RESPIRATION AND THE SYNTHESIS OF ORGANIC MOLECULES

We have seen that all living cells respire and that respiration provides chemical energy and certain intermediate compounds, both needed for many cell functions. This intimate relationship between respiration and other cellular reactions can readily be seen when one considers the formation of carbohydrates, fats, amino acids, and proteins in the cell. All these compounds are ultimately derived from the sugars produced during photosynthesis.

During development of a fat-storing seed such as squash, the seed may store as much as 20% of its dry weight as fat. However, it is sugar and not fat that is translocated through the phloem into the seed. Obviously, a mechanism exists in the cells of the seed to convert sugar into fat. How is this done?

Fat molecules are made up of two parts: (*a*) a three-carbon-atom alcohol (glycerol), combined with (*b*) one or more organic (fatty) acids (Fig. 5.6). These two parts of a fat molecule are synthesized from respiration intermediates; glycerol may be synthesized from the cleavage products of glycolysis, and fatty acids are formed by a complicated series of enzymatic reactions acting on the acetyl-CoA formed when pyruvic acid loses CO_2 (Fig. 14.7). The production of fat is further tied to respiration, since the energy from ATP as well as from NADP is needed for fat synthesis.

Similarly, amino acid synthesis is linked to respiration. Some of the same organic acids that are formed as intermediates during the cycle of respiration are used in amino acid synthesis. Again, the ATP produced by respiration is used for the necessary activation of the amino acids used in protein synthesis.

Only a few of the many sugars and other carbohydrates found in a plant cell are directly products of photosynthesis. The interconversion of one sugar into another or the polymerization of a sugar in the synthesis of a high-molecular-weight carbohydrate such as starch or cellulose involves the activation of the sugar. Energy for these sugar interconversions is again derived from respiration.

RESPIRATION AND CELL ULTRASTRUCTURE

We know that photosynthesis occurs within the chloroplasts. We may ask ourselves, "Where in the cell does respiration take place?" In a very active cell, such as a meristematic cell, the rate of respiration is very high. If the cell is in an inactive tissue, such as in a dried seed, the rate of respiration is very slow indeed. Just as investigators of photosynthesis have literally taken the cell apart in order to study its photosynthetic apparatus, the cell has been broken apart, and each part (soluble phase, nucleus, chloroplasts, dictyosomes, microbodies, and mitochondria) has been studied to determine its role in cellular metabolism.

We have seen that the complete oxidation of a sugar molecule involves a whole series of reactions. Some of these, particularly those of glycolysis, and the PPP takes place in the soluble phase of the cytoplasm. These reactions do not appear to be strongly associated with cell organelles or membranes. Others, particularly those involving the aerobic oxidation of pyruvic acid (the acid cycle, terminal oxidation, and oxidative phosphorylation steps of respiration), are carried out by mitochondria.

Mitochondria have been isolated from plant as well as animal cells and when properly prepared will break down pyruvic acid in the presence of oxygen. Carbon dioxide is produced, and ATP may be synthesized during the oxidation reactions. Thus, respiration can be studied outside the cell in the test tube, with preparations from living cells.

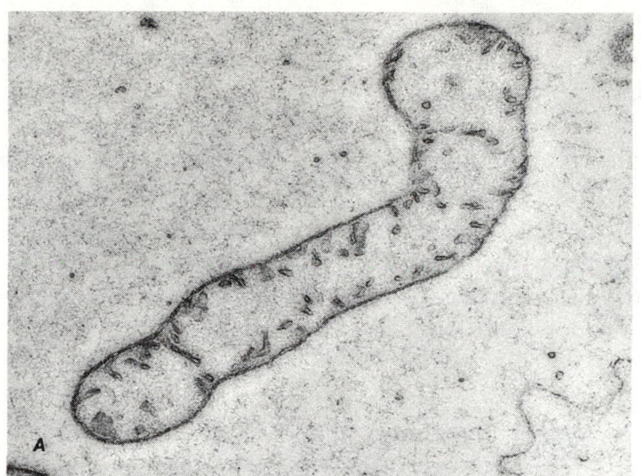

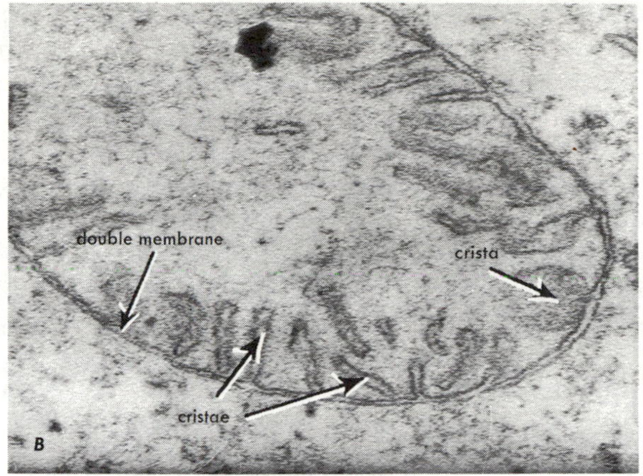

Figure 14.10
A, electronmicrograph of a thin section of a mitochondrion, the cellular site of aerobic respiration. Mitochondrion is from a companion cell in phloem of squash (*Cucurbita*), ×8000. *B*, portion of mitochondrion from *Carteria* showing the double mitochondrial membrane and the inward extension of the inner component of the membrane to form the cristae, ×43,000. (*A*, courtesy of K. Esau; *B*, courtesy of Carole Lembi.)

Recall the details of mitochondrial structure (see page 47). Figure 14.10 shows a mitochondrion from a companion cell of squash (*Cucurbita*) and from the alga *Carteria*. Here, the characteristic double boundary membrane that separates the mitochondrion from the rest of the cytoplasm is clearly seen. The circular and elongated dark bodies inside the mitochondrion are sections through an internal membrane system that is made up of fingerlike protuberances extending inward from the inner mitochondrial membrane. It is believed that many of the enzymes necessary for respiration are located on this membrane system in a definite relationship to each other. We see that the respiratory activity of mitochondria is closely associated with its structure in a manner similar to that in which the photosynthetic activity of chloroplasts is related to their structure. We see also that an integration of activities exists between the cytoplasm, where the early stages of food breakdown occur, and the mitochondria, where the final stages of aerobic respiration take place.

During the life of the cell, there is a continuous traffic of molecules across the mitochondrial membranes. Pyruvic acid, inorganic phosphate ions, and ADP molecules are three important substances moving into the mitochondria from the cytoplasm.

Just as the H^+ gradient produced during electron transport in photosynthesis is believed to be involved in ATP synthesis, in respiration also a H^+ gradient is produced across the mitochondrial envelope during electron transport. In this case, H^+ is pumped out of the mitochondrion and ATP is produced by ATPase activity in the inner membrane of the mitochondrial envelope. The newly formed ATP may move from the mitochondria to some other place in the cell. As work is done there, the ATP is hydrolyzed to ADP and inorganic phosphate. The ADP may again enter a mitochondrion. In this way, chemical energy in the substrate (pyruvic acid) molecules is continuously moving into the mitochondria, where it is transformed into a more readily usable form of energy (ATP). This in turn is constantly moving from the mitochondria to all parts of the cell.

THE ENVIRONMENT AND RESPIRATION

Since respiration is the process by which some chemically stored energy in foods is transformed into a form immediately available to do work in the cell, any conditions either inside or outside the cell that affect the rate of respiration will influence cellular activites.

Cell Hydration

Many seeds remain viable for long periods of time when stored in relatively dry air. The water content of the cells at this time may be less than 10% of the seed weight. In contrast, the water content of active protoplasm may be as high as 90% or more. Under low moisture conditions, respiration and some other cellular activities slow to a very low rate; others stop completely. Respiration goes on at a very slow rate, so little oxygen is consumed, little carbon dioxide is given off, and only very minute amounts of heat are released. Energy is needed only to maintain a certain little-understood steady state in the quiescent protoplasm. Growth has ceased, mineral salts are not being absorbed, and cell divisions have stopped, as have many, if not all, reactions involving synthesis of new materials. Because of this dormant state, seeds may be kept in large storage bins such as grain elevators for long periods of time.

If even a small amount of water is added to the seeds, imbibition occurs, the seeds swell, respiration increases rapidly and, if the seeds are closely confined, as in a grain elevator, the temperature inside the mass of seeds that are insulated from the outside may increase to a point where the seeds will be killed. In this example, the rate of respiration is greatly influenced by the water content of the cell. A number of other factors, both external and internal, influence the rate at which a cell respires and consequently affect all the cell's activities that depend upon the use of energy released in respiration. *In addition to the degree of hydration of the cell, factors such as temperature, oxygen supply, food availability, carbon dioxide concentration, the presence of certain chemicals, and the age of the cell commonly influence cell respiration.*

Temperature

Temperature has a particularly marked effect upon most biological reactions, especially those that are controlled by enzymes. In the temperature range from near freezing (0°C) to about 30°C, an increase of 10°C approximately doubles the rate of respiration. Above 30°C the harmful effects of high temperature on the cell may become marked. At these higher temperatures, cellular enzymes progressively become inactivated and respiration decreases.

Nevertheless, over the long time of evolution, certain organisms have evolved characteristics that have enabled them to survive in otherwise hostile environments. In this way, certain species of algae and bacteria are adapted to respire and grow under temperature extremes that would kill unadapted species. Particularly impressive is the presence of some of these organisms in hot springs and streams where temperatures may exceed 60°C.

Food

Since respiration involves the utilization of food, the presence of easily available food is necessary if a cell is to continue to respire and stay alive. Photosynthesizing leaf cells, of course, are able to produce their own food in the light. Under normal conditions, they must also form enough food to supply the needs of all the nonphotosynthetic cells in the plant. If a plant is kept in the dark, continued food usage without food synthesis will rapidly deplete the available food. Stored food reserves, particularly starch and sugars, may be drawn upon by all living cells. If the plant is prevented from carrying out photosynthesis and if food is not supplied, the plant will eventually starve to death. In fact, any factor that limits photosynthesis and thus food availability must indirectly influenced respiration.

Oxygen

Although most plant cells can continue for a time to oxidize foods even in the absence of gaseous oxygen (anaerobic respiration), molecular oxygen is normally necessary for the health of higher plant cells. Rarely does the concentration of oxygen in the atmosphere deviate enough from the normal 21% to appreciably affect the rate of respiration. However, underground stems, seeds, and roots may be in an oxygen-poor environment, since the microorganisms in the soil as well as the plant parts themselves may use the oxygen in the soil atmosphere faster than it is replaced from the air. Under these conditions, respiration in the roots may be decreased. The effect of aeration on root growth is shown in Fig. 11.5. Similar conditions of low oxygen level and high carbon dioxide concentration may occur in the internal cells in bulky plant organs such as large fleshy fruit.

A knowledge of the effects of the gas environment on respiration of stored fruits and vegetables has been of great practical importance to shippers and handlers of produce. Although the absence of oxygen is detrimental to the tissues of most plants, modified atmospheres made of low concentrations of oxygen, which inhibit both aerobic and anaerobic respiration, and higher levels of inert gases have been effective in prolonging storage life and in improving fruit and vegetable quality after shipping.

COMPARISON OF PHOTOSYNTHESIS AND RESPIRATION

Photosynthesis	Respiration
1. CO_2 and H_2O are used	1. O_2 and food are used
2. Food (carbohydrate) and O_2 are produced	2. CO_2 and H_2O are produced
3. Energy from light is trapped in chlorophyll and in food	3. Energy in food may be temporarily stored in ATP or lost as heat
4. ATP is produced by use of light energy (photosynthetic phosphorylation)	4. ATP is produced by oxidation of food (oxidative phosphorylation)
5. Hydrogen is transferred from H_2O to NADP to form NADPH	5. Hydrogen is transferred from food to NAD or NADP to form NADH or NADPH
6. ATP and NADPH are used primarily to drive reactions involving sugar synthesis	6. ATP and NADH or NADPH are available to do many types of work in the cell
7. Only chlorophyll-containing cells carry out photosynthesis	7. Every living cell carries out respiration
8. Occurs only in light	8. Occurs both in light and in darkness
9. Occurs in chloroplasts in eukaryotic plants	9. Glycolysis occurs in cytoplasm, while the final steps of aerobic respiration occur in mitochondria
10. Total photosynthesis must exceed total respiration for growth to occur	

SUMMARY OF RESPIRATION

1. Sugar is prepared for oxidation through the process of phosphorylation during which phosphorus is transferred from an organic phosphorus donor, ATP, to the sugar molecule. Sugar cleavage then occurs, resulting in the production of two three-carbon-atom phosphate intermediates. These in turn are oxidized to the common intermediate, pyruvic acid.
2. If molecular oxygen is absent, pyruvic acid usually is changed to ethyl alcohol and carbon dioxide in the final stages of anaerobic respiration in plant cells. Only small amounts of energy are released in this process.
3. If molecular oxygen is present, the pyruvic acid becomes further oxidized to carbon dioxide and water through the organic acid cycle. In this process (aerobic respiration), large amounts of energy are stored in energy-rich compounds that are used when work is done in the cell.
4. The two major functions of respiration are (*a*) the transformation of potential energy stored in food to an energy form readily available to do work (ATP) and (*b*) the production of intermediate products that are used in the synthetic reactions of the cell.
5. Energy-yielding oxidative reactions may be coupled to energy-requiring reactions.
6. Respiration is carried out through an integration of reactions going on in the cytoplasm (glycolytic reactions) and the mitochondria (Krebs cycle and terminal oxidation reactions).
7. Alternate pathways of respiration, such as the pentose phosphate pathway, occur in some cells.
8. Cell hydration, temperature, oxygen supply, food availability, carbon dioxide concentration, and cell age all affect the rate of respiration.

CHAPTER 15

THE FLOWER AND THE FLOWERING PLANT LIFE CYCLE

THE FLOWER

The flower initiates the sexual reproductive cycle in all Anthophyta (the flowering plants) and, in so doing, it usually terminates the growth of the shoot bearing the flower. With some plants—annuals, biennials, and some perennials—death follows flowering and seed set. With most perennials the capacity for vegetative growth is regained after sexual reproduction and the development of the next generation. In woody perennials, provision is always made for continued vegetative growth, through mixed buds or associated shoot buds following flower production. The function of the flower is to facilitate the important events of gamete (haploid reproductive cell) formation and fusion.

Of all the characteristics of flowering plants, the flower and fruit are the least affected by changes in the environment. For instance, we have seen that leaf shape is influenced by age, light, water, and nutrition. The basic morphology and anatomy of flowers and fruits are not affected quite so much by environmental factors and are therefore important parts for angiosperm classification. In addition, flowers have a high esthetic value, and the resulting fruits and seeds are of major importance in food production.

The essential steps of sexual reproduction, meiosis and fertilization, take place in the flower. The complete sexual cycle involves: (*a*) the production of special reproductive cells, accompanied by meiosis, (*b*) pollination, (*c*) fertilization, (*d*) fruit and seed development, (*e*) seed and fruit dissemination, and (*f*) seed germination. The

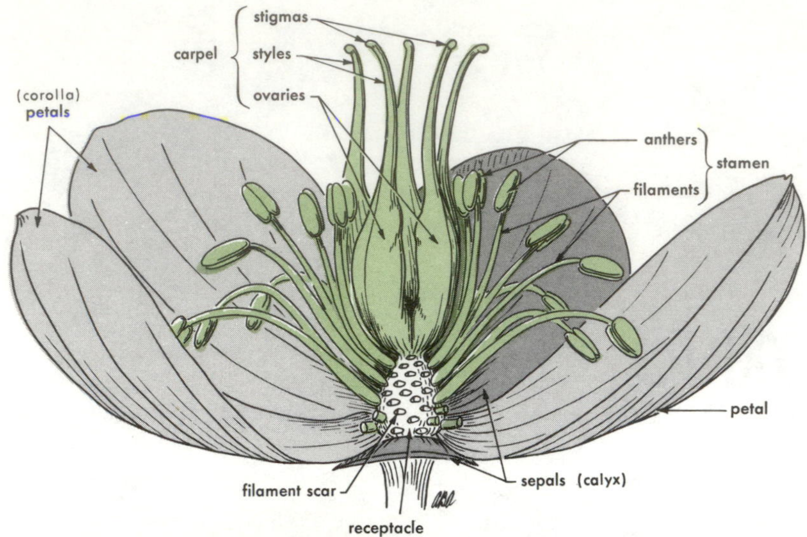

Figure 15.1
A diagram of a longitudinal section of a flower of Christmas rose (*Helleborus*). The perianth consists of two similar whorls; there are numerous stamens arranged in a spiral on a cone-shaped receptacle. Five separate carpels form the central whorl of floral parts.

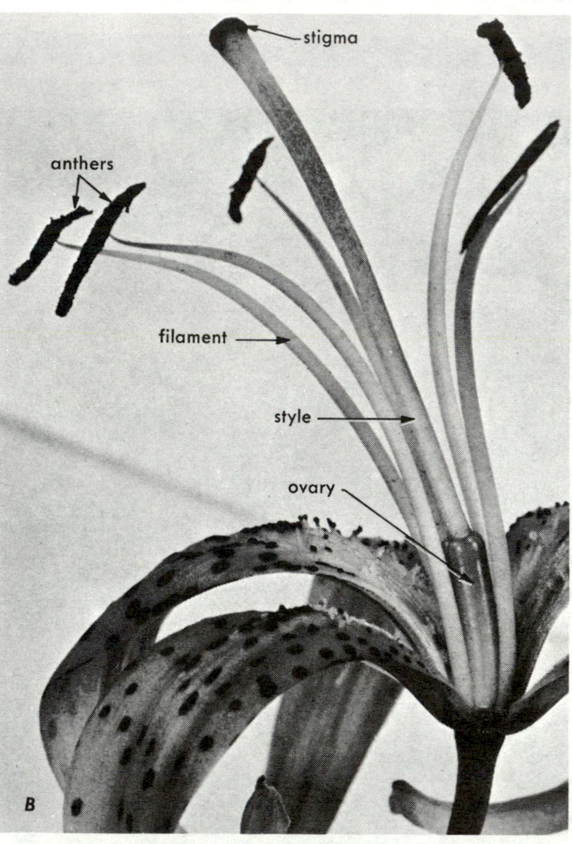

Figure 15.2
The essential floral organs. *A, Magnolia,* showing many separate stamens and separate carpels arranged in a spiral on the receptacle. *B,* regal lily (*Lilium regale*), showing five of the six stamens with anthers and filaments, and three carpels in a single pistil with stigma, style, and ovary.

seed completes the process of sexual reproduction in the angiosperms, and the embryo in the seed is the first stage in the life cycle of new individuals.

FLOWER MORPHOLOGY

Flowers are of many different forms. They also vary greatly in size, color, number of parts, and arrangement of parts. There are flowers so small that their organs are scarcely visible to the unaided eye; for example, the flowers of the duck-weeds, free-floating plants common in ponds throughout the world. Then there are the flowers of a plant (*Rafflesia*), growing on the floor of dark tropical forests of the Malay archipelago, that are 3 to 4 ft in diameter.

A typical flower is composed of four whorls of modified leaves: (*a*) **sepals,** (*b*) **petals,** (*c*) **stamens,** and (*d*) a **carpel** or **carpels,** all attached to the **receptacle,** the modified stem end that supports these structures (Fig. 15.1, **Fig. 15.5,** page 289).

The **sepals** enclose the other flower parts in the bud. Generally, they are green. All the sepals taken collectively constitute the **calyx;** that is, the calyx of a flower is composed of more or less distinct parts, the sepals.

The **petals** are usually the conspicuous, colored, attractive flower parts. Taken together, the petals constitute the **corolla.**

The **stamens** form a whorl, lying inside the corolla. Each stamen has a slender stalk or **filament** at the top of which is an **anther,** the pollen-bearing organ. The whorl or grouping of stamens is called the **androecium** (Fig. 15.2*A*).

The **carpel** or **carpels** comprise the central whorl of modified floral leaves (Fig. 15.1, Figs. 15.2, **15.5**). Collectively, the carpels are spoken of as the **gynoecium.** Each *individual structure* in the gynoecium is commonly referred to as a **pistil.** A pistil may be composed of a single carpel, or of several united carpels in the center of the flower. Thus, a pistil may be formed from *one* or *more* carpels. The pistil may be dissected from a flower as a unit. When one carpel forms one pistil, a carpel may be physically separated from a flower as a whole unit (Figs. 15.2, **15.5,** 15.6*D,F*). When several carpels form a single pistil, the pistil must be cut apart in order to separate out the carpels (Fig. 15.6*H*). The name "pistil" is ancient and is derived from the resemblance of numerous commonly occurring pistils to the pestle of the mortar-and-pestle set always present in the medieval pharmacy, where it

Figure 15.3
Primitive pistil and stamen from living plants in southeast Asia. *A*, pistil of *Drimys piperita*. *B*, cross section of pistil (dotted line). *C*, pistil laid open. *D*, stamen of *Degeneria vitiensis*. *E*, cross section of stamen (*A*, *B*, *C*, redrawn from I. W. Bailey and B. G. L. Swamy, *Am. J. Bot.* **38,** 373. *D*, *E*, redrawn from J. E. Canright, *Am. J. Bot.* **39,** 484. Also I. W. Bailey and A. C. Smith, J. Arnold Arboretum **23,** 356 © 1942 by Arnold Arboretum. Reprinted by permission of publishers.

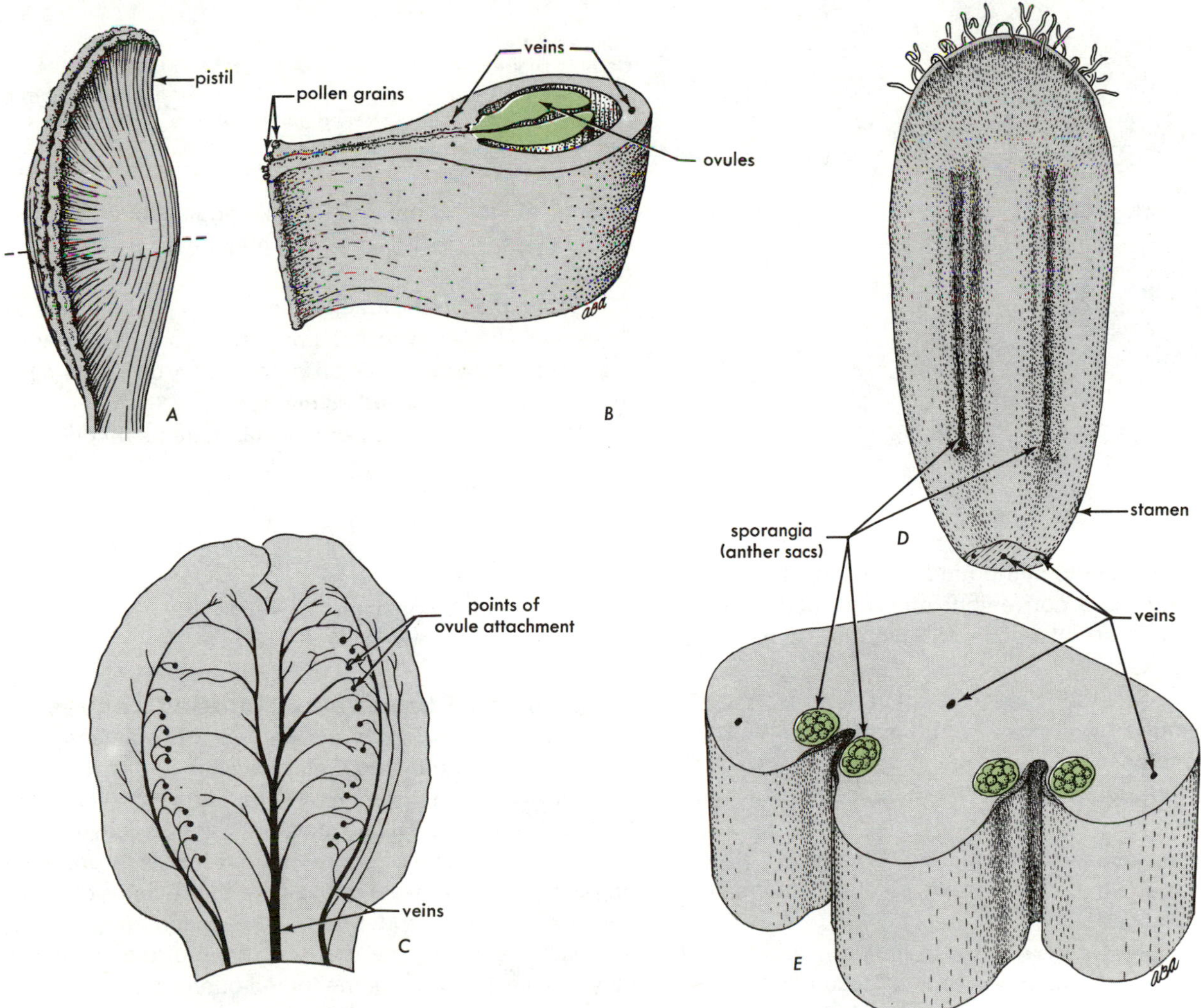

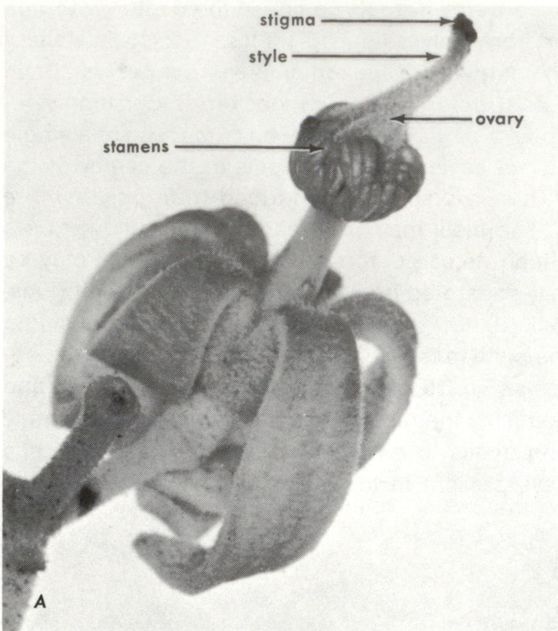

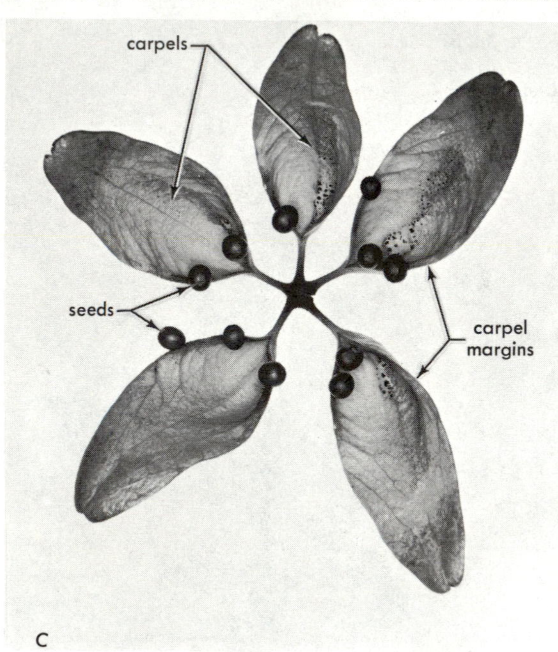

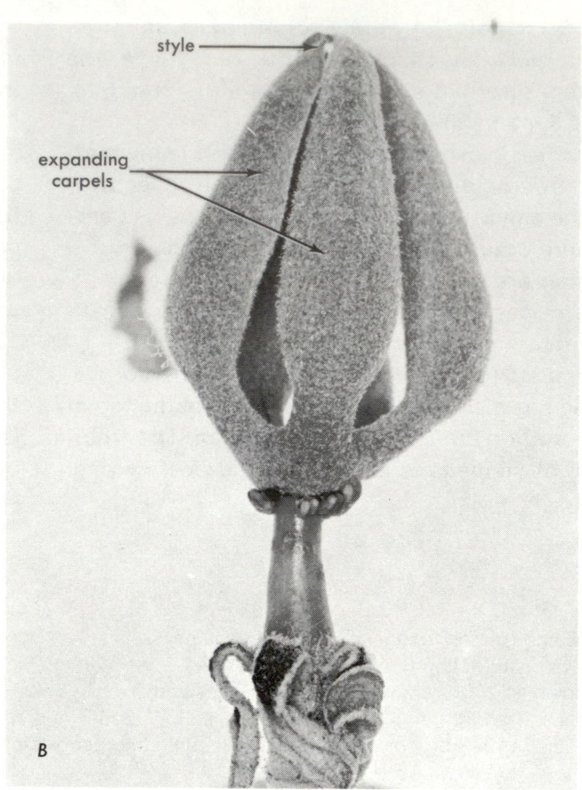

Figure 15.4
Carpels suggestive of foliage leaves. *Sterculia plantanifolia*. *A*, flower showing a single gynoecium, ×5. *B*, after pollination, carpels separated except at expanding tips, ×3¼. *C*, dehiscence of matured ovary showing seed attached to the margins of the five leaflike carpels, ×5/8.

was frequently used to compound medicines from floral parts. The term is a convenient one, but modern concepts of flower structure have rendered its meaning somewhat inexact.

There are generally three distinct parts to each pistil: (*a*) an expanded basal portion, the **ovary,** in which are borne the **ovules;** and (*b*) the **style,** a slender stalk supporting (*c*) the **stigma** (Fig. 15.1). The pollen is deposited on the stigma.

The **receptacle** is the enlarged end of the flower stem or stalk to which the sepals, petals, stamens, and pistils are attached.

The term **perianth** is applied to the calyx and corolla collectively. It is frequently used to describe flowers, such as the tulip, in which the two outer whorls, though present, are morphologically indistinguishable. The individual parts of such a perianth may be referred to as a **tepal.**

Sometimes individual flowers or compact clusters of flowers will have a whorl of small leaves or **bracts** standing close below them. Such a collection of bracts subtending flowers is called as **involucre.**

The parts of the flower in outline are as follows:

Receptacle

Calyx, consisting of sepals } Perianth
Corolla, consisting of petals }

Androecium, consisting of stamens
Gynoecium, consisting of carpels

Carpels and Stamens as Modified Leaves

The question arises, why are carpels and stamens considered modified leaves? One reason is because the early developmental stages of floral parts closely resemble those of leaves. The second, although circumstantial, concerns the shape of spore-bearing leaves in primitive angiosperms. If we were to examine the stamens of primitive flowers such as *Magnolia* or *Degeneria* (Fig. 15.2*A;* 15.3*D,E*), we would see that the shape of the stamen is very leaflike. In addition, the folded open carpels of primitive angiosperms are leaflike in appearance (Fig. 15.3*A*

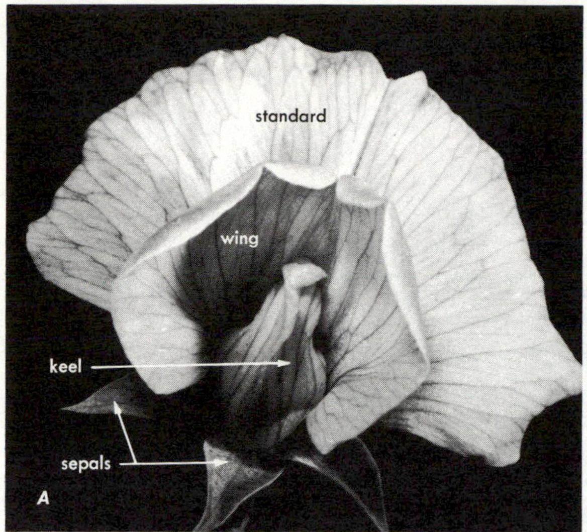

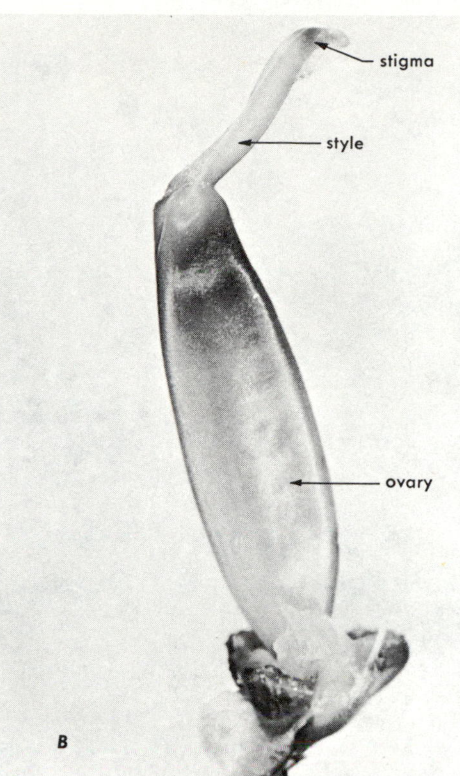

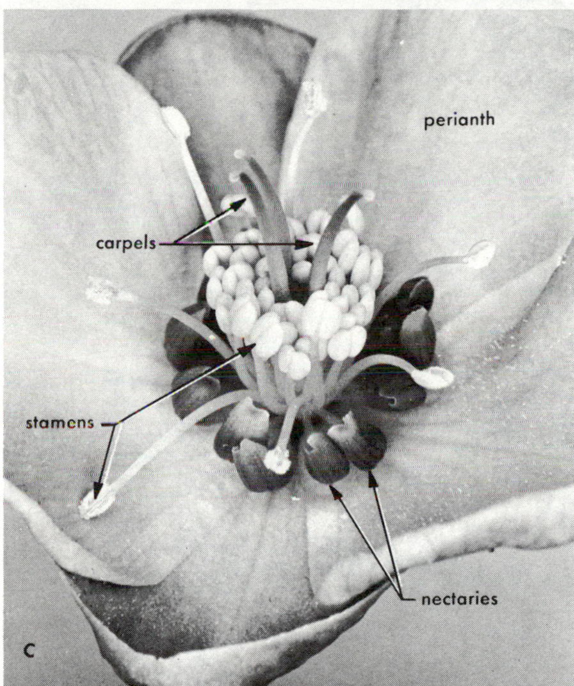

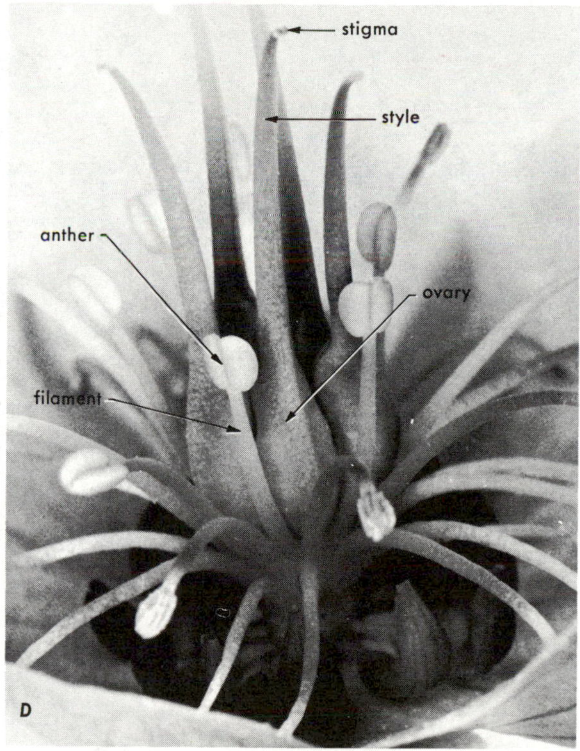

Figure 15.6
Flowers and essential organs. *A*, flower of garden pea (*Pisum sativum*). *B*, gynoecium of garden pea consisting of a single carpel. *C*, flower of Christmas rose (*Helleborus*) with stamens and carpels. *D*, maturing gynoecium of Christmas rose consisting of several loosely associated carpels. *E*, flower of *Cotyledon*. *F*, opened flower of *Cotyledon*, showing stamens adnate to corolla tube and five separate compact carpels. *G*, flower of *Tulipa*. *H*, enlarged view of stamens and compound pistil of *Tulipa*. *G*, ×½; *B*, ×2; *A*, *C*, *E*, *F*, *H*, ×1, *D*, ×2. (E-H on page 274).

to *C*). If a young pea pod, which is a single carpel, is opened carefully along its ventral suture (toward the axis), the margins may be folded back, showing the seeds, which have developed from ovules, along the margin of a leaflike carpel (Fig. 16.1*D*).

The relationship between carpels and stamens is even more striking in *Sterculia plantanifolla,* because in this plant there are five simple pistils united only by their stigmas (Fig. 15.4*A* to *C*). When these stigmas mature, they open to show five very leaflike carpels that bear seeds along their margins. Carpels may thus be compared with leaves. The evolutionary steps in carpel development still remain speculative.

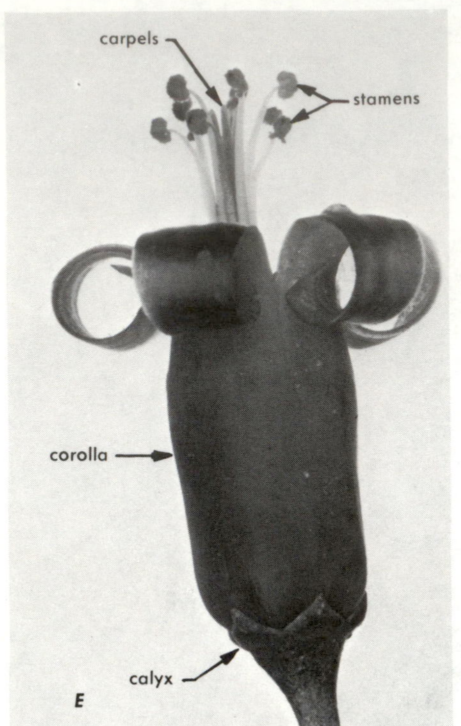

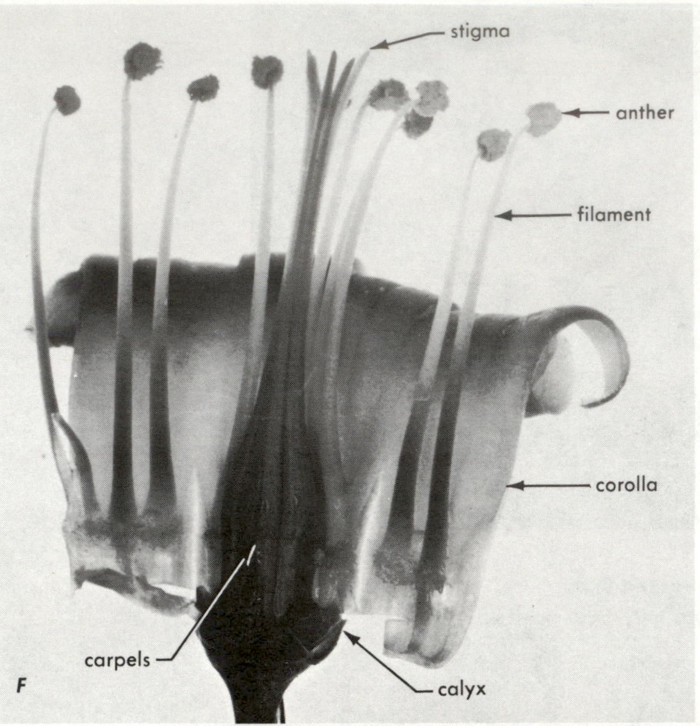

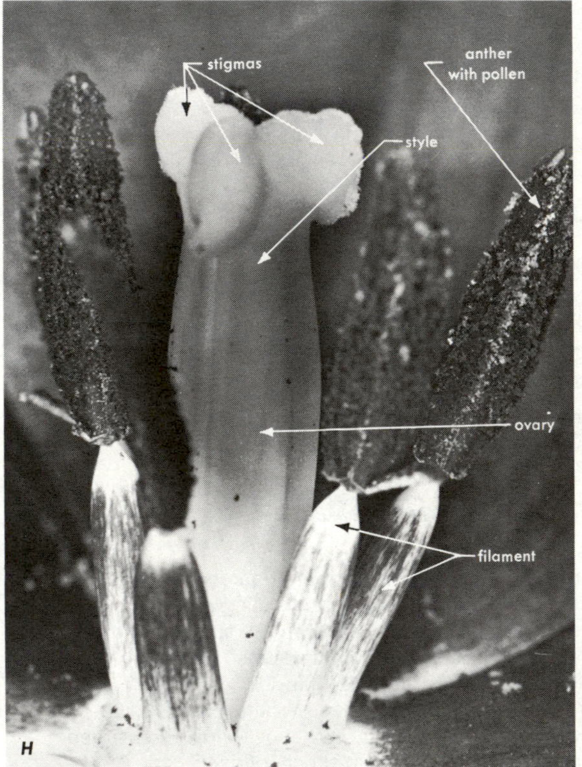

Figure 15.6 (continued)

VARIATIONS IN FLORAL STRUCTURE

General Description

Complete and Incomplete Flowers

The parts of a typical flower, sepals, petals, stamens, and carpels, are all attached to the receptacle. A flower with all four sets of floral leaves is said to be a **complete flower** (Fig. **15.5,** 15.6). An **incomplete flower** is one in which one or more of the four sets are lacking. For example, there are (*a*) flowers that have no perianth, (*b*) flowers that have a calyx but no corolla, (*c*) flowers with carpels but no stamens, and (*d*) flowers with stamens but no carpels. The flowers of the calla lily are examples of flowers without a perianth (Fig. 15.7*A*). In the goosefoot family, which includes spinach and beet, the flowers

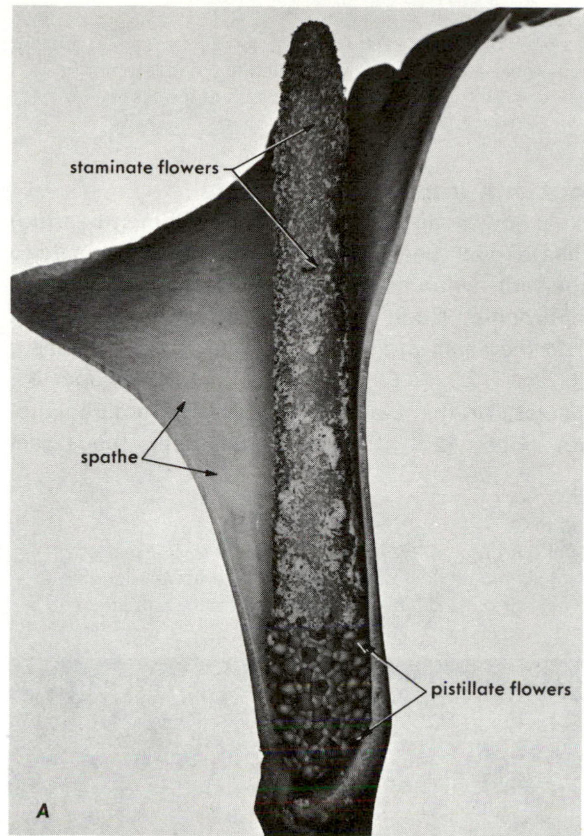

Figure 15.7
Flowers with perianth parts lacking. *A*, spathe of calla Lily (*Zantedeschia aethiopica*), sepals and calyx absent. *B*, flower of *Clematis*, petals absent and sepals prominent.

Figure 15.8
Imperfect flower. *A*, pistillate and *B*, staminate flowers of squash (*Cucurbita*), ×2.

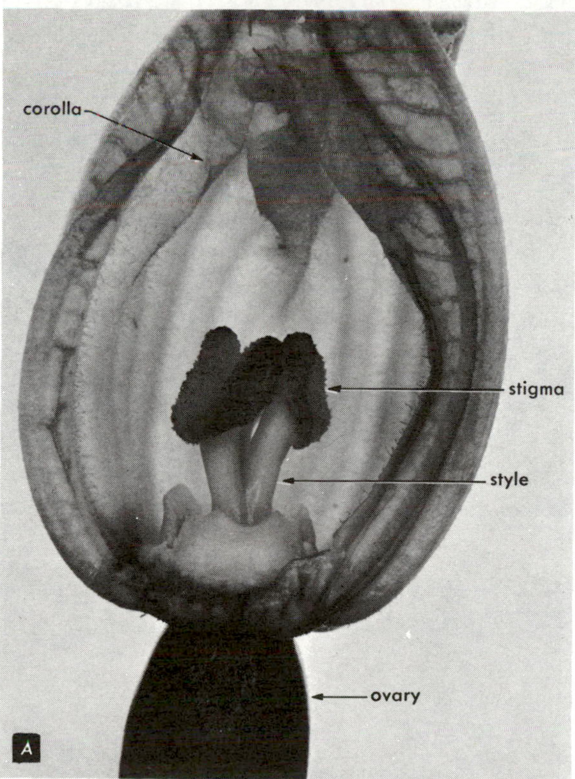

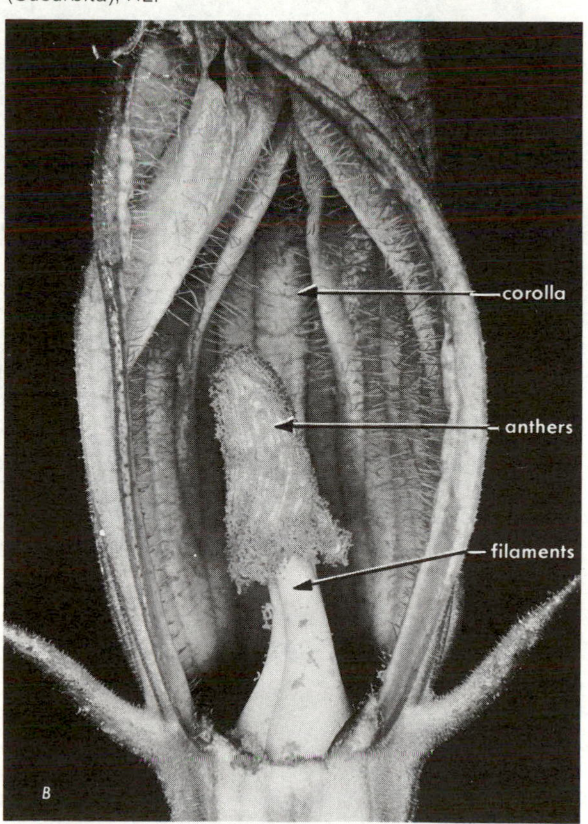

have greenish sepals but no petals; but in *Clematis* and a number of other plants, although there are no petals, the sepals are colored and petal-like (Fig. 15.7*B*). Unisexual flowers (with either carpels or stamens but not both) are fairly common in the plant kingdom. Examples are corn (*Zea mays*, Figs. 15.9, 15.10), oak, walnut (Fig. 15.11), willow (Fig. 15.12), poplar, squash (Fig. 15.8), hemp, asparagus, hop, and date palm (Fig. 15.13).

Figure 15.9
Inflorescences of corn (*Zea mays*). Left: long silks (styles) arising from the kernels (ovaries), many of which make up the ear (inflorescence). Right: stamen-bearing flowers of the tassel, ×1/6.

Perfect and Imperfect Flowers

Unisexual flowers are either **staminate** (stamen-bearing) or **pistillate** (pistil-bearing). Unisexual flowers are said to be **imperfect,** whereas bisexual flowers are **perfect** or **hermaphroditic.** When staminate and pistillate flowers occur *on the same individual plant,* as in corn, squash, walnut (Figs. 15.8, 15.9, 15.10, 15.11), and other species, the species, or the plant, is said to be **monoecious.** In corn (Figs. 15.9, 15.10), for example, the tassel

Figure 15.10
Flowers of corn (*Zea mays*). *A,* the tassel, ×1/5. *B,* exserted anthers of staminate flowers, ×1. *C,* pistillate flowers forming a very young ear, ×½. *D,* several pistillate flowers with attached styles, ×1.

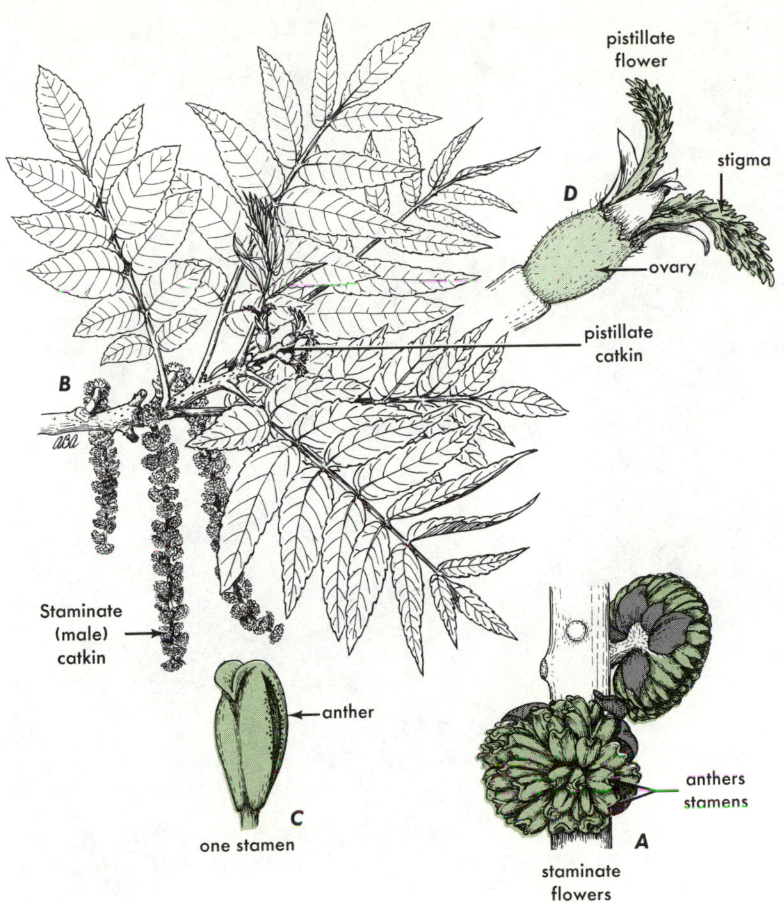

Figure 15.11
English walnut (*Juglans regia*) flowers. *A*, staminate flowers. *B*, staminate inflorescence (catkin). *C*, dehiscing anther. *D*, pistillate flower.

Figure 15.12
Willow (*Salix*) inflorescence and individual flowers. *A*, male catkin, ×1. *B*, single male flowers, ×15. *C*, single female flowers, ×15.

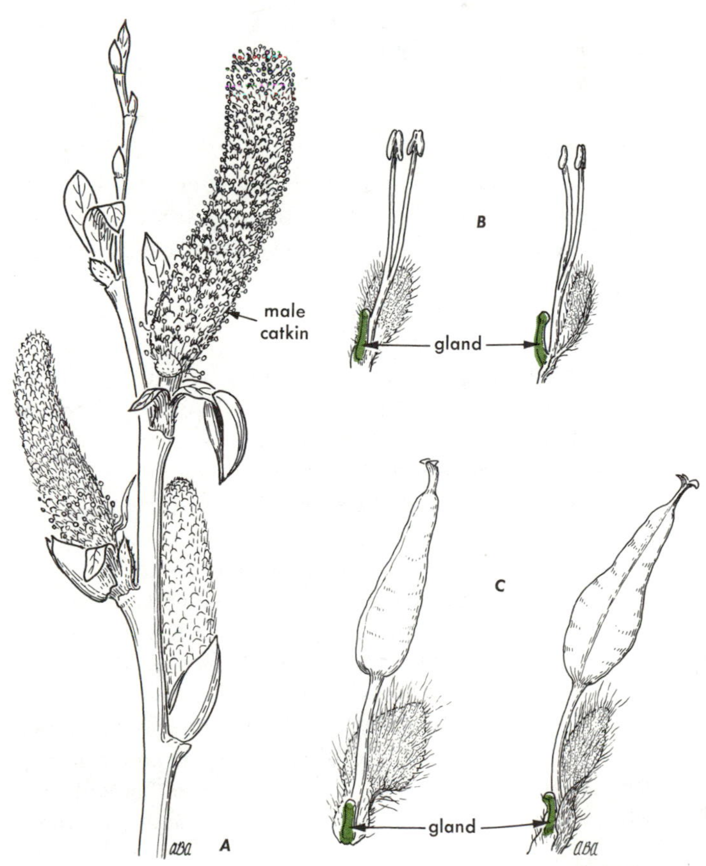

(borne at the top of the stalk) consist of a group of staminate flowers, and the young ear is a group of pistillate flowers. When staminate and pistillate flowers are borne *on separate individual plants,* as in asparagus, willow, and many other species, the species, or the plant, is said to be **dioecious.** For example, in a commercial asparagus field, approximately half the individual plants bear only staminate flowers, and half bear only pistillate flowers. In such circumstances, we speak of staminate or "male" plants and of pistillate or "female" plants. Only pistillate (female) plants bear fruit. Another example of dioecism is the date palm (Fig. 15.13). Some individual palms are staminate; others are pistillate. The edible fruit (date) is produced only by pistillate palms. In such instances, there will be no fruit production unless staminate plants and pistillate plants are growing near enough together for the pollen to be transferred. In commercial plantings of dates, most of the individual palms are pistillate, that is, fruit-bearing. Dates are propagated by offshoots, which arise chiefly near the base of the

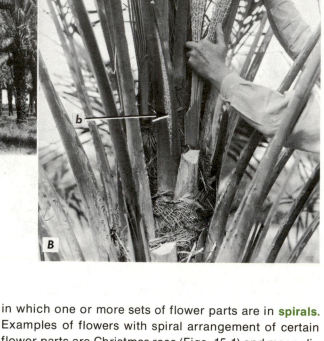

Figure 15.13
Date palms (*Phoenix dactylifera*). *A*, commercial grove. *B*, artificial pollination of date flowers: *a* strands of male flowers being placed in the center of the female cluster; *b*, freshly opened flower cluster ready for pollination; *c*, flower cluster after pollination, with the strands tied to hold male flowers in place. (*A*, courtesy of California Date Administrative Committee. *B*, courtesy of Nixon.)

stem in the early years of the palm's life. Offshoots from a staminate palm grow into staminate palms, and offshoots from a pistillate palm into pistillate palms. In a commercial date garden, it is economically desirable to have as many pistillate or fruit-bearing individuals as possible, and this is secured by vegetative propagation. A relatively few staminate palms are scattered throughout the garden—only enough to supply pollen. Pollen may be carried naturally by wind or it may be collected and dusted by hand on the pistils (Fig. 15.13*B*)

Some species of plants produce three kinds of flowers: staminate, pistillate, and perfect (hermaphroditic). For example, in red maple, one may find both unisexual and bisexual flowers in the same flower cluster.

Often, staminate flowers have rudiments of a pistil, and pistillate flowers have rudiments of stamens.

Whorled and Spiral Arrangements

In most flowers, sepals, petals, stamens, and carpels are in **whorls** or circles on the axis of the flower. For example, in *Sedum* **(Fig. 15.5*B*)** and *Cotyledon* the sepals arise in a whorl at the outer lowest level on the receptacle; slightly above this whorl are the petals; above these is a circle of stamens attached to the petals; and in the center are two united carpels. In contrast with this whorled arrangement of flower parts is the arrangement in which one or more sets of flower parts are in **spirals.** Examples of flowers with spiral arrangement of certain flower parts are Christmas rose (Figs. 15.1) and magnolia (Fig. 15.2).

Floral Symmetry

In many flowers such as those of *Liriodendron* **(Fig. 15.5*A*)**, lily (Fig. 15.2*B*), and *Sedum* **(Fig. 15.5*B*)**, the corolla is made up of petals of similar shape that radiate from the center of the flower and are equidistant from each other. Other flower parts have a similar arrangement. Such flowers are said to be **regular.** In these cases, even though there may be an uneven number of parts in the perianth, any line (**Figs. 15.5*B*,** 15.14) drawn through the center of the flower will divide the flower into two similar halves. They may be exact duplicates or mirror images of each other.

Irregular flowers, such as sweet pea (Fig. 15.6*A*) and mints (Fig. 15.14*B*), have whorls either (*a*) with dissimilar flower parts, (*b*) with parts that do not radiate from the center, or (*c*) not equidistant from one another. In most of these flowers, only one line will divide the flower in equal halves; the halves are usually mirror images of each other. Some flowers (bleeding heart, Dutchman's breeches), though irregular, may be bisected by any number of lines into similar mirror images.

Figure 15.14
Floral symmetry. *A*, regular flower of columbine (*Aquilegia*), ×1. *B*, irregular flower of a mint (*Salvia*), ×2.

The irregular bean or pea flower has a corolla composed of the following (Fig. 15.6*A*): one broad conspicuous petal (the **banner** or **standard**); two narrower petals (**wings**), one on each side; and, opposite the banner, two smaller petals that are united along their edges to form the **keel.** Other examples of irregular flowers are mints, violets, orchids, and snapdragons.

Union of Flower Parts

In the flower of *Sedum,* illustrated in **Fig. 15.5*B*** all parts of the flower are separate and distinct; that is, each sepal, petal, stamen, and carpel is attached at its base to the receptacle. In many flowers, however, members of one or more whorls are to some degree united with one another, or are attached to members of other whorls. A precise terminology useful to plant taxonomists has been developed to indicate these conditions. Union with other members of a given whorl is termed **coalescence** (or **connation**). Partial or complete union of the sepals along their edges occurs in the flowers of mints (Fig. 15.14*B*), violets, evening primroses, peas (Fig. 15.6*A*), and many other plants. This condition is referred to as **synsepaly.** Petals may also be attached to one another (Figs. 15.6*F*, **15.16*C*** page 290) and they are said to be **sympetalous** (the noun would be **sympetaly**).

Stamens may coalesce, as is seen in the legume, squash, and cotton families (Figs. 2.4, 15.8*B*). This coalescence is **synandry.** In the cotton flower, the filaments of many stamens are united to form a sheath, whereas the anthers are separate. In the flower of orange, there are 20 to 60 stamens, united at their bases to form groups. And in the thistle family the anthers are united into a tube, whereas the filaments are distinct.

When the carpels of the gynoecium coalesce, the situation is called **syncarpy,** and one may say that the pistil is compound. This condition occurs in many families of plants and may be seen to advantage such flowers as those of lily (Fig. 15.3*B*), tulip (Figs. **15.5;** 15.6*H*), and squash (Fig. 15.8).

When parts of a flower have not joined, the prefix **apo-** (separate) may be used to describe the flower; for example, **apopetalous.** These situations are outlined below:

Whorl	No coalescene	Coalescence
Sepals	Aposepalous	Synsepalous, synsepaly
Petals	Apopetalous	Sympetalous, sympetaly
Stamens	Apoandrous	Synandrous, synandry
Carpels	Apocarpous	Syncarpous, syncarpy

Union of members of two different whorls also occurs, **adnation.** For example, in such flowers as *Cotyledon* (Fig. 15.6*F*), almond **(Fig. 15.16*B*),** snapdragons, and honeysuckle, the stamens are attached to the corolla rather than to the receptacle. In these cases, the stamens are said to be adnate to the corolla.

Elevation of Flower Parts

In flowers of magnolia (Fig. 15.2*A*), lily (Fig. 15.2*B*), and tulip (**Fig. 15.16*A*,** page 290, Fig. 15.15*A*), the receptacle is convex or conical and the different flower parts are arranged one above another. They occur in the following order (beginning with the lowest): sepals, petals, stamens, and carpels (Figs. 15.1, 15.15*A*). The gynoecium is thus situated on the receptacle above the points of origin of the perianth parts and androecium. An ovary in this position is said to be **superior.** In the daffodil (Fig. 15.15*C*) or sunflower (Fig. 15.23) the ovary appears to be below the apparent points of attachment of the perianth parts and the stamens. This is an **inferior** ovary. With an inferior ovary, the anatomy of the flower parts indicate that the lower portions of the three outer whorls, calyx, corolla, and androecium have fused to form a tube or **hypanthium** (Fig. 15.15*B*). The ovary in the daffodil is completely adnate with the hypanthium.

In a flower with a superior ovary, sepals, petals, and

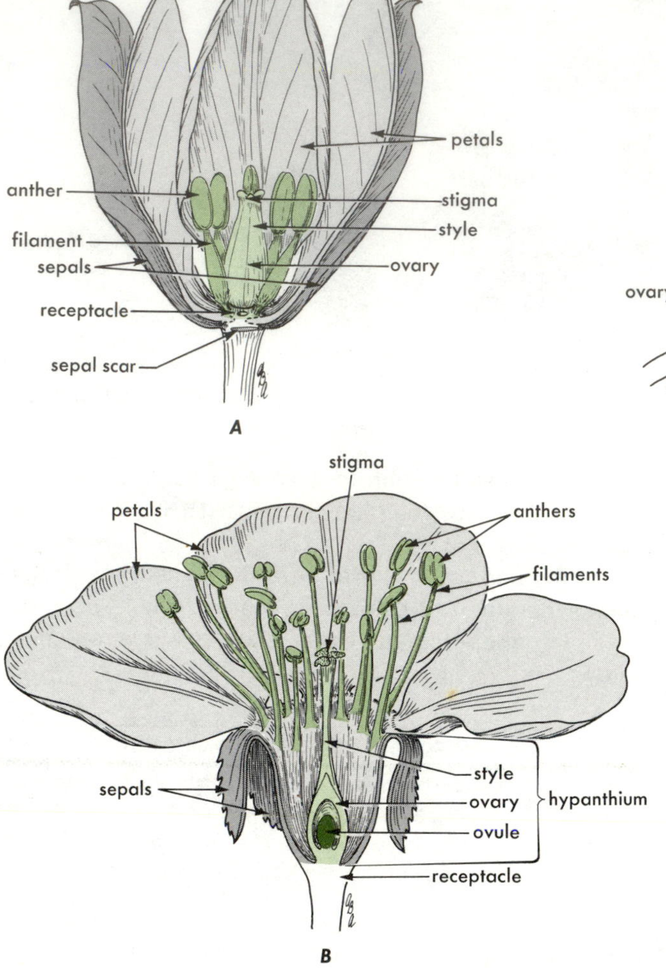

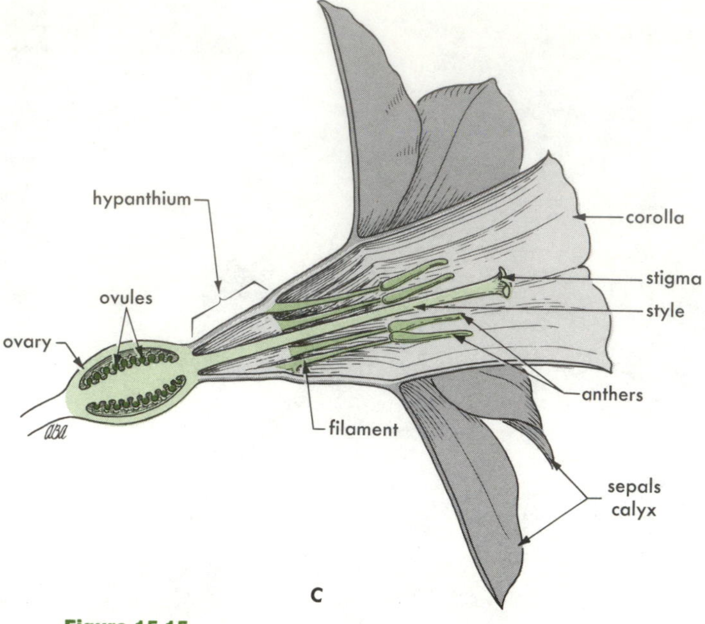

Figure 15.15
Diagram of elevation of floral parts. *A*, hypogyny (*Tulipa*). *B*, perigyny in cherry flower (*Prunus avium*). *C*, epigyny in daffodil flower (*Narcissus pseudonarcissus*).

stamens arise from the outer, lower portion of the concave receptacle, below the point of origin of the carpels (Fig. 15.1). The perianth and stamens are **hypogynous,** or with reference to the three outer whorls, a condition of **hypogyny** exist (Figs. 15.15*A*, **15.16*A***). In a flower with an inferior ovary, the perianth and stamens appear to arise from the top of the ovary, and they are **epigynous** (Figs. 15.15*C*, **15.16*C***).

In some flowers such as the plum and apricot, the hypanthium does not become adnate to the ovary. The perianth parts and stamens arising from the rim of cuplike hypanthium around the ovary are **perigynous** (Figs. 15.15*B*, **15.16*B***).

Inflorescences

In most flowering plants, flowers are borne in clusters or groups. Morphologically, an **inflorescence** is a flower-bearing branch or system of branches. In manuals for the identification of flowering plants, there are many different terms descriptive of various kinds of inflorescence. Only the most common ones are discussed and illustrated here (Fig. 15.17).

A very simple type of inflorescence may be found in such plants as currant (*Ribes*) and radish (*Raphanus*). It is called a **raceme.** In this type, the main axis has short branches, each of which terminates in a flower. Each flower is on a short branch stem called the **pedicel.** The main axis of a raceme continues to grow in length more or less indefinitely; the apical meristem persists. Primordia of leaves arise in the usual manner along the margin of this apical meristem, and in the axil of each leaf a flower is borne. The oldest flowers are at the base of the inflorescence and the youngest are at the apex.

In a simple raceme, the flowers are on pedicels that are about equal in length. In a **spike** (Fig. 15.17), the main axis of the inflorescence is elongated, but the flowers, each in the axil of a bract, are sessile (without a pedicel). The **catkin** is a spike that usually bears only pistillate or staminate flowers. The inflorescence as a whole is shed later. Examples are willow and walnut (Figs. 15.11 and 15.12).

In all the inflorescences just mentioned, the flowering axis is elongated. If it is short, flowers appear to be arising umbrellalike from approximately the same level. An inflorescence of this kind, in which pedicels are of nearly equal length, is called an **umbel** (Fig. 15.17). The onion (*Allium*) is a good example. The **head** is an inflorescence in which the flowers are sessile and crowded together on a very short axis. Members of the family Asteraceae (Compositae), including thistle (*Cirsium*) and sunflower, have this type of inflorescence (Figs. 15.17 and 15.21).

Inflorescences of the raceme type may be compound, that is, branched. A branched raceme is called a **panicle** (Fig. 15.17). Compound spikes occur in wheat, rye, and certain other grasses.

In contrast to the raceme types of inflorescences,

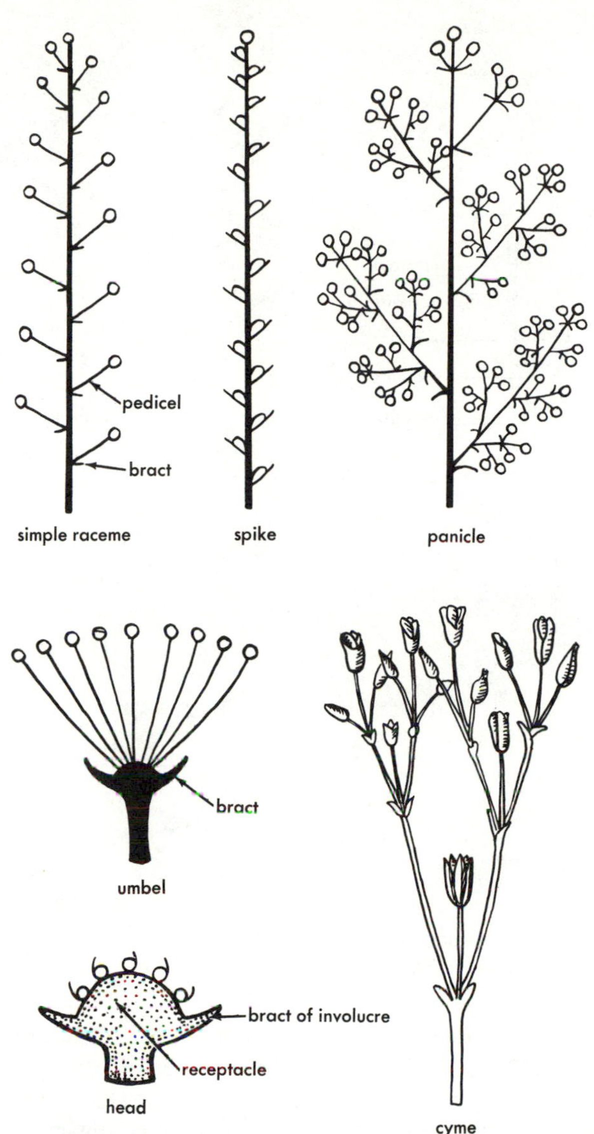

Figure 15.17
Types of inflorescences.

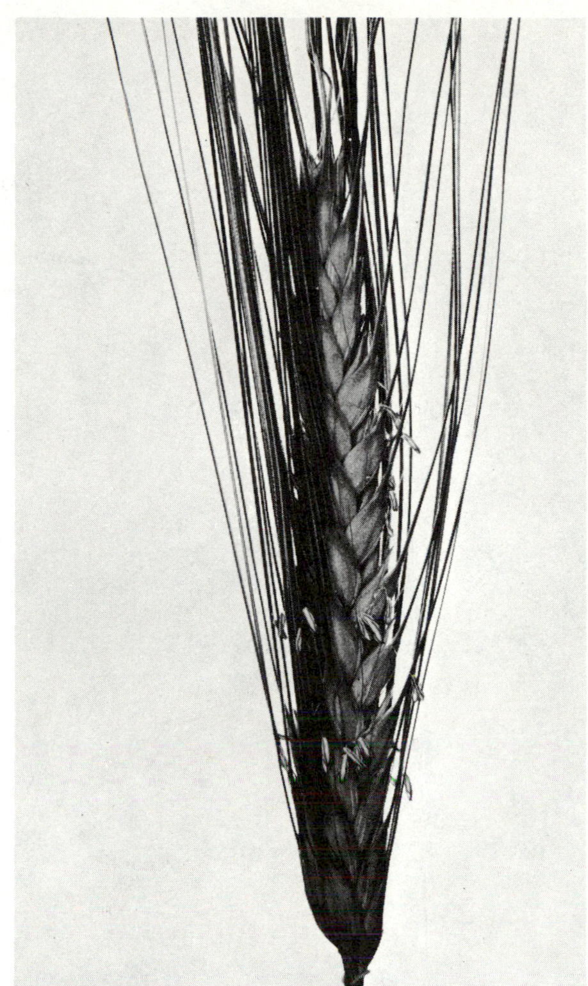

Figure 15.18
Spike of barley (*Hordeum vulgare*), ×1.

just described, is the **cyme** (Fig. 15.17). In the cyme, the apex of the main axis produces a flower that involves the entire apical meristem; hence, that particular axis ceases to elongate. Other flowers arise on lateral branches farther down the axis of the inflorescence and, thus, usually the youngest of the flowers in any cluster occurs, farthest from the tip of the main stalk. The flower cluster of chickweed (*Cerastium*) is an example of the cyme.

We have discussed in some detail the various aspects of flower structure as it occurs throughout the groups of flowering plants. We shall now proceed to describe the specialized flowers of two very important families. The Poaceae (Gramineae) (grasses) and the Asteraceae (the sunflower family). Several members of the Poaceae are of great economic importance, supplying our single largest source of food directly in the form of flour and indirectly as feed for cattle, sheep, and hogs. The Asteraceae family, while not a major source of food products, is made up of a large number of species with a wide distribution and colorful flowers; zinnias, marigolds, dahlias (Fig. 15.20), dandelions, and many similar flower types belong to this family.

The Grass Flower

Grass flowers generally grow in a **spike** (Fig. 15.17, 15.18), as in barley, or are loosely grouped in a **panicle** (Fig. 15.19), typical of oats. In all instances, individual small flowers or florets are associated in groups known as **spikelets** (Fig. 15.19*B*). Each spikelet is separated from its neighbors by two small modified leaves or bracts known as **glumes.** They are found at the base of the spikelet and in some cases (wheat, oats, barley), they may fairly well enclose it. A separation of the glumes reveals the individual florets of the spikelet attached to a slender stalk, the **rachilla.**

Each floret is in turn protected by two additional bracts, the **lemma** and the **palea** (Fig. 15.19*C*). In wheat, barley, oats, and other grains, the lemma is large and may have a long slender **awn** attached to it. The palea is

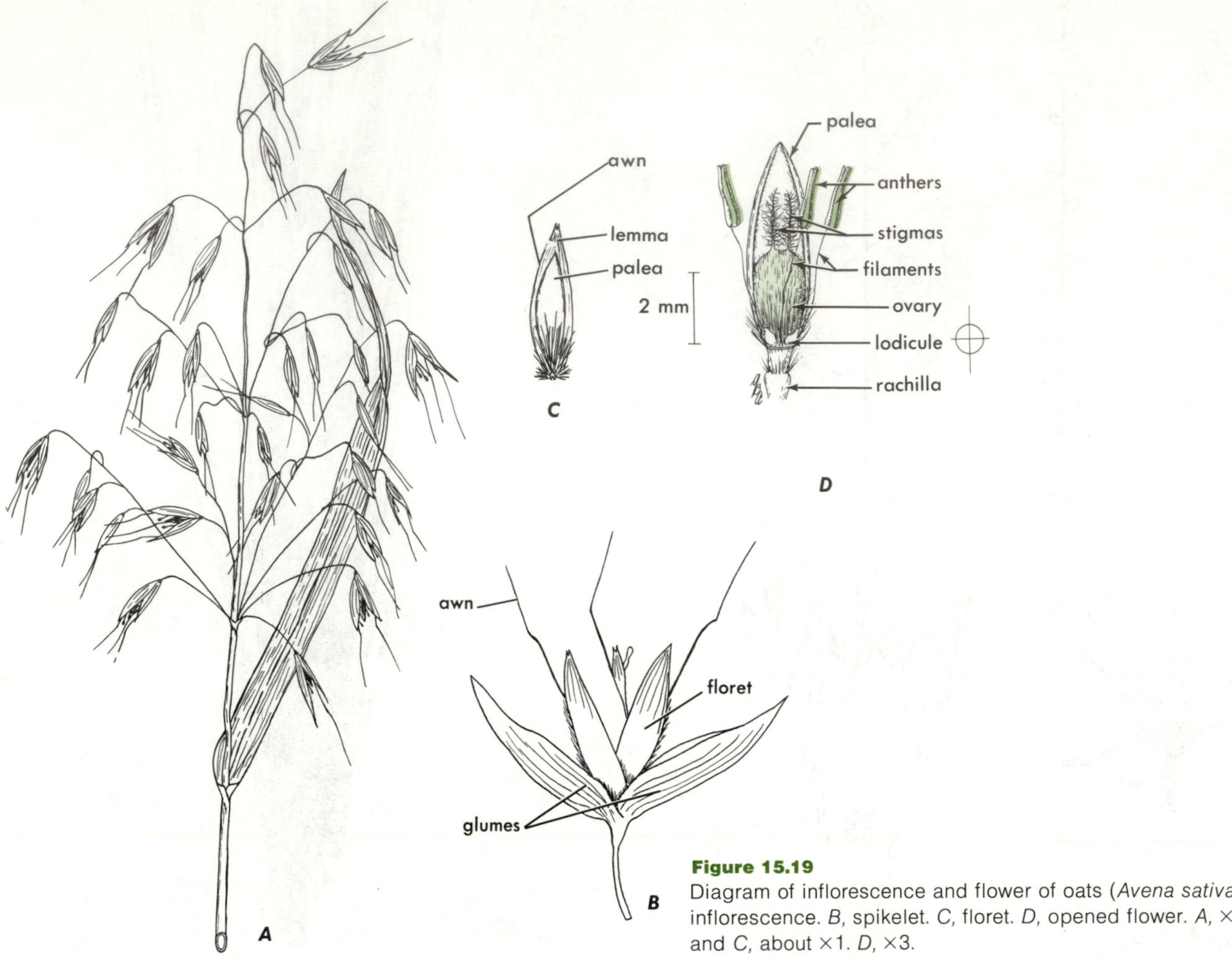

Figure 15.19
Diagram of inflorescence and flower of oats (*Avena sativa*). *A*, inflorescence. *B*, spikelet. *C*, floret. *D*, opened flower. *A*, ×1/10; *B* and *C*, about ×1. *D*, ×3.

small and frequently enclosed by the lemma. The essential parts consist of the androecium, composed of three stamens, and the gynoecium, composed of two fused carpels (Fig. 15.19*D*). When flowers are mature, the palea and lemma separate slightly, exposing two feathery stigmas and allowing the anthers to hang free from the spikelet on greatly lengthened filaments (Figs. 15.18, 15.10*B*). At the base of the floral parts, two small protuberances, the **lodicules,** may be distinguished. These are thought to be greatly modified perianth parts and are supposed to function in the separation of lemma and palea at pollination time (Fig. 15.19*D*).

In corn (Fig. 15.10), the flowers are imperfect, with staminate flowers grouped in the tassel, and pistillate flowers eventually developing into the ear. The peculiar and very important difference between corn flowers and those of other grasses is the absence of bracts of any sort around pistillate flowers.

The Composite Flower

As the name implies, the composite flower is a group of small flowers arranged to give the appearance of a single typical flower. Several types of composite flowers are shown in Fig. 15.20. This characteristic grouping of flowers is called a **head.** In dahlia (Fig. 15.20*B*), sunflower (Figs. 15.21*B*, 15.22), and many other composite flowers, there are two distinct types of individual flowers, **ray flowers** and **disk flowers.** Some species may, however, have all ray flowers (cactus dahlia, Fig. 15.20*A*), and others all disk flowers (globe thistle, Fig. 15.20*C*). Ray flowers are frequently sterile or pistillate.

The disk flower is generally complete. It is epigynous; the small tubular corolla arises from the top of the inferior ovary (Figs. 15.21, 15.22, 15.23). There are two carpels, as evidenced by the forked stigma (Figs. 15.22*C*, 15.23), which, when receptive to pollen, protrude from the corolla tube. The stamens, five in number, have separate filaments, but their anthers are coalesced to form a tube around the style. In sunflower, stamens usually mature before the stigma and protrude from the corolla tube. These details are shown in Figs. 15.21 and 15.23*A*. Referring to the figures, note that flowers on the periphery of the flower head have been pollinated, and both stigmas and stamens have withered. Adjacent flowers show protruding stigmas dusted with pollen. Next come several rows of dark anthers capped by white masses of pollen (Fig. 15.22*C*). The flowers in the center have not yet opened. A calyx is not present as such, but many composite flowers are surrounded by a conspicu-

Figure 15.20
Three types of composite flowers. *A,* cactus dahlia (*Dahlia juarezii*), all ray flowers. *B,* dahlia (*Dahlia pinnata*), both ray and disk flowers. *C,* globe thistle (*Echinops exaltatus*), all disk flowers, ×½.

ous tuft of hairs or otherwise modified calyx parts. This is known as **pappus.** In sunflower, these are simply two small sepal-like structures (Fig. 15.23). In dandelion, the pappus is feathery and aids in the dispersal of the fruit.

Two kinds of **floral bracts** sometimes occur in composite flowers. The whole head may be surrounded by green **involucral bracts.** Each individual flower may have its own **receptacular bract** arising from the receptacle near the base of the flower (Fig. 15.21*B*)

THE DEVELOPMENT OF THE FLOWER

As noted previously, the floral apex loses its ability to elongate, and all of it eventually differentiates into floral leaves. Moreover, the floral leaves are very much crowded together, not being separated by long internodes (as are foliage leaves), and they lack axillary bud primordia. Compare **Fig. 7.7** page, 117 which shows the shoot tip of bean, with Fig. 15.24*A,* which is a floral apex of *Ranunculus* (buttercup). As a rule the sepals are in a whorl; that is, they arise on the stem axis at approxi-

Figure 15.21
The composite inflorescence. *A,* head of golden glow (*Rudbeckia laciniata*), ×½. *B,* median section of head of sunflower (*Helianthus annuus*), ×1. In all figures, fertilized flowers are outermost, the stigmas reflexed; the next inner row shows exserted recurved stigmas; next lies a circle of flowers with stamens ready to shed pollen; and in the center are the young unopened flowers of the head.

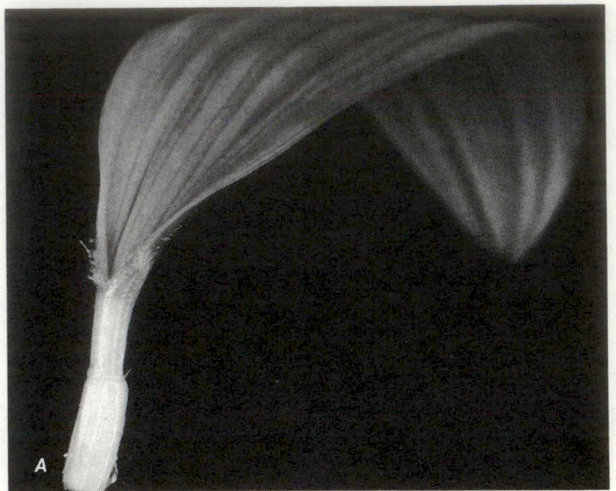

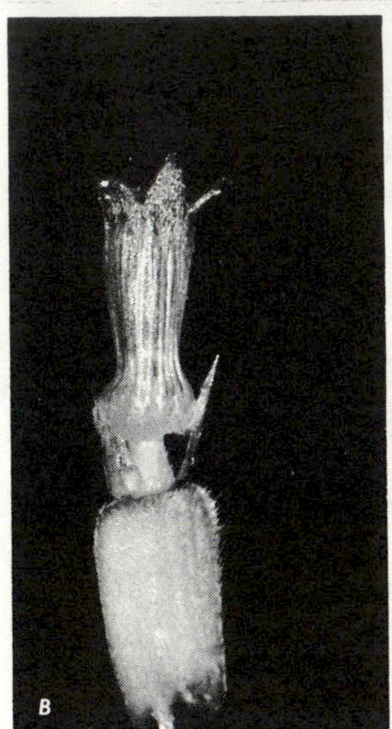

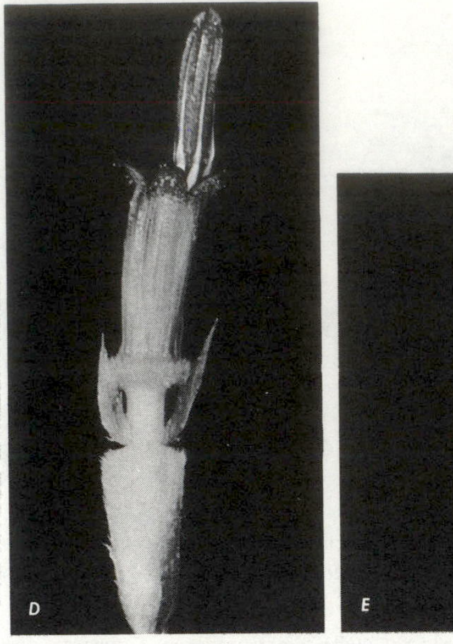

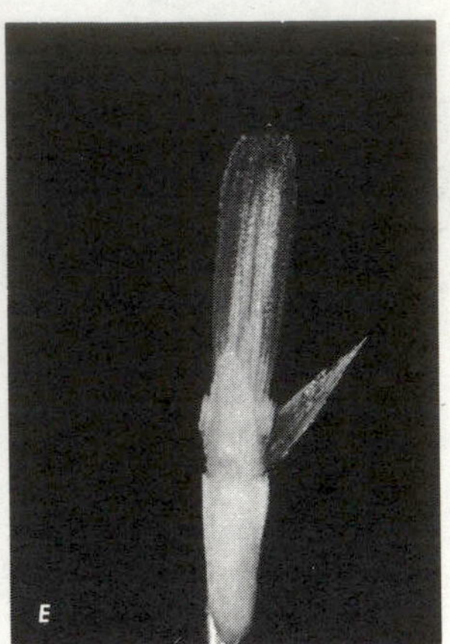

Figure 15.22
Individual flowers of a sunflower head (*Helianthus annuus*). *A*, a ray flower; *B* through *E*, disk flowers. *B*, flower after fertilization, both stigmas and stamens have withered. *C*, flower with stigmas exserted and recurved. *D*, flower with staminal column exserted. *E*, unopened flower, ×2.

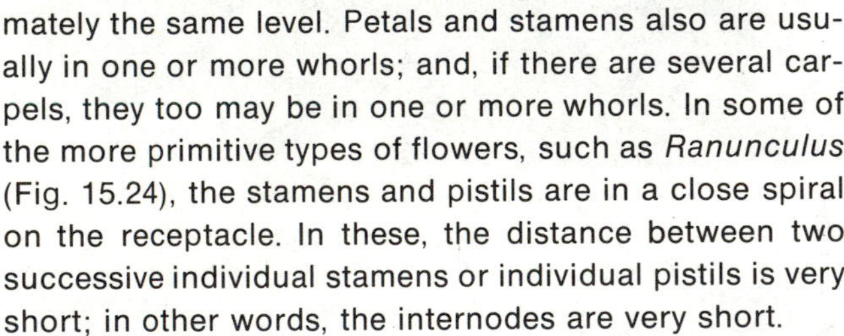

mately the same level. Petals and stamens also are usually in one or more whorls; and, if there are several carpels, they too may be in one or more whorls. In some of the more primitive types of flowers, such as *Ranunculus* (Fig. 15.24), the stamens and pistils are in a close spiral on the receptacle. In these, the distance between two successive individual stamens or individual pistils is very short; in other words, the internodes are very short.

The development of lettuce flowers (*Lactuca sativa*) illustrates, in a general way, some of the steps of floral development (Figs. 15.24*B*, *C*, *D*, *E*). The floral apex of lettuce, a composite flower, supports many flower primordia (Fig. 15.24*B*). Examination of these primordia shows that petals appear first and rise above the developing stamens and carpels (Fig. 15.24*C*). Lettuce is epigynous, and the two carpels form a depression in the center of the flower primordium (Fig. 15.24*C*, *D*). However, regardless of flower type, the carpels always originate as hollow flasks in the center of the flower primor-

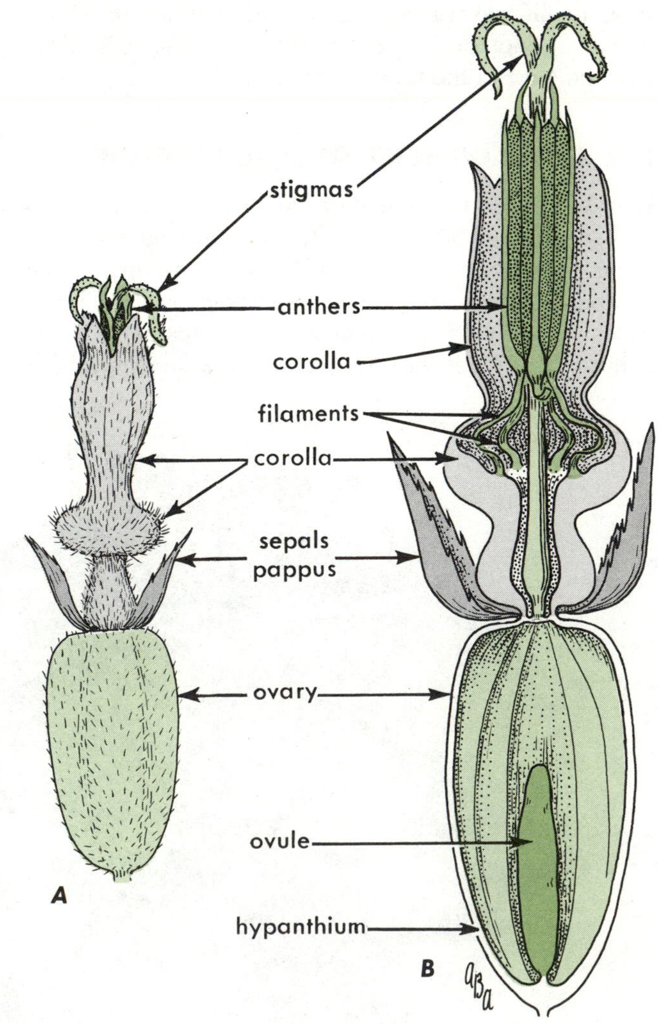

Figure 15.23
Disk flowers from a sunflower head. The pollen has been shed, and stigmas are raised and exserted. *A*, external view ×2½. *B*, enlarged longitudinal cross section, ×4.

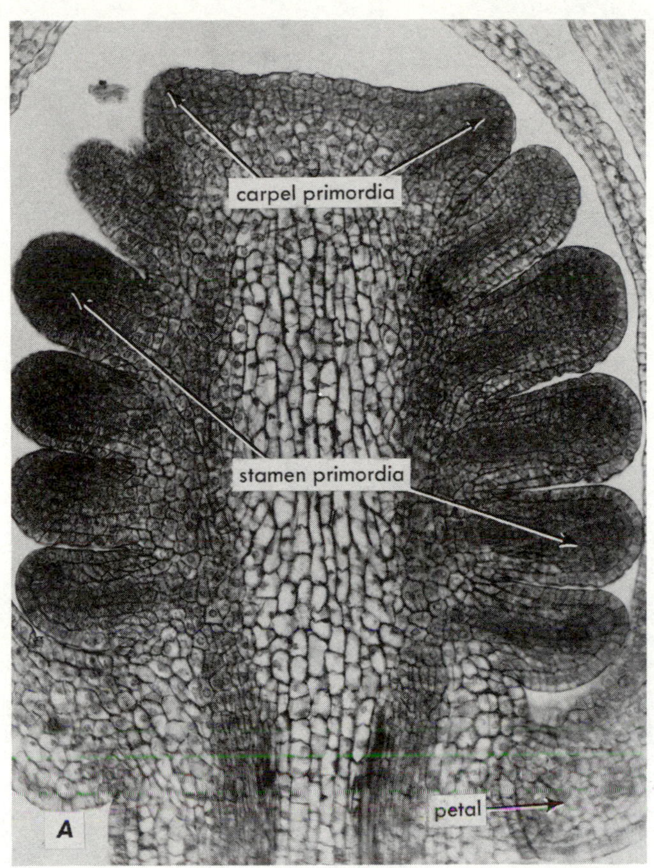

dia. As development continues, the petals and stamens elongate. The carpels also elongate, and their margins grow toward each other, finally fusing. The result is the hollow ovary which, at this early stage, may already contain the primordium of an ovule (Fig. 15.24*E*). Upward growth of the carpel margins continues and results in the style and stigma.

ANGIOSPERM LIFE CYCLE

The floral organs essential for sexual reproduction are the stamens (androecium) and carpels (gynoecium) (Figs. 15.1, **15.26*A*,** page 291). The perianth, composed of calyx and corolla, is a protective covering of the stamens and pistils and when it is present it also serves to attract and guide the movements of insects and other pollinators.

The Androecium

Each stamen consists of an anther supported by a filament (Fig. 15.25*A*). The anther usually consists of four elongated and connected lobes called pollen sacs. Early in the development of the anther, the pollen sacs each contain a mass of dividing cells called **microsporocytes** (microspore mother cells) (Fig. 15.25*B*, **15.26*B***). Each microsporocyte divides by meiosis to form four haploid (1*n*) microspores **(Fig. 15.26*C,D,E*).** The nucleus of each

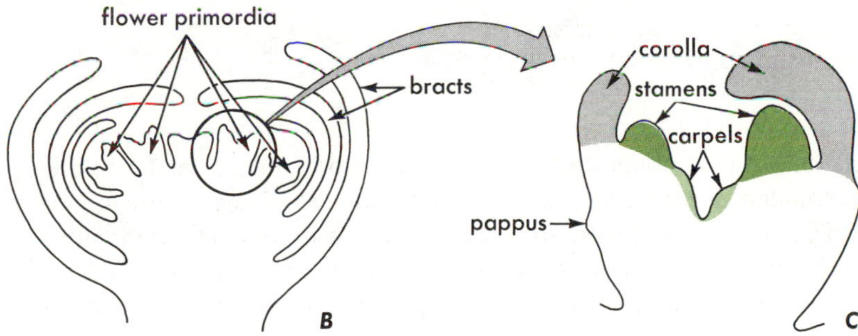

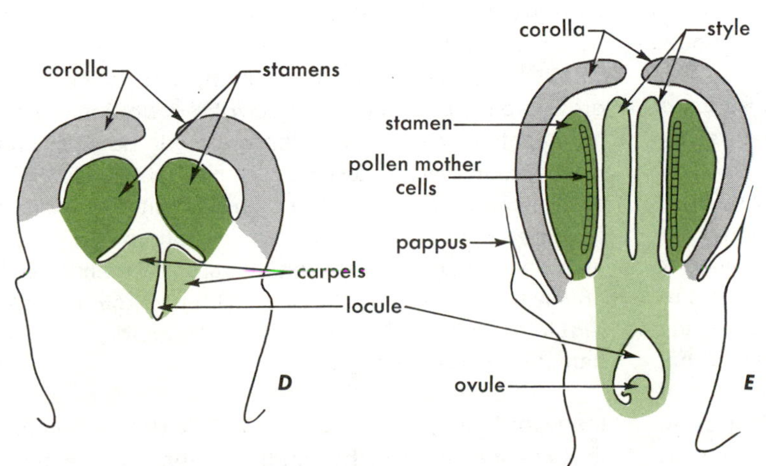

Figure 15.24
Development of the floral apex. *A*, photomicrograph of floral apex of *Ranunculus* showing numerous stamen primordia and the earliest carpel primordia, ×200. *B* through *E*, diagram of the developing floral apex of lettuce (*Lactuca sativa*). *B*, young head showing several flower primordia. *C*, young flower showing carpel, stamen, and corolla primordia. *D*, slightly more advanced stage showing carpel just prior to fusion of the margins. *E*, carpel margins have fused, forming the hollow ovary in which occurs a primordium of an ovule; corolla tube, stamens, and stigma are in early stages of differentiation. (*A*, courtesy of S. Tepfer; *B* through *E* redrawn after H. A. Jones, *Hilgardia* **2,** 454.)

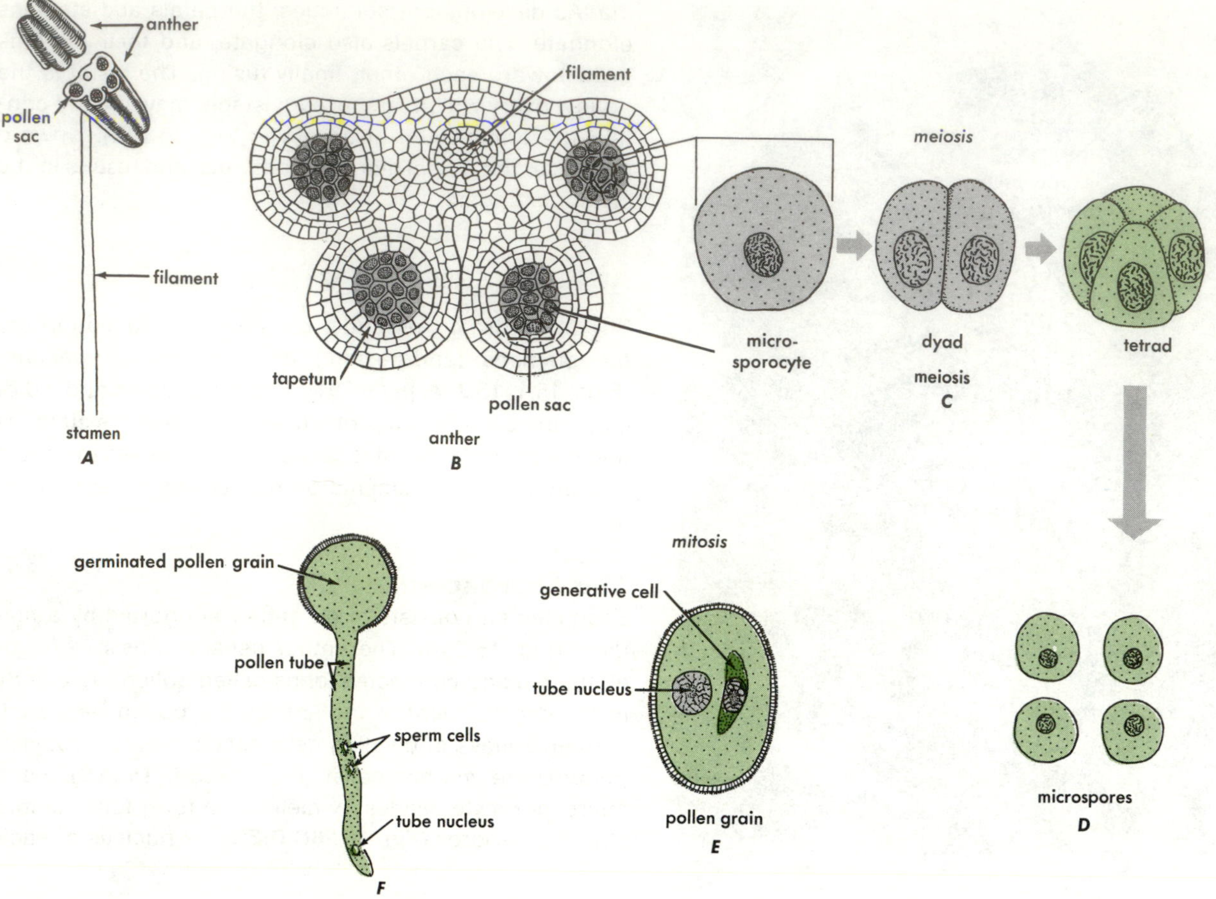

Figure 15.25
Development of pollen from a microsporocyte to the pollen grain. *A*, stamen. *B*, cross section of anther. *C*, development of tetrad of cells from the microsporocyte by meiosis. *D*, four microspores. *E*, pollen grain, *F*, germination of pollen grain.

microspore then divides by mitosis to form a two-celled **pollen grain** that contains a **tube cell** as well as a smaller **generative cell** (Fig. 15.25*E*, **15.26*E***). The role of this two-cell, haploid, **male gametophyte** (gamete-producing plant) is to produce sperm nuclei necessary later for fertilization.

The pollen grain is surrounded by an elaborate, ornate cell wall. The pattern of the pollen wall can be intricate and beautiful; at the same time it is useful both to the plant and to plant taxonomists. The wall is composed of an extremely hard material called sporopollenin. This material is so hard that pollen grains many thousands of years old still retain their same pollen wall texture and pattern (see Chapter 31).

Remember that floral structures are relatively unaffected by environmental changes. Consequently, the form of such structures as pollen grains can be used to make comparisons between plant species. Plant taxonomists study pollen wall patterns and are able to make intelligent guesses about relationships in their evolution.

After the pollen grains are mature, the anther wall splits open and the pollen are shed **(Fig. 15.26*F*).** In some manner (which will be discussed later), the pollen are transported to the stigma of adjacent or distant flowers. This process is called **pollination.** Once on the stigma, if the pollen and the stigma are genetically compatible, it will germinate to form an elongate **pollen tube** (Fig. 15.25*F*). The ornate pattern of the pollen grain wall serves as the means for receptive stigmas to recognize compatible pollen. The troughs and ridges of the pollen wall (Fig. 15.27) apparently contain proteins in specific patterns and concentrations. If the patterns correlate with those of the stigma, metabolic events are triggered within the pollen grain to stimulate the pollen tube to grow into the pistil's tissue. Lack of recognition indicates incompatibility and the pollen grain will not germinate. Recognition systems also exist at different levels. For example, some compatibility/incompatibility reactions are triggered after pollen tube germination. In this case, an incompatibility reaction would actually inhibit further pollen tube growth in the style. Other reaction systems regulate pollen tube entry into the embryo sac and gamete release.

The elaborate cell walls of pollen grains present many intriguing problems (Fig. 15.27). Their intricate details are species-specific and their chemical constitution

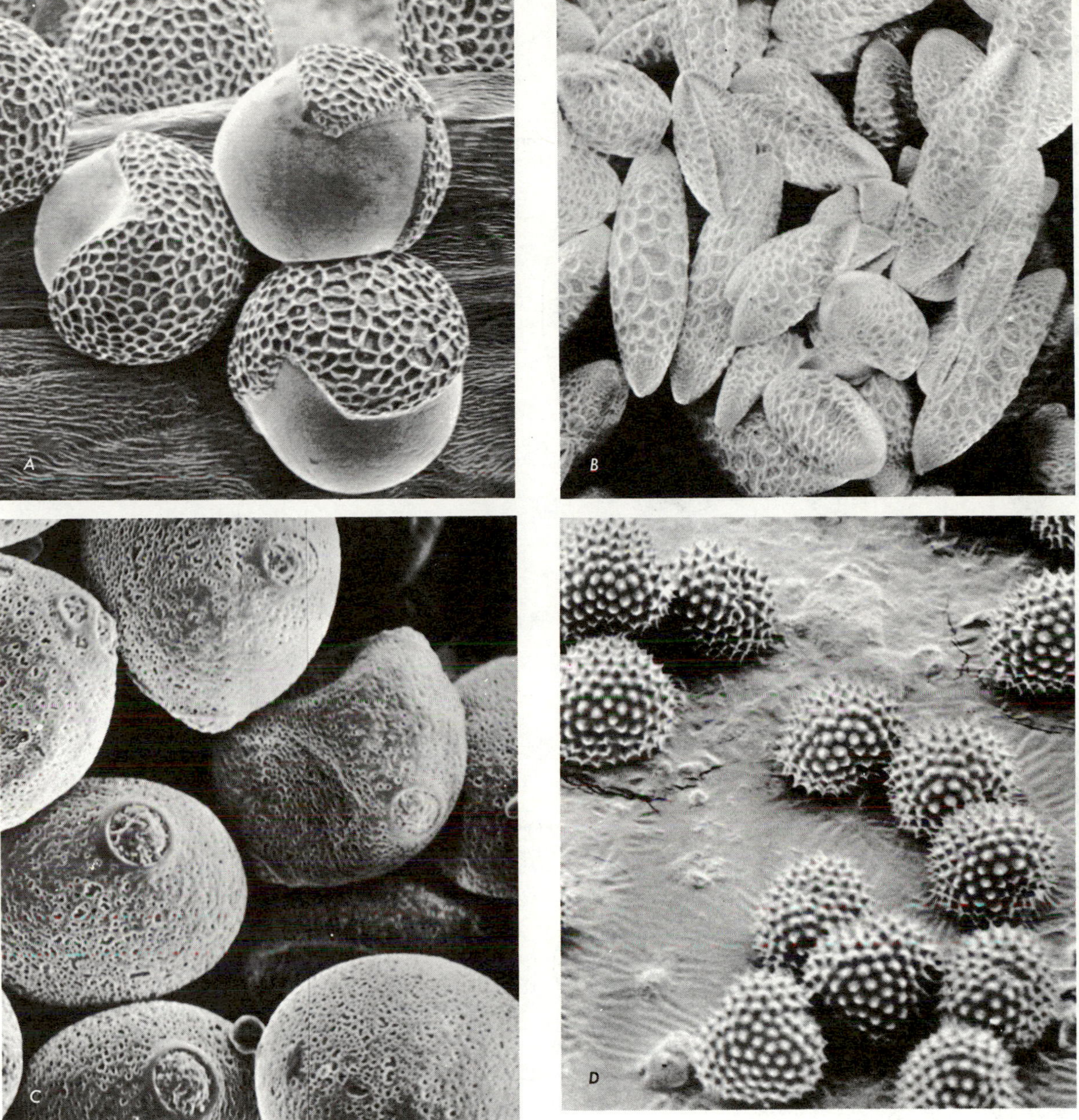

Figure 15.27
Pollen grains as seen with the scanning electron microscope. *A, Iris,* ×210. *B,* day lily (*Hemerocallis fulva*), ×275. *C,* cucumber (*Cucumis sativus*), ×870. *D,* ragweed (*Ambrosia*), ×1000.

is so resistant to decay that pollen may still be present in fossil and subfossil plant deposits when all other traces of biological material have been lost (Fig. 18.16). How do genes exercise control over the deposition of such inert materials outside the protoplast? The deposition of the wall begins during meiosis and passes through a number of steps, some of which occur after developing pollen grains lie free in the pollen sac in the anther. There is evidence of some change in ribosomes in developing pollen and of ribosome association with the plasmalemma. There is also evidence that anther cells surrounding pollen sacs may be influential in wall pattern formation. Sculpturing of pollen walls is so precise and consistent in its major features that it forms the basis for a well-developed field of **pollen taxonomy.**

The Gynoecium

The structure of the gynoecium depends upon the number and arrangement of carpels comprising it (Fig. 15.6). In the pea flower (Figs. 15.6*A, B*), there is a single carpel forming the gynoecium (Fig. 15.28). It consists of three parts: (*a*) the **ovary,** an expanded basal portion, and (*b*) the **style,** a slender stalk that ends in (*c*) a hairy irregular portion, the **stigma.** In the flower of the Christmas rose (Figs. 15.6*C, D*) there are five separate and distinct carpels comprising the gynoecium, and each carpel has its

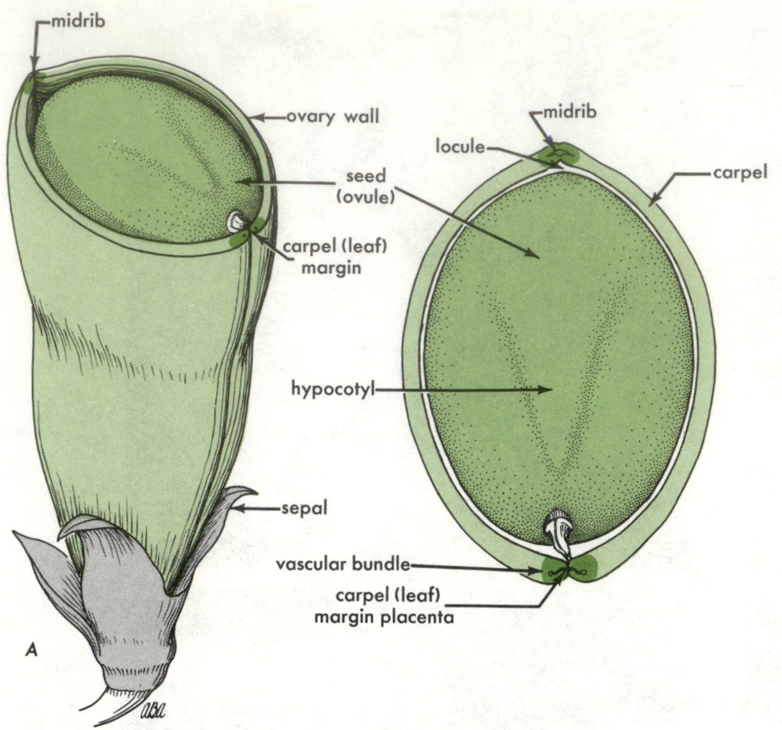

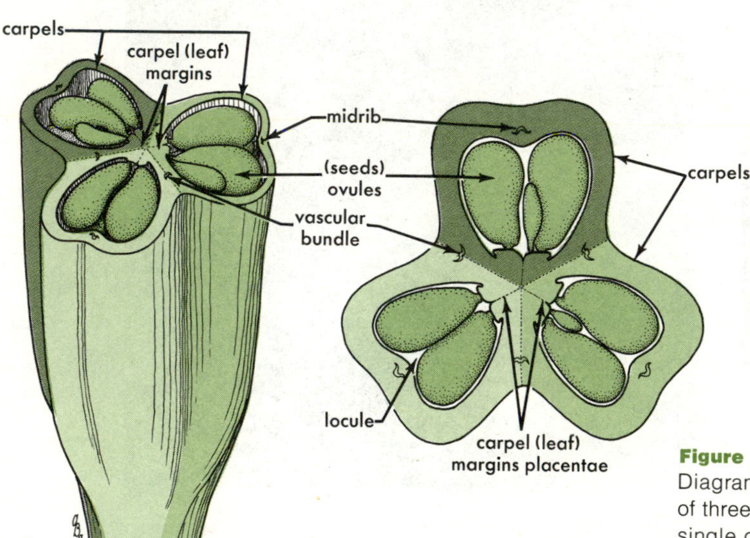

Figure 15.28
Diagram comparing gynoecia consisting of a single carpel and of three united carpels. *A*, section of ovary of pea consisting of a single carpel. *B*, section of ovary of *Tulipa* showing three united carpels.

own ovary, style, and stigma (Figs. 15.1, 15.6*D*). In the flower of *Cotyledon* **(Fig. 15.5*F*)** there are again five carpels which, while not actually fused together, grow so compactly as to appear on casual observation to form a single central pistil. In the tulip (Figs. 15.6*G, H*) there are three carpels so completely fused that only a single structure is present to represent the central whorl of floral leaves. A single pistil composed of one carpel is called a **simple pistil** (Fig. 15.1). When two or more carpels fuse to form a pistil, it is known as a **compound pistil** (Figs. 15.6*H*, 15.28*B*).

The ovary is a hollow structure having from one to several chambers, or **locules** (Fig. 15.28). The number of carpels in a compound pistil is generally related to the number of stigmas (Fig. 15.6*G, H*), the number of locules (Fig. 15.28), and, sometimes, the number of faces of the ovary (Fig. 15.28*B*). The pea has a single stigma and the ovary has one locule (Figs. 15.6*A, B*, 15.28*A*). A similar situation holds for each pistil in the gynoecia of Christmas rose (*Helleborus*) and *Cotyledon* (Figs. 15.1, 15.6*D, F*). There are three stigmas in the tulip gynoecium, the ovary is three-sided (Fig. 15.6*H*) and contains three locules (Fig. 15.28*B*); it is composed of three carpels.

Cross sections of the gynoecia of the garden pea and of the tulip are shown in Fig. 15.28. If one assumes that the carpel is a modified leaf, the structure of these two gynoecia may be interpreted as follows: The pea gynoecium would be a modified leaf folded along its midrib

Figure 15.5
Flowers showing three different arrangements of flower parts. *A, Liriodendron tulipifera,* three sepals, six petals, numerous stamens, many carpels compressed in inner whorl. *B, Sedum* sp., five sepals, five petals, ten stamens, five carpels. *C, Lilium tigrinum,* perianth consists of two whorls, each with three similar sepals and petals, six stamens, and three carpels coalesced in one pistil.

Figure 15.16
Elevation of floral parts. *A*, hypogyny in tulip flower *(Tulipa)*. *B*, perigyny in almond flower *(Prunus amygdalus)*. *C*, epigyny in daffodil flower *(Narcissus pseudonarcissus)*. *A* and *C*, about ×⅓; *B*, ×1.

Figure 15.32
Floral guides for insects collecting nectar or pollen. *A*, *Viola*, splashes of color and guide lines leading to throat of corolla. *B*, section of nasturtium flower showing tubular spur for nectar and guide hairs leading to entrance to tube. *C*, mimicry—the flower of the orchid *Ophrys* resembles a female wasp. (*C*, courtesy of R. Norris.)

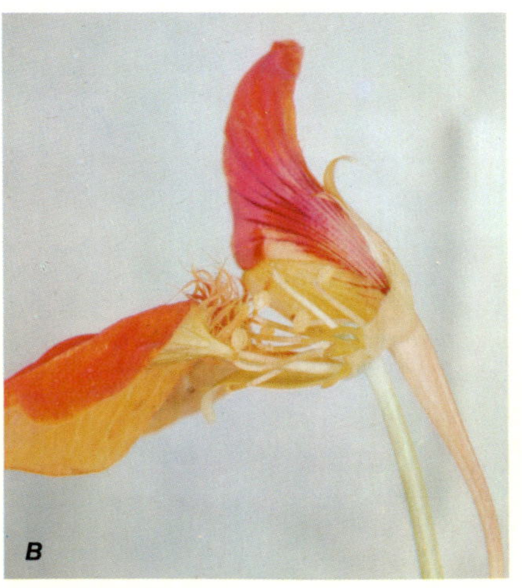

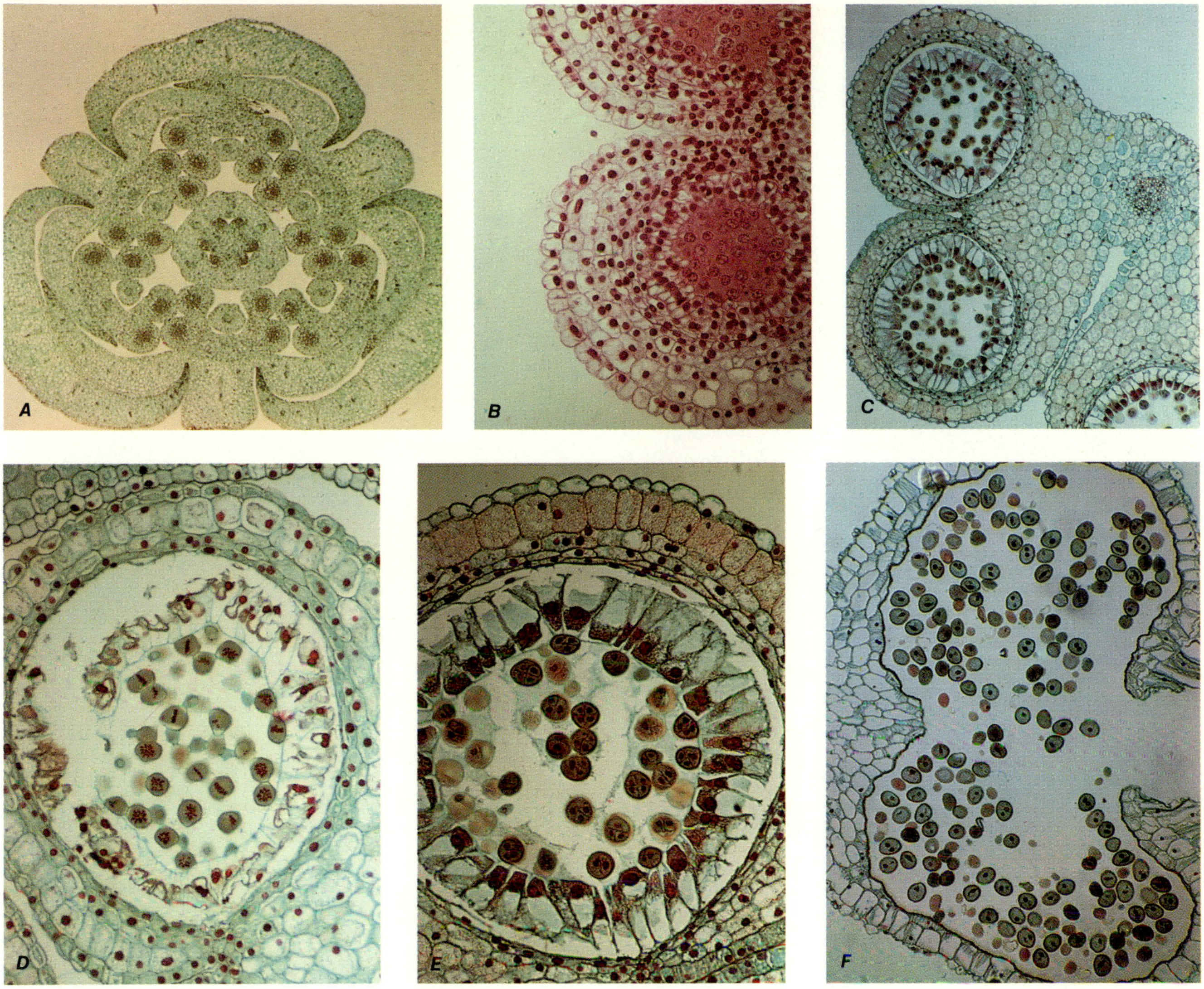

Figure 15.26
Pollen formation. *A*, cross section of lily *(Lilium* sp.) floral bud, three carpels in the center are surrounded by six anthers. *B*, cross section of two locules of an anther containing microsporocytes. *C*, microsporocytes undergoing meiosis. *D*, anther with closer view of same stage. *E*, microspore tetrad stage. *F*, microspore dividing mitotically to form a pollen grain. Anther wall (right side) has split to release pollen grains.

Figure 15.35
Tissue culture propagation. *A*, direct method from shoot tips with jojoba *(Simmondsia chinensis)*. *B*, jojoba culture showing new root initiation.

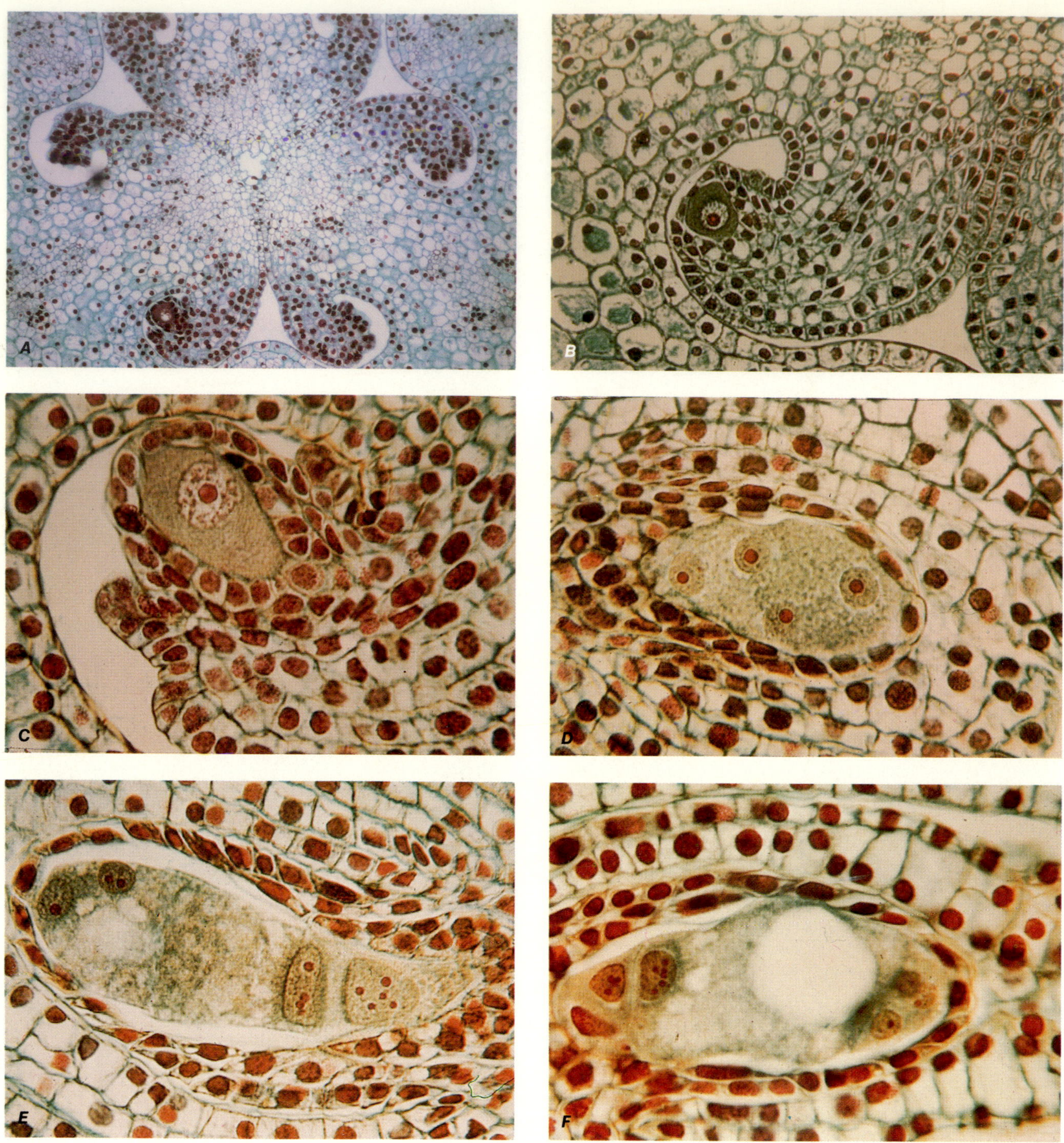

Figure 15.30
Embryo sac development. *A*, cross section of lily ovary showing location of ovules in three locules. *B*, ovule with a large megasporocyte. *C*, later stage, the integuments and single layered nucellus shown surrounding the megasporocyte. *D*, four megaspore stage. *E*, mature embryo sac. *F*, second four nucleate stage.

(Fig. 15.28*A*) with its margins fused. There is a single vascular bundle for each of the fused margins. The mature peapod will open by splitting both along the midrib, dorsal suture, and between the margins, ventral suture. The region where the ovules are attached is the placenta (Figs. 15.28*A*, 15.31). In the tulip, the margins of the three carpels are fused into a central placental region. Thus, in this interpretation, here too, the ovules are borne on the margins of infolded modified leaves.

The table below shows the relationship between the parts of the gynoecia in the flowers shown in Fig. 15.5.

Plant	Carpels	Pistil
Pea	1 carpel	1 simple pistil
Christmas rose	5 single and separate carpels	5 simple pistils
Cotyledon	5 distinct but closely appressed carpels	5 simple pistils
Tulip	3 fused carpels	1 compound pistil

In all these cases the term gynoecium refers to the carpels taken collectively, without reference to their number or their manner of association with each other. The term "simple pistil" refers to a single carpel, while "compound pistil" denotes the union of several carpels to form a single structure.

Placentation

The tissues within the ovary to which the ovules are attached are called **placentae** (singular, **placenta,** Fig. 15.28). The manner in which the placentae are distributed in the ovary is termed **placentation,** and its determination is helpful in classification: When the placentae are on the ovary wall, as in the violet, currant, gosseberry, and bleeding heart (*Dicentra,* Fig. 15.31*D*), the placentation is **parietal.** When they arise on the axis of the ovary that has several locules, as in lilies and fuchsia, the placentation is **axile** (Fig. 15.31*C*). Less frequently, the ovules are on the axis of a one-loculed ovary, in which event the placentation is **central,** as in the primrose family (Fig. 15.31*A*, *B*).

Style and Stigma

The style is a slender stalk that terminates in the stigma. It is through stylar tissue that the pollen tube grows. In some flowers, the style is very short or entirely lacking; in others, it is long. In corn (*Zea mays*), the corn silks are the styles (Figs. 15.9, 15.10*C*, *D*). As a rule, the style withers after pollination, but in some plants (for example, *Clematis*) it persists and becomes a structure that aids in the dispersal of the fruit (Fig. 15.7*B*). The stigmatic surface often has short cellular outgrowths that aid in holding the pollen grains; and sometimes it secretes a sugary and sticky solution, the **stigmatic fluid.** In many wind-pollinated plants, such as the grasses, the stigma is much branched, or plumelike.

The Ovule

The ovule, the structure that will eventually become the seed, arises as a dome-shaped mass of cells on the surface of the placenta. The outer cells of the dome develop to form one or two protective layers, the **integuments** (Figs. 15.29*A* to *C*, page 294, **15.30*B*,** page 292). The integuments do not fuse at the apex of the ovule, thus leaving a small opening, the **micropyle** (Fig. 15.29*C*). While this development is taking place, one of the internal dividing cells of the ovule, the **megasporocyte,** is enlarging in preparation for meiosis (**Fig. 15.30*A*,*B***). The tissue within which the megasporocyte has differentiated is known as the **nucellus.** The ovule then is composed of one or two outer protecting integuments, along with the micropyle, nucellus, and megasporocyte (Fig. 15.29*E*).

Embryo Sac Development

Meiotic division of the megasporeocyte may be considered as the first step in the development of a seed. As a result of these divisions, a row of four cells called the **megaspores** is produced in the nucellus. Although there is a wide variation in the fate of the megaspores among different members of the Anthophyta (see Chapter 19), as a rule, three of the cells (the ones nearest the micropyle) disintegrate and disappear, whereas the one farthest from the micropyle enlarges greatly (Fig. 15.29*E*). This megaspore, with its one nucleus and cytoplasm, now develops into the mature embryo sac. The usual stages are as follows: (*a*) *a series of mitotic divisions;* the megaspore nucleus divides, forming a two-nucleate embryo sac, each of these two nuclei divides, forming a four-nucleate embryo sac, and then each of the four nuclei divides, forming an eight-nucleate embryo sac (Figs. 15.29*F*, *G*, *H*) (*b*) *migration of nuclei;* after the first division the daughter nuclei migrate to opposite poles of the embryo sac, and after the last division, one nucleus from the sets of four at the opposite poles of the embryo sac migrates toward the center (Fig. 15.29); (*c*) *cell formation about nuclei;* the nuclei, with their surrounding cytoplasm, form cells. Six of these nuclei, each with its associated cytoplasm, become complete individual cells. The two polar nuclei form a single, binucleate cell. This results in a seven-celled embryo sac. These cells have names as follows: At the micropylar end of the embryo sac there is one **egg cell** associated with two **synergid cells.** Since it is frequently difficult to differentiate the egg cell from the other two, these three cells are sometimes referred to as the **egg apparatus.** The two nuclei that migrated centerward approach each other; they are called **polar nuclei** (Fig. 15.29), and form a binucleate cell called the **endosperm mother cell.** The three nuclei remaining at the end of the embryo sac opposite the micropyle form the **antipodal cells.** The em-

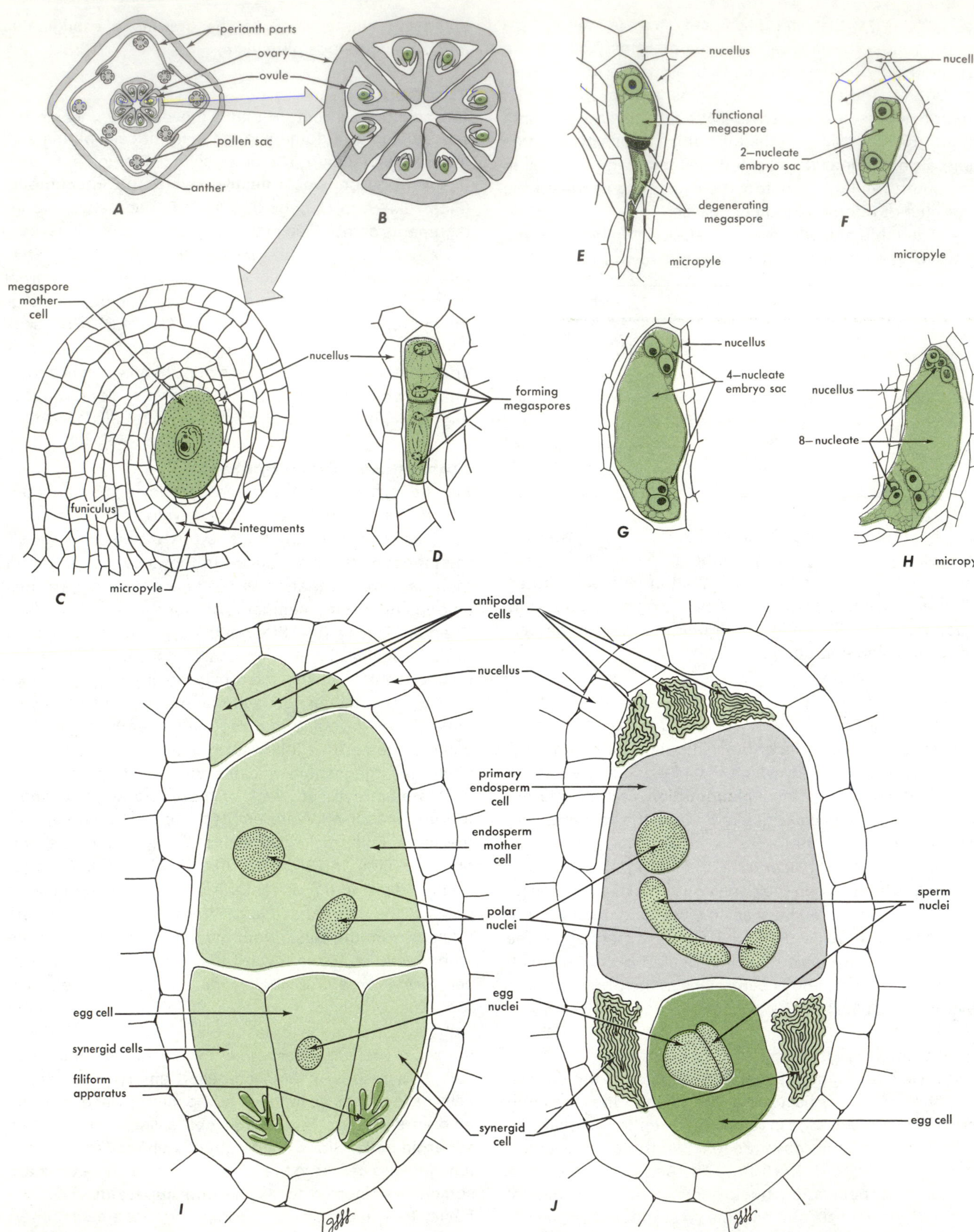

Figure 15.29
Embryo sac development. *A*, cross section of a flower bud showing the four whorls, four-carpelate ovary in center. *B*, enlarged view of ovary showing four carpels and ovules. *C*, an ovule, megasporocyte in prophase of meiosis. *D*, late telophase of second meiotic division. *E*, four megaspores, three degenerating. *F*, two-nucleate embryo sac. *G*, four-nucleate embryo sac. *H*, eight-celled embryo sac. *I*, mature embryo sac. *J*. fertilization, synergids degenerating.

bryo sac, which may be considered to be a seven-celled haploid plant, is now mature and ready for fertilization (Fig. 15.29*I*).

Embryo sac development is generally illustrated by slides of lily ovaries; however the lily type of embryo sac development is not common, and a similar sequence of stages occurs in only a relatively few species. In lily and other species having this type of development, all four megaspores, rather than a single megaspore, function in the subsequent development of the embryo sac. The sequence of stages results in polyploid endosperm tissue, a change that has genetic significance and means that the type of embryo sac development assumes practical importance for the improvement of foods derived from endosperm. The sequence of stages in embryo sac development as it occurs in lily is shown in **Fig. 15.30** page 292.

In lily the four megaspores function, and three move to the antipodal end of the developing embryo sac while the fourth passes to the micropylar end **(Fig. 15.30*D*).** Mitosis ensues. The three antipodal nuclei share a common spindle. The separate sets of anaphase chromosomes are thus triploid. The two large nuclei at the antipodal end each have three sets of chromosomes. Those at the micropylar end have one set of chromosomes each **(Fig. 15.30*E*).** A second normal mitosis ensues in each of the four nuclei, resulting in four nuclei at each end of the embryo sac. One haploid and one triploid nucleus migrate to the center of the sac and make up the polar apparatus. This leaves two synergid nuclei and the egg nucleus at the micropylar end and three antipodal nuclei at the other end of the sac. **Figure 15.30*F*** shows the mature embryo sac. Because the embryo sac is so large, it is difficult to get all eight nuclei of the mature embryo sac in a single section. **Figure 15.30*F*** illustrates the breakdown of the antipodal cells; it shows one polar nucleus and what are probably two synergids and the egg cell near the micropylar end.

Germination of the Pollen Grain

Pollen grains adhere to stigmatic surfaces. In many flowers, the stigmatic surface has short outgrowths to which pollen grains adhere. In some species, the stigma produces a sticky secretion, the **stigmatic fluid.** The hybridizer soon learns to recognize a receptive stigma and thus the proper time to apply pollen artificially. The pollen grain germinates on the surface of the stigma. The protoplast of the grain absorbs water and swells, breaking the outer wall and forming a protoplasm-lined tube. The pollen tube is a slender threadlike growth; it penetrates the tissue of the stigma, grows down through the style, and enters the ovary. Styles of some flowers are hollow, but most of them are not, and so passage of the tube apparently involves the secretion of tissue-dissolving enzymes by the advancing tip of the tube.

In short-styled flowers, pollen tubes need to grow only a short distance. For example, in pea and tulip, the style has a length of only 2 to 3 mm (Figs. 15.6*B, H*). On the other hand, in common corn, the distance from the stigmatic surface at the end of the corn silk to the young corn grain (ovary) may be as much as 50 cm (Fig. 15.10*C*). In a long style, the pollen tube is not a continuous, unbroken tube extending from stigmatic surface to ovary. It consists only of a very short, advancing apical portion containing two sperm cells. The older tube, through which the sperms have passed, gradually disintegrates.

The rate of growth of pollen tubes varies widely. In wheat, for example, fertilization normally occurs 1 to 2 days after pollination. Under favorable conditions, pollen tubes of corn may grow 15 cm in 24 hr. In some oaks, the tube takes almost a year to grow 2 to 3 mm.

As a rule, when the pollen grain is shed from the anther, it is a two-celled haploid plant, containing a generative cell and a tube cell (Fig. 15.25*E*). Before or during growth of the pollen tube, the generative cell undergoes division, forming two sperm cells (Fig. 15.25*F*). The two sperms, and in some species, the tube nucleus, move to the tip of the pollen tube, retaining that position as the tube grows. In other species, the tube nucleus degenerates either before the tube starts to grow or shortly thereafter. The two sperm nuclei bear the hereditary characters of the male parent.

The pollen of many plants will germinate in water. That of other plants requires a nutrient medium, such as a solution of cane sugar. For example, it was found that sugar beet pollen germinates abundantly on culture media containing 1.5% agar and 40% cane sugar. But the pollen of some species will not germinate in any artificial medium thus far tried.

Pollen varies considerably in the length of time it will remain viable, depending particularly upon moisture and temperature conditions surrounding the grains. Plant breeders may desire to keep pollen viable for a considerable length of time in order to make use of it in cross-pollination. It is sometimes shipped long distances for special hydridizing purposes. Pollen of date palms will retain its viability for several months, if kept dry. Experiments show that sugar beet pollen will germinate fairly well after 50 days if kept at low temperature and low humidity. Apple pollen has been successfully stored for 4½ years at 50% relative humidity and at a temperature of 2 to 8°C. Pollen kept dry will withstand greater temperature extremes than pollen kept moist.

Fertilization

Reaching the ovary, the pollen tube grows toward one of the ovules, usually enters the micropyle, penetrates one or more layers of nucellar cells, and enters the embryo sac (Fig. 15.29), approaching, in one manner or another, the egg cell and the endosperm mother cell. These stages are not easy to study, but it appears that the tip of the tube ruptures, one sperm cell enters the egg, and the other sperm enters the endosperm mother cell. Within the egg, sperm and egg nuclei fuse. Within the endosperm mother cell, the other sperm nucleus and two polar nuclei fuse (Fig. 15.29*I*, *J*). This double fusion of egg with sperm and polar nuclei with sperm is sometimes called **double fertilization.** The **zygote** and **primary endosperm cell** result from these fusions, the antipodal cells and synergid cells will degenerate, and conditions are set for further development of seed and fruit.

Only one pollen tube functions in fertilization of each embryo sac. If an ovary contains many ovules, each of course with one embryo sac, one pollen tube normally enters each ovule. Sometimes, however, fertilization of the egg is not effected, in which event it does not develop into an embryo. The presence of 100 or more mature seeds in a watermelon fruit is evidence that at least as many individual pollen tubes grew down the style of the flower and discharged their contents into separate embryo sacs. As a matter of fact, however, usually many pollen tubes disintegrate at some point along their path of growth from the stigma to the embryo sac.

The immediate external evidence of fertilization is withering of stigma and style and, in many flowers, dropping of petals. If flowers are bagged and pollination is prevented, petals usually remain fresh for a much longer time than in pollinated flowers.

One interesting question concerns how the pollen tube is directed into the embryo sac. The answer has been worked out partially for cotton (*Gossypium*). In the embryo sac of cotton, one or both of the synergid cells begin to shrivel and die before the pollen tube enters. It has been suggested that this activity results in the flow of some chemical substance from the synergid that possibly influences the direction in which the pollen tube grows. Two observations favor this hypothesis. First, at the base of both synergids is found a highly convoluted cell wall, known as the filiform apparatus (Fig. 15.29*I*). Cell walls in this configuration are commonly believed to be associated with active chemical secretion. The second bit of evidence is that the pollen tube actually penetrates one of the synergids and empties its contents into it. The two sperm then pass through the incomplete upper cell wall of the synergid; subsequently, one of them moves through the incomplete side wall of the egg to fertilize the egg cell, and the other moves deeper into the embryo sac to fuse with the endosperm mother cell.

The fusions that occur in the embryo sac initiate a number of changes in the entire ovary:

1. Development of zygote to form the embryo plant.
2. Development of primary endosperm cell to form endosperm (reserve food supply of the seed).
3. Development of integuments to form seed coat.
4. Absorption or disintegration of nucellar tissue. In some plants, however, a portion of the nucellus may become storage tissue of the seed rather than the endosperm. Such tissue of nucellar origin is called **perisperm.**
5. Development of ovary tissue to which ovule or ovules are attached to form the fruit.
6. Possible stimulation of accessory flower parts, such as the receptacle, sepals, or petals, to increased growth and incorporation into the fruit.

Thus, the importance of fertilization is to stimulate growth of certain floral parts and to bring about a withering of others. The net result is the development of the ovary wall into the fruit and the ovule into the seed.

Pollination

Pollination may be brought about by either **wind** or **insects,** and occasionally by water or birds. Flower structure and pollen type is generally adapted to one or the other of these pollinating vectors (Table 15.1). In the case of insect pollination, adaptation may be very complex, indicating a long evolutionary association between insect vector and plant.

Figure 15.31
Three types of placentation. *A* and *B*, free central (*Primula*). *C*, axile (*Fushia*). *D*, parietal (*Dicentra*). (Redrawn from Priestley and Scott, *An Introduction to Botany*. © 1949 by Longmans Green. Reprinted with permission of the Longmans Group. Ltd.)

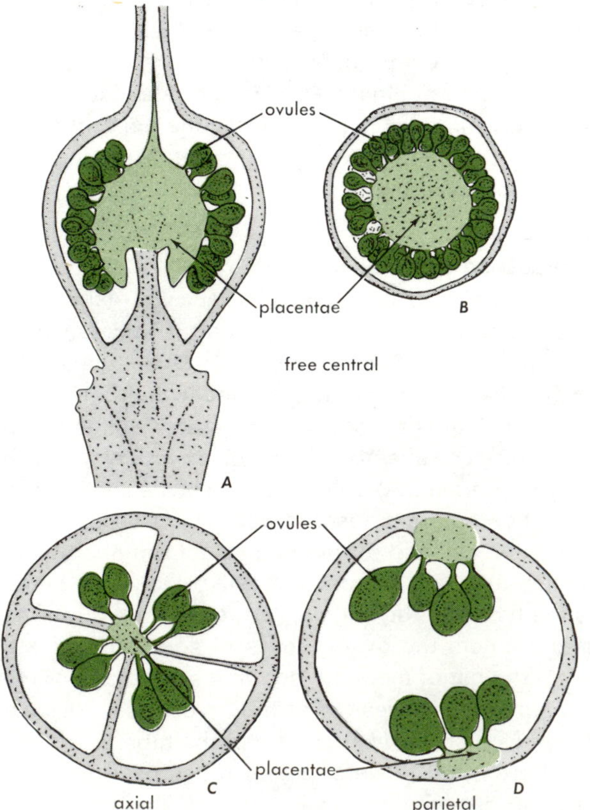

Pollinating Vectors

Wind. Pollen is carried chiefly by wind and insects. Rarely, water and birds are agents of pollination. **Wind pollination** is the common type in plants with inconspicuous flowers, as in grasses (Figs. 15.10, 15.19), poplars, walnuts (Figs. 15.11, 15.12), alders, birches, oaks, ragweeds, and sage. Such plants usually produce pollen in enormous quantities. Flowers of grasses are inconspicuous, and most of them lack odor and nectar and hence are unattractive to insects. Furthermore, their pollen is light and dry and is easily blown. Their stigmas, in many cases, are feathery and expose a large surface to flying pollen. In cottonwoods, alders, birches, oaks, walnuts, and hickories, the flowers are in **catkins** (Fig. 15.12*A*). Staminate catkins are pendulous and move easily in wind. The light pollen is shaken from anthers and is readily carried away by breezes. In many catkin-bearing trees, flowers open before leaves unfold so that pollen movement is unhampered.

Living Vectors. Living vectors include a variety of insects, a few species of birds, and even some bats. Plants that are pollinated by living vectors usually share certain characteristics. They possess a colorful perianth that, together with other floral parts, may be arranged into a complex architecture (**Fig. 15.32,** page 290). They also produce both a sweet-tasting fluid, called nectar, and volatile compounds having unique odors. Work of many investigators has shown that these characteristics are highly adapted for attraction of pollinators.

Bees, for example, are able to detect and distinguish between many odors and colors and degrees of sweetness. Bees can distinguish at least three colors in addition to ultraviolet, and these colors include the most common perianth shades (yellow and blue). Bees are directed to nectar-producing glands by splashes or lines of contrasting color **(Fig. 15.32*A*).** When the insect attempts to reach nectar or collect pollen, floral architectures ensures transfer by the insect of pollen to the stigma. In nasturtium **(Fig. 15.32*B*),** for example, nectar is held in a long narrow tube or spur. A foraging bee directs his proboscis down the tube, rubbing the stigma. Pollen, which may have been picked up in earlier visits to other flowers, is deposited on the stigma.

TABLE 15.1
TYPICAL TRAITS OF FLOWERS POLLINATED BY DIFFERENT VECTORS

	Pollination Vector				
Trait	Wind	Beetle	Hummingbird	Bee	Moth
Color	Green, brown, dull; petals often absent	White to off-white, dull	Red or orange	Bright white, blue, yellow; may reflect ultraviolet light	White or pale
Odor	None	None to strong, fruity, or fetid	None	Often mild, pleasant	Strong and sweet, often pronounced at night
Nectar and nectar guides	None	If present, easily accessible; guides absent	Abundant, often at base of corolla tube or in spurs; guides absent	Commonly present but may be hidden; guides present	Moderate to abundant; guides absent
Flower shape	Regular, small, often unisexual; anthers and stigmas exserted; copious production of pollen	Large, open, regular; many parts	Regular or irregular, but petals form a long tube without a landing platform at edge; often pendent	Often irregular or tubular; few parts; landing platform present; flowers erect	Typically regular, but often tubular with spurs at base; may open only at night; few parts; no landing platform; flowers horizontal or pendent
Rewards for pollinator	None	Pollen, nectar, all flower parts may be chewed	Nectar	Pollen or nectar	Nectar
Examples	Walnut, grasses	*Magnolia,* dogwood	*Fuschia,* hibiscus	Larkspur, snapdragon, violet	Tobacco, Easter lily, some cacti

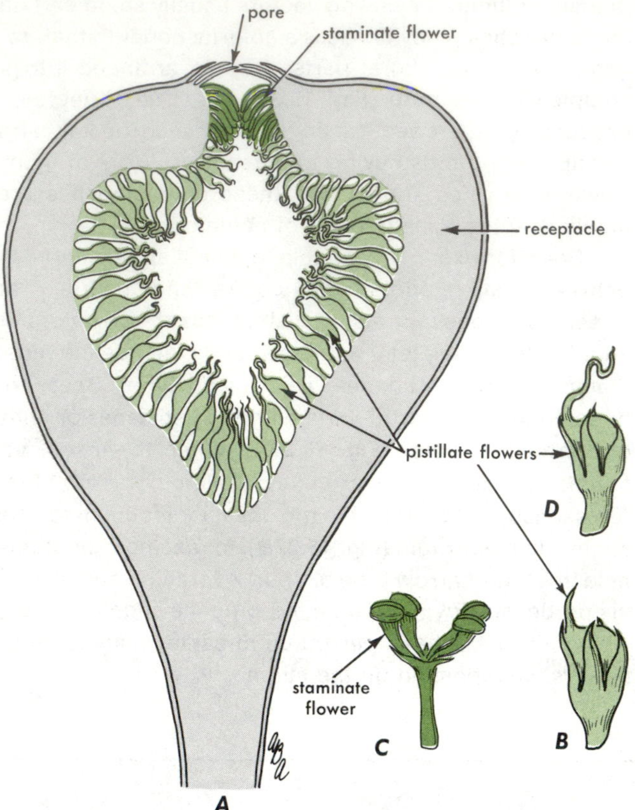

Figure 15.33
Diagram of a fig. *A, B,* and *C,* inedible caprifig. *A,* section of receptacle showing location of staminate and pistillate flowers, ×2. *B,* short-styled abortive pistillate flower, ×3. *C,* staminate flower, ×3. *D,* fertile, long-styled pistillate flower of the edible Calimyrna fig, ×3.

Orchid flowers have many very specialized mechanisms to ensure pollination. The petal shape and color of species of *Ophrys* closely resemble the appearance of a female wasp or fly **(Fig. 15.32*C*).** Male wasps or flies, emerging from the pupal stage before the females, mistake the *Ophrys* flower for females. The male lands on the flower and attempts copulation; repeated pseudocopulation results in pollination.

There is a very close interdependence between the life cycle of the commercial fig (*Ficus carica*) and that of its pollinator, the fig wasp (*Blastophaga psenes*). There are two types of fig trees, one that produces edible fig fruit and one that produces only small, hard, inedible fruit, which is nothing more than an incubator of fig wasp larvae.

An immature, inedible fig is an inflorescence. Small, unisexual flowers line the interior of a hollow, flask-shaped structure that has a pore at the enlarged end. Male flowers lie near the pore and female flowers line the rest of the cavity (Fig. 15-33*A*). The female flowers have pistils with short styles (Fig. 15.33*B*), and female wasps are able to enter the inflorescence through the pore and deposit eggs in each ovary. The eggs hatch into larvae and later the larvae change to wasps, eat their way out of the ovaries, and mate inside the immature fig. The males usually remain inside and die, but the females escape through the pore, becoming dusted with pollen from the male flowers (Fig. 15.33*C*) as they do so.

Searching for ovaries in which to deposit eggs, the female may enter the inflorescence of an edible fig. This inflorescence differs from that of an inedible fig by having only female flowers, and the pistils have much longer styles (Fig. 15.33*D*). The wasp is unable to oviposit because of the elongated styles, but she succeeds in pollinating flowers. Seeds and an edible fig fruit will later result. Eventually, she will find an inedible fig inflorescence, lay her eggs, and start the cycle over again. A single branch of an inedible fig can be grafted to an edible fig tree, and it will provide sufficient numbers of pollinators for the rest of the tree.

Types of Pollination

There are essentially two different kinds of pollination, determined by the genetic similarity of the plants involved. If anthers and stigmas, essential organs in pollination, have the same genetic constitution (that is, have chromosomes bearing identical genes, whether or not they are produced on the same or different individual plants), transfer of pollen from anther to stigma is **self-pollination.** If two parent plants with different genetic constitutions are involved, transfer of pollen from the anther of one to the stigma of the other is **cross-pollination.**

Two individual plants may belong to the same pure line. In wheat, for example, the stigma of a flower usually receives pollen from its own anthers. Following fertilization, the ovary of the self-pollinated flower develops into a grain. The plant that develops from this grain in turn produces many flowers. If these flowers are self-pollinated, the many individual plants that develop from the grains constitute a **pure line.** If pollen from one individual of a pure line is placed on the stigma of a flower of another individual of the same pure line, the result, genetically, is the same as when pollen is placed on the stigma of the flower that furnished the pollen.

Forms of Self-Pollination.

1. Transfer of pollen from an anther to a stigma of the same flower. This is the normal method in cereals (except rye and corn), garden peas, and some other plants.
2. Transfer of pollen from anthers to stigmas of other flowers on the same plant. This occurs in many plants with perfect flowers and is usual in monoecious plants.
3. Transfer of pollen from anthers to stigmas of flowers on other genetically identical plants.

Cross-Pollination. Cross-pollination always involves two plants with different genetic constitutions. The differences may involve one or many characters. The two parents may be of the same variety or of the same species; they may be of different species or even of different genera.

Asexual Reproduction

Apomixis

This term refers to a type of reproduction in which no sexual fusion occurs but in which structures usually concerned in sexual reproduction are involved. Normally, egg nuclei will not start on the series of changes that result in embryo plants unless fertilization has occurred.

Occasionally an embryo develops from an *unfertilized egg.* This occurrence is called **parthenogenesis.** Two types of parthenogenesis are as follows: that in which the embryo develops from an *n,* or haploid, egg cell (the individuals are usually sterile), and that in which the embryo develops from a 2*n,* or diploid, egg cell. In the latter type, meiosis is omitted in the development of the embryo sac. Another form of apomixis is that in which the embryo plant arises from tissue surrounding the embryo sac. These "adventitious" embryos (diploid) occur in *Citrus, Rubus,* and other plants.

Apomixis has variations other than those given above, but all forms of this phenomenon involve the origin of new individuals without nuclear or cellular fusion.

In some plants, deposition of pollen on the stigma is a prerequisite to apomictic embryo development, even though a pollen tube does not grow down the style and nuclear fusion does not take place. In such instances, there is evidence that hormones formed in the stigma, or furnished by the pollen, are transferred to the unfertilized egg cell, initiating there the changes that result in embryo development.

Vegetative Propagation

For centuries many horticultural plants have been reproduced vegetatively. For example, potatoes are usually propagated by cutting tubers into a number of sections. Obviously, all plants growing from tubers or sections of tubers from one plant have the same genetic composition. Such a group of asexually reproducing plants is called a **clone.** Another example: All individual pear trees in an orchard may be parts of a single parent tree; that is, they may have been propagated by cuttings or buds from one parent tree. In this event, these individual pear trees have the same genetic composition. Pollination within clones is essentially self-pollination in that, although pollen goes from one plant to another plant, both plants have the same genetic constitution because they are reproduced by vegetative means.

Natural strategies for asexual propagation of plants with identical genetic characteristics (clones) exist in numerous plant families. Members of the *Crassulaceae* are particularly rich in these mechanisms and will be used as examples. Vegetative propagation can be categorized into two classes (1) plants in which propagation is induced only after a tissue or plant organ is wounded; and (2) plants that maintain a residual meristem from which plantlets develop.

Propagation from leaves of the jade plant (*Crassula argentea*) is a classic example of the wound response type. In this case, when a leaf is removed from the plant, a wound repair mechanism is automatically initiated to seal the cut-off tissues by cell division and deposition of fatty substances (tannins and suberin) in the cut cells to protect against water loss and pathogen entry. As part of the process a new plantlet is developed from a vascular bundle just inside the wounded surface of a petiole after leaf fall (Fig. 15.34*A*). This same response can be observed by sterilizing a leaf and placing a segment of it on sterile agar in a closed flask (Fig. 15.34*B*). An interesting observation is that the new plantlet that forms always develops on the petiole end of the leaf segment. A polarity exists that internally controls the onset and location of the new plantlet.

Other plants, such as *Bryophyllum,* have evolved a propagation mechanism whereby small patches of cells (residual meristems), located in the notches of the leaf margin, are induced to divide and differentiate into new attached plantlets (Fig. 15.34*C*). These plantlets are apparently stimulated by some environmental cue, such as day length. Once formed, they eventually fall to the ground and develop into new plants.

Tissue Culture Propagation

Vegetative propagation from excised plant parts can also be induced experimentally in the laboratory. This procedure, **tissue culture,** capitalizes on a property of plant cells called **totipotency.** That is, any plant cell not already irreversibly differentiated, has the potential under ideal conditions to form into an entire new plant. Plant cells do not normally express this character because they are under the constraints of their neighboring cells, and the influences of their internal and external environment. Studies by many scientists since the 1940s have demonstrated that it is possible to manipulate individual plant cells to form new plants, even if a particular species is not prone to vegetatively reproduce under natural conditions. Think of the importance of this discovery from an agricultural viewpoint. In this manner it is potentially possible to propagate thousands of identical individuals from a single parent without sexual recombination of genes that could result in the loss of desirable characteristics. All of the new individuals would belong to the same genetic clone and maintain the same favorable characteristics such as high yield in a crop plant, or large showy flowers in a horticultural variety.

This type of induced propagation has been accomplished with several important crop plants including, corn, rice, wheat, pea, flax and others. It is done directly from shoot tip explants, or indirectly from various plant parts. The direct method involves removing the shoot tip

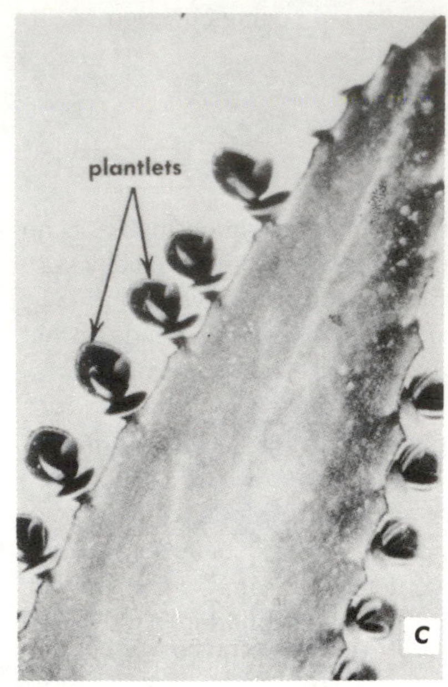

Figure 15.34
Examples of vegetative propagation. *A, Crassula argentea* leaf removed from plant forms new plantlet at the petiole end. *B,* plantlets also form from *C. argentea* leaf pieces placed in sterile culture—always on the petiole—end of the leaf piece. *C, Bryophyllum* is an example of the residual meristem type where plantlets form directly on the leaf from previously existing meristems.

from the source plant, sterilizing it in a dilute bleach solution, rinsing it in sterile water, and transferring it into a sterile flask containing a growth medium composed of agar, mineral salts, vitamins, and plant hormones (see Chapter 20). Depending upon the temperature, light condition, and the relative mixture of medium components, the transplanted shoot tip will elongate, branch, and produce roots (**Fig. 15.35,** page 291). This method has been used successfully for many plants, but is not universally applicable, because generally only one or a few plantlets can be produced from one explant shoot tip, making the procedure an expensive process if one is interested in propagating many plantlets.

The indirect method makes it potentially possible to propagate hundreds, perhaps thousands of identical plantlets from one explant! An explant source, perhaps a leaf, or a stem, is sterilized, rinsed, segmented, and transferred to culture medium (Fig. 15.36). The cells of these organ segments can be induced to divide and to form a large mass of cells, called a ***callus.*** Upon further manipulation of the medium and the environment, roots, shoots, and whole new plantlets can be induced to develop that can be later transplanted to soil.

SUMMARY

1. The flower, the distinguishing structure of the Anthophyta, is formed of four whorls of parts specialized to carry out sexual reproduction, including pollination, fertilization, and seed production.
2. A floral apex has lost its ability to elongate and its potentiality for vegetative growth. The ability for vegetative growth is not regained until the completion of the two nuclear phenomena, meiosis and fertilization.
3. The four whorls of floral leaves are (*a*) the calyx, composed of sepals; (*b*) the corolla, composed of petals; (*c*) the androecium, composed of stamens; and (*d*) the gynoecium, composed of carpels.
4. There are two parts to a stamen, the pollen-producing anther and the filament.
5. A single unit of the gynoecium is frequently called a pistil. If it consists of one carpel, it is a simple pistil; if it consists of two or more fused carpels, it is a compound pistil.
6. A pistil consists of three parts, the stigma, the style,

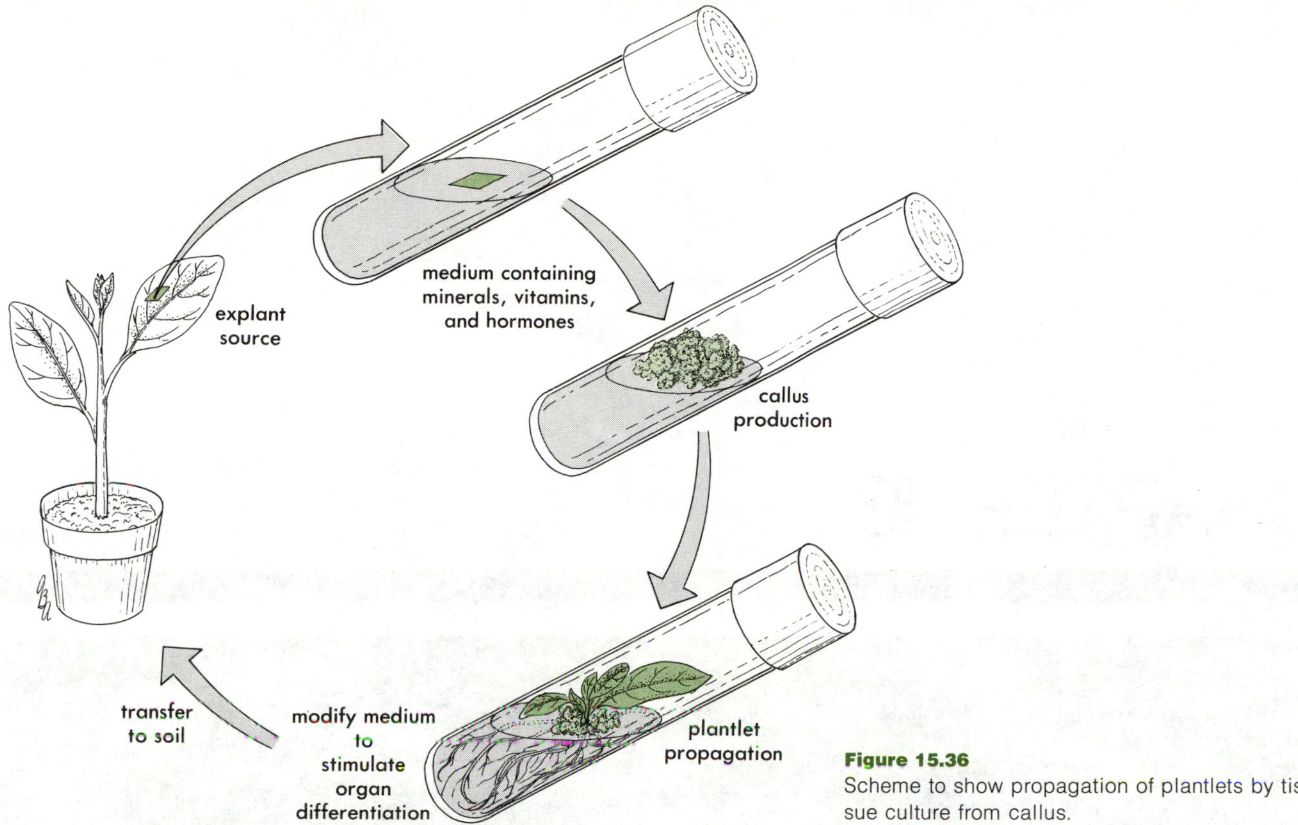

Figure 15.36
Scheme to show propagation of plantlets by tissue culture from callus.

and the ovary. The stigma is receptive to pollen, and the ovary encloses the ovules. At the time of fertilization, the ovules consist of integuments, nucellus, and embryo sac.

7. Meiosis takes place both in the anthers and in the ovule.
8. Pollen may be considered a two-celled haploid plant or male gametophyte protected by a cell wall, which is frequently elaborately sculptured. Two sperms are produced by the generative cell.
9. The embryo sac may be considered a seven-celled haploid plant or female gametophyte cells.
10. Pollination is the transfer of pollen from anthers to stigmas. A cellular recognition system regulates the germination of compatible pollen. A pollen tube growing down the style to the embryo sac serves as a pathway by which the two sperms reach the embryo sac.
11. Floral architecture facilitates pollination, which is most frequently carried out by wind or insect vectors.
12. Double fertilization, the union of one sperm with the egg cell and of a second sperm with the endosperm mother cell, occurs only in the Anthophyta.
13. The embryo plant develops from the zygote. The endosperm develops from the primary endosperm nucleus and supplies some nourishment for embryo and sometimes for seed development.
14. Variation in floral architecture arises (*a*) from variation in number of parts in a given whorl; (*b*) from symmetry in floral parts; (*c*) by connation of floral parts; (*d*) by adnation of floral parts; (*e*) by elevation of floral parts; and (*f*) by the presence or absence of certain whorls.
15. Flowers are either solitary on flower stalks or grouped in various inflorescences such as heads, spikes, catkins, umbels, panicles, racemes, or cymes.
16. There are essentially two types of pollination: cross-pollination, which results in much variation in the progeny, and self-pollination, which results in progeny having great similarities.
17. Apomixis is a type of reproduction in which an unfertilized egg develops into an embryo without sexual fusion.
18. Other means of asexual reproduction are propagation by cuttings, and propagation of plantlets naturally, in certain species, either by a wound response or a residual meristem system.
19. Tissue culture is a relatively new method of asexual reproduction where new plants are formed from tissue or organ explants grown in artificial medium. New plantlets are grown directly from shoot tip explants, or indirectly from callus. Roots, shoots, or whole plantlets may be induced by modification of the growth medium or environmental parameters.

CHAPTER 16

THE FRUIT, THE SEED, AND GERMINATION

THE FRUIT

fruit is the ripened ovary of a flower; it may include other floral parts. In many cases, fruit development usually depends on pollination and the activity of indoleacetic acid, or other growth substances.

Fruits are auxiliary structures in the sexual life cycle of only the Anthophyta. Their development has unquestionably been an important factor in the successful evolution of flowering plants. Fruits protect seeds, aid in their dissemination, and may be a factor in timing their germination. Fruits are highly constant in structure and are thus very important plant parts in the classification of the Anthophyta. In everyday usage, the term ''fruit'' usually refers to a juicy and edible structure, such as an apple, plum, peach, cherry, orange (*Citrus*), or grape. Structures that are not popularly thought of as fruits such as string beans, eggplant (*Solanum melongena*), okra (*Hibiscus esculentus*), squash, and cucumbers, which are commonly called ''vegetables,'' and ''grains'' of corn, oats, wheat (*Triticum*), and other cereals are all fruits in a botanical sense.

Development of the Fruit

Development of the fruit (pod) of pea and bean is described as an example of the changes that may occur during transformation of the ovary into a fruit. Pistils of bean and pea flowers are each composed of one carpel (Fig. 16.1); that is, one ovule-bearing leaf. Recall that the ovary of bean is formed by fusion of margins of the carpel

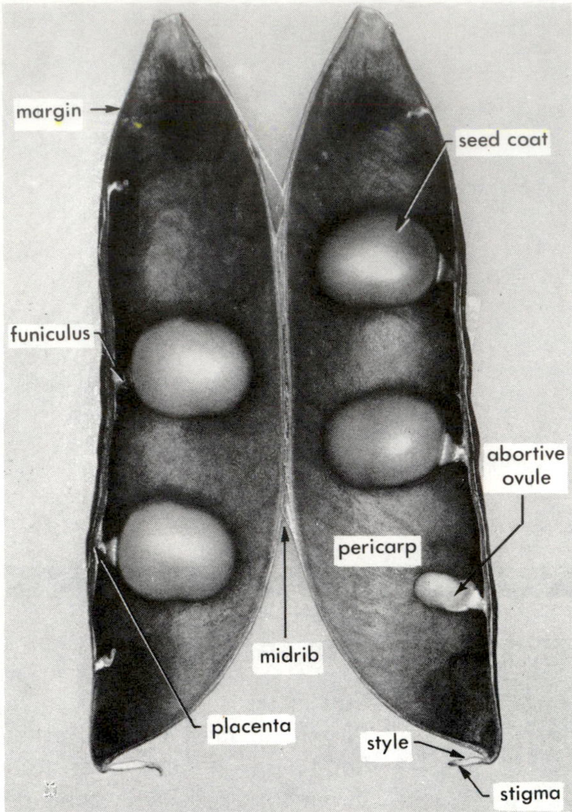

Figure 16.1
Opened pea pod (*Pisum salivum*) showing developing seeds attached to carpel margins, ×1.

and that ovules are attached to these margins. A cross section of the bean ovary (Fig. 16.2) shows ovary wall, ovule, locule. The ovary wall may be divided into three distinct layers, (*a*) an outer epidermis, (*b*) an inner epidermis, and (*c*) a middle zone consisting of several layers of cells. There are three carpellary bundles, one for each margin (ventral carpellary bundles) and one for the midrib opposite them (dorsal carpellary bundle).

Fertilization initiates a series of changes in the embryo sac and other tissues of the ovule that lead to the development of the fruit and seed. Tissues of the ovary wall undergo marked changes into three layers. The fruit wall (developed from the ovary wall) is called the pericarp, and the three more or less distinct parts, in order beginning with the outermost, are **exocarp, mesocarp,** and **endocarp** (Fig. 16.3, shown for almond). When the pod is mature, floral structures such as pedicel, calyx, withered stamens, style and stigma, and even remnants of the corolla may also be present.

Mature pods usually show the presence of small, underdeveloped (abortive) ovules. It is probable that these ovules were not fertilized (Fig. 16.1). If none of the ovules within the bean ovary is fertilized, the ovary does not enlarge. In most plants, normal fruit development takes place only if pollination is followed by fertilization, and if fertilization is followed by seed development.

Parthenocarpy

The stimulus of fertilization usually exerts an influence on all parts of the flower, particularly on the ovary and its contained ovules. In Fig. 16.4, normal and seedless fruits of tomato are compared. Note the greatly increased development of the fruit that accompanies normal seed development. In some plants, *normal fruit development may take place without fertilization.* Such unfertilized fruits may or may not be seedless, depending upon the occurrence of apomixis and the subsequent development of embryos. On the other hand, fertilization may occur, but ovules may fail to develop into mature seeds even though fruit may form in a normal fashion. In a strict sense, only fruits that develop without fertilization are called **parthenocarpic fruits.** Practically, it is not always possible to know without extensive study, whether a given fruit is parthenocarpic. For instance, Thompson seedless grapes were thought to be parthenocarpic until it was shown that fertilization takes place, but ovules fail to mature into seeds. In many citrus fruits, seeds formed by fertilization are accompanied by seeds with adventitious embryos. To complicate definitions further, sexually formed seeds are not as robust as adventitious seeds and frequently do not develop. This situation has led to a loose use of the word parthenocarpy to mean simply seedless fruits.

Parthenocarpic (seedless) fruits are quite regularly produced in such cultivated plants as English forcing cucumber, certain varieties of eggplant, navel orange, banana, pineapple, and some varieties of apple and pear.

In certain plants, seedless fruits may be induced by pollen that is incapable of fertilizing the ovules. For example, in some orchids, the placing of dead pollen or a water extract of pollen upon the stigma may start fruit development.

Recently, parthenocarpy has been induced in some plants by spraying the blossoms with dilute aqueous solutions of certain growth substances. Commercial application of this technique is being made.

Kinds of Fruits

In classifying the different kinds of fruits, the following criteria are taken into account:

(a) The structure of the flower from which the fruit develop.

(b) The number of ovaries involved in fruit formation.

(c) The number of carpels in each ovary.

(d) The nature of the mature pericarp (dry or fleshy).

(e) Whether or not the pericarp splits **(dehisces)** at maturity.

(f) If dehiscent, the manner of splitting.

(g) The possible role that sepals or receptacle may play in formation of the mature fruit.

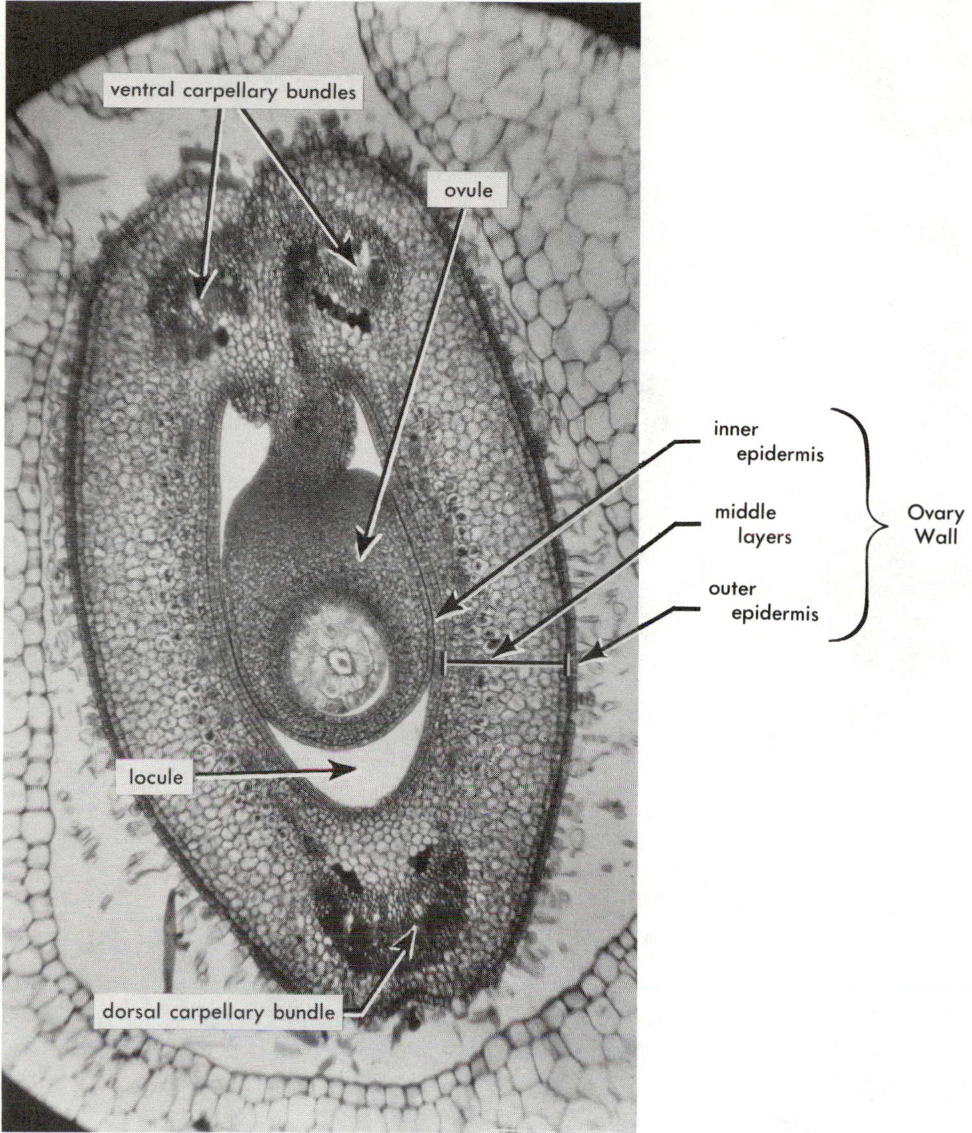

Figure 16.2
Cross section of bean ovary after fertilization, ×20.

Simple fruits are derived from a *single ovary.* They may be dry or fleshy; the ovary may be composed of one carpel or of two or more; and the mature fruit may be dehiscent or indehiscent.

Aggregate fruits and **multiple fruits** are formed from *clusters* of simple fruits. The difference between these types of fruits depends on the number of flowers involved in their formation. In strawberry (*Fragaria*) (Fig. 16.5) and blackberry (*Rubus*) (Fig. 16.6), the aggregate fruit is derived from the many ovaries of a single flower. These matured ovaries or fruits are all attached to a common receptacle. Such groupings constitute aggregate fruits, and the simple fruits comprising them may be classified according to the scheme of classification of simple fruits given below. Mulberry, (*Morus*), fig (*Ficus*), and pineapple (*Ananas comosus*) (Fig. 16.7) are multiple fruits. Multiple fruits consist of the enlarged ovaries of several flowers, more or less grown together into one mass. In some plants—mulberry, for example—associated floral structures form a part of the fruit, and in the fig the receptacle enlarges to become the sweet edible portion (Figs. 16.8, 15.33).

Simple Fruits

Pericarp Dry and Dehiscent

Legume or pod. This type of fruit, characteristic of nearly all members of the pea family, Fabaceae (Leguminosae), arises from a single carpel, which at maturity generally dehisces along both sutures (Figs. 16.1 and 16.9). In the bean or pea pod, the "shell" is pericarp, and the "beans" or "peas" are seeds. Pods may be spirally twisted or curved as in alfalfa or vetch (*Vicia sativa*) (Fig. 16.25*H*). A number of legumes, however,

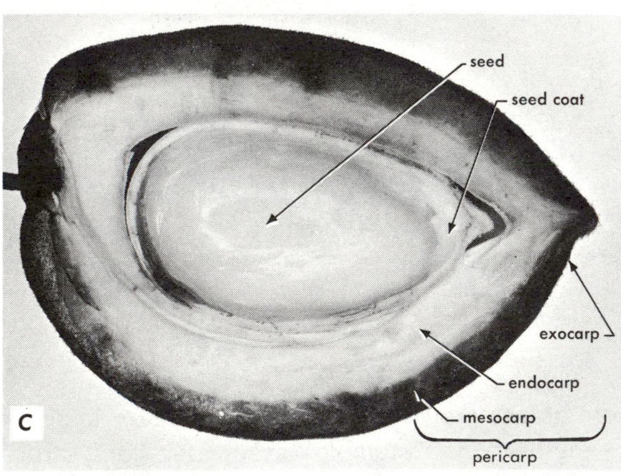

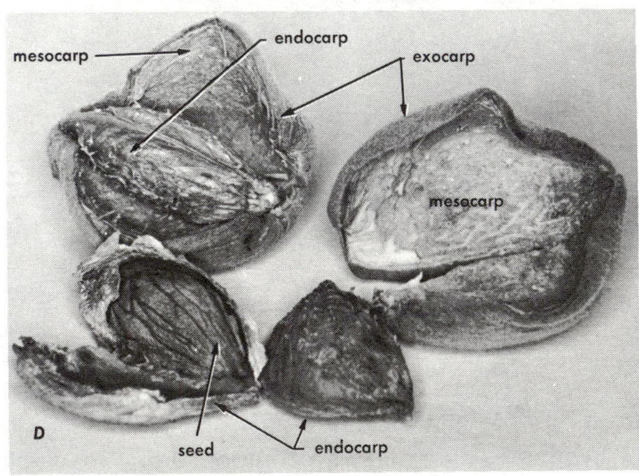

Figure 16.3
A fruit is a ripened ovary. The ovary of an almond flower (*Prunis amygdalus*) develops into the almond. *A,* almond blossom, ×1. *B,* developing ovary bursting calyx tube, ×1. *C,* section of a maturing fruit showing division of pericarp and seed, ×1. *D,* mature almonds showing split outer portion of ovary walls (shucks) and intact inner portion (shell), ×½.

such as honey locust, alfalfa, and bur clover have pods that do not dehisce.

Follicle. An example of follicle is the fruit of magnolia (Figs. 16.10). The follicle develops from a single carpel, and opens along one suture, thus differing from the pod, which opens along both sutures.

Capsule. Capsules are derived from compound ovaries, that is, an ovary composed of two or more united carpels. Each carpel produces a few to many seeds. Capsules dehisce in various ways (Figs. 16.11*A* to *I*).

Silique. This is the type of fruit (Fig. 16.12*A* to *D*) characteristic of members of the mustard family, Brassicaceae (Cruciferae). The silique is a dry fruit derived from a superior ovary consisting of two locules. At maturity, the dry pericarp separates into three portions; the

Figure 16.4
Tomato (*Lycopersicon esculentum*) fruits. *A* and *B*, normal development after fertilization. *B*, cross section. *C* and *D*, seedless or parthenocarpic fruits. *D*, cross section, ×¾.

Figure 16.5
Aggregate fruit of the strawberry (*Fragaria*). *A*, flower. *B*, fruit, achenes with enlarged receptacle, ×2. (Courtesy of R. S. Bringhurst.)

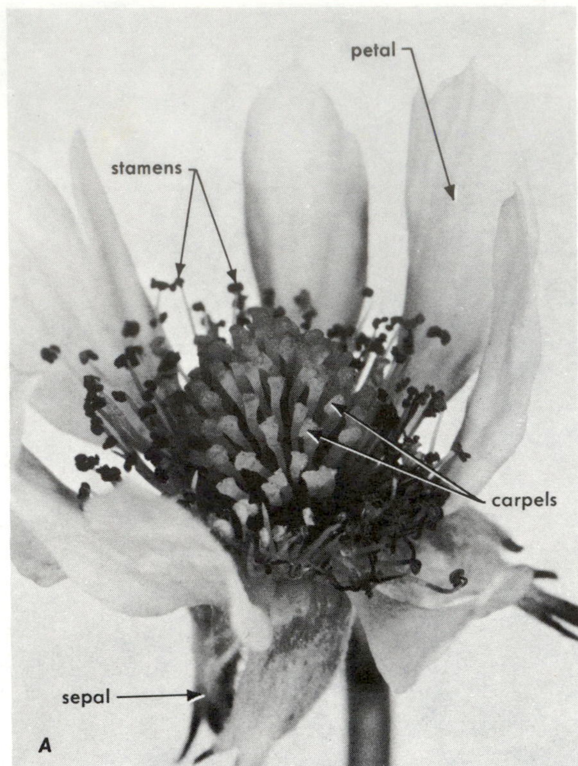

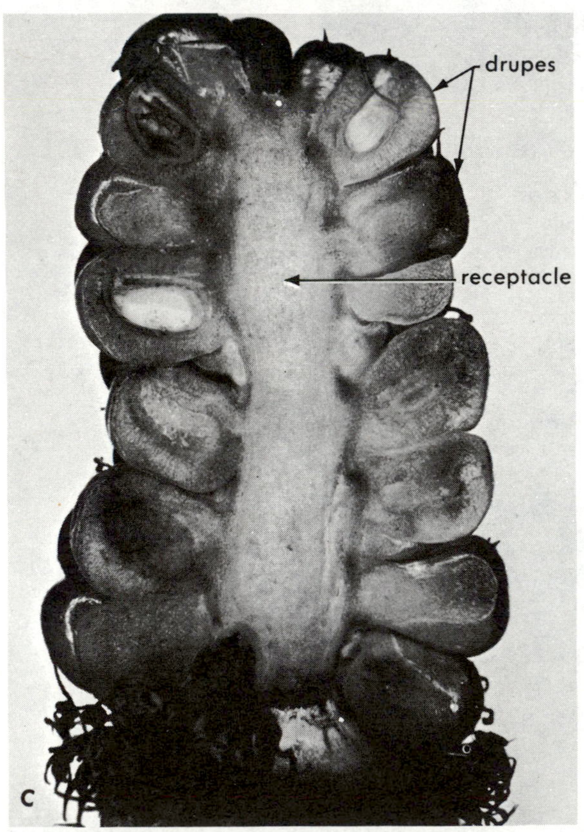

Figure 16.6
Aggregate fruit of the blackberry (*Rubus ursinus*). *A*, flower. *B*, flower shortly after petals have fallen. *C*, section of fruit, ×3.

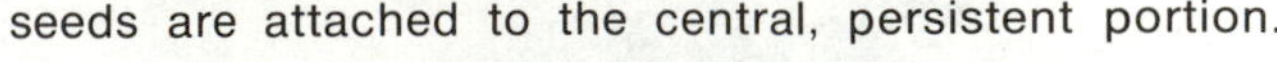

seeds are attached to the central, persistent portion.

Pericarp Dry and Indehiscent

Achene. Sunflower is an example of this type of fruit. These fruits are commonly called "seeds," but as in sunflower (Fig. 16.13), a carefully broken pericarp reveals that the seed within is attached to the placenta by its stalk. This pericarp may be separated easily from the seed coat, that is, from the layer of cells just beneath it. Achenes are indehiscent.

Grain or caryopsis. This is the fruit of the grass family, Poaceae (Gramineae), which includes such important plants as wheat, corn, and rice (*Oryza*). Like the achene, the grain is a dry, one-seeded, indehiscent fruit (Fig. 16.14). It differs from the achene, however, in that pericarp and seed coat are firmly united all the way around.

Samara. This is a dry, indehiscent fruit, which may be one-seeded, as in the elm (*Ulmus*), or two-seeded, as in maple (*Acer*) (Fig. 16.15*A*). These fruits are typified by an outgrowth of the ovary wall, which forms a winglike structure.

Schizocarp. This is the fruit characteristic of the carrot family, Apiaceae (Umbelliferae), which includes such common plants as carrot (*Daucus*), celery (*Apium*), and parsley (*Petroselinum*). The schizocarp is a dry fruit consisting of two carpels that split, when mature, along the midline into two one-seeded indehiscent halves (Fig. 16.15*B*).

Nut. The term nut is popularly applied to a number of hard-shelled fruits and seeds. A typical nut, botanically

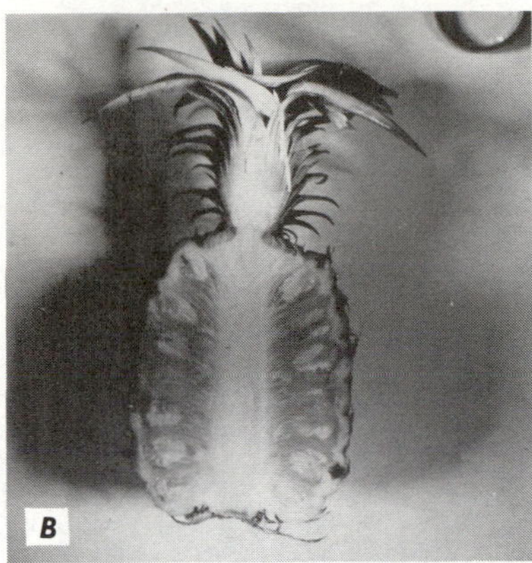

Figure 16.7
Pineapple (*Ananas comosus*) a multiple fruit. The flower is inferior, the resulting fruits are berries, sterile in the cultivated species, which become embedded in the swollen edible rachis. A jagged-edged bract extends out over the fruit. The fruits are spirally arranged. *A*, exterior view. *B*, section through center showing central rachis surrounded by coalesced fruits. $\times\frac{1}{10}$.

Figure 16.8
The fruit of the fig (*Ficus carica*). *A*, end of fleshy receptacle, showing pore, which is lined with staminate flowers. *B*, receptacle turned inside out, showing pistillate flowers that have matured into small drupes, ×1.

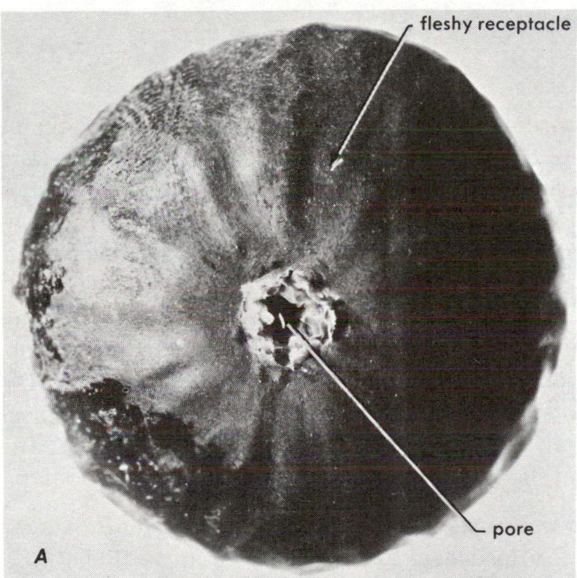

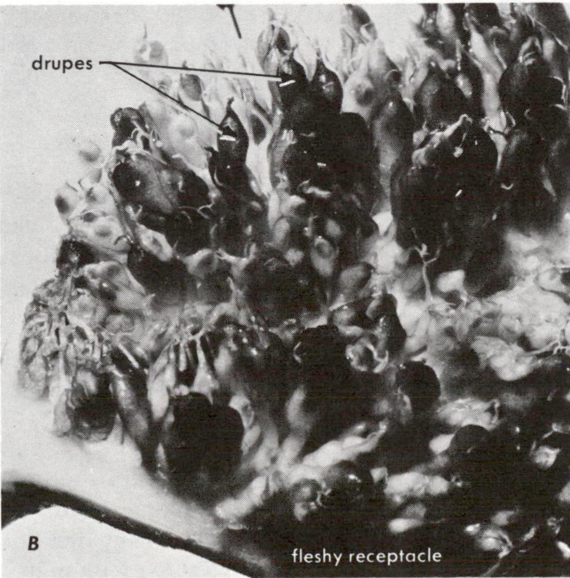

speaking, is a one-seeded, indehiscent dry fruit with a hard or stony pericarp (the shell). Examples are the chestnut (*Castanea*), walnut, and acorn (Fig. 16.16*A*,*B*). An acorn, the fruit of the oak, is partially enclosed by a hardened involucral cup. The husk or shuck of the walnut, which has been removed before the product reaches the market, is composed of involucral bracts, perianth, and the outer layer of the pericarp. The hard shell is the remainder of the pericarp. Unshelled almonds are drupes from which the hulls (exocarp and mesocarp) have been removed. Brazil nuts are seeds, not fruits. And the un-

Figure 16.9
Pod of Scotch broom (*Cytissus scoparius*). *A*, unopened. *B*, partially split. *C*, dehisced down both sutures.

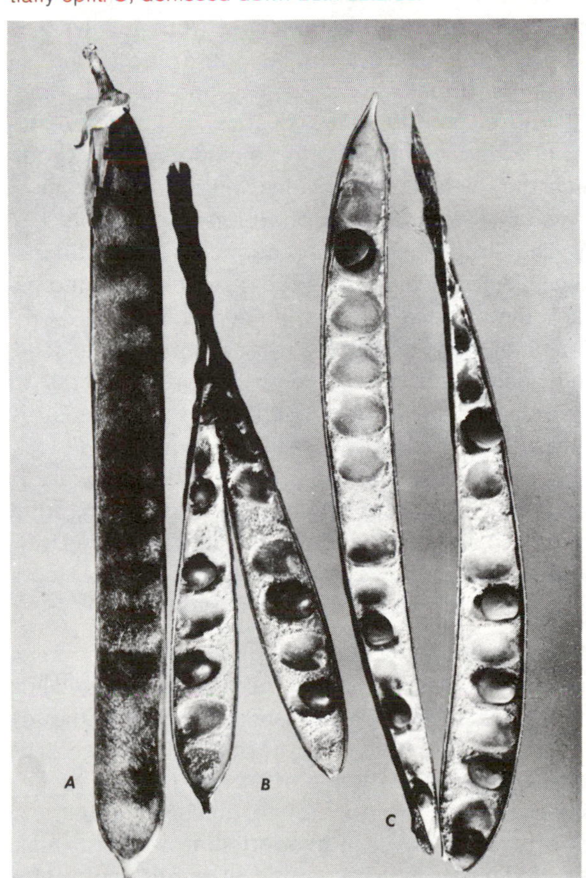

Figure 16.10
Follicles. *A, Magnolia,* each carpel shown in Fig. 15.2. *A* has now developed into a follicle, ×½. *B,* dehiscing follicle of a South American milkweed (*Oxypetalum caeruleum*).

shelled peanut is a pod, corresponding to the fruit of bean or pea; the edible portions are seeds within the pod.

Pericarp Fleshy

Drupe. Examples of this type of fleshy fruit are cherry, almond, peach, and apricot, all of which are members of a genus in the Rosaceae (rose family). The olive (*Olea*) fruit is also a drupe. The drupe is derived from a single carpel, and is usually one-seeded. However, if one examines young ovaries of flowers of almonds, cherries, plums, or other members of the group to which they belong, two ovules will be found, one of which usually aborts (fails to develop into a seed, Fig. 16.17). The drupe has a hard endocarp consisting of thick-walled stone cells. The exocarp is thin, forming the skin, and the mesocarp forms the edible flesh.

The pit or stone of a cherry, for example, is one seed, with thin seed coats, surrounded by a stony endocarp. In other words, the pit is composed of the seed plus the stony inner layer (endocarp) of the ovary wall.

In almond fruits (Figs. 16.3, 16.17), the mesocarp is fleshy like a typical drupe when the fruit is young, but it becomes hard and dry as it develops, and forms the hull. The shell of the almond is endocarp. In coconut, also a drupe, the pericarp tissues become dry at maturity.

Berry. This fleshy type of fruit is derived from a compound ovary. Usually, many seeds are embedded in a flesh, which is both endocarp and mesocarp, although the line of demarcation may be difficult to see. The tomato is a common example of a berry (Fig. 16.4).

The citrus fruit (lemon, orange, lime, and grapefruit) is a type of berry called a **hesperidium.** It has a thick, leathery rind (peel), with numerous oil glands, and a thick juicy portion composed of several wedge-shaped locules (Fig. 16.18). The peel of a citrus fruit is exocarp and mesocarp; the pulp segments are endocarp. The juice is in pulp sacs or vesicles; they are outgrowths from the endocarp walls (Fig. 16.18*C*), and each mature vesicle is composed of many living cells filled with juice.

Another berrylike fruit is that of members of the family Cucurbitaceae, which includes watermelon, squash, pumpkin, and cucumber. It is called a **pepo** (Fig. 30.11). The outer wall (rind) of the fruit consists of receptacle tissue that surrounds and is fused with the exocarp. The flesh of the fruit is principally mesocarp and endocarp.

Pome. This type of fruit is characteristic of a subfamily of Rosaceae, to which belong apple (Fig. 16.19) and pear. The pome is derived from an inferior ovary (Figs. 16.19*A,B*). The flesh is enlarged hypanthium, (fleshy floral tube) and the core has come from the ovary (Fig. 16.19*C,D*).

Compound Fruits – Formed From Several Ovaries

Aggregate Fruits. An aggregate fruit is one formed from numerous carpels of one individual flower. These fruits are made up of many simple fruits that may be classified as types of simple fruits. The strawberry flower has numerous separate carpels on a single receptacle (Fig. 16.5). The ovary of each carpel has one ovule, and the ovary develops into a one-seeded dry fruit (achene). The receptacle to which these fruits are attached becomes fleshy; the whole structure, which we call a strawberry, is an aggregate of simple fruits, each an achene. The receptacle is stem tissue and consists of a fleshy pith and cortex with vascular bundles between them. The hull of

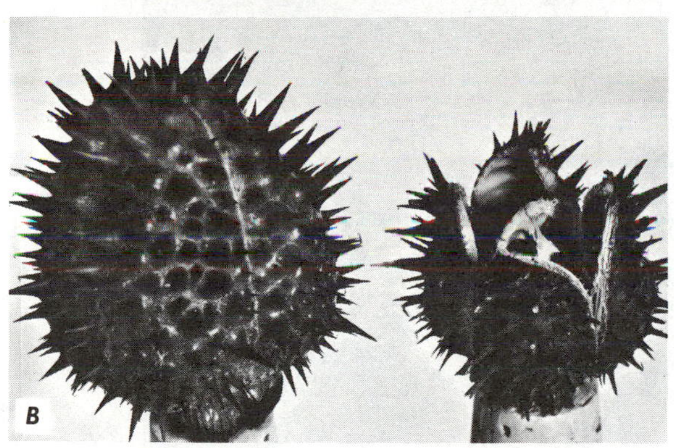

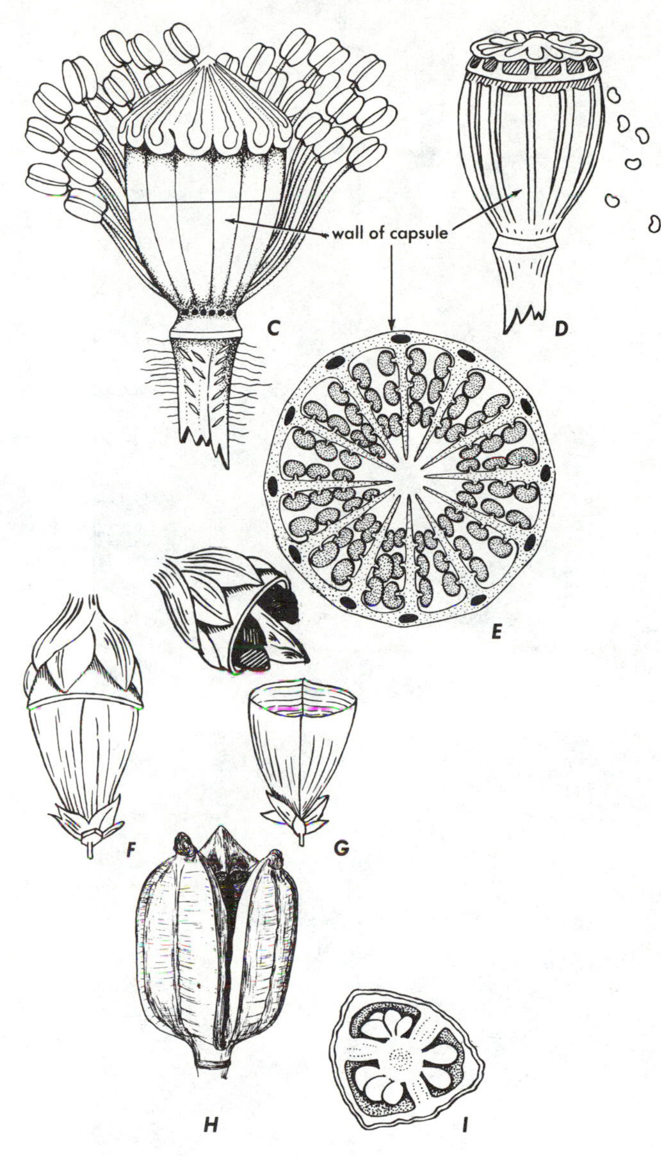

Figure 16.11
Capsules. *A, Amaryllis. B, Datura. C,* poppy capsule, side view before dehiscence. *D,* mature poppy capsule dehiscing by pores at the top. *E,* cross section of poppy capsule, showing its many carpels. *F,* mature capsule of plantain. *G,* capsule of plantain opening by transverse lid. *H,* capsule of tulip dehiscing lengthwise. *I,* cross section of tulip ovary. All approximately ×1. (*A* and *B,* courtesy of W. Russell; *D, E,* and *F,* redrawn from E. Krosmo. *Unkraut im Ackerbau der Neuzeit.* © 1930 by Springer-Verlag.)

the strawberry fruit is composed of persistent calyx and withered stamens. The achenes are usually spoken of as seeds.

Flowers of raspberry and blackberry, and other species of *Rubus* have essentially the same structure as those of strawberry. In these flowers, many separate carpels attached to one receptacle develop into small drupes (Fig. 16.6).

Multiple Fruits. A multiple fruit is one formed from individual ovaries of several flowers. These fruits, considered individually, may be classified as types of simple fruits. The fig (Fig. 16.8) and pineapple (Fig. 16.7) are examples of multiple fruits; the individual fruits composing them are nutlets in fig and parthenocarpic berries in pineapple.

The fig fruit we eat is an enlarged fleshy receptacle. Its flowers are small and attached to the inner wall of the receptacle. Both staminate and pistillate flowers occur and may be borne in the same or in different receptacles.

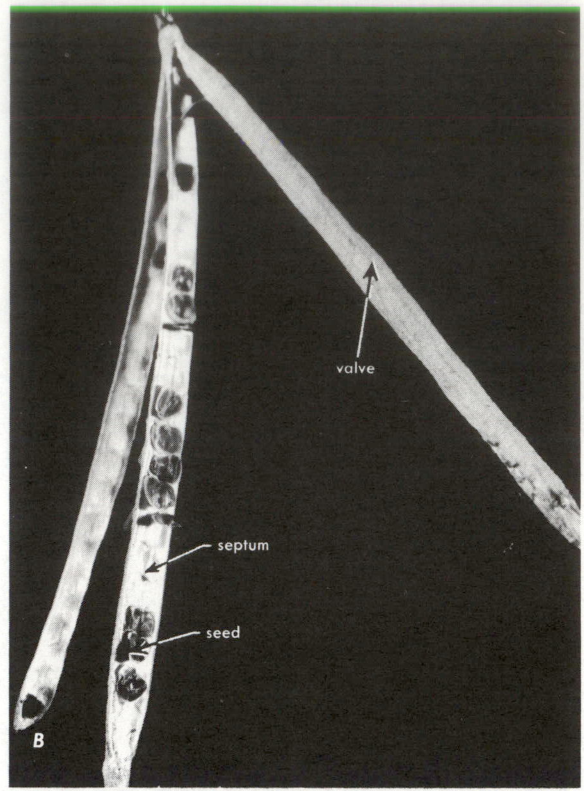

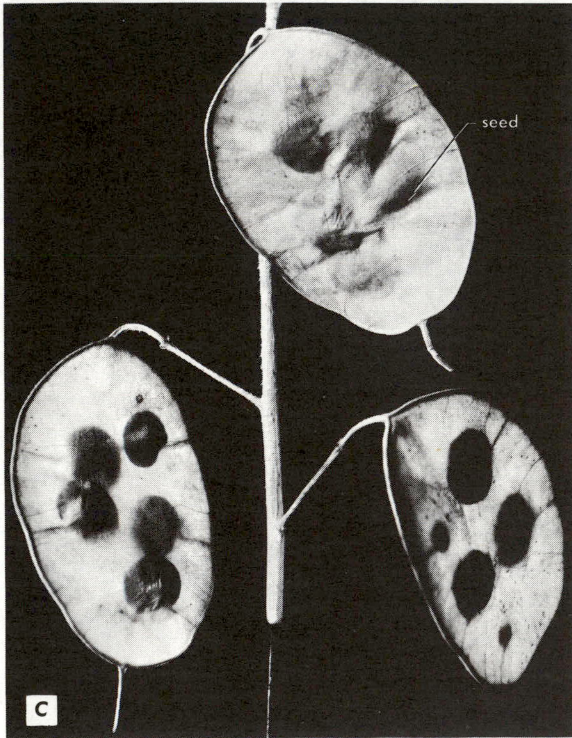

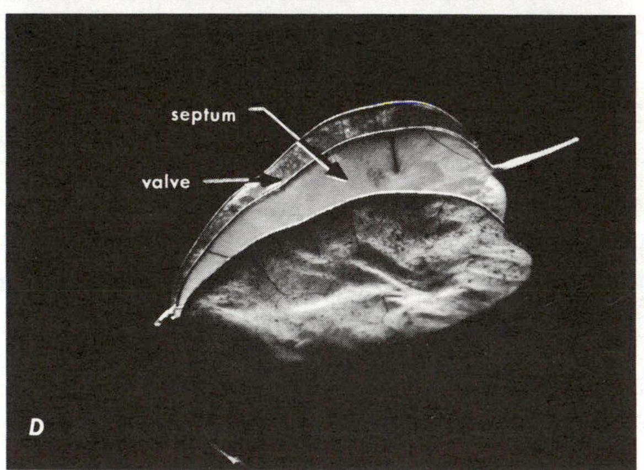

Figure 16.12
Siliques. *A* and *B*, of stock (*Matthiola incana*). *A*, unopened, ×¾. *B*, partially split to show orientation of valves and septum, ×¾. *C* and *D*, of moonwort (*Lunaria annua*). *C*, unopened, ×¾. *D*, opened to show valves and septum, ×¾. (Courtesy of W. Russell.)

Each ovary develops into a nutlet that is embedded in the wall of the receptacle. Thus, the fig is derived from many flowers, all attached to the same receptacle. Common edible figs (Mission and Kadota) are parthenocarpic. Smyrna figs and several other varieties do not mature their fruits unless fertilization and seed formation have taken place (Fig. 15.33).

Accessory Parts of Fruits. In the strawberry, the edible portion is a thickened pulpy central stem in which achenes are embedded. Tissues other than the ovary wall that form part of a fruit, are referred to as **accessory.** Thus, much of the fleshy fruit of pineapple, apple, and strawberry is accessory.

Summary of Fruits

1. A fruit is a ripened ovary, plus any other closely associated floral parts.

2. The initiation of fruit development generally depends upon pollination and fertilization.
3. Fruit development is accompanied by the development of the ovule into the seed.
4. There are *three different kinds of fruits*, classified on the basis of the number of ovaries and flowers involved in their formation:

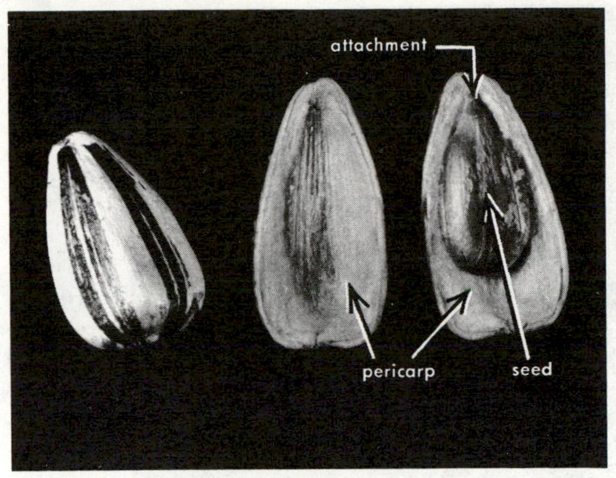

Figure 16.13
Achene of sunflower (*Helianthus annuus*), unopened, and opened to show attachment of seed, ×1.

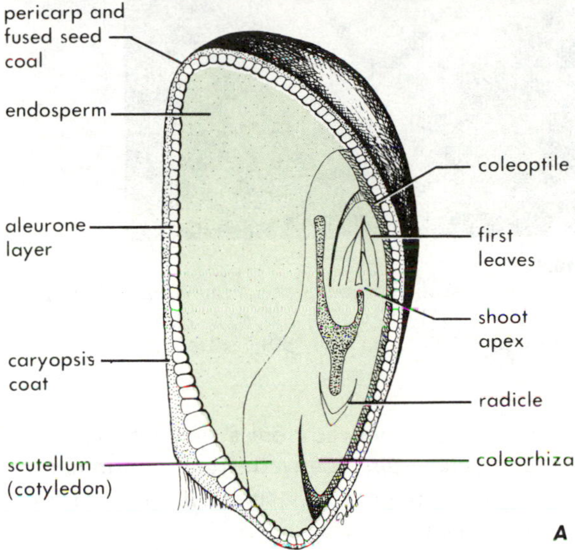

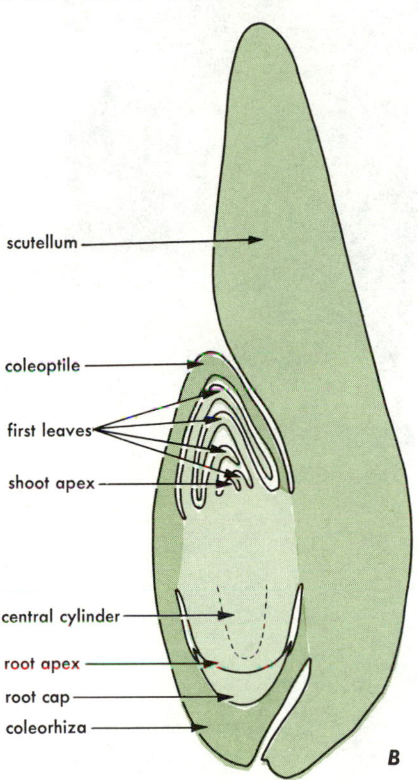

Figure 16.14
A, median section of caryopsis of yellow foxtail (*Setaria lutescens*). *B*, longitudinal section of an embryo of corn (*Zea mays*).

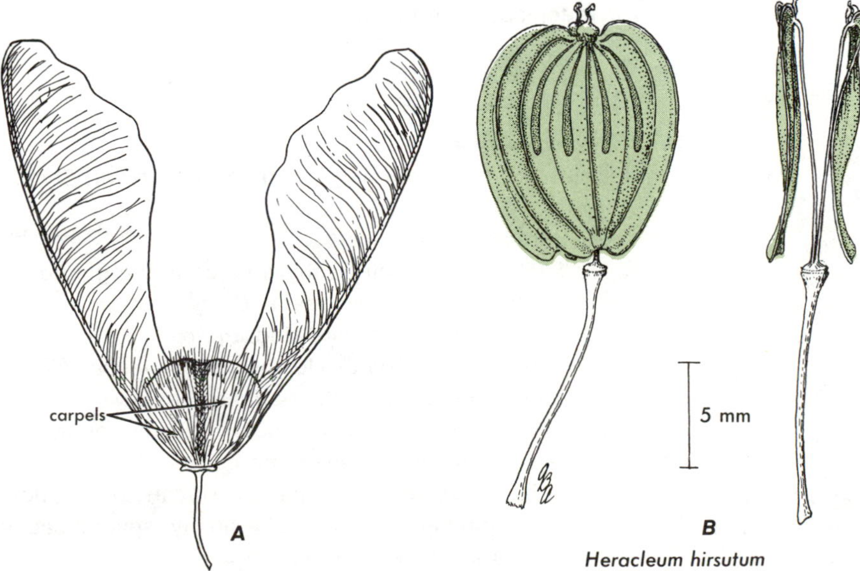

Figure 16.15
A, two-seeded samara of maple (*Acer*). *B*, example of schizocarp (*Heracleum hirsutum*); left is face view and right is side view.

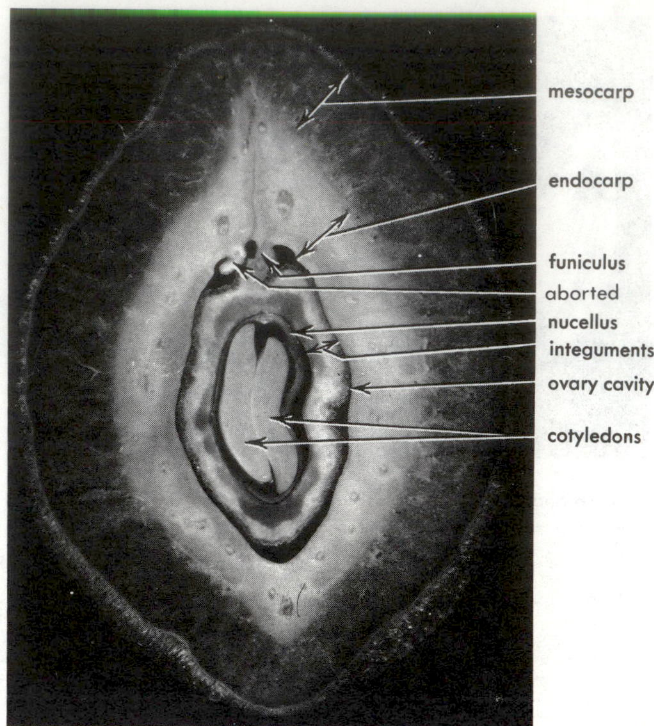

Figure 16.17
Median transverse section of young almond fruit (*Prunus amygdalus*) (drupe). (From R. M. Holman and W. W. Robbins, *A Textbook of General Botany*, John Wiley & Sons, New York.)

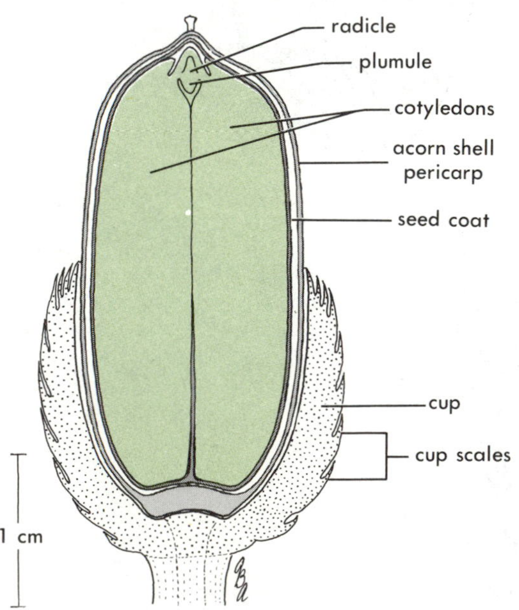

Figure 16.16
Acorns of oak (*Quercus*). *A*, the fruits are a type of nut and the cups are fused involucrol bracts, ×1. *B*, internal view of acorn of *Quercus Kellogii*.

(*a*) *simple fruits*, derived from a single ovary;
(*b*) *aggregate fruits*, derived from a number of ovaries belonging to a single flower and on a single receptacle;
(*c*) *multiple fruits*, derived from a number of ovaries of several flowers more or less grown together into one mass.

5. Simple fruits may have a dry pericarp or a fleshy pericarp. If the pericarp is dry, it may be dehiscent (splitting at maturity, allowing seeds to escape) or indehiscent (not splitting).

Key to Fruits

I. Fruit formed from a single ovary of one flower. *Simple fruits.*
- **A.** Pericarp fleshy.
  - **1.** The ovary wall fleshy and containing one or more carpels and seeds. *Berry.*
    - **a.** Ovary wall a hard rind. *Pepo.*
    - **b.** Ovary wall with a leathery rind. *Hesperidium.*
  - **2.** Only a portion of the pericarp fleshy.
    - **a.** Exocarp thin; mesocarp fleshy; endocarp stony; single seed and carpel. *Drupe.*
    - **b.** Outer portion of pericarp fleshy, inner portion papery, floral tube fleshy; several seeds and carpels. *Pome.*

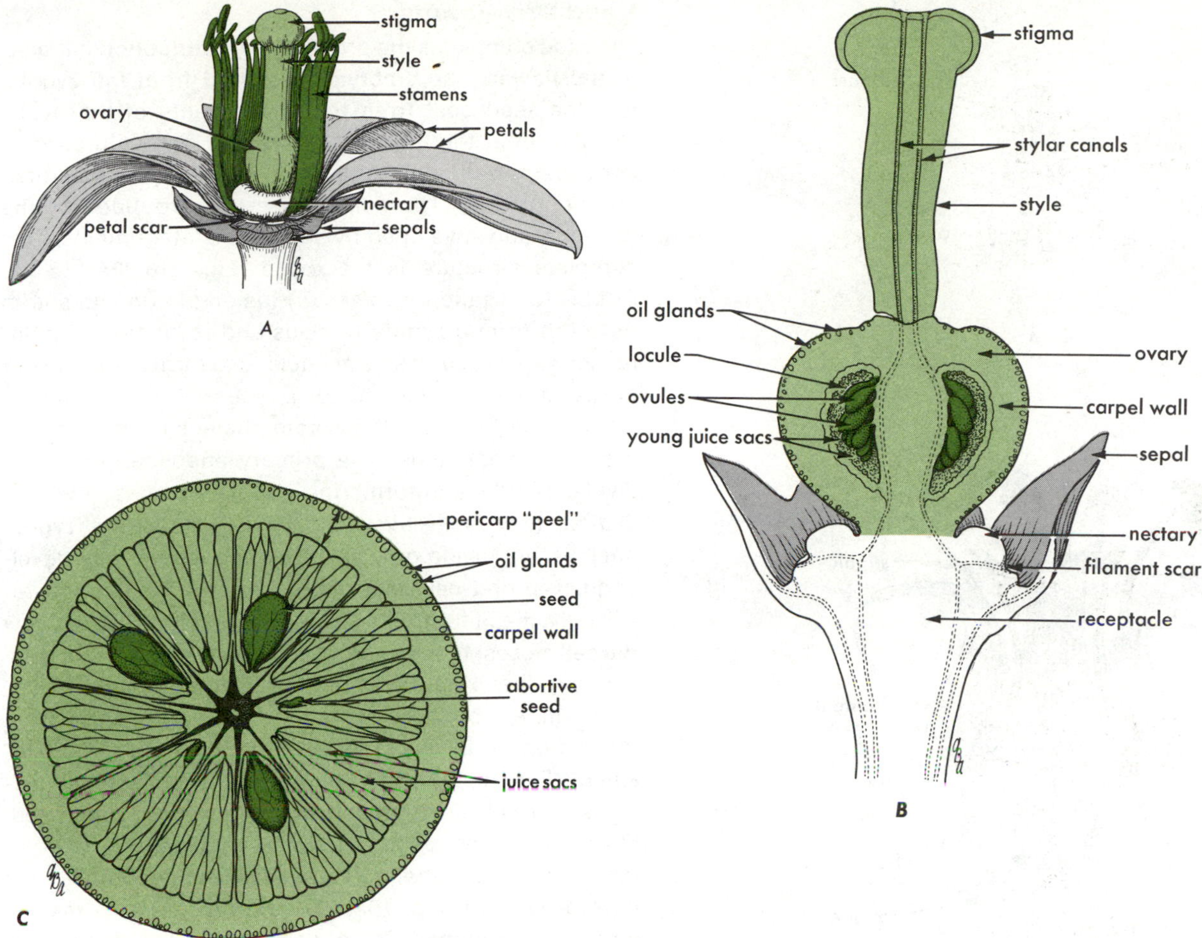

Figure 16.18
Flower and fruit (hesperidium) of orange (*Citrus sinensis*). *A*, flower, dissected to show pistil and stamens. *B*, lengthwise section of maturing ovary. *C*, cross section of mature fruit.

B. Pericarp dry.
  1. Dehiscent fruits.
    a. Composed of one carpel.
      (1) Splitting along two sutures. *Legume.*
      (2) Splitting along one suture. *Follicle.*
    b. Composed of two or more carpels.
      (1) Dehiscing in one of four different ways. *Capsule.*
      (2) Separating at maturity, leaving a persistent partition wall. *Silique.*
  2. Indehiscent fruits.
    a. Pericarp bearing a winglike growth. *Samara.*
    b. Pericarp not bearing a winglike growth.
      (1) Carpels, two to many, united when immature, splitting apart at maturity. *Schizocarp.*
      (2) Carpels one, if more, not splitting apart at maturity; one-seeded fruits.
        (a) Seed united to the pericarp all around. *Caryopsis* or *grain.*
        (b) Seed not united to the pericarp all around.
          Fruit large, with thick, stony wall. *Nut.*
          Fruit small, with thin wall. *Achene.*

II. Fruits formed from several ovaries.
  A. Fruits developing from one flower. *Aggregate fruits* (classify the individual fruits in key for simple fruits).
  B. Fruits developing from several flowers. *Multiple fruits* (classify individual fruits in key for simple fruits).

THE SEED

The seed provides a highly efficient mechanism to: (*a*) carry plants over periods unfavorable for growth; (*b*) maintain a reserve supply of potential plants in case of successive unfavorable seasons; and (*c*) disseminate plants. The tremendous production of seeds ensures the renewal or continuance of plant populations. Many animals, including man, cooperate in this population renewal and dissemination.

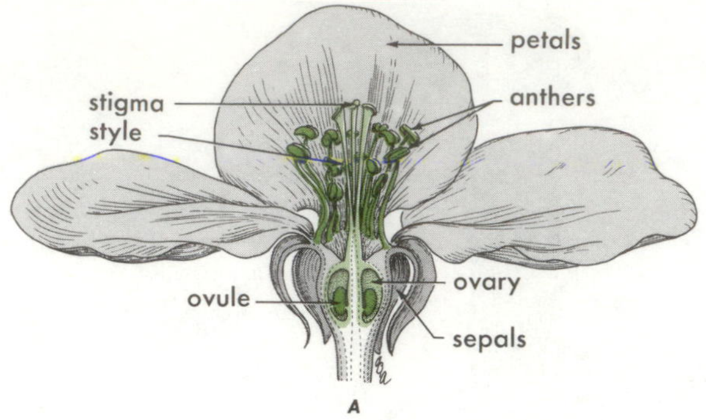

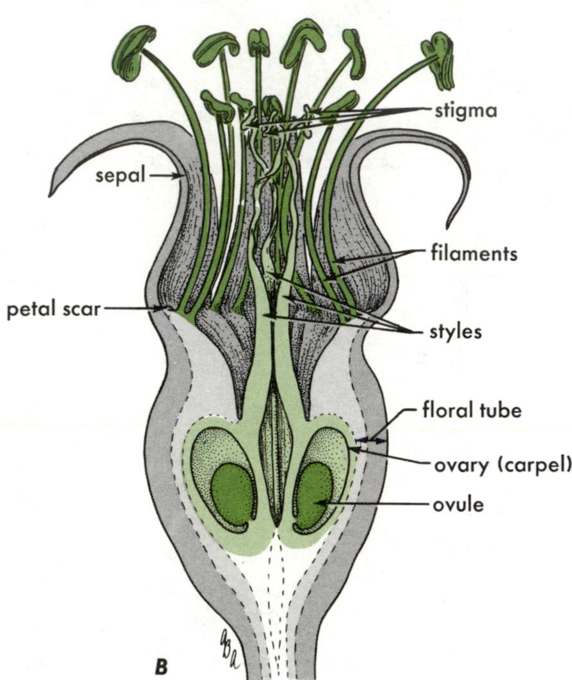

Seed Development

The seed completes the process of reproduction initiated in the flower. The embryo developed from the zygote, and the seed coat from the integuments of the ovule. Recall that at the time of fertilization, egg cell, primary endosperm cell, synergids, and antipodals constituted the embryo sac. The embryo sac is surrounded by the nucellus, and enveloped by one or two integuments. This complete structure is the ovule (Figs. 16.20*A*, 15.29). Double fertilization involves (*a*) fusion of egg and sperm nuclei to form a zygote nucleus and (*b*) fusion of polar nuclei with a second sperm nucleus to form a primary endosperm nucleus (Fig. 16.20*B*).

After fertilization, the zygote nucleus remains quiescent for a short time. The primary endosperm nucleus divides rapidly and forms the **endosperm tissue** which, at first, may lack cell walls (Fig. 16.20*C*). The zygote nucleus will divide only after the endosperm has developed. The first cell divisions of the new generation results in a file of from four to eight cells (Fig. 16.20*D*). Now the cell closest to the micropyle elongates (the basal cell), pushing the other cells of the filament into the endosperm. At about the same time, the cell furthest from the micropyle divides at right angles to the axis of the filament (Fig. 16.20*E*). This is the first in a series of divisions that will finally result in the embryo of the seed. Further divisions result in a globular stage (Fig. 16.20*F*), then into a heart-shaped stage after the two cotyledons have developed (Fig. 16.20*G*), and finally into a mature embryo surrounded by a seed coat, the seed (Fig. 16.20*H*).

Figure 16.19
Flower and fruit (pome) of apple (*Malus sylvestris*). *A*, median longitudinal section of flower. *B*, median longitudinal section of a maturing fruit. *C*, median section of fruit. *D*, cross section of fruit. (*C* and *D*, redrawn from W. W. Robbins, *Botany of Crop Plants*. © 1931 by Blakiston. Reprinted with permission of the publisher.)

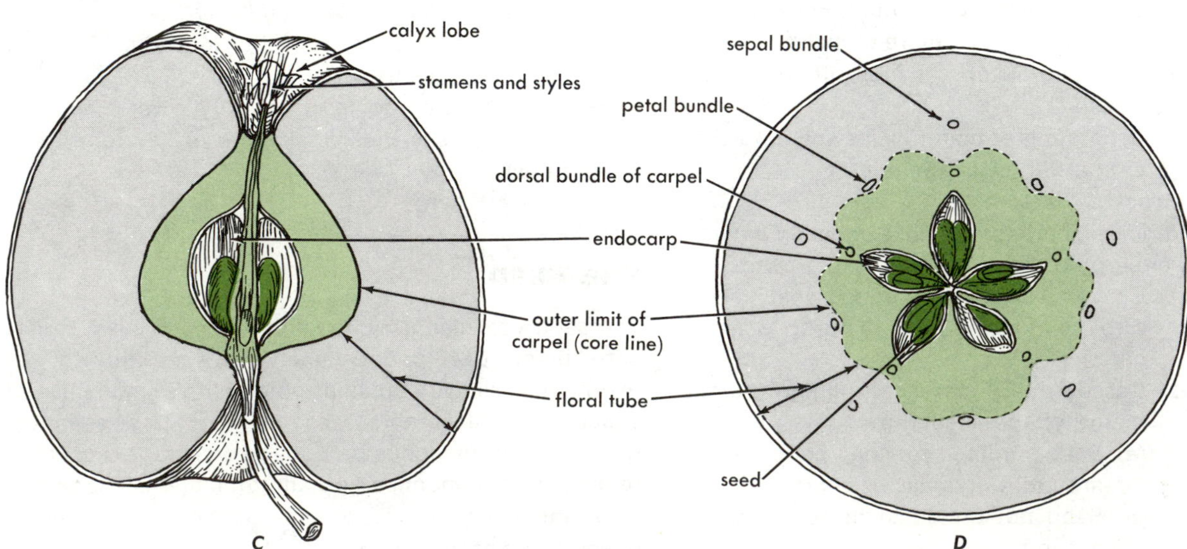

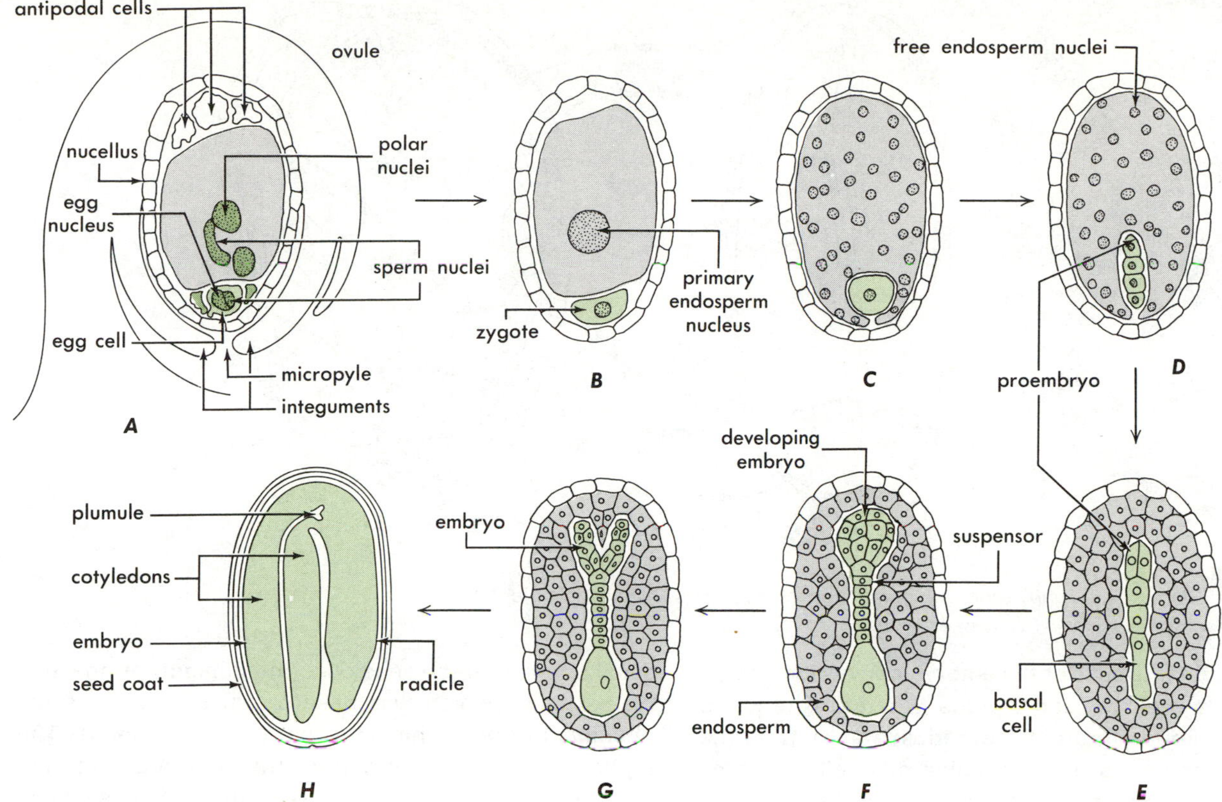

Figure 16.20
Diagram of the development of dicotyledonous angiosperm seed. *A*, ovule after fertilization. *B*, after fusion of gamete nuclei to form zygote and endosperm. *C*, free nuclear divisions in the endosperm. *D*, filamentous stage of the proembryo. *E*, elongation of suspensor cell and division of proembryo cell. *F*, globular stage of the embryo. *G*, heart-shaped stage of embryo development. *H*, mature seed.

The course of events just described is very general, for there are many variations in details of embryo development. Note too that the major distinction between seeds of monocotyledonous and dicotyledonous plants is the occurrence of one or two cotyledons, respectively (Figs. 16.20*H*, 16.21).

While development of the embryo is taking place, the nucellus, endosperm, and integuments are also undergoing changes that are characteristic of the group of plants to which the seeds belong. In the great majority of plants, nucellus and endosperm are required only for the initial stages of development. This is particularly true of the nucellus, which is generally used up as a nutritive source in early stages. It persists as a food storage tissue, the **perisperm,** in seeds of sugar beet (*Beta*), and a few other species.

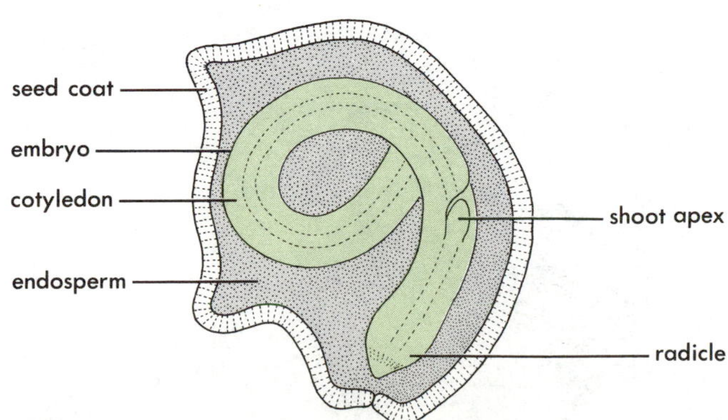

Figure 16.21
A longitudinal section through an onion seed (*Allium cepa*) showing the embryo coiled within the endosperm.

The persistence of endosperm as a food reserve occurs in relatively few dicotyledonous seeds; the castor bean (*Ricinus communis*) is a good example (Fig. 16.22). In monocotyledonous plants, however, the endosperm often persists in seeds as the primary food source during germination. Important grasses, such as corn, oats, wheat, and barley are exploited agriculturally because of this stored food.

In most seeds, integuments of the ovule become hard and horny seed coats in the mature seed. The scanning electron microscope shows the seed coats to be variously and sometimes beautifully sculptured (Fig. 16.23).

Kinds of Seeds

Seeds of angiosperms differ in two ways: (*a*) they have one or two cotyledons and (*b*) they store food either in the embryo, in the endosperm, or more rarely, in nucellar tissue, the perisperm, which lies between embryo and seed coat.

When food storage occurs within the embryo, the

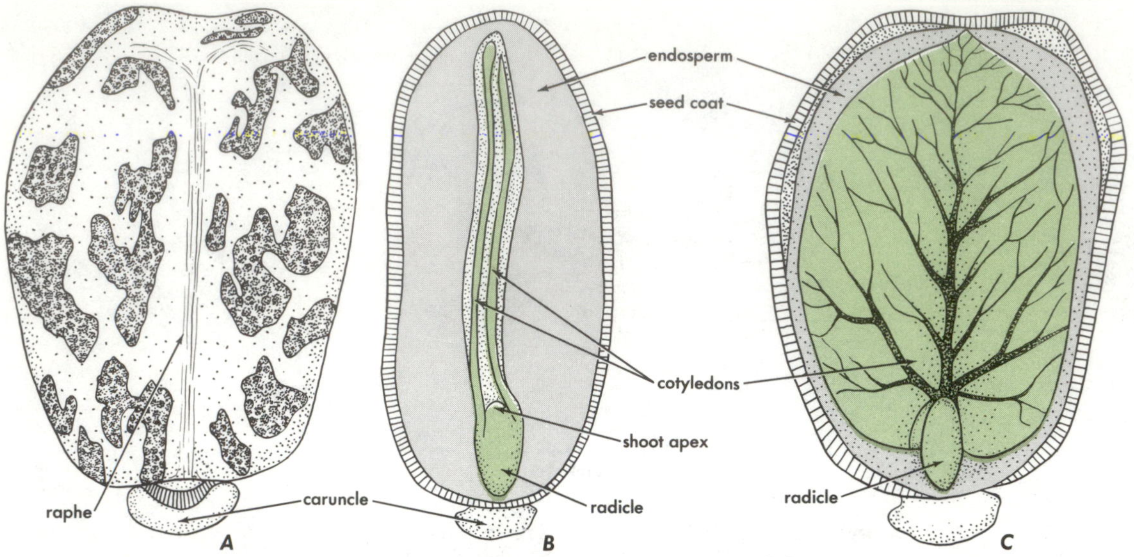

Figure 16.22
Castor bean (*Ricinus communis*) seed. *A*, external appearances. *B*, section showing edge view of embryo. *C*, section showing flat view of embryo.

normal vascular tissues of the embryo convey the solubilized food to meristematic regions, where it is required for growth. Food stored in the endosperm, outside the embryo, is absorbed through epidermal cells. In some cases, the cotyledons become specialized to act as absorbing organs (Fig. 16.28). In most grasses, the entire cotyledon has become a highly specialized absorbing organ known as the scutellum (Fig. 16.14).

Common Bean

The fruit of the bean plant is a pod; within it are seeds. The external characters of the bean seed are more clearly seen after soaking it in water. Points of interest are the **hilum, micropyle,** and **raphe** (Fig. 16.24). The hilum is a large oval scar, near the middle of one edge, left where the seed broke from the stalk or **funiculus** when the beans were harvested (see Fig. 16.1). The micropyle is a small opening in the seed coat (integument) at one side of the hilum, which was observed in the ovule as the opening through which the pollen tube entered. The raphe is a ridge at the side of the hilum opposite the micropyle. It represents the base of the funiculus, which is fused with the integuments. Conducting tissue present in the funiculus will be continued in the raphe. At the end of the raphe, this conducting tissue may fan out over the ovule or seed and lose its identity as conducting tissue. This region is known as the **chalaza.** When present, it occurs at the end of the ovule or seed opposite the micropyle. An elevation or bulge of the seed coat adjacent to the micropyle marks the position of the radicle (embryo root) within the seed.

When the seed coat of a soaked bean is removed, the entire structure remaining is embryo; no endosperm is present. The following parts of the embryo can be ob-

Figure 16.23
The coats of many seeds are characteristically sculptured. *A*, morning glory (*Convolvulus arvensis*), ×23. *B*, California poppy (*Eschscholtzia californica*), ×62.

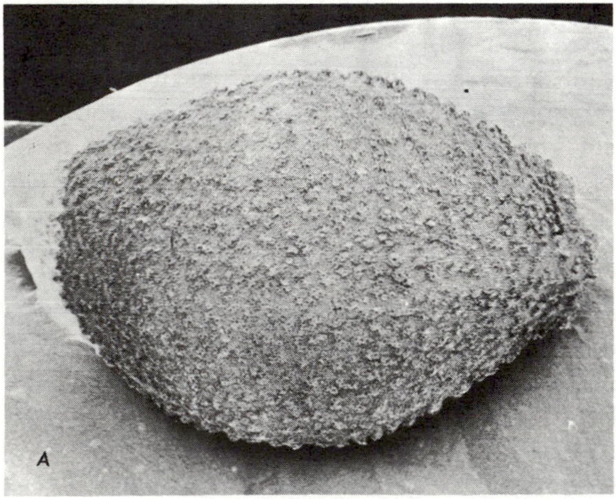

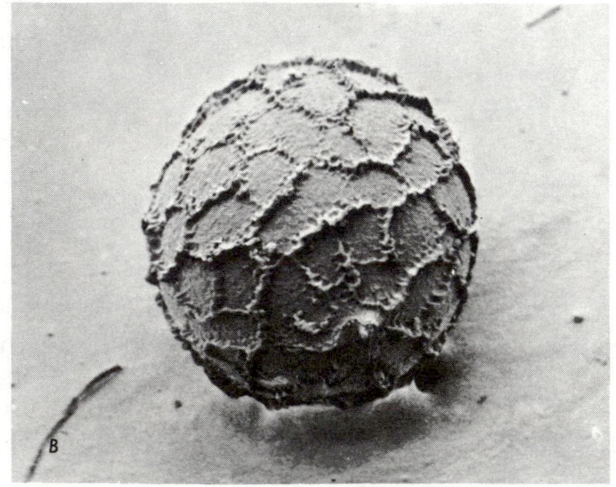

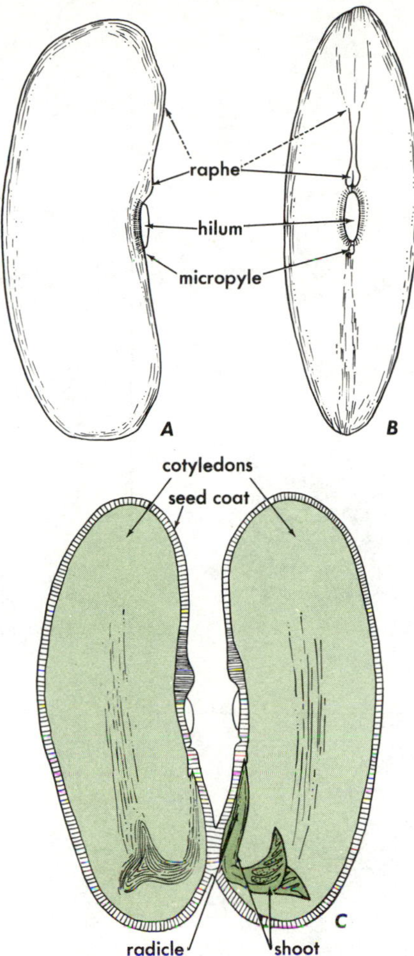

Figure 16.24
Bean (*Phaseolus vulgaris*) seed. *A*, external side view. *B*, external face or edge view. *C*, embryo opened.

served: (*a*) a shoot consisting of two fleshy cotyledons, a short axis (the hypocotyl) below the cotyledons, and a short axis (the epicotyl) above the cotyledons, bearing several minute foliage leaves and terminating in a shoot tip; and (*b*) the root or radicle (Fig. 16.24).

Castor Beans

The external points of interest of this seed are: (*a*) **caruncle,** a spongy structure, an outgrowth of the outer seed coat; (*b*) hilum and (*c*) micropyle, which are covered by the caruncle; and (*d*) the raphe, which runs the full length of the seed (Fig. 16.22).

The chalaza region is marked, in the castor bean, by a protuberance at the end of the seed opposite the caruncle. In contrast to the bean seed, the castor bean seed has a massive endosperm in which the embryo is embedded. The shoot of the embryo consist of (*a*) two thin cotyledons with conspicuous veins, (*b*) a very short hypocotyl (shoot axis below the cotyledons), and (*c*) a minute epicotyl (shoot axis above the cotyledons). The root portion of the embryo axis consists of a small radicle.

Grasses

The so-called "seed" of corn, or of any other grass, is in reality a fruit, the caryopsis. It is a one-seeded, dry, indehiscent fruit with the pericarp (ovary wall) firmly attached to the seed. Pericarp and seed coats are so firmly attached to each other and to other tissues of the grain that it is impossible to peel them away.

Longitudinal sections of a caryopsis of yellow foxtail (*Setaria lutescens*) and an embryo of corn (*Zea mays* are shown in Figs. 16.14*A* and *B*). The endosperm constitutes the bulk of the caryopsis and is composed of (*a*) an outer layer (single row of cells) the **aleurone layer** and (*b*) a starchy endosperm. Cells of the aleurone layer contain proteins and fats but little or no starch.

Like all embryos of other seed plants, the grass embryo has an axis with a **shoot apex** and a **root apex** (Fig. 16.14). The shoot apex, together with several rudimentary leaves, are surrounded by a sheath, the **coleoptile.** The rudimentary root (radicle) is also surrounded by a sheath, the **coleorhiza.** At the juncture of shoot and root is a very short stemlike region. A relatively large part of the grass embryo is a cotyledonlike structure called the **scutellum.** It is a shield-shaped structure that lies in contact with the endosperm. The outer cells of the scutellum secrete enzymes that digest the stored foods of the endosperm. These digested foods move from endosperm cells through the scutellum to the growing parts of the embryo.

It is noteworthy that the process of milling to make polished white rice removes the outside caryopsis coat, plus the entire protein-rich aleurone layer, and the outer layers of the endosperm. This means that most of the nutritious protein is removed before the rice is eaten! The removed material, composed of the outside layers and the embryo axis, is sold as bran.

Onion

The onion seed consists of seed coats enclosing a small amount of endosperm. The embryo is a simple axis, the radicle and single cotyledon being quite prominent. The shoot apex is located close to the midpoint of the axis and appears as a simple notch. The embryo is coiled with the radicle usually pointing toward the micropyle (Fig. 16.21).

Dissemination of Seeds and Fruits

Agents in Seed and Fruit Dispersal

The chief agents in seed and fruit dispersal are **wind, water,** and **animals.**

Wind. The structural modifications of seeds and fruits that aid in dissemination by wind are of several types. The most common of these are winged types (Fig. 16.25*C*) and plumed types (Fig. 16.25*A,B*). Some seeds are simply dispersed by the violent dehiscence of the pericarp (Fig. 16.25*H,I*).

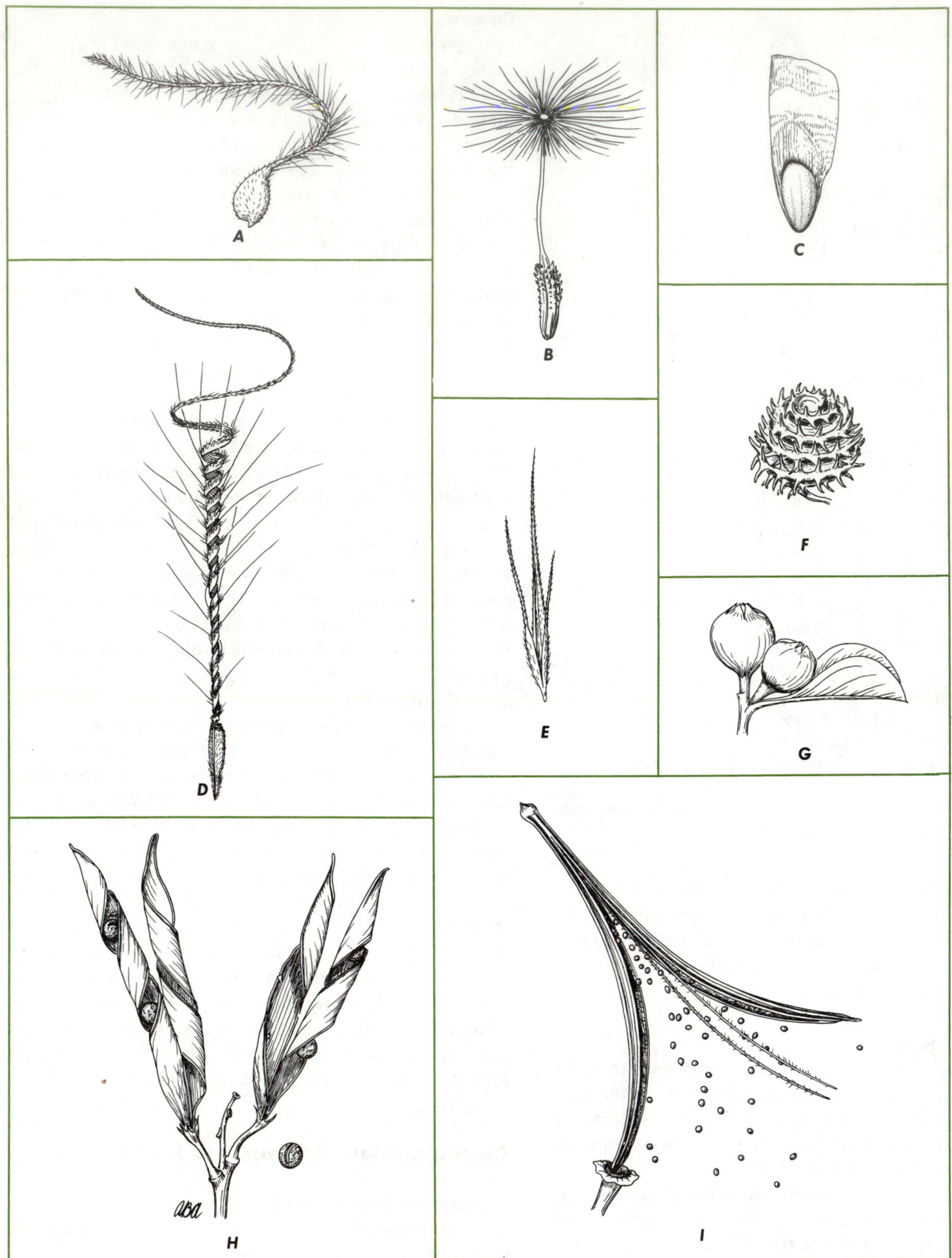

Figure 16.25
Fruits and seeds showing various devices aiding in dissemination. **Wind:** *A, Clematis,* ×4. *B,* dandelion (*Taraxacum vulgare*), ×5. *C,* seed of Coulter's big-cone pine (*Pinus coulteri*), ×1. **Attachment:** *D,* cranesbitt (*Geranium*), ×3. *E,* foxtail (*Hordeum hispida*), ×3. *F,* bur clover (*Medicago denticulata*), ×4. *G,* fleshy **edible** fruit of *Cottoneaster,* ×2. Violent **dehiscence** of pericarp. *H,* vetch (*Vicia sativa*), ×2. *I,* California poppy (*Eschscholtzia californica*), ×2.

Animals. Many seeds and fruits are carried by animals, both wild and domesticated. Seeds with beards, spines, hooks, or barbs adhere to hair of animals (Figs. 16.25*D,E,F*). An example is bur clover (*Medicago denticulata*)

Seeds of many plants pass through the digestive tract of animals without having their viability impaired. Fleshy, edible fruits may be eaten by birds (Fig. 16.25*G*) and carried by them long distances, and then the seeds regurgitated or discharged with the excrement. Squirrels carry nuts, such as those of walnut and hickory (*Carya*), and the seeds of pines. Seeds of some aquatic and marsh plants and of mistletoe (*Phoradendron*), which are covered with a sticky material, are carred on the feet of birds.

Water. Fruits with a membranous envelope containing air like those of sedges, or with a coarse, loose, fibrous outer coat, as in the coconut (*Cocos*), are well adapted for dispersal by water. A great variety of fruits and seeds float in water, even though they lack special adaptations to ensure buoyancy, and are readily transported long distances by moving water. In irrigated districts, irrigation water is a very important means by which weed seeds are distributed.

SEED DORMANCY

Seeds can remain viable for remarkably long periods. In one study, begun in 1878, Dr. W. J. Beal of Michigan State University buried jars containing seeds from several plant species. At 5- and 10-year intervals a jar was opened and the seeds tested for germination. Most species remained viable for at least 10 years, and one species, the moth mullein (*Verbascum blattaria*), still germinated after more than 90 years. This, however, is no record for seed longevity. Seeds from the Oriental lotus (*Nelumbo nucifera*) have been removed, still viable, from archeological diggings known to be more than 1000 years old! Seeds of a few other species have been recovered from cold and anaerobic deposits, dated at over 1000 years, and have also proven viable. It is truly remarkable that seeds can remain living over such long periods.

Many seeds will not germinate even when supplied with water, oxygen, and a favorable temperature. There are various factors in different plants that are associated with the breaking of dormancy. For instance, light is necessary for the germination of some grass seed and for some strains of lettuce (*Lactuca*) and perhaps other plants. Water and gases must pass through the seed coats. If seed coats restrict the movement of water and gases or are impermeable to one or the other or to both, the seed coats must decay, be broken, or be scratched, allowing water and oxygen to reach the embryo before germination can start. In some seeds, the embryo itself fails to germinate when provided with favorable conditions. In orchids, the seeds are disseminated while embryos are rudimentary and the embryos must develop further before germination.

Many kinds of seeds are known that will not germinate unless the seeds, while on a moist substrate, are subjected for a time to temperatures close to freezing. At the other extreme, some seeds will not germinate unless they have been subjected to the rather high heat of a fire. Another type of dormancy is produced by the presence of natural chemical inhibitors. These occur in many fruits and serve to keep the seeds dormant while they are enclosed by the fruit. In other cases, inhibitors are present in the seeds themselves or may be produced by decaying leaves, or forest litter. The influence of the plant growth substances on dormancy and seed germination is still unclear.

GERMINATION

What induces germination? Many seeds will germinate when provided with moisture, oxygen, and a favorable temperature. The water content of seeds is low, between 5 and 10%. The cytoplasm with contained organelles is scarcely recognizable because it is crowded between the large amount of reserved food materials (Fig. 16.26*A*). Water is, at first, imbibed very rapidly, which results in the swelling of the protoplasm with a reappearance of organelles (Fig. 16.26*B*). Increase in water content makes possible an increased metabolic activity. From this point in time the seed must have an adequate water supply for survival.

The first indication that the processes of germination have begun is generally the swelling of the radicle. In all cases, the radicle imbibes water rapidly and, bursting the seed coats and other coverings that may be present, starts to grow downward into the soil. This helps to assure that the young seedling has an adequate supply of water and nutrients when the shoot breaks through the surface of the soil. Although the succeeding steps of germination are essentially similar, there are variations. For instance, in germination of beans, peas, castor beans, and onions, a structure with a sharp bend or **hook** is first forced through the soil (Figs. 16.27, 16.28, 16.29, 16.30), but the structure forming the hook is different in each case. In bean (*Phaseolus vulgaris*), the hypocotyl (the part of the shoot below the cotyledons) elongates (Fig. 16.27). In pea (*Pisum sativum*), the epicotyl (the shoot above the cotyledons) elongates (Fig. 16.29). In both cases, cotyledons and shoot apex remain below the ground for a short time. Above the ground, the hook now straightens. In the case of bean, a straightening of the hypocotyl raises cotyledons and shoot apex above ground (**epigeal germination,** Fig. 16.27); a lengthening and straightening of the epicotyl will now pull the shoot apex away from the cotyledons (Fig. 16.27).

Upon straightening out of the pea epicotyl, however, the cotyledons will remain in the ground, and only the

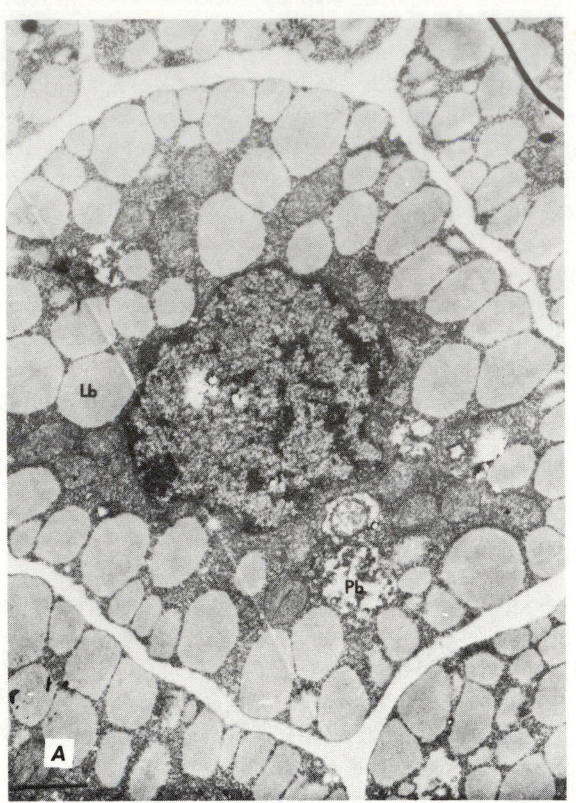

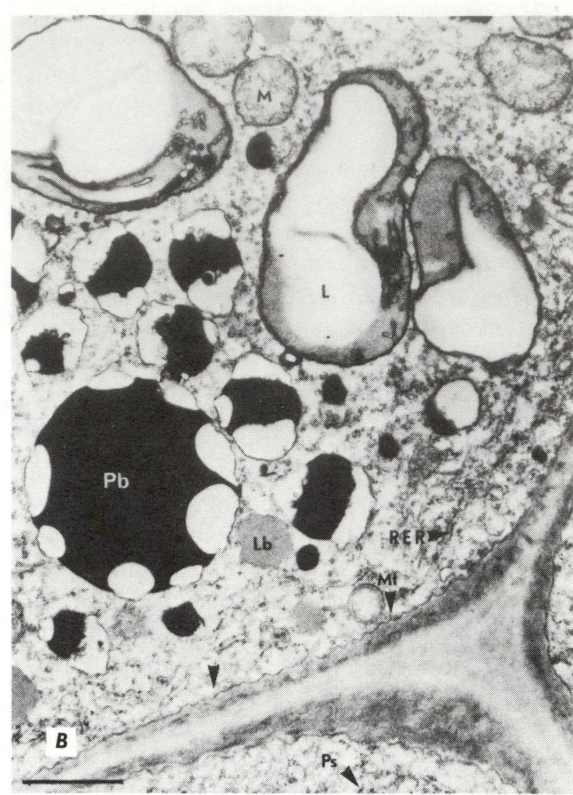

Figure 16.26
Yellow foxtail grass (*Setaria lutescens*). Electron micrographs. *A,* scutellum cell in dry caryopsis before germination; note the abundance of storage organelles (Lb, lipid bodies and Pb, protein bodies). *B,* root cell after 65 hours of germination; the cell now has fewer storage organelles and a greater number of organelles involved in metabolism (*M,* mitochondria and *L,* leucoplasts). *A,* ×10,3000; *B,* ×13,3000.

Figure 16.27
Stages in the germination of a bean (*Phaseolus vulgaris*) seed.

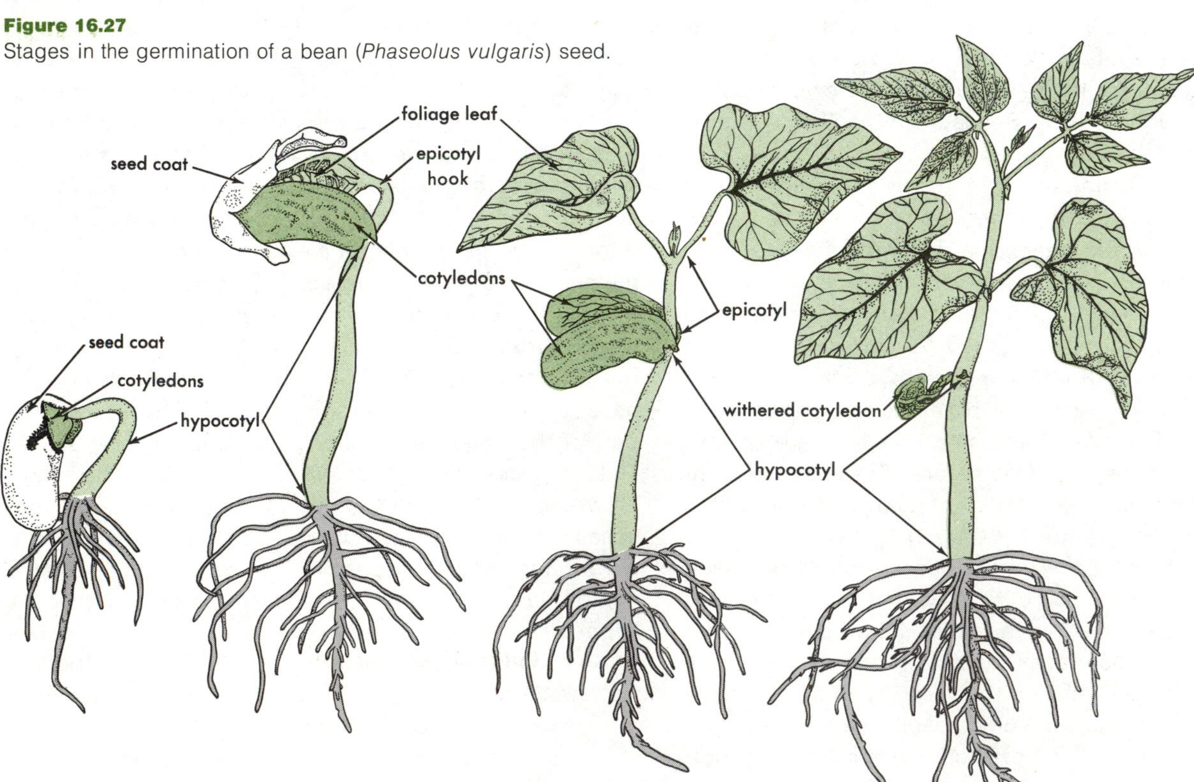

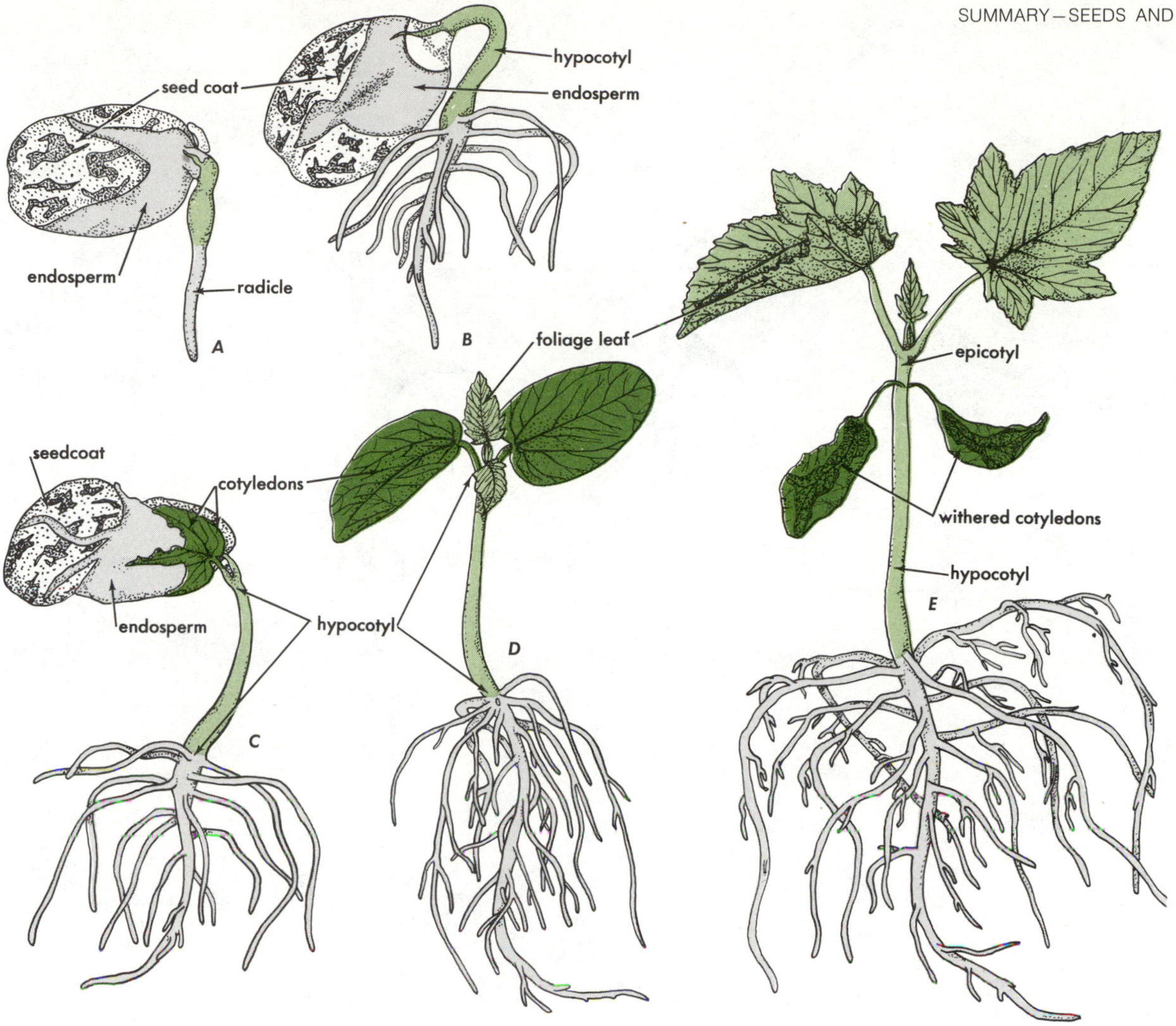

Figure 16.28
Stages in the germination of a castor bean (*Ricinus communis*) seed.

apex and first leaves will be raised upward (**hypogeal germination,** Fig. 16.29). In most cases, cotyledons, whether raised above ground or remaining below, do not carry on effective photosynthesis for a long time.

However, the straightening of the hypocotyl hook of the castor bean seedling lifts the cotyledons and the endosperm above the ground (Fig. 16.28). The cotyledons first function as absorbing organs facilitating the transfer of food from the endosperm to the rest of the seedling. When the reserve food supply in the endosperm is exhausted, the cotyledons become green, similar to foliage leaves, and carry out photosynthesis.

In onion, the primary root first penetrates the soil. Then a sharply bent cotyledon breaks the soil surface, and slowly straightens out. The cotyledon of the onion is tubular and its base encloses the shoot apex (Fig. 16.30). There is a small opening at the base of the cotyledon through which the first leaf finally emerges. In grasses, the situation is more complex. The shoot and root apices are enveloped by tubular sheaths known as the coleoptile and coleorhiza, respectively (Fig. 16.31). The primary root rapidly pushes through the coleorhiza. The root continues to grow, but is eventually replaced by adventitious roots that arise from the lower nodes of the stem (Fig. 16.31). The coleoptile elongates and emerges above ground, becoming 2 to 4 cm long. At this time, the uppermost leaf pushes its way through the coleoptile and, growing rapidly, becomes part of the photosynthesizing shoot.

SUMMARY—SEEDS AND GERMINATION

1. Seeds may store food within or outside the embryo. In most dicotyledons such as bean, food is stored in the two cotyledons. Cotyledons may also serve as absorbing and, later, as photosynthesizing organs. Food may be stored in an endosperm rather than in the cotyledons.
2. In monocotyledonous seeds, food is usually stored in an endosperm. In grasses, such as corn, a single cotyledon-like structure (scutellum) has become a highly specialized absorption organ that does not emerge from the seed. In seeds like those of onion, the cotyledon emerges from the seed coats and be-

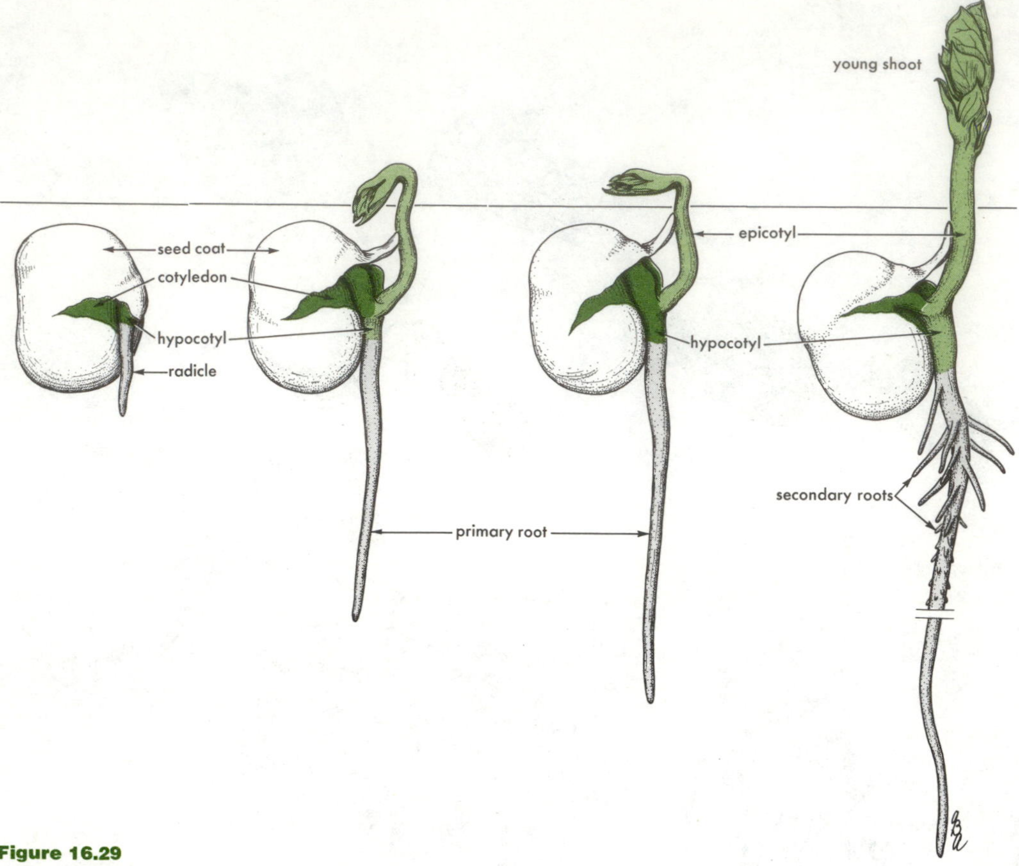

Figure 16.29
Stages in the germination of a pea (*Pisum sativum*) seed.

Figure 16.30
Germination of an onion (*Allium cepa*) seed.

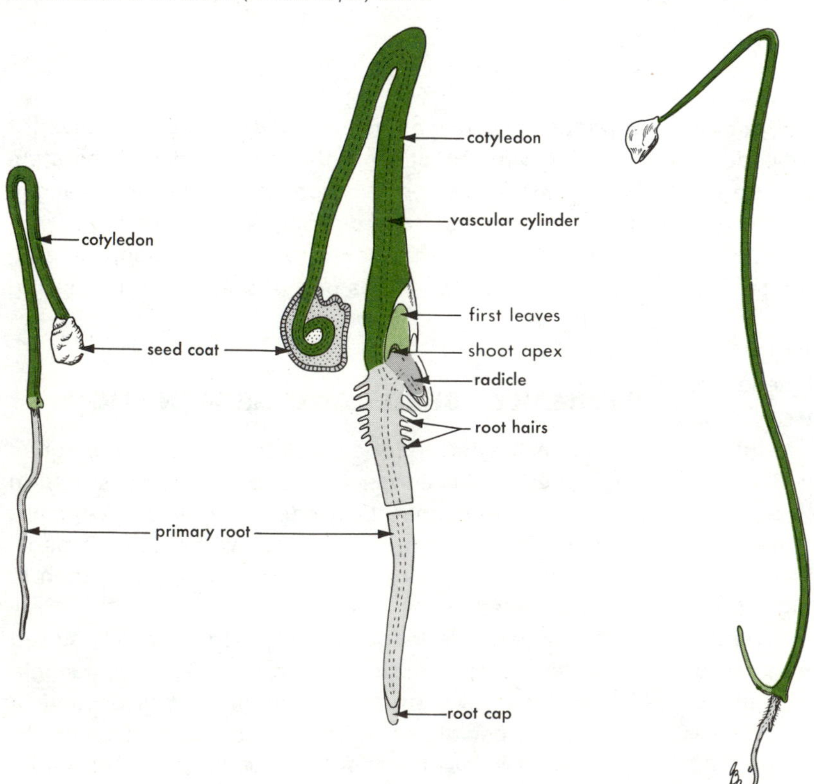

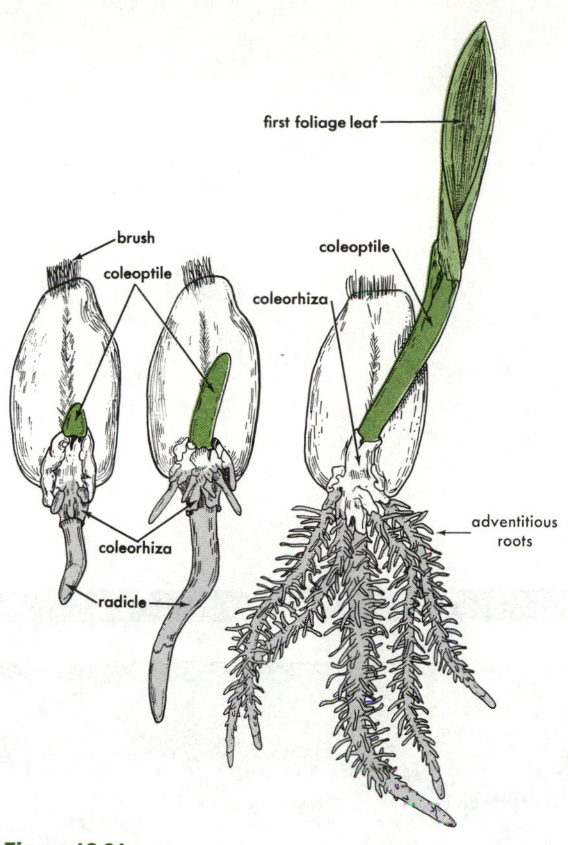

Figure 16.31
Germination of a grain of wheat. (*Triticum*).

comes green, but its tip continues to absorb food from the endosperm.

3. Mature seeds may be dormant and, depending on the species and the immediate environment of the seeds, they may remain viable and dormant from a few months to many years.
4. In germination, the cotyledons may be elevated above the ground, sometimes to become photosynthetically active, or in the case of some monocotyledons, to continue absorption of the endosperm. In other cases, the cotyledons remain below the ground.
5. Dormancy is usually broken by the provision of moisture, oxygen, and a favorable temperature. Other factors, such as light, the removal of chemical inhibitors, or the destruction of the seed coats, may be required in some instances.

CHAPTER 17

MEIOSIS AND BASIC GENETICS

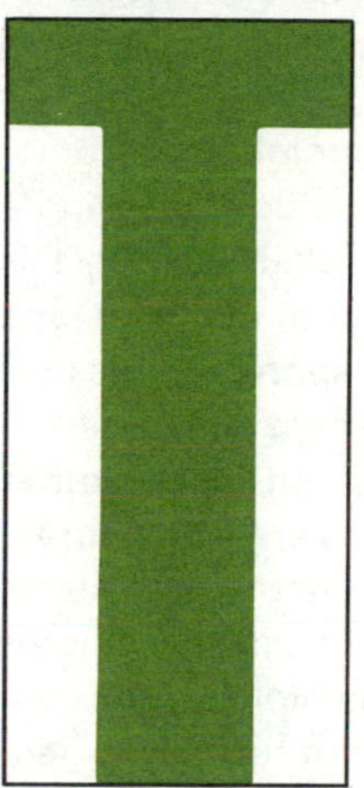

he simplest type of life cycle is the division of a single-celled plant to form two new plants. Most individuals, plant or animal, are not so simply constructed; specialized cells are set aside for reproductive processes. For example, in flowering plants, flowers and seeds possess such specialized cells. Even in these plants, the special reproductive cells are in a sense immortal, for one can trace their history back to the beginning of life through a series constituted mainly of meristematic and reproductive cells. Every plant and animal living today must have a chain of ancestors stretching back to the beginning of life. During this long period, the characteristics of present-day plant and animal life have evolved. How has this happened? Are species of animals and plants constant? Do they change? How similar to each other are individuals of a given species? If individuals of a given species vary, what are the causes of variation? Can this variation be controlled or predicted? How do the specialized reproductive cells function? What is the importance of sex? These and many other questions arise when one considers the reproductive cycle of plants and animals. Partial answers may be given, but to no question stated above is there, as yet, a final and complete answer.

THE SEXUAL REPRODUCTIVE CYCLE

We have seen in Chapter 15 that during the sexual life cycle of an angiosperm, microsporocytes and megasporocytes undergo meiosis leading eventually to the development of pollen and embryo sacs, respectively. In both instances, meiosis results in the formation of four cells each with half the number of chromosomes as the parent cells. Subsequently during fertilization, the union of gametes and fusion of their nuclei results in a doubling of the number of chromosomes in the resulting zygote. Of very great importance in this cycle is the behavior of chromosomes, the carriers of **genes,** which are the units responsible for the morphological and physiological characteristics of every individual. Recall (Chapter 6) that the genes are specific sequences of nucleotide pairs in the chromosomal DNA molecule.

The pairs of genes at the same position or **locus** in an homologous pair of chromosomes, though affecting the same trait, need not be identical. For example, a pair of genes may affect the stature of a plant, one carrying information for tallness and the other for dwarfism. Such specific forms of genes affecting the same characteristic and located at the same locus in a chromosome are called **alleomorphs** or **alleles.** Different alleles have slightly different sequences of nucleotides. However, only one allele can occupy one locus in an individual chromosome at any one time.

During meiosis the two sets of homologous chromosomes found in the**diploid** parent are reduced to a single set of chromosomes, and the cell or plant bearing this single set is said to be **haploid.** The cells in which meiosis occurs are **meiocytes.** If they produce **spores** – cells able to produce new plants directly – as generally happens in plants, the meiocytes are sometimes called **spore mother cells.** The spores resulting from meiosis are **meiospores.** Since meiospores give rise to a new generation of plants directly, it follows that plants arising from meiospores must be haploid. In order to regain the diploid state, two haploid cells must fuse. Cells capable of fusing are generally specialized cells and are called **gametes.** The diploid cell resulting from the gamete fusion is a **zygote** and will give rise by mitosis to a diploid, or $2n$, plant body.

Figure 17.1
Diagram of a corn cell showing the 10 pairs of chromosomes and the nucleolus. Homologous chromosomes are tightly paired. Kinetochores and common knobs are shown on the chromosomes. Redrawn from Morgan and Rhoades.

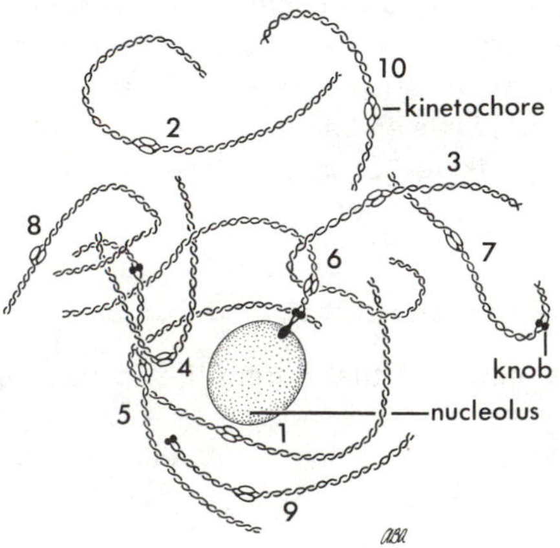

Sometimes conditions arise that result in three, four, or more sets of homologous chromosomes per cell. This condition is known as **polyploidy** and, depending upon the number of homologous chromosomes, the plants are **triploid** ($3n$), **tetraploid** ($4n$), **hexaploid** ($6n$), etc.

The two phenomena, **meiosis** and **fertilization,** involving, as they do, the assortment and combination of chromosomes, form the basis of the sexual reproductive cycle. Indeed, the great diversity of plant and animal life is due largely to changes that occur in a chromosome set during meiosis and fertilization. In many plants (asparagus and date palm, for example), two separate parents are involved in sexual reproduction; in others, such as pea and wheat, the reproductive cells are produced on the same parent. Thus, chromosome behavior at meiosis and fertilization constitutes a mechanical basis for inheritance, and inheritance itself may be placed on a firm mathematical foundation.

REVIEW OF MITOSIS

The precise and coordinated series of events involving DNA replication and chromosomal behavior during mitosis were described in detail in Chapter 6. We shall briefly review mitosis here in preparation of our study of meiosis.

In prophase of mitosis, extended chromosomes divide and contract to form two chromatids per chromosome. The kinetochores then become aligned at the equator of the cell; during anaphase, the two chromatids derived from one chromosome separate and move to opposite poles of the cell; during telophase, derivative nuclei are reconstituted; and the development of a cell wall completes the formation of two identical cells. In this process, there has been no change in the chromosome number, and no reassortment of chromosomes; each derivative nucleus has the same complement of genes. It should be emphasized, however, that all cells containing $2n$ chromosomes have two sets of chromosomes. In corn, for instance, there are 20 chromosomes; two sets of 10 each. In all $2n$ cells of corn, there would be two sets of each of the 10 different chromosomes. In certain stages of cell division individual chromosomes of many plants can be identified by characteristic structural features (Fig. 17.1).

CHARACTERISTICS OF MEIOSIS

The division stages of chromosomes during meiosis are similar to the division stages of chromosomes undergoing mitosis. Meiosis begins with the chromosomes present as long, slender uncoiled threads. They proceed to shorten, thicken, and split into chromatids as in mitosis. The chromatids separate at anaphase. Replication of DNA occurs in the interphase preceding meiosis and again in the interphase after the completion of meiosis, when one chromatid is present in each chromosome. The basic difference between mitosis and meiosis lies in the *pairing of homologous chromosomes* in meiosis so that four chromatids, two derived from each of the pairing homologous chromosomes are associated at mid-prophase. In mitosis, only the two sister chromatids, derived from a single chromosome, are associated at this point. This introduces a complication in meiosis, for the chromatids of paired homologous chromosomes are able to exchange partners with each other. Such exchanges between chromatids derived from opposite members of a homologous pair of chromosomes (nonsister chromatids) may be expressed as visible changes in the offspring. There are two divisions in meiosis, which result in four cells from the initial mother cell. Each of these four cells has only one complete set of chromosomes.

Since two divisions are required for the completion of meiosis, it is common to designate the two divisions by the numerals I and II and the phases as prophase I, metaphase I, anaphase I, telophase I, prophase II, metaphase II, anaphase II, and telophase II.

Meiosis—First Division

Prophase I

During prophase I of meiosis (Figs. 17.2*A,B*), the nuclear threads of the vegetative nucleus, present in the diploid number, contract to form chromosomes, each with two chromatids; in this respect the process resembles mitosis.

As in mitosis, the nucleolus disappears. The nuclear membrane also disappears during the later stages of prophase I (Fig. 17.2*B*). The process is complicated, however, because before the chromatids become apparent, the homologous chromosomes pair. In so doing, they approach and coil about each other (Figs. 17.3*A,B*). The two chromatids of a chromosome, although present in interphase, appear only after the pairing of the homologous chromosomes is well-advanced.

Figure 17.2
Early prophase I. *A*, this photomicrograph and those in Figs. 17.3 through 17.11 show meiosis in lily anther. Note paired threads.

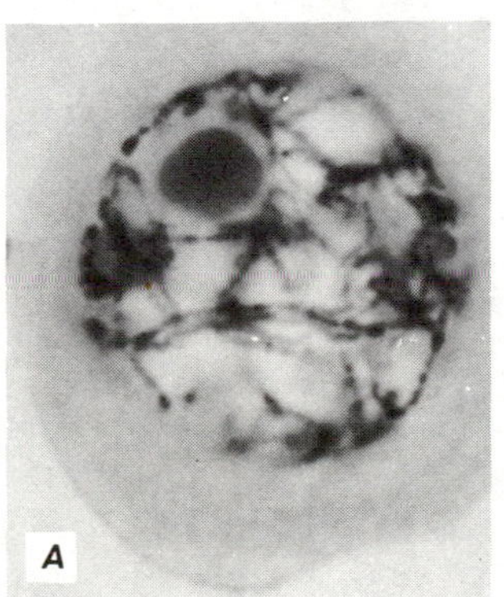

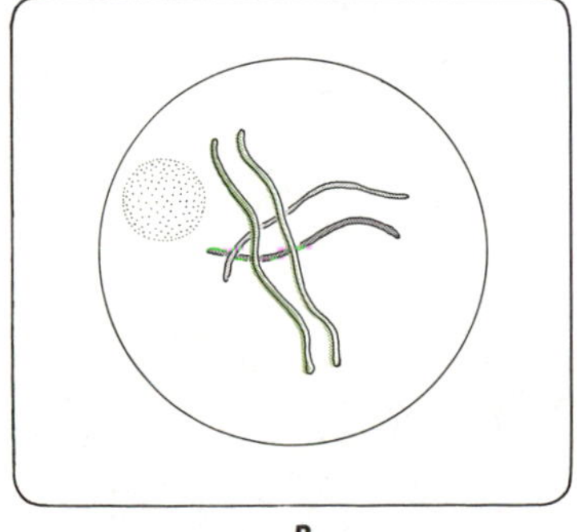

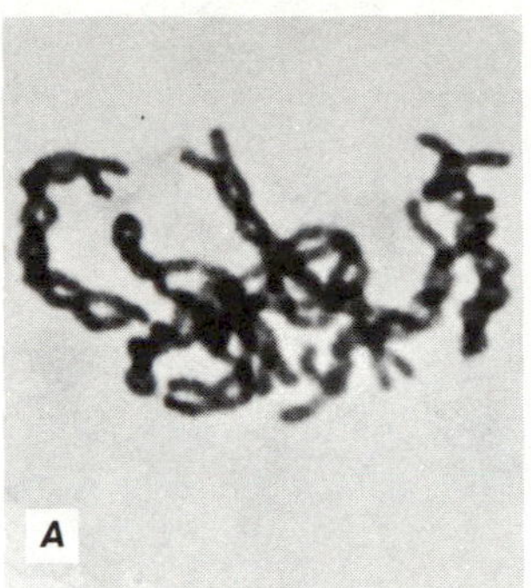

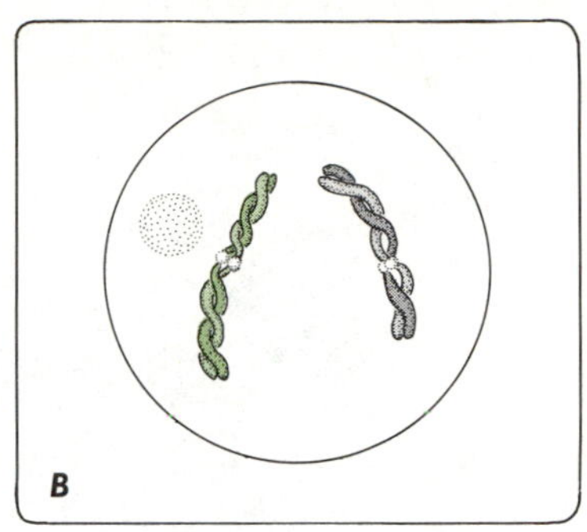

Figure 17.3
Late prophase I. (*A*, courtesy of R. Gankin.) Each body represents two paired chromosomes; note chiasmata.

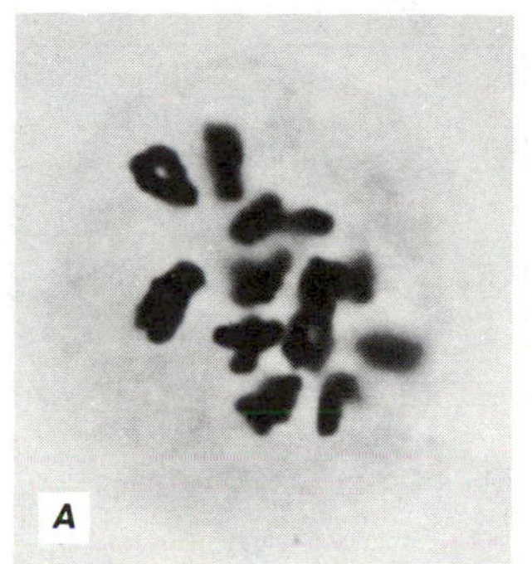

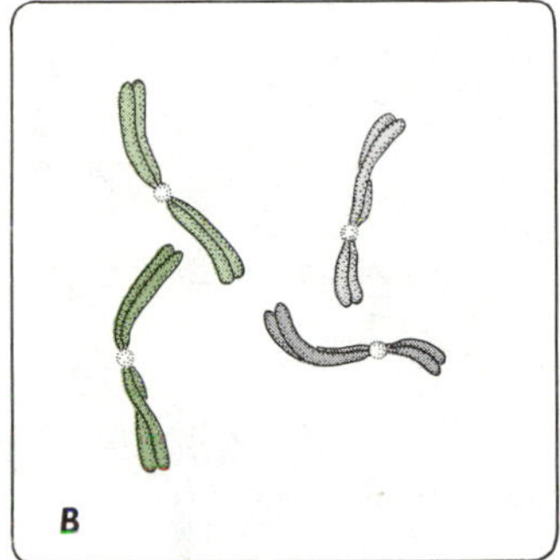

Figure 17.4
Late prophase I.

The resulting figure is composed of two paired homologous chromosomes and four chromatids (Figs. 17.4*A,B*). This stage of prophase I involves two phenomena: (*a*) the pairing of homologous chromosomes and (*b*) the formation of chromosomes with two chromatids each.

Chromatids thus formed do not remain unchanged; they normally break at several points and rejoin in such a way that a given reconstituted chromatid may be composed of parts of four chromatids. This breaking and rejoining of the chromatids is called **crossing over,** and the cross formed by the chromatids involved in the interchange is known as a **chiasma** (**chiasmata,** plural). Chiasmata may occur at any point along the paired chromosomes.

Thus, a third step in the prophase I of meiosis is (*c*) the formation of chiasmata due to the breaking and rejoining of chromatids from homologous chromosomes.

The mechanism and consequences of crossing over will be discussed in a later section.

Metaphase I

At metaphase I the kinetochores of the paired homologous chromosomes pass to the equator of the cell. The spindle apparatus develops and spindle fibers grow toward the poles (Fig. 17.5).

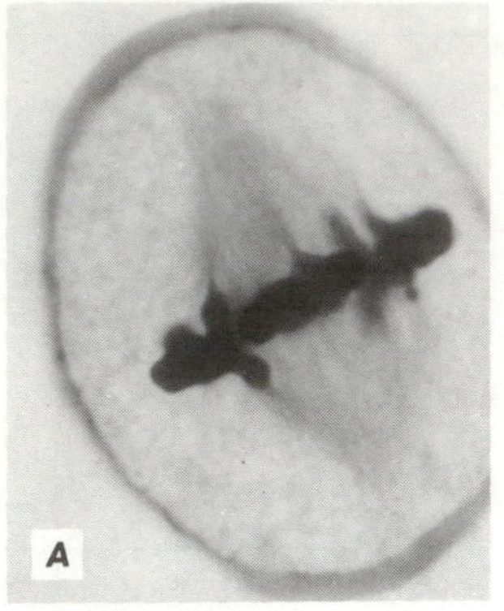

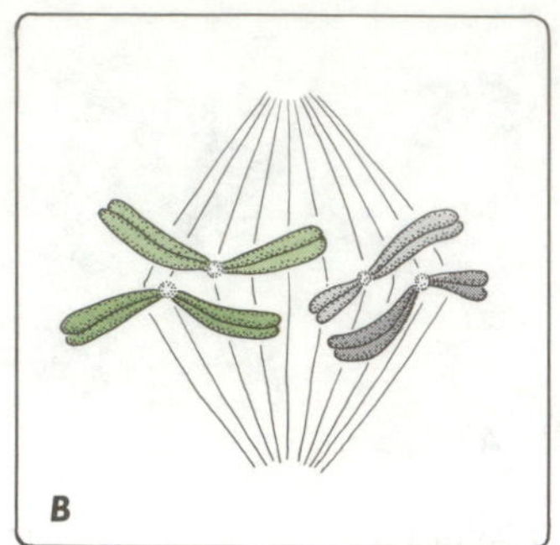

Figure 17.5
Metaphase I.

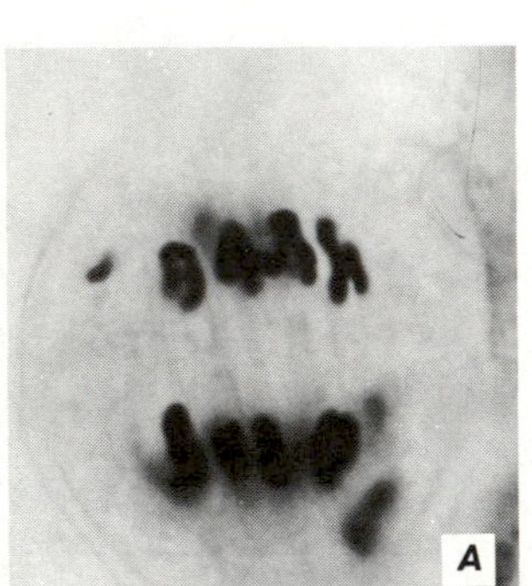

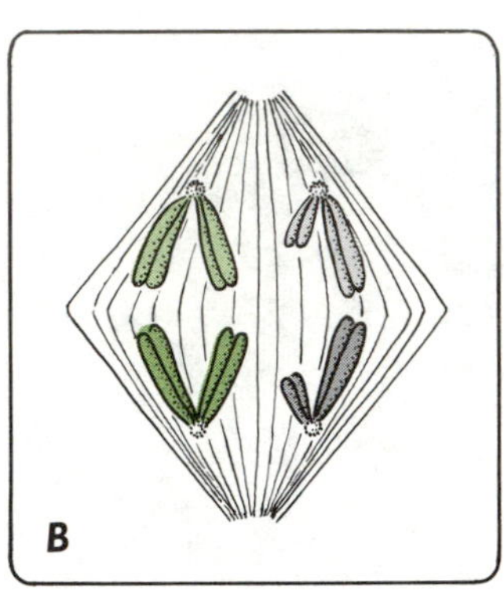

Figure 17.6
Anaphase I.

Anaphase I

In anaphase I, the kinetochores of *whole chromosomes* separate from each other and, with their associated chromosomes, move to opposite poles of the cell (Figs. 17.6*A*,*B*).

Telophase I

Following anaphase I, the chromatids group together at opposite poles of the cell (Figs. 17.7*A*,*B*) and immediately prepare for the second meiotic division. Each telophase chromosome consists, as usual, of two chromatids. In this sense, the *n*, or haploid number of chromosomes, is present.

Meiosis—Second Division

Prophase II

A second meiotic division now follows, which separates these rearranged chromatids. During prophase II (Figs. 17.8*A*,*B*), chromosomes again form, each with two composite chromatids.

Metaphase II

The kinetochores approach the equatorial plate, forming metaphase II (Figs. 17.9*A*,*B*).

Anaphase II

As in mitosis, the kinetochores now split and separate. In anaphase II, single chromatids move to opposite poles of the cell (Figs. 17.10*A*,*B*) and are reconstituted in telophase II into nuclei (Figs. 17.11*A*,*B*)

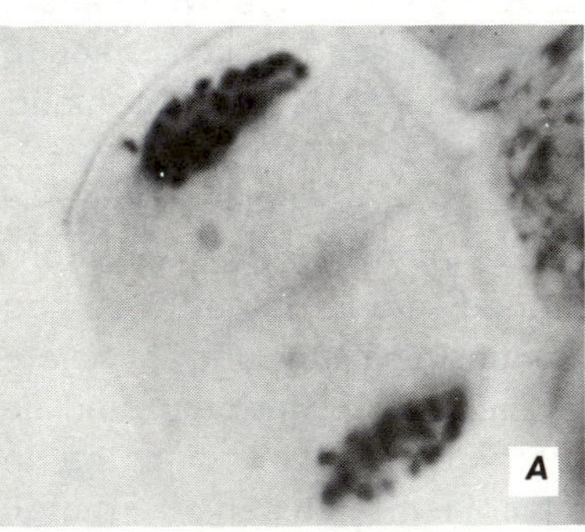

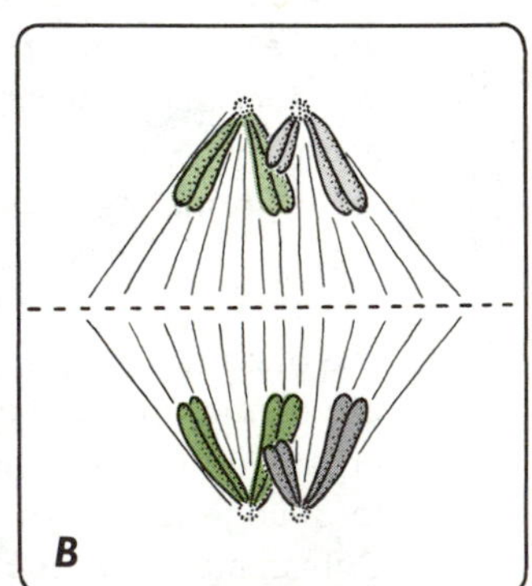

Figure 17.7
Telophase I.

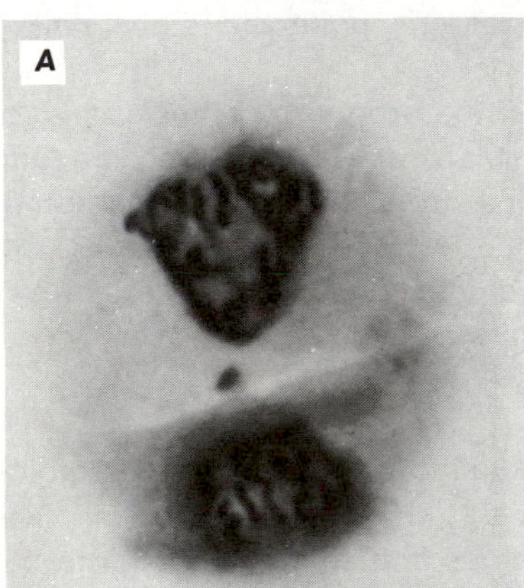

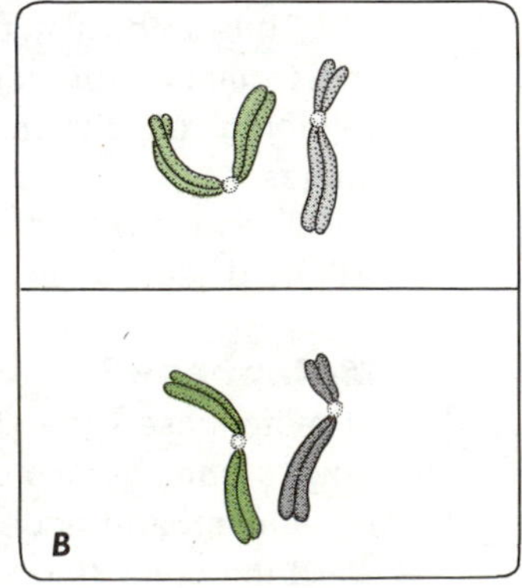

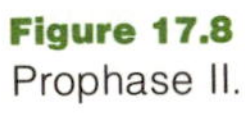

Figure 17.8
Prophase II.

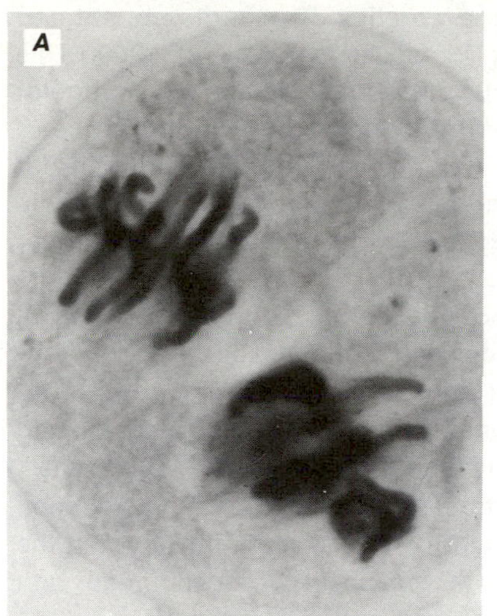

B

Figure 17.9
Metaphase II.

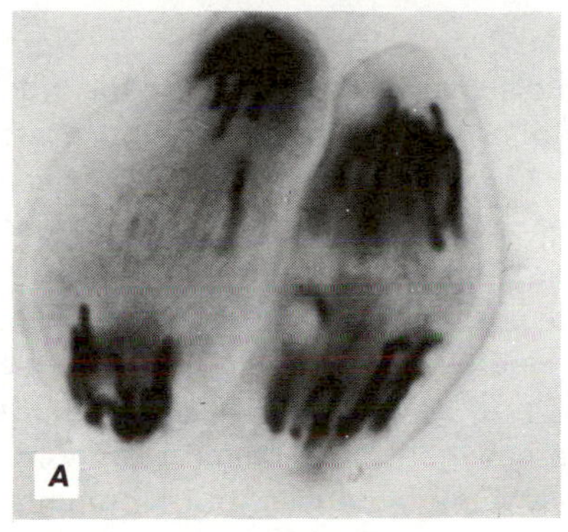

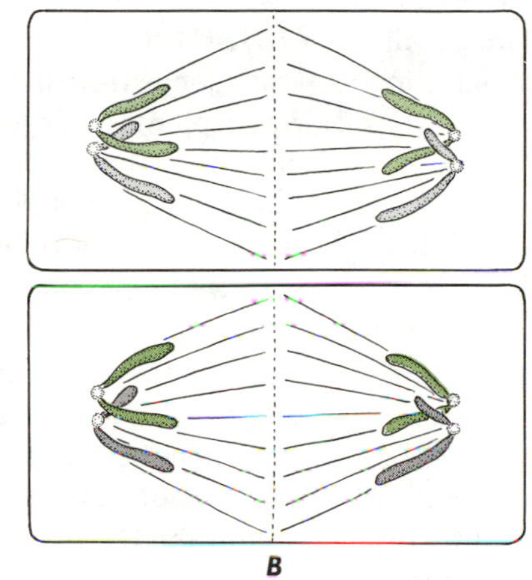

B

Figure 17.10
Anaphase II.

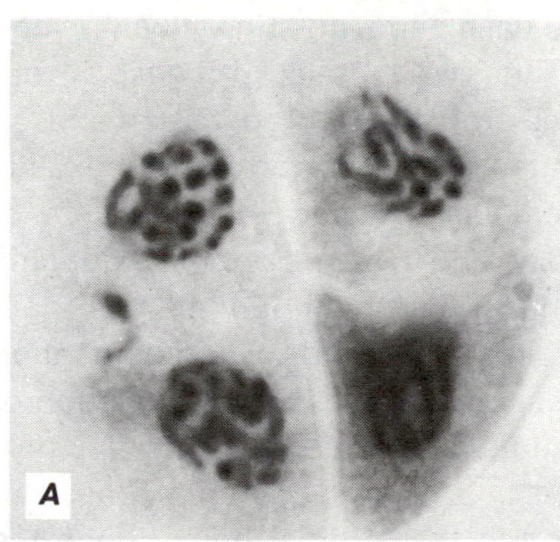

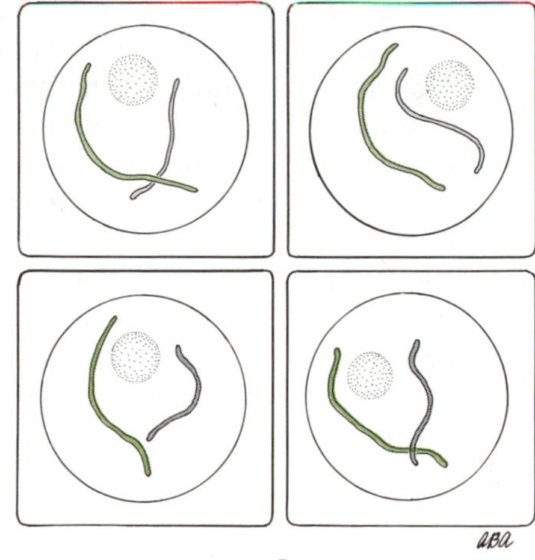

B

Figure 17.11
Telophase II.

Telophase II

Each nucleus thus formed contains one of the four chromatids derived from the pairing of homologous chromosomes in prophase (Fig. 17.11).

Walls develop about each new nucleus and associated cytoplasm, thus forming cells with the *n* or haploid number of chromosomes.

FERTILIZATION

Let us assume that two alleles affecting plant height occur in a cross-pollinated plant. Let *T* represent the allele carrying the information for tall stature and *t* the allele for short stature. Recall that meiosis occurs in the formation of pollen and also in the development of the embryo sac. One-half the pollen grains and eggs formed will carry the allele *T;* and the other half will carry the allele *t.* Since large numbers of both pollen and egg cells are formed, and since the fusion of any two may occur, we should expect the following combination:

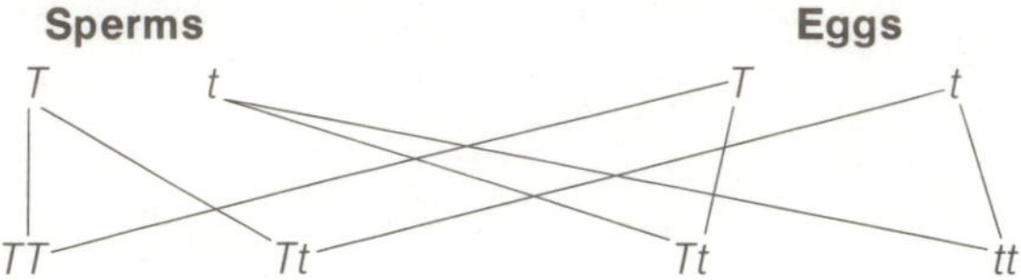

The many hundreds of sperm cells produced in hundreds of pollen tubes will carry either *T* or *t* alleles in approximately equal numbers. These sperms will unite with eggs carrying either the *T* or *t* allele.

Notice that there is one chance for the formation of zygote *TT,* one chance for the formation of zygote *tt,* and two chances for the formation of zygote *Tt.* If the genes *Tt* governed height, we should expect to find plants of different heights occurring in certain very definite ratios in a breeding plot. That this situation actually does occur has been well-established by a long series of experiments.

BREEDING SYSTEMS IN PLANTS

The art of plant and animal improvement is very ancient. It has been practiced by many different races of men for thousands of years. Wheat, corn, and rice are examples of plants that have been improved by men of three different continents. The domestication of these grains was accomplished before the beginning of recorded history and ranks as an accomplishment of major importance. The task was so well done that the tribes that grew these grains were assured of a food supply and could turn their energies to activites other than getting food. Civilization resulted. Attempts to increase crop yields and the nutritive values of plants, or to introduce new and better food plants, have been continued down through the ages to the present time. The last 50 years has seen a great increase in plant and animal improvement and in a knowledge of the mechanism of inheritance in plants and animals.

Within the past decade some dwarf varieties of wheat, developed in Mexico under the direction of Victor Borlaug, have doubled the yields of wheat in many countries. Similar advances have been made with rice. High-yielding dwarf strains of this grain have been introduced into almost all the tropical countries of Asia. The introduction of these two crops has so enlarged the food production of the developing countries, that we are in the midst of a "Green Revolution."

Production of hybrid corn in the United States led the way to the development of these high-yielding crops. Only a few genes account for high yields, and the great majority of high-yielding strains carry the same genes. Thus, only a handful of varieties of some important crops are cultivated. For instance, in the United States, two types of peas and nine varieties of peanuts constituted 95% or more of their respective crops in 1969. This means a great genetic uniformity of certain particular genes. Most hybrid corn previous to 1970 carried the *T* gene which resulted in a male sterile plant. It turned out that this gene also rendered the plants very susceptible to the fungus causing corn blight. In 1970, about 15% of the corn crop was lost to this fungus. This occurrence points up the hazard of genetic uniformity in agricultural plants. Very little is currently known about the interactions between wheat and its parasites, and there definitely exists the probability that the new wheats, and rices too, are very vulnerable to damage by disease. Thus, while great increases in production have been achieved, genetic similarity has resulted, introducing the very great possibility of widespread crop damage with the danger of famine. It has occurred before.

Mutation

Most genes have been discovered by examining the individual plants in large uniform populations to find the rare individual that differs from its neighbors. It has been found that such differences occur in rather characteristic ratios. When these new plant types are bred to the standard **wild type** (the gene normally occurring in the wild population), they are generally recessive to the wild type and the new traits seem to be deleterious to the plants that bear them. These new genes are called **mutations.** They may be due to a variety of causes, but a common explanation is that an error has occurred in the duplication of the DNA helix. How might this occur?

In our discussion of the DNA molecule and its replication, it was pointed out that gene specificity resided in the arrangement of the base pairs, adenine–thymine and cytosine–guanine, in the DNA helix. It was further pointed out that the helix uncoiled in replication and, as the two strands separated, a new strand formed exactly like the disjoining member. The perfect replication of thymine opposite adenine and cytosine opposite guanine results in the new identical DNA helix. Sometimes there may arise a **copy error** caused perhaps by a modification of the nitrogenous bases, perhaps by a small deletion of a portion of a chromosome. Such a copy error would give the replicating DNA helix a slightly different base sequence than that found in the original helix. Since the DNA molecule is involved in the production of specific cellular protein through *m*RNA, we would expect the change of sequence in the base pairs to be reflected in all events leading to the final form of the trait related to the gene in question. Such a change in the replication of the DNA molecule is one cause of mutations.

Variation

Variation in a population of plants or animals may be due to a number of conditions. Variation is probably always related to genes. The number of genes involved in the expression of a trait, and the relationships between

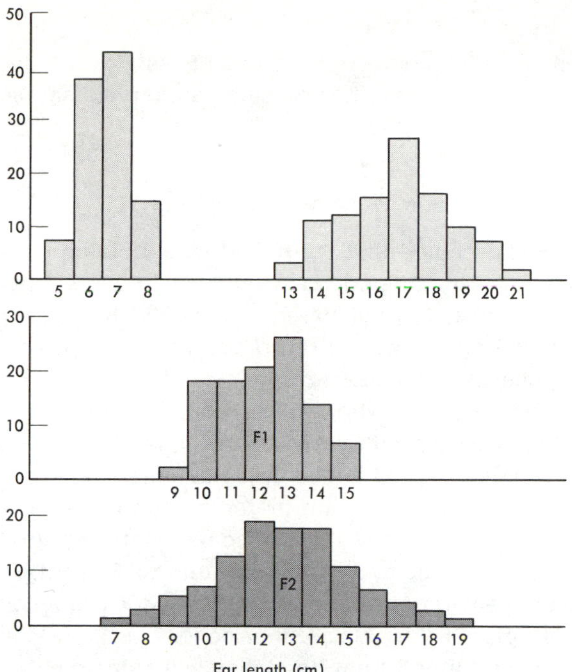

Figure 17.12
Continuous variation. Two varieties of corn are crossed. Black Mexican has long ears and Tom Thumb has short ears. The lengths of the ears of the F_1 show continuous variation between the ear lengths of the parents; there is continuous variation over a greater range in the F_2. (Courtesy of J. L. Brewbaker, *Agricultural Genetics,* p. 57. © 1969. Reprinted by permission of Prentice-Hall, Inc., Englewood Cliffs, New Jersey.)

Figure 17.13
Gregor Mendel. (Courtesy of The Bettmann Archive, Inc.)

genes, play an important part in the expression of the trait and in the relative importance of the environment upon the type and amount of variation. Mendel's revolutionary approach to plant breeding lay in his plan to follow the numerical presence or absence of single traits through a number of generations of carefully controlled pollinations and seed collections. The differences between the progeny were large and easily separated the progeny into distinct categories. This is **discontinuous variation.** It may be contrasted with **continuous variation,** in which the variation is evenly and widely distributed about a mean (Fig. 17.12). Continuous variation is a very important aspect of modern genetics. However, the principles established through studies on discontinuous variation form the foundation for explaining the mechanism underlying continuous variation as well as other breeding systems.

Mendel's Contributions

Modern **plant breeding** and **genetics** date from 1900, when Gregor Mendel's (Fig. 17.13) experiments on hybridizing garden peas were first appreciated. The importance of his experiments lies in the introduction to plant breeding of several procedures that, in 1900, were new and revolutionary. They are as follows:

1. The study of the inheritance of single or **unit** characters.
2. The keeping of accurate records of the number of times a given unit character appears in the offspring of selected parents.
3. The maintenance of the pollination of the experimental plants under the complete control of the investigator at all times.

While these procedures are accepted today as self-evident, they were so revolutionary in 1865 when Mendel's work was done that the standard scientific magazines of the time refused to publish Mendel's results. Instead, his experiments were presented before a small society of men in his home town. This society was not greatly unlike some of the present-day luncheon clubs and, like some of them, it published its proceedings. Thus, Mendel's work was placed on record, where it lay unnoticed for 35 years. In 1900, it was discovered by three prominent European plant breeders and it has become the cornerstone of all modern work in plant and animal genetics.

The Monohybird Cross

Mendel selected garden peas for his experimental plants. He carefully took pollen from the anthers of a dwarf-growing variety and dusted it on the stigma of a tall-growing variety. The seeds resulting from this cross-pollination were collected and planted the following season. All the plants that grew from these seeds were tall. The same results were obtained when the tall variety supplied the pollen:

tall plant × dwarf plant (pollen parent)
↓
all tall plants

or

dwarf plant × tall plant (pollen parent)
↓
all tall plants

Then flowers of the tall progeny were self-pollinated. Peas are normally self-pollinated, and so in order to obtain this seed Mendel merely had to keep stray pollen from reaching the stigmas of his experimental plants. The seeds resulting from these self-pollinated flowers were collected and planted the following spring. Upon counting the plants that grew from these seeds, he found that 787 were tall plants and 277 were dwarf plants. In other words, about three-fourths of these pea plants resembled one of the original pair of parents and about one-fourth resembled the other parent.

It has been found convenient to give designations to these generations of plants. The two original parents are designated by **P.** The first generation, comprising only tall pea plants, is the **first filial generation,** or the **F_1**; the second generation of three-quarters tall and one-quarter dwarf pea plants is the **second filial generation,** or the **F_2**. Subsequent generations of self-pollinated plants would be consecutively the F_3, F_4, F_5, and the like.

Note in the cross that, even though we are dealing with two different plants, we are concerned with only **one character**—height of growth. A cross dealing with a *single unit character is a* **monohybrid cross.** Mendel studied, in all, seven monohybrid crosses. The results were similar to those just described for height. They were as follows:

1. The F_1 always resembled one of the parents: tall × dwarf gave tall; wrinkled × round seeds gave round seeds; and the like. From this result, Mendel concluded that one expression of a given character was **dominant** (tall, or round seeds), whereas the other aspect of the same character was **recessive** (dwarf, or wrinkled seeds).
2. Selfing of the F_1 resulted in an F_2 generation in which the original forms of a given character **segregated** to give three times as many dominant plants as recessive plants.

Mendel further postulated a possible mechanism for this genetic behavior. He knew that the F_1 plants must contain factors, now called **genes,** responsible for both tallness and dwarfness, even though all the plants were tall, because (*a*) tall and dwarf plants were crossed to produce the tall F_1, and (*b*) upon selfing, the tall F_1 gave rise to both tall and dwarf individuals. If we then represent the gene for tallness by *T* and the gene for dwarfness by *t*, we may diagram the genes determining height that are present in the tall F_1 by *Tt:*

Tt = tall growing F_1 pea plants

Since the F_1 hybrid contains two genes, *T* and *t*, determining height, each true breeding parent plant should contain two similar genes, thus:

TT = tall-growing parent
tt = dwarf-growing parent

Mendel assumed that the pollen would contain but one gene, and that the eggs similarly would contain but one gene. Thus, the tall parent, *TT*, would produce *T* pollen and *T* eggs, and the dwarf parent, *tt*, would produce *t* pollen and *t* eggs.

The parent plants whose nuclei contain the homologous chromosomes with the identical genes *TT* or *tt* are said to be **homozygous** for either the dominant or recessive genes. The F_1 plant with the homologous chromosomes carrying the genes *Tt* is said to be **heterozygous.**

Pollen produced by the heterozygous tall F_1 plant, *Tt*, will contain either gene, *T* or *t*. Likewise, eggs produced by this F_1 plant will contain either gene, *T* or *t*, and the sperm nuclei formed from the pollen will contain either gene, *T* or *t*. Many thousands of pollen grains, sperms, and eggs will be formed by the tall heterozygous F_1 plant. A sperm containing the gene *T* may unite with either a *T* or a *t* egg, as follows:

T sperm + *T* egg = *TT* zygote
T sperm + *t* egg = *Tt* zygote

A sperm containing the gene *t* may unite with a *T* egg or a *t* egg, as follows:

t sperm + *T* egg = *Tt* zygote
t sperm + *t* egg = *tt* zygote

These four possible combinations of genes will occur at fertilization and may be diagrammed as follows:

		Possible sperms	
		T	*t*
Possible eggs	*T*	Zygote 1 *TT*	Zygote 2 *Tt*
	t	Zygote 3 *Tt*	Zygote 4 *tt*

The possible types of eggs are arranged along the left of the checkerboard, and the possible types of sperms at the top. The squares of the checkerboard represent the possible zygotes, and they are derived by combining the gametes opposite the squares. For instance, zygote 1 is derived from sperm *T* and egg *T;* zygote 2 is derived from sperm *t* and egg *T*, and so on. The genes, *Tt*, or *TT*, or *tt*, for example, present in the plant, constitute the **genotype** of the plant. Reference to the diagram will show that there are three different genotypes: *TT*, *Tt*, and *tt*. The appearance of the plant is the **phenotype.** For instance, the appearance or the phenotype of the plant *TT*

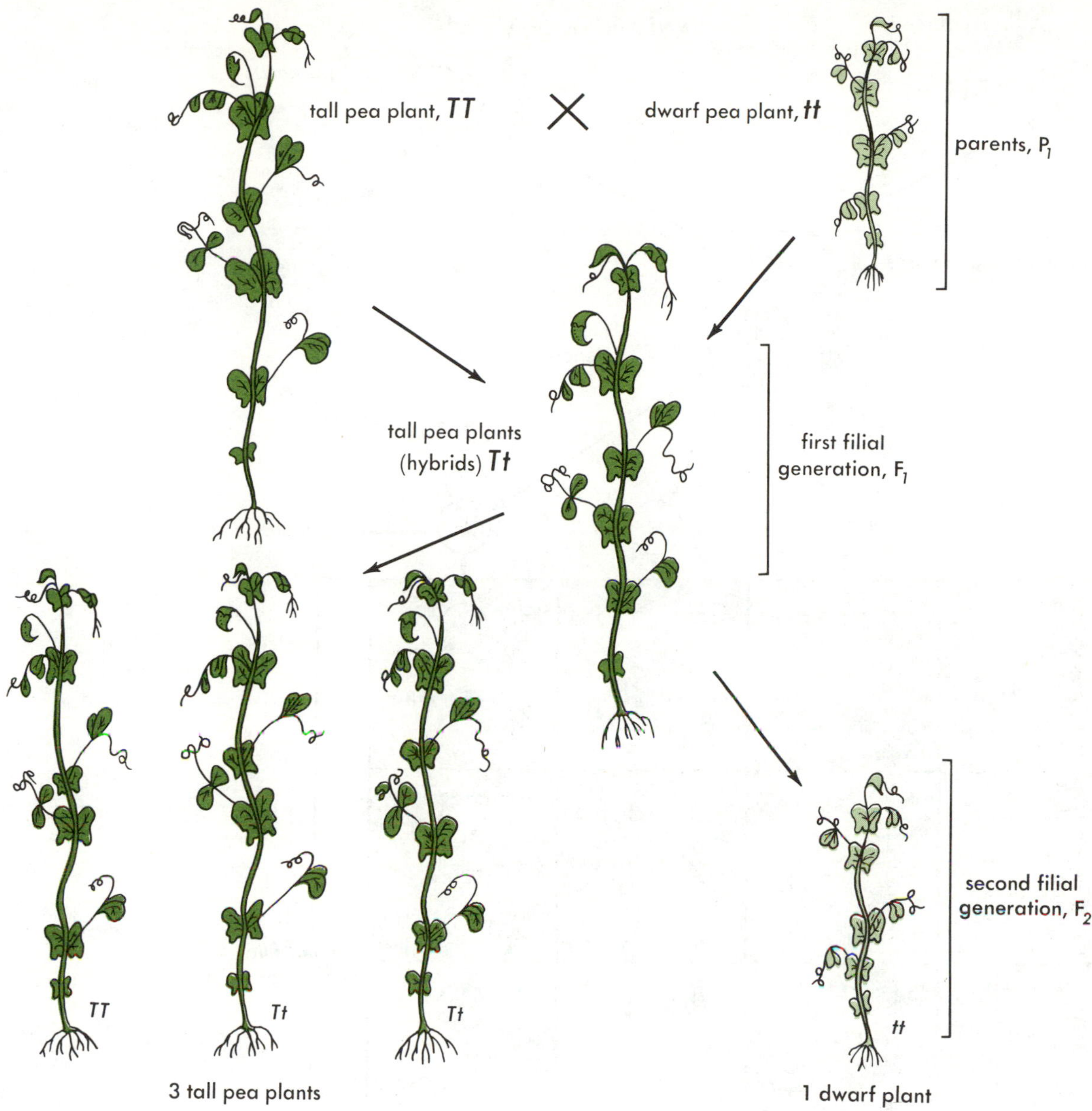

Figure 17.14
Diagram of monohybrid cross between tall (*TT*) and dwarf (*tt*) pea plants.

is tall; of the plant *Tt* is tall; and of the plant *tt* is dwarf. Note that there are three times as many tall plants represented on the checkerboard as there are dwarf plants. Furthermore, if Mendel's assumption regarding these factors or genes is correct, the homozygous tall plant *TT*, when self-pollinated, should give only tall offspring; the homozygous dwarf plant *tt*, when self-pollinated, should give only dwarf offspring; the tall heterozygous *Tt* plants, when self-pollinated, should give tall and dwarf plants in the ratio of 3 tall plants to 1 dwarf plant. This is what actually happens. A complete monohybrid cross is shown in Fig. 17.14.

This cross gives us additional information about these genes. It might be thought that they could interact with each other to bring about some modification in their genotypes, either in the zygote nucleus or in the line of vegetative nuclei leading to the next generation of spore or pollen mother cells, or that the cytoplasm could act upon them as they are replicated at each somatic division in the interval between fertilization and meiosis. But none of these things happens. Genes *T* and *t* retain their identity and separate from each other inviolately at each succeeding meiosis, generation after generation. Another important fact about them is that they occupy equivalent spots or loci on homologous chromosomes. They are alleles.

The Dihybrid Cross

Let us consider a cross concerning two characters, height of the plant and form of the seeds. This is a **dihybrid cross,** that is, a cross involving *two unit characters* (Fig. 17.15). One parent plant is tall and has round

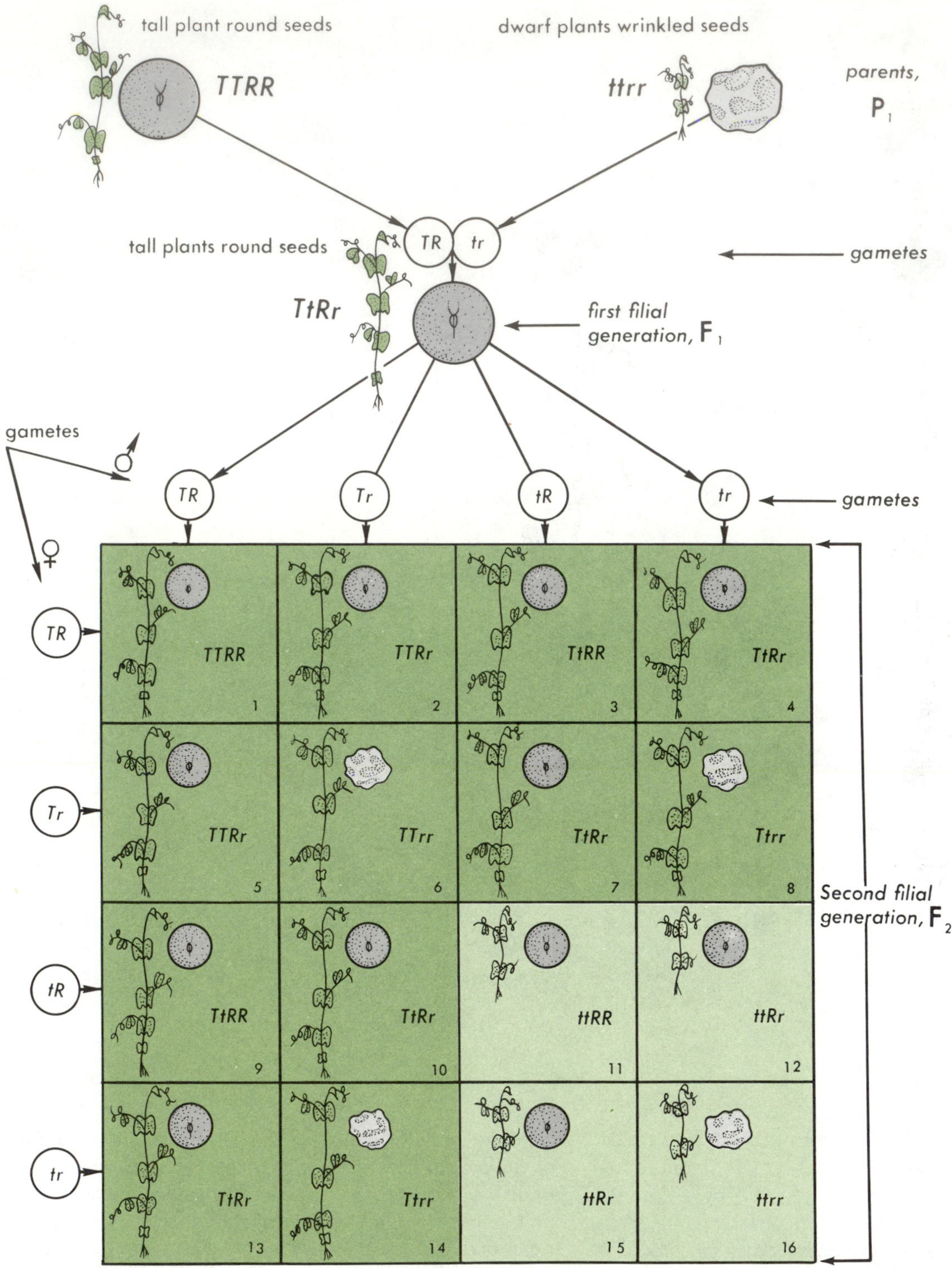

Figure 17.15
Diagram of a dihybrid cross between a tall pea plant with round seeds (*TTRR*) and a dwarf pea plant with wrinkled seeds (*ttrr*).

seeds; the other parent plant is dwarf and has wrinkled seeds. We know that both these parent plants are homozygous for these particular characters because upon self-pollination they breed true. When cross-pollinated, all the resulting F_1 plants are **tall** and have **round** seeds. We know from this outcome that tallness is dominant over dwarfness and that roundness is dominant over wrinkledness. We may therefore write the genotype of the parent plants as follows:

The tall plant with round seeds, *TTRR*
The dwarf plant with wrinkled seeds, *ttrr*

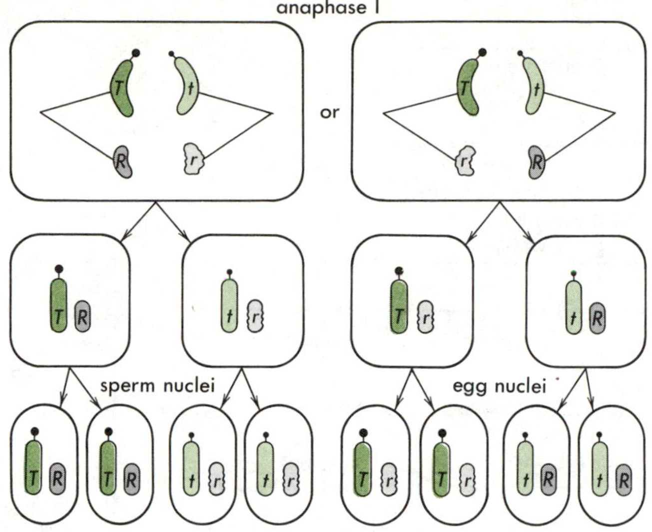

Figure 17.16
Types of sperms and eggs that may be formed from a plant having the heterozygous genotype *TtRr*.

Since all plants possess the characters of height and seed form, factors determining these characters must always be present in every plant.

After meiosis, the pollen and eggs produced by the tall plant with round seeds will contain the genes *TR*. The pollen and eggs produced by the dwarf plant with wrinkled seeds will be *tr*. Union of sperms from a pollen grain *TR* with an egg *tr* (or a pollen grain *tr* with an egg *TR*) will result in a zygote having the genotype *TtRr* (Fig. 17.15). This plant will be tall and have round seeds. It is heterozygous for both height and seed shape.

The four possible types of eggs and sperms that may be formed during meiosis from a plant with genotype *TtRr* are shown in Fig. 17.16. The sperms and eggs that result are:

Sperms	*TR, Tr, tR, tr*
Eggs	*TR, Tr, tR, tr*

Any one of the four types of sperms may unite with any one of the four types of eggs. The number of possible combinations is best shown on the checkerboard of Fig. 17.15. The four types of sperms are listed along the top of the checkerboard; the eggs are placed at the left. Each square represents a possible zygote. Examination of these squares will show that four phenotypes are possible:

1. Nine tall plants with round seeds, squares numbered 1, 2, 3, 4, 5, 7, 9, 10, 13.
2. Three tall plants with wrinkled seeds, squares numbered 6, 8, 14.
3. Three dwarf plants with round seeds, squares numbered 11, 12, 15.
4. One dwarf plant with wrinkled seeds, square number 16.

A dihybrid cross carried through the F_2 generation may be simply diagrammed as follows:

Parent genotypes	*TTRR*	×	*ttrr*
Parent phenotypes	Tall plants with round seeds		Dwarf plants with wrinkled seeds
Gametes	*TR* ↘		↙ *tr*
F_1 genotype		*TtRr*	
F_1	All plants heterozygous, tall with round seeds		
Self-polination of F_1	*TtRr*	×	*TtRr*
Gametes	*TR, Tr, tR, tr* ↘		↙ *TR, Tr, tR, tr*
F_2 phenotypes	9 tall plants with round seeds 3 tall plants with wrinkled seeds 3 dwarf plants with round seeds 1 dwarf plant with wrinkled seeds		

In this cross, it is important to note that the genes for height and seed shape separate or *segregate independently of each other* during the formation of meiospores from the megasporocyte.

Soon after Mendel's work was discovered, it was noticed that his results could be explained by assuming that the genes were located in the chromosomes. It has now been well-demonstrated that the genes are carried in the chromosomes, and maps have been prepared showing their relative positions in the chromosomes of many plants.

The Backcross

In a backcross offspring are crossed to one of the parents. Usually, a parent having homozygous recessive alleles is selected, because such a cross yields much information regarding the genotype of an offspring. This type of a backcross is called a **testcross.** For instance, there are four different phenotypes in a dihybrid cross and nine different genotypes. There would be an average of nine tall F_2 plants with round seeds for every 16 offsprings produced, and they may have any one of four genotypes: *TTRR, TTRr, TtRR, TtRr* (Fig. 17.15). If the progeny of a backcross between one of these F_2 plants and a homozygous recessive parent (*ttrr*) contains equal numbers of tall plants with round seeds, tall plants with wrinkled seeds, short plants with round seeds, and short plants with wrinkled seeds, the genotype of the F_2 plant must have been *TtRr* (Table 17.1).

LINKAGE AND CROSSOVER

We have seen that during prophase I of meiosis, homologous chromosomes pair, and there may be a breaking

Table 17.1
PROGENY OF BACKCROSS: *TtRr* × *ttrr*

Gametes of *TtRr*	Gametes of *trtr*	Genotype in Progeny of Backcross	Phenotype of Progeny of Backcross
TR	*tr*	*TtRr*	Tall, round seeds
Tr	*tr*	*Ttrr*	Tall, wrinkled seeds
tR	*tr*	*ttRr*	Dwarf, round seeds
tr	*tr*	*ttrr*	Dwarf, wrinkled seeds

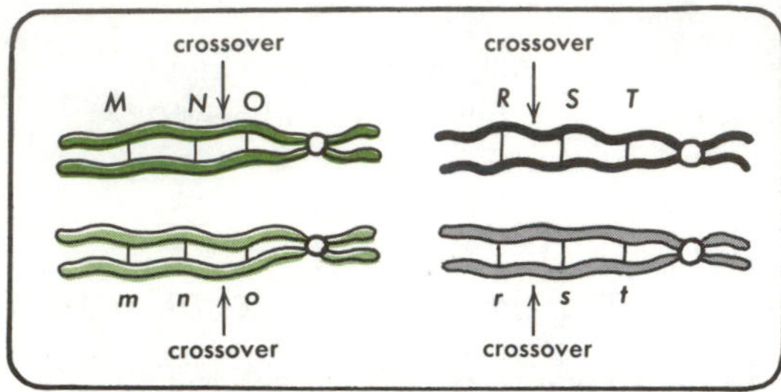

Figure 17.17
Location of genes on chromatids. Future location of crossover indicated by arrows.

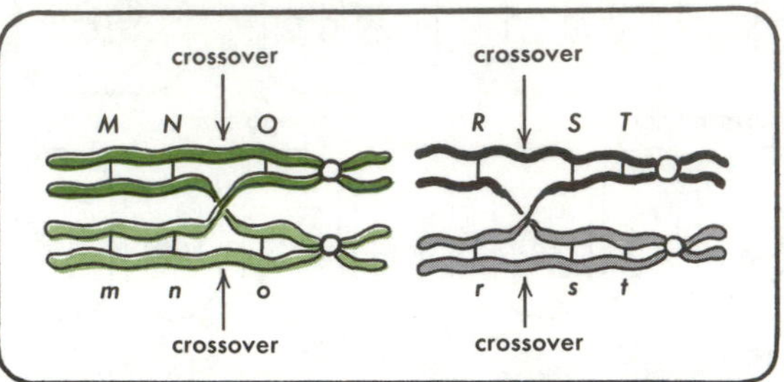

Figure 17.19
Metaphase I. Nonsister chromatids of homologous chromosomes have broken and, at the crossover region have exchanged pieces with each other.

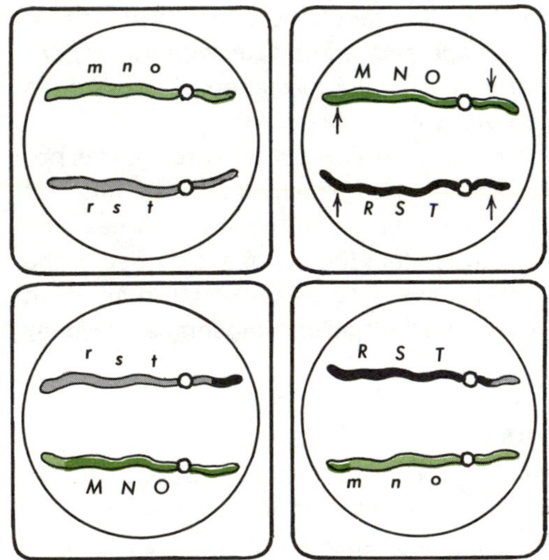

Figure 17.18
Meiospores that would arise if the crossover had occurred distally to the third marker or in the short arm of the chromosome; the three markers in each chromosome would have remained linked in their original order.

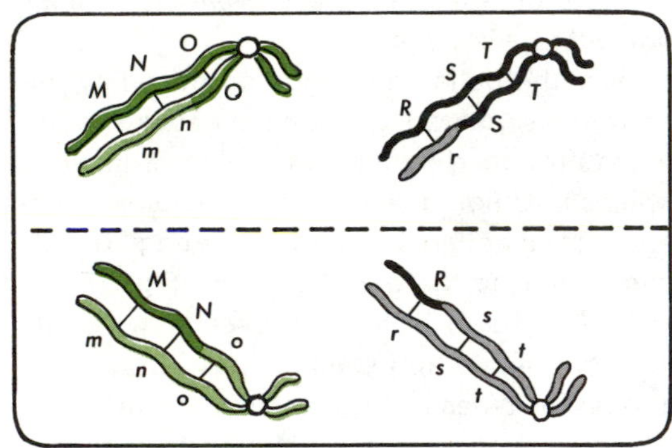

Figure 17.20
Early anaphase I. Kinetochores separate. Sister chromatids are still associated at the kinetochore but, because of the crossover, nonsister chromatids are associated distally to the kinetochore.

and reconsititution of sister chromatids at chiasmata. The distribution of genes during meiosis depends upon the separation of the chromatids, and the location and amount of interchange taking place between them. This is illustrated in the following diagrams. Figure 17.17 represents two pairs of homologous chromosomes, each with three pairs of genes to serve as markers; *Mm, Nn,* and *Oo* on the green homologs and *Rr, Ss,* and *Tt* on the gray homologs. If the chiasmata occurred between *Oo* and kinetochore, or to the left of *Mm,* the relationship between these three pairs would not change: *M, N,* and *O* would remain **linked** together as would *m, n,* and *o* on the chromatids of the original chromosomes.

At metaphase II and anaphase II, the chromatids, bearing *M, N, O* and *m, n, o,* would separate and the three markers *M, N, O, m, n, o,* would remain linked together so that only two types of meiospores *M, N, O* and *m, n, o* (Fig. 17.18) would result.

The same would happen if the exchange occurred to the left of *Rr. T, S,* and *R* would remain linked, as would *t, s,* and *r.* The number of different kinds of meiospores to be derived from one to several linked allelic pairs is two or four, depending upon whether crossing over has occurred between the alleles. Without crossing over, two of the meiospores would be similar; with crossing over, all four meiospores would be different.

Thus, in Fig. 17.19, chiasmata are shown between *Nn* and *Oo* on the green chromosome and between *Rr* and *Ss* on the gray chromosome.

In this case, because of chiasmata formation, whole, complete chromosomes are not separated from each

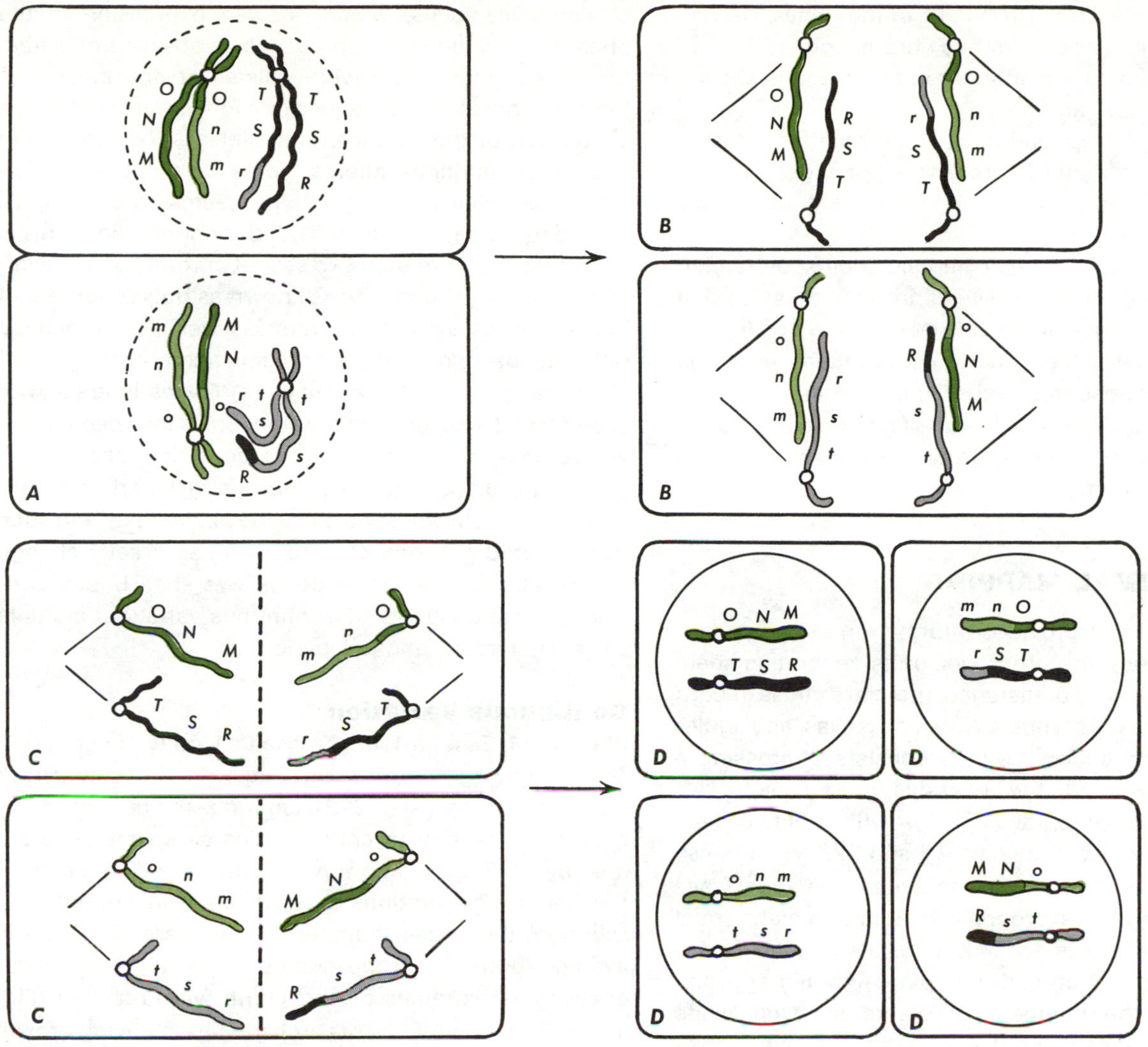

Figure 17.21
A, interphase. Two sister chromatids attached at the kinetochores. *B,* early anaphase II. The chromosomes have moved to the metaphase II plate and the kinetochores have split. Here chromatids are moving to opposite poles of the cell. Note the disposition of the markers. *C,* early telophase II. *D,* four meiospores are formed.

other, for on the side of the chiasma away from the kinetochore, sister chromatids will separate just as in mitosis (Fig. 17.20). Anaphase I of meiosis is thus the separation of two chromatids. Since the chromatids have become variously modified, the separation involves both a separation of whole chromosomes (at the kinetochores) and a separation of sister chromatids as in mitosis (across the chiasma, Fig. 17.20).

In this example, the separation of chromatids during the second division of meiosis would result in the formation of four cells each with a different genetic composition (Figs. 17.21*A,B,C,D*). The chromatids within each of these four cells have been variously modified by crossing over.

The proportion of meiospores having crossover chromosomes in a large number of meiotic divisions will depend upon the number and location of interchanges. For example, if exchanges occur between *Oo* and *Nn* in four out of 100 meiocytes, eight of the resulting 400 meiospores would show interchanges. This is because one meiocyte gives rise to four meiospores, and because a single interchange results in two crossover and two noncrossover chromatids. In this case, 2% of the resulting meiospores would show interchanges. The allelomorphs *Mm* are further removed from *Oo* than are *Nn* (Fig. 17.19). It turns out that the greater the distance separating genes, the greater the number of chromatid exchanges. We might expect that exchanges would occur between *Mm* and *Oo* in eight of 100 meiocytes. This is twice as many as occurred in the shorter distance between *Oo* and *Nn.* Sixteen of the 400 meiospores would show exchanges resulting from crossing over between *Oo* and *Mm.* The percentage of exchanges between genes linked together on the same chromosome is taken to indicate the relative distances separating them and is expressed as **crossover units.**

Now, consider the distribution of the genes, *Oo* and *Tt*, which are on nonhomologous chromosomes. Each of the four meiospores are different in the example shown; they are *OT*, *Ot*, *oT* and *ot*.

Combining segregation of chromatids during meiosis with chromatid interchanges, or crossing over, brings about variation of the genetic constitution of the resulting meiospores.

Meiospores give rise to haploid plants of varying genetic constitutions; they will all produce gametes that fuse to give rise, once again, to a new diploid generation. The gametes carry the genetic constitutions endowed them by the meiospores, and if meiospores differ, gametes must also differ. The fusion of gametes having different genetic constitutions further greatly enhances the possibility for variation.

CHROMOSOMAL MAPPING

The backcross also provides information as to whether genes are linked. The crossover units separating them may be determined. For instance, there are genes in corn that are expressed as white sheaths on leaves, and liguleless blades. The backcross here consists of crossing a plant homozygous for the recessive white sheaths (*ws ws*) and liguleless blades (*lg lg*) with a plant heterozygous for these traits: *Lg lg* and *Ws ws*. If these alleles are on different chromosomes, the four types should appear in the progeny in approximately equal numbers. If the alleles are linked on the same chromosome, the numbers of wild-type plants should approximately equal the numbers of plants showing white sheaths and blades without ligules. A small number of plants with the dominant ligule leaves, but with the recessive white sheath, accompanied by an equal number of plants with the complementary traits, would indicate some crossing over between linked genes. It turns out that about 46% of the progeny of such a cross resemble the heterozygous parent and another 46% have white sheaths and ligule blades. About 8% were crossover types. Therefore, these genes, *lg* and *ws*, are linked and separated by eight crossover units.

OTHER BREEDING SYSTEMS

The breeding experiments so far discussed have involved one or two genes directly related to one or two traits. Height was controlled by one gene, leaf color or flower color, by another, and so on. These genes were independent of each other in activity, segregating at meiosis, and combining at fertilization. A variety of other more complicated breeding systems are known: (*a*) Numerous genes may play a part in determining the expression of a trait. This results in **continuous variation,** rather than in the **discontinuous variation** just discussed. (*b*) We have considered the presence of a single pair of alleles, at one chromosome locus, which showed expression in the phenotype as the simple presence or absence of a trait. Sometimes there are several alleles at one locus and their action may show a series of slight differences in the expression of the trait. Several alleles at one locus are known as **multiple alleles.** (*c*) The number of heterozygous alleles in a genotype seems to be directly related to the vigor of an individual or population. This is **heterosis.** (*d*) More than two sets of chromosomes occur in many plant groups. This is known as **polyploidy,** and it influences the vigor of a plant, as well as the genetic ratios to be expected in breeding experiments. Polyploidy rarely occurs in animals. (*e*) Genes interact with each other. (*f*) Portions of chromosomes may become **inverted** and, in the heterozygous condition, show interesting and unexpected expressions. (*g*) Parts on nonhomologous chromosomes may be exchanged, and such **translocated** portions of chromosomes greatly change the genetic expression of genes. We shall briefly consider selected aspects of continuous variation, multiple alleles, heterosis, and polyploidy.

Continuous Variation

When E. M. East (in 1905) crossed a variety of corn (Tom Thumb) having ears 5–8 cm long with a variety (Black Mexican) having ears 13–21 cm long, the ears of the F_1 generation were 9–15 cm long. The ears were 7–19 cm long in the F_2 generation. A classification of the ears of the F_1 and F_2 generations by length showed a normal distribution: the largest number of ears were close to the average length, with the number of ears gradually decreasing as extremes of ear length were reached (Fig. 17.12). Most traits of plants and animals involving growth rates, vigor, coloring, quality and amount of oil or starch, or quality of fiber or flour, show continuous variation. Traits showing continuous variation must be studied by statistical methods. It is necessary to count many individuals and to take many measurements, weights, and evaluations of color on a colorimeter. The collected figures are then processed by computers. These are **metrical traits.** They are influenced by numerous genes in different locations in the genotype. As the number of genes contributing to the expression of a trait increases, the importance of the contribution of an individual gene to the trait decreases. The many genes thus contributing to the expressing of a trait are **polygenes.** Polygenes are no different from other genes; some may show dominance, and they may interact with each other or mutate. Traits controlled by polygenes more easily respond to the environment than do traits controlled by a single or a few genes. These factors make continuous variation difficult to study, but in the age of the computer, investigations on continuous variation are providing significant information on the mechanism of inheritance.

Multiple Alleles and Pseudoalleles

On p. 328 we defined alleles as a pair of genes that retained their identity and separated inviolate at each succeeding meiosis. While it is true that genes do not appear to react with each other or to be greatly influenced by their environment in their normal process of replication, we are no longer sure that a gene can be strictly defined as a unit of crossing over. This is because all alleles do not neatly determine the expression of a character.

Sometimes a given locus may not consist of a single allele, but may apparently contain an allelic series in which one allele is dominant over a series of recessive alleles. Such alleles are frequently called **multiple alleles.** It is important to remember, however, that only two alleles are present at a time, one on each homologous chromosome.

Alleles are paired genes and, by definition, at a given locus only two, one on each chromosome, can be present at a given time. Generally, one of these is dominant, the other recessive, and in an allelic series this same relationship holds. One gene is generally dominat over all others, but the remainder show varying degrees of dominance over each other.

In corn, at least eight alleles forming an allelic series control the expression of pigment in the leaves, stem, aleurone, and pericarp. They are all located on a chromosome at a point known as the *A locus.* At least three different types of pigments are influenced by the activity at this locus; purple anthocyanins, yellow-brown anthocyanins, and a red-brown pericarp pigment. The allele A^{st} is dominant over all other alleles with respect to plant and aleurone pigmentation, giving purple pigmentation in the leaves and aleurone and red in the pericarp. However, several in the series tend to dilute these colors when present in a heterozygous condition with A^{st}.

A type of recombination may rarely occur between the alleles of such a series, or in situations resembling such a series. In *Drosophila,* about once in 500,000 crosses, crossing over between multiple alleles occurs. This means that these multiple alleles are not strictly alleles, for they behave like separate genes in that crossing over occurs between them at rare intervals. They are pseudoalleles, and they prove that the classical gene may itself be subdivided into physical units. Indeed, in very rapidly multiplying organisms such as bacteria, it is possible to study chromosome interchange within the confines of a gene. In corn, pseudoalleles are known that are separated by 0.0015 of a map unit, a distance that matches the space occupied by about 1,500,000 of the nucleotide pairs of the DNA molecule. In bacteria, pseudoalleles separated by only 0.000001 of a map unit are known, and this figures out to be about 10 nucleotide pairs in the DNA molecule.

Heterosis

Heterosis refers to the presence of a large number of heterozygous alleles in a genotype. The recessive alleles are generally less desirable. Therefore, an increase in the vigor of a plant should occur when one member of a pair of homozygous recessive alleles is replaced by a dominant allele. Thus, an increase in heterozygosity should increase vigor. An increase in heterozygosity is easily obtained. Inbreeding rapidly increases the homozygosity of a genotype. By crossing carefully selected inbred lines, it is possible to attain a very high degree of heterozygous alleles, and the progeny of such a cross exhibit great vigor, being larger and more robust than the initial parents of the pure lines. If the increased vigor were due simply to the dominant genes, genotypes having large numbers of homozygous dominant alleles should be as vigorous as plants with an equal number of heterozygous alleles. This is not the case. Vigor seems to depend upon the heterozygous condition and not upon the presence of many dominant alleles. A number of hypotheses have been developed to account for the relationship between heterozygosity and vigor, but the reason for the relationship remains unclear.

How then may vigor be fixed in a species? Given lines of hybrid corn certainly produce year after year in a given predictable way. Asexual reproduction will of course, fix the number of heterozygous alleles in a population. This does seem to be of importance because a great many plants do reproduce extensively by asexual means.

Strains of hybrid corn are obtained first by establishing pure lines of known constitution. This step requires about seven years of self-pollination, resulting in the development of many homozygous strains. Any two homozygous strains are then crossed to form F_1 hybrids, known to commercial corn breeders as *single-cross hybrids.* Since there are many homozygous strains differing in some respects, it is possible to obtain a large number of different single-cross hybrids, and, therefore, the selection of the best homozygous strains requires a thorough knowledge of the traits desired in the commercial strain. The seed produced in a field of such single-cross hybrids, $A \times B$ (Fig. 17.22), for instance, would give an F_2 generation with its resulting segregation of characters. This is impractical commercially and necessitates a further step. Two carefully selected single-cross hybrids, $A \times B$ and $C \times D$ (Fig. 17.22), are crossed, given what is known as *double-cross* seed. This is sold commercially and produces *double-cross* hybrid plants of remarkable uniformity, capable of yielding large amounts of grain. The steps in the production of double-cross hybrid corn seed are shown in Fig. 17.22 and are outlined in the diagram on page 343.

Polyploidy

Polyploidy is the increase in the number of chromosomes, or of chromosome sets, above two. Polyploidy may arise in two ways: by the doubling of an homologous set of chromosomes **(autopolyploidy)**; or by combining two complete sets of chromosomes from different species of plants **(allopolyploidy).**

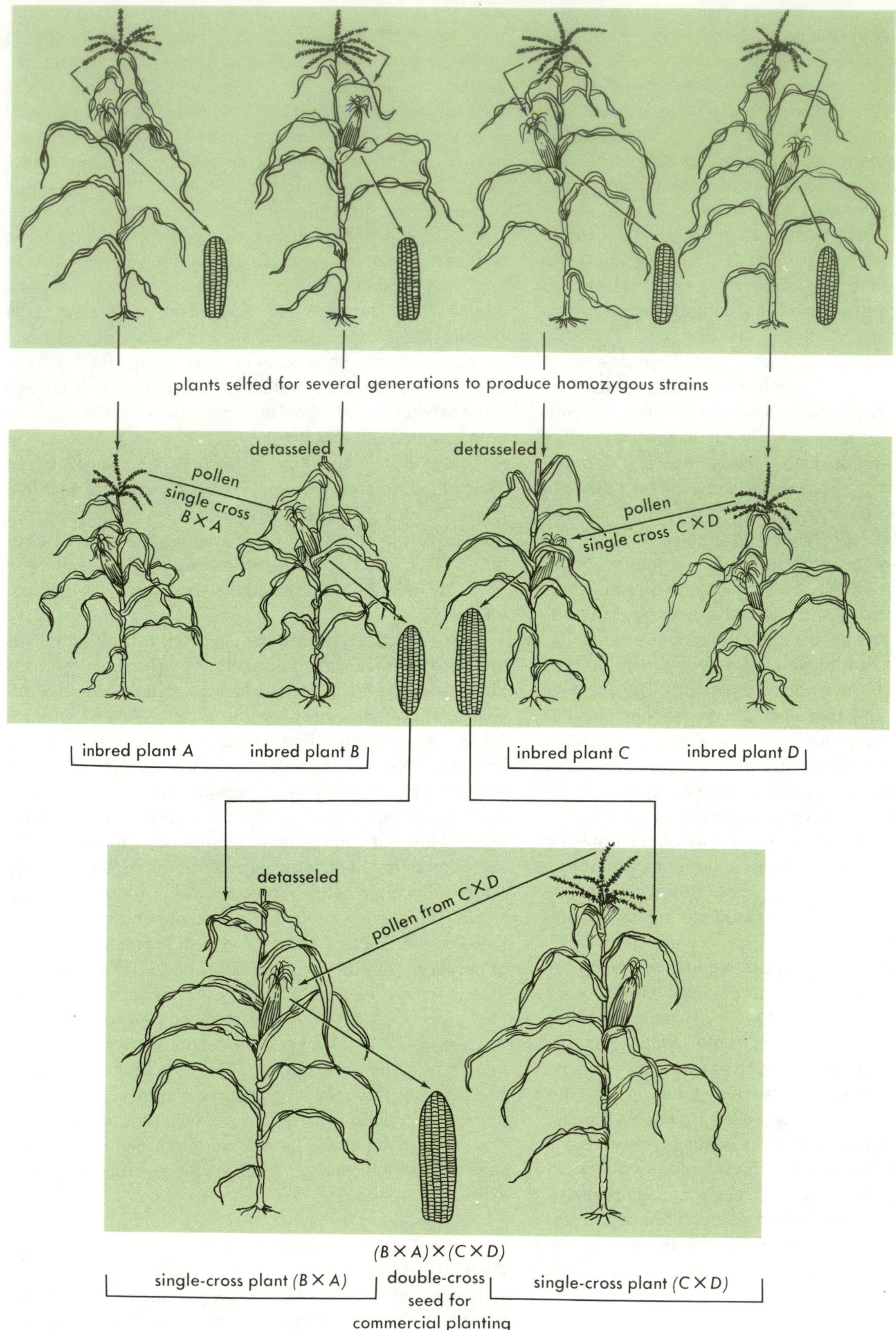

Figure 17.22
The manner in which commercial hybrid corn is produced. (Redrawn from *Farmers' Bulletin* 1744, U.S. Department of Agriculture, Bureau of Plant Industry, Soils and Agricultural Engineering.)

Steps in Production of Hybrid Corn Seed

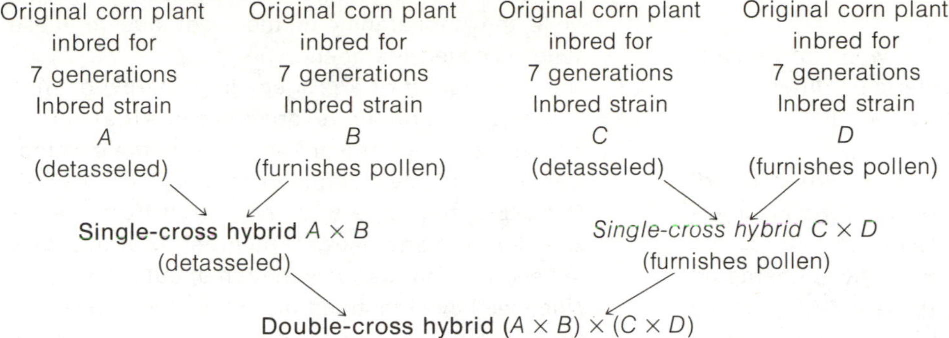

Autopolyploidy

The drug colchicine, dervied from a crocus growing in the Middle East, inhibits spindle formation when applied to cells during mitosis. The separation of chromatids occurs at anaphase, but they remain within the nuclear envelope. When the nucleus is reconstituted, it contains four sets of homologous chromosomes. This is a **tetraploid,** and since its four sets of chromosomes are homologous, it is an **autotetraploid.** Such a doubling of the chromosome number in a zygote would result in an autotetraploid plant, and such plants are known. This plant would have four chromosomes with the *A* gene. If it is homozygous, there would be four *A* genes or four *a* genes. There would be varying degrees of heterozygosity, *Aaaa, AAaa,* or *AAAa.* Pairing between four homologous chromosomes is irregular, and since gametic ratios depend upon the amount of pairing and crossing over, these ratios vary considerably. If a particular autotetraploid has the constitution *AAAa,* eight chromatids would arise during meiosis and they could show random segregation. Thus the gametic ratios of *AA, Aa,* and *aa* could range from the extremes of 3*AA*:3*Aa*:O*aa* to 15*AA*:12*Aa*:1*aa.* In addition, one gamete might receive three *AAA* chromosomes and its counterpart one *a* chromosome. Such irregularity renders autotetraploids unstable genetically, and unless they reproduce asexually, they seldom become permanently established. Sugar cane is an autopolyploid that is reproduced asexually. Breeding experiments require the annual screening of some million seedlings derived from crosses between plants having different genotypes.

Allopolyploids

Hybrids between species are usually sterile. They do not have sets of homologous chromosomes; thus, pairing is very irregular, and viable seed is seldom set. However, should a doubling of chromosomes occur in a zygote having one complete set of chromosomes from each parent, the resulting tetraploid cell would have only two sets of homologous chromosomes. The plant arising from this kind of doubling of chromosome numbers is an **allotetraploid.** Pairing and crossing over are now normal; the parental chromosomes pair with each other as they would in the parent plants. This allotetraploid is a fertile plant. It will not successfully backcross with its parents and does not look like them. It is essentially a new species or even a new genus.

Aneuploids

It is not necessary that whole sets of chromosomes be added to those already present. One, two, three, or four chromosomes may be added to a plant with a diploid number. If one chromosome is added, a pair of homologous chromosomes will become a triplet of three homologous chromosomes. Instead of two *A* chromosomes (*AA*), there will be three *A* chromosomes (*AAA*). This plant is a **trisomic** plant. A diploid chromosome number of 30 indicates 15 pairs of homologous chromosomes. Since a single chromosome may be added to each of these homologous pairs, a plant having 15 pairs of chromosomes will have 15 possible different trisomic plant types. Numerous cases are known in which all the possible trisomics have been obtained. Two or more extra chromosomes may be present, and different arrangements are possible. Plants having extra chromosomes are **aneuploids.**

More information on polyploidy is given in Chapter 31.

SUMMARY

Summarizing, the following important facts have been established regarding the mechanism of inheritance:

1. The factors responsible for inheritance are located in small, definite regions of chromosomes. In other words, inheritance is due to small regions of chromosomes. These factors or chromosome regions are called **genes.**
2. Most higher plants have the diploid number of chromosomes, the vegetative cells containing two sets of homologous chromosomes.
3. Two genes thus interact in the development of many characters in the diploid plant.

4. The gametes are haploid and consequently carry only one member of each pair of homologous chromosomes.
5. Members of a gene pair, *TT* and *tt*, while influencing a single character (height), determine different aspects of that character, such as tallness and dwarfness.
6. When different forms of a character are involved, such as tallness and dwarfness, one form is generally dominant, and the other recessive.
7. When a plant, heterozygous for a given character, such as *Tt*, is self-pollinated, the resulting progeny will number three times as many dominant plants as recessive plants.
8. When a plant, heterozygous for two given characters, such as *TtRr*, is self-pollinated, the resulting progeny will be of four different types, in the following ratio: nine plants will show only the dominant characters (tall and round), three will show one dominant and one recessive (tall and wrinkled), three will have the second dominant and second recessive (dwarf and round), and one plant will have both recessives (dwarf and wrinkled).
9. The chromosomes carrying the genes segregate independently of each other; thus, *T* may go with *R* or with *r*.
10. Genes are linked together on chromosomes. While such genes normally segregate as a unit, crossing over involving a recombination of characters occurs regularly between them.
11. Mutations may arise by an error in the replication of the nucleotide pairs in the DNA molecule.
12. The sum total of genes (particularly those under discussion) within a nucleus consititute the genotype, while the appearance of the plant due to these genes is called the phenotype.
13. The segregation of a progeny into sharply defined classes is discontinuous variation and is the result of the expression of one or two genes in the development of a precisely definable trait.
14. The expression of a trait may result from the interaction of many genes. The progeny in this case will show a continuous variation about a mean.
15. When we have identified more than two forms of a gene (or more than two variations in the nucleotide sequence of a locus), we call those multiple alleles.
16. At rare intervals, recombination takes place between what appear to be genes in an allelic series.
17. A high degree of heterosis or a large number of heterozygous alleles in a genotype greatly increases vigor.
18. The diploid number of chromosomes is the optimum condition for the normal steps of meiosis. Polyploid individuals of groups of plants have numbers of chromosomes other than the diploid numbers. An increase in the number of sets of homologous chromosomes is autopolyploidy. Allopolyploidy occurs when the chromosome number is changed by the addition of sets of nonhomologous chromosomes. Aneuploid refers to any deviation from complete sets of chromosomes.
19. Changes in chromosome number by polyploidy may result in increased vigor and great disturbance in meiosis and fertilization.

CHAPTER 18

PLANT ECOLOGY

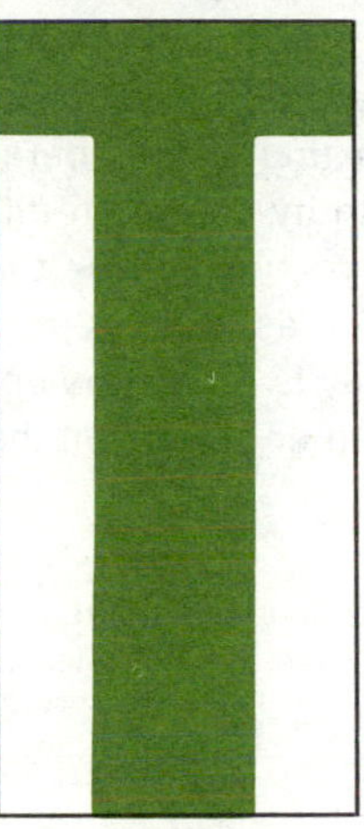

hrough evolution, existing plant species have achieved a balance with their environment. In each type of habitat, certain species group together. Fossil records indicate that similar groups of species have lived together for millions of years. The species of these groups share incoming solar radiation, soil, water, and nutrients to produce a relatively constant amount of living matter. They recycle nutrients from the soil to living tissue and back to the soil again; and they divide the environment's resources. Plant ecology attempts to explain this balance between plants and their environment. For example, what stresses does the environment put on plants, and how do plants respond to them? Plant ecology also attempts to determine how this natural balance may be artificially imitated or manipulated to improve human life.

Plant ecology deals with levels of biological organization beyond the organism level: the population, the community, the ecosystem. A **population** is a group of closely related organisms that normally interbreed. A **community** is composed of all the populations in a given habitat. An **ecosystem** consists of the living community plus the nonliving factors of its environment.

COMPONENTS OF THE ENVIRONMENT

The environment of an organism includes all of the living and nonliving things around it. Some important environmental components are moisture, temperature, light, soil, and living organisms, and all of them work separately and jointly to influence plant distribution and behavior. Species differ in their range of tolerance to these environmental factors. For example (Fig. 18.1), coast redwood (*Sequoia sempervirens*) is restricted to a narrow strip of California and adjacent southwestern Oregon, about 9500 square km, that experiences heavy summer fog. Coast redwood is not tolerant of variations in moisture or temperature. Red maple (*Acer rubrum*), however, occurs over about 4,000,000 square km of eastern North America in wet, dry, cold, or warm environments. It is tolerant of wide variations in moisture and temperature.

There are two types of environment. The **macroenvironment** is influenced by the general climate, elevation, and latitude of the region. Weather Bureau data on rainfall, wind speed, and temperature are measures of the macroenvironment. Measurements are taken at a standard height of 1.5 m above ground in a clear area away from buildings or trees.

The **microenvironment** is the environment that is close enough to the surface of an organism or object to be influenced by it. For example, bare soil tends to absorb heat; consequently, the temperature just above or below the soil surface is much higher than air temperature (Fig. 18.2). Light quality and quantity are much different for herbs beneath a forest canopy than for the leaves of the canopy itself (Fig.18.3). Air as far as 10 mm from the surface of a leaf on a still day is less turbulent and higher in humidity than free air further away from the leaf. In marshy areas, where the water table is close to the surface, minor dips or rises in the topography greatly influence the root microenvironment. Plants growing in shallow depressions are often subject to more frequent frost than those growing on higher ground, because cold air will settle in the depressions. The microenvironment is as important for plant growth as the macroenvironment.

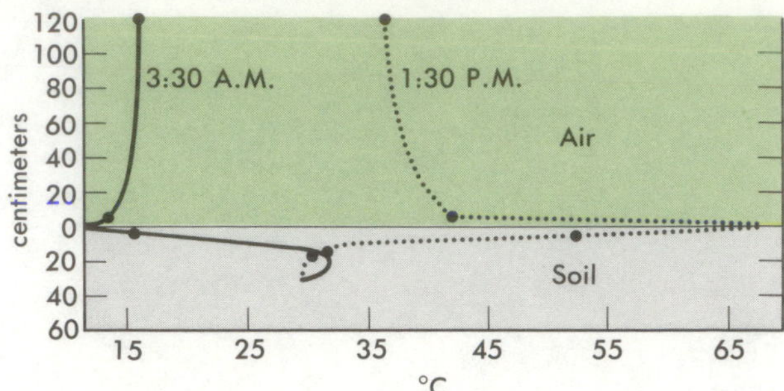

Figure 18.2
Air and soil temperatures at several positions just above and just below the soil surface. Temperatures were measured during the warmest and coolest parts of the day. (Redrawn from W. D. Billings, *Plants and the Ecosystem,* © Wadsworth Publishing Company, Belmont, Ca. Reprinted with permission of the publisher.)

Moisture

Water is a most important factor in determining both the

Figure 18.1
Distribution of coast redwood (*Sequoia sempervirens*) and red maple (*Acer rubrum*) in the United States. (Courtesy of *Silvics of Forest Trees of the United States.* Department of Agriculture Handbook No. **271,** Washington, D.C.)

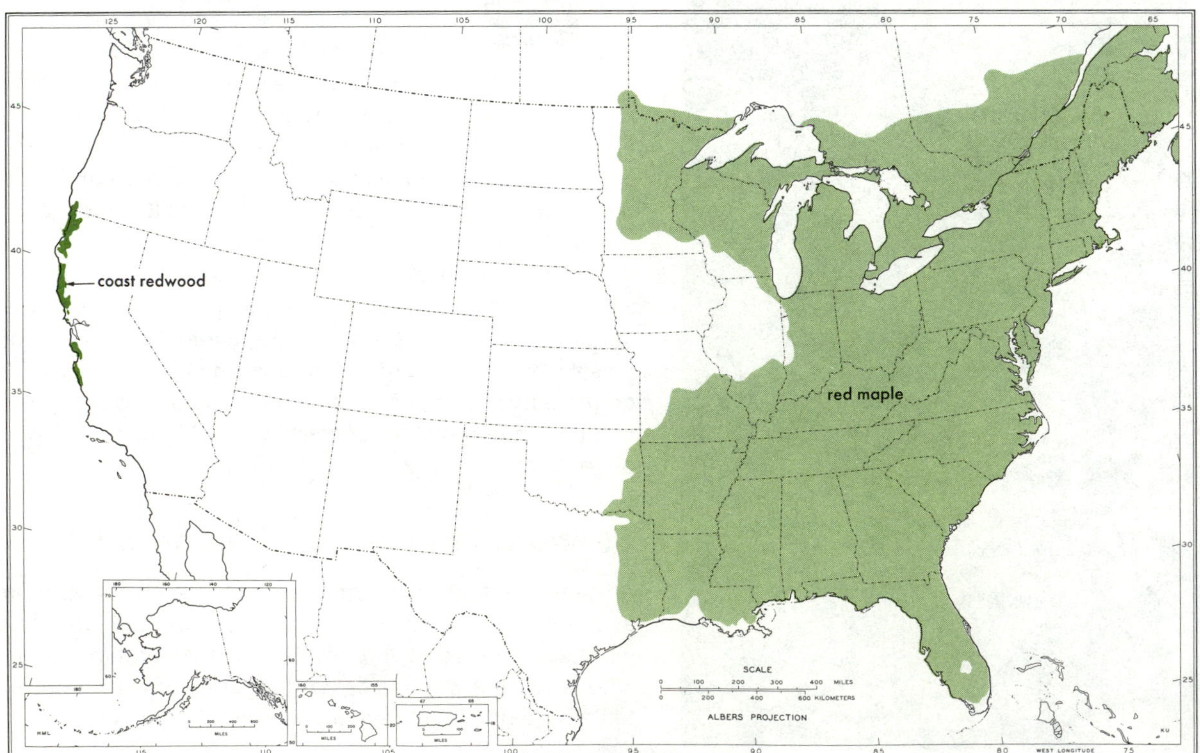

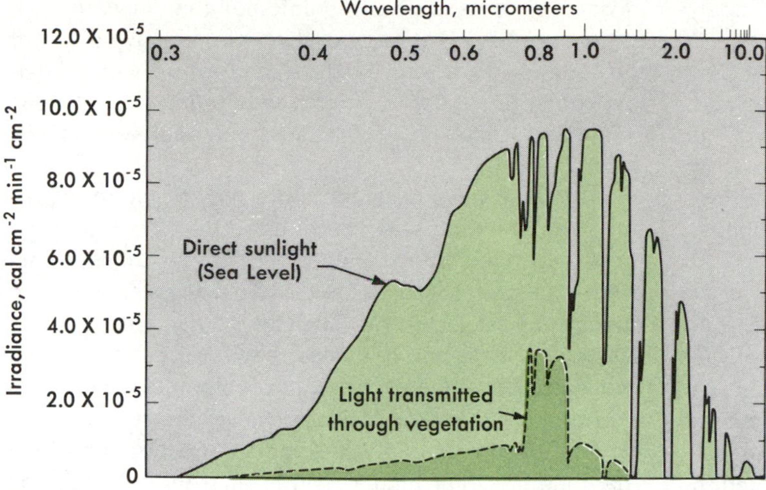

Figure 18.3
Distribution of solar radiation energy by wavelength. Energy was measured in direct sunlight and beneath a forest canopy. (Redrawn from D. M. Gates, *Ecology* **46,** 1. © 1965 by Ecological Society of America.)

distribution of plants over the earth's surface and the character of an individual plant. Probably no single factor is so largely responsible for the abundance of plants in a habitat as is the supply of water. So important are the water relationships of plants that various attempts have been made to classify plants on the basis of these relationships. One such classification divides plants into (*a*) **xerophytes,** which are able to live in very dry places, (*b*) **hydrophytes,** which live in water or in very wet soil, and (*c*) **mesophytes,** which thrive best with a moderate water supply.

Plants with xerophytic characteristics, which limit transpiration or in other ways closely control water balance, occur in different climatic zones, especially deserts. Xerophytic plants do not necessarily have a lower transpiration rate than do mesophytes when water is ample, but they do possess one or more characteristics that enable them to survive periods of drought. The most effective of them are a thick cuticle, early or daytime stomatal closure, reduction of the transpiring surface, and water storage tissue.

The time of precipitation is as important to plant growth as the total annual amount. For example, tropical rain forests and tropical savannahs may both receive the same total yearly rainfall—let's say 200 cm—but the forest receives an equal share of the total each month, while the savannah has pronounced dry and wet seasons. The result is that the two vegetation types are quite different: tropical rain forests support tall, evergreen trees and vines in profusion, but tropical savannahs support grasses, shrubs, or short trees, which are often deciduous.

Plants adapt to dry climates in several ways. Cactus plants develop water storage tissue in the stem and reduce transpiration by the loss of leaves; all their photosynthetic activity is performed in stem tissue. Mesquite shrubs (*Prosopis*) and salt-cedar trees (*Tamarix*) possess long roots that tap ground water, sometimes at depths below 53 m. **Epiphytes,** growing on tree trunks above the soil (Fig. 18.4), trap falling rain in leaf axils or in dead cells on the surface of roots. These epiphytes are not parasites; they depend on trapped water and debris for all growth requirements. Annual herbs adapt to drought by completing a brief life cycle only during periods of sufficient soil moisture. *Boerhaavia repens* of the Sahara Desert is a small annual that can go from seed

Figure 18.4
Tillandsia, an epiphyte native to Central America and Ecuador. (Courtesy of the New York Botanical Garden.)

to seed in 10 to 14 days, but most annuals have a life span of 3 to 8 months.

There is some evidence to show that plants can absorb dew through their leaves, but the amount of water obtained in this way is not great. Usually, dew, fog, or high humidity are of more value in reducing the rate of transpiration, than by physically entering the plant. (Exceptions include small epiphytes, certain annual plants, mosses, and lichens.)

Plants may also utilize fog by condensing the mist into large drops that trickle down the stem or drip from the branches, increasing soil moisture considerably. The amount of water added to soil in this way by trees on the San Francisco Peninsula was measured as 5 to 74 cm in one summer, the exact amount within that range depending on the type of exposure. Normal precipitation (rain) for the area is only 64 cm a year, with almost no rain in summer.

Temperature

Temperature influences moisture availability and the rate at which chemical reactions occur. At freezing temperatures, water changes to ice and is unavailable for plant growth; consequently, in cold climates one of the principal factors limiting growth may be drought. Absorption of water by roots is slower in cold soil than in warm soil. At high temperatures, evaporation may remove much of the surface soil moisture before the new growth of shallow roots can reach it, and transpiration will also be stimulated, resulting in further rapid depletion of soil water.

Brief extremes in temperature may be more important in determining plant distribution than are long-term moderate temperatures. Palms and many cactus plants seem excluded from areas that experience a specific frequency of frost. California lilac (*Ceanothus megacarpus*) seeds germinate poorly unless they have been exposed briefly to high temperatures. Many temperate zone plants are not frost-resistant until they have been exposed to moderately low temperatures for a short time and thus "hardened." Other plants require an alternation of day and night temperatures (a **thermoperiod**) for best growth.

Some deciduous plants, which have evolved in latitudes characterized by cold winters, require significant periods of near-freezing temperatures when they are dormant. Without such exposure their seeds may fail to germinate, vegetative growth the next season may be delayed or poor, and flowering may be delayed. Thus, for example, certain peach varieties cannot be grown successfully in areas of relatively warm winters.

Figure 18.5
Distribution of mountain hemlock (*Tsuga mertensiana*), showing the relationship between altitude and latitude.

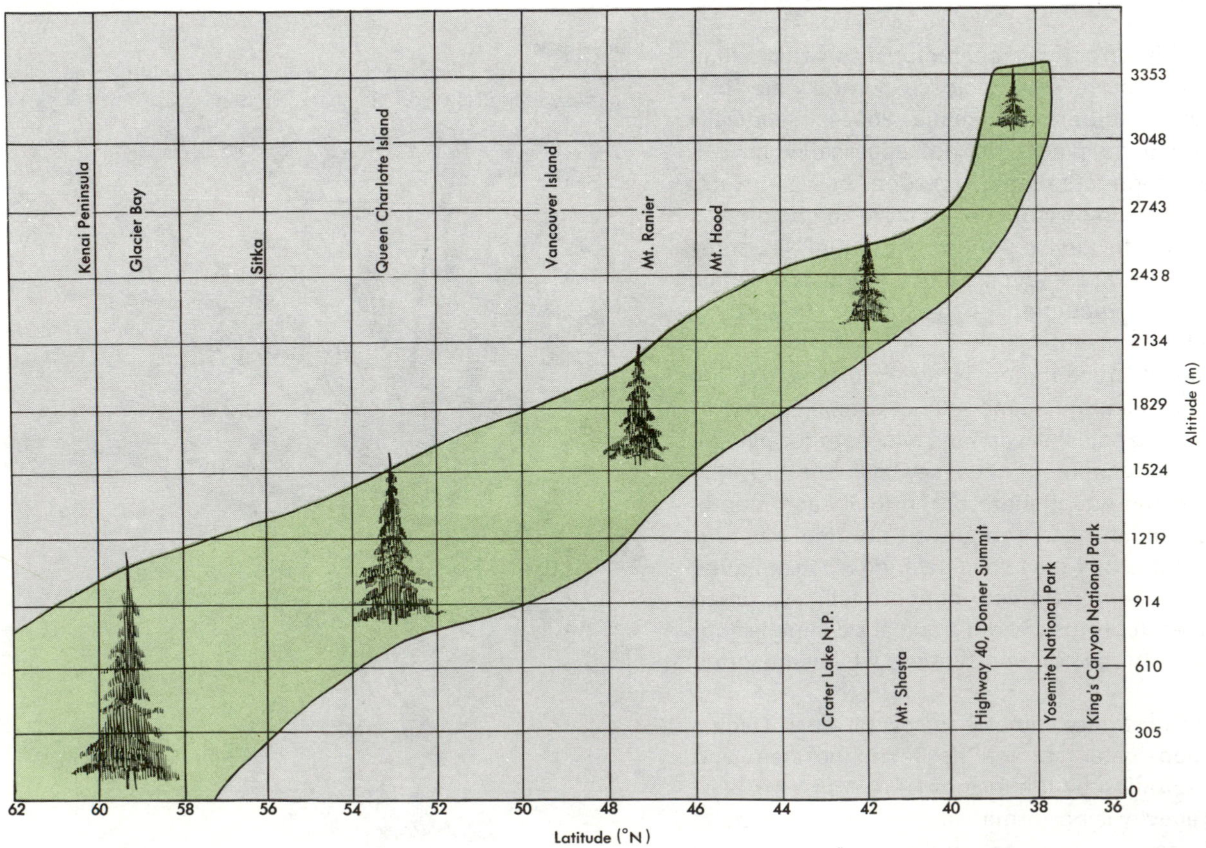

Topography greatly influences soil and air temperature. In mountains latitude and elevation combine to restrict species to certain belts (as shown for mountain hemlock in Fig. 18.5). Even at the same latitude and elevation, minor differences in the direction of slope **(aspect)** create different microenvironments. A gorge running east-west through an Indiana forest was studied to illustrate the effect of aspect on plant growth. The gorge was 65 m wide at the top and 45 m deep; its sides supported scattered trees, shrubs, and herbs. Meteorological instruments were placed 15 cm above and below the soil surface midway down each side. During spring, the south-facing slope was found to exhibit a greater daily range of topsoil temperature, a higher mean air temperature, a higher rate of soil water evaporation, and lower relative air humidity than the north-facing slope. Of nine spring-flowering species present on both banks, the average flowering time was six days earlier on the south-facing bank. A similar difference in flowering time could be achieved on level ground only over a distance in latitude of 180 km.

Temperature is only an estimate of the heat energy available from solar radiation. Solar radiation at the limits of our atmosphere is equivalent to about 2 cal per cm² per min. Much of this radiation is absorbed, scattered, or reflected within the atmosphere, and only half may reach the ground and heat it. In turn, the warm earth reradiates energy back to space—terrestrial radiation. The difference between solar (incoming) radiation and terrestrial (outgoing) radiation is termed **net radiation.** During a clear day at the equator, solar radiation is greater than terrestrial and, thus, net radiation is positive; during the night, the reverse is true and net radiation is negative. But, over a 24 hr period, a month, or a year, net radiation may be positive or negative, depending on the location. For example (Fig. 18.6), at Aswan, Egypt—a warm desert climate—net radiation is positive each month, with a peak in June. Net radiation is positive only five months of the year at Turukhansk, Siberia—a cold subartic climate—and the annual net radiation is very close to zero.

Figure 18.6
Monthly net radiation of Aswan, Egypt, and Turukhansk, Siberia. (Redrawn from D. M. Gates, *Energy Exchange in the Biosphere.* © 1962 by Harper and Row, Publishers. Reprinted with permission of the publisher.)

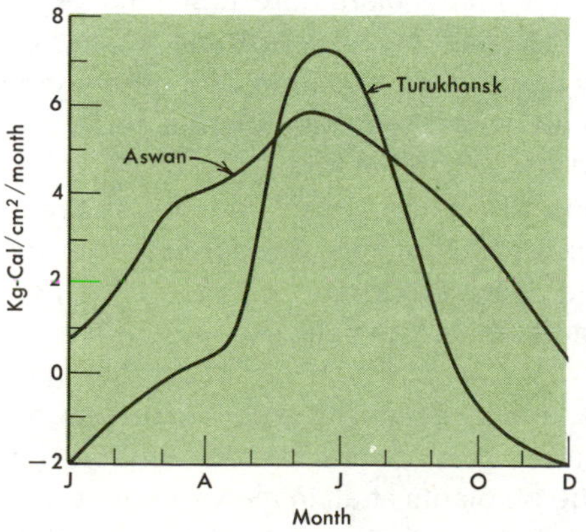

Light

Figure 18.3 shows how solar radiation is changed in quality and quantity by its passage through green leaves. Light on the forest floor may be only 1 to 5% as intense as that of full sunlight and it is primarily green in color. Some herbs of the deciduous forest complete their life cycle in early spring, before the trees above them develop their complete foliage and reduce light intensity. Leaves of plants that develop in shade exhibit different morphology, anatomy, and physiology than those that develop in sunlight, even when both are attached to the same plant. Shade leaves are larger, thinner, contain less chlorophyll per gram of tissue, have less well-defined palisade and spongy mesophyll layers, and achieve maximum rates of photosynthesis at much lower light intensities than do sun leaves (Fig. 18.7). Recent biochemical evidence has shown that the difference in photosynthetic rates is due to a difference in the activity of the enzyme responsible for CO_2 fixation.

When a large tree dies in a mature forest, the canopy is opened and increased light intensity reaches the ground beneath. This increased light radiation induces seedlings there to begin rapid growth. Seedlings must be able to survive many years of slow growth in shade before such an opening "releases" them. Eastern hemlock (*Tsuga canadensis*) is remarkably shade-tolerant. Mature trees of this species live for 1000 years. They can retain the capacity for rapid growth when released from shade to an age of 400 years. Saplings only 2 m tall and 2 to 3 cm in diameter may have a ring count indicating an age of 60 years.

Trees that are not shade-tolerant may become established and grow to maturity if some disturbance such as fire or logging first removes the shade. Seedlings of shade-tolerant species grow beneath them, however, and replace them as they die, while their own seedlings do not remain alive. For this reason, it is difficult to maintain pure stands of valuable shade-intolerant species

Figure 18.7
Photosynthetic response to different light intensities for sun and shade leaves of beech (*Fagus sylvatica*). (Redrawn from P. Boysen-Jensen and D. Muller, *Jahrb. Wiss. Bot.* **70,** 493, 1929.)

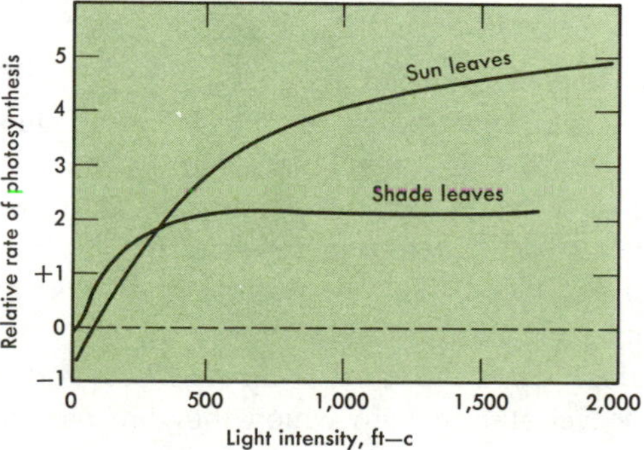

such as teak, mahogany, Douglas fir, or the southeastern yellow pines. These species only invade disturbed sites and do not long maintain themselves.

Light is also reduced in intensity and quality by passage through water. Algae are not commonly found below a depth of 60 m except in very clear water. A record depth for algae may have been found in Lake Tahoe, California. A sample of water and bottom mud from a 140 m depth contained the alga *Chara*. Light intensity at that depth, despite the water clarity, is only 0.1% of full sunlight (see Chapter 23).

Brief exposure to light may be as important for some plants as long-term exposure. Recall from Chapter 16 that seeds of some species can be stimulated to germinate in the dark if they are first exposed briefly to full sunlight; this is done by simply shaking them for a few seconds in soil held in a glass container. This treatment could be duplicated in nature by a plow turning the soil and seeds in it, briefly exposing the seeds to sunlight, then reburying them. Plowing aerates the soil, makes it more permeable for root growth, and removes plants at the surface—all conditions of advantage to an emerging seedling.

Plants also respond to the length of day (photoperiod) by flowering, going dormant, or emerging from dormancy (see Chapter 20).

Soil

Soil is the part of the earth's crust that has been changed by contact with the living and nonliving parts of the environment. It is a weathered, superficial layer typically 1 to 3 m thick, made up of decomposed and partly decomposed parent rock material with associated organic matter in various stages of decomposition. There are many kinds of soils and many different soil conditions. The character of natural plant covering and the behavior of crops depend on soil conditions as well as climatic conditions. Environmental influences that operate through soil are called **edaphic factors;** those that act on the plant through the atmosphere are called **climatic factors.**

Each environment creates a unique soil type. These soil types have their own history of development, morphology, and chemical attributes. In the United States alone, there are 14,000 soil types, each more formally called a **soil series** by soil taxonomists. Similar series are grouped into families, and families into orders.

Soil Profile Development

The profile of a spodosol (old name = podzol) was illustrated and described in Chapter 11 (see Fig. 11.1). It was characteristic of wet temperate regions, such as those that support dense coniferous forests in Canada. **Spodosol** is one of 10 **soil orders** currently recognized by soil taxonomists (Table 18.1). To emphasize their different attributes, we shall briefly discuss four other orders and the vegetation with which they are associated.

Alfisols and **ultisols** lie beneath the deciduous forests of the eastern United States. They are related to spodosols in that they are leached and acidic, but not as extremely as spodosals.

Mollisols (old name = chernozem) develop beneath grasslands that experience moderate rainfall and high summer temperatures. Moisture is no longer sufficient to leach the soil severely; the pH is 7 to 8. Fibrous root systems of grasses thoroughly permeate this soil, and their decay evenly enriches it with organic matter. An *A* horizon, containing about 10% organic matter by weight, occupies the top 60 cm. It may be almost black in the upper part, but becomes dark brown below. Just beneath the *A* horizon is a calcium carbonate accumulation, leached from above. The layer is sometimes called the A_c horizon. Below it is the parent material; there may be no *B* horizon.

Oxisols (old name = laterites) form in regions of high rainfall with continuously warm temperatures, such as those that support tropical rain forests. Mineral cycling is rapid because of fast plant growth and fast litter decomposition. Soil is acidic in pH. Removal of the forest canopy quickly leads to soil deterioration because litter decomposition no longer compensates for leaching. Excessive amounts of fertilizer must be added if the land is farmed. Silicates (sand) are leached, but clay and iron are not; the iron in clay is oxidized to Fe_2O_3, giving the topsoil a brilliant reddish color.

Azonal Soils

Some soils support unusual vegetation, or the soils fail to develop a characteristic profile because of unusual parent material, an accumulation of salts, or poor drainage that leads to frequent flooding.

Depending on the type of parent material, soils may be abnormally low or high in particular ions. For example, **serpentine** soils, among other characteristics, are very low in calcium, and many plants will not grow on them unless calcium is added. Some plants which do grow on serpentine are **endemic** to that soil (*restricted* to it and not normally found growing on other soils). Transplant tests, however, have shown that the serpentine endemics will grow perfectly well—in fact better—on normal soil *if* the surrounding plants are first removed. It appears, then, that the endemics are restricted to serpentine because they are poor competitors with other plants in nonserpentine areas, but they can tolerate serpentine soil and here find fewer competitors.

In arid regions, short and violent rainstorms create streams of excess water that run off and collect in low basins. In time, the ponds evaporate and leave the topsoil permeated with salts which had been in solution. The soil in the center of the basin (which lay under the most water) is highest in salt content; the soil at the edge is lowest. Plant distribution parallels salt distribution (Fig. 18.8): there may be no plants at all in the very saline center, but salt-tolerant plants such as salt grass (*Distichlis*)

Table 18.1
THE DERIVATION AND GENERAL SIGNIFICANCE OF THE NAMES OF THE 10 SOIL ORDERS

Order Name	Derivation of Name	Characteristics
Entisol	Artificially created	Soils of very limited development. Profile properties largely inherited from the parent material.
Inceptisol	L. *inceptum*, beginning	Soils exhibiting moderate development. Identifying horizons are of types that form relatively quickly.
Aridisol	L. *aridus*, dry	Soils of arid regions. Limited change in parent material because of low climatic intensity.
Mollisol	L. *mollis*, soft	Soils with friable surface horizons noticeably darkened by organic matter. The degree of base saturation is high.
Spodosol	L. *spodus*, wood ash	Soils with illuvial *B2* horizons having significant accumulations of free Fe and Al oxides, noncrystalline clays, and humus; usually without structure and often partially cemented.
Alfisol	Artificially created	Soils containing a *B* horizon with significant amounts of crystalline clays and generally a moderate to high degree of base saturation.
Ultisol	L. *ultimus*, last	Soils containing a *B* horizon with significant amounts of crystalline clays but generally with a low degree of base saturation.
Oxisol	Gr. *oxid*, acid, sharp	Highly weathered soils containing a *B* horizon consisting primarily of sesquioxides or 1:1 clays.
Vertisol	L. *verto*, to turn	Soils that form large cracks on drying; self-plowing.
Histosol	Gr. *histos*, tissue	Organic soils.

Figure 18.8
Zonation of plants around a saline basin, Grand Coulee, Washington. The saltiest soil in the center of the basin is devoid of plants. Surrounding this is a broad belt of salt grass (*Distichlis stricta*), then a narrow belt of dark shrubs (greasewood, *Sarcobatus vermiculatus*), and finally sagebrush (*Artemisia tridentata*). From R. F. Daubernmire, *A Textbook of Plant Autecology*, second edition, John Wiley & Sons, New York.)

grow in a zone further out; less tolerant shrubs such as greasewood (*Sarcobatus*) grow still further out; and the usual shrubs of the area, such as sagebrush (*Artemisia*), surround the sink. Transplant experiments indicate that seeds of salt-tolerant plants will germinate and grow very well in nonsaline soil, and they appear to be restricted to saline soils only because they are poor competitors with other plants growing on nonsaline soil.

Very few plants are adapted to soils that are frequently flooded. A lack of soil oxygen impedes normal respiration and nutrient uptake; furthermore, some reduced ions, such as ferrous ion, are toxic. Among the adaptations that plants exhibit in flooded soils are: **aerenchyma** tissue in stems and larger roots (porous parenchyma tissue that permits the passage of air from above-ground organs to below-ground organs); secretion of oxygen out the roots into nearby soil, so that potentially toxic ions are oxidized; and the accumulation of malate during respiration, instead of potentially toxic ethanol.

Fire

In the past 40 years, fire has been rediscovered as a major ecological factor. We say "rediscovered" because primitive human beings were well aware of its effect and learned to use it for their own purposes. It may well be that their most important food crops, their domestic animals, their routes of migration, and some of their cultural attributes had been molded by natural and man-made fires. Certainly, many animals are today attracted to fire and exhibit behavioral patterns in relation to fire that imply thousands of years of evolution in response to it.

Most natural fires are started by lighting, and in North America many vegetation types, including grassland, chaparral, and at least some conifer forests, owe their distribution and community structure in large measure to lightning fires. On U.S. National Forest land, during the 22-year period 1945 to 1966, lightning fires accounted for 64% of all fires, an average of 5000 fires each year. In addition, early explorers and settlers invariably commented on the frequency of fires in forests and grasslands. In the words of E. V. Komerek, a recent investigator: "Lightning fires are an integral part of our environment and though they may vary in both time and space, they are rhythmically in tune with global weather patterns. Our environment can be called a fire environment."*

One of the first vegetation types shown to be dependent on frequent fire for its maintenance in this country is the pine savannah along the southeastern coastal plain (Fig. 18.9*A*). Tall, scattered loblolly, slash, shortleaf, and longleaf pines dominate the area, and a thick growth of grasses (mainly *Andropogon* and *Aristida*), with some herbs, cover the ground beneath. During the nineteenth century when this vegetation was protected from fire, it was noticed that the pines gave way to oak; in time the result was a thick oak forest with little grass. As the pines were valuable for lumber and turpentine, and the grass for forage, landowners were concerned about this trend. But the importance of fire was not conclusively demonstrated until 1930, with longleaf pine (*Pinus palustris*).

Figure 18.9
A, the pine savannah of the southeastern coastal plain. Pine shown is longleaf (*Pinus palustris*); area is in North Carolina. *B*, close-up of longleaf pine seedling in "grass" stage.

Longleaf pine is tolerant to fire, and dependent on it. The seeds germinate in the fall soon after they drop to the ground from the cones. During their first year of growth, the seedlings are very sensitive to even the slightest fire. However, during the next 2 to 4 years,

* E. V. Komarek, Sr. 1968, The nature of lightning fires, pp. 5–41. In *Proceedings of the 7th Tall Timbers Fire Ecology Conference,* Tall Timbers Research Station, Tallahassee, Florida.

longleaf pine seedlings are in the "grass" stage (Fig. 18.9*B*). Most of the growing is done by the roots, and the stem apex remains close to the ground. The terminal bud is covered with a dense mat of hairs that protects it from surface fires that may sweep the area during this time. Fire is even beneficial at this stage, because it kills a particular fungal disease that parasitizes the long needles.

By the time the seedling is 3 to 5 years old, a surface fire is desirable to remove grasses that may have covered the seedlings and shaded it from the sun. The seedling will then make a spurt of stem growth so rapid, that by the age of 8 to 9 years that young tree's canopy will be high enough above the ground to be out of reach of surface fires. In addition, its thick bark can endure the heat of a fire without permitting damage to the cambium. If fire is kept from such an area for 15 years, grasses and young oak and pine saplings of other species become so dense that reproduction of longleaf pine is suppressed. If fire continues to be kept out beyond that point, oak trees gradually replace the pines and form a closed forest.

The standard practice today is to burn the grass and pine debris about every four years. The grass is quickly able to regenerate itself, and pine reproduction is favored, while oak seedlings are killed. If longer periods go by without fire, too much dry litter collects on the ground, increasing the hazard of setting a fire so hot that mature trees are damaged. OF course, no matter how "cool" the fire, great care is taken to control it. The cost for control burning amounts to considerably less than $5 per acre.

On the west coast recognition of the importance of controlled burning lagged behind its perception on the east coast. In California, the absence of fire in some areas has produced potentially catastrophic conditions, especially in the mixed-conifer forest of the Sierra Nevada mountains. Before settlement by Europeans, natural fires swept these forests about once every eight years (as revealed when fire scars in tree trunks are dated by ring count). However, those lightning fires did not create raging, rampaging, extremely hot fires, because there was only a relatively small amount of litter on the ground. The appearance of those fire-adapted forests was stately and open, according to early reports by travelers. As shown in Fig. 18.10, however, much of the mixed-conifer belt of the Sierra Nevada (about 1500 m elevation) today is no longer stately and open. We now have evidence that in the absence of fire, these forests are transformed into crowded fir (*Abies*) and incense-cedar (*Calocedrus*) forests, and that the longer fire is excluded the more catastrophic is the eventual, inevitable fire.

California experiences a high frequency of lightning strikes during dry, late summer, and it is impossible to prevent fires from starting. A considerable amount of litter—needles, twigs, and bark—is shed each year. So long as fires move through an area frequently, say every 8 to 25 years, the fire does not become hot enough to reach the upper canopy and significantly damage the trees. The native trees are adapted to such frequent, light burns. The seedlings of many of them require contact with mineral soil in order to become established because litter does not retain moisture. Fire, of course, clears the ground of litter. In controlled experiments, seedlings of ponderosa pine, Jeffery pine, and big tree are much more abundant on burned plots than on unburned plots. If the seedlings are protected from fire for 10 to 15 years, they become tall enough and develop a bark that is thick enough to withstand a light surface fire. However, should fire be excluded for much longer periods, so much brush,

Figure 18.10
Thick understory of shade-tolerant trees that has developed in a Sierran mixed-conifer forest as a result of fire exclusion. (Courtesy of H. Biswell.)

Figure 18.11
A grove of giant sequoia (*Sequoiadendron gigantea*) in Yosemite National Park, California.

white fir, and litter collects that even mature trees cannot withstand the flames that eventually come. Many magnificent stands of big tree (*Sequoiadendron,* Fig. 18.11)—the most massive tree in the world, and found nowhere else in the world—are now, for this reason, in great danger of being destroyed by fire.

Biological Factors

It is typical for several plant and animal species to coexist in a given habitat. Very rarely does a single species, to the exclusion of all others, control a habitat. If all the species in a group are individually examined, they will be seen to utilize different parts of the environment or alternate with each other in time. Some plants (green plants) are producers of carbohydrates, but others (parasitic and saprophytic fungi) are consumers or decomposers of it; some animals feed on plants, others feed on insects, and others feed on small mammals; some plants are trees that utilize full sunlight, others like ferns utilize weak light; some animals are nocturnal, others are diurnal; some plants are evergreen, others deciduous; and so on. The portion of the environment utilized by each species is called its **niche.** The niche has often been called the "occupational address" of a species. In a particular area at a given time it is thought that one and only one species can occupy or fill each niche. Two species that occupy similar niches compete strongly with each other.

Competition

Competition may be defined as the decreased growth of two species of plants because of an insufficient supply of some necessary factor(s) (Table 18.2). Sometimes only a single factor, such as a mineral element, is lacking, but usually two or more factors are limiting and it is difficult to separate them and determine which one creates the greater competition stress. Competition is very important in determining plant distribution. Many species restricted to saline, dry, or nutritionally poor soils would actually grow better on "normal" soil if other plants were removed. These species are restricted to their niches because they are poor competitors in comparison with the numerous plants that populate more moderate niches. They will die because the other plants have faster root and shoot growth rates, and remove water and sunlight. Sometimes, introduced plants become widespread pests and actually replace native species because they are better competitors than the native species. This seems to be the reason for the enormous increase of cheatgrass (*Bromus tectorum*) in the intermountain region of the United States (Chapter 9).

Amensalism

Amensalism may be defined as the inhibition of one species by another. Whereas competition results from the removal of a resource, amensalism results from the addition of something to the environment. In the coastal hills of southern California, sage shrubs (*Salvia*) cover the slopes, and grasses carpet the valleys. Occasional pockets of shrubs occur in the grassland. Figure 18.12 is an aerial view of these pockets of shrubs. The ground beneath and about the shrubs is bare of grass, and the grass is stunted as far as 9 m from the shrubs. Since the zone of stunting is well beyond the limits of shrub root growth, competition between shrubs and grass for soil moisture is not the cause of stunting. If clumps of grass are transplanted to the bare zone, their growth is severely retarded. Analysis of the soil beneath the shrubs, the bare zone, or the grasses, shows no major differences in physical or chemical attributes. The reason for this distribution is that many volatile oils are emitted from sage leaves, and two of the oils in particular—cineole and camphor—are very toxic to grass seedlings. When both fresh sage leaves and grass seeds are placed in a closed

Table 18.2
SOME CATEGORIES OF BIOLOGICAL INTERACTION IN TERMS OF THE EFFECT THE RELATIONSHIP HAS ON EACH PARTNER. "OFF" INDICATES THAT THE TWO ARE NOT IN CONTACT; "ON" INDICATES CONTACT. NO EFFECT = 0, STIMULATION = +, DEPRESSION = −

	Effect			
	On		Off	
Name of Interaction	#1	#2	#3	#4
Neutralism	0	0	0	0
Competition	−	−	0	0
Mutualism	+	+	−	−
Protocooperation	+	+	0	0
Commensalism	+	0	−	0
Amensalism	0	−	0	0
Parasitism/herbivory	+	−	−	0

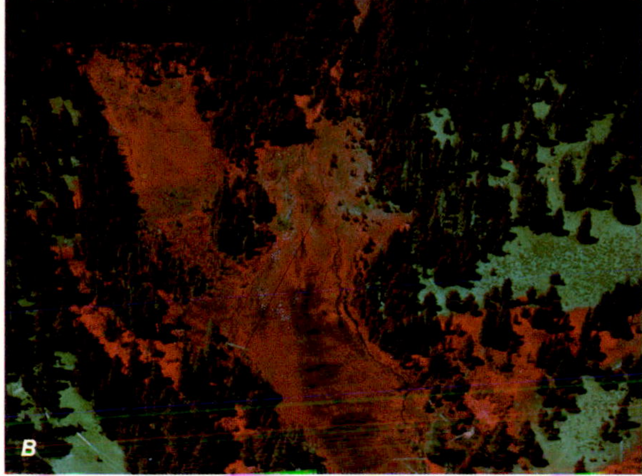

Figure 18.18
A, normal and *B*, infrared aerial photographs of the same area, showing mountain bog (red), sagebrush (green), and pine forest (dark brown) communities. Notice how different species show up in contrasting colors with infrared, and how rivulets of water in the bog become more obvious. (Courtesy of S. Rae.)

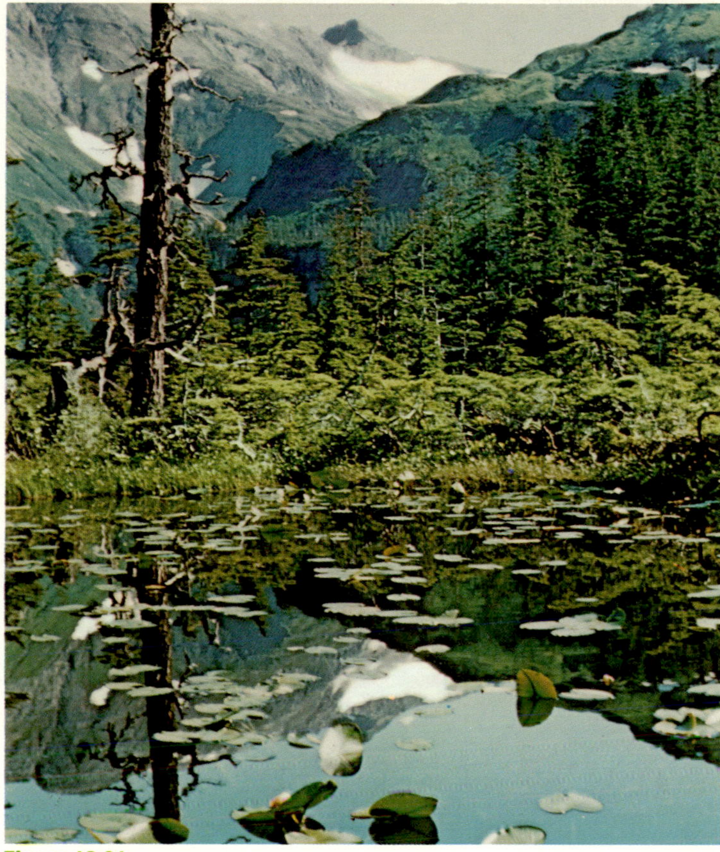

Figure 18.21
Succession of communities about a pond near Juneau, Alaska. Standing water in the pond supports leaves of cow lily *(Nuphar)*; mats of *Sphagnum* moss and sedge *(Carex)* encroach on the pond along its edge; beyond is a shrub-dominated community *(Empetrum, Cassiope, Kalmia)* with some mountain hemlock *(Tsuga mertensiana)*; finally, sitka spruce *(Picea sitchensis)*, the tall, dark-green trees, dominate the rest of the area. (Courtesy of J. Major.)

Figure 18.23
Taiga, northern Canada. (Courtesy of W. D. Billings.)

Figure 18.24
Coniferous forest, Olympic Peninsula, Washington. Notice the moss- and lichen-covered trunks and branches and the dense carpet of ground vegetation.

Figure 18.25
Change of leaf color in the deciduous forest during autumn.

Figure 18.26
A, tropical rain forest, New Guinea. The dense lower vegetation is absent farther into the forest. Notice the hanging lianas. *B*, rain forest in New Zealand. Note tree ferns. (*A*, courtesy of G. Webster.)

Figure 18.27
A, African savannah in Kenya. The flat-topped trees are *Commiphora,* source of the biblical myrrh. (Courtesy of O. A. Leonard.) *B,* the original steppe, or prairie vegetation of central California, dominated by the bunch grass *Stipa pulchra.* (Courtesy of J. Major.)

Figure 18.29
A, desert scrub in southern New Mexico, dominated by creosote bush *(Larrea divaricata). B,* desert scrub in southern Arizona with succulents (cactus species).

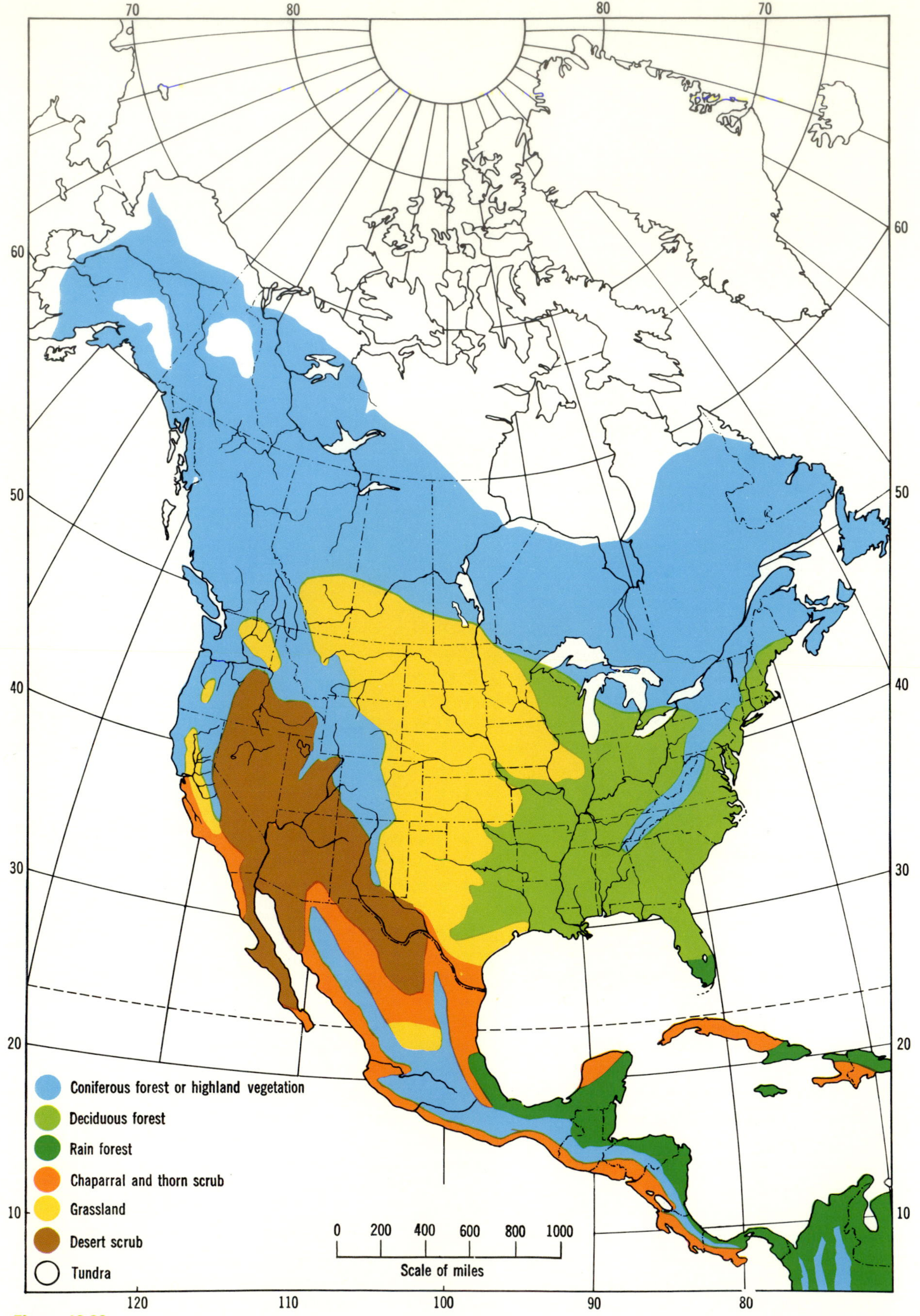

Figure 18.30
Map of the major vegetation types of North America.

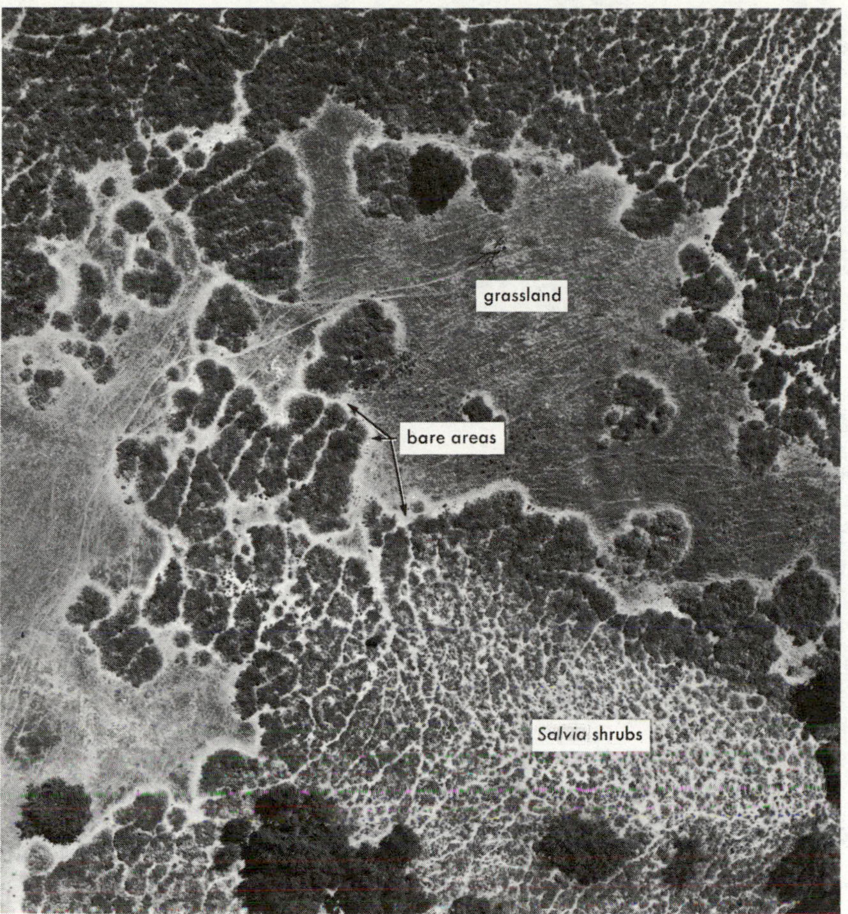

Figure 18.12
Aerial photo of *Salvia* shrubs adjoining grasslands. The light bands between and around the shrubs are bare of grass. (Courtesy of C. H. Muller.)

chamber, germination is often reduced to zero, and radicle growth is considerably reduced. These same toxins can be isolated from the air around sage shrubs, from the natural soil, and from seed coats of grasses in the soil. The conclusion of some ecologists is that the bare areas are caused by toxins emitted by *Salvia* that inhibit the germination and growth of grasses. More recent studies indicate that a high intensity of foraging by birds and small mammals, which nest in or beneath the shrubs, also contribute to maintaining the bare zones.

Amensalism by chemical means may be very common in nature. This happens because plants are leaky systems, passively contributing all sorts of substances to their environment. One investigator grew seedlings of 150 species in a nutrient culture that included several radioactive elements as markers. The plants took up the isotopes through their roots and transported them to all parts of the plant, including the leaves. The plants were then exposed to a mist, and the water that condensed and ran off the leaves was collected for analysis. He found that 14 elements, 7 sugars, 23 amino acids, and 15 organic acids—all radioactive—had been leached from the plants. In nature, these substances would have been leached by rain from the leaves and would have accumulated in the soil. Other studies have shown that roots are similarly leaky.

Herbivory

Herbivory, another form of interaction, plays an obvious part in plant distribution. A glance along pasture fences reveals that normally abundant but palatable plants are absent from the pasture, yet are common outside it; weedy species, which are not palatable, are common in the pasture, yet are rare outside it. Grazing animals are responsible for this difference. Plant defenses against herbivores include: (1) a spatial distribution that makes it difficult for herbivores to locate and damage host plants; (2) a life cycle timing, such as seasonal leaf drop, that makes the host plant unavailable to herbivores; and (3) the manufacture of substances that are distasteful or otherwise inhibit herbivores from feeding on host plants. Parasitism is herbivory by a plant rather than by an animal, but the type of interaction is the same.

Mutualism

Animal pollinators form a **mutualistic** relationship with plants. The bee, moth, and beetle, which cross-pollinate flowers while feeding on nectar or pollen, are familiar examples. Lichens are mutualistic associations of fungi and algae, and mycorrhizae are mutualistic associations of fungi and higher plants (Chapter 26). Epiphytes, mentioned earlier, form a **commensal** relationship with the host tree. **Protocooperation** is exhibited by adjacent members of some species that form natural root grafts, thus sharing soil resources.

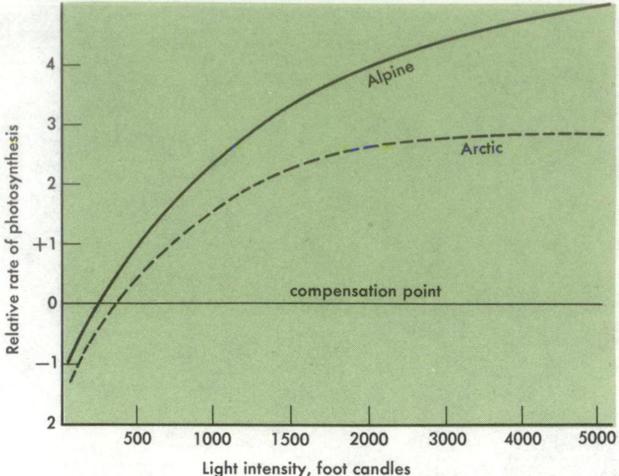

Figure 18.13
Photosynthetic response to different light intensities for alpine and arctic ecotypes of alpine sorrel (*Oxyria digyna*). (Redrawn from H. A. Mooney and W. D. Billings, *Ecological Monographs* **31,** 1. © 1961, Ecological Society of America.)

ECOLOGY AND PLANT POPULATIONS

Ecologists would like to use species as indicators of particular environments. They would like to know, for example, that a certain species only grows in a certain range of temperature or moisture. Then the temperature and moisture patterns of an area could be accurately predicted by just noting whether that plant species grows there. Unfortunately, most species are tolerant of a wide range of environments and they are not genetically uniform. Individuals of species *X* that grow in one environment are slightly different from others of species *X* that grow in another environment. Their genetic variation makes it possible for them to compete successfully in several different environments.

In the early twentieth century, the Swedish botanist Göte Turesson collected seeds of species that had a wide range of habitats—from lowland, southern, and central Europe to northeastern Russia in the Ural Mountains. When he germinated the seed and grew the plants in a uniform garden in Akarp, Sweden, he found that members of widespread species were not uniform. Plants that grew from seeds collected in warm, lowland sites often were taller and flowered later in the year than those that grew from seeds collected in cold, northern sites. Yet these variants had very similar flower morphology and were interfertile, traits indicating that the variants were all part of one species. Turesson called these variants within species **ecotypes.** He hypothesized that they were ecologically and genetically adapted to particular environments.

Evidence that has since accumulated indicates that most wide-ranging species are composed of a continuum of ecotypes, each differing slightly in morphology and/or physiology. For example, alpine sorrel (*Oxyria digyna*) is a small perennial herb that grows in rocky places above timberline in mountains and north of timberline in the arctic. Both the alpine and arctic environments experience long periods of freezing weather, but alpine areas receive much higher light intensities during the growing season than do arctic areas. Physiological ecotypes have resulted: the optimum light intensity for photosynthesis in alpine plants is much higher than that for arctic plants (Fig. 18.13). The two ecotypes also differ morphologically in leaf color, leaf shape, and the frequency of rhizome production.

Applications of the ecotype concept have been especially useful in forestry. The U.S. Forest Service conducts tests to determine the best ecotype of a species to be used for reforestation in a given area. Seed is collected throughout a species' range and is germinated in a uniform garden; seedling growth is carefully watched, and the ecotype best suited for the garden climate is selected for nearby reforestation programs.

ECOLOGY AND PLANT COMMUNITIES

Some environments support only one species. Most environments, however, support groups of species associated together in **communities.**

Provided that distances are not too great, a given community repeats itself wherever the environment is suitable. For example, high peaks in the Smoky Mountains of North Carolina all exhibit red spruce (*Picea rubens*), mountain ash (*Sorbus americana*), Fraser fir (*Abies fraseri*), and a ground carpeted with several moss and fern species (Fig. 18.14). The spruce, fir, ash, mosses, and ferns are important members (but not the only ones) of this high-elevation forest. High elevations at different latitudes or on different continents support other species, so there are other types of high-elevation forest communities. There are hundreds of nonforest communities as well. Along slopes of the California Coast Range at moderate elevation, a shrubby community of scrub oak (*Quercus dumosa*), manzanita (*Arctostaphylos*), and *Ceanothus* repeats itself (see Fig. 18.28). Black spruce (*Picea mariana*), *Sphagnum* moss, and the shrub Labrador tea (*Ledum palustre*) are associated in virtually every bog across Canada (**Fig. 18.21,** page 357).

Communities are named after their most common

Figure 18.14
Subalpine forest in the mountains of North Carolina. The trees are red spruce (*Picea rubens*) and Fraser fir (*Abies fraseri*); ferns carpet the floor of a local clearing caused by wind-throw of the trees. (Courtesy of U.S. Forest Service.)

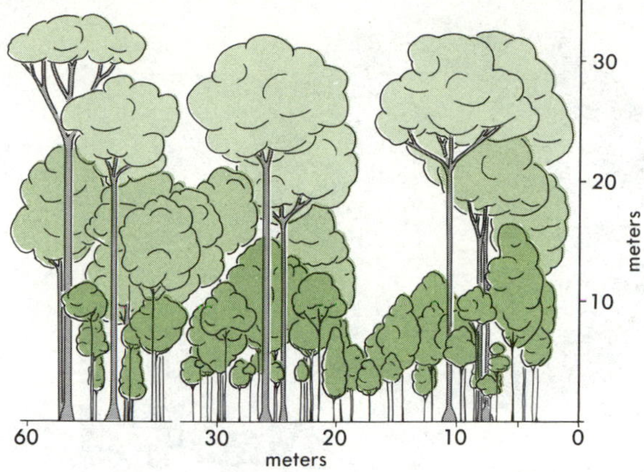

Figure 18.15
Profile diagram of the architecture of a tropical rain forest in British Guiana. There are three tree strata, at approximately 35 m, 20 m, and 10 m; shrub and herb species are uncommon. Small saplings are included in the 10 m stratum. (Redrawn from P. W. Richards, *Tropical Rain Forest*. © 1952 by Cambridge University Press. Reprinted by permission of the publisher.)

large species—the plants that receive the full force of the macroenvironment. The spruce and fir of the high-mountain community in the Smoky Mountains occur in greater numbers than does the mountain ash, and their leaf canopies influence the microenvironment in which the carpet of ferns and mosses grow. The spruce and fir are said to **dominate** the community, and the community name is **spruce-fir.** For similar reasons, we have **black spruce** communities and **scrub oak** communities.

Communities exhibit a unique architecture or structure, produced by the leaf canopies of their dominant and subordinate species. In the tropical rain forest of British Guiana, one group of species forms a canopy at 30 m, a second group at 18 m, and a third at 9 m (Fig. 18.15). The ground floor is almost bare of shrubs or herbs. In deciduous forest communities of the United States, there is one tree canopy layer at about 18 m, a layer of short trees (like dogwood) at 3 m, a rich shrub layer at 1 m, and a seasonally present layer of herbs on the ground. Some desert communities exhibit only a single (shrub) layer.

Paleoecology

Fossil impressions millions of years old sometimes reveal communities very similar to those that exist today. These fossil communities are so similar to modern communities that we are able to reconstruct past climates from them. A fossil community near the town of Goshen in central Oregon, dated to the Eocene epoch (50 million years ago), included the following genera: *Drimys, Magnolia,* holly (*Ilex*), oak (*Quercus*), *Octea,* and fig (*Ficus*). The closest living approximation of this fossil community today occurs 2400 km south of it in the temperate uplands of Central America. Apparently, the climate in North America is now much colder than it was in the Eocene. The reconstruction of past vegetation types and climates by the use of fossil evidence is called **paleoecology** (see also Chapter 31).

Another form of fossil evidence is pollen that has been trapped in lake sediments. As pollen is shed near a lake, some sinks to the bottom and is incorporated with the silt and organic matter that forms sediment. A chronological sequence of pollen is thus preserved; the deeper the pollen occurs in the sediment, the older the pollen. Pollen of many species is resistant to decay under such anaerobic conditions, and may remain intact for thousands of years. Cores of sediments can be examined under the microscope and the pollen identified as to genus (Fig. 18.16) and sometimes even to species. Within limits, paleoecologists assume that the abundance of pollen in the core is proportional to the abundance of that species in the past.

Figure 18.17 summarizes the pollen profile of sediment from a lake in Nova Scotia. The present vegetation consists of a deciduous forest on hilltops with sugar maple (*Acer saccharum*), beech (*Fagus grandifolia*), and yellow birch (*Betula lutea*); and a coniferous forest in the valleys, with balsam fir (*Abies balsamea*) and white spruce (*Picea glauca*). But Fig. 18.17 shows that this pat-

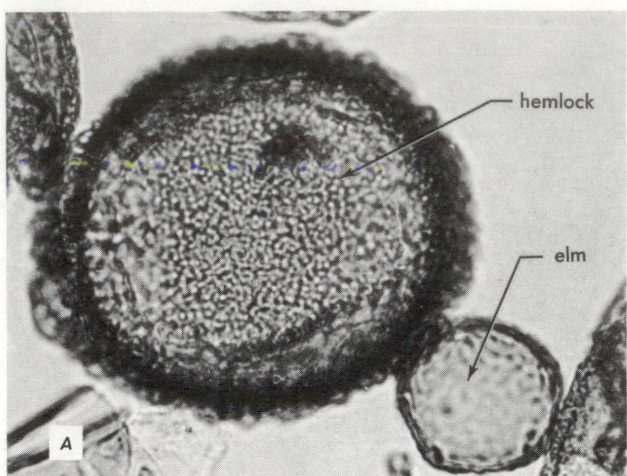

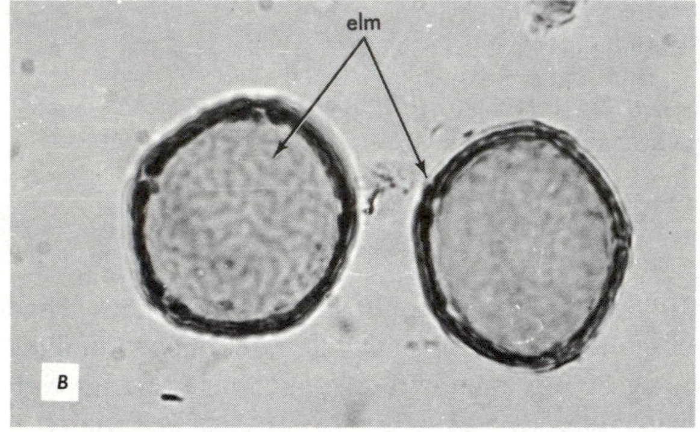

Figure 18.16
Fossil (*A*) and modern (*B*) pollen of the same genus.

tern of vegetation is relatively recent. Sediments at a depth of 5 to 6 meters, corresponding to an age of about 9000 years, reveal pollen mostly from herbs and shrubs, which are species characteristic of meadow vegetation near permanent ice. The nearest such vegetation today occurs 1000 km to the north of the lake. Shallower depths, corresponding to an age of about 6000 years, show peaks in spruce and fir, indicating dominance of typical northern coniferous forest. In yet shallower depths, deciduous species of birch and maple increase in abundance and coniferous species decline, indicating a continual warming trend in the climate.

Aerial photographs, using infrared-sensitive film, can reveal the species composition of communities. Different species stand out vividly because they reflect different amounts of infrared (**Fig. 18.18,** page 357). Such photographs can be used to map the distribution of species or communities over large areas.

How May Communities Be Measured?

If communities are to be useful as tools with which to estimate the nature of the environment, they must be described more completely than just by their name, because minor variations in species composition indicate different environments. For example, in areas where spruce and fir are both dominant and very abundant, the environment is probably different from where they are still dominant but less abundant. How can the abundance of a species or the composition of a community be measured?

1. One way to characterize a community is to measure the amount of ground covered by each species. The amount of ground cover is important because it affects the microenvironment of the community. To sample the ground cover, one could stretch a tape measure across the ground; the length of tape covered by each plant canopy is proportional to the total ground cover of each species. This method of determining ground cover is called a **line transect.** Numerical information from line transects often contradicts first impressions. In desert shrub communities, for example, a person looking across the landscape would think that the shrubs covered much of the ground (see **Fig. 18.29,** page 359), but such an estimate is in error because of the low

Figure 18.17
Pollen profile of sediment from a lake in Nova Scotia. Numbers along horizontal axis refer to relative abundance of each type of pollen (From D. A. Livingstone, *Ecological Monographs* **38,** 87. © 1968 by Ecological Society of America.

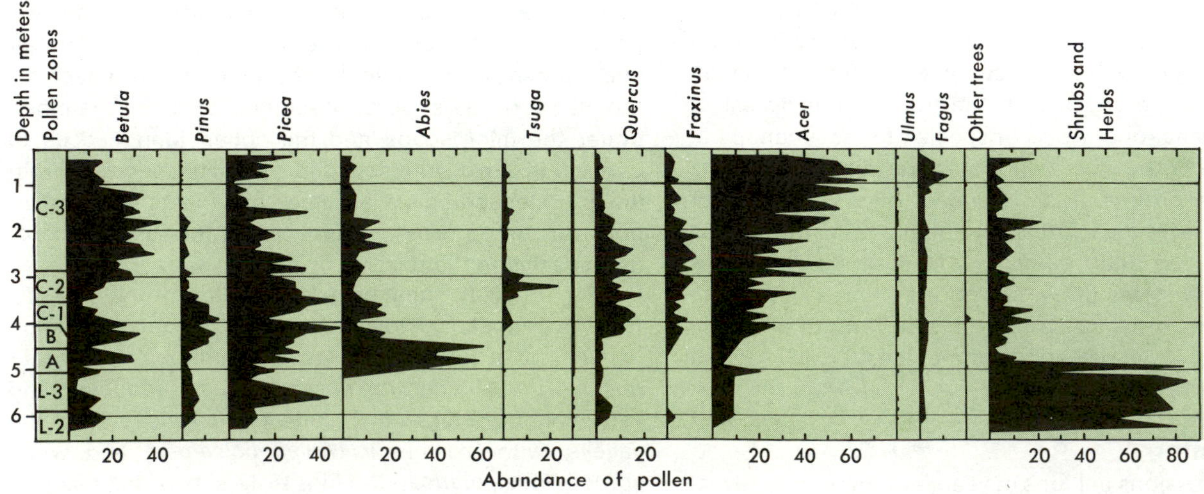

perspective. If seen from above, when the canopy edges can be projected perpendicularly to the ground, it is apparent that actually only 10 to 20% of the ground is covered. Forest communities, on the other hand, exhibit more than 100% cover because of overlap in canopy layers at different levels.

2. Another method of estimating cover, which is especially useful when plants are low to the ground, is by the use of quadrats. **Quadrats** are variously shaped frames that can be placed on the ground directly over the plants. The amount of ground covered by each species can be estimated as a percentage of the area enclosed by the quadrat frame. A community can be sampled by placing quadrats at randon or at regularly spaced locations throughout it. Usually, not more than 10% of the total community area is included in the quadrat samples.

3. Other methods of sampling, based on mathematical assumptions, can be used to determine whether individuals are distributed at random, regularly (like trees in an orchard), or in clumps (Fig. 18.19). Nonrandom distribution (regular or clumped) may result from propagation by asexual means, the method of seed disperal, competition, or some pattern (like terracing) in the microenvironment. Most temperate zone species are distributed in clumps.

Succession

As we have already mentioned, the microenvironment beneath a plant community is very different from that in the open. Temperature, humidity, soil moisture, and light are all affected by the canopy. A stable community consists of species whose seedlings can survive in its unique microenvironment; seedlings of other species do not survive. If the stable community is removed by some catastrophe, leaving the soil exposed to full sunlight, the first species that colonize the site are not those of the old community. They are seedlings of species adapted to growth in full light intensity.

In parts of the southeastern United States where oak-hickory communities are stable, the first species to invade following disturbance are annual and perennial herbs, which grow best in high light intensity, such as horseweed (*Conyza canadensis*), *Aster,* and broomsedge (*Andropogon virginicus*). These make up the **pioneer** community. Pine seeds blow in from neighboring areas during the first several years after the disturbance, and within five years there are many pine seedlings. As the seedlings grow, they compete with the herbs for moisture, and their expanding canopies begin to change the microenvironment. Within 30 years of the disturbance, a tall stand of pine results. Examination of the forest floor, however, shows many oak and hickory seedlings and very few pine seedlings. Pine seedlings grow poorly in shade and under conditions of root competition, but those of oak and hickory do well. Within 50 years of the disturbance, the hardwood seedlings form a well-defined understory beneath the pines (Fig. 18.20). Whenever a pine tree dies, it is replaced in the upper canopy by an oak or a hickory. Within 200 years of the disturbance, the forest again consists only of mature oak and hickory trees, and the forest floor is covered with many seedlings of the same trees. The shrubs and herbs associated with an oak-hickory community are also present. Such a sequence of change in plant communities is called a **succession** or **sere,** and the stable community that is capable of indefinitely maintaining itself is called a **climax community.** In this case, oak-hickory is the climax community. All the intermediate, temporary communities are called **seral stages.**

Sometimes, climax communities are prevented from becoming established because of a recurring disturbance. There is some evidence that part of the prairie of

Figure 18.19
Diagrammatic representation of three types of plant distribution: *A,* clumped; *B,* regular; *C,* random.

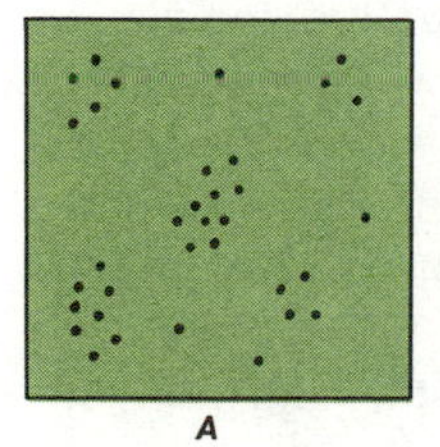

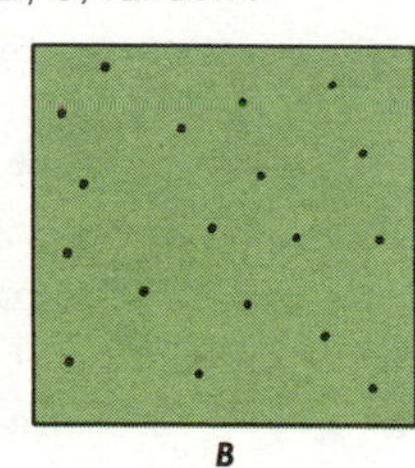

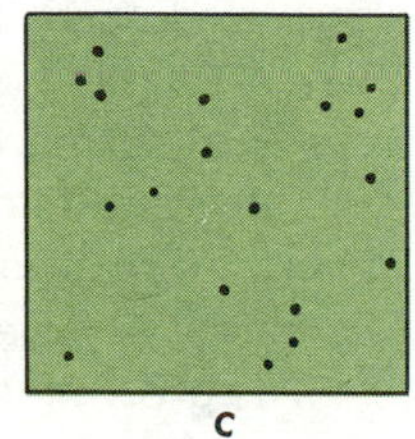

Figure 18.20
A stand of longleaf pine (*Pinus palustris*), about 50 years old, showing a well-developed understory of broad-leaved species of oak and hickory. In time, the pines will die and be replaced by maturing members of this understory. (Courtesy of U.S. Forest Service.)

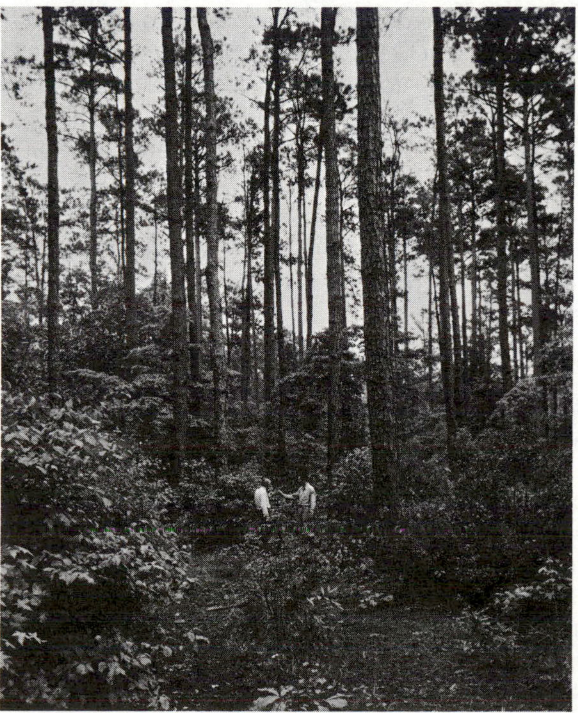

the central United States that greeted pioneers could have been a forest were it not for recurring prairie fires. The fires destroyed slow growing tree seedlings, but grasses quickly regenerated or germinated. When fire is controlled, prairie near the forest edge soon supports many free seedlings.

The first ecologists to elaborate successional pathways—men like Josef Paczoski in Europe and Frederick Clements in the United States—had no experimental evidence for documentation. They could not record changes as time passed, for most successions take hundreds of years to complete. They determined the existence of successional pathways by comparing a climax community with adjacent small areas that had been disturbed at known times in the past. They examined freshly disturbed or denuded areas (as a result of, for example, ice, landslides, fire, logging, and wind-throw), which revealed the likely pioneer community, or the first stage of succession. Sites that had been disturbed at some time earlier (verified from records of fires, grazing, or logging) exhibited later successional communities because the passage of time was greater. Complete sequences of succession were sometimes represented as bands of different communities about a gradually filling bog or behind a retreating glacier (**Fig. 18.21,** page 357). The community closest to the lake or glacier was the most recently established and represented the pioneer community. The next closest community occupied soil that had been exposed for a longer time; at some earlier time it had supported the pioneer community, but enough time had passed for the next successional stage to appear. Beyond the second community was a third, representing the third successional stage, and so on until the climax community was reached. Ring counts of trees closest to the pond or glacier, coupled with estimates of the rate of silting or of glacial movement, gave clues as to the time necessary for each successional stage to be reached.

The conclusions of these early workers have yet to be experimentally or empirically verified, and only recently have some of the causes of succession (such as competition, amensalism, and growth requirements of seedlings) been experimentally dealt with.

VEGETATION TYPES OF THE WORLD

The term **vegetation** refers to the life form of plants dominating a community: trees, shrubs, or herbs; deciduous or evergreen; coniferous or broadleaved. A community dominated by evergreen angiosperms belongs to a different vegetation type than one dominated by shrubs or one dominated by conifers. On the other hand, many different communities dominated by different conifers belong to the same vegetation type. We will briefly examine six major vegetation types: tundra, taiga, deciduous forest, tropical rain forest, savannah-prairie, and scrub.

Tundra

If one were to travel north or south toward the poles, it would be evident that the trees gradually become stunted and less common, and finally disappear completely. At the same time, low shrubs, perennial herbs, and grasses become dominant. The resulting meadowlike vegetation is called **tundra.** Annual plants are very rare, and the reproduction of perennials is chiefly by vegetative means such as rhizomes. Shrubs are dwarfed, gaining normal height only in the protection of the lee of boulders or small hills. The winter wind, carrying ice, acts like a sand blast and prunes back stems wherever they project beyond the boulder or hill.

The warmest month has an average daily mean temperature of 10°C or less and the growing season is short. Only 2 to 4 months of the year have average daily temperatures above freezing. Approximately the top 30 cm of soil thaws during the growing season and roots may freely penetrate it, but below this level soil water may be permanently frozen; this soil is called **permafrost.** Annual precipitation is about 25 cm, very little falling as snow. The terrain is generally flat, but small depressions are common and water tends to stand in them during the growing season. Day length fluctuates from 24 hr in June (northern hemisphere) to none at all in December.

A similar tundra vegetation occurs in mountains at elevations above treeline (Fig. 18.22). The climate, however, is slightly different. Solar radiation is more intense, day length does not fluctuate so extremely, and frosts are more common during the growing season. Many plants are tinged with red from abundant anthocyanin. This substance absorbs some of the high light intensities that may damage the plant's photosynthetic apparatus.

The lower limit of these alpine areas depends on latitude. At 60°N in Alaska, treeline is at 900 m; at 45°N in the Rocky Mountains, it is at 3000 m; at 20°N in central Mexico, it is at 4000 m; at 5°S in the Andes of South America, it is back to 3600 m; and at 50°S in Chile, it is back to 900 m. Because summer temperatures are cooler near the coast, timberline also depends on distance from the ocean. Thus, timberline is lower in the Cascades than in the Rockies.

Taiga

Just south of the tundra is the **taiga,** a broad belt of communities dominated by conifers. A similar belt occurs in mountains, just below the alpine zone (**Fig. 18.23,** page 357). Trees are slender, short (15-20 m), and relatively short-lived (less than 300 years), but the forests are dense. The taiga is made up of a patchwork of forests on upland sites that alternate with bogs formed in depressions. Soils are acidic and relatively infertile spodosols; in northernmost or uppermost parts of the taiga the soil has permafrost below a surface layer.

The growing season is 3 to 5 months long, and tem-

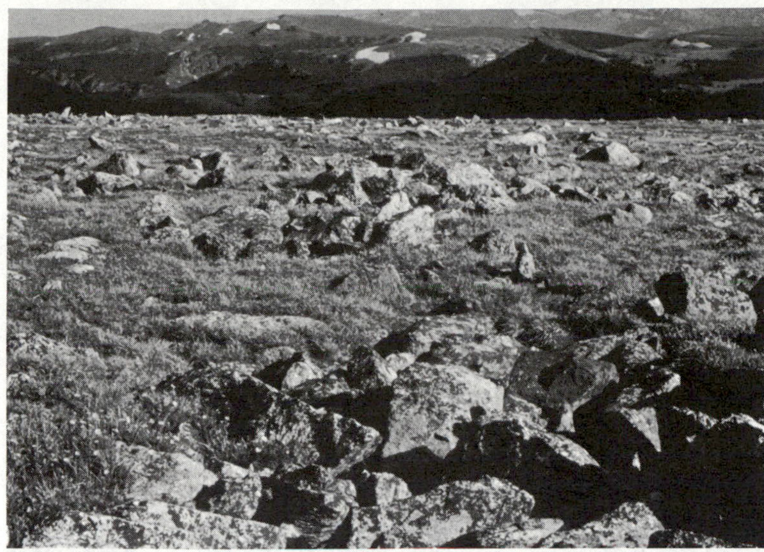

Figure 18.22
Alpine tundra, Beartooth Plateau, Wyoming, 3,300 m elevation. (Courtesy of J. Major.)

peratures above 30°C are occasionally reached. Winter temperatures are very severe; the mean daily temperature is below freezing for six months of the year, and at Yakutsk, Siberia, the average January daily temperature is −43°C. Annual precipitation is less than 50 cm, and the air is exceptionally dry. The taiga is monotonous and silent; animal life is not abundant.

At lower elevations in some mountains, and along the coast (where temperature extremes are not so severe), coniferous forests are much more luxurious. Trees are taller and the undergrowth denser. Many favorite recreation areas of the western United States are situated in these richer forests. The Olympic Peninsula of Washington receives over 200 cm of annual precipitation and supports the most spectacular coniferous forest in the world (**Fig. 18.24,** page 358). Douglas fir, the most heavily cut lumber species of the United States, reaches its best growth in this area.

Deciduous Forest

In contrast to the taiga, the **deciduous forest** supports a diversity of plant life and is vibrant with animal activity. Deciduous angiosperm trees dominate communities of this forest. When day length becomes shorter and nights cooler in late summer, hormone concentrations in such trees change, and the destruction of chlorophyll follows, allowing other pigments to show through in brilliant fall colors of red, yellow, and orange (**Fig. 18.25,** page 358). In spring, before new leaves have fully expanded from buds, beautiful and distinctive herbaceous perennials come into flower. Shrubs are common, but not dense.

These forests occur in areas with a cold (though not severe) winter, and a warm, humid summer. Snowfall may be heavy, but most of the annual precipitation falls as summer rain. Annual precipitation ranges from 75 to 125 cm. Since the growing season may be as long as 200 days, it is not surprising that much of the deciduous forest in Europe, Asia, and North America has been cut and replaced with cropland.

Soils are alfisols in the north and west, ultisols in the south. (The latter have striking yellow or red colors in the *B* horizon, which is often exposed if plowed.)

Tropical Rain Forest

No other vegetation can quite equal the **rain forest** in sheer diversity of species and complexity of community structure (Figs. 18.15, **18.26,** page 358). The deciduous forest may exhibit 5 to 25 different tree species per acre, but the tropical rain forest supports 20 to 50. The tree canopies typically form three layers, and beneath this dense overstory there are very few shrubs and herbs. Light intensity on the ground is less than 1% of full sun. Most of the herbaceous plants grow as epiphytes on trunks and branches of trees. Vines cling to tree trunks and wind their way up to the canopies. Seeds of a few species, such as the well-named strangler fig, germinate in the canopy and grow down to the soil. Many of the tree trunks flare out at the base in peculiar buttresses.

The species are evergreen. A given leaf may be shed or a given bud may break at any time of the year. The leaves tend to be very large, with long, tapering tips that may serve to drain the surface of moisture. Rainfall is high, often over 250 cm a year, but it is evenly distributed throughout the year. Temperature is also uniform throughout the year, averaging 27°C. Warm temperatures and high humidity couple to produce a climate oppressive to humans. Afternoon showers, which reduce the temperature, offer only short relief.

Farming peoples of the rain forest have for centuries practiced agriculture on a cut-burn-cultivate-abandon plan. The vegetation is removed in a small area by cutting and burning, cultivation follows for a few years, and the plot is abandoned. Abandonment follows because crop growth declines each year. The rain leaches nutrients from the exposed topsoil and the nutrients are not replaced by litter as they are beneath the normal forest canopy, as mentioned earlier in this chapter. Much of the rain forest of Asia and Africa has been disturbed in this way. More recent disturbance involves bulldozing and clear-cutting, especially in South America; the rain forest may be incapable of reinvading these areas. The vegetation that invades these abandoned plots is shrubby, dense, and often includes bamboo. It is this successional stage, which eventually will give way to rain forest, that serves as the "jungle" of television and films. Soils are typically in the oxisol order.

Savannah and Prairie

Forests give way to grassland as rainfall decreases and becomes more seasonably distributed. At first, trees remain close enough to retain a closed canopy, but gradually they become more widely spaced, and eventually they drop out altogether except along river banks and canyons. **Savannah** is the term applied to grassland with widely spaced trees whose canopies cover less than 30 % of the ground (**Fig. 18.27*A*,** page 359); **steppe** or **prairie** refers to grassland without trees (**Fig. 18.27*B*).**

Grasslands of the United States are heavily used for agriculture and grazing purposes, and little of the original vegetation or animal life remains today. Only from the records of explorers and early settlers are ecologists able to reconstruct in their imagination what these thousands of square miles once looked like.

The eastern part, with relatively high rainfall, supported a dense "sea" of grass some 1 to 2 m tall. Here is a description from an anonymous journal dated 1837:

> The view from this mound . . . beggars all description. An ocean of prairie surrounds the spectator whose vision is not limited to less than 30 or 40 miles. This great sea of verdure is interspersed with delightfully varying undulations, like the vast swells of ocean, and every here and there, sinking in the hollows or cresting the swells, appears spots of trees, as if planted by the hand of Art for the purpose of ornamenting this naturally splendid scene.

This is now the corn belt and is probably one of the most agriculturally fertile areas of the world. Productivity is so high that it permits us to export grain to many countries around the world. Its mollisol soils are dark brown in color from the rich amount of organic matter left by decaying fibrous grass roots.

The drier, western part supported shorter and sparser stands of grass, but there was still enough growth to allow graziers to cut and bale it as hay. Today, the western prairie supports a shorter, sparser cover of grass. Reasons for such a decline include too much grazing pressure by herds of cattle and a slight drying trend in the climate of the area over the past 75 years. Shrubs from the desert and junipers from the foothills have invaded (possibly because of fire suppression), and they have further reduced grass cover.

Grasslands are the home of the largest terrestrial herbivores and carnivores. Vast herds of bison that roamed the North American prairie are gone, but animal preserves in African savannahs (**Fig. 18.27*A*,** page 359) still excite visitors by a spectacular animal cast that includes eland, wildebeest, giraffe, elephant, and lion.

Scrub

Scrub is vegetation dominated by shrubs, and it occurs in areas with hot, dry summers and moderate winters. In areas with high, but seasonal, precipitation, the shrubs are tall and thorny, but if rainfall is only 40 to 50 cm a year, scrub can be a nearly impenetrable thicket of shrubs with evergreen, hard, spiny, small leaves (Fig. 18.28). Scrub of this sort is found around the Mediterranean Sea, on coastal hills of California and Chile, in southwestern Australia, and at the tip of South Africa. Even though the species of plants in these widely separate areas are entirely different, the vegetation has come to look strikingly similar. In the western U.S., this type of scrub is called **chaparral.**

Desert scrub (**Fig. 18.29,** page 359) grows in areas of rainfall below 25 cm a year. Shrubs are widely spaced

Figure 18.28
Chaparral in the foothills of California.

and have far-reaching root systems (much of which runs just beneath the soil surface). The shrubs may be associated with **succulents,** plants that store water in stems or leaves, like the common cactus of the southwestern United States. Some shrubs have deep roots that tap permanent supplies of ground water, while others reduce transpiration by dropping their leaves every dry period and developing new ones every rainy period. Seeds of desert annuals may remain viable for long periods in the soil until a favorable year, when rainfall is sufficient to trigger germination and a colorful show of flowers a few weeks or months later. Although desert scrub seems to be made up mainly of perennial species, an actual count of all species appearing throughout the year reveals nearly as many annuals as perennials.

A Final Note

We have described major vegetation types as if they have sharp boundaries with, and differ greatly from, all other types. Actually, vegetation types typically grade into one another or form complex mosaics that may extend for thousands of hectares. Meeting zones of two or more vegetation types are called **ecotones.**

A summary map of the major vegetation types of North America is shown in **Fig. 18.30,** page 360.

ECOLOGY AND ECOSYSTEMS

Energy Flow

We have examined plant ecology at the population, community, and vegetation levels. A final level of organization is the ecosystem. The ecosystem of a particular habitat includes the plant and animal communities, the physical (nonliving) environment, and all the interactions between them.

The organisms of an ecosystem can be divided into three categories: producers, consumers, and decomposers. **Producers** are green plants and certain bacteria that perform photosynthesis. **Consumers** are parasitic, herbivorous, or carnivorous plants and animals that feed on other plants or animals. **Decomposers** are saprophytic bacteria, fungi, and certain animals that obtain nu-

Figure 18.31
Loss of energy in western rangeland. Of 700,000 kcal of radiant energy that reach each square meter of surface, only 0.1% becomes available as plant food for the cattle grazing the rangeland. Only 0.01% of the plant food consumed during an eight month period is converted to meat available for human consumption, that is, only 0.001% of the original incident solar energy. Numbers are calories per square meter. (Redrawn from W. A. Williams, *Journal of Range Management* **19,** 29, 1966.)

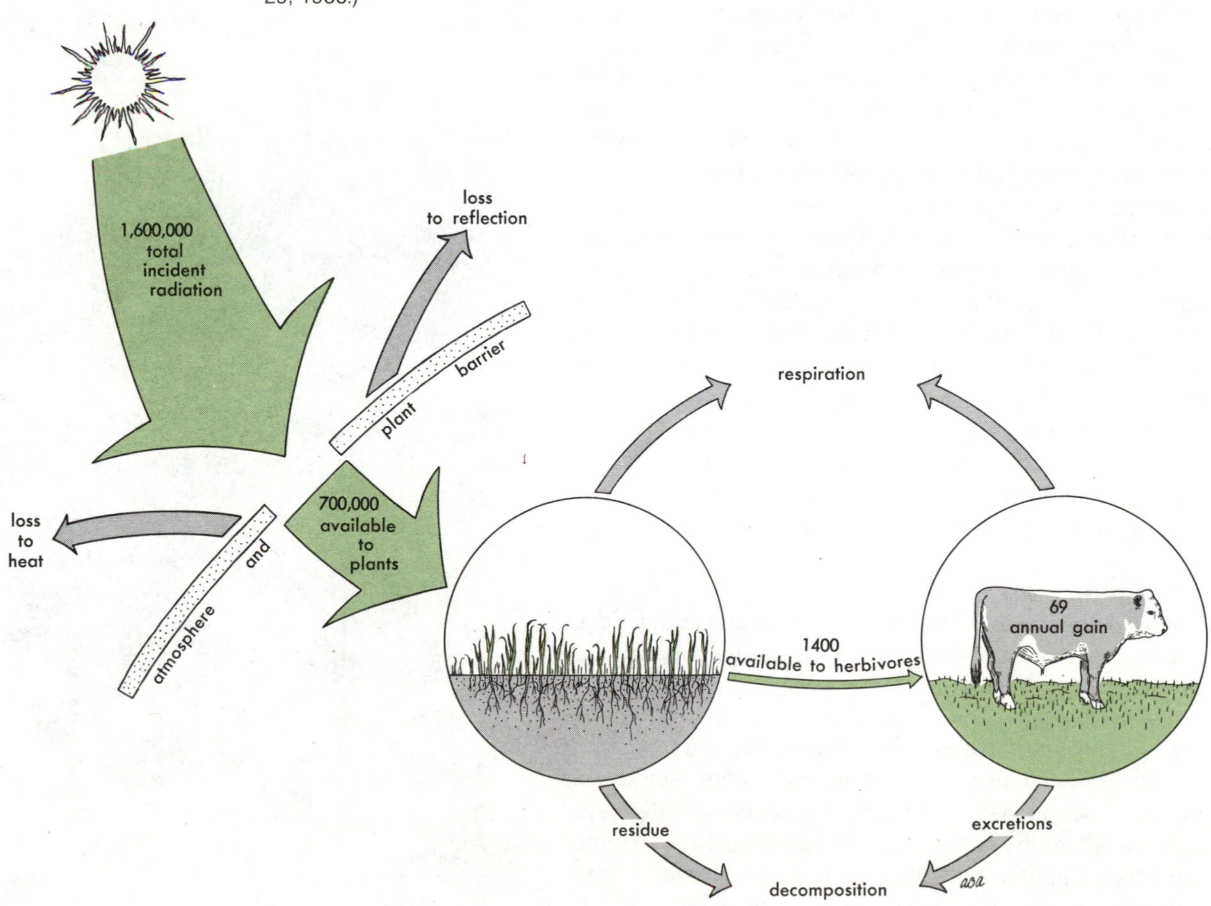

trition by breaking down dead organic matter. The three groups form a complex web of interdependence: consumers feed on producers, and decomposers feed on the organic remains of both and convert them to lower-energy, smaller molecules that are taken up again by producers.

Caloric energy is also passed through the ecosystem, and in this transfer organisms are linked to each other by a **food chain.** Food chains are short in the arctic tundra: meadow plants trap light energy and convert it to chemical energy (e.g., proteins or carbohydrates); caribou transfer some of this energy to themselves, retaining some of it, excreting some, and respiring some; and human beings transfer some of the caribou caloric energy by eating a fraction of the total caribou. Aquatic ecosystems exhibit longer food chains: algae (plankton especially) produce the caloric energy; some is passed to the small animals that feed on plankton, then to the small fish that eat the plankton feeders, then to larger fish, and then to shore birds or land animals like bear or man.

There is tremendous variation in net productivity from one ecosystem to another. **Net productivity** equals calories produced in photosynthesis minus calories lost in respiration. Desert ecosystems, for example, exhibit less than 0.5 gram of dry matter produced per square meter of land per day; grasslands, 0.5 to 3.0; deciduous and coniferous forest, 3 to 10; agricultural areas utilized all year (as in sugar cane production), 10 to 25.

Energy transfer through the ecosystem involves losses each time energy passes from one type of organism to another. Figure 18.31 summarizes a very simple food chain consisting of California rangeland on which steers are grazed. The range receives 700,000 kilocalories (kcal) of sunlight energy per square meter per year. After deducting the energy lost by reflecting and heating, the fraction spent by plants on their own respiration, and other energy channeled into roots and other unconsumable parts of the plants, only 800 kilocalories are taken in by the grazing animals. The animals in turn spend nearly all their energy unproductively (from the point of view of human beings) in respiration, waste matter, and inedible bones and hide, leaving a net of only 69 kilocalories per square meter, of the original 700,000, that goes into meat. This represents about 139 kg of edible beef on the hoof per ha per year.

Pollution

Human beings add to the environment many substances whose effect on plants and animals is poorly understood. If organisms in food chains or ecosystems shared by humans are damaged by these substances, then humans may be indirectly, but nevertheless severely, damaged.

These substances are called **pollutants.** Some are liquids or solids that have been introduced to waterways and may affect the metabolism of aquatic plants, such as algae (see Chapter 23). Others are gases released to the atmosphere; they may affect the growth of wild or cultivated terrestrial plants. Among the most important of these airborne pollutants are sulfur dioxide (SO_2), ozone (O_3), and peroxyacetyl nitrate ($C_2H_3O_5N$, also called PAN).

Sulfur is present in iron, copper, lead, and zinc ores; during refining it is released as SO_2. It is also released from the burning of coal. The SO_2 enters leaves through stomata and goes into solution on the wet mesophyll cell walls as sulfurous acid. Inside the cells, such vital metabolic processes as protein synthesis are disrupted because of the reducing nature of this acid. Ultimately, cells shrink and collapse, producing visible chlorotic lesions on the leaves. Chronic exposure of such sensitive crops as alfalfa, barley, and lettuce to only 0.1 to 0.5 ppm (parts per million) of SO_2 causes visible damage and reduced yield. The effect of SO_2 on natural vegetation is dramatically revealed in Copper Basin, Tennessee, where early in this century a refinery freely released SO_2 before its toxic nature was understood. Over an area of perhaps 40 km^2 the original forest was destroyed (Fig. 18.32). The barren land was so severely eroded that subsequent attempts to revegetate the area have been unsuccessful.

Ozone and PAN are important components of urban smog. In large part they are a result of the incomplete combustion of automobile fuel (Fig. 18.33). Our chemical

Figure 18.32
Copper Basin, Tennessee. The normal eastern deciduous forest was destroyed by sulfur fumes from a smelter in the early twentieth century and revegetation attempts have been unsuccessful.

understanding of smog and its affect on plants started less than 30 years ago. Ozone is naturally present in air, but smog contains 10 to 100 times the natural concentration, and this concentration is sufficient to cause millions of dollars of crop damage a year. Ozone enters through the stomata, and its oxidizing nature causes many metabolic malfunctions, including chloroplast breakdown. In the mountains near Los Angeles, air circulation takes sufficient quantities of ozone to a height of 1700 m to cause needle drop, reduced growth, and the ultimate death of trees in ponderosa pine forests (Fig. 18.34).

Other experiments in the Los Angeles Basin have shown the effect of PAN on citrus. Established lemon and orange trees were encased in miniature greenhouses. Some greenhouses circulated normal, smoggy, outside air; others circulated air that had first been filtered of PAN. Over the period of six years, the average yearly fruit yield of lemon trees was reduced 39% by PAN, and the orange tree yield was reduced 64%. Evidently, PAN interferes with the light reactions of photosynthesis and also with respiration.

It is hoped that ecologists will eventually show us how ecosystems can be rationally manipulated to simultaneously satisfy all our needs: industry, urbanization, agriculture, conservation, and recreation. Knowledge must be gained and compromises made between our needs and what the environment can support.

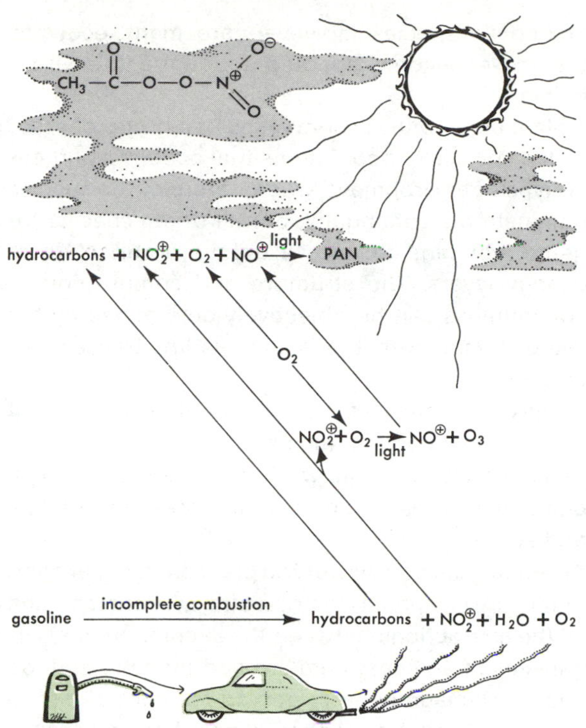

Figure 18.33
The formation of ozone (O_3) and peroxyacetal nitrate (PAN) in smog. The formula for PAN is given at the top of the diagram.

Figure 18.34
The effect of smog on ponderosa pine growth in a mixed-conifer forest near the Los Angeles Basin. *A,* cross section of a normal 30-year-old pine. *B,* the same age tree exposed to smog-polluted air. (Courtesy of J. R. McBride et al., *California Agriculture* **29,** 8.)

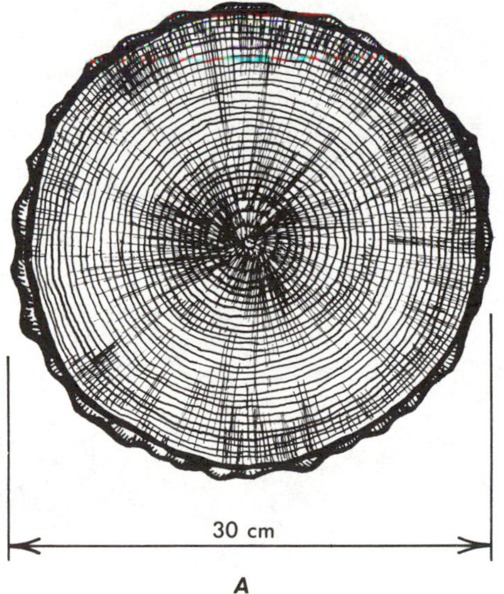

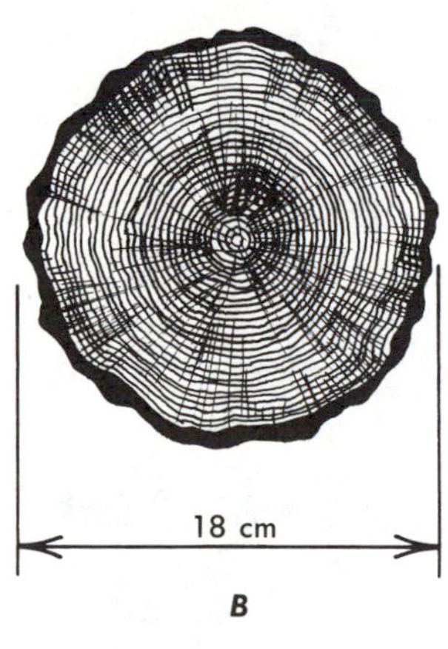

SUMMARY

1. Plant ecology is the study of how the environment influences plant distribution and behavior. It is also the study of the structure and function of nature. Ecology deals with the population, community, and ecosystem levels of organization.
2. The components of the environment include moisture, temperature, light, soil, fire, and living organisms. Each species utilizes a different part of the environment: its niche.
3. The seasonal timing of rainfall may be as important as the annual total in determining the kind of vegetation present.
4. Slope aspect affects temperature in the microenvironment. It also affects the amount of net radiation that reaches the plants.
5. Light quantity and quality is changed as light passes through a leaf canopy or through water.
6. Soil formation proceeds through mechanical and chemical weathering and through plant activity. Through time, soils in different regions develop their own characteristic profiles. Soils can be classified by profile characteristics into 10 orders and thousands of soil types.
7. Fire is an important, natural environmental factor that affects the distribution and architecture of many vegetation types.
8. Forms of biological interaction include competition, amensalism, predation, and mutualism.
9. A taxonomic species includes genetically similar plants that are interfertile. Within any wide-ranging taxonomic species, however, are many ecotypes that have become adapted to different environments.
10. A plant community is composed of a group of associated species or populations that occupy a particular type of environment. One or more of the species dominate the community, but all of the species together combine to form a unique architecture of canopy layers. The structure and composition of communities can be objectively determined by the use of sampling methods such as line transects or quadrats.
11. Climax communities perpetuate themselves, but successional communities are transitory.
12. Vegetation types of the world include tundra, taiga, deciduous forest, tropical rain forest, grassland, and scrub.
13. An ecosystem consists of the plant and animal communities of a habitat, their abiotic environment, and all the interactions between these components. Organisms of an ecosystem are tied together by food chains. Damage done to one component of a food chain (as through the action of pollution) may cause indirect but serious damage to the entire ecosystem.
14. Important atmospheric pollutants include sulfur dioxide, ozone, and peroxyacetyl nitrate. The latter two are components of smog and cause millions of dollars of crop damage a year as well as ecological damage to natural vegetation.

CHAPTER 19

PRINCIPLES OF PLANT TAXONOMY

Plant taxonomy deals with the **identification** and **classification** of plants. The term identification implies that many plant groups exist, and each group possessing unique characteristics can be separated from all other groups. These characteristics, as we shall see, involve morphology, anatomy, physiology, cytology, biochemistry, and geographic distribution. The term classification implies that plant groups can be ranked in some hierarchical relationship, on the basis of similarities in their characteristics. This ranking attempts to reflect real genetic relationships.

As mentioned earlier (Chapter 2), taxonomy is building a library of the world's plant resources. Taxonomists are making a census of the unique characteristics of each of the half-million species that make up the world's flora. This census is far from complete for two reasons. First, many areas of the world have not been botanically explored and their floras can only be estimated; second, the process of evolution will continue to produce new species even in well-known areas and human activities will continue to produce plant extinctions.

The important, exciting goal of the census is not identification, but rather classification. Taxonomists hope that, as the census becomes complete, patterns of similarity will become apparent within this maze of diversity, and that these patterns will enable them to reconstruct the evolutionary history of the present flora. They hope it will be possible to determine which characteristics—out of the hundreds that plants express—are most reliable in showing pathways of evolution. Are they characteristics of flower morphology, wood anatomy, chemical composition, geographic distribution, breeding behavior, or chromosome number? Which traits in each category are **primitive** (imitating those of early, now-extinct plants)? Which traits are **advanced** (derived in time from primitive traits)? When the answers to such questions are known, it will be possible to construct a **phylogenetic classification**—one based on genetic, evolutionary relationships. Such a classification, in addition to summarizing the past, might prove useful in predicting future pathways of evolution.

HISTORICAL ASPECTS

Pre-Linnaean Period

The history of taxonomy can be very roughly divided into a pre-Linnaean (300 BC–1753 AD) period and a post-Linnaean (1753–present) period.

Before the time of written history, of course, man had learned to identify many plants at what we now consider the genus or species level. The plants named were those particularly useful as spices, medicines, foods, drugs, or religious symbols. The written history of botany begins with the Greek Theophrastus (ca. 300 BC), a student of Aristotle. He described and classified almost 500 kinds of plants in the book *De Historia Plantarum.* He utilized morphological characters like habit (tree, shrub, herb), length of life (annual, biennial, perennial), corolla form (petals fused together or free), and ovary position (superior, inferior) to distinguish among his plants.

Many of the taxonomists who followed the lead of Theophrastus through the dark ages and Renaissance are typified by Otto Brunfels (ca. 1500 AD) of Germany. He too classified plants on the basis of gross morphology, but he contributed to the record by expanding the list of known plants and accompanying many of the descriptions with drawings. Hans Weiditz was Brunfels' illustrator, and he did a magnificent job of carving accurate and intricate woodcuts from which prints were made (Fig. 19.1). The descriptions and drawings of many species were compiled in books referred to as **herbals.** Herbaceous plants used for medicinal purposes were often emphasized in the herbals. It was common to assign medicinal properties to a plant on the basis of its form: If the plant or parts of it imitated the shape of an organ, then it was used for correcting ailments of that organ. For example, the thallus of liverworts, which resembles the shape of a liver, was used to treat liver ailments.

Linnaeus

Carolus Linnaeus (ca. 1750, Fig. 2.15) represents the culmination of this period. Born Carl Linné in southern Sweden, he attended the University of Uppsala and received an M.D. degree from the University of Haderwijk in the Netherlands. During his early days at Uppsala, however, he had become interested in the classification of

CXCVI Contrafayt

Storckenschnabel.

Von dem Nammen.

Torckenschnabel/Gottes gnad/Kranch/

Figure 19.1
Part of one page from Brunfels' sixteenth century herbal. The plant illustrated is called stork's bill (probably *Erodium cicutarium*).

plants and had developed his own system, called the sexual system. It was based almost entirely on the morphology and number of stamens and pistils, and proved very efficient in classifying the many new plants then being brought to Europe from Africa, America, and Asia and propagated in botanic gardens. After working on the collections of many large botanic gardens, Linnaeus published the book *Species Plantarum* in 1753. The work included about 6000 species and 1000 genera.

Each species was described in Latin by a sentence limited to twelve words which began with the genus name. Linnaeus considered this sentence (a **polynomial**) to be the official, scientific species name, but he also coined a shortened form consisting of the genus name and one additional word from the polynomial. This shortened form (a **binomial**) served as his common name for the plant. He described spiderwort (Fig. 19.2) by this polynomial: *Tradescantia ephemerum phalangoides tripetalum non repens virginianum gramineum,* common name *Tradescantia virginiana.* Loosely translated, the polynomial means: The annual, upright Tradescantia from Virginia which has a grasslike habit, three petals, and stamens with hairs like spider legs, common name Tradescantia of Virginia. The genus name was in honor of John Tradescant, a gardener to King Charles I.

Not surprisingly, the binomial became favored over the polynomial by later taxonomists, and was accepted as the official, scientific name. Common names were left to individual choice and not restricted to Latin. The usual English common name for *Tradescantia virginiana* is spiderwort (spider plant), which comes from the jointed

Figure 19.2
Tradescantia virginiana (spiderwort). Note the jointed staminal hairs, from which the common name derives, ×2.

hairs on stamen filaments that resemble the jointed legs of a spider (Fig. 19.2).

Table 19.1
ASSUMPTIONS FOR THE BESSEYAN PHYLOGENETIC SCHEME

Primitive Characters	Advanced Characters
Plants woody	Plants herbaceous
Flowers bisexual	Flowers unisexual
Floral axis elongated	Floral axis short
Foral parts spirally arranged	Floral parts whorled
Floral parts numerous	Floral parts few
Sepals or petals free	Sepals or petals fused
Floral symmetry radial	Floral symmetry bilateral
Fruit single	Fruit aggregated

Table 19.2
ARRANGEMENT OF SOME FLOWERING PLANT FAMILIES ACCORDING TO BESSEY AND ENGLER

Bessey's System	Engler's System	Gradient
DICOTYLEDONAE	MONOCOTYLEDONAE	Primitive
Magnoliaceae	Gramineae	↓
Ranunculaceae	Liliaceae	
Malvaceae	Iridaceae	
Cruciferae	Orchidaceae	
Solanaceae	DICOTYLEDONAE	
Labiatae	Salicaceae	
Rosaceae	Ranunculaceae	
Leguminosae	Magnoliaceae	
Cucurbitaceae	Cruciferae	
Salicaceae	Rosaceae	
Umbelliferae	Leguminosae	
Compositae	Malvaceae	
MONOCOTYLEDONAE	Umbelliferae	
Liliaceae	Labiatae	
Gramineae	Solanaceae	
Iridaceae	Cucurbitaceae	
Orchidaceae	Compositae	Advanced

Post-Linnaean Period

After Darwin published *On the Origin of Species* in 1859, emphasis in taxonomy shifted to (*a*) a search for characters which reflected genetic (evolutionary) relationships, and (*b*) the construction of a phylogenetic classification scheme. Construction of phylogenetic schemes demanded that decisions be made as to which traits are primitive and which are advanced. As there was little or no fossil evidence to support any of several hypotheses, early schemes were both numerous and often in conflict. For example, the American C. E. Bessey started with the assumptions listed in Table 19.1.

Although Bessey's assumptions are still widely accepted today, some 60 years after he proposed them, disagreements continue to arise over specific interpretations. For example, some angiosperms such as willow (see Chapter 30) have flowers that lack petals or sepals. Are these plants primitive, never having evolved a complete flower, or did they once possess flowers with petals and sepals but have since lost them, hence really being advanced? Bessy thought they were advanced and consequently regarded plants like willow (*Salix*), birch (*Betula*), and maple (*Acer*) as closely related to advanced families. In contrast, equally competent taxonomists like Adolph Engler and Karl Prantl, living at about the same time, placed these plants among the most primitive angiosperm families (Table 19.2).

Whether the absence of parts represents a primitive or derived condition can sometimes be determined by examination of the flower's vascular system. Vascular bundles that branch in the receptacle and extend to petals or sepals may still show the original branching even though the petals or sepals have been lost through time. Apparently, the vascular system is very **conservative** (does not readily change). In other cases, the missing parts have not been lost altogether and are represented by small projections of glandular or apparently functionless tissue. Evidence of this sort, gathered since the time of Bessey, has confirmed his hypothesis that plants like *Salix* are indeed advanced.

Another source of disagreement arises from answers to this question: Do species that now resemble each other represent progeny from a common ancestor, or could it be possible that they started from quite different sources and simply evolved to look alike? These two possibilities are diagrammed in Fig. 19.3. It is seen that species *A* and *B*, which are very similar today, actually arose from quite different stocks. Similar environments, however, have shaped them along convergent paths. This pattern is termed **convergent evolution.** A striking example of convergent evolution is the present similarity between cactus plants of southwestern United States and certain *Euphorbia* species of Africa. Both groups grow in hot, arid climates, and as seen in Fig. 19.4, both have a similar morphology: They have no leaves, their stems are ribbed and swollen and contain water-storage

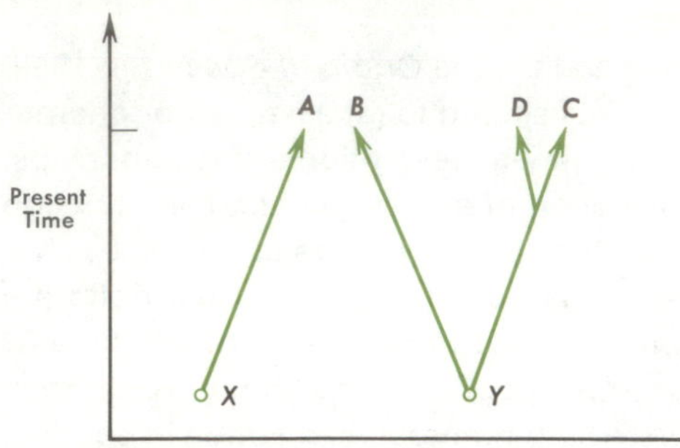

Figure 19.3
Hypothetical pathways of species evolution. *X* and *Y* represent ancestral forms from which *A, B, C,* and *D* have since evolved. The paths leading to *A* and *B* represent convergent evolution; the paths leading to *B* and *C* represent divergent evolution.

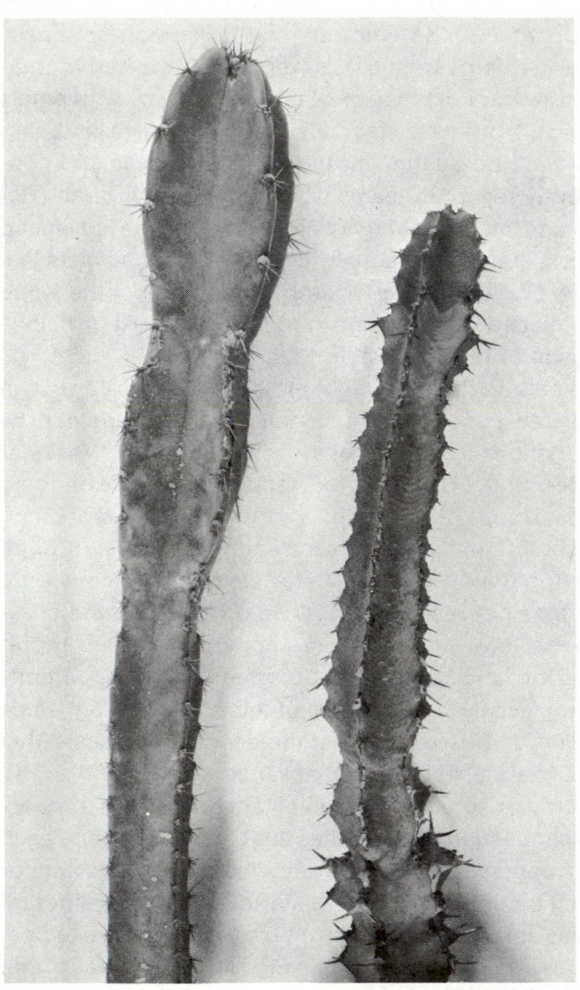

Figure 19.4
Convergent evolution of members of the Cactaceae (left) of the southwestern United States and members of the Euphorbiaceae (right) of Africa, ×1/10.

tissue, and they are armed with spines. Yet, other traits, like flower morphology and geographic distribution, are so different that it is clear they must have arisen from different stocks.

On the other hand, *B* and *C* (Fig. 19.3), which are today less similar than *A* and *B,* actually did evolve from a common ancestor. This pattern is called **divergent evolution.** As time passes, *B* and *C* will become more and more different. The same pattern will be repeated by *C* and *D,* but at this point in time they are still quite similar.

If a certain group of plants arose from a single ancestral type, the group is referred to as **monophyletic** (Chapter 2). If a group arose from several different ancestral types, the group is referred to as **polyphyletic.** Bessey considered each separate group of seed plants (the Anthophyta, Coniferophyta, Ginkgophyta, etc.) to be monophyletic; some taxonomists considered all the seed plants together to be monophyletic; still others did not think that even one group, the Anthophyta, was monophyletic. Unless fossil evidence is at hand, it is very difficult to separate the effects of convergent and divergent evolution. At present, most taxonomists agree that the largest group of plants which is *definitely* monophyletic is the family.

Rules of Nomenclature

Considering the rich variety of hypotheses and plant characters upon which one could base a classification scheme, it is not surprising that the names of many plants have changed with time, being taken from one genus to another or being split into varieties or subspecies. *Tradescantia virginiana* L. is still an accepted name, more than 200 years after it was coined, but others that Linnaeus selected have been changed. For example, in 1753 Linnaeus described a plant from America as Canadian pine, *Pinus canadensis.* Later, the taxonomist Carriere realized it was a hemlock, not a pine, and transferred the species to the genus *Tsuga.* The new name was *Tsuga canadensis,* Canadian or eastern hemlock. How does this affect the old, valulable herbarium collections of the same plant which had been labeled *Pinus canadensis?* How would later taxonomists realize the connection between the two? There would be references, in older books, to something called *Pinus canadensis* which no longer is a name found in any taxonomic reference book. Who would know what plant was meant? There was a need, then, to allow for changes in name but at the same time preserve continuity and prevent a haphazard proliferation of new names.

The answer to the need was a set of international rules on botanical nomenclature, adopted by the International Botanical Congress of 1930, held in Cambridge, England. These rules were explicit guidelines for the determination and coining of valid plant names. Some rules dealt with the categories of classification available, others with the procedure of identifying new species, and others with the procedure for changing names.

One rule provided that, to be valid, a plant name must be published in Latin, accompanied by a short, descriptive paragraph which sets the new species off from related species, and that this description must be published in some recognized journal or **flora,** which is a book that summarizes all the plant species of a region.

Another rule provided that "each **taxon** (a general word for any plant group; pl. = taxa) . . . can bear only one valid name, the earliest that is in accordance with the Rules of Nomenclature." If two or more people independently proposed names for the same plant, then the earliest one proposed would be the valid one. Common sense indicated that there should be some limitation to this principle of priority, and for vascular plants the year 1753 (date of publication of *Species Plantarum*) was selected. Names coined before that year were not valid, and the principle of priority only applied to names coined after that year.

Other rules dealt with changes of name. One of these is quite important, for it helps solve the problem of *Pinus* (*Tsuga*) *canadensis*. It provides that when a genus or species is changed, the name of the original author be placed in parentheses and be followed by the name of the author who has now changed it. Thus, *Pinus canadensis* L. becomes *Tsuga canadenis* (L). *Carr.* It is now immediately apparent that *Tsuga canadensis* is not the original name for the plant and Linnaeus had given it another name. The original name can be found in references like the Kew or Gray Herbarium Indexes that list all the previous names (sometimes there are several) for presently named species. Portions of pages of *Index Kewensis* are shown in Fig. 19.5.

METHODS OF TAXONOMIC RESEARCH

Traditional (Morphological)

About 20 years ago, the geneticist C. H. Waddington said it is "an empirical fact that living organisms do not vary continuously over the whole range, but they fall into more or less well-defined groups, which are commonly called species." This philosophy is quite appropriate for a traditional taxonomist. He realizes that man's love of classifying nature often puts boundary lines between things where boundaries simply do not exist; but when it comes to species, he is convinced from extensive field observations that discrete species do exist and that classification at this level is not arbitrary or artificial.

A traditional taxonomist primarily examines plant morphology, searching for a few traits that consistently enable him to separate plants into "well-defined groups." In pre-Darwin days, a variety of traits were utilized, but now the choice is limited to those that presumably are (*a*) genetically controlled (rather than environmentally controlled) and (*b*) conservative in the evolutionary sense. Flower and fruit morphology, for example, meet these criteria, but leaf size may not.

It should not be supposed that the traditional taxonomist limits himself at the start of a study to the examination of only a few traits. He examines many characters, measures many plants, and keeps records of all these. He eventually, however, subjectively selects only a few char-

Figure 19.5
Part of two pages of *Index Kewensis*. Note that *Pinus canadensis* L. is shown to be a defunct synonym (by the = sign) for *Tsuga canadensis* Carr. The abbreviations refer to the publications where the original species description occurs. For example, *Pinus canadensis* is described in *Species Planatarum*, p. 1421.

PINUS:—
Boursieri, Carr. Conif. ed. I. 398 = contorta.
brachyphylla, Parl. in DC. Prod. xvi. II. 424 = Abies brachyphylla.
brachyptera, Engelm. in Wisliz. Tour N. Mexico, 89 = ponderosa.
bracteata, D. Don, in Trans. Linn. Soc. xvii. (1836) 443 = Abies bracteata.
Brunoniana, Wall. Pl. As. Rar. iii. 24. t. 247 = Tsuga Brunoniana.
bruttia, Tenore, Prod. Fl. Nap. p. lix = pyrenaica.
bullata, Roezl, Cat. Sem. Conif. Mexic. (1857) 16 = Montezumae.
Bungeana, *Zucc. in Endl. Syn. Conif.* 166.—China.
Buonapartea, Roezl, ex Gord. Pinet. 218 = P. Ayacahuite.
cairica, D. Don, ex Gord. l. c. 166 = halepensis.
calabrica, Hort. ex Gord. l. c. 168 = P. Laricio.
californica, Loisel. in Duham. Arb. ed. Nov. v. 243; Hartw. in Journ. Hort. Soc. ii. (1847) 189 = insignis.
Calocote, Roezl, ex Gord. Pinet. Suppl. 73 = P. Teocote.
canadensis, Bong. in Mém. Acad. Pétersb. Sér. VI. ii. (1833) 163 = Tsuga Mertensiana.
canadensis, Duroi, Obs. Bot. 38 = Picea alba.
canadensis, Linn. Sp. Pl. ed. II. 1421 = Tsuga canadensis.
canaliculata, *Miq. in Journ. Bot. Néerl.* i. (1861) 86. —China.
canariensis, *C. Sm. in Buch, Beschr. Canar. Ins.* 159. —Ins. Canar.

TSUBAKI, Adans. Fam. ii. 399 (1763) = **Camellia**, Linn. (Ternstr.).

TSUGA, Carr. Conif. ed. I. 185 (1855). *CONIFERAE*, Benth. & Hook. f. iii. 440.
MICROPEUCE, Gord. Pinet. Suppl. 13 (1862).
Balfouriana, *McNab, in Journ. Linn. Soc.* xix. (1882) 211.—Am. bor. occ.
Brunoniana, *Carr. Conif.* ed. I. 188.—Reg. Himal.
canadensis, *Carr. l. c.* 189.—Am. bor.
caroliniana, *Engelm. in Coult. Bot. Gaz.* vi. (1881) 223.—Am. bor.
diversifolia, *Mast. in Journ. Linn. Soc.* xviii. (1881) 514.—Japon.
Douglasii, Carr. Conif. ed. I. 192 = Pseudotsuga Douglasii.
Hookeriana, *Carr. l. c.* ed. II. 252.—Calif.
Lindleyana, Roezl, Cat. Grain. Conif. Mex. 8 = Pseudotsuga Douglasii.
Mertensiana, *Carr. Conif.* ed. II. 250.—Am. bor. occ.
Pattoniana, *Engelm. in Gard. Chron.* (1879) II. 756.—Am. bor. occ.
Roezlii, *Carr. in Rev. Hortic.* (1870) 217.—Calif.
Sieboldii, *Carr. Conif.* ed. I. 186.—Japon.
Tsuja, A. Murr. in Proc. Hort. Soc. Lond. ii. (1862) 508 = Sieboldii.

TSUSIOPHYLLUM, Maxim. in Mém. Acad. Pétersb. Sér. VII. xvi. (1871) n. IX. 12. *ERICACEAE*, Benth. & Hook. f. ii. 602.
Tanakae, *Maxim. l. c.*—Japon.

acters to serve in determining the number of taxa. The characters he selects are those that show discontinuities and are most helpful in separating related taxa.

Anatomical

We have already seen, with *Salix*, that the vascular system of a plant is very conservative and may be used to show genetic relationship between taxa. Anatomical studies of particular cell types within the vascular system reveal evolutionary patterns. Vessel elements, for example, show the changes summarized in Table 19.3 (see also Fig. 19.6).

Evidence for this set of hypotheses comes from examination of fossils of lower plants that have vessels (such as *Selaginella*), and of *correlations* between changes of vessel characters with changes in other characters such as flower morphology (that is, plants with primitive flower morphology also show primitive vessel characters). Similar hypotheses of evolution have been applied to other cell types, such as fibers.

Differences in anatomy are usually minor at the level of genus or species; nevertheless, these small differences can show evolutionary relationships. Creosote bush (*Larrea tridentata*) is a common desert plant in the southwestern United States **(Fig. 18.29,** page 359). Although members of the species are very similar in morphology across its range from California to Texas, the species is not genetically homogeneous. Populations of creosote bush occur in three separate desert regions. The plants in each region have different numbers of chromosomes.

Each desert, then, has its own subspecies of *L. tridentata.* Fiber length shows the same pattern. Table 19.4 summarizes average fiber length of *Larrea* plants grown from seed in a uniform garden, all tested at the same age. The standard hypothesis for direction of fiber evolution is that advanced fibers are longer than primitive ones. Hence, the data in Table 19.4 indicate that western subspecies of *L. tridentata* have evolved from eastern ones.

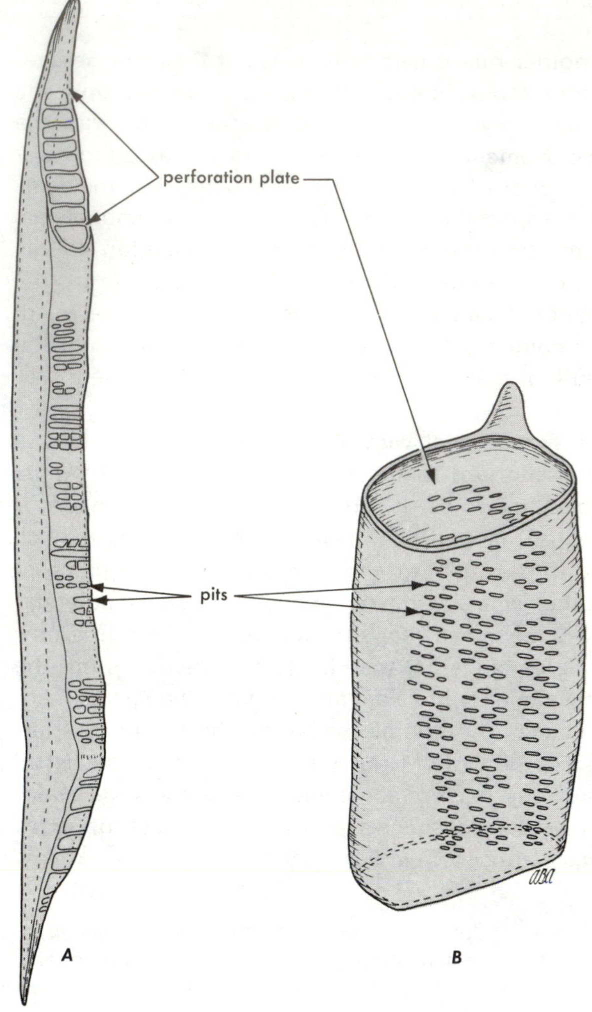

Figure 19.6
Advanced and primitive vessel elements: *A,* primitive type, from tulip tree (*Liriodendron tulipifera*). *B,* advanced type, from oak (*Quercus* sp.), ×100.

Table 19.3
CHARACTERISTICS OF PRIMITIVE AND ADVANCED VESSEL ELEMENTS

Primitive	Advanced
Similar in appearance to a tracheid	Not similar to a tracheid
Slender in cross section (100 μ)	Broad in cross section (250 μ)
Long about (800 μ)	Short (about 300 μ)
Tapered ends	Truncated ends
Angular in cross section	Round in cross section
End wall multiperforated	End wall with single perforation

Table 19.4
LENGTH OF FIBERS IN CREOSOTE BUSH SEEDLINGS

Area Where Seed Collected (Desert)	Average Fiber Length (μ)
California, Nevada (Mojave)	873
Arizona (Sonoran)	799
New Mexico, Texas (Chihuahuan)	628

Embryological

The pattern of embryo development is thought by some to be extremely conservative and to actually reflect in some measure the evolutionary history of the organism. For example, the embryo of man passes through early stages that are almost identical to those of lower groups of chordates (such as fish), and this is taken by some people as evidence of the evolutionary relationship between man and these other chordates.

Taxonomists have found that the pattern of embryo development in the Anthophyta is not uniform, and sometimes differences in the pattern correlate with differences in plant morphology. The position of the ovule within the ovary can be very distinctive and characteristic

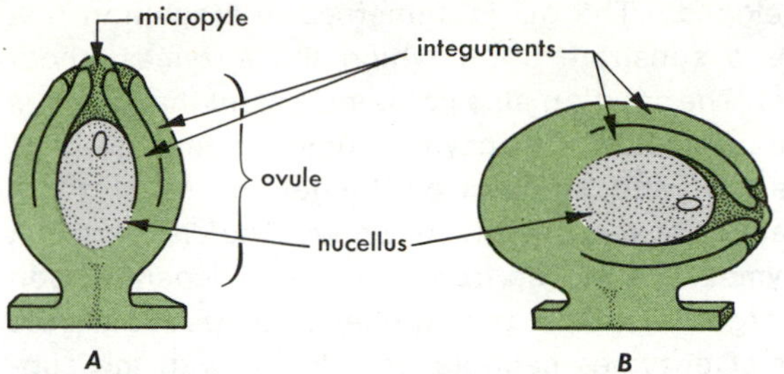

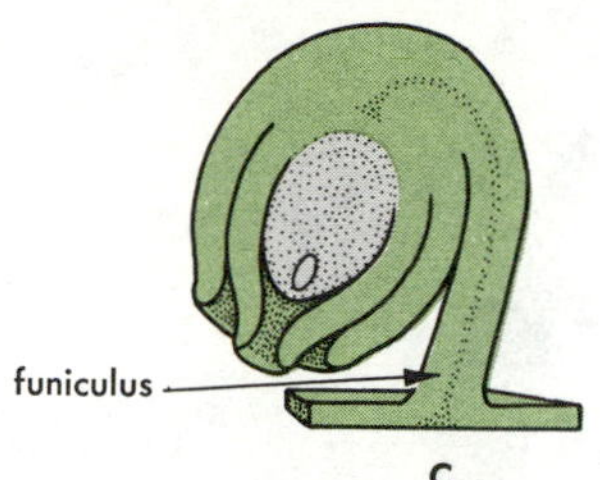

Figure 19.7
Three types of ovule position within the ovary: *A,* orthotropous; *B,* campylotropous; *C,* anatropous.

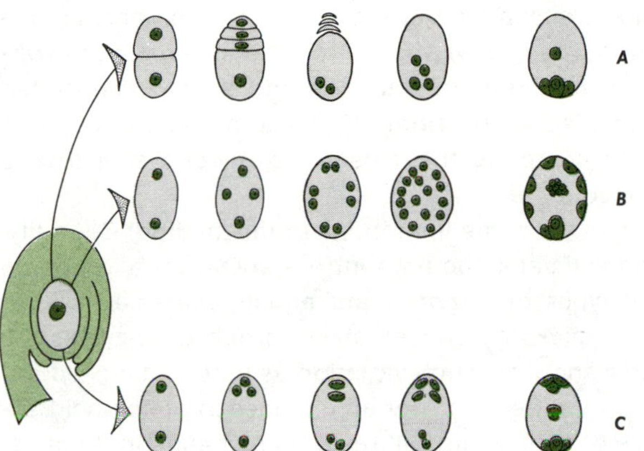

Figure 19.8
Number and position of cells within the mature embryo sac of three species: *A,* evening primrose (*Oenothera*); *B, Peperomia; C,* lily (*Lilium*). (After B. M. Johri in P. Maleshwari (ed.), *Recent Advances in the Embryology of Angiosperms,* 1963, University of Delhi.)

for entire families or orders; the ovule can be shaped so that the micropyle is directly opposite the point of ovule attachment to the placenta **(orthotropous),** or the micropyle can be facing the placenta **(anatropous),** or be somewhere between the two extremes **(campylotropous)** (Fig. 19.7). The number and position of cells in the embryo sac (prior to fertilization) can vary tremendously; there are eight cells in the embryo sac of lily (*Lilium*), four in evening primrose (*Oenothera*), and sixteen in *Peperomia* (Fig. 19.8). The number of cotyledons, of course, is one of the criteria for separating members of the Anthophyta into two classes, Monocotyledoneae and Dicotyledoneae. The degree of embryo differentiation at the time of seed dispersal may vary (members of the orchid family exhibit very rudimentary and incompletely formed embryos). The amount of endosperm may also be characteristic of large groups, for example, the endosperm of grains such as wheat, rice, and corn is copious.

Biochemical

We have already seen (Chapter 2) that algal divisions are separated by biochemical characteristics, such as the type of chlorophyll or accessory pigments, the major component of the cell wall, or the principal form of food stored. Only within the past years, however, has biochemistry been regularly applied as a taxonomic tool for higher groups of plants. Members of closely related taxa are chemically treated to extract particular compounds and the amounts (or presence versus absence) of the compounds extracted are compared. The more similar the taxa are chemically, the closer their genetic relationship is presumed to be.

Almost every category of biochemical compound has been used at one time or another in these studies—sugars, amino acids, fats, oils, alkaloids, alcohols, terpenes, phenols—and they have been isolated by paper chromatography, electrophoresis, fractionation, or, more recently, gas chromatography. Many of these compounds are high in molecular weight and are structurally complex. It is reasoned by biochemical taxonomists that these complicated molecules probably evolved at only one time and in only one original group of plants; therefore, species or genera or families that possess the same complicated molecules are undoubtedly related to this original group of plants.

The existence of hybrids can sometimes be dramatically shown from biochemical evidence. Natural hybrids of two *Eucalyptus* species (Fig. 19.9 and Table 19.5) were suspected to occur in parts of Australia. The hybrids were hypothesized because unidentifiable *Eucalyptus* plants, somewhat intermediate in morphology, existed near the parent species. Leaf samples were collected from random members of each category and particular oils were extracted. The biochemical evidence of hybridization supported morphological evidence, because the suspected hybrids were intermediate in oil composition.

Gel Electrophoresis

One of the most widely used methods for comparing taxonomic similarity of organisms at the molecular level, is

Table 19.5
OIL CONTENT (IN RELATIVE UNITS) OF TWO *EUCALYPTUS* SPECIES AND SUSPECTED HYBRIDS

Category	Geranyl Acetate	Cineole	Eudesmol
E. macarthuri	4.0	0.0	3.0
E. cinerea	0.0	4.0	0.5
Suspected hybrids	0.8	1.6	1.4

From Pryor and Bruant, *Proc. Linn. Soc. N.S. Wales* **83,** 55, 1966.

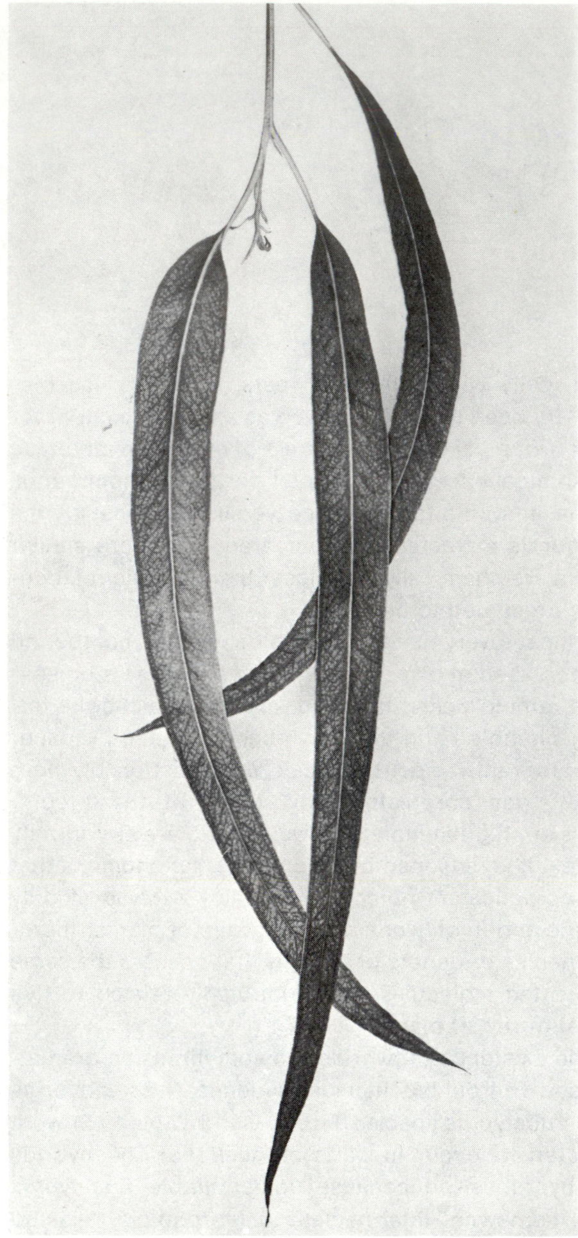

Figure 19.9
Branch and foliage of a typical *Eucalyptus* tree, ×½.

gel electrophoresis. This method detects differences between enzymes (or other proteins) on the basis of differences in their charge and shape. Such differences are the result of changes in the amino acid sequence of the molecule (see Chapter 5). Since amino acids are coded for by small segments of the DNA molecule, amino acid differences may be equivalent to single gene differences.

A small amount of liquid extract of crushed leaves (or any other plant organ desired) is placed in a notch cut into a gelatinous material (Fig. 19.10). This gel, often made from potato starch, either fills a tube or coats a glass plate. The extract is placed at one end of the gel, and an electric potential difference is set up across it. After several hours, the current is turned off, and the gel is "developed." The gel is immersed in a solution that contains a substrate upon which the enzymes under study act. The solution also contains a stain that couples with the product of the enzyme–substrate reaction. The result is that bands of color on the gel (Fig. 19.11) mark the locations of very specific enzymes. The distance that the enzymes have moved from their origin depends upon their charge and shape; the number of bands reveals the number of enzymes capable of reacting with the substrate.

When members of a species are analyzed in this way, many of the bands are identical, revealing a high degree of genetic similarity, but there still may be considerable variation. The degree of genetic relationship between different species, genera, families, and even divisions, can be estimated in the same way: The closer the banding patterns compare, the closer the genetic similarity is presumed to be.

There are some limitations to the method. Basically, it underestimates the amount of genetic variation. First, not all types of enzymes are equally variable. If a researcher picks a "conservative" group of enzymes, he will now show as much variation as if he had picked another group. Second, not all changes in amino acid sequence of an enzyme will result in a change in the electrical charge or shape of the molecule. This means that two slightly different molecules will still migrate to the same point on the gel, and in those cases, they will be scored by the experimenter as identical, even though they actually are different.

Despite these drawbacks, gel electrophoresis allows taxonomists to estimate the degree of genetic similarity between organisms quickly and conveniently. Such an estimate is much more accurate than one based on morphological similarities (like leaf shape and flower structure). Most morphological traits are controlled by many genes, not single genes, and can be strongly affected by the environment. But with amino acids, we are dealing with single genes, and environmental influences are largely eliminated.

Biological

Biological taxonomists are interested in determining what the **natural biotic units** are. By natural biotic units, they mean populations of plants which maintain their distinctiveness because of **biological barriers** that genetically isolate them from other populations. These isolating barriers may be due to breeding behavior (time of flowering, type of pollinator), habitat or geographic isolation, or inability to form fertile hybrids with closely related groups (sometimes because of differences in chromosome number or chromosome morphology).

Biological taxonomists often deal with taxa at, and below, the rank of species because they are vitally concerned with detecting divergent evolution at an early stage. They are interested in discovering how species

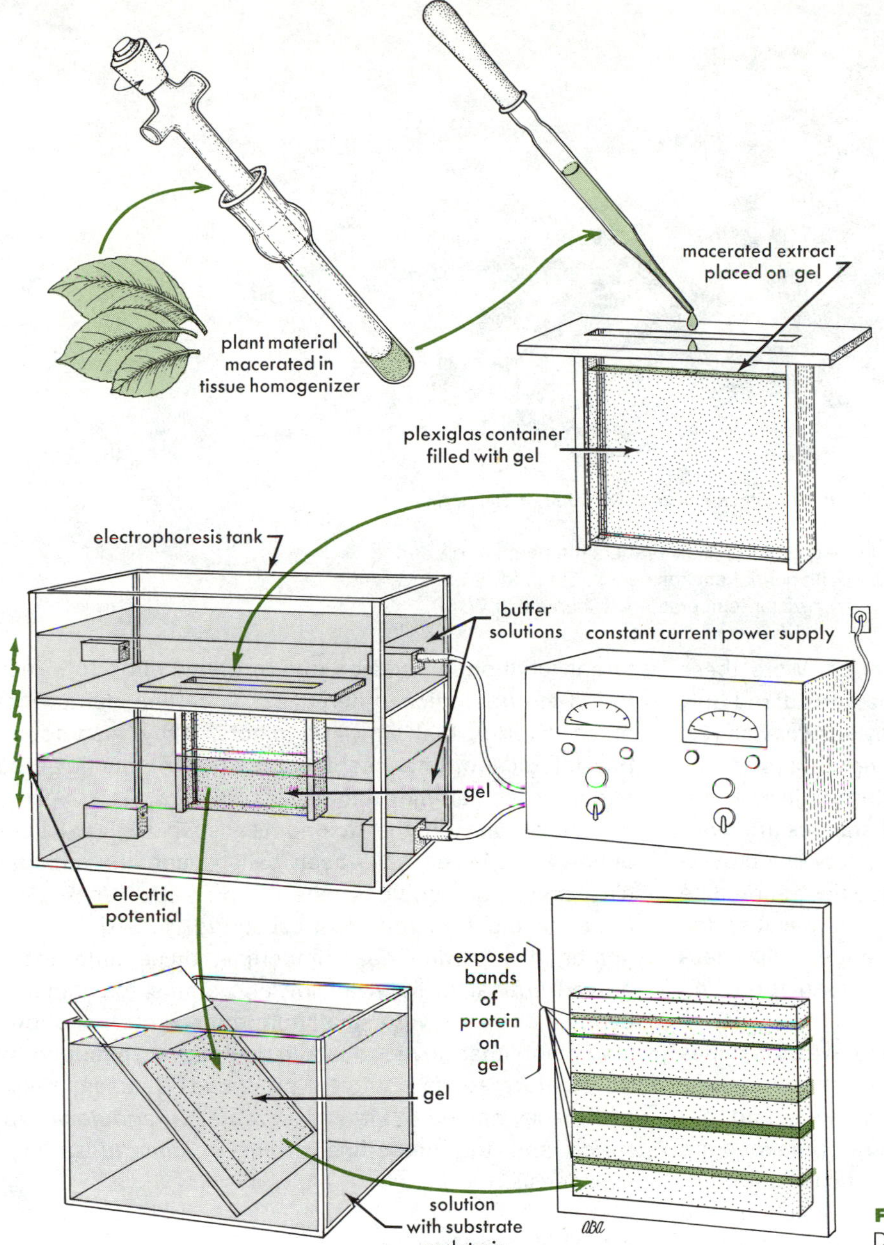

Figure 19.10
Diagram of a gel electrophoresis experiment.

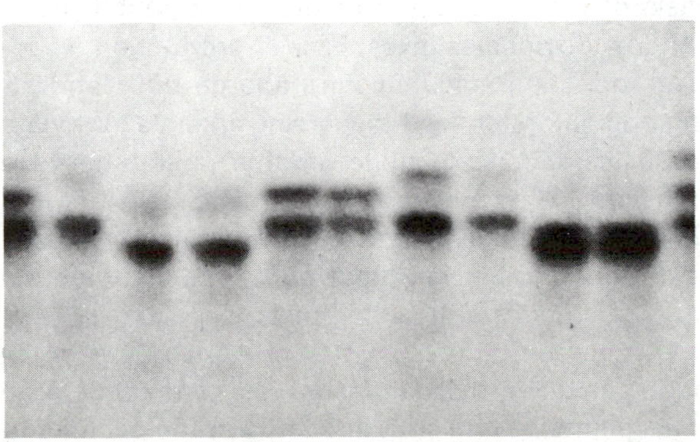

A "developed" gel of 11 *Clarkia* genotypes. The enzyme bands have migrated from the "top" of the gel shown, downward, (Courtesy of L. D. Gottlieb.)

become distinct and how they are able to maintain their distinctiveness even though living in close proximity or in similar habitats. Biological taxonomists are often referred to as **biosystematists,** and their field as **biosystematics.**

Conflicts continue to arise between biosystematics and traditional taxonomy because their conclusions are not always the same. The "natural biotic units" do not always correspond to "well-defined groups" of morphological distinctiveness. The conflicts have, in part, a historical cause. As a result of world explorations in the eighteenth and nineteenth centuries, it became clear that some species within a genus differed only slightly in morphology and habitat from region to region. One species might be in the mountains, another in the lowlands; one species might be along the coast of Europe, another

Figure 19.12

Growth of *Potentilla glandulosa* ssp. *typica* after one growing season at each transplant garden. This coastal subspecies, or ecotype, grew best in its normal environment at Stanford; it grew very poorly at timberline and was killed by the following winter temperatures. (Courtesy of W. M. Hiesey.)

along the coast of eastern North America. Were these really distinct species or should they be considered only as variants of a single species? The early answer was yes, they were distinct species. Much later, biosystematic research sometimes showed that fertile hybrids can result (or at least potentially can, once distances are overcome) from crosses between these species. The biosystematists then concluded that these "species" should be classed as subspecies of a single species because the morphological and geographic differences were less important indicators of relationship than breeding behavior.

On the other hand, biosystematists have sometimes discovered that breeding barriers exist between nearby populations of what traditional taxonomists had described as a single species (that is, the populations are morphologically identical). Yet, the genetic isolation undoubtedly is the first step of divergent evolution, and these populations represent distinct species at a very early stage. They are referred to as **sibling species.**

Subspecies, Varieties, Races, and Ecotypes

Many species that are widespread and occupy a variety of habitats may be composed of several genetically distinct subspecies. The subspecies are all interfertile, but hybrids are not common because hybrids are less suited to the habitats than are parental types. Clausen, Keck, and Hiesey of Stanford University have shown the existence of subspecies in plants such as five-finger (*Potentilla glandulosa,* Fig. 19.12), which grows in California from the coast near Stanford to high elevations near timberline in the Sierra Nevada mountains. Clausen, Keck, and Hiesey dug up plants of five-finger that were growing all along this range and transplanted them together in uniform gardens at Stanford and elsewhere. In the uniform garden, any differences among the plants from different locations would be due to genetic control. They found morphological differences; timberline plants were much shorter than lowland plants. They also found physiological differences; the timberline plants flowered earlier in the summer than lowland plants.

Depending on the taxonomist, such regional subunits of a species have been called varieties, subspecies, races, or ecotypes (see Chapter 18). When five-finger plants from different locations were grown together in a garden near timberline, these differences proved critical to survival: lowland plants that did not become dormant were killed during the winter, and lowland plants that flowered late in summer were nipped by frost before seeds could be produced. Clausen, Keck, and Hiesey named the lowland group *P. glandulosa* ssp. *typica* and the timberline group *P. glandulosa* ssp. *nevadensis.*

r and K Strategies

Biosystematists are also interested in comparing taxa on the basis of their survival strategies. How much metabolic energy do species invest in seed production, vegetative reproduction, accumulating a large bulk, and attracting pollinators or seed-dispersing animals? Why are some species genetically quite variable while others are very constant? What mechanisms do plants have to avoid or minimize competition and ensure survival? Experiments designed to answer these questions combine the talents of ecologists and taxonomists.

One example of different strategies can be seen by comparing an oak with a dandelion. Oaks and other large perennial plants devote a lot of energy to the accumulation of great bulk that can withstand many environmental stresses. The price paid is a long juvenile period of development when reproduction is nil, and even after that the amount of energy put into flower and seed production is low, compared to the total mass of the plant. Seeds (fruits) are usually large and are disseminated by gravity, water, or animals. Seed production and seedling establishment can be very low for many years in succession due to poor environmental factors, but the population density remains constant because of the long life span of individuals. This is called a **K** strategy by ecologists. Dandelions and most herbaceous annuals, on the other hand, invest a high proportion of their energy in seed production and have a short life cycle. Each seed is small and often wind-disseminated, which means there is very little food reserve to maintain the young seedling during establishment. The result is that most seedlings perish. But, because the reproductive capacity of the few survivors is so high, the population density is maintained. This is called an **r** strategy. Obviously, both K and r strategies can lead to success in an evolutionary sense. The question is, under what set of environmental conditions will each strategy be the most successful?

Numerical

Another approach to taxonomy is to consider all forms of evidence equally; all characteristics, whether morphological, anatomical, biological, or biochemical, have equal weight in the decision-making process. This approach, which relies to a great extent on multivariate statistical methods, is called **numerical taxonomy.** A basic tenet of numerical taxonomists is that a great deal of evidence is needed to objectively separate taxa. Recall C. H. Waddington's comment (page 379) that living organisms ". . . fall into more or less well defined groups, which are commonly called species." The problem that many taxonomists encounter is that the species they work with are less well defined; hence attempts to establish boundaries between them become more arbitrary. The aim of the numerical taxonomist is to erect, by way of simultaneous analysis of many characteristics (from 20 to 300 or more), objectively defined boundaries that may be reproduced by later taxonomists studying the same sets of information.

An example of the different interpretations produced by subjective traditional evaluations may be found in the genus *Quercus* (oaks). *Quercus falcata* is the southern red oak native to the southern half of the eastern United States. This oak has been variously defined as a single species with four varieties (*falcata, triloba, pagodaefolia,* and *leucophylla*) or as two species (*Quercus falcata* and *Quercus pagoda*). It is interesting that a recent numerical taxonomic investigation, based on simultaneous analysis of 15 equally weighted morphological characteristics, provided evidence to support the latter interpretation. The numerical analysis was based on data from 20 trees, 5 of each supposed variety, growing in a 20 km^2 area. The analysis demonstrated that this group of 20 trees consisted of two statistically independent subgroups, each with 10 trees. The varities *falcata* and *triloba* comprised one subgroup and varieties *pagodaefolia* and *leucophylla* comprised the other. While it is evident that there are not four independent varieties, the question of the taxonomic status of the two subgroups is still debatable, that is, are they two species or one species with two varieties? The answer depends upon the taxonomist's opinion of the distinctiveness of the two groups.

Numerical taxonomic analyses may be categorized generally as either **phenetic** or **cladistic** analyses. The basic difference is that phenetic analyses, exemplified by the oak problem presented above, are meant to reflect overall patterns of phenotypic similarity while cladistic analyses are meant to reflect patterns of evolutionary relationship. Although the two methods may produce identical patterns of relationship among the entities being classified, they are based on somewhat different assumptions. For either analysis, the numerical taxonomist must decide on operational species or taxonomic units (OTUs) that will form the basis for the analysis. OTUs may be individual plants, separate populations of the same species, separate species within a genus, separate genera, etc. In addition, a suitable number of characteristics must be selected to allow comparison of the OTUs. Phenetic analyses are based on overall similarity, that is, pairs of OTUs are compared over all available evidence and a coefficient of similarity is determined, while cladistic analyses are based on evidence of similar patterns of specialization in different characteristics (recall Bessey's assumptions on primitive and advanced characteristics).

One method of depicting the results of a phenetic analysis is by way of a branching diagram called a **dendrogram** (Fig. 19.13) produced by **cluster analysis.** Cluster analysis is a method whereby groups, or clusters, of OTUs that have high coefficients of similarity are erected and then joined to other such groups. The result is a branching diagram that connects all of the OTUs and OTU clusters at levels corresponding to their degree of similarity. By selecting an appropriate range of similarity to represent a given level in the taxonomic hierarchy, the taxonomist may then recognize species, genera, etc. For example in Fig. 19.13, groups in which all OTUs have similarities between 85 to 100% might be recognized as part of the same species, while a 65% criterion might be used for genera. The ultimate interpretation of the dendrogram is dependent upon the taxonomist's knowledge of the included OTUs.

A cladistic analysis will also result in a branching diagram, called a cladogram, but in this case the branching pattern is designed to depict not only similarity but also evolutionary relatedness. Such a diagram is illustrated in Fig. 19.14, which shows supposed evolutionary relationships of species, OTUs *A* to *E*, in a hypothetical genus. The cladogram depicts modern species as the end points of different pathways of evolution. For example, in Fig. 19.14, *D* and *E* represent the modern end points of two evolutionary lines departing from a hypothetical common ancestor. All the species in the genus are considered to have evolved from a single ancestor

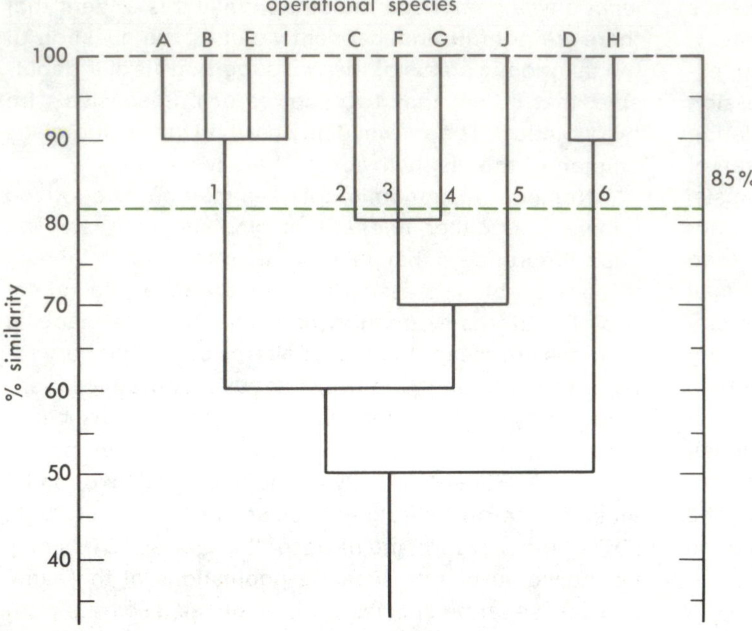

Figure 19.13
Dendrogram, summarizing percentage similarity between pairs and groups of operational species. If members of the same species were defined as showing at least 85% similarity (dotted line), then the 10 operational species would be lumped into six species as indicated by the intersections of the lines, 1 to 6, with the dotted 85% line.

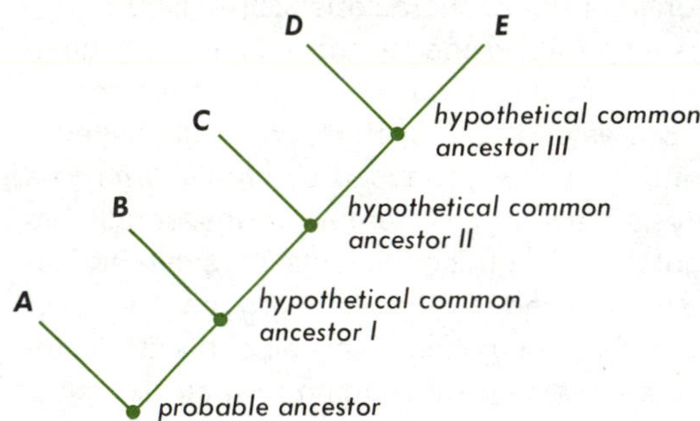

Figure 19.14
A cladogram, showing supposed evolutionary relationships between species (or OTU's) *A-E*. Intermediate ancestors are shown as dots but no longer exist.

(that is, the genus is considered to be monophyletic) indicated as the probable ancestor in the figure. The positions of each of the species are then determined by comparing patterns of character changes; the characters of the probable ancestor are viewed as primitive and those of the modern species are advanced. The position of each species is then determined by the number of characters in which it differs from the probable ancestor. In Fig. 19.14, *A* differs very little from the probable ancestor whereas *D* and *E* demonstrate the greatest degree of difference from the probable ancestor.

The point of emphasis in numerical taxonomic studies is that the methods employed, which usually have a rigorous mathematical basis, are truly objective and will always produce the same results given the same data. This is not true of traditional taxonomy, however, in that two taxonomists examining the same collection of plants often have produced quite different conclusions. Yet, the interpretation of the results of a numerical taxonomic analysis is also dependent on the taxonomist's view of the OTUs in the analysis. While the clusters of OTUs produced may be the same, one taxonomist may call the clusters species and another may call them genera. This points out that the computer cannot give us the answers. The interpretation of results is dependent upon the trained taxonomist or systematist who is aware of the nature of the OTUs analyzed.

The numerical taxonomist, then, suggests that when one considers many traits instead of a few, one will see a continuum of variation. Boundaries between species can be erected by traditional or numerical taxonomists. These boundaries should be viewed as arbitrary and artificial in the final analysis. The viewpoints of traditional and numerical taxonomists, then, are quite different.

SUMMARY

1. As a floral survey of the world becomes more complete, the goal of taxonomy becomes the construction of a phylogenetic classification scheme that summarizes the evolutionary history of our flora.
2. The history of taxonomy can be divided into a pre-Linnaean period and a post-Linnaean period. Linnaeus culminated the pre-Darwinian period (ca. 1753) and organized plant nomenclature; nearly 200

years later, the rules of plant nomenclature were further elaborated.

3. Taxonomic research may take any of several directions, utilizing patterns of morphology, anatomy, embryology, biochemistry, and breeding behavior, any of which may be used to show genetic relationships between taxa. The numerical taxonomist utilizes information from all fields and does not give extra weight to any particular type of information. This is in contrast to other taxonomists who do choose to subjectively weight particular types of information.

CHAPTER 20

PLANT GROWTH AND DEVELOPMENT

All the genetic information required to build and operate a plant is carried in each individual cell encoded in the DNA of the nucleus, mitochondria, and plastids (Chapter 5). During the life of a plant, this genetic information is used to direct increases in size **(growth)** and changes in form **(development).** As we have seen in looking at the architecture of cells, tissues, and organs in previous chapters, this growth and development requires a division of labor among cells, tissues, and organs. The formation of these differences is called **differentiation.**

The challenge of plant growth and development is to understand how particular bits of that genetic information are selected to chart a cell's growth and differentiation into a mature form that may differ in many ways from the undifferentiated meristematic cell from whence it ultimately derived. How is the reading of that genetic information accomplished so as to result in a red-colored cell in a rose petal, a green leaf cell, or a tracheid? Why doesn't a buried seed, upon germination, form enlarged leaves underground? While we may at times emphasize the processes of differentiation at the cellular level, because cells are the basic functional unit, it should be understood that differentiation can be coordinated at the tissue and organ level as well.

Stored in this genetic information there must exist a library of programmed sequences of differentiation and development. One obvious example of such a sequence is the process of mitosis where, upon some signal, there is initiated a precisely orchestrated series of biosynthetic events, organelle movements, and membrane transformations. The choices of which program to follow, of when to do what, are made in response to appropriate cues. What constitutes an appropriate cue? Appropriate cues may be (*a*) signals from the environment, (*b*) hormonal signals from other parts of the plant or neighboring cells, (*c*) activities of neighboring cells or tissues, (*d*) position factors expressing the location within the plant body, and (*e*) nutritional factors, to name a few. A cue does not have to carry much information in itself in order to release a complex sequence of development. A period of low temperatures may trigger the change in the shoot apex that stops leaf initiation and allows the flower-making program to be read. The same low-temperature cue may be used to initiate the metabolic changes making a frost-sensitive bud that would be killed by −5°C into a frost-hardy bud capable of surviving temperatures of −30°C.

The interval between perception of the signal and full development of the response can vary from seconds to months. Certain wavelengths of light can alter the movement of ions across membranes to a detectable degree in 10 sec. Application of auxin, a plant hormone (p. 394), to some tissues can result in increased growth within 10 min. The effects of low temperature on growth, however, may require weeks or months to complete. In the case of very rapid responses, the response may be so set up that there is no need to involve the direct intervention of RNA or protein synthesis. In other cases, new RNA molecules are doubtless formed, and corresponding new proteins appear (see Chapter 6). In others, cell divisions form new cells and even new organs are initiated. Thus, the extent of involvement of the metabolic machinery of the cell varies from one programmed sequence to another.

Variation in the numbers and enzyme complement of cell organelles constitutes an important aspect of differentiation. A leaf mesophyll cell, a root parenchyma cell, and a tomato fruit mesocarp cell all have the information in the nucleus and proplastids necessary to make a green chloroplast. Yet, the leaf mesophyll cell usually forms the complete photosynthetic apparatus (a mature fully green chloroplast) only if it is exposed to light; the root cells normally do not differentiate chloroplasts even when exposed to light; and the tomato fruit cell makes functioning green chloroplasts early in the life of the fruit. Later, during ripening of the fruit, a new course of differentiation besets these chloroplasts whereby the thylakoid membranes are rearranged, chlorophylls disappear, and red carotenoid pigments are synthesized in large amounts (Fig. 4.16*D*). In the fruit, this second differentiation of the chloroplasts into red chromoplasts is only one aspect of the whole complex differentiation process that is programmed to change a fruit from a seed-manufacturing organ into a seed-dispersing organ, attractive to the eye and taste of birds and animals.

ENVIRONMENTAL CUES AND THE SELECTION OF PROGRAMMED PATTERNS OF RESPONSE

The development of a typical plant is much more the result of interaction with the environment than is the morphological development of one of our animal friends. The control or regulation of development may be considered at three levels: (*a*) within the individual cell, **intracellular,** (*b*) between cells, **intercellular,** and (*c*) from the outside of the organism, **environmental.** This is represented schematically in Fig. 20.1.

The next few paragraphs outline the kinds of problems and choices faced by a plant and how a variety of environmental cues elicit programmed patterns of response that maximize chances of survival of the species. In later sections, we will examine the internal mechanisms of control and their relationship to these environmental factors.

A wide array of blocks to seed germination are programmed to enhance the probability of germination under favorable conditions. These blocks can start in the fruit before seed maturity. Fleshy fruits frequently provide the seed with an inhibitor of germination such as abscisic acid (see later pages in this chapter) that prevents germination in the fruit prior to dispersal, fruit breakdown, and leaching away of the inhibitor. Occasionally, one can see a malfunctioning of blocks to germination in grapefruit and certain new tomato varieties where seeds do germinate within the fruit.

Temperature

A requirement for exposure to a period of low temperature may favor germination in the spring. Some seeds will germinate only after sufficient rain, when the accompanying leaching action has washed out an inhibitor from the seed coat. This appears, for instance, to delay germination of some desert annuals until they have had a rainfall of adequate amount to support a few weeks' growth. Other seeds germinate only when buried (light inhibits germination), while some, especially small seeds with limited food reserves, may germinate only near or at the surface (light is required for germination). This light requirement is often the block removed when soil is disturbed and formerly buried seeds are stimulated to germinate, a characteristic of certain weeds familiar to gardeners.

Gravity

The force of gravity acting upon organelles within cells gives directional information to the root enabling it to grow downward as soon as it emerges from the seed. In response to the same stimulus, the shoot is programmed to grow upward, the most likely location of light. Rhizomes may respond to the gravity vector by growing horizontally. Branches on trees often grow at a particular

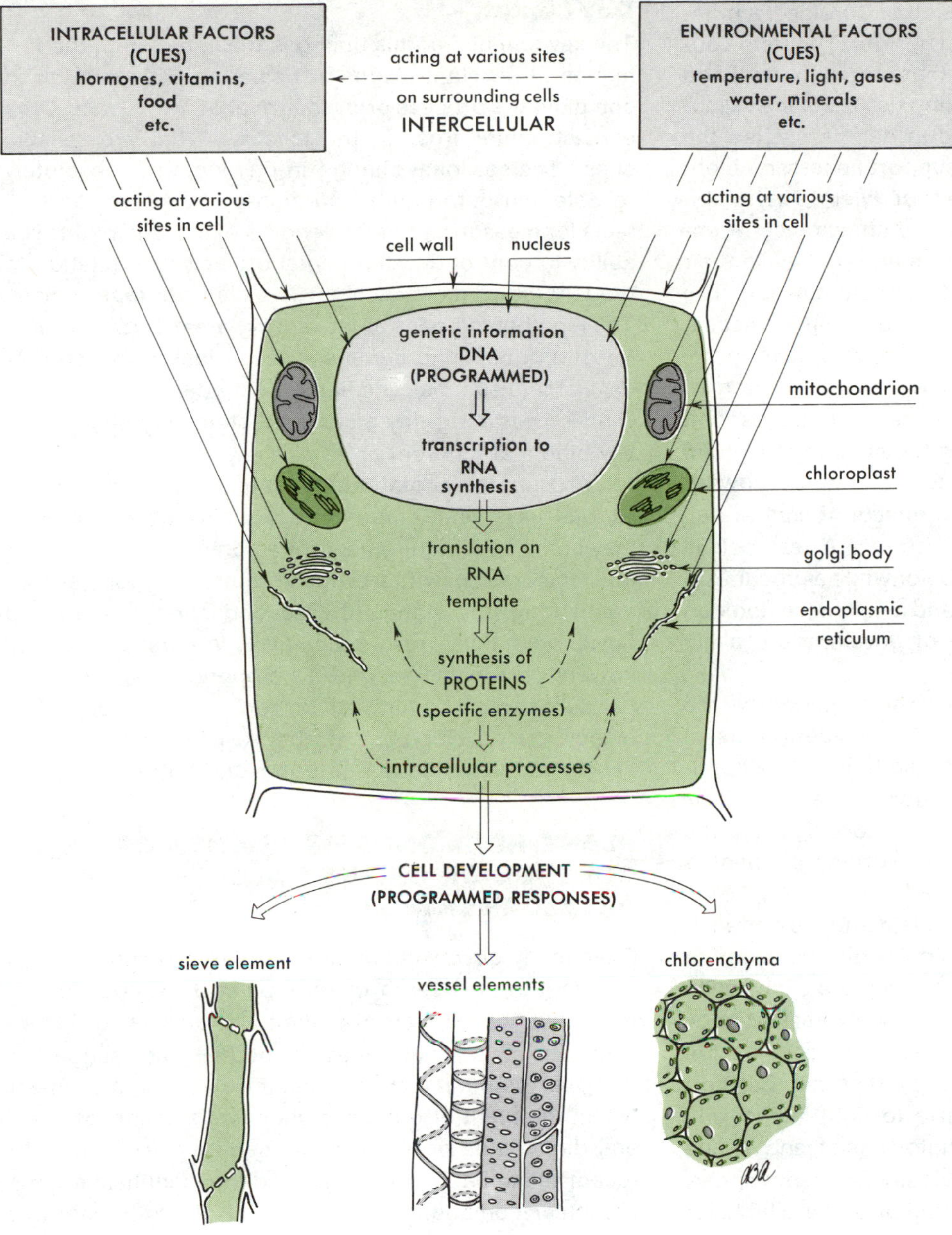

Figure 20.1
Diagram showing different levels of regulation of plant development such as intracellular, intercellular, and environmental.

angle to the vertical. The direction of gravity is a very reliable cue on which to base these growth responses.

Light

Why don't leaves unfurl and enlarge beneath the soil? The shoot system as it emerges from a germinating seed is a quite delicate structure that must push its way through soil of varying composition. It is a characteristic that dicotyledonous plants do not appreciably expand their leaves until they reach the soil surface (**Figs. 20.2, 20.3,** page 409). The cluster of leaf primordia and apical meristem are usually inverted and *pulled* up through the soil behind the recurved portion of the hypocotyl or epicotyl (the **hook**). The delicate leaves and apex are thus protected. In monocotyledonous plants, the young leaves are rolled up and completely ensheathed in a protective tube, the **coleoptile.** When the coleoptile reaches the surface, the leaves push out of the splitting coleoptile and unroll. The cue that signals arrival at the soil surface is light received by the pigment, **phytochrome.** Phytochrome is a special light-absorbing molecule that serves as a light-sensing system, making metabolic and developmental processes responsive to the presence or absence of light. Growing a seedling in absolute darkness results in the kind of growth pattern adapted for movement through the soil (**Figs. 20.4, 20.5,** page 409). The seedling is **etiolated.**

In addition to the leaf expansion or unrolling, a number of other changes are initiated by light. The stem usually elongates very rapidly in darkness (or in the soil) and light inhibits elongation. Supporting tissues of the vascular system are weakly developed in darkness when the soil would normally give all the support necessary. Light stimulates increased development of xylem. The inhibition of growth can be related to the appearance of enzymes that catalyze lignin synthesis and that make their appearance 2 to 3 hr after the start of illumination. The growth of the stem is programmed to develop the necessary strength to support itself and its leaves when growing above ground. While light slows the elongation of much of the stem, certain cells inside the hook respond by increased elongation, causing the stem to straighten out and the developing leaves to be oriented upward **(Fig. 20.2).** These morphogenetic effects of light are all initiated by activation of phytochrome. A leaf cell is triggered into one pattern of behavior while a procambial cell takes off on another course and a cell in the hook on another programmed sequence of development, all in response to the same stimulus.

It may happen that the first light perceived by the emerging plant is coming from only one side. The usual response is for the stem or coleoptile to initiate a bending growth that orients the tip toward the light. This bending is initiated by absorption of more light on one side than on the other. The light-absorbing pigment is distinct from phytochrome.

The cells of the mesophyll of a leaf and the cortex of a young stem are programmed to delay differentiation of the photosynthetic apparatus until they are exposed to light **(Figs. 20.3, 20.5).** Then chlorophyll synthesis starts, membranes are organized into thylakoids, and the full battery of chloroplast enzymes is synthesized. Furthermore, the regulation is responsive to light intensity in many cases so that leaf morphology (sun and shade leaves) and enzyme content (ribulose bisphosphate carboxylase) are adjusted to the amount of light available for photosynthesis (see Chapters 10, 13).

Light filtered through the leaves of other plants is of lowered intensity and different spectral composition (Chapter 18). This appears to constitute a cue or cues to increase stem elongation, a response raising the chances of growing out of the shade of other plants and icnreasing light capture for photosynthesis.

On much of the land surface of the earth the suitability of conditions for plant growth varies seasonally with respect to temperature, water availability, and total sunlight. Whatever the unfavorable season, it is typical that plants native to the area characteristically manage to time their activities in such a way as to complete the formation of seeds or resistant stages prior to the onset of the unfavorable season. The unfavorable season for one plant need not be the same for another species in the same area.

Day Length

The key to much of this timing is the initiation of the formation of resistant stages or flowering early enough to complete the process prior to arrival of, for instance, the earliest winter frost in the locality. What are reliable cues? The seasonal change in day length is absolutely reliable and both plants and animals have evolved systems for measuring relative lengths of day and night. The ability to control developmental processes in relation to the lengths of day and night is called **photoperiodism.** The lengthening days of late spring are a certain indicator of an upcoming summer season just as the shorter days of late summer and early fall predict the oncoming winter. Thus, **long-day plants** and **short-day plants** usually bloom at different times of the year.

Frequently, photoperiodic signals are combined in various ways with temperature signals. A biennial plant may combine a requirement for a considerable period of low temperature (winter) with a requirement for a subsequent long day signal (the second spring) to initiate flowering. It thus grows vegetatively the first season and flowers only in the second. Photoperiodic and/or temperature signals can initiate flowering, the onset of dormancy, the development of frost resistance, leaf senescence, and the initiation of tubers and bulbs.

PLANT HORMONES AS AGENTS OF CONTROL OF GROWTH AND DEVELOPMENT

From many observations and some experiments, it was apparent to nineteenth century plant physiologists that the growth of one part of a plant was closely correlated with the growth or activities of another. They suggested that this communication must involve the movement within the plant of unknown chemical substances. Several direct lines of experiments can be traced down to recent years. Each line has yielded the identification of a particular compound that occurs in plants in very low concentrations. When these compounds have been purified carefully and applied to plants, they exert powerful controls on functions and plant development.

A set of five types of plant growth hormones seems to be nearly universal in seed plants. These are the most important agents involved in coordinating the growth of the plant as a whole. They frequently constitute the cues that determine the course of development and differentiation. Many of the effects of the external environment on development are mediated by changing the synthesis or distribution of these hormones within the plant. The five plant growth hormones are **auxins, cytokinins, gibberellins, abscisic acid,** and **ethylene** gas (Fig. 20.6). Depending upon the system, they may act essentially alone or in some sort of balance with one or another.

There are enough occurrences of hormones and hormone responses among more primitive plants to

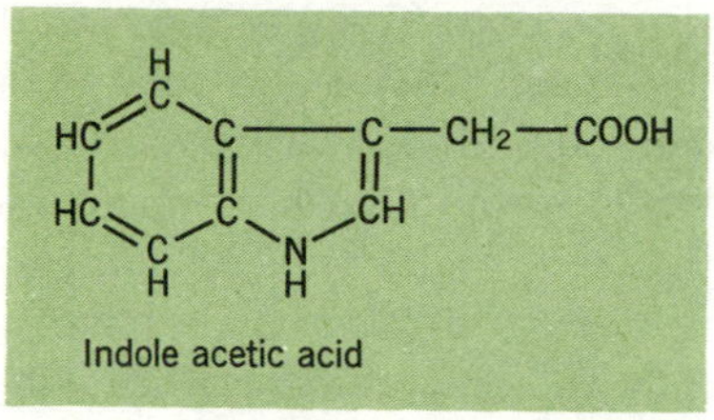

Indole acetic acid

A: Molecule of the auxin, indole acetic acid (IAA), made up of two rings and a side chain

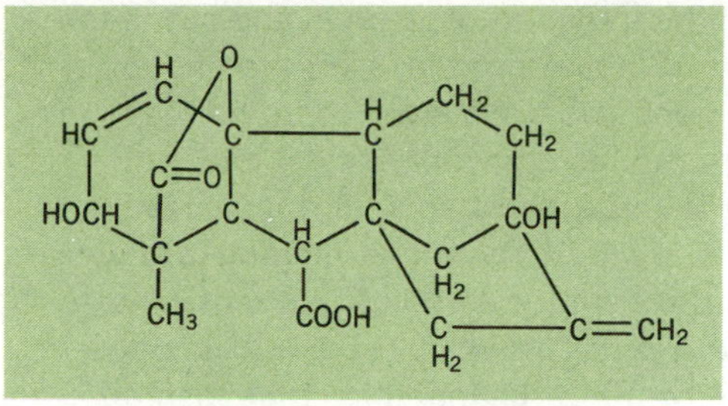

B: Molecule of gibberellic acid

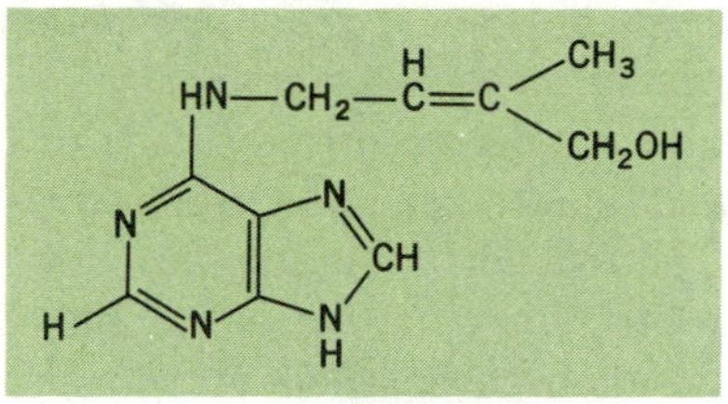

C: Zeatin

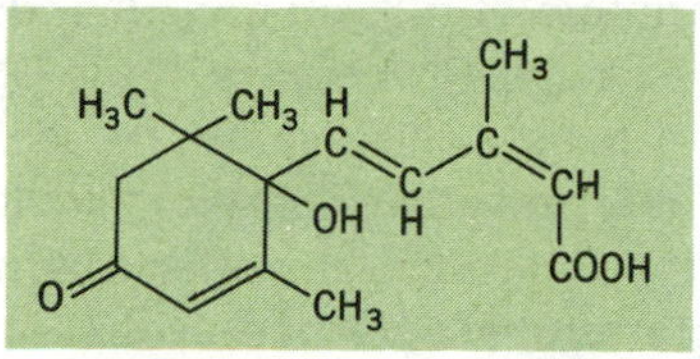

D: The structure of abscisic acid (also called dormin or abscisin II)

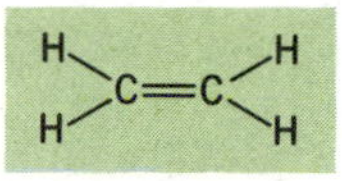

E: Ethylene

Figure 20.6
Plant hormones. *A,* indole acetic acid, IAA, the most common naturally occurring auxin. *B,* gibberellic acid, GA_3, one of more than 40 very similar natural gibberellins. *C,* zeatin, one of several natural cytokinins. *D,* abscisic acid, ABA. *E,* ethylene, a gaseous growth regulator.

suggest that the evolutionary progenitors of seed plants had at least several of the important control systems. It is very likely that increasing importance and diversification of hormonal control mechanisms was an important component of the evolution of the seed plants.

Some hormones conform to the animal physiology definition of a hormone: a substance produced in one place in the organism that moves to another site, where it exerts its action. But that definition is unnecessarily restrictive for plant hormones. For example, one hormone, auxin, is generally produced in a shoot apex and then transported by a specific transport system to act on the growth of the stem and buds further down. However, there is a reasonable basis for saying that sometimes auxin must function in the same cells that synthesize it. Ethylene, another hormone, is a gas and is probably not transported in the plant in the dissolved state in any specific way. However, a precursor to ethylene may be transported from the roots of flooded plants to the shoots where it is converted to ethylene. The leaves then curl in response to excess ethylene.

Plant hormones sometimes initiate a sequence of developmental changes such that they must certainly be acting to alter the pattern of gene expression. In other cases, plant hormones appear to be controlling the rate of some enzymatic activity without directly involving the gene-directed synthesis of RNA and the RNA coding of protein synthesis (Chapter 6). In some cases, we can identify specific proteins synthesized as a result of application of the hormone to certain tissues. On the other hand, the response elicited in one plant system by a given hormone may be quite different from that produced by the same agent in another plant or in a different part of the same plant. So one hormone may initiate a number of different seemingly unrelated responses. One might take as a convenient working idea the proposition that the concentration of a given hormone is a meaningful cue about the internal environment of the plant and one that might initiate numerous programmed responses. This postpones the necessity of trying to decide whether or not there is one mechanism of action of a given hormone before any one mechanism is known. It is easier to consider that these hormones are like radio tubes or transistors in that the kind of control function they exert depends on the kind of circuit they are plugged into.

It is most probable that these hormones bind with protein receptor sites in the cell to carry out their actions. Studies relating the molecular architecture and activity of natural and synthetic hormones show that receptor sites are nearly as discriminating as are the sites on enzymes that bind specific substrate molecules. Recently several proteins that bind auxin molecules have been identified. One type seems to be on the plasmalemma and is involved in the transport of auxin in and out of the cell. Another auxin-binding protein is on the endoplasmic reticulum and appears to be involved in the reaction whereby auxin promotes growth. A cytokinin-binding protein has been found associated with ribosomes. The recent discoveries of these hormone-binding proteins lends support to the hypothesis that hormones bind to particular receptor proteins whose activity is consequently turned on or off, or changed in some fashion.

One actively pursued proposal is that hormones bind to specific binding proteins that are involved in activating or inhibiting the transcription of a given gene (controlling the synthesis of messenger RNA copies of the DNA encoded gene) or set of genes. This hypothesis seems tantalyzingly close to being proven in a case or two. It is also clear that some of the most important effects of hormones and their receptor proteins do not involve such direct control of gene expression.

AUXINS*

Control of Cell Enlargement by Auxins

Cell enlargement is a process critical to the life of almost every cell in a plant, and the ability of the plant to control this process precisely is central to growth and morphogenesis. A plant cell may be thought of as an inflatable bag (the protoplast) surrounded by a supporting cover (the cell wall) that may be rigid or may be expanded under pressure, depending upon its makeup. By using respiratory energy and carrier systems, ions such as K^+, Na^+, and Cl^- are pumped into the vacuole, and water follows, osmotically creating a turgor pressure. The turgor pressure in a typical cell might be maintained at 5 bars or about 5 kilograms per cm^2. Is the cell wall going to expand while under this internal pressure? Obviously, some cells do expand under this condition. Does the plant control the elongation rate by changing the turgor pressure against walls of unchanging properties? Or does the cell use a more or less constant force against a wall whose extensibility can be changed either to allow expansion or not? Plants predominantly use changes in wall extensibility to regulate growth, but they frequently use turgor changes to accomplish readily reversible swelling and shrinking of cells that underlie, for instance, the sleep movements of leaves, the opening and closing of the leaflets of the sensitive plant (*Mimosa pudica*) and, of course, the guard cell swelling and shrinking that controls stomatal opening. Auxin plays a role in changing wall extensibility.

The coleoptile of grasses has been considered since Charles and Francis Darwin's time as especially favorable material for studying this elongation process *per se* and the systems that control it. We now see that the coleoptile is used in these studies at a point in development where no more cell divisions will occur, there is little response to growth-promoting hormones other than auxin, and the cells are pretty well synchronized in a state of readiness to expand. Other plant materials tend to have cell divisions and enlargement proceeding at the same time and to be sensitive to the action of other hormones, simultaneously making them more complex and difficult to analyze. Although not properly a stem, the coleoptile at this stage grows in much the same way as the rapidly elongating portion of a stem.

Under normal turgor pressure, the cells of the coleoptile will expand rapidly if an optimal concentration of auxin is present, and slowly if it is too low or very high. Elongation is dependent upon an active metabolism. Inhibition of respiration causes nearly complete cessation of growth in a few minutes. In order to continue extension, the walls must constantly be worked upon by the protoplast through the plasmalemma. New materials are secreted through the membrane and deposited around the inside surface of the wall (cellulose microfibrils, p. 74) or well into the interior of the wall (other polysaccharide and protein components). There must be a breaking and reformation of cross-linking bonds between various of these macromolecules, making the framework of the wall (Fig. 5.5). How the cell controls these changes that occur outside its plasmalemma is one of the most intriguing problems in plant physiology. Auxin regulates some aspects of this wall metabolism to promote the loosening of the structure to allow the turgor force to stretch the wall.

Whether a cell expands in all directions equally to form an isodiametrical cell as in cortical parenchyma or whether the expansion is restricted to one axis to give an elongate, fusiform or tubelike cell depends upon the wall. If the supporting cellulose microfibrils are oriented randomly around the cell surface, expansion occurs uniformly, forming a larger sphere. If they are predominantly oriented more or less perpendicularly to one axis as are the hoops on a barrel, the wall will expand along one axis toward the ends but the hoops prevent increase in diameter. The result is formation of a tubular cell.

Another important point is that the growth process is very tightly controlled. Auxin presented to an oat coleoptile section deficient in auxin causes a growth rate increase in 10 minutes. Removal of the auxin supply results in a slowing of elongation in less than 30 minutes. Therefore there must be a continuous supply of auxin in the growing tissue to maintain growth. The rapidity of the response and other experiments are taken together to indicate that in this kind of effect on the cell wall, auxin does not act primarily through some direct effect on RNA or protein synthesis. There is, however, a requirement for protein synthesis if a cell is to continue growth over any period of time.

* Auxin is a general term that applies to a group of compounds all of which affect plants in the same way. It includes the naturally occurring compounds indoleacetic acid (IAA) and phenylacetic acid as well as many synthetic compounds. In this chapter, the term auxin will usually mean IAA.

Control of Auxin Concentration

The demonstration that elongation rates respond so quickly to removal or resupply of auxin makes it obvious how important the factors must be that control the concentration of auxin. The coleoptile and the growing stem generally separate the site of auxin synthesis from the zone of rapid elongation. The tip of the coleoptile or the stem tip with its cluster of young leaves synthesize indoleacetic acid (the major natural auxin) from the amino acid, tryptophan. The auxin is then moved down the stem through the elongating region (Fig. 20.7). The movement is called **polar transport** because it is a one-way, energy-requiring, active movement away from the tip. Typical rates of movement are 10 to 15 mm per hour, much slower than movement of materials in the xylem and phloem that may exceed 1 meter per hour. Furthermore, polar transport in the youngest stem regions is counter to the direction of movement in the phloem and xylem

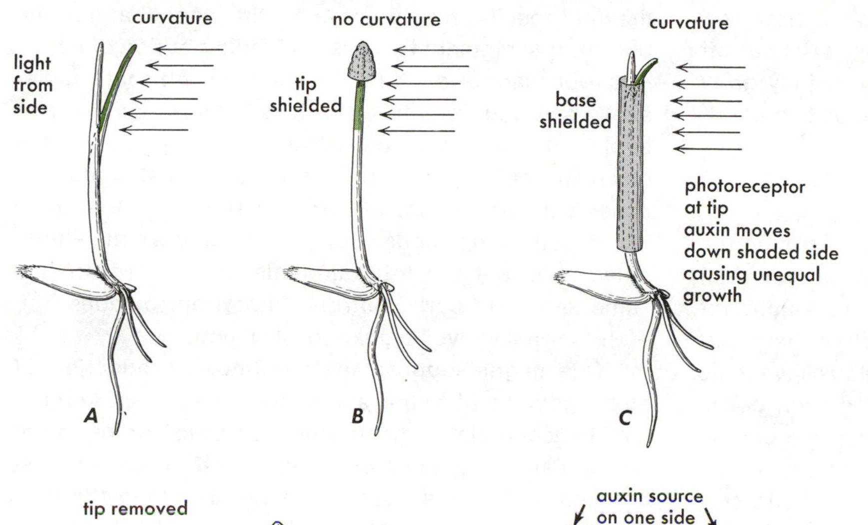

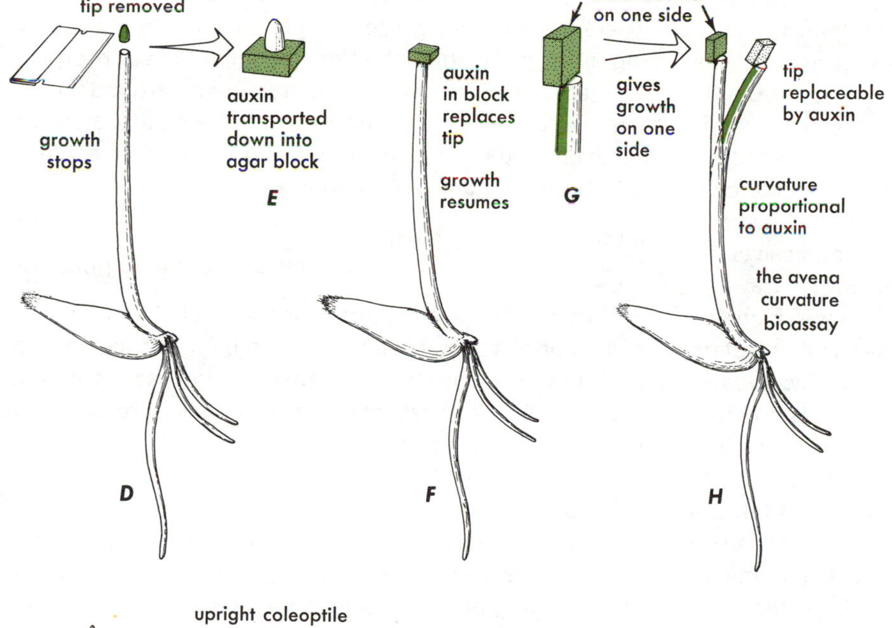

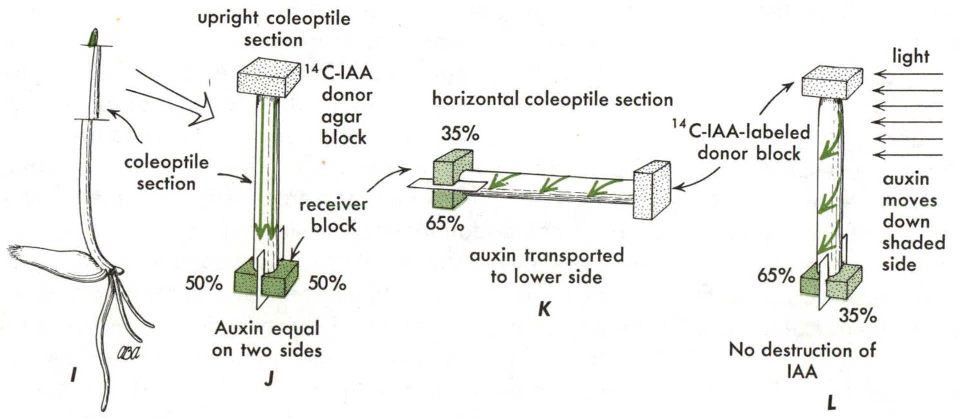

Figure 20.7
Diagram of experiments showing how regulation of auxin transport can result in phototropic curvature (*A, B, C, L*) or in geotropic curvature (*I, J, K*), and the *Avena* curvature bioassay used in estimating minute amounts of auxin (*D, E, F, G, H*). Experiments in which IAA labeled with radioactive carbon was used (*I, J, K, L*) show how geotropic and phototropic stimuli cause auxin to be transported laterally. Unequal amounts of IAA accumulate in the receiver blocks.

and has been shown to occur even in stem tissue from which all vascular elements were removed.

Thus, auxin is swept from the tip through the growing zone, and the concentration in this stream sets the growth rate of the tissue. Remove the tip of a coleoptile, the auxin source, and the remaining auxin is rapidly drained away from the growing zone. The growth rate drops in a few minutes. Put back the tip or replace the auxin supply with a block of agar containing indoleacetic acid and growth is soon restored (Fig. 20.7). If, however, the auxin source is placed on only one side of the cut-off stump, the cells lined up below the auxin supply grow more than those on the other side. This leads to a bending of the coleoptile or stem.

Tropic Curvatures

Plants use this system of unequal supply of auxin on the two sides of a stem to orient the direction of growth in response to environmental cues. These curvature responses are called **tropic curvatures.** When the shoot and root emerge from the seed, it is of obvious advantage to direct shoot growth upward and root growth down. The direction of gravitational force is a reliable cue. In growing regions in both shoots and roots, there are particles heavy enough to sink to the bottom of the cell. These sedimenting particles are called **statoliths,** and in higher plants much evidence indicates that amyloplasts perform this function. On reorientation of a cell from its original position (e.g., put a plant on its side), experiments have shown the amyloplasts to be deposited on the new bottom side in a few minutes. In some unknown manner, this reorientation of particles within the cell changes the direction of auxin transport progressively to the lower side so that by the time the auxin has traveled from the tip to the main growing zone there may be twice as much auxin moving through the lower half of the stem compared with the upper side (Fig. 20.7). The greater growth on the lower side pushes the tip until it is pointing up. The statoliths settle back to their original location, and auxin transport becomes uniform on both sides again. The original change in statolith positions takes a few minutes, and by 15 minutes there is enough auxin concentration difference to measure the beginning of curvature. In some rapidly growing seedlings, the stem may turn upright within an hour from the time of putting it horizontal. This tropic curvature in response to gravity is called **geotropism.**

The geotropic curvature of roots downward has many aspects in common with the shoot response, except that there is considerable doubt that auxin is the hormone involved. It is clear that the root cap produces some controlling factor, probably abscisic acid, which moves to the zone of elongation to cause curvature by inhibition of growth of the lower side. The role of auxin in root geotropism is not clear.

Another stimulus that achieves lateral redistribution of auxin is light striking one side of the stem (unilateral illumination). The process of growing toward or away from unilateral light is called **phototropism** (Fig. 20.7). A pigment is relatively concentrated in cells just below the tip of coleoptiles and stems. Absorption of blue wavelengths of light somehow alters the pigment in a way that is linked to auxin transport. Dim light coming from one side of an axis results in more pigment activation on the lighted side than on the shaded side. Less auxin is moved down the lighted side, and a curvature of the tip toward the light results. We do not know the steps that link the phototropic pigment to the lateral redistribution of auxin. However, careful experiments have shown that indoleacetic acid (auxin) labeled with radioactive carbon and applied to the tips of coleoptiles moves symmetrically down the coleoptile in the dark. Light treatments that cause curvature result in as much as double the auxin moving down the shaded side as moves down the illuminated side, but the total amounts transported are the same as in the dark controls. Phototropism does not, then, normally involve auxin destruction.

One might suppose that continued production of auxin at the tip of a stem and its transport down ought to lead to accumulation somewhere. However, no region of accumulation has been detected; therefore, auxin must be destroyed or inactivated somewhere along the line. There are enzymes present in plants capable of destroying much more auxin than the plant could ever produce. The real question is how is auxin protected from destructive oxidation long enough to act. There are also enzymes that inactivate auxin by tying another molecule to it to form an inactive compound:

indoleacetic acid + aspartic acid →
indoleacetylaspartic acid (inactive)

This latter system functions especially to prevent auxin concentration from rising too high. In fact, supplying high concentrations of auxin to tissues frequently results in the synthesis of the enzyme system responsible for the inactivation of IAA.

Apical Dominance

In the typical pattern of growth, the growing tip of the stem with its young leaves exerts an inhibitory effect on the sprouting of lateral buds on the stem below the apex and on subsequent growth of lateral branches that do sprout. This phenomenon is called **apical dominance.** The relative effectiveness of this dominance of apical growth over lateral growth varies with the distance from the tip of the stem, the age of the plant, the genotype, and nutrition and other environmental factors. A plant with strong apical dominance has little or no branching, for example, the sunflower. Weak apical dominance leads to a bushy appearance with numbers of side branches, as in tomato plants. In grasses, branching occurs at the very basal nodes of the stem. These branch shoots are called **tillers** and the process tillering. Relative rates of tillering are important characteristics of cereal grain varieties.

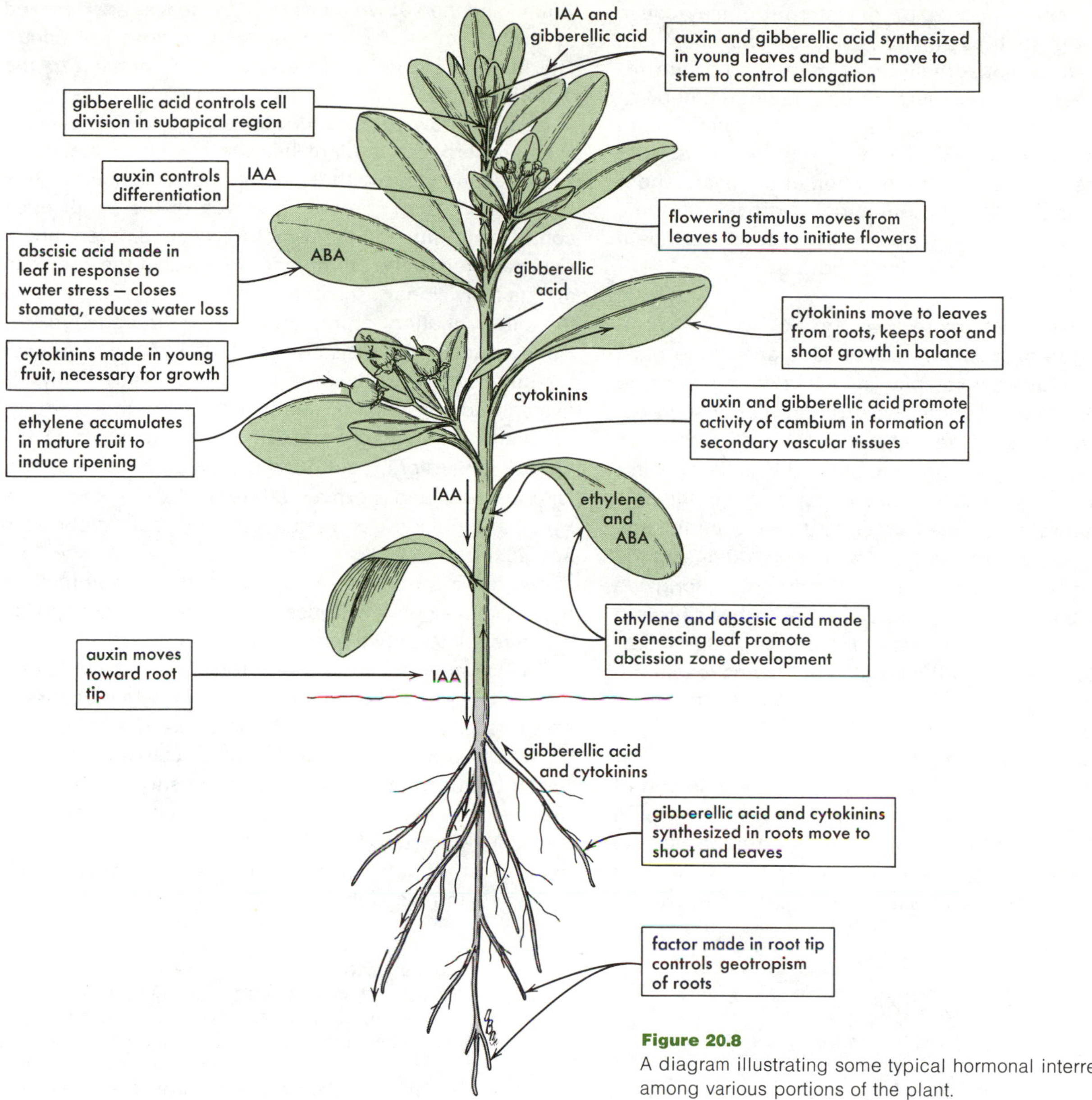
IAA and
gibberellic acid
auxin and gibberellic acid synthesized
in young leaves and bud – move to
stem to control elongation
gibberellic acid controls cell
division in subapical region
auxin controls
differentiation
IAA
flowering stimulus moves from
leaves to buds to initiate flowers
abscisic acid made in
leaf in response to
water stress – closes
stomata, reduces water loss
ABA
gibberellic
acid
cytokinins move to leaves
from roots, keeps root and
shoot growth in balance
cytokinins made in young
fruit, necessary for growth
cytokinins
ethylene accumulates
in mature fruit to
induce ripening
auxin and gibberellic acid promote
activity of cambium in formation of
secondary vascular tissues
IAA
ethylene
and
ABA
ethylene and abscisic acid made
in senescing leaf promote
abcission zone development
auxin moves
toward root
tip
IAA
gibberellic acid
and cytokinins
gibberellic acid and cytokinins
synthesized in roots move to
shoot and leaves
factor made in root tip
controls geotropism
of roots

Figure 20.8
A diagram illustrating some typical hormonal interrelationships among various portions of the plant.

The existence of a reservoir of suppressed lateral buds provides performed young shoot tips that are capable of rapid assumption of growth to replace a damaged stem tip. Experiments with pea seedlings have shown that by about 4 hr after removal of the growing stem tip there is the beginning of increased metabolic activity in the lateral bud next below the tip. One of the early events in the activation of lateral buds is the differentiation of the xylem to form a connection between the bud trace and the xylem in the bud itself. Completion of vascular connections to the stem is followed by growth of this new branch.

Removal of the apical bud or shoot tip removes the source of an inhibiting influence that passes down the stem. If the tip is replaced by a supply of auxin in a lanolin paste or agar block, the lateral buds remain inhibited. Thus, auxin can "*replace the step tip.*" If a cytokinin is applied directly to the lateral bud that is suppressed by the presence of an active shoot tip, outgrowth of the lateral bud is stimulated. After the branch has started to grow in response to cytokinin, its continued elongation is dependent on its own supply of auxin in the branch tip. Application of auxin to the recently sprouted lateral bud increases its growth. Thus, auxin functions in different ways before and after sprouting of the bud.

Auxin coming from the stem tip has been shown to inhibit the formation of vascular connections to the lateral bud. Cytokinins, on the other hand, promote the vascular differentiation. It is probable that the cytokinins originate in the roots and move up in the xylem sap. While we know some of the major aspects of the control system, it is not certain that differentiation of vascular connections is the first event that occurs in the bud as it is released from the influence of the apex.

There are many other "growth correlations" where the growth and development of one plant part is related to that of another (Fig. 20.8). Apical control may cause lateral organs such as branches, leaves, rhizomes, and stolons to grow horizontally or at some angle with respect to gravity that is different from the upright habit of the leader shoot. Removal of the dominant leader is often followed by bending of a nearby branch into an upright direction as it takes over the function of dominant shoot (**Fig. 20.9,** page 409). The direction of growth of some rhizomes and stolons has been shown to be dependent on hormones coming from the shoot tip.

Cell Differentiation in the Xylem

Differentiation of tracheary elements is one of the most extensively studied examples of cell differentiation in plants. The results of differentiation are easily seen because the changes in the cell are extreme. They result in extensive wall thickenings enabling the cells to withstand water stress, and they also result in loss of the protoplast to allow unimpeded water flow. One experimental approach has been to perform delicate operations on the shoot tip to study how young leaf primordia alongside the apex influence differentiation of procambial cells into tracheids or vessel elements. A leaf primordium promotes differentiation in the procambial strand leading to it and this promotory effect of the primordium can be replaced by auxin (Fig. 20.10).

In *Coleus* stems, a wound that severs a vascular bundle is followed by cell division and differentiation of tracheary elements from parenchyma cells in a path around the wound to reconnect the interrupted bundle. This regeneration process requires a supply of auxin that is normally transported polarly out of leaves above the wound and then down the stem. If the leaves are removed at the time of wounding, regeneration does not occur. The leaves can be "replaced" by auxin applied to the petiole stump.

Further down the stem where elongation has ceased, even an herbaceous plant like the sunflower may have some secondary growth from a vascular cambium. Here again, auxin from the young leaves of the shoot apex continues to function in promoting xylem differentiation. Auxin applied to the stump after removal of the apex can maintain divisions in the cambium and can also maintain the differentiation of the interfascicular cambium. In woody plants, cambial activation in the spring as growth is resumed involves the production of auxin by newly activated buds. A wave of cambial divisions moves down the stems stimulated by auxin moving down from the buds. Furthermore, continued cell division in the vascular cambium and normal differentiation of secondary xylem and phloem requires a supply of both gibberellin and auxin.

Further advances concerning the differentiation process have come with tissue cultures. It is now possible to regulate growth conditions, principally with auxin and cytokinins, to induce pea root cortex cells to differentiate into tracheids in tissue culture. As it occurs in these cells, the process includes an early synthesis of DNA, formation of polyploid nuclei, cell division, and finally rapid differentiation into tracheary elements. In these experiments, 16% of the new cells formed were tracheary elements. Doubling of the chromosome number is a part of differentiation of some other species and cell types, too, as is the case, for instance, in some root

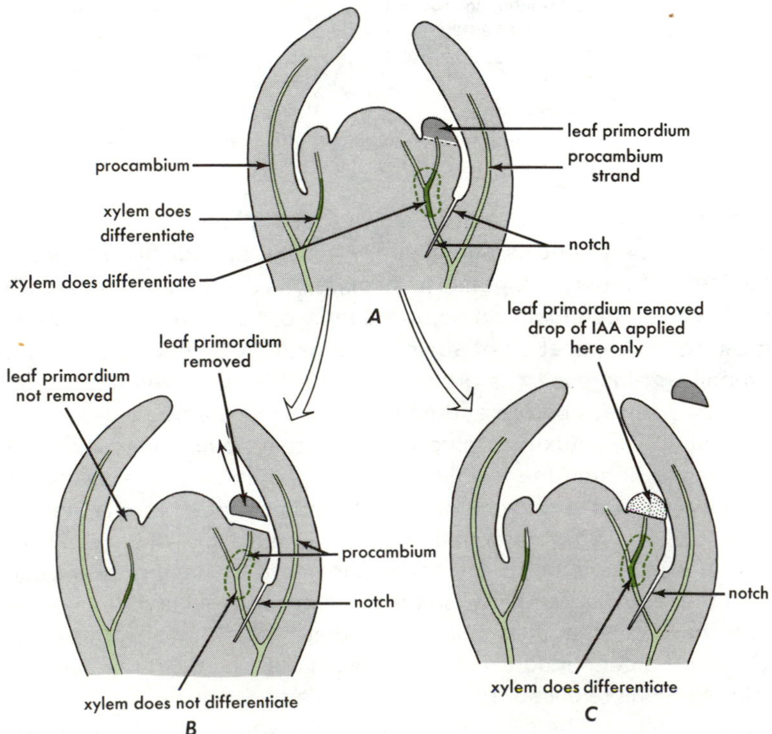

Figure 20.10
Diagram of an experiment showing how a leaf primordium provides a stimulus for xylem differentiation in the procambium. The notch severing the procambium isolates the tissue of interest above it. *A*, control with intact leaf primordium. *B*, leaf primordium removed. *C*, leaf primordium removed and a drop of auxin substituted. Auxin is effective replacement for stimulus from the primordium.

hair cells. The ability to induce specific cell differentiation at will in large quantities of cells provides a powerful means of studying differentiation and the role of hormones in the process.

GIBBERELLINS

Gibberellins and Stem Elongation

The shoot tip with its cluster of leaf primordia exerts control over growth activities below it in more ways than just supplying auxin. In sunflower, for instance, **gibberellins** (Fig. 20.6) can pretty well substitute for the shoot tip if it is removed. Removal of the shoot tip stops growth; gibberellins restore growth whereas auxin does not. Gibberellins are synthesized in the young petiolate leaves that surround the shoot tip. Sometimes they are made also in root tips, but it is not yet clear just what is the role of the gibberellins from the roots. Here we are dealing with the activities of a stem over a period of days. Gibberellins have been shown to promote cell division in the stem immediately below the apical meristem. It is in this region that enough divisions occur to provide a large fraction of the total cells involved in primary growth of a stem. There are synthetic chemicals, some of commercial agricultural importance, that act in the plant quite specifically to inhibit the synthesis of gibberellins. When these growth retardants (for example, CCC, AMO-1618) are applied to plants, cell division in the subapical region practically stops, whereas the leaf-initiating activity of the apical meristem continues. The result is a stem that essentially stops elongating but continues to make leaves separated by extremely short internodes. A plant of this form is called a **rosette,** whereas plants with elongate internodes are called **caulescent** (see both forms in Figs. 20.20, 20.23). The application of gibberellin to a retardant-treated plant causes the resumption of both cell division and elongation in the internodes and subapical region. Thus, gibberellins are an important factor in stem elongation.

Many plants naturally grow in a rosette form for at least part of their growth cycle. They typically respond to gibberellin application by rapid stem elongation, **bolting,** that is a part of flowering (Figs. 20.20, 20.23). As will be discussed later, the bolting response can be caused by an environmental factor such as cold treatment (winter) or long days, through triggering an increase in gibberellin.

As to the relation between auxins and gibberellins in stem elongation, it is apparent that the two frequently function simultaneously in the same stem. In some cases, gibberellins may stimulate cell division, thus providing more cells on which auxin can act. In the coleoptile, gibberellins act at an early stage of development, and the auxin-sensitive stage described above comes later in the life of the cells.

In Europe, commercial use is made of the growth retardant CCC or Cycocel (2-chloroethyltrimethyl ammonium chloride) to inhibit stem elongation in wheat plants. Shorter, stronger stems result and the plants are much more resistant to lodging, being knocked down by wind and rain. Lodging makes harvesting difficult. Short stems in grains are also achieved genetically in breeding programs and are important factors in Mexican wheats and Philippine IR8 rice, plants of the "green revolution," that have raised crop productivity in Asia and Central America.

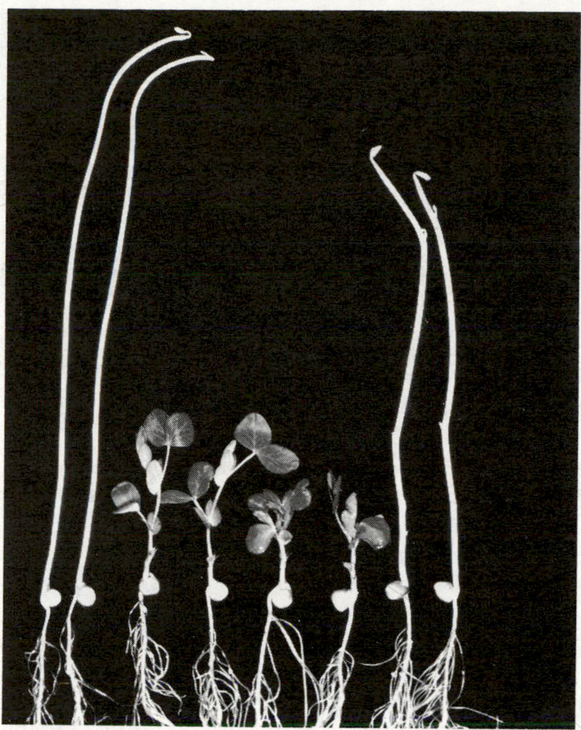

Figure 20.11
Dark- and light-grown pea seedlings of tall (left four plants) and dwarf (right four) varieties 8 days old. The etiolated plants were grown in the dark; the middle four were grown in the light.

"Dwarf" forms in a variety of plants are frequently due to a diminished gibberellin synthesis or to an enhanced production of a natural antagonist of gibberellin action. Dwarfism is sometimes dependent on light for expression, as in peas where both dwarf and tall varieties are equally tall in darkness (Fig. 20.11). Dwarf forms of corn and peas are used in estimating the concentration of gibberellins in extracts of plants or plant parts. Fig. 20.12 shows such a bioassay wherein quantities of gibberellins are estimated by comparison to the ability of known concentrations of the hormone to stimulate stem elongation. Bioassays have been instrumental in the purification and identification of all the plant growth hormones. The appropriate plant material may respond in hours or days to minute quantities of hormones that would be impossible to identify chemically.

Figure 20.12
Dwarf peas (*Pisum sativum*) showing the promotion of stem elongation by gibberellic acid applied 7 days prior to photograph. This response is used to bioassay the gibberellin contents of plant extracts. (Courtesy of L. Rappaport.)

Gibberellins and Enzyme Synthesis

As a grain such as barley or corn starts to germinate, the embryo begins to grow but has limited food reserves in itself. The main reserves are in the starchy endosperm, a collection of cells loaded with starch, reserve proteins, and some nucleic acids. A special layer of living cells, the aleurone layer, surrounds the main endosperm tissue and is instrumental in digesting the stored materials to soluble forms that can diffuse to the embryo. Early in germination, the embryo synthesizes gibberellin (GA) which diffuses to the aleurone layer where the GA acts as a signal to activate the synthesis and secretion of the enzymes necessary to digest stored materials in the endosperm. If the embryo is removed from the seed prior to germination, very little digestion of the remaining endosperm takes place. Adding gibberellin in minute quantities to the embryo-less endosperm induces the synthesis and secretion of enzymes just as if the embryo had provided the stimulus (Fig. 20.13).

The barley aleurone layer can be isolated and studied independently of the embryo and storage endosperm. It is a collection of cells with no growth activities and no division, but with a very active ability to synthesize a few proteins and to secrete them. As such it has been used to study the mechanism of action of gibberellin, and it has yielded much useful information, however, without as yet yielding the molecular basis for GA action. About 6 hr after adding gibberellin, the aleurone cells start to make α-amylase, an enzyme that breaks starch into soluble sugars, and other enzymes involved in breakdown of proteins and nucleic acids. The appearance of α-amylase is due to *de novo* synthesis (the enzyme is assembled from amino acids). This increased protein synthesis is due to the appearance of the specific messenger RNA coding for the α-amylase enzyme protein. This example of hormonal action via control of enzyme synthesis still is missing the identity of the original receptor molecule for the gibberellin. It is likely that the GA receptor interacts with the particular gene coding for the α-amylase messenger RNA. The proof or disproof of that hypothesis should come soon. Gibberellin applications causes a number of dramatic changes in the aleurone cells prior to the appearance of amylase. New ribosomes and endoplasmic reticulum membranes appear, as well as enzymes involved in the synthesis of membrane lipids. The addition of GA to the aleurone unleashes the development of the endosperm digesting system. We understand only parts of how it does this.

In the manufacture of beer, the starch stored in the grain endosperm must be hydrolyzed to a soluble form before its conversion into alcohol by the glycolytic enzymes of yeast. The natural production of amylase occurs in the malting barley. The addition of extra GA speeds up amylase synthesis by the aleurone enough to be used commercially.

CYTOKININS

Cytokinins, Cell Division, Organ Initiation, and Senescence

It has long been recognized that cell division in many tissues is blocked or limited by insufficient concentrations of some hormonal factor. Coconut milk was shown to be a rich source of a cell division-inducing factor. A suitable bioassay was developed in which parenchyma cells in chunks of tobacco pith tissue responded to the factor by division and growth. First, a synthetic, and later, naturally occurring cell division-stimulating compounds of a class called cytokinins were isolated, identified, and studied. Zeatin (Fig. 20.6) and isopentenyladenine (IPA) are two

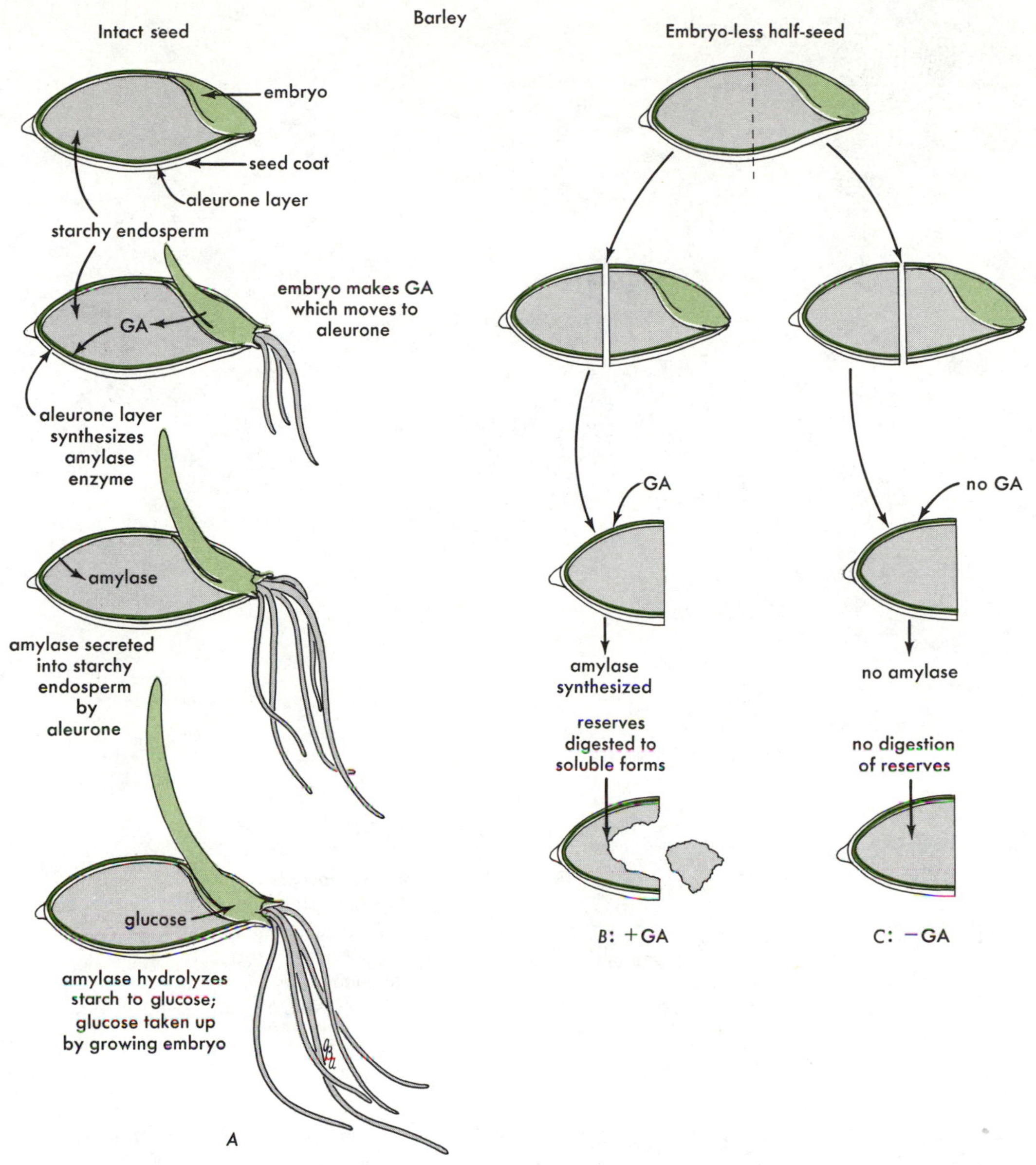

Figure 20.13
Diagram illustrating how gibberellin from the embryo induces the synthesis of the starch-degrading enzyme, α-amylase, in the aleurone layer.

naturally occurring cytokinins. Kinetin is a synthetic cytokinin. Actually, cytokinins exert control over many processes, and some do not involve cell division.

Cytokinins have been proved to occur in high concentration in rapidly dividing tissues, especially young fruits where they are apparently synthesized. Thus, corn grains at the milk stage are a rich source of zeatin, and young apples, plums, and other fruits contain high concentrations of cytokinins at the time when cell divisions are maximum.

The tobacco pith tissue culture has yielded much information on the hormonal control of morphogenesis. Freshly isolated cells will enlarge somewhat if supplied with nutrients and auxin, but will not divide unless minute amounts of a cytokinin are added. Thus, cytokinins can release the process of cell division. Moreover, by varying the balance between auxin and cytokinin, it is possible selectively to initiate the development of root or shoot primordia (Fig. 20.14). High auxin-to-cytokinin ratios cause root initials to differentiate; a suitably treated block of pith tissue can be covered with roots. A low auxin-to-cytokinin ratio causes differentiation of clumps of cells into apical meristems; the resulting shoots can cover the original block of pith. Intermediate concentration ratios stimulate a growth of relatively undifferentiated masses of cells, a callus tissue. Shoots can be removed, rooted, and grown to mature plants. Even single cells from the pith have been grown into mature plants, showing that a parenchyma cell contains all the genetic information necessary to form a plant: The origi-

Figure 20.14
Diagram of the control of differentiation exerted by interaction of auxin and cytokinin. Pieces of tobacco pith tissues were aseptically grown on nutrient medium (tissue culture) supplemented with various levels of the two hormones.

nal differentiation of the parenchyma cells did not involve a loss of genetic material.

Many observations point to a close balance between shoot and root growth. Certainly, a part of cytokinin physiology must involve the supply of cytokinins to the shoot through the xylem from the root tip where at least a portion of the plant's cytokinins are made. Cytokinins have been found in the xylem sap exuding from cut-off stumps of several species. If a leaf is removed from a stem, a sequence of changes is initiated. The changes can be called **senescence,** and they may include the breakdown of storage carbohydrates, cessation of protein synthesis, and an increase in the breakdown of proteins, nucleic acids, and chlorophyll. These changes resemble somewhat those typically occurring during natural leaf senescence. If the leaf is put under conditions where it forms adventitious roots, the senescence is halted. The roots can thus substitute for attachment to the plant. Furthermore, cytokinin application to the leaf can substitute for the roots, too, in preventing or delaying much of the senescence. In practice, the delay or prevention of chlorophyll loss in isolated leaves or disks cut from leaves is a sensitive bioassay for cytokinins.

The maintenance of active RNA and protein synthesis appears to be an essential part of cytokinin action in delaying senescence. It is easy to visualize a strong correlation between root and top growth involving cytokinins. Treatments that slow root growth, such as waterlogging or flooding the root zone, cut down the cytokinin concentration in the xylem sap.

A fascinating puzzle has developed concerning the mechanism of cytokinin action. Cytokinins occur free or simply combined with ribose or ribose phosphate, and they also occur as a functional part of *t*RNA molecules that carry certain of the amino acids to the ribosomes for assembly of proteins. It would be attractive to hypothesize that the dramatic effects one sees by adding cytokinins to the tobacco pith tissue or to isolated leaves are due to cytokinin action as part of *t*RNA involved in protein synthesis. But the available biochemistry says that one cannot add a preformed cytokinin and expect it to get into *t*RNA. The cytokinin in *t*RNA is made after the *t*RNA is assembled. One proposal is that while a cytokinin as part of certain *t*RNAs has a certain function to perform, the role of added cytokinins is to act in another way. This proposal suggests that free cytokinins are made in cells by breakdown of *t*RNA. This is just one example of the probability that successful application of the concepts of molecular biology to problems of plant development requires more skill than did the original development of the concepts with bacteria.

ETHYLENE

Ethylene as a Natural Growth Regulator

Ethylene is a simple gas, $CH_2{=}CH_2$ (Fig. 20.6), that is made in small quantities by many plant tissues in which it serves as a powerful natural regulator of growth and development. It is also a common product of combustion and a frequent air pollutant. It can create havoc in a florist's greenhouse, it is routinely used to bring about the uniform ripening of bananas, and in uncontrolled application it can lead to fruit spoilage. Since it is naturally involved in triggering the fruit-ripening process in so many fruits, it has been called the fruit-ripening hormone. More recent work has established many roles for ethylene in normal vegetative growth, too, so it has broad importance in growth regulation as do the other hormones we consider in this chapter.

The fact that ethylene is a gas requires a few explanations about how it normally functions and how it can be controlled or manipulated to modify its concentration in tissues for experimental or practical purposes. It is synthesized in the plant from the amino acid, methionine, a constituent of all cells. Since ethylene has only limited solubility in the aqueous phase of a cell, it volatilizes into the intercellular air spaces and diffuses away into the surrounding atmosphere as does carbon dioxide formed in respiration. Natural ethylene action in a tissue is dependent upon its continued synthesis. Its concentration in a tissue is determined by the relative rates at which it is synthesized and at which it diffuses away. If ethylene is introduced into the air surrounding a tissue, it diffuses in and raises the internal concentration to equal the external concentration. Enclosing plants or especially fruits in a container or space with restricted air circulation can cause the concentration of ethylene made by the plant itself to build up to the point that it has drastic effects on growth and development. Effective levels are frequently in the range of 0.1 to 1 part per million of air. Even the levels of ethylene present in urban air pollution are sometimes sufficient to cause measurable biochemical changes in plants. One successful method of achieving ethylene application under field conditions is to spray plants with a compound that spontaneously breaks down in the tissue to release ethylene. One such compound, chloroethylphosphonic acid, yields ethylene, chloride, and phosphate, which are all natural plant constituents.

An example of ethylene action in controlling vegetative growth is illustrated by a dicotyledonous seedling such as the pea during its early growth. As the plumule emerges from the germinating seed in the soil, the tip of the seedling shoot maintains a mode of growth that protects the delicate apical meristem and prevents young leaves from being damaged or torn off as the hook, the recurved upper portion of the stem, is pushed through the soil. A high rate of ethylene synthesis in the young plumule in darkness maintains the hook and prevents leaves from enlarging appreciably (Figs. 20.2, 20.3). The transition from the underground mode of growth to the above-ground mode is triggered by absorption of light by phytochrome. One of the changes induced by the activation of phytochrome is the shutdown of ethylene synthesis to a much lower level. The leaves are thus relieved of ethylene inhibition of expansion. The hook "opens" or straightens out to present the expanding leaves to sunlight, which further stimulates leaf development (**Fig. 20.3,** page 409).

In many ways, growth in darkness is essentially equivalent to growth underground. However, growth underground may provide additional problems—environmental cues and plant responses—that are not illustrated by simply growth in darkness. In peas if a restricting obstacle such as a soil crust or stone is encountered by the hook, pressures are generated in the tissue, and within a few hours a dramatic rise in ethylene synthesis can be measured. The result is an ethylene-induced swelling of the stem and an inhibition of elongation. Enlargement of the stem diameter enables it to exert a greater upward force against the obstacle. If the stimulus is prolonged, the stem geotropic response is modified and it starts to grow almost horizontally. These responses to extra-high ethylene increase the likelihood of penetration through, or growth around, a restriction and thus, successful emergence from the soil. They can easily be induced in dark-grown plants by gassing with ethylene (Fig. 20.15).

Ethylene may promote the germination of certain seeds.

In the growth of young trees, the relative height and diameter of the trunk is very much a function of the motion it undergoes as a result of wind. If a young tree stem is tied to a rigid stake, as is frequently the case in container-grown trees, or is supported by other plants in a crowded nursery, the usual response is a tall, slender, weak stem. If the stake is flexible enough to allow some movement in response to gentle winds, the plant has a sturdier form more like one growth without support. The plant that grows without support from a young stage typically has a shorter stem, larger in diameter, and with a tapered form coming from a wide base at the soil level. Routine exposure to high winds can greatly increase this response, as can be seen in trees on windy sea coasts or

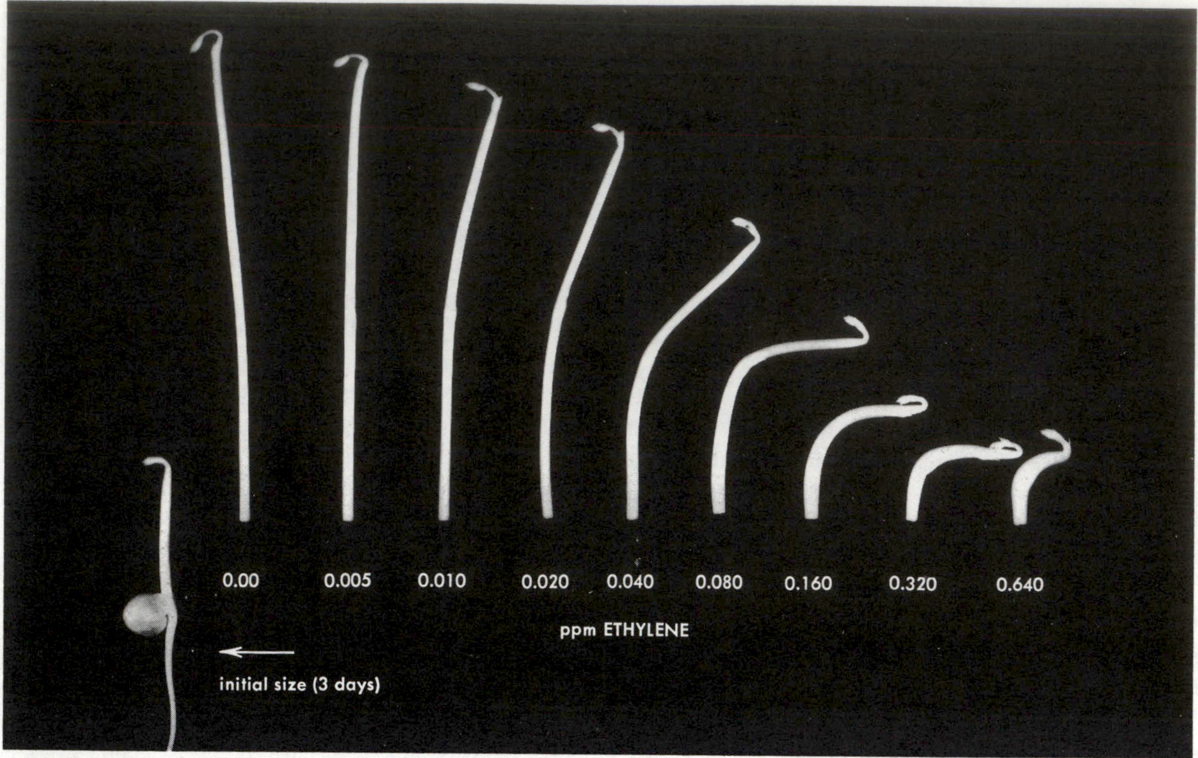

Figure 20.15
The response of dark-grown pea seedlings to various levels of ethylene during 4 days (ppm = parts per million ethylene in air). Internally generated ethylene under stress can induce similar responses. (Courtesy of J. Goeschel and H. Pratt.)

Figure 20.16
The effect of 28 consecutive days of brief, periodic shaking on the growth of "Bonny Best" tomato *(Lycopescicon esculentum)*. Left: unstressed control. Center: 30 sec of stress once daily. Right: 30 sec of stress twice daily. (Courtesy of C. Mitchell. *J. Am. Soc. Hort. Sci.* **100,** No. 1, 161. © 1975, American Society of Horticulture Science.)

mountain ridges. Controlled gassing of tree trunks with ethylene stimulates enlargement, and it is thought that stress-induced ethylene synthesis in the trunk contributes to this growth response.

Herbaceous plants may also be profoundly affected by mechanical stimulation. For example, tomato plants stimulated by shaking, flexing, or rubbing the plant axis, when compared to unstimulated control plants, may show reduction in node and leaf number, shortening of internodes, nodal swelling, epinasty of leaves, a deep greening of the leaves, and an increased lateral branch development (Fig. 20.16). An auxin–ethylene interaction appears to be involved in this mechanical alteration of growth. One possible explanation is that mechanical manipulation of the stem tissue stimulates ethylene production that in turn interferes with the auxin transport system.

Ethylene and Auxin Activity

Ethylene and auxin physiology are often intertwined. It seems to be nearly universal that high concentrations of auxins induce the synthesis of ethylene in tissues where ethylene synthesis is low. It is not easy to tell how frequently the endogenous production of ethylene is controlled by endogenous auxin, but many of the responses to the auxin type of herbicides are in reality responses to ethylene. In turn, ethylene inhibits the transport of auxin in many tissues. The result is an elaborate control of the effective auxin level in a tissue.

Two examples of the relatedness of some (but certainly not all) auxin and ethylene responses are the control of flower morphogenesis in cucurbits (cucumbers, squash, melons), and in the initiation of flowers in bromeliads (e.g., pineapple). A typical cucurbit plant first forms male (staminate) flowers, and at later nodes it forms perfect flowers. There are some strains that have only female (pistillate) flowers. It has been shown that applied auxin or some other treatment to raise internal ethylene favors the initiation of female flower parts on flowers that would normally lack them, or it tends to make pistillate flowers from normally perfect flowers. Gibberellin, either naturally high concentrations or artificially applied, promotes maleness, the initiation of anthers. There appears to be an ethylene–gibberellin balance that determines, at the time of flower initiation, which organs the flower will have. Being able to control the sex of flowers chemically has simplified the commercial production of hybrid seed by restricting the parentage of seed to the desired cross.

Pineapple plants can be induced to flower and thus form fruit on a precise schedule by applying either auxin or ethylene or an ethylene-releasing substance to the plant. This is of great commercial importance in coordinating harvesting and processing or shipping activities. On a somewhat more humble scale is the trick of inducing ornamental bromeliad houseplants to flower by putting them in a transparent airtight container with a ripe apple as an ethylene source. Bromeliads are seemingly unique in that ethylene is not generally an initiator of flowering.

Abscission of leaves and the role of abscisic acid in the process were discussed earlier, in Chapter 10. Ethylene treatment can also be very effective in causing abscission of leaves and fruits. Ethylene can interfere with the normal transport of auxin out of leaves. A regular auxin supply to the petiole is a necessity to prevent the formation or maturation of the abscission layer at the base of the petiole. When high ethylene concentrations inhibit the polar transport of auxin down the petiole, the abscission layer forms and the leaf soon abscises. An ethylene-releasing chemical is under investigation for promoting abscission of fruits for mechanical harvesting. Under certain conditions, it is successful for some fruits without producing too many undesirable side-effects.

Orchid flowers are in part valued for their longevity in cultivation as well as their beauty. In cultivation they are isolated from their natural pollinators and this contributes to their longevity. The packets of pollen grains, pollinia, normally carried by pollinators, are very rich in auxin, and when a pollinium or auxin is applied to the orchid stigma a rapid production of ethylene ensues that triggers the further development of the flower and the rapid senescence of the petals. This can be most disappointing to the orchid fancier.

Ethylene and Fruit Maturation

During the normal development of a flower, on through to a ripened fruit with mature seeds, there is a period that requires extensive growth and development of the embryo in the seeds. During this time, dispersal would be premature. There is in many fruits a programmed pattern of changes that convert the fruit from a seed-manufacturing to a seed-dispersal organ. One change frequently is a color change from green to yellow, orange, red, or blue; this increases its visibility to potential animal dispersal agents. Chlorophylls are broken down and other pigments such as the red and blue anthocyanins are synthesized. Another change is the conversion of starch and organic acids to sugar so that the fruit become sweet. Some of the materials in the cell walls are broken down so that the cells become more loosely bound to each other and the fruit is softer. Volatile flavor components are synthesized, and these contribute much of the flavor we value. One important link in this grand plan is the rise in ethylene to a critical concentration within the fruit. When this critical concentration is reached, the whole pattern of changes we call **ripening** is initiated. In many fruits, ripening can be hastened or at least made to occur uniformly by applying ethylene to fruits that are fully enough developed but have not attained the critical internal level of ethylene to trigger the ripening process.

The whole battery of changes requires extensive

synthetic activities by the fruit. This requires the expenditure of considerable metabolic energy, as is shown by a large increase in respiration while these changes are being made. This rise is called the **climacteric rise** in respiration. It can be measured as CO_2 production, O_2 consumption, and a sudden rise in ethylene production well over the rise that triggered the process initially.

It is standard commercial practice to pick bananas green in the tropics, transport them to storage rooms near the point of use and then treat them with ethylene gas to ensure that all the bananas in the bunch ripen at the same rate. They will ripen in about 5 days.

Ethylene may cause ripening when it is not wanted. In fruit held in cold storage, elaborate precautions are used to prevent ethylene from accumulating in the atmosphere to a point where it would trigger ripening. When a fruit goes through the phase of rapid ripening, the production of ethylene may increase many times over the rate that was initially necessary to reach the critical triggering level. Thus, "one bad apple in a barrel" producing ethylene may trigger the ripening of the rest. Or the ethylene produced by the mold on an orange may be enough to trigger undesirable changes in the rest of the oranges.

ABSCISIC ACID

Abscisic Acid and Stomatal Closure

The role of abscisic acid (ABA) (Fig. 20.6) in regulating abscission was outlined in Chapter 10. There are other roles played by ABA that make it a part of a rapidly responding system sensitive to water stress. When transpiration exceeds water absorption, the water content of leaves decreases. Turgor is lost and wilting ensues. It has been shown that in less than half an hour, under water stress, the leaf may begin to make and accumulate abscisic acid. Within a few hours, a manyfold increase in ABA concentration has been observed in a number of species. It has also been shown that ABA fed to leaves with open stomata can start stomatal closure in times as short as 5 minutes. Abscisic acid appears to act by interfering with the uptake or retention of potassium (or sodium) in guard cells. Since the high guard cell turgor required to maintain open stomata cannot be maintained without potassium ions, the stomata close. When water loss by transpiration is thus slowed, turgor in the leaf as a whole is regained, and the leaf recovers from wilting. The high level of ABA is metabolized away in a day or two and stomata begin to open normally.

Abscisic Acid as an Inhibitor

Another role for abscisic acid probably accounts for its frequent occurrence in high concentration in fleshy fruits. Abscisic acid is a potent inhibitor of seed germination, without being injurious to the seed. It probably functions in preventing germination in the fruit. It may carry over in the seed after release from the fruit and serve as part of the mechanism that prevents premature germination. It may be removed from seeds by the leaching action of rains.

Abscisic acid is a potent inhibitor of growth in many systems, but it differs from many other inhibitors in being nontoxic and that its inhibitory action is easily reversible with GA or auxin, depending upon the system.

THE USE OF GROWTH REGULATORS IN THE CONTROL OF GROWTH

We have discussed some specific characteristics of five types of plant hormones. This knowledge of hormone action has been very effectively applied to the practical control of plant growth. A very few of many possible examples in which growth-regulating chemicals are successfully used are: the selective killing of dandelions in a grass lawn; the inhibition of fruit ripening; the stimulation of growth of larger fruit. In addition to the examples we have discussed already, we will consider a few more examples in detail.

Control of Sprouting

We have seen how auxin can suppress the growth of lateral buds. This property may be used to advantage in controlling sprouting. Undesirable bud development may be delayed or inhibited by the application of certain auxins. For example, potato tubers exposed to the vapors of a synthetic growth regulator do not sprout in storage.

Rooting of Cuttings

Cuttings of certain species normally root slowly or produce a very small number of roots. It is now possible in some of these species to secure vigorous root production (Fig. 20.17) by treating the cuttings with various auxins, such as indoleacetic acid, indolebutyric acid, β-naphthoxyacetic acid, and numerous substituted phenoxy and benzoic acids. These growth regulators in proper concentration initiate cell divisions and stimulate the production of root primordia in the cuttings. The bases of the cuttings are immersed in the growth-regulator solution for a period of 12 to 24 hr, the strength of the solution ranging from 1 part of the substance in 5000 to 100,000 parts of water. The length of treatment and strength of solution that give optimum rooting vary with the species.

Root formation in cuttings of *Hibiscus* was found to depend upon a combination of indoleacetic acid and sugars and nitrogenous substances. The sugars and nitrogenous substances are contributed by the leaves, which explains the fact that rooting is very slight, even when cuttings are treated with a growth regulator, unless leaves are present.

Of significance is the fact that certain growth regula-

Figure 20.17
The promotion of root initiation by the synthetic auxin, indolebutyric acid, on American holly *(Ilex opaca)* cuttings. Row *A*, cuttings that stood in an aqueous solution of 0.01% indolebutyric acid for 17 hours before being placed in sandy rooting medium. Photographed after 21 days. Row *B*, untreated controls. Courtesy of U.S. Department of Agriculture.)

tors affect the course of differentiation. For example, short segments of the tap root of dandelion normally develop buds from the upper cut surface, and callus and roots from the lower cut surface. If the auxin, indolebutyric acid, for example, is applied to these segments, roots differentiate from both cut surfaces; or, if the growth regulator in tissues is decreased by proper treatments, leaves can be caused to differentiate from both ends of the segments.

Tissue Culture and Plant Propagation

Very recently there has been a dramatic practical application of the knowledge of how to initiate cell division and organ formation by treating pieces of plant tissue with cytokinins and auxins. Using tissue culture methods (see Chapter 15), desirable ornamental varieties, difficult to propagate fruit tree rootstocks, virus-free plants, and even the most desirable douglas fir timber trees have been multiplied in large numbers. Although horticulturists have vegetatively propagated plants for hundreds of years, there are many plants for which the classical methods did not work at all or were so slow as to be impractical. It is now possible to initiate rapid growth in tissue cultures (as in Fig. 20.14), manipulate hormone concentrations to cause the initiation of many little shoots, root cuttings of these shoots, plant the rooted cuttings in soil, all in the space of a few months on a scale of thousands or even a million plants.

Control of Fruit Development

The development of fruits, particularly tomatoes, without pollination has been induced by application of certain growth regulators. A high percentage of seedless fruit results when the flowers are treated before pollination; but flowers treated after pollination set fruit, most of which have seed. Substances employed for this purpose include the auxins indoleacetic acid, β-naphthoxyacetic acid, and 4-chlorophenoxyacetic acid. They are readily available synthetic compounds.

Control of Abscission

Certain substances are effective in retarding the preharvest drop of fruit such as apples, pears, and citrus. After treatment with these chemicals, the fruits cling to the trees for a number of days longer than normal and may attain more satisfactory color and maturity. Growth regulators effective for this purpose are the auxins, naphthaleneacetic acid, and 2,4-dichlorophenoxyacetic acid (2,4-D). The growth regulator apparently retards the processes that result in the formation of the abscission zone at the base of the petiole (p. 195).

Growth Regulators as Herbicides

Certain synthetic growth regulators, although stimulative in extremely small quantities, seriously disturb physiological processes in plants when added in larger amounts. Some of these compounds have turned out to be very potent weedkillers. Although not all growth regulators can be used to kill weeds, the so-called "phenoxy" compounds are particularly effective. Because these compounds are selective, killing broad-leaved plants and leaving the grasses relatively unharmed, they are particularly useful for destroying many kinds of weeds growing in combination with various grasses or crops of grain. Thus, it is possible to spray a lawn with 2,4-D (2,4-dichlorophenoxyacetic acid) and, without injuring the grass, kill dandelions growing in it. The use of a related compound, 2,4,5-T (2,4,5-trichloropheoxyacetic acid) has been prohibited because of possible harmful side-effects to humans. It appears that it is an impurity, dioxin (a powerful poison), in some samples of these chemicals, rather than the chemicals themselves, that is harmful to man.

The fact that growth regulators are translocated throughout the plant in the phloem makes these herbicides particularly effective in killing deep-rooted perennial plants. Studies of the factors affecting their movement in plants have been greatly facilitated through the use of radioactive isotopes. Figure 20.18 shows the movement of amitrol (ATA) containing radioactive carbon through a plant.

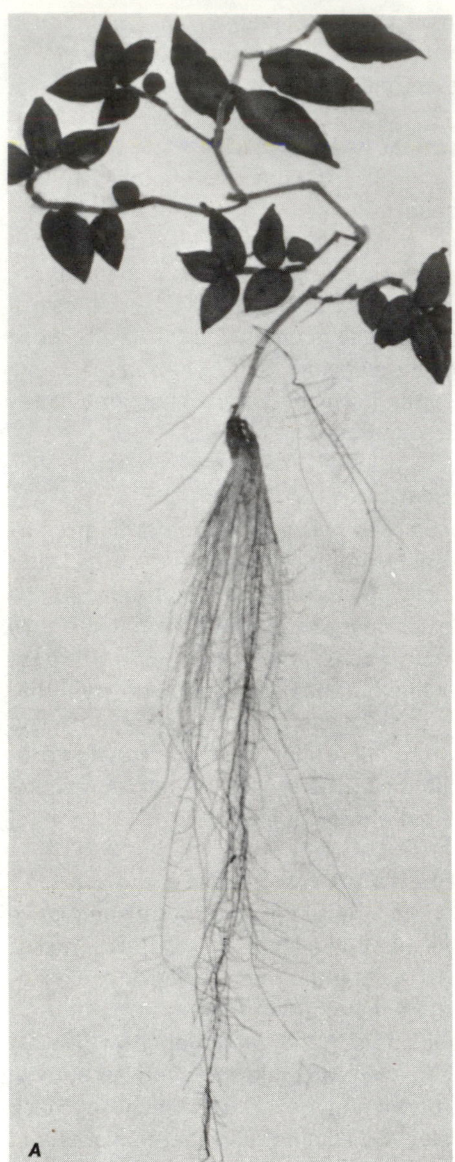

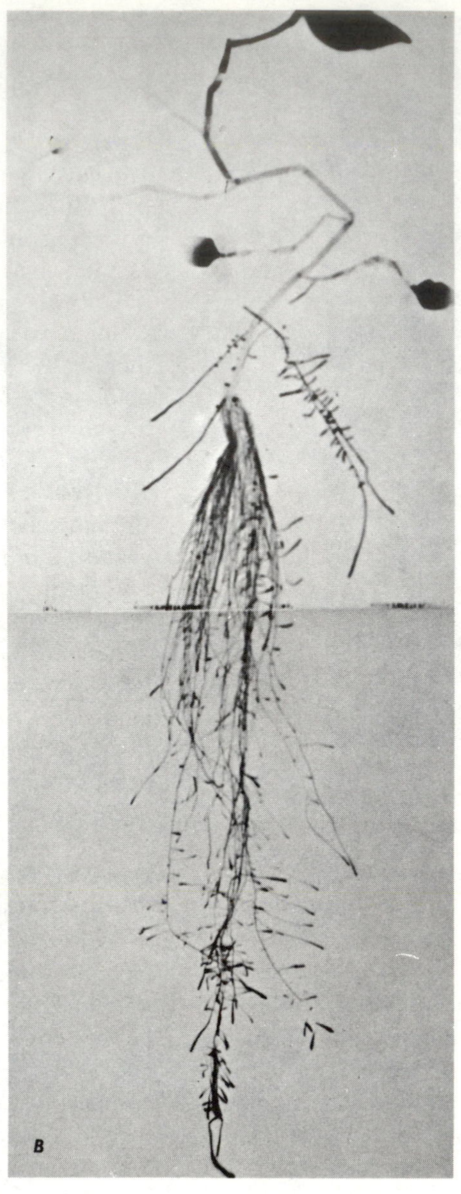

Figure 20.18
A, silhouette of the entire *Zebrina* plant; *B*, autoradiogram of the *Zebrina* plant showing the path of movement of ATA (amitrol) containing radioactive carbon. ATA was applied to the leaf at the upper right. (Courtesy of A. S. Crafts.)

There are numbers of mechanisms through which herbicides kill plants or parts of plants. A few mechanisms can be described in some detail, while others cannot. Effective doses of auxin-type herbicides, such as 2,4-D, produce growth changes that indicate they are doing many of the same things natural auxin can do. This includes the stimulation of abnormally high rates of ethylene synthesis and/or stimulation of cell division in the phloem region with blockage of phloem transport.

Other herbicides may block specific metabolic reactions. The phenyl urea type of compounds reach a specific site in the electron transport chain, within chloroplasts, resulting in a blockage of the oxygen-evolving reactions of photosynthesis. Other nonselective herbicides, such as Paraquat or Diquat, are in themselves not very damaging, but when they get into chloroplasts they are reduced and the resulting free radicals do great damage to cell constituents. These compounds require that the plant be exposed to sunlight to exert their effect.

There are many bases of selectivity whereby one plant is killed and another survives application of herbicides. Again, some modes of resistance are well-understood while many more are not. The most important method of achieving selectivity is probably by applying precisely controlled amounts of herbicide that have been found, by trial and error, to be effective in killing some, and not other, plants. The reasons it takes less of a compound to kill one plant than it does another may have to

Figure 20.2
Pea seedlings grown in light and in darkness. Note minimum stem elongation and maximum leaf development in light. The center two plants have had only 1 day in light, which caused straightening of the stem (hook opening) and start of leaf expansion and greening, as compared to the fully etiolated seedlings on the left.

Figure 20.5
Corn *(Zea mays)* leaf development. A series showing the leaf enclosed in the coleoptile, emergence from the coleoptile in darkness where unrolling is prevented, an artificially unrolled leaf, and a similar leaf grown in light, fully unrolled with greening completed.

Figure 20.3
Pea leaf development as a function of light. Fully developed green leaves from plants in continuous light; early stage of development after 1 day in light (center); etiolated leaf from dark-grown plant protected by plumular hook. All 8 days old.

Figure 20.4
Corn seedlings grown in light or in darkness showing the long mesocotyl growth that pushes the coleoptile and its enclosed first leaf and shoot apex to the soil surface. In the light-grown seedlings the mesocotyls are only a few millimeters long.

Figure 20.9
Apical dominance and negative geotropism in shoots of *Photinia* sp. Right: upright shoot showing apical dominance and suppression of lateral bud growth. Left: shoot that has been bent. Note development of lateral shoots and their negative geotropic response.

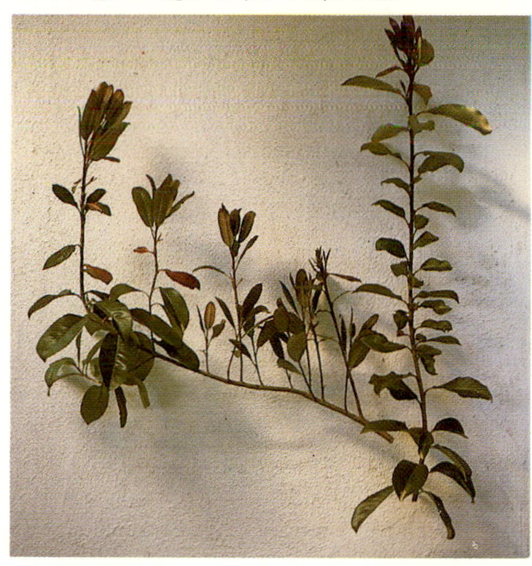

Figure 20.22
"Sleep" movements of silk tree *(Albizia julibrissin)* leaves. *A,* 3:00 p.m. *B,* 7:30 p.m.

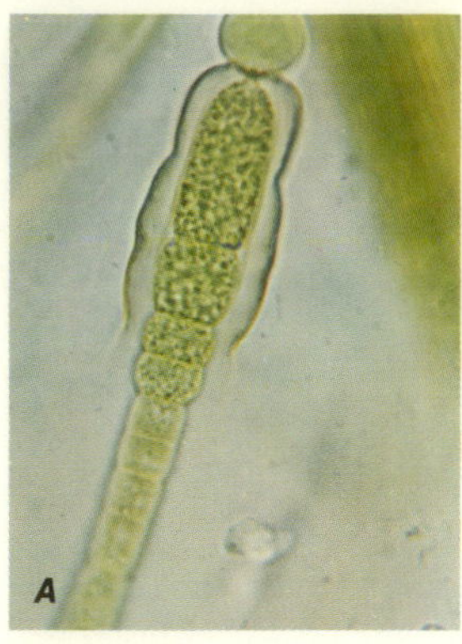

Figure 21.22
A, Gloeotrichia, $\times 400$. *B*, colony of *Oscillatoria*, $\times 125$. (*A*, courtesy of N. Lang.)

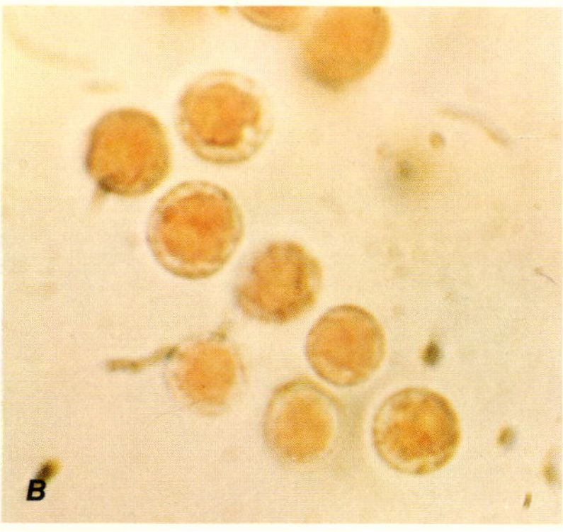

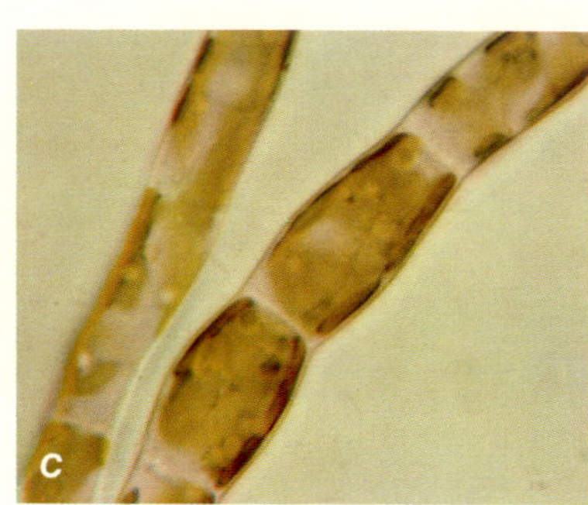

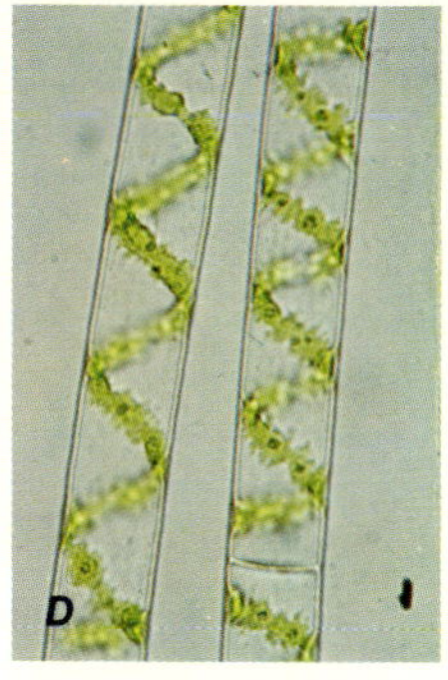

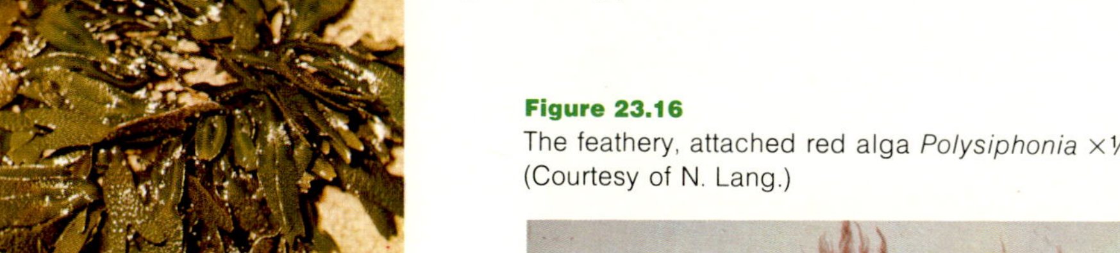

Figure 23.2
Algal diversity. *A*, mat of *Cladophora* in a shallow pool, $\times \frac{1}{10}$. *B*, *Porphyridium*, $\times 1000$. *C*, several cells of the brown alga *Pylaiella*, $\times 200$. *D*, section of filament of the green alga *Spirogyra*, $\times 400$. *E*, the brown alga *Fucus*, $\times \frac{1}{5}$. (*C*, courtesy of David Brown; *D*, courtesy of N. Lang.)

Figure 23.16
The feathery, attached red alga *Polysiphonia* $\times \frac{1}{2}$. (Courtesy of N. Lang.)

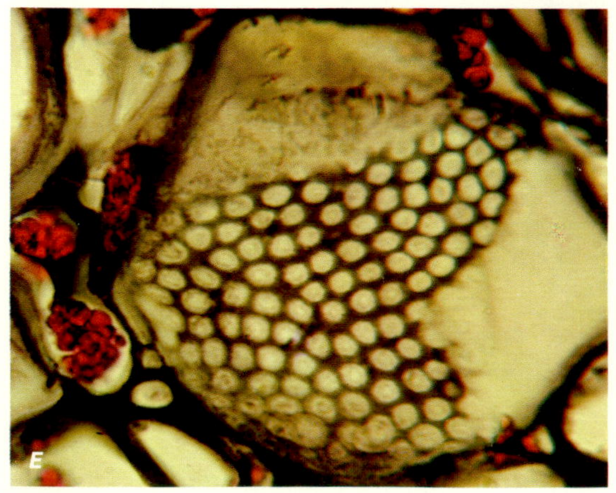

Figure 23.20
E, detail of the end wall of a sievetube-like cell, showing pores through which mannitol and other substances can flow in the stipe of the brown alga, *Macrocystis*, $\times 1000$. (Courtesy of Knute Fisher.)

do with the amounts sticking on the surface of two plants, the amounts penetrating the cuticle and entering cells, the amount and pattern of translocation, the relative rates of breakdown or detoxification in the plants, and the abilities of different plants to recover after portions of the plant have been severely injured. Successful use of herbicides requires highly trained people, and it is most effective in crop situations, where one is concerned only with the enhanced survival of one species.

VITAMINS AND THE REGULATION OF GROWTH

Extensive research in the field of animal nutrition has shown that animals, including man, may suffer from certain so-called vitamin deficiencies. The diet may contain water, mineral salts, carbohydrates, fats, and proteins in sufficient quantities to supply their energy requirements, but something may be lacking which is necessary for normal development and good health. If this something, which we call a **vitamin,** is added, even in the minutest quantities, the individual recovers from the deficiency symptoms.

Today, a number of different vitamins have been discovered, each with specific physiological roles. Not only do vitamins function in animal metabolism, but also, many of these same chemicals have been found to regulate physiological processes in plants as well. One of the chief difficulties in studying the functions of vitamins in plants is that, in contrast to animals that are unable to synthesize vitamins, green plants in general produce their own supply. Thus, it is difficult for a plant physiologist to produce a vitamin deficiency in a plant. If the plant manufactures the vitamin, the physiologists cannot easily deprive it of the vitamin.

Certain organs in plants, however, are unable to produce all the vitamins they need. The culture of these organs isolated from the plant has led to a better understanding of the role that vitamins have in plant growth. As an example, let us look at the influence of vitamins on root growth.

Being devoid of chlorophyll, roots are unable to synthesize sugar, and hence are dependent upon the green organs of the plant. We may ask: Do the leaves produce substances, other than sugar, essential for the normal growth of roots? If growing root tips are cut off and placed in a medium containing water, essential mineral salts, and sugar, the isolated roots will not continue to grow normally. They will grow for a prolonged period of time, however, if an extract of yeast cells is added to this medium. The growth substances in yeast extract are shown to be **thiamine, nicotinic acid,** and **pyridoxine,** all part of the vitamin B complex. The living cells of the roots of some species of plants cannot synthesize thiamine, nicotinic acid, or pyridoxine. These substances are manufactured in the leaves under the influence of light and move from them to the roots. Roots of other kinds of plants can produce pyridoxine but not nicotinic acid or thiamine.

Thiamine is widely distributed in the plant kingdom. It has been found in many kinds of bacteria, in yeasts and other fungi, in both fresh-water and marine algae, and in mosses, liverworts, ferns, and the higher plants. Some bacteria, yeasts, and other fungi are able to synthesize thiamine, whereas others are not. Almost all the thiamine in a wheat grain is in the bran. It is now common practice to enrich white flour with thiamine.

Young plant tissues of higher plants, principally the leaves, are particularly active in synthesizing thiamine. From these tissues it moves to other parts of the plant. That thiamine is produced in leaves and translocated through the phloem to roots may be shown by removing a girdle of bark from the stem of a plant. It will be found that thiamine accumulates above the girdle. If the girdle is made on the petiole between the leaf and stem, thiamine accumulates in the leaf. As stated, thiamine is not produced in roots (at least in most plants that have been studied), but it is essential for root growth. Thus, thiamine has some of the characteristics of plant hormones in that it is produced in one part of a plant, and translocated to another part, where in small amounts it is very effective in influencing growth. However, it differs from the hormones discussed above in that thiamine, nicotinic acid, and pyridoxine all are cofactors (p. 72) of known enzyme reactions. Thus, thiamine may be classed as a root-growth hormone; it plays a role in the normal formation of roots in many seed plants.

Carefully controlled experiments lead us to believe that most ordinary plants growing with their roots in a soil containing the required mineral nutrients, and with their leaves in light, synthesize sufficient quantities of thiamine in the leaves to meet all requirements of the plant. Seeds usually are relatively rich in thiamine. Thus, the seedling probably has an adequate supply of the vitamin to support normal growth until such time as newly formed leaves are synthesizing the substance. Cuttings, however, may be deficient in thiamine. A few leaves on stem cuttings promote root growth, not only because they contribute carbohydrates to the roots, but also because they may contribute other growth-influencing factors.

LIGHT AND THE CONTROL OF DEVELOPMENT

It is not particularly surprising that green plants should have evolved mechanisms to ensure placement of their leaves to maximize light capture. What is remarkable, however, is the variety of developmental processes into which are woven light-mediated controls. These controls are sensitive to the presence or absence, direction, daily duration, and spectral composition of the light in the plants' environment. Any light-mediated event requires a pigment to absorb the light and to be altered by that light absorption long enough to react in some biochemical system in the cell. Only a very small fraction of the kinds of chemicals making up a living plant cell absorb visible light. These light-absorbing chemicals are called pigments. Photosynthetic pigments are present in high concentrations and contribute color to tissue, but growth-controlling pigments are low in concentration and do not impart a noticeable color to tissues. There are two well-characterized growth-controlling pigments, **phytochrome** and **protochlorophyll,** plus a third, a **phototropic pigment** that is not chemically identified, and possibly several other distinct but less well-known pigments. Most of the remaining pages in this chapter will be devoted to phytochrome. We will discuss protochlorophyll in relation to the greening process at the end of the chapter.

Phytochrome is a remarkable pigment that seems essentially universal in vascular plants. It is a protein that exists in two stable forms, active and inactive, and the two forms are different in color. The inactive form absorbs red light and is called P_{Red} or P_R, and the active form is called $P_{Far\text{-}red}$ or P_{FR}, because it absorbs far-red light just at the extreme of human visual sensitivity. When either form absorbs light, it is converted to the other form. Since P_R absorbs red light most efficiently, red light (660 nm wavelength) is most efficient in converting P_R to P_{FR}. Conversely, far-red (730 nm) light converts P_{FR} to P_R.

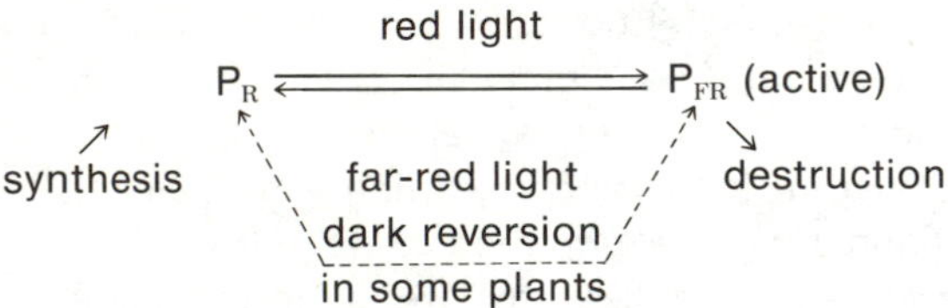

When chlorophyll absorbs light, the change induced in the molecule lasts about 1 billionth of a second (10^{-9} sec). When phytochrome absorbs light, the ultimate effect is long-lasting. P_R is stable indefinitely in the dark, and P_{FR} may exist and exert its effect for many hours, even in darkness, before being inactivated or spontaneously converted back to P_R. Thus, a little bit of red or far-red light can essentially switch phytochrome activity on or off. In contrast to chlorophylls and phytochrome, pigments of another class, the red- or blue-colored **anthocyanins** do not do anything with the light energy they absorb other than convert it to heat.

Phytochrome absorbs the various wavelengths of sunlight to different extents, the net result being that in sunlight a mixture of P_R and P_{FR} is reached. When exposed to artificial light, incandescent or fluorescent, the relative amounts of P_R and P_{FR} are quite different from that achieved by sunlight. Growth of plants is different in artificial light unless a mixture of incandescent and fluorescent light that approximates sunlight is used to achieve a mixture of P_R and P_{FR} similar to that in sunlight.

As mentioned earlier, some seeds do not germinate in darkness. After seeds are soaked, a brief exposure to sunlight or, more particularly, red light will activate their germination. Red light acts by converting phytochrome to the active P_{FR} form. P_{FR} relieves some as yet unknown block in metabolism, and the seed can go on to germinate. If far-red light is given immediately following the red treatment, P_{FR} is converted back to P_R before P_{FR} has enough time to act. Now the seeds do not germinate. The far-red reverses the red treatment. If, after the red treatment, a delay is introduced prior to the far-red reversing treatment, the P_{FR} is given more time to act, and it may do enough in a short time to enable germination to proceed. In that case, the far-red reversal treatment is ineffective as far as the physiological response goes, although it is still effective in converting the pigment back to P_R.

Thus, phytochrome functions in many seeds as an indicator of the presence or absence of light. Some seeds that are buried are prevented from germination by the accumulation in them of metabolites as they sit in the restricted air supply of the soil environment. Though initially they may not have had a light requirement for germination, they can acquire one under these conditions. That light requirement for germination is only satisfied when the soil is disturbed naturally or by cultivation.

The main signal to a seedling that it has emerged from the soil is the formation of P_{FR} by light. This P_{FR} causes changes in many growth processes in the stem and leaf. In dicotyledonous plants, ethylene synthesis is shut down to a low level in the apical meristem or terminal bud and leaf enlargement starts and the hook opens (**Fig. 20.3,** page 409). Stem elongation is generally inhibited by P_{FR}, and changes in the differentiation of xylem tissue lead to added strength necessary to raise and support leaves above ground. In grasses, light promotes unrolling of the first leaf as it breaks out of the ensheathing coleoptile (**Figs. 20.4, 20.5,** page 409). But the rolled-up leaf can attain considerable size in grasses in the dark, so that leaf expansion is not an important initial response in a typical grass. It appears that formation of P_{FR} rapidly leads to a release of active gibberellin from an inactive form and that increased GA is the more immediate cause of leaf unrolling. Applied GA can substitute for light in this particular response.

In some tissues, phytochrome-controlled differentiation can be followed at the biochemical level. For instance, irradiation of etiolated seedlings may induce the synthesis of the enzyme phenylalanine ammonia lyase that catalyzes an important step in the biosynthesis of lignin deposited in secondary cell walls. P_{FR} may also stimulate the synthesis of anthocyanin pigments that give apples their red color.

When a plant finds itself almost continuously in the shade of other plants, the quality of light falling on the plant is changed in a way that can markedly affect the phytochrome status of the tissue. As light passes through successive layers of leaves, the red wavelengths are filtered out by chlorophyll, while the far-red wavelengths pass through. In the shade created by other plants, light rich in far-red tends to markedly lower the P_{FR} concentration. The result is more rapid stem elongation and an increased probability of growing out from under the shade of other plants. Even in light-grown plants, phytochrome is present, and experiments can be devised to show how it exerts subtle controls on stem elongation. In this example, a change in light quality can trigger a response that has obvious survival advantages.

While the mechanism of action of phytochrome has not yet been established, some work suggests that it may be attached to membranes and exert some control over membrane function.

PHOTOPERIODISM

It was mentioned earlier (p. 392) that the photoperiod frequently provides the timing signal for flowering and other activities that prepare a plant to survive a season unfavorable for growth. The formation of dormant buds, resistance to freezing temperatures, formation of tubers and bulbs may be responsive to photoperiodic stimuli.

Photoperiodic control of flowering involves the perception of day length in the leaves. Under the right conditions, a chemical stimulus is made in the leaves, and moved out through the phloem to terminal or axillary buds, where it induces a change in the kind of organs being initiated, a change from initiation of leaf primordia to the initiation of flower primordia. In the case of photoperiodic control of bud dormancy, the leaves produce something that moves to the shoot apex, causing it to stop elongation and to make bud scales and the beginnings of next season's shoot. Similarly, a factor from the leaves may initiate tuber formation.

Photoperiod and Flowering

Plants that generally are induced to flower during the lengthening days of spring respond to day lengths longer than a certain minimum or critical photoperiod. These are called **long-day plants** (Figs. 20.19, 20.20). A long-day plant might flower after having been exposed to day lengths longer than 11, 12, or 13 hr. The exact critical photoperiod is a characteristic of each species. Examination of Fig. 20.17 will show that to initiate flowering around May 1 requires progressively longer critical day lengths at higher latitudes. Conversely, if a plant is stimulated to flower by 13 hr days in its native habitat, that critical day length will come earlier in the year the farther away from the equator the plant is moved. Such a plant might be poorly adapted to the season at more northerly (or southerly) latitudes because of flowering prematurely.

Figure 20.19
Diagram illustrating the response of photoperiodically sensitive plants to the seasonal progression of daily light and dark periods and how the day lengths vary with latitude. Approximate latitudes: Mexico City, 19°; New Orleans, 30°; San Francisco, 28°; New York, 41°; Chicago, 42°; Montreal, 45°; Paris, 49°; London, 51°; Oslo, 60°; Leningrad, 60°.

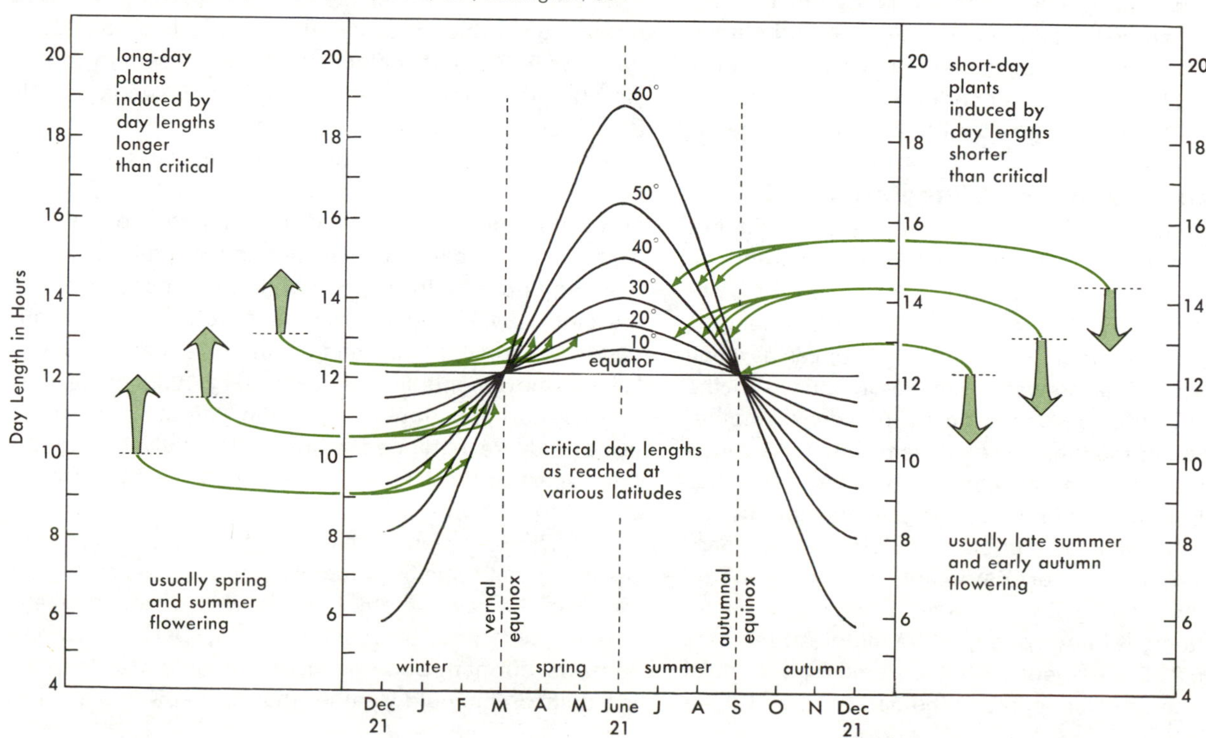

Figure 20.20
Effect of day length on behavior of the long-day plant henbane *(Hyoscyamus)*. Plants were grown with 8 hour photoperiod until they were about a month old and were then subjected to 24 photoperiods of 10, 11, 12, 13, 14, or 16 hours. Initiation of flowers and accompanying stem elongation (bolting) occurs on days longer than 12 hours. (Courtesy of Plant Industry Station, Corps Research Division, Agricultural Research Service, U.S. Department of Agriculture.)

Short-day plants are those whose flowering is initiated by days shorter than a critical length. They are generally induced by the shortening days of late summer or early fall. Short-day plants may be thought of as long-night plants, because it is the length of an uninterrupted dark period that is critical to timing (Fig. 20.21). Thus, a plant requiring days of less than 13 hr in reality is responding to nights of more than 11 hr.

The photoperiod may be either an absolute requirement or a partial requirement, causing flowering at an earlier time or with greater intensity. Many other plants indifferent to photoperiod are called **day-neutral plants.** Their flowering may be controlled either by other environmental factors such as temperature and water stress, or by a genetically determined pattern of development.

Phytochrome and the "Biological Clock"

By what mechanism does a plant measure time? First of all, the time-measuring system can be very accurate and quite independent of temperature. Some plants commonly respond quite differently to day (or night) lengths differing by half an hour and even to differences as small as 10 minutes. This precision enables response to the progression of the seasons to occur in 1 week. Another characteristic of the timing mechanism is that it involves phytochrome as the photoreceptor that indicates whether it is light (day) or dark (night). A long dark period that would otherwise be inductive (initiate changes leading to flowering) to short-day plants can be rendered ineffective by forming phytochrome P_{FR} with a short light treatment (a night interruption) in the middle of the dark period (Fig. 20.21). Evidently, phytochrome P_{FR} present at the end of the day disappears in the early hours of darkness as it spontaneously reverts to P_R. A short-day plant requires more than a minimum dark period without the active phytochrome P_{FR}. Conversely, a long-day plant is prevented from flowering if it has less than a minimum daily period with P_{FR} (light). So phytochrome P_{FR} is inhibitory to flowering during the night in short-day plants and is promotory to flowering in long-day plants.

The least well-known component of the time-measuring system is the "biological clock." The biological clock is a metabolic system that goes through a cycle once approximately every 24 hr. It serves to synchronize physiological, biochemical, and activity functions on a daily cycle. Biological clocks must be present in organisms ranging from single-celled algae to flowering plants and in animals including humans. It is clear that phytochrome and the "clock" interact to measure time, and under the proper circumstances the block to making the flowering stimulus is removed. The functioning of the biological clock can be observed independently of the flowering process in plants that link another process to the clock. Even under continuous light or continuous darkness, some plants can maintain a daily cycle of "sleep" movements of their leaves (**Fig. 20.22,** page 409). The leaves hang down during the "night" part of the cycle and are extended more horizontally during the "day" part of the cycle as determined by their internal clock.

Thus, many plants are able to time various activities on the annual cycle by virtue of accurately measuring the length of the day or night. Flowers are initiated in time to mature seeds prior to frost or drought. During favorable seasons, dormant buds are made capable of withstanding unfavorable cold weather that probably lies ahead.

Figure 20.21
Effect of day length on behavior of *Chrysanthemum*, a short-day plant. *A*, plant that received light of natural short days of autumn and blossomed at the usual time. *B*, plant that received an hour of light near the middle of each night for several weeks beginning just before flower buds would normally have been initiated; thus, each long dark period was divided into two short ones. This interruption was sufficient to delay flowering. (Courtesy of Plant Industry Station, Crops Research Division, Agricultural Research Service, U.S. Department of Agriculture.)

Supplemental Factors Controlling Flowering

One might inquire why a short-day plant does not respond to the short days of late winter or early spring prior to the time the days get too long to be inductive. There are generally supplemental factors controlling the development of plants in their natural habitat that keep a short-day plant from blooming in the early spring when the days might otherwise be short enough. For instance, many annual plants are not sensitive to the photoperiod until they have reached a certain size. The normal season for seed germination may preclude the plants attaining the necessary maturity prior to the arrival of the long days of spring. Thus, the only season in which it would be capable of responding under normal conditions would be late summer or early fall.

The Nature of the Flowering Stimulus

We have spoken of the flowering stimulus without indicating more of its nature other than it is something that is made in the leaves and moves through the phloem to the buds. Some experiments strongly suggest that the same flowering stimulus serves long-day, short-day, and day-neutral plants. They apparently differ only in the conditions required to bring about its synthesis. Much work has gone into attempts to extract and identify the promotory substance without substantial success. Recent work has shown that when the leaves of certain long day plants are held under short days, they produce a second controlling substance that is inhibitory to flowering. There thus appears to be a more complex control by both stimulatory and inhibitory factors that has to be unravelled before we understand flowering very well. Empirical testing has, however, uncovered procedures whereby known plant hormones and metabolites can, in some plants, be used to cause flowering under noninductive photoperiods or to prevent flowering under otherwise inductive conditions.

Gibberellins applied to many long-day rosette plants will produce flowering under short days and bring about a response similar to long days in the plants shown in Fig. 20.20. Gibberellins are, on the other, inhibitory to flower initiation in many other plants, especially woody plants. In woody plants, it is often possible to stimulate flowering by treating with growth retardants that inhibit gibberellin synthesis. Recall also that auxin or ethylene are effective in initiating flowering in pineapple. In some plants, abscisic acid application has induced flowering. The potential practical importance of being able to control flowering time in more crops cannot be overestimated.

VERNALIZATION AND FLOWERING

Another mechanism for timing flowering makes the passage of winter a prerequisite to flowering. Plants having this type of control include both winter annuals and biennials. Winter annuals typically germinate just prior to or during the winter so that the young plant passes through a period of weeks of exposure to low temperatures prior to the warmer weather of spring and summer when they flower.

Biennials generally have a full season of vegetative growth the first year, exposure to cold weather in a winter, and then flower the second spring and summer. They need to achieve a larger size before the cold treatment is effective. Without exposure to the low temperatures, neither group will flower, or flowering will be delayed (Fig. 20.23). Gibberellins can sometimes substitute for the cold requirement. This promotion of flowering in response to cold treatment is called **vernalization.** The cold treatment generally requires temperatures of 10°C or less for a period of weeks. Frequently, the cold

Figure 20.23
Substitution of gibberellic acid for the cold requirement (winter) in the flowering of the biennial carrot *(Dacus carota)*. *A*, control plant under long days only. *B*, long days plus gibberellic acid, no cold treatment. *C*, long days plus cold treatment, no gibberellic acid. (Courtesy of A. Lang.)

requirement is followed by a long-day photoperiod requirement that prevents the flowering from occurring too early in the spring. In vernalization, the growing tip may be the site of perception of the cold stimulus, and no transportable factor is involved.

"Winter" cereals, wheat and rye, are normally planted in the fall so that they germinate prior to winter and go on to flower promptly the next spring and summer after making a minimum number of leaves. If they are planted in the spring and are not vernalized, flowering is delayed until many more leaves are formed. "Spring" cereal strains are normally planted in the spring and flower promptly in response to long days. Winter cereal strains can be artificially vernalized prior to a spring planting date by moistening the germinated grains and holding them at about 1°C for a period of weeks.

DORMANCY, PHOTOPERIODISM, AND TEMPERATURE

In the discussion of photoperiodism, we indicated that many woody perennial plants use the photoperiod to induce a dormant condition in the bud prior to the onset of unfavorable weather. The bud typically goes through a period of increasing dormancy until it can no longer be rapidly reactivated by favorable environmental conditions or a shock treatment such as defoliation. By fall it may be in a state of rest wherein no treatment will activate it other than the passage of time at low temperature, a process called **chilling.** The satisfaction of a chilling requirement requires temperatures a little above freezing but generally below 10°C. Normally, the requirement is met by midwinter and at that time a branch brought into the greenhouse will have buds bursting in a few weeks. The buds are no longer in a condition of rest. Their chilling requirement has been satisfied. They are able to respond to the warming temperatures of spring with renewed growth. The photoperiod induced the dormant condition and the chilling process signaled that it was safe to respond to spring-like weather.

The chilling requirement for many deciduous fruit trees has been extensively studied. Typical varieties may require 250 to 1000 hr of exposure to chilling temperatures to produce a rapid uniform burst of flower and leaf buds in the spring. A variety planted in an area of insufficient chilling will have limited and sporadic bud opening and will set a poor crop of fruit. The lack of sufficient cold weather sets the southern limit for crops such as peaches, apricots, cherries, plums, and apples, and for effective flowering of ornamentals such as lilac (*Syringa vulgaris*).

Another temperature-related phenomenon is the development of **frost hardiness.** In the summertime, a few degrees below freezing will kill leaves and buds. During the fall, there can be (if the genotype provides the capacity to so react) development of resistance to temperatures below freezing. Short photoperiods and exposure to temperatures just above freezing lower the temperature at which buds and even leaves are irreversibly damaged. This cold resistance may progressively increase from −5°C early in the fall to −30° or even −50°C in the buds and twigs by late fall. This protects the buds and stems from the extreme cold and drying effects of winter weather.

Olive trees are evergreen and live in Mediterranean climates with mild winters. They make use of a chilling requirement for a different purpose, the initiation of flowers, rather than for the termination of rest. Thus, one environmental signal may be linked to the control of a variety of aspects of development. The pattern of linkage of signals and responses is one of the characteristics of an ecotype (Chapter 18) or population.

THE GREENING PROCESS

The process of greening, the formation of chlorophyll in leaves and young stems, etc., is precisely controlled to prevent the use of food reserves and energy in constructing the photosynthetic apparatus unless light is available for photosynthesis. Angiospermous plants do not make chlorophyll in the dark. Each molecule of chlorophyll is made by a reaction requiring the absorption of light by protochlorophyll hooked up to a special catalytic protein, the photochlorophyllide holochrome. In darkness, a small amount of this photoreceptor system is maintained in plastids with a minimum of structural and enzymatic differentiation. Without light there is very little development of this etioplast (Fig. 4.16*B*). When light is absorbed by the protochlorophyllide holochrome, chlorophyll is formed and some new protochlorophyll is made to replace the protochlorophyll that was used up. In a period of 36 or 48 hr an etiolated leaf can be converted to a fully green leaf (**Fig. 20.3, 20.4,** page 409). New membranes are formed in the plastid and the enzymes involved in the electron transport and carbon cycles (Chapters 14, 15) accumulate. Phytochrome frequently participates in this light-activated greening process, too. In a few hours after the start of illumination, photosynthesis can be observed.

SUMMARY

1. Plants are able to respond to a wide variety of environmental situations because they have systems for monitoring important environmental factors and they have programmed patterns of response to these stimuli.
2. The influence of one part of a plant on the growth and development of another part is important in keeping the plant growing as a coordinated unit. The correlation of activities in one part with another may be through food, mineral nutrient or water supplies, or by the production and distribution of specific hormones.
3. Plant hormones are natural compounds, usually transported from a site of synthesis to a site of action, which act in small quantities to regulate growth and development in many ways.
4. Auxin, gibberellins, cytokinins, abscisic acid, and ethylene are the important plant growth hormones.
5. Auxins promote cell enlargement by altering cell walls, making them more extensible. Growth is controlled by controlling the synthesis, transport, and inactivation of auxin. Phototropic and geotropic curvatures in the shoot are regulated by changing the transport of auxin to the growing cells.
6. Auxins also help control other developmental processes such as apical dominance, cell division, and differentiation of specific cell types in the xylem.
7. Gibberellins also control stem elongation through effects on both cell elongation and on division in the subapical region. Other developmental effects of gibberellins include promotion of the synthesis of specific enzymes.
8. Cytokinins are active in control of cell division in specific tissues, in maintenance of active nucleic acid and protein metabolism in leaves, and in correlating root and shoot growth.
9. Ethylene is important in controlling stem elongation along with auxin, in maintaining the etiolated growth habit in darkness, and in regulating the ripening process in many fruits.
10. Abscisic acid is important in regulating the leaf's response to water stress through stomatal closure, in control of seed germination, and in the development of the abscission layer.
11. Light is absorbed by specific growth-active photoreceptors in addition to the photosynthetic pigments. These receptor systems include phytochrome, the phototropic pigment, protochlorophyll, and others.
12. Phytochrome exists in active and inactive forms that are interconvertible by red and far-red light. It is used to sense whether or not light is present, and when activated by light it is involved in starting the development of the above-ground mode of growth. It may inhibit stem elongation, promote leaf expansion, promote the germination of light-sensitive seeds, and many other developmental events.
13. Flowering is frequently precisely timed with the seasons through photoperiodism. A system in the leaf can measure the length of the day or night and regulate the synthesis of a flowering stimulus. The flowering stimulus moves from the leaf to receptive lateral or terminal buds, where it causes the meristem to make flowers rather than leaves.
14. There are two main types of photoperiodically controlled plants, long-day and short-day plants, responding respectively to day lengths longer than critical or to night lengths longer than critical. Phytochrome is the photoreceptor involved in sensing day or night. Many plants are photoperiodically unresponsive, or day-neutral.
15. Buds may be induced to go dormant by a photoperiodic system and only activated again by favorable conditions after passing through the winter season, a prolonged period of low temperatures, the chilling period.
16. A prolonged period of low temperatures may vernalize certain plants, that is, remove a metabolic block to the formation of flowers by the apex. This serves as a signal denoting the passage of winter.
17. Light is necessary in flowering plants for the conversion of protochlorophyll to chlorophyll. This requirement for light prevents the development of the photosynthetic apparatus in plant parts not exposed to light.

CHAPTER 21

PROKARYOTES

Certain organisms exhibit a uniquely simple internal cell structure. Although the taxonomic diversity of these primitive plants is very small, accounting for only 1 to 2% of all species in the plant kingdom, they have a vast ecological and economic importance. Many are important as decomposers, contributing to the cycling of elements in the ecosystem. Those which are photosynthetic may be the only producers in some of the harshest environments on earth. Others are parasitic on plants, causing millions of dollars of crop damage each year, or are parasitic on man and other animals, causing many deaths each year and uncountable cases of illness. These are the **prokaryotes.**

Cells of these simple organisms contain genetic material, but they lack the organized nuclei of eukaryotes and the nature of their chromosome is unique. Most are unicellular, with a cell volume 1 to 2 orders of magnitude smaller than that of typical eukaryotes. These cells respire and some may photosynthesize, but they lack the mitochondria and chloroplasts of eukaryotes. Those prokaryotes that are colonial or filamentous represent an extreme of thallus complexity; most are unicellular.

Prokaryotes are also chemically unique, differing from eukaryotes in cell wall composition, nature of storage products, and details of certain metabolic pathways. In fact, they are unique in virtually every aspect: cell size, complexity, ecological function, cytology, chemistry, life cycles, genetics. Prokaryotes were present in the fossil record 1 to 2 billion years prior to eukaryotes, and it is possible that they were ancestral to eukaryotes.

The prokaryotes (sometimes spelled procaryotes) can be subdivided into: (1) the blue-green algae and (2) the bacteria. The taxonomic relationships of prokaryotes are not well understood, consequently there is a confusion of terms above the genus level. The blue-green algae, for example, have been put with the algae by some workers, and called the Cyanophyta division, while others have placed them with the bacteria and called them the Cyanobacteria. The bacteria have been put into the division Eubacteriae by some, to emphasize their uniqueness, and into the Schizomycophyta by others, to emphasize links with the plant kingdom. We will follow the latest edition of *Bergey's Manual of Determinative Bacteriology* and avoid technical names for taxonomic levels above the order.

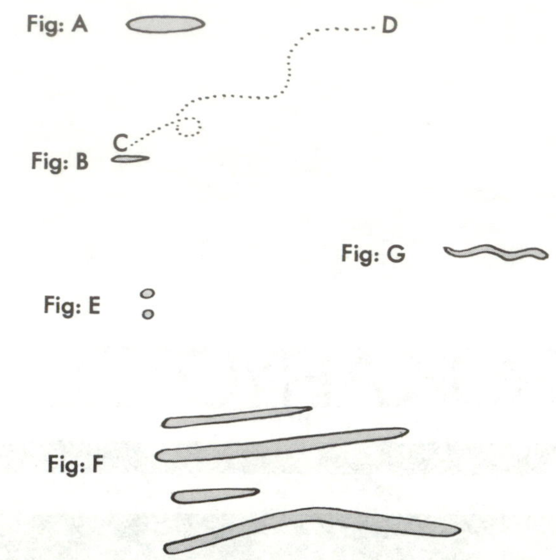

Figure 21.1
A copy of van Leeuwenhoek's drawings of the bacteria he obtained from his mouth. He described them thus: "I then most always saw, with great wonder, that in said matter there were many very little animalcules very prettily a-moving." (From *Arcana Natura Detecta Delphis Batovorum,* 1695.)

BACTERIA

Bacteria are of worldwide distribution, comprising a large number of species, mostly single-celled or simple filamentous forms between 0.5 and 5 μm diameter and between 1 and 20 μm long. Multiplication results from the simple division of single cells, a process known as **fission.** Some bacteria form spores, but the spores are more a means of carrying the species over periods unfavorable to growth than a means of multiplication.

Bacteria play an important role in the decomposition of organic matter, are indispensable in maintaining soil fertility, affect the quality of the water and milk we drink and of the food we eat, and cause diseases of animals and plants. We are likely to gain the impression that all bacteria are harmful, because we associate them with such dread human diseases as typhoid fever, tuberculosis, and other types of infections. But many kinds of bacteria are beneficial, in fact, indispensable in the life of the world.

Discovery of Bacteria

Our knowledge of the shapes and morphology of bacteria depends largely on the type of microscopes with which they are studied. Antony van Leeuwenhoek, of the Netherlands, first saw them in the summer of 1676. He was curious as to what made pepper hot, so he set some peppercorns aside in a cup of water and several days later examined this water under a simple single-lens microscope at a magnification of about 400 times. To his amazement he saw many actively moving, very small organisms that he called animalcules. He did not prepare any drawings of these small organisms until several years later, and these drawings, when finally prepared, showed organisms taken from his own mouth. These drawings, which are shown in Fig. 21.1 clearly indicate that he was describing bacteria for the first time.

While van Leeuwenhoek suggested that these small organisms might be related to fermentation, decay, and disease, his suggestion was not taken seriously. Their relation to these processes was finally discovered quite accidentally. Pasteur, a young French chemist, was retained in 1860 by the French government to improve the quality of French beers. A chemist was considered the proper person for the task because it was then thought that fermentation was a purely chemical process. Pasteur, however, firmly established the part played by living organisms in fermentation. He definitely showed for the first time that yeasts were necessary for the production of wine from grapes. Pasteurization, the flash heating of liquids to kill undesirable organisms, was first used to preserve French wines destined for export.

During the 1890s the French Academy of Sciences, of which Pasteur was a member, was hotly debating whether or not life could arise spontaneously. The outcome of this debate was the definitive demonstration by Pasteur that microorganisms always arise from preexisting organisms. Living organisms, whether bacteria or insect larvae, never arose *de novo* from decaying matter. A long series of experiments in Pasteur's laboratory succeeded in establishing the very definite relationship between bacteria (and yeasts) and the processes of fermentation, decay, and disease.

The role of bacteria in disease was also first shown in the last 25 years of the nineteenth century. In 1876 Robert Koch, a German country doctor, proved that anthrax was caused by a bacterium (*Bacillus anthracis*). He was able to isolate the bacterium, to inject it into test mice and to cause them to be diseased, then to isolate the bacterium again from the diseased mice. He took pictures of the organism through the microscope two years later, providing the first photographic evidence of bacteria. At about the same time, Burrill first recognized a bacterium as a plant pathogen.

Forms of Bacteria

Bacteria are of three general shapes: **spherical, rod-shaped,** and **spiral.** Spherical bacteria are called **cocci** (**coccus,** singular, Fig. 21.2), rod-shaped ones are called **bacilli** (**bacillus,** singular), and spiral forms are known as **spirilla** (**spirillum,** singular). There is some intergrading between these forms. It is sometimes difficult, for instance, to distinguish between a very short rod and a coccus that has elongated in preparation for cell division. Furthermore, shape depends to a certain extent upon the age of the bacteria and upon the environment.

Figure 21.2
First electron micrographs of bacteria. *A, Pneumococcus. B, Streptococcus. C. Staphylococcus aureus. D, Bacillus anthrax.* A11 ×10,000. (*A,* courtesy of Charles Pfizer and Company; *B, C, D,* courtesy of J. Hillier.)

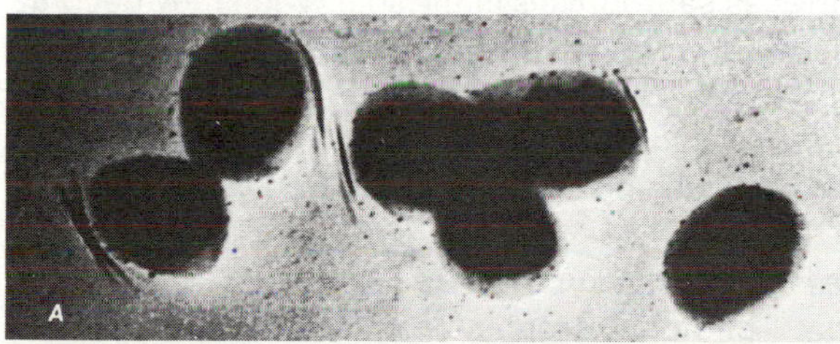

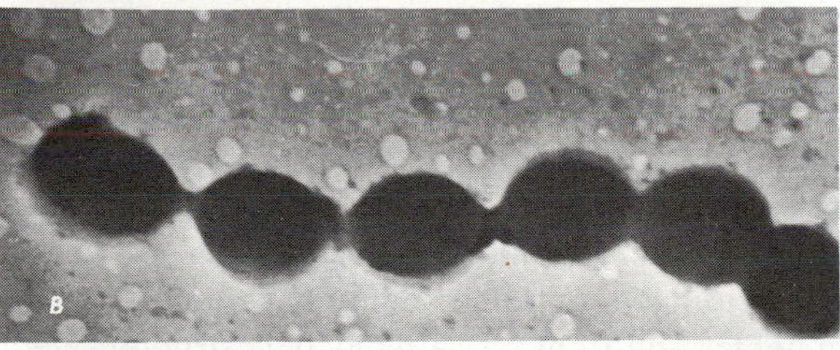

Bacteria of the coccus form, in chains resembling strings of pearls, are called streptococci (Fig. 21.2). Several species of the genus *Streptococcus* are very important and cause scarlet fever, erysipelas, mastitis of cows, and sinus infections. *Streptococcus lactis* is an agent in the souring of milk. Spherical bacteria may sometimes remain associated in irregular flattened masses; these forms are frequently referred to as **staphylococci** (Fig. 21.2). One species of *Staphylococcus* is the usual causal agent of boils and abscesses. Some species of spherical bacteria remain associated in cubical packets of 8, 64, or even more cells; these species belong to the genus *Sarcina.* Rods sometime form chains of cells (Fig. 21.2).

Bacteria are among the smallest living organisms. Some of them are close to the limit of visibility with the most powerful light microscopes. Several billion bacteria may be present in a cubic centimeter of soil. The very smallness of bacteria is a factor of major importance in their life. All the nourishment, all gases and inorganic salts that a bacterium requires, must be taken in through its cell wall. All its waste products, all toxins or poisons, certain enzymes, and other special substances that are associated with bacterial activity must pass outward through the surface of the bacterial cell. The more surface a bacterium possesses in relation to its volume, the more readily will these substances diffuse.

Distribution of Bacteria

These minute organisms are ever-present about us. They occur in the air, water, and soil. They exist on the surfaces of all animal and plant bodies and on the surfaces of almost everything we touch. Live bacteria are present in milk and in all foods that have not been sterilized. Even foods that have undergone the usual process of sterilization will sometimes contain viable spores of bacteria. Spores have been subjected to temperatures near absolute zero without being killed. Bacteria occur naturally in the digestive tracts of animals. Nitrogen-fixing bacteria are closely associated with the living cells in the roots of flowering plants, such as legumes, alders, bitterbrush, and others.

Water from deep, cold wells or springs is usually devoid of bacteria. Whereas the surface few centimeters of most soils teem with bacteria, the number decreases as the depth increases.

Although bacteria are almost universally distributed, it is possible, by taking great precautions, to keep rooms relatively free of these organisms. Such precautions are observed in hospital operating rooms and in certain types of biological and bacteriological laboratories. Surfaces are washed in disinfecting fluid, sterilized gowns are worn, and only instruments that have been placed in boiling water are used. Even the air may be purified by steam or ultraviolet radiation. This condition of freedom from bacteria is known as **asepsis. Sterilization** is the destruction of all living forms, including bacteria, that may be within or upon a particular object.

Movement

Bacteria are so small that many can be carried passively for long distances in water or air. But some forms actively move, and can propel themselves at rates of 5 to 50 μm per second. In proportion to body size, this is a rate comparable to 4 to 40 mph by a deer.

Locomotion is provided by gliding, flexing, or swimming mechanisms. Swimming is accomplished with flagella that may be attached to one end of the cell (polar) or all along the cell (peritrichous, Fig. 21.3). Each flagellum passes through the cell wall and plasmalemma and is anchored in the outer cytoplasm. Flagella are also present in some eukaryotes, but their ultrastructure is more complex. In the spirochetes, groups of flagella extend the length of the cell within the layers of the wall and form an axial filament; movement is caused by a flexing of the cell (Fig. 21.4). The mechanism of gliding is not well understood, but it requires a solid medium, and the cells leave a trail of slime behind.

Figure 21.3
Caryophanon latum showing many flagella, ×5000. (Courtesy of C. F. Robinow.)

Nutrition and Energy

Like all other living organisms, bacteria need a source of carbon and a source of energy. Based on these two parameters, we can divide the bacteria into four nutritional groups: (1) **photoautotrophs,** which use light as an energy source and CO_2 as a carbon source; (2) **photoheterotrophs,** which use light as an energy source and various organic compounds (e.g., acetic acid, $C_2H_4O_2$) as carbon sources; (3) **chemoautotrophs,** which oxidize reduced inorganic compounds (e.g., NH_3, H_2, H_2S, Fe^{++}) to obtain energy and which use CO_2 as a carbon source; and (4) **chemoheterotrophs,** which oxidize or reduce a variety of organic compounds to obtain both energy and carbon. Most bacteria are chemoheterotrophs, and the energy/carbon source is taken up through the plasmalemma.

A number of bacteria can shift from one trophic group to another, depending on the environment. Some of the metabolic pathways that yield energy and carbon can only occur in the presence of free oxygen, and others can only occur in the absence of oxygen. Finally, some chemoheterotrophs are free living **saprophytes,** able to break down dead organic remains, such as the litter in soil, while others are **parasitic.** Among the parasites are the disease-causing bacteria of plants and animals. Sometimes the distinction between parasite and saprophyte is not clear-cut. For example, certain types of bacteria thrive as saprophytes in the soil. When they become embedded in wounded tissue, however, they may become parasites. Such are the bacteria causing tetanus and gaseous gangrene in man, and blackleg in cattle, and bacteria causing soft rot of fruit and vegetables.

Bacterial Photosynthesis

Bacterial photoautotrophs do not perform photosynthesis in exactly the same way that blue-green algae and eukaryotes do. Photosystem II (see Chapter 13) is lacking, and as a result free oxygen is not liberated. Bacterial photosynthesis is strictly anaerobic. A second difference is that chlorophyll *a* is not involved; instead a series of **bacteriochlorophylls (Bchl** *a, b, c, d,* and *e*) and accessory pigments (carotenoids) are the light-harvesting pigments. Any Bchl has a molecular structure very similar to that of chlorophyll *a,* but certain side groups are unique and the absorption spectrum is also slightly different (Table 21.1).

The purple and green bacterial groups use CO_2 or various organic compounds (fatty acids, organic acids,

Table 21.1
ABSORPTION PEAKS OF PURIFIED CHLOROPHYLLS IN ETHER SOLUTION[a]

	Spectral Region				
Chlorophyll	Ultraviolet	Blue	Green	Red	Far-Red
a		430(100)		615(13)	662(77)
Bchl *a*	358(100) 391(68)		577(29)		773(100)
Bchl *b*	368(100) 407(87)		582(30)		795(96)
Bchl *c*		428(100)		622(29)	660(63)
Bchl *d*		424(100)		608(17)	654(61)
Bchl *e*		458(100)		593(12)	647(32)

[a] Wavelengths are in nm, and the relative amount of light absorbed is shown in parentheses (e.g., 100 = maximum absorption) (from Stanier et al., 1976).

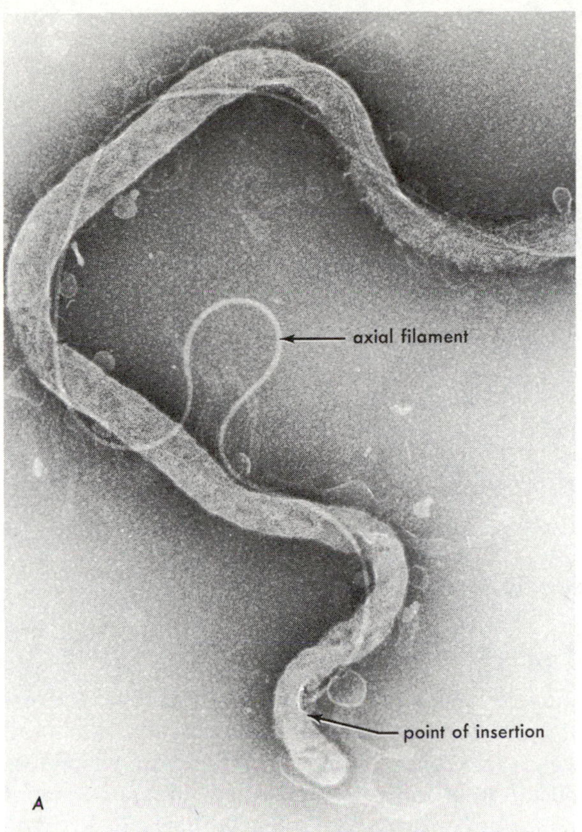

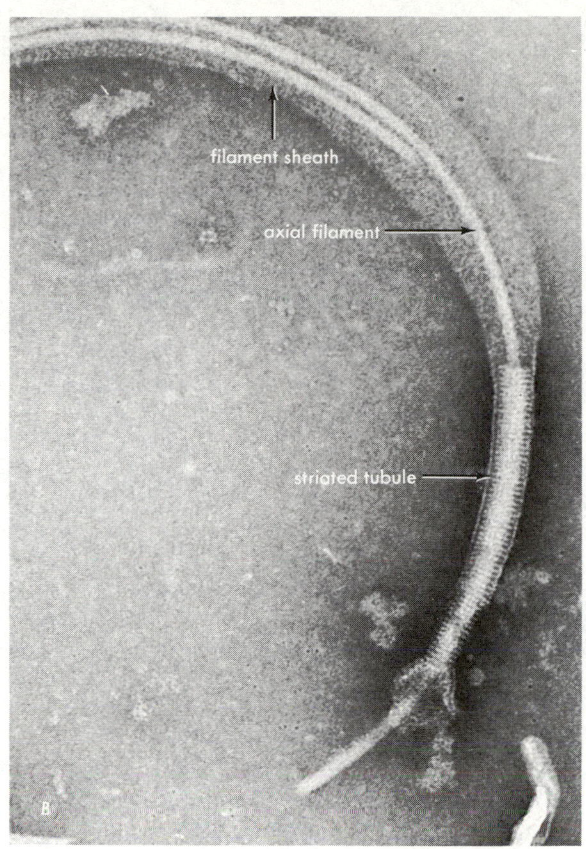

Figure 21.4
Axial filaments of *Spirochaeta stenostrepta*. *A*, about ⅔ of the cell, × 43,000. *B*, detail of the other end of the cell, ×120,000. The loop of axial filament shown in *A* is atypical for it is normally embedded in the layers of the cell wall. (Courtesy of S. C. Hold and E. Canale-Parola, *J. Bacteriol.* **96,** 882, © by American Society of Microbiology.)

carbohydrates, alcohols) as the source of carbon, and instead of water they use H_2S, H_2, or organic compounds as the source of H. Ribulose bisphosphate carboxylase (Chapter 13) is the carbon-fixing enzyme and the Calvin cycle appears to be the metabolic pathway of dark reactions. For example, the photosynthetic reaction in the purple sulfur bacteria is approximately:

$$2CO_2 + H_2S + 2H_2O \xrightarrow[\text{Bchl}]{\text{light}} 2(CH_2O) + H_2SO_4$$

Although photosynthetic bacteria lack chloroplasts, they do contain specialized membranes that function as the site of photosynthesis. In the green bacteria, a line of cigar-shaped vesicles lies just beneath the outer membrane. Each is a sac, about 50 nm wide and 2 to 3 times as long, surrounded by a single layered membrane only 3 to 5 nm thick. Light-harvesting accessory pigments are found only within these sacs. Bchl is in the plasmalemma. Light reactions in the purple bacteria take place on complex infoldings of the plasmalemma, which sometimes appear as vesicles, sometimes as parallel layers reminiscent of the grana of chloroplasts (Fig. 21.5). These membranes, however, are simply extensions of the outer membrane and are continuous with it. Only the green bacteria have a separate membrane system within the cytoplasm. Some purple bacteria also contain polyhedral bodies within a single-layered membrane (Fig. 21.6). The granular contents are rich in RuBP carboxylase, hence probably are the site of dark reactions. These bodies have been called **carboxysomes.**

Figure 21.5
Section of *Rhodomicrobium vannielii* showing array of peripheral photosynthetic membranes, ×60,000. (Courtesy of E. S. Boatman and H. C. Douglas, *J. Biophys. Biochem. Cyto.* **11,** 469. © 1961 by The Rockefeller University Press. Reprinted with permission of the publisher.)

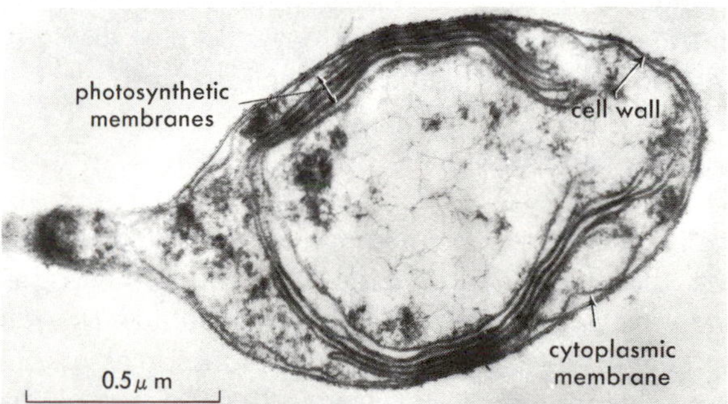

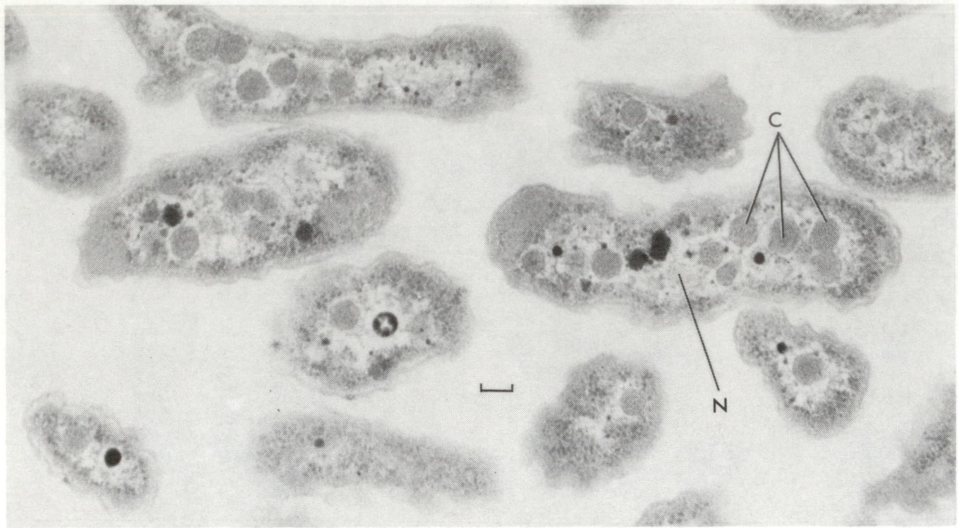

Figure 21.6
Electron micrograph of a section of *Thiobacillus*, containing numerous carboxysomes *(c);* N designates nucleoplasm. Bar indicates 0.1 μm. (Courtesy of J. M. Shiveley.)

Bacterial Respiration

Bacteria, like all living organisms, require energy to carry on living processes. Bacteria that generally use atmospheric oxygen in respiration are called **aerobes.** Many bacteria, however, are able to **ferment:** to respire without atmospheric oxygen and without the subsequent oxidation of food materials to carbon dioxide and water (see Chapter 14). They are able to obtain sufficient energy from glycolysis, the oxidation of food to organic acids. They do not absolutely require the greater amount of energy that is released when oxygen combines with alcohol, pyruvic acid, or some other simple organic compound to form carbon dioxide and water. Indeed, a few species of bacteria cannot grow in the presence of oxygen. Bacteria that grow in the absence of oxygen are called **anaerobes.** If even small amounts of oxygen hinder their growth, they are known as **obligate anaerobes.**

The enzymes involved in respiration and ATP generation are bound to the plasmalemma that has a function analogous to that of the mitochondria of eukaryotes. About 10 to 20% of all bacterial protein (enzymes) is contained in the cell membrane. The surface area of the membrane is increased by regions of infolding, called **mesosomes** (Fig. 21.7), and respiration activity may be concentrated there. Mesosomes may also play a role in the creation of a cross wall during cell division.

Nitrogen Fixation

Some bacteria and blue-green algae are capable of absorbing and fixing N_2 into ammonium (NH_4^+), which can then be metabolized and used to meet the nitrogen demands of the cell. Most plants cannot utilize N_2, and instead take up nitrogen from the soil solution as nitrate (NO_3^-) or ammonium (NH_4^+). Nitrogen fixation is ecologically and economically important, for it improves the fertility of water and soil. The application of manufactured ammonium as fertilizer is energy demanding. Nonindustrialized countries rely on animal waste, including that of humans for much of their nitrogen fertilizer needs. The world applies 50 million metric tons of nitrogen fertilizer to crops every year; biological nitrogen fixation provides more than 3 times that amount, mostly to natural, unmanaged ecosystems. It is likely that biological nitrogen fixation will become more important to agriculture in the near future as human beings learn to regulate and control this process.

The basic equation for nitrogen fixation in blue-green algae is:

$$2N_2 + 6H_2O \xrightarrow[\text{nitrogenase}]{\text{low } O_2\text{, ATP}} 4NH_3 + 3O_2$$

The reaction is inhibited by atmospheric concentrations of oxygen; thus nitrogen-fixing organisms either inhabit anaerobic environments or possess structural or physiological barriers that exclude oxygen from the cytoplasm. The enzyme nitrogenase is of major importance in the pathway, and the presence of small amounts of molybdenum seem important to its activity. The reaction requires an input of energy in the form of ATP, which comes from fermentation or photosynthesis. Some nitrogen-fixing bacteria are free living and must respire their own glucose (derived from organic matter in the environment) to provide the energy; it is a very inefficient process, requiring about 100 g of glucose for every 1 g of N_2 fixed. Other bacteria infect the roots of higher plants and obtain glucose from the host. This is a symbiotic relationship, however, for in return the higher plant absorbs excess fixed nitrogen from the bacteria. It is estimated that 2 to 5% of the higher plant's photosynthetic product is used

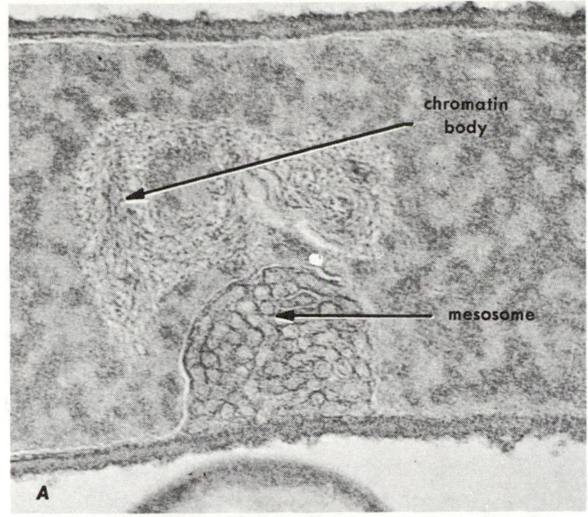

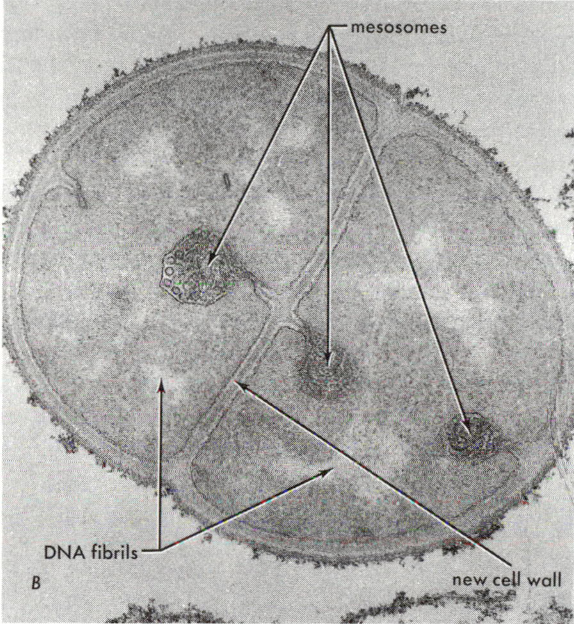

Figure 21.7
Mesosomes. *A*, the array of coiled membranes in the cytoplasm of *Bacillus subtilis* constitutes a mesosome. *B*, thin section of a vegetative cell of *Sarcina ureae*. Note the distinct association of the mesosomes with the areas of cross-wall formation. Nuclear material is present in the less dense ares throughout the cytoplasm, ×80,000. (*A*, courtesy of A. Ryter, *Bact. Rev*. **33,** 45. *B*, courtesy of S. C. Holt and E. R. Leadbetter, *Bact. Rev*. **33,** 346. © 1969 by American Society of Microbiology. Reprinted with permission of publisher.)

by the bacteria in nitrogen fixation.

Legumes (beans, peas, alfalfa, vetch, soybeans) are commonly infected by the nitrogen-fixing bacterium *Rhizobium*. *Rhizobium* cells are present in many soils and they enter the root through a root hair, causing the host cell's membrane to invaginate. The bacteria multiply and move through this infection thread from cell to cell into the cortex. They proliferate in cortical cells and they induce the root cells to divide and to enclose the infected cells in a small gall, called a **nodule.** There may be many nodules, each about 1 mm in diameter, on a single root system (Fig. 21.8). Cortical cells inside the nodule are crowded with *Rhizobium* cells. As a result of the symbiosis, legumes contain 2 to 3 times as much protein as grains, and they are an important protein source to the human diet.

The actinomycetes, a group of bacteria that form filaments rather than single cells, also include species that fix nitrogen and create nodules on higher plants. The plants infected are not legumes. They are generally woody and often grow on nutritionally poor soils, such as bogs, sand, gravel, and mine spoil. Many of these plants

Figure 21.8
Nodules produced by *Rhizobium leguminosarum* on the roots of a legume. Scale is in inches. (Courtesy of Francis Smith.)

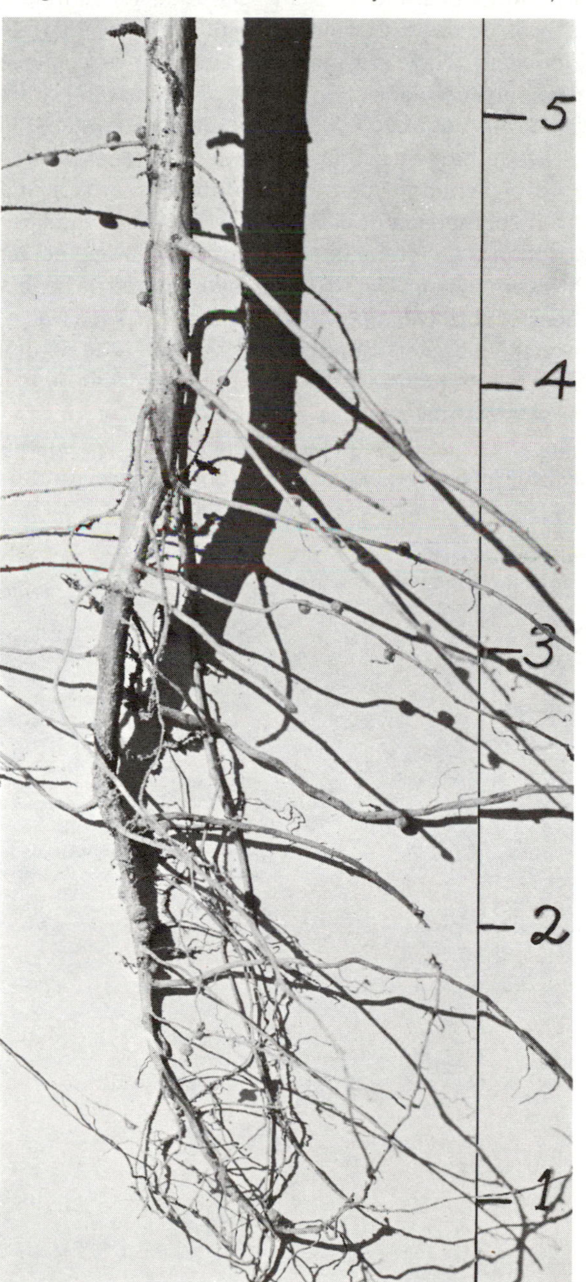

are pioneers in succession. Many species of alder (*Alnus*), *Ceanothus,* wax myrtle (*Myrica*), *Casuarina,* and bitterbrush (*Purshia*) among other genera, exhibit such nodules. The role of several bacteria in the cycle of nitrogen through a typical ecosystem is shown in Fig. 21.9.

Cell Structure

Cell Wall

Not all bacteria possess walls, but in those which do, the wall is a bag-shaped macromolecule, called peptidoglycan, of enormous dimensions that surrounds the protoplast. The basic, repeating unit of this molecule is essentially two glucose units with a side chain of several amino acids attached (Fig. 21.10). Cross-linkages between the amino acid chains hold the subunits together. **Gram-positive** bacteria have a thick, relatively homogeneous wall composed mainly (40–90% dry weight) of peptidoglycan. They are called Gram-positive because they retain a blue-black stain invented by Christian Gram in 1884. **Gram-negative** bacteria, which do not retain the stain, have a thinner, two layered wall: the inner layer is made up of peptidoglycan and the outer layer is a unit membrane composed of protein, phospholipid, and lipopolysaccharides (complex molecules of molecular weight over 10,000). The two wall layers together have a thickness of about 10 nm, while the homogeneous wall of Gram-positive bacteria have a thickness of 10 to 50 nm. Penicillin prevents the synthesis of peptidoglycan, which explains its effectiveness as an antibiotic.

These wall features have a great taxonomic significance, and all bacteria can be broken into three groups solely on this basis: Gram-positive, Gram-negative, and the **Mycoplasmas,** which do not have a wall. There are a few exceptions, such as the genera *Halobacterium* and *Halococcus:* their walls lack peptidoglycan, but they maintain their structural integrity only so long as the cells are in strongly saline solutions.

Protoplast

The prokaryote cell is characterized by its simplicity of internal architecture, but this does not mean that the protoplast is homogeneous. The cytoplasm has a granular appearance due to the presence of many ribosomes (Fig. 21.11). It may be penetrated by invaginations of the cell membrane. The green bacteria are exceptional in that vesicles surrounded by a membrane are separate from (but tightly appressed to) the cell membrane. Some photosynthetic bacteria have membrane-bound carboxysomes in the cytoplasm. Gas vesicles—hollow cylinders bounded by a 2 nm thick layer of protein (not a unit membrane)—are present in some bacteria. Food is stored in the cytoplasm as glycogen or as poly-β-hydroxybutyrate.

The nucleoplasm has an irregular shape and a sharp boundary, but a membrane does not separate it from the cytoplasm. During cell division, the nucleoplasm pinches in two, and the eukaryotic pattern of chromosome separation on a spindle is not evident. Bacteria are haploid organisms, with but a single chromosome. The chromosome consists of a circular, double helical DNA molecule. One or more small, extrachromosomal circular bits of DNA may also be present. These pieces replicate

Figure 21.9
Diagrammatic path of nitrogen through an ecosystem.

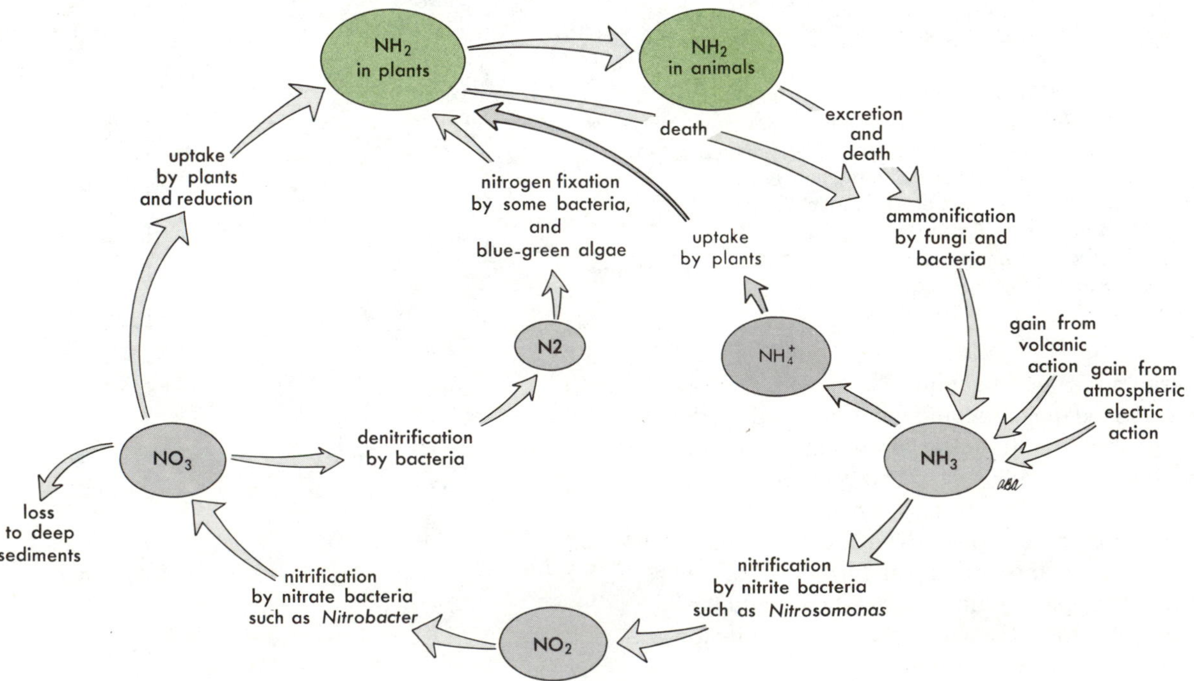

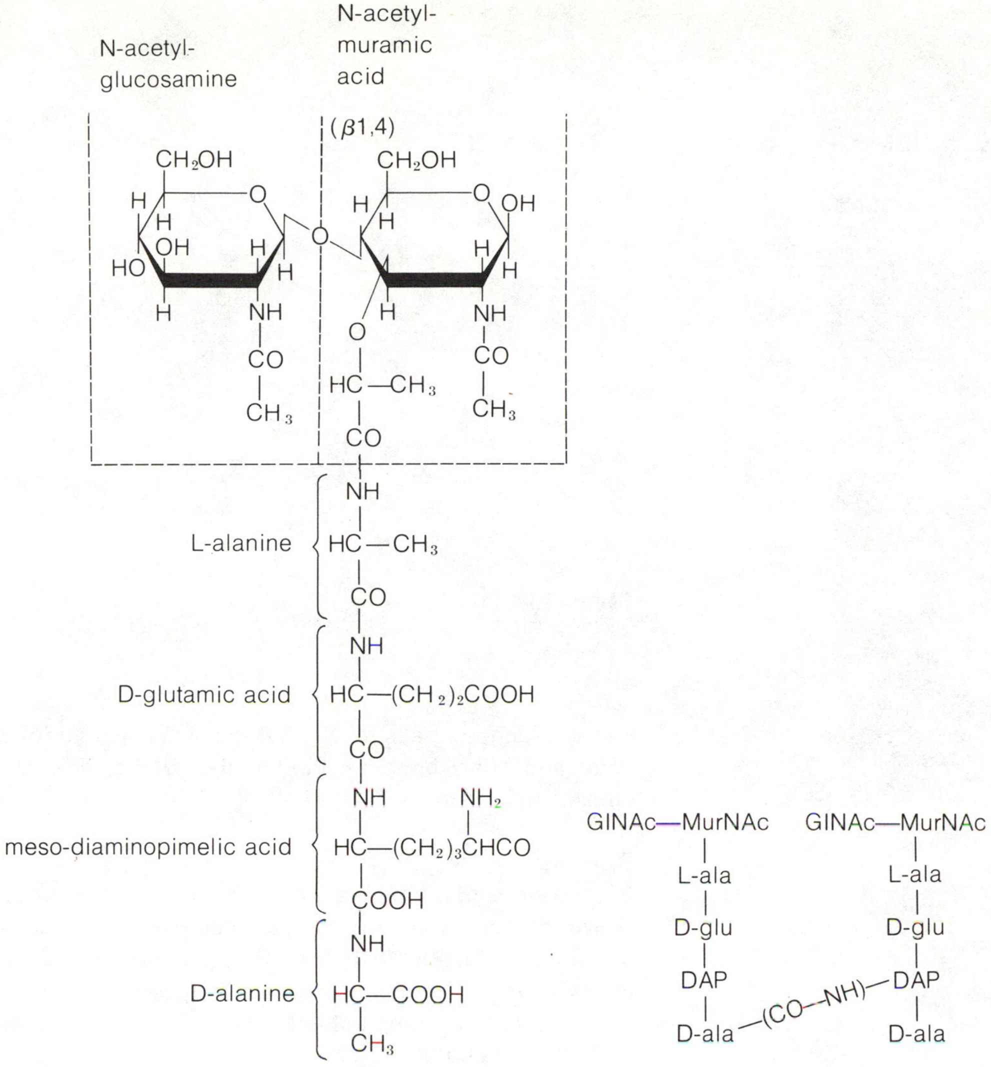

N-acetyl-glucosamine

N-acetyl-muramic acid

(β1,4)

CH_2OH H O H OH HO H H H NH CO CH_3

CH_2OH H O OH H H H O H NH CO CH_3 HC—CH_3 CO

L-alanine: NH, HC—CH_3, CO

D-glutamic acid: NH, HC—$(CH_2)_2COOH$, CO

meso-diaminopimelic acid: NH, NH_2, HC—$(CH_2)_3CHCO$, COOH

D-alanine: NH, HC—COOH, CH_3

A

GlNAc—MurNAc | L-ala | D-glu | DAP | D-ala—(CO—NH)—DAP

GlNAc—MurNAc | L-ala | D-glu | DAP | D-ala

B

Figure 21.10
General structure of a peptidoglycan. *A*, Complete structure of a single subunit, showing the linkage between the two amino sugars that make up the glycan strand, and between muramic acid and the four amino acids in the short peptide chain. *B*, Schematic, simplified representation of structure shown in *A*. Representation of the mode of cross-linking between the terminal carboxyl group of D-alanine on one subunit and the free amino group of the diamino acid (diaminopimelic acid) on an adjacent subunit.

when the cell replicates, and they carry genes that may confer resistance to drugs, induce the production of toxins, or permit sexual conjugation. These pieces of DNA are called **plasmids.**

Asexual Reproduction

Cell Division

Most bacteria divide by simple fission. The cell is "pinched" in two—much as a sausage might be cut by the tightening of a slip knot. As the process nears completion, the two cells pull apart, stretching the cell wall and plasma membrane. This thin, stretched portion finally breaks and the ends of the two cells round up. The nucleoplasm apparently divides before the division of the cell (Fig. 21.12). Some bacteria also divide by budding, in which a cell protuberance grows to equal the volume of the parent cell, then becomes detached.

Under optimum conditions, a bacterium may divide and its progeny grow to full size, ready for another

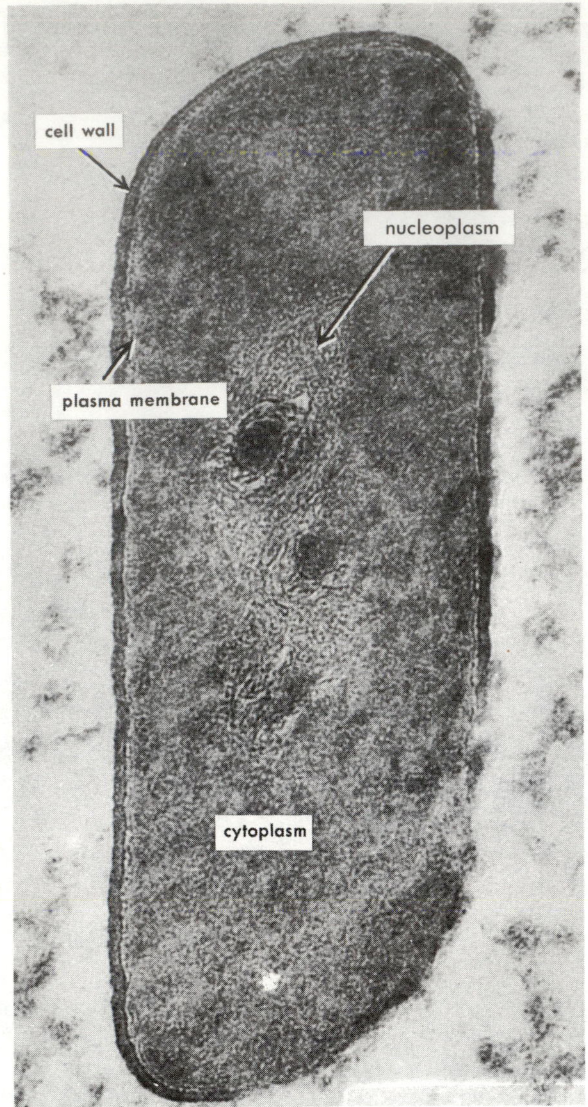

Figure 21.11
A longitudinal section through a cell of *Bacillus subtilis*, × 80,000. (Courtesy of A. M. Glauert, *et al. Exptl. Cell Res.* **22,** 73. © 1961 by Academic Press Reprinted with permission of the publisher.)

division, in 10 minutes. If such conditions were to prevail for 1 day, 200 trucks of 5 ton capacity would be required to haul away the progeny of a single cell. Obviously, these conditions have never prevailed for so long a period. Nevertheless, the reproductive capacity of bacteria is enormous. Bacterial growth is everywhere held in check. Probably the most important checks are: (*a*) lack of food; (*b*) the production of substances unfavorable to growth by the bacteria themselves, by other bacteria, by fungi, by higher plants, or by the host in which the bacteria may be living; and (*c*) competition for oxygen or other nutrients. The rate of increase of a bacterial population is, in general, very similar to that of other organisms, including yeasts, fruit flies, and man. There is a short initial period during which increase in numbers is very slow; then the bacteria increase very rapidly and the population reaches a peak at which it remains constant for a time; and after the peak the number of bacteria decreases rather slowly (Fig. 21.13).

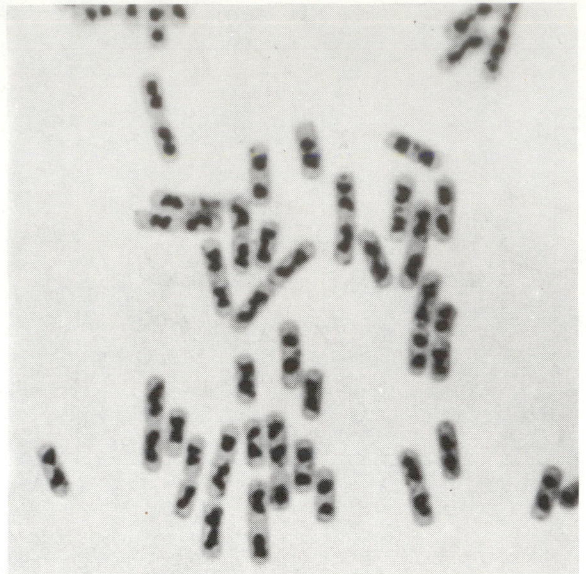

Figure 21.12
Dividing chromatin bodies in *Bacillus cereus*. (Courtesy of C. F. Robinow.)

Spores

Several kinds of bacteria are able to form resistant cells, known as **spores,** at the end of their period of active growth (Fig. 21.14). Each spore is protected by several layers of wall. Inside, the protoplast may be highly dehydrated and show no detectable metabolic activity (especially true of endospores).

Endospores are spores formed within the cell wall of a parent cell. First, the chromosome replicates, creating two nucleoplasm bodies, which then coalesce into a rod-shaped region. A small end of that nucleoplasm is

Figure 21.13
A growth curve showing an initial slow rate of growth followed by a rapid logarithmic rate or growth, and finally by the establishment of a colony with little or no growth.

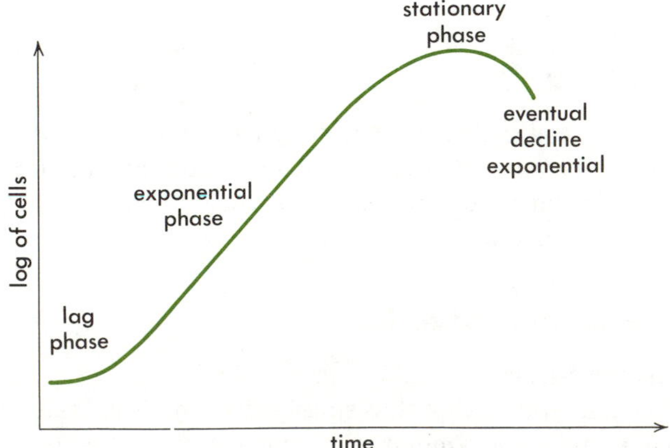

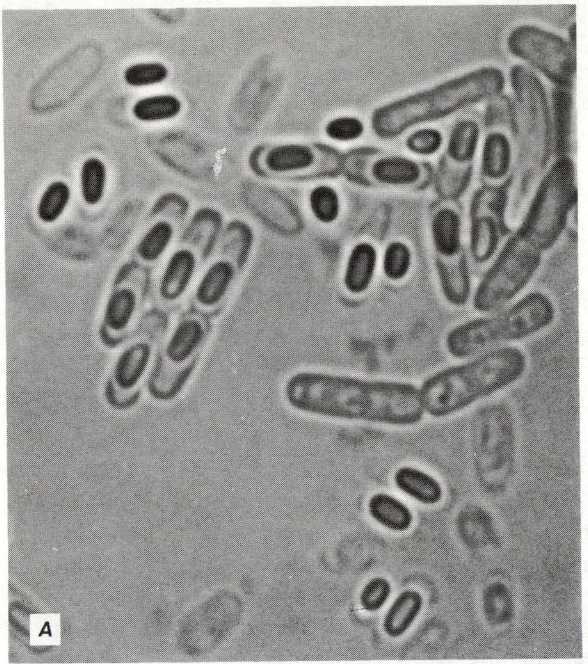

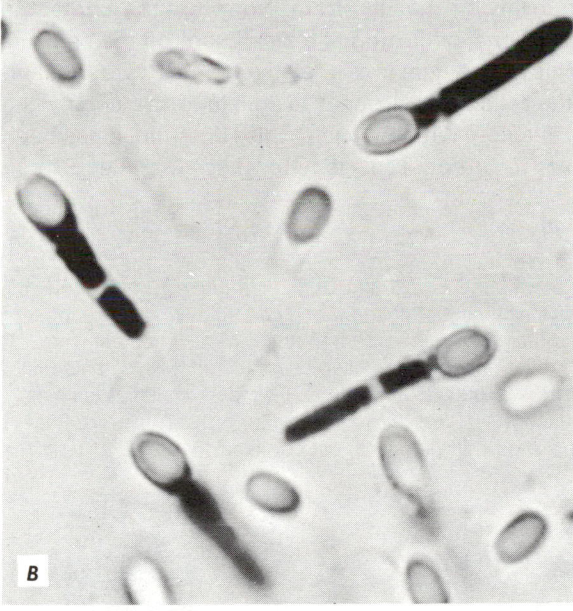

Figure 21.14
A, vegetative forms, sporulating cells, and free spores of *Bacillus cereus*. (Courtesy of C. F. Robinow, *The Bacteria*, Vol. **1**, I. C. Gunsalus and R. Y. Stanier, (eds.). © 1960 by Academic Press. Reprinted with permission of the publisher.) *B*, sporulating cells and free spores of *Clostridium pectinovorum* (Courtesy of C. F. Robinow, *The Bacteria*, Vol. **1**, I. C. Gunsalus and R. Y. Stanier, eds., Academic Press, New York. Reprinted with permission of publisher.)

pinched off and becomes surrounded by its own membrane and cell wall. The cell wall is composed of several layers and is mainly protein, but the outermost layer is highly resistant to treatments that solubilize most proteins. Ultimately the spore is liberated from the parent cell. Generally the spore will remain dormant for a period of time before it can be induced to germinate.

Spores may remain alive for many years under conditions quite unfavorable to growth. They resist desiccation, are not killed by long exposures to temperatures as high as 80°C, and may be suspended in boiling water for 20 minutes or longer without being killed. Their resistance to heat, coupled with their ability to become active immediately upon being placed in a suitable environment, makes them of great interest to food technologists, doctors, and also students of living cells. Most of the spore-forming bacteria are rod-shaped forms belonging to the genera *Bacillus* and *Clostridium.* While most bacteria form but one spore per cell, a few are known that may form two or more spores per cell. Such spores occur very commonly in soil bacteria. Fortunately, few disease-producing **virulent** (or **pathogenic**) bacteria form spores.

Transformation

Sexual reproduction is definitively indicated by crossing-over or gene recombination. It was through gene recombination that sexual reproduction was first demonstrated in the bacteria. The first suggestion of gene recombination occurred quite unexpectedly in studies on antibody formation and virulent and nonvirulent colonies of *Streptococcus pneumoniae.* Here genetic information was passed from one living bacterium to another through the medium of dead bacteria.

Pneumonia is caused by *Streptococcus pneumoniae,* and the virulent bacteria in the sputum of a patient are surrounded by large capsules formed by a carbohydate. When grown on agar these virulent bacteria form large smooth colonies. It has been shown that the carbohydrates of the capsules are able to stimulate the formation of antibodies in the host, and that four strains of bacteria occur as indicated by the type of carbohydrate present in the capsule and its antibody response. These four types are known simply as Type I, II, III, and IV. Avirulent bacteria producing rough colonies on agar (due to their inability to produce capsules) occur as frequent mutations of the virulent smooth form. Once obtained, these mutants do not revert to the virulent smooth form but may be carried unchanged through hundreds of culture transfers and tens of thousands of generations.

We have thus established for study three traits of *Streptococcus pneumoniae:* (*a*) virulence, (*b*) type of colony formed, and (*c*) type of carbohydrate in the capsule. Each of these traits may mutate, virulent to avirulent, rough to smooth colony form, and one type of capsule carbohydrate to another. We may study these characters just as we did the various characters of peas (Chapter 17). Thus, avirulent bacteria producing rough colonies on agar may mutate to a virulent, smooth form.

In some studies on the activity of these bacteria,

living, avirulent, rough colony bacteria derived from ancestors with Type I capsules were injected into a mouse, together with a large number of *heat-killed,* virulent, smooth colony, Type II bacteria.

These injections should have been harmless because only *living* virulent bacteria can induce the disease. However, the mouse died, and *living,* virulent Type II cells were recovered from its dead body, although *no living* virulent cells of *S. pneumoniae* were injected into the mouse. We know that the dead bacteria cannot cause disease, and we do not believe that dead bacteria can become alive. The only other explanation is that somehow or other genetic information must have passed from the dead, virulent, smooth, Type II bacteria into the living, avirulent, rough Type I bacteria. This hypothesis was confirmed when it was shown that an extract taken from the dead, virulent form could indeed transform the living, avirulent bacteria into a lethal culture. When this extract was purified, DNA turned out to be the material that brought about the change. This change of the genetic information in a bacterial cell by the addition of DNA from another culture of bacteria is known as **transformation.** This experiment may be outlined as follows:

Living bacteria		**Dead bacteria**		**Living bacteria**
avirulent	+	virulent	→	virulent
rough colony		smooth colony		smooth colony
no capsule		Type II capsule		Type II capsule
Living bacteria		**DNA from**		**Living bacteria**
avirulent	+	virulent	→	virulent
rough colony		smooth colony		smooth colony
no capsule		Type II capsule		Type II capsule

Sexual Reproduction

Transformation indicated that gene recombination (Chapter 17) occurred in bacterial cells. It further stimulated a search for the presence of naturally occurring gene recombination in bacterial cells. Evidence for gene recombination was first obtained by Lederberg in a brilliant, yet simply designed, experiment. *Escherichia coli,* a common bacillus of the large intestine, needs no specific growth factors. It is able to synthesize all its protoplasm from a simple broth of glucose and inorganic ions. Lederberg obtained two mutant strains that were unable to synthesize certain compounds that were required for their metabolism and hence could not grow in a mineral medium. In order to obtain growth, biotin (B) and methionine (M) had to be added to the growth medium of one strain and threonine (T) and leucine (L) had to be added to that of the second strain. Since each strain could synthesize the metabolite that the other could not, *gene recombination, if it occurred,* should result in some progeny that would grow on the minimal broth medium, thus:

$B^-M^-T^+L^+$ cannot grow because it cannot synthesize biotin (B^-) or methionine (M^-)	×	$B^+M^+T^-L^-$ cannot grow because it cannot synthesize threonine (T^-) or leucine (L^-)

↓

after gene recombination
$B^+M^+T^+L^+$
can grow because it can synthesize
all four compounds.

A bacterium that now has all the plus genes should give rise to a colony in the minimal culture to which these compounds have not been added, because it will be able to synthesize them. When this experiment was set up, a strain that was able to synthesize all four compounds was actually found. This demonstrated that recombination of genes did actually occur in bacteria. As it took place only once in about a million cells, its occurrence was too infrequent to allow for detailed study.

Eventually a strain of *E. coli* was found in which recombinations were sufficiently frequent to make a genetic analysis of inheritance in bacteria possible. This strain is known as the Hfr strain, and Fig. 21.15 shows a mating involving the Hfr strain. Note that the contact is rather tenuous, occurring at a small spot on the cell surface. This attachment is of a relatively short duration, and when the cells separate, each is able to form a colony in a normal fashion. With the use of this Hfr strain, it has been possible to prepare a genetic map for the *E. coli*

Figure 21.15
Mating between mutant strains of *Escherirchia coli* (Hfr and F^-). Note the attachment of phage particles to the F^- mutant (upper bacterium) and their absence from the lower, Hfr mutant. The DNA from the phage heads has been injected into the bacterial protoplast, ×95,000. (Courtesy of T. F. Anderson, E. L. Wollman, and F. Jacob, *Ann. Inst. Pasteur* **93,** 450. © 1957, Institute Pasteur. Reprinted with permission of publisher.)

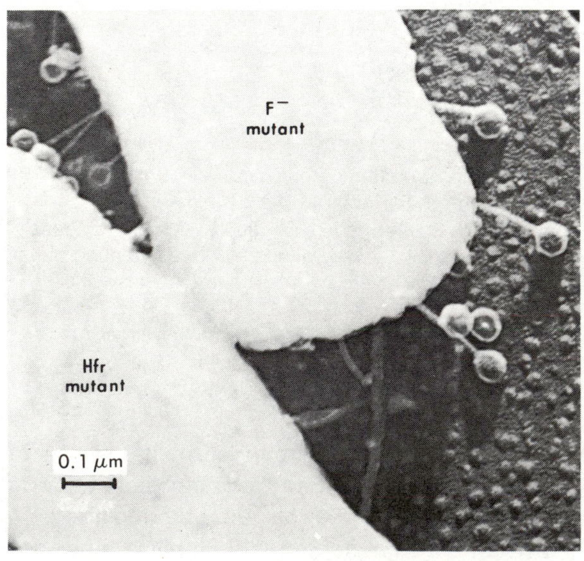

chromosome just as for corn or *Drosophila,* although the methods are of necessity somewhat different from those used for higher forms.

It must first be demonstrated that a true mating has occurred and that a transforming principle is not involved. This is easily accomplished by placing a fine sintered glass disk in the base of a U-tube and growing the different strains of the bacteria in the opposite arms (Fig. 21.16). A transforming principle (DNA) can diffuse through the sintered glass disk, but bacteria cannot pass it. If a recombination occurs when strains are mixed in the same culture medium but does not occur when they are kept separated by a filter, the recombination must be the result of a mating. This experiment proves that a transforming principle was not present and that a mating occurred.

As with higher plants, it is possible to observe certain phenotypic changes. Hfr bacteria genetically marked for motility and shape may easily be distinguished in a mixed culture from a strain (F^-) possessing the opposite characters (Fig. 21.17). The attachment between Hfr and F^- cells, which normally lasts less than 90 minutes, may be observed. When the cells separate, each produces a clone, but only the F^- progeny show any gene recombination. The progeny of the Hfr bacterium resemble the parent cell. Thus, transfer of genetic material is only from the donor Hfr cell to the recipient F^- cell.

The ability to conjugate and to donate DNA to another cell is conferred by a plasmid, called an **F** (for fertility) **plasmid.** Cells with an F plasmid will attach to other cells. The nuclear chromosome and the plasmid will duplicate, and one of the duplicated F plasmids will pass into the recipient cell. Generally, that is the only DNA transferred, the contact is broken, and the only result is that both cells now have the F plasmid. The original

Figure 21.16
Diagram of an experiment to test whether a bacterial mating has taken place. If genetic information is not transferred across the sintered glass disk, a mating is necessary to bring about gene recombination.

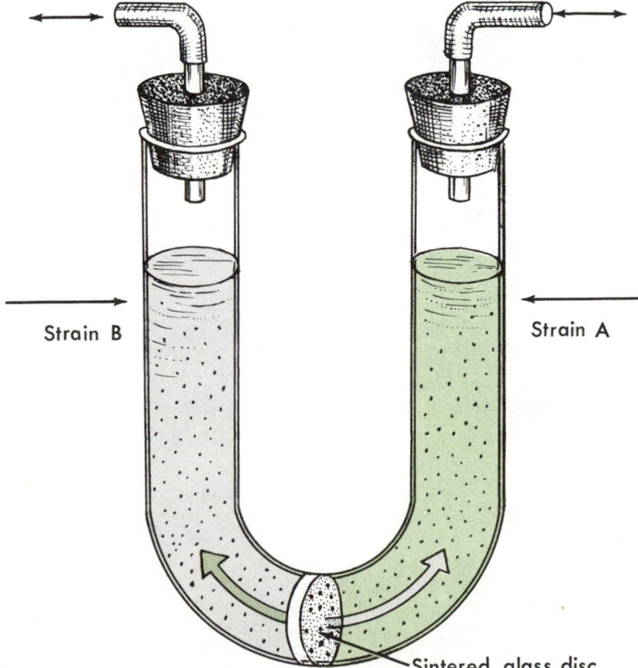

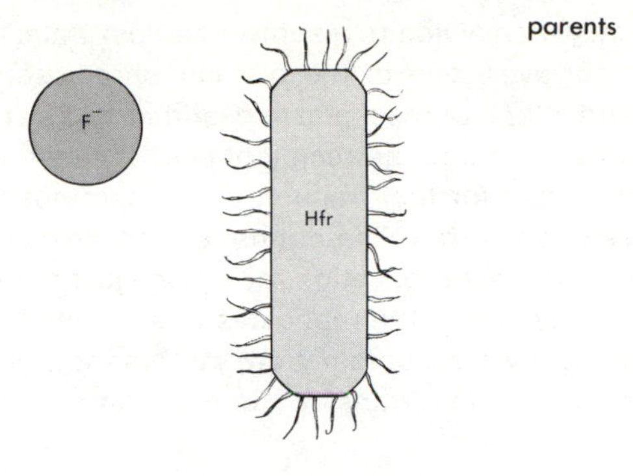

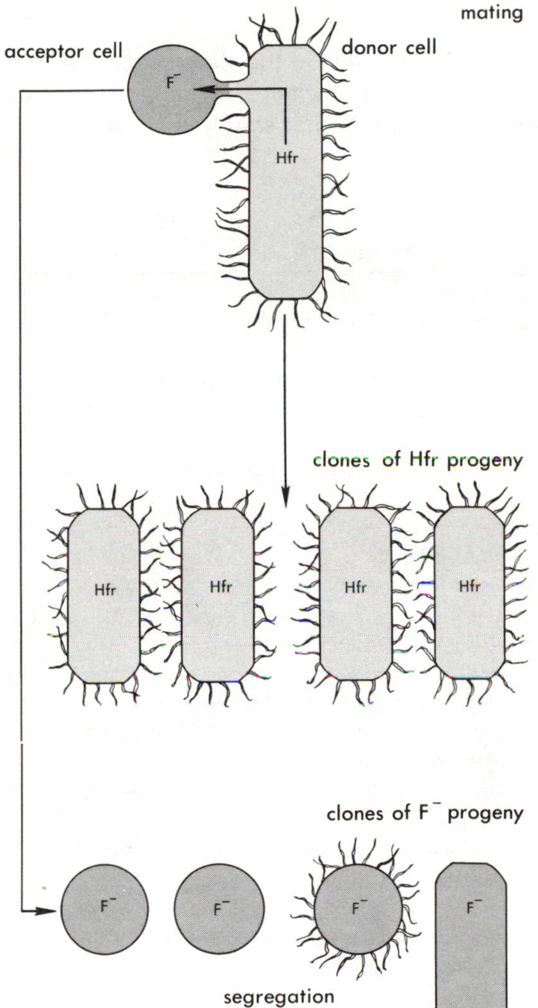

Figure 21.17
Diagram to show how gene recombination may be traced in bacterial matings. Experiment starts with mating of Hfr and F^- mutant strains of *E. coli*. After mating, these two bacteria are allowed to produce clones; the appearance of the individuals of the clones is shown.

donor cell may then divide to resume a haploid state. Occasionally, however, one of the nuclear chromosomes will break and attach to the F plasmid, and all or a part of it will be transferred into the recipient cell. The recipient will now be diploid for that fraction of the chromosome that has been transferred. The donor cell remains viable because chromosome replication took place just prior to mating, and only one of the replicates was donated. **Hfr** (for high frequency of recombination) **strains** show chromosome breakage and transfer more frequently than other strains.

Classification

Cell function is as important in the classification of bacteria as cell shape. Many bacteria, in fact, can only be identified by what they do, not simply by how they appear. The 3000 or so species of bacteria can be divided into a number of major groups based on wall traits, source of energy and carbon, cell shape, form of motility, requirement for oxygen, and whether they produce spores. In the most recent edition of *Bergey's Manual of Determinative Bacteriology,* the bacteria are classified into 19 "parts" (Table 21.2). Only brief comments will be made for a few of those groups, in our attempt to show the range of diversity within the bacteria.

Photosynthetic bacteria (Part 1) are Gram-negative rods, cocci, or spirilla. Some have flagella. Sometimes the cells occur in chains. Cell color (purple, green, yellow, orange, brown) is due to various carotenoids. Some cells contain gas vesicles. They are limited to relatively few aquatic habitats, such as shallow, eutrophic ponds, deep (25 m) anaerobic strata in lakes, or hot springs rich

Table 21.2

MAJOR GROUPS ("PARTS") WITHIN THE BACTERIA, AS RECOGNIZED IN *BERGEY'S MANUAL OF DETERMINATIVE BACTERIOLOGY* (1974)

Part 1.	Phototrophic Bacteria. One order, 3 families, 18 genera
Part 2.	Gliding Bacteria Two orders, 8 families, 21 genera; also 6 genera of uncertain affiliation
Part 3.	Sheathed Bacteria 7 genera
Part 4.	Budding and/or Appendaged Bacteria 17 genera
Part 5.	Spirochaetes One order, 1 family, 5 genera
Part 6.	Spiral and Curved Bacteria One family, 2 genera; also 4 genera of uncertain affiliation
Part 7.	Gram-Negative Aerobic Rods and Cocci Five families, 14 genera; also 6 genera of uncertain affiliation
Part 8.	Gram-Negative Facultatively Anaerobic Rods Two families, 17 genera; also 9 genera of uncertain affiliation
Part 9.	Gram-Negative Anaerobic Bacteria One family, 3 genera; also 6 genera of uncertain affiliation
Part 10.	Gram-Negative Cocci and Coccobacilli One family, 4 genera; also 2 genera of uncertain affiliation
Part 11.	Gram-Negative Anaerobic Cocci One family, 3 genera
Part 12.	Gram-Negative Chemolithotrophic Bacteria Two families, 17 genera
Part 13.	Methane-Producing Bacteria One family, 3 genera
Part 14.	Gram-Positive Cocci Three families, 12 genera
Part 15.	Endospore-Forming Rods and Cocci One family, 5 genera; also 1 genus of uncertain affiliation
Part 16.	Gram-Positive Asporogenous Rod-Shaped Bacteria One family, 1 genus; also 3 genera of uncertain affiliation
Part 17.	Actinomycetes and Related Organisms Four genera not assigned to a family; 1 family with 2 genera; 1 order with 8 families and 31 genera
Part 18.	Rickettsias Two orders, 4 families, 18 genera
Part 19.	Mycoplasmas One class, 1 order, 2 families, 2 genera; also 2 genera of uncertain affiliation.

in sulfide. Many species can fix N_2. *Chloroflexus* can form masses of orange to green filaments of cells several millimeters thick that resemble mats of blue-green algae. *Halococcus* and *Halobacterium* occur only in highly saline water and are able to generate ATP from light absorbed by a protein called **bacteriorhodopsin,** and they do not contain Bchl. These genera are also unusual in that their cell walls lack peptidoglycans.

Gliding bacteria (Part 2) contain several Gram-negative groups. Despite the name, this part does not contain all gliding bacteria (one genus of green photosynthetic bacteria, *Chloroflexus,* has gliding motility). The **myxobacteria** is one group in Part 2. They are rod-shaped cells that form a colony in a matrix of tough slime. They produce spores that may be borne singly, in chains, or in clusters on stalks (Fig. 21.18). They are commonly found in soil, and they secrete enzymes that lyse (split) other bacteria or that enable them to digest cellulose.

Spirochetes (Part 5) are long, flexible Gram-negative cells capable of swimming by axial flagella. Some are free-living in anaerobic mud or water, and others are parasites of mollusks or vertebrates, including man. *Treponema* causes syphilis and yaws, *Borrelia* causes relapsing fever, and *Leptospira* causes a form of jaundice.

Gram-negative aerobic bacteria (Part 7) include some economically important genera, such as *Rhizobium,* which fixes N_2 symbolically with legumes, and the acetic acid bacteria that produce vinegar. Some are chains of cells surrounded by a **sheath** that lies outside the wall. The sheath is laid down by the terminal cell and consists of protein, polysaccharides, and lipids. Other Gram-negative cells may secrete a layer of slime, called a **capsule,** around the cell. Cells of *Caulobacter* are stalked, and holdfast at the base of the stalk attaches the cell to a substrate.

Gram-negative facultatively anaerobic rods (Part 8) are also known as the enteric bacteria (enteron = intestine). The well-known *Escherichia coli,* a common bacterium in the human gut, is an example. Others include *Salmonella,* which causes typhoid and food poisoning, *Shigella,* which causes bacterial dysentery, *Erwinia,* which causes fire blight of pears and soft rot of fruit, and *Yersinia,* which causes bubonic plague.

Gram-negative chemoautotrophs (= chemolithotrophs, Part 12) are very important in the cycling of nutrients through the biosphere. They are widespread in soil and water, and some occur in hot springs. Cell shape ranges from round to rod to spiral; some are motile by flagella, some by gliding. Plasmalemma infoldings may be extensive (Fig. 21.19). There are four groups, based on the oxidation reaction that provides their energy:

nitrifying bacteria	$NH_3 \rightarrow NO_2^- \rightarrow NO_3^-$
sulfur bacteria	H_2S, S, or $S_2O_3^{--} \rightarrow SO_4^{--}$
iron bacteria	$Fe^{++} \rightarrow Fe^{+++}$, $Mn^{++} \rightarrow Mn^{+++}$
hydrogen bacteria	$H_2 \rightarrow H_2O$

Figure 21.18
Scanning electron micrograph of the fruiting bodies of the myxobacterium *Chondromyces crocatus* (magnification ×820). (Courtesy of J. Pangborn and P. Grilione. *J. Bacteriology,* **124,** 1558. © 1975 by American Society of Microbiology. Reprinted with permission of publisher.)

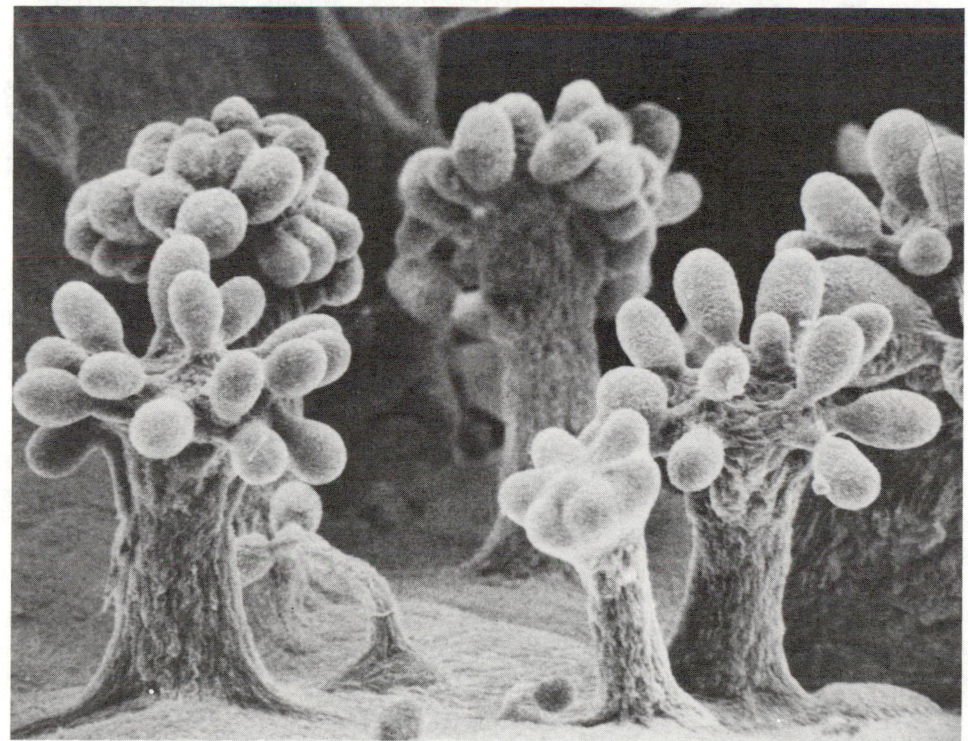

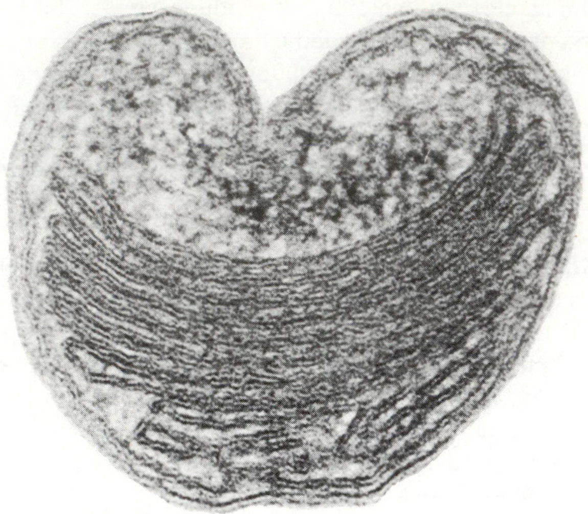

Figure 21.19
The chemotroph, *Nitrosocystis,* oxidizing NO_2^- to NO_3^- anaerobically, has an array of parallel membranes, ×90,000. (Courtesy of R. G. E. Murray.)

Gram-positive endospore-forming rods and cocci (Part 15) are also widespread in soil. If soil containing *Clostridium* enters a deep wound, it may cause tetanus or gangrene. *Clostridium* may also occur on vegetables, such as corn and beans. If *Clostridium botulinum* is not killed in the process of canning, it grows and multiplies in the canned product, producing one of the most powerful poisons known. Food poisoning from **botulinus toxin,** however, has become quite rare. Commercial canning processes are adequate to kill all spores, and home canners have been so well-educated that home-canned food is almost as unlikely to contain viable *Clostridium botulinum* as is commercially canned food.

Actinomycetes (Part 17) are Gram-positive bacteria that form bodies resembling those of molds and other fungi. Cells grow and divide in long filaments that branch and intertwine. Cross walls are generally absent or widely spaced. Spores (sometimes endospores) are produced at the ends of these filaments. *Streptomyces* is the largest genus in this group. It is common in soil and the odor of damp soil in fact comes from this organism. It is an important source of antibiotics.

Rickettsias (Part 18) are short, Gram-negative rods that are obligate parasites and cannot be cultured on nonliving media. They cause Rocky Mountain spotted fever, Q fever, and scrub typhus in man, and their presence in insects that suck plant juices, for example, aphids, suggested that they might be found in plants. A rickettsia-like organism was isolated in 1970 from dodder, an angiosperm parasite. Pierce's disease has long been a very serious and mysterious disease of grapes. Pierce first described it in 1880 and noted bacteria in relation to diseased tissue. He was unable to cultivate bacteria from his diseased grapes, and the causative organism has eluded many investigators for close to 100 years. In 1972, Coheen, working at the University of California in Davis, discovered with the electron microscope the presence of rickettsia-like organisms in grape vines having the symptoms of Pierce's disease.

Mycoplasmas (Part 19) include the smallest known cellular organisms. Cell volume is less than 0.1 μm^3, or about one tenth the volume of a typical bacterium such as *E. coli.* A cell wall is lacking, and the cell shape is plastic. They contain about one quarter the amount of DNA present in an *E. coli* cell. Some are plant pathogens that are disseminated by leaf hoppers or by grafting, Disease symptoms range from yellowing, growth abnormalities like witches broom, and phloem necrosis (Fig. 21.20).

Mycoplasmas apparently have a rather broad distribution. They have been isolated from sewage and decaying matter as well as from dogs, rats, mice, and other animals. They will grow on an enriched artificial medium containing a beef heart infusion with serum or other similar additions.

Virus yellows in plants were long described as virus diseases of plants. Recently, over 30 of these virus yellows diseases have been shown to be caused by a mycoplasma. Other long-baffling plant diseases once thought to be due to a virus infection but now known to be caused by mycoplasmas are pear decline, citrus stubborn disease, and Western X disease of peaches and cherries. Correct knowledge of the true causative organ-

Figure 21.20
Mycoplasma organisms in a sieve tube of a peach tree showing symptoms of peach-x disease, ×10,000. (Courtesy of G. Nyland.)

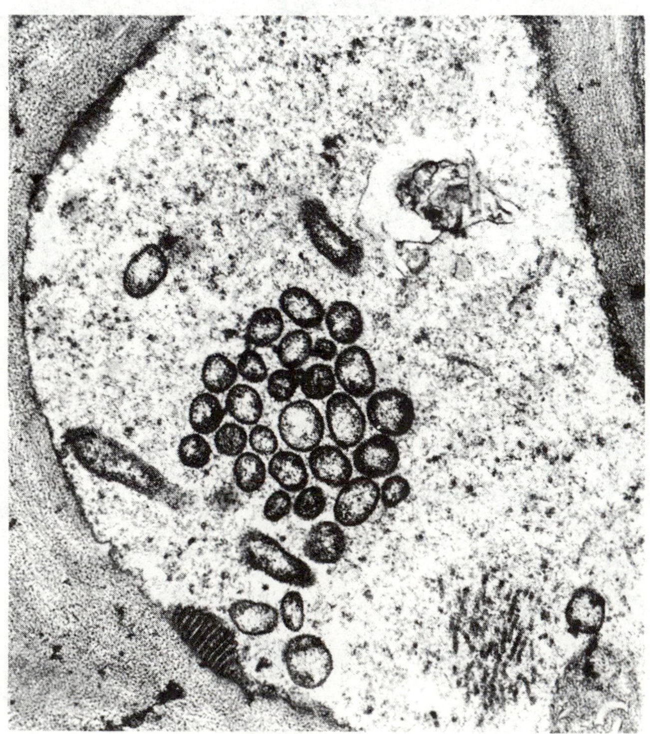

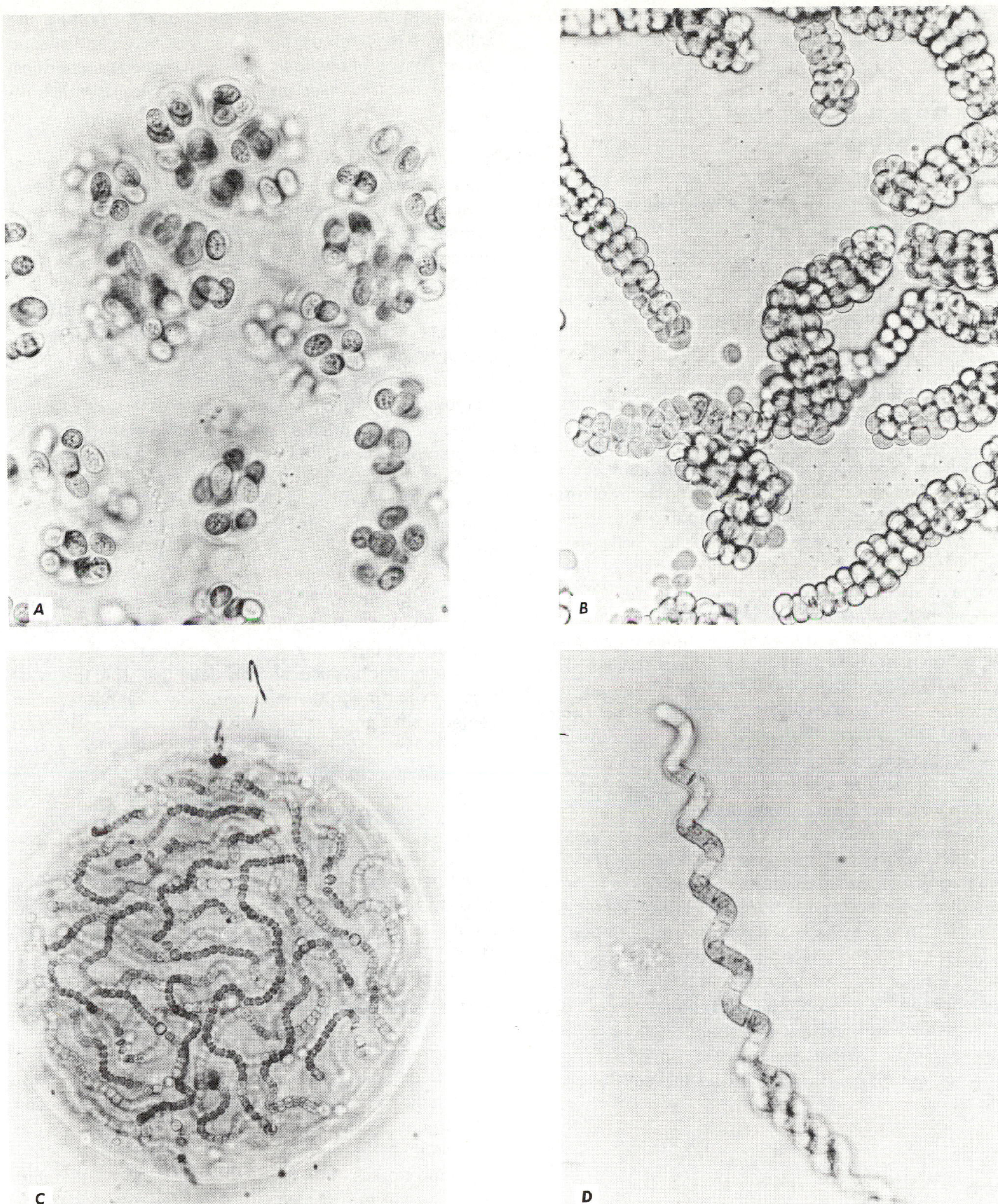

Figure 21.21
Diversity of form in the Cyano bacteria. *A* and *B*, loosely organized colonies of *Gloeocapsa* sp. and Fischerella musicola, both ×800; note the pronounced gelatinous sheath around *Gloeocapsa*. *C*, a colony of the filamentous *Nostoc* sp., ×880, with enlarged heterocysts visible, all embedded in a gelatinous matrix. *D*, the spiral filamentous *Arthrospira* sp., two filaments are intertwined, along one portion of the figure, ×880.

isms is a great step to devising effective means of controlling these baffling plant diseases.

BLUE-GREEN ALGAE, OR CYANOBACTERIA

These prokaryotes differ from the bacteria in chemical, cytological, metabolic, and morphological ways. All are photosynthetic and possess chlorophyll *a;* furthermore, they have unique accessory pigments and photosynthesis occurs on a series of internal membranes (not simply invaginations of the plasmalemma as in most bacteria). Cell volume ranges from 5 to 50 μm^3, in contrast to 0.01 to 5 μm^3 for bacteria. They have about twice as much DNA as does *E. coli.*

Morphological diversity ranges from unicells, to small colonies of cells, to simple and branched filaments (Figs. 21.21, **21.22,** page 410). Some of the filamentous forms show a small degree of differentiation: certain cells are modified for holding the filament to a substrate, some for reproduction, some for such specialized functions as nitrogen fixation. None of the cells exhibit flagella, but some are capable of a gliding or flexing movement.

Blue-green algae are commonly found in bodies of fresh and salt water, and they may at times be the major photosynthetic organisms in aquatic food chains. They are especially abundant in shallow, warm, nutrient-rich, or polluted water low in oxygen, and cell density can be great enough to color the water, creating a **bloom.** Accessory pigments may give such blooms a red color; for example, *Trichodesmium erythraeum* may have given the Red Sea its name. Most blooms disappear in a few days, but the cells can release nerve toxins lethal to animals that swim in or drink the water. *Microcystis, Nostoc, Aphanizomenon, Gloeotrichia,* and *Anabaena* all are capable of producing toxins. Some 0.5 mg of *Microcystis* toxin will cause the death of a mouse within an hour.

Other blue-green algae may be in the surface layers of soil, in the spray zone on rocky coasts just above the intertidal zone, on snow at high elevations, in acid bogs, and in alkaline hot springs at temperatures up to 75°C. Some species live symbiotically in the tissues of lichens, liverworts, cycads, and ferns, and in the cells of some protozoans.

CELL STRUCTURE AND FUNCTION

The cell wall is multilayered and is indistinguishable from the wall of Gram-negative bacteria. Its major component is a peptidoglycan. Outside of the 28 nm thick wall is a sheath equally as thick, containing fibrils loosely embedded in an amorphous matrix that may be colored yellow, brown, red, or violet (Fig. 21.23). A slime shroud, analogous to the capsule of bacterial cells, may extend beyond the sheath; its presence can be shown by putting the cells in an India ink background. The sheath and shroud are composed of pectic acids and mucopolysaccharides, and the total thickness is affected by environmental conditions.

The cytoplasm contains many ribosomes and appears granular. In filamentous forms, fine plasmodesmata connect adjacent cells. The plasmalemma may form invaginations, but in addition there are a series of parallel membranes within the cytoplasm that are separate from the plasmalemma. The light reactions of photosynthesis occur on these membranes, which contain chlorophyll *a* and several accessory pigments, phycoerythrin, phycocyanin, and allophycocyanin. The accessory pigments are grouped in rods or discs (called **phycobilisomes**) that are attached to both sides of the membranes and that are about twice the size of ribosomes. These pigments harvest light of wavelength 550 to 650 nm, and pass the energy on the chlorophyll *a.*

Other cytoplasmic inclusions are gas vesicles, granules of glycogen (the carbohydrate storage product), lipid droplets, granules or arginine and aspartic acid polymers (a nitrogen storage product), and polyhedral carboxysomes (which may be the site of CO_2 fixation in the dark reactions). Gas vesicles are especially prominent in floating, aquatic unicells, and it is likely that they contribute to buoyancy.

The nucleoplasm is sharply delimited from the cytoplasm, even though there is no nuclear membrane. As in bacterial cells, there is only one chromosome, composed of a circular, double-stranded molecule of DNA. A histone coating is absent.

NITROGEN FIXATION

About one-third of all blue-green algal species are able to fix nitrogen. In most of these cases, nitrogen fixation occurs in specialized cells called **heterocysts** (Fig. 21.24). These are enlarged, thick-walled cells, with a dense cytoplasm. The internal membranes no longer lie in parallel arrays, and these cells may have lost photosystem II, hence do not generate O_2. Plasmodesmata connect the heterocysts to adjacent cells within a filament. It is possible that the thick wall helps maintain an anaerobic condition in the cytoplasm, necessary for the activity of nitrogenase. The metabolic pathway of fixation is the same as that in bacteria.

In the tropics, blue-green algae such as *Nostoc* and *Anabaena* are purposely cultured in rice paddies to increase the fertility of the soil. Some experiments showed the *Anabaena cylindrica* could fix 3 to 400 kg of N per hectare per year. This organic nitrogen is liberated to the rice when the algae are mixed into the soil and acted on by decomposers. Rice yield increases dramatically when blue-green algae are cultivated this way, as a "free" form of fertilizer.

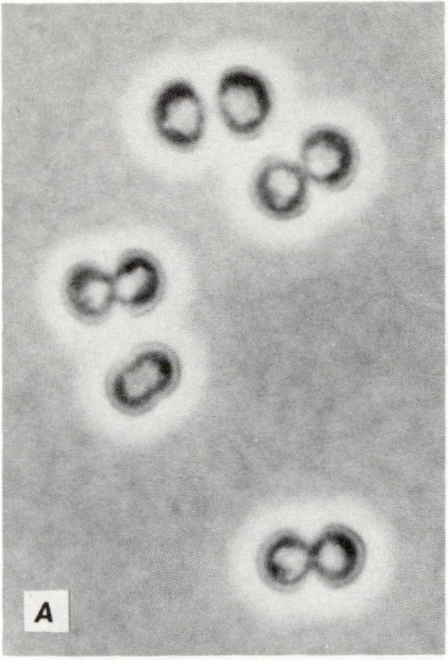

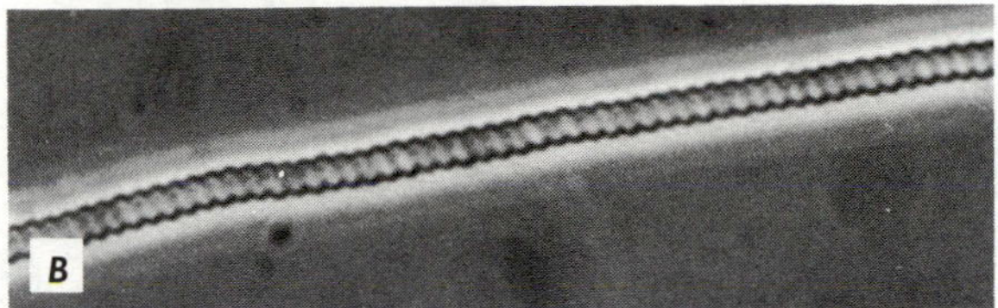

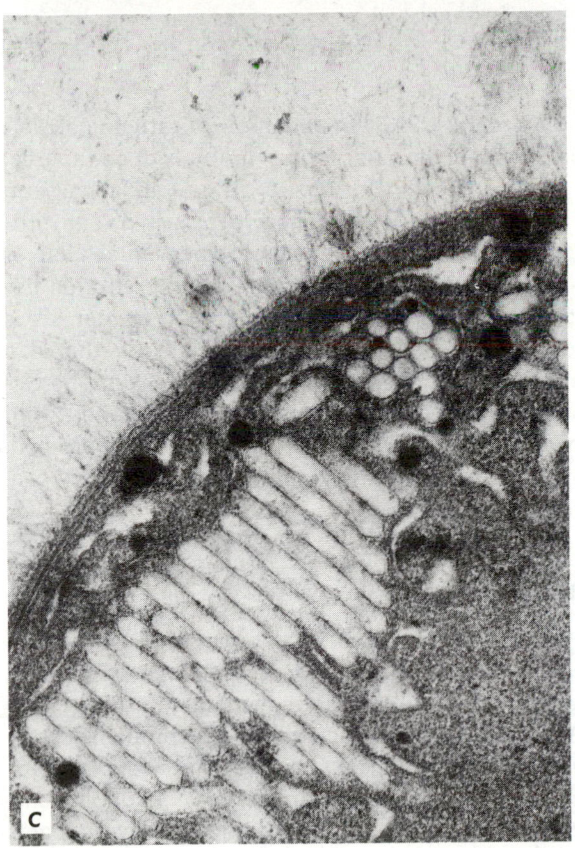

Figure 21.23
Sheath and wall of blue-green algal cells. *A*, sheath around *Snyechococcus*, ×1000. *B*, sheath around the filamentous *Spirulina versicolor*, ×1250. *C*, ultrastructure of *Anabaena flos-aquae*, ×73,500, showing microfibrils in the outer sheath, a four-layered cell wall, gas vesicles in longitudinal section (rods) and cross section (circles), photosynthetic membranes, dark-staining lipid bodies, and many granular ribosomes.

Some filamentous and unicellular species (e.g., *Gloeocapsa*) can fix nitrogen even though they lack heterocysts. It is not understood how they maintain a low internal oxygen concentration.

REPRODUCTION

Sexual reproduction has never been documented for the blue-green algae. Several forms of asexual reproduction, however, do occur. These forms involve cell division, fragmentation, akinetes, and spores.

Unicells reproduce themselves by cell division, which proceeds much as it does in the bacteria. That is, the long circular DNA strand replicates itself and separates into two equal circular strands. The nucleoplasm and cytoplasm divide in two without the appearance of chromosomes, a spindle apparatus, or a cell plate. Colonies enlarge by cell division and filaments elongate by cell division.

Filaments may fragment, often at weak points where a cell has collapsed, or sometimes next to a heterocyst. The segments that are released are called **hormogonia** (sing. = **hormogonium**), and they can produce a new individual by continued cell division at the apex.

Some filamentous species form **akinetes**, which are enlarged cells with a thickened wall, many internal food reserves, and a large amount of DNA. After a resting stage, cell division occurs within the cell, the parent wall ruptures, and a small, short filament of cells is released. Heterocysts may also function in this way, but their major function is nitrogen fixation.

Spore formation is not common. Some unicellular, epiphytic species are capable of pinching off an apical part of the cell and releasing it as an expospore, whereas others can divide within the parent wall and produce many endospores.

SYMBIOSES

Lichens

Lichens are **symbiotic, mutualistic** (see Chapter 18) **associations** between certain fungi and algae. Many biologists classify them with the fungi (see Chapter 26). Each lichen takes on a shape and color and habitat that depends on the species of fungus and alga present. Some lichens are firmly appressed to a substrate **(crustose),** others are attached at one point and appear as exfoliating "leaves" (**foliose** lichens), and others are cylindrical and freestanding or pendulous and epiphytic **(fruticose).**

Most lichens contain green algae, but those with blue-green algae may nevertheless be very common and abundant in a given habitat. *Chroococcus, Gloeocapsa,* and *Nostoc* are the usual blue-green algae involved, thus these lichens are capable of nitrogen fixation. Lichens are more completely described in Chapter 26.

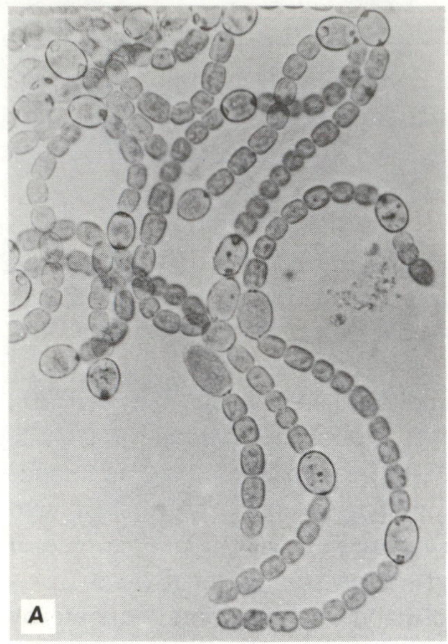

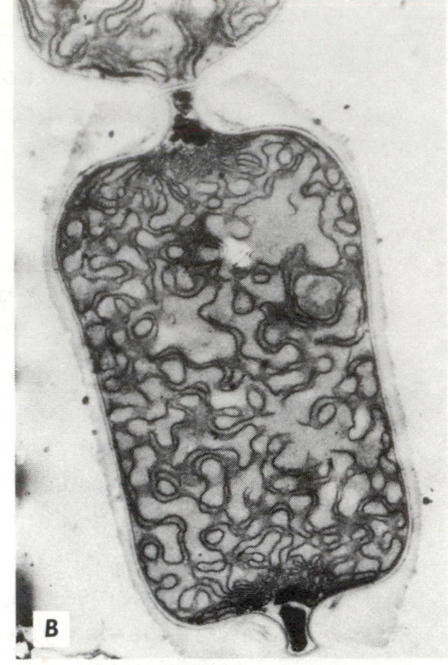

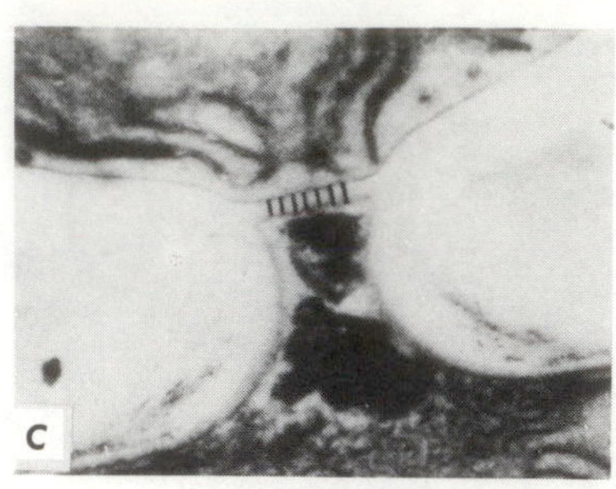

Figure 21.24
Ultrastructure of the heterocyst of the blue-green alga *Anabaena*. *A*, filaments of A. *azollae*, ×880, showing enlarged heterocysts in different stages of development. *B*, electron micrograph of a heterocyst of A. *cylindrica*, ×22,000, showing constructions at heterocyst poles, three layers outside the cell wall, and contorted thylakoids (photosynthetic membranes) in the cytoplasm. *C*, an enlargement of *B*, showing microplasmodesmata connecting the cytoplasm of the heterocyst with that of the adjoining vegetative cell. The microplasmodesmata have been artificially darkened in this photo to enhance their visibility.

Symbioses with Other Organisms

Some blue-green algae live within the cells of other algae. For example, *Richelia* lives in the diatom *Rhizosolenia.* Others live in the cells of protozoans. For example, *Synechococcus* lives in the fresh water flagellate *Cyanophora.* The blue-green algal cells within protozoans have lost their cell walls and they function as chloroplasts. They divide prior to host cell division and are passed into the progeny cells.

Other blue-green algae live symbiotically in the tissue of the liverworts *Blasia* and *Anthoseros,* in the water fern *Azolla,* and in the cycads *Zamia* and *Cycas. Blasia pusilla* lacks chloroplasts of its own and it is nutritionally dependent upon *Nostoc* colonies scattered in the thallus. The gametophyte thallus of *Anthoceros* has mucilaginous cavities, some of which harbor *Nostoc* filaments. *Anabaena* is present in *Ricciocarpus,* and also in leaf cavities of the small, floating, aquatic fern *Azolla* (Fig. 21.25). Research on the cultivation of *Azolla* in rice paddies is becoming very important in California and elsewhere. The nitrogen fixed by the blue-green symbiont is released to the soil and water, improving rice yield without the application of artificial fertilizer. *Anabaena* infects root cortex cells of certain cycads, much as *Rhizobium* bacteria infect legume roots, and it is possible that the symbiotic relationships are analogous.

CLASSIFICATION

The Cyanophyta or Cyanobacteria may be broken down into three orders. The Chamaesiphonales are unicellular, colonial, or filamentous forms that produce spores. Some of the unicellular species grow epiphytically on other aquatic plants. The other two orders do not produce spores. The Chroococcales include unicellular or colonial forms. *Gloeocapsa,* which can fix nitrogen despite the absence of heterocysts, is in this order, as is *Microcystis,* a member of the phytoplankton that can create blooms and release a potent neurotoxin. The Oscillatoriales are filamentous and include all genera with heterocysts (*Nostoc, Anabaena, Rivularia, Gloeotrichia,* and others), and a few without (*Oscillatoria, Lyngbya, Spirulina,* and others).

SUMMARY

1. Prokaryotes (procaryotes) account for 1 to 2% of all species in the plant kingdom; they include the bacteria and the blue-green algae.
2. Bacteria may be unicellular or filamentous. They are ecologically important as decomposers (saprophytes) or consumers (parasites). Economically, they have enormous positive and negative impacts through their roles as plant-animal-human disease agents, in fermentation, and in nitrogen fixation.
3. Bacteria are widely distributed and some forms are tolerant of many environmental extremes. They may be carried passively by air or water, or are capable of gliding, flexing, or swimming motions (the latter two by flagella with a uniquely simple ultrastructure).

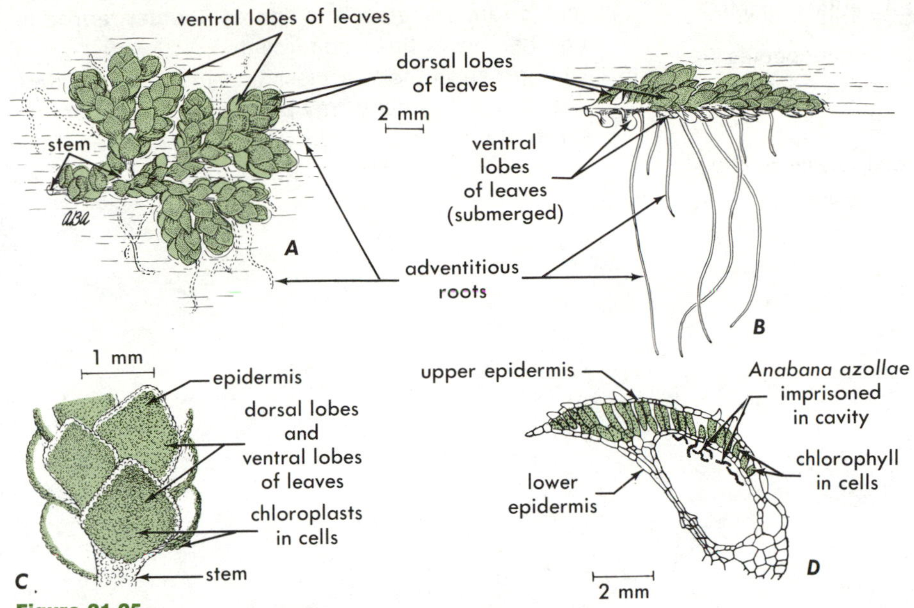

Figure 21.25
The floating fern *Azolla,* which has a symbiotic relationship with the nitrogen fixing blue-green alga *Anabaena.* A and B, whole plant views from above and from the side; C, details of the small, clasping leaves; D, leaf anatomy, showing blue-green alga *Anabaena.*

4. Based on carbon source and energy source, bacteria are placed in the following trophic categories: photoautotrophs, photoheterotrophs, chemoautotrophs, or chemoheterotrophs. In photosynthetic forms (photoautotrophs), photosystem II is absent and O_2 is not evolved, anaerobic conditions are required, and bacteriochlorophyll replaces chlorophyll *a.* Chloroplasts are absent. Dark reactions follow the Calvin cycle.
5. Gram-positive bacteria possess a 10 to 50 nm thick cell wall largely composed of peptidoglycan. Gram-negative bacteria have a thinner wall composed of peptidoglycan to the inside, a unit membrane of protein and lipids to the outside. Very small bacteria called mycoplasmas and certain halophilic bacteria lack any kind of wall.
6. Bacterial cytoplasm contains ribosomes, invaginations of the outer membrane (which function in photosynthesis and respiration), carboxysomes, gas vesicles, and glycogen granules. The nucleoplasm is not bounded by a membrane.
7. During cell division the nucleoplasm pinches in two. A single, circular chromosome replicates. Two daughter cells pull apart until the stretched point of connection ruptures. Some bacteria are capable of budding, and others can produce resistant cells (spores).
8. Sexual reproduction rarely occurs except for an Hfr strain of *Escherichia coli.* Two cells touch and fuse at a point, then DNA passes from a donor cell to a recipient cell. Generally, only bits of extrachromosomal DNA called plasmids are passed, but sometimes part or all of a chromosome will also be passed.
9. Classification of bacteria is based on cell function (metabolism), cell shape, wall traits, form of motility, ability to form spores, requirement for oxygen, and trophic category. There are approximately 3000 species.
10. Blue-green algal cells are about 10 times the volume of bacterial cells, they contain twice the amount of DNA, all are photosynthetic, internal membranes are present, and chlorophyll *a* is present. They may be unicellular, colonial, or filamentous, and some filaments show differentiation among the cells. Flagella are absent.
11. Blue-green algae are especially abundant in shallow, warm, nutrient-rich or polluted water that is low in oxygen. Some species are tolerant of high elevations, cold temperatures, alkaline hot springs, acid soils, or salt spray. A few live symbiotically with other plants or protozoans.
12. Blue-green algal cytoplasm contains ribosomes, parallel arrays of membranes, gas vesicles, glycogen granules, lipid droplets, and carboxysomes. The nucleoplasm is not bounded by a membrane. A single, circular chromosome is present. The cell wall is like that of gram-negative bacteria; however, it is thicker and may also be enveloped by an outer sheath of fibrils and a slime shroud, both containing pectic acids and mucopolysaccharides. Plasmodes-

mata penetrate the wall, connecting adjacent cells.

13. About one-third of all blue-green algal species fix nitrogen. Most of them fix nitrogen in specialized cells called heterocysts.
14. Asexual reproduction involves cell division, fragmentation, akinetes, and spores. Sexual reproduction has never been documented.
15. Blue-green algae are classified into three orders, based on capacity to produce spores or heterocysts, Chamaesiphonales, Chroococcales, Oscillatoriales, and form of the plant body.

CHAPTER 22

VIRUSES

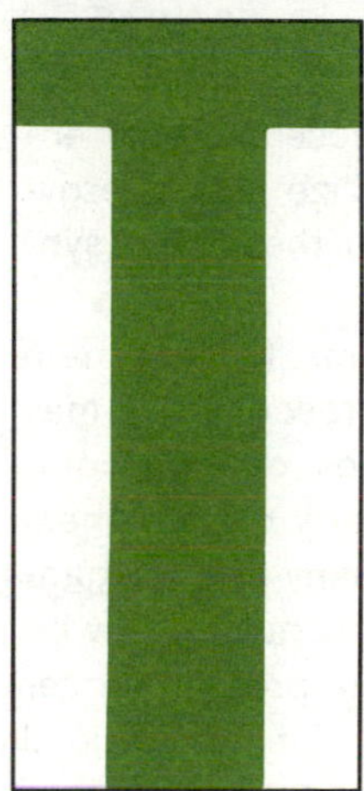

oward the close of the nineteenth century, D. J. Iwanowski, a Russian plant scientist, carefully pressed some sap from a stunted and diseased tobacco plant. With a small amount of this sap he inoculated a healthy tobacco plant. The healthy plant soon developed signs of a disease similar to that of the plant from which the sap was taken. The experiment was repeated, but this time the sap from the diseased plant was forced through a filter with pores so small that bacteria could not pass through them. The healthy plants inoculated with this filtered sap soon contracted the disease of the original plant. This was a new discovery—a disease could be transmitted from one plant to another by a filtered sap that was free of visible living bodies even when viewed by the most powerful light microscopes. The sap contained something (a **virus**) that caused an infectious disease.

THE NATURE OF VIRUSES

Viruses are a third major category of life, distinct from prokaryotes and eukaryotes by being acellular. A virus has two phases in its life cycle, and in neither phase is its basic unit a cell. Viruses are small particles, less than 0.2 μm in one dimension and thus below the resolving power of the light microscope. They are all, as far as known, unable to multiply outside a host. Outside the host viruses exist as inert, infectious particles called **virions.** A virion consists of DNA or RNA surrounded by a protein coat. There is enough genetic information to contain 10 to several hundred genes. The DNA may be single-stranded or double-stranded, circular or linear. The RNA may be single-stranded or double-stranded. Inside a host cell, a virus consists of replicating DNA or RNA alone, and the host cell provides all other material necessary to produce virions once again and to continue the cycle.

Viruses appear to become a part of the host cell. They do not possess energy-yielding or synthetic enzyme systems. Viral constituents, particularly their nucleic acids, are similar to those of their hosts and induce the host cell to elaborate viral, rather than host, materials. Therefore, viruses cannot multiply outside cells. Due to this close association with living cells, we know viruses only because of their direct effect upon either bacterial, animal, or plant cells. Indeed, viruses may be more logically compared to cellular constituents than to whole cells. Viruses, as cell constituents, offer a logical explanation of their infectious nature. The similar cell and viral constituents are the nucleic acids that carry genetic information to specify cell functions. Since the host cell cannot distinguish viral nucleic acid from its own, it utilizes these new genetic messages to obediently synthesize constituents for the virus. Some of the new virus-specified constituents are enzymes. These and host enzymes synthesize new virus and utilize the energy-providing pathways of the host cell in these new syntheses.

All viruses are small infectious agents. Most of them can be seen only with the electron microscope. We may catalog them as rods, spherical particles, or more complicated particles, but we cannot properly classify them by their morphology alone as we do many living organisms. Table 25.1 gives a list of several plant viruses with shapes and sizes. For convenience, the present procedure is to group them as bacterial, plant, or animal viruses. This is an artificial classification, and we may find that some viruses are infectious for a wide range of host organisms.

Smallpox, poliomyelitis, and measles are examples of viral diseases of man. Dog distemper is caused by a virus. The so-called mosaic diseases of plants, characterized by a mottling of the leaves, are also viral diseases.

Table 22.1

SOME PLANT VIRUSES WITH THEIR DIMENSIONS AND VECTORS

Virus	Size (Nanometers)	Vector
	Spherical viruses	
Tobacco necrosis	20	Fungus
Bean mosaic	30	Aphid
Squash mosaic	30	Beetle
Potato yellow dwarf	110	Leafhopper
	Short-rod viruses	
Hydrangia ringspot	44 × 16	Unknown
Tobacco mosaic	280 × 15	Unknown
	Long-rod viruses	
Bean yellow mosaic	750 × 12	Aphid
Beet yellows	1250 × 10	Aphid

Even though the actual agent causing these viral diseases was long unknown, much has been learned of methods to combat the diseases.

For instance, a disease known as curly top practically eliminated sugar beet fields in California by 1920. Careful study of the disease showed that it was caused by a virus and that it was transmitted by leaf hoppers. The leaf hopper overwinters on weedy plants adjacent to sugar beet fields and rapidly infects the young beet seedlings in the spring. It is still impossible to cure a viral infection once it has become established in a host, and it is not practical to develop a vaccine to inoculate sugar beets against the virus. Two pathways are open to plant pathologists interested in viral disease control: (*a*) develop sugar beet plants resistant to the virus, or (*b*) control the population of leaf hoppers. Sugar beets resistant to the virus have been developed and the leaf hoppers are controlled by insecticides and herbicides that keep down the overwintering weeds.

Recall from Chapter 21 that some diseases of plants long thought to be virus-induced (e.g., some "virus" yellow diseases) are now known to be caused by bacteria (mycoplasmas).

BACTERIAL VIRUSES

Bacterial viruses were first known as **bacteriophages,** a term that has been shortened to **phage** and is now used to designate a bacterial virus. Phages were discovered independently by two investigators, Twort (1915) and d'Herelle (1917), both of whom were curious over the sudden dissolution, or **lysis,** of bacterial colonies or portions of colonies. They were able to show that the destruction of the bacteria was due to an infectious agent, (*a*) too small to be seen with the light microscope, (*b*) capable of passing through a fine filter, and (*c*) capable of being transmitted in bacterial cells. Its characteristics thus conformed to the definition of a virus.

It is now known that there are at least two developmental pathways that a phage may take once it gains entry to a bacterial cell. If the phage is **virulent,** or **lytic,** it will induce the host to produce many virus particles and to extrude those particles (in the case of some animal viruses), or, more typically, to lyse (split open) and to release the particles all at once. If the phage is **nonvirulent** (synonyms include **temperate** or **prophage**), the viral DNA or RNA will be repressed. Ultimately the repression may be released and nonvirulent phage will become virulent. We will discuss each of these cycles in turn.

Virulent Phage

Bacterial phage often consists of a head and a hollow tail region (Fig. 22.1). Fibrils extend from the tail and these play a role in attaching the virion to the host cell wall. Some virions can attach to flagella. A contractile sheath may be present around the tail. Head, tail, sheath, and fibrils are protein in nature. In the T-even phage shown in Fig. 22.1, the sheath then contracts and the DNA within the head is injected into the host cell, much as a syringe would force liquid through a needle inserted in skin. The penetration process may be different, and unexplained as yet, in other phage. The protein head, about 70 nm in diameter, and the tail usually remain outside the bacterium (Fig. 22.2).

Immediately upon the entry of the viral DNA, replication of bacterial DNA stops. Synthesis of new bacterial enzymes also stops, but those present continue to function for a time. Under the influence of viral DNA, bacterial enzyme systems either synthesize, or are themselves converted into systems capable of elaborating necessary phage constituents. New viral RNA synthesis is necessary to initiate and direct the synthesis of new viral constituents. Phage DNA is now replicated in the bacterial cell. This activity results in the elaboration of 100 or more new phage particles, complete with protein tails and head coating, the latter enclosing a viral DNA core. The bacterium now disintegrates or undergoes lysis and viral particles are released to infect more bacteria (Fig. 22.3). The whole process takes about 1 hour and is shown diagrammatically in Fig. 22.4.

Viral DNA does not contain the base cytosine as does bacterial DNA, but another base, hydroxymethylcytosine (HMC). This is a convenient circumstance, as it makes it possible for an investigator to distinguish viral and bacterial DNAs.

Nonvirulent Phage

If the injected RNA or DNA is repressed by the bacterial cell, the host cell is termed lysogenic. The virus nucleic acid may become incorporated into the DNA of the host chromosome or may become attached to the plasmalemma; in either case it replicates with the host and is passed on to progeny cells. In this state, the phage is known as a **prophage** and it confers immunity upon its host against invasion by other similar phage particles. Rarely, a progeny cell is formed that does not contain the prophage. Metabolism of these phages is so subtly linked to that of the host bacterium that both may exist through

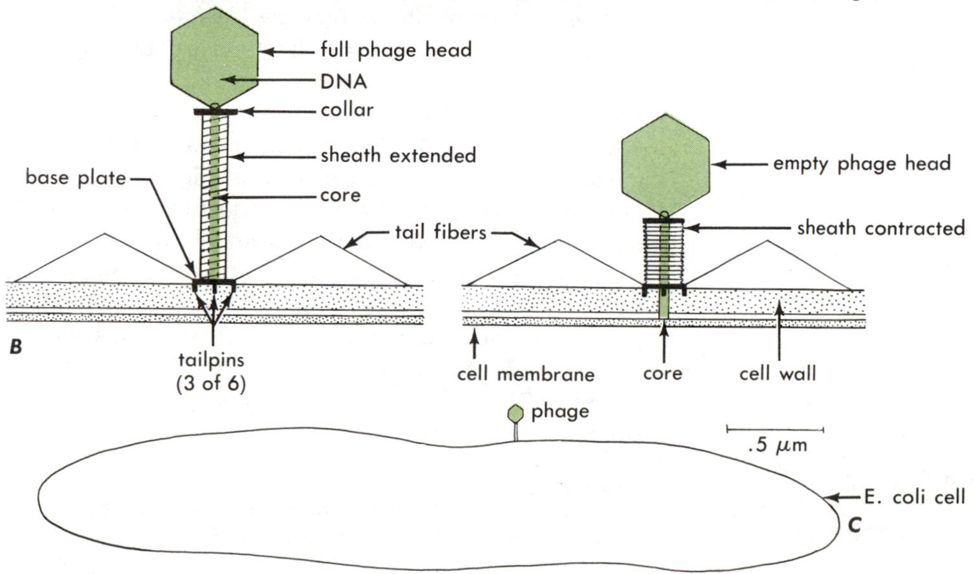

Figure 22.1
A, an electron micrograph of a T-even phage particle, ×,60,000. *B*, diagrammatic representation of phage before (left) and after (right) injection of DNA through the bacterial wall. *C*, relative size of phage and bacteria. (*A*, courtesy of A. S. Brenner, et al., *J. Mol Biol.* **1,** 281. © by Academic Press Inc. London Ltd. *B*, redrawn from R. Y. Stanier, E. A. Adelbert, J. Ingraham. *The Microbial World*, pp. 378. © 1976 4th ed., Reprinted by permission Prentice-Hall, Inc., Englewood Cliffs. New Jersey.)

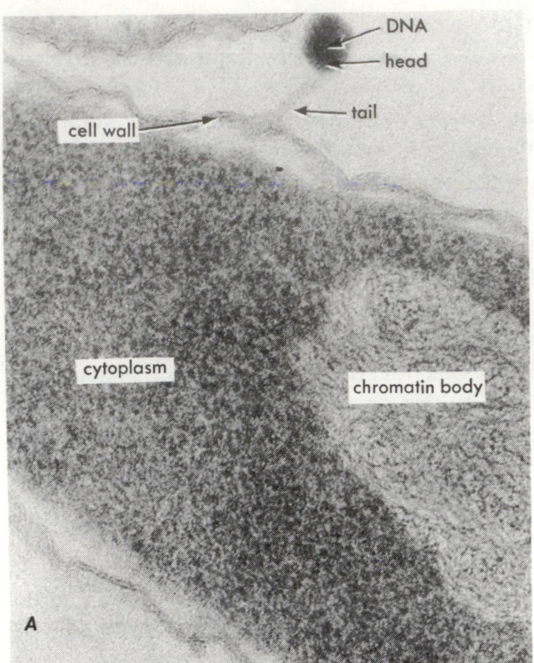

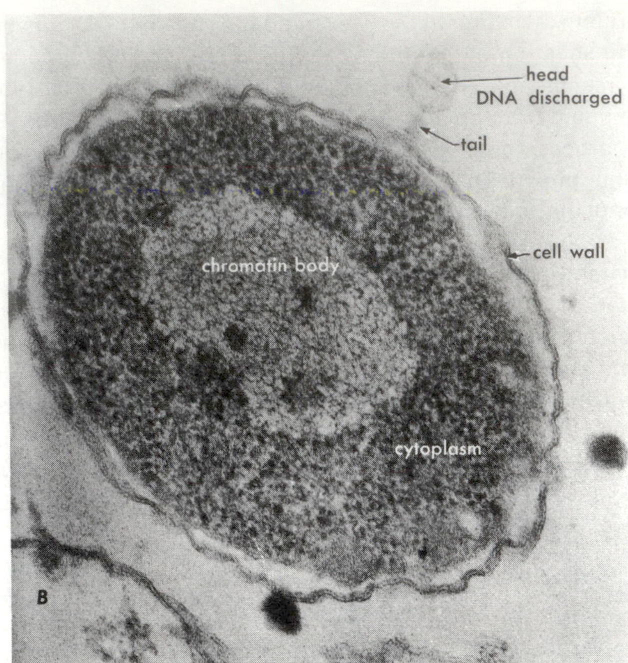

Figure 22.2
Phage T2 attached to the surface of an *E. coli* cell. *A,* before discharge of DNA from the head, ×20,000. *B,* after discharge of DNA from head, ×22,000. (Courtesy of E. H. Cota-Robbles.)

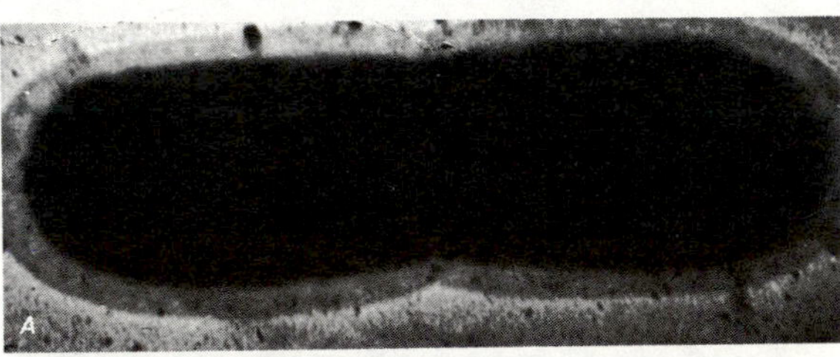

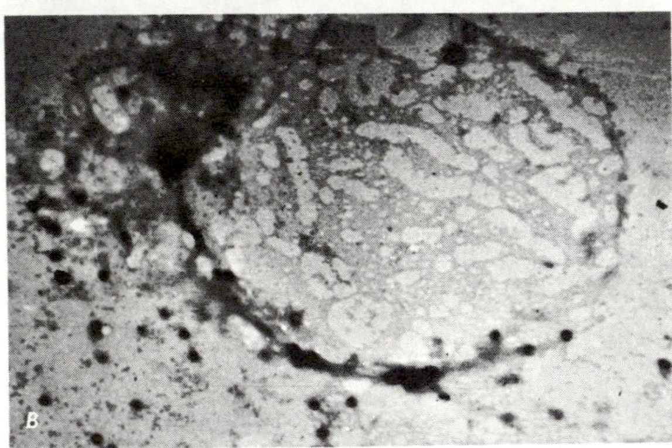

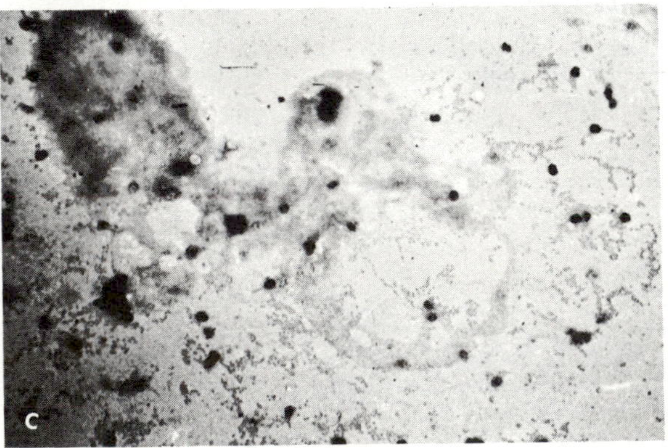

Figure 22.3
Disintegration of a bacterium and multiplication of virus particles. (Courtesy of S. E. Luria.)

many generations without apparent injury to the host. The prophage changes to an active phage when its DNA becomes separated from that of the host. At this juncture, phage DNA directs the bacterial cell to elaborate phage constituents; the phage multiplies and eventually matures; infectious, temperate phage particles are released from lysing bacterial hosts.

During replication of phage, a temperate phage particle can excise a segment of bacterial DNA and carry it into another bacterium; the new host may be influenced by bacterial DNA carried by the phage. This process is known as **transduction** and occurs as described below.

Transduction

A strain of *Salmonella typhinurium* is known that is resistant to streptomycin; it may be designated as str-r. Phages may be reproduced in bacteria resistant to streptomycin. A phage becomes attached to the bacterial cell wall and phage DNA enters the bacterial protoplast (Fig. 22.5*A*). The bacterial DNA disintegrates and there is a multiplication of the vegetative virus (Fig. 22.5*B*). Disintegration of bacterial DNA does not destroy the individual bits of genetic information but seems to distribute it at random in the bacterial cell (Fig. 22.5*C*). When phage DNA becomes incorporated in protein heads of newly forming phage particles, replicated phage DNA carries with it some original bacterial DNA (Fig. 22.5*D*). This original bacterial DNA is known as the **transducing fragment.** Upon lysis, phage particles are released into the medium,

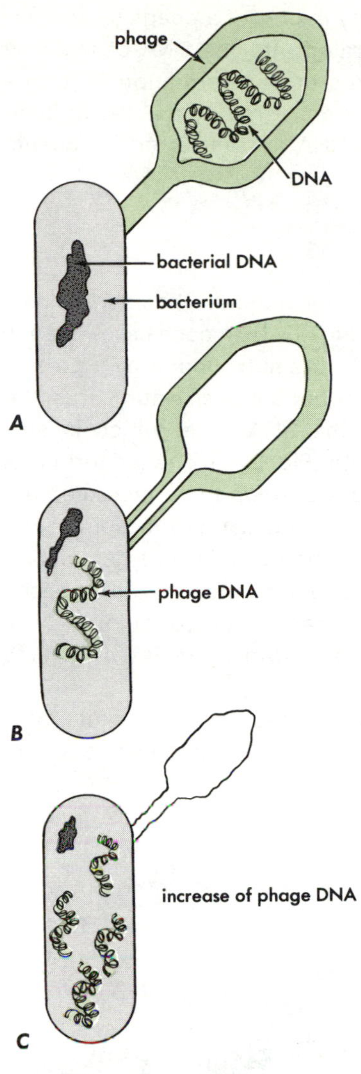

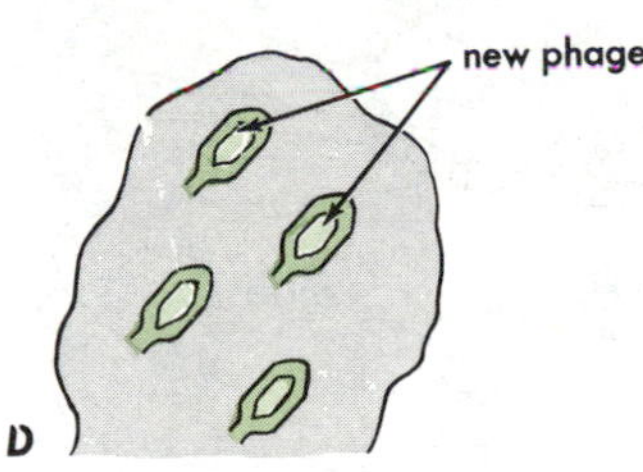

Figure 22.4
Diagram of a life cycle of a virulent phage. The bacterial cell and virion have not been drawn to scale.

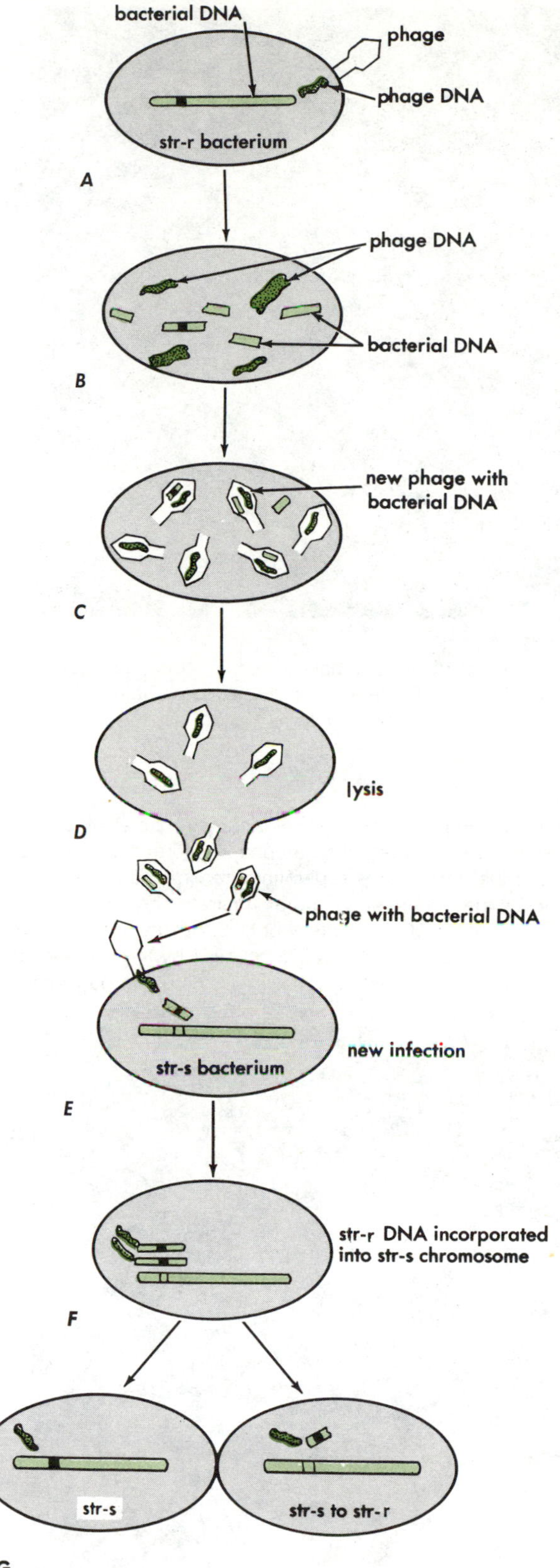

Figure 22.5
Diagram to show probable steps in the transduction of genetic information from one bacterium to another by way of a transducing virus. (Redrawn after R. Sager and F. J. Ryan, *Cell Heredity*, John Wiley & Sons, New York.)

where they may be collected and transferred to a culture of streptomycin-sensitive bacteria (str-s). If this culture is now grown on a medium containing streptomycin, about one in a million bacterial cells will be str-r.

What has happened is this: When phage DNA is injected into the bacterium, the tranducing fragment enters also (Fig. 22.5*D*) and is replicated along with phage DNA. Since this is a temperate phage, lysis will not occur immediately, and upon division of the bacterial cell the str-r gene in the tranducing fragment somehow takes the place of the str-s gene in the bacterial DNA (Fig. 22.5*F*). Note that in this case the transducing fragment merely goes along with the viral DNA. There is also evidence that the host str-r gene may become incorporated in the viral DNA by gene recombination. **Transduction** is

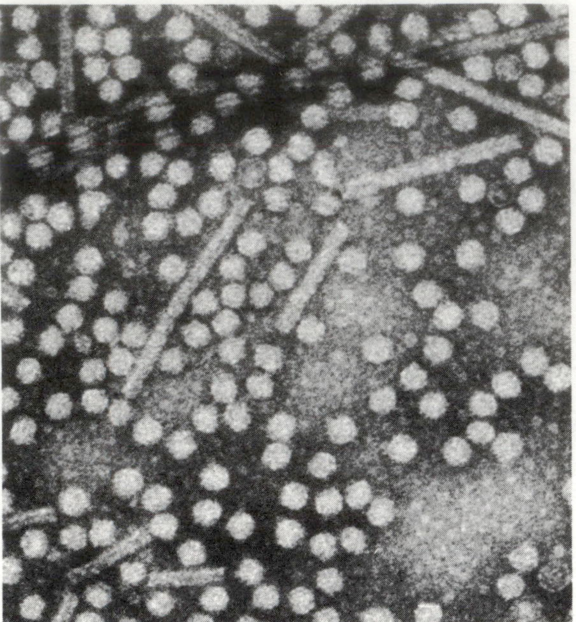

Figure 22.6
Rod-shaped particles of tobacco mosaic virus and isodiametric particles of grape fan leaf virus. TMV virus particle is 18 nm wide. (Courtesy of Robert Taylor.)

Figure 22.7
Bushy stunt virus stained to show the central core, which is composed of RNA. The clear area around the core is a layer of protein, and the dark circles separating the virus particles are formed by the precipitation of uranyl nitrate absorbed on the protein sheath, ×150,000. (Courtesy of H. E. Huxley and G. Zubay, *J. Biophys. Biochem. Cytol.* **11,** 273. © 1961 by the Rockefeller University Press. Reprinted with permission of the publisher.)

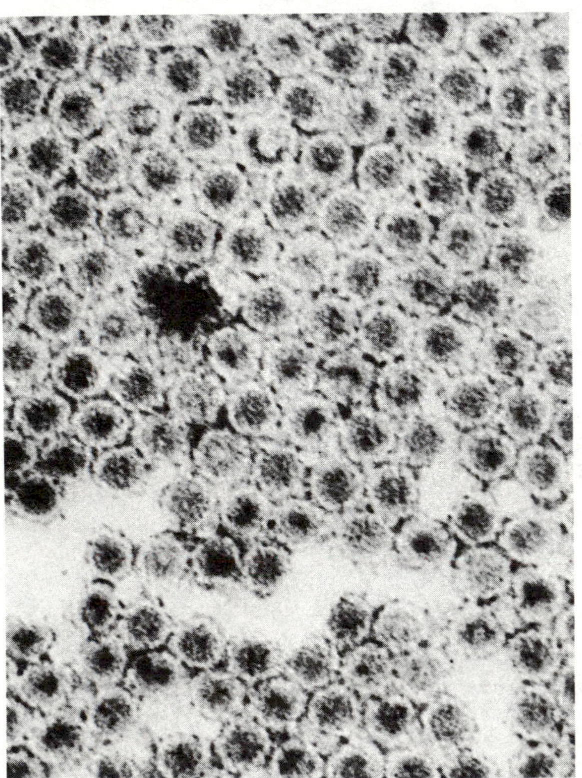

the transfer of host DNA by means of a phage to another host where it is incorporated into the DNA of the new host. It might be compared to transformation: in transduction, bacterial DNA is transferred by a virus, while in transformation, bacterial DNA is transferred by agents other than phage.

PLANT VIRUSES

Plant viruses occur as either spheres, bacilliform particles, or elongated rods (Fig. 22.6). In each case, there is some sort of RNA or DNA core surrounded by a protein sheath. In the case of the spherical bushy stunt virus, the center of the sphere is the RNA core, which is surrounded by a protein sheath (Fig. 22.7). The rod forms, as illustrated by tobacco mosaic, seem to be more complicated, for example, twisted into a uniform spiral (Fig. 22.8). The presence of virus particles may be detected in the chloroplasts of tobacco plants infected with tobacco mosaic virus (Fig. 22.9). Under certain conditions, these particles may become arranged in a precise crystalline array (Fig. 22.10).

Many angiosperms are susceptible to virus infection. So far, no viral diseases of mosses has been reported.

Figure 22.8
Diagrammatic interpretation of the structure of tobacco mosaic virion.

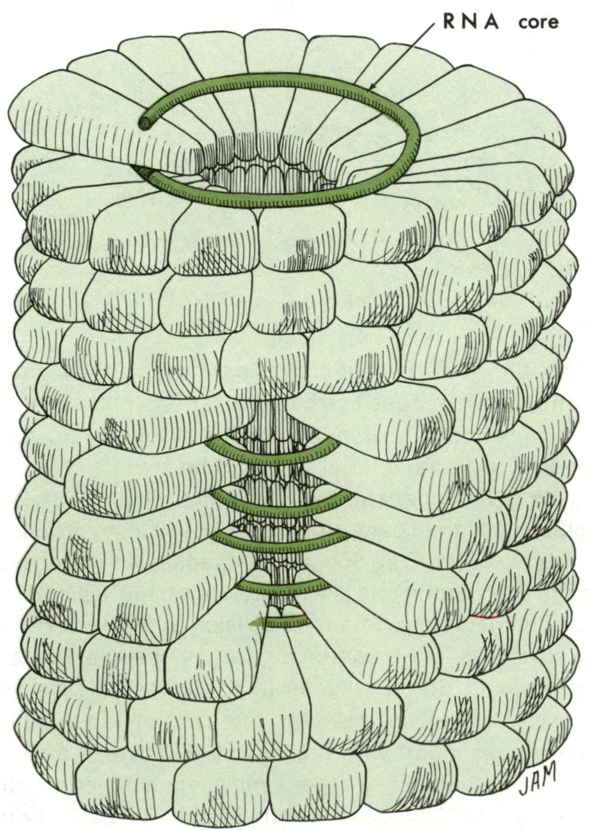

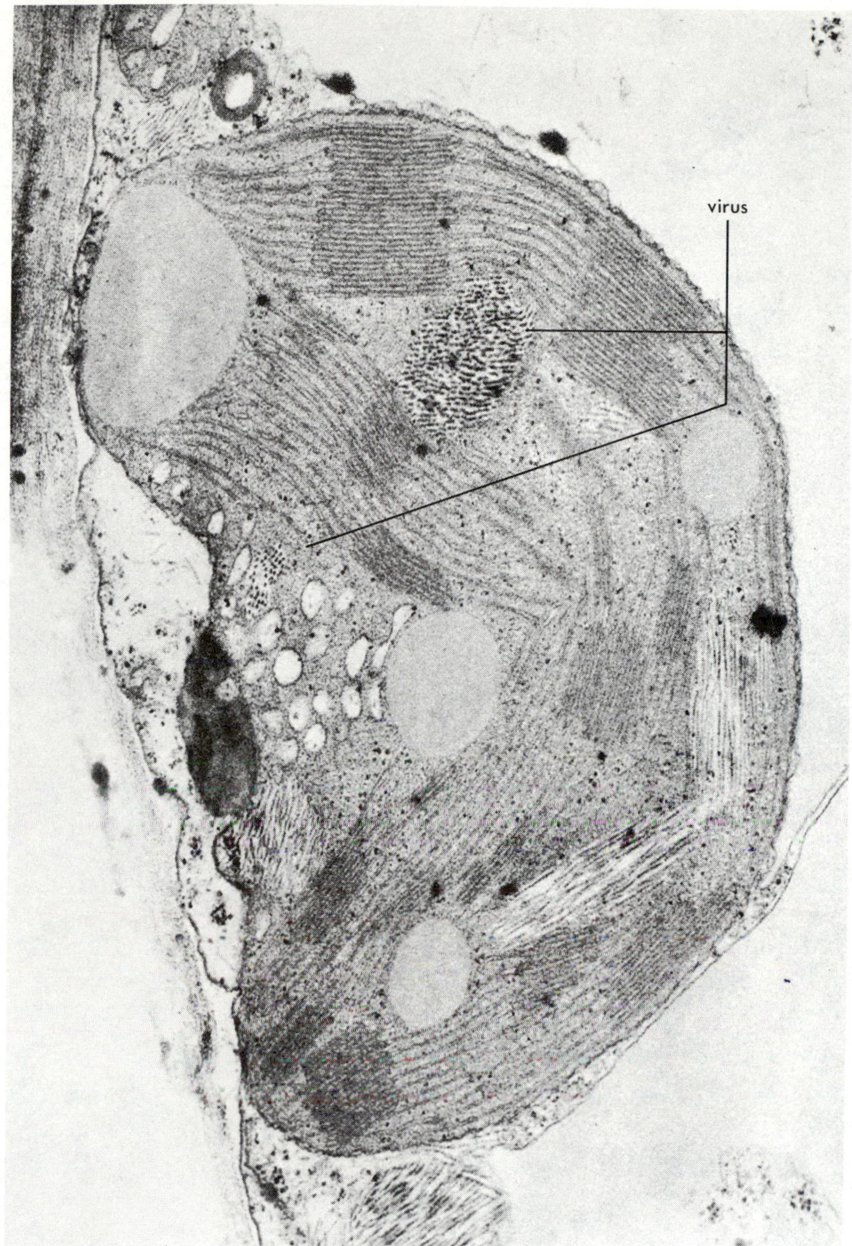

Figure 22.9
Chloroplast from a tobacco plant infected with tobacco mosaic virus (U-5 strain). Note array of crystalline like particles, ×25,000. (Courtesy of T. A. Shalla.)

Tobacco itself may suffer from 15 different viruses, and potatoes from 18. As early as 1935, viral diseases were known for 1100 species of angiosperms. Furthermore, a single virus may exist as several distinct strains. Fifty-four strains have been described for the common tobacco mosaic virus.

Symptoms of Viral Diseases of Plants

One of the most common symptoms of viral infection in plants is the occurrence of light-green or yellow areas on leaves (Fig. 22.11). These areas range from small circles to large irregular patches, to stripes, or irregular bands. In some instances, the entire diseased plant may be a much lighter green than the healthy one. Many variegated horticultural varieties of plants undoubtedly are infected with some mild acting virus. Diseases characterized by mottling or variegation of leaves are referred to as "mosaics," and almost all are regarded as viral diseases. We have, for instance, tobacco mosaic, potato mosaic, and sugar cane mosaic.

The yellow or lighter areas of diseased leaves are usually thinner in cross section than are those of normal healthy leaves. Chloroplasts may be absent or very reduced in size (Fig. 22.11) and in number, or they may be

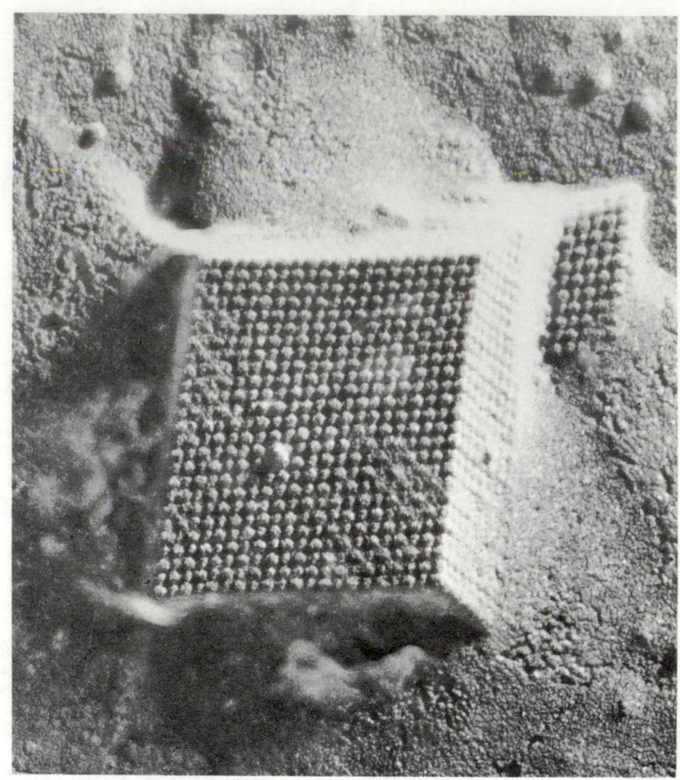

Figure 22.10
A crystalline particle of the tobacco necrotic virus, ×84,000. (Courtesy of L. W. Lablow, *J. Ultrastruct. Res.* **2,** 177. © 1958 by Academic Press. Reprinted by permission of the publisher)

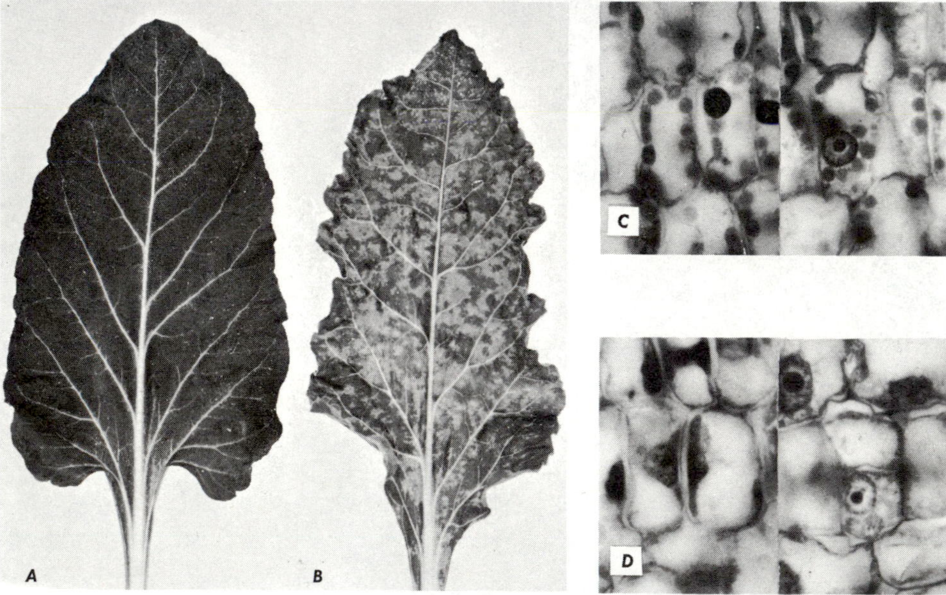

Figure 22.11
A, healthy sugar beet leaf. *B,* sugar beet leaf infected with mosaic virus. *C,* cells from healthy leaf. *D,* cells from diseased leaf. (Courtesy of K. Esau and C. W. Bennett.)

yellow in color. This reduction in green pigment, and therefore in the color of the leaf, is called **chlorosis.** A chlorotic leaf is one that has light green or yellow areas. Not all chlorotic leaves, however, harbor a mosaic virus; chlorosis may be due to other factors, including mineral deficiencies.

In addition to being chlorotic, leaves of virus-infected plants may be irregular in shape with a rough or puckered surface, may be rolled in along the edges, may be reduced in width, or may turn yellow and drop prematurely (Fig. 22.12).

Flowers of infected plants are frequently dwarfed, mottled, or streaked. Streaking of flowers occasionally produces a highly prized horticultural variety; many hundreds of dollars have been paid for tulip bulbs infected with a virus that causes a streaking of the flowers

Figure 22.12
Healthy plants compared with those infected with a virus. *A*, healthy bean. *B*, diseased bean. *C*, healthy grape. *D*, diseased grape. (Courtesy of Plant Pathology Department, University of California.)

(Fig. 22.13). Fruits may be small, distorted, and mottled. Tomatoes, peppers, melons, and beans infected with certain viruses show a distinct mottling; tomatoes and cucumbers may be distorted and much reduced in size. Stems may show discoloration of various sorts; frequently, they are dwarfed and contain cankers or dead areas.

Figure 22.13
Tulip flower infected with a virus disease.

Viruses belonging to the mosaic group induce changes primarily in parenchyma tissue. Vascular tissues, however, also may be injured as the result of a primary injury to the parenchyma tissue of leaves and stems. Other viruses primarily harm the conducting tissue, particularly the phloem. The phloem may also serve to translocate virus particles. The sieve tubes are injured and frequently killed, forming a mass of dead cells on the outer edge of the vascular bundle. This mass impairs the movements of food in the plant and results in injury to other plant tissues.

In many plants, viruses seem to interfere with the tissues responsible for production, as well as translocation, of food (Figs. 22.14, 22.15). This damage may result in a stunting of all organs of the plant (Fig. 22.16). Viral diseases of plants claim a large toll from worldwide agriculture. Swollen shoot virus infects cacao trees in West Africa; over 1 million cacao trees were lost in one western province of the Gold Coast in 1950, and in one five-year period 74% of the 30 to 40 year old trees were killed. In the 1950s, another viral disease, tristeyo, caused the loss of 7 million citrus trees in São Paulo State, Brazil. Although the curly top disease of sugar beets is under control, another viral disease of sugar beets, virus yellows, probably causes a 35% reduction in sugar yields over large areas of the western United States each year. Unfortunately, many plant viral diseases, particularly of the mosaic type, are more insidious. They induce less conspicuous symptoms and cause less dramatic losses, and hence are tolerated more willingly by growers. Over a period of years, their cumulative effect represents an enormous loss of food production. In developing coun-

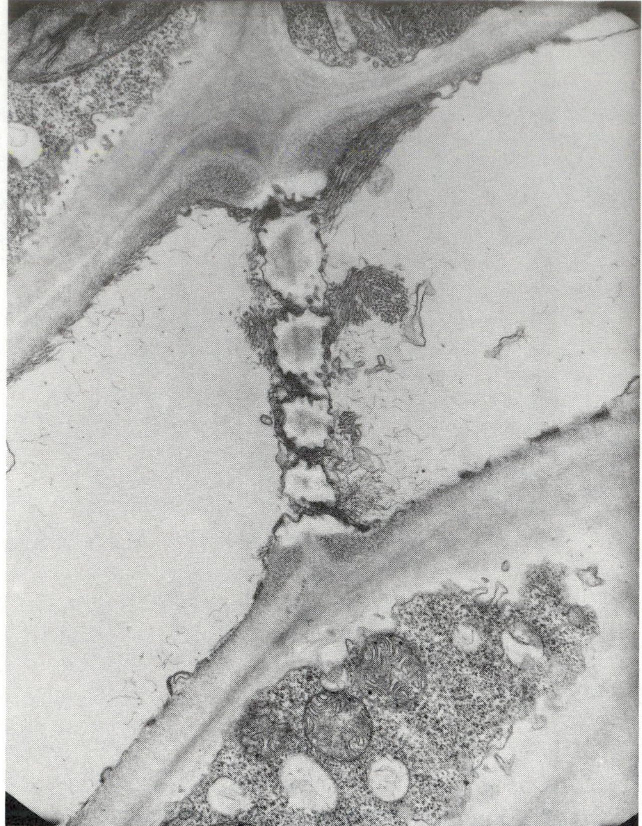

Figure 22.14
Virus particles present at the sieve plate in a sieve tube of a plant infected with beet-yellow virus. (Courtesy of K. Esau, *Viruses in Plant Hosts: From Distribution, and Pathologic Effects.* © 1968 by the University of Wisconsin Press, Board of Regents of the University of Wisconsin System.)

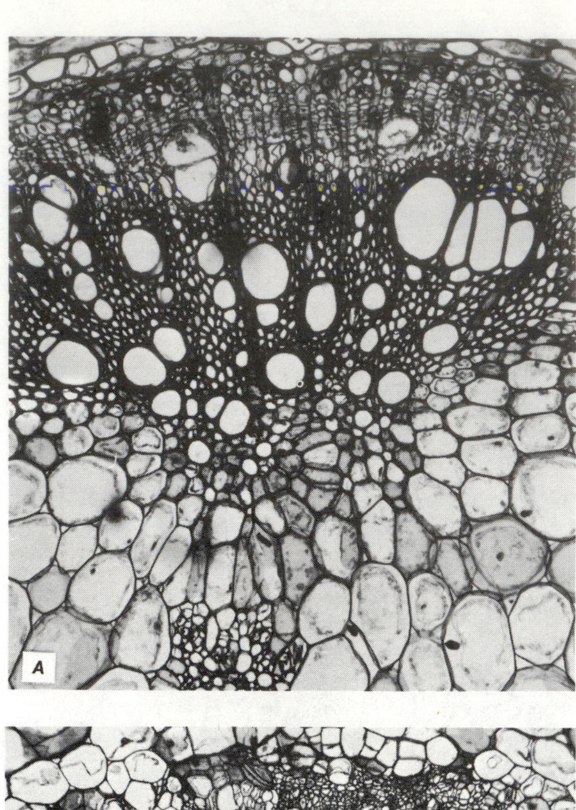

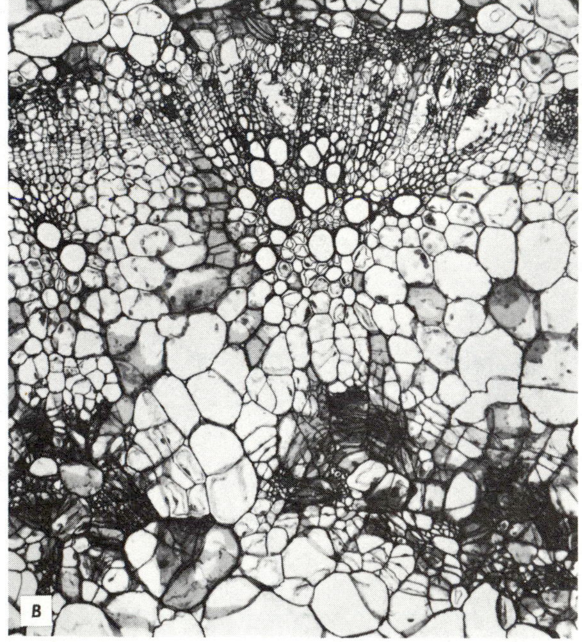

Figure 22.15
Healthy and diseased vascular tissues of tobacco. *A*, healthy. *B*, infected with curly top mosaic. (Courtesy of I. Esau.)

tries of the world these losses are frequently of crucial importance.

Some viruses are so highly infectious that they may be carried from one plant to another by almost any sort of contact. Workers in the tobacco field may carry tobacco virus from one plant to another on their hands or clothing. The tobacco virus is not destroyed in the manufacture of smoking tobacco. Thus, should cigarettes made from virus-infected leaves be smoked by a field worker, he may easily carry the virus on his hands to healthy plants. Tobacco virus may also cause a disease of tomato plants.

A few of the more common viral diseases of plants are alfalfa mosaic, beet yellows, and beet mosaic. Mosaic of legumes and potato mosaic are caused by several viruses. Wheat mosaic is transmitted by root-inhabiting fungi common to many soils. Tobacco ring spot and similar viruses are transmitted by soil-inhabiting nematodes that feed on plant roots.

The Reproductive Cycle of a Plant Virus

Transmission of Viruses

Any organism that carries a disease-causing agent from one living thing to another is called a **vector.** Insects, particularly those that feed on plant juices, are the most important vectors of viruses of plants. Certain types of aphids, thrips, and leafhoppers account for the transmission of viruses from one plant to another. The aphids feed by injecting a long stylet into plant leaves or stems and sucking out plant juices. They obtain sap mostly from phloem tissues, from parenchyma cells of the cortex of stems, and from the mesophyll of leaves. The stylet may wind its way between cells of the cortex until it

Figure 22.16
A, healthy tobacco plant. *B*, tobacco plant infected with curly top virus. (Courtesy of K. Esau.)

reaches the phloem, or it may pass directly through all cells in its path. Thrips have large mandibles with which they chisel their way into epidermal cells. They then suck out the cell contents.

When insects feed on plants infected with virus diseases, they ingest virus along with cell sap. The virus passes into the intestine of the aphid or thrip, is absorbed into the bloodstream, and is carried to the salivary glands. Saliva secreted by feeding insects then carries virus, and plants upon which aphids or thrips feed may become infected with virus. There is some reason to believe that virus may undergo development in the insect. With certain diseases, the insect cannot infect a healthy plant immediately after feeding upon a diseased plant; an incubation period (within the insect) of several days must intervene before the virus is capable of causing infection when given off with saliva. Some plant viruses have a more casual relationship with their insect vectors. Though virus may be ingested by the vector, this is irrelevant, as only a small amount of virus that adheres to the stylet of the insect participates in transmission.

Some viruses can be transmitted from one host plant to another only through specific insects. Often, however, the relationship between insect, virus, and plant is not specific. One insect may carry several virus diseases, several different insects may carry the same virus, and several species of plants may be infected with the same virus. Cucumber mosaic virus is transmissible by more than 60 species of aphids, and the virus has the capacity to infect plants in more than 40 dicotyledonous and monocotyledonous families.

Replication of Virus RNA

In the plant cell, an invading virus particle soon sheds its protein coat to release the viral RNA or DNA. In the case of viruses that have a chromosome consisting of single-stranded RNA, the viral nucleic acid combines directly with host ribosomes that now synthesize virus-specified proteins. The viral RNA behaves like *m*RNA. Some of these newly synthesized viral proteins are enzymatically active, and they induce viral nucleic acid replication. This occurs in several stages.

In the first step, a complementary strand of viral nucleic acid, the plus strand, is synthesized using the invading viral RNA as a template. This is **transcription.** This new strand is composed of the matching complementary nucleotide bases. This new strand and the template from which it was formed combine spontaneously to produce a double-stranded helical molecule with many of the properties of DNA. Additional virus-elicited enzymes, or perhaps the same enzyme that constructed the new nucleic acid strand, now proceed to copy both strands simultaneously to produce 100 to 200 of the double-helical DNA-like molecules. These new molecules are a replication form of the virus and can be isolated from the infected plant; they now instigate the replication of many new plus RNA strands identical to the original invading strand. They possess the same property of inducing the host ribosomes to produce new virus-specified proteins.

Late in the replication process, some virus-directed enzymes activate host cell metabolism to synthesize viral coat proteins. These proteins now combine with the plus strands of RNA to produce mature virus particles. This is **reconstitution.**

SUMMARY

1. Viruses are acellular, hence, unique from prokaryotes and eukaryotes. Outside a host cell, a virus exists as an inert infectious particle (a virion) composed of DNA or RNA surrounded by a protein coat. Inside a host cell, the viral DNA or RNA induces the host cell to produce more virions, which are released when the cell dies.
2. Viruses are smaller than 0.2 μm in diameter, hence, can only be seen with the electron microscope.

They are rod-shaped, spherical, or of complex shape.

3. Many diseases of plants, animals, man, and bacteria are caused by viruses. Plant viruses may be spread by insects, field workers, soil fungi, or soil animals. Recently, some diseases thought to be viral have been shown to be caused by bacterial mycoplasmas.
4. Viruses that infect bacteria are called phage. Some phage is virulent, causing host cell death, but others are nonvirulent and become incorporated into the apparently healthy progeny of the infected parent cell.
5. Transduction is the incorporation of host DNA into virions and its transfer to another host, where it may be expressed (provided the virus is nonvirulent). Transduction is a type of transformation.
6. Symptoms of virus disease in plants include variegation in color of leaves or petals, irregular leaf shape, premature leaf crop, and dwarfing of flowers, fruits, or entire plants.

CHAPTER 23

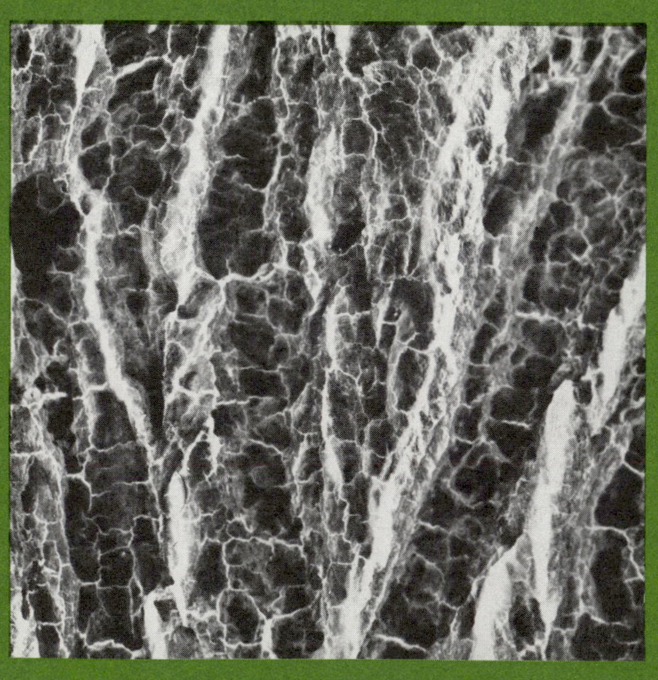

ALGAE: MORPHOLOGICAL DIVERSITY

Algae are photosynthetic, nonvascular plants. They may exist as single cells, as loosely organized clumps of cells **(colonies),** as flat, leaflike sheets, or as intricately branched and intertwined **filaments** (Figs. 23.1, **23.2,** page 410). Only rarely, in the kelps, do algal bodies become large and complex enough to differentiate their tissues into organs that resemble roots, stems, and leaves (Fig. 23.1*K*). Approximately 25,000 species of algae have been identified, but many more, especially those in the oceans, remain to be described for the first time. It is estimated that there may be 10 times this number of species living today.

Grouping all these species under the general classification of algae is highly artificial because there does exist a far greater diversity of form and metabolism among them than among the bryophytes and the vascular plants. It is difficult to find definitive traits that are common to all algae yet absent from all other divisions. Possibly the only structures possessed by all higher plants that never occur in the algae are: a protective jacket of sterile cells surrounding the developing gametes, and an embryo stage in the life cycle. In addition, most algal bodies are relatively undifferentiated. Their aquatic and semi-aquatic environments lack survival pressure for the evolution of vascular and supporting tissue. The surrounding water eliminates the need for water transport and strengthening tissue. Their small size and the general presence of photosynthesizing cells throughout the plant body greatly reduces (or eliminates) the need for food transport. Thus vascular and supporting

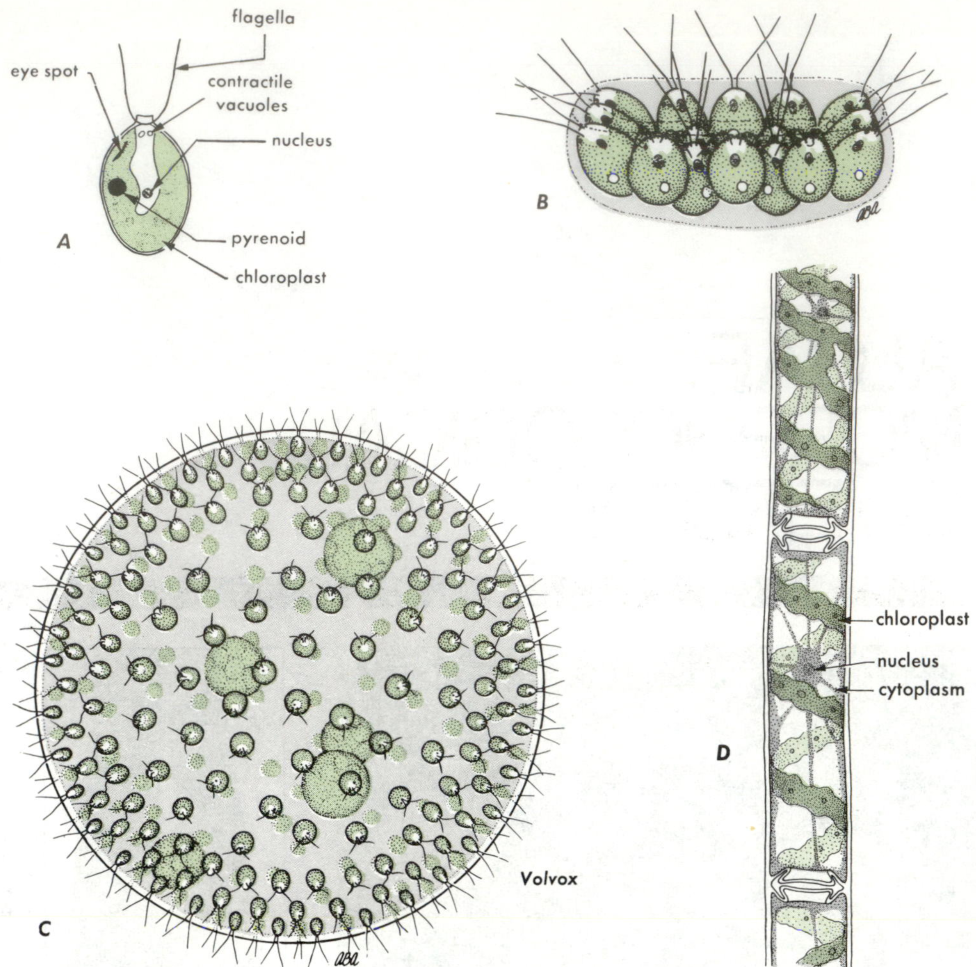

Figure 23.1
Algal diversity. Unicellular forms include A, *Chlamydomonas* in the Chlorophyta. Colonial forms include B, *Gonium* and C, *Volvox,* both in the Chlorophyta. Filamentous forms can be septate and relatively simple in differentiation, as D, Spirogyra, or E, *Hydrodictyon,* or nonseptate and differentiated into prostrate, anchoring filaments and upright, photosynthetic filaments as F, *Caulerpa,* all in the Chlorophyta. Sheetlike forms include G, *Porphyra* in the Rhodophyta. More differentiated, larger forms with leaflike fronds include seaweeds and kelps, in the Phaeophyta and Rhodophyta, such as H, *Fucus,* I, J, *Ptilota filicina,* and K, (on page 460) *Laminaria andersonii* (about 1-2 m tall).

tissues have not evolved in the algae. All the cells of most algae can carry on photosynthesis and obtain water and nutrients directly from their surroundings. Algae lack true roots, stems, and leaves (with vascular tissue); such a body is called a **thallus.** Plants with such a body are called **thallophytes.**

Algae can be found nearly everywhere. They float in air or water. They are attached to tree trunks or branches, to the bottom of streams, to soil particles, or to rocky intertidal cliffs battered by surf. They are also found growing symbiotically with other plants or animals. Algae occur in the most severe habitats on earth. Some species grow on snow in perpetually freezing temperatures; others thrive in hot springs at temperatures of 70°C or more; a few live in extremely saline bodies of water, such as the Great Salt Lake in Utah; yet others survive the pressure and low light intensity conditions of 100 m or more below the surface of lakes or seas. They have even been detected 1 km from ground zero, 6 months after a 20 kiloton atomic bomb tower test in Nevada, and they were considered to be survivors of the explosion.

Probably the most commonly noticed natural urban habitats for algae are the sides of glass fish tanks. They may also generally be found around leaking faucets and in garden or park pools that are not kept "pure" with chemicals. The "bloom" occurring during the summer on many lakes or the scum found on ponds is actually algae. Microscopic forms occur in most natural waters, especially the top 75 m of the ocean. Here they constitute a primary food source for marine animals of all types and account for about 50% of the oxygen released into the atmosphere through photosynthesis.

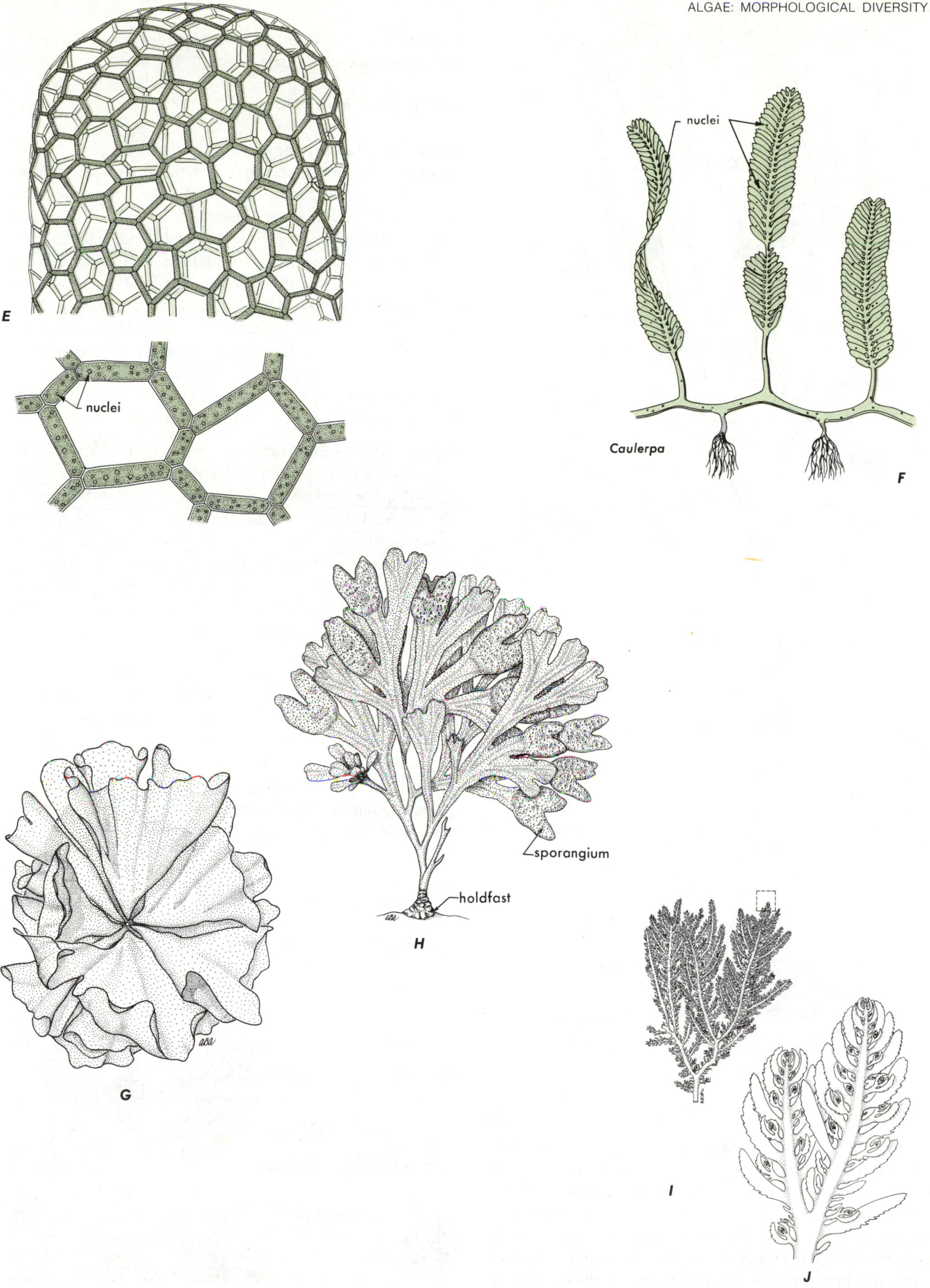
E
nuclei
nuclei
Caulerpa
F
sporangium
holdfast
H
G
I
J

Figure 23.1 *(Continued)*

IMPORTANCE OF ALGAE

Algae are important in two basic, but quite different ways. They are important to the entire biosphere because of the ecologically vital functions they perform. These functions include: the production of carbohydrates, which places the algae at the base of food chains, the fixation of nitrogen, and reef building. They are also economically important to people because they serve as food, fodder, and fertilizer and also have many industrial and pharmaceutical uses.

Ecological Functions

Algae as Producers

In aquatic ecosystems, algae are the major producers. Shallow bodies of water may depend on productivity by higher plants such as rushes, water lilies, duckweed, or the like (see Fig. 23.9) or they may be fueled by a constant supply of decomposing plant material **(detritus)** washed down from the surrounding land. About 70% of the globe is nonterrestrial and most of that is covered by deep water. Here algae are the only producers. They have been called the grasses of the oceans, converting solar energy into chemical energy that is passed up through the rich marine food chain, and releasing oxygen as an important by-product to both water and atmosphere.

Algae in Marine Ecosystems. Algae (and other life) occupy several distinct habitats in aquatic ecosystems. One habitat, along rocky oceanic coasts, is the **intertidal** zone (Fig. 23.3). This zone is especially severe for the growth of algae, because the plants are alternately exposed by low tide and inundated by high tide. When exposed, the plants are beset by desiccation, high temperatures, high light intensity, and increased salinity. When the tide comes in, all environmental factors change abruptly. In winter the plants may be locked in ice.

The **neritic** zone is below the intertidal, but is still relatively shallow and near shore. Species of attached algae in the intertidal and neritic zones have their own particular distribution: some occur at the upper end, others occur in the middle, and others in the lower region (Fig. 23.4). Their distribution probably reflects different degrees of adaptation to the stresses of exposure, for those species at the upper part of the intertidal zone are ex-

Figure 23.3
The intertidal zone, exposed at low tide, along the central coast of California.

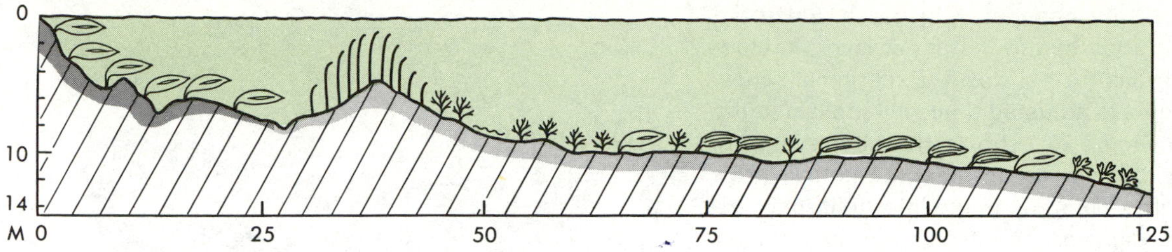

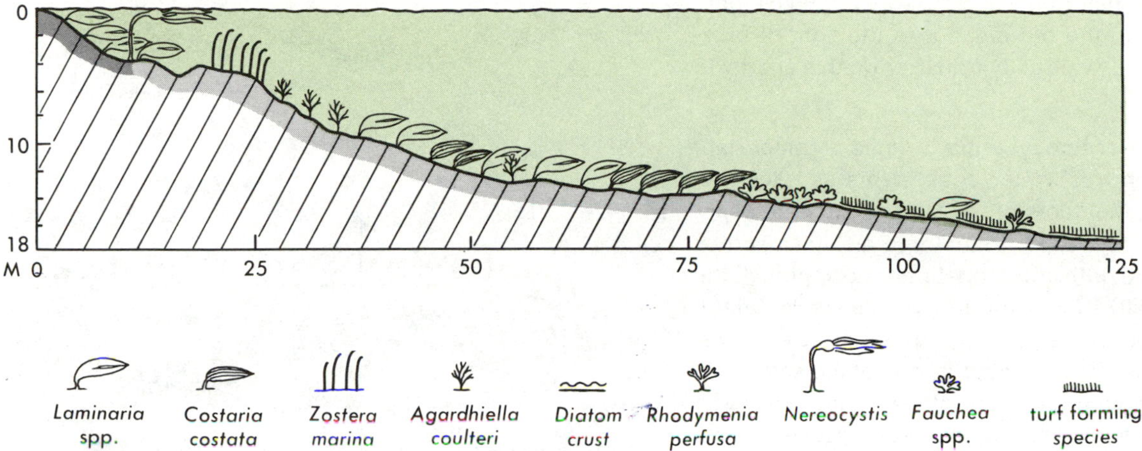

Figure 23.4
Representative transects of benthic (bottom attached) algae in the near-shore region of Puget Sound, Washington. Dark substrate is rocky.

posed more frequently and for longer periods of time than species farther down. Experiments tend to confirm this hypothesis: algae that grow in the upper part of the intertidal zone are actually more resistant to drying, and to fluctuations in temperature, light intensity, and salinity.

Gradually, with increasing depth, the quality and quantity of penetrating light becomes less and less favorable for photosynthesis. The **compensation depth** or **compensation light intensity** is the point where growth is no longer possible (Fig. 23.5). Respiration equals photosynthesis at this point. In deep bodies of water, the habitat between the surfaces and the compensation depth is dominated by floating algae, the **phytoplankton.**

Phytoplankters are usually microscopic, but they do include the large, floating *Sargassum* seaweed, namesake of the Sargasso Sea off Bermuda. The microscopic forms are typically unicellular and maintain buoyancy by storing oil droplets or by developing fine projections that extend out from the cell wall (Fig. 23.6). The average life span of a given cell is probably measured in hours or days; if it does not reproduce the protoplast dies and the cell wall remains will sink to the bottom. Most phytoplankters are members of the Bacillariphyta (diatoms) and Pyrrhophyta (dinoflagellates) divisions, but the Chlorophyta (greens), Cyanophyta or Cyanobacteria (blue-greens), and Euglenophyta are also represented.

Figure 23.5
The penetration of sunlight into the clearest ocean water. The compensation intensity for most phytoplankton, where respiration = photosynthesis, is 1/100th to 1/1000th the intensity of full sunlight at the ocean surface; compensation depth is about 550 ft (170m).

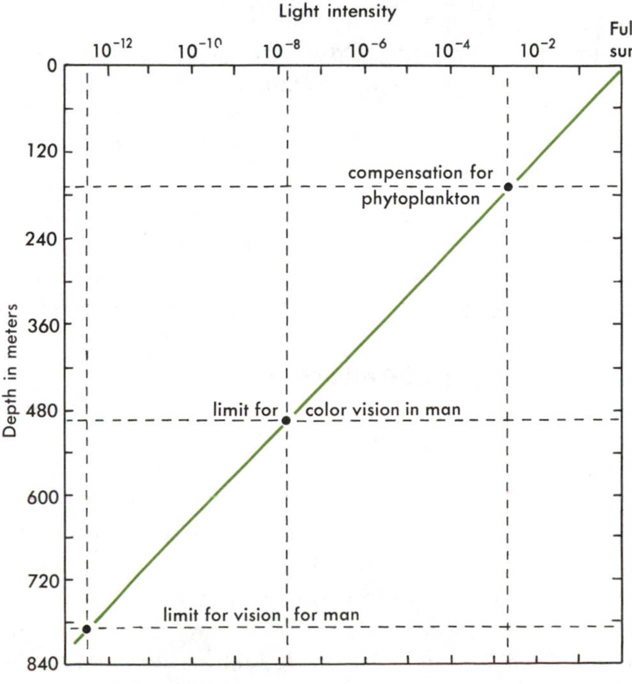

The quality of light changes with depth. Since red light is completely absorbed in the upper layers, a blue-green twilight prevails further down. Experiments show that aquatic algae have adjusted their metabolism to the light at this depth. Figure 23.7 shows the action spectrum of photosynthesis for the green alga sea lettuce (*Ulva taeniata*), which grows high up in the intertidal zone; superimposed on the same graph is the action spectrum for the red alga *Myriogramme spectabilis,* which grows at much greater depths. You can see that the peak has shifted and condensed to center around the blue-green region, 480 to 580 nm. Special accessory pigments (phycobilins) present in the red algae trap the light of blue-green wavelengths, eventually transferring this energy to cholorophyll.

Algae in Freshwater Ecosystems. Algae are important producers in such freshwater ecosystems as lakes, rivers, and streams. Members of the Chlorophyta are especially prominent (Fig. 23.8). Planktonic and filamentous attached forms are both common, but none approach the size and complexity of red and brown seaweeds. *Chara,* in the Charophyta, is perhaps the most complex (Figs. 23.9*F,G*), and it is differentiated into rootlike and stemlike portions; the stem consists of node and internode regions. At the shallow margins of lakes, vascular plants such as sedge (*Scirpus*), cattail (*Typha*), and floating or submerged pondweeds and water lilies (Fig. 23.9) are the major producers.

If lakes are deep enough, their water may stagnate into three zones in summertime. An upper zone, the **epilimnion,** is relatively warm and sunlight is intense enough to support large phytoplankton populations. Oxygen level in the epilimnion, consequently, is high. Below the epilimnion is a narrow zone where temperature drops rapidly with depth, on the order of 4°C per meter. This is the **thermocline,** and it serves as a barrier to the mixing of the epilimnion with colder, darker water at the lake bottom—the **hypolimnion.** Often, but not always, the thermocline marks the light compensation depth for phytoplankton, and when this is the case, the hypolimnion is low in oxygen. In the fall, as air temperature drops, the temperature of the epilimnion also drops and may finally equal the temperature of water in the hypolimnion. At that time the lake is no longer stagnant and stratified, and oxygen levels near the bottom can be recharged.

Productivity and Pollution

The density of phytoplankton is not very high in the open ocean, perhaps a few thousand cells per liter, but the oceanic expanse is enormous and this results in a high annual rate of gross productivity: 32.6×10^{16} kcal (3 quintillion, 260 quadrillion). This is about three times the annual productivity of all the world's grassland and pasture, and four times that of all cropland.

As great as this productivity is, however, it is limited by many environmental factors such as light, temperature, and nutrients. If these limiting factors are ameliorated, productivity will increase manyfold. A natural process of increasing productivity and a changing environment takes place very slowly in most bodies of water because silt accumulates, making the depth shallower and the water uniformly warmer, and vegetation encroaches along the edges and contributes increasing amounts of detritus. This process is called **eutrophication,** and it ordinarily takes centuries, millennia, or eons. Human activities, however, mainly through the disposal of industrial, agricultural, and residential wastes, can telescope eutrophication into decades. This is called **cultural eutrophication.**

As nutrients increase, algal and bacterial activity increase, resulting in more turbidity and lower light. Oxygen becomes limiting near the bottom; this limitation plus siltation prove fatal to fish eggs of some species. Algal species replace one another in an aquatic succession: generally greens are replaced by blue-greens that may impart disagreeable odors or tastes to the water (Fig. 23.10). As the base of the food chain changes, so

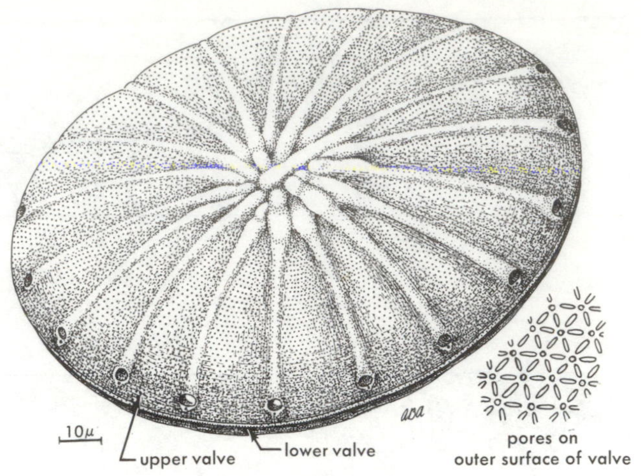

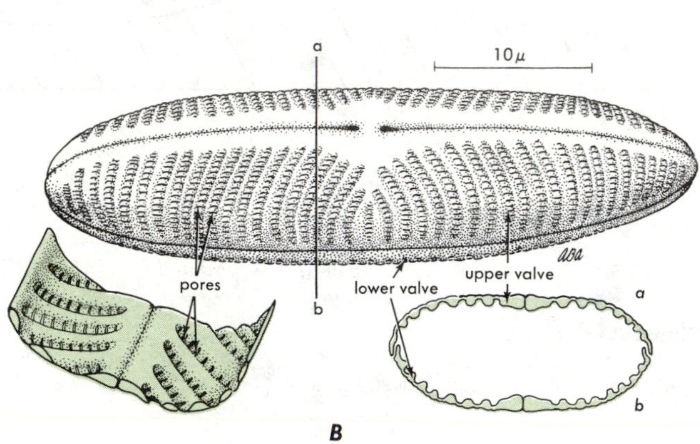

Figure 23.6
Examples of phytoplankters. *A* to *D,* diatoms *Asteromphalus elegans, Navicula digitoradiata, Asterionella formosa,* and *Biddulphia biddulphia. E* and *F,* dinoflagellates *Ceratocorys aultii* and *Ceratium* sp.

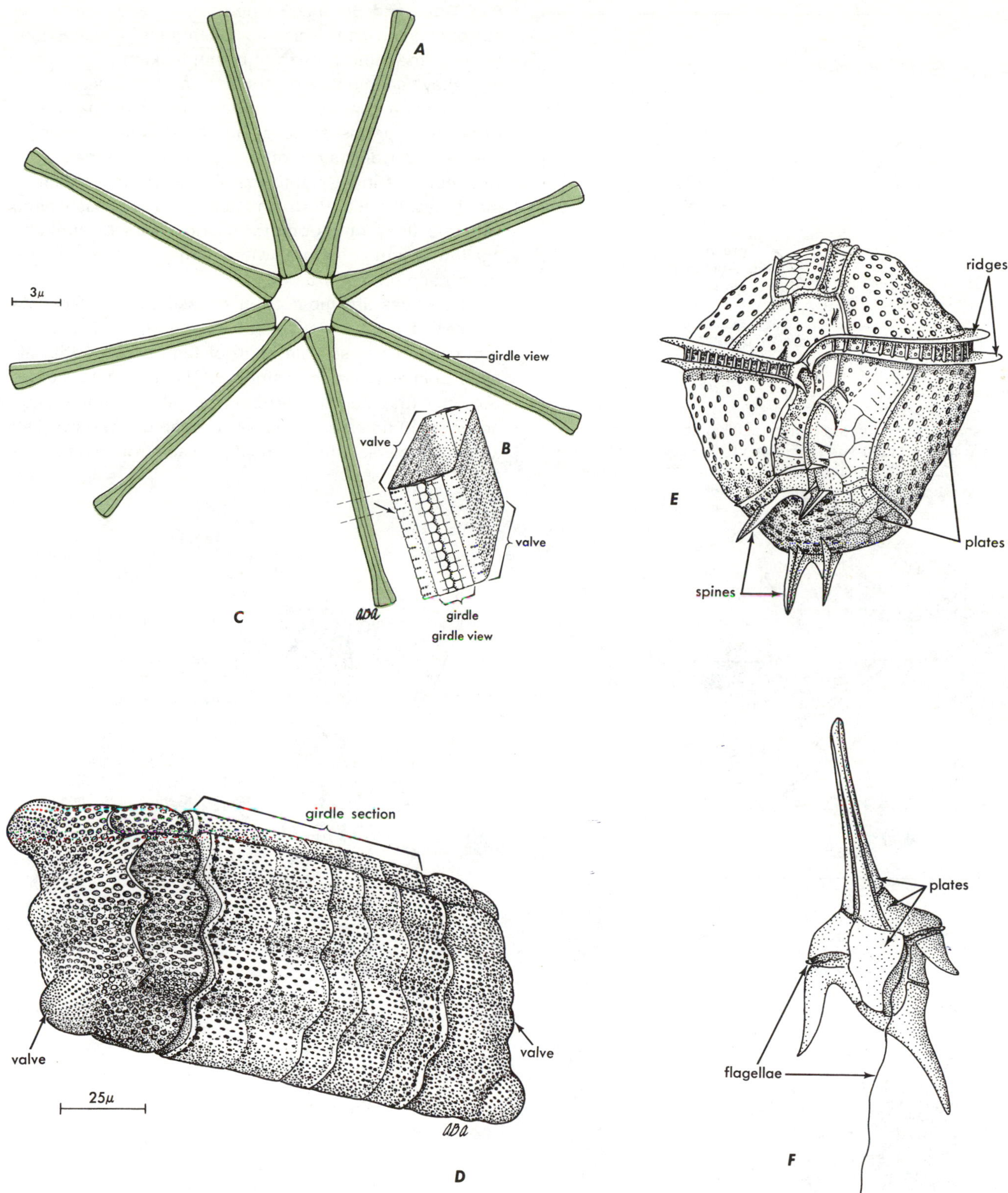

does the entire chain. After 100 years of accelerating industrial and residential development along its shores, Lake Erie had changed from a relatively clean lake with important fishing and recreational values to an aesthetic and economic loss. Plankton and other algae increased six- or seven-fold in some places, and the composition shifted toward the noxious blue-greens. Oxygen content of bottom waters declined by 67%. Commercial catches of trout, whitefish, herring, sauger, walleye, and blue pike declined by 99.9%, while such fish as carp, shad, alewives, and smelt increased. Recently, the rate of pollution has declined and water quality has improved.

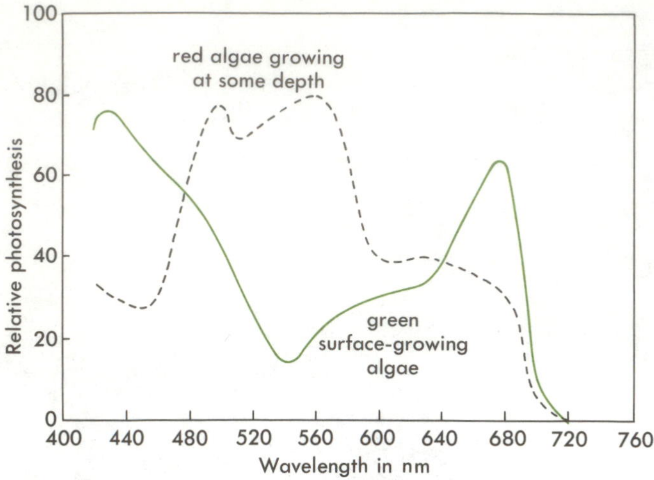

Figure 23.7
Action spectrum for photosynthesis in a green, surface-growing alga (solid line) and in a red alga growing at some depth (dashed line). (Data from T. Haxo and L. R. Blinks. *J. Gen. Physiol.* **33,** 389. © 1950 by The Rockefeller University Press. Reprinted with permission of the publisher.)

Red Tides and Blooms. One of the results of cultural eutrophication is the episodic appearance of **red tides** or **blooms:** the rapid growth of phytoplankton populations until they become dense enough to color the water. This phenomenon can occur in bodies of water that range from small ponds to lakes to large coastal regions. Marine occurrences are often reported along the southwest coast of India, southwest Africa, southern California, Texas, Florida, Peru, and Japan. Factors associated with red tides or blooms are long hours of sunlight, shallow, warm, offshore water, and high levels of nitrogen and phosphorus.

Red tides are most often caused by dinoflagellate species of *Gymnodinium* or *Gonyaulax.* Both species produce a water-soluble toxin of high potency that affects animal nervous systems. It is related to curare poison, extracted from certain tropical flowering plants, and it is 10 times as effective as cyanide. Massive fish kills and the poisoning of shellfish result from red tides. A

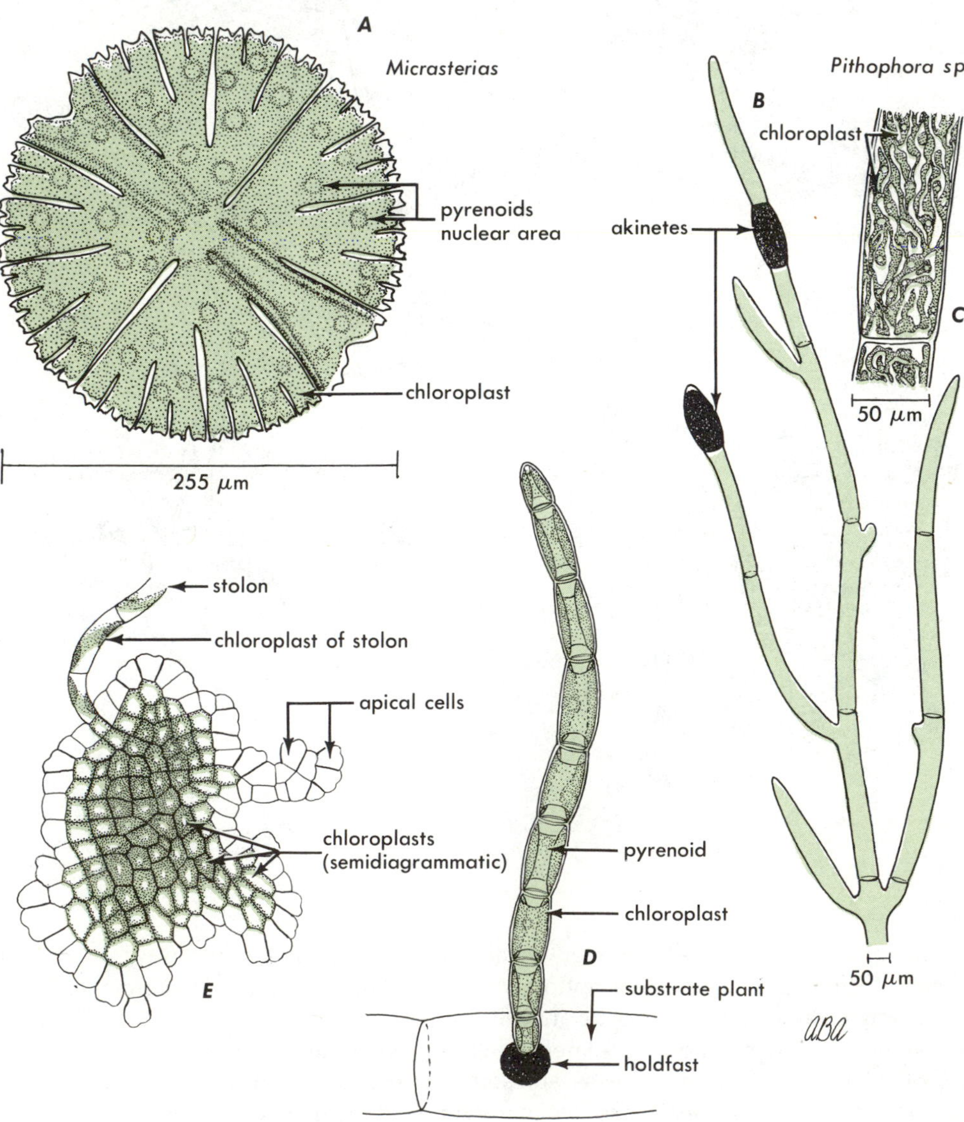

Figure 23.8
Representative algae of fresh, clean water. *A, Micrasterias,* a green desmid. *B, Pithophora,* a filamentous green alga. *C,* detail of netlike chloroplast in *Pithophora* cells. *D, Ulothrix,* an epiphytic green alga. *E,* a young colony of the red alga *Hildebrandia rivularis.* Colonies are flat, and connected by filamentous "stolons." (*D* is redrawn from H. W. Nichols, *A. J. Bot.* **52,** 9. © 1965 by Botanical Society of America.)

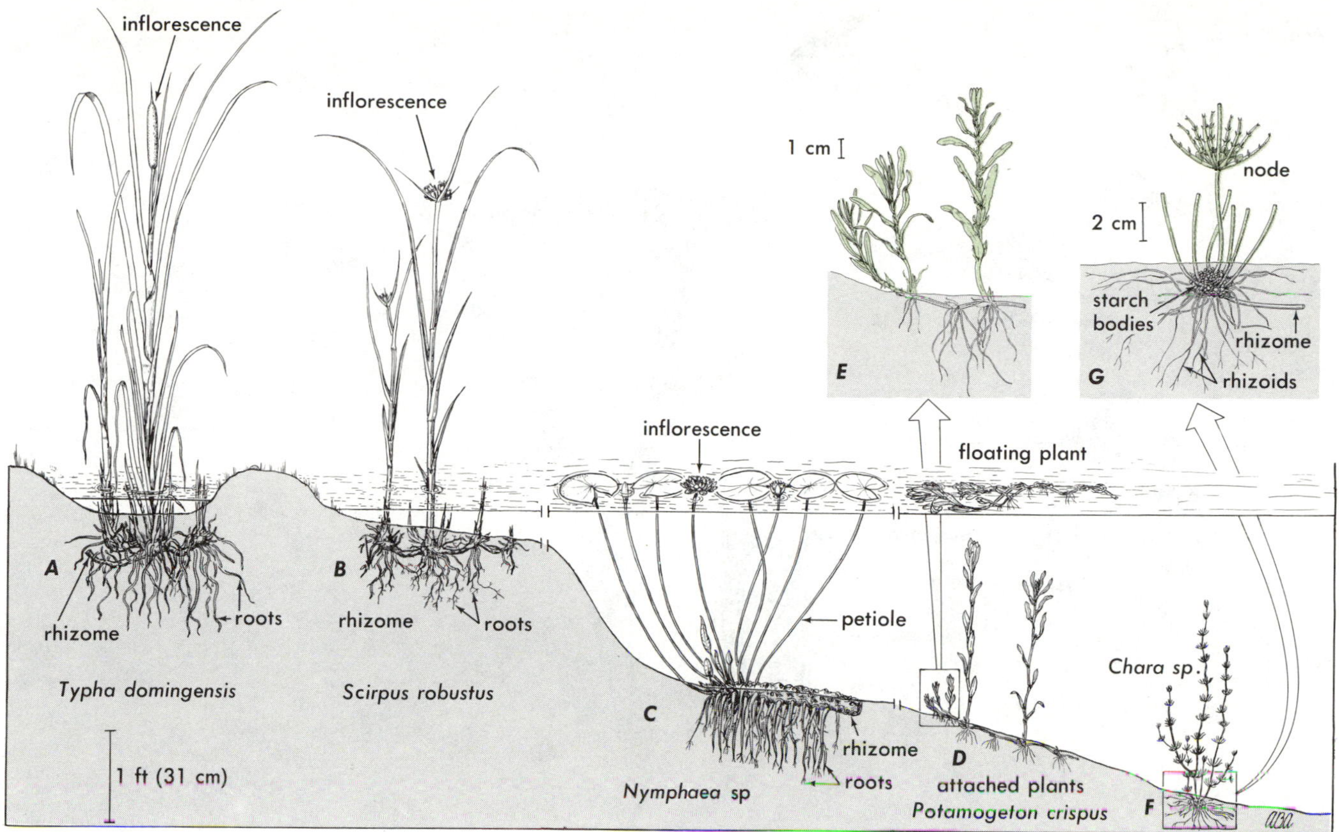

Figure 23.9
Important producers in fresh water systems. *A* to *E*, vascular plants. *A*, emergent cattail *(Typha domigensis)*. *B*, emergent sedge *(Scirpus robustus)*, more tolerant of long-term flooding than cattail. *C*, submerged waterlily *(Nymphaea* sp.*)*, with floating leaves and inflorescences. *D* and *E*, submerged and floating segments of pond weed *(Potamogeton crispus)*. *F* and *G*, the green alga *Chara*, differentiated into rhizoids, "rhizome," nodes, and internodes.

1947 episode off Florida killed an estimated 500 million fish.

Other blooms are caused by the chrysophyte *Prymnesium parvum* and blue-green species of *Microcystis*, *Anabaena* (Fig. 23.10*A*), *Nostoc*, *Aphanizomenon*, *Gloeotrichia*, and *Oscillatoria* (**Fig. 21.22*B*,** page 410). The blooms are not necessarily blue-green in color, however; a species of *Trichodesmium* causes red blooms, which give the Red Sea its name. Some of the plants above also produce toxins.

Economic Uses

Algae as Food or Medicine

Seaweed—marine algae of moderate size, which usually grow in the intertidal or neritic zones—forms an important part of the human diet and medicine chest in several parts of the world. In the orient, seaweed harvesting has been known for 5000 years. Shen Nong, the legendary Chinese "father of medicine," prescribed seaweed for certain ailments in 3600 BC. Some 3000 years later, Confucius praised its curative value. For centuries, the Japanese have used algae as a healthful, tasty supplement to their rice diet. The demand for *nori* (the red alga *Porphyra*) has grown to such an extent in Japan that it is cultivated in shallow, intertidal bays (Fig. 23.11). The Polynesians in Hawaii were known to have utilized and named at least 75 species of *limu* (algae) as food sources. Some rare species were cultivated in marine fish ponds for the nobility. Dulce (the red seaweed *Rhodymenia palmata*) has been known as a food for 12 centuries in the British Isles. It was the Irish who discovered that small quantities of Irish moss (the red alga *Chondrus crispus*), when boiled with milk, would produce a gelatin dessert that the French later called *blanc mange.* Brown algae off the coast of California (mainly *Macrocystic pyrifera*, Fig. 23.12), at one time were harvested for their content of iodine, which is added in trace amounts to the diet to prevent goiter, an enlargement of the thyroid gland in the neck.

It is not as food or medicine, however, that algae are most important today. With some exception, they do not have much nutritive value—in fact, their major constitu-

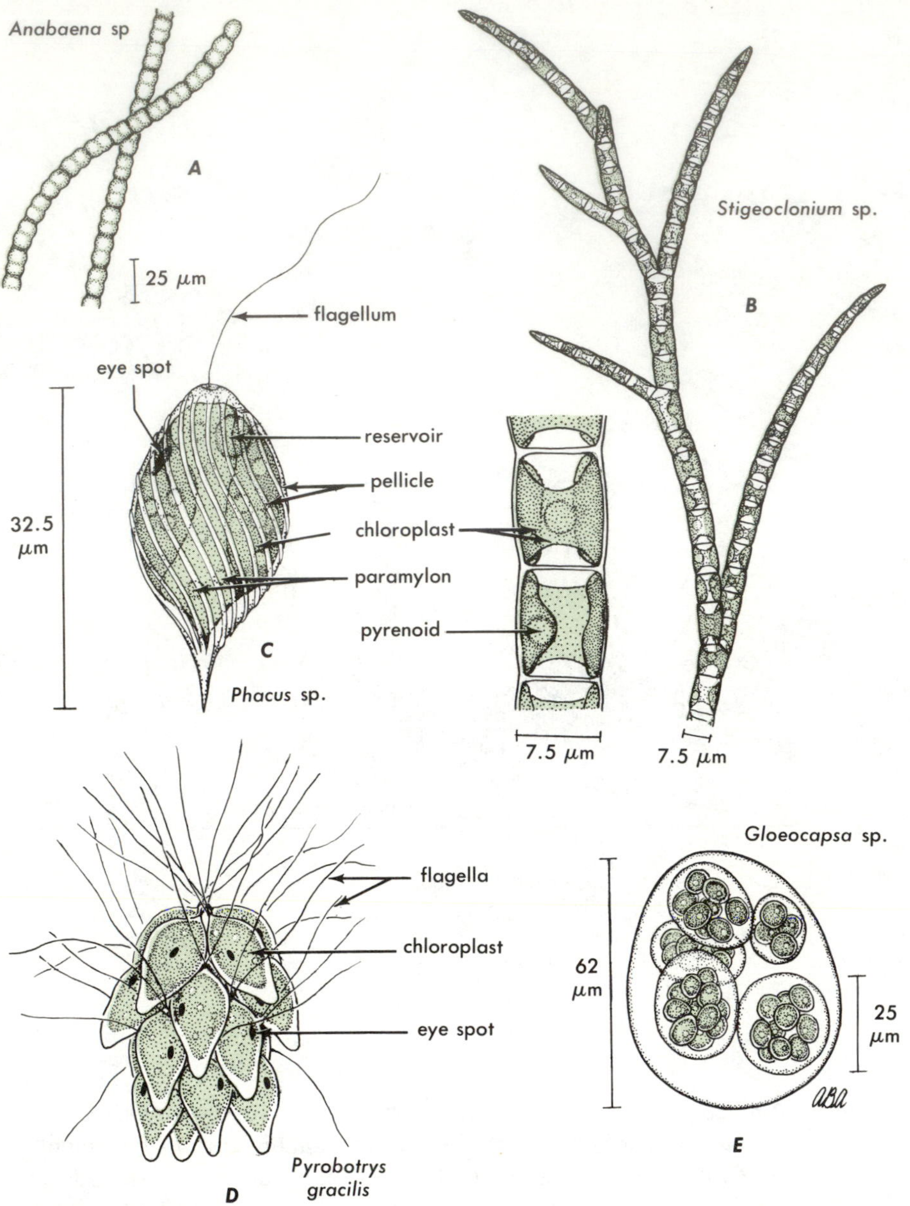

Figure 23.10
Representative algae of fresh, but polluted water. *A*, the blue-green alga *Anabaena*. *B*, the green alga *Stigeoclonium*, with a detail of girdle-shaped chloroplasts. *C*, *Phacus*, in the Euglenophyta. *D*, the colonial green alga *Probotrys gracilis*. *E*, the colonial blue-green alga *Gloeocapsa*. (*D*, redrawn from P. C. Silva and G. F. Papenfuss, 1953 Pub. No. 7. State Water Pollution Control Board, Sacramento.)

ents are largely indigestible. Algae are used more as condiments, garnishes, or desserts than as staple foods, much as we use lettuce, watercress, celery, or whipped cream. Iodine now is regularly obtained from other sources and added to table salt. Some claims as to the health-giving value of seaweed do not have much foundation in fact.

Fodder and Fertilizer

Seaweeds not only contain such important trace elements as iodine, but they contain large amounts of potash (potassium), nitrogen, phosphorus, and other components of good fertilizer or cattle feed supplements. In historic times, the North American Indians and the Scotch-Irish used Irish moss as a fertilizer to build up poor soils for such diverse crops as corn and potatoes. Seaweed has more recently been shown to compare favorably in nutrition to barnyard manure. *Macrocystis* (Fig. 23.12) continues to be harvested off the European coast for use as a cattle feed supplement.

Cell Walls and Cell Wall Extracts

Peculiar characteristics of the cell walls of diatoms, brown algae, and red algae have led to many recent industrial, pharmaceutical, and dietary advances. It is in these areas that the algae have their greatest economic

Figure 23.11
Nori (*Porphhyra tenera*, Rhodophyta) culture, Sendai Prefecture, Honshu Island, Japan. *A*, distance view, showing the extent of the hibi nets in January, about the time of harvest. *B*, closer view, showing the hibi nets about 30 cm above mean low water in September, at the beginning of nori cultivation.

value. These characteristics result in products such as diatomite, agar, carrageenan, and algin, each of which we shall consider.

Diatomite. Diatomite, also called diatomaceous earth, is a chalky, sedimentary rock composed of the cell wall remains of unicellular algae called **diatoms**. Diatoms, you recall, are important members of the phytoplankton (Fig. 23.6). One of the richest deposits of diatomite is located near Lompoc, California (Fig. 23.13). About 15 million years ago, that area was submerged beneath a warm, shallow sea. As diatoms flourished and died, their remains accumulated in bottom sediments. These particular remains persisted intact because the wall is not composed of carbohydrates such as cellulose; rather it is 95 percent silica.

The walls of a diatom fit together like two halves of a Petri dish and the glasslike case is perforated with hundreds of microscopic pores (Fig. 23.14). Each half is called a **frustule** and the region of overlap is the **girdle.** Under the electron microscope, it is possible to see that even the pores have pores. The perforations form exquisite, symmetrical designs, each design peculiar to a given species. Diatoms extract the opaline silica from the surrounding water by a process that is still not understood. In more recent geologic time, the Lompoc area was uplifted above sea level and the diatom deposit was revealed by erosion. Today, such companies as General Refractories and Johns-Manville mine the diatomite for industrial and pharmaceutical use.

Diatomite makes a superior filter or clarifying material because the microscopic pores create a large surface area (0.5 lb—230 g—has a surface equal to the area of a football field), and the rigid walls are incompressible. Diatomite is inert and can be added to many materials to provide bulk, improve flow, and increase stability. In those ways it is used in cement, stucco, plaster, grouting, dental impressions, paper, asphalt, paint, and pesticides. Diatomite is also used as an abrasive.

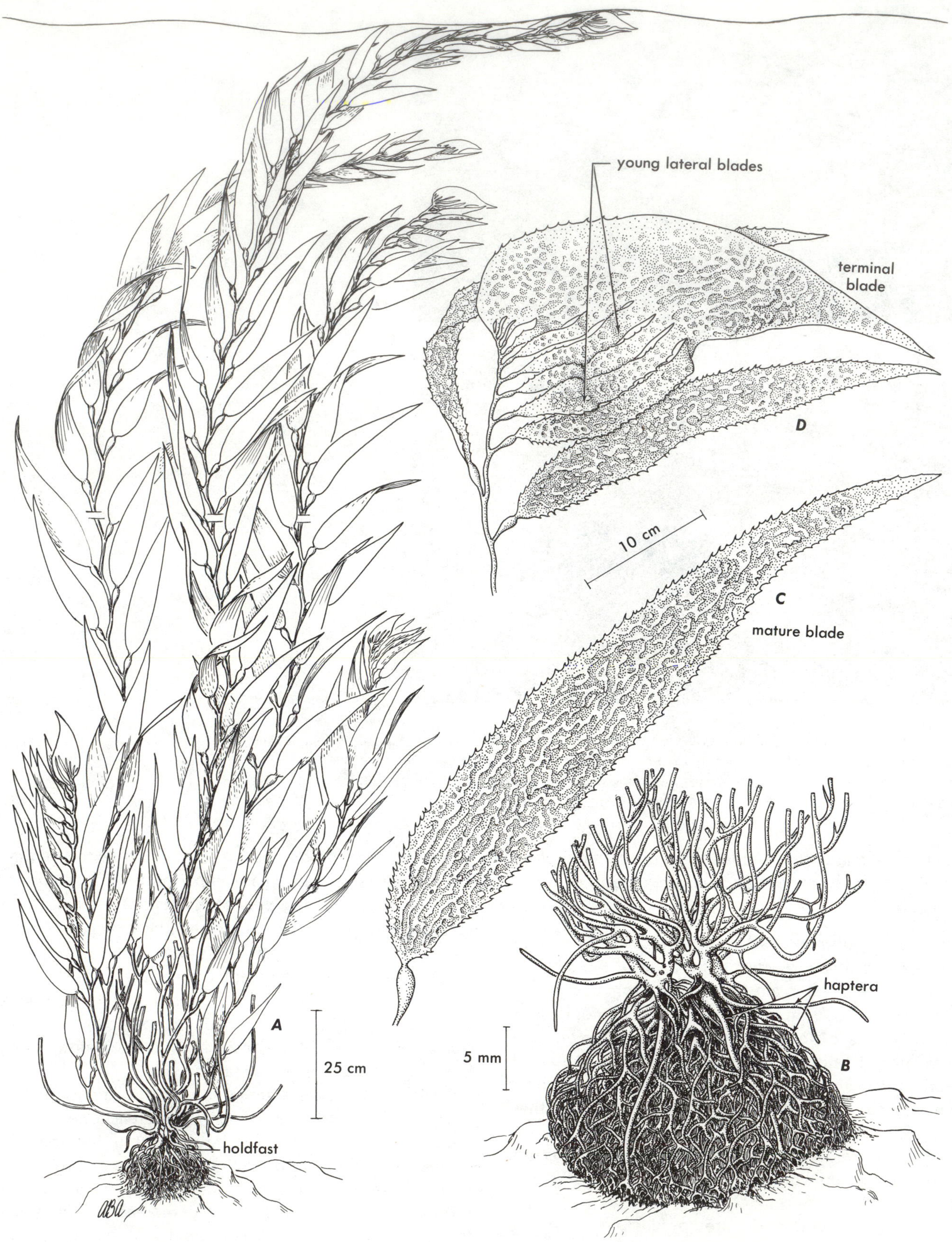

Figure 23.12
The kelp *Macrocystis pyrifera,* including details of the blades (*C* and *D*) and holdfast *(B)*.

Figure 23.13
Aerial view of diatomite deposits near Lompoc, California.

Agar

Less than 100 years ago, a physician's wife, Frau Fanny Eilshemius Hesse, discovered the use of agar as a bacterial culture medium. She passed the information on to her husband, and he to Robert Koch, and Koch to the world via his important scientific writings in the late nineteenth century.

Agar (or agar-agar) is a polysaccharide, analogous to starch or cellulose but chemically quite different. It is found in the walls of certain red algae. Mainly species of *Gelidium* and *Gracilaria* seaweeds are commercially harvested, divers descending 3 to 12 m in the warm near-shore waters off Japan, China, California, Mexico, South America, Australia, and the southeastern United States.

In addition to its bacteriological use, agar is important in the bakery trade. When added to icing, it retards drying in the open air or prevents running in cellophane packages. Because it is virtually indigestible, agar is also used medically as a bulk-type laxative.

Carrageenan

This is another polysaccharide from the walls of red algae. It is mainly harvested from Irish moss and the substance takes its name from the town of Carragheen, County Cork, along the south shore of Ireland, where its properties were first discovered.

Carrageenan reacts with the proteins in milk to make a stable, creamy, thick solution or gel. Consequently, it is used in ice cream, whipped cream, fruit syrups, chocolate milk, custard, evaporated milk, bread, and even macaroni. It is added to dietetic, low-calorie foods to bring back the appropriate mouth "feel" of nondietetic foods. Carrageenan is also used in toothpaste, pharmaceutical jellies, lotions of many sorts, and as a whiskey chaser for a cold "cure." Irish moss is commercially harvested in this country in Maine, principally by two companies: Kraft Foods and Marine Colloids.

Algin

Since this long-chain polymer is made up of repeating organic acid units, it is also called **alginic acid.** It is the principal wall component of brown algae, constituting up to 40% of the middle lamella and primary wall by weight. Water is strongly absorbed to algin, resulting in a thick solution. For example, one tablespoon of algin added to a quart (one liter) of pure water will increase the viscosity

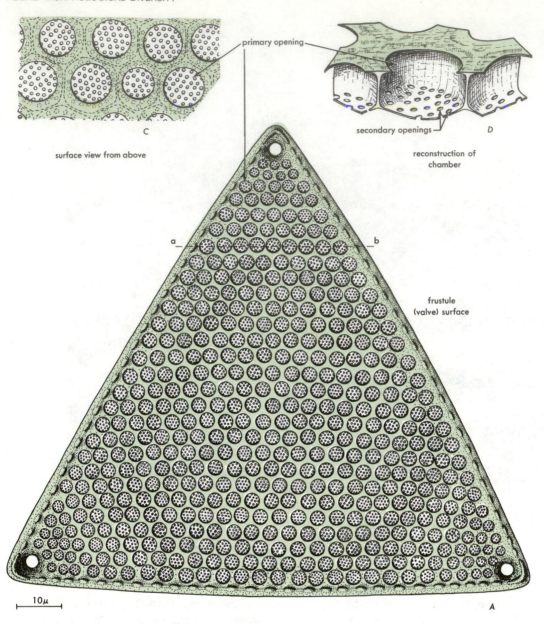

primary opening

C

surface view from above

secondary openings

D

reconstruction of chamber

a
b

frustule (valve) surface

10μ

A

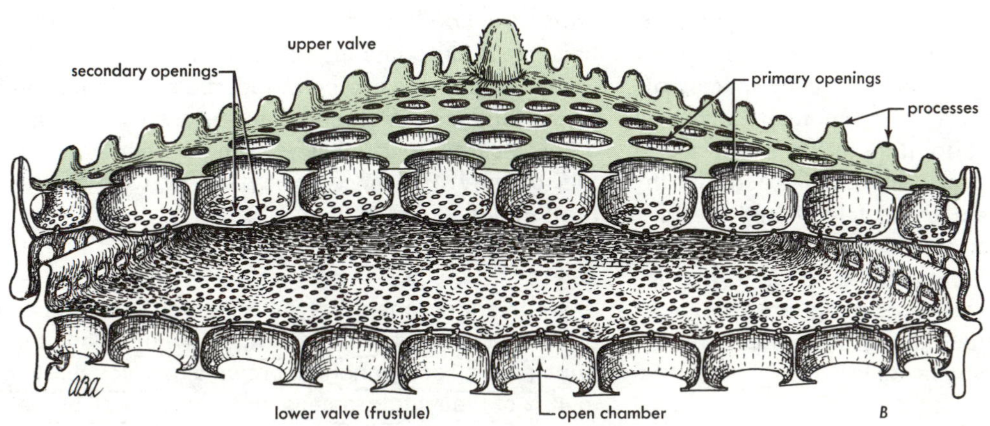

Figure 23.14
Cell wall details of the triangular diatom *Triceratium favus*. *A*, face view. *B*, cross sectional view, taken at *a-b* in *A*; *C*, and *D*, details of primary and secondary openings in the cell wall.

to approximately that of honey. In nature, algin may be valuable to intertidal browns during exposure to air by enabling water to be retained in and on the plant. Commercially, however, it has many uses (Table 23.1).

Hundreds of species possess algin, but only a few are commerically harvested: *Macrocystis pyrifera* along the California coast, and species of *Laminaria* (Fig. 23.1*K*), *Fucus* (Fig. 23.1*H*), and *Ascophyllum* off England. *Macrocystis* is the largest **kelp** (a general term for large, marine, brown algae) known, and it has the fastest growth rate of any multicellular plant in the world. From a single-celled zygote, it grows into a young plant rooted on a rocky bottom 6 to 30 m below the surface, and then into a 60 m long mature giant—much of it floating on the surface—in the course of a single growing season (Fig. 23.12). At this stage it is differentiated into a rootlike region (the **holdfast** composed of many haptera) that anchors the plant to bottom substrate, a stemlike region (the **stripe**), and numerous leaflike **fronds** that arise all along the stipe and possess gas-filled **bladders** at their base that increase buoyancy.

There have been recent attempts to attach young *Macrocystis* plants to artificial, floating supports some distance offshore in southern California. If these test trials work well, kelp could be farmed in this manner.

Table 23.1

SOME OF THE PRODUCTS IN WHICH ALGIN IS USED

Food Products	Industrial
Bakery icings and meringues	Water base paints
Salad dressings	Wall joint cements
French dressings	Welding rod coatings
Dietetic dressings	Textile print paste
Pickle relish	Textile sizing
Meat sauces and pepper sauces	Latex creaming and thickening
Orange concentrates	Adhesives
Fruit drinks	Paper coatings
Dietetic beverages and drinks	Corrugated paper
Beer	Paper sizing
Ice cream and related frozen desserts	Paper cartons for foods, soaps, and detergents
Egg nog	Food wrappers
Creamed cottage cheese	Waxed cardboard cartons
Cream cheese	Boiler compounds
Pasteurized cheese spreads	Can sealing compounds
Canned buttered vegetables	Battery plate separators
Canned chow mein	Mold release coatings
Canned meat stews	Wax cleaner polishes
Cake mixes	Ceramic glaze
Puddings, pie and cake fillings	Sugar beet clarification
Fountain syrups	Finger paints
Buttered pancake syrups	Antibiotic tablets and suspensions
Berry syrups	Dental impression compounds
Chip dips and mixes	Toothpaste
Instant dessert mixes	Surgical jellies
Chocolate drink	Mineral oil emulsions
Delicatessen salads	Medicated rubbing ointments
Candy	Tranquilizing tablets
Dessert and salad gels	Hand lotion
Breading batters	Facial beauty masque

ALGAL CLASSIFICATION

As shown in Table 23.2, there are many algal divisions that differ in biochemistry, cytology, and habitat. Differences in cell wall construction, number of placement of flagellae, and type of chlorophyll are quite fundamental and significant. We shall briefly survey seven of the divisions.

Rhodophyta: The Red Algae

Cyanobacteria and Rhodophyta have a few characteristics in common. Both posses only chlorophyll *a*, though some reds have also chlorophyll *d*, and identical accessory pigments called **phycobilins**, neither possesses flagellated cells, and both form a nonstarch storage product related to amylopectin. Chloroplasts are present, in the Rhodophyta, but they lack grana (Fig. 23.15) and carbohydrate accumulates in the cytoplasm, not in the stroma of the chloroplast as it does in the green algae and higher plants.

The Rhodophyta are eukaryotic, possess cellulose walls, and exhibit some of the most complex life cycles, incorporating both sexual and asexual reproduction, of any group of plants. Their range of morphological diversity is also greater than that of the blue-greens, including sheetlike forms and complex, finely branched filamentous forms. In some species, pits connect the cytoplasm of adjacent cells. A few reds are unicellular.

The reds are almost exclusively marine, and are most abundant in warm water, often extending to some depth. Again like the Cyanobacteria, some forms are lime-encrusted and participate in tropical reef building.

Morphological diversity in the subclass Florideophycidae is attained by an aggregation of filaments. The simplest algae in the subclass are much branched, small, rather delicate forms. In more complex forms, filaments are compacted into a parenchymalike mass of cells, and the plant body consists of a dissected blade, stipe, and holdfast such as exhibited by *Gigartina*.

Growth is almost always apical and, as in all other species, morphological form is related, among other things, to cellular polarity. This is well-demonstrated in *Polysiphonia* (**Fig. 23.16,** page 410). *Polysiphonia* has a single apical cell. Cell division with the mitotic spindle parallel to the long axis of the filament results in an increase in length of the filament (Fig. 23.17*A*). Cell division at right angles to the axis of the filament produces an array of five pericentral cells, precisely arranged around the central axial cell (Fig. 23.17*B*) This means that the planes of division giving rise to the pericentral cells are also precisely oriented to each other. Furthermore, the order of production of the new pericentral cells is the same in all plants. Cells two and three are formed on opposite sides of cell one. Cell four forms beside two, and cell five forms between cells three and four.

Polarity at an angle to the main axis of the filament results in lateral branches (Fig. 23.17*C*). In this case, the daughter nuclei in the dividing apical cell end up in opposite corners of the cell, and the cell plate cuts the dividing cell at an angle to the axis of the filament. Nor-

Table 23.2
MAJOR CHARACTERISTICS OF 9 ALGAL DIVISIONS[a]

Division	Chlorophylls and Accessory Pigments	Cell Wall	Storage Product	Flagella	Habitats	Number of Species	Notes
Rhodophyta (reds)	*a* (+ *d* in some) phycobilins	Cellulose + sometimes agar or carrageenan	Floridean starch (= amplopectin)	none	98% marine; often warm, deep water	4000	Mostly seaweeds; complex life cycles
Chrysophyta (golden algae, class Chrysophyceae only)	*a* + *c* + fucoxanthin	Cellulose	Fats, oils, chrysolaminaran	2, unequal, anterior	Mainly fresh water	300	
Xanthophyta (yellow-greens)	*a* (+ *c* in some)	Cellulose or none; sometimes with Si and diatomlike	Oil or chrysolaminaran	2, various	Mainly fresh water	400	Sometimes placed in Chrysophyta
Euglenophyta	*a* + *b*	None	Paramylon (paramylum)	1–3, equal, anterior	Mainly polluted fresh water	450	Many animal-like properties
Chlorophyta (greens)	*a* + *b*	Cellulose	Starch	2, equal, anterior	Widest distribution of any division	7000	Possible precursor of higher plants
Charophyta (stoneworts)	*a* + *b*	Cellulose	Starch	2, equal, anterior	Often bottom dwellers in fresh water	250	Differentiated into nodes and internodes; gametangia with sterile jacket, sometimes put in Chlorophyta
Bacillariophyta (diatoms)	*a* + *c* + fucoxanthin	Silica + pectin	Fats, oils, chrysolaminaran	generally none	Mainly aquatic	8000+	Prominent in phytoplankton; sometimes put in Chrysophyta
Pyrrhophyta (dinoflagellates, class Dinophyceae only)	*a* + *c* + peridinin	None or plates of cellulose	Starch(?), oil in some	2, unequal, lateral	93% marine	1000	Prominent in phytoplankton
Phaeophyta (browns)	*a* + *c* + fucoxanthin	Cellulose + algin	Laminaran, mannitol	2, unequal, lateral	99.7% marine; often, shallow water	1500	No unicellular forms; includes the kelps (another name for brown seaweeds)

[a]A few small groups have not been included; see Chapter 21 for blue-green algae.

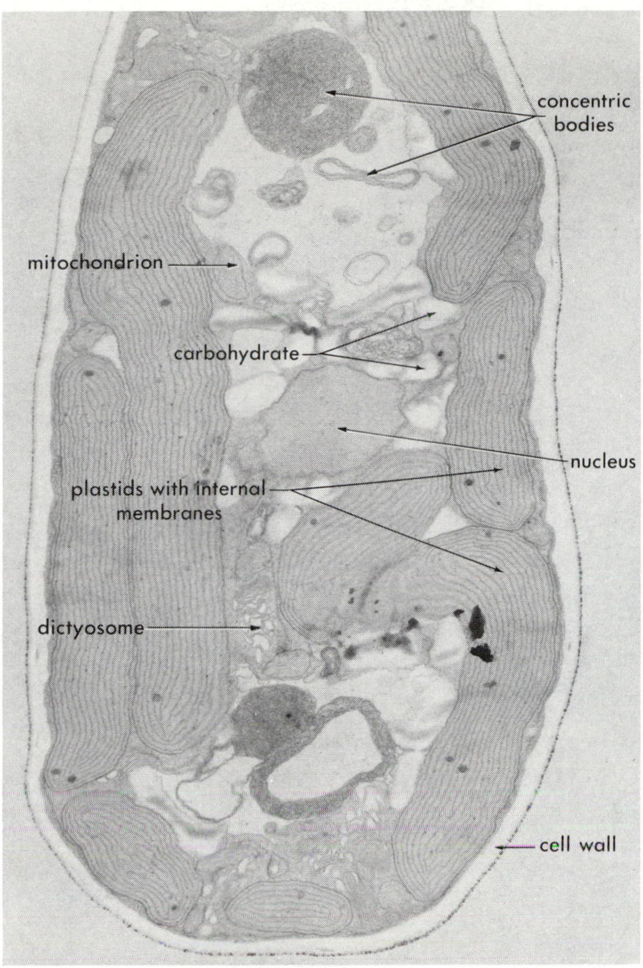

Figure 23.15
A vegetative cell from the freshwater red alga *Batrachospermum moniliforme*. Chloroplast membranes do not form grana, and the carbohydrate is stored free in the cytoplasm, ×20,000. (Courtesy of David Brown.)

mal polarity is restored in the apical cell. Polarity in the second cell is now established at an angle to the main filament, and a protuberance forms from this cell in line with this newly established polarity. Subsequent cell divisions are at right angles to this polarity, cutting off successive cells to form a lateral branch.

Thus, the mature thallus of *Polysiphonia* is a branching thallus formed by a line of central or axial cells surrounded by a regular number of pericentral cells. Axial and pericentral cells are the same length. A comparable structure occurs in many of the Florideophycidae. Morphological variation, which is considerable in the subclass, results from the number of cells forming the axis, variation in the size of cells, and the degrees of branching. In any event, the parenchymalike body is basically filamentous in origin.

Euglenophyta

This is a small division consisting mainly of unicellular, motile species (Fig. 23.18). Members of this division have several protozoan features, such as the lack of a cell wall (although the outer cytoplasm may be modified for rigidity), rapid movement, and the ability to produce a nonstarch food reserve called paramylon. Some Euglenophyta, in addition, lack chloroplasts and either engulf or absorb food. Exposure to ultraviolet radiation can halt chloroplast division but not cell division: the result is that within a few generations some progeny cells will lack chloroplasts. These cells will continue to live as heterotrophs and produce progeny just like themselves.

Chlorophyta: The Green Algae

The green algae are predominately fresh-water forms, but they also exist in salt water, on snow, in hot springs, on soil, on branches, and on the leaves of terrestrial plants. They form the second largest division within the algae. (Diatoms are the largest.)

Cytologically, this group shares several important traits with higher plants: chlorophyll *a* and *b*, similar accessory pigments, starch as a storage product, and cellulose cell walls. For this reason, some scientists hypothesize that ancient Chlorophyta served as the ancestors of all higher plants. The most complex greens have well-developed prostrate portions and upright portions, indicating a high degree of differentiation. This type of habit is called **heterotrichy.** Other forms as shown in Figs. 23.1*A* to *F*, 23-8*A* to *D*, 23.10*B* and *D*) are unicellular, filamentous (with or without cross walls), colonial, and sheetlike.

Bacillariophyta: The Diatoms

We have already discussed these organisms earlier in the chapter as one of the major components of phytoplankton. They exist as single cells (Figs. 23.6*A* to *D*, 23.14)—sometimes stalked and sedentary, sometimes floating free—or as filaments. As seen in a face view, their silica walls are either circular or elongate, giving rise to two main orders, Centrales and Pennales. Some diatoms are capable of a gliding motion even though cilia and flagella are absent.

Diatoms contain chlorophyll *a* and *c*, store food reserves as oil, and their unusual silicate wall is embedded in a pectin matrix. They are mainly aquatic, but they are also found in soil.

Pyrrhophyta: The Dinoflagellates

Together with the diatoms, this group dominates the phytoplankton. Most of its species are unicellular, flagellated, and marine (Figs. 23.6*E,F*), but a few are colonial or filamentous. These organisms in great density cause "red tides." The cells are either naked, that is, with only a thickened membrane around them as in Euglenophyta, or they are enclosed by a unique cellulose wall that resembles patches of armor plating stuck together. Usually two flagella are present, both emerging from the same pore, but otherwise different. One flagellum is flat

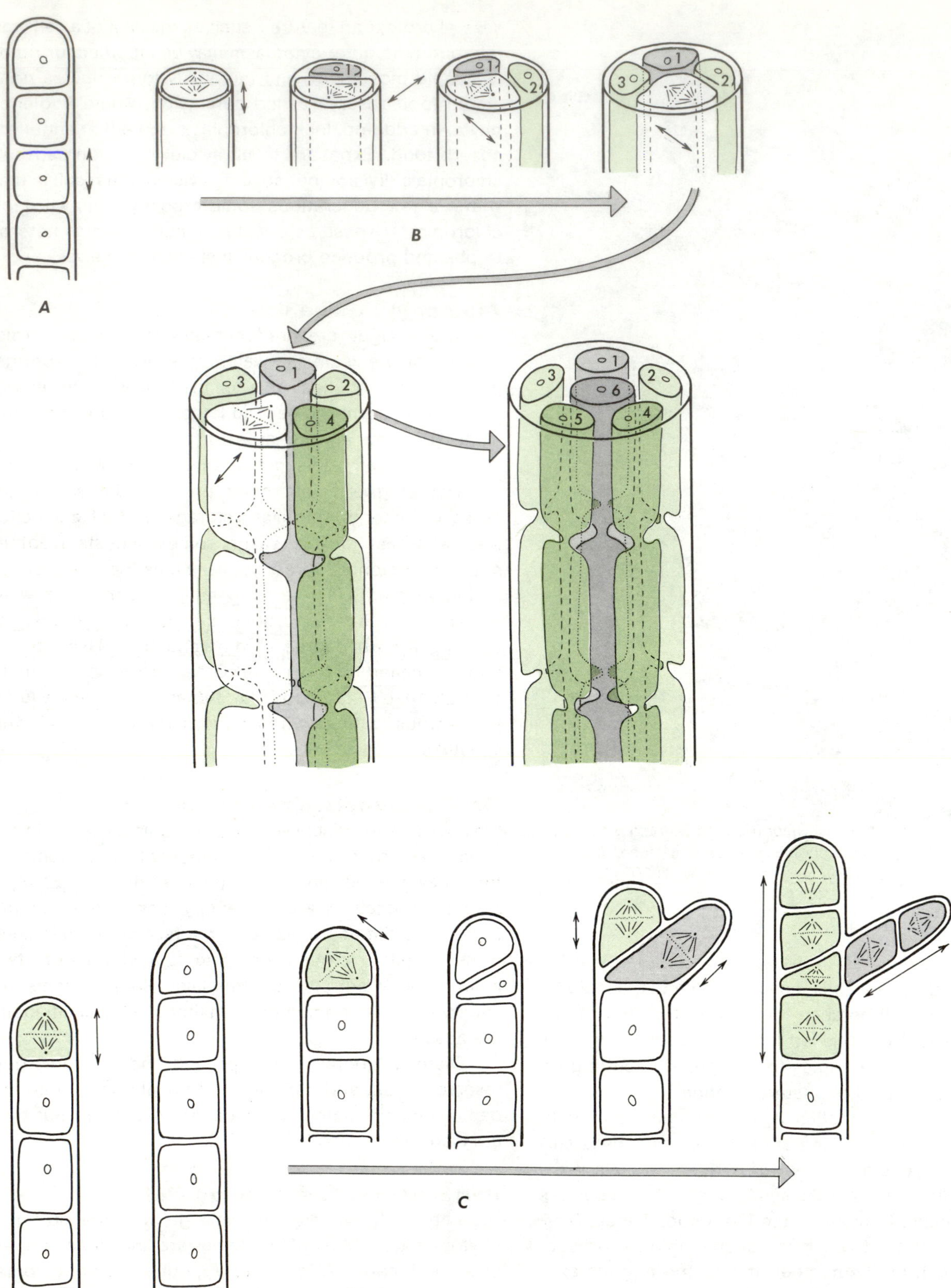

Figure 23.17
The early cell divisions in *Polysiphonia* are very precisely ordered and easily followed. They are responsible for the organization of the mature thallus. *A*, cell divisions only parallel to the axis of the filament would result in a uniseriate filament. Regularly spaced divisions at an angle to the axis produce regular branches. *B*, divisions at right angles to the axis of the filament, always in the order indicated, result in a thallus formed of a central cell surrounded by five pericentral cells. *C*, the origin of a branch. Arrows indicate polarity of cell divisions.

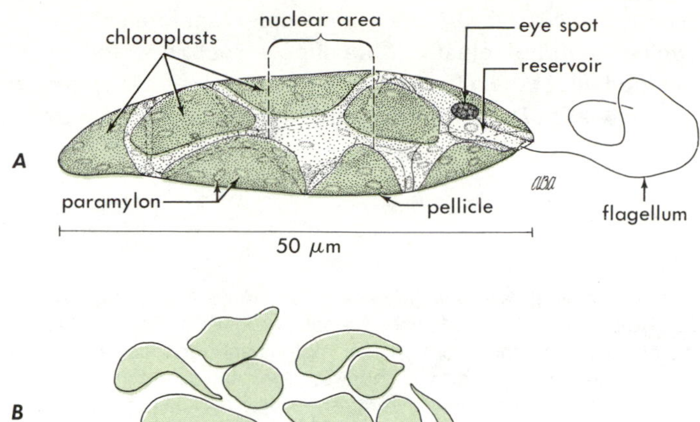

Figure 23.18
A, detail of *Euglena gracilis,* showing several thin, curved chloroplasts, storage granules of paramylon, the more rigid exterior cytoplasm and plasmalemma (the pellicle—not a cell wall), eyespot, and flagellum. *B,* various cell shapes a single cell of *Euglena* may fake over a short period of time.

and ribbonlike and encircles the cell in a transverse groove; it is responsible for rotation and some forward movement. A second flagellum trails behind and provides forward movement while at the same time acting as a rudder.

Dinoflagellates contain chlorophyll *a* and *c* and a brown pigment, peridinin, which gives the group its green-brown or orange-brown color. True starch appears to be stored. Some forms are luminescent and contribute to the glow of water when it is disturbed at night, as in the wake of a ship.

Phaeophyta: The Brown Algae

The brown algae include filamentous forms, sheetlike forms, and large kelps with more complex differentiation in anatomy and morphology than any other algae (**Fig. 23.2,** page 410, 23.12). Unicellular forms are unknown. Chlorophyll *a* and *c* are present, plus the pigment fucoxanthin, which gives these plants their characteristic brownish color (though color can range from olive-brown to golden-brown to practically black). Carbohydrate is stored as mannitol or laminaran, not as starch, and it accumulates as granules in the cytoplasm. As in the Rhodophyta, the chloroplasts do not contain grana (Fig. 23.19). The cell walls are composed of cellulose and often a great deal of algin as well.

The most primitive Phaeophyta, the Ectocarpales, are small-branched filamentous forms (Fig. 24.6). The heterotrichous habit is developed in most of them. The prostrate system consists of a few filaments and serves to anchor an upright system of branching photosynthetic and reproductive filaments to other algae or to rocks. This prostrate system of filaments exhibits apical growth. The upright system appears as feathers or compact brownish tufts. Cell division is diffusely distributed along the branching filaments. It is rarely apical or confined to special regions.

Figure 23.19
A chloroplast and a mitochondrion from a cell of the brown alga *Egregia menziesii.* Note clear areas at opposite ends with DNA fibrils and the chloroplast membranes in parallel bands, ×26,000. (Courtesy of T. Bisalputra and A. Bisalputra, *J. Cell. Biol.* **33,** 511. ©1967 by the Rockefeller University Press. Reprinted with permission of the publisher.)

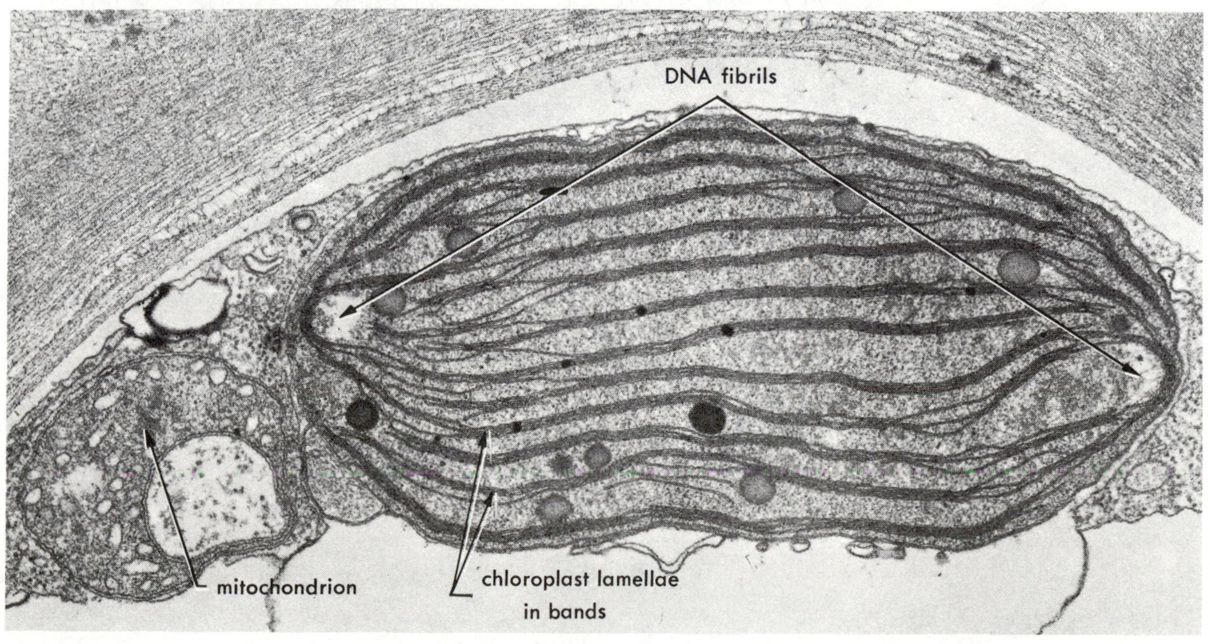

The most advanced Phaeophyta are in the Laminariales. They are large plants called **kelps,** with regions specialized for anchorage, conduction, support, and photosynthesis. Kelps such as *Macrocystis, Laminaria,* and *Postelsia* (Fig. 23.20), are complex anatomically as well as morphologically. If the stipe is sectioned and examined under the microscope, several regions are apparent (Fig. 23.20*D*). Cells in the outermost layer not only are

Figure 23.20
A, an example of a kelp, or brown seaweed: *Postelsia palmaeformis* (sea palm), showing organs that resemble roots, stems, and leaves, about 25 cm tall. *B,* young plant. Anatomical details of the stipe of *Postelsia*. *C,* diagrammatic cross section showing the major regions and the location of detail drawings. *D,* details of the meristoderm region, cortex with muscilage canals, and inner cortex with sieve-tubelike cells in a transition region near the central medulla. *E,* detail of the end wall of a sieve-tubelike cell, showing pores through which mannitol and other substances can flow in the stipe of the brown alga, *Macrocystis,* ×1000. (Courtesy of Knute Fisher.) For color version of Figure 20.23*E*, see page 410.

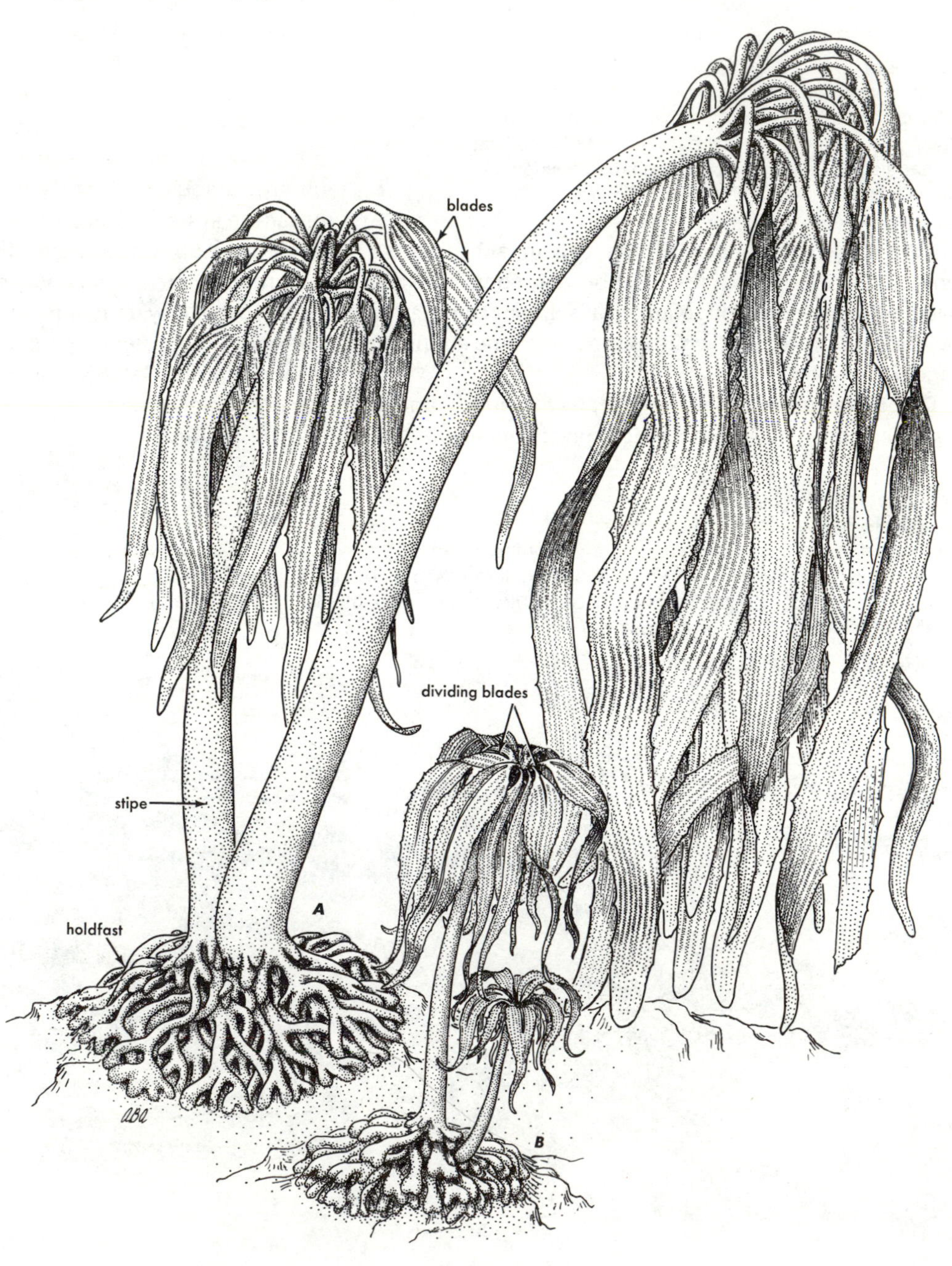

protective, but they remain meristematic and contain chloroplasts as well. To distinguish this diverse tissue from mere epidermis, it is given the name **meristoderm**.

A broad cortical region is composed of parenchyma-like cells. Mucilage-secreting cells form definite canals through the cortex. Loosely packed filaments of cells fill the central region, the **medulla.** Some inner cortex cells next to the medulla appear to function as sieve-tube members; they have sieve plates, form callose, adjoin one another to form continuous tubes, and are known to translocate mannitol by a mechanism resembling that found in vascular plants (Figs. 23.21, 23.20*E*). However, these kelps contain no tissue that resembles xylem. The value of a photosynthate-conducting system in these large plants is easy to perceive. The mass of floating fronds considerably reduces the penetration of light into the water below; consequently the lower part of the stipe and the holdfast may be at such a low light intensity that respiration exceeds photosynthesis. They must be nourished by food translocated from above.

Unlike many of the other algae, the dominant generation of the kelps is diploid. The dominant generation is also larger and more complex than that of any other alga.

DIVERSITY OF FORM WITHIN THE GREEN ALGAE

Within divisions, evolutionary relationships are demonstrated by thallus morphology. It is widely accepted that unicells are the most primitive and that evolutionary advance proceeded, in general, along similar lines in all divisions. There were three tendencies in this evolu-

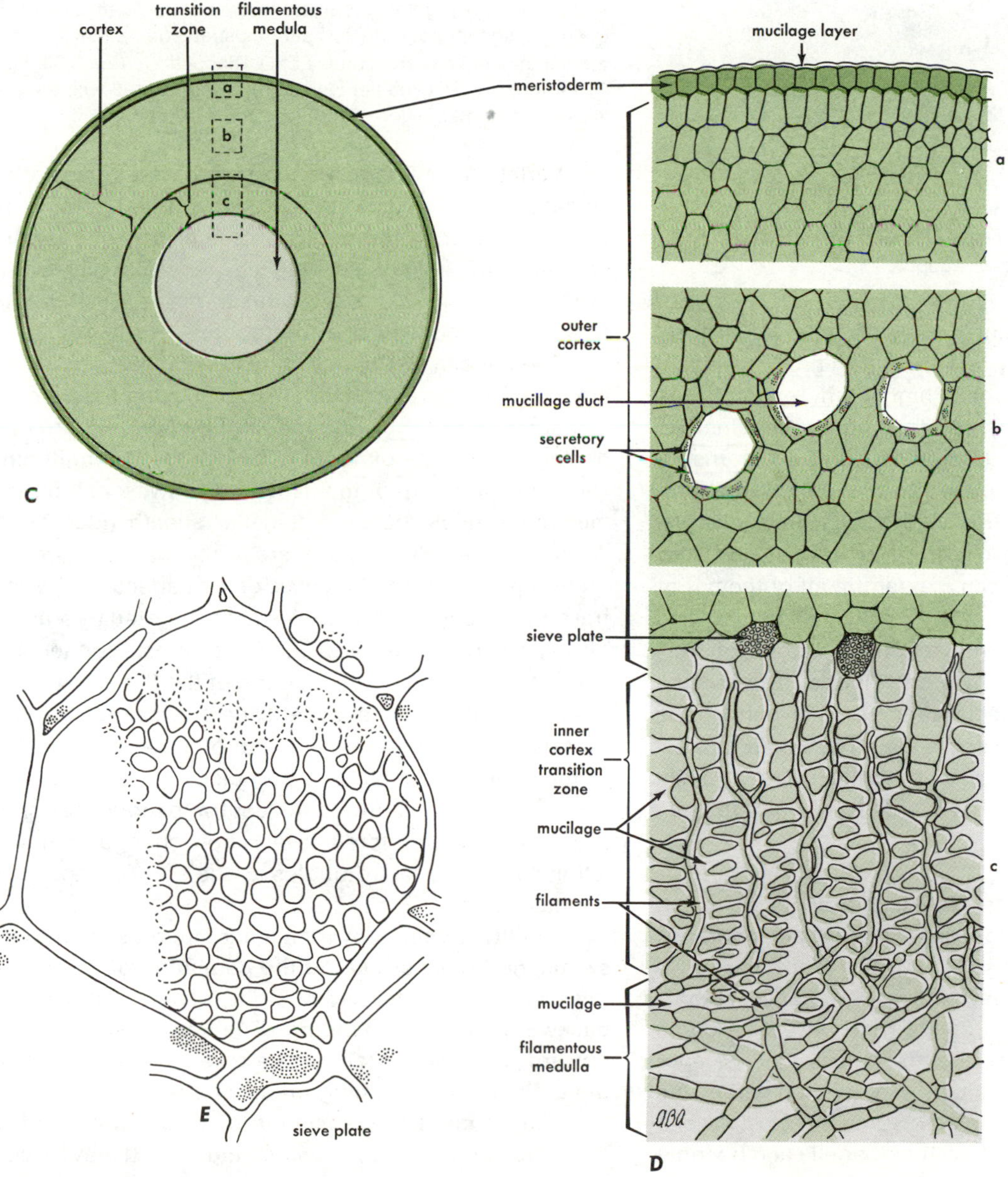

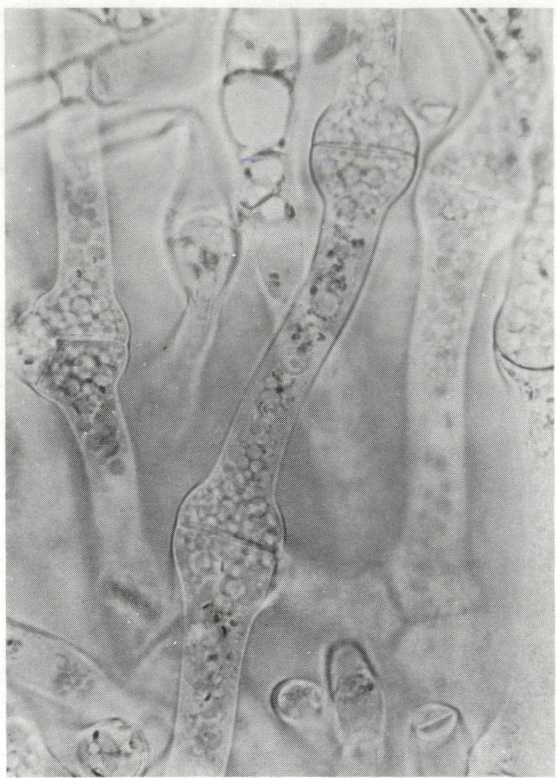

Figure 23.21
Photograph of sieve-tubelike cells, with swollen junctures, from the kelp *Laminaria groenlandica*, about ×900.

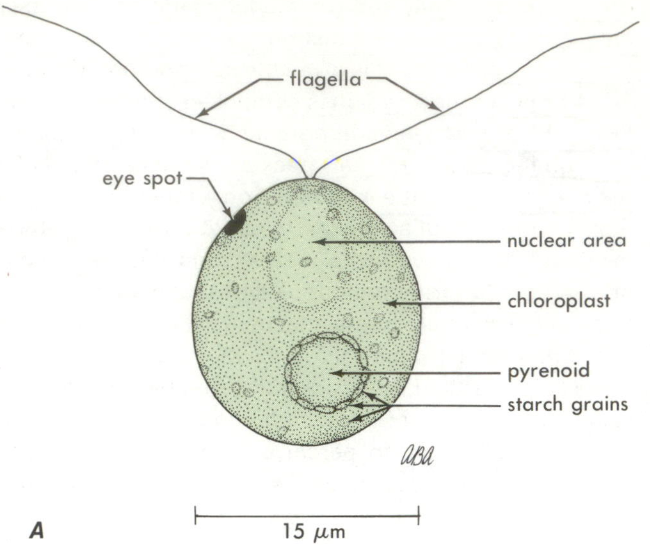

Figure 23.22
A, drawing of a vegetative cell of the gree alga *Chlamydomonas*. *B*, an electron micrograph, to better illustrate the relationship of the cell organelles. (*B*, courtesy of D. Ringo. *J. Cell Biol*. **33,** 543. © 1967 by The Rockefeller University Press. Printed with the permission of the publisher.)

tionary expression: (*a*) single cells cooperated to form rather complex colonies (the **volvocine** line, Fig. 23.1*C*); (*b*) single cells became locally highly differentiated (the **siphonous** line, Fig. 23.1*F*); and (*c*) single cells remained attached together after division to form filaments or simple sheets of parenchymalike tissue (the **tetrasporine** line, Figs. 23.1*D,E,* 23.8*B,D,* 23.10*B*). While parallel changes have apparently occurred in all divisions, they have not occurred to the same extent in all of them. For instance, there are only unicells in Euglenophyta, and unicellular forms are not known in the Phaeophyta. Furthermore, thalli of Phaeophyta and Rhodophyta show a much higher degree of specialization than is found in the other divisions. We shall discuss the morphology of algal thalli, proceeding from less complex to more complex arrangements as illustrated by examples of Chlorophyta.

Volvocine Line

An example of this line is the order Volvocales in the Chlorophyta. Most individuals in the *Chlamydomonas* genus are oval or oblong (Fig. 23.22). They have two flagella at their anterior end and a cup-shaped chloroplast. Most species have a single red body of carotene with a characteristic ultrastructure; this is the **eye spot** (Fig. 23.23). The nucleus is centrally located within the cup formed by the chloroplast. Some species have a stage when the individuals lose their flagella and become embedded in a common mucilaginous matrix.

Variations in flagellation are important in algal classification, but details will not be considered here. Of interest, however, is the remarkable similarity of the ultrastructure of eukaryotic flagella. The flagella of all organisms form *Chlamydomonas* to man have a uniform internal arrangement of microtubules.

The flagella of *Chlamydomonas* will serve as an example. Here, there are two anterior flagella entering the cell at an angle. They end in **basal bodies** that are connected by a bridge of fibers (Fig. 23.22*B*). A longitudinal section near the tip of the flagellum shows a unit plasma membrane enclosed by a flagellar sheath (Fig. 23.24). Within the flagellum there are three bands of microtubules parallel to the long axis of the flagellum. The central band is longer. The tubules are separated by a matrix of some sort. An array of indefinite cross connections may also be seen. A cross section of the flagellum at this level shows the flagellar sheath and plasmalemma. It also reveals that the outer circle of microtubules is composed of nine sets of double microtubules and that a pair of single microtubules forms the central band. Other cross sections taken at different levels show changes in the relationship of the microtubules as the sections approach the basal body.

Motility is gained by a flagellar motion similar to the swimming breast stroke. At the beginning of the stroke, the flagella are extended straight forward. They sweep backwards without bending. The return stroke is an upward wave motion and again resembles the return of the arms in the swimming breast stroke.

The permanent association of chlamydomonadlike cells into curved plates (*Gonium*, Fig. 23.1*B*) may be considered the first step in advance in the volvocine line.

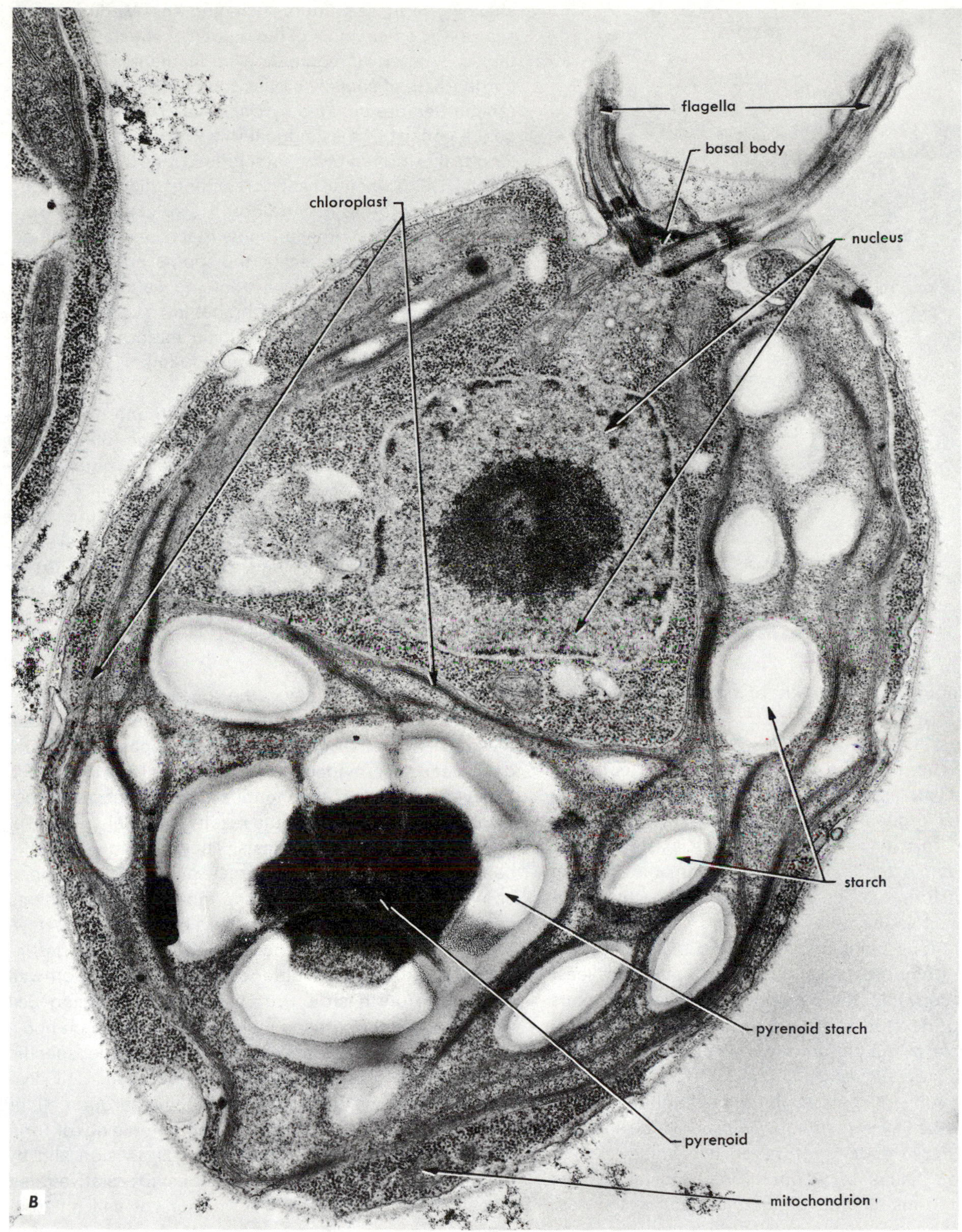

Each new colony is formed within a parent cell and released from the parent cell as a small but fully formed new colony.

Further advance is occasioned by the formation of a hollow sphere of 32 similar individual cells. They are connected to each other by delicate strands. Increase in the number of cells forming a colony and specialization of some cells for reproduction culminates in the genus *Volvox* (Fig. 23.1*C*). The colonies contain from 500 to 20,000 cells, only a few of which are reproductive cells confined to a small area. The vegetative cells of *Volvox* still resemble *Chlamydomonas*.

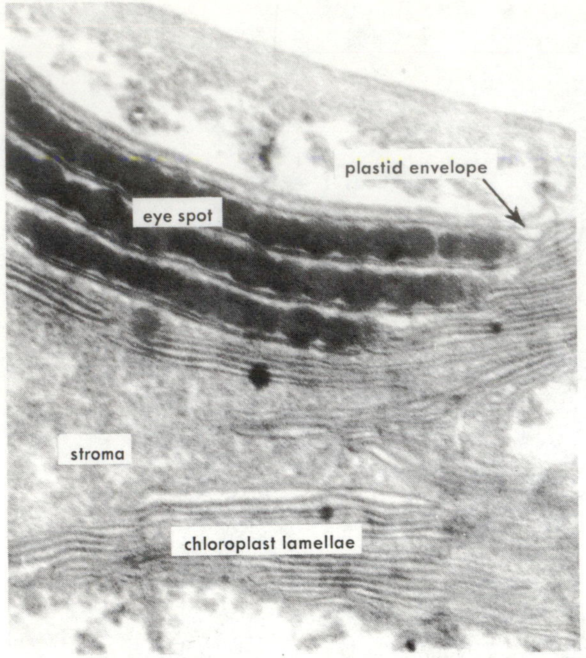

Figure 23.23
The eye spot in *Chlamydomonas* consists of several rows of granules containing carotene. It lies wholly within the chloroplast, ×25,000. (Courtesy of Carole Lembi.)

Siphonous Line

The outstanding characteristic of this line is the complete lack, or rare production, of cross walls **(septa).** The body of most siphonous species is thus multinucleate and coenocytic. In forms completely lacking septa, the peripheral cytoplasm, with embedded nuclei, plastids, and all other organelles, is continuous throughout the plant body. This also means that there exists a single central vacuole extending through all filaments. Such an evolution of form may be envisioned with a chlamydomonadlike cell as the primitive starting point. Evolution of the siphonous plant body occurred several distinct times in the Chlorophyta and has been paralleled at least once (*Vaucheria*) in the Xanthophyta. The siphonous habit did not develop in other divisions.

The most primitive forms consist only of an aerial green, multinucleate vesicle anchored to the soil by colorless branching rhizoids (*Botrydium,* Xanthophyta) or by a single rhizoid (*Protosiphon,* Chlorophyta). The irregular branching or budding of this vesicle would result in the irregular thallus of *Valonia* (Chlorophyta). Simple filamentous forms such as *Vaucheria* occur in the Xanthophyta. Two types of multinucleate forms occur in the Chlorophyta. Some, like *Hydrodictyon* (Fig. 23.1*E*), have cross walls separating multinucleate cells; others are strictly coenocytic. These coenocytic forms, belonging to the order Codiales, branch extensively and with a high degree of regularity, conferring considerable diversity of form on the adult thalli.

The genus *Caulerpa,* which forms extensive mats on shallow warm sea floors including the Mediterranean sea, is a good example of the degree of specialization attained by intertwining coenocytic filaments (Fig. 23.1*F*). Large tracts of shallow areas of the floor of the Mediterranean Sea are covered by *Caulerpa*. The plants of this genus consist of a cylindrical rhizome up to 3 meters in length. It is formed by a characteristic intertwining of filaments. This rhizome bears branching rhizoids growing downward into the ocean floor. It also sends upward a number of photosynthetic shoots that take a great variety of shapes. In some species, the shoots are featherlike, while in other species they are flat and leaflike. They, like the rhizome, are formed of branching coenocytic filaments arranged in a characteristic pattern. Growth in these forms is at the apex of each filament, where concentration of protoplasm also occurs.

In contrast to this extensive growth and differentiation of coenocytic filaments, stands the uninucleate single-celled *Acetabularia* (Fig. 4.21). Here, a high degree of specialization occurs within the limits of a single nucleated cell about 25 mm long and divided into three regions of specialization: (*a*) a branched rhizoid system, one branch of which contains the single nucleus; (*b*) an upright stalk; and (*c*) an expanded upper portion, the cap.

Tetrasporine Line

Although a *Volvox* colony in the volvocine line may consist of 20,000 cells with some degree of specialization, all these cells were formed within a single parent cell. The daughter cells provided their own cell wall at a later stage in development. In the siphonous line, mitoses occur, but cytokinesis does not, and the resulting nuclei are not set apart in individual protoplasts. They remain within the parent cell wall, which enlarges and may take on a complex shape. In the tetrasporine line, cell divisions occur; each mitosis is accompanied by cytokinesis. Now, the parent cell wall is retained on three sides as an essential part of the two daughter cells. A new common wall forms the fourth side. Protoplasts and enclosing cell walls become cooperating units in a tissue or plant body. The need to confine cell division to a single specific period of the life cycle has been eliminated. Cell divisions, mitosis and cytokinesis may occur through all, or much, of the life of a plant. Size and complexity of form may now be attained by continued cell division and by controlling: (*a*) polarity of cell divisions, (*b*) relative rates of cell divisions, and (*c*) relative rates of cell enlargement.

Gametes of the green alga *Ulothrix* resemble mature chlamydomonad cells, as do the gametes of many other algae. When zoospores of *Ulothrix* germinate, the new cells adhere to each other and the cells divide with spindles aligned in a straight line, so that repeated division of all cells results in a filament formed by a single row of cells (Fig. 23.8*D*, 23.25). This is a **uniseriate** fila-

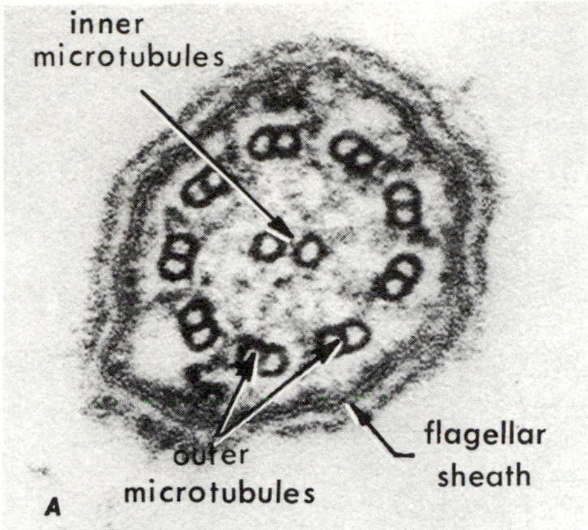

Figure 23.24
Flagella of *Chlamydomonas reinhardi*. *A*, cross section, taken near tip of flagellum. There is an outer circle of nine paired microtubules (doublets) and two central single microtubules. The microtubules are embedded in a matrix that is surrounded by a flagellar membrane and a flagellar sheath, ×100,000. *B*, a longitudinal section of a flagellum at its tip end; the central microtubules are longer than those composing the outer circle of doublets, ×68,000. (Courtesy of D. Ringo, *J. Cell. Biol.* **33,** 543. © 1967 by the Rockfeller University Press. Reprinted with permission of the publisher.)

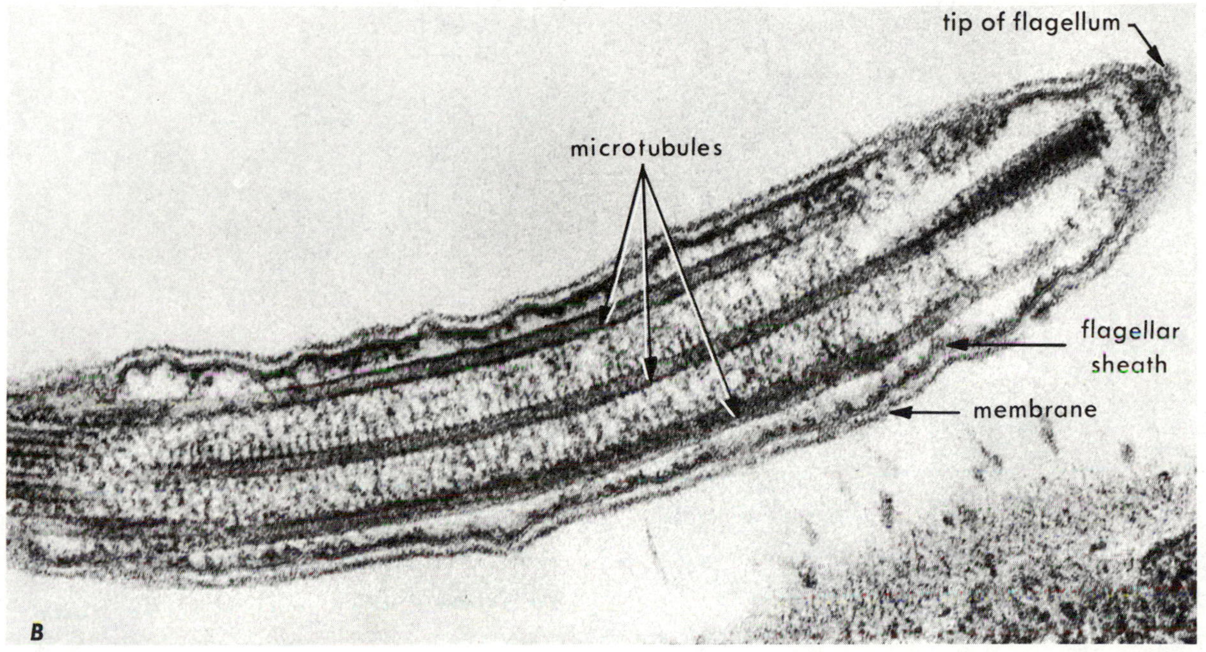

ment. Two very important things have happened: (*a*) cytokinesis results in adherent, cooperating daughter cells and (*b*) a definite polarity for all cells has been established for this simple multicellular plant. Adhering cells and genetic control of polarity of cell division are two basic requirements for attainment of great diversity of plant form.

In *Ulothrix* and other simple unbranched forms, all spindles are oriented in the same direction and all cells in the filament may divide. Upon establishing a second pair of poles at right angles to the axis of the first pair, a diversity of thalli will arise, depending upon the timing of cell divisions in the two planes. If all cells divided regularly at right angles to each other, but in the same plane, a thin tissue, one cell in thickness, would form a **monostromatic thallus** (Fig. 23.25*C*). This occurs in the green alga *Monostroma*. If, instead, every tenth cell divided, and continued to divide, at right angles to the line of divisions forming the filament, a flat filament with regular branches would result (Fig. 23.25*B*). If every twentieth cell could now divide in a plane at right angles to the plane of the flat branching filament, a branching filament with its branches at right angles to each other would result. If the rates and locations of cell divisions are confined to specific cells of the main filament, lateral branches will arise at definite points as in *Draparnaldia* (Fig. 23.25*B*). Note that the cells of lateral branches are much smaller than those of the central branch.

If all cells early in the development of a monostromatic thallus divided but once in a plane at right angles to the plane of the monostromatic thallus, a thallus two cells in thickness **(distromatic thallus)** would result. Sea lettuce, the marine green alga, *Ulva*, has a distromatic thallus (Fig. 23.25*D*). With division possible in three planes, form is determined by the relationship between the rates of division, the location of the divisions, the rates of cell expansion, and the kinds of differentiation that occur in the three planes. Cell divisions under these

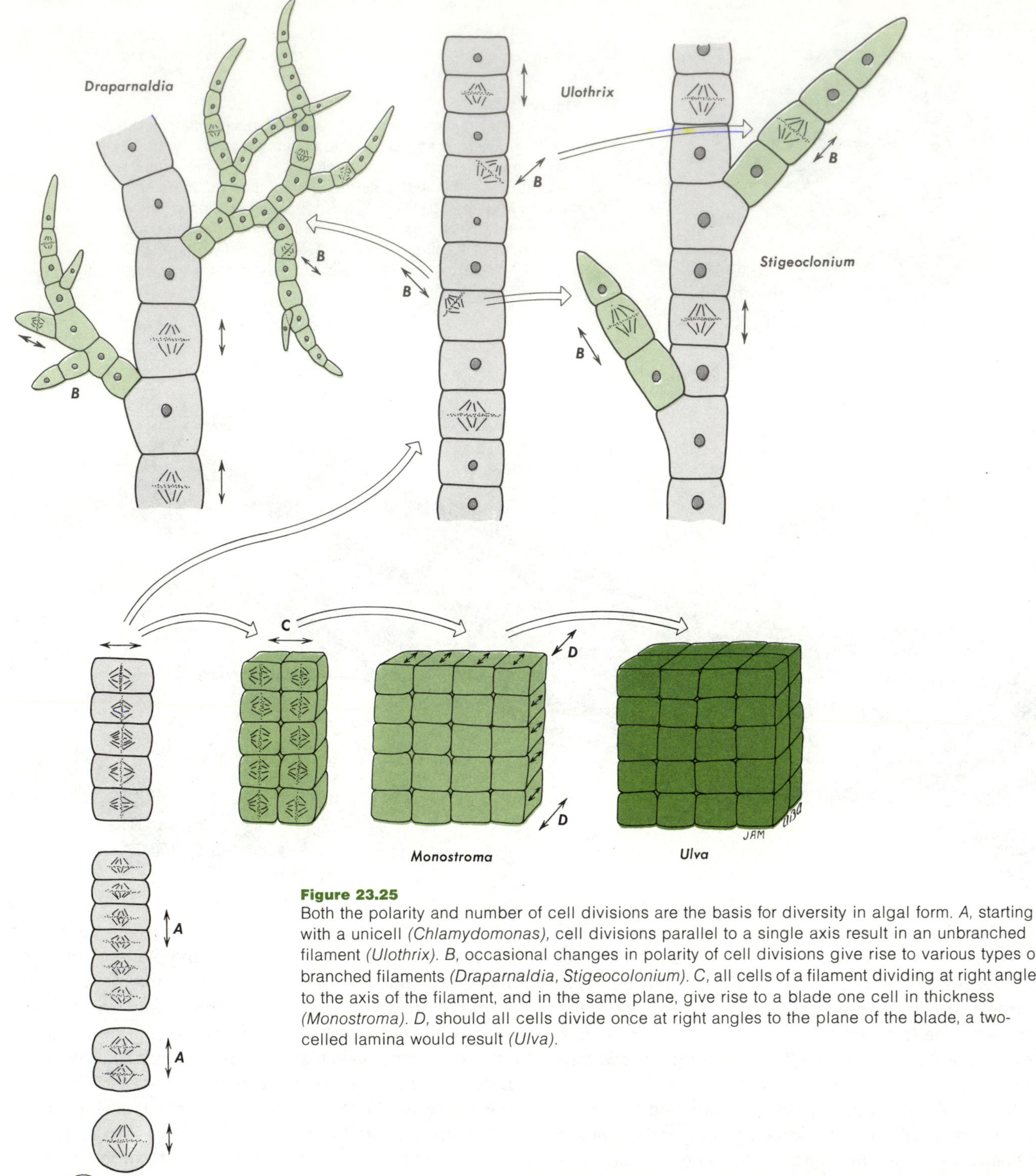

Figure 23.25
Both the polarity and number of cell divisions are the basis for diversity in algal form. *A*, starting with a unicell *(Chlamydomonas)*, cell divisions parallel to a single axis result in an unbranched filament *(Ulothrix)*. *B*, occasional changes in polarity of cell divisions give rise to various types of branched filaments *(Draparnaldia, Stigeocolonium)*. *C*, all cells of a filament dividing at right angles to the axis of the filament, and in the same plane, give rise to a blade one cell in thickness *(Monostroma)*. *D*, should all cells divide once at right angles to the plane of the blade, a two-celled lamina would result *(Ulva)*.

conditions are generally confined to specific locations or meristems at the apex of the thallus or at definite intercalary points.

While polarity of cell division is of importance in determining plant forms, environmental factors may also be involved. Nevertheless, it would be exciting to discover the genetic mechanism that determines which cells are to divide at right angles to their neighbors. Just what type of change in base pairs in DNA brings about the change from a filament to the monostromatic thallus of *Monostroma* or the distromatic thallus of *Ulva?*

SUMMARY

1. Algae lack true roots, stems, and leaves (i.e., with vascular tissue). Their thallus may be unicellular, sheetlike, colonial, filamentous, or large and complex as in the kelps. Gametes develop in unicellular gametangia; thus there is no sterile, protective jacket of cells around the gametes as there is in higher plants.
2. Algae are widely distributed. In aquatic ecosystems, they may be the major producers and form the base of the food chain. Floating algae (generally microscopic and unicellular) comprise the phytoplankton. Cultural eutrophication causes rapid growth of phytoplankton and can lead to blooms or red tides with toxic by-products.
3. Some algae have human use as food, sources of essential nutrients, fodder for domesticated animals, and fertilizer for crops.
4. The porous, silicious walls of diatoms make them valuable as industrial filters, extenders, or additives that improve flow or stability of other materials, and as abrasives.
5. Agar and carrageenan, polysaccharides in the walls of red algae, and algin from the walls of brown algae have many human uses.
6. The algae can be taxonomically divided into 10 or more divisions, based on storage products, wall chemistry, types of chlorophylls, and types of flagella. Seven of the divisions were briefly discussed in the text.
7. Members of the Rhodophyta (red algae) are almost exclusively marine, especially in tropical water. Some secrete lime incrustations and are important contributors to reef buildings. They are often delicately filamentous, and prominant pits may connect the cytoplasm of adjacent cells.
8. Members of the Euglenophyta have several protozoan features, such as lack of a cell wall, flagellar movement, and the accumulation of a nonstarch storage product (paramylon).
9. Members of the Chlorophyta (green algae) predominantly occupy fresh water habitats. Cytologically, this group shares several traits with higher plants; thus this group may be phylogenetically ancestral to higher plants.
10. The Bacillariophyta (diatoms) is the largest division of algae. Its members dominate the phytoplankton, especially in fresh water systems. Members of the Pyrrophyta (dinoflagellates) are also abundant in the phytoplankton, and their numbers may become dense enough to produce red tides and luminescent surface water in marine systems.
11. Members of the Phaeophyta (brown algae) include large kelps with organs resembling roots, shoots and leaves. Anatomical specializations include a meristoderm that permits growth in girth, and a phloemlike tissue that permits translocation of food reserves (mannitol). Kelps are diploid.
12. It is generally accepted that unicellular forms are most primitive in the algae, and that evolution proceeded in three lines of increasing complexity: the volvocine line, leading to colonies; the siphonous line, leading to enlarged cells or to nonseptate filaments of many cells; and the tetrasporine line, leading to septate filaments, sheets of cells, or parenchymalike tissue.

CHAPTER 24

ALGAE: REPRODUCTION

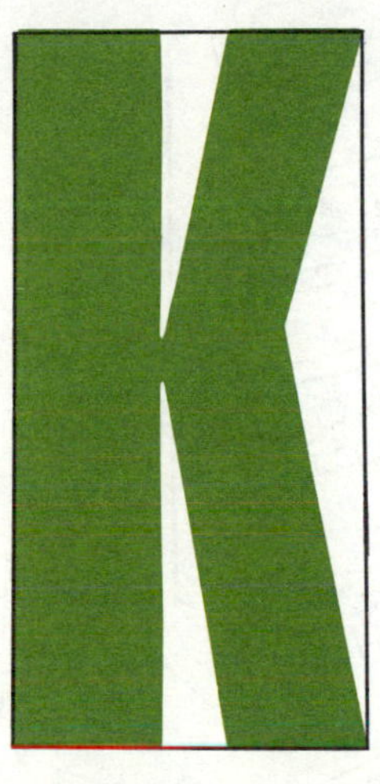

nowledge of the reproductive processes in the algae is incomplete and largely confined to the Rhodophyta, the Phaeophyta, and the Chlorophyta. In some other divisions, detailed knowledge is known only for a very few species, genera, or families. Asexual reproduction in the algae commonly occurs under conditions favorable for growth, and sexual reproduction commonly when conditions are less favorable. There is no typical life cycle for the algae.

ASEXUAL REPRODUCTION: VEGETATIVE

Vegetative reproduction is widespread in many forms and is probably used in all divisions. Indeed, in those algal groups where sexual reproduction is unknown or rare, asexual reproduction may be the only form of reproduction.

The longitudinal splitting of a biflagellate motile *Euglena* into two, still motile, but uniflagellate daughter cells is simple vegetative reproduction. The diatoms may divide vegetatively for up to five years. Recall that the cells have two sections that fit into each other like the top and bottom of a Petri dish. After division, a new wall forms within the old wall (Fig. 24.1*A*). This means that the new cell receiving the lower and smaller wall will be smaller than the parent cell. Eventually, a minimum size is reached and division stops. Full size is regained at the time of sexual reproduction (Fig. 24.1*B*).

Vegetative reproduction in filamentous forms occurs by a simple fragmentation of the filament. In some species, specialized cells seem to be associated with the

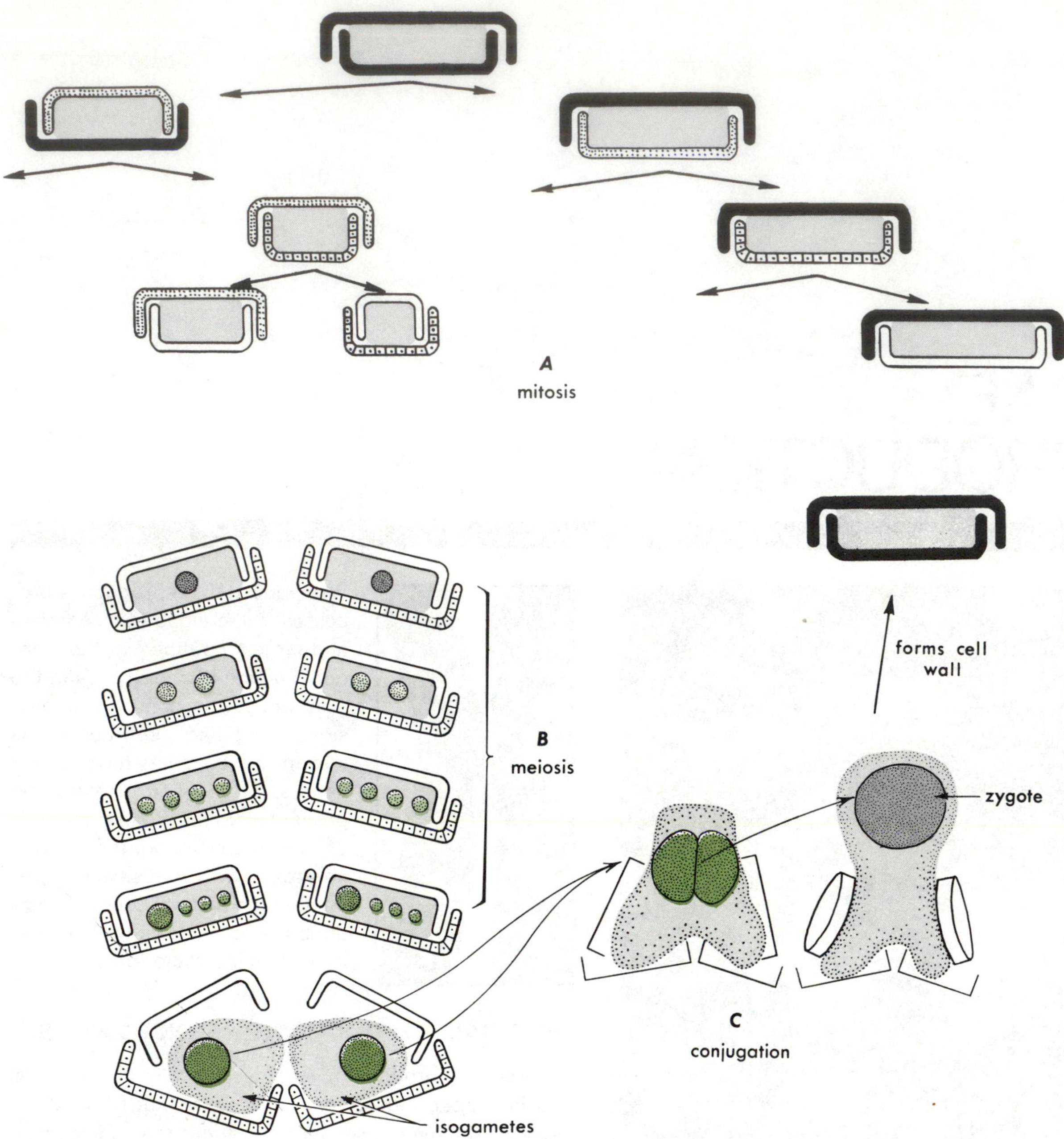

Figure 24.1
A gametic life cycle as represented by a diatom. *A*, the progeny cells continually become smaller if, after cell division, the new cell forms within the silica shell of the old cell. *B*, meiosis occurs; three of the nuclei with the *n* number of chromosomes degenerate and the single remaining haploid cell becomes a gamete. *C*, fusion of the two gametes to form a zygote. (*A* and *C*, redrawn after G. M. Smith, *Freshwater Algae of the United States*. © 1933 by McGraw Hill Book Company. Reprinted with permission of the publisher.)

breaking of the thallus. In species of the Cyanophyta, notably *Oscillatoria* (see Chapter 21), the point at which fragmentation occurs is marked by a dead cell. Separation at these dead cells results in short filaments. Other species may modify somatic cells by adding a thick wall, and these cells may function as **resting spores,** living through a hostile period of time in dormancy when all other somatic cells die.

ASEXUAL REPRODUCTION: MITOSPORES

Asexual spores are always preceded by mitosis and may be called **mitospores** to distinguish them from spores preceded by meiosis, called **meiospores.** The latter constitute a stage in sexual reproduction and will be considered later. Since only mitotic divisions are involved, the genotypes of all asexual spores arising from one and the same parent are identical with each other and with the parent plant. They will produce a population of genetically identical individuals. When a population has the genotype best fitted to a given growing condition, or locality, it will multiply by asexual reproduction and populate that locality. Such a population of genetically identical individuals is a **clone.**

Mitospores may be either motile or nonmotile. Motile spores are called **zoospores** and nonmotile spores are frequently called **aplanospores.** Mitospores are produced in specialized cells called **sporangia.** In many algae, the sporangia show little, if any, difference in appearance from ordinary vegetative cells. The number of spores produced depends on the species, but it is typically 16 to 64.

Most zoospores are pear-shaped or spherical. Depending on the species, they have two, four, or many flagella. After liberation from the sporangium, zoospores are actively motile for as short a time as a few minutes or for as long as three days. Their movement and periods of activity are frequently affected by light. After their period of activity they settle to the bottom of a pond or culture tank, lose their flagella, and by cell division begin to develop a new thallus.

SEXUAL REPRODUCTION

Sexual reproduction is responsible for variation within a population. All of the new plants in a given population formed by sexual reproduction will have different genotypes from the parents; those plants best adapted to the growing conditions will become established.

The cells that fuse in sexual reproduction are known as **gametes.** Gametes have the haploid or n number of chromosomes. The single cell resulting from the fusion is a **zygote,** and has a **diploid** or $2n$ number of chromosomes. By cell division, the originally one-celled zygote can produce a diploid plant (the **sporophyte** generation). The number of chromosomes is later returned to the haploid, or n number by meiosis. Reproductive cells resulting from meiosis, with the n number of chromosomes, you recall, are meiospores.

Meiosis occurs at different places in the sexual cycle, and the cells in which it occurs have been called **meiocytes.** Meiospores can be motile or nonmotile. If they lodge in an appropriate environment, they germinate and by cell division produce a haploid plant (the **gametophyte** generation). The gametophyte generation produces the sexual cells called gametes. A generalized life cycle, within which any algal or higher plant can be placed, is shown in Fig. 24.2. It shows the relationship between the sporophyte and gemetophyte generations, and the role that meiospores and gametes play.

Gametes are produced in cells called **gametangia.** A gametangium, in the most specialized case, may be indistinguishable from an ordinary vegetative cell, except

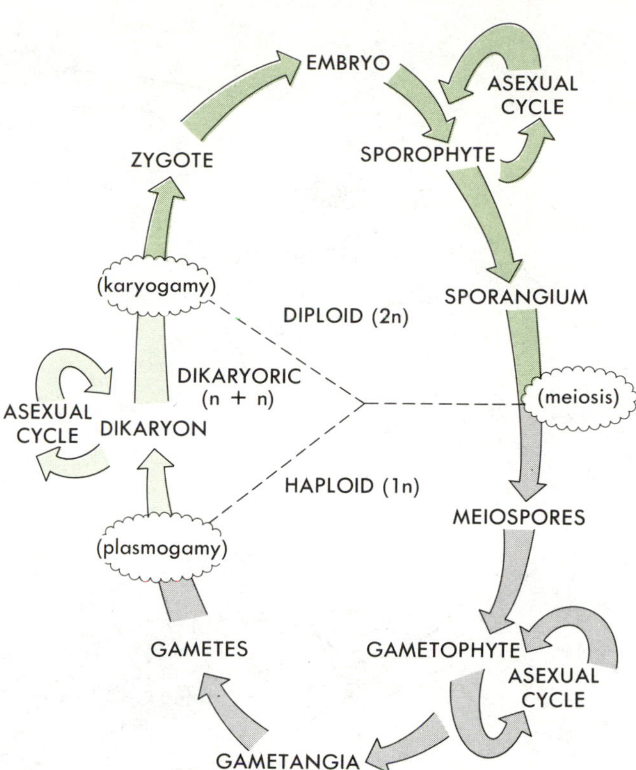

Figure 24.2
A generalized life cycle, within which the life cycle of any one particular plant can be placed. No single plant has a cycle that includes all the sexual and asexual loops shown here. The dikaryotic phase is sometimes called the heterokaryotic phase in other texts.

that several additional mitoses occur within it, giving rise to 16 to 32 cells. In the unicellular, motile green alga *Chlamydomonas*, these cells strongly resemble the vegetative cells, but they may be smaller (Fig. 24.3).

Sometimes gametes and spores are difficult to tell apart. In the simple filamentous species of the green alga *Ulothrix* (Fig. 24.4), an ordinary vegetative cell gives rise to 8 to 64 flagellated cells. When only 8 motile cells are formed, they act as spores, germinating directly and forming a new (haploid) plant on their own. Since these spores are motile, a synonym is zoospore. The cell in which they were formed is a **sporangium.** However, when a greater number of cells is produced, the cells behave as gametes, fusing in pairs to produce a diploid cell (the zygote). In this case, the cell in which the flagellated cells were produced is a gametangium.

When gametes resemble each other, they are called **isogametes** and the species that produces them is **isogamous** (Fig. 24.5). Thus, *Chlamydomonas* and *Ulothrix* are isogamous.

In some species, the gametes differ in size, one being slightly smaller than the other. Gametes with only a slight difference in size are said to be **anisogametes** (heterogametes), and species expressing this state are **anisogamous species** or exhibit **anisogamy** (heterogamy) (Fig. 24.5).

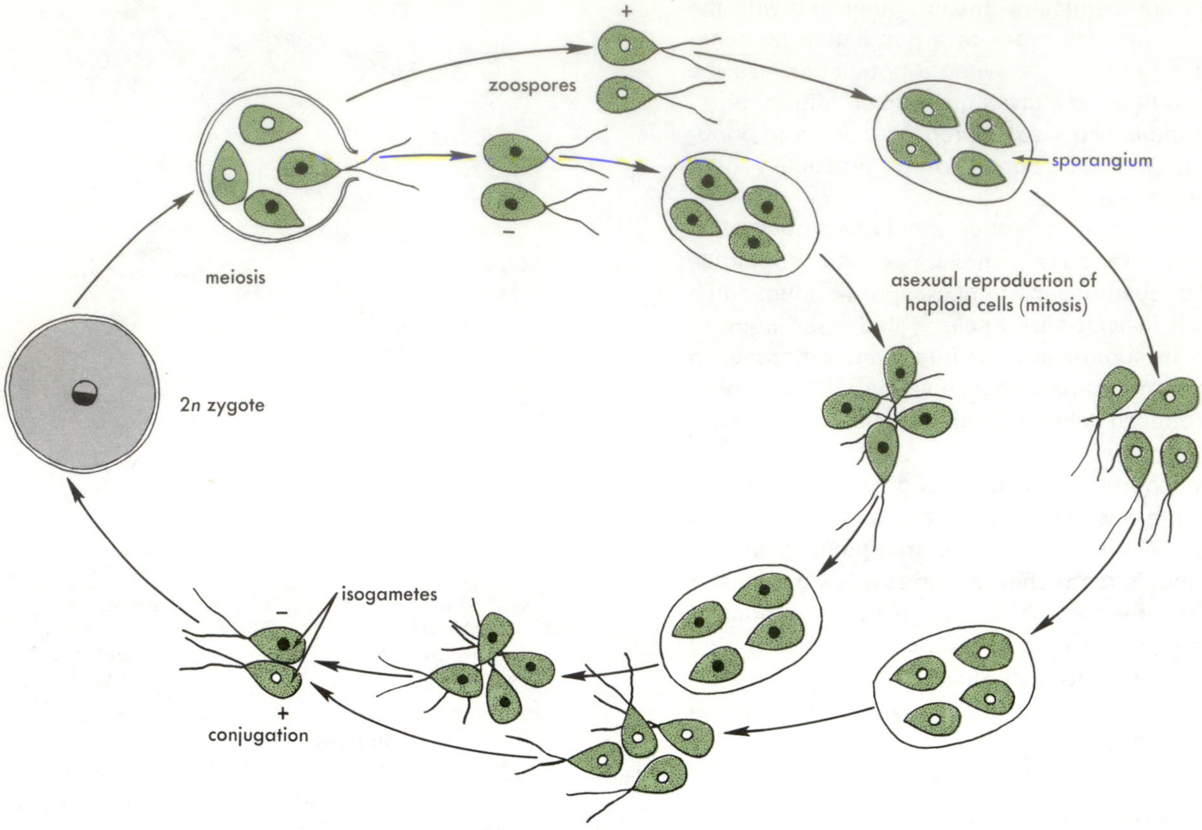

Figure 24.3
A zygotic life cycle is represented by the green alga *Chlamydomonas*. Meiosis occurs during the germination of the zygote; thus the zygote is the only diploid (sporophytic) cell. (Redrawn after F. Moewus.)

In other species, the gametes not only differ in size but in degree of motility. One type of gametangium produces many small, motile gametes called **sperm.** A sperm-producing gametangium is known as an **antheridium.** Another vegetative cell may become quite enlarged, sometimes becoming flask-shaped. The protoplast retracts slightly from the cell wall. It has become an **egg** cell, and the cell in which it is produced is known as an **oogonium.** The sperms are liberated from the antheridium and swim toward the oogonium. One sperm enters the oogonium, either through the funnel end or by a breaking down of the oogonial cell wall. Union of egg and sperm now ensues; this process is known as **fertilization.** When gametes differ in size and activity as they do in *Laminaria, Fucus,* and *Polysiphonia,* the plants exhibit **oogamy** (Fig. 24.5).

How does a sperm cell find an egg cell? It could happen by chance, much as wind-blown pollen lands on stigmas of flowers of its own species. However, there is considerable evidence that the female gametangia or gametes in plants produce chemical agents that attract the sperms to the close proximity of the egg cells. For instance, a solution prepared from the mature female filaments of the alga *Oedogonium* may be drawn up into a capillary tube. When the tube is placed in a suspension of sperms, the sperms will collect at the tip of the capillary tube and some of them will eventually enter the tube. This attraction seems to be species-specific for *Oedogonium.* Once the gametes fuse, a diploid zygote cell is formed. Further cell division will produce the sporophyte generation.

Alternation of Generations

Fertilization constitutes one of the two critical steps of a complete sexual life cycle; meiosis is the second critical step. We may distinguish three types of life cycles, depending on the relation of meiosis to fertilization.

In most animals, meiosis results directly in gamete formation. In a sense, the entire gametophyte is composed of gametes. In this case, the mature individuals are diploid and the haploid stage is limited to the gametes. This is a **gametic life cycle.** It occurs in the diatoms and in the order Codiales of the Chlorophyta.

In many of the more primitive algae, meiosis accompanies the germination of the zygote. In this case, the mature plants are haploid and the diploid stage is limited to the zygote. This is a **zygotic life cycle.** It occurs in many Chlorophyta and in a few primitive Rhodophyta.

In most plants, meiosis and fertilization take place in distinct generations. One generation is haploid and pro-

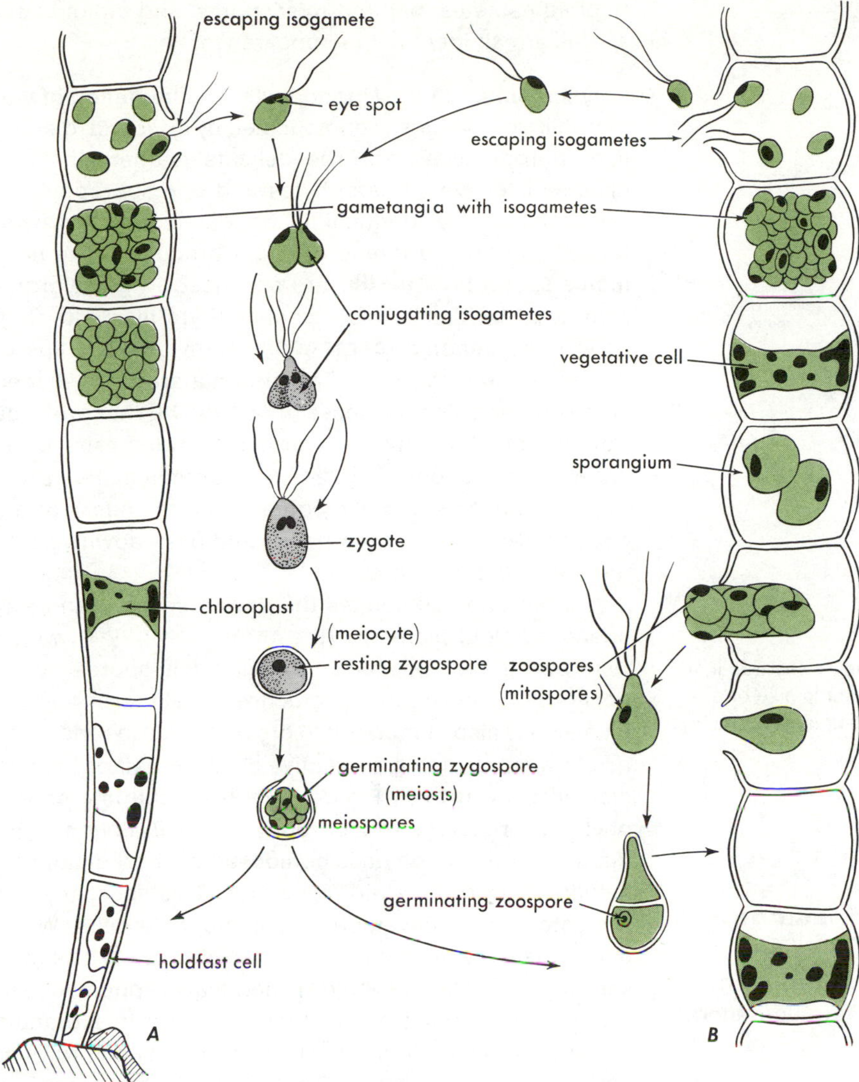

Figure 24.4
A zygotic life cycle as represented by the filamentous green alga *Ulothrix.* As with *Chlamydomonas,* only the zygote is diploid.

duces gametes; the other is diploid and produces meiospores. This is a **sporic life cycle.** The plants of the sporophyte and gametophyte generations may be identical in appearance; if so, there is an **alternation of isomorphic generations,** as occurs in some Chlorophyta, primitive Phaeophyta, and many higher Rhodophyta. In **alternation of heteromorphic generations,** the haploid plants of the gametophyte generation have different forms from the diploid plants of the sporophyte generation. This occurs in a few Chlorophyta and in all of the more advanced Phaeophyta, and in the rest of the plant kingdom beyond the algae.

Gametic Life Cycle

The diatoms are unicellular forms that may divide by cell division for long periods. With each division, one of the cells is smaller than the parent cell (Fig. 24.1*B*). At the end of a period of mitotic divisions, meiosis occurs and four gametes are produced, not all of which may function (Fig. 24.1*C*). Depending on the species, these most generally are isogametes, but they may be anisogametes or even eggs and sperm. Ater fusion, the zygote increases greatly in size and, again depending on the species, may begin a rest period. The germination of the zygote is by mitotic cell division. Note that in diatoms, the products of meiosis are gametes, not meiospores. The gametes themselves are the only haploid cells in the life cycle.

Zygotic Life Cycle

Chlamydomonas and *Ulothrix* are examples of a zygotic life cycle (Figs. 24.3, 24.4). In each of them, two haploid gametes fuse to produce a diploid zygote. Meiosis occurs in the germination of the zygote, which is the only diploid cell in the life cycle.

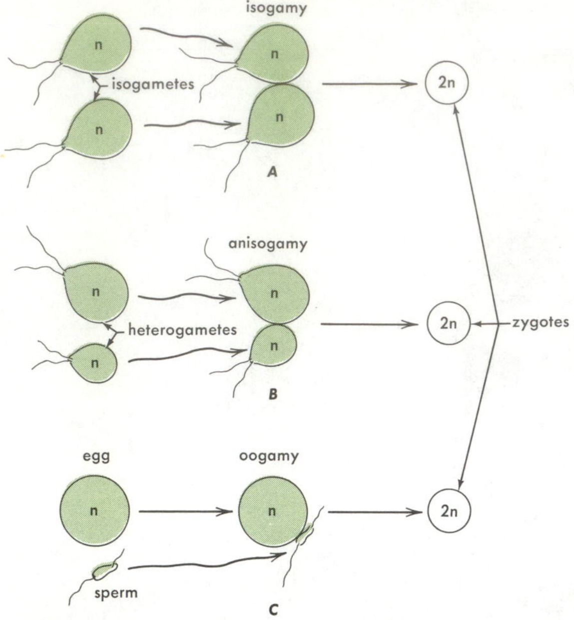

Figure 24.5
Differences in gametes. *A*, in isogamy, the gametes are identical. *B*, in anisogamy, the gametes are slightly different in size or activity. *C*, in oogamy, the gametes are very different in size and activity and are called sperms and eggs.

Sporic Life Cycle: Alternation of Isomorphic Generations

Ectocarpus (Phaeophyta) and *Polysiphonia* (Rhodophyta) will serve to illustrate alternation of isomorphic generations.

Ectocarpus. In the haploid generation of *Ectocarpus* (Fig. 24.6), specialized reproductive cells (gametangia) are formed. The gametangia are grouped together, forming elongated, slightly curved structures. Each cell in this structure is a gametangium that has resulted from mitotic divisions and will produce one motile gamete with the haploid chromosome number. Two isogametes conjugate to form the usual diploid zygote. The zygote, upon germination, will grow into a diploid sporophyte (Fig. 24.6C) whose thallus is identical in appearance to that of the gametophyte thallus. Two different kinds of sporangia are formed on this sporophyte plant. One of them resembles exactly the group of gametangia found on the gametophyte plant. Like the gametangia, all sporangium cells have been formed through mitotic divisions, and each cell will produce a motile diploid zoospore that can germinate directly into a new diploid plant (Fig. 24.6*B*). The other type of sporangium found on the sporophyte plant is spherical; after meiosis, it will produce numerous motile zoospores or meiospores that have the haploid chromosome number. These meiospores grow into new gametophyte plants (Fig. 24.6*A*). Haploid gametes, haploid meiospores, and diploid zoospores are all identical in appearance.

Polysiphonia. In the Rhodophyta, an alternation of isomorphic generations is complicated by a second, distinctive sporophyte phase. The delicate feathery thalli of male and female gametophytes and one sporophyte of *Polysiphonia* are identical in appearance and growth (Figs. 24.7*A*,*B*). No motile cells are produced. The nonmotile sperm **(spermatia)** are produced in great profusion near the tips of male gametophyte filaments (Fig. 24.7*C*) The oogonia **(carpogonia)** form near the tips of the female gametophyte. The oogonia are flask-shaped cells with long slender necks **(trichogynes)** that protrude from a protecting envelope formed from the cells at the bases of the oogonia (Fig. 24.7*D*). Spermatia are carried by water currents to make contact with the necks of the oogonia. Nuclei enter the necks and pass downward to unite with the female nuclei.

Fertilization stimulates the development of an array of short diploid filaments (the **carposporophyte**), which will eventually produce diploid mitospores **(carpospores).** The original protective envelope of haploid filaments is also stimulated to grow. It enlarges and gives rise to a surrounding protective case formed of haploid filaments (**cystocarp,** Fig. 24.7*E*). Thus, the first sporophyte generation (carposporophyte) in *Polysiphonia* and many other Rhodophyta is composed of short diploid filaments that produce diploid spores. This sporophyte is protected by an envelope of gametophyte cells. When discharged, the diploid carpospores give rise to a second sporophyte **(tetrasporophyte),** identical in appearance to the two gametophyte plants. Meiosis occurs in sporangia in the sporophyte plant in cells formed between rows of axial and pericentral cells (Fig. 24.7*G*). Each sporangium produces four meiospores **(tetraspores)** that on germination give rise to male or female gametophyte plants similar in appearance to the sporophyte plants.

Sporic Life Cycle: Alternation of Heteromorphic Generations

Kelp. Alternation of heteromorphic generations occurs in both the Chlorophyta and Phaeophyta but is most highly developed in the Phaeophyta, particularly in the kelps of the order Laminariales. The sporophyte generation of the kelps is large and well known; its vegetative characteristics have been discussed in some detail in the previous chapter. The sporangia arise from meristematic cells, of the meristoderm, that form the outermost layer of cells on the fronds (Fig. 24.8). The sporangia usually arise in groups **(sori)** and are accompanied by an elongation of adjacent cells that serve to protect the developing sporangia. Depending on the species, each sporangium produces 8 to 64 motile meiospores.

The meiospores germinate into male or female ga-

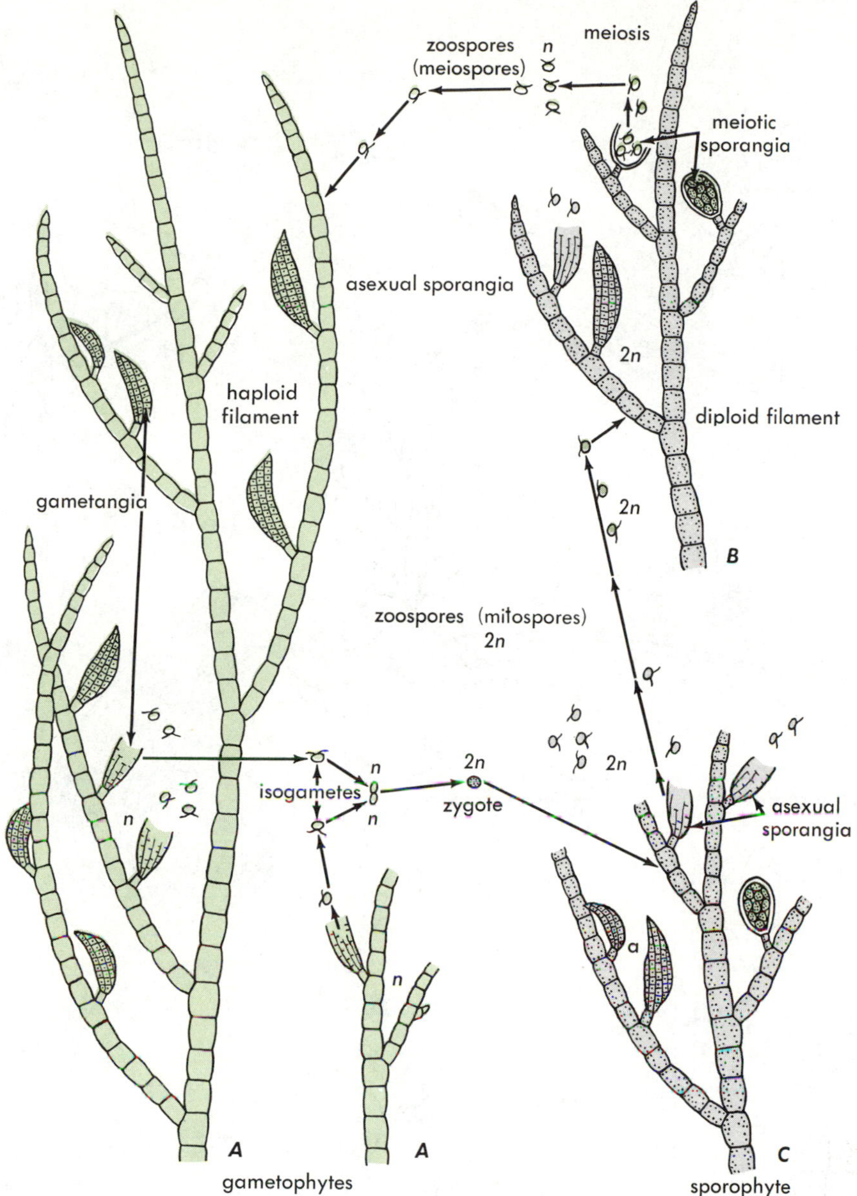

Figure 24.6
Sporic life cycle, alternation of isomorphic generations differing only in having diploid and haploid chromosome numbers, as represented by the filamentous brown alga *Ectocarpus*.

metophyte plants, both of which consist of a small branched filament (Figs. 24.8*C,D*). The antheridia are produced in large numbers, either singly or in groups at the ends of short branches. There appear to be no specialized oogonia, as eggs may be produced in any of the cells of the female gametophyte (Fig. 24.8*E*). The eggs are usually discharged from the oogonium before fertilization. The egg is thus fertilized outside of the oogonium in open sea water.

Fucus. Alternation of heteromorphic generations has reached an extreme in *Fucus*. The diploid generation is the conspicuous (Figs. 23.1*H*, **23.2*E*,** page 410) and the haploid phase has been reduced to a few haploid cells. These do not develop into free-living vegetative gametophytes but become transformed into sperm and eggs and fuse to produce a new diploid generation.

The ends of the diploid strap-shaped thallus of *Fucus* are swollen and notched. They bear small raised areas (Fig. 24.9*A*). A section through these areas shows them to be small cavities **(conceptacles)** that open by small pores to the sea water (Fig. 24.9*B*). Unilocular sporangia arise in these cavities, and they are of two types. Microsporangia are formed profusely at the ends

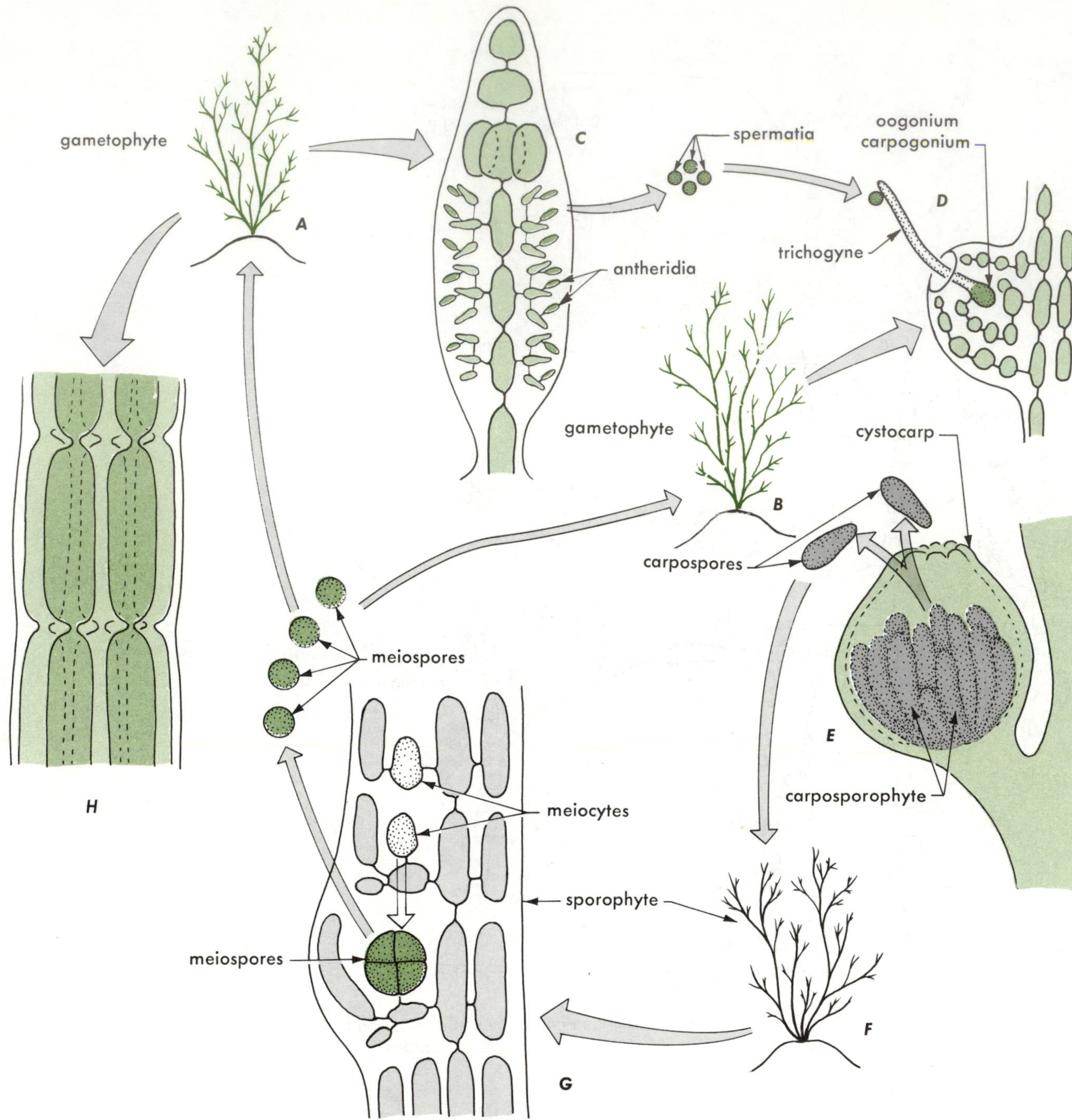

Figure 24.7
Sporic life cycle, alternation of isomorphic generations, as represented by the red alga *Polysiphonia*. *A*, male gametophyte plant. *B*, female gamotophyte plant. *C*, production of spermatia at the tip of male gamotophyte. *D*, the carpogonium (oogonium) is protected by filaments and spermata become attached to the protruding trichogyne. *E*, the zygote produces a body of 2*n* carpospores, and the gametophytic cystocarp enlarges and surrounds the carposporophyte. *F*, a sporophyte, identical in appearance to the gametophyte, arises from the carpospores. *G*, some pericentral cells give rise to meiocytes which, through meiosis, produce four meiospores (tetraspores). *H*, the vegetative structure of both gamotophytes and the sporophyte are identical. (Redrawn after R. S. Scagel et al. *An Evolutionary Survey of the Plant Kingdom*. © 1965 by Wadsworth Publishing Company, Belmont, California. Reprinted with permission of the publisher.)

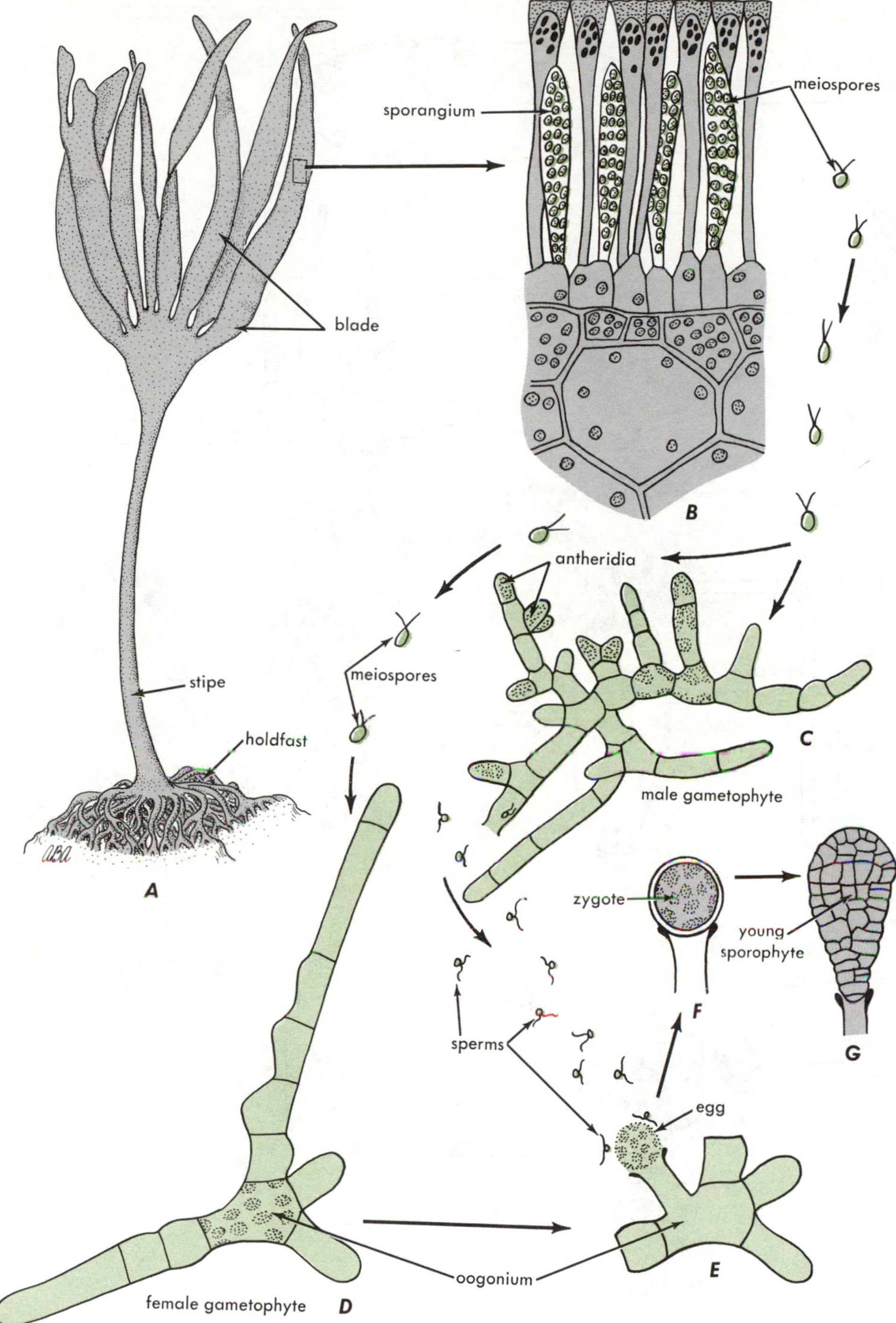

Figure 24.8
A sporic life cycle, with alternation of heteromorphic generations, as shown by the brown alga *Laminaria*. A, B sporophyte; C-G gametophyte details.

of short branching filaments, and megasporangia are large spherical cells (Fig. 24.9*C*) Sterile hairs are numerous, surround the sporangia, and may protrude outward through the pore. In Pacific Coast species of *Fucus,* both types of sporangia occur in the same cavity, but in Atlantic Coast species they occur on separate plants.

The microsporangium contains a single cell—a microsporocyte (Fig. 24.9*F*). Meiosis takes place, and four haploid cells are produced. Each can be simultaneously considered a meiospore and a microgametophyte. Each microgametophyte will rapidly divide by four mitoses, giving rise to 16 cells (a total of 64 cells within the old

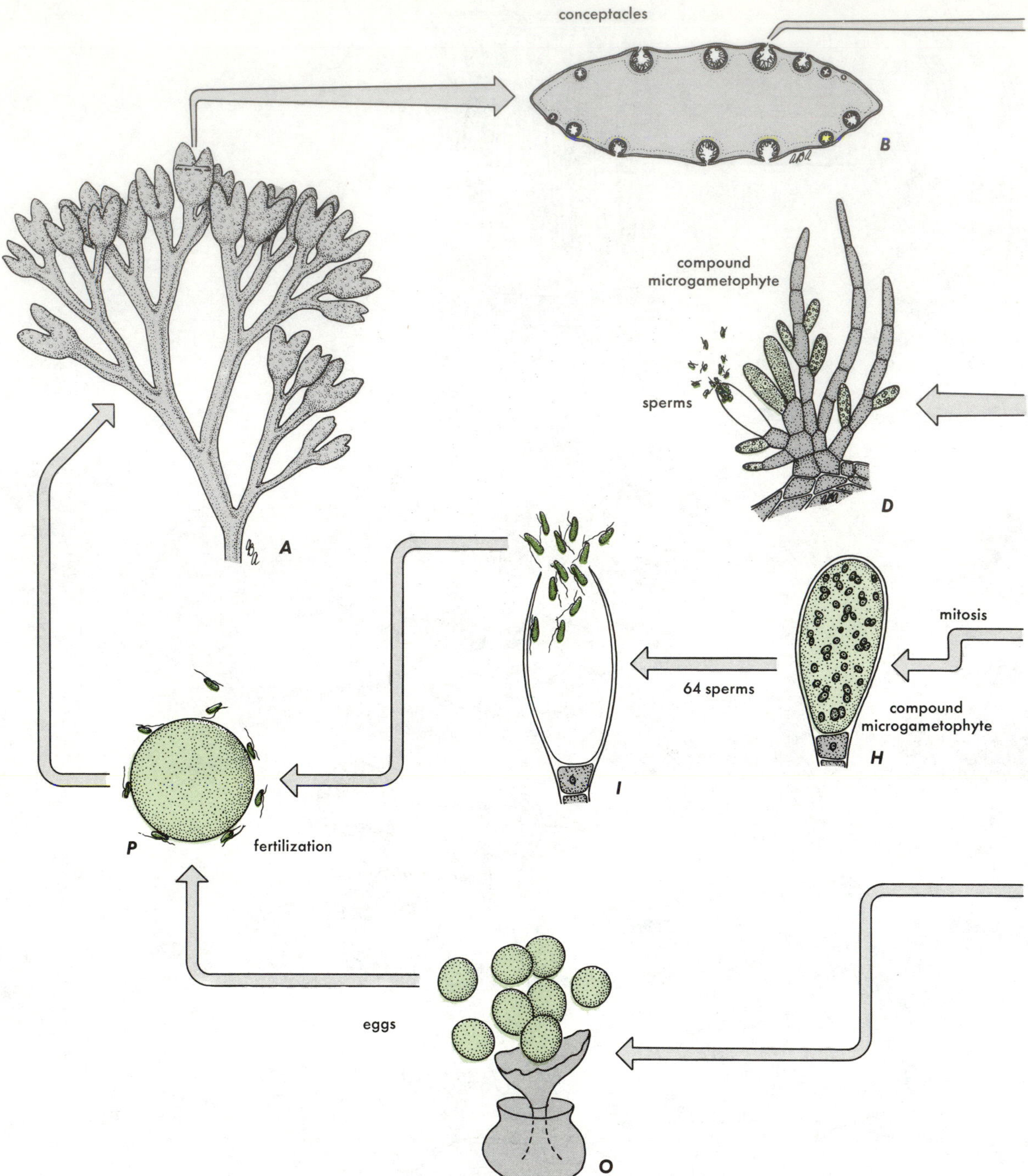

Figure 24.9
A sporic life cycle, as shown by *Fucus*. *A*, dichotomously branching blade with swollen tips. *B*, cross section of tip showing cavities (conceptacles). *C*, the conceptacles contain sterile hairs (paraphyses), shorter filaments bearing many small antheridial meiocytes, and large oogonial meiocytes on short stalks. *D*, detail of filaments bearing sperm. *E*, detail of oogonial meiocyte after meiosis. *F* and *G* meiosis in male meiocyte. *H*, mitotic division resulting in a compound microgametophyte. *I*, each cell functions as a sperm, shown leaving the microsporangium, *J*, *K*, and *L*, meiosis in the oogonial meiocyte. *M*, a single mitosis following meiosis. *N*, an eight-celled megagametophyte forms. *O*, the eight cells leave the megasporangium and function as eggs. *P*, fertilization takes place in the open water. (*F* through *P* adapted from R. R. Scagel et al., *An Evolutionary Survey of The Plant Kingdom*. © 1965 by Wadsworth Publishing Company, Belmont, California. Reprinted with permission of the publisher.

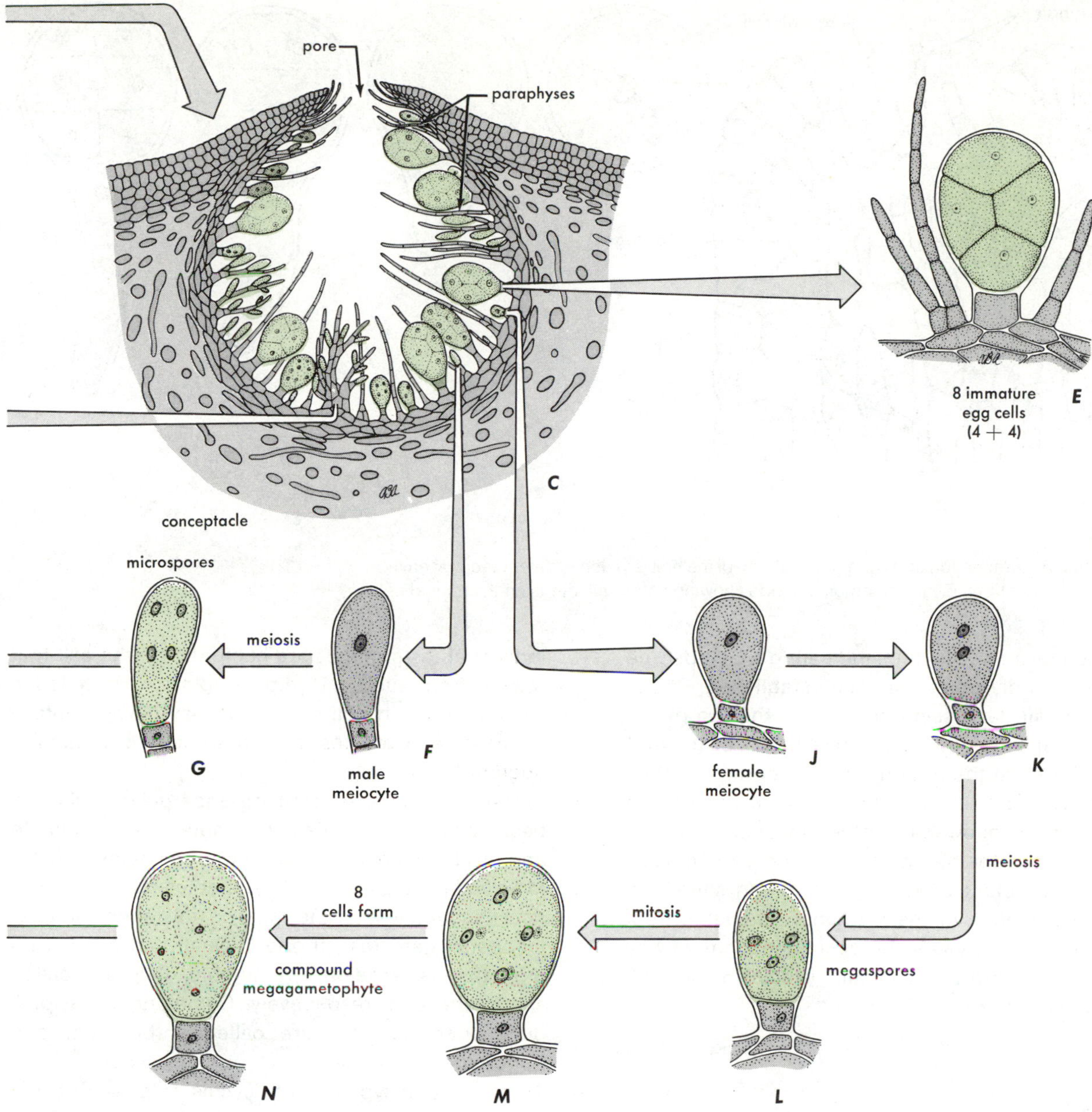

sporangium walls). Each of the 16 (64) cells will differentiate into a sperm and be discharged into the cavity.

The megasporangium also contains a single large cell—a megasporocyte. Meiosis produces four meiospores, and each meiospore undergoes a mitosis without cell division; the result is four binucleate cells, each one of which is a megagametophyte. Ultimately, cell walls do form separating each nucleus and some cytoplasm, resulting in eight cells that further differentiate into eight eggs (Figs. 24.9*E,O*). The unfertilized eggs are discharged from the old megasporangium walls and they pass, with the sperms, outside the cavity into the open sea where fertilization takes place.

The polarity of cell division in the Phaeophyta has been studied experimentally in the germination of the zygote of *Fucus*. The first division of the zygote is preceded by formation of a small protuberance in the zygote cell wall. The plane of the first division is at right angles to this protuberance (Fig. 24.10*A*). This division is unequal; the larger of the two cells formed receives most of the chloroplasts and develops into a vegetative and reproductive thallus. The smaller cell, with fewer chloroplasts, first forms a slender rhizoid that later develops into a holdfast (Fig. 24.10*A*).

The protuberance in the zygote cell wall develops in response to blue light, lower pH, or warmth. The plane of the first cell division of the zygote appears to be determined by purely physical factors. It does not seem to be under genetic or nuclear control. Polarity does not seem to play a prominent part in the next few cell divisions, for the successors of the larger cell simply form a spherical cell mass. However, this sphere soon elongates into a

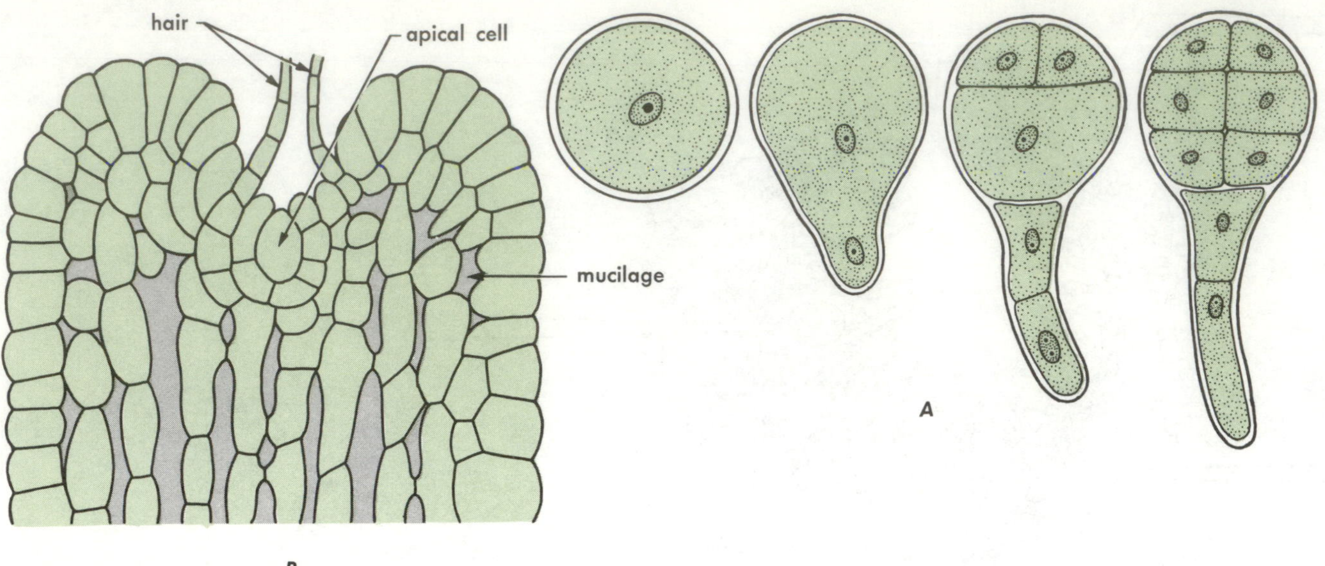

Figure 24.10
Growth in *Fucus*. *A*, germination of zygote. The polarity of the first division is determined by environmental factors such as light. *B*, apical region of blade showing notch with meristematic apical cell.

pear-shaped mass of cells, thus indicating that a degree of polarity in cell divisions has been established.

Soon a small depression appears in the top of the young pear-shaped thallus. A group of from one to eight apical cells forms at the bottom of this depression (Fig. 24.10*B*). The young thallus now becomes flat, and cells in the apical region follow each other in an orderly sequence to produce the flat strap-shaped dichotomously branching *Fucus* plant body (Fig. 23.1*H*). In spite of a fairly good knowledge of the sequence of events from zygote to the mature thallus, the basic questions of why and how cells come to divide in such precise planes still require explanation.

SUMMARY

1. There is no typical life cycle for the algae as a group. In some divisions, detailed knowledge of reproduction is limited to a few species, genera, or families. Asexual reproduction may be the only form of reproduction in some algae.
2. Vegetative reproduction includes cell division, fragmentation of filaments, the production of resting spores, and the production of mitospores. Mitospores arise from preceding mitotic cell divisions that occur in a specialized cell called a sporangium. Mature spores may be motile or nonmotile.
3. Spores can each produce a new individual plant, but gametes must fuse in pairs in order to produce a new individual. As gametes are haploid ($1n$), the cell that results from fusion is diploid ($2n$); it is called a zygote. Through cell division it may develop into a sporophyte. Gametes are produced in a gametangium.
4. Certain cells in a sporophyte undergo meiosis and become meiospores. Each meiospore may germinate and produce a haploid individual, the gametophyte.
5. Male and female (or + and −) gametes may look identical (isogametes), or they may be unlike (anisogametes, heterogametes). If one is small and motile and the other is large and nonmotile, they are called sperm and egg, respectively. Gametangia that produce such gametes are called antheridium and oogonium, respectively.
6. Three general types of life cycles exist among the algae (and other organisms): gametic, zygotic, and sporic. In the gametic life cycle, the haploid stage is confined to the gametes; all mature individuals are diploid. In the zygotic life cycle, the diploid stage is restricted to the zygote; all mature individuals are haploid. In the sporic life cycle, there is an alternation of generations; one generation of mature plants is haploid and it produces gametes, the other is diploid and it produces meiospores. The generations may look alike or appear so different that they seem unrelated.
7. Life cycles of diatoms, *Chlamydomonas, Ulothrix, Ectocarpus, Polysiphonia,* a large kelp, and *Fucus* were described.

CHAPTER 25

MYXOMYCOTA AND OOMYCOTA

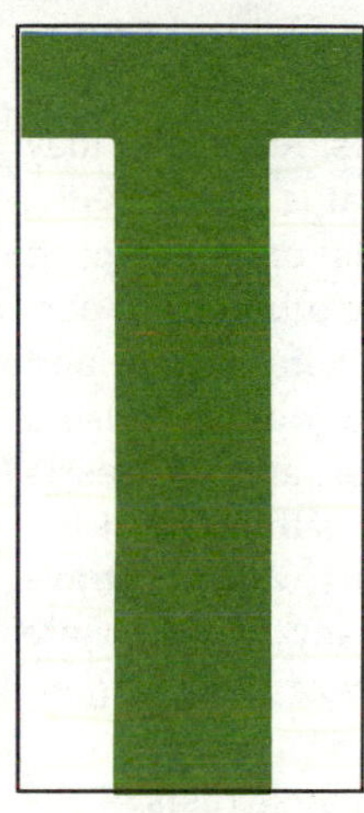

LOWER FUNGI

The fungi (sing., fungus) are eukaryotic organisms that lack chlorophyll and so cannot perform photosynthesis. They differ from animals in that they usually possess a cell wall, they are usually nonmotile, and they reproduce asexually by spores. Some biologists place the fungi in a separate kingdom. In this book, we are following tradition by placing the fungi in the plant kingdom, and we will divide the fungi into three divisions.

There are more than 200,000 species of fungi. We commonly meet the fungi as mushrooms, mildews, molds, and the organism that causes athlete's foot. A few fungi provide food for humans, but many more compete with us for food sources and several of them use us as their own living food supply. The baking and brewing industries have an ancient partnership with the fungus known as yeast (*Saccharomyces*), and the medical profession has the makings of another long partnership with fungi such as *Penicillium chrysogenum* for their production of penicillin and other antibiotics. It is doubtful whether any other group of organisms is associated with human beings in as widely varied an array of relationships. Some of those relationships are deleterious to humans.

It is well-known that the ravages of human diseases have changed or modified the history of peoples. Similarly, epidemics of diseases that lay waste important food plants may influence greatly the course of events. From 1843 to 1847, a fungus known scientifically as *Phytophthora infestans* spread rapidly through the potato fields of Ireland. This fungus causes a severe disease of potatoes known as **late blight.** It not only kills the foliage but also infects the tubers, causing them to rot rapidly. In addition, blighted potato tubers contain an alkaloid that causes abortion and birth defects. Children born to undernourished mothers may be retarded.

In the middle of the nineteenth century, Irish peasants depended chiefly on the potato as a source of food. Consequently, from 1843 to 1847, when weather conditions were just right for the rapid development of *Phytophthora infestans,* and when the potato disease that it caused attained epidemic proportions, the Irish experienced a disastrous food famine. During these years 2 million died of starvation and related effects of the blight, and 1 million others moved across the ocean to New York City. They and their descendants have left an imprint upon American life.

Fungi obtain the energy necessary for respiration either by absorbing nutrients from dead organic remains (saprophytism), or by absorbing nutrients from living tissue (parasitism). Common examples of parasitic fungi are those that cause mildew of roses, late blight of potatoes, and rust of cereals. The common field mushrooms are good examples of saprophytic forms. A parasite may or may not be "disease-producing," that is, **pathogenic.** The fungus *Phytophthora infestans* is pathogenic; it produces a disease of the potato. On the other hand, certain bacteria subsist on the roots of alfalfa, beans, and other members of the legume family but do not cause a disease. The organism upon which the parasite lives is the **host.** For example, the potato plant is the host of the parasitic fungus *Phytophthora infestans.* **Plant pathology** is the study of plant diseases, many of which are caused by fungi. **Mycology** is the study of fungi in general.

Some parasites, such as those that cause rusts of cereals, are obliged to secure their nourishment from living tissues. They are known as **obligate parasites.** And certain saprophytes, like some mushrooms, thrive only on nonliving organic materials. They are known as **obligate saprophytes.** On the other hand, some fungi have the faculty of growing as either a parasite or a saprophyte and are referred to as **facultative parasites** or **facultative saprophytes.** The most common plant pathogens are facultative parasites.

CLASSIFICATION OF FUNGI

There are three fungal divisions. The first, Myxomycota, comprises the **slime molds.** It is a small division whose members have a vegetative plant body lacking a firm wall and obtaining nutrients by ingestion. The other two divisions exhibit a cell wall, but the structural component of the wall differs: it is cellulose in the Oomycota but it is **chitin** in the Eumycota. Chitin is a polymer identical to that found in the cuticle of insects and crustaceans (Fig. 25.1). A further difference between the latter two divisions is that asexual reproduction is by spores in Oomycota, but typically by **conidia** in Eumycota. Spores are produced in specialized cells or chambers called sporangia, but conidia are not formed within sporangia. End cells of a fungal filament may round up and be cut off, forming conidia (Fig. 25.2), or portions of a fungal filament may fragment, also forming conidia. A synonym for conidium is **conidiospore.** We will postpone further discussion of the Eumycota until Chapter 26.

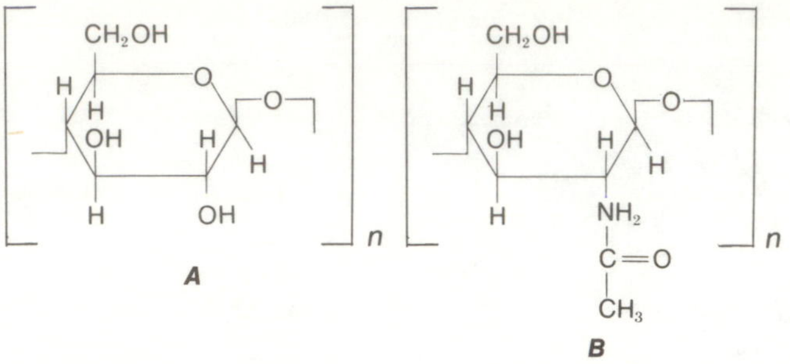

Figure 25.1
A, the repeating unit glucose in cellulose. *B,* the repeating unit N-acetylglucosamine in chitin.

MYXOMYCOTA

The slime molds are of scientific interest because they combine the characteristics of both plants and animals and their level of organization shifts from unicellular to multicellular. There are approximately 500 species, divided into two groups: the acellular slime molds, with a coenocytic vegetative stage; and the cellular slime molds, with a cellular vegetative stage.

Acellular Slime Molds

The vegetative body consists of a slimy mass of naked protoplasm in which there are many nuclei without separating walls. The vegetative body is called a **plasmodium** (plural, **plasmodia**) (Fig. 25.3). This plasmodium has no definite shape; it creeps slowly by amoeboid movement, usually within rotting tree trunks, across leaves, or in crevices, engulfing solid particles of food as it goes. All slime fungi are saprophytes. The absence of cell walls, amoeboid movement, and the ability to take solid food particles into the protoplasm are characteristics that are usually associated with animals. When slime fungi reproduce, however, they form spores with cellulose walls; thus, they have reproductive characteristics that are definitely those of plants.

Prior to the reproductive stage, the plasmodium moves to a drier substratum. After a time, the plasmodium ceases moving and forms one or more spore-producing structures (**sporangia** or fructifications, Fig. 25.3). There is great variation in the form and color of the sporangia among the different species. Some have a very delicate structure and brilliant coloring. The spores

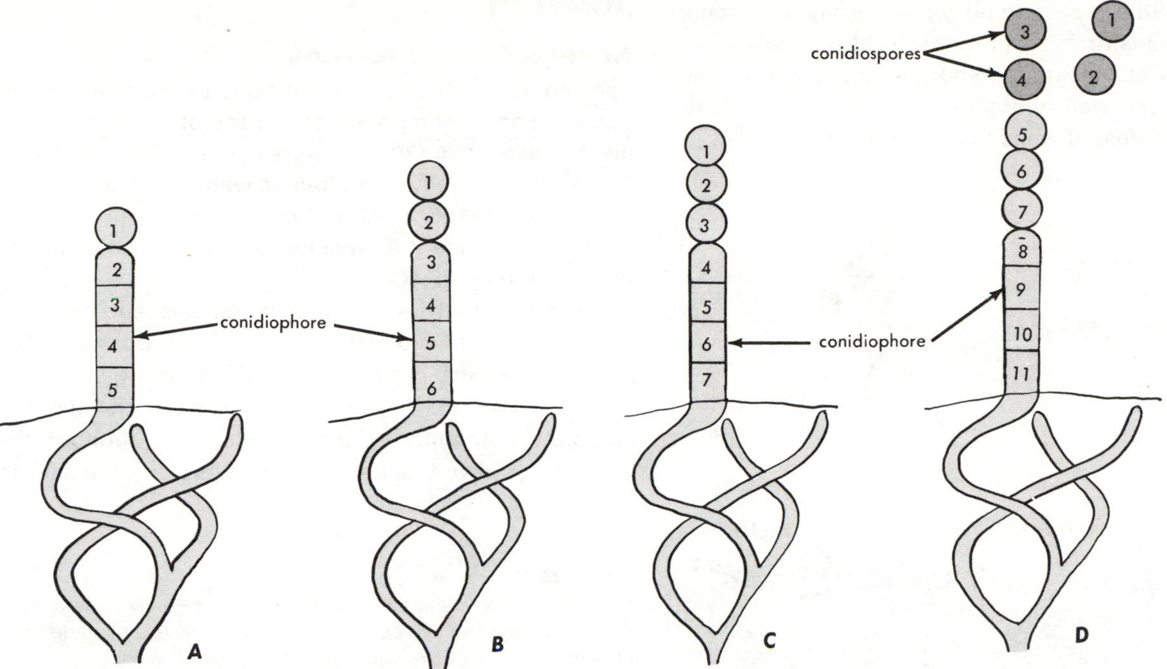

Figure 25.2
Diagram showing formation of conidiospores. *A,* tip of fungal filament (hypha or conidiophore) divided into cells of similar sizes, the outermost cell of which, cell 1, has rounded up. *B, C,* and *D,* sequential cells 2 to 7 have rounded up and some are released, to be carried by air, water, or living vectors to other areas.

Figure 25.3
Life cycle of a myxomycete, *Physarum.* (Modified from C. J. Alexopoulous, *Introductory Mycology,* John Wiley & Sons, New York.)

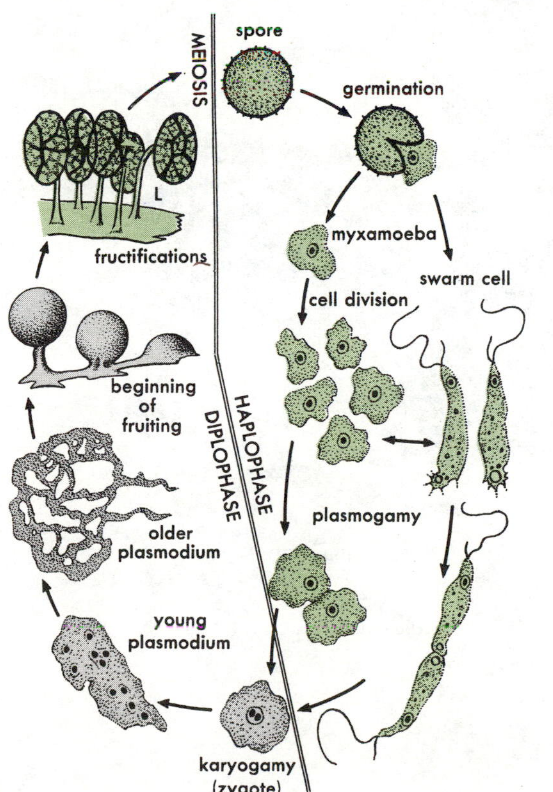

formed within the sporangia are uninucleate and are surrounded by a cellulose wall. They are released from the sporangia and spread by wind. In the presence of water, they germinate; the wall is ruptured, and the contents escape in the form of a flagellated or amoeboid naked mass of protoplasm that may later multiply by fission. All acellular slime molds that have been carefully studied have a sexual stage. Pairs of these bodies may fuse to form a $2n$ zygote. Each zygote may grow into a $2n$ plasmodium, or zygotes may coalesce to form a $2n$ plasmodium.

Cellular Slime Molds

The cellular slime molds differ from the true slime molds in that the plasmodium of a cellular slime mold is clearly multicellular—the cells that mass together to form the plasmodium never lose their outer membrane, thus they remain discrete. For this reason, the vegetative body is often called a pseudoplasmodium.

Dictyostelium was the first member of this group to be discovered and described. It is commonly found in soil, and is relatively easy to isolate. The life cycle begins with germination of spores. Each spore liberates a single amoeboid cell. The amoebae multiply by mitosis and produce a large population of individual cells.

These amoeboid cells lack walls, are capable of amoeboid movement, and engulf food particles, such as bacteria, just as true amoebae in the Protozoa. For some reason, certain cells begin secreting a chemical attractant that causes nearby cells to migrate to these central cells. The result is a plasmodium that migrates for some time, then transforms itself into a base, stalk, and sporangium. Cells inside the sporangium round up, secrete a cellulose wall about themselves, and are later released as spores. All cells appear to be 1 n.

OOMYCOTA

General Characteristics

The vegetative thallus or plant body of the members of this division is composed of a mass of threads of filaments called **hyphae** (**hypha,** singular) (Figs. 25.4*A*,*B*). The vegetative body may often be webby and delicate, as in common bread mold and other molds. The mass of hyphae forming the vegetative body is called the **mycelium** (Fig. 25.4*C*).

The nuclei divide normally by mitosis, and cell division follows immediately after nuclear division. The vegetative hyphae usually have no cross walls (**septa,** singular **septum**). They are multinucleate, and there is complete continuity of the cytoplasm throughout the whole vegetative thallus. Hyphae of this sort are said to

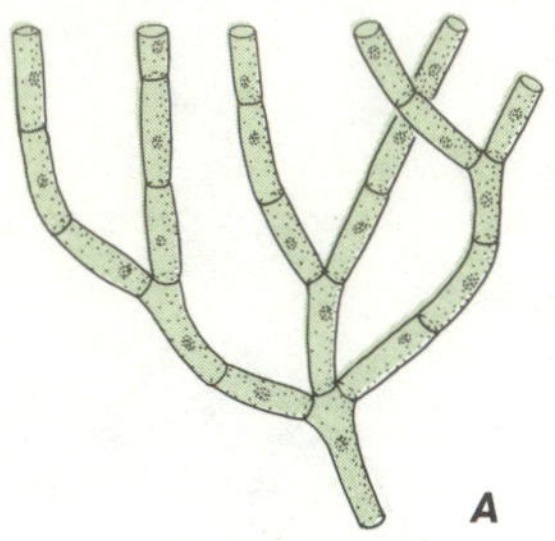

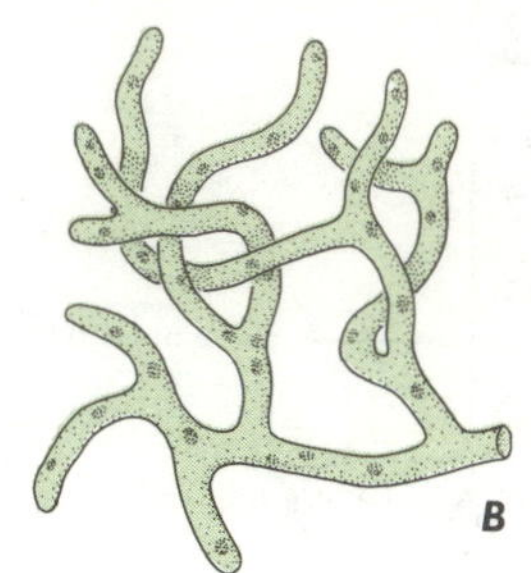

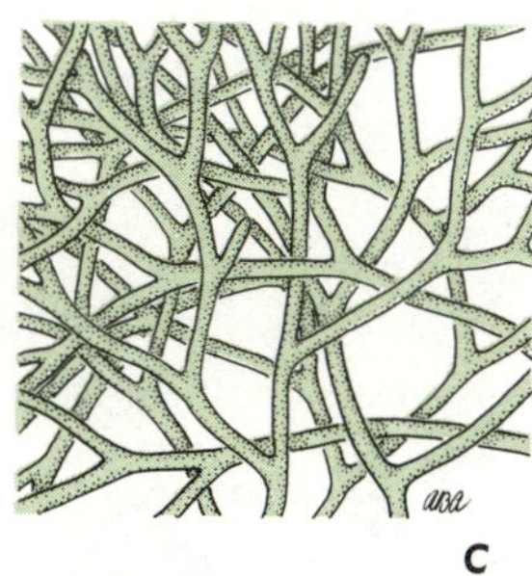

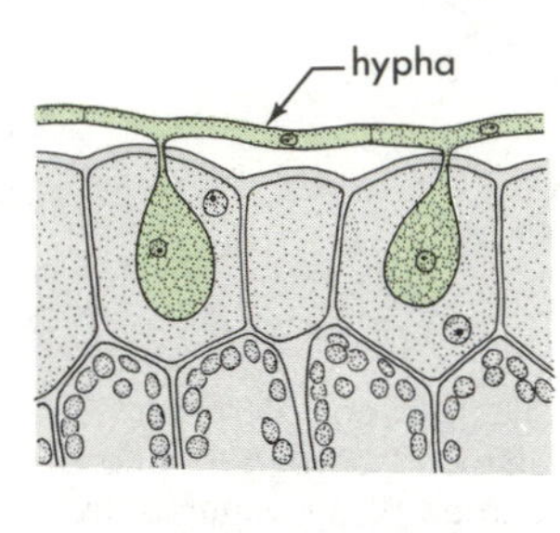

Figure 25.4
Various aspects of the fungal plant body. *A,* septate hyphae. *B,* nonseptate hyphae, typical of Oomycota. *C,* mycelium, made up of many hyphae. *D,* diagrammatic view of two haustoria. *E,* electron micrograph of an haustorium of *Erysiphe* in the Eumycota. The haustorium is within an epidermal cell of barley. Note that haustoria are separated from the cytoplasm of host cells by a membrane and that haustoria are thus in a vacuolelike space. Cellular organelles are prominent in haustoria. (*E,* ×11,200, courtesy of C. E. Bracker.)

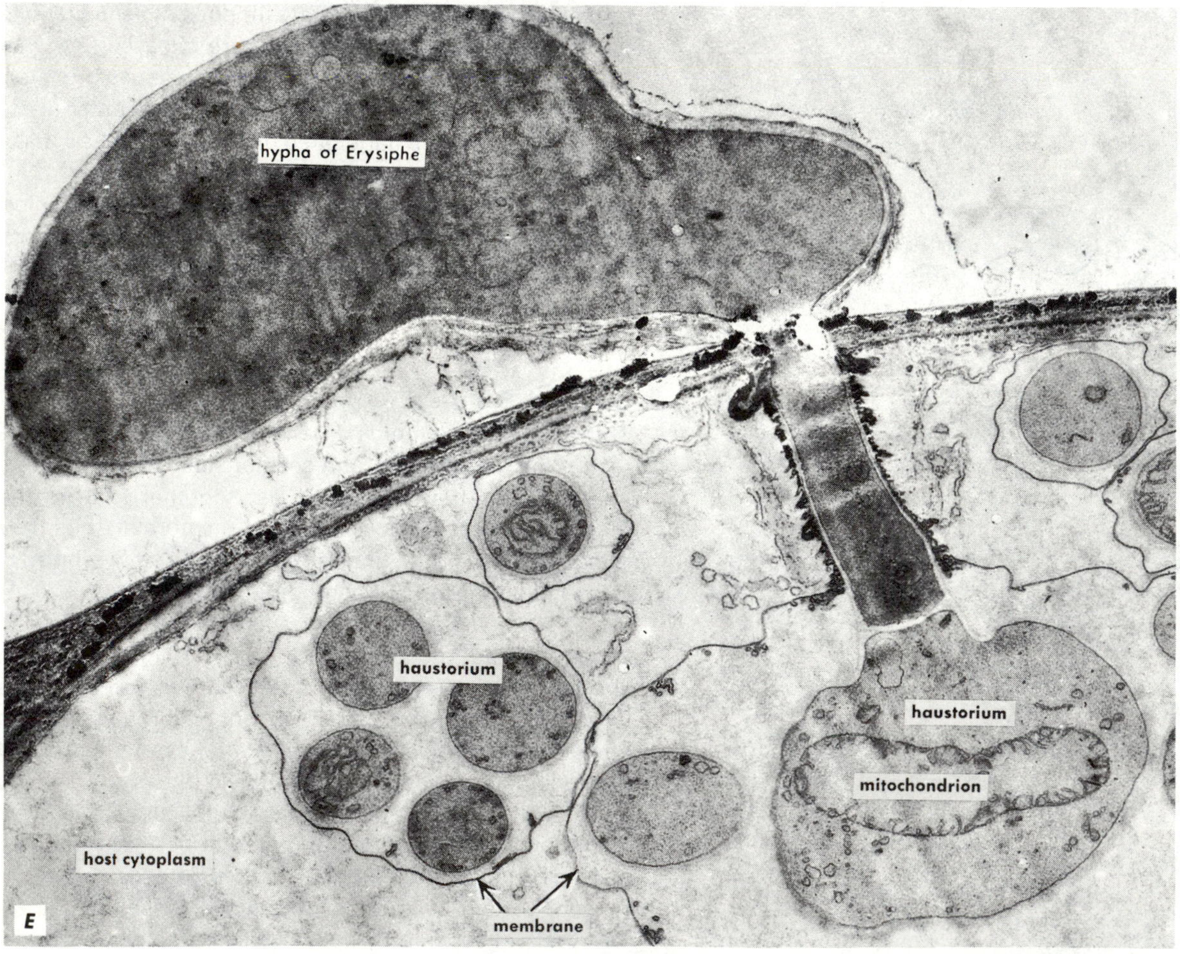

be **coenocytic** and, since no cross walls or septa are formed in such filaments, they are also called **nonseptate.** When cross walls are present, the filaments are said to be **septate.**

Hyphae grow only at their tips; there the walls are soft and extensible under the turger pressure from the protoplast. As a region of wall is left behind the growing tip, it becomes thicker and thus highly resistant to further stretching. Branches originate as bulges in the sides of the growing hyphal tips, where the walls are still soft. The branches behave exactly as their parent hyphae. Hence the system of hyphae progressively extends and ramifies as it absorbs nutrients and water from the substratum. The hyphae possess a "chemical sense" and can orient their growth and branching toward sources of raw materials such as sugars, amino acids, water, and minerals. The orientation of growth in response to chemical signals is **chemotropism.** Its effect here is to cause the fungus progressively to invade food sources such as moist slices of bread and decaying stored fruits.

There is some cooperation between neighboring hyphae, but the degree of cooperation rapidly declines with distance. Hence as the mycelium grows and spreads, its advancing hyphae tend to form local groups that function independently of one another, as if they belonged to entirely separate mycelia. Older parts of the mycelium may be abandoned, their usable contents withdrawn, and their empty walls sealed off by cross walls that resemble abandoned tunnels. Such an event, or the accidental shearing of mycelia into parts by outside forces, readily converts a single mycelium into many mycelia. Thus, **fragmentation** is a real and useful mode of vegetative reproduction that is available to nearly all fungi.

The cytoplasm of fungal hyphae resembles that of other eukaryotic plant cells in having mitochondria and an endoplasmic reticulum. The mitochondria are long, cylindrical, circular, or branching, as may be seen in Fig. 25.5. The endoplasmic reticulum is in the form of slender paired membranes. The small profiles (S) may be similar to bodies seen in the cytoplasm of higher plants that have no generally accepted name, but are sometimes called **microbodies.**

Function of the Mycelium

The vegetative mycelium carries on the general activities of fungal metabolism, such as absorption, digestion, respiration, and secretion, but not photosynthesis. Since it is incapable of synthesizing its own foods, it must obtain nourishment either from nonliving organic matter or from living plants or animals. The food must be rendered soluble so that it may diffuse through the walls of the hyphae and reach the protoplast. Some fungi obtain food from even the hardest of woods and from solids of many other sorts. These solid substances are liquefied or otherwise rendered diffusible by enzymes that are secreted by the hyphae. Some fungi have specialized hyphae known as **haustoria** that can penetrate living cells and obtain nourishment therefrom (Figs. 25.4*D,E*).

True fungi are solution feeders: they can take in only small molecules. In many cases their food requirements are very simple, consisting of no more than an organic carbon compound such as a sugar, plus water, minerals, and perhaps a few vitamins (enzyme cofactors, required

Figure 25.5
Portion of a hypha of *Rhizoctonia solani*, ×24,000. (Courtesy of C. E. Bracker.)

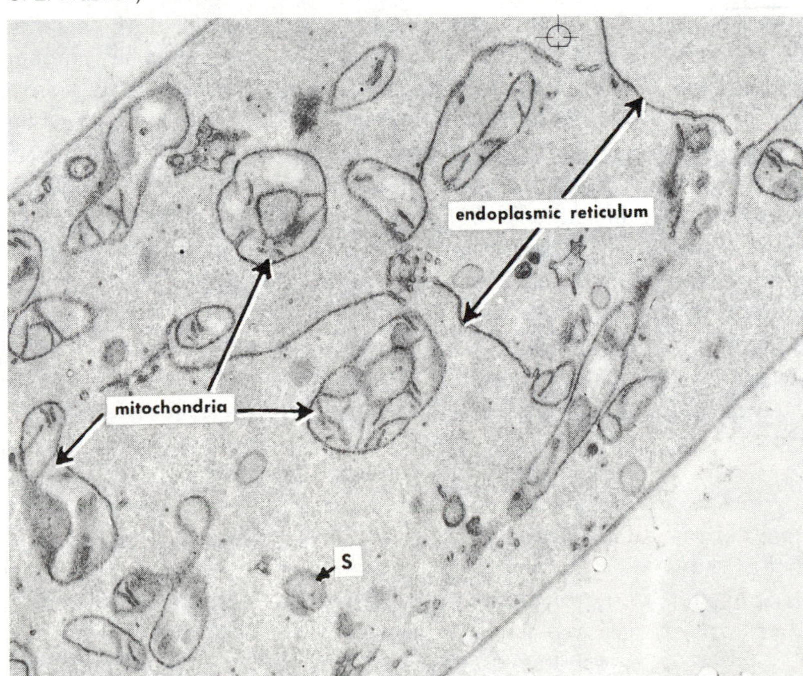

only in small amounts, that the fungus cannot build for itself). The vitamin most often needed by fungi is the same vitamin B_1 which is essential in human nutrition.

Since the principal function of the mycelium is to obtain nutritive materials, it is usually found in close association with a source of food. It may be growing inside a living tree, in a dead stump, on or in a leaf, in aging cheese, in a manure pile, or in many other kinds of organic substances. The mycelium of most fungi is not adapted to withstand much drying; hence it seldom grows openly exposed to the atmosphere, unless the relative humidity is high.

Reproduction

The sporangia, spores, gametangia, and gametes of the Oomycota show a great range of structure and resemble in some respects the reproductive structures found in algae. Simple cell division and fragmentation occur in some fungi. Both motile and nonmotile spores are produced in various types of sporangia. Isogamy is characteristic of several genera. Heterogamy, involving gametangia resembling those of *Vaucheria,* occurs in several families.

Although the vegetative mycelium of most fungi is the actively destructive portion, the reproductive structures are of more importance from the disease-control standpoint and as a basis of classification. Usually, fungi multiply and are disseminated by abundant spores; it is therefore important to prevent the dissemination of spores, or to stop their germination, or to kill the young hyphae after germination. It is seldom possible to rid the infected plant of a vigorous mycelium.

Spores may be borne in specialized cells called **sporangia** (spore cases). The sporangia (**sporangium,** singular) may occur either on typical filaments on special upright hyphae, or on highly branched hyphae. Aerial sporangia are formed as in bread mold. In other species, the sporangia are submerged and germinate in thin films of water. The spores produced by such sporangia are motile and are called **zoospores.** The hyphae that bear sporangia are called **sporangiophores.**

Many Oomycota are severe plant pathogens. Some, however, may infect fish or insects or cause diseases in man. Many members of this division, like the common water mold, are normally saprophytes or weak parasites. Some are minute forms of one to several cells, and a few of these smaller sorts are reported to attack small aquatic animals. Most representatives of the Oomycota develop a more or less extensive mycelium of indefinite form (Fig. 25.6).

The Oomycota are divided into several orders. Specialists do not agree completely on the best manner of classifying the plants and, therefore, the number of recognized orders depends on the authority one is consulting. We shall consider several forms distributed in the orders Saprolegniales and Peronosporales.

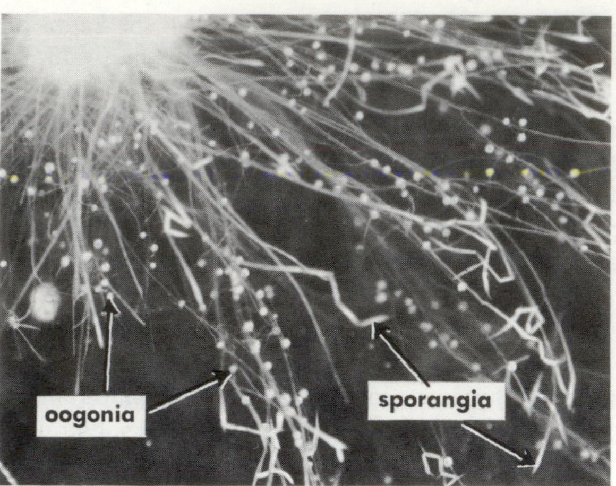

Figure 25.6
Water mold growing on a hemp seed, ×3.

Saprolegniales

Fungi of this order live in fresh or salt water and are generally saprophytic. Even the species that attack the gills of fish grow on tissues that have been weakened or suffered injury. Because of the aquatic habitat of many members of this order, they are frequently called **water molds.** Some species can be easily cultivated by placing small pieces of meat, egg albumin, radish seeds, or dead flies in a dish of pond water (Fig. 25.6). After the mycelium has ramified through the dead fly or other substratum, hyphae grow outward into the water. A small ball of white hyphae, 1 to 3 cm in diameter, may thus be formed. Reproductive cells are produced by these hyphae.

Asexual Reproduction. With ample food supply, the mycelium increases in size with but little tendency to produce reproductive cells. If, however, a well-developed mycelium is transferred to distilled water in which a food supply is lacking, sporangia will usually appear. Sporangia are formed at the ends of hyphae by a cross wall cutting off the tip of a hypha from the rest of the mycelium as in *Saprolegnia* (Fig. 25.7*A*). The sporangium is a multinucleate structure. After a time, the protoplasmic contents of the sporangium divide into a large number of spores, each with one nucleus (Figs. 25.7*B,C*). Upon maturity, the spores are discharged from the sporangium. In *Saprolegnia,* each zoospore has two flagella attached to its anterior end that enable it to swim actively (Fig. 25.7*D*). After a time, the zoospores settle down, lose their flagella, develop a cellulose wall, and pass through a resting period. Upon resuming activity, they escape from the wall; the two newly developed flagella are now attached laterally, and the zoospores, thus equipped, swim about for a period. If they come to rest on a suitable substance, each sends out a tubular outgrowth that penetrates and infects this substance.

In a few species of water molds, one or even both zoospores stages are suppressed. When zoospores

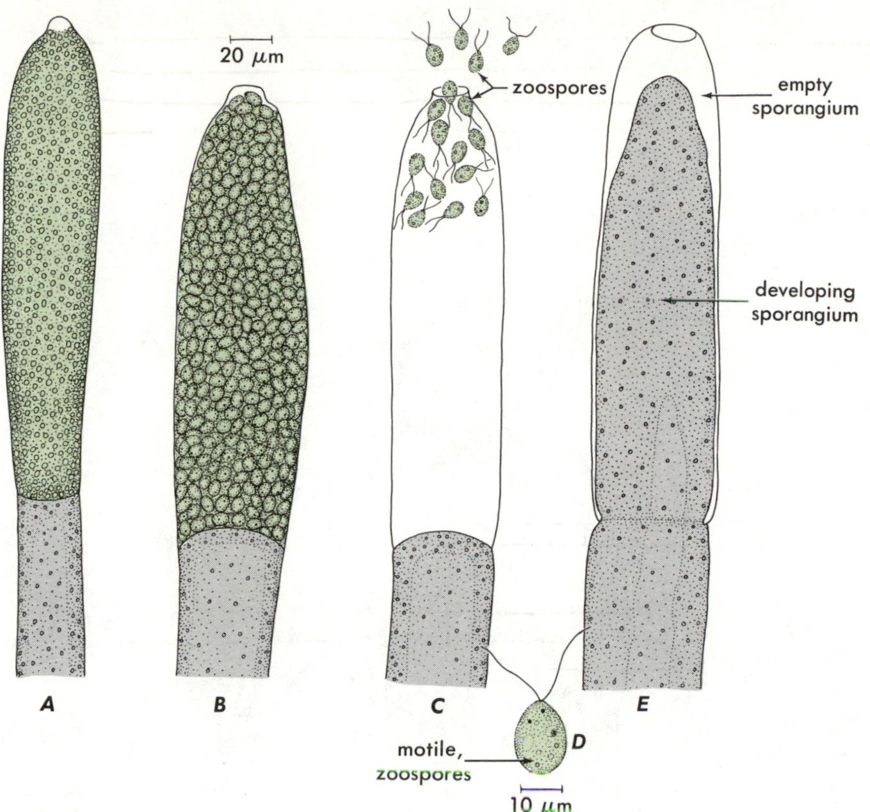

Figure 25.7
Zoosporangia of *Saprolegnia*. *A*, immature sporangium. *B*, mature sporangium. *C*, discharge of zoospores. *D*, one zoospore. *E*, empty sporangium being renewed. (Based on W. C. Coker, *The Saprolegniaceae*. © 1923 by the University of North Carolina Press. Reprinted with permission of publisher.)

emerge from the sporangia, they may germinate directly into a new mycelium. In certain other species, the spores never leave the sporangia but germinate while still enclosed within it, and the germ tubes pierce the old sporangial wall. In still other species, spores are not even formed; the sporangia germinate directly into a coenocytic mycelium.

Sexual Reproduction. The **oogonia** are spherical cells, formed at the tips of short side branches (Fig. 25.8). When mature, they are three to four times the diameter of ordinary hyphae. The cytoplasm in the swollen tip becomes denser than that in regular hyphae. Meiosis occurs, and from one to twenty eggs may be formed from the protoplasmic contents in each oogonium, depending upon the species. The eggs are spherical, dense bodies of protoplasm, each containing one nucleus.

The **antheridia** are also formed at the tips of branches, in some species near the oogonia. Each is separated from the main filament by a cross wall. The antheridia are usually curved and not much greater in diameter than the hyphae from which they arise. The contents of a mature antheridium consist of several nonflagellated male gametes. The antheridium comes in contact with an oogonium. In some instances, one oogonium may have several antheridia attached to it. A short slender hypha, the **fertilization tube,** grows from the side of an antheridium adjacent to an oogonium (Fig. 25.8*B*), penetrates the oogonial wall, and comes in contact with one or more eggs. If the oogonium contains several eggs, the fertilization tube usually branches, sending a branch to each egg. Nuclei (male gametes) from the antheridium migrate into the tube and any branches present. One nucleus and possibly some cytoplasm pass into each egg, and fertilization ensues. The fertilized egg or zygote develops a heavy wall, becoming an **oospore** (Fig. 25.8*C*), and usually will not germinate for several months, even under favorable conditions; hence, it is well-adapted to survive unfavorable conditions. Upon germination, the oospore sends out new diploid hyphae, which, if they find a source of food, rapidly grow into a typical mycelium. If food is scarce, the formation of zoospores follows soon after germination. Not uncommonly, eggs may develop into new hyphae without fertilization.

Peronosporales

Nearly all the Peronosporales are obligate parasites. In general, heavy-walled oospores carry them over unfavorable periods, and various sorts of asexual spores bring about rapid multiplication under suitable conditions.

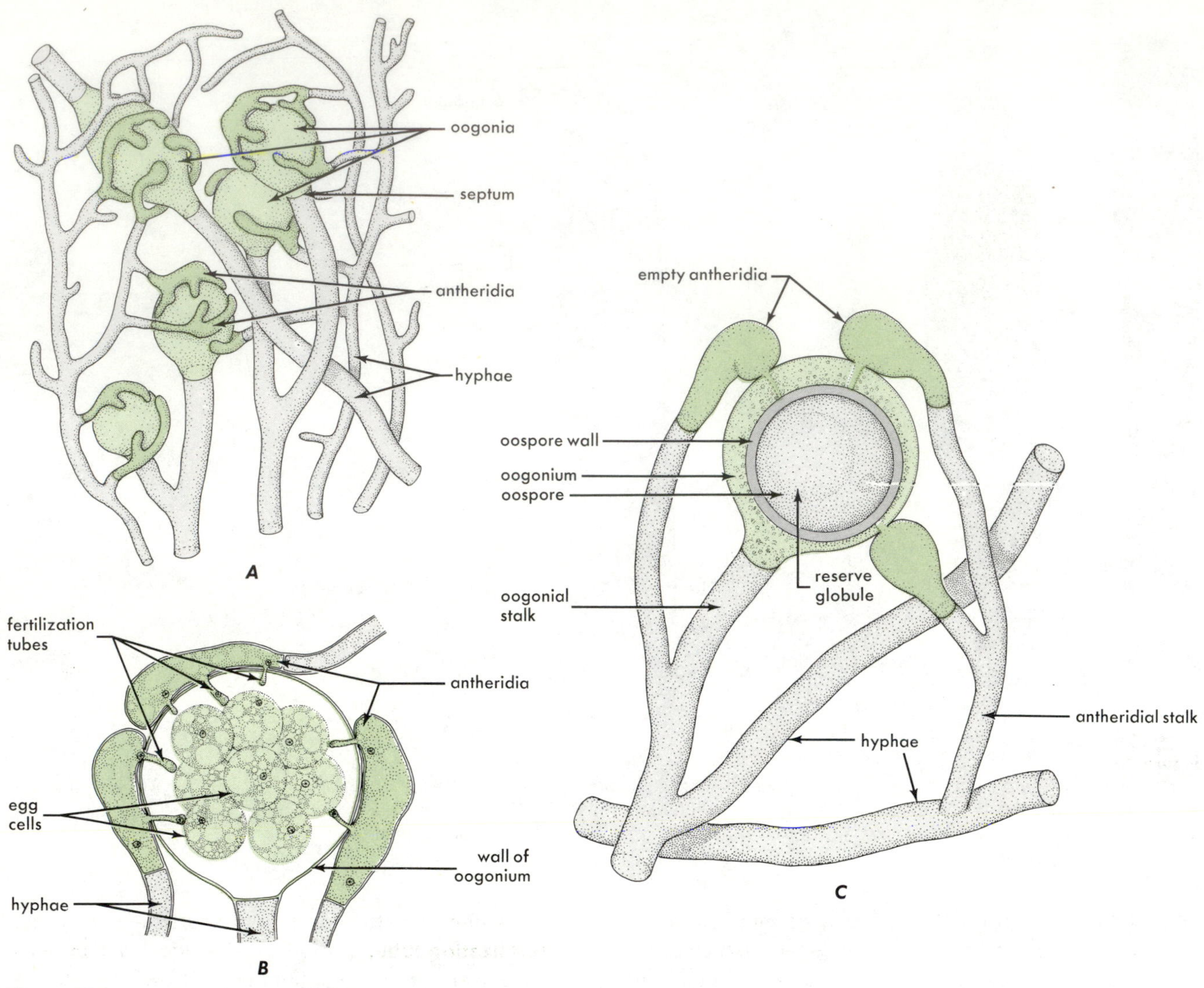

Figure 25.8
Gametangia in water molds. *A, Saprolegnia* oogonia and antheridia with associated hyphae. *B, Saprolegnia* antheridia, oogonium with many eggs, fertilization tubes in place. *C, Pythium* oospore formation following fertilization. (*B*, based on W. C. Coker, *The Saprolegniaceae.*© 1923 by the University of North Carolina Press. *C*, based on J. T. Middleton, *Memoirs of the Torrey Botanical Club,* **20,** 1. © 1943 by the *Torrey Botanical Club,* and V. D. Mathews, *Studies on the Genus Pythium.* © 1931 by University of North Carolina Press.)

Because many of them form a downy growth on the surface of the host or substrate, they are sometimes called **downy mildews.**

Asexual Reproduction. The asexual reproductive structures of the different genera are adapted to various habitats and show more variation than do the sexual reproductive bodies.

A few Peronosporales, for example, *Pythium,* may live in water or on moist soil as saprophytes; they may infect aquatic plants and are frequently responsible for the "damping off" of the seedlings of many farm and hothouse plants. In keeping with their moist habitat, these fungi are propagated by zoospores that strongly resemble those of the water molds. Sporangia are formed at the tips of the hyphae (Fig. 25.9*A*). When proper conditions prevail, the sporangia open and the zoospores emerge. The zoospores are small and are able to swim in the water or wet soil for a time. They may then round up and become encysted, but eventually, they germinate and develop a new mycelium. Seedlings may be infected at or just below the soil surface and are quickly killed.

Other Peronosporales (*Phytophthora, Plasmopara, Albugo*) require a moist habitat and grow best in rainy or humid weather, but they are strictly terrestrial. Many of them are parasites on flowering plants. The mycelia of downy mildews grow within leaves or stems of plants between the cells. Haustoria penetrate the cells.

Aerial sporangia can be formed on long sporangiophores that extend out through the stomata of the in-

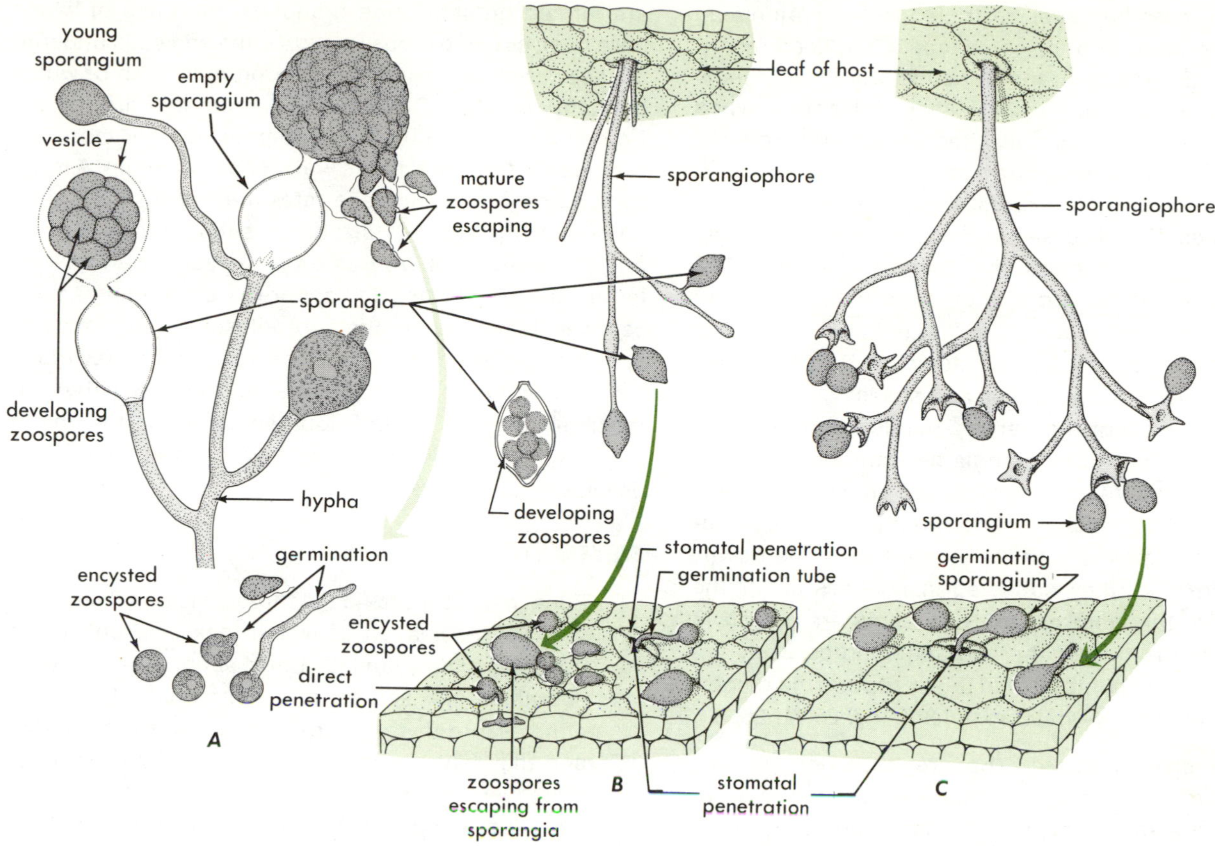

Figure 25.9
Asexual reproduction in the Peronosporales. *A, Pythium; B, Phytophthora; C, Bremia.*

fected plant (Figs. 25.9*B,C*). The sporangia are disjoined from the special hyphae producing them and are disseminated by air currents.

Some sporangia eventually come to rest on the leaves of susceptible plants. When moisture on the leaf is sufficient, zoospores escape from the sporangia. The zoospores are very small and are able to move about in the film of water on the leaf surface. They soon send out small hyphae called germ tubes that penetrate the host tissues and bring about new infections (Fig. 25.9*B*). *Phytophthora infestans* reproduces asexually in this manner. One fact that the potato growers of England and Ireland learned from the great potato famine of 1846 was that the disease was most severe in the dampest areas and almost nonexistent in dry areas. In other species, the sporangia may germinate directly into germ tubes (Fig. 25.9*C*).

Downy mildew is produced by a fungus having asexual reproduction similar to that of *Phytophthora.* Downy mildews are very common infections on many cultivated and wild plants (Fig. 25.10). They are frequently found on hops, beans, grasses, melons, alfalfa, peas, sugar beets, grapes, and other plants. They are easily identified by the sporangiophores that may, in severe cases, nearly cover the leaf.

Downy mildew of grape, *Plasmopora viticola,* nearly destroyed French vineyards in the late nineteenth cen-

Figure 25.10
Downy mildew *(Bremia)* on lettuce leaf. (Courtesy of Plant Pathology Department, University of California.)

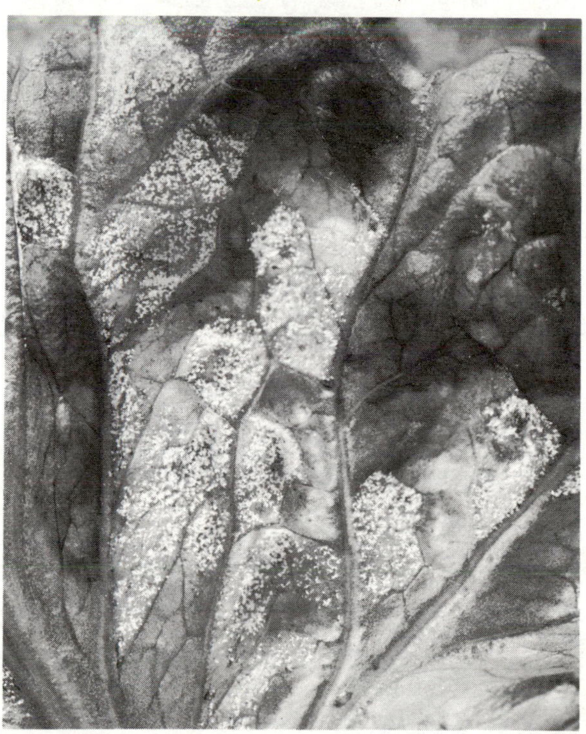

tury. The disease had been restricted to North American species of grapes, and for some time it had been recognized that European grape varieties were particularly susceptible to the fungus, for they did poorly when brought to the northeastern United States. Inadvertently, *Plasmopora viticola* was introduced to a nursery in Bordeaux in 1878 on grape rootstocks imported from the United States. The disease spread rapidly, nearly devastating vineyards throughout the Mediterranean area. The mycologist Alexis Millardet is credited with saving the wine industry by developing the first fungicide in 1882, a mixture of lime and copper sulfate called Bordeaux Mix, applied as a dust to the leaves and stems.

A small group of the Peronosporales, known as the "white rusts," develop sporangia beneath the epidermis of such crop plants as mustards and spinach. There is but one genus (*Albugo*) in this group. The sporangiophores do not grow out of the stomata as in the downy mildews. Instead, they collected in pustules under the epidermis of the stem or leaf. Sporangia are cut off in chains from the tips of the sporangiophores. They accumulate in large numbers and finally rupture the epidermis, forming creamy-white pustules (Figs. 25.11, 25.12).

Sexual Reproduction. Sexual reproduction in the Peronosporales is similar to that observed in the Saprolegniales. In many forms, it immediately precedes death of the host plant. Gametangia (oogonia and antheridia) are formed on short side branches, the ends of which have been cut off by cross walls. Some end cells undergo meiosis, swell to form spherical oogonia, each of which contains one egg. Close by, the antheridia are formed from end cells of other side branches. One or more antheridia press closely to the wall of an oogonium. A fertilization tube forms and penetrates the oogonial wall until it contacts the egg. Fertilization results by the fusing of one sperm nucleus with an egg nucleus. The resulting fertilized egg becomes an oospore by developing a thick cell wall that protects the protoplasm against adverse conditions (Fig. 25.13). When conditions are favorable for growth, the oospore germinates, forming, either immediately or after the development of a short hypha, a large number of diploid zoospores, each of which may develop into a hypha.

SUMMARY

1. Fungi may be classified in three divisions. They total 200,000 species, all of which lack chlorophyll but possess some plantlike traits such as a cell wall, sporangia and spores, and nonmotility. Fungi are either saprophytes or parasites (as are bacteria), and they have large positive and negative impacts on human affairs. Some fungi may be both parasitic and saprophytic, depending on environmental conditions.

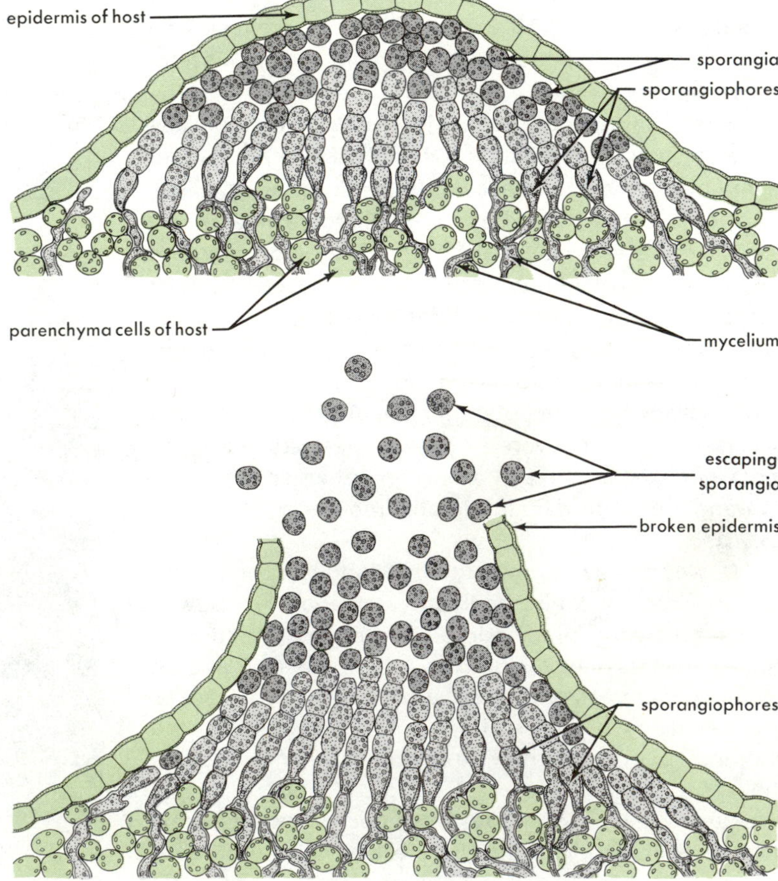

Figure 25.11
Pustules on shepherd's purse caused by sporangia of *Albugo*. (Redrawn from G. M. Smith, *Cryptogamic Botany*, Vol. 1, *Algae and Fungi*. © 1955 by McGraw-Hill Book Company, New York. Reprinted with permission of the publisher.)

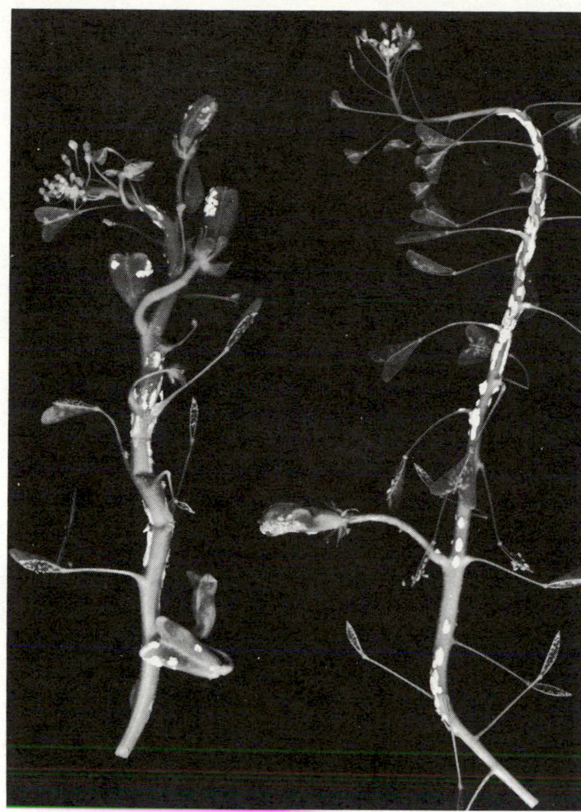

Figure 25.12
Albugo on shephard's purse. (From R. M. Holman and W. W. Robbins, *A Textbook of General Botany*, John Wiley & Sons, New York.)

2. Two of the three fungal divisions are discussed in this chapter: Myxomycota and Oomycota. Together they may be called the "lower fungi" and they differ from the third division (Eumycota, "higher fungi") by lacking chitin and conidia.
3. Myxomycota (slime molds) have life cycles with unicellular and multicellular stages. Acellular slime molds have a diploid vegetative body consisting of a naked mass of protoplasm (no cell wall) with many nuclei. This plasmodium moves in a slow, amoeboid way, engulfing solid bits of food and bacteria. It has a gametic life cycle. Cellular slime molds have much smaller plasmodia, with membranes separating the nuclei, and a sexual stage is unknown (the plasmodium may be haploid). Their life cycles are summarized on page 500–502. Spores have cellulose walls.
4. Oomycota differ from Myxomycota in having cellulose cell walls throughout the life cycle. The basic unit of the plant body is a nonseptate filament called a hypha. Hyphae grow from the tip, produce many side branches, and ramify into a network called a mycelium.
5. Nutrients are absorbed through hyphal walls and membranes from the surrounding medium. Enzymes that solubilize the nutrients are secreted into the medium. Some parasitic hyphae are specialized as haustoria that invade the cytoplasm of host cells.
6. Members of the Oomycota can be divided into several orders. Only two are discussed in the text: Saprolegniales (water molds) and Peronosporales (downy mildews).
7. Members of the Saprolegniales are aquatic and generally saprophytic. Asexual reproduction is typi-

Figure 25.13
Oospores of *Albugo*, ×150.

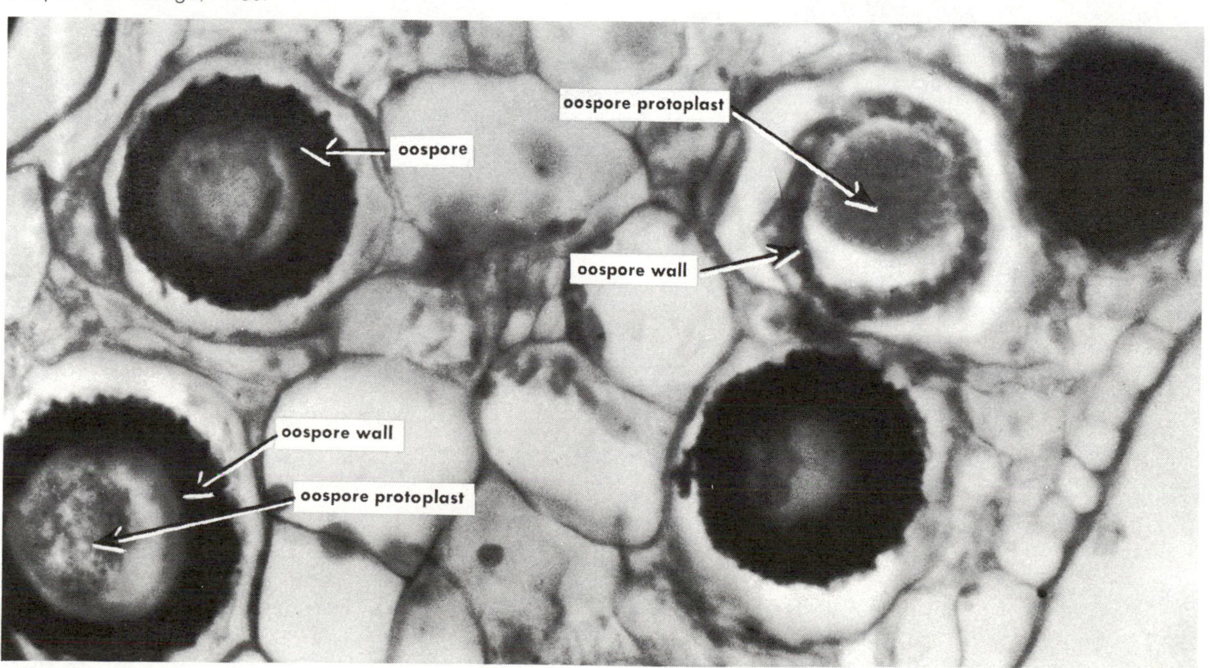

cally by motile spores. Sexual reproduction involves eggs and sperms. The zygote has a resistant, resting stage before germinating into a diploid mycelium. These plants have a gametic life cycle (see pages 504–506).

8. Members of the Peronosporales are obligate parasites. The group includes damping-off fungi (*Pythium*), late blight of potato (*Phytophthora*), downy mildew of grape (*Plasmopora*), and white rusts (*Albugo*). Life cycles are similar to those in the Saprolegniales (see pages 505–508).

CLASSIFICATION

Division	Myxomycota
Genera	*Physarum*
	Dictyostelium
Division	Oomycota
Order	Saprolegniales
Genus	*Saprolegnia*
Order	Peronosporales
Genera	*Pythium*
	Phytophthora
	Plasmopara
	Albugo
	Bremia
	Peronospora

CHAPTER 26

EUMYCOTA

The Eumycota consists of three subdivisions. With rare exception (e.g., the yeasts), the vegetative thallus is a mycelium composed of intertwining hyphae. The cell walls are of chitin, and asexual reproduction is often by means of conidia. The mycelium is nonseptate in the Zygomycotina (bread molds and related species), but it is septate in the other two subdivisions. The cross walls, however, are perforated with complex pores (Fig. 26.1) that enable protoplasm and sometimes organelles to flow from cell to cell.

We are including the form group Fungi Imperfecti in this chapter, because there is reason to believe that most of these species are Ascomycotina whose sexual cycles have been lost or never observed. The fungal component of most **lichens** is also typically an ascomycete, and the lichens are often classified as one order within the Ascomycotina; consequently they too are included in this chapter. Another symbiotic association of fungi with other plants is the **mycorrhiza,** which typically involves the roots of higher plants and soil-inhabiting Ascomycotina or Basidiomycotina, and they will be discussed in this chapter.

SUBDIVISION ZYGOMYCOTINA

The name of this subdivision refers to the production of a resting zygote (a zygospore) that is the result of the fusion of gametes. This is the only diploid structure in the life cycle, in contrast to many of the fungi discussed in Chapter 25. No motile cells of any sort are found in this subdivision; all its members produce typical aerial

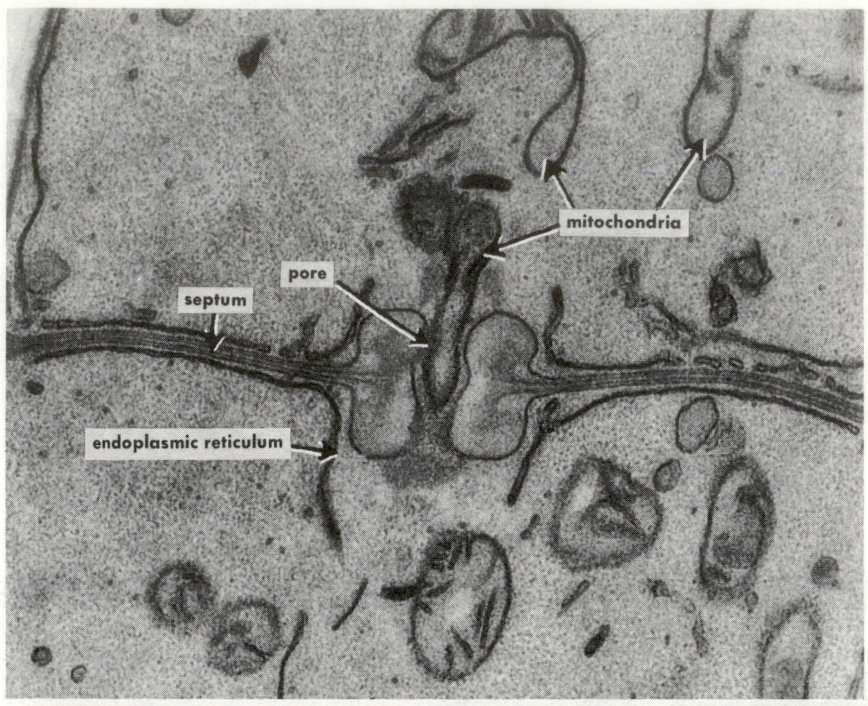

Figure 26.1
A small portion of a hypha of *Rhizoctonia solani* showing a septum, with pore, through which a mitochondrion appears to be passing, ×24,000. (Courtesy of C. E. Bracker and E. E. Butler.)

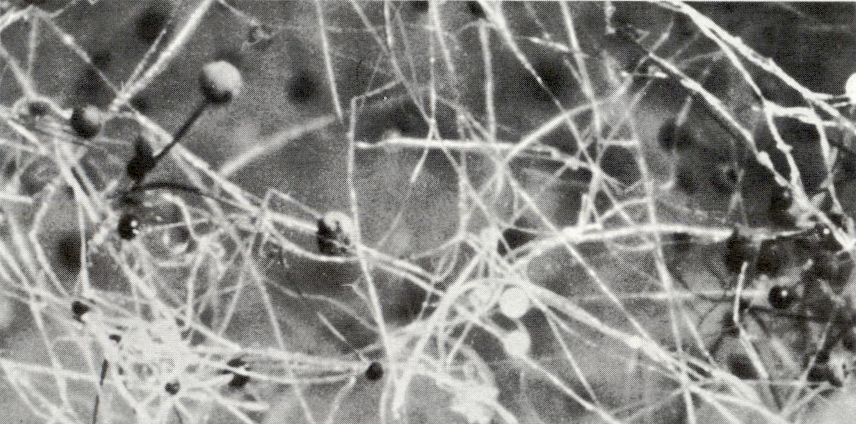

Figure 26.2
Hyphae, mature sporangia (black), and immature sporangia (white) of *Rhizopus stolonifer*, ×4.

sporangia, and they regularly have a coenocytic mycelium of indefinite form. Most species are saprophytic, some are weak parasites of plants, a few are specialized parasites of animals, and some are actually obligate parasites on other Zygomycotina. We shall consider only the common bread mold.

The mycelium at first grows chiefly within the substrate, which may be composed of various kinds of organic matter. Eventually, aerial hyphae develop so that the surface of the substrate may become covered with a mass of hyphae (Fig. 26.2).

Common bread mold, *Rhizopus stolonifer*, is a member of the order Mucorales and is of worldwide distribution. It grows well on a large variety of organic substances. Although it is mainly a saprophyte, it does attack and considerably damage sweet potatoes, berries, and fruit while in transit or in storage. It grows luxuriantly on sweet potatoes and bread, and these foods may be utilized as media for the growth of laboratory cultures.

Asexual Reproduction

After the mycelium has become well established upon and in a substrate, certain aerial hyphae, usually of larger diameter than those in the substrate, grow just above its surface for a short distance and then come in contact again with the substrate. They are known as **stolons.** At the point of contact with the substrate, new hyphae form. Hyphae are of three types: (*a*) stolons; (*b*) branching hyphae, called **rhizoids,** that penetrate the substrate and serve to anchor the mycelium and to absorb nutrients; and (*c*) hyphae that grow upright and produce sporangia at their tips and therefore are **sporangiophores** (Fig. 26.3).

The development of a sporangium is as follows: The tip of the sporangiophore swells, and a bulging wall cuts off an apical cell, which becomes the sporangium. The dome-shaped wall separating the sporangium from the parent hypha is called the **columella.** Numerous spores are formed from the protoplasm within the sporangium,

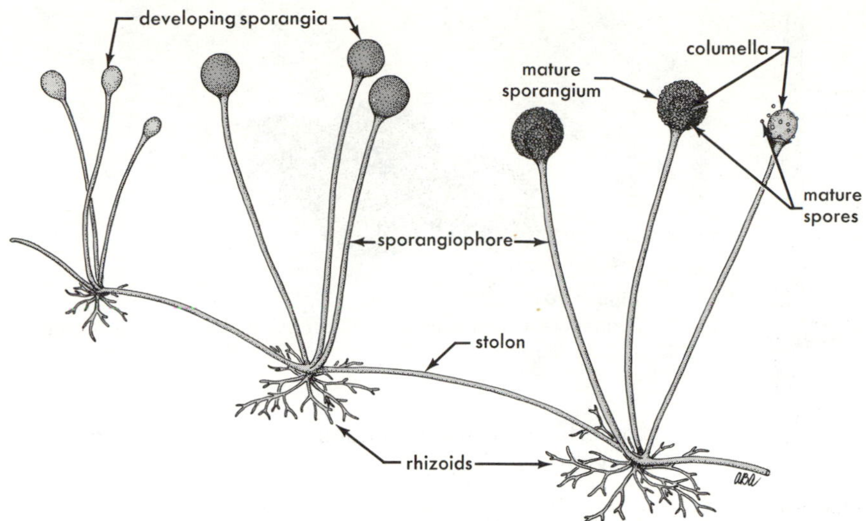

Figure 26.3
Asexual reproduction in *Rhizopus stolonifer.* Rhizoids usually develop more extensively than shown.

Figure 26.4
Stereoscan micrograph of spores of *Rhizopus stolonifer,* ×400. (Courtesy of S. Cook and D. Hess.)

and when the spores are ripe (Fig. 26.4), the outer wall of the sporangium falls apart and the spores are dispersed by air currents. On a suitable substrate, the spores germinate and develop new hyphae. Immature sporangia are white; mature ones are black.

Sexual Reproduction

The characteristic type of sexual reproduction is conjugation. The first step is the contact of the tips of short club-shaped hyphal branches, brought about by chemical attraction (Fig. 26.5). Once in contact, the ends of these two short hyphae swell and elongate slightly. A cross wall forms back from the tip of each hypha, separating a terminal cell with many nuclei from the parent hypha. The two tip cells thus formed are gametangia. The walls of the gametangia that are in contact dissolve, permitting the two protoplasts to fuse. The zygote resulting from this union develops into a **zygospore** (Fig. 26.6), which has a thick wall and is quite resistant to unfavorable conditions. The zygospore will later germinate, undergo meiosis, produce a sporangium, and release meiospores. Each meiospore can germinate and continue the life cycle.

There is some variation in the way different species develop gametangia and zygospores. Let us assume that in a given community no bread mold has ever been found. By chance, one bread mold spore is brought in and grows, producing many millions of spores that spread the fungus throughout the community. We now have a strain of bread mold growing in this community that has arisen from a single spore. The probabilities are that sexual union will never occur, because many species, although morphologically similar, are physiologically differentiated into **sexual strains.** Their similar appearance and behavior make it impossible to designate them as male and female. Instead they are called, for convenience, **plus** (+) and **minus** (−) **strains.** The bread mold spore introduced under these conditions into this community would be either plus or minus. A plus strain will not conjugate with a plus strain, nor a minus strain with a minus strain and, hence, no sexual reproduction will occur. Asexual reproduction, however, will be normal. If the opposite strain were now introduced and established, isogametes and zygospores would form. Species of fungi that are differentiated into mating strains are said to be **heterothallic.**

SUBDIVISION ASCOMYCOTINA

The fine flavors of some cheeses are due to ascomycetes, and some edible species of fungi are members of this group. Ascomycetes include several severe, though uncommon, human pathogens, and certain other representatives are known to the medical profession mainly for the beneficial drugs derived from them. Ergot, a drug

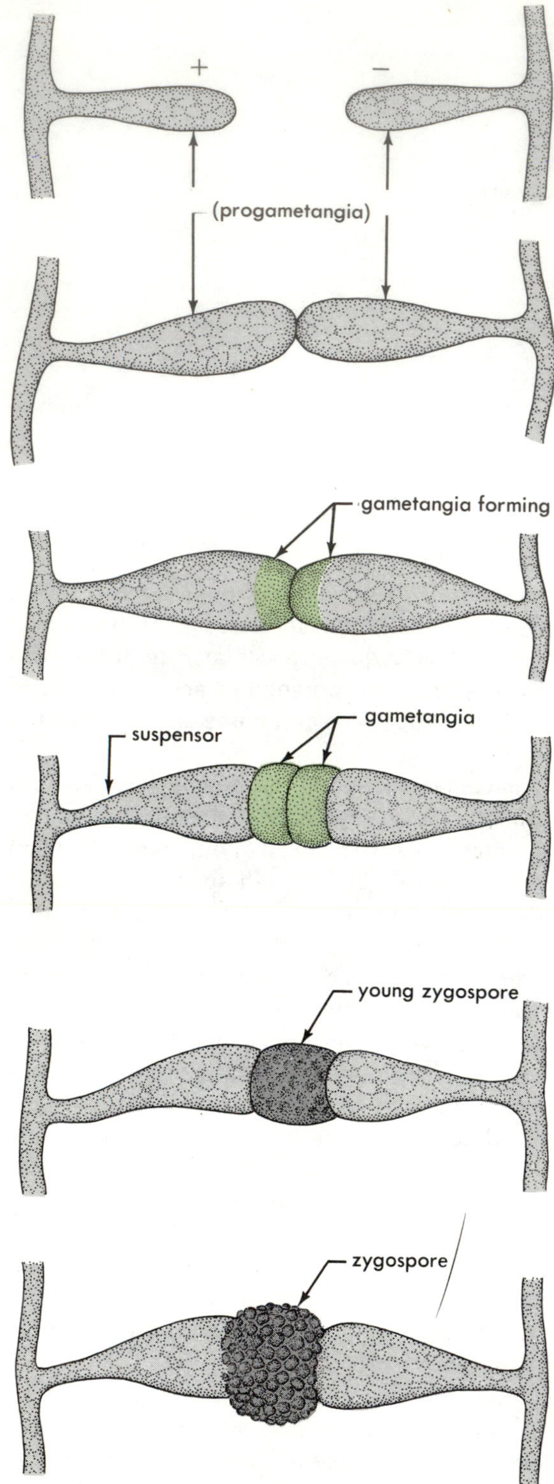

Figure 26.5
Sexual reproduction in *Rhizopus stolonifer.* (Based on G. M. Smith, *Cryptogamic Botany*, Vol. **I,** *Algae and Fungi.* © 1955 by McGraw-Hill Book Company. Reprinted with permission of the publisher.)

widely used to control bleeding, is derived from an ascomycete that infects grasses. Yeasts, also members of this class, are a source of many vitamins and are of great

Figure 26.6
Zygospore with suspensors, *Rhizopus stolonifer,* ×300. (Courtesy of R. Riding.)

importance in the production of alcohol and in breadmaking.

Reproduction

Asexual Reproduction

Various types of asexual spores are chiefly responsible for the dissemination of Ascomycotina during their period of active growth. In mild climates asexual spores may survive the unfavorable seasons, but they are usually killed by cold or by very hot and dry weather. Conidia are commonly produced, and in many species the arrangement of conidia and conidiophores is characteristic and of taxonomic importance.

Sexual Reproduction

Meiosis followed by mitosis in most Ascomycotina results in eight meiospores in a terminal cell (Fig. 26.7*A*). The appearance of these spores suggests beans or marbles within a cellophane sac; thus the common name **sac fungi.** The Greek word for sac is **ascus** (plural, **asci**), from which is derived the subdivision name. The meiospores are generally called **ascospores.** The asci are always the terminal cells of special hyphae. They are usually located in a reproductive structure, the **ascocarp** (Fig. 26.7*B*), and have developed by a specialized modification of fertilization from the female gametangium, the oogonium, or as it is known in this group, the **ascogonium.** The male gametangium is called an **antheridium,** as in preceding groups.

The ascocarp, composed of both vegetative and ascus-bearing hyphae, is characteristic of the species. It may be microscopic or as much as 10 to 15 cm in diameter. There are mainly three general types of ascocarps:

1. **Cleistothecium:** hollow, completely closed sphere (Figs. 26.8*A,B*).
2. **Perithecium:** hollow, flask-shaped body with narrow opening (Figs. 26.8*C,D*).
3. **Apothecium:** open, cup-shaped body (Figs. 26.8*E,F*).

The end cells (asci) of the ascus-bearing hyphae, in many forms, line the inner surface of the ascocarp. This surface layer is the **hymenium** or fertile layer. Sterile

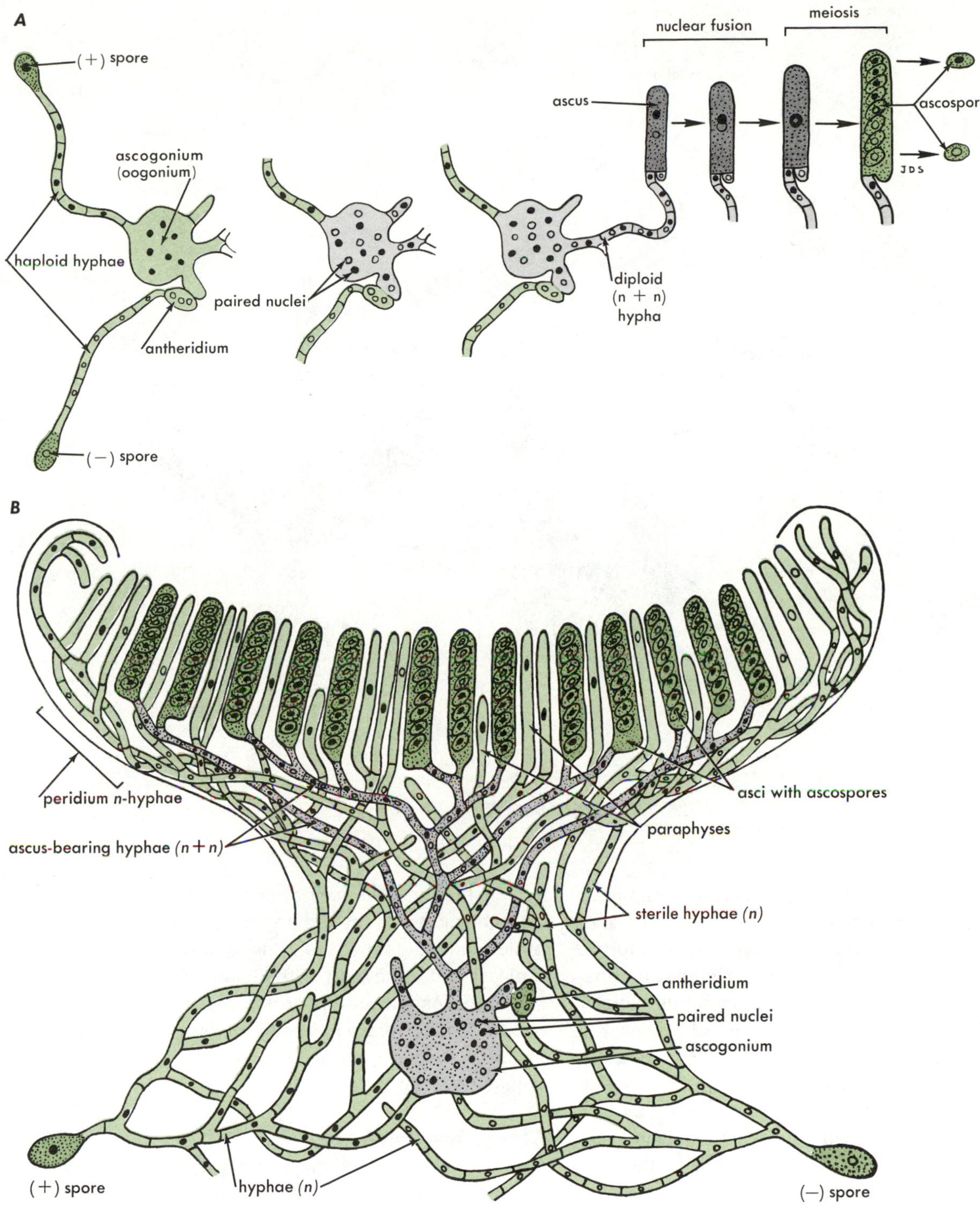

Figure 26.7
A, diagram showing sexual life cycle of an Ascomycete. *B,* diagram of a cross section of an apothecium type of ascocarp. (Redrawn from L. W. Sharp, *Fundamentals of Cytology.* © 1943 by McGraw-Hill Book Company. Reprinted with permission of the publisher.)

cells, called **paraphyses,** also arise in the hymenium (Figs. 26.7*B*, 26.8*D*) and are more numerous and generally longer than asci.

In most Ascomycotina, the ascocarps are the direct result of fusion. The description that follows is a generalized account of gametangial fusion as it occurs in many Ascomycotina. The gametangia develop from the haploid mycelium growing within the host or substrate. The female gametangium, or oogonium, is a single cell with many nuclei. Antheridia are elongated cells borne on short side branches of adjacent filaments. Both gametangia may have special accessory cells. After an antheridium establishes contact with an oogonium, the male nuclei pass into it. Male and female nuclei pair, but

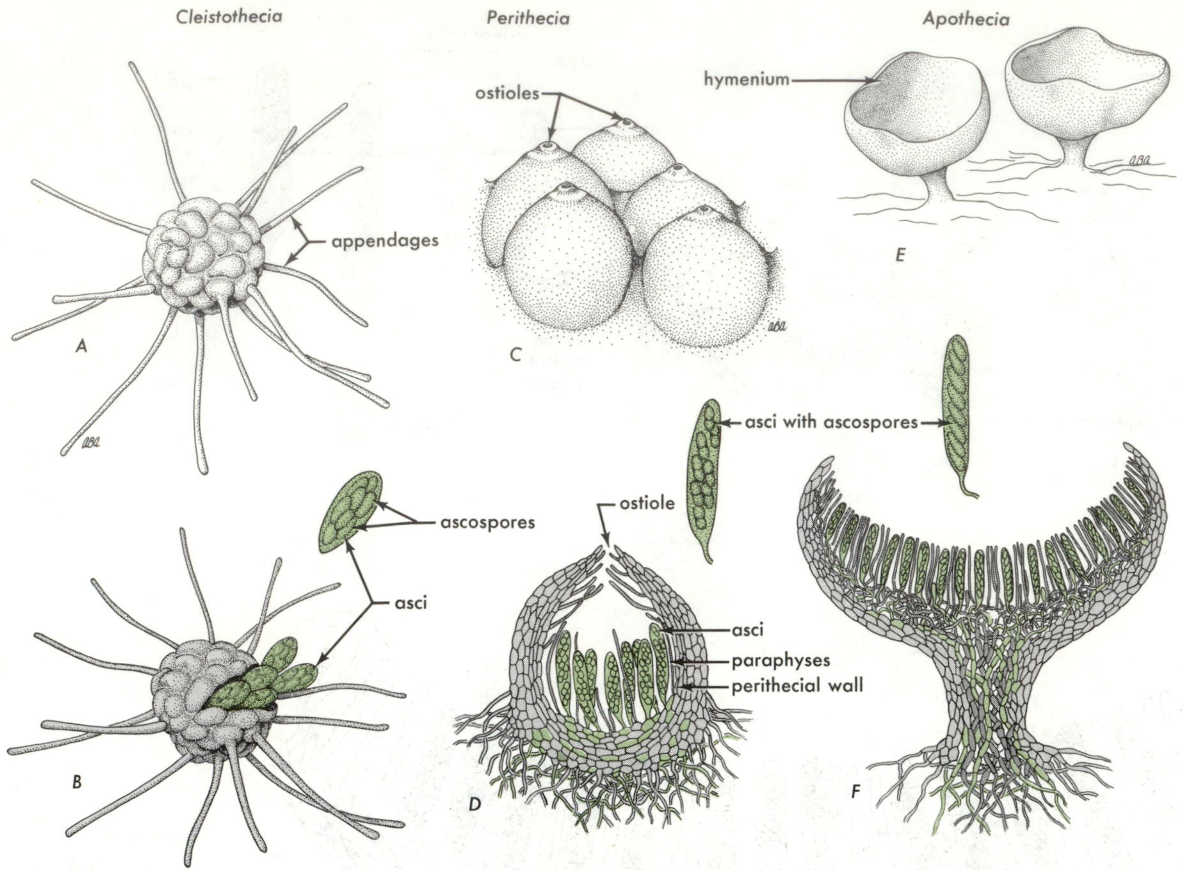

Figure 26.8
Diagram showing three types of ascocarps. *A* and *B*, cleistothecia, ×500; *1C* and *D*, perithecia, ×150; *E* and *F*, apothecia, *E* ×¼, *F* ×10. (*D*, redrawn from E. A. Gäuman, *The Fungi*. © 1952 by Hafner Publishing Company and E. A. Gäuman, *Comparative Morphology of Fungi*. © 1928 by McGraw-Hill Book Company. Reprinted with permission of the publisher.)

do not fuse as is normally the case. This process now stimulates the growth of hyphae from the oogonium and from the surrounding haploid mycelium. The cells forming from the oogonium are binucleate, each having one male and one female nucleus. This is not a true diploid condition; it may be designated by $n + n$ rather than $2n$, and the paired nuclei are frequently referred to as a **dikaryon.** Since it is from these hyphae that the asci will eventually develop, they may be called **ascogenous hyphae.** The coordinated growth of the haploid and ascogenous hyphae results in the formation of the **ascocarp.** The terminal cells of the very much branched ascogenous hyphae, together with the ends of some haploid hyphae, form the **hymenial layer.** The hymenial layer is more or less surrounded on the outside and protected by the vegetative haploid hyphae that form a layer called the **peridium** (Fig. 26.7*B*). It is in the young ascus that the nuclei of the dikaryon finally fuse to form a true diploid cell. Meiosis occurs immediately after the nuclear fusion and is followed, in most species, by a mitotic division giving rise to eight ascospores.

The sexual phase of most Ascomycotina is thus complicated by the fact that the male and female nuclei do not fuse immediately after the union of protoplasts from the antheridium and oogonium. This gives rise to a prolonged stage between the union of the sex protoplasts and the fusion of the sperm and egg nuclei.

In the Ascomycotina so far described, the development of the ascocarp has been preceded by gametangial fusion. In some forms, an ascocarp may develop without gametangial fusion.

Classification

Species within this subdivision may be grouped into two classes and a dozen orders. Representative genera and common names of eight orders are shown in Table 26.1. Notice that *Eurotium* and *Talaromyces* have other generic names in parentheses. The parenthetical names are synonyms dating back to the time when the sexual stage of the plants was unknown and they were placed in the form class Fungi Imperfecti. When the sexual stages were finally seen, it was clear that these plants were Ascomycotina, and the form genera were replaced with new generic names.

Each of the representative genera in Table 26.1 will be briefly discussed.

Table 26.1
CLASSIFICATION OF ASCOMYCOTINA GENERA DISCUSSED

Class	Order	Genus	Common name
Hemiascomycetes	Endomycetales	*Saccharomyces*	Yeasts
	Taphrinales	*Taphrina*	Leaf-curl fungi
Euascomycetes	Eurotiales	*Eurotium* (*Aspergillus*)	Green mold
		Talaromyces (*Penicillium*)	Blue mold
	Erysiphales	*Erysiphe*	Powdery mildews
	Clavicipitales	*Claviceps*	Ergot
	Helotiales	*Monilinia*	Brown rot fungus
	Pezizales	*Peziza*	Cup fungus
	Tuberales	*Tuber*	Truffles

Saccharomyces (Yeasts)

Yeast cells may be spherical, ellipsoid, more or less rectangular, or, in vigorously growing cultures, sometimes hyphalike (Figs. 26.9*A,C*). Their rate of growth seems to influence their shape to some extent. Under ordinary microscopic magnification, living cells appear lacking in much structural detail, but under the electron microscope, nuclei, vacuoles, storage granules, and mitochondria can be seen (Fig. 26.9*D*).

Reproduction. Yeasts are normally single-celled. In old cultures, the cells may remain attached, forming short, branched chains. Some yeasts divide by fission. In most yeasts, however, new cells grow out from the mother cell, much as a small bubble would form if a piece of thin rubber were made to expand through a small opening in some heavier material. The small "bubbles" formed from the mother yeast cell are called **buds.** They enlarge and finally separate from the parent cell. This process of vegetative reproduction is termed **budding** (Fig. 26.9).

Most yeasts form asci, each containing from one to eight ascospores, the number being constant for a given species. As in other Ascomycotina, the production of asci usually is associated with a sexual cycle. In the formation of ascospores, the nucleus of a diploid cell divides into several nuclei, and each, with some associated cytoplasm, becomes delimited as a spore. The spores lie within the parent cell, which is essentially an **ascus** (Fig. 26.10). During the formation of ascospores, a reduction division occurs. Thus, in yeast, the parent cell is diploid, and the ascospores are haploid. In some yeasts, the nuclei of adjoining haploid vegetative cells fuse, the resulting zygote dividing to produce ascospores that multiply by vegetative division. In certain other yeasts, a fusion of haploid ascospores takes place.

The life cycle of *Saccharomyces cerevisiae,* the common yeast of commerce, is of considerable interest because it may reproduce asexually by budding in both haploid and diploid phases. Its life history is of such a nature as to make this species particularly suited for studies of fundamental biological significance.

Economic Importance. The main sources of energy for many yeasts are sugars, which are oxidized in respiratory processes within the yeast cell. Oxygen is not required for the first steps of the oxidation. The sugar, nevertheless, is broken down to a simple organic acid. Energy is released by glycolysis (see Chapter 14). If the oxygen supply is ample, this acid is oxidized, with the release of relatively large amounts of energy, to carbon dioxide and water. If the oxygen supply is deficient, the organic acid will be changed to carbon dioxide and alcohol. Thus, yeasts growing in a sugar solution well supplied with oxygen will produce carbon dioxide and water, and the yeast cells will multiply rapidly. On the other hand, yeasts growing in a sugar solution poorly supplied with oxygen will tend to form carbon dioxide and alcohol and will multiply slowly. The oxidation of sugars to carbon dioxide and alcohol by yeasts without the presence of oxygen is known as **alcoholic fermentation** (see Chapter 14). This process is utilized in the production of industrial alcohol, winemaking, brewing, and bread-making.

Yeasts have played a very important part in vitamin research. Several vitamins are synthesized by yeast plants, which are an important commercial source of these highly valued substances.

Taphrina

All members of this genus are highly pathogenic to plants, especially fruit trees. They infect leaves, flowers, fruits, and young shoots. They induce unequal growth in leaf cells, causing diseases frequently referred to as **leaf curls.** The fungus causing peach leaf curl is well-known. The mycelium of this fungus is intercellular in the palisade tissue of leaves. Here, it stimulates some cells to more rapid growth, which results in wrinkling of the leaf. Eventually, the hyphae push their way between the epidermis and the cuticle, where the individual cells swell, round off and thicken their walls, and form a compact layer. These cells then elongate, rupturing the cuticle, and become asci, each one of which contains ascospores. Thus, the asci are produced at the surface of the infected organ. The opening buds as they start to grow in the spring are infected by spores that have overwintered between the bud scales. A copper fungicide applied just before the buds swell is a very effective control. Typical gametangia are not formed.

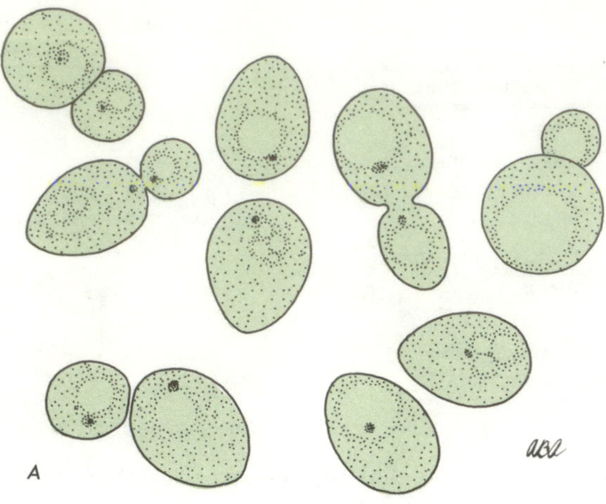

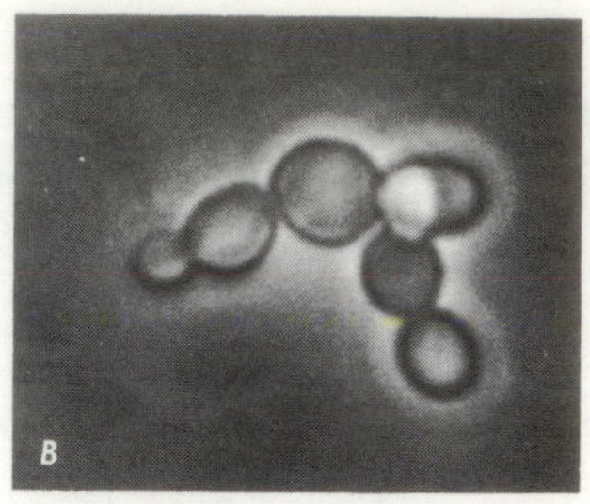

Figure 26.9
Yeasts in various stages of budding. *A, Saccharomyces cerevisiae. B, Torulopsis apicola. C, Candida tropicalis. D,* electron micrograph of budding in *Saccharomyces cervisiae. A* to *C,* about ×1000; *D,* ×25,000. (*B, C,* courtesy of H. Pfaff, *D,* courtesy of E. Vitols, R. J. North, and A. W. Linnane, *J. Biophys. Biochem. Cytol, 9, 689.* © 1961 by The Rockefeller University Press. Reprinted with permission of the publisher.)

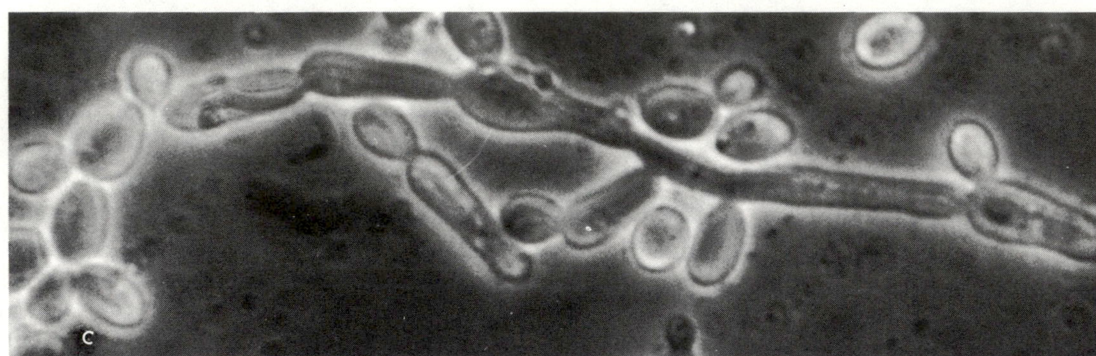

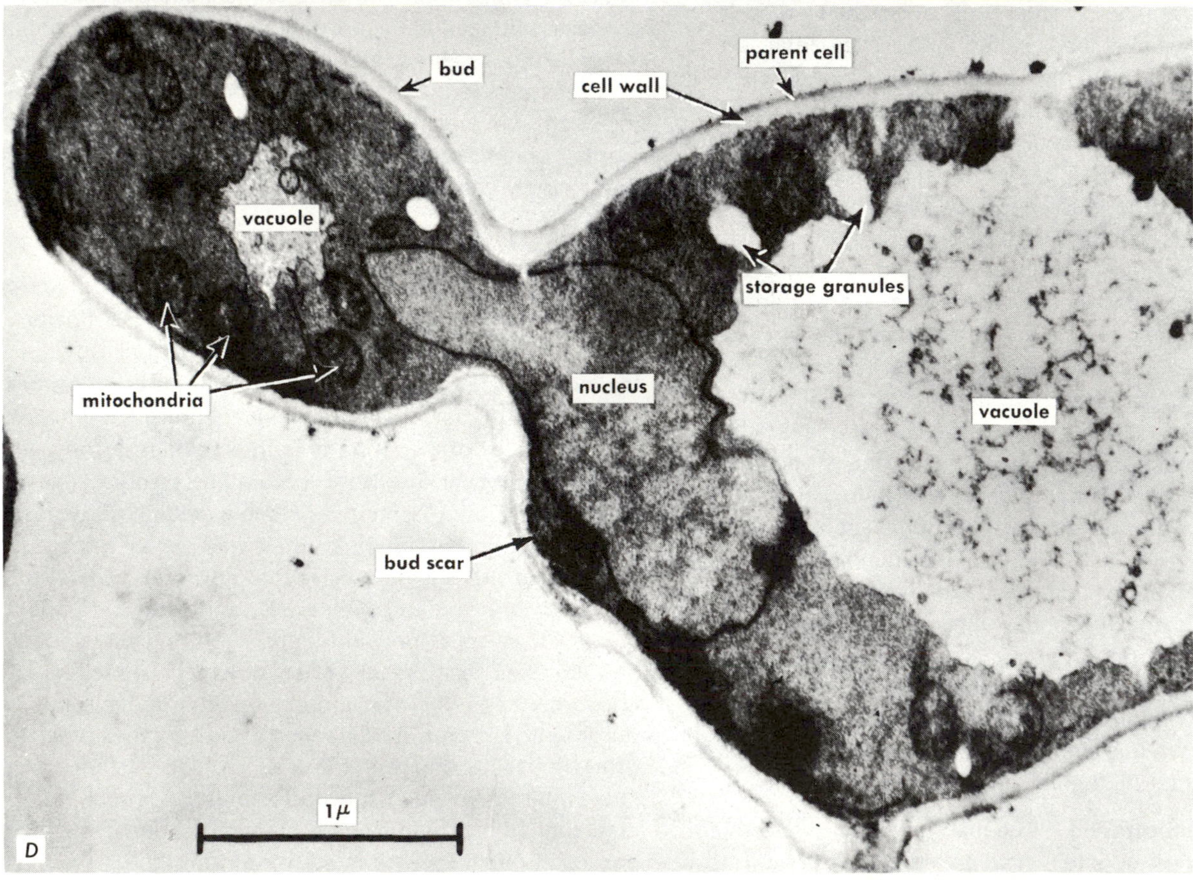

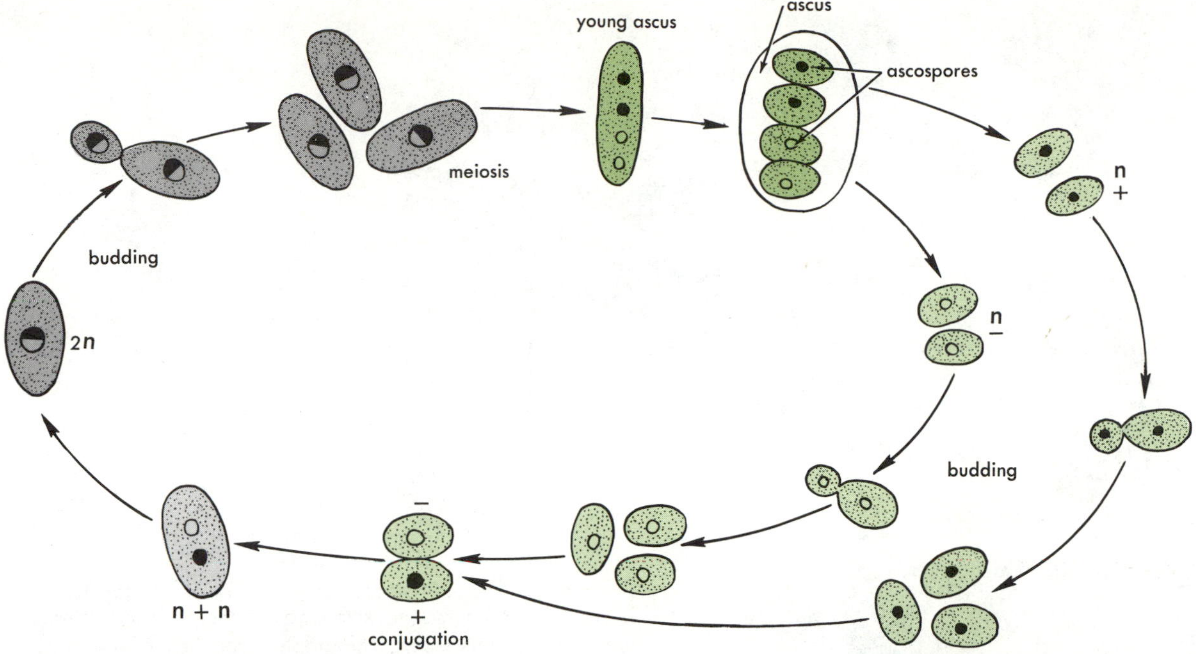

Figure 26.10
Life cycle of *Saccharomyces cerevisiae.*

Form Genera *Penicillium* and *Aspergillus*

Asexual reproduction in *Penicillium* and *Aspergillus* is effected by means of conidia. In *Penicillium,* the spores are formed on profusely branched conidiophores (Figs. 26.11*A,D*). In *Aspergillus,* the tip of the conidiophore swells and conidiospores form in long chains radiating from this swollen tip (Figs. 26.11*B,C*).

The ascogonium of *Talaromyces vermiculatum* (*Penicillium*) is an elongated slender multinucleate cell, and the antheridium is simply a club-shaped, swollen hypha. Details of fertilization and meiosis are obscure, but eventually cleistothecia develop, containing scattered asci without an organized hymenial layer (Fig. 26.12). Asexual reproduction is by means of conidia. It must be emphasized that the asexual stage is the usual and greatly predominant mode of reproduction, the sexual stage being rare indeed.

Penicillium and *Aspergillus* are probably the most widespread fungi. They are the common blue, green, and black molds that occur on citrus fruits, jellies, and preserves. Their conidia are everywhere in the air and soil, and in the biological laboratory they are frequent contaminations in culture media. Enzymes secreted by these fungi are particularly active in digesting starch and other carbohydrates. When purified, these enzymes are important industrial preparations. *Aspergillus oryzae* is used in the preparation of rice wine and soybean sauces. Several species of *Aspergillus* are important in cheese manufacture. A disease resembling tuberculosis is caused by *Aspergillus fumigatus,* and a number of other *Aspergillus* species cause diseases in plants. Strains of *Aspergillus flavus* form aflotoxins that are toxic to animals.

Erysiphe

The powdery appearance of the surface of leaves infected with many members of *Erysiphe* and related genera suggests the common name, **powdery mildews.** All the powdery mildews are obligate plant parasites (**Fig. 26.13,** page 539). Their food requirements are closely integrated with the metabolism of the host plants, and frequently the host plant is not killed. This relationship ensures the fungus a continued food supply. Many can live only on a special host, whereas others have a wide host range. They do not grow in artificial culture media. These characters indicate a high degree of specialization and a long-standing relationship between parasite and host.

The mycelium is generally confined to the surface of the leaves, flowers, or fruits of apples, grains, rose, grapes, cherries, and other plants (Fig. 26.14). Haustoria penetrate epidermal and parenchyma cells, from which they secure nourishment. At first the mycelium on the surface of the leaf appears like a delicate cobweb. Eventually, it assumes a white powdery or dusty appearance owing to the development of numerous conidiospores.

Asexual Reproduction. This type of reproduction is effected by conidiospores and usually accounts for the rapid propagation of the fungus during the growing season. The conidiophores are short filaments that stand outward from the mycelium on the surface of the host (Fig. 26.14). The spores themselves may remain attached to each other and form long characteristic chains. Sulfur dusted on the host plants at this stage of the fungal life cycle is an effective control against infections from these spores.

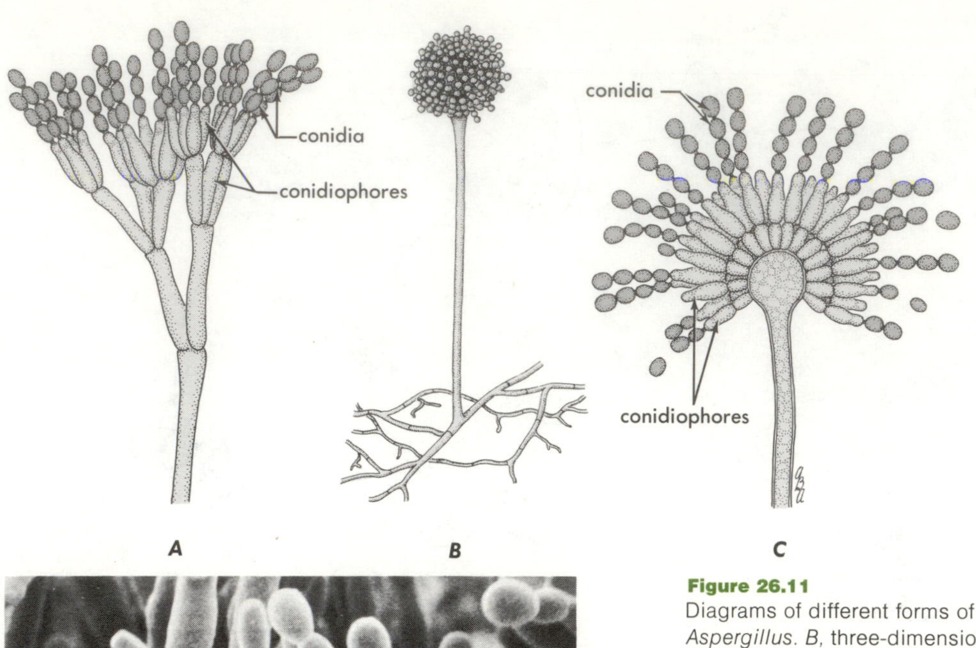

Figure 26.11
Diagrams of different forms of conidiophores. *A, Penicillium. B, C, Aspergillus. B,* three-dimensional appearance. *C,* optical section through *B. D,* stereoscan micrograph of conidiophores with conidia of *Penicillium digitatum,* ×3500. (Courtesy of E. Butler.)

Sexual Reproduction. The ascogonium is always located at the end of a hypha and may be slightly swollen. The antheridia are small cells, also occurring at the tips of hyphae. An antheridium becomes closely appressed to an ascogonium; an opening appears in the wall separating the gametangia; and the male nucleus migrates through it into the ascogonium. Binucleate hyphae develop from the binucleate zygote, and simultaneously ad-

Figure 26.12
Sexual stages in *Talaromyces vermiculatum (Penicillium vermicalatum). A,* vegetative hypha. *B,* formation of antheridium and ascogonium. *C, D,* and *E,* development of antheridia and ascogonia with first stage of fertilization. *F,* cross section of the loosely formed primitive ascocarp. *G,* ascospores. *H.* vegetative hyphae.

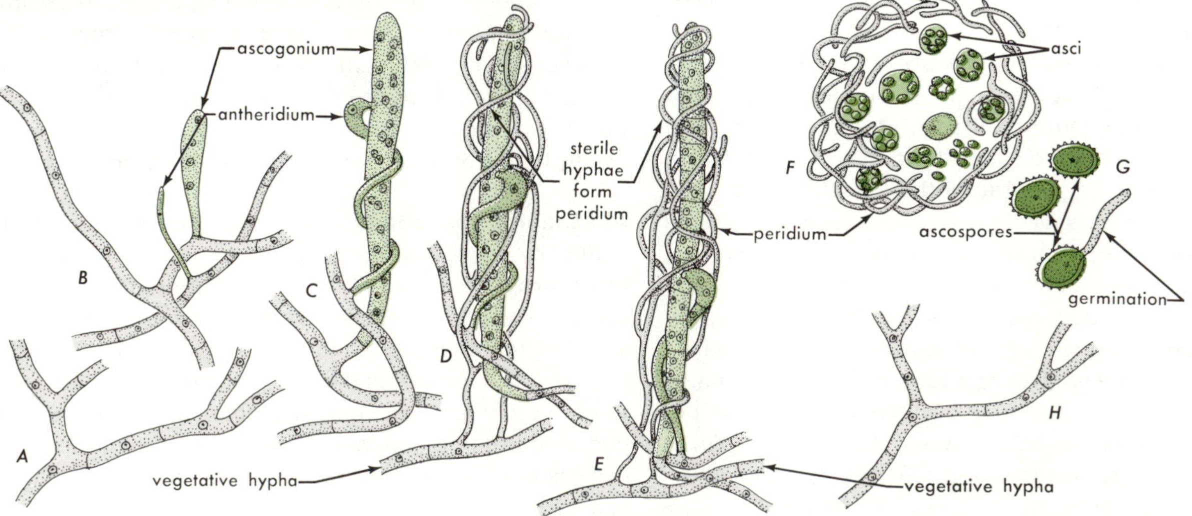

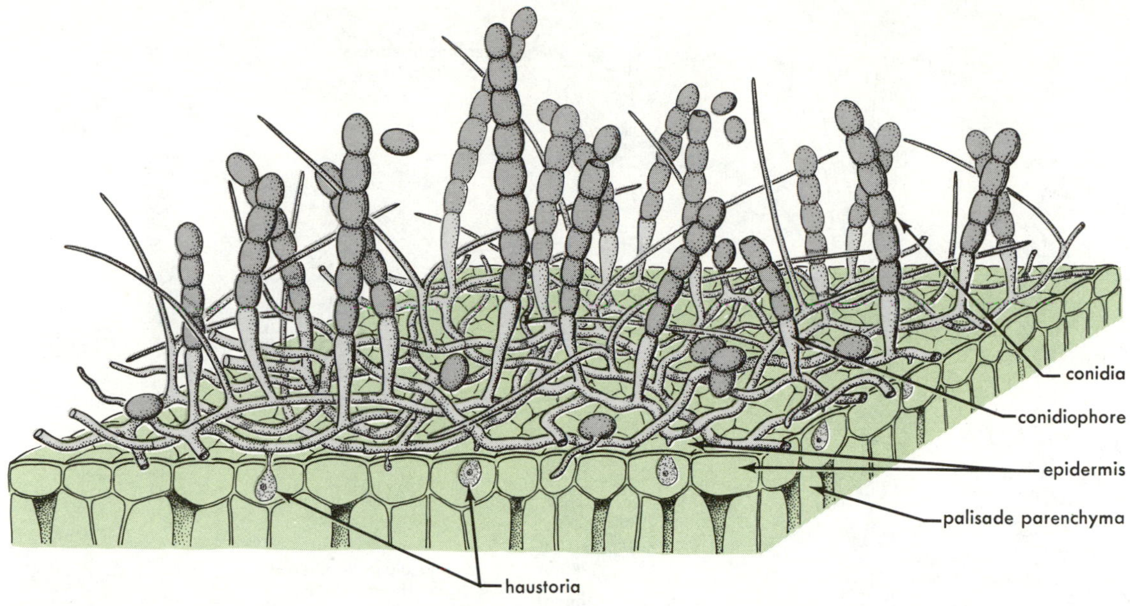

Figure 26.14
Powdery mildew on leaf surface. (Redrawn from *Selecta Fungorum, Fungorum Carpologia of the Brothers Tulasne, L. R.* and *C,* Vol. I. 1861-65. Translated into English by W. B. Grove; A. H. R. Buller and C. L. Shear, eds., 1931. Published by Clarendon Press. Reprinted by permission of the publisher.

jacent haploid vegetative hyphae also develop. A small black **cleistothecium** is formed and encloses one or several asci **(Fig. 26.13*A*).** Nuclear fusion and reduction division take place in the ascus, and ascospores develop. The ascospores are usually discharged by force from the cleistothecium.

Appendages, characteristic of the genera, extend outward from the cleistothecia, and these may aid in the dispersal of the cleistothecia and in attaching them to a new host.

Claviceps

The genus *Claviceps* is parasitic on grasses, including grains. A dormant mycelium that replaces the mature grain is known as **ergot.** It possesses several alkaloids that have medicinal properties. Ergot constricts the blood vessels, particularly those that pass into the hands and feet, thus depriving the extremities of a normal blood supply. In humid summers in Central Europe, *Claviceps* may infect rye heavily. In the Middle Ages before the nature of the fungus was understood, ergot would be milled along with the grains of rye. The contaminated flour, which might contain as much as 10% of powdered mycelium, would be baked in bread. A continued diet of bread from this flour resulted in much misery. Madness, abortion, a burning sensation in the extremities, gangrene, loss of limbs, and hallucinations are some results of ergot poisoning. In 944 in France, 40,000 people died from ergot, and fatalities continue to be recorded up to the present. The disease was known as "Holy Fire" or St. Anthony's fire. Today, ergot is a valued drug used to control hemorrhage, particularly during childbirth. It also is used in the treatment of migraine headaches. One of the alkaloids in ergot is closely related to LSD (lysergic acid diethylamide), and it is possible that this is the basis of hallucinations induced by ergot.

With the knowledge of the poisonous nature of ergot, diseased grain is no longer milled and *Claviceps* is not of concern in human diet. However, cattle may occasionally feed on infected grasses having ergots in place of seeds and thus be poisoned.

Life History of *Claviceps* (Fig. 26.15)

1. Ascospores are mature when the first flowers of rye, wheat, or other host grasses open. A given race of *Claviceps* infects certain species of cereals or grasses. The mature ascospores infect the young flowers.

2. A mycelium develops throughout the ovary of the infected flower. Finally it completely replaces the ovary, assuming its general shape.

3. Conidia appear on the surface of this mycelium, together with a sticky, sweet secretion. Insects collect the secretion and, in doing so, disperse the conidia.

4. The compact mycelium replacing the ovary grows upward and becomes a hard and horny body. It is somewhat longer than the mature grain and is purple in color. This dormant, elongated mycelium is the **ergot.** It is composed of a compact mass of tough hyphae that, in a dormant condition, survives periods unfavorable to growth.

5. When conditions become suitable, the hyphae of the ergot produce ascogonia and antheridia. After fertilization the ascogenous hyphae, accompanied by adjacent vegetative hyphae, develop a short, upright stalk. The stalk supports a number of small flask-shaped perithecia.

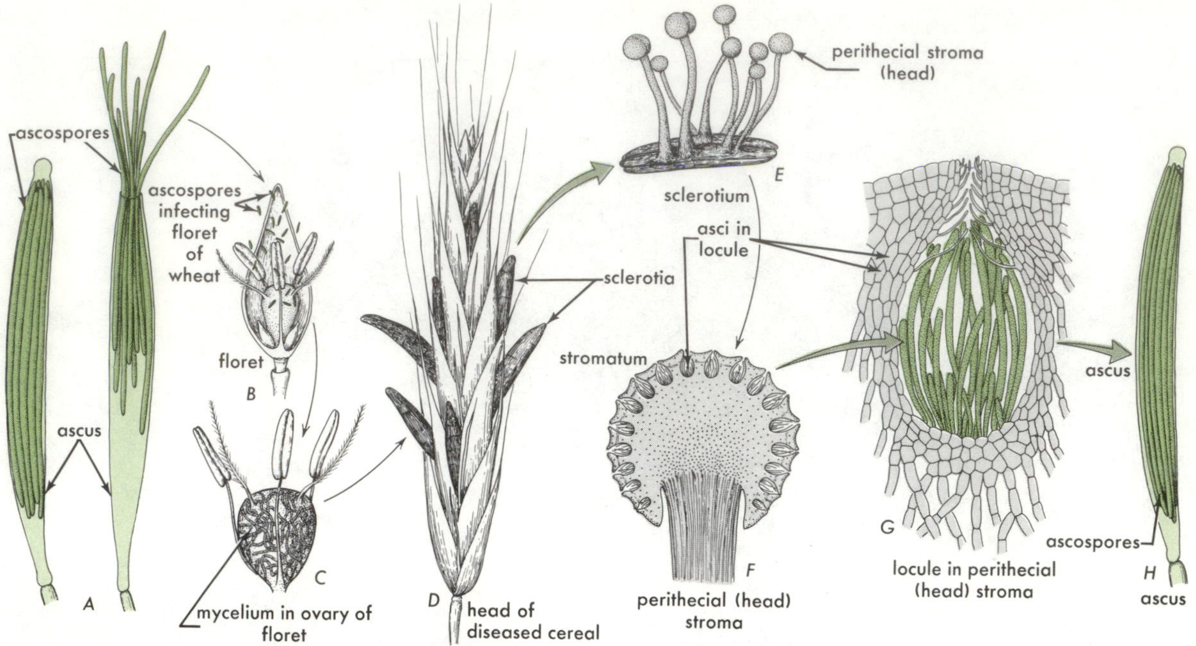

Figure 26.15
Life cycle of *Claviceps*. (*A, D, E, F,* and *H* from E. A. Gäuman, *Comparative Morphology of Fungi*. © 1928 by McGraw-Hill Book Company. Redrawn from G. M. Smith, 1955. *Cryptogamic Botany*, Vol. I, *Algae and Fungi*. © 1955 McGraw-Hill Book Company. Reprinted with permission of the publisher.)

6. Meiosis takes place; asci and ascospores are formed within the perithecia.

7. As the asci mature, they extend out of the perithecial opening. Pressure develops in the asci, shooting the ascospores into the air.

Monilinia and *Peziza*

The two orders to which these genera belong, the Helotiales and Pezizales, are commonly called the cup fungi, because the asci are borne in open, saucer-shaped fruiting bodies (**apothecia,** Figs. 26.8*E*, page 518, **26.16,** page 539). The majority of lichen fungi are cup fungi, but they are placed in a separate order, the Lecanorales, and we shall discuss them later in the chapter. The basic taxonomic difference between the cup fungi orders is whether or not the asci have a trapdoor opening at the tip (an operculum), for aid in spore discharge. Helotiales lacks such an opening, while Pezizales possesses it. The apothecia may be as much as 10 cm in diameter and they are often brightly colored.

Some, such as *Peziza,* are saprophytes. Others are economically important parasites, such as *Monilinia,* brown rot of stone fruits. *Morchella* (morel; Fig. 26.17), in the Pezizales, is a highly prized edible fungus. Its fruiting body is intricately sculpted into many small cups.

Life History of *Monilinia fructicola* (Fig. 26.18)

1. Ascospores formed during spring, or conidia formed on a mummy, infect blossoms. Conidia form on blighted blossoms.

2. Conidiospores produced in blossoms rapidly infect healthy fruits under favorable weather conditions.

3. An extensive mycelium develops within the fruit, causing it to rot (**Figs. 26.16*B,C*)**.

4. The rotted fruit dries, becoming a mummy. It may drop to the ground or remain on the tree.

5. In early spring, gametangia form in the mummied fruits that have fallen to the ground.

Figure 26.17
The common edible morel, *Morchella esculenta*.

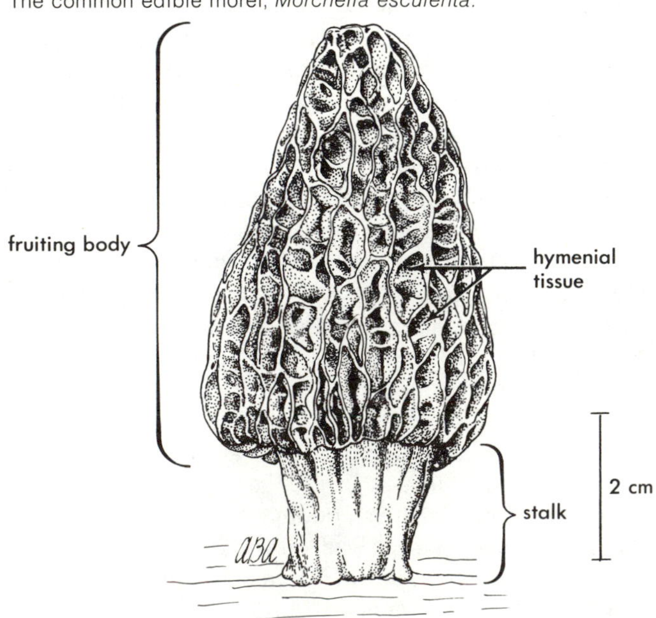

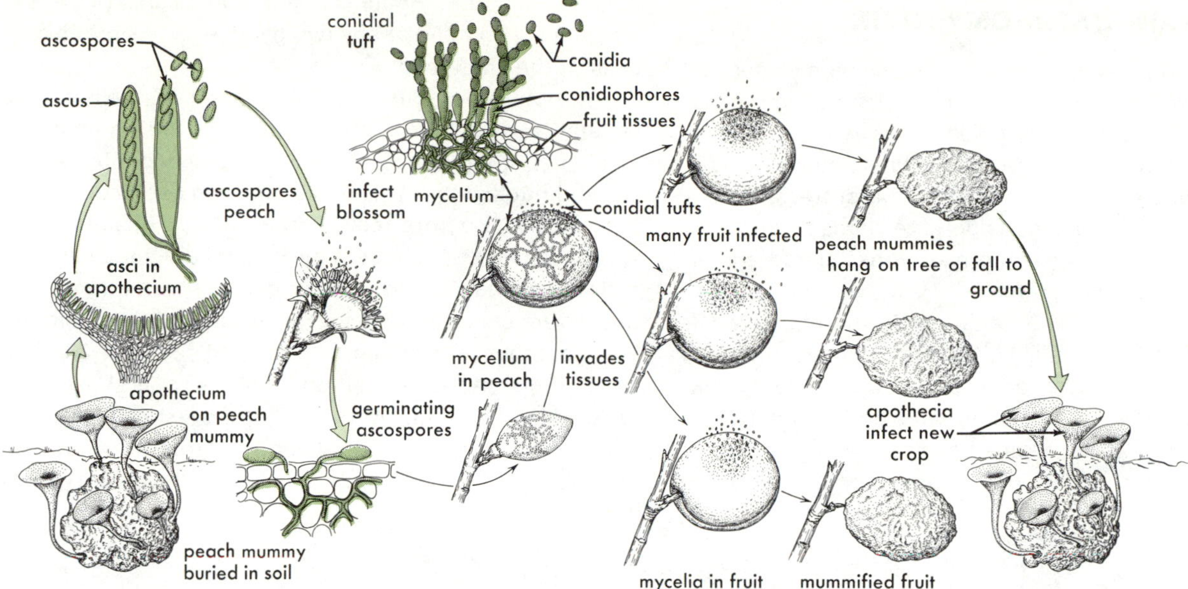

Figure 26.18
The life cycle of the brown rot fungus *Monilinia fructicola*. (Parts based on G. N. Agrios, *Plant Pathology*. © 1969 by Academic Press. Reprinted with permission of the publisher.)

6. Fertilization ensues.

7. Apothecia develop from ascogenous and vegetative hyphae.

8. The resulting asci shoot ascospores several inches into the air and reinfection occurs.

Tuber

These ascomycetes have totally enclosed, subterranean fruiting bodies. The ascocarps are essentially spherical, ranging in diameter from less than 1 to more than 10 cm, and some are highly prized by gourmets for the particular flavor that they impart to food. They are generally produced by mycorrhizal fungi, and so are found near the base of trees. Commercially valuable species are associated with European species of oak and beech. They are so valuable that they are hunted with trained dogs or pigs who scent the ascocarps (called **truffles** in the market) and lead their owner to them. Then the animal is restrained while the truffle is dug up. There have been attempts to cultivate them in Italy and France.

Significant Features of the Ascomycotina

General Characteristics

1. The mycelium is septate. The cells of the main vegetative hyphae have one to several haploid nuclei. The ascogenous hyphae have two haploid nuclei ($n + n$) in each cell.

2. Many are severe plant pathogens, causing diseases of fruit and nut trees and of grains.

3. Some saprophytic forms, such as the yeast, are of considerable economic importance.

4. Yeasts are simple Ascomycotina, consisting of a single cell.

Reproduction

1. Asexual
   a. Usually by means of conidiospores.
   b. By budding (in the yeasts).
2. Sexual
   a. Definite gametangia are not developed in the more primitive species; typical ascogonia and antheridia form in the more advanced types.
   b. Fertilization generally involves:
      (*1*) The union of two protoplasts without the fusion of the nuclei
      (*2*) Stimulation to growth of ascogenous hyphae ($n + n$) and of haploid vegetative hyphae.
      (*3*) Union of two haploid nuclei in ascus ($n + n \rightarrow 2n$).
   c. An ascocarp develops as a result of the growth of ascogenous and haploid vegetative hyphae.
      (*1*) The inner layer (hymenium) of the ascocarp gives rise to asci.
      (*2*) The outer layer (peridium) of the ascocarp is composed of haploid vegetative filaments.
   d. There are three types of ascocarps: (*1*) cleistothecia, (*2*) perithecia, and (*3*) apothecia.
   e. Meiosis occurs in the formation of the ascospores in the ascus. Eight ascospores are generally produced in each ascus.

SUBDIVISION BASIDIOMYCOTINA

Well-known representatives of the basidiomycetes include mushrooms, toadstools, bracket fungi, puffballs, and species causing such diseases as wheat rust and corn smut. Many of them are saprophytes, being particularly important in the decay of dead forest trees. Others are parasitic and cause considerable damage to forest and orchard trees, wheat, corn, onions, snapdragons, roses, and many other plants.

No specialized sex organs are formed; nearly all, however, reproduce sexually. Conjugation can occur (*a*) between ordinary hyphal cells, (*b*) between two special cells, or (*c*) between special spermlike bodies and receptive hyphae (Fig. 26.19). Some species are **heterothallic;** that is, only plus and minus strains conjugate. Other species are **homothallic,** conjugation taking place between any two cells of any two hyphae, or even between cells of the same hypha.

In both types, prior to conjugation the cells of the vegetative mycelium contain unassociated haploid nuclei; after conjugation the resulting cell contains two haploid nuclei ($n + n$). The two nuclei in each cell do not fuse until just before meiosis (Fig. 26.19). Fusion of the two nuclei and subsequent meiosis take place in special cells known as **basidia.** The meiospores formed by these basidia are called **basidiospores.** Generally, the basidium may be either (*a*) a single club-shaped cell (Fig. 26.19*A*), (*b*) a single short, filamentous cell, or (*c*) a short four-celled filament (Fig. 26.19*B*). The club-shaped basidia occur on special spore-bearing structures, as in mushrooms, bracket fungi, and puffballs. This type of

Figure 26.19
Generalized life cycles of Basidiomycotina showing fusion of protoplasts, fusion of nuclei, and meiosis (production of basidiospores). *A*, common mushroom. *B*, fungus causing wheat rust. *A*, redrawn from G. M. Smith, *Cryptogamic Botany*, Vol. 1. *Algae and Fungi*. © 1955 by McGraw-Hill Book Company; H. Kniep and A. H. R. Buller, *Researches on Fungi*. Vol. VII, Royal Society of Canada. University of Toronto Press; *B*, based on Kniep and Buller (see *A*) and R. F. Allen, *Phytopath*, **23,** 574. © 1933 by the American Phytopathological Society. Reprinted with permission of the publisher.)

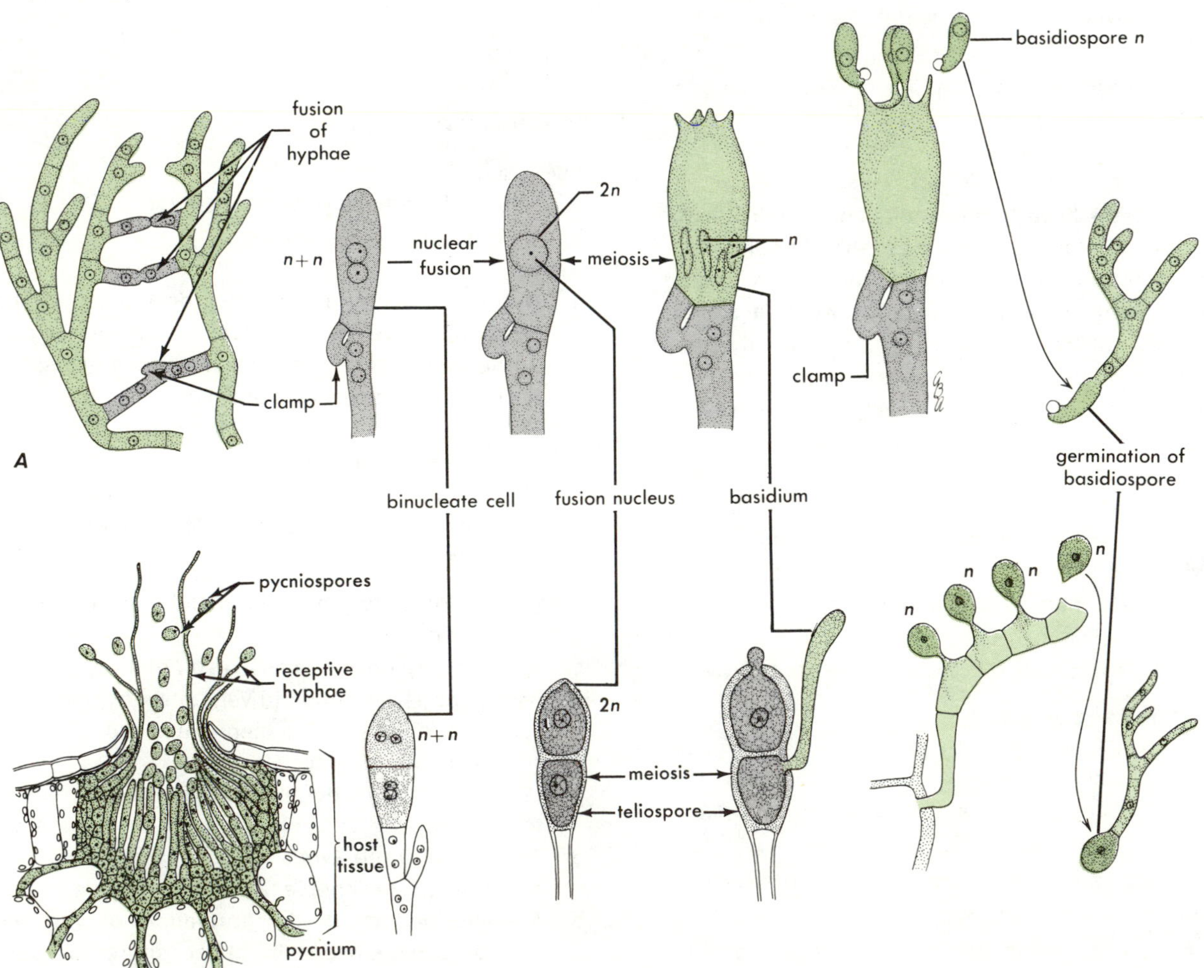

basidium characterizes the class Homobasidiomycetes. When the basidium is a single short, filamentous cell or a short four-celled filament, it is characteristic of the subclass Heterobasidiomycetes. In both types, the basidiospores are attached to the basidia by short stalks, the **sterigmata** (singular, **sterigma**).

Class Homobasidiomycetes

The mushrooms, bracket fungi, and puffballs are Homobasidiomycetes. Only a few are parasitic, but these few do extensive damage to forest trees, some orchard trees, and a small number of garden crops. Several Homobasidiomycetes are always found growing in association with certain species of trees. It is believed that some of them may form mycorrhizae. Certain Homobasidiomycetes obtain their carbohydrates from wood and may even cause extensive decay of timbers in mines and buildings.

Although both haploid and $n + n$ mycelia develop extensively, the $n + n$ phase is of special interest because it is dominant in the life cycle and it gives rise to the spore-bearing body, or **basidiocarp,** in which the basidia are developed. The hyphae form a tangled mass in the substrate or host and emerge to form a basidiocarp—the mushroom, puffball, or bracket fungus. As the basidiocarp grows, the ends of certain hyphae generally become aligned in a layer that will bear spores when mature. This is the **hymenium layer,** some cells of which gradually enlarge to become basidia.

The development of the basidiospore is shown in Fig. 26.20. The young basidium contains the $n + n$ nuclei, or the dikaryon (Fig. 26.20*A*). These two nuclei fuse to give the true diploid nucleus shown in Fig. 26.20*B*. Meiosis gives rise to the four haploid nuclei destined to pass into the basidiospores (Fig. 26.20*C*). When this stage is reached, the sterigmata have extended outward from the tip of the basidium and are ready to receive the basidiospore nuclei. The nuclei present in the basidium will pass through the sterigmata. The basidiospore of Fig. 26.20*D* is mature and ready for dispersal as the haploid nucleus has moved into it. Note that one nuclear division has been completed here.

There are several orders of Homobasidiomycetes. We shall consider mainly the order Agaricales, of which common mushrooms are members, and the order Polyporales, which contains the bracket fungi.

The Mushroom Basidiocarp

The basidiocarp of a typical mushroom such as the commercially grown *Agaricus campestris* (Fig. 26.21) consists of a short upright stalk or **stipe,** attached at its base to a mass of mycelium and expanding on top into a broad cap or **pileus.** Many mushrooms have a ring of tissue around the stipe rather close to the pileus. This ring occurs because when it is young, the pileus was attached to the stipe at this point. The underside of the pileus of *Agaricus* is formed by thin **gills** radiating outward from the stipe. These gills are lined with a hymenium or spore-bearing layer. If a young gill is sectioned, most basidiospores will still be attached to the basidia. Four basidiospores are attached to each basidium. Figure 26.21*B* shows a section through three gills. Note two stages of development of basidia and the loose tangle of hyphae forming the center of the gill. When a basidiospore is discharged, it is shot horizontally from its position on the basidium straight into the space between two gills. A basidiospore is somewhat like a toy rubber balloon, having a large volume for its weight. It is shot from the basidium with a force sufficient to carry it about midway between the two gills. It then falls straight downward. It has been estimated that some basidiocarps may discharge as many as a million spores a minute for several days.

If the cap of a mushroom is placed on a piece of paper and carefully protected from wind, the spores will fall straight downward and form a **spore print.** Spore prints are important in the identification of mushrooms, since they indicate the color of the spores and the pattern of the gills.

The spores of *Agaricus compestris* are brown, whereas those of other Agaricales may be white, black, red, yellow, or ochre. The spores of *Amanita muscaria,* the "fly-mushroom," are white. *Amanita muscaria* has other distinguishing features: When the basidiocarp is in the button stage, it is completely surrounded by a thin tissue known as the **universal veil** (Fig. 26.22*A*). As the basidiocarp expands, this veil is torn. Part of it remains attached to the cap, where ruptured fragments of it may be seen in the mature mushroom; the remaining part stays underground, forming a layer around the base of the stalk (Fig. 26.22*E*). The enlarged basal structure is referred to as the **volva** or "death cup." The fragments of white on the reddish yellow cap are remnants of the universal veil. The white gills and ring are visible.

Amanita phalloides is one of the most deadly poisonous of all mushrooms. As a general precaution, one should avoid all white-spored mushrooms with a death cup and a ring on the stipe.

Some species of the Agaricales are hallucinogenic, inducing euphoria, visions, and intoxication. *Amanita muscaria,* found in deciduous and coniferous forests of Europe and Asia, is an hallucinogenic mushroom. The mushrooms are gathered, dried, then chewed. It is possible that the Indian sacred plant soma, venerated in poetry and hymns that date back 3000 years is *A. muscaria.* Mushroom worship goes back equally far in the new world. Aztecs and Mayans practiced religious cults that used mushrooms of the genus *Psilocybe* (and others) to induce visions and prophecy. Extreme care must be taken in chewing hallucinogenic mushrooms, because they are toxic and can lead to death or illness if consumed in an inappropriate quantity. Some American Indians still use mushrooms in religious rites.

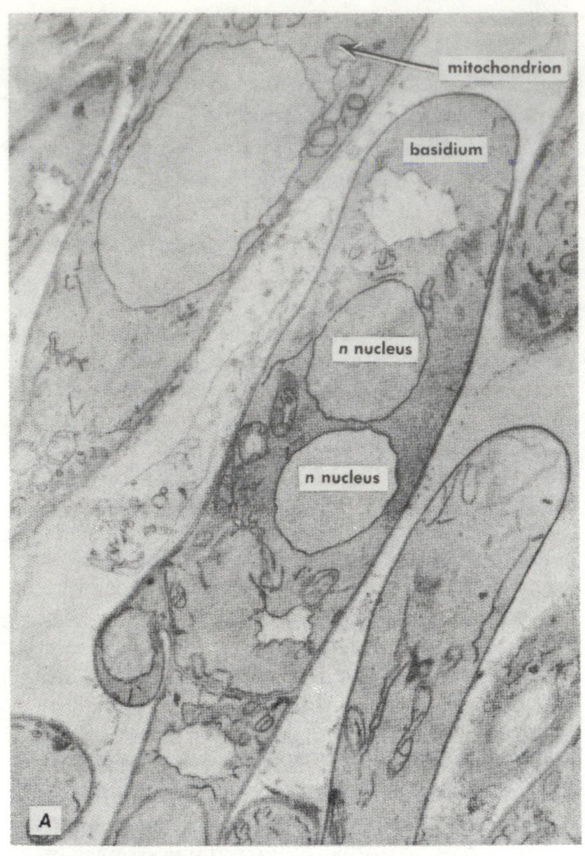

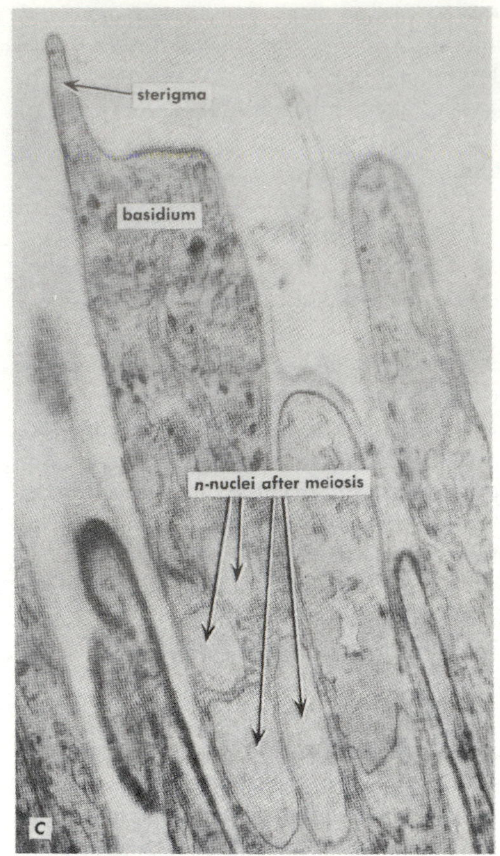

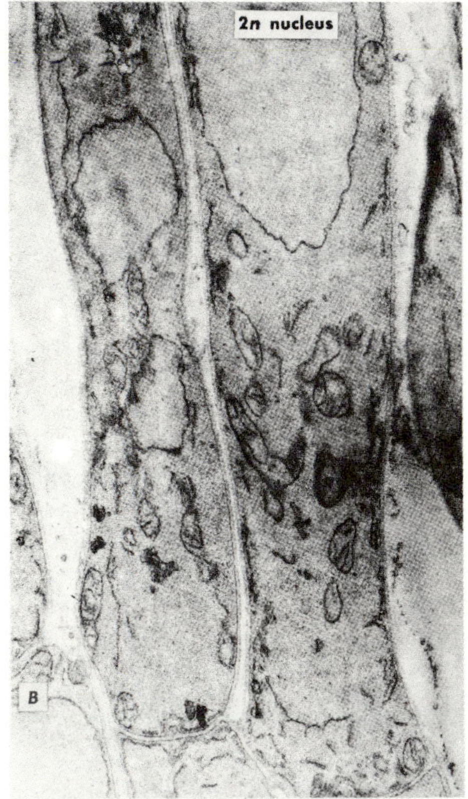

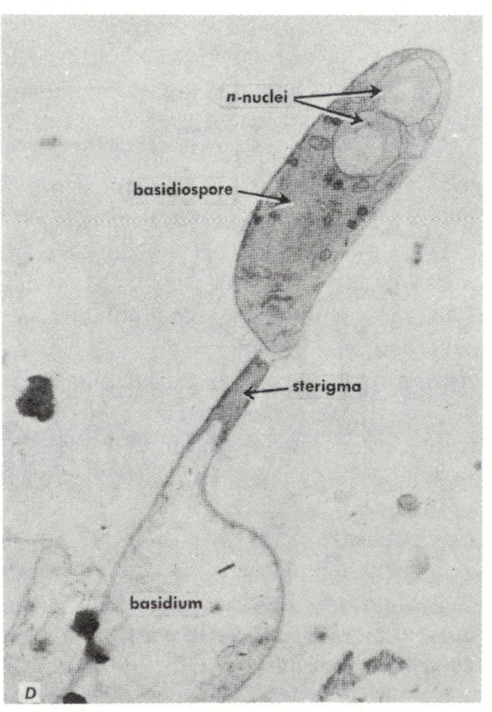

Figure 26.20
Basidiospore formation in *Schizophyllum commune*. Above: *A*, young basidium with $n + n$ nuclei, ×3000. *B*, basidium with $2n$ nucleus, ×5000. *C*, four n-nuclei after meiosis, ×3000. Note development of sterigma. *D*, basidiospore attached to sterigma; the n-nucleus is completing its first division. (Courtesy of K. Wells.)

Figure 26.21
Structure of the basidiocarp of a common mushroom. *A, Agaricus campestris,* the common edible mushroom, ×1. Note the gills. *B,* section showing three gills of a mushroom. *C,* section of a gill cut parallel with plane of basidia. It shows the production of basidiospores. (*B* and *C* redrawn from A. H. R. Buller. *Researches on Fungi.* © 1905-1950 by Longman Green. Reprinted with permission of Longman Green, Ltd.)

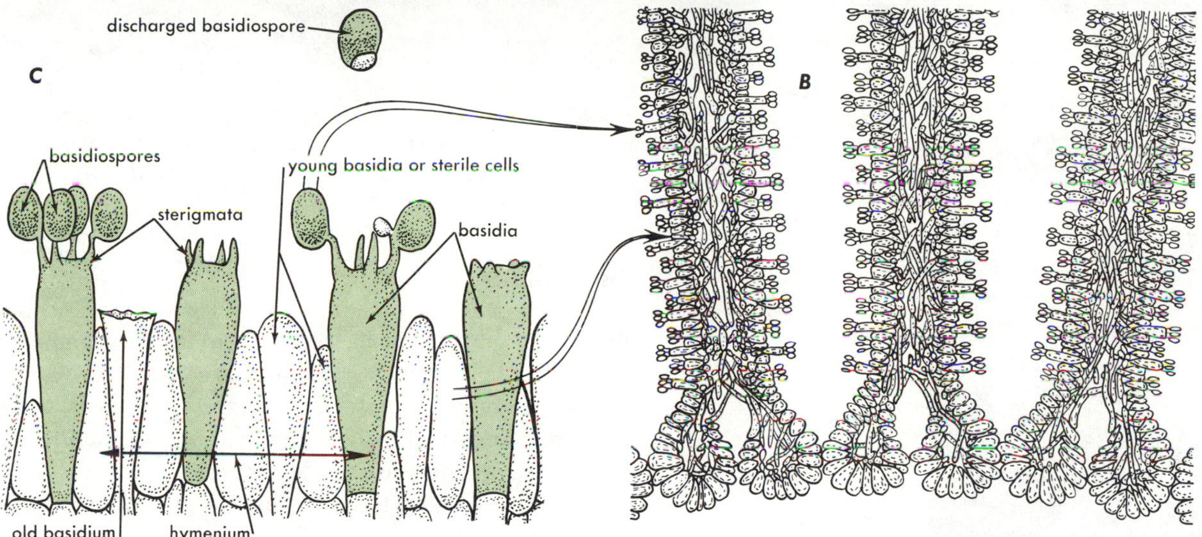

Basidiocarps of Bracket Fungi

These fungi lack gills. The lower surface of the cap or bracket has many small pores that extend upward to form small tubules (Fig. 26.23*A*). The hymenium lines the tubules. For this reason, bracket fungi are also called *pore fungi.*

Some bracket fungi cause serious damage to forest trees and lumber in mines and wooden buildings. During World War II, much green lumber had to be used, and destruction by fungi of improperly cured lumber resulted in much loss. One of the best-known fungi in this group is *Ganoderma applanatum* (Fig. 26.23*B*). It causes a disease of forest trees known as *white-mottled rot,* to which many hardwoods, such as poplar, birch, maple, basswood, oak, and elm are susceptible. True firs, Douglas fir, spruce, and hemlock may also suffer from this disease.

The basidiocarp of *Ganoderma* is a very elegant shelf-shaped bracket or **conk.** The conks are perennial and grow to a large size. They are gray on top, and the lower surface has millions of pores lined with a creamy-white hymenium.

Most of the bracket fungi of the forest are facultative parasites. They normally live in the dead heartwood, to which they gain entrance by wounds. They are, however, able to invade the living cells of sapwood, which they may destroy. Some trees, such as cedars and redwoods, are more rot-resistant than others, and this may be due to the accumulation of certain metabolic by-products that inhibit fungal growth.

Class Heterobasidiomycetes

Many members of the class are parasites on vascular plants. Like the Homobasidiomycetes, they possess, as far as is known, both haploid and $n + n$ phases, and reduction division takes place in the development of the

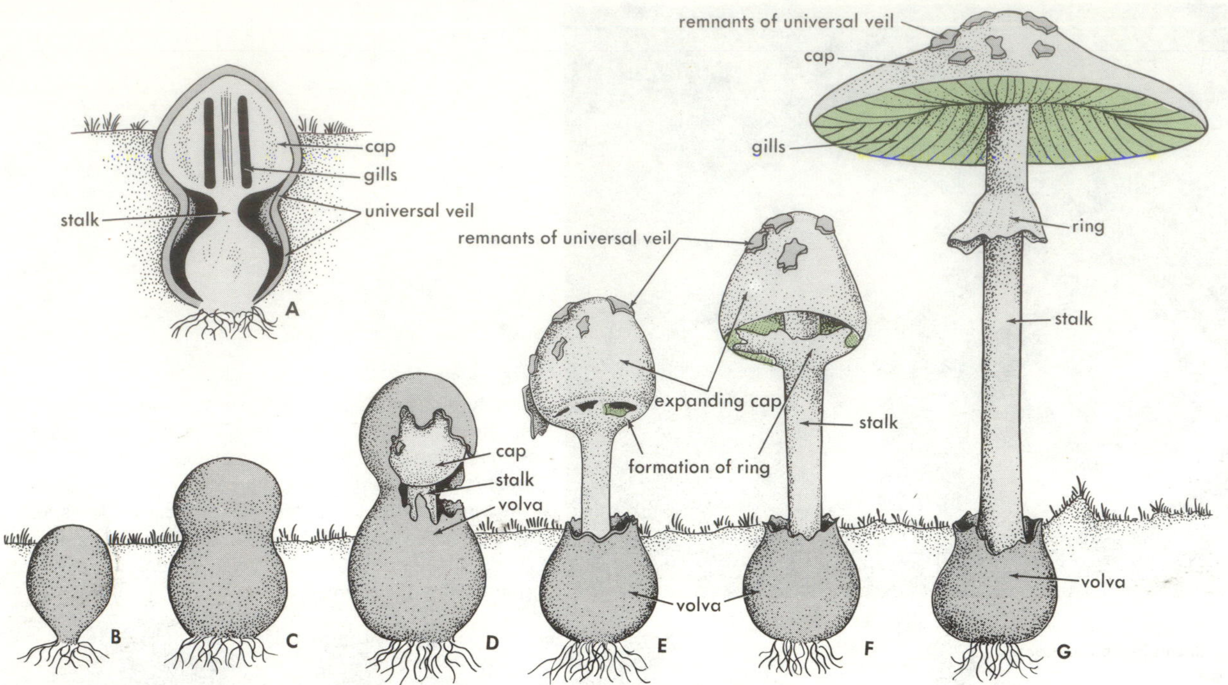

Figure 26.22
The basidiocarp of a common poisonous mushroom *(Amanita)*. *A* through *G*, developmental stages showing formation of volva and universal veil. *H*, mature mushroom; note remnants of veil on cap (*A* through *G*, redrawn from Gibson, *Edible Mushrooms and Toadstools*, Harper and Brothers, 1899.)

basidia or basidiospores. Asexual reproduction may occur in either the haploid or $n+n$ phase or in both. Conjugation may take place between two nonspecialized cells, as in the Homobasidiomycetes. In some forms, however, special cells conjugate or a spermlike cell may conjugate with a receptive hypha.

Some of the Heterobasidiomycetes may have two hosts; that is, separate phases of their life history are passed on different plants. Such forms are **heteroecious.** If only one host is required to complete their life history, they are said to be **autoecious.**

There are three orders in this class. The Tremellales have waxy, gelatinous, or membranous basidiocarps, and this order includes the "jelly fungi." We shall discuss only the two other, more economically important orders, Uredinales (rusts) and Ustilaginales (smuts).

Uredinales

The Uredinales, or rust fungi, are of worldwide distribution. They are known to infect a very large number of higher plants, and all are obligate parasites. Black or red spores develop in pustules on the leaves of many grasses, roses, and mallows, and they (particularly the red spores) suggested the name of **rust.**

There are many types of rust fungi and they show marked variation in their life histories. In nearly all of them, however, basidiospores are produced on short four-celled basidia that usually develop from an overwintering spore, known as a **teliospore.** Fusion of the two haploid nuclei and reduction division occur during the development of the basidia and basidiospores. The details of conjugation are obscure in many species.

Common wheat rust, caused by *Puccinia graminis,* is of great economic importance, and its life history is well known. It produces all the different types of spores common to the order. We shall follow its life history in detail. This life cycle was first described by Anton De Bary in the middle of the nineteenth century. De Bary is often called the "father" of plant pathology.

Life History of *Puccinia graminis* (Figs. 26.19*B*, 26.24, **26.25, 26.26,** pages 540–541, 26.27.) In the early spring, the overwintering teliospores germinate on the soil or stems of wheat and, in fact, any place where there is moisture.

Teliospores of *Puccinia graminis* are two-celled spores. A short four-celled basidium develops from each

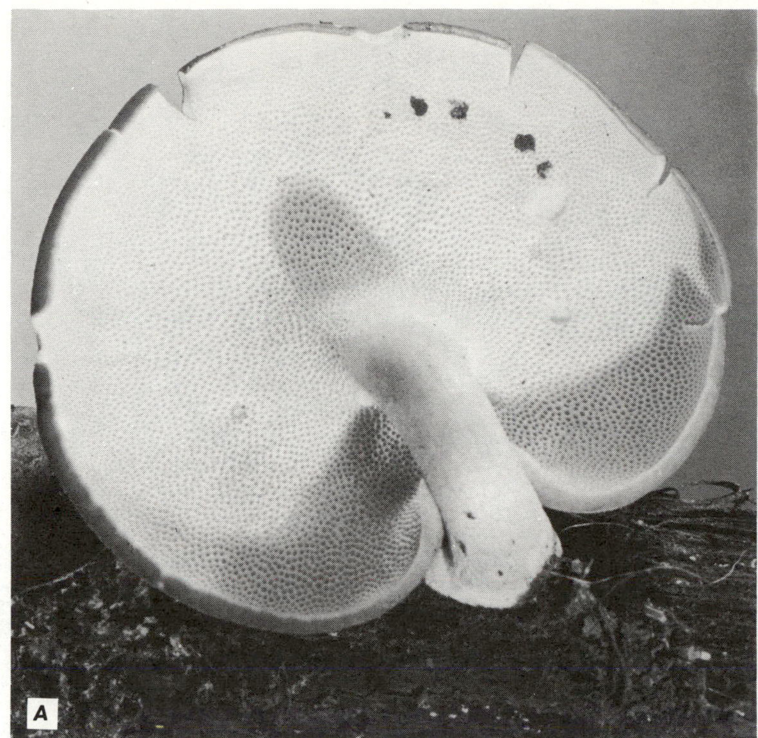

Figure 26.23

Basidiocarps of Polyporales. *A,* the underside of the pore fungus *Polyporus brumalis,* $\times 1/3$. *B,* basidiocarp of *Ganoderma applantatum,* $\times 1/3$ (*A,* courtesy of Boche. *B,* courtesy of Brownell.)

cell; only one is shown in Fig. 26.19*B.* A single haploid basidiospore is formed from each cell, so that four basidiospores eventually develop from each cell of the teliospore. Two of these four basidiospores are plus strain (+), two are minus strain (−).

The basidiospores are carried by air currents and can infect only the young leaves, fruit, or twigs of common barberry bushes. They germinate chiefly on barberry leaf surfaces and produce a germ tube, which penetrates the epidermis (Figs. 26.24, **26.25**). Soon a mycelium develops in the tissues of the host. Then, after a time, pustules, called **spermagonia,** appear on the upper surface of leaves. In section a spermagonium is pear-shaped and has a small pore opening through the upper epidermis to the exterior of the leaf. Some hyphae extend upward through the pore; others, lining the interior of the spermagonium, produce large numbers of spermlike cells, called **spermatia.** These are forced out of the spermagonia through the pores.

Puccina graminis is heterothallic; both plus and minus spermagonia and spermatia appear on barberry leaves. These develop from plus or minus basidiospores, both of which are haploid. Spermatia are transferred by wind or insects to adjacent spermagonia, where they come in contact with the **receptive hyphae,** which are essentially female gametangia. A plus (+) spermatium fuses with a cell of a minus (−) receptive hypha, and vice versa. An $n + n$ diploid mycelium results; its cells each contain two haploid nuclei.

Shortly after fusion, a mass of $n + n$ hyphae appears in the spongy parenchyma of the barberry leaf close to the lower epidermis. Here, a second series of pustules called **aecia** are formed (Figs. 26.24, **26.25**). The aecia face downward, and the upper side toward the leaf mesophyll becomes lined with $n + n$ hyphae. The growth of the aecia eventually brings about the rupture of the lower epidermis, and binucleate **aeciospores,** cut off from the ends of the numerous hyphae lining the base of the aecium, are exposed to the atmosphere. The aeciospores thus released are disseminated by wind and they can infect only wheat plants. They remain viable for only a few weeks.

The hyphae of germinating aeciospores gain entrance to wheat plants by growing through the stomata. A delicate mycelium develops and the hyphae form haustoria that draw nourishment from surrounding cells of the wheat plant but do not kill them. About 10 days after infection, red binucleate spores, **uredospores,** are produced by the mycelium. The epidermis of the wheat plant is ruptured by the mass of spores, forming open **uredia** (Figs. 26.24, ***26.25B,C,*** **26.26**). The uredospores thus produced are carried by wind currents to other wheat plants. They are capable of remaining viable for many months, provided that they are not exposed to freezing temperatures. The fungus is spread throughout a grain field during the growing season by the red uredospores. The progress of the disease throughout a wheat-growing area may be so rapid as to attain epidemic proportions.

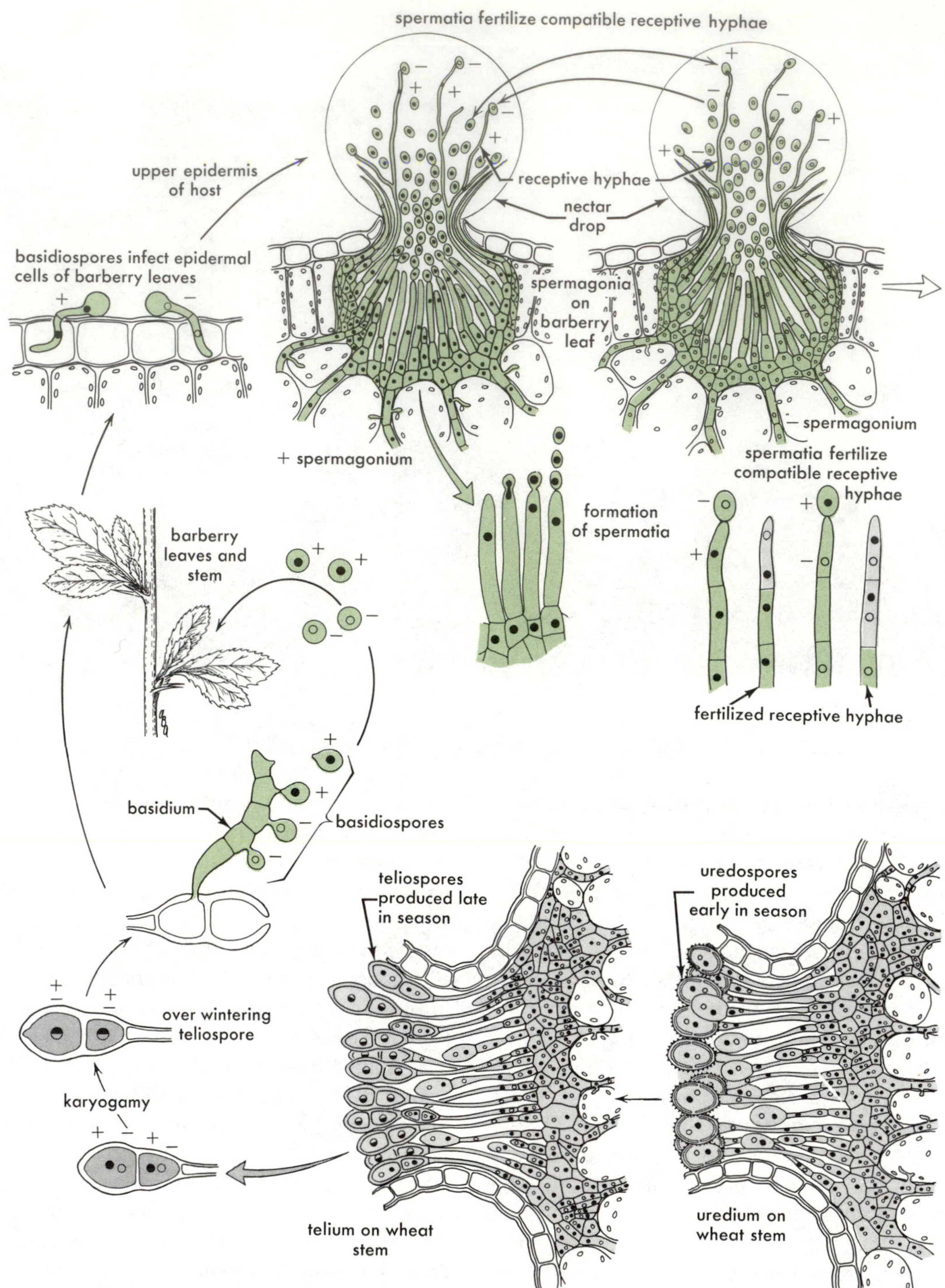

Figure 26.24
The life cycle of *Puccinia graminis*. (Portions based on A. H. R. Buller, *Researches on Fungi,* Vol. VII, Roy. Soc. Canada. © 1950 by University of Toronto Press; R. F. Allen, *Phytopath.* **23,** 572. © 1933 by the American Phytological Society; and E. A. Bessey, *Morphology and Taxonomy of Fungi.* © 1950 by the Blakeston Company. Reprinted with the permission of the publishers.

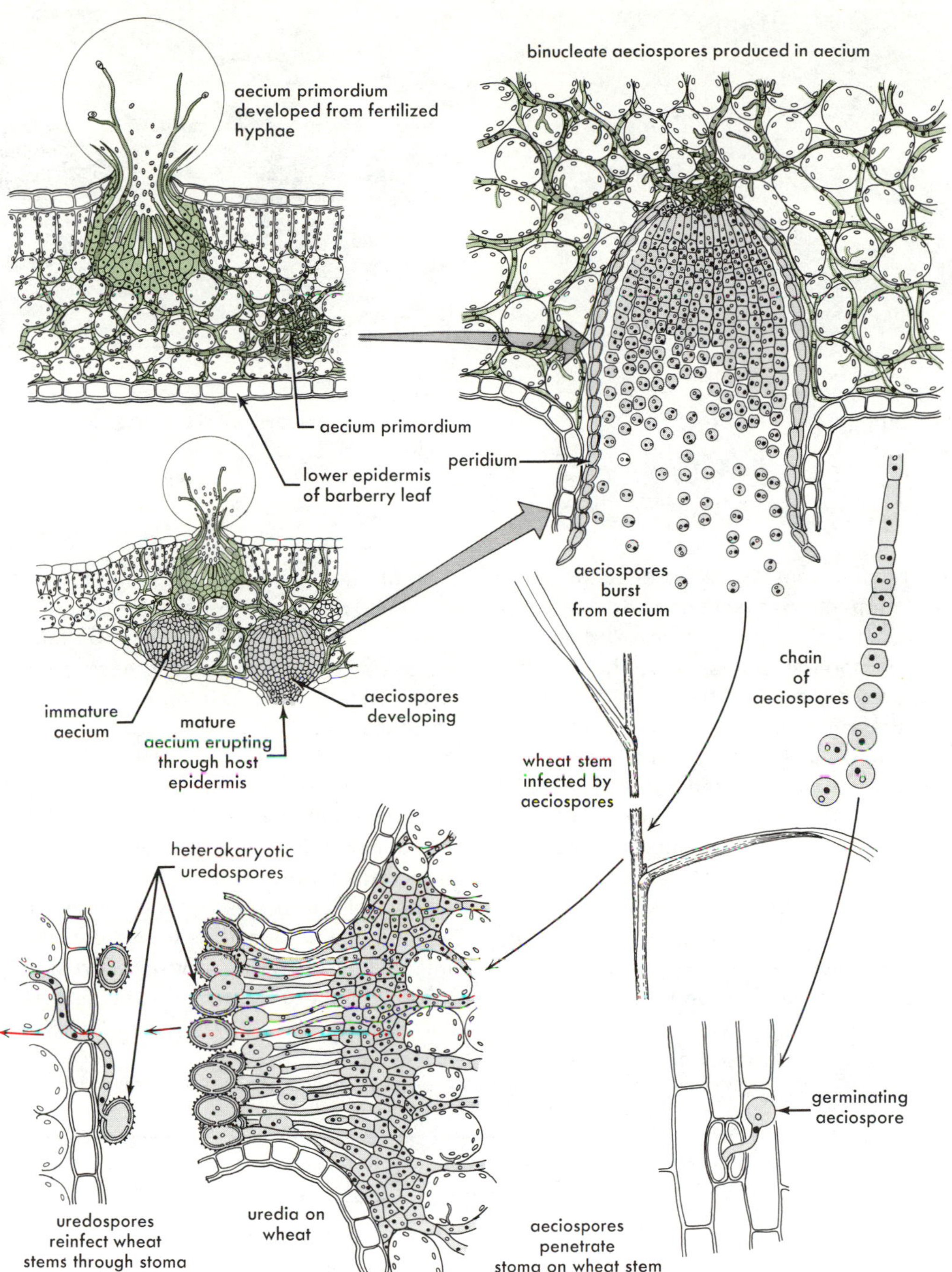

aecium primordium
developed from fertilized
hyphae
binucleate aeciospores produced in aecium
aecium primordium
lower epidermis
of barberry leaf
peridium
aeciospores
burst
from aecium
chain
of
aeciospores
immature
aecium
mature
aecium erupting
through host
epidermis
aeciospores
developing
wheat stem
infected by
aeciospores
heterokaryotic
uredospores
germinating
aeciospore
uredospores
reinfect wheat
stems through stoma
uredia on
wheat
aeciospores
penetrate
stoma on wheat stem

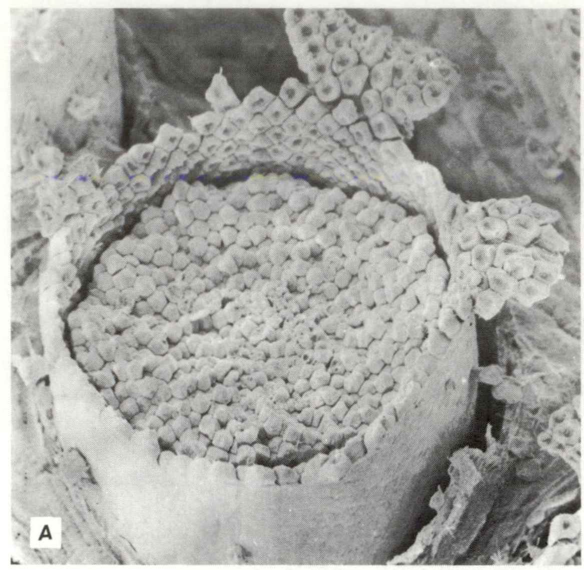

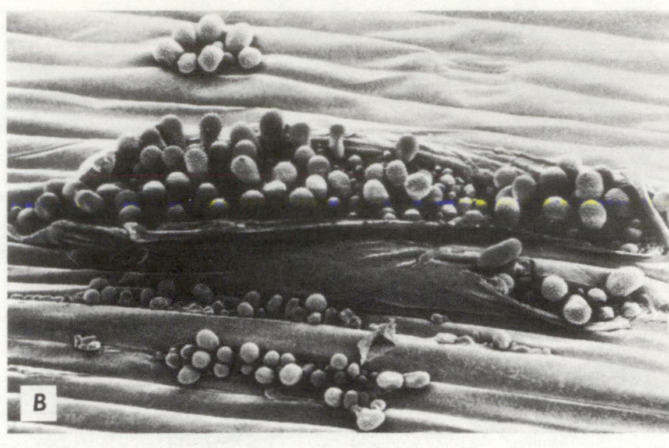

Figure 26.27
A, stereoscan view of aecium with aeciospores, ×250. *B,* stereoscan view of a uredial sorus on wheat, ×500. (*B,* courtesy of E. Butler.)

As the wheat begins to mature, the production of red uredospores by the infecting mycelium gives way to the production by the same mycelium of heavy-walled two-celled overwintering spores– **teliospores.** The pustules producing teliospores are called **telia.**

Each cell of the teliospore is, at the outset, binucleate; the two haploid nuclei, one (+) and one (−), are still present and distinct. Early in winter, the two haploid nuclei fuse, the true diploid condition resulting. The teliospores overwinter in the $2n$ condition. If they have been exposed to freezing temperatures, meiosis will occur the next spring. In the rusts, a short basidium grows from each cell of the teliospore, the four haploid nuclei resulting from meiosis migrate into it, and cell walls develop. Basidiospores now form, one from each cell of the basidium, thus starting the cycle again.

Puccinia graminis is heteroecious; it requires two hosts to complete its life cycle. Certain other rusts are autoecious, needing but one host to complete their life cycle. Not all rusts develop all the spore types found in *Puccinia graminis,* nor do they all have an identical life history.

Wheat rust may be controlled in cold climates by eradicating all barberry bushes (the alternate host) near wheat fields. Barberry will not be present for new infection in the spring, and uredospores will be killed by the winter cold; thus the rust will die. In regions with warm winters, the uredospores may overwinter, wheat may be infected in the spring, and wheat rust may therefore be more difficult to control. Figure 26.28 shows how over-

Figure 26.28
Northward progression of uredospores of wheat rust, showing date of first infection for one particular year. (Courtesy of U.S. Department of Agriculture.)

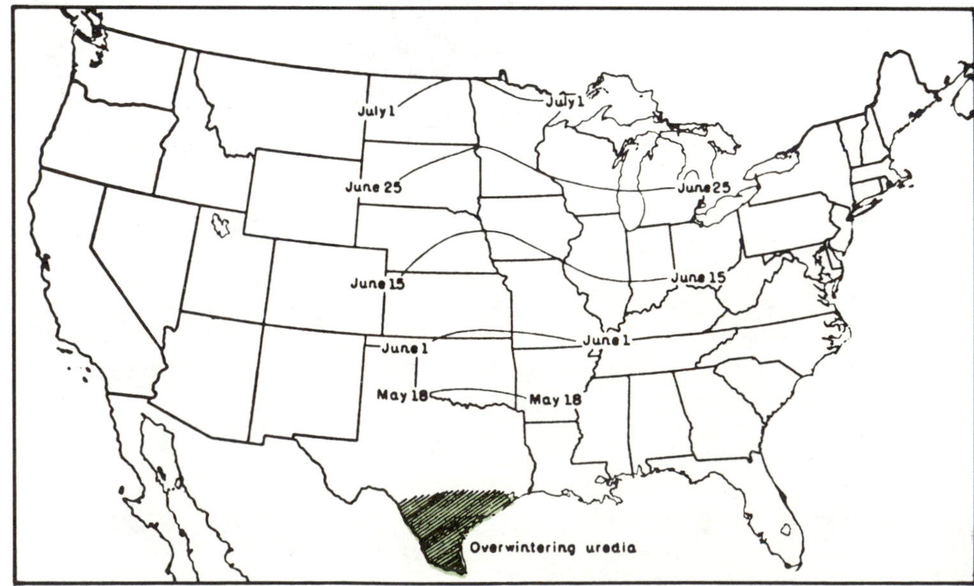

wintering uredospores in Southern Texas infect wheat the next spring in a progressive wave moving north, thanks to prevailing winds from the south.

Different varieties of *Puccinia graminis* are able to attack other grains. Variety *avenae* attacks oats; variety *tritici* attacks wheat. Breeding work with wheat, however, has shown that many biological strains exist even within the variety *tritici*. Indeed, new rust strains seem to appear from time to time, and so it becomes necessary to continue the breeding of wheat varieties that will be relatively resistant to them. More than 300 different strains of wheat rust are known.

Significant Steps in the Life History of *Puccinia graminis*

1. Two hosts, wheat and barberry, are necessary to complete the life history. In other words, *Puccinia graminis* is heteroecious.

2. Uredospores (one-celled, red) and teliospores (two-celled, black) are produced on the wheat plant. Uredospores can infect other wheat plants; they are killed by vigorous winters. Teliospores cannot infect any plant; they are overwintering spores, germinating in the spring on the soil or wherever else there is moisture.

3. A short four-celled basidium develops from each of the two cells of a teliospore. Meiosis takes place in this development. One basidiospore (haploid) forms from each of the four cells of the basidium. Basidiospores, either plus or minus, infect young tissues of the barberry.

4. Plus or minus spermagonia (both kinds, haploid) develop in the upper portion of the barberry leaf. The spermagonia produce spermatia and receptive hyphae.

5. Fertilization results from union of a spermatium with a receptive hypha of the opposite strain.

6. The resulting $n + n$ mycelium forms an aecium that releases binucleate aeciospores from openings in the lower surface of the leaf.

7. Aeciospores infect young wheat plants, producing an $n + n$ mycelium that eventually produces uredospores and teliospores.

Ustilaginales

All members of the Ustilaginales are plant parasites causing a group of diseases known generally as the **smuts.** Two families are generally recognized: Ustilaginaceae and Tilletiaceae. The life histories of the fungi causing cereal smuts vary, particularly in the manner of infection of the hosts: there may be **local infection, blossom infection,** or **seedling infection.** We shall discuss several genera under these headings.

Local Infection. *Ustilago maydis*, a member of the ustilaginaceae, causes common corn smut (Fig. 26.29*A*). Infection by spores of U. *maydis* is said to be **local,** meaning that infections remain localized, producing pustules or large tumors. Perhaps the most noticeable tumors are those that occur in the corn ear. The kernels become much enlarged due to the development within them of an extensive mycelium that eventually gives rise to a mass of black, heavy-walled, overwintering spores. Since haploid nuclei fuse and meiosis occurs within them, they are teliospores. These spores may lie dormant in the soil, on old corn stalks, or in manure.

A four-celled basidium develops when the teliospore germinates. Two basidiospores of one mating type, and two of another, are usually found on each basidium (Fig. 26.30*A*). The basidiospores may infect corn plants directly or they may multiply by budding. The budding cells are able to bring about infection. At any rate, a small haploid mycelium develops soon after infection (Fig. 26.30*B*). Mycelia of opposite mating types develop closely together, and conjugation occurs by the fusion of cells from small haploid mycelia of opposite strains. After conjugation, the $n + n$ mycelium, depending upon its location in the corn plant, may form a small pustule or develop into a tumor as large as a baseball. Most of the hyphal cells in the pustule or tumor change into teliospores.

Blossom Infection. Spores of *Ustilago tritici* and *Ustilago avenae* can infect *only the pistil* of wheat and oat flowers, respectively. They are known as **blossom-infecting smuts,** and the disease they produce is called the loose smut of wheat or oats. Teliospores mature in heads of diseased plants at the time healthy plants are in blossom (Fig. 26.31). The young pistil is directly invaded and a small mycelium eventually develops in the embryo. The infection has no injurious effect on the developing grain, which matures normally. The mycelium within the embryo remains small and dormant, thus carrying the fungus over seasons unfavorable to growth.

When the grain germinates, the mycelium within resumes activity. It grows best in or near meristematic tissue and, thus, keeps pace with the development of the wheat or barley plant. The presence of the mycelium close to the meristematic tissue has an accelerating influence on the growth of the host, which matures rapidly, producing flower heads. The mycelium invades the developing ovary, eventually replacing the host tissue completely. The entire head of grain becomes a black, loose mass of teliospores (Fig. 26.29*B*) and, when the spores blow away, nothing is left but the axis of the spike.

Meanwhile, healthy plants have been growing more slowly and developing normal florets. Teliospores, and florets ready for pollination, occur at the same time. Teliospores are carried by wind to the healthy florets, where the ovaries are eventually invaded.

Seedling Infection. *Tilletia tritici*, belonging to the family Tilletiaceae, causes **bunt** or **stinking smut** of wheat. Only young seedlings can be infected. The smuts of this family are consequently referred to as the **seedling-infecting smuts.**

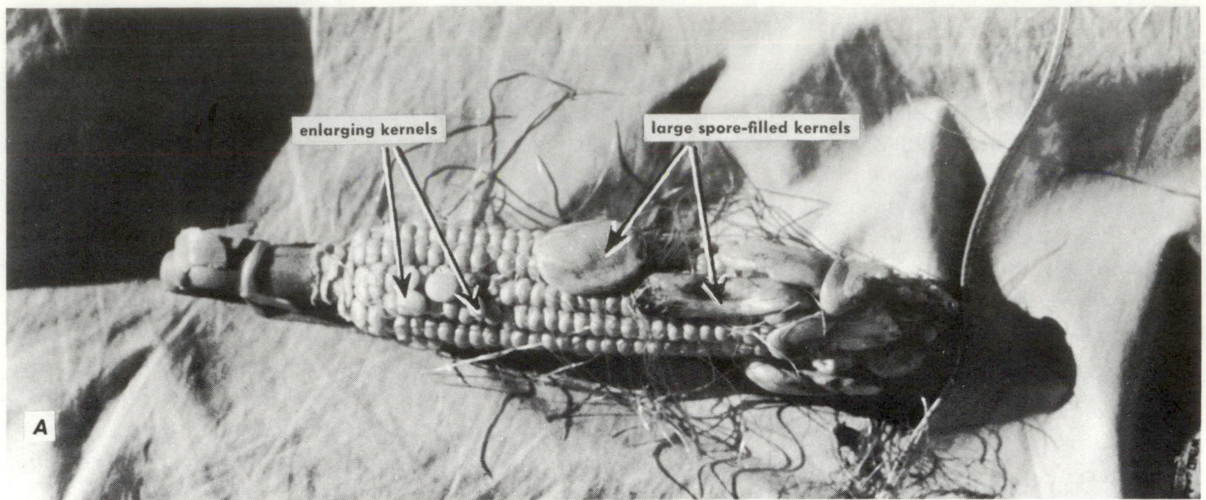

Figure 26.29
Smut *(Ustilago)*. *A*, ear of corn with smutted kernels. *B*, healthy (left) and smutted (right) panicles of oats.

As in the blossom-infecting smuts, masses of smut spores (teliospores) replace the mature grain of wheat (Fig. 26.32). In contrast to loose smuts, however, the grain coats remain intact, and infected grains resemble normal ones. At harvest time, the infected grains are broken and the teliospores are thoroughly mixed with the healthy grains, to which they firmly adhere. As the name stinking smut indicates, the teliospores have a very unpleasant odor. Grain mixed with them is unfit for flour or animal feed. The presence of large numbers of smut spores as dust during threshing greatly increases the fire hazard; explosions and serious fires were not infrequent when heavily smutted grain was threshed.

The teliospores carry the fungus over the season unfavorable for growth. Adhering to the grain, they are sown with it. Meiosis occurs in the teliospore, and frequently several mitotic divisions follow, producing several haploid nuclei. Upon germination of the teliospore, a short basidium is formed that gives rise to numerous basidiospores. The basidiospores conjugate by means of conjugation tubes, either while still attached to the haploid mycelium or after falling away. After conjugation, a short $n + n$ mycelium is formed that produces conidia. Seedlings may be infected by these conidia, or a second short mycelium may form that produces another crop of conidia. The young mycelium grows best in, or adjacent to, meristematic tissue. It keeps pace with the developing wheat plant. When the grain starts to head out, the mycelium enters the young ovaries of developing flowers. Hyphal filaments may eventually replace all normal cells of the grain, except the coats. Teliospores are formed by these hyphal cells, and the cycle is repeated.

Significant Features of the Basidiomycotina

A. Vegetative characteristics.
   1. Hyphae septate; extensive haploid mycelium; extensive $n + n$ mycelium, two haploid nuclei in each cell.
   2. Homobasidiomycetes: mostly saprophytic or mycorrhizal forms, a few species parasitic on trees.
   3. Heterobasidiomycetes: many parasitic on plants; some, such as the heteroecious rusts with a complicated life cycle.

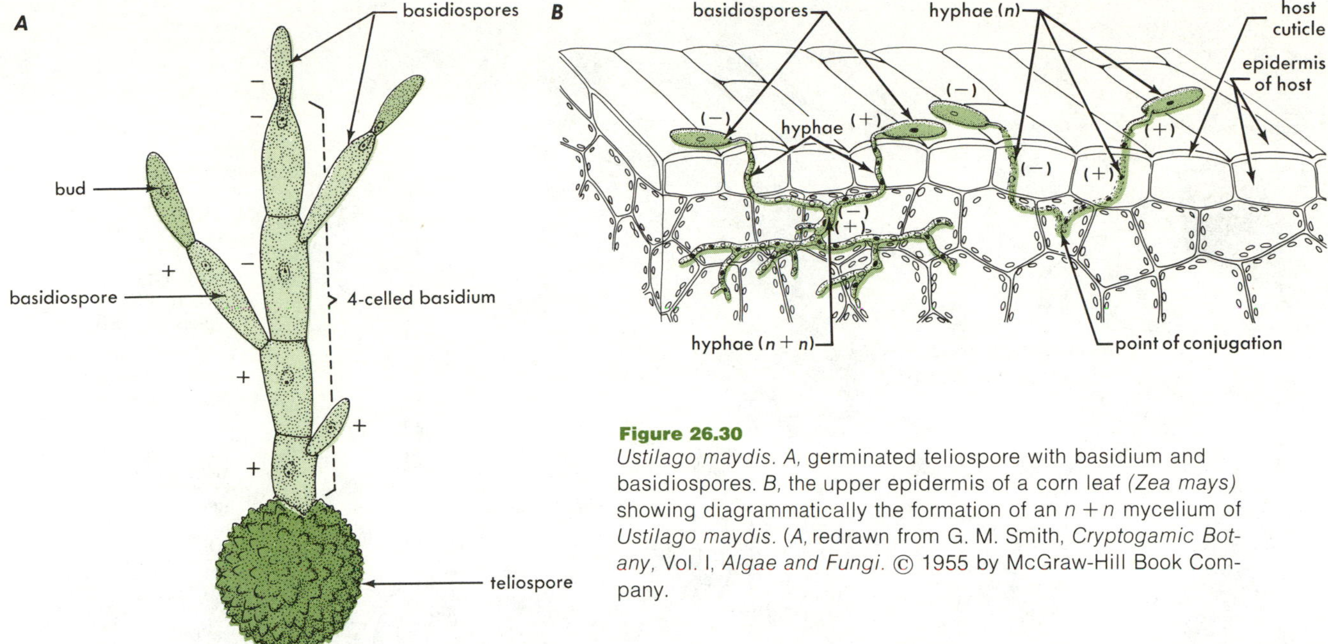

Figure 26.30
Ustilago maydis. A, germinated teliospore with basidium and basidiospores. *B,* the upper epidermis of a corn leaf *(Zea mays)* showing diagrammatically the formation of an $n + n$ mycelium of *Ustilago maydis.* (*A,* redrawn from G. M. Smith, *Cryptogamic Botany,* Vol. I, *Algae and Fungi.* © 1955 by McGraw-Hill Book Company.

Figure 26.31
Life cycle of *Ustilago tritici,* a blossom-infecting smut. (Parts based on G. N. Agrios, *Plant Pathology.* © 1969 by Academic Press. Reprinted with permission of the publisher.)

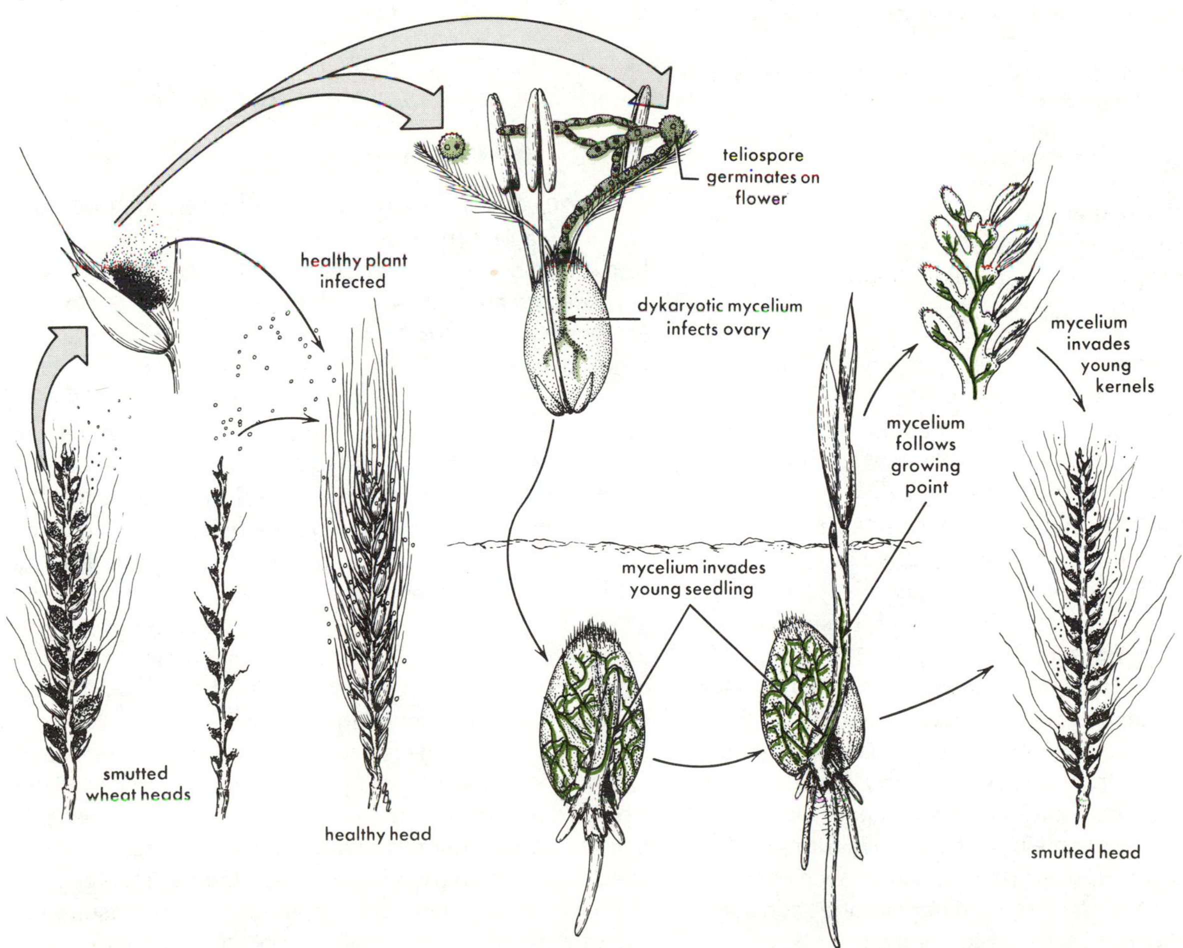

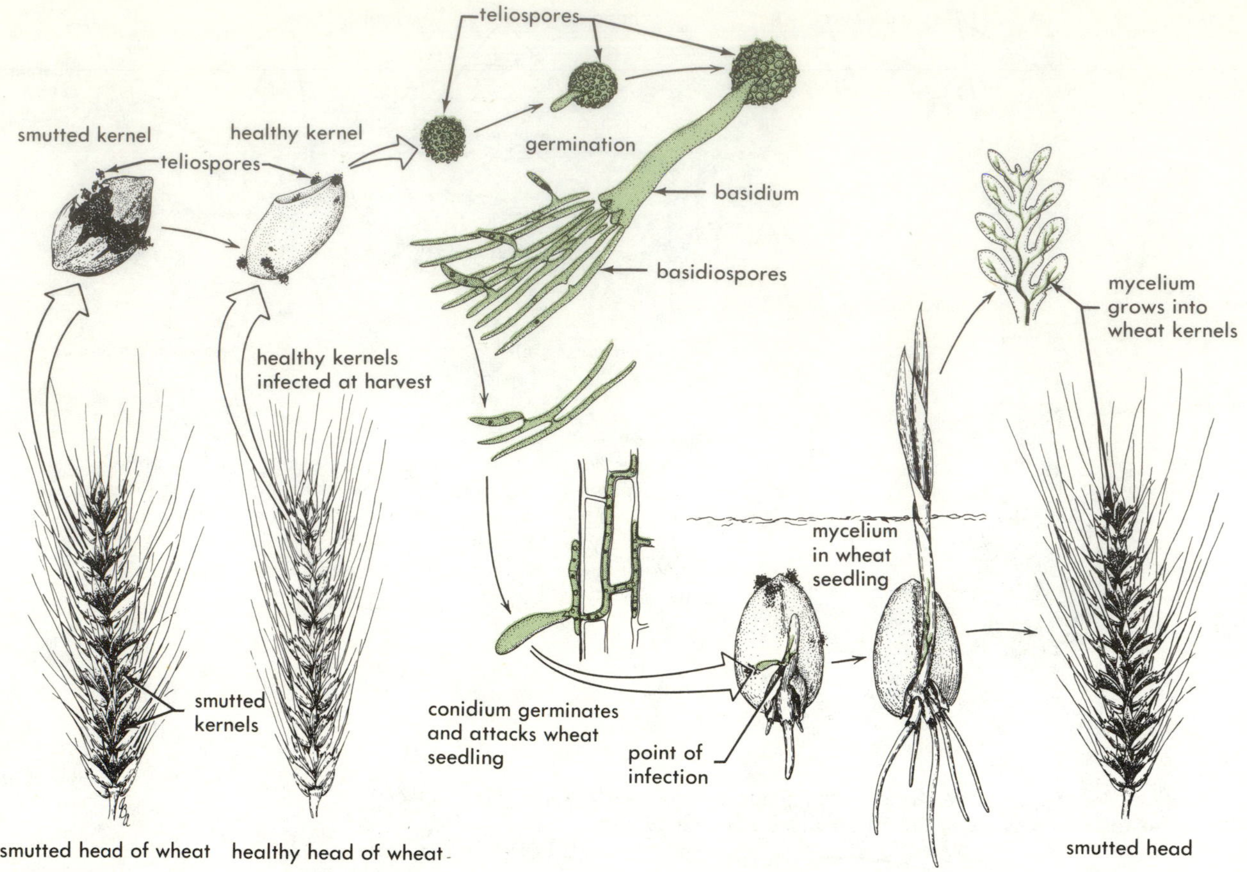

Figure 26.32
Life cycle of *Tilletia tritici*, a seeding-infecting smut. (Parts based on G. N. Angrios, *Plant Pathology*. © 1969 by Academic Press. Reprinted with permission of the publisher.)

B. Reproductive characteristics.

1. Asexual: asexual spores formed by a few species of Homobasidiomycetes; various sorts formed by different species of Heterobasidiomycetes; uredospores and aeciospores formed by the rusts.

2. Sexual: no motile sex cells or gametangia.
   (*a*) Homobasidiomycetes: (*1*) union of two hyphal cells containing haploid nuclei; (*2*) $n + n$ mycelium with two haploid nuclei in each cell developed from the union of the two haploid hyphal cells; (*3*) basidiocarp (mushroom, puffballs, bracket fungi) produced by the $n + n$ mycelium; (*4*) hymenium consisting of club-shaped basidia developed on the basidiocarp; (*5*) two haploid nuclei fuse in each basidium; (*6*) meiosis results in four basidiospores.
   (*b*) Heterobasidiomycetes: (*1*) basidia often develop from resistant teliospores; may be short filamentous cells or short four-celled filaments; meiosis occurs in formation of basidiospores; (*2*) fertilization with eventual fusion of two haploid nuclei varies considerably from species to species.

FUNGI IMPERFECTI

About 20,000 species of fungi are known only by their asexual stages. The class name, Fungi Imperfecti, arises from the custom of calling the sexual stages of fungi **perfect stages** and the asexual stages **imperfect stages.** Since only the imperfect stages of this large group of fungi are known, they are called the Fungi Imperfecti. Obviously, the classification is an artificial one.

In general, the structure of the hyphae, which are septate, and of spores, suggests that many imperfect fungi may be Ascomycotina; others may be Basidiomy cotina. Their sexual states have not been observed or no longer exist. The classification of these fungi is fraught with great difficulties, largely because a natural classification is based upon sexual states and the morphology of sexual and asexual stages is by no means coordinated. For instance, two taxa may have very similar conidial stages but very different sexual stages. The various categories of classification in this class are designated as **form genera** or **form families** to show that the members of the genera or families do not necessarily have a natural or family relationship. For instance, some species have been named simply on the basis of the host upon which they were found, a procedure that resulted in naming hundreds of literally nonexistent "species."

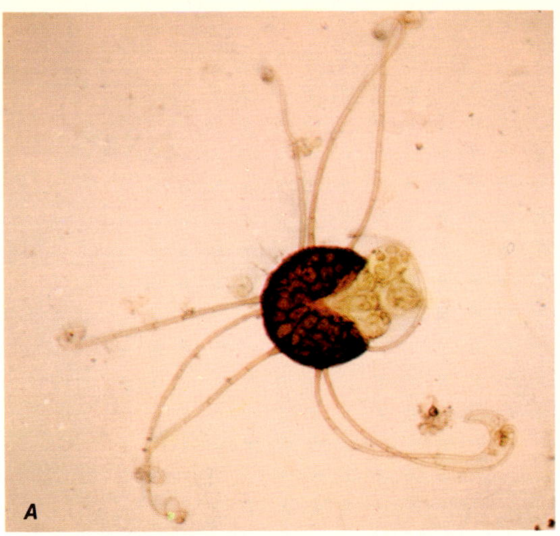

Figure 26.13
Powdery mildews. *A*, cleistothecium with asci of powdery mildew *(Uncinula)*, ×5. *B*, whitish areas on grape leaf are mycelium of *Uncinula*, ×⅓. *C*, grapes heavily infected with powdery mildew, ×½. (Courtesy of J. Ogawa.)

Figure 26.16
A, apothecia of the brown rot fungus *(Monilinia fructicola)*, ×½. *B*, brown rot on cherry compared with a healthy fruit, × ½. *C*, brown rot on peach, × ⅕. (Courtesy of J. Ogawa.)

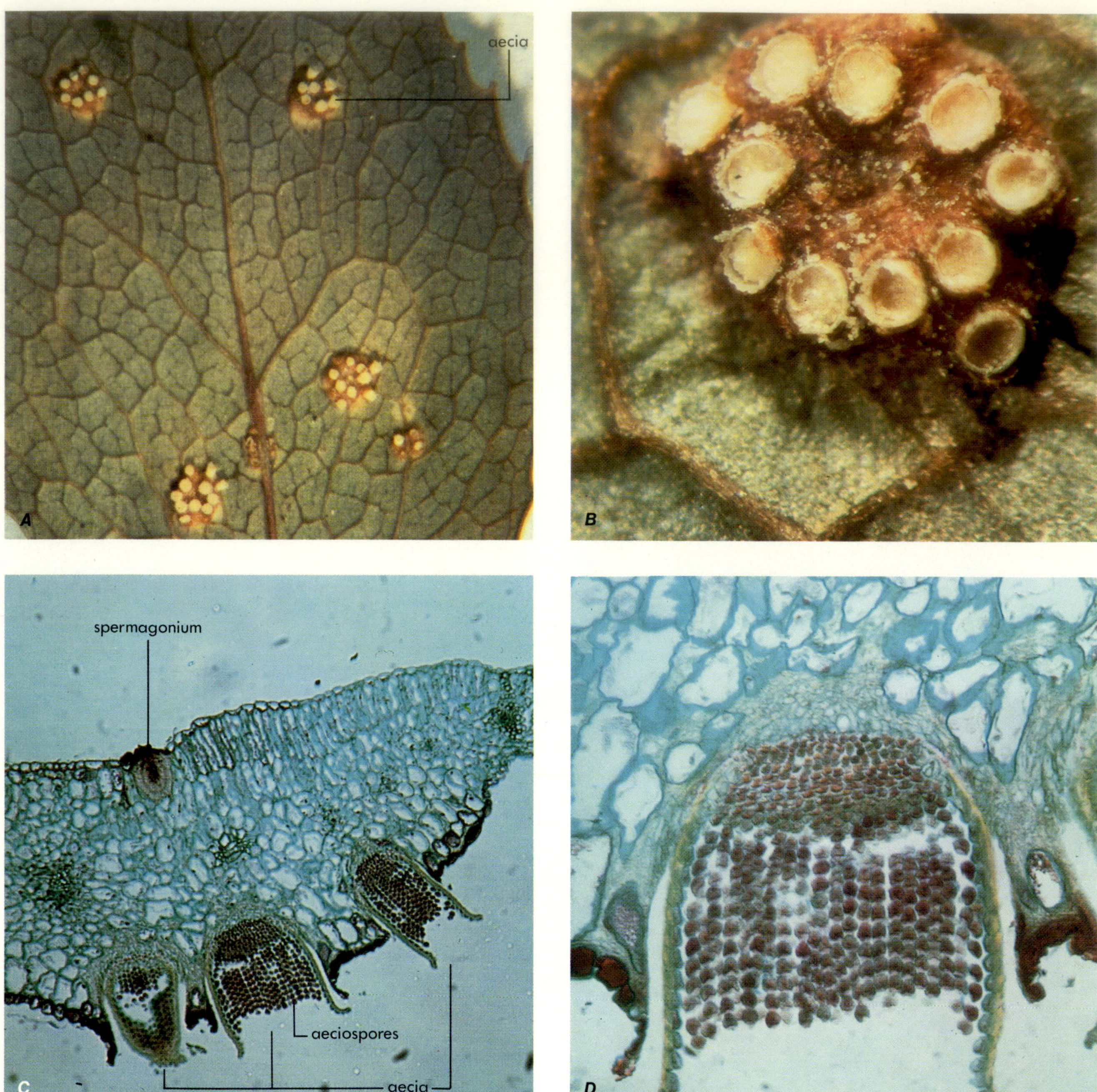

Figure 26.25
Stages in the life cycle of *Puccinia graminis* on barberry leaf. *A,* lower surface of barberry leaf *(Berberis vulgaris)* showing groups of aecia, ×1. *B,* enlarged view of groups of aecia, ×150. *C,* cross section of barberry leaf showing spermagonium on upper surface, aecia on lower surface, ×150. *D,* section of an aecium, ×1000.

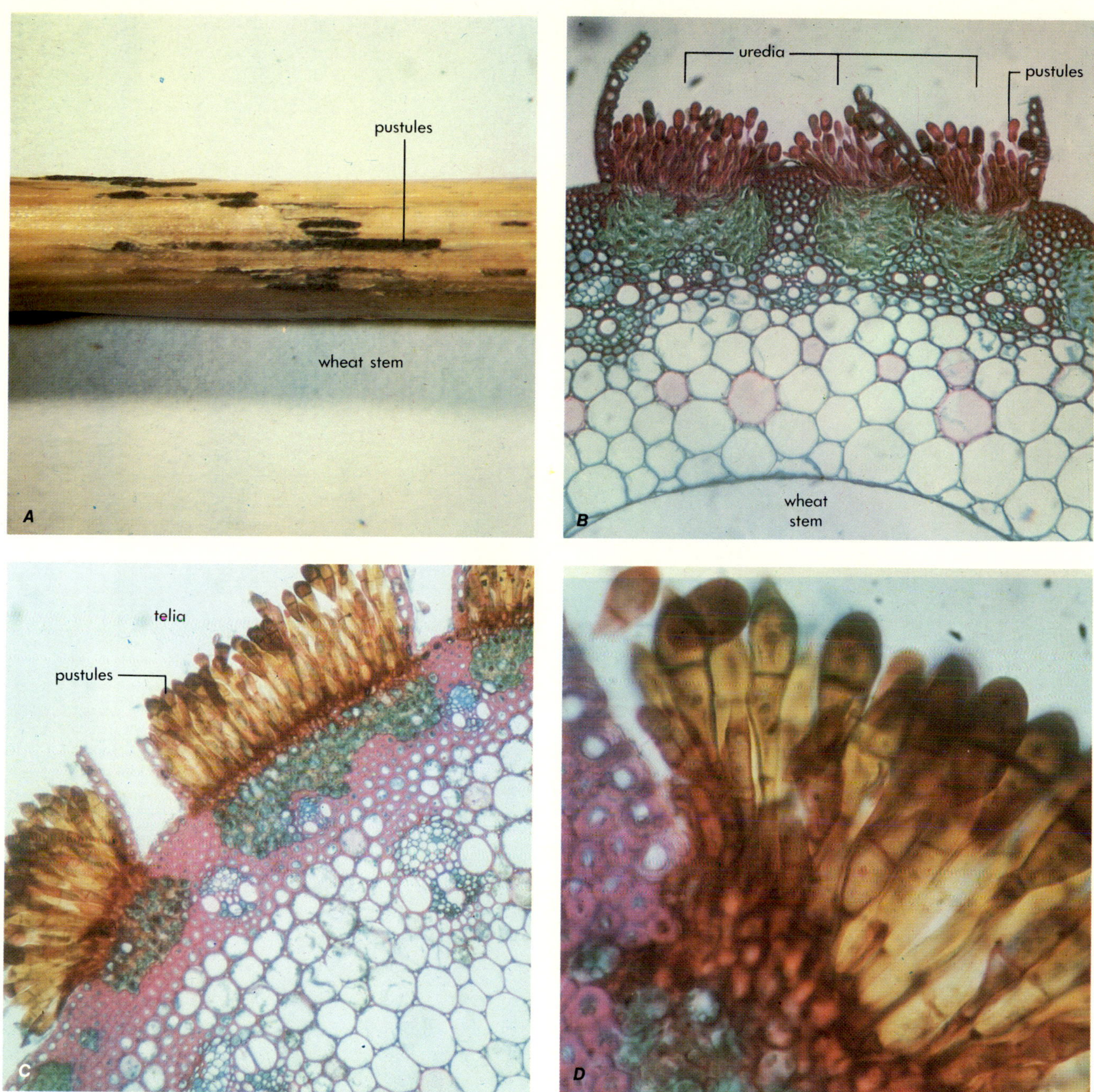

Figure 26.26
Stages in life cycle of *Puccinia graminis* on stem of wheat. *A*, sori on wheat stem, ×1. *B*, cross section of uredia, ×100. *C*, cross section of telia, ×100. *D*, enlarged view of telium showing teliospores, ×1000.

Figure 26.36
Various types of lichens. *A,* the conspicuous green foliose lichen is *Peltigera aphthosa,* and it is surrounded by the white branches of the fruticose lichen *Cladonia* sp. *B,* crustose lichen *Acarospora chlorophana* colors a rock on the skyline of a Sierra Nevada ridge. *C, Ramalina reticulata,* a fruticose lichen, clothes the branches of an oak in the foothills of the Coast Ranges along the Pacific Coast.

Only a single form group will be mentioned, the form order Moniliales; it is the largest in the class and has over 10,000 form species. Most of the fungal pathogens of man belong here. Some of them have been demonstrated to be causal agents of various types of allergy. *Penicillium* and *Aspergillus* are genera frequently placed in this order because there are many species belonging to these genera whose sexual stages are unknown.

Penicillium

Penicillium chrysogenum has won deserved fame because penicillin, a drug derived from it, will inhibit the growth of bacteria without injuring human tissue. Substances such as penicillin, formed by one organism that inhibits the growth of other organisms, are called **antibiotics.** Our knowledge of them dates back to 1870, but their significance was not fully appreciated until the discovery of penicillin.

In 1928, Alexander Fleming investigated extracts of *Penicillium notatum* and after demonstrating their antibiotic properties gave the antibiotic the name penicillin. It was not until after 1940 that a concentrated effort was made to produce penicillin on a commercial scale, using a different species. The reason for its antibacterial nature was not discovered until the 1950s. As mentioned in Chapter 21, it prevents the synthesis of peptidoglycan, the major cell wall material of bacteria.

Many thousands of kilograms of the mycelium of *Penicillium chrysogenum* have been grown, and the species has been subjected to intensive study, yet no sexual stages have ever been observed. The absence of sexual reproduction presents the geneticists who are studying *Penicillium chrysogenum* with a difficult breeding problem. Nevertheless, they have succeeded in growing strains that produce many times more penicillin per kilogram than do the original strains.

Predacious Fungi

Several forms in the Moniliales are adapted to capture and destroy microscopic animals. A few actually trap nematode worms by forming small rings that quickly constrict when stimulated by the contact of a nematode crawling through them. *Dactylaria,* a nematode-trapping form, is shown in Fig. 26.33.

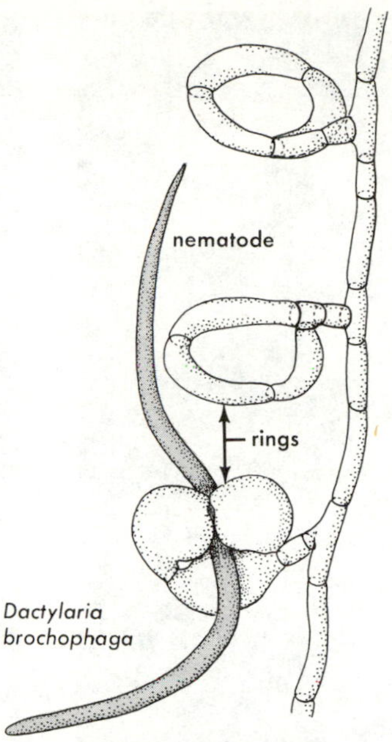

Figure 26.33
Nematode trapping loops of *Dactylaria*. Redrawn from E. A. Gäumann, *Fungi*. © 1952 by Hafner Publishing Company. Reprinted with permission of the publisher.)

The Parasexual Cycle

Since the sexual process has not been observed to occur in the Fungi Imperfecti and genetic variation is not thought to be a part of asexual reproduction, the question may be asked: Is genetic variation possible in the Fungi Imperfecti? The answer is, apparently, yes, at least in the laboratory. Careful genetic analysis has produced evidence for two possible sources of variation. One is very similar to heterosis; a selective advantage appears to result from certain stable combinations of nuclei in heterokaryon. A **heterokaryon** is a hypha, spore, or mycelium with genetically unlike nuclei. The second observation yields results similar to those observed in a true sexual life cycle. The two nuclei of a heterokaryon fuse to form a diploid nucleus. Recombination of genes takes place in some manner. Eventually, in the fungal filaments a somatic reduction takes place, the haploid condition is again attained, and the crossing over of selected traits may be observed. This process is called the **parasexual cycle.** It is well-established in laboratory cultures where it may account for the great variability of some Fungi Imperfecti, but its significance in the world is not known.

FUNGI IN SYMBIOTIC ASSOCIATIONS WITH OTHER PLANTS

Mycorrhizae

Mycorrhizae (singular, mycorrhiza) are fungal associations with the roots of higher plants. There are at least two types of association: **ectomycorrhizae** and **endomycorrhizae.** Endomycorrhizal fungi are unicellular and live within individual root cells, but ectomycorrhizal fungi have typical hyphae that cover the root tips in a thick mat (Fig. 26.34) and penetrate between cortical cells (Fig. 26.35). These hyphae do not form haustoria, but contact with root cells is nevertheless very close and metabolites are transferred in both directions. Radioactive carbon in carbon dioxide, for example, can be fixed in photosynthesis by the higher plant and later be detected in the fungus. Radioactive isotopes of phosphorus, calcium, and potassium can also be shown to be taken up in greater amounts by plants with mycorrhizae than by plants without them.

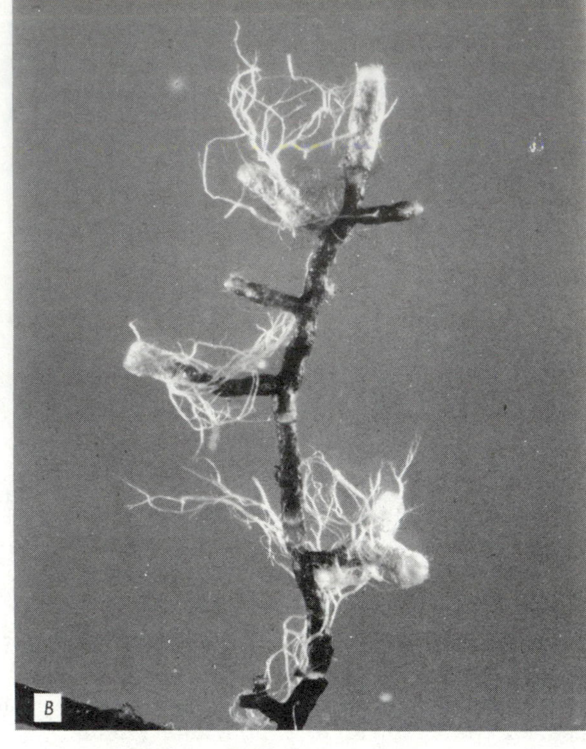

Figure 26.34
Various shapes that mycorrhizae may take. Note the short club-shaped lateral roots that are infected with fungi; their white covering is a mantle of hyphae. (Courtesy of B. Zak, *Ectomycorrhizae.* © 1973 by Academic Press. Reprinted by permission of the publisher.)

Mycorrhizal associations occur throughout the plant kingdom. They have been found on roots of many trees, shrubs, and herbs of flowering plants; on conifers, ferns, mosses, and lower vascular plants. Possibly more than 80% of all angiosperms exhibit them. In some cases, the presence of mycorrhizae is essential to normal plant development. Seedlings of orchids and heaths fail to survive if the fungi are absent, and plantings of pines in areas where they do not normally grow (such as Australia) do poorly unless the mycorrhizal fungi are introduced to the soil. These fungi may also contribute to plant development by secreting hormones, particularly IAA.

Fungi of all types are involved in these associations, but Basidiomycotina are by far the most common. The diversity of pairings between species of fungus and species of higher plants is very broad. One root system may harbor several different mycorrhizal fungi on different roots or even along the length of one primary root segment, and the hyphae of one mycorrhizal fungus can branch from the root system of one species to a nearby root system of another species. In general, ectomycorrhizal species are common among temperate broad-leaved and needle-leaved trees, while endomycorrhizal species are common among grasses and tropical plants of all life forms.

Lichenized Fungi – Lecanorales

There are about 20,000 species of lichens, most of them belonging to the order Lecanorales in the class Euascomycetes. They constitute, numerically, the largest order in the Mycota. Furthermore, they are probably the most visible and the least known. They grow on stones (**Fig. 26.36*B*,** page 542), painting cliffs in vivid colors. They hang in festoons from trees **(Fig. 26.36*C*)** and clothe

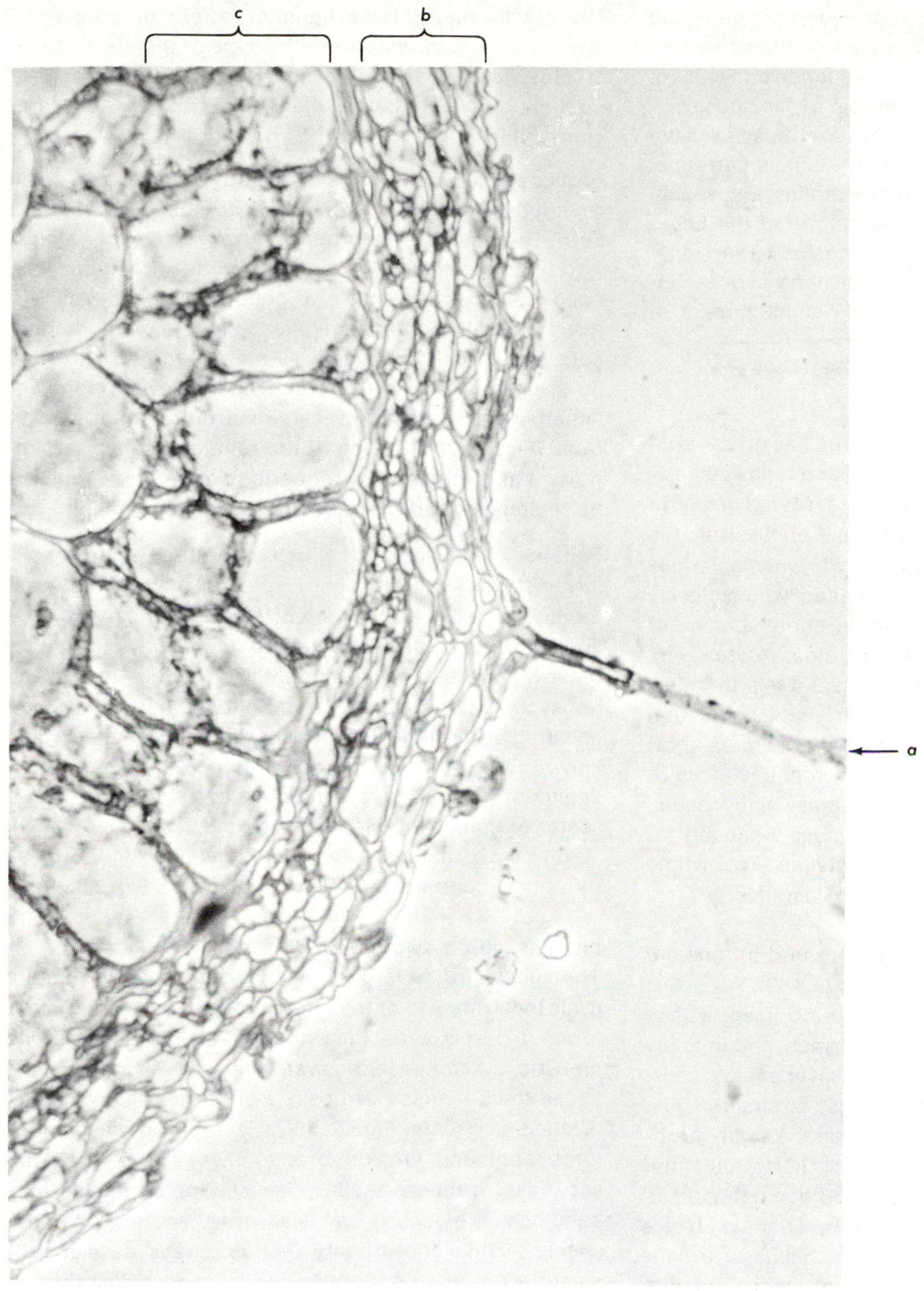

Figure 26.35
Cross section of a mycorrhiza, showing hyphae branching off into the soil *(a)*, the fungal mantle outside the root *(b)*, and the cortex with hyphae forming an intercullular net between root cells *(c)*. (Courtesy of F. H. Meyer, *Ectomycorrhizae*. © 1973 by Academic Press. Reprinted by permission of the publisher.)

their branches with thalli having intricate patterns. They grow under water and on rocks wet with salt spray of oceans. They cover square miles of Arctic tundra and the floor of cool forests **(Fig. 26.36*A*)** and stabilize the soil of deserts. Some lichens you may know are British soldiers and reindeer "moss"; few other lichens have common names.

What is a lichen? It is a stabilized thallus of two plants, an alga and a fungus (**Fig. 26.37,** page 559). Each has its own identity. They grow separately from each other with difficulty, if at all. Neither, alone, is able to form the characteristic thallus of their association. In a few cases it has been possible for the partners to be disassociated and to be cultured separately. The algae grow poorly, may need a sugar supplement, even when grown in the light. It has proven difficult to reassociate the separated partners. In no other fungi does the vegetative mycelium form a vegetative thallus characteristic of the species.

It turns out that only about 20 algal species have been identified as the algal partner. The fungal partner is usually responsible for the form of the thallus, and sexual reproductive stages are entirely fungal. All of the Lecanorales have ascocarps (**Fig. 26.39,** page 559), and most of them are apothecia. There are perhaps a dozen lichenized Basidiomycotina. Only a very small number of Fungi Imperfecti are lichenized.

Thallus Structure—Internal

While the general thallus form of lichens has great variability, the internal arrangement of alga and fungus is relatively constant. The simplest arrangement, occurring in only a few species is found in the genus *Ephebe.* Here the alga, a branched minute brushy form, *Stigonema,* largely determines the shape of the lichen thallus, which closely resembles that of the brushy alga. In some species of *Collema* (**Fig. 26.37*A*),** the blue-green alga, *Nostoc* (Fig. 21.21) with its usual mucilaginous envelope is the algal partner **(phycobiont).** Instead, however, of the usual long intertwining filaments, the *Nostoc* here occurs as short filaments, sometimes, as only a single cell. It is more or less evenly distributed throughout the mass of mycelium. Hyphae and algae are usually more closely packed together toward the surfaces of the thallus. The whole thallus is embedded in mucilage which becomes soft and swollen when wet.

In most species of lichens the alga and fungus are associated in a regular fashion **(Fig. 26.37*B*).** A **cortex,** composed of variously compacted and modified hyphae always forms on the upper or outer surface, and in many species on the lower surface as well. A layer of algal cells, frequently in characteristic groupings, is situated just below the upper cortex. Hyphae enmesh the algae, probably introducing haustoria into many of them. Below the **algal layer** there occurs a variously modified network, or felt, of mycelium known as the **medulla.** There is always some manner of attaching the lichen thallus to a substrate. If a lower cortex is lacking, the mycelium extends downward into the substrate **(Fig. 26.37*B*).** When a lower cortex is present, one or another of a variety of attachment stalks may occur.

Thallus Form—External

There are four thallus forms: **crustose, fruticose, foliose,** and **two-fold thalli.**

Crustose. Collectors frequently refer to these lichens as crusts **(Figs. 26.36*B*, 26.38*A*).** And so they are, a thin layer of lichen tissue more or less completely covers the substrate. The thallus may be a smooth and continuous growth, or it may be marked by irregular cracks **(rimose).** The cracks may form a definite pattern of **areoles.** A lower cortex is missing from this type of thallus. In some crustose lichens a peripheral lower cortex develops. The margin of the areoles then turn upward to form **squamules.**

Fruticose. This thallus is formed by a network of small rounded or flattened branching strands **(Figs. 26.36*C*, 38*B*).** There is always an outer cortex enclosing an algal layer and medulla. The center of the strand may be hollow, filled with a mass of cottony hyphae, or with a mass of parallel stiffened hyphae to form a strengthening core. Such thalli may occur as minute shrubby growths covering rocks like a fur coat. Others develop long reticulated nets hanging in festoons from the branches of trees **(Fig. 26.36C).** Much of the reindeer "moss" so important as a food supply for caribou and reindeer in Arctic regions is *Cladonia alpestris,* a fruticose lichen.

Foliose. This leaflike thallus has both an upper and a lower cortex with an algal layer and medulla between them. The upper and lower cortex differ in construction and color. An extensive system of ridges may develop on the lower surface giving strength to the lobe just as the honeycomb pattern gives strength to the thin metal sheet covering an airplane wing. Foliose lichens are the largest known. The lobes of *Lobaria pulmonaria* may have a diameter of 40 cm. They are attached to a substrate along a sector of thallus margin. Other foliose lichens are small, about 1 cm in diameter, and are attached to rock surfaces by a single central strand (*Dermatocarpon,* **Fig. 26.38*D***).

Lichens with a Twofold Thallus

The lichens in the large family Cladoniaceae have a life cycle involving two of the above types of thalli. There first forms a purely vegetative primary thallus that may be areolate, squamulose, or even foliose. A secondary fruticose thallus arises vertically from this primary thallus (*Cladonia crispata,* **Fig. 26.38C**). The reproductive structures (apothecia and/or soredia) always develop on the secondary fruticose thallus. The secondary thallus may become very extensive and is always characteristic of the species. While the primary thallus always develops, in some few species such as reindeer moss (*Cladonia alpestris*) it consists only of small granules and soon disappears.

Reproduction

The sexual reproductive stage of lichens is entirely that of the fungal partner. Asexual reproduction involves both partners and is therefore a lichen trait only.

Vegetative

Fragmentation. Probably most lichens may be dispersed by fragmentation of the thallus. In many crustose and some foliose forms the thallus spreads slowly over the substrate. Sections die and disintegrate. New thalli spread from the remaining living sections. In some fructicose and foliose forms thallus fragments may be dispersed to initiate new thalli.

Lichen Propagules. There are two types of thalline structures adapted for vegetative reproduction; **soredia** and **isidia.** These propagules are common on both fructicose and foliose thalli but are less common on the crustose lichens.

Soredia. Soredia are groups of algal cells enveloped in a mass of mycelium. The whole individual structure ranges from 25 to 100 mμ in diameter. They are formed in the algal and medulla layers and erupt through pores or cracks in the cortex. Like masses of spores, they form a powdery covering to the thallus surface, from which they are dispersed by wind currents. Consisting, as they do, of both mycelium and algae, they are highly adapted for vegetative dispersal of the lichen species that produce them.

Isidia. Isidia are small protuberances of the thallus. They originate from the medulla and incorporate algae and cortex as they grow upon the lichen surface. They are always covered with a cortex. Isidia are larger than soredia, 0.01 to 0.03 mm, in diameter and appear as masses of small growths on the thallus surface. It has been postulated that in some species they are adapted to facilitate photosynthesis by increasing the surface area of the thallus. However, they are easily broken from the thallus and lichens that produce them in abundance may not produce ascocarps.

Sexual Reproduction

As far as is known the development of perithecia and apothecia is similar to that observed for the nonlichenized ascomycetes (Fig. 26.8). Cleistothecia are unknown. However, the complete sexual life cycle is poorly understood. Many lichens produce pustules resembling the spermatogonia of wheat rust. Spermatialike cells are produced within these pustules, but their function has not been established. Not since 1888 has their germination been reported. They have been observed a number of times associated with hyphae similar to the adhesion of spermatia to the receptive hyphae in the wheat rusts. They are thus considered by some lichenologists to be spermatia, but which, at least, in some cases no longer function as sperms. In many lichen keys the pustules are referred to as **pycnidia** and the spermatia variously designated as **pycnidiospores, microspores,** or **microconidia.**

In the family Lecidiaceae and a few other cases the apothecia closely resemble those found in the nonlichenized ascomycetes, except that the peridium is invariably black and brittle **(Figs. 26.39*A,B*).** It is referred to as a **proper margin.** All other apothecia in the order are surrounded by the thallus, the peridium is inconspicuous or even lacking, and the apothecium is now surrounded by a **thalline margin (Fig. 26.39*C*).**

In general the perithecia are more or less embedded within the thallus and are similar in structure to the perithecia found in the nonlichenized ascomycetes.

The ascospores are discharged as in the ascomycetes in general. Their germination in a number of species has been studied. The association of a young germling mycelium with algae has not been observed in nature. A complicating and unexplained fact is, the most frequent algal phycobiont, *Trebouxia,* has not yet been shown to be a free living alga. So the manner in which a germling becomes associated with the most common algal partner, *Trebouxia,* remains a puzzle.

Lichen Summary

1. Lichens form a symbiotic (mutualistic) relationship between an alga and a fungus.
2. There are about 20,000 species of lichens, but only about 15 different species of algae enter the relationship.
3. The fungal partner is usually responsible for the form of the thallus.
4. Asexual reproduction is by means of soredia and isidia, reproductive structures containing both alga and fungus and so unique to the lichens.
5. Sexual reproduction involves ascocarps and in general resembles the process as observed in nonlichenized ascomycetes.

SUMMARY

1. Members of the Eumycota have chitinous cell walls and usually reproduce asexually by conidia (conidiospores). This division includes the subdivisions Zygomycotina (bread molds), Ascomycotina (sac fungi), and Basidiomycotina (club fungi), as well as the Fungi Imperfecti, mycorrhizal fungi, and lichen fungi.
2. Zygomycotina species exhibit a zygotic life cycle, in contrast to many Myxomycota. The mycelium is nonseptate. Most species are saprophytic.
3. The life cycle of *Rhizopus stolonifer* is described (see pages 514–515). The gametophyte generation is heterothallic: that is, it has + and − sexual strains, which are morphologically identical but genetically distinct. This species has a zygotic life cycle.
4. Ascomycotina species have septate mycelia and reproduce asexually with conidia. Heterothallic, haploid hyphae fuse to produce a dikaryon($n + n$) cell. The dikaryon divides by mitosis, producing ascogenous hyphae that terminate with an ascus cell. In the ascus, fusion of the two haploid nuclei occurs, followed by meiosis, generally resulting in eight meiospores. The asci line a fruiting body of various shape (the ascocarp).

5. Yeasts (*Saccharomyces*) are members of the Ascomycotina. They are generally unicellular and may reproduce asexually by budding. Yeasts perform alcoholic fermentation and are cultured for use in winemaking, brewing, and breadmaking.
6. Important pathogens in the Ascomycotina include leaf curls (*Taphrina*), powdery mildews (*Erysiphe*), and ergot (*Claviceps*). *Claviceps* parasitizes grasses; its fruiting body lies in the grain and it may accidentally be eaten. Alkaloids within it cause a burning sensation in the limbs, gangrene, halucinations, madness, even death. The life cycle of *Claviceps* is summarized on pages 523–524.
7. The cup fungi *Monilinia* and *Peziza* are also members of the Ascomycotina; the former is parasitic on stone fruits where it causes the disease called brown rot, while the latter is a saprophyte. The life cycle of *Monilinia* is shown on pages 524–525. Edible morels (*Morchella*) and truffles (*Tuber*) are also in this subdivision.
8. Members of the subdivision Basidiomycotina also form $n + n$ hyphae, but these hyphae play a larger role in the life cycle than they do in the sac fungi: they may form independent mycelia and they may form fruiting bodies. Eventually, specialized cells at their tips, called basidia, undergo fusion and meiosis, resulting in four meiospores. The meiospores are extruded from the basidium and ultimately released. There are two classes within this subdivision: Homobasidiomycetes and Heterobasidiomycetes.
9. Mushrooms, puffballs, bracket fungi, and some mycorrhizal fungi are all in the class Homobasidiomycetes. Most are saprophytes. The genus *Amanita* contains both poisonous and hallucinogenic species. The life cycle of mushroom (*Agaricus*) is described on pages 527–529.
10. Jelly fungi, rusts, and smuts are in the class Heterobasidiomycetes. Most are parasitic, some infecting different hosts at different states of the life cycle (e.g., they are heteroecious). The life cycle of wheat rust (*Puccinia graminis* var. *tritici*) is described on pages 530–535. Life cycles of species of the smut *Ustilago* that infect blossoms and seedlings are described on pages 535–538.
11. About 10% of all fungal species are placed in the form class Fungi Imperfecti. The "perfect" or sexual stages of these fungi have never been observed. Based on vegetative traits, however, most of these species are sac fungi. *Penicillium* is in this group. It produces an antibiotic that prevents the synthesis of peptidoglycan, the major wall material of bacteria.
12. Mycorrhizae are fungal associations with the roots of higher plants. They appear to be mutualistic (symbiotic), in that the "host" receives additional mineral nutrients through the enlarged absorbing surface of the mycelium, and the fungus obtains photosynthate from the "host." Hormones may also be passed from fungus to "host."
13. Lichens are mutualistic associations between fungi (usually Ascomycotina) and algae (greens or blue-greens). The shape, color, and anatomy of a lichen depends upon the species of fungus and alga present; hence lichen "species" can be recognized and named. Lichens can reproduce asexually by liberating bits of tissue (soredia, isidia) that contain algal and fungal cells.

CLASSIFICATION

Division	Eumycota
Subdivision	Zygomycotina
Order	Mucorales
Genus Species	*Rhizopus stolonifer*
Subdivision	Ascomycotina
Class	Hemiascomycetes
Order	Endomycetales
Genus Species	*Saccharomyces cervisiae*
Order	Taphrinales
Genus	*Taphrina*
Class	Euascomycetes
Order	Eurotiales
Genera	*Eurotium* (*Aspergillus*)
	Talaromyces (*Penicillium*)
Order	Erysiphales
Genus	*Erysiphe*
Order	Clavicipitales
Genus	*Claviceps*
Order	Helotiales
Genus	*Monilinia*
Order	Pezizales
Genus	*Peziza*
Order	Lecanorales
Order	Tuberales
Genus	*Tuber*
Subdivision	Basidiomycotina
Class	Homobasidiomycetes
Order	Agaricales
Genus Species	*Agaricus campestris*
	Amanita muscaria
	A. phalloides
Order	Polyporales
Genus Species	*Ganoderma applanatum*
Class	Heterobasidiomycetes
Order	Tremellales
Order	Uredinales
Genus Species	*Puccinia graminis*
Order	Ustilaginales
Family	Ustilaginaceae
Genus Species	*Ustilago maydis*
	U. tritici
	U. nuda
Family	Tilletiaceae
Genus Species	*Tilletia tritici*
Form subdivision	Fungi imperfecti
Form order	Moniliales
Form Genus	*Penicillim notatum*
Species	*P. chrysogenum*
	Aspergillus sp.
	Dactylaria brachophaga

CHAPTER 27

BRYOPHYTA

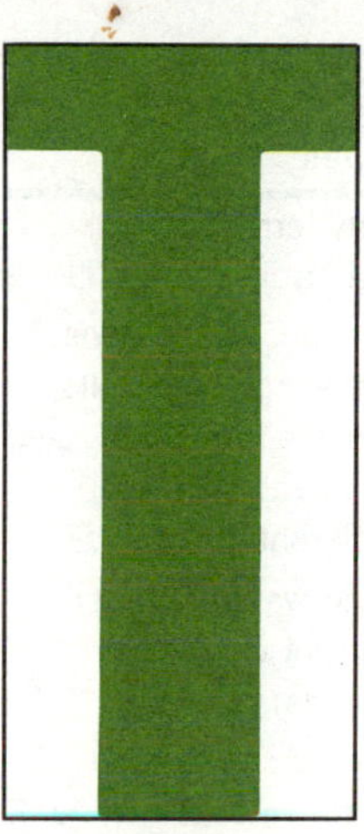

The division Bryophyta includes **hornworts, mosses,** and **liverworts.** Although all bryophytes are morphologically distinct from thallophytes, it is difficult to separate them from algae in general on the basis of any single character. Perhaps the two greatest differences between algae and bryophytes are the general structure of the plant bodies and gametangia.

It will be recalled that the plant body of thallophytes is generally composed either of single cells (or filaments of cells), or of intertwining filaments, resulting in a more or less complex body. Layers of parenchyma tissue occur in some forms. With the exception of one stage in the life history of mosses, the bryophyte plant body is never filamentous. It is composed of blocks or sheets of cells forming a parenchymatous tissue.

Gametangia of thallophytes are unicellular or, if multicellular as in *Ectocarpus,* lack a protective jacket of sterile cells. Each and every cell of the multicellular structure of *Ectocarpus* produces gametes. However, multicellular female gametangia characteristic of all members of the Bryophyta produce but a single gamete, the egg, which is surrounded by a flask-shaped layer of protecting cells. This characteristic female gametangium is also found in all the higher plants except for the flowering plants. It is called an **archegonium.** Male gametangia of both *Ectocarpus* and the bryophytes produce numerous sperms, but sperm-producing cells of the bryophytes are always surrounded by a protective layer of parenchyma cells. This male gametangium is called an **antheridium.**

Water is required by most algae and bryophytes to effect fertilization. A definite alternation of generations occurs in all members of the Bryophyta, and the sporophyte (diploid) generation is always more or less dependent on the gametophyte (haploid) generation. In all Bryophyta, an **embryo sporophyte,** consisting of a spherical or elongated mass of tissue, is formed directly after the germination of the zygote and is retained and nourished by the gametophyte generation.

Asexual spores are not formed in Bryophyta. Members of this division reproduce asexually either by fragmentation of the gametophyte or by special multicellular bodies known as **gemmae.**

Bryophytes occur on soil, trunks of trees, and rocks, and many are truly aquatic. These and other traits of the Bryophyta are contrasted to those of the thallophytes in Table 27.1.

Table 27.1
TABULATION OF DIFFERENCES BETWEEN THALLOPHYTES AND BRYOPHYTES

Thallophytes	Bryophytes
1. Mostly aquatic	**1.** Mostly terrestrial, but moist habitat
2. A few of the kelps have elements that resemble sieve cells; otherwise, no conduction cells	**2.** Some mosses have cells suggesting sieve cells; otherwise food conducted in relatively undifferentiated cells
3. None have water-conducting elements	**3.** Simple water-conducting cells (not vessels or tracheids) in some
4. In general, no specialized water- and nutrient-absorbing tissue; rhizoids and haustoria may occur in some fungi	**4.** Rhizoids anchor and absorb water and mineral nutrients
5. Mostly filamentous or a lacework of intertwining filaments; parenchyma in a few	**5.** Only one stage of mosses is filamentous; all others are formed of parenchyma cells
6. An alternation of hetero or isomorphic generations in many forms	**6.** All have an alternation of heteromorphic generations
7. Both sporophytes and gametophytes nutritionally independent	**7.** Gametophyte independent; sporophyte nutritionally dependent
8. Gametangia are either single cells or groups of single cells not accompanied by a jacket of sterile vegetative cells	**8.** Gametangia are always composed of gamete-producing cells covered by a jacket of sterile vegetative cells
9. Asexual reproduction by spores or conidia	**9.** Asexual reproduction by gemmae
10. Embryo absent	**10.** Embryo present

ALTERNATION OF GENERATIONS

As just mentioned, all Bryophyta have a distinct alternation of generations. A haploid plant, producing gametes, alternates with a diploid plant; meiosis occurs in the diploid plant, and haploid spores containing *n* sets of chromosomes in each nucleus are formed. Thus, a haploid gamete-producing plant—the **gametophyte**—regularly alternates with a diploid spore-producing plant—the **sporophyte,** as shown in the diagram.

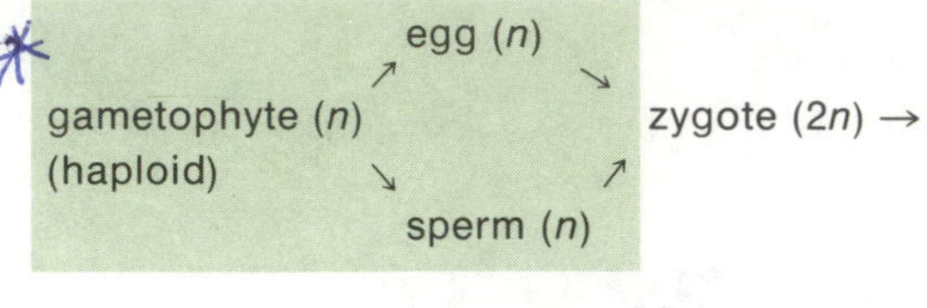

sporophyte (2*n*) →
(diploid)

meiospores (*n*) → gametophyte (*n*)
(haploid)

Sporophyte and gametophyte plants are but two phases in a complete sexual life history.

CLASSIFICATION

The Bryophyta are divided into three classes, **Hepaticae (liverworts), Anthocerotae (hornworts),** and **Musci (mosses).** The Hepaticae are the most primitive bryophytes. The simplest liverworts consist of a flat, ribbonlike, green thallus (Fig. 27.1) that produces gametes and a meiospore-producing capsule or sporangium embedded within the gametophyte thallus. The name liverwort is very old, having been used in the ninth century. It was probably applied to these plants because of their fancied resemblance to the liver and the belief that plants resembling human organs would cure diseases of the organs they resembled. At any rate, a prescription for a liver complaint in the 1500s called for "liverworts soaked in wine."

Less primitive Hepaticae have simple "stems" and "leaves," and sporophytes are borne above green gametophytes. Mosses are probably the best-known class of Bryophyta. They have either an upright or prostrate leafy gametophyte, and may form extensive mats on moist, shady soil (Fig. 27.2). The sporophyte is raised above the leafy gametophyte. *Anthoceros* has the simplest gametophyte of the group but its sporophyte has a meristematic region, which is a type of tissue more characteristic of higher forms than of Bryophyta.

ECONOMIC IMPORTANCE

Bryophytes are not extremely important economically, but they do have an interesting ethnohistory. Native

Figure 27.1
Liverworts. *A, Ricciocarpus,* ×1. *B, Riccia,* ×1. (Courtesy of W. Russell.)

American Indians in Utah, for instance, ground up mosses, such as *Mnium* and *Bryum,* into a paste and applied it as a poultice to treat burns and bruises. In Europe, the growth of mosses such as *Dicranoweisia* was encouraged on shingle roofs to make them water-tight.

Sphagnum is by far the most important moss economically. The surface layers of "stems" and "leaves" of *Sphagnum* are composed of alternating files of living green cells and large empty hyaline cells (Fig. 27.3). The hyaline cells are capable of absorbing water or other fluids in great quantities as high as 20 times their own weight. In the 1880s in Germany it was discovered, quite by accident, that *Sphagnum* moss pads make excellent absorbent bandages. A workman in Kiel lacerated his arm while working in a peat moor. Since he had no other bandage material, he packed his wound with dry peat (*Sphagnum*). Ten days later, when he was able to obtain treatment, the wound had healed. After that, disinfected bags filled with dry *Sphagnum* became a common wound dressing in Germany. The use of *Sphagnum* as wound dressings was not popular outside Germany until World War I. At that time they were used on a large scale by the Allied Armies as well. The British, for example, used 1,000,000 *Sphagnum* dressings each month. *Sphagnum* dressings were used very little in the United States, but the American Red Cross did publish inquiries and instructions on finding useful *Sphagnum.* More recently, *Sphagnum* is still used as packing for vegetables, in potting mix to increase the water holding capacity of the soil, and old compressed *Sphagnum* is used as a fuel in the form of a low grade coal known as peat.

Figure 27.2
Moss gametophytes and sporophytes of the hair cap moss, *Polytrichum.*

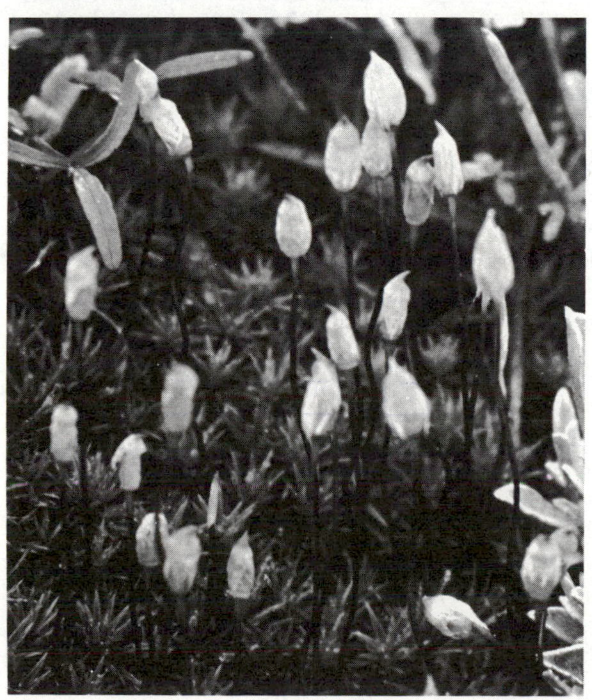

PHYLOGENY

Although mosses and liverworts appear to lie somewhere between thallophytes and vascular plants, there is no evidence that they are the ancestral precursors of vascular plants. The fossil record of bryophytes is rather meager. One liverwort fossil dates from the Devonian, but no moss fossil dates back further than the Carboniferous (see Chapter 31); the first fossils of vascular plants ap-

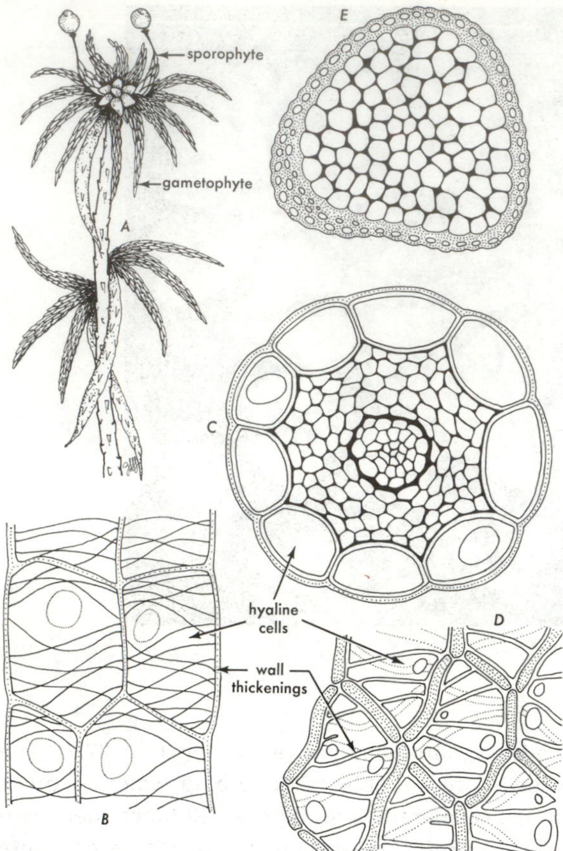

Figure 27.3
A, habit of sporophyte plant of *Sphagnum papillosum*, ×2. *B*, external view of "stem" showing large hyaline cells, ×450. *C*, cross section of branched "stem" showing external hyaline cells and thick-walled cells of the central axis, ×590. *D*, "leaf" cells: note the narrow chlorophyll containing cells and the larger hyaline cells with peculiar bandlike thickenings, ×300. *E*, cross section of *Tetraphis pellucida* "stem" to show its simple structure, ×70. (Redrawn from R. F. Scagel et al., *An Evolutionary Survey of the Plant Kingdom*, 1965. Wordsworth Publishing Company, Inc., Belmont, California. Reprinted by permission of the publisher.)

pear 50 to 100 million years *earlier* in the record. It is more likely that bryophytes evolved from primitive vascular plants, than the other way around, and that the several classes of bryophytes may have evolved independently rather than from some one common ancestor. There are sufficient morphological differences between mosses and liverworts to lead some botanists to classify them as separate divisions (Hepatophyta and Bryophyta).

Most fossil bryophytes date from the last 65 million years, and they are very similar to modern (living) forms. Consequently, we have no idea of what primitive, ancestral forms may have looked like, nor how mosses and liverworts may have been closer linked in the past. The most common guess, however, is that bryophytes represent an evolutionary dead end, an historic branch that did not lead from thallophytes to higher plants, but which instead lead from vascular plants to simpler, modified, degenerate forms.

CLASS HEPATICAE

Hepaticae are commonly known as liverworts. There are some 9,000 species of liverworts. The great majority of them grow in moist, shady localities. The gametophyte is the prominent plant, and the sporophyte is usually partially dependent on the gametophyte. The gametophyte is green and may grow either as a flat ribbon or as a leafy shoot. In either event, the plant body is frequently called a thallus, even though its internal structure does not correspond to that of the thallophytes. Depending on the taxonomist, this class may be subdivided into four, five, or six orders. We shall discuss the characteristics of two genera in the Marchantiales, *Riccia* and *Marchantia*, and of one genus in the Jungermanniales, *Porella*.

Marchantiales

Gametophytes are small, green, ribbon-shaped plants that branch regularly by a simple forking at the growing tip, resulting in a number of Y-shaped branches (Fig. 27.1). In some species, a rosette may be formed. The upper surface of the thallus is composed of cells adapted for photosynthesis and arranged to form air chambers that open to the surface by a pore (Fig. 27.4*A*). Several types of storage cells generally make up the lower surface. **Rhizoids,** specialized elongated cells performing the functions of roots (anchorage, and absorption of water and mineral nutrients), extend downward from the lowermost layer of cells. The thallus is thus differentiated into distinct upper and lower portions. Scales, frequently brown or red, are formed on the lower surface.

Riccia

Riccia is a widely distributed genus, and, although it requires water for active growth, most species will withstand considerable drought. Several species are aquatic, growing either on mud or on the surface of small ponds.

Gametophyte. The gametophyte is a small green thallus, frequently forming a rosette (Fig. 27.1*B*). Tissue on the lower side is composed of colorless cells that may contain starch. Tissue on the upper side consists of vertical rows of chlorophyll-bearing cells, between which are air chambers (Fig. 27.4*A*). The gametangia are embedded in deep, lengthwise depressions or furrows, in the dorsal surface of the thallus. Antheridia and archegonia are usually found on the same gametophyte.

Antheridia. Antheridia of different representatives of the Bryophyta are similar in structure, though they may vary somewhat in shape. In *Riccia*, they are pear-shaped and composed of two types of cells, (*a*) fertile and (*b*) sterile (Fig. 27.4*B*). Fertile cells give rise to sperms, which are relatively numerous, small, and dense with protoplasm. Sterile cells form a protective jacket, one cell in thickness, around the fertile cells.

Mature sperms consist mainly of an elongated nucleus with two long flagella. They may be shot from the mature antheridium with considerable force, or they may be extruded slowly in a single mucilagenous mass. In any event, they do not leave the antheridium until enough moisture is present to allow them to swim about.

Archegonia. The archegonium of *Riccia* is a flask-shaped structure consisting of two parts, (*a*) an expanded basal portion, the **venter,** and (*b*) an elongated **neck** (Fig. 27.4*C*). Four **cover cells** are located at the top of the neck. Each archegonium contains a single egg cell, which is located in the venter. A short stalk attaches the archegonium to the gametophyte. Archegonia of a similar structure are found in other Bryophyta.

Shortly before the egg cell is mature, the cover cells separate. At the same time, the cells in the center of the neck dissolve, so that an open canal connects the venter with moisture outside the archegonium. Free-swimming sperms move toward certain chemical substances formed by the archegonium, for they swim to it and then down the canal opened by dissolution of the neck canal cells. Several sperms may enter the archegonium, but only one fertilizes the egg in the venter.

Sporophyte. As a result of fusion of sperm and egg nuclei, the zygote contains the diploid, or $2n$, set of chromosomes. Mitosis now proceeds in a more or less orderly manner until a spherical mass of some 30 or more similar cells is formed (Fig. 27.4*D*). These cells are partially dependent upon the gametophyte. This mass of undifferentiated, cells comprising the young sporophyte is an **embryo.** It is located within the venter of the archegonium.

Further development of the embryo involves differentiation of an outer layer of cells to form a protective jacket of sterile tissue surrounding a mass of cells that are capable of forming spores. This fertile or spore-forming tissue is called **sporogenous tissue.** The sporogenous cells continue to divide by ordinary cell division until many have formed. All these cells are **sporocytes,** and will undergo meiosis, producing spores with the haploid number of chromosomes (Figs. 27.4*E*,*F*,*G*,*H*).

Meiosis consists of two cell divisions, during which the number of chromosomes is halved. Since two divisions are involved, four spores are formed from each sporocyte. Each group of four spores, formed from a single sporocyte, is usually referred to as a **tetrad** or quartet of spores.

The mature sporophyte of *Riccia* (Fig. 27.4*E*) is called a **sporangium** or **capsule.** It is composed of a jacket of sterile cells enclosing a mass of spores. The sporophyte wall breaks down when the spores are mature. These spores remain embedded in the gametophyte thallus and are not released until after the death and decay of the gametophyte. When spores germinate, each grows into a typical gametophyte plant.

Significant Steps in the Life History of *Riccia*

1. The gametophyte, a small, green, flat plant, absorbs water and mineral salts from the soil and carbon dioxide from the air. It contains chlorophyll and can synthesize food. All gametophytic nuclei are haploid (contain n sets of chromosomes).
2. Gametangia (antheridia and archegonia) develop in a furrow on the upper surface of the gametophyte.
3. Each antheridium produces thousands of sperms.
4. A single egg is formed in the venter of each archegonium.
5. Gametes (eggs and sperms) contain n sets of chromosomes.
6. When sufficient moisture is present, mature sperms are extruded from antheridia. The neck of the mature archegonium is opened and sperms swim down to the egg in the venter.
7. Each egg is fertilized by a single sperm.
8. The zygote, as a result of the union of two haploid gametic nuclei, is diploid.
9. An embryo sporophyte, consisting of a spherical mass of undifferentiated cells, partially dependent upon the gametophyte, develops from the zygote. Each nucleus of the embryo sporophyte is diploid ($2n$).
10. The cells of the embryo differentiate into a sporangium consisting of a jacket of sterile cells enclosing a mass of fertile cells.
11. Sporocytes, each with $2n$ chromosomes, form in sporangia.
12. Sporocytes undergo meiosis. A quartet of spores forms from each sporocyte. All sporocytes contain $2n$ chromosomes, and spores contain n chromosomes.
13. Upon death and decay of old gametophytes, spores are liberated and new gametophytes are developed from them.

Marchantia

Marchantia may grow in large mats on moist rocks and soil in shady locations. It is widely distributed and fairly common on banks of cool streams. It is somewhat better adapted to growing on land than is *Riccia,* but considerable moisture is still required for active growth and for fertilization.

Gametophyte. The gametophyte of *Marchantia,* and of other members of the family to which it belongs, differs from that of *Riccia* in that gametangia are on special disks raised some distance above the vegetative thallus (**Fig. 27.5,** page 560). The thallus of *Marchantia* is similar to that of *Riccia,* but much coarser. It is strap-shaped with a prominent midrib **(Fig. 27.5A).** It shows dichotomous branching. The tips of the branches are notched, indicating growth by one or several apical cells. An individual thallus may be 1 to 2 cm broad. Their length and

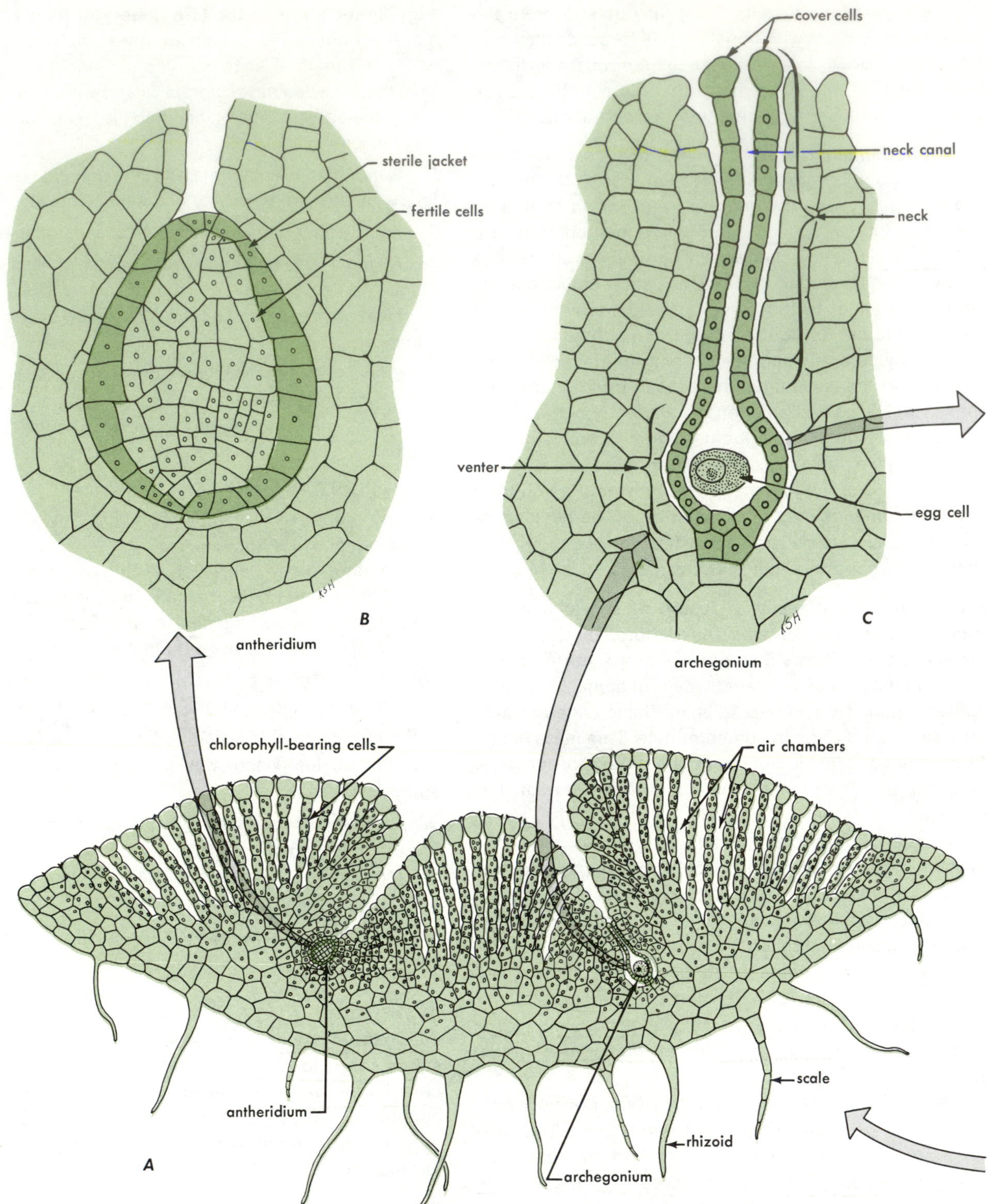

Figure 27.4
Riccia. A, cross section through a thallus; gametangia are on the upper surface of the thallus between the photosynthetic filaments of in a notch in the thallus, ×8. *B,* an antheridium; note jacket of sterile cells surrounding the developing sperm cells. *C,* archegonium on thallus, neck canal open, egg cell in venter. *D,* young sporophyte developing in venter of old archegonium. *E,* sporophyte with sporocytes still enclosed in venter of old archegonium. *f,* meiosis occurs within the sporophyte. *G,* the venter now contains fragments of the old spore case (sterile cells of the sporophyte) and spores. *H,* a spore. *B* through *G,* about ×50; *H,* ×100.

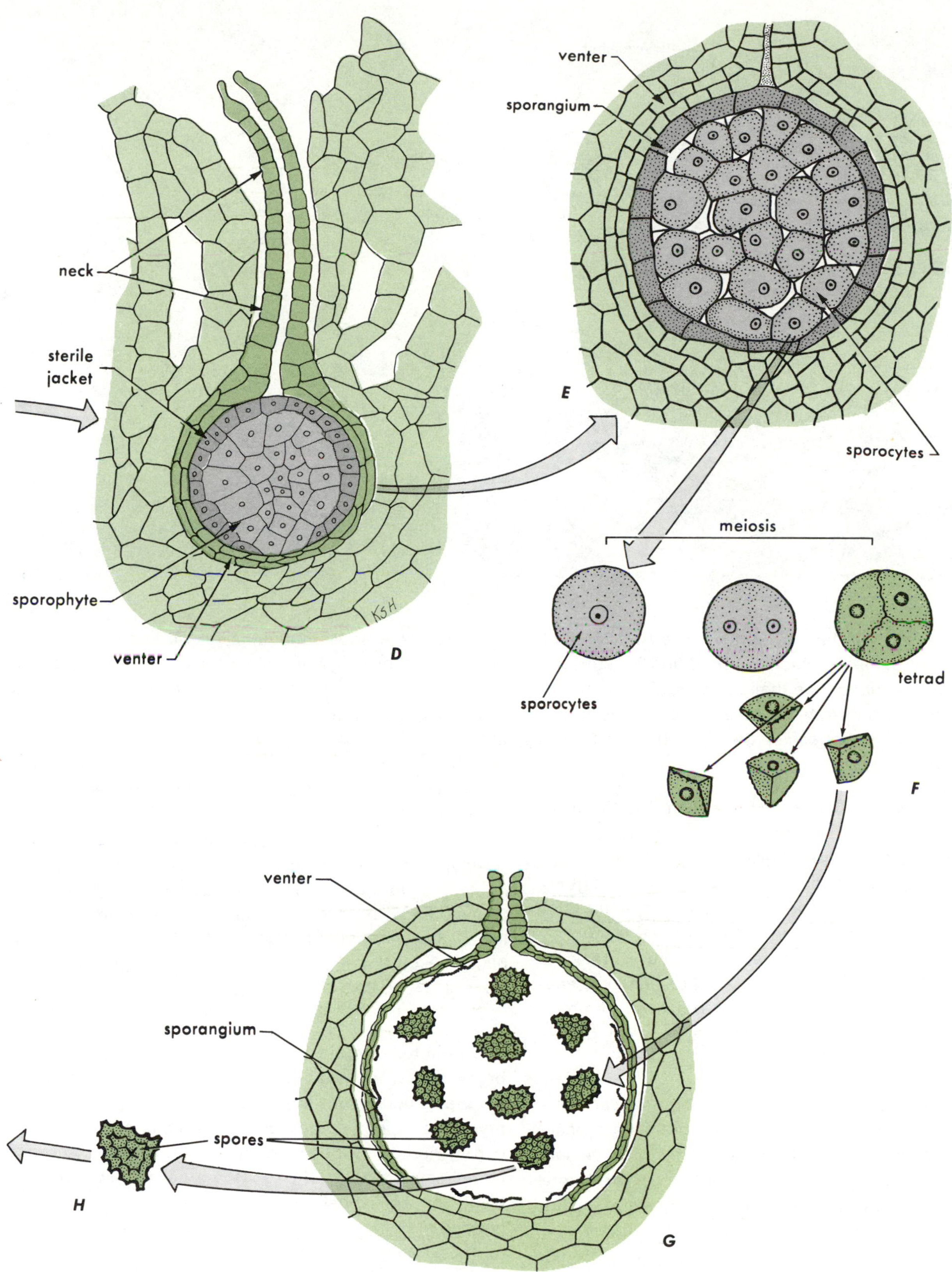

the degree of branching depend upon growth conditions. The thalli are frequently crowded and overlap each other to some extent. It is usual for new growth to overlap the older decaying ends of adjacent thalli. On its upper surface are polygonal areas, with a small but conspicuous pore in the center. These areas, with their air pores, mark the outlines of air chambers, each filled with short filaments of cells containing chloroplasts (Fig. 27.6). As

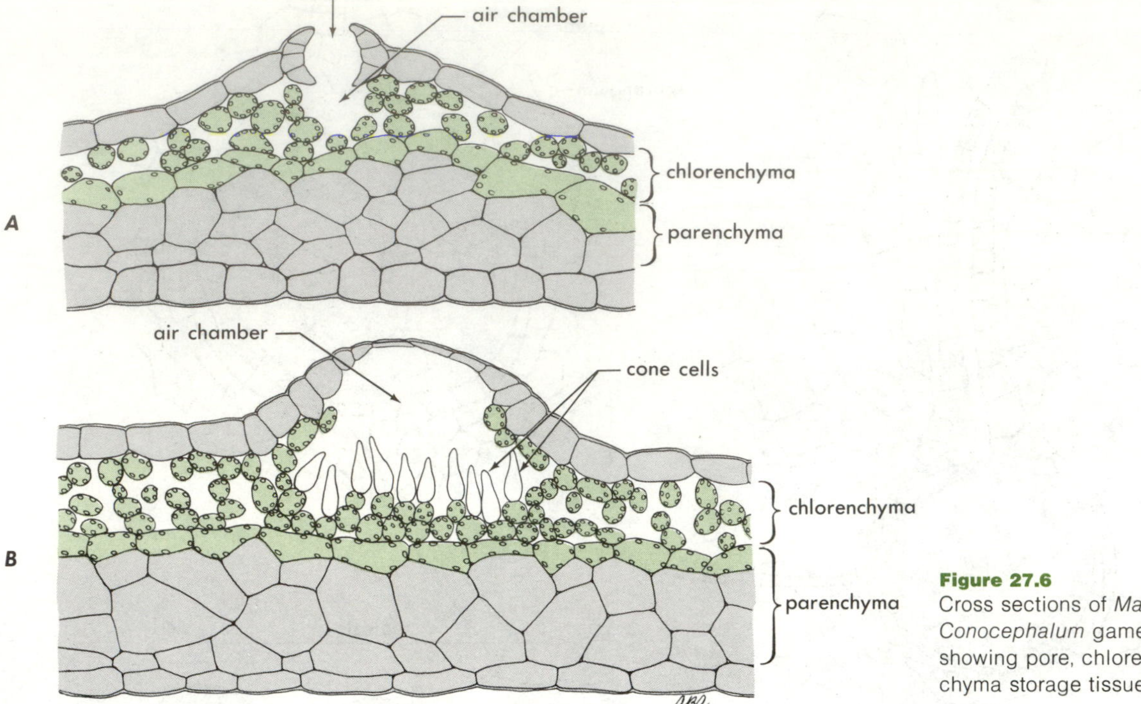

Figure 27.6
Cross sections of *Marchantia, A,* and *B, Conocephalum* gametophyte thalli showing pore, chlorenchyma, parenchyma storage tissue, and epidermis.

in *Riccia,* the lower surface of the thallus is composed of colorless cells, some of which are modified for storage. Rhizoids, which anchor the thallus and serve as organs for absorption, grow from the cells covering the lower surface. Several rows of scales are also attached to this lower surface.

Asexual reproduction. *Marchantia* reproduces asexually in two ways: (*a*) Older parts of the thallus die and younger portions, no longer attached, develop into new individual plants; and (*b*) small cups, known as **gemmae cups,** form on the upper surface, and small disks of green tissue, called **gemmae,** grow from the bottom of these cups **(Fig. 27.5*B*)**. Gemmae, when mature, break off from the thallus and are distributed into nearby areas. Raindrops are often the agents that break off gemmae and scatter them away from the thallus. New gametophyte plants grow from gemmae.

Sexual reproduction. The gametangia are very similar in structure to those of *Riccia.* Antheridia are pear-shaped bodies composed of a jacket of sterile tissue surrounding sperms (**Fig. 27.7*A*,** page 561). They are borne on disks raised above the thallus on slender stalks, called **antheridiophores (Fig. 27.5*C*).** Antheridiophores are modified portions of the thallus having furrows, rhizoids, and air chambers. Flask-shaped antheridia develop in cavities on the upper surface of the disk, the youngest antheridia being close to the other margin of the disk. Mature sperms are extruded in a mucilagenous mass.

Archegonia are structurally similar to those of *Riccia* with venter, neck, and cover cells **(Figs. 27.7*B*,*C*).**

Antheridia and archegonia of *Marchantia* and heterothallic; that is, gametophytes are either male or female. Archegonia are borne on specialized branches called **archegoniophores.** The disk at the top of the archegoniophore is frequently an eight-lobed structure. In development, the lobes bend downward and then grow inward toward the stalk. Thus, the youngest portion of an older disk bearing archegonia is on the underside and close to the stalk **(Figs. 27.5*D*, 27.7*B*).** Archegonia develop in the lower surface of the disk but in tissue similar to that found on the upper surface of the thallus. Striking fingerlike processes grow out from between the lobes. The archegonia mature, and fertilization takes place when the disk of the archegoniophore is but slightly elevated above the thallus.

Sporophyte. Zygotes develop, as in *Riccia,* into embryos, which are spherical masses of undifferentiated, colorless, dependent tissue (**Fig. 27.8*A*,** page 561). Embryos are diploid, the nuclei of all cells containing $2n$ chromosomes. Subsequent growth, however, is more complicated than in *Riccia.* Some cells of the embryo divide to form a mushroomlike growth that becomes embedded in gametophyte tissue. This growth is called the **foot.** Cells in the central portion of the embryo form a stalk or **seta.** As the seta elongates, the lower cells develop into a **sporangium (Fig. 27.8*B*).** Lengthening of the seta suspends the sporangium below the disk. While this development is taking place, the stalk of the archegoniophore elongates, lifting the disk with its archegonia well above the thallus **(Figs. 27.5*D*,*E*).**

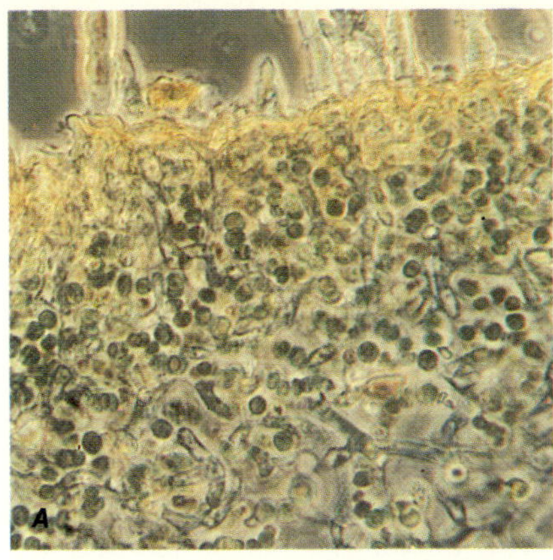
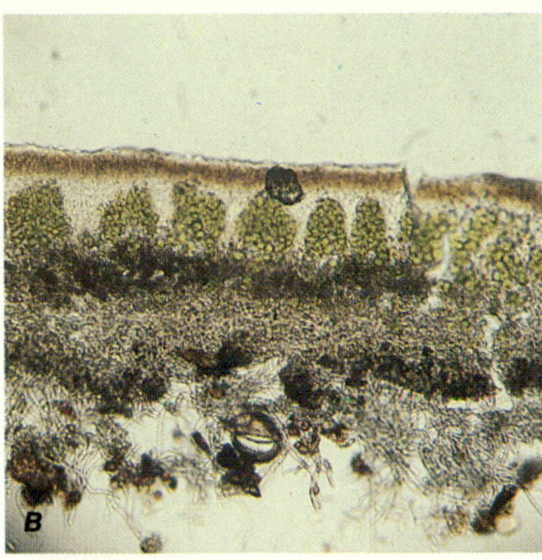

Figure 26.37
Cross sections of lichen thalli showing association of alga and fungus. *A*, *Collema* sp. with blue green alga, *(Nostoc)*, in short chains distributed throughout the lichen thallus, ×30. *B*, *Lecanora muralis* with the green alga *Trebouxia* in a definite layer. Section also shows the structure of a lichen thallus: upper cortex, algal layer, medulla, and hyphae extending downward into the substrate, ×10.

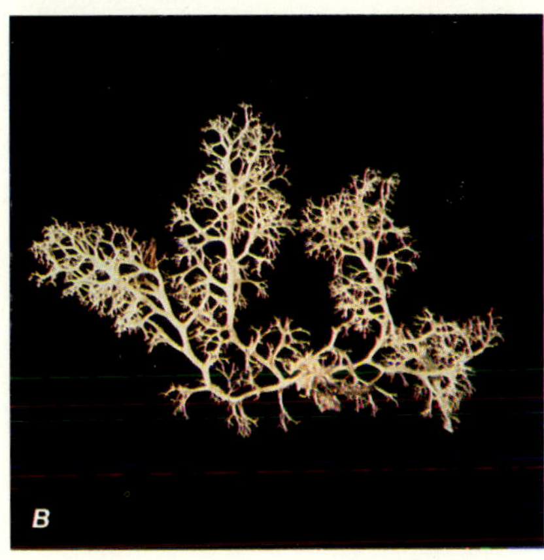
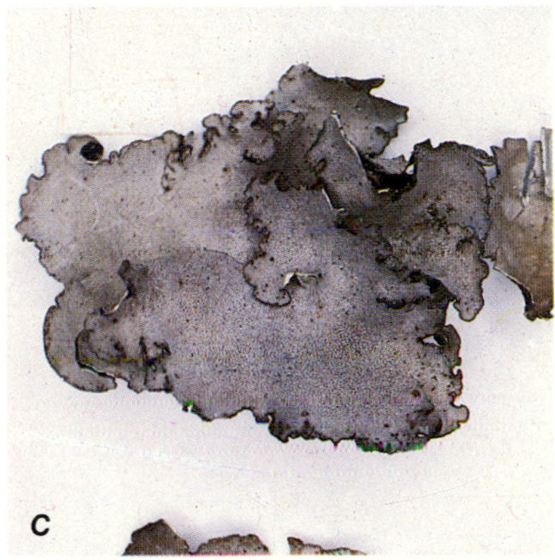

Figure 26.38
Various types of lichen thalli. *A*, crustose, *Lecidea atrobrunnea* with thallus of closely packed squamules, ×10. *B*, fruticose, *Cladonia impexa*, ×0.5. *C*, foliose, *Dermatocarpon reticulata* is a small foliose form growing on rocks, ×0.5. *D*, twofold thallus of *Cladonia crispata*. Note leaflike primary thallus and the upright secondary thallus, ×0.4.

Figure 26.39
Ascocarps. *A*, apothecium with proper margin, *Lecidea atrobrunnea*, ×30. *B*, section of apothecium of *Rhizocarpon bolanderi*, ×100. *C*, apothecium with a thalline margin, *Lecanora chrysoleuca*, ×20.

Figure 27.5
Marchantia. *A,* habit, shows dichotomously branching thallus with prominent midrib, gemmae cups, antheridial heads, and archegonial heads with mature (yellow) sporophytes, ×1/6. *B,* gemmae cups, ×1. *C,* antheridial heads, ×2. *D,* archegonial heads, ×1.5. *E,* undersurface of an archegonial head; the sporophytes are enveloped by a protective perianth. Immature sporophytes are green; ripening sporophytes show yellow, ×3. *F, G,* and *H,* time sequence showing opening of sporophyte capsule. (arrows), ×3. *I,* mass of spores being extruded from sporophyte capsule, ×10.

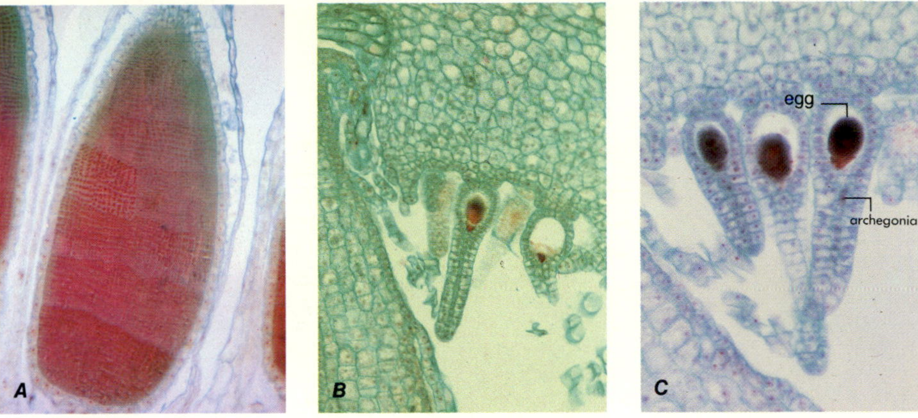

Figure 27.7
Light micrographs ot gametangia of *Marchantia*. *A*, antheridium in antheridial head, *B*, archegonium suspended from lower surface of archegonial head, *C*, details of archegonia.

Figure 27.8
Light micrographs of developing sporophyte of *Marchantia*. *A*, embryo sporophyte, *B*, elongating sporophyte showing three regions. *C*, mature sporophyte. *D*, higher magnification of mature sporophyte showing spores and elaters.

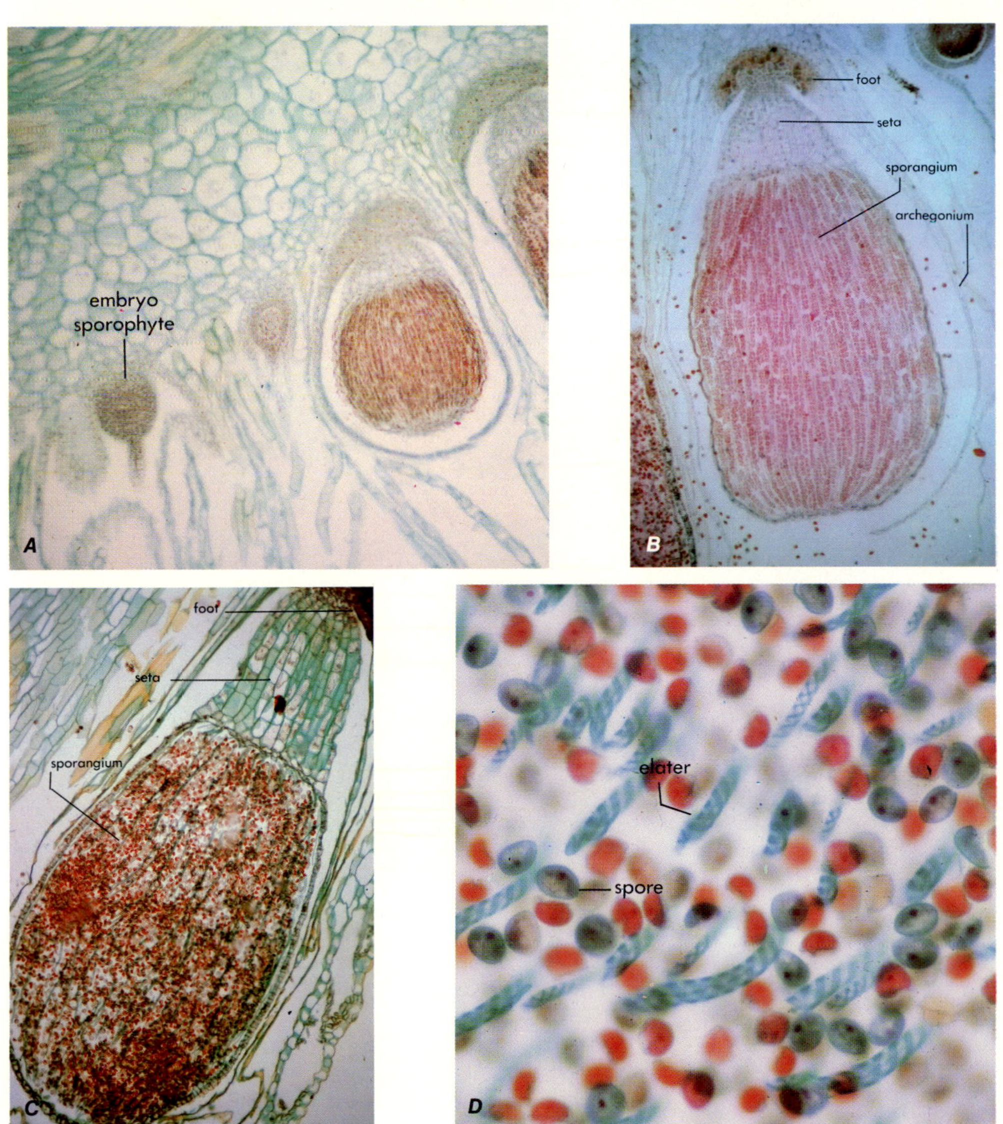

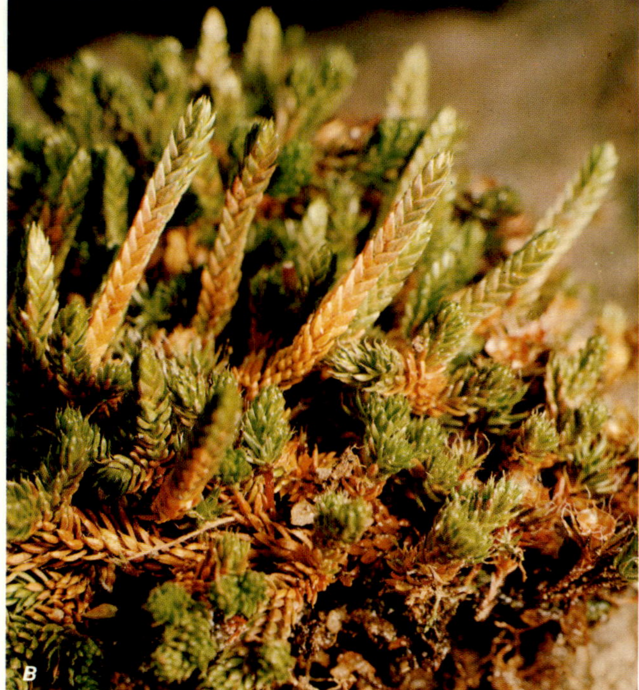

Figure 28.9
Selaginella. A, Selaginella watsonii, a montane species growing in crevices in granite. *B,* strobili of *S. watsonii. C, S. emmeliana,* native to tropical America and requiring a moister, warmer climate, ×⅓.

Fertilization stimulates enlargement of the old archegonium, which keeps pace with the enlargement of the developing sporophyte. As a result, the sporophyte is continually enclosed within the archegonium, which, because of its increase in size and change in function and shape, is now known as the **calyptra.** In addition, the surrounding gametophyte tissue produces two other envelopes that protect the sporophyte.

The mature sporophyte **(Fig. 27.8C)** is composed of three parts: **foot, seta,** and **sporangium.** The foot is an absorbing organ. The seta serves to lower the sporangium away from the archegonial disk and thus to facilitate distribution of spores. Before meiosis, the sporangium or capsule is composed of a jacket of sterile cells surrounding a mass of sporocytes, among which are a number of sterile elongated cells. Four spores, each containing *n* chromosomes, develop from each sporocyte. The elongated sterile cells are transformed into spiral elements that change shape under the influence of varying moisture conditions. These cells are called **elaters** and they aid in dispersal of spores **(Fig. 27.8*D*).**

Dissemination of spores is aided by two structural features not found in *Riccia:* (*a*) the sporangium of the sporophyte hangs from the lower side of the raised archegonial disk and (*b*) elaters help to empty the spore case. Spores develop immediately into new gametophytes if they land in an appropriate habitat.

Significant Steps in the Life History of *Marchantia*

1. The gametophyte thallus absorbs water and mineral salts from soil and carbon dioxide from air. It is green and can synthesize foods. All nuclei of the gametophyte are haploid (*n*).

2. Gametangia (antheridia and archegonia) are borne on upright branches called antheridiophores and archegoniophores.

3. The zygote is diploid (2*n*).

4. The embryo sporophyte, consisting of a spherical mass of undifferentiated cells, is formed from the zygote.

5. Cells of the embryo sporophyte differentiate into foot, seta, and sporangium.

6. The foot is an absorbing organ. It is embedded in the gametophyte, from which it receives nourishment.

7. The sporangium consists of a jacket of sterile cells surrounding a mass of sporogenous tissue interspersed with elongated sterile cells.

8. The seta is a stalk that, in lengthening, lowers the sporangium below the surface of the archegonial disk.

9. Repeated mitotic cell divisions in sporogenous tissue result in sporocytes, each containing 2*n* chromosomes.

10. Sporocytes divide by meiosis into quartets of spores, each containing *n* chromosomes.

11. The elongated sterile cells are transformed into spiral elaters.

12. The presence of elaters and the hanging position of the sporophyte aid in disseminating spores.

13. Spores germinate immediately into new gametophyte plants.

Jungermanniales

This is the largest order of liverworts containing two-thirds of all species in the Hepaticae. The gametophytes of these liverworts are leafy, in contrast to thallus plants; that is, the gametophytes have stemlike and leaflike organs that may be held somewhat erect, off the ground. Rhizoids anchor the above-ground portion. A very common, widespread genus in this order is *Porella* (Fig. 27.9).

Porella plants occur in dense mats. Shaded, older portions of the stems lack leaves, and rhizoids arise in those regions. Some species, such as the one illustrated, are nearly devoid of rhizoids, so their major function is not the absorption of water or nutrients. Younger portions of stems are densely clothed with leaves that arise in three ranks. The lower (ventral) rank of leaves is much smaller than the upper two (Fig. 27.9*C*). The leaves are one cell thick and lack a cuticle.

Antheridia occur in the axils of leaf and stem, and archegonia occur on short, leafy branches. Sporophytes resemble those of *Marchantia.*

CLASS ANTHOCEROTAE

The Anthocerotae have the simplest gametophytes of the Bryophyta. They are small, green thallus plants with little internal differentiation of vegetative tissues. They are slightly lobed, with numerous rhizoids growing from the lower surface (Fig. 27.10). The antheridia are similar in structure to those encountered among the Hepaticae. They are located in roofed chambers in the upper portion of the thallus (Fig. 27.11*A*). The archegonia are embedded within the thallus and in direct contact with the vegetative cells surrounding them (Fig. 27.11*B*).

The sporophyte (Figs. 27.10, 27.12) of the Anthocerotae is in striking contrast to those of the Hepaticae. The subepidermal cells contain chloroplasts, and typical stomata are found in the epidermis. A foot embedded in the thallus serves as an absorbing organ. The sporangium is an upright elongated structure. Sporogenous tissue forms a cylinder parallel with the elongated axis of the sporangium. Spores mature in progression from top down. A meristematic region lies just above the foot and continually adds new cells to the base of the sporangium. Its presence means that spores may be produced over long periods.

Under exceptionally favorable growing conditions, the sporophyte may lengthen greatly. Some sporogenous tissue at the base of the sporangium may be replaced by a conspicuous conducting strand. The foot enlarges and, through decay of the gametophyte, comes into more or less direct contact with the soil. Such

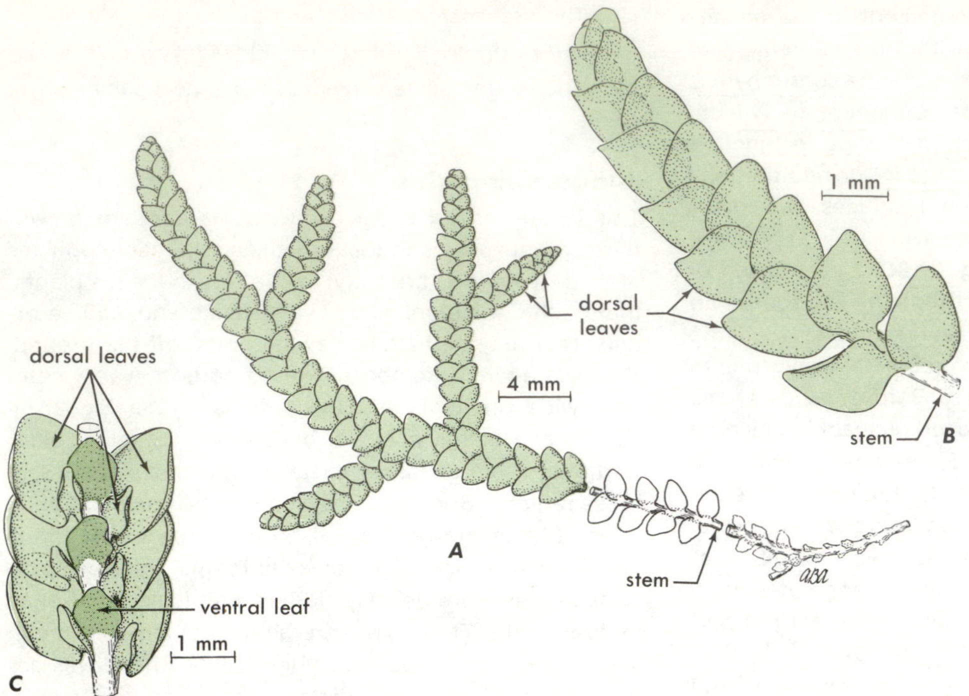

Figure 27.9
Porella, a leafy liverwort. *A*, habit sketch. *B*, detail of two dorsal ranks of leaves. *C*, detail of smaller, ventral leaves.

sporophytes are capable of maintaining themselves independently for some time, and they represent the most primitive sporophytic terrestrial plants.

Figure 27.10
Anthoceros gametophyte with upright dependent sporophytes, ×3.

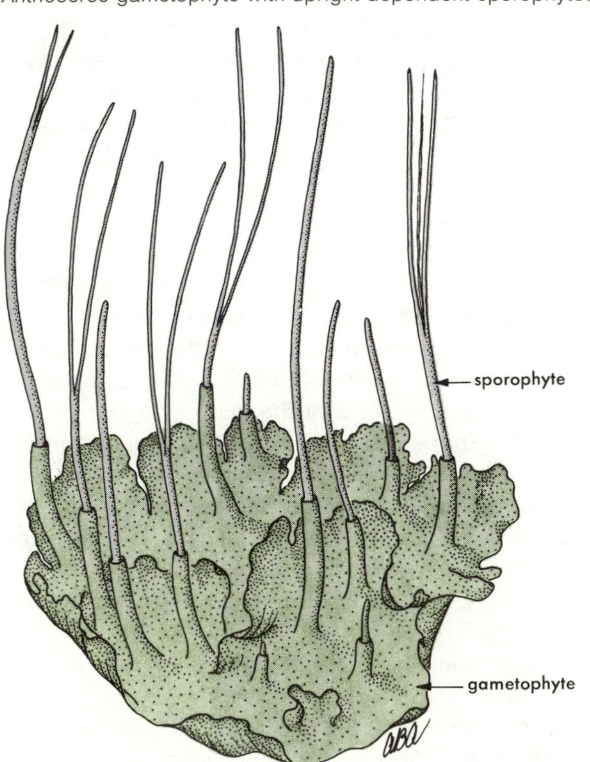

CLASS MUSCI

Although the mosses are small plants, they are nevertheless conspicuous. They frequently cover rather large areas of stream banks. They grow on rocks and trees, and are sometimes submerged in streams. Although some mosses are able to resist considerable drought, all require moisture for active growth and reproduction.

Mosses as a class show great structural uniformity. Gametophytes of nearly all 20,000 species have two growth stages: (*a*) a creeping, filamentous stage (the protonema), from which is developed (*b*) the moss plant with an upright or horizontal "stem" bearing small, spirally arranged green "leaves."

Gametophyte

Germinating spores of mosses do not develop directly into a leafy gametophyte but first become a filamentous structure. This early stage of gametophyte is called the **protonema** (Fig. 27.13). The protonema is not a permanent structure, although it may branch considerably under favorable conditions and cover a rather large area of soil, sometimes forming a green coating resembling algal growth. Cells composing the protonema contain numerous chloroplasts. At various locations, cell divisions along the protonema produce swollen regions called "buds". From them develop the upright gametophyte stage. It is interesting that although a moss protonema is gametophytic tissue of a lower plant, cytokinins, which are growth hormones that stimulate cell

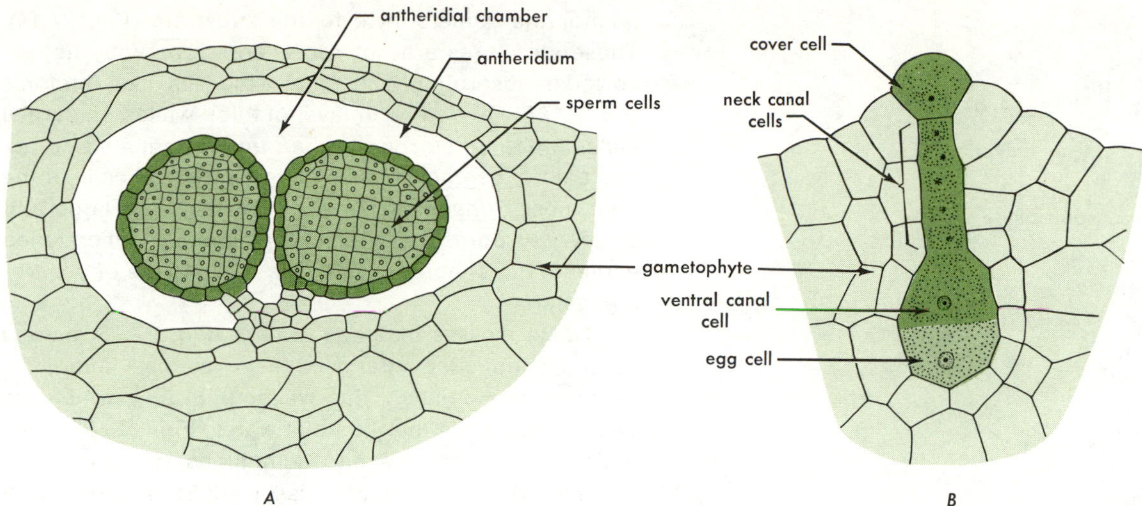

Figure 27.11
Section through an antheridium (*A*) and an archegonium (*B*) of *Anthoceros*, ×50. (Redrawn from G. M. Smith et al., *A Textbook of General Botany*. © 1953 by the Macmillan Company. Reprinted with permission of the publisher.)

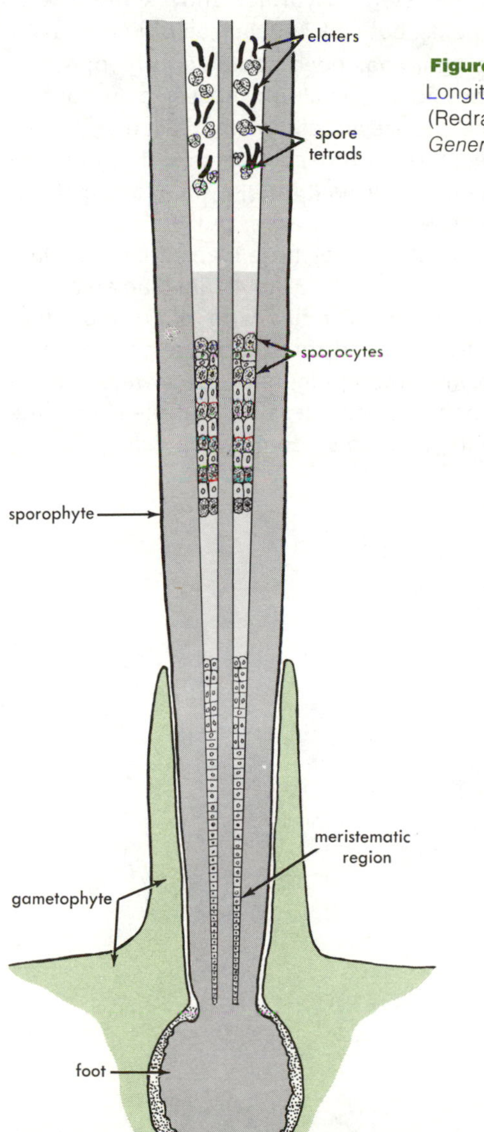

Figure 27.12
Longitudinal median section through sporophyte of *Anthoceros*. (Redrawn from R. M. Holman and W. W. Robbins, *A Textbook of General Botany*, John Wiley & Sons, New York.)

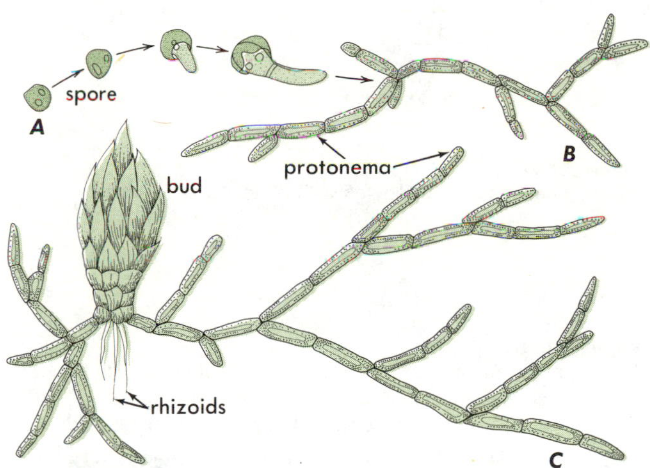

Figure 27.13
Funaria. *A*, germination of meiospores. *B*, protonema. *C*, protonema with "bud." (Redrawn from W. P. Schimper, *Recherches sur les mousses*, Strasbourg, 1848.)

division and bud development in higher plants (p. 400), also induce bud formation in moss protonemata. In fact, a rapid biological test for cytokinin activity has been developed using moss protonemata. In this test, 20-day-old protonemata of the moss *Funaria hygrometrica* are treated with the sample being tested for cytokinin activity. If cytokinin is present in the test solution, buds develop on the protonemata within 48 hours.

The mature "stem" of the leafy gametophyta shows a wide range of differentiation, depending on the species, age, and environment. Elongate cells called rhizoids

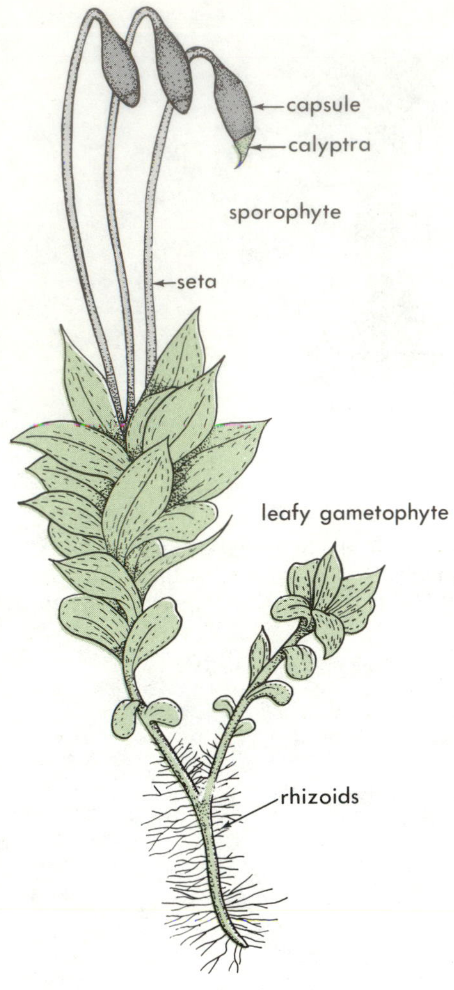

Figure 27.14
Mnium, showing gametophytes with attached sporophytes.

anchor the gametophyte to the substrate (Fig. 27.14). These structures are not roots, and apparently not involved in absorption. The simpler mosses, like *Tetraphis* (Fig. 27.3*E*) have an outer layer of thick-walled epidermal cells **(stereids)**, surrounding an undifferentiated cortex made up of parenchymalike cells. Epidermal cells of the peat moss *Sphagnum* (Figs. 27.3*A* to *D*) are large and empty, with pores that open to the outside. As mentioned earlier in the chapter, these cells are capable of absorbing water.

Some mosses, such as *Mnium* (Fig. 27.15) have a central conducting strand in their "stems." *Mnium* is made up of elongated, thin-walled **hydroids**, which are dead, empty cells that conduct water (Fig. 27.16). Their end walls are oblique and sometimes very thin, perforated with pores, or partly dissolved. Experiments with dyes show that translocation of water can occur in these cells. Some moss "leaf" midribs also contain hydroids, but only in one genus are these known to connect with the hydroids of the "stem." Hydroids show some resemblances to tracheids, but lack specialized pitting and lignified walls. No lignin has been detected in bryophytes.

A few of the most specialized mosses also contain cells resembling the sieve cells of the vascular plants. Between the epidermis and central strand, elongated cells with oblique end walls **(leptoids)** may occur (Figs. 27.16, 27.17). These cells are alive, but their nuclei are degenerate and inactive. They have many enlarged plasmodesmata in their end walls (35 to 40 nm diameter), and callose may be present. Studies with $^{14}CO_2$ show that sugars may be translocated through these cells at rates of 0.3 to 50 cm/hr. Although leptoids are always associated with parenchyma cells, it is not yet clear that the parenchyma cells function as companion cells.

Figure 27.15
Gametophyte "stems" of mosses. *A*, longitudinal section through "stem" of *Mnium* showing central strand of sclerenchymalike cells, ×100. *B*, cross section of *Mnium* showing thick-walled stereids; thin-walled cortex; central region of partially collapsed hydroids, ×72.

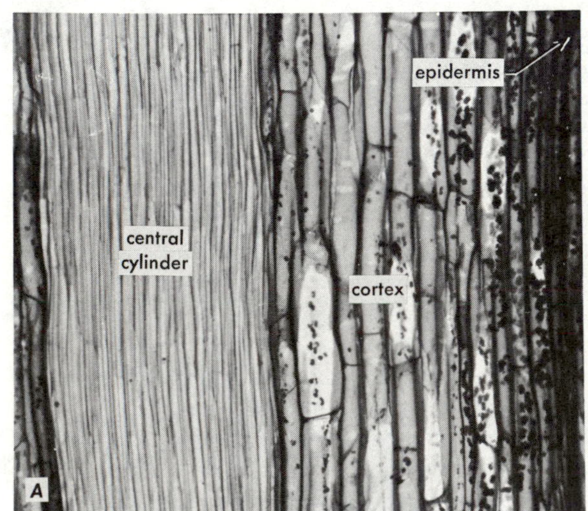

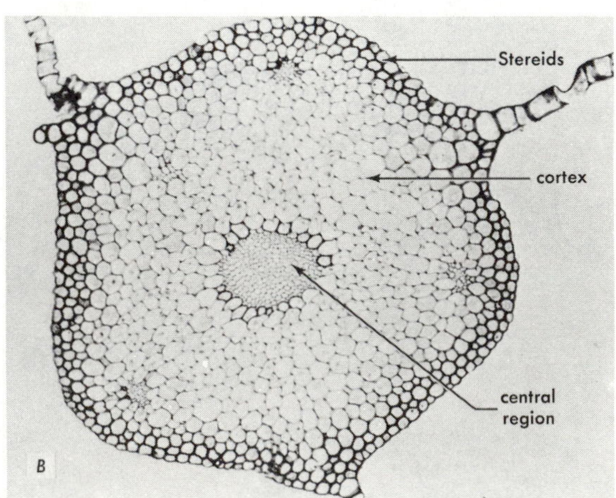

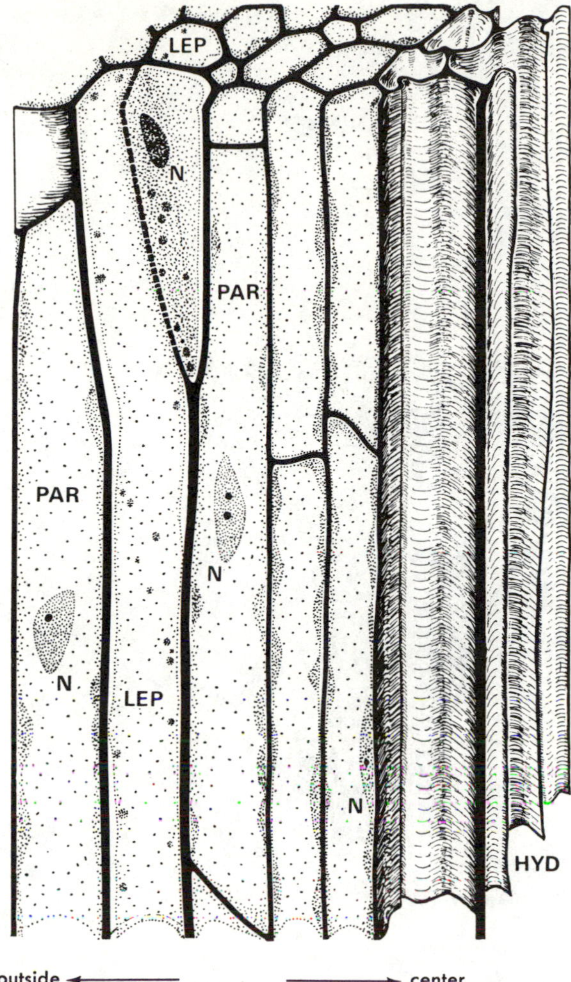

Figure 27.16
Partial "stem" portion of a moss showing "phloemlike" leptoids (LEP) and "xylemlike" hydroids (HYD). PAR=parenchyma, N=nucleus.

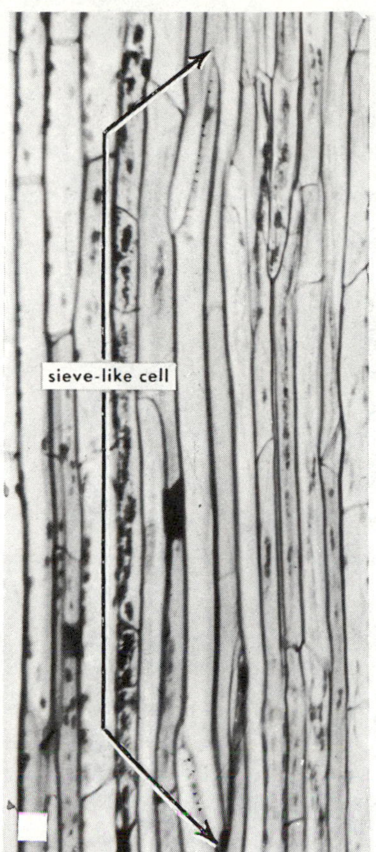

Figure 27.17
Longitudinal section through gametophyte "stem" of *Polytrichum* with phloemlike cells, ×100. (Courtesy of E. M. Gifford, Jr.)

It is important to emphasize that: (1) most mosses do not contain leptoids; (2) there are important differences between hydroids and tracheids, and between leptoids and sieve cells; (3) there is as yet no firm evidence to show these cells are phylogenetically related to conducting cells of vascular plants. If there is a relationship, it is that these cells are degenerate forms, derived from primitive vascular plants. That is, mosses were probably not the precursors of vascular plants, but rather the other way around.

Many mosses reproduce asexually with gemmae that form on the tips of leaves or in clusters, at the end of stalks that branch from the main stem.

Sexual Reproduction

The gametophytes of many mosses are monoecious (homothallic); that is, both antheridia (sperm-producing) and archegonia (egg-producing) are produced by the same gametophyte (Fig. 27.18). Some mosses are **dioecious** (heterothallic); in other words, antheridia and archegonia are produced on separate gametophytes. In *Mnium,* shoots-bearing antheridia are easily recognized, by the surrounding leaves that spread around them somewhat like petals of a flower. The group of antheridia appears as an orange spot in the center of the terminal cluster of "leaves" (Fig. 27.19).

Antheridia of most mosses (Fig. 27.19) are enclosed by cells forming a **sterile jacket.** These cells contain chloroplasts that become orange-red when the antheri-

Figure 27.18
Mnium, showing location of gametangia, ×1.

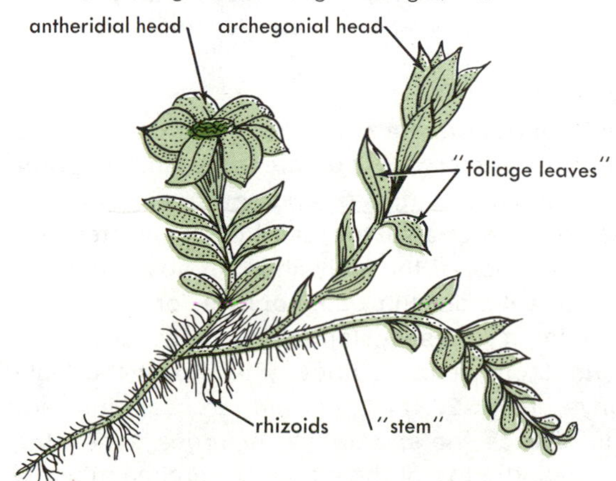

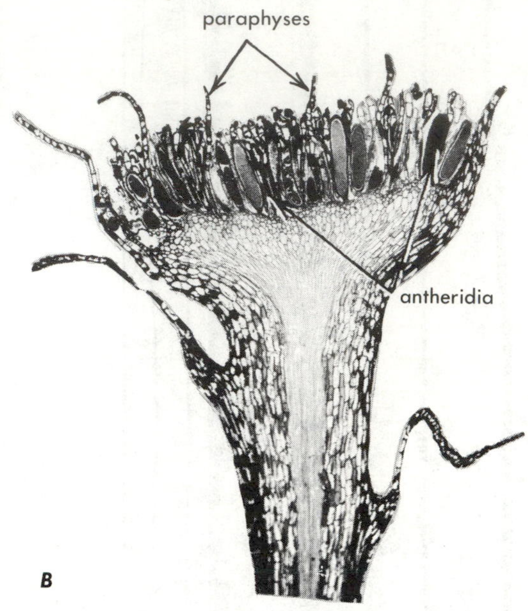

Figure 27.19
Antheridial heads of mosses. *A, Polytrichum. B, Mnium.* In some mosses, archegonia may be found mixed with antheridia on the same head.

dium ripens. The sperms consists mainly of an elongated nucleus and each has two flagella. The antheridia are surrounded by club-shaped, multicellular, sterile hairs **(paraphyses)** with conspicuous chloroplasts.

The archegonia have a long **neck,** a thickened middle portion called the **venter,** which contains the egg, and a long **stalk** (Fig. 27.20). When sufficient moisture is present, sperms are extruded from the antheridium. The neck cells of the archegonium separate, opening a passage to the egg. Sperms swim down, possibly responding to a chemical signal, and one fuses with the egg. Mosses are dependent on free water for fertilization.

Sporophyte

Soon after fertilization, the zygote begins to develop into a spindle-shaped embryo that differentiates into a sporophyte consisting of a **foot, seta,** and **sporangium** or **capsule.** The foot penetrates the base of the venter and grows into the apex of the leafy shoot. It absorbs water and nourishment from the gametophyte for its growth and development. The seta elongates rapidly, raising the sporangium 1 cm or more above the top of the leafy gametophyte (Fig. 27.14). The old archegonium increases in size as the sporophyte enlarges. When the seta elongates, the top of the expanded archegonium is torn from its point of attachment to the gametophyte and elevated with the sporangium. The upper end of the old archegonium, now known as the **calyptra,** remains for a time as a covering for the sporangium (Fig. 27.14).

The sporangium of mosses may measure 1 to 3 mm in diameter and 2 to 6 mm long. It is surrounded by an epidermal layer composed of cells similar to those in the epidermis of higher plants. Stomata occur in the epidermis covering the lower half of the sporangium. The sterile tissue forming the inner portion of the sporangium may conveniently be divided into three regions, each of which may be recognized by the type of cells comprising it. The fourth region, composed of sporogenous cells, forms a layer around the columella (Fig. 27.21). The cells of the sporogenous tissue (sporocytes) may increase in number by mitosis; each cell contains the diploid number of chromosomes. Meiosis takes place next, and haploid spores containing $1n$ chromosomes result. They are the first cells of the gametophyte generation.

The columella projects upward, forming a small dome above the main mass of the sporangium. The four or five outer layers of cells of this dome differentiate into a dry, brittle cap called the **operculum.** The cells immediately beneath the operculum form a double row of triangular **peristome teeth** (Fig. 27.22). The broad bases of

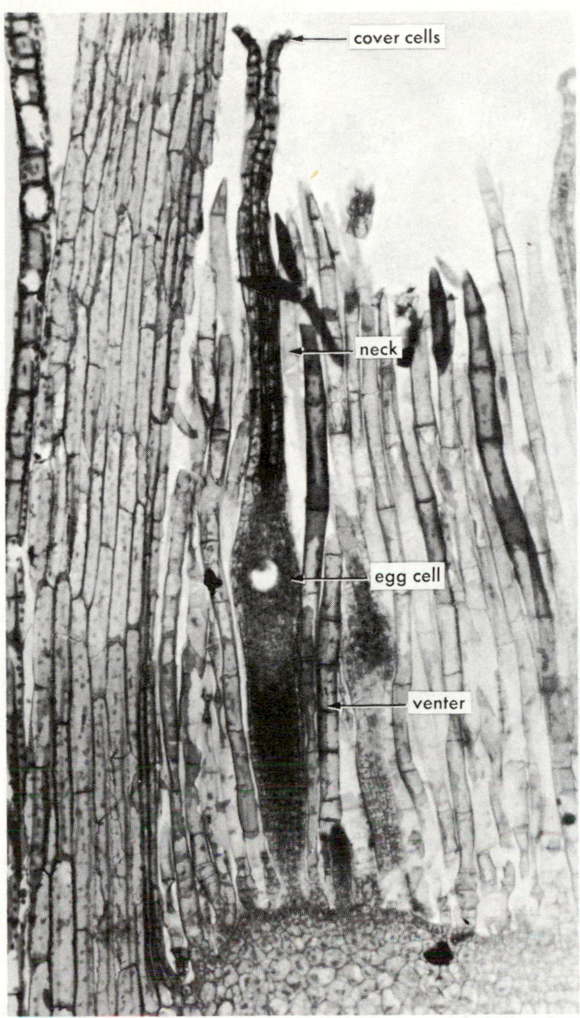

Figure 27.20
Archegonium surrounded by paraphyses *(Mnium)*, ×50.

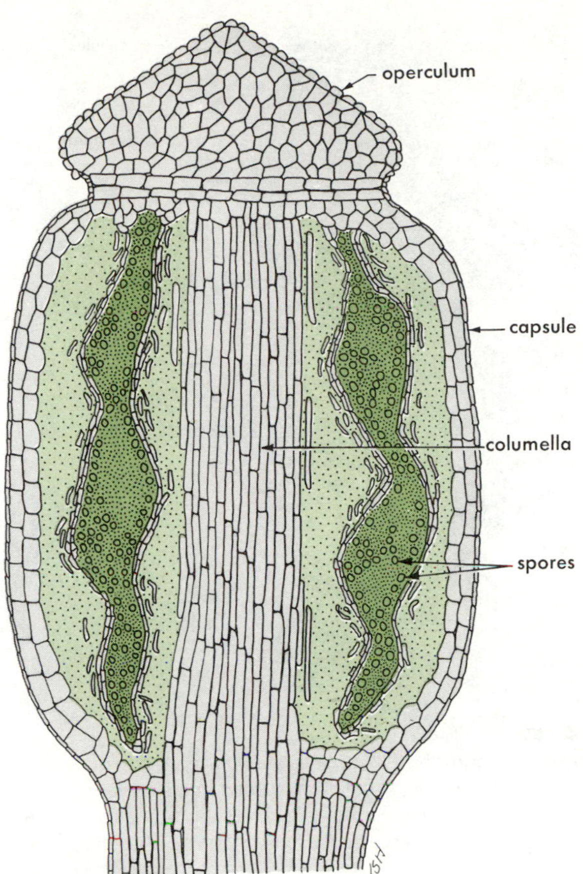

Figure 27.21
Median section through a mature but unopened capsule of *Mnium*, ×25.

the teeth are attached to the thick-walled deciduous cells that form the **annulus** around the upper end of the sporangium. When the sporangium matures and becomes dry, thin-walled cells of the annulus break down, allowing the operculum to fall away and expose the peristome teeth. By this time, most of the thin-walled cells within the columella have collapsed and the cavity thus formed is filled with a loose mass of spores. Peristome teeth are rough and are very sensitive to the amount of moisture in the air. When they are wet or the atmospheric humidity is very high, they bend into the cavity of the sporangium; when dry, they straighten and lift out some of the spores, which are then disseminated by air movements. If a spore comes to rest on moist soil and if illumination and temperature are favorable, it germinates and grows into a protonema.

SUMMARY OF MOSS LIFE HISTORY AND CHARACTERISTICS

1. The gametophyte consists of (*a*) a filamentous, branched, algallike structure called a protonema, and (*b*) leafy shoots that develop from buds on the protonema. The shoots consist of a stalk bearing rhizoids at its lower end and "leaves" throughout its length. The gametophyte is green and able to synthesize food. All gametophyte nuclei are haploid.
2. Antheridia and archegonia develop at the apex of the leafy gametophyte.
3. Each antheridium produces hundreds of sperms.
4. A single egg is formed in the venter of each archegonium.
5. Gametes (eggs and sperms) each contain 1*n* chromosomes.
6. When sufficient moisture is present, mature sperms are extruded from the antheridium. The neck of the mature archegonium is open and sperms swim down it to the egg in the venter of the archegonium.
7. Each egg is fertilized by one sperm.
8. The zygote, as a result of the union of gametic nuclei, is diploid.
9. An embryo sporophyte, consisting of a spindle-shaped mass of undifferentiated cells, partially dependent on the gametophyte, develops from the zygote.
10. The embryo develops into a sporophyte consisting of a foot, seta, and sporangium. The sporangium

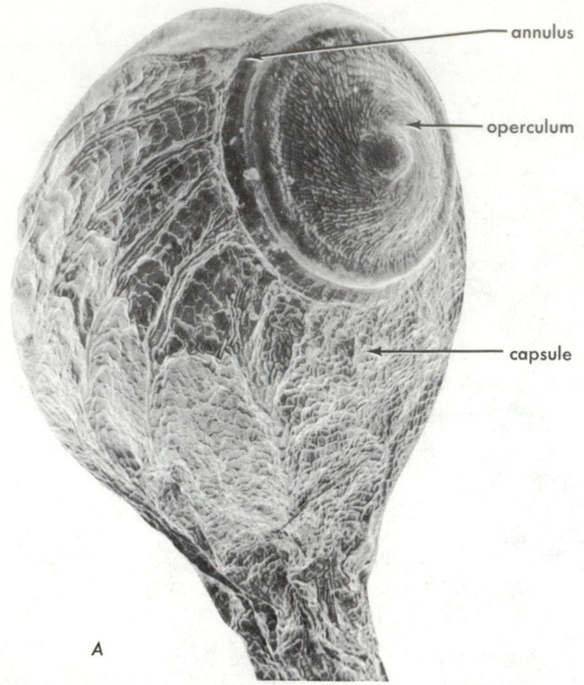

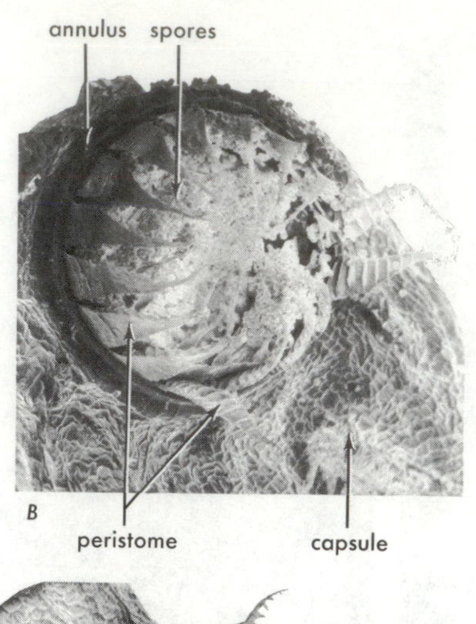

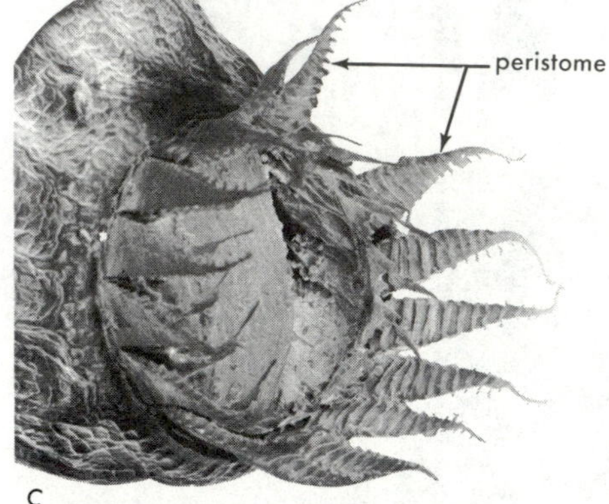

Figure 27.22
Stereoscan micrograph of a capsule of *Funaria*. *A*, capsule with operculum in place. *B*, operculum has been shed, peristome teeth partially open, spores attached to them. *C*, peristome teeth fully open, capsule empty, ×300. (Courtesy of W. Russell and D. Hess.)

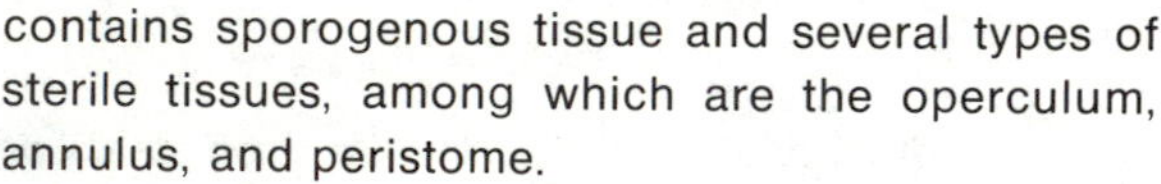

contains sporogenous tissue and several types of sterile tissues, among which are the operculum, annulus, and peristome.

11. Sporocytes, each containing $2n$ chromosomes, form from sporogenous cells.
12. Sporocytes undergo meiosis. Four spores result from each sporocyte.
13. When spores germinate, they form the protonema.
14. In the central strand of some mosses, for example, *Sphagnum* and *Mnium*, there are specialized cells called hydroids that may be involved in water conduction.
15. Cells specialized for sugar transport, called leptoids, are also found in some mosses.

SUMMARY

1. Members of the Bryophyta differ from thallophytes in having: (1) a sterile, protective jacket around developing gametes (i.e., the gametangium is multicellular); (2) in generally being parenchymatous rather than filamentous; and (3) in possessing an embryo stage (i.e., the zygote is protected and nurtured by the parent, rather than being formed in the open environment). The female gametangium is flask-shaped and contains a single egg; it is called an archegonium. The male gametangium is still called an antheridium, as in the thallophytes. There is a sporic life cycle with unlike generations. The gametophyte generation is dominant and independent, the sporophyte is shorter-lived and partially or entirely parasitic on the gametophyte.
2. Asexual spores are not formed. Asexual reproduction occurs by fragmentation or by dispersal of bits of special tissue called gemmae.
3. Bryophytes are divided into three classes: Hepaticae (liverworts), Musci (mosses), and Anthocerotae (hornworts). Although bryophytes are small plants that grow in moist habitats, some species exhibit traits that adapt them to terrestrial, even dry, conditions: stomata, cuticle, vascular tissue. These plants may represent simplified, derived offspring from primitive vascular plants, rather than being evolutionary precursors of vascular plants.
4. Three genera of Hepaticae are described: *Riccia, Marchantia,* and *Porella* (the latter is a "leafy" liverwort that resembles a moss). The life cycle of *Riccia* is summarized on pages 554–555 and that of *Marchantia* on pages 555–563.
5. The life cycle of *Anthoceros* is described on pages 563–564. Some sporophytes are capable of rooting in the soil and thus become independent of the gametophyte, in contrast to other members of the division.

6. Musci is the largest class in the Bryophyta with 20,000 species. A typical moss life cycle is described on pages 564–570. Some mosses, such as *Mnium,* have a central conducting strand in the gametophyte axis consisting of tracheidlike hydroids and sieve-tubelike leptoids. As with other members of this division, the sporophyte consists of a parasitic foot, a seta, and a capsule that produces meiospores.

CLASSIFICATION

Division	Bryophyta
Class	Hepaticae
Order	Marchantiales
Genera	*Riccia*
	Ricciocarpus
	Marchantia
Order	Jungermanniales
Genus	*Porella*
Class	Anthocerotae
Genus	*Anthoceros*
Class	Musci
Genera	*Funaria*
	Mnium
	Polytrichum
	Sphagnum

CHAPTER 28

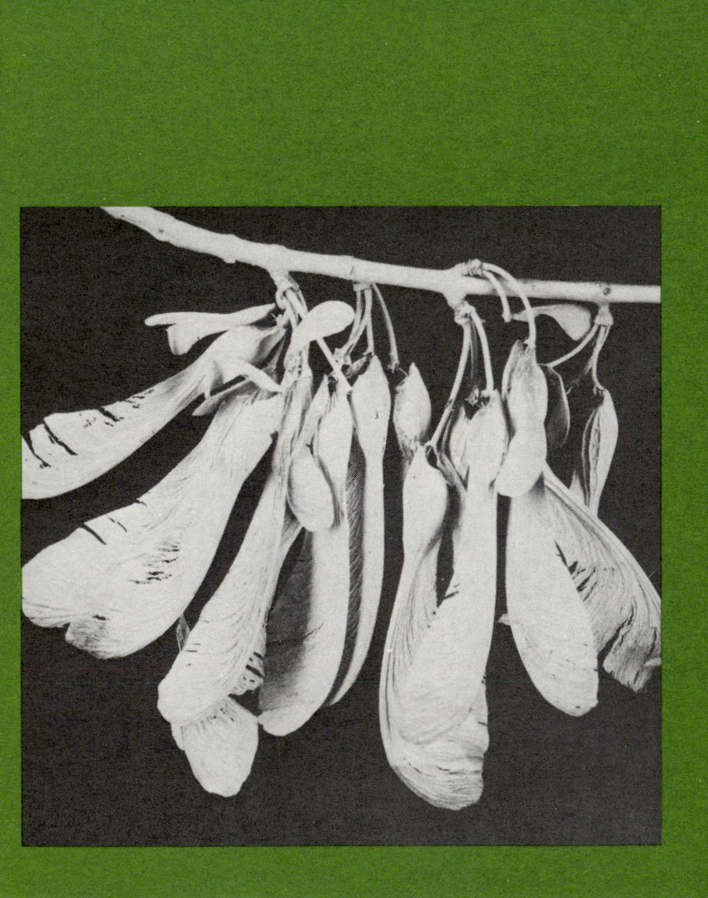

LOWER VASCULAR PLANTS

In contrast to algae, fungi, liverworts, and mosses, the sporophytes of vascular plants possess a well-developed vascular system that serves to conduct water, mineral salts, and food. There are 10 divisions of vascular plants; six are seed-bearing and four do not bear seeds. The latter four make up the lower vascular plants: **Psilophyta, Lycophyta, Sphenophyta,*** and **Pterophyta.**

The number of families and genera in these divisions is small, and their members do not form a very conspicuous part of the current land flora. Only the Pterophyta (ferns) has many species and they are prominent in certain cool, moist habitats. However, plant fossils indicate that there was a period in the earth's history when the members of the Lycophyta, Sphenophyta, and Pterophyta formed a large and dominant flora (Chapter 31).

Although they are land plants, they still require free water for fertilization. The sporophyte is the dominant generation and—except for the youngest stages in the formation of an embryo—it is independent of the gametophyte. Gametophytes are small, and some may lack chlorophyll. In its simplest form, the plant body of the sporophyte is an axis. Meristematic tissue terminates the shoots and roots. Most lower vascular plants are herbaceous perennials, but some ferns reach tree stature and have some woody tissue (Fig. 28.1).

DIVISION PSILOPHYTA

This division is represented in the existing flora by a single family, the Psilotaceae, with two genera, *Psilotum*

* We choose to use the old designation, Sphenophyta, for the division to which the horsetails belong, because an alternative designation, Arthrophyta, might be confused with Anthrophyta or Arthropoda.

Figure 28.1
A tree fern *(Cyathea)* growing in Golden Gate Park, San Francisco, $\times 1/50$.

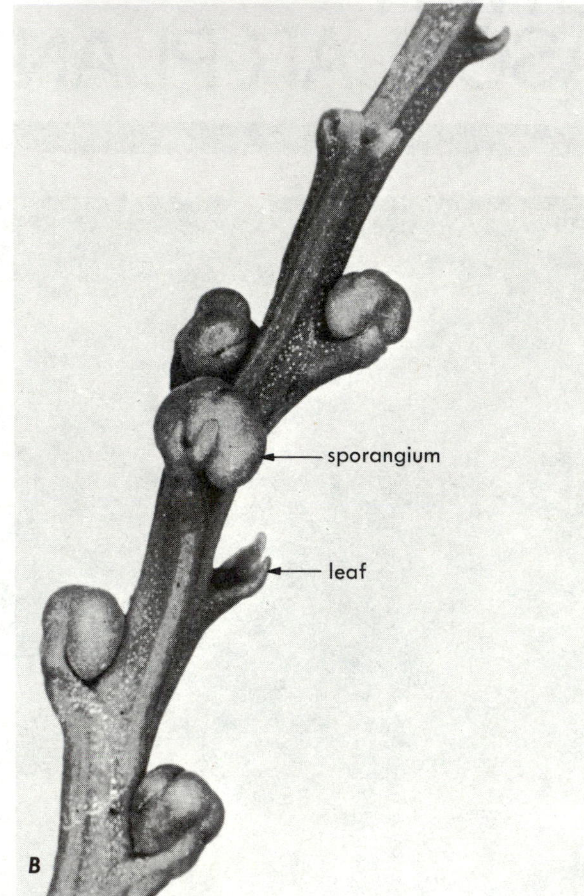

Figure 28.2
Psilotum nodum. A, plant. Notice sporangia on branches at left, $\times 1/5$. *B,* end of branch showing scale leaves and sporangia at nodelike swellings, $\times 2$.

and *Tmesipteris.* They are rare plants found mainly in the tropics, one species of *Psilotum* grows as far north as Florida. Figure 28.2 shows the simple plant body of *Psilotum nudum.* Roots are not present and leaves (microphylls) are small and scalelike. These are not "true" leaves (megaphylls); they are instead more like epidermal outgrowths. The branching, upright stem is slightly flattened and contains chlorophyll. A fungus is always associated with the branched rhizome, which is clothed with many rhizoids. The vascular tissue is simple (Fig. 28.3), consisting of poorly developed phloem and xylem. An endodermis, with a conspicuous Casparian strip, separates the central vascular tissue from the outer cortex. The xylem contains spiral and scalariform tracheids.

Sporangia are borne in axils of some of the scalelike leaves (Fig. 28.2*B*). Meiosis occurs in the sporangia, forming tetrads of meiospores. When the meiospores are ultimately released from the sporangia, they may land in a suitable habitat, germinate, and divide by mitosis repeatedly to form a gametophyte.

Psilophyta gametophytes are little more than a nonphotosynthetic axis of parenchymatous tissue several millimeters long (Fig. 28.4). Xylem and phloem are rarely present. Gametophytes may grow on the surface of soil, rock, or bark, or may grow within the soil. Nutrients are provided via a symbiotic or parasitic relationship with a fungus, which ramifies throughout the tissue.

Gametangia are scattered over the surface of the gametophyte: flask-shaped archegonia with eggs, and

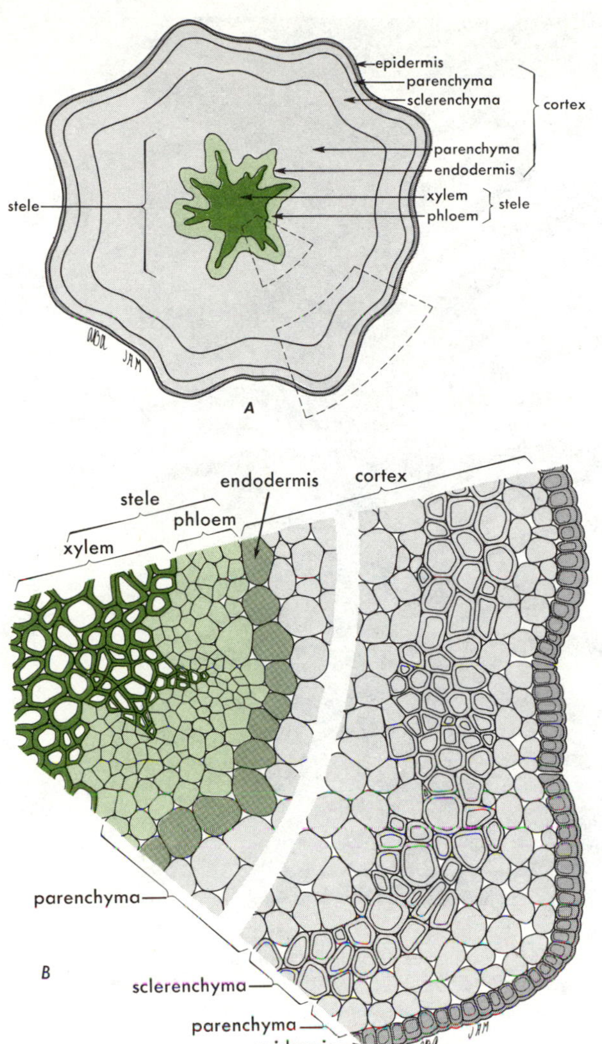

Figure 28.3
Cross section of *Psilotum* stem. *A,* diagram to show arrangement of tissues, ×8. *B,* enlarged sector showing cellular detail, ×450.

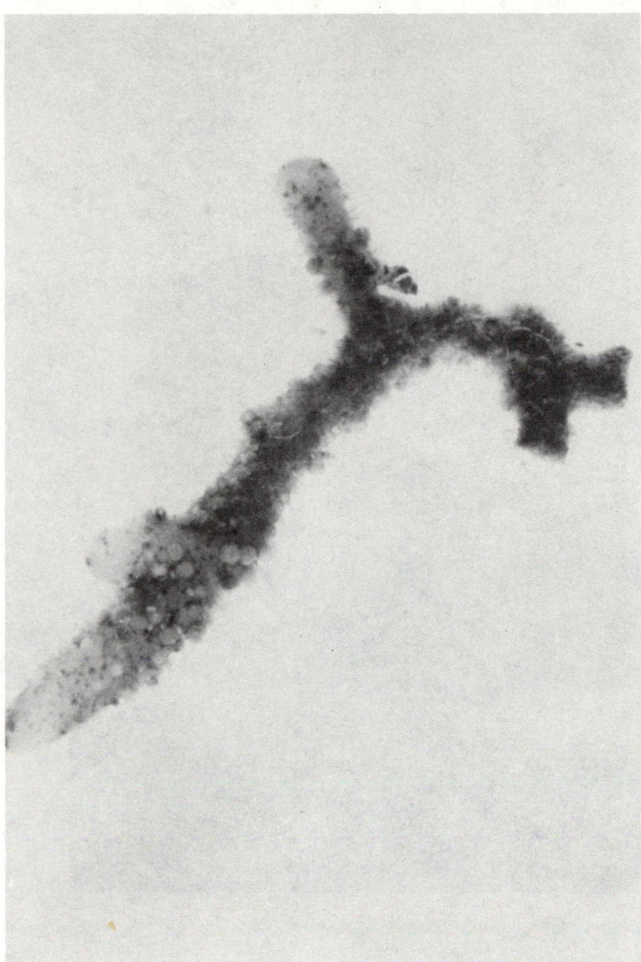

Figure 28.4
Gametophyte of *Psilotum*. Spherical objects are antherida. (Courtesy of D. Brandon.)

spherical antheridia with spirally coiled, multiflagellate sperm.

After a sperm swims through free water and reaches an egg, fertilization occurs and a diploid zygote cell begins dividing. The young sporophyte (embryo) consists of a shoot-root axis attached to the gametophyte by a **foot.** While the shoot-root portion is assuming form, the foot enlarges by repeated cell divisions, sending haustorial outgrowths into the gametophytic tissue. The foot, by virtue of its position and organization, is well suited for the functions of anchorage and absorption until the shoot-root axis becomes physiologically independent and a mature sporophyte results.

DIVISION LYCOPHYTA

The Lycophyta are well-represented in the northern hemisphere by three genera: *Lycopodium, Selaginella,* and *Isoetes.* We shall consider only the first two genera in detail.

There are many fossil Lycophyta belonging to several orders, Lepidodendrales being one of the best known. Now-extinct representatives of this order were forest trees, some of which bore seeds and possessed megaphylls (Chapter 31).

The leaves of the living members of this division are generally small and usually arranged in a spiral. Leaf gaps are never formed at the junction of stem and leaf in the Lycophyta, so that a stem of a member of this division has an unbroken central core of xylem. The leaves are microphylls; the central vein does not branch.

Lycopodium

Approximately 400 species are in this genus. Most are trailing plants, many forming short, upright branches that loosely resemble pine seedlings (Fig. 28.5). They are frequently called "ground pine" or "club moss." Although widely distributed, they are most abundant in subtropical and tropical forests. They cannot grow in arid habitats. Several species grow in the eastern and north-

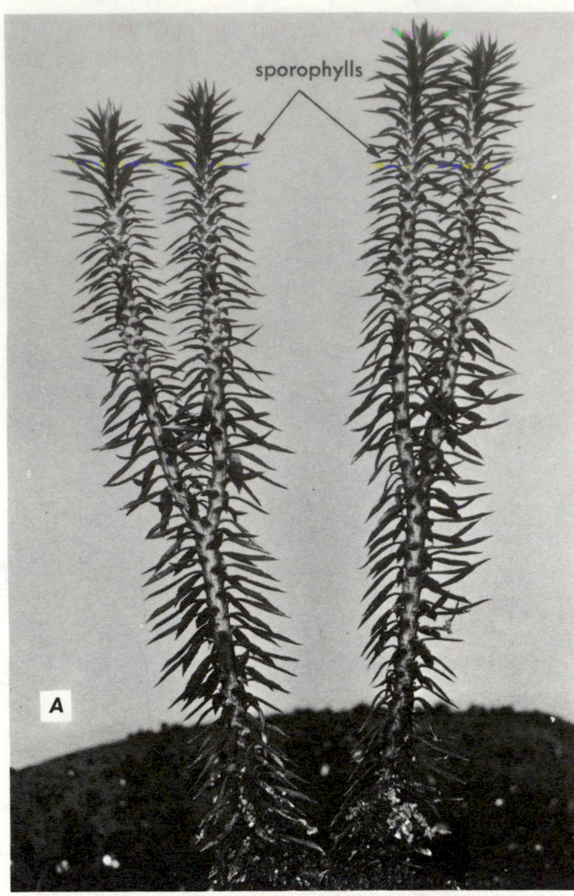

Figure 28.5
Sporophylls in *Lycopodium*. *A* and *B*, sporophylls similar to vegetative leaves in *L. selago*. *C*, sporophylls different from vegetative leaves and grouped in terminal strobili in L. obscurum.

western United States, but none occur in the more arid states of the southwest. Some eastern species are in danger of extinction because they are popular as Christmas decorations.

Mature Sporophyte

The main stem branches freely and is prostrate. Upright stems, approximately 20 cm in height, grow from the horizontal stem. Both types of stems are sheathed with small green leaves. Small but well-developed adventitious roots arise irregularly from the underside of the horizontal stem. As in many higher plants, the primary root, which grows from the embryo, is not long-lived.

The xylem in a *Lycopodium* stem occurs as a complex system of anastomosing strands, so that its distribution varies greatly in different cross sections of the stem (Fig. 28.6). The phloem is present between the strands of xylem and thus, too, forms a complex system of anastomosing strands. The xylem is composed of tracheids, whereas the phloem contains sieve cells and some parenchyma cells. An endodermis encircles the vascular cylinder. *Lycopodium* stems also lack pith, as do roots of most higher vascular plants.

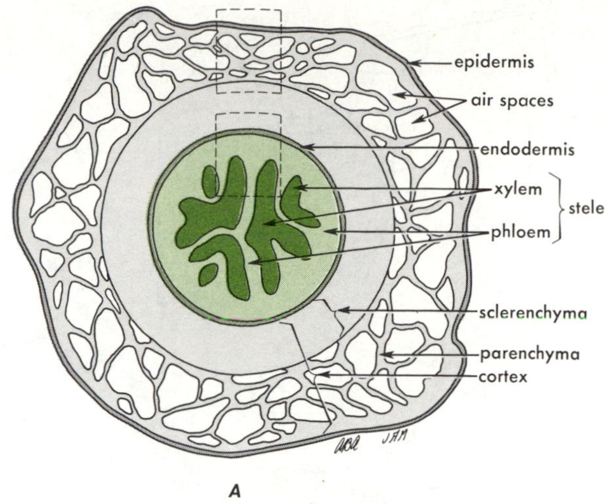

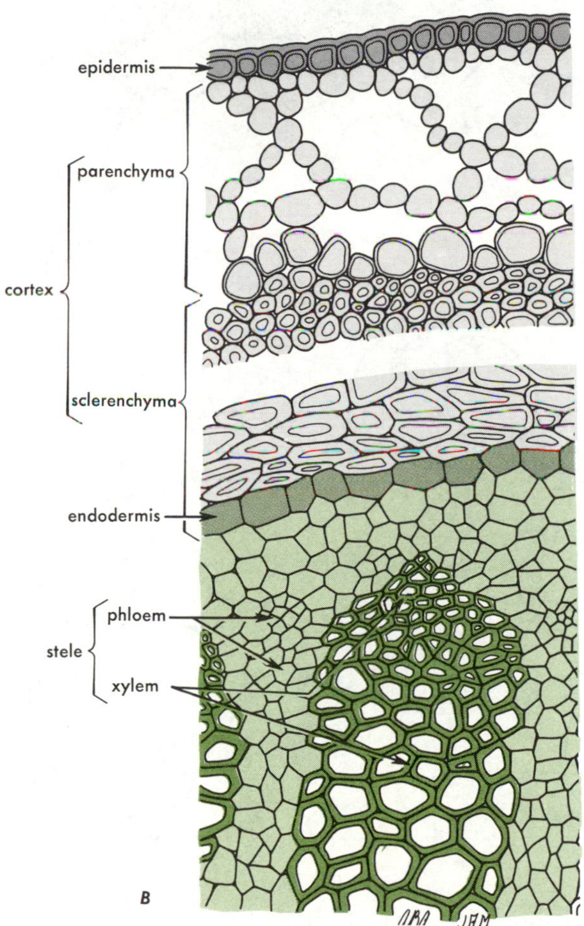

Figure 28.6
Cross section of *Lycopodium* stem. *A,* showing irregular distribution of xylem and phloem, ×65. *B,* enlarged sector showing cellular detail, ×260.

Leaves of *Lycopodium* show a well-developed epidermis with stomata, a mesophyll with many air spaces, and a midvein (Fig. 28.7).

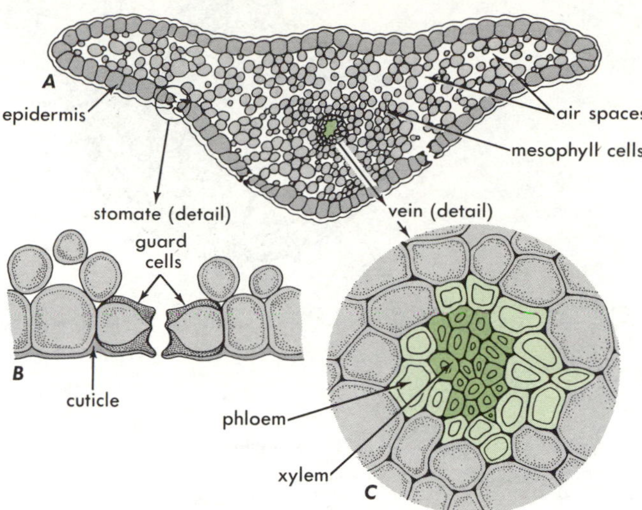

Figure 28.7
A, cross section of *Lycopodium* leaf. *B,* stomate detail. *C,* vein detail.

Reproduction

Lycopodium may reproduce asexually or through an alternation of generations involving distinctive gametophytic and sporophytic generations. Both generations are independent and the sporophyte generation is the dominant generation. Meiospores are borne in sporangia in or near the leaf axils (Figs. 28.5*B*, 28.8*A,B,C*), sheathing the stem. Spores and sporangia of a given species are all alike. Leaves bearing sporangia are called spore-bearing leaves, or **sporophylls.** In some species, they closely resemble ordinary nonspore-bearing leaves in their structure and appearance. In other species, they differ from sterile leaves in size, shape, position, and color. Such modified sporophylls are grouped together closely at the ends of stems, forming a **cone** or **strobilus** (Figs. 28.5*C*, 28.8*A*).

Gametophyte. The meiospores germinate and develop into gametophytes. Gametophytes in some species grow above ground and are green; in other species, gametophytes are subterranean and lack chlorophyll. The gametophytes are always associated with a fungus. Both male and female gametangia are found on the same gametophyte. Gametangia are borne on the upper portion of the gametophyte (Fig. 28.8*E*). Fertilization occurs when sufficient free water is present to allow sperms to swim to mature archegonia.

Sporophyte Embryo. The embryo sporophyte possesses (*a*) a well-developed foot, (*b*) rudiments of a short primary root, (*c*) leaf primordia, and (*d*) a short shoot apex (Fig. 28.8*I*). The embryo grows directly into the mature sporophyte plant.

Selaginella

Selaginella species resemble those of *Lycopodium* in their general appearance but are smaller and more nu-

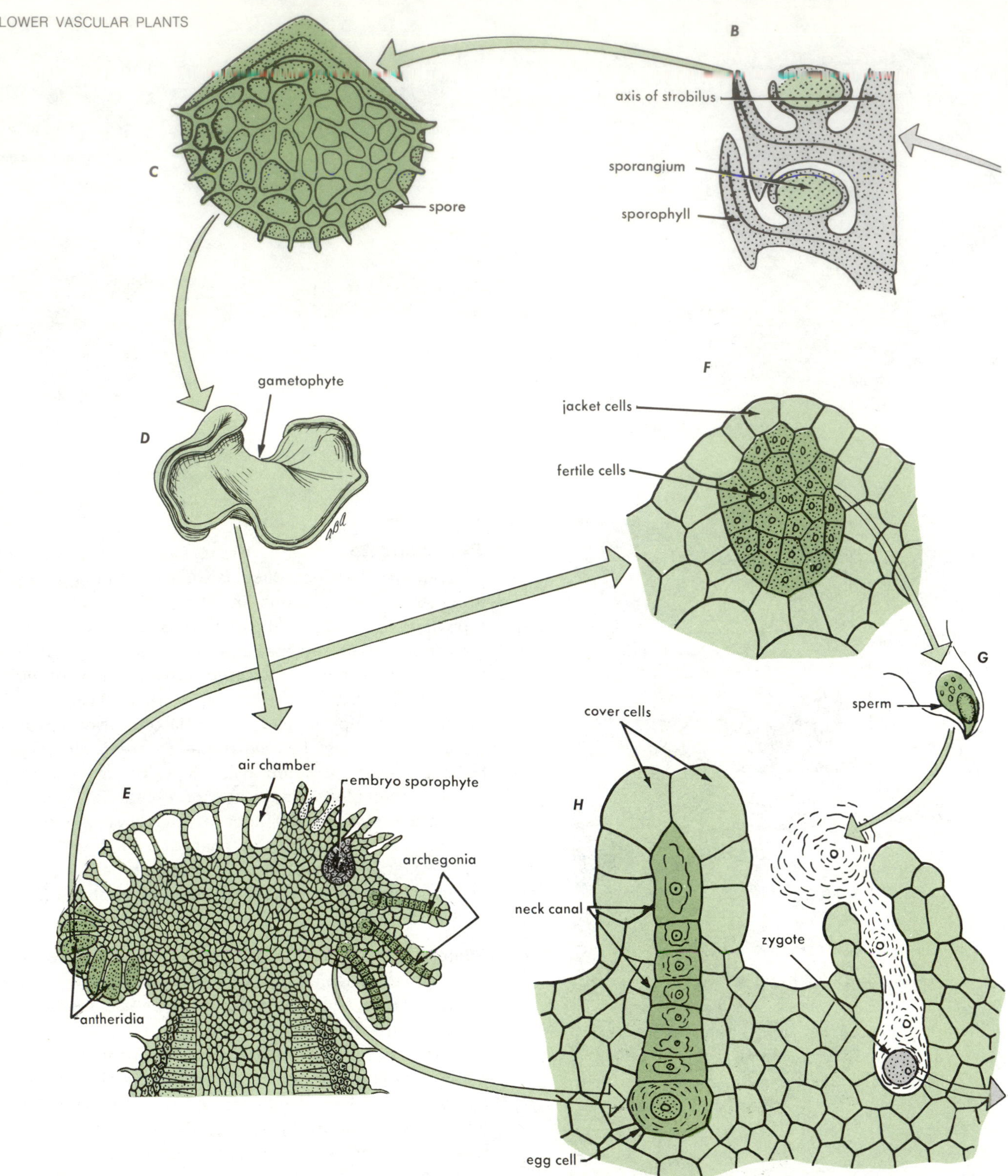

merous. They number more than 700. Although they are widely distributed, most of them are tropical; a few grow in temperate zones. Some species are adapted to withstand periods of drought and hence may grow in relatively dry localities (**Fig. 28.9,** page 562).

Mature Sporophyte

As in *Lycopodium,* the sporophyte of *Selaginella* generally consists of a branched, prostrate stem with short, upright branches, usually only 10 cm high. In some species, the stem is upright and slender and taller. Both horizontal and upright stems are sheathed with small leaves in four longitudinal rows or ranks.

Two species of *Selanginella* are shown in **Fig. 28.9.** One of them, *Selanginella watsonii,* grows in exposed rock crevices in the higher Sierra Nevada and is able to withstand periods of drought. The more delicate *Selanginella emmeliana* grows best in a humid environ-

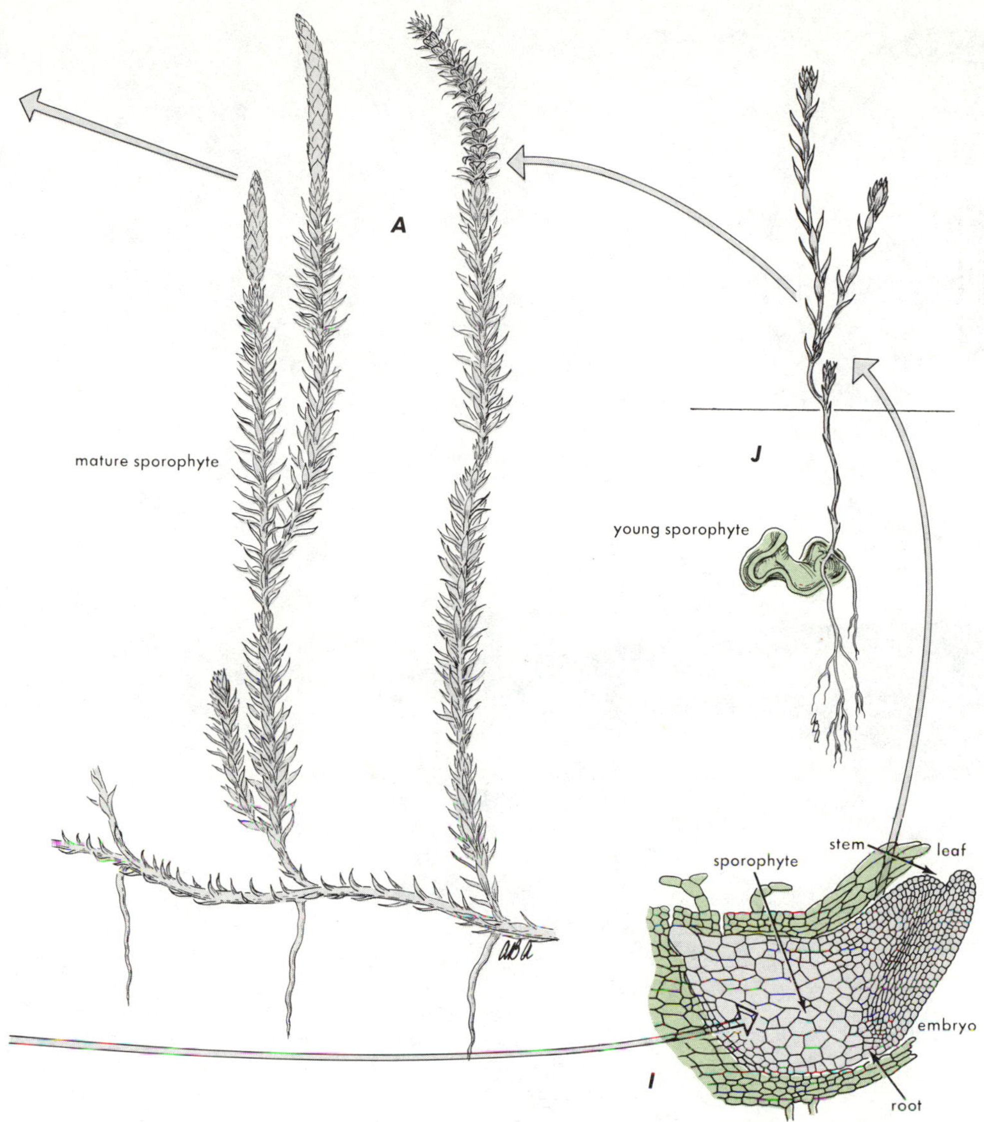

Figure 28.8
Stages in the life cycle of *Lycopodium*. *A, Lycopodium annotinum,* ×1. *B,* section of sporophyll showing location of sporangia, ×3. *C,* spore, ×50. *D,* gametophyte, ×10. *E.* section of gametophyte showing location of gametangia, ×25. *F,* antheridium, ×100. *G,* sperm, ×200. *H,* archegonium, ×100. *I,* section of embryo, ×50. *J,* underground gametophyte with sporeling, ×3. (*B,* after M. G. Sykes, *Ann. Botany (London)* **22,** 63. *C,* after E. Pritzel, in A. Engel and K. Prantl, *Die näturlichen Pflanzinfamilien. E. G, H,* after H. Bruchmann, *Flora* **101,** 220. *J,* material supplied by A. J. Eames.)

ment; it is frequently grown as an ornamental plant.

The vascular system in stems of *Selaginella* consists of one to several branching strands, each having a central core of xylem surrounded by phloem. The cross sections in Fig. 28.10 show a single strand in a stem of *Selaginella.* It occurs in a large air space and is supported by strands of radially arranged endodermal cells. Vessels are present in the xylem of several species of *Selaginella.* Parenchyma and sclerenchyma tissue form a cortex, which is bounded externally by an epidermis.

Reproduction

As in *Lycopodium,* spores are borne in sporangia, which grow in or near axils of sporophylls. Although sporophylls do not differ greatly in appearance from sterile leaves, they are always grouped to form cones, or strobili, at the ends of upright branches.

Two types of sporangia are formed: **megasporangia** (Greek, *mega,* large) and **microsporangia** (Greek, *micro,* small). As the names indicate, megasporangia produce larger spores. A single strobilus usually contains both types or sporangia (Fig. 28.11 *B,C.D*).

Within a developing megasporangium, all but one of the **megasporocytes** (potential spore-producing cells) degenerate. This remaining megasporocyte, nourished in part by fluid resulting from degenerating sporocytes

epidermis
cortex
stele
xylem
phloem
endodermis
A

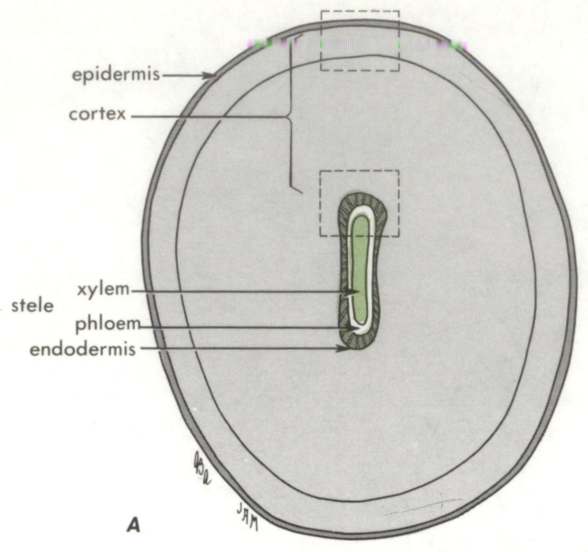

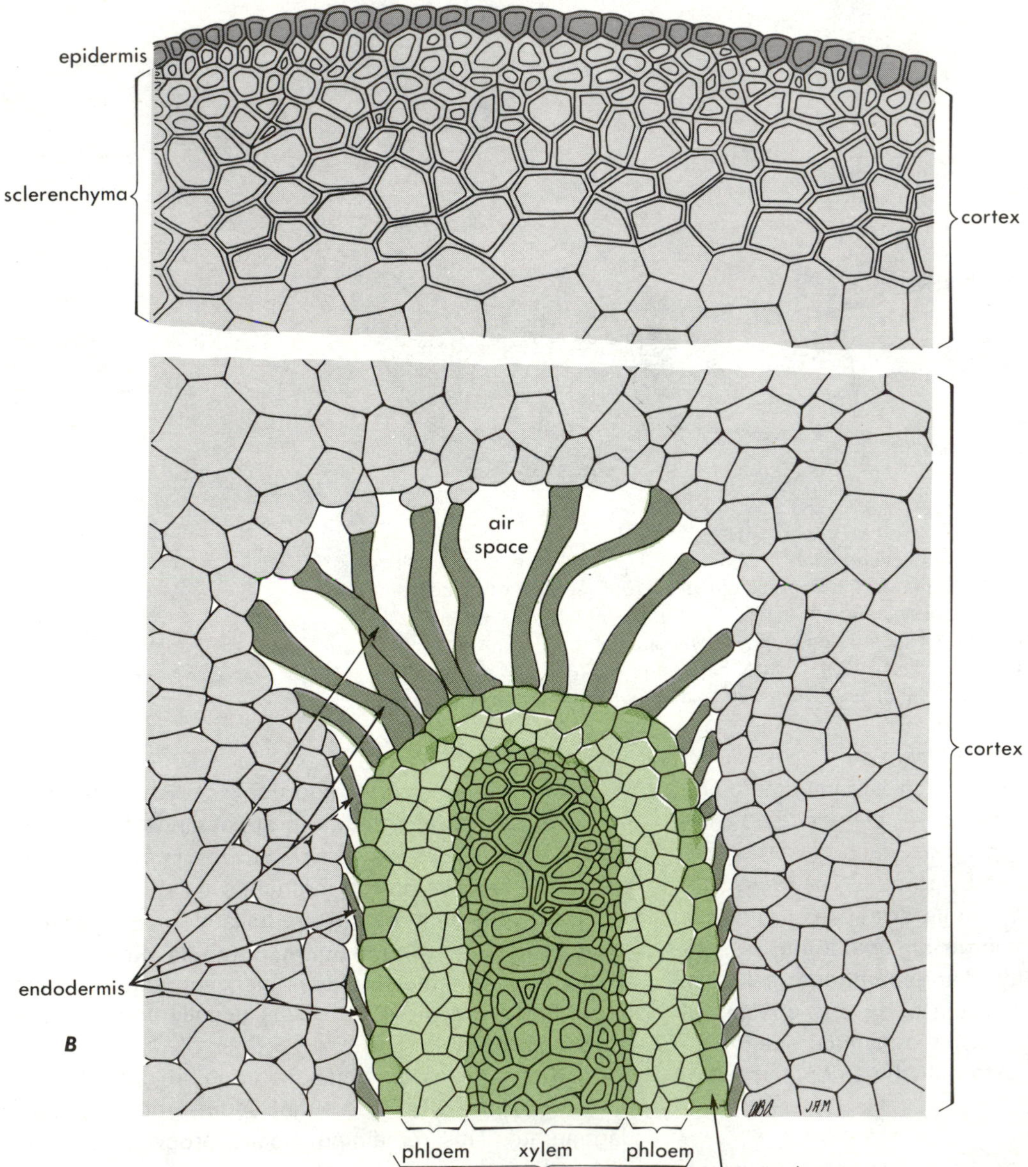

epidermis
sclerenchyma
cortex
air
space
cortex
endodermis
B
phloem
xylem
phloem
stele
pericycle

Figure 28.10
Cross section of a stem of *Selaginella*. *A*, diagram showing distribution of stem tissues; a single flattened stele is supported in an air space by filaments of endodermal cells. The cortex is composed of parenchyma and sclerenchyma cells. *B*, detail showing cellular structure.

and in part by cells surrounding the spore cavity, increases greatly in size. During meiosis, four large spores, called **megaspores,** are formed from the megasporocyte. Each megaspore may germinate and give rise to a female gametophyte (**megagametophyte;** Fig. 28.11*M*).

Only a few microsporocytes within the developing microsporangium degenerate. The approximately 250 microsporocytes that remain undergo meiosis, each forming four small spores, or **microspores** (Fig. 28.11*E,I*).

The production by a given species of two distinct types of meiospores—megaspores and microspores—is called **heterospory.** Thus, *Selaginella* is heterosporous. *Lycopodium,* which produces but one spore type, is said to be **homosporous.**

Female Gametophyte. Repeated cell divisions within the megaspore result in the female gametophyte, which is contained within the megaspore wall until it nears maturity. Its increase in size eventually ruptures the megaspore wall, and a small cushion of colorless gametophytic tissue protrudes from the megaspore along the lines of rupture (Fig. 28.11*M,N*). Archegonia develop on the protruding cushion.

The megaspore (containing the female gametophyte) may be shed from the cone or strobilus at almost any stage of the development of the gametophyte. In some species, it may be retained in the strobilus until well after fertilization. In any event, fertilization occurs only when sufficient water, either from rain or dew, is present, allowing sperm to swim to archegonia.

Male Gametophyte. Upon germination, microspores divide into two cells. One of them, the **prothallial cell,** is small and does not divide further; it represents the vegetative portion of the male gametophyte. The other cell, by repeated divisions, develops into an antheridium composed of a jacket of sterile cells enclosing 128 or 256 bilflagellated sperms. This development takes place within the microspore wall. Microspores are shed from the microsporangium midway in the development of the male gametophyte and grow to maturity without direct connection with parent sporophyte or soil. Usually, microspores sift down to the bases of the megasporophylls below the microsporophylls. In this position, they are close to the developing female gametophytes. The sperms escape when the microspore wall ruptures.

Sperms swim to the archegonia, which grow on that portion of the female gametophyte protruding from the megaspore. Fertilization ensues, and the resulting zygote initiates the diploid or sporophyte generation.

Sporophyte Embryo. Of the two cells formed by the first division of the zygote, only one develops into an embryo. The other cell grows into an elongated structure, the **suspensor,** which pushes the developing embryo into gametophytic tissue, where there is a food supply (Figs. 28.11*N,O*).

The embryo is a structure with (*a*) a foot, (*b*) a root, (*c*) two embryonic leaves, and (*d*) a shoot. In certain species of *Selaginella,* the embryo is held by the megaspore and retained within the strobilus. It does not pass into a dormant state, as do embryos of seeds, but continues to grow. Young sporophytes may be found extending from the strobilus of parent sporophytes. If the developing embryo were to pass into a period of dormancy while being held by the parent sporophyte, a condition would arise that approaches the seed habit.

Isoetes

The genus *Isoetes,* commonly called "quillwort" in America and Merlin's Grass in Europe, includes more than 60 species. It may be found submerged in shallow water, at least part of the year. In the Sierra, it frequently forms dense mats covering large areas of shallow lake bottoms. All species are small plants. The plant body resembles some monocotyledonous plants and consists of a tuft of leaves growing from a greatly shortened axis, (corm), from which roots arise (Fig. 28.12).

Lepidodendrales

The Lepidodendrales is an important fossil order of the division Lycophyta. Many representatives were heterosporous, and true seeds were formed by some species. No members of the order are in existence today. It was, however, a dominant order during the period when the great deposits of coal were being formed. If one can imagine a lush forest of *Selaginella*-like trees, 45 m tall and 2 m in diameter, a fair picture of this forest may be obtained. There were also smaller, shrubby and herbaceous Lepidodendrales. Since two spore types, and seeds, were formed by these ancient plants, we must believe that both heterospory and the seed habit are very old plant characters and not characters confined to present-day plants (see also Chapter 31).

DIVISION SPHENOPHYTA

Members of this division once grew very abundantly, as their good representation in the fossil record shows. Today, the division is represented by a single genus (*Equisentum*) of about 25 species (Fig. 28.13). Many species inhabit cool, moist places, but *Equisetum arvense* also grows in dry habitats. Most species are characterized by the presence of silica in the epidermis of stems. Because of this trait, they were used in colonial days to scour pots and pans and hence were called scouring rushes. The genus is also commonly known as the horsetails.

Mature Sporophyte

The sporophyte of *Equisetum* is the dominant phase of its life history. In one tropical species, the sporophyte is vinelike and may reach 7 m in length. Usually, 1 m represents the maximum upright growth. All species are perennials and have a branched rhizome from which

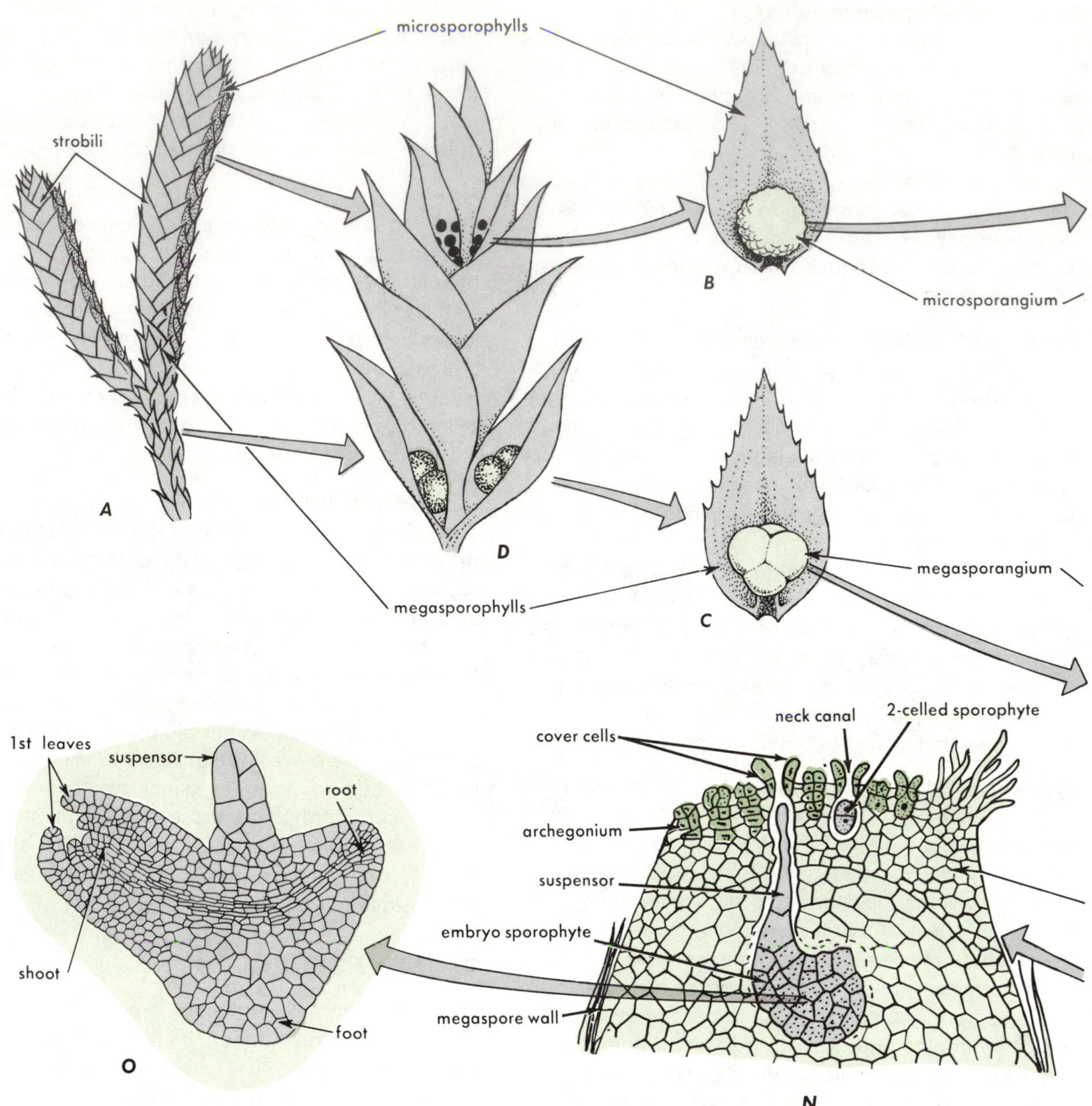

Figure 28.11

Stages in the life cycle of *Selaginella*. *A*, shoot of *Selaginella watsonii* showing two strobili, ×2. *B*, microsporophyll with microsporangium, ×5. *C*, megasporophyll with megasporangium, ×5. *D*, strobili showing sporophylls with sporangia, ×2. *E*, microsporangium discharging microspores, ×50. *F*, megasporangium discharging megaspores, ×50. *G*, microspores sift downward and lodge in axils of megasporophylls close to megaspores. *H*, a microspore, ×100. *I*, section of a microspore, ×150, *J*, male gametophyte within microspore, ×150. *K*, antheridium with developing sperm cells. *L*, mature sperm cells, ×100. *M*, female gametophyte within megaspore, ×100. *N*, section of female gametophyte showing location of archegonia in a young embryo pushed, by its suspensor, into the vegetative tissue of the female gametophyte, ×100. *O*, the young embryo sporophyte generally remains embedded in vegetative gametophyte tissue, ×100. (*E, H, I, J, K, L*, redrawn from R. A. Slagg, *Am. J. Botany* **19,** 106, © 1932 by Botanical Society of America, *I, J, O,* after H. Bruchmann, *Flora* **104,** 180. Reprinted with permission of the publishers.)

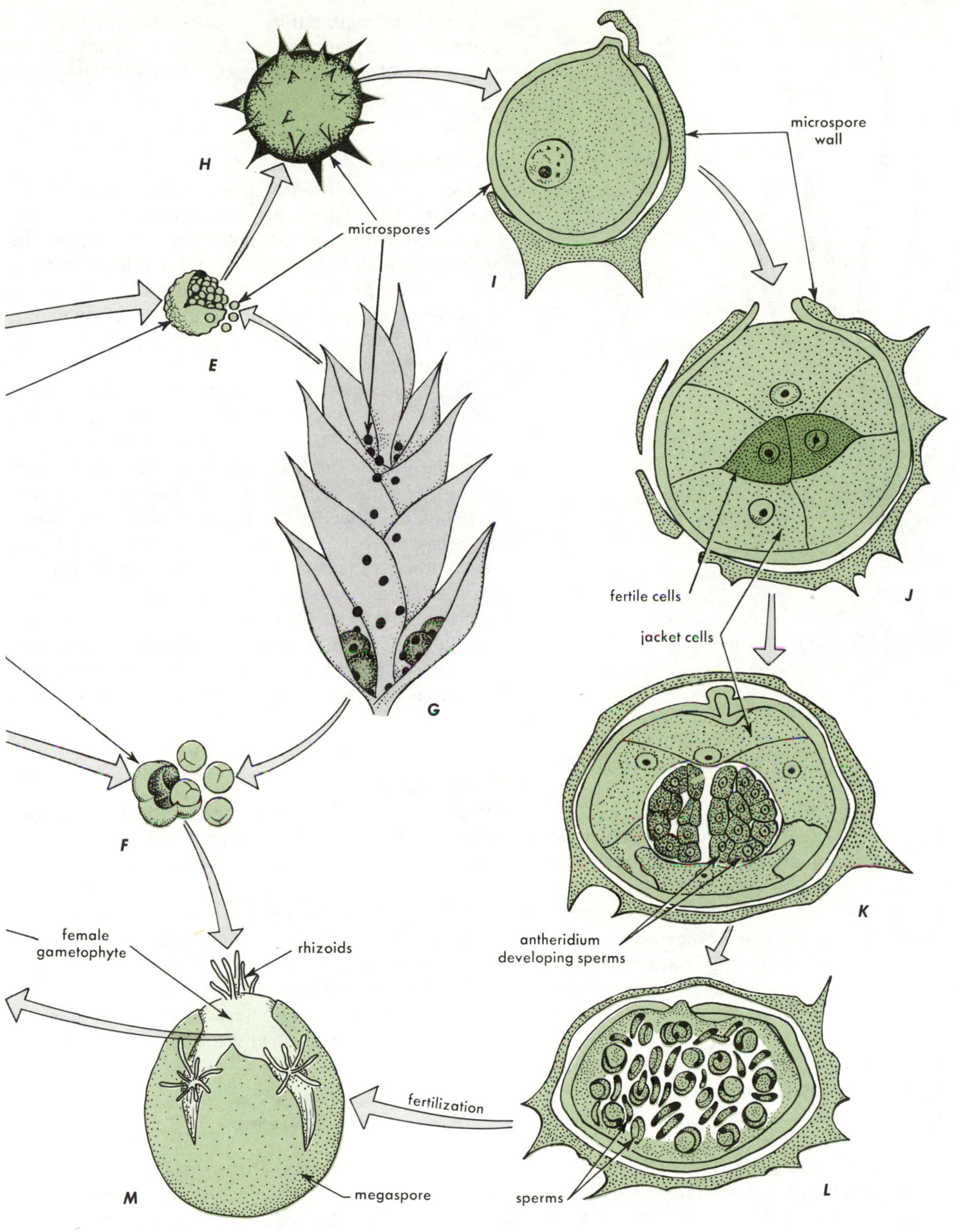
H
microspores
I
microspore
wall
E
G
fertile cells
J
jacket cells
F
K
female
gametophyte
rhizoids
antheridium
developing sperms
fertilization
M
megaspore
sperms
L

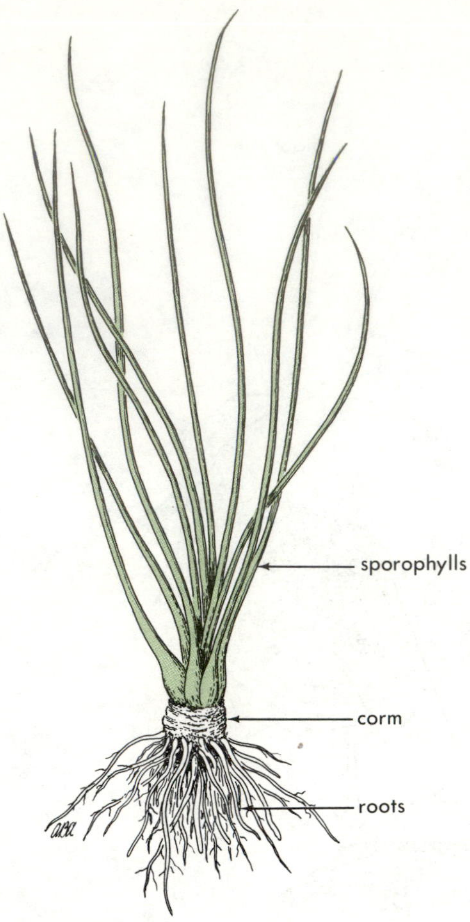

Figure 28.12
Sporophyte of *Isoetes*.

upright stems arise. Depending on the species, stems may branch either profusely or sparingly. In either case they are straight and marked by vertical ridges and distinct nodes (Fig. 28.13*D*). Bases of nodes are sheathed by whorls of small simple leaves. Many branches may also form at each node. The microphyllous leaves are much reduced in size, nongreen and, in many species, short-lived. The stems are green and are, therefore, the organs of food manufacture. Roots occur at the nodes of the rhizomes, but they may also develop from stem nodes if the stem becomes procumbent on a moist surface.

Except at nodes, stems of *Equisetum* are hollow and the ribs make prominent markings on their outer circumferences. The ridges are formed of sclerenchyma tissue and not only project outward, but extend inward almost to the small vascular bundles (Fig. 28.14). There are air spaces, or vallecular canals, between vascular bundles. Photosynthetic tissue lies between these air canals and the thin, outer layer of sclerenchyma tissue and epidermis. Vascular strands are also marked by a smaller, carinal canal (Fig. 28.14*B*). Arms of xylem extend outward from this canal, and phloem tissue lying outward from the canal is present between radial arms of xylem. Some species contain vessels in the xylem. An endodermis is present, but its location varies from species to species; it may surround each vascular bundle, or encircle a ring of vascular bundles. In some species, there is also an inner endodermis.

Reproduction

In all species of *Equisetum,* the sporangium-bearing organs, **sporangiophores,** are specialized structures very different from ordinary leaves. They are grouped together in strobili, or cones (Fig. 28.13) at the summit of main upright branches and occasionally on lateral branches. In most species, strobili are borne on ordinary vegetative shoots; in a few species, they are formed only on special fertile shoots.

The sporangiophores are stalked, shield-shaped structures borne at right angles to the main axis of the cone (Fig. 28.13*C*). The cone may be compared to a pole to which open umbrellas have been fastened, the handles of the umbrellas being at right angles to the pole. Sporangia are attached to the underside of the shield (umbrella), close to its edge. They extend horizontally inward, toward the axis of the cone. Four meiospores with the haploid number of chromosomes are formed from each sporocyte. All meiospores are morphologically alike. *Equisetum,* therefore, is homosporous. Upon maturity the meiospores are discharged from the sporangia. The spores are fragile and normally live but a few days (Fig. 28.15).

Gametophyte

Gametophytes are small green bodies about the size of a pinhead and consist of a cushionlike base with many erect, delicate lobes (Fig. 28.16). They are easily cultured in the laboratory.

Gametangia are similar to those of *Lycopodium.* Fertilization takes place only when there is free water. Sperms are spiral shaped and multicilate.

The zygote develops directly into an embryo. No suspensor is formed. The embryo is similar to those already described, except that the foot is small or lacking.

DIVISION PTEROPHYTA

The division Pterophyta comprises the ferns, some 10,000 species, most of which are shade-loving plants of small size, their upright leaves, or **fronds,** generally being their most prominent feature. Most species native to temperate zones have an underground rhizome with leaves and roots arising at nodes (Fig. 28.17); a few are small, floating aquatics with or without rhizomes (Figs. 28.18, 21.25). Some ferns of tropical regions do grow into fairly large trees. All Pterophyta have a definite alternation of generations, with both sporophyte and gametophyte being autotrophic plants.

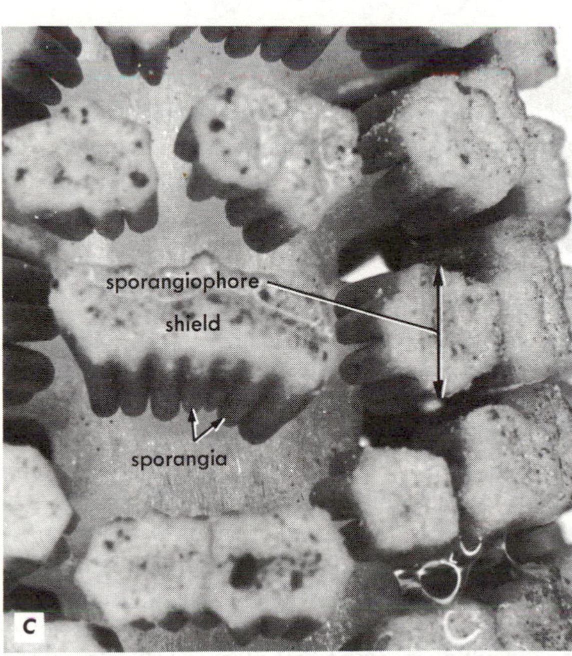

Figure 28.13
Strobili of *Equisetum telmateia*. *A,* fertile stems with strobili; also note much-branched sterile stems, ×¼. *B,* strobilus with sporangiophores, ×2. *C,* enlarged view of several sporangiophores, ×10. *D,* enlarged view of stem showing three nodes, ×2.

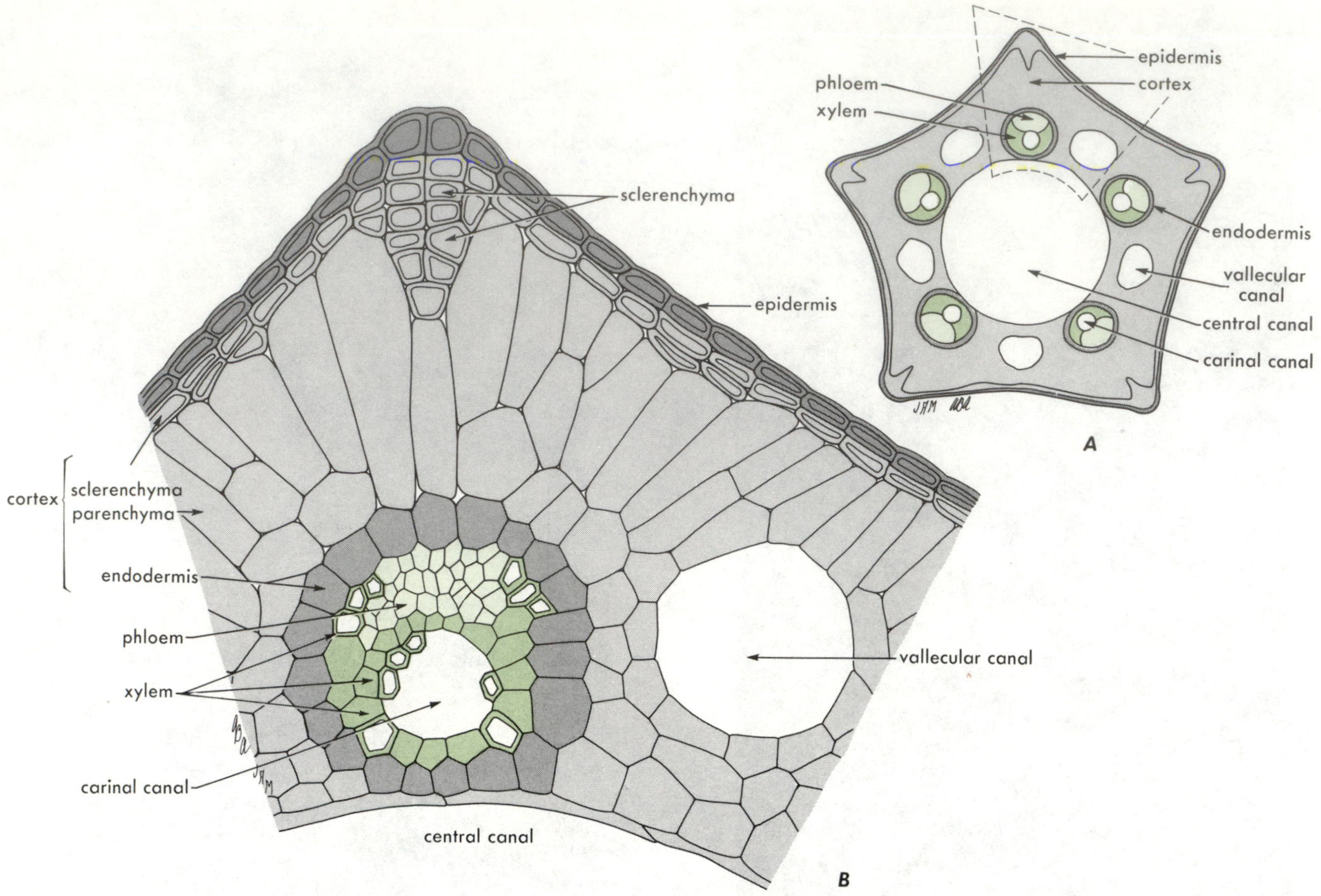

Figure 28.14
Cross section of *Equisetum* stem. *A,* diagram showing arrangement of stem tissues, ×70. *B,* enlarged view showing details of cellular structure, ×400.

Figure 28.15
Spores of *Equisetum* showing the hygroscopic ribbons, ×100. *A,* ribbon coiled; *B,* ribbon open. (Redrawn from R. von Wettstein.)

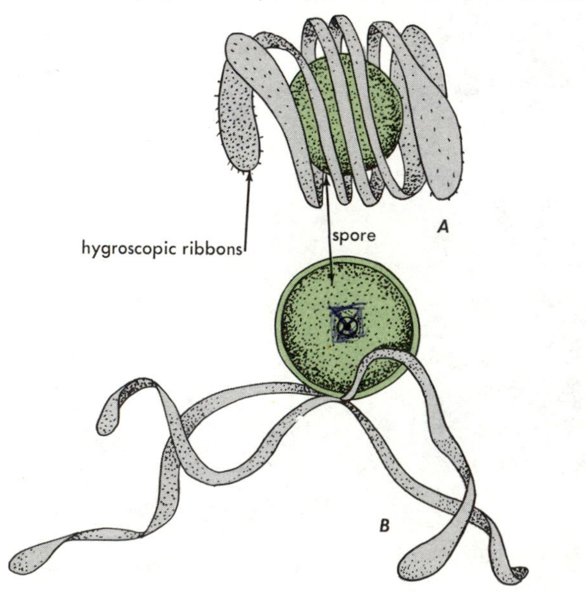

The Pterophyta comprise four orders, the largest of which, the Filicales, includes the **true ferns.** The Filicales are divided into 11 or more families, of which the largest and best-known in the United States is the Polypodiaceae. Most of the discussion below deals with representatives of this family.

Mature Sporophyte

The sporophyte, which is the dominant generation of all ferns, possesses an underground stem or rhizome from which leaves and adventitious roots arise. Leaves are the most prominent part of the fern plant and they vary greatly in size and form. Young fern leaves are rolled into tight spirals and consist chiefly of meristematic tissue (Fig. 28.19). As the leaf matures, it unwinds from the base upwards. The upright, expanded portion is mature, but the leaf continues to grow by cell division at its coiled tip. In certain species, the apical meristem may remain active for years, resulting in leaves that are nearly 3 m long. The uncoiled, fully expanded fern leaf lacks any residual meristematic tissue.

Most fern leaves are compound, although simple

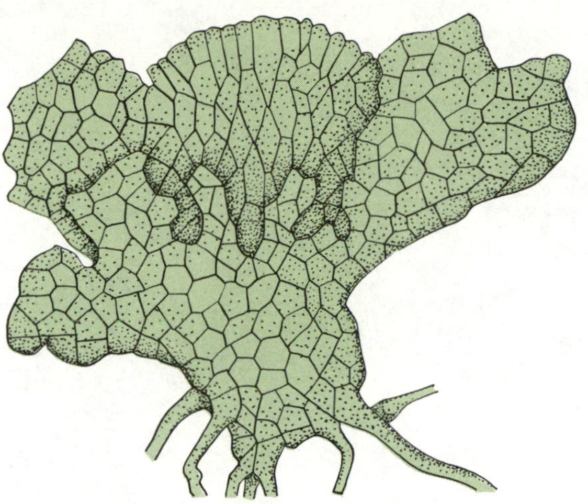

Figure 28.16
Gametophyte of *Equisetum*, ×20. (Redrawn from E. R. Walker, *Bot. Gaz.* **92,** 1. © 1931 by the University of Chicago Press. Reprinted with permission of the publisher.)

Figure 28.17
A typical, mature fern sporophyte, showing rhizome below ground and fronds above ground. Lady fern, *Athyrium felix-femina*.

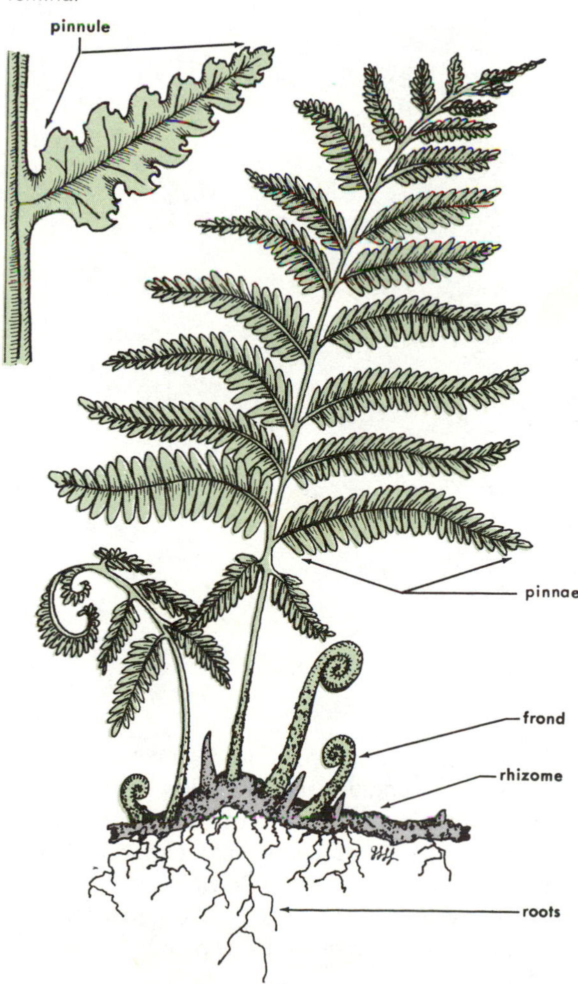

types exist in all groups. The most common type of fern leaf has a stout or rigid petiole, which is prolonged to form a rachis from which leaflets (pinnae and pinnules) arise (Fig. 28.20).

The vascular tissue of ferns is organized into one or more vascular strands, each having a core of xylem, surrounded by phloem and separated from the ground parenchyma by an endodermis. These vascular bundles are interconnecting, so that, as in *Lycopodium,* the precise arrangement will vary from section to section. However, there are arrangements characteristic of various genera. For instance, in *Polypodium* there is a ring of small bundles placed well out from the center of the rhizome (**Fig. 28.21**, page 595). Each strand is formed of a central core of xylem, surrounded by phloem, and the whole has a bounding endodermis. The xylem consists largely of tracheids, vessel elements being known in only two genera. Sieve cells occur in the phloem.

A typical fern leaf has a well-developed epidermis with chloroplasts in the epidermal cells and stomata on the lower surface. The mesophyll may be relatively undifferentiated or divided into palisade and spongy parenchyma layers **(Fig. 28.21*C*).**

Reproduction

Vegetative Reproduction

Vegetative reproduction may occur in one or two ways: (*a*) by death and decay of the older portions of the rhizome and the subsequent separation of the younger growing ends; and (*b*) by the formation of leaf-borne "buds," which detach and grow into new plants or produce new plantlets on the frond (Fig. 28.20*F*). Such buds occur in only a few genera. Some clumps of ferns are thought to be hundreds of years old.

Sexual Reproduction

In the sexual life cycle, independent sporophyte and gametophyte generations alternate with each other. The vegetative structure of the sporophyte has been described above. Spores are borne in sporangia, which ordinarily develop on the lower surface or margins of fronds (**Fig. 28.22,** page 596). Not all leaves are fertile (i.e., spore-producing), and the fertile leaves in some species are dissimilar in structure to the sterile (non-spore producing) leaves. The distribution of sporangia on the leaf surface varies considerably in different genera and species. Sporangia may (*a*) cover much of the lower surface, (*b*) be grouped in **sori** (singular, **sorus**) and grow in a definite relationship with veins, or (*c*) grow only along margins of the leaf.

When sporangia are grouped together in sori, a structure called the **indusium** (**Figs. 28.22*D*,** 28.24*B*), which is sometimes umbrellalike, may be present, thereby protecting young and developing sporangia. Fre-

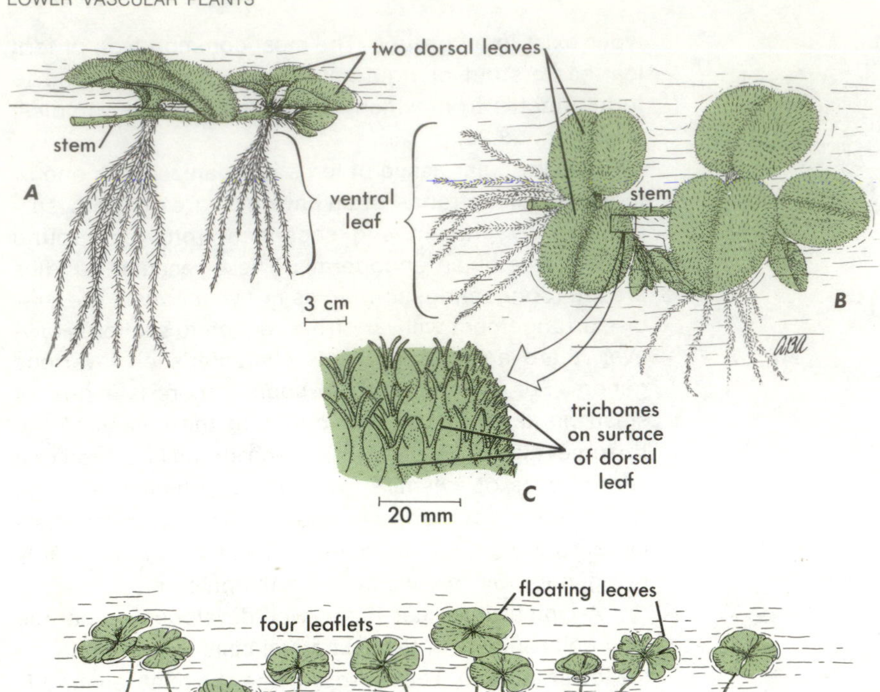

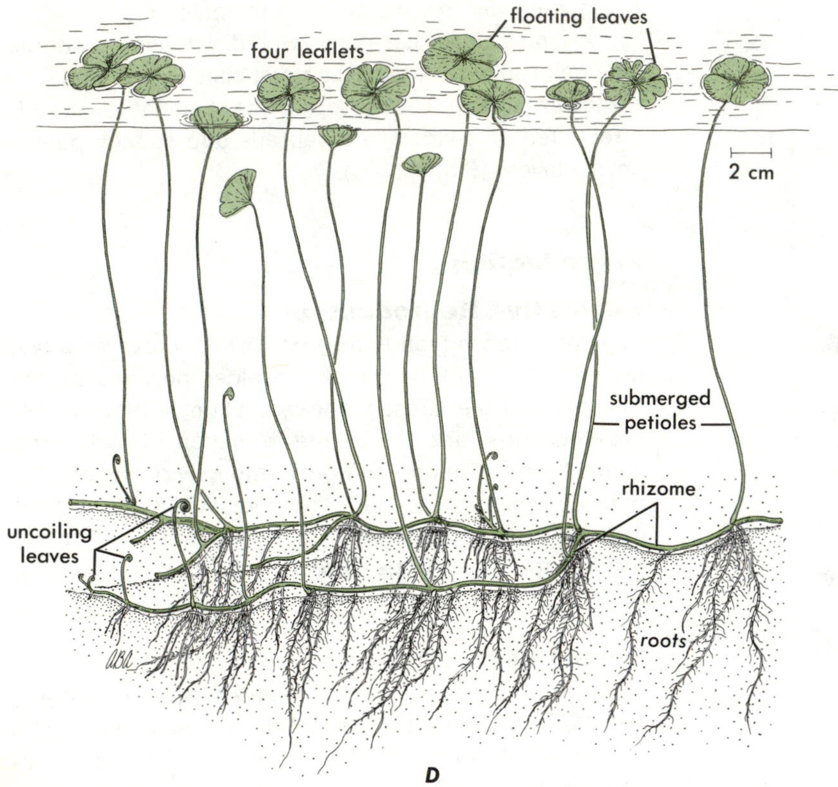

Figure 28.18
Aquatic ferns. *A, B, C,* details of *Salvinia; D, Marsilia.*

quently, marginal sporangia are protected by the curled edge of the leaf itself, which forms a **false indusium.**

The sporangium is a delicate, watch-shaped case (**Fig. 28.22*F***), consisting of a single layer of epidermal cells, only one row of which possesses heavy walls. This row, which nearly encircles the sporangium in the Polypodiaceae, is the **annulus;** it functions in opening dried mature sporangia and aids in dispersal of ripe spores by opening, then snapping back (Fig. 28.24*B*).

The young sporangium is filled with sporocytes. Meiosis results, as always, in spores with a reduced number of chromosomes. Ferns, with the exception of two families, the Marsileaceae and the Salviniaceae, are homosporous. These two families are widely distributed, small, aquatic, and unfernlike in appearance (water ferns, see Fig. 28.18).

Figure 28.19
Fern fronds retain meristematic tissue at their tightly rolled tips. As differentiation proceeds, the tip uncoils. *A*, this "fiddle head" is the tightly rolled young frond of the tree fern. *Cyathea*, ×¼. *B*, maturing frond of the bracken fern (*Pteridium aquilinum*); the tips are still tightly rolled, ×⅓.

Gametophyte

Gametophytes, or **prothallia,** of the Polypodiaceae are small, flat, green, heart-shaped structures (Fig. 28.23) with rhizoids on their lower surface (Fig. 28.24). In most species, they apparently mature rapidly and are not long-lived. Antheridia and archegonia are borne on the same prothallus. Antheridia are formed when the prothallus is very young and are scattered over its lower surface. Normally, 32 sperms develop within each antheridium.

Archegonia form later than antheridia and are usually clustered close to the notch, also on the undersurface of the gametophyte (Fig. 28.24*H*).

Fertilization occurs when free water is present and sperms are thus able to swim to the neck of the archegonium. The resulting zygote is diploid and rapidly develops into an embryo sporophyte comprising a foot, root, stem, and leaf (Fig. 28.24). The embryo develops directly into a young sporophyte, without a dormancy period.

SUMMARY

1. Lower vascular plants is a general name for vascular plants that do not produce seeds. This group consists of four divisions, the Psilophyta, Lycophyta, Sphenophyta, and Pterophyta (see summary table 28.1). The latter division, the ferns, contains the most species and is the most widespread.
2. The Psilophyta is represented by only two genera, *Psilotum* and *Tmesiptris*. Sporophytes of *Psilotum* consist of a branched, underground rhizome with rhizoids (no true roots) and an upright, green, branched stem with scalelike microphylls. It is homosporous and the meiospores germinate to produce nonphotosynthetic, small gametophytes. Free water is required for fertilization. The embryo sporophyte is at first attached to the gametophyte by an absorbing organ (the foot), and it depends for nourishment on the gametophyte.
3. The Lycophyta is represented by only three genera in this book: *Lycopodium, Selaginella,* and *Isoetes* (the quillworts). Two other genera, *Stylites* and *Phylloglossum* were not discussed.
4. Sporophytes of *Lycopodium* (club mosses) are less than 20 cm tall and exhibit true roots, stems, and leaves. The stems are sheathed with small leaves that contain an epidermis with stomata, a mesophyll with intercellular air spaces, and a central vein.

Sporangia are borne in the axils of leaves; they may be clustered near the tips of branches in a strobilus (cone). The genus is homosporous, and the meiospores produce small gametophytes that may or may not have chlorophyll. Free water is required for fertilization, and the embryo sporophyte is at first attached to the gametophyte by a foot, as in *Psilotum*.

5. Sporophytes of *Selaginella* are similar in appearance to those of *Lycopodium*. Vessels are present in the xylem of some species. Sporangia are borne in the axils of leaves and they are clustered into strobili. Sporangia and spores are of two kinds. Megasporangia produce megaspores that germinate and grow (within the ruptured spore wall) into female gametophytes. Microsporangia produce microspores that germinate and grow (within the spore wall) into male gametophytes. Microspores may sift down the strobilus to lie near megaspores, then they rupture to release sperm; if free water is

Table 28.1

COMPARISON OF PSILOPHYTA, LYCOPHYTA, SPENOPHYTA, AND PTEROPHYTA

Generation	Psilotum	Lycopodium	Selaginella	Equisetum	Ferns
Sporophyte	A branching stem, scale leaves, no roots	Prostate branching stem with upright branches, roots, and leaves	Prostate branching stem with upright branches, roots, and leaves, also scales	Rhizome, upright, jointed stem, small leaves	Rhizome, roots, and leaves
	Herbs or epiphytes	Herbs	Herbs	Herbs	Herbs or trees
	Simple vascular system	Simple vascular system	Simple vascular system, but vessels are present	Simple vascular system but vessels may be present	Stem: several strands of vascular tissue each with xylem surrounded by phloem; vessels: only in two genera; sieve cells: no companion cells in phloem
	One type of sporangium	One type of sporangium	Megasporangia and microsporangia	One type of sporangium	One type of sporangium (mostly)
	Homospory	Homospory	Heterospory	Homospory	Homospory (mostly)
	No sporophylls	Sporophylls present	Sporophylls present	Sporangiophores	Leaves bear spores in sporangia
	No cone	A cone in some species	Cone	Cone composed of sporangiophores	No cone
	Embryo develops attached to gametophyte	Embryo develops attached to gametophyte	Embryo develops within female gametophyte on sporophyll	Embryo develops attached to gametophyte	Embryo develops attached to gametophyte
Gametophyte	Irregular subterranean structure, associated with fungus	Irregular to tapered subterranean structure, associated with fungus	Male gametophyte: one prothallial cell, and an antheridium within microspore	A single green thallus resembling *Anthoceros*	Heart-shaped, small, green, completely independent thallus
	Gametangia embedded in thallus	Gametangia embedded in thallus	Female gametophyte: small amount of tissue within megaspore, cushion protrudes in which gametangia are embedded	Gametangia embedded in thallus, neck of archegonium protruding	Antheridium of approximately 6 to 10 cells, archegonia partially embedded in gametophyte
	Motile sperms	Motile sperms	Motile sperms	Motile sperms	Motile sperms

Figure 28.20
Diversity of form in fern fronds. *A,* simple entire frond of hart's-tongue fern *(Phyllitis scolopendrium),* $\times \frac{1}{10}$. *B,* pinnate frond of *Polypodium,* $\times \frac{1}{2}$. *C,* bipinnate frond of *Pellaea. D,* bipinnate frond, with serrate leaflets, of the rock brake fern, *Cryptogramma acrostichoides,* $\times 1$. *E,* fertile leaf of *Cryptogramma,* $\times 1$. *F,* much-dissected frond of *Asplenium;* young plants form at meristematic leaf tips $\times \frac{1}{2}$.

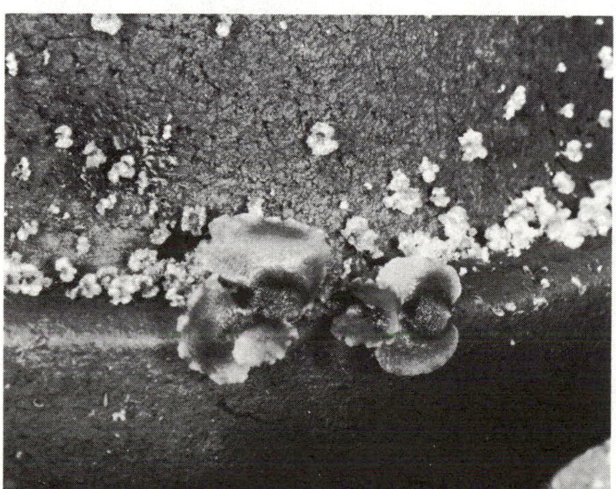

Figure 28.23
Fern gametophytes about 1-10mm in diameter growing on a flower pot, $\times 2$. (Courtesy of E. M. Gifford.)

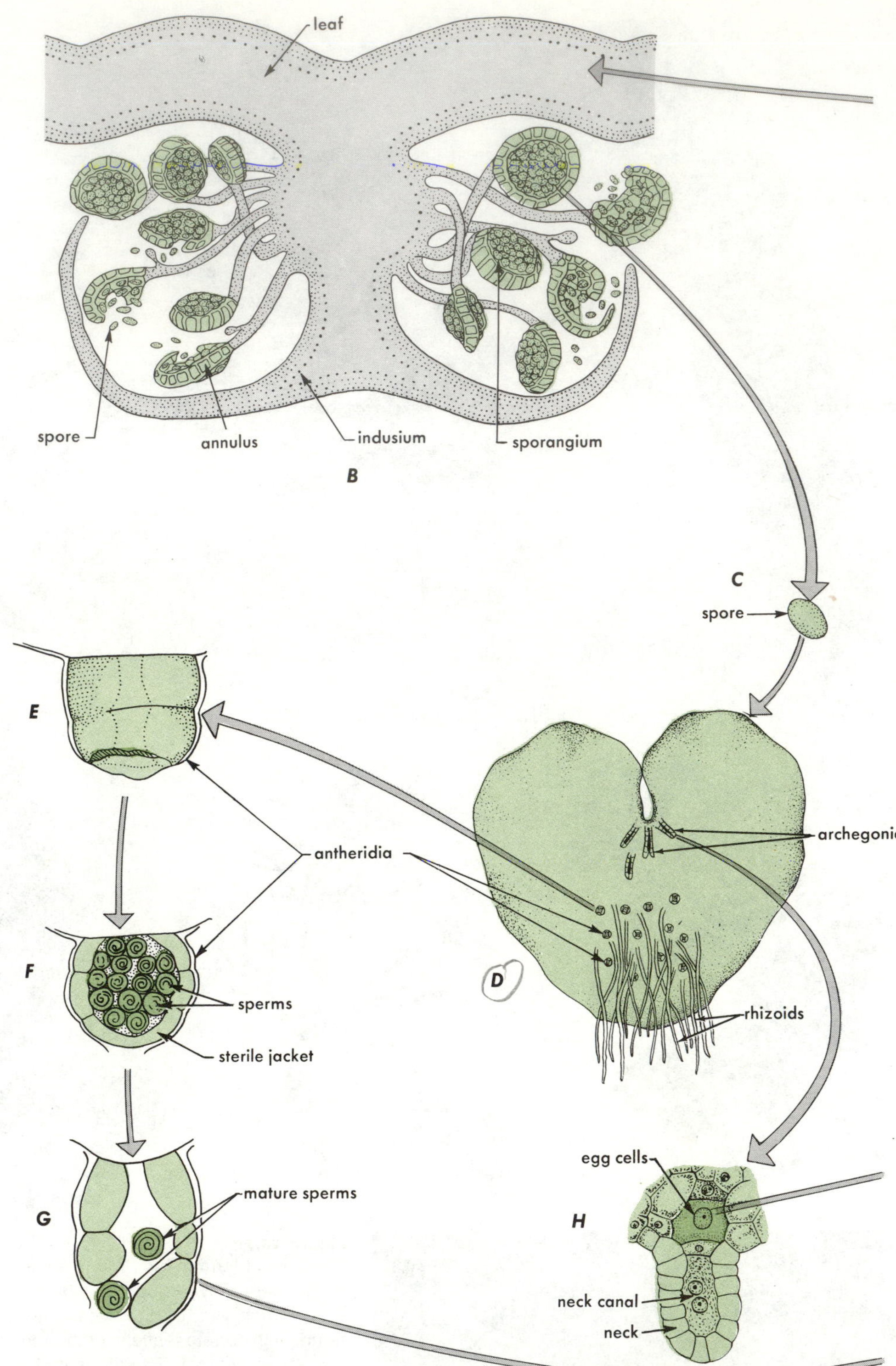

Figure 28.24
Stages in the life cycle of a fern. *A,* a frond of a fern, ×1. *B,* section through a sorus, showing indusium, sporangia, and spores, ×50. *C,* spore, ×100. *D,* gametophyte, ×5. *E,* antheridium, ×100. *F,* section of antheridium showing sperm, ×100. *G,* open antheridium discharging sperm, ×100. *H,* archegonium, ×100. *I,* archegonium receptive to sperm, ×100. *J,* section of gametophyte showing zygote within the venter of the archegonium, ×100. *K,* section of gametophyte and embryo, ×100. *L,* gametophyte with attached young sporophyte, ×10. *(E, F, G* redrawn from M. E. Hartman, *Bot. Gaz.* **91,** 252. © 1931 by The University of Chicago Press. *H* and *J* after D. H. Campbell, *Mosses and Ferns*. © 1905 by The Macmillan Company. *K* and *L* after R. M. Holman and W. W. Robins, *A Textbook of General Botany*. © 1939 by John Wiley and Sons. Reprinted with permission of the publishers.)

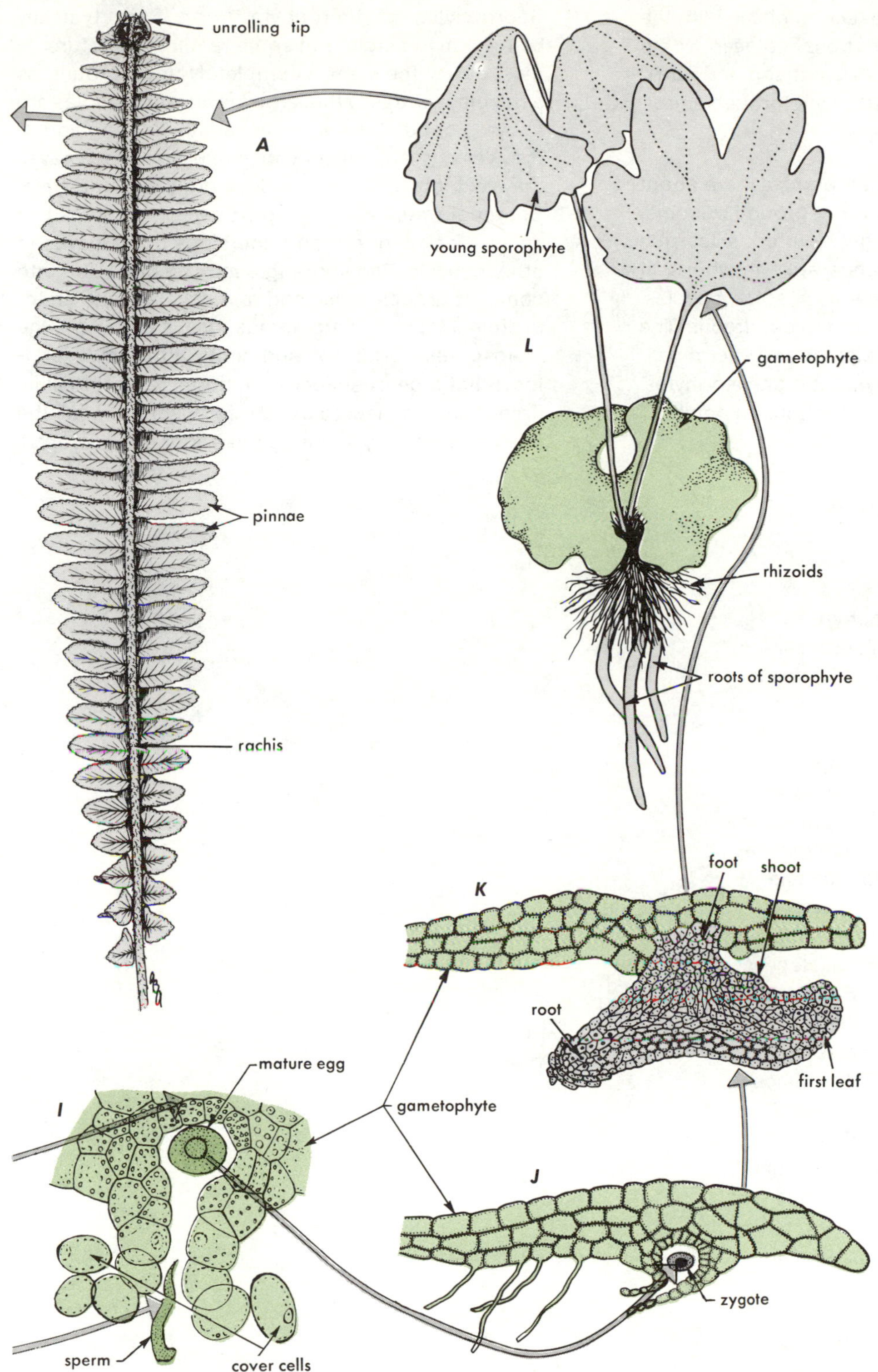
unrolling tip
A
young sporophyte
L
gametophyte
pinnae
rhizoids
roots of sporophyte
rachis
foot
shoot
K
root
first leaf
mature egg
I
gametophyte
J
zygote
sperm
cover cells

present fertilization can be accomplished in the strobilus. The embryo is at first pushed deep within the female gametophyte by a suspensor, and nutrition is gained from the gametophyte through a foot. *Selaginella* is heterosporous.

6. Sporophytes of Sphenophyta (horsetails) are about 0.5 m tall and consist of an underground rhizome, with true roots, and an upright, jointed, silica-rich stem. Microphylls and branches arise in whorls at the nodes. Vessels are present in some species. Sporangia are grouped into complex strobili. The division is homosporous. Meiospores germinate and grow into small, photosynthetic gametophytes. Free water is required for fertilization. The embryo stage is similar to that of *Psilotum.*

7. Sporophytes of Pterophyta (ferns) are typically below 1 m in height, but some reach tree stature. In most cases, the stem is completely underground as a rhizome, and only the large leaves and petioles are above ground. A few ferns are small, floating aquatics. Vessels are present in some species. Leaves may exhibit palisade and spongy parenchyma. Sporangia are produced on specialized leaves in some species, or on the undersides of all leaves in other species. The sporangia may be clustered into compact groups called sori and covered with an indusium. Most ferns are homosporous and, if so, the meiospores germinate and grow into small, photosynthetic, heart-shaped gametophytes (prothallia). Free water is necessary for fertilization, and the embryo sporophyte is much like that of *Psilotum.*

CLASSIFICATION

Division	Psilophyta
Family	Psilotaceae
Genera	*Tmesipteris*
	Psilotum nudum
Division	Lycophyta
Family	Lycopodiaceae
Genus	Lycopodium
Family	Selaginellaceae
Genus	*Selaginella*
Family	Isoetaceae
Genus	*Isoetes*
Division	Sphenophyta
Family	Equisetaceae
Genus	*Equisetum*
Division	Pterophyta
Order	Filicales
Family	Polypodiaceae
Genera	*Polypodium*
	Pteris
Family	Cyatheaceae
Genus	*Cyathea*
Order	Marsileales
Genus	*Marsilea*
Order	Salviniales
Genera	*Azolla*
	Salvinia

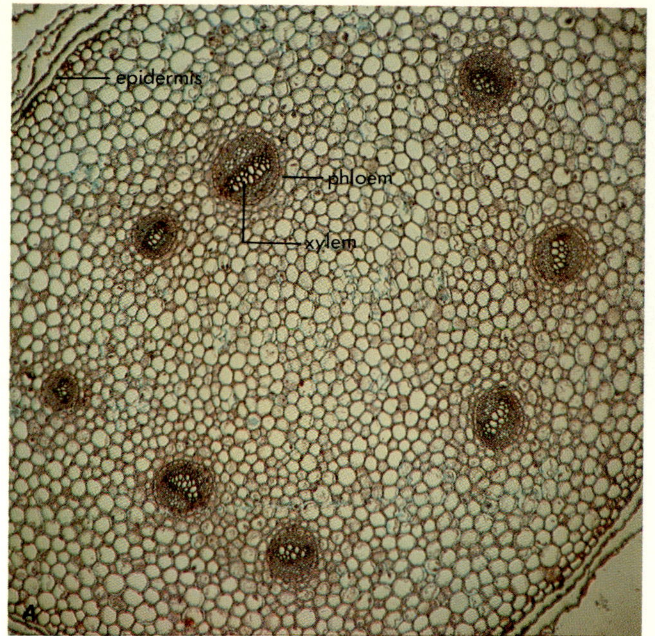

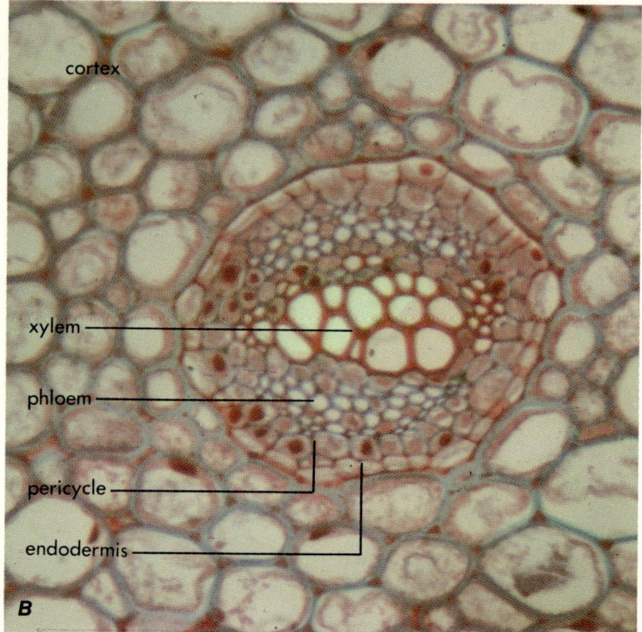

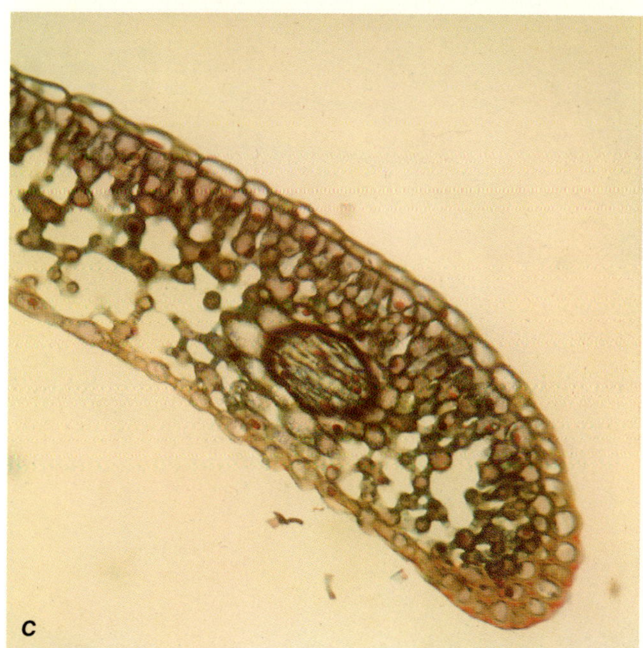

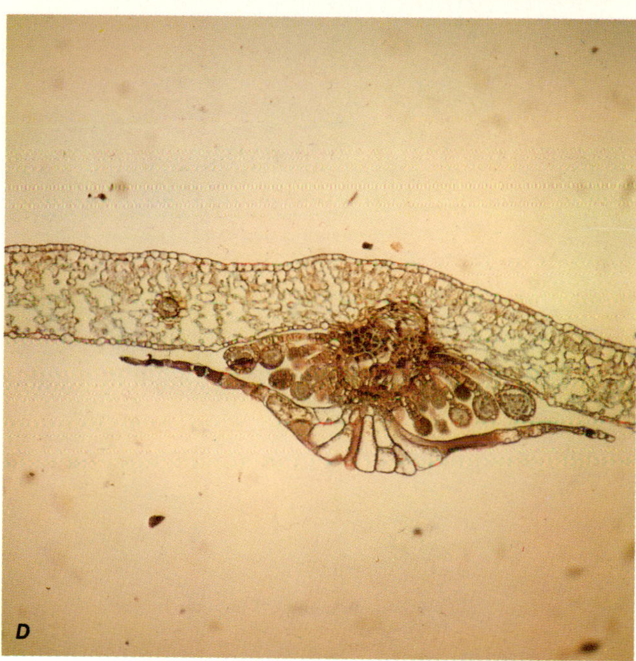

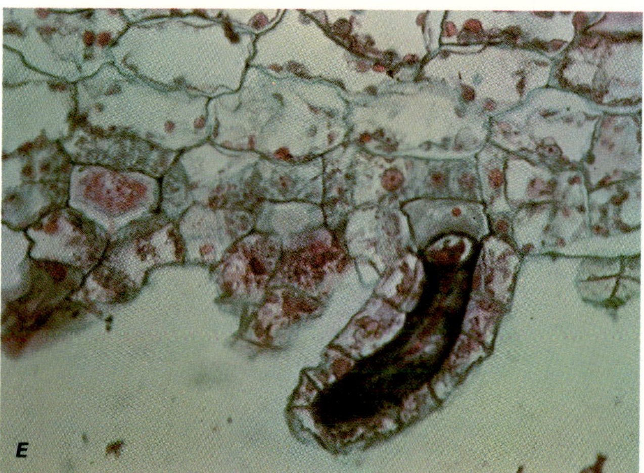

Figure 28.21
Fern anatomy. *A, Polypodium* rhizome cross section showing arrangement of vascular strands, ×25. *B,* a single vascular strand, ×160. *C,* cross section of a frond of *Polystichum munitium,* ×100. *D,* cross section of a sorus of *Crotomium,* ×30. *E,* archegonium, ×100.

Figure 28.22
Fern sori. *A*, immature marginal sori with indusium, *Adiantum* sp., ×10. *B*, false marginal indusium on sorus of *Pellaea rotundifolia*, ×5. *C*, opened sporangia in sorus of *Pellaea rotundifolia*, ×35. *D*, sori of *Polystichum* with peltate indusium, ×5. *E*, hooded indusium of *Nephrolepsis* attached by a broad base to a veinlet, ×10. *F*, elongated sorus of *Asplenium* ×5. *G*, sori of *Phanerophlebia* with indusia, ×10.

CHAPTER 29

GYMNOSPERMS

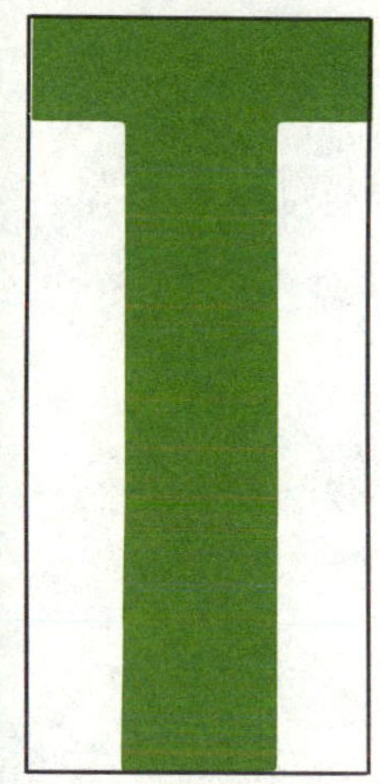

he **higher vascular plants** possess both vascular tissue and seeds. **Gymnosperms** (conifers and their relatives) and **angiosperms** (flowering plants) together make up the higher vascular plant group. These plants dominated the world's vegetation during the Cenozoic era (Chapter 31).

Gymnosperms are a major source of lumber and paper pulp in North America. This group is represented by such common trees as pines (*Pinus*), spruces (*Picea*), firs (*Abies*), and true cedars (*Cedrus*), all of which possess well-developed seed-bearing cones (Fig. 29.1). Other gymnosperms, such as *Ginkgo,* yews (*Taxus*), and Mormon tea or joint-fir (*Ephedra*), do not bear cones.

A trait common to all gymnospermous plants is the absence of a protecting case, that is, an ovary wall, around the seeds. In pines and other cone-bearers, seeds are borne on the surface of scales that comprise the cone and, though well-protected by the scales, they are not surrounded by floral parts. In yews, seeds are partially surrounded by a red berrylike structure (Fig. 29.2*A*). In *Ephedra,* ovules are borne in axils of short bracts (**Fig. 29.19,** page 616) and in *Ginkgo,* naked ovules are attached to ends of short branches (Fig. 29.2*B*). By contrast, the seed of an angiosperm is surrounded by a matured ovary wall.

Seeds lacking protection of an ovary wall are said to be ''naked''—thus the name gymnosperm, derived from two Greek words *gymnos* (naked) and *sperm* (seed). Fossil Lepidodendrales, belonging to the Lycophyta, bore seeds more than 200 million years ago, thus indicating that the seed habit is neither of recent origin, nor indica-

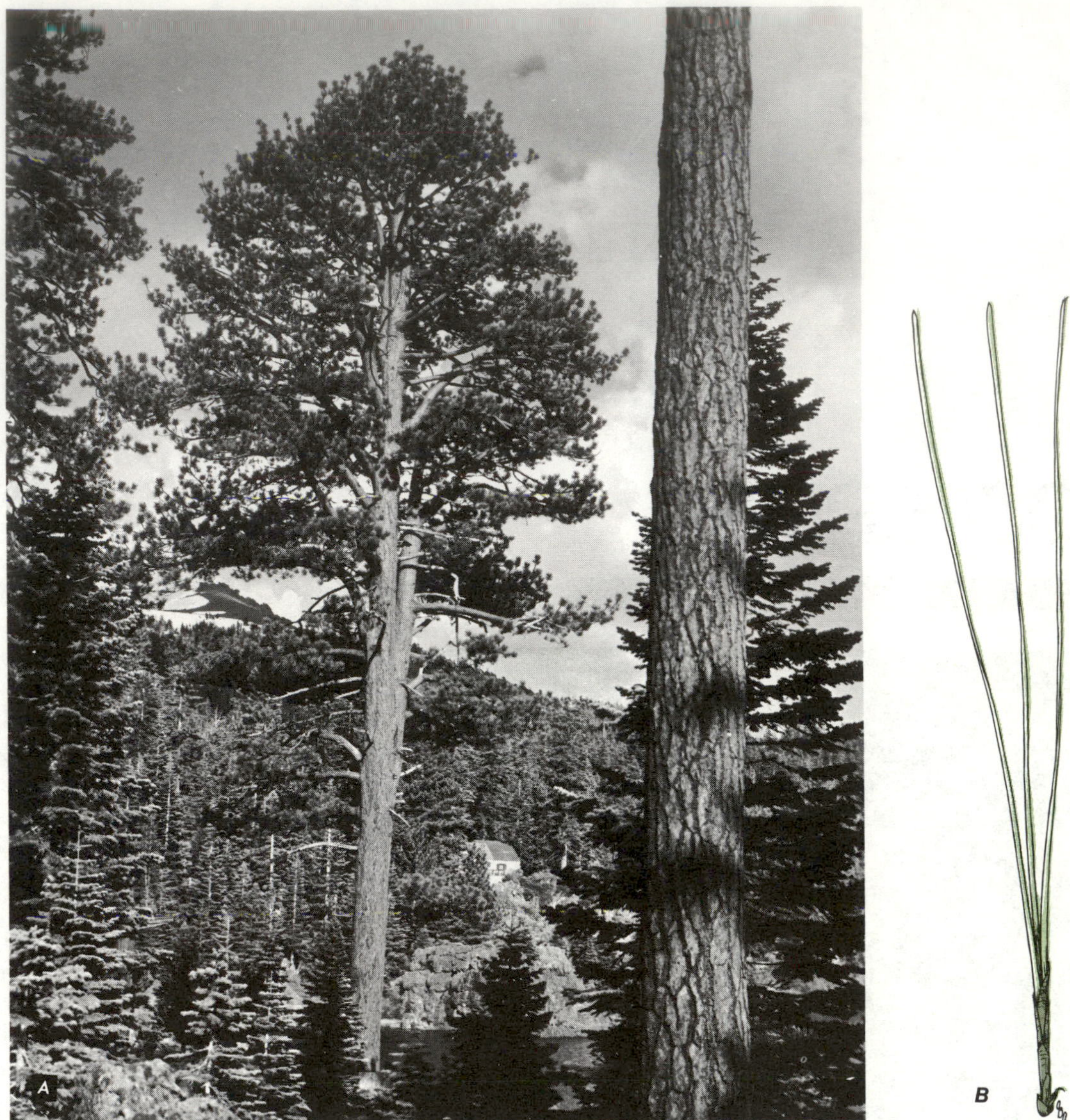

Figure 29.1
Pines. *A, Pinus jeffreyi* (Jeffrey pine), showing the trunk of one tree and the crown of a second tree. *B,* fascicle of ponderosa pine needles.

tive of a single evolutionary line.

There are four divisions of living gymnosperms, encompassing more than 600 species. Of these divisions, we shall only briefly consider: (*a*) the **Cycadophyta,** or **cycads;** (*b*) the **Ginkgophyta,** composed of only one species, the maiden-hair tree; and (*c*) the **Gnetophyta,** an artificial grouping of three unusual genera, *Gnetum, Ephedra,* and *Welwitschia.* We will discuss a fourth division, the **Coniferophyta** or conifers, in much more detail, and we begin with that group.

DIVISION CONIFEROPHYTA

Classification

Without exception, the well-known and economically important gymnosperms of temperate zones belong to the Coniferophyta. There are seven families and 550 species. They comprise pines, hemlocks, firs, spruces, junipers, yews, redwoods, and many others. Not all of them bear cones, yet the cone is such a conspicuous feature of many of them that the division has been named the Coniferophyta or cone bearers. Most conifers form true cones. Juniper berries are in reality cones with fleshy adhering scales. In yews, seeds are surrounded at the base by a more or less pulpy, berrylike body and are not borne in cones. In either case, the seeds are naked, not being surrounded by an ovary wall.

The following brief descriptions of several common genera enable us to introduce the more important conifers in some of the seven families.

Family Pinaceae

***Pinus* (Pines).** Pines are usually large trees. Some

Figure 29.2
Gymnosperms without typical cones. *A*, branches of *Taxus* with seeds; note berrylike aril surrounding seed. *B*, branch and seeds of *Ginkgo biloba*.

members of bristlecone pine (*Pinus longaeva*) are the oldest living things, over 5000 years old (see Fig. 29.9). The leaves are needlelike, two or more growing together (except in *Pinus monophylla*) in a **fascicle** or group, which is sheathed at the base (Fig. 29.1*B*). The cones vary greatly in size and shape and are very characteristic of the species to which they belong.

***Abies* (Firs).** Firs are stately trees of a symmetrical, cylindrical or pyramidal shape. The leaves are flat and linear; in cross section they are relatively broad, without marked angles. The cones are erect on the branches and shatter at maturity (Fig. 29.3).

***Picea* (Spruces).** These trees closely resemble the firs, but they can be distinguished by the position of the leaves on the branchlets, by the angular appearance of the leaves in cross section, and by the cones that are hanging and do not shatter at maturity (Fig. 29.4).

***Tsuga* (Hemlocks).** The trees of this genus are pyramidal with slender horizontal branches. The leaves are usually two-ranked, linear, flat, and with a short petiole. They resemble the leaves of firs but are much shorter. The cones are small.

***Pseudotsuga* (Douglas Fir).** Only two species occur in the United States. One of them, the Douglas fir (*Pseudotsuga menziesii*), is the most important timber tree of the United States. It may grow to a height of 60 m, with a trunk diameter of 3 m. Its leaves are flat, like those of the true firs, but have white lines on either margin and a groove along the upper surface. The cones are 5 to 11 cm long, pendulous, and easily recognized by bracts extending outward below each scale.

***Larix* (Larches).** Trees belonging to this genus grow in the cooler portions of the Northern Hemisphere. They differ from most other members of the Pinaceae in that

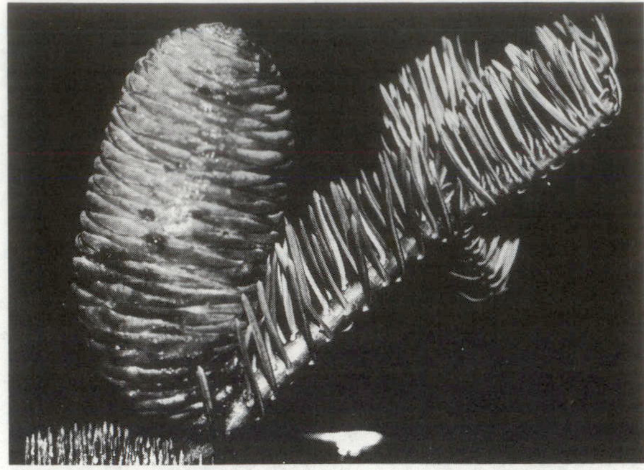

Figure 29.3
Abies, ovulate cone and branch. Note upright arrangement of needles. Branches lower on the tree, in the shade, exhibit leaves that lie flat.

they are deciduous. The American larch, or tamarack, is a tall tree frequently found in bogs. The needles are short, linear, and grouped in crowded clusters on short spurs. On leading shoots, however, the needles are arranged spirally. The cones are small and persistent (Fig. 29.5).

Family Cupressaceae

This family contains the junipers, cypresses, and false cedars (true cedars, *Cedrus* species, are in the Pinaceae). The leaves are borne singly, but are usually small and scalelike and are crowded closely around the stems (Fig. 29.6*B*). The cones may be woody, as in *Cupressus,* or fleshy, as in *Juniperus.* The cone of juniper is often called a "berry," but it is a modified cone, composed of fleshy scales that completely enclose the seeds (Fig. 29.6*C*).

Family Taxodiaceae

This family contains the bald cypress, dawn and coast redwoods, and Sierra bigtree. Coast redwood (*Sequoia sempervirens*) may be the tallest tree in the world, individuals over 60 m being common and the record height being 112 m. Redwood has outstanding lumber qualities, not the least of which is its resistance to decay. Consequently, logging has removed 90% of all virgin redwood forest in the past 150 years. Only half of the remaining acreage is protected from future logging. Sierra bigtree (*Sequoiadendron giganteum*) is the most massive of any plant or animal species in the world; its trunk at breast height may reach 9.8 m in diameter and its height above 82 m. The trunk alone of the General Sherman tree weighs 625,000 kg (over 680 tons).

Although some members of this family, such as redwood, are currently very restricted in their range, this was not always the case. At the end of the Mesozoic era, about 65 million years ago, coast redwood (*Sequoia sempervirens*) and dawn redwood (*Metasequoia glyptos-*

Figure 29.4
Picea. *A,* branch of *P. alba* with ovulate cones. *B,* leaves of *Picea.* (*A,* courtesy of the American Museum of Natural History. *B,* redrawn from L. H. Bailey, *The Cultivated Conifers in North America.* © 1933 by The Macmillan Company. Reprinted with permission of the publisher.)

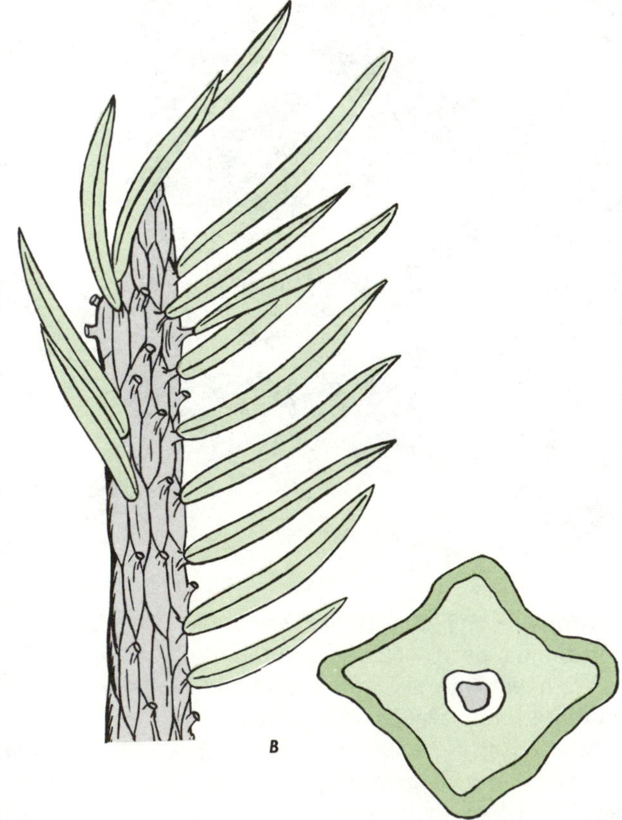

Figure 29.5
Larix lyallii. A, branch with needles. *B,* mature ovulate cones.

troboides) (**Fig. 29.7,** page 613) were widely distributed in the north temperate zone (Fig. 29.8). Today, coast redwood is restricted to a narrow, coastal strip of California less than 13,000 km^2 in area, and dawn redwood is restricted to a small area in central China. Dawn redwood, in fact, is so narrowly distributed that it was thought to be extinct, and was known only from the fossil record until it was rediscovered in 1941. Seedlings have since been distributed all over the world.

Family Taxaceae (Yews)

With two exceptions, ovulate cones are not borne by members of the yew family. The seed of this family so strongly resembles a drupe or a nut that it is usually called a fruit. The embryo is protected by an outer flesh called an **aril,** and an inner hard pit. Both these tissues may be derived from the integuments of the ovule; hence, the structure is morphologically a naked seed. As shown in Fig. 29.2*A*, the outer red aril of ground pine (*Taxus canadensis*) almost encloses the seed. It drops away when the seed is mature.

Native yews are common only in a few places in the United States. Possibly the best-known yew is the English yew, which, because of the excellent bows that were made from its wood, is closely linked with English history and folklore. Yews are now widely cultivated.

Life History of a Conifer

The following outline of a gymnosperm life cycle is based largely on the genus *Pinus.* It differs from life cycles of other genera mainly in requiring three summers for completion. Significant stages of pollination, fertilization, and maturation of the seed occur within a decimeter of each other on a given branch, thus making it an ideal subject for class study.

Sporophyte

All species of pines are trees. They range in size and age from scrubby, 70-year-old fire-adapted species such as knobcone pine (*Pinus attenuata*) to large, straight trunked ponderosa pine (*Pinus ponderosa*) 500 years old, to twisted, slow-growing bristlecone pine (Fig. 29.9) 5000 years old.

Conifer wood has no vessels. A three-dimensional reconstruction from a small piece of *Sequoia* is diagrammed in Fig. 29.10. It is extemely regular. Tracheids are elongated cells that appear in cross section as square, hollow cells. Those formed in the spring are largest in diameter; as the season progresses, newly formed tracheids are smaller in diameter with thicker walls. The heavy-walled cell is called a fiber-tracheid. Note, xylem rays are composed of brick-shaped parenchyma cells, only one cell wide, and sometimes with dark-staining contents. Other parenchyma cells may run in vertical columns; these are called axial parenchyma.

Many gymnosperms produce resin. Typically, resin occurs in long **resin ducts** that are surrounded by parenchyma cells (Fig. 29.11). Resin may play a role by inhibiting the activity of wood-boring insects. *Sequoia* does not produce resin.

In the phloem, companion cells are missing, but **sieve cells,** fibers, and parenchyma cells are present.

The needle leaves of *Pinus* show considerable adaptation to drought (Fig. 29.12): sunken stomata, thick cuticle, fibrous epidermis, close-packed mesophyll, and veins only in the center of a thick leaf. Such modifications are important in winter, when soil moisture is frozen and the evergreen leaves are subjected to drying winds.

All conifers produce two kinds of spores—**microspores** and **megaspores**—borne in cones that are morphologically distinct. The two types of cones are known, respectively, as **staminate** (or pollen) and **ovulate** (or seed) **cones**.

Staminate (Pollen) Cone

Staminate cones average 10 mm or less in length, by 5

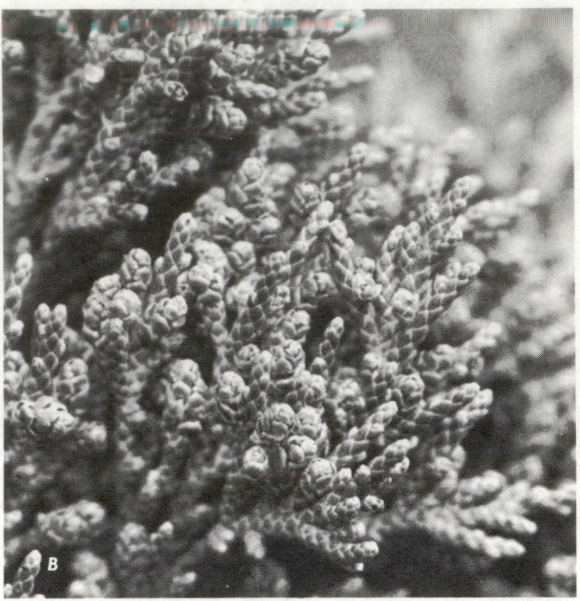

Figure 29.6
Juniperus occidentalis growing at 2200 m in the Sierra Nevada Mountains of California. *A*, tree. *B*, branch with staminate cones. *C*, branch with ovulate cones.

mm in diameter. They are borne in groups, usually on lower branches of trees (Fig. 29.13*A*). Each cone is composed of a large number of small scales (microsporophylls) arranged spirally on the axis of the cone. Two microsporangia develop on the under surface of each scale.

Stages of microspore development are quite similar to those of *Selaginella* and to spores of mosses and ferns. Microsporocytes, which are surrounded by a nutritive cell layer, the **tapetum,** undergo meiosis, and four microspores result. As usual, each microspore contains the haploid number of chromosomes. The nucleus within the newly formed microspore divides several times by mitosis, forming a **pollen grain** that contains two viable, haploid nuclei and vestiges of several vegetative cells (Fig. 29.14*H*). Enormous numbers of pollen grains are finally shed from the microsporangium. They are light in weight and bear two wings that may facilitate dispersal by wind.

Ovulate (Seed) Cone

The ovulate cone, when mature, is the well-known cone of pines and other conifers. Each cone is composed of an axis upon which are borne several woody scales in a spiral fashion. Ovulate cones develop at tips of young branches in early spring (**Fig. 29.15,** page 614). Two ovules, each enclosing a single megasporangium, develop on the upper surface of **ovuliferous scales** (**Fig. 29.16*B*,** page 615).

The ovules first appear as small protuberances on the upper surface of this scale, close to the axis of the cone. A protective layer of cells, the **integument,** develops early on the outer surface of the ovule. In the end of each ovule near the axis of the cone, there is a small opening, the **micropyle,** through which pollen grains may enter.

In pine, one megasporocyte lies in the center of each ovule (Figs. 29.14*C*, **29.16*C***); several are contained in ovules of redwoods and cypresses. Like microsporocytes, the megasporocyte is surrounded by a nutritive tissue called the **nucellus**. The nucellus is in reality a megasporangium because it surrounds the area where megaspores arise.

Female Gametophyte. The megasporocyte soon di-

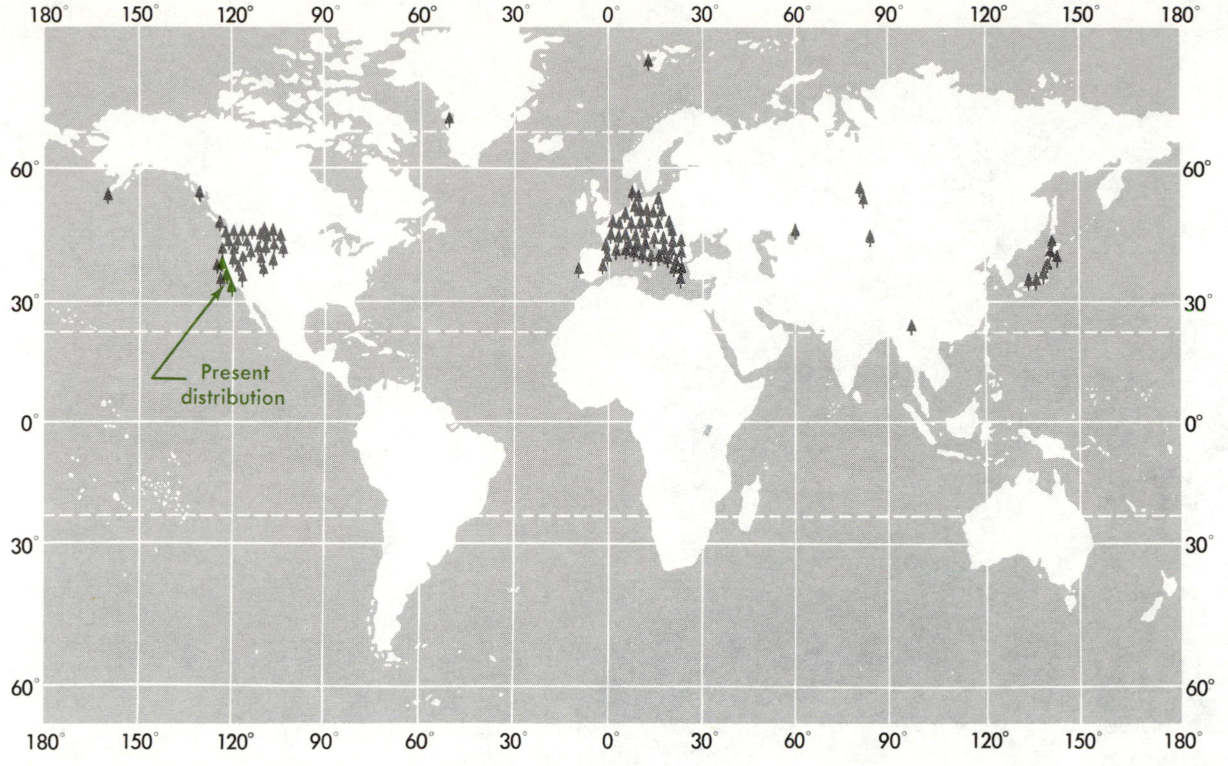

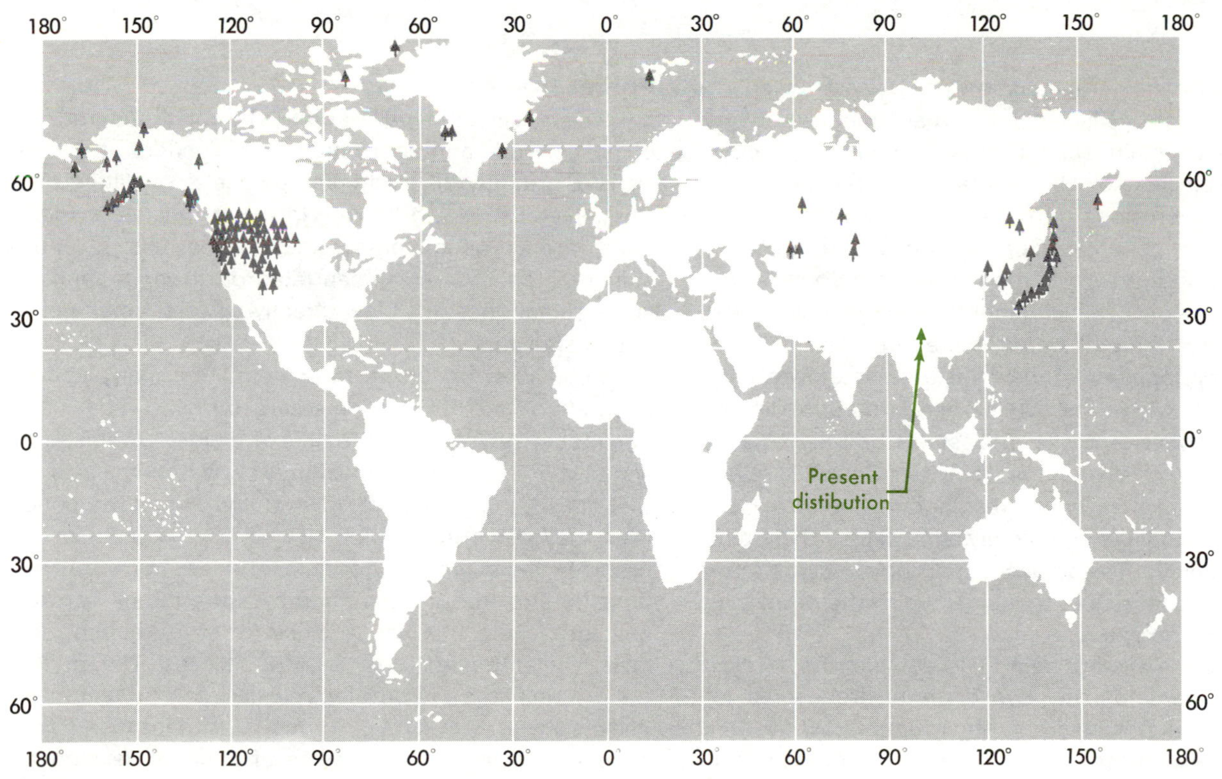

Figure 29.8

Distribution of fossil and living redwoods, *Metasequoia* (below) and *Sequoia* (above). (From A. Florin, *Acta Horti Bergiani,* **20.**)

Figure 29.9
Pinus longaeva (bristlecone pine) in the White Mountains of California, growing near timberline.

vides by meiosis. Four megaspores, usually arranged in a single row of four cells, result from each megasporocyte. The nucleus within each megaspore has the haploid number of chromosomes. Generally, only one of the four megaspores develops into a female gametophyte; the other three degenerate. Germination of the remaining megaspore and growth of the female gametophyte progresses very slowly. Several months are required in most conifers, and 13 months are required in pine.

The development of the female gametophyte takes place entirely within the ovule. There are approximately 11 mitotic divisions of the megaspore before cell walls begin to appear between the newly formed nuclei. Walls do gradually form, however, resulting in a small mass of gametophytic tissue, completely enclosed by the diploid cells of the ovule. The adjacent sporophytic tissue of the ovule, the nucellus, is in part digested by the developing female gametophyte.

While cell walls are being laid down in the developing female gametophyte, two or more archegonia differentiate at its micropylar end. Figures 29.14*K* and **29.16*E*** show that the ovule at this stage consists of integuments, nucellus, and female gametophyte containing several archegonia, each with its enclosed egg. Directly beneath the micropyle is a space, the **micropylar chamber.** Nucellar tissue lies between the micropylar chamber and the archegonia.

Pollination

It will be recalled that **pollination** in flowers is the transfer of pollen from anther to stigma; in conifers, it is the transfer of pollen from staminate cone to ovulate cone. Conifer pollen is windblown. In many species, ovulate cones are borne on higher branches of the tree and staminate cones are concentrated on lower branches; since pollen does not blow directly upward, cross pollination is usual.

Pollination occurs in most conifers in early spring soon after the ovulate cone emerges from the dormant bud (Table 29.1)—about the time of meiosis. At this age, scales of young cones turn slightly away from the axis so that pollen grains can sift down to the axis of the cone. Here they come in contact with a sticky substance secreted by the ovule. As this material dries, it draws some pollen grains through the micropyle into the micropylar chamber where they lodge close to the developing female gametophyte (Figs. 29.14*K*, **29.16*E***). The pollen grain germinates and develops slowly into a tubular male gametophyte as it grows through the nucellus. Thus, male and female gametophytes of conifers develop to maturity in close proximity within the ovule. They are both dependent on nucellar tissue for nourishment and protection.

Several nuclear divisions occur in the tube, but no cell walls are formed. Two of the last-formed nuclei are

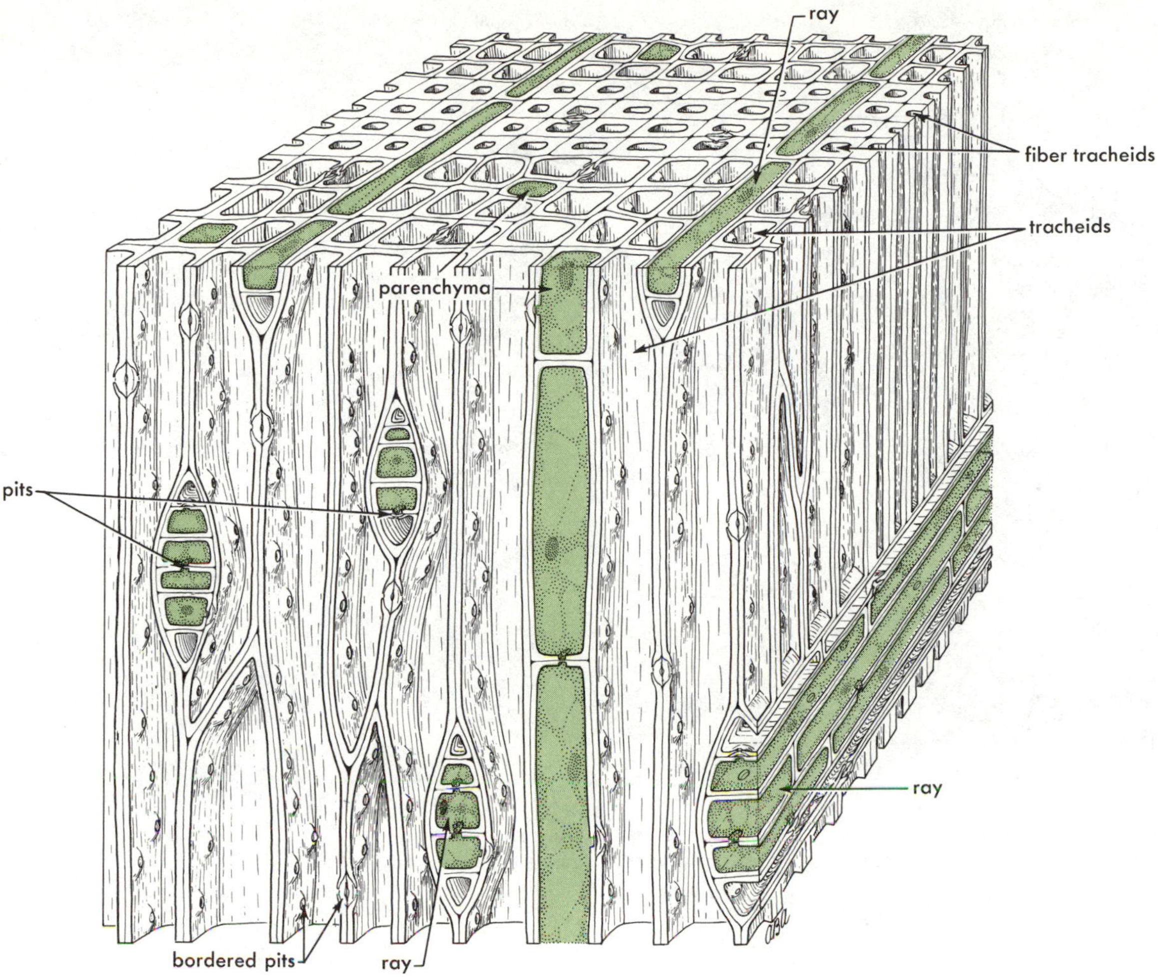

Figure 29.10
Block diagram of secondary xylem of redwood *(Sequoia sempervirens).*

sperm nuclei. This branched **pollen tube,** containing two sperm nuclei and several vegetative nuclei, is the male gametophyte (Fig. 29.14*I*).

Fertilization and Embryo Formation

Development of male and female gametophytes is so co-ordinated that the egg is formed and ready for fertilization when the pollen tube, containing two sperm nuclei, has reached the archegonium. In pine, this development requires about one year. At the time of fertilization, pine cones are generally green and the scales tightly closed, showing a spiral pattern of arrangement **(Fig. 29.15*D*).** The sperm nuclei, together with other protoplasmic contents of the pollen tube, are discharged directly into the egg cell. Sperm nuclei do not possess cilia and hence are not actively motile. One sperm nucleus comes in contact with the egg nucleus and unites with it. The nonfunctioning sperm nucleus and the other protoplasmic material discharged into the egg cells soon undergo disorganization.

Figure 29.11
Resin duct in pine wood, as seen in cross section.

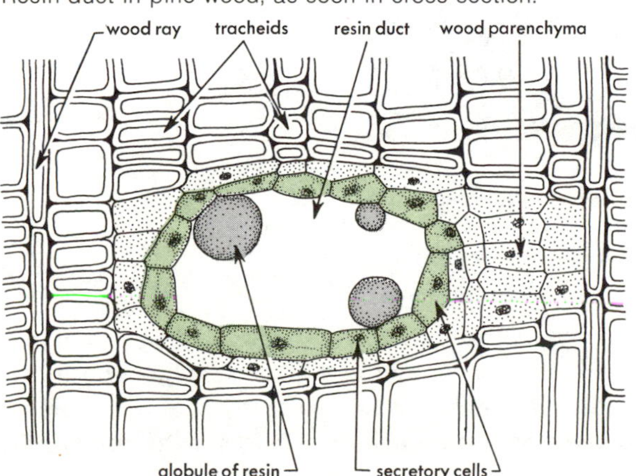

The formation of the embryo is preceded by the development of a relatively elaborate **proembryo.** This structure consists of four tiers of four cells each (Fig. 29.14*M*). The four cells farthest from the micropylar end of the proembryo may each develop into an embryo. The intermediate cells are the suspensor cells; they elongate greatly and push the embryo cells deep into the female

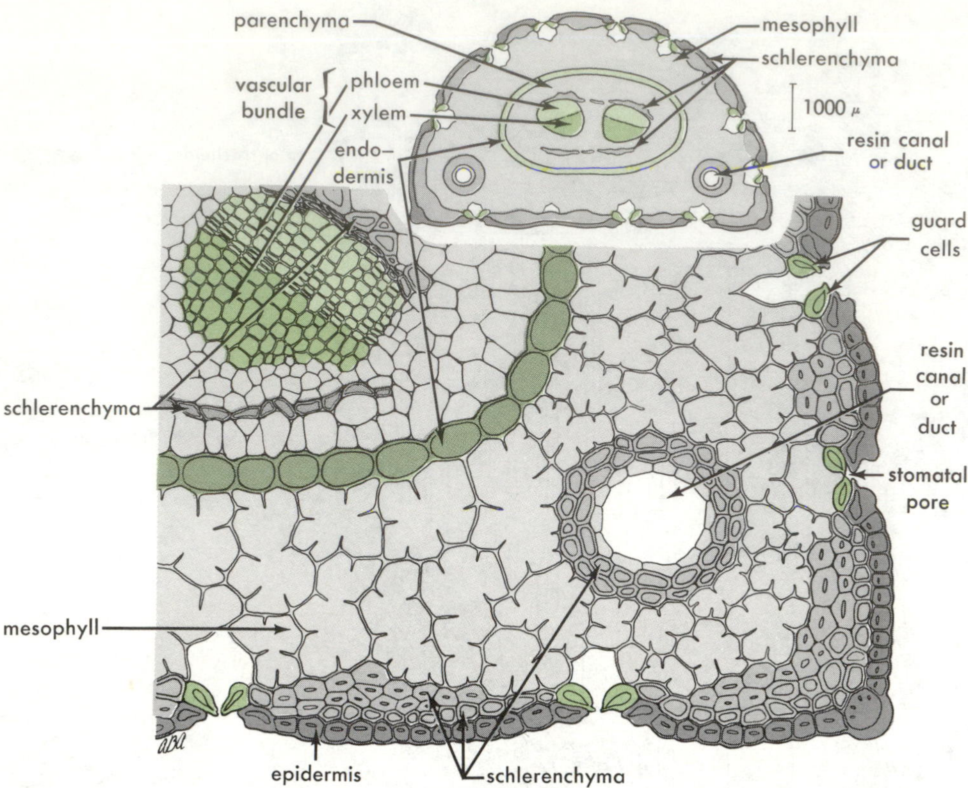

Figure 29.12
Cross section of a two-needled pine leaf. The inset (above) shows the entire cross section, with major tissues outlined; the large drawing shows cellular detail of a wedge from the epidermis in to one of the vascular bundles at the center.

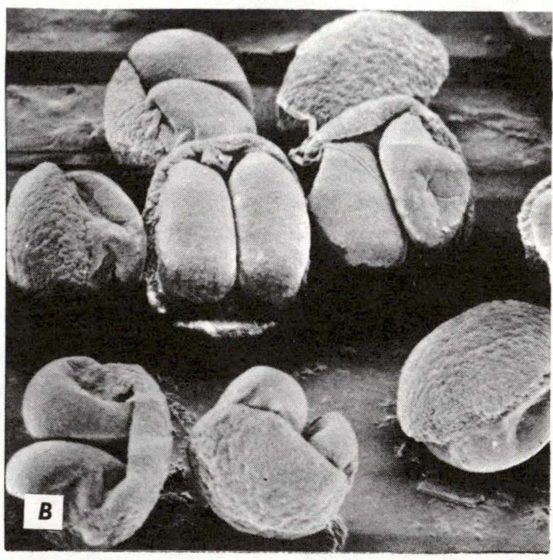

Figure 29.13
Staminate cones of *Pinus*. *A*, end of a branch with a cluster of staminate cones holding ripe pollen. *B*, scanning electron micrograph of pollen grains (probably shrunken from treatment in the SEM).

gametophyte. While this development is taking place, the female gametophyte continues to grow, digesting most of the rest of the nucellus. It enlarges and becomes packed with food to be used not only for the growth of the embryo but also as a reserve in the seed.

It will be recalled that the female gametophyte usually contains two or more archegonia. Since the egg in each archegonium may be fertilized and since each of the four embryo-forming cells may give rise to an embryo, eight or more embryos may develop in every seed. Normally, however, only one embryo survives, but seeds with two well-formed embryos are not rare.

The mature embryo consists of several **cotyledons** or seed leaves, **epicotyl, hypocotyl,** and **radicle** or rudimentary root (Fig. 29.14*N*). The embryo is embedded, as previously mentioned, in the enlarged female gametophyte. Both embryo and female gametophyte are surrounded by a papery shell and a hard protective **seed coat,** formed from the nucellus and the integument of the ovule. The whole structure is the **seed.** In many pines, the seeds are winged.

Normally, in pines, the seed matures some 12 months after fertilization; two years intervene between initiation of ovules and formation of seeds (Table 29.1). Young pine seedlings (Fig. 29.14*P*) have several cotyledons.

Gymnosperm seeds may remain dormant for many years, and some may remain embedded in the old mature cones for six years or more. Heat causes the resin coating the cones of some species to melt, allowing the cones to open and release the seeds. This behavior has considerable survival value, since large numbers of seeds are released after fires, and injured trees are thus replaced by the young ones. In many species, however, seeds are shed soon after they are mature.

The Conifer Life Cycle In Perspective

The pine life cycle, representative of the Coniferophyta, is part of a general life cycle trend from Bryophyta to Anthophyta. Quite simply, that trend is: an increasing importance and complexity of the sporophyte generation and at the same time a decreasing size and degree of independence of the gametophyte generation (Fig. 29.17).

As a generalization, thallophytes are part of this trend. In the algae and fungi, the gametophyte generation tends to be dominant; that is, a given species tends to spend most of its active life span in the haploid state. Furthermore when sperm and egg (or + and − gametes)

Table 29.1

A GENERALIZED LIFE HISTORY OF A PINE

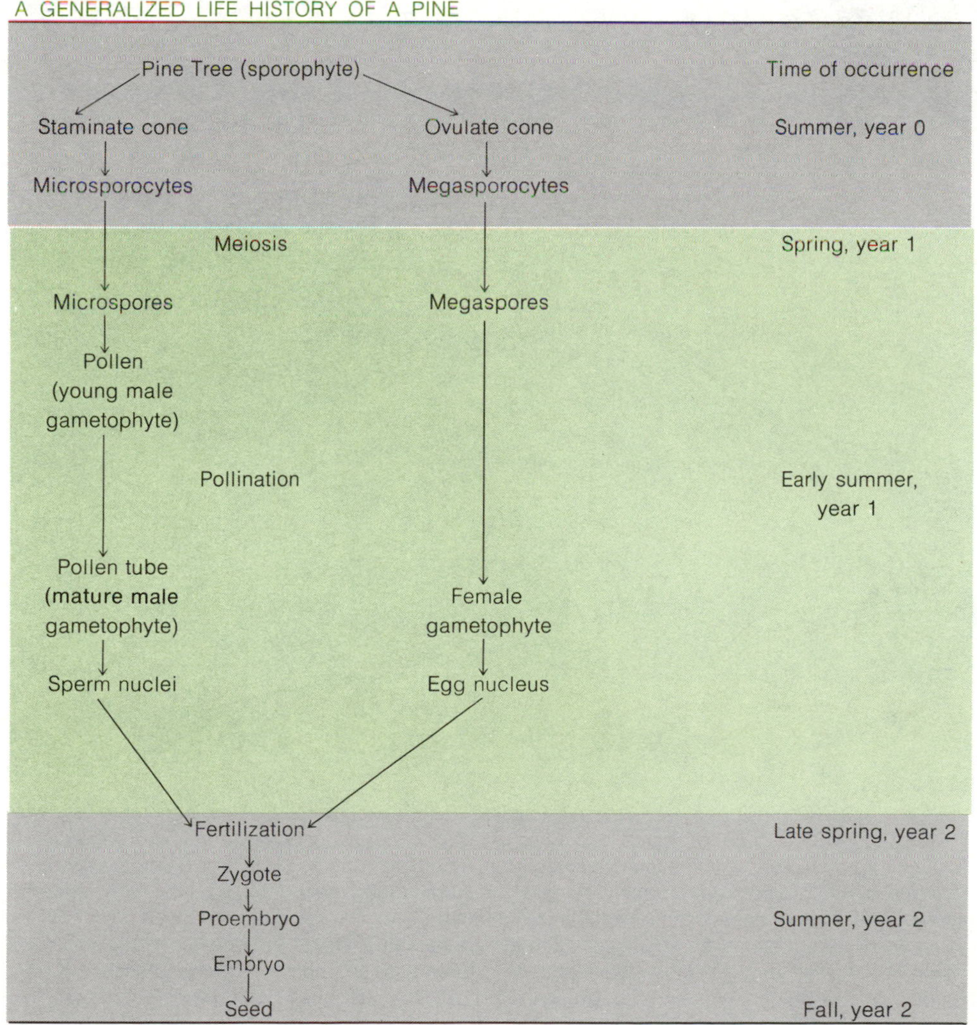

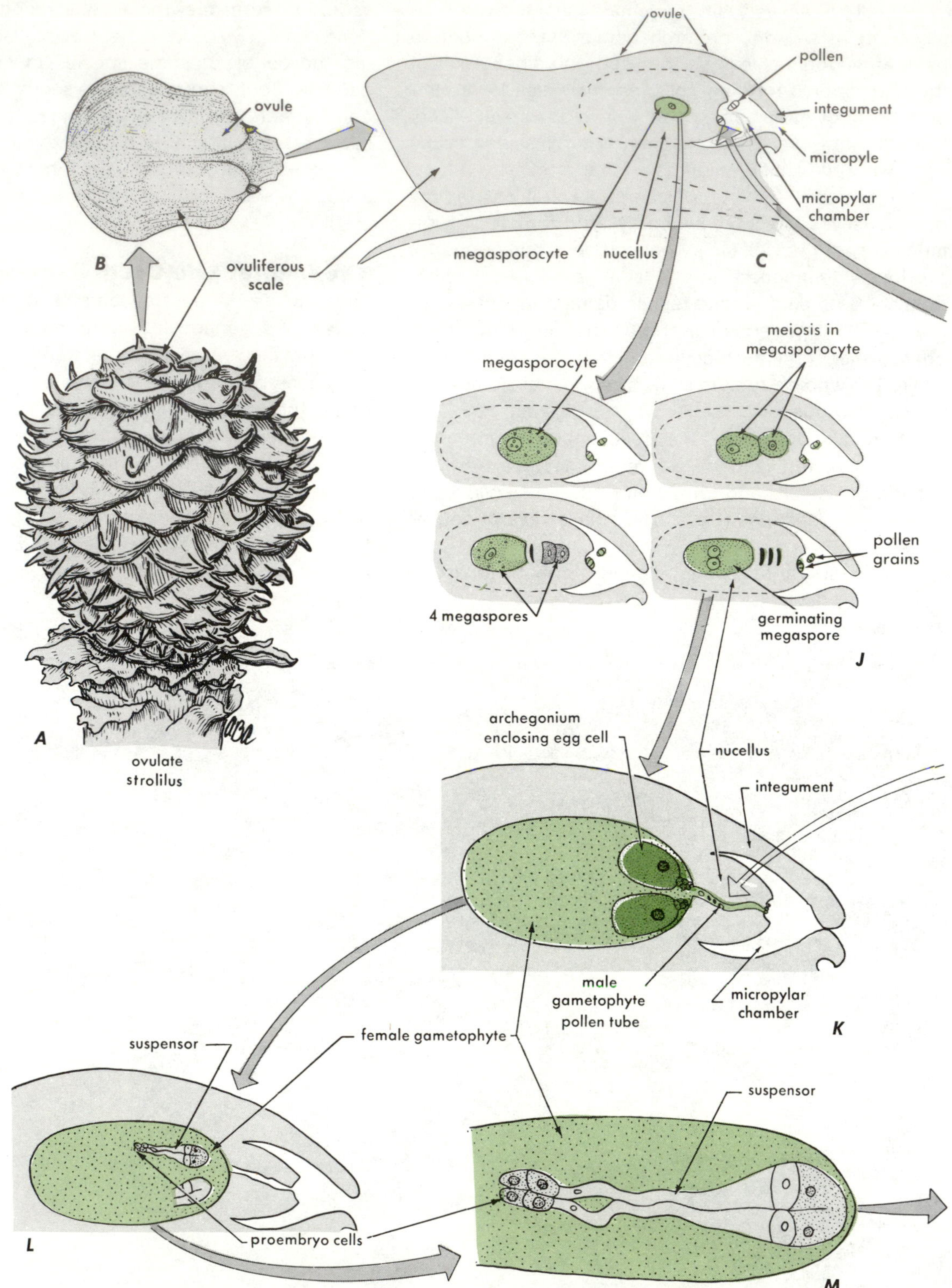

Figure 29.14

Stages in the life cycle of a pine. *A*, young ovulate cone. *B*, scale from ovulate cone showing two ovules on upper surface. *C*, section of ovulate cone showing megasporocyte cell and pollen in pollen chamber. *D*, staminate cone. *E*, scale from staminate cone. *F*, section of staminate scale. *G*, winged pollen grain. *H*, development of male gametophyte in pollen grain. *I*, pollen tube. *J*, formation of linear tetrad on megaspores. *K*, female gametophyte with egg cell and pollen tube. *L*, suspensor and proembryo sporophyte within female gametophyte. *M*, proembryo. *N*, section of seed showing seed coats, female gametophyte, and embryo sporophyte. *O*, winged seed. *P*, seedlings. (*C, G, I* redrawn after J. J, Coulter and C. J. Chamberlain, *Morphology of the Gymnosperms*. © 1917 by The University of Chicago Press.)

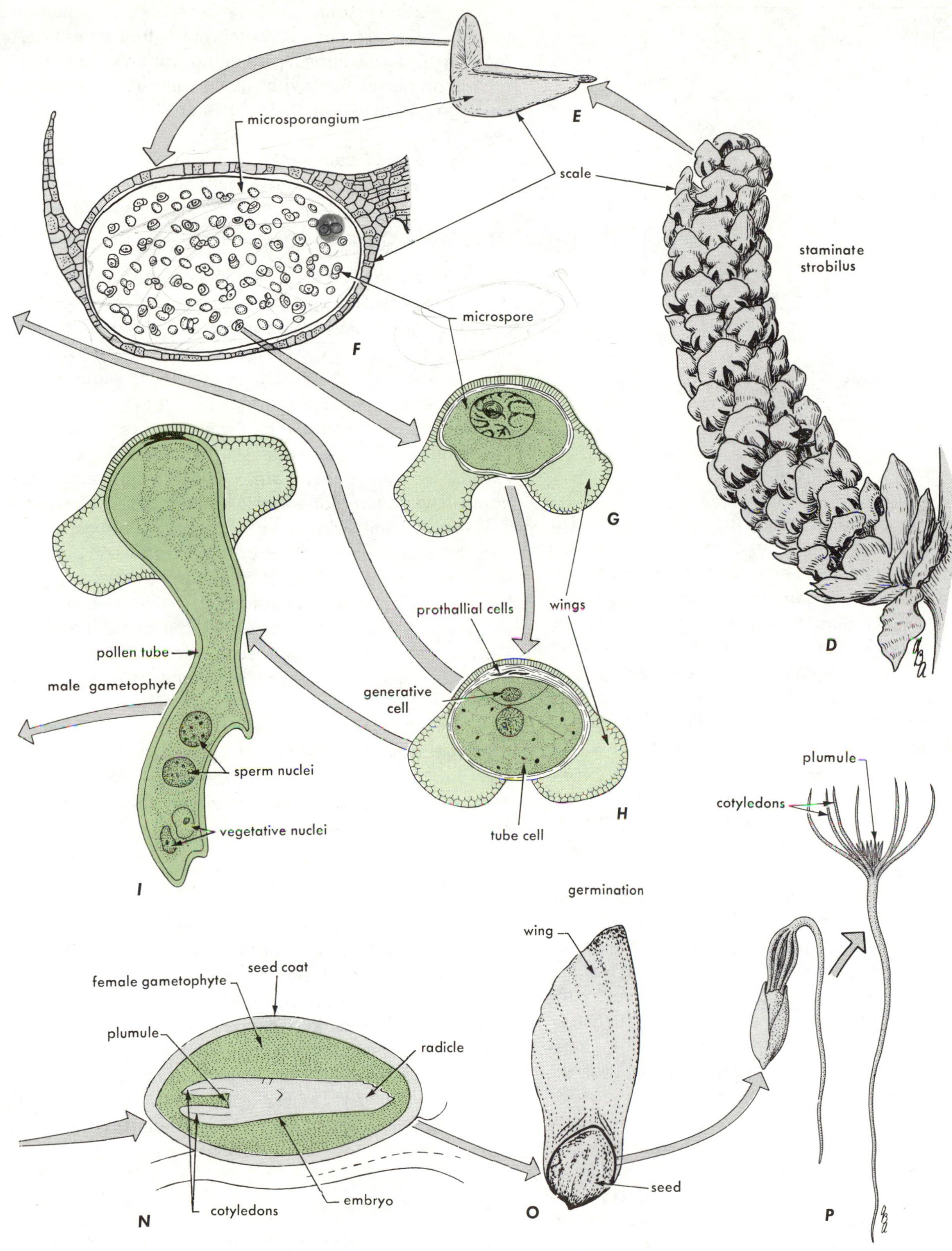

join to produce a zygote, that cell is not retained or protected within parental tissue. Recall the life cycle of *Fucus,* where the eggs and sperm are released from gametangia and meet in the open water; the zygote then begins dividing into a young plant while still in the open water.

In the Bryophyta, the gametophyte generation is still dominant and it tends to be more anatomically and

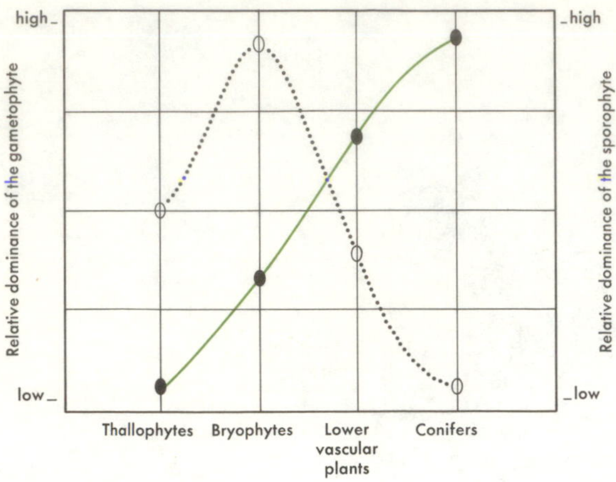

Figure 29.17
Life cycle trend from thallophytes to conifers, highly diagrammatic. Sporophyte, solid green line; gametophyte, dotted line.

morphologically complex than the sporophyte. The sporophyte is still not an independent entity, because it partly relies on the gametophyte for nutrition during its entire life. In contrast, the gametophyte is photosynthetic and completely independent and can reproduce asexually to continue growth indefinitely (indeed, in some mosses, the sporophyte generation is rarely seen). Cells that resemble tracheids and sieve cells are present in some species. There is more protection given to the young sporophyte (embryo) that in thallophytes: the zygote is retained in the tissue of the gametophyte parent and its early growth takes place within this tissue. There is also additional protection given to developing gametes: the gametangia have a sterile jacket of cells around the gametes.

In the lower vascular plants, the sporophyte becomes the dominant generation—dominant in terms of life span, size, and anatomical complexity. It also becomes an independent phase. During the embryo stage, the sporophyte is still attached to and nourished by the gametophyte, but ultimately it becomes established in the soil and is from then on independent of the gametophyte. Multicellular gametangia are retained, as is the requirement for free water for fertilization.

In the conifers, the size, complexity, and longevity of the sporophyte generation become even more extreme, and the gametophyte generation becomes dependent on the sporophyte for nutrition and protection. Both the male gametophyte (pollen tube) and the female gametophyte are parasitic on the nucellus ($2n$ tissue) and incapable of manufacturing their own food or of existing in the open. They exhibit very little differentiation, and antheridia are not recognizable. The gametes are protected to a greater degree than in lower vascular plants in that the egg and the pollen tube that carry sperm to egg always lie within surrounding tissue.

Pollination is traditionally regarded as an adaptation to a dry, terrestrial habitat. Furthermore, free water is not required for fertilization. Another innovation is additional protection for the embryo by its enclosure in a dispersal package (the seed) complete with stored food and a hard seed coat.

Figure 29.17 summarizes the shift in life cycle patterns that we have seen from thallophytes to conifers.

DIVISION CYCADOPHYTA

Some 200 million years ago members of the Cycadophyta formed an extensive portion of the earth's flora, probably constituting the food supply for at least some of the herbivorous dinosaurs. Today, there remain 9 to 10 genera of cycads with about 100 species growing in widely separated areas of the earth's surface but largely confined to the tropics. Only one genus, *Zamia,* occurs naturally in the continental United States; it is found in southern Florida. Various genera are, however, cultivated outdoors in warmer regions, and in greenhouses elsewhere.

Two cycades, *Cycas revoluta* and *Dioön spinulosum,* are shown in Fig. 29.18. Both were grown in the conservatory at Golden Gate Park, in San Francisco. These trees are palmlike in appearance; in fact, *Cycas revoluta* is known as Sago palm, In *Zamia,* the genus native to Florida, the trunk is largely subterranean (Fig. 29.18*C*), but in *Cycas* it forms a single straight bole. The trees grow slowly, a 2 m high specimen being perhaps as much as 1000 years old.

That these trees are really gymnosperms and not palms is readily evident from their large cones (Figs. 29.18*A,B*). A primitive feature of the division is the presence of flagellated sperm.

DIVISION GINKGOPHYTA

Only one living representative, the maiden-hair tree (*Ginkgo biloba*), remains of this very ancient division of plants. It has been reported as growing wild today in forests of remote western China. *Ginkgo* has, however, been grown for centuries on Chinese and Japanese temple grounds and is now cultivated in many countries. It is a large tree with characteristic small, fan-shaped leaves (Fig. 29.2*B*) that are divided into two lobes. The sperms are flagellated. In the United States the *Ginkgo* is commonly planted as a city sidewalk tree.

DIVISION GNETOPHYTA

This division consists of only three genera, and in some respects their traits place them as intermediate between the gymnosperms and the angiosperms. For example, they contain vessels in the xylem, the ovules appear to be surrounded by two integuments, and the pollen-producing structures superficially resemble stamens. All these traits are shared by angiosperms and are absent in other

Figure 29.7
Comparison of the new dawn redwood *(Metasequoia)* with coast redwood *(Sequoia)* trees about 30 years old: *A*, coast redwood; *B*, dawn redwood. Coast redwood: *C*, branches; *E*, seed cones; *G*, pollen cones. Dawn redwood; *D*, branches; *F*, seed cones; *H*, pollen cones.

Figure 29.15

Reproductive stages of *Pinus*. *A*, *Pinus monticola* (western white pine). *B*, staminate cones of *Pinus canariensis*, $\times 1/10$. *C*, six month old ovulate cones of *Pinus halapensis*, $\times 2/10$. *D*, year-old cones of *Pinus halapensis*, $\times 3/10$. *E*, mature cones with seeds of *Pinus halapensis*, $\times 3/10$.

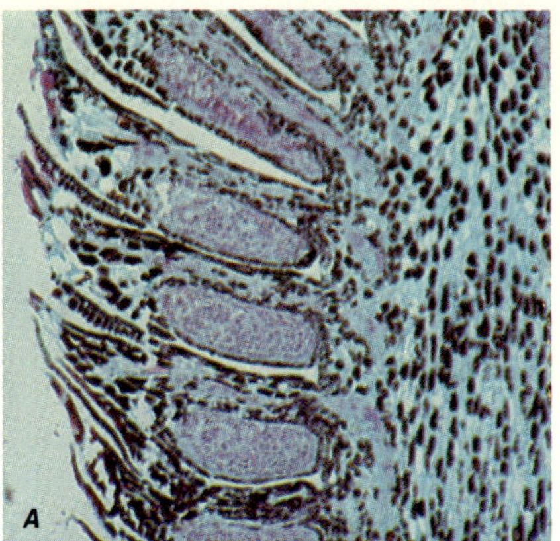

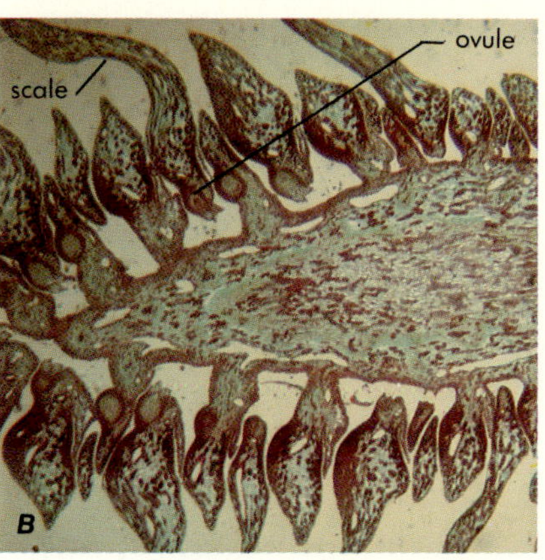

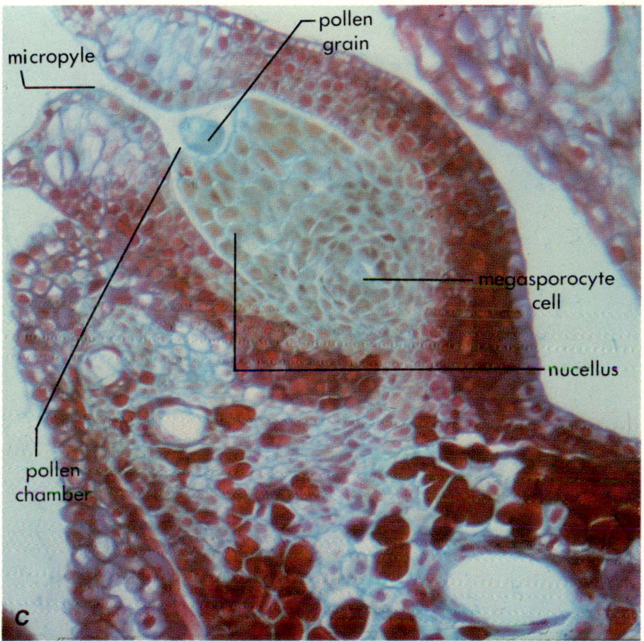

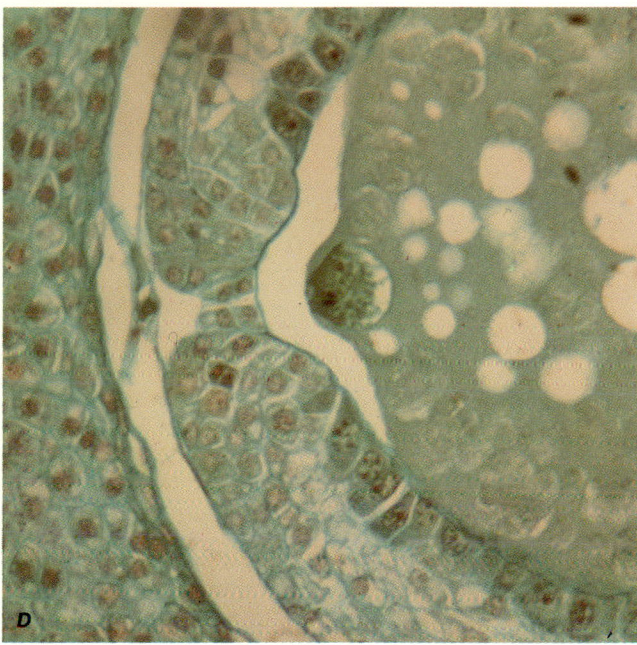

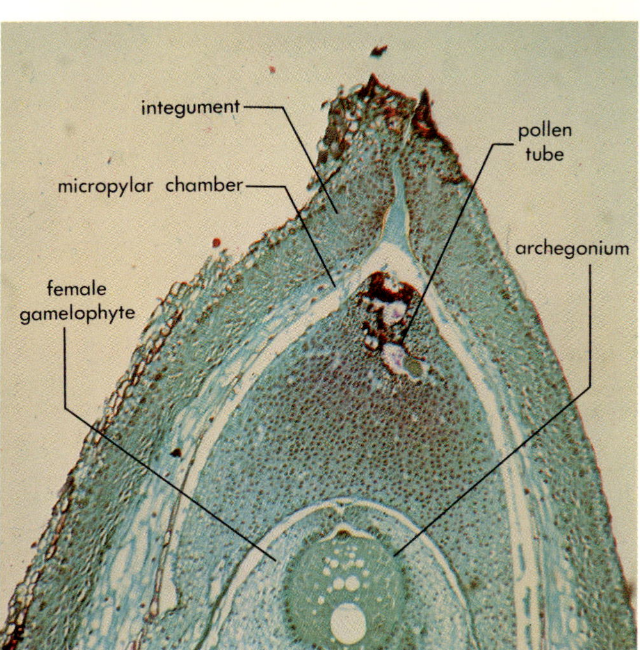

Figure 29.16

Reproductive stages of *Pinus*. *A*, section through a staminate cone, ×2.5. *B*, section through a six-month-old ovulate cone, ×2.5. *C*, ovule at time of pollination, ×25. *D*, tip of ovule at time of fertilization, ×25. *E*, archegonium in female gametophyte, ×100.

Figure 29.19
Gnetales. *A, Welwitschia. B,* two species of *Ephedra,* the larger from China, the smaller from California. *C,* node from China species, ×4. *D,* flower, note droplet at apex of the naked ovule, ×5. *E,* ovules of California species. *F,* mature ovule partially enclosed by red arilles, ×5.

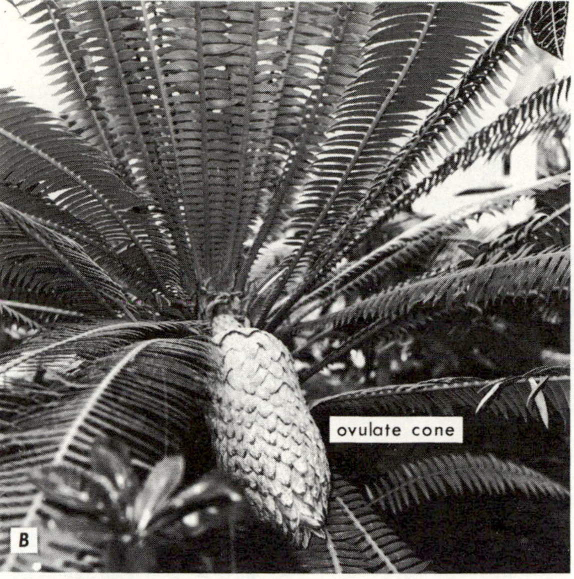

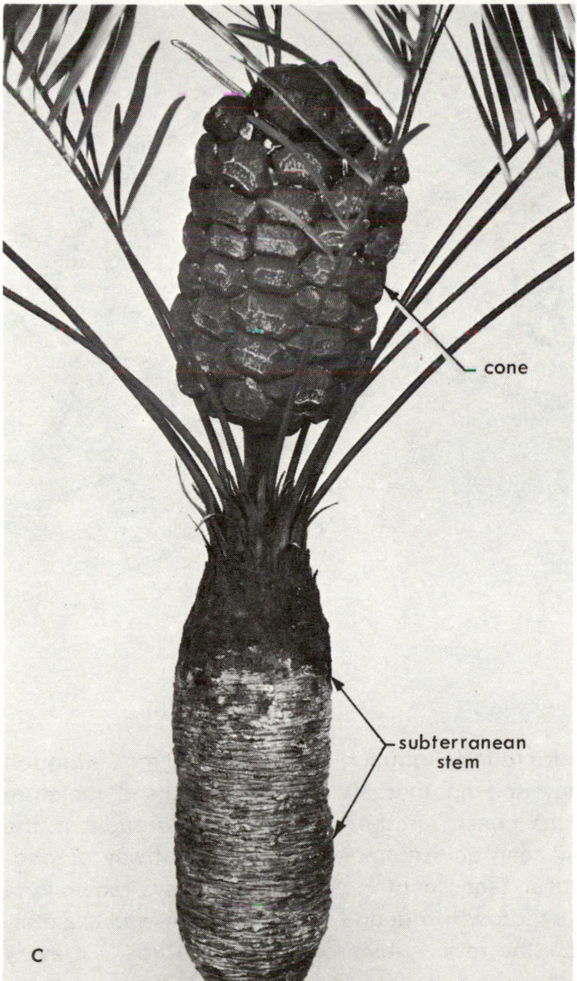

Figure 29.18
Cycads. *A, Cycas revoluta* showing a microsporangiate cone. *B, Dioön spinulosum* showing ovulate cone. *C, Zamia* showing ovulate cone and subterranean stem. (Courtesy of the Field Museum of Natural History.)

gymnosperms, yet true flowers and fruit are absent—seeds are still borne naked.

Only one genus, *Ephedra* (joint-fir, Mormon tea), is found in North America. Its 40 species are distributed in warm temperate parts of the Mediterranean region, India, China, the southwestern deserts of the United States, and mountainous parts of South America. It is a shrubby xerophyte, with whorls of small deciduous leaves at prominent joints; most photosynthesis is accomplished in the green stems (**Fig. 29.19,** page 616).

The other genera look quite different from *Ephedra* and exhibit important life cycle differences. Their inclusion into one division is a matter of convenience for classification and the resulting division is quite artificial. *Gnetum* is a tropical genus of 30 species, mainly climbing lianas. The leaves look very much like typical dicot leaves. *Welwitschia* is a xerophyte, distributed along the southwest coast of Africa. It has two long, straplike leaves that trail along the soil surface (**Fig. 29.19**).

Welwitschia mirabilis

As detailed in Figs. 29.20*B*, 29.21*E* to *I*, individual units of the male cone bear a striking resemblance to flowers. Each "flower" consists of several scales or bracts, six anthers, and a central, sterile structure that resembles a pistil. The "pistil" is an outgrowth of the integument of an ovule; the integument is drawn out into a stylelike tube with a stigmalike disc at the apex. A drop of fluid, similar to the pollination drop of the pine, may be secreted on that stigmalike disc. The ovule is sterile, however, and the cone is functionally male.

Female cones are borne on separate plants. A naked ovule is present inside each cone scale. The integuments

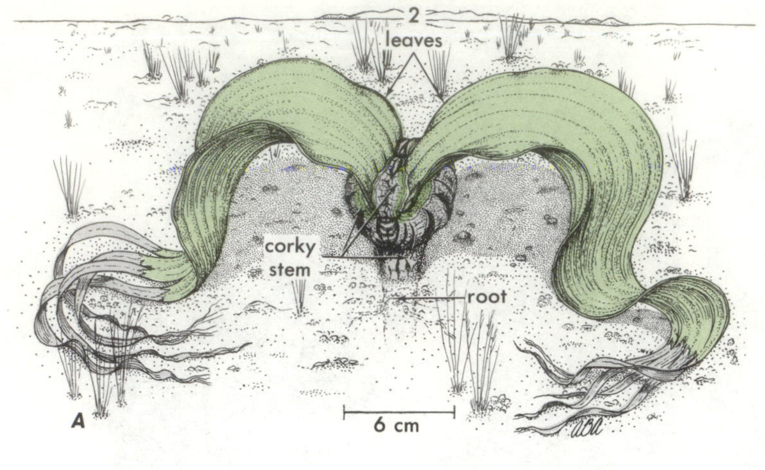

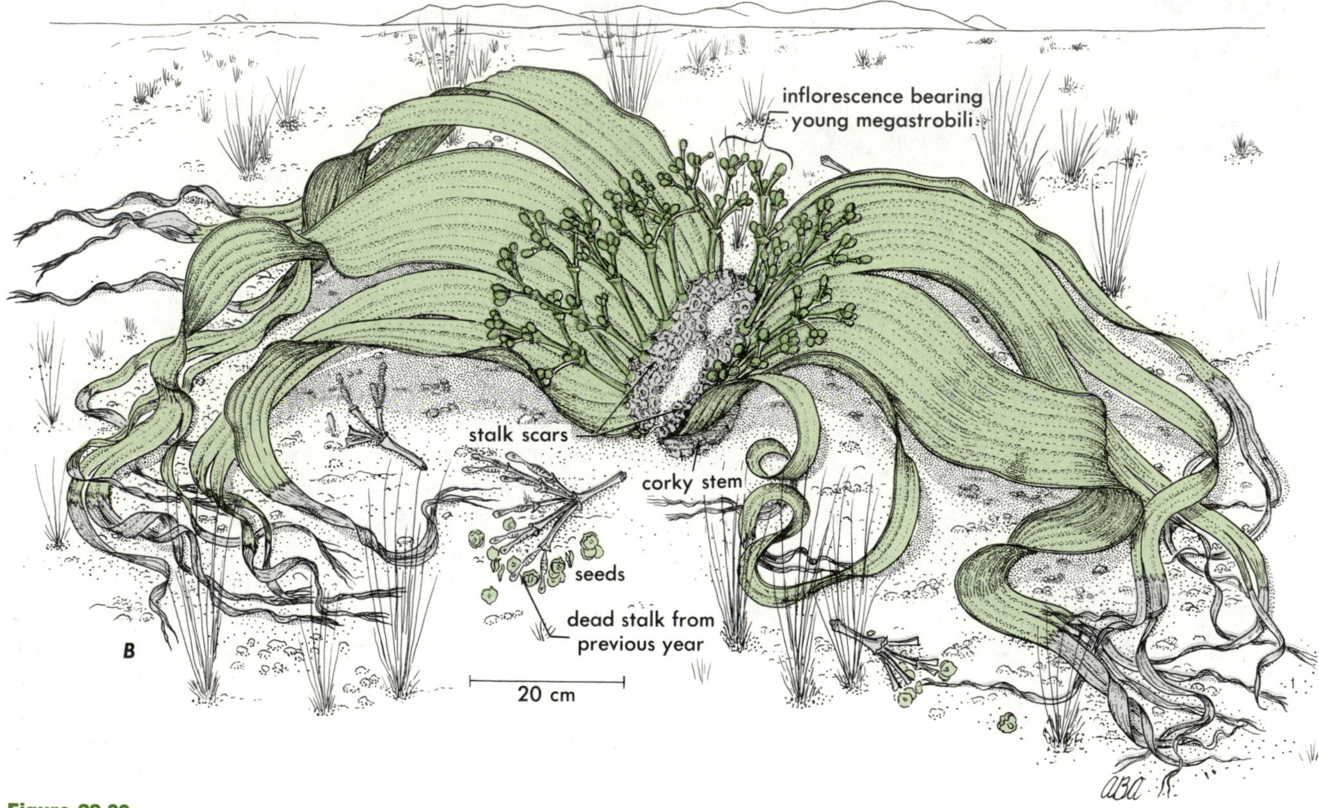

Figure 29.20
Habit sketches of *Welwitschia mirabilis* growing in Namib Desert. *A*, young (10- to 20-year-old) plant. *B*, maturing (100- to 200-year-old) female plant.

are expanded into four wings (perianth, bracteoles) (Fig. 29.21*B*). Pollen is carried by wind from the anthers of the male cone to the ovules of the female cone.

The plants are slow growing, most photosynthate going into an exceptionally well-developed tap root. The above-ground portion consists of two leaves that become frayed at the ends. The leaves are leathery, with many sclereids, considerable lignin, and numerous crystals of calcium oxalate. Stomata are sunken, which is probably an adaptation to the arid habitat in which these plants are found. Annual rainfall is only 20 mm, although fog may add another 50 mm of condensed moisture along the coast. Another adaptation to drought is the crassulacean acid metabolism (CAM) pathway of photosynthesis (see Chapter 13). CAM plants are known for a very slow growth rate and for a very high water-use efficiency (the ratio, grams of water vapor lost for every gram of carbon dioxide fixed, is very low, as much as one-tenth the ratio of mesophytic C_3 plants). Plant ages of 1000 to 2000 years are not uncommon.

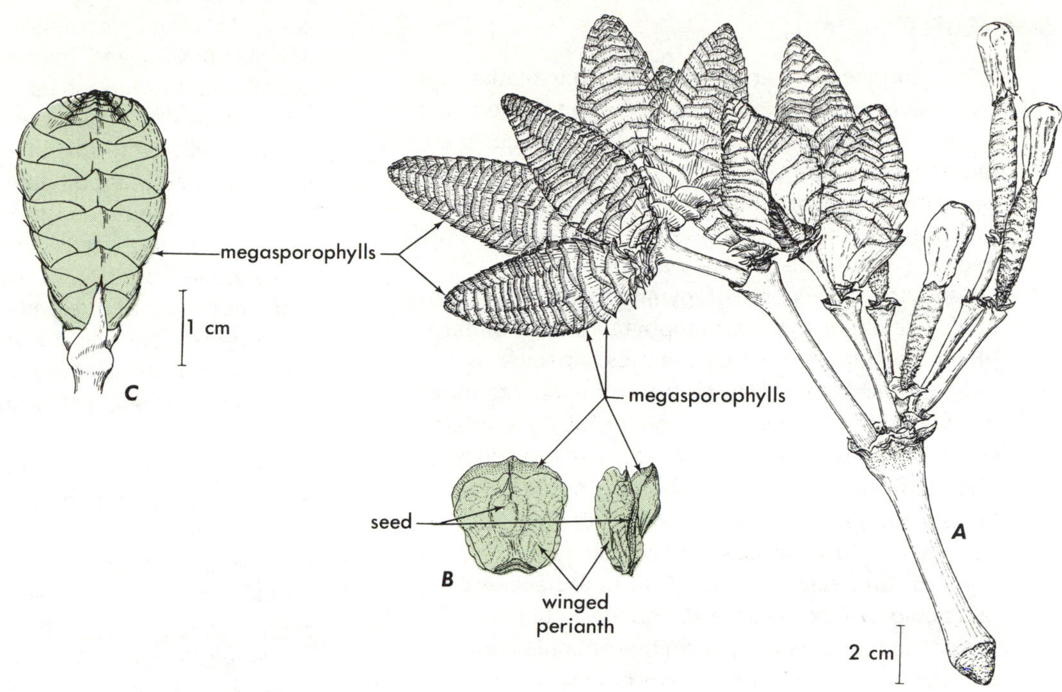

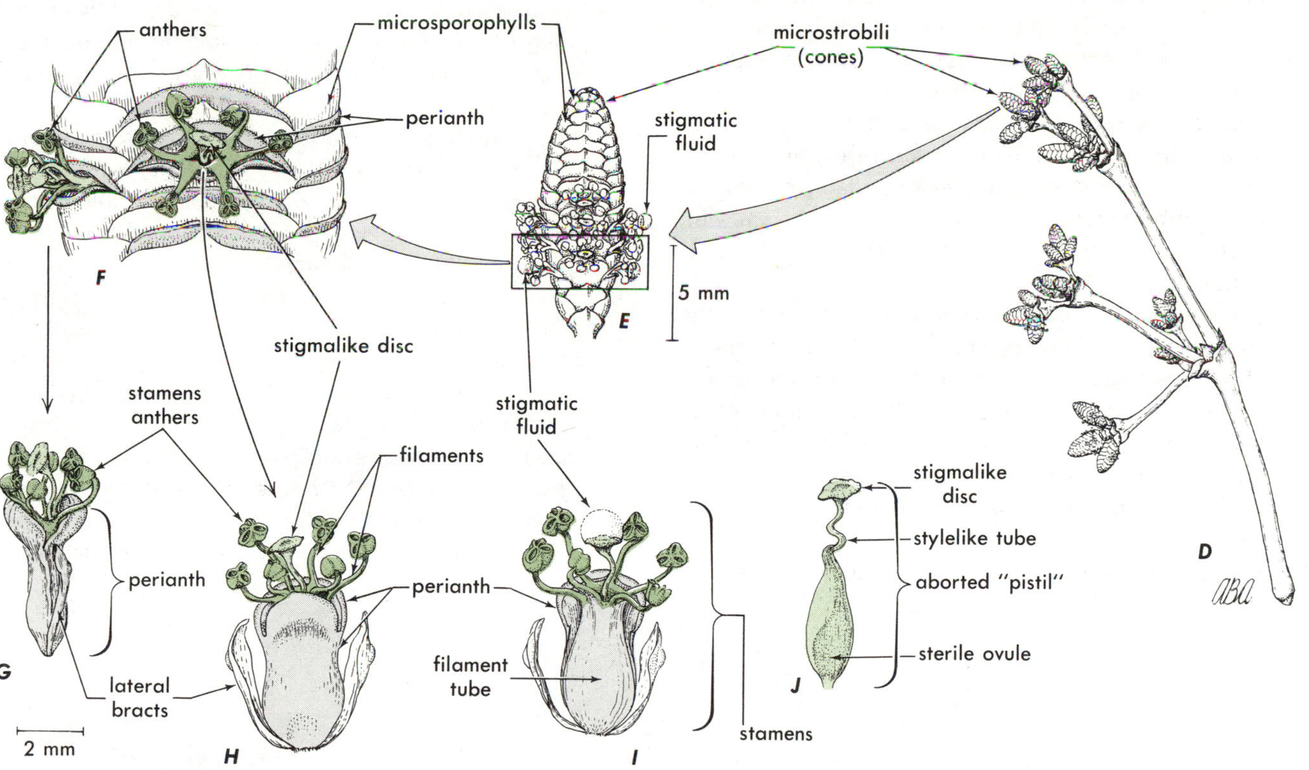

Figure 29.21
Details of female *(A-C)* and male *(D-J)* cones of *Welwitschia*. *A*, megasporangiate inflorescence bearing mature female cones. *B*, each bract (megasporophyll) encloses one ovule that develops into a seed. The integuments develop into winged outgrowths. *C*, immature female cone. *D*, microsporangiate inflorescence bearing male cones. *E*, portion of a male cone at anthesis, showing exerted stamens. *F*, one male "flower" with six stamens and a nonfunctional stigmalike disc. *G-I*, different views of a male "flower" showing several subtending bracts. *J*, detail of the nonfunctional pistil-like portion, which is an outgrowth of the integument of a sterile ovule. (Labeling with assistance of Dr. C. H. Bornman, University of Lund, Sweden.)

SUMMARY

1. The gymnosperms are higher vascular plants and possess both vascular tissue and seeds. Seeds aid plant dispersal, protection of the embryo, and establishment of the young sporophyte. A second advancement is pollination, which permits sperms to reach eggs while surrounded by protecting tissue and in the absence of free water.
2. Of the four divisions of gymnosperms, Cycadophyta, Ginkgophyta, Gnetophyta, and Coniferophyta, the latter contains the most species, is the most widespread, and contributes the most to modern vegetation types. The Coniferophyta contains seven families, several of which were discussed. The Pinaceae contains pines, firs, spruces, hemlocks, Douglas fir, larch and true cedars (*Cedrus* sp.). The Cupressaceae contains junipers, cypresses, and false cedars. The Taxodiaceae contains bald cypress, redwood, and Sierra bigtree. The Taxaceae contains the yews. Other families are better represented in the southern hemisphere.
3. *Pinus* (pine) life cycle was examined as representative of the Coniferophyta. The sporophyte is large and anatomically complex. Vessels are absent, but considerable secondary xylem is produced. Leaves are modified for arid conditions. Pine is heterosporous, the microspores being produced in small (staminate) cones on lower branches. Megaspores are produced in other (ovulate) cones on upper branches; these cones are at first small but over a two-year period mature into large, woody cones.
4. The microspores begin dividing while still within the spore wall. After two divisions, and the elaboration of air-filled wings on the outside of the thickened spore wall, growth ceases and the immature male gametophytes are liberated to the wind. These are pollen grains. Some, by chance, sift into young ovulate cones and are sealed in when the cone scales close.
5. Two ovules are present on each scale in the ovulate cone. An ovule consists of an outer integument, nucellar tissue, and the female gametophyte. One nucellar cell functions as a megasporocyte, and, of the four resulting megaspores, only one survives. It begins to divide to form the female gametophyte at the same time that pollination (the transfer of pollen from male to female cone) occurs. At maturity, the female gametophyte is a relatively undifferentiated, multicellular, nonphotosynthetic thallus embedded in the nucellus. At one end are several archegonia.
6. If the pollen grain is drawn through the micropyle (an opening in the integument), it comes to lodge against the nucellus. It germinates and a pollen tube begins to parasitically grow toward the female gametophyte through the nucellus. Finally, a third cell division produces two sperm and the formation of the male gametophyte is complete. No antheridium is recognizable.
7. Fertilization is accomplished when the pollen tube ruptures near an archegonium and one sperm unites with an egg. The resulting zygote divides and differentiates into an embryo. A mature embryo lies surrounded by female gametophyte tissue, some remnants of the nucellus, and a seed coat (from the integument). The embryo is dormant until the seed is shed and comes to lie in an appropriate microenvironment.
8. In general, life cycles from the thallophytes to conifers have shown a definite trend or pattern: the dominance and complexity of the sporophyte have increased and the dominance and complexity of the gametophyte have declined.
9. The Cycadophyta contain about 100 species of tropical, slow-growing, palmlike plants. The Ginkgophyta consists of a single species that may now be extinct in the wild but is widely cultivated. The Gnetophyta exhibit some traits that link them more closely to the flowering plants than to other gymnosperm groups. Only one of its three genera, *Ephedra,* is found in North America. This latter division is quite artificial in terms of classification.

CLASSIFICATION

Division	Cycadophyta
Genera	*Cycas*
	Zamia
	Dioön
Division	Ginkgophyta
Genus and species	*Ginkgo biloba*
Division	Coniferophyta
Family	Pinaceae
Genus	*Pinus*
Genus	*Abies*
Genus	*Picea*
Genus	*Tsuga*
Genus	*Pseudotsuga*
Genus	*Larix*
Family	Cupressaceae
Genus	*Cupressus*
Genus	*Juniperus*
Family	Taxodiaceae
Genus	*Taxodium*
Genus	*Sequoia*
Genus	*Sequoiadendron*
Genus	*Metasequoia*
Family	Taxaceae
Genus	*Taxus*
Division	Gnetophyta
Genus	*Ephedra*
Genus	*Welwitschia*
Genus	*Gnetum*

CHAPTER 30

ANGIOSPERMS

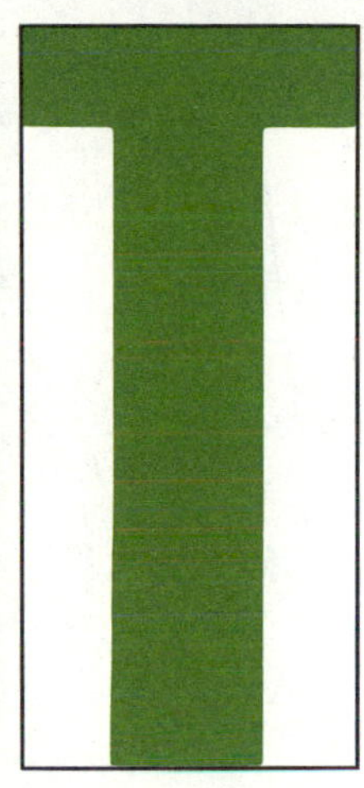

he flowering plants, **Angiosperms,** are the dominant plants of the world today. They include nearly all the crop plants in orchards, gardens, and fields. Hardwood forests, shrublands, grasslands, and deserts are composed chiefly of flowering plants. They show great variation in form, from simple stemless, freefloating duckweed *(Lemna)* through a whole series of herbaceous types to shrubs and trees such as oaks (*Quercus*) and beeches (*Fagus*). We have already studied the structure and physiology of the angiosperm plant body in considerable detail. Angiosperms constitute the **Anthophyta,** which is divided into two subdivisions, **Monocotyledonae** and **Dicotyledonae.**

The outstanding and unique structure of the angiosperms is the *flower* (Fig. 30.1). The flower is a compressed shoot bearing floral leaves. In a complete flower the floral leaves are sepals, petals, stamens, and carpels. Some flowers have only the essential reproductive structures, stamens and carpels. Other flowers are unisexual and have either stamens or carpels.

All flowering plants produce seeds except some commercial seedless varieties (e.g., grapes and bananas). In angiosperms, seeds are borne within a closed structure, the ovary, which eventually becomes the fruit.

REVIEW OF LIFE CYCLE

The life cycle of the angiosperms was thoroughly discussed in Chapter 15. At this point let us simply review its most important aspects.

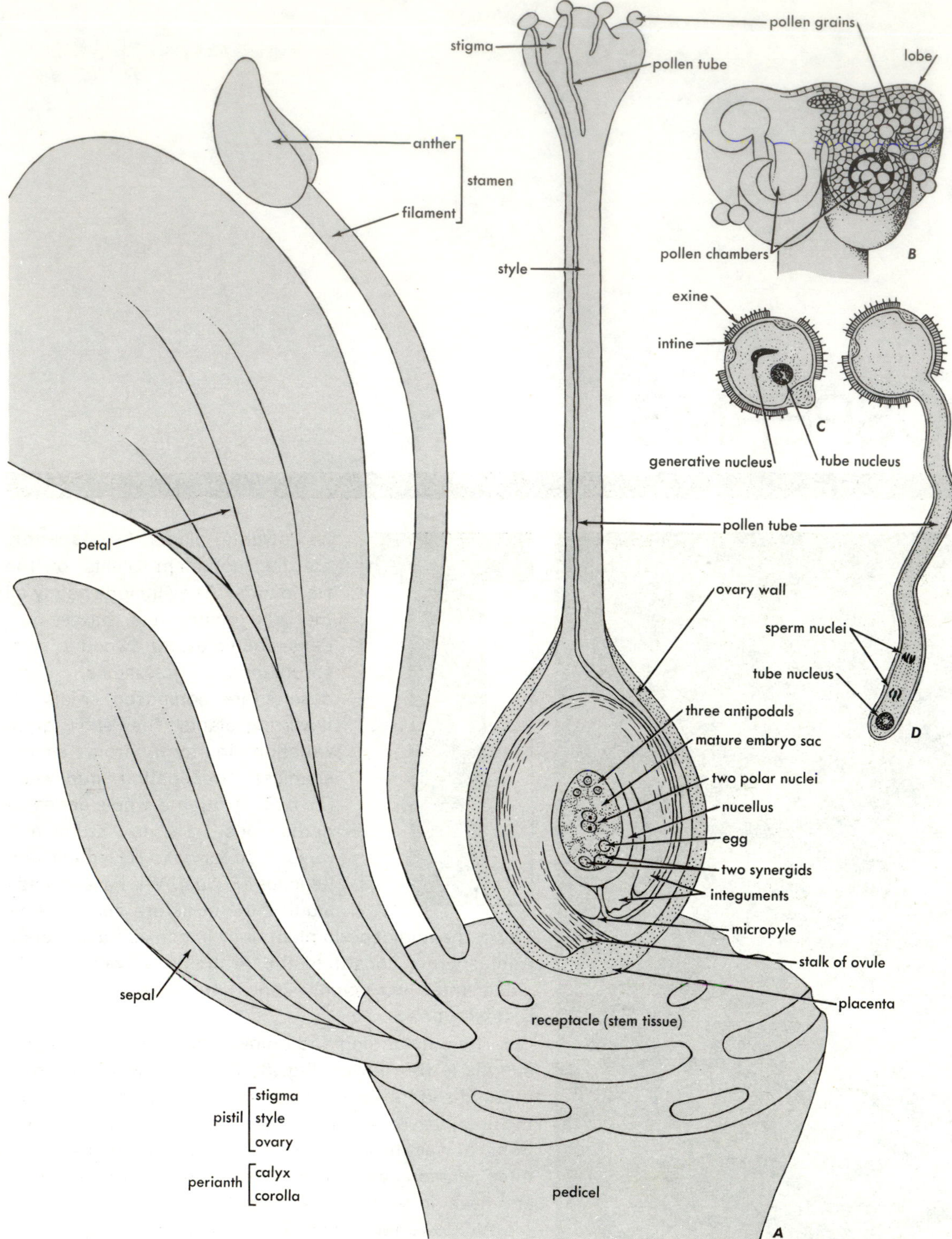

Figure 30.1
A, diagram of a flower. *B*, cross section of an anther. *C*, mature pollen grain. *D*, germinating pollen grain. (*C* and *D*, redrawn from Bonnier and Sablon, *Cours de Botanique*, Librairie Generale de l'Enseignement.)

Male Gametophyte

The anther is the part of the stamen responsible for production of pollen. Microsporocytes form within the pollen sac (Fig. 30.1). They divide by meiosis to form haploid ($1n$) microspores, which in turn divide mitotically to form a generative and a tube nucleus (Fig. 30.1*C*). A heavy, sculptured wall forms around the microspore and a mature pollen grain results. The pollen is now shed and conveyed in one fashion or another to a stigma where it germinates. A pollen tube penetrates the stigma (Figs. 30.1*A,D*) and grows through the style, down to the ovule within the ovary. The generative nucleus usually divides within the pollen tube into two sperm nuclei (Fig. 30.1*D*). Each sperm nucleus, with its associated cytoplasm, is a sperm cell.

Free water is not needed for fertilization in angiosperms; motile, ciliated sperms are not produced by any members of this division. The pollen tube with sperm cells and tube nucleus, if present, constitute the mature male gametophyte.

Female Gametophyte

An enlarged cell, the megasporocyte, within nucellar tissue of a young ovule undergoes meiosis and forms four megaspores arranged in a row. While this process is occurring, integuments form about the nucellar tissue, resulting in the formation of an ovule. The three megaspores closest to the micropyle generally disintegrate. The remaining megaspore undergoes three successive mitotic divisions, resulting in seven cells of the embryo sac or female gametophyte: one egg cell, two synergid cells, one endosperm mother cell with two polar nuclei, and three antipodal cells (Fig. 30.1*A*).

Fertilization and Seed Development

With penetration of the pollen tube into the mature embryo sac, the stage is set for fertilization. In angiosperms, fertilization involves not only the union of the egg cell with the sperm cell but, in addition, the union of a second sperm cell with the endosperm mother cell to form the primary endosperm cell. Since there are two nuclei in the endosperm mother cell, the nucleus of the primary endosperm cell will have three sets of chromosomes. The fusions of two sperm cells, one with the egg, the second with the endosperm mother cell, is **double fertilization.**

The resulting zygote generally develops directly into a small proembryo, from one end of which a typical embryo develops. The fate of the primary endosperm cell depends on the species. This cell gives rise to an endosperm that plays some role in nourishing the developing embryo. In a few genera, notably the grasses, the endosperm enlarges and persists in the seed to form the source of nourishment for the young seedling. A seed always includes an embryo and a food supply to enable the young seedling to establish itself, except in rare cases like orchids where the embryo remains quite rudimentary. Food is generally stored either in cotyledons of the embryo or in endosperm.

With the germination of a seed and the development of the seedling into a flowering plant, the life cycle of an angiosperm is completed. The essential steps are shown in Table 30.1

Table 30.1
GENERALIZED LIFE HISTORY OF AN ANGIOSPERM

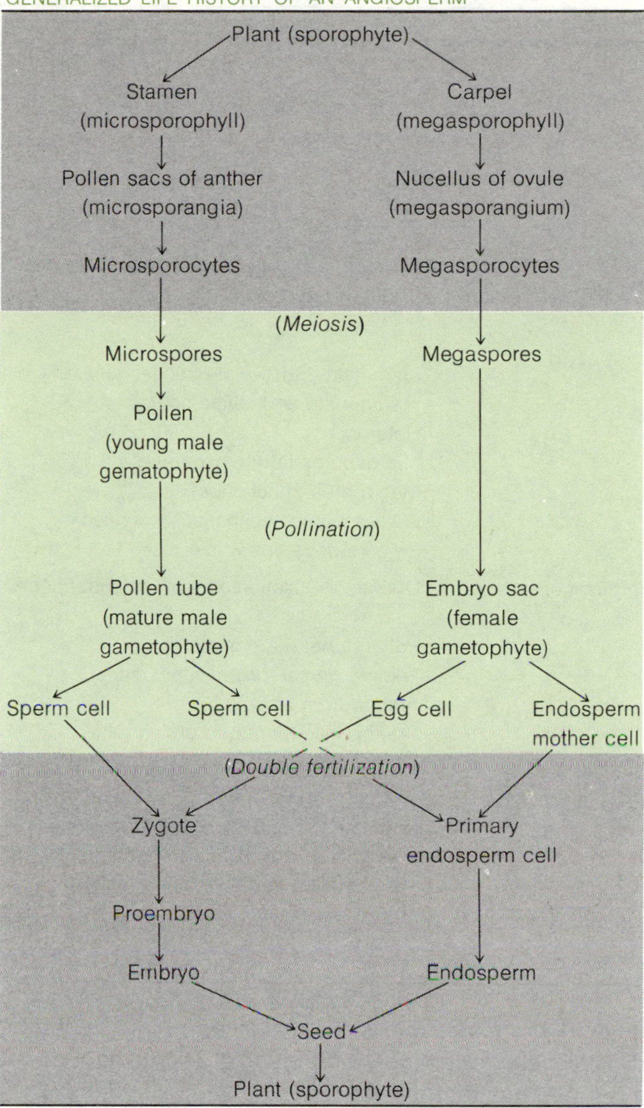

COMPARISON OF ANGIOSPERM LIFE CYCLE WITH MORE PRIMITIVE PLANTS

Certain comparative details between structures and life cycles of members of the gymnosperms and angiosperms are given in Table 30.2. Details concerning the Psilophyta, Lycophyta, Sphenophyta, and Pterophyta are given in Table 28.1. A general comparison of life cycles in thallophytes and bryophytes, appear in Table 27.1.

In all Pteropnyta the sporophyte is the dominant generation, and both gametophyte and sporophyte generations are independent. In gymnosperms and angiosperms, there occurs a further reduction in size and complexity of the gametophyte to a condition of complete dependency on the parent sporophyte. This is accompanied by an overall increase in the general complexity of the sporophyte. It can be stated that, as a

Table 30.2
COMPARISON OF THE CONIFERS AND FLOWERING PLANTS

Generation	Gymnosperms	Angiosperms
	Roots, stems, and leaves present	Roots, stems, and leaves present
	Trees and shrubs, no herbs	Trees, shrubs, and herbs
	Stem: one core of xylem surrounded by phloem, vessels in one small division only	Stem: one core of xylem surrounded by phloem or several vascular strands with phloem exterior, vessels present
	Sieve cells without companion cells	Companion cells and sieve-tube members
	Pith present in stems, not in roots	Pith in stems, not in roots, of dicots; in some monocot roots
Sporophyte	Cambium develops in all species	Cambium in perennial forms; absent generally in annuals
	Staminate and ovulate cones, generally, with staminate and ovlate scales	Flowers, stamens, and carpels
	Heterospory	Heterospory
	Microspores in microsporangia	Microspores in anther sacs (microsporangia)
	Megapores in nucellus	Megaspores in nucellus
	Ovules present, exposed on scales	Ovules present, enclosed within carpel
	Embryo develops within a seed	Embryo develops within a seed
Fertilization	Double fertilization absent; no endosperm	Double fertilization, resulting in zygote and primary endosperm cell
	Pollen tube: male gametophyte	Pollen tube: male gametophyte
	Female gametophyte within ovule	Female gametophyte (embryo sac) within ovule
	Antheridium absent	Antheridium absent
	Sperm cells formed in pollen tube	Sperm cells formed in pollen tube
Gametophyte	Motile sperms in several primitive genera, but free water not needed for fertilization	
	Nonmotile sperms in all other forms	Nonmotile sperms
	A very greatly reduced archegonium completely embedded in the female gametophyte	Archegonium only suggested by synergid cells of the embryo sac
	Embryo within female gametophyte and seed coat	Embryo within seed coats, remains of nucellus, and endosperm

guiding principle, the more advanced forms in these divisions are more highly adapted to grow and flourish in a dry land habitat than are the more primitive forms. The most obvious developments designed to accomplish this are: (*a*) adaptations that remove any dependence of the plants on free moisture for fertilization, and (*b*) adaptations, both vegetative and reproductive, that enable the plant to grow and reproduce on dry land. Examples are given below.

Protection of the Female Gametophyte

The gynoecium of the angiosperm flower is essentially a modified leaf, or leaves, enclosing an ovule, or ovules, in which the female gametophyte is to be found. The ovary, as we know, eventually develops into a fruit. This enclosure of ovules and, subsequently, seeds by an ovary wall is a new development that sets angiosperms apart from all other groups of plants. The female gametophyte (embryo sac) has gained, in the ovary wall, an added protective barrier. However, this has necessitated further adaptations to enable sperm to pass through this added protection: (*a*) development of a receptive stigma and (*b*) growth of a pollen tube through the style.

Size of Gametophytes

Reduction in size of gametophytes has proceeded to a point in which the male gametophyte, the pollen tube, consists of one vegetative nucleus and two sperm cells. The female gametophyte is a seven-celled structure, one of whose cells is an egg cell. A second cell, the endosperm mother cell, is an evolutionary innovation in that it, too, is receptive to a sperm cell. The remaining five cells are vegetative cells. It should be pointed out that embryo sacs of some angiosperms have more cells, while a few of them have less, but all have an egg cell and an endosperm mother cell.

Nourishment for the Developing Embryo

Double fertilization is still another difference between the life cycle of the angiosperms and the more primitive forms. In the latter, nourishment for the young developing sporophyte is generally provided by the female gametophyte (Figs. 28.8), or, in the case of many ferns, by the simple prothallus, or gametophyte (Fig. 28.24). In angiosperms, food for the developing seedling is stored either in the endosperm or in the cotyledons of the embryo itself.

Fruit

Fertilization stimulates tissues around the zygote to begin further development. In angiosperms, the ovary wall is stimulated to produce a fruit, and the fruit is a development originating with, and characteristic of, angiosperms. The protection of the seed from its surroundings by the ovary wall is accompanied by various devices for bringing it under conditions favorable for germination.

EVOLUTION IN THE ANGIOSPERMS

If angiosperms represent the culmination of an extensive evolutionary development, it is likely that within the division Anthophyta, evolution has been and still is active. In other words, families of angiosperms must be related and it should be possible to arrange the families in a sequence that would give some indication of their relationships. Some families will be found to be more primitive than others; and from the more primitive types the more advanced and specialized forms have presumably arisen (Chapter 15).

The Besseyan System

Many systems of classifying plants have been proposed. A system that has found much favor with American botanists was developed by Charles E. Bessey (1900s). His system regards the order Ranales as being the most primitive. In this order, the Magnoliaceae is one of the most primitive families, with the Ranunculaceae somewhat more advanced. The Christmas rose *(Helleborus)*, magnolias (Fig. 30.3), and buttercups *(Ranunculus)* are representative of these families.

If we consider the arrangement of flower parts in magnolias and buttercups to be most primitive, then it is possible to compare them to other families to determine their apparent relationships. The principal tendencies in evolution of the flower, according to the Besseyan system, are as follows: (*Note:* Refer back to Chapter 15 to review flower terminology.)

1. From an elongated to a shortened floral axis.
2. From a spiral to a whorled condition of floral parts.
3. From numerous and separate stamens and carpels to few and connate stamens and carpels.
4. From numerous and separate sepals and petals to few and connate sepals and petals.
5. From complete and perfect flowers to incomplete and imperfect flowers.
6. From regular to irregular flowers.
7. From hypogyny to epigyny.

According to the Besseyan view, there were at least three main lines of advance from the primitive Ranalian type of flower. These lines culminated in (*a*) mints (Lamiaceae), *(b)* asters (Asteraceae), and *(c)* orchids (Orchidaceae) (Fig. 30.2).

SELECTED FAMILIES OF ANGIOSPERMS

Let us now examine selected families of angiosperms with two objectives in mind: (*a*) to learn the characteristics of some important angiosperm families and (*b*) to discover how these families fit into the Besseyan system of angiosperm classification. In our discussion of these matters so far, floral characteristics have been emphasized. This is because their form and structure are little influenced by the external environment. They also are structurally more constant from generation to generation than are vegetative parts. Because of this constancy in structure, floral parts have a much greater value in both plant identification and in tracing evolutionary relationships than do vegetative characters. However, the latter cannot be neglected.

Since the magnolia and buttercups are thought to be primitive forms, we shall begin our discussion with these families, then proceed to follow through a line of ascent ending in the mint family. We shall consider the line culminating in the Asteraceae (Compositae)* and complete our discussion with the line of monocotyledonous families. It should be further noted that lines of development are not straight but branched and treelike. The arrangement of the families discussed here as they occur in the Besseyan system is shown in Fig. 30.2.

Magnoliaceae to Lamiaceae

In this line of ascent (Fig. 30.2), flowers in all families are hypogynous. Syncarpy (the fusion of carpels) occurred very early, as did a reduction in number of floral parts. Other types of connation and the change to irregular flowers appeared somewhat later but still early in the line of ascent. Based on their floral characteristics, primarily, it is possible to show an evolutionary connection between the following families: Magnoliaceae, Ranunculaceae, Papaveraceae, Malvaceae, Brassicaceae, Solanaceae, Salicaceae, and Lamiaceae (Fig. 30.2).

The Malvaceae is a clearly marked family with relatively primitive characters as shown by its regular, perfect, and hypogynous flowers. There are numerous stamens with connate anthers, and the five or more carpels are connate. In Solanaceae, the flowers, while perfect and regular, show connation of sepals, petals, and carpels, with stamens adnate to the corolla tube. The Salicaceae is a family in which reduction in both number and size of floral parts has resulted in its advanced condition. The flowers are imperfect and grouped in catkins, and both calyx and corolla are lacking. The pistillate flowers are hypogynous and, in staminate flowers, the number of stamens had been reduced to one or two. The Lamiaceae, or mints, represent the most advanced family in

*According to rules of plant nomenclature family names should end in -ceae. Traditionally, the names of certain families such as Cruciferae and the Compositae have not ended this way. In this chapter the currently more acceptable synonym names ending in -ceae will be used.

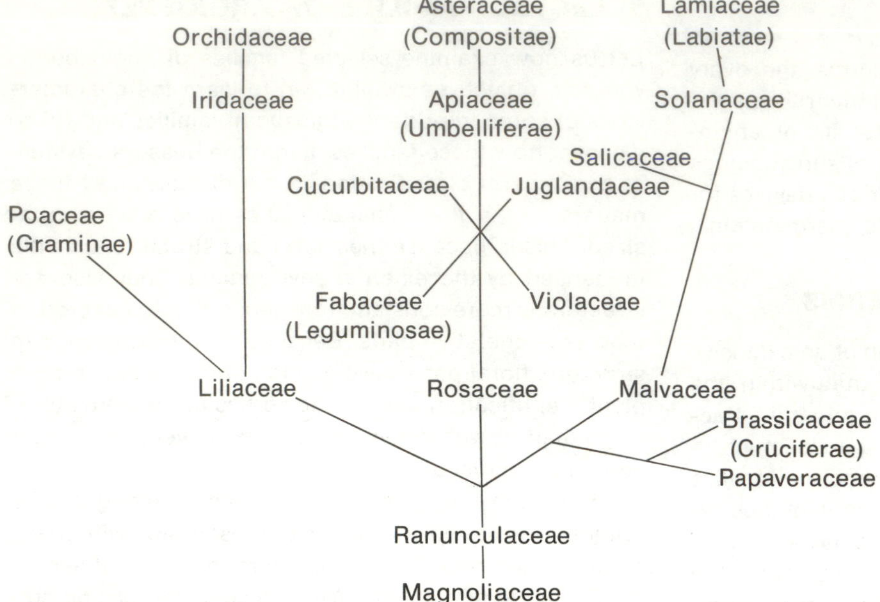

Figure 30.2
The evolutionary tendencies in the Angiospermae according to the Bessayan system. (Courtesy of McMinn.)

this line of ascent. The flowers are hypogynous, with an irregular symmetry; there is a reduction in number of parts to four or two; and sepals, petals, and carpels are connate. The stamens are not only adnate to the corolla tube but are also highly specialized for insect pollination. Let us consider some of the general characteristics of these families.

Magnoliaceae

This group shares with the Ranunculaceae the distinction of being very ancient. The tulip tree, *Liriodendron,* as shown by fossil remains, once had a very wide distribution. Most species of the Magnoliaceae are trees or fairly large shrubs. Magnolias themselves are magnificent trees, *Magnolia grandiflora* grows to 30 m or more with stiff, large, evergreen leaves and large white blossoms 18 to 20 cm across. The magnolias are largely warm-climate species, but tulip trees will tolerate northern winters (Fig. 30.3). There are some 12 genera and 230 species* in the family.

Ranunculaceae

Most of these are herbs but there are a few small shrubs and woody climbers. Leaves are frequently compound, a feature that has led to the common name of crowfoot family. There are about 50 genera and approximately 1900 species growing largely in the North temperate and arctic regions. Many of our common garden flowers belong to this family. Among these are *Delphinium, Aquilegia, Paeonia, Ranunuculus* (Fig. 30.4), *Anemone,* and *Clematis.* The marsh marigold (*Caltha*) and *Hepatica,* common spring flowers, are members of this family. One species, *Aconitum napellus,* yields the drug aconite, a cardiac and respiratory sedative, and some species of *Delphinium* are poisonous to livestock.

Papaveraceae

This is a small family of only 250 species, but it is of great economic importance because it contains the source of opium. The opium poppy (*Papaver somniferum*) is native to the Middle East, but is is now widely cultivated in India, Southeast Asia, and the Mediterranean region. Incisions are made in the fruit of the opium poppy when it is still immature, and the milky latex (opium) is collected. It contains a mixture of many alkaloids, including morphine, codeine, and papaverine. Perhaps only 5% of the opium crop is sold legally. Most species of this family are annual or perennial herbs. The flowers are typically solitary, with 4 to 12 free, often crumpled petals, numerous stamens, and a single large pistil that lacks a style (Fig. 30.5). Many species have ornamental value.

Brassicaceae (Cruciferae)

This is a large, distinct family of many cultivated forms, as well as some that are noxious weeds. Cabbage, cauliflower, broccoli, brussels sprouts, kohlrabi, and kale are

*The estimates of genus-species numbers were taken from Willis, J.C. *A Dictionary of the Flowering Plants and Ferns,* 8th edition, revised by H. H. Airy Shaw, 1973, Cambridge University Press. Note that species numbers from other authors may differ depending on the system used to calculate their estimates.

Figure 30.3
A, Magnolia grandiflora flower and *B,* fruit with hanging seeds.

all horticultural varieties of a single species, *Brassica oleracea.* Wild mustard, another *Brassica* species, (Fig. 30.6) is a sometimes noxious weed in grain fields, though sometimes used as cover crop in orchards. Radishes, turnips, and stocks are other members of this family, as are many garden herbs. The family has a worldwide distribution in temperate and subarctic zones; all are herbs. There are about 375 genera and 3200 species.

Malvaceae

This is a large family of 95 genera and 1000 species. Cotton, taken from seed coats of various members of the genus *Gossypium,* makes this family important from the viewpoint of politics, agriculture, and industry (Fig. 30.7). The cotton of commerce occurs as long hairs on seeds, which are borne in large capsules. Herbs, shrubs and trees, and such ornamentals as *Hibiscus,* okra, hollyhocks, and numerous weeds are members of this family.

Solanaceae

A large family with many tropical forms, it is also well-represented in temperate regions. There are about 90 genera and over 2000 species, some 1700 of the latter being in a single genus, *Solanum* (Fig. 30.8). To this family belong tomatoes, potatoes, tobacco, eggplant, peppers, *Petunias,* and *Salpiglossis.* Although the family is of worldwide distribution, most of the cultivated forms were brought under domestication in the Western Hemisphere. Many plants in the family produce poisons and drugs, such as belladonna and atropine. Even such a

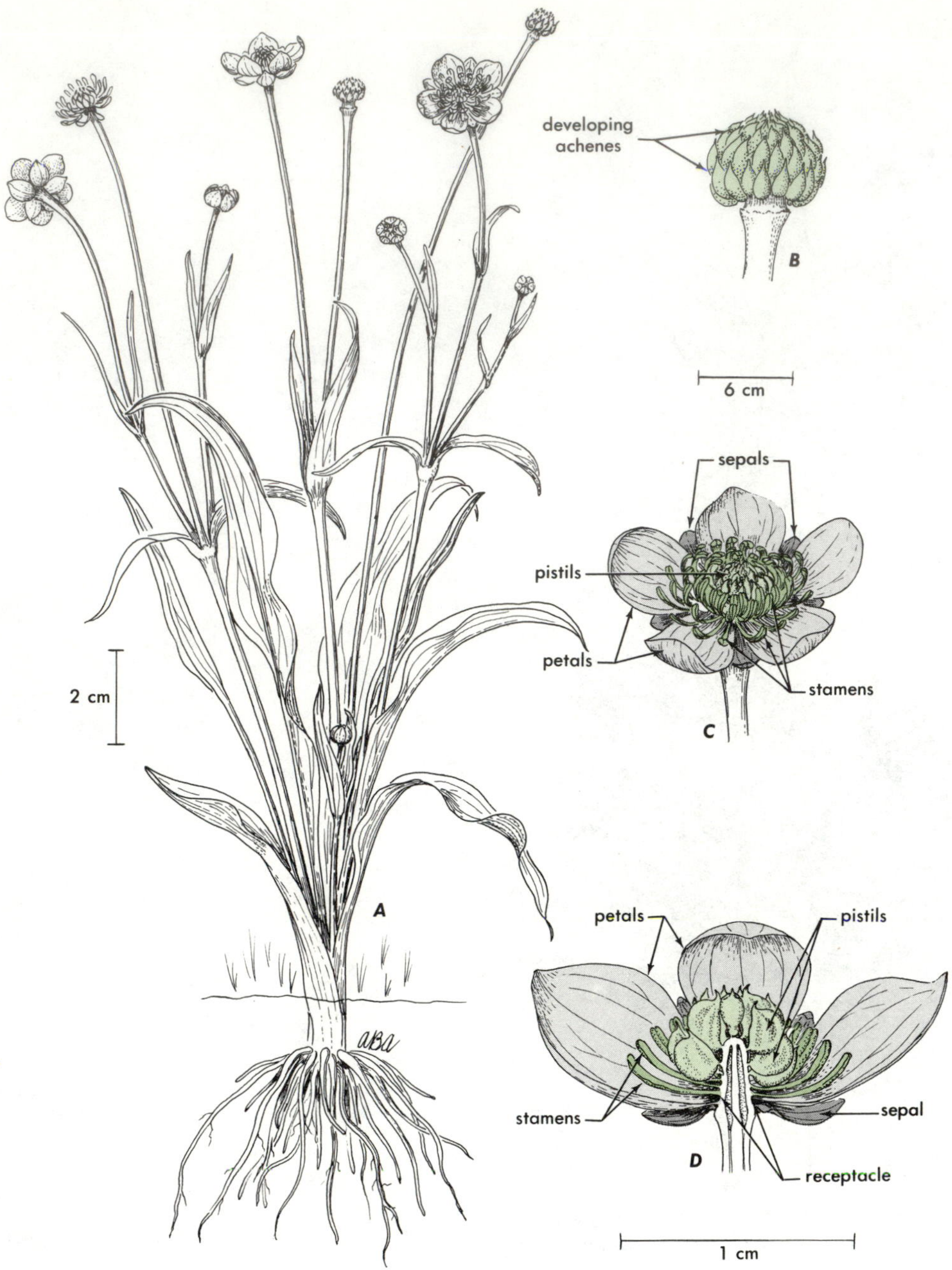

Figure 30.4
A, Ranunuculus alismaefolius entire plant. *B,* fruit head. *C,* flower. *D,* flower longitudinal section.

common plant as the tomato was long supposed to be poisonous, as its scientific name *Lycopersicon,* which means "wolf peach," seems to indicate. There are many erect and climbing herbs in the family, also some shrubs and small trees.

Salicaceae

The willow family is comprised of three genera (*Salix,* willows; *Populus,* poplars; *Chosenia,* an asiatic shrub), with about 530 species. They are mostly trees, but some shrubby forms are known. They are very abundant in the Northern Hemisphere, mostly in temperate zones. Baskets are woven from branches of the basket willow, and paper pulp is obtained from trunks of one species. Willows are common along water courses (Fig. 15.12), frequently overhanging or even choking mountain streams, to the ill comfort of fishermen. Pussy willow is a familiar example of this family.

Lamiaceae (Labiatae)

Members of this family, such as mint and sage represent the supposed highest advance of one of the three lines of angiosperm evolution. It is a large family of considerable economic importance, largely because of volatile oils

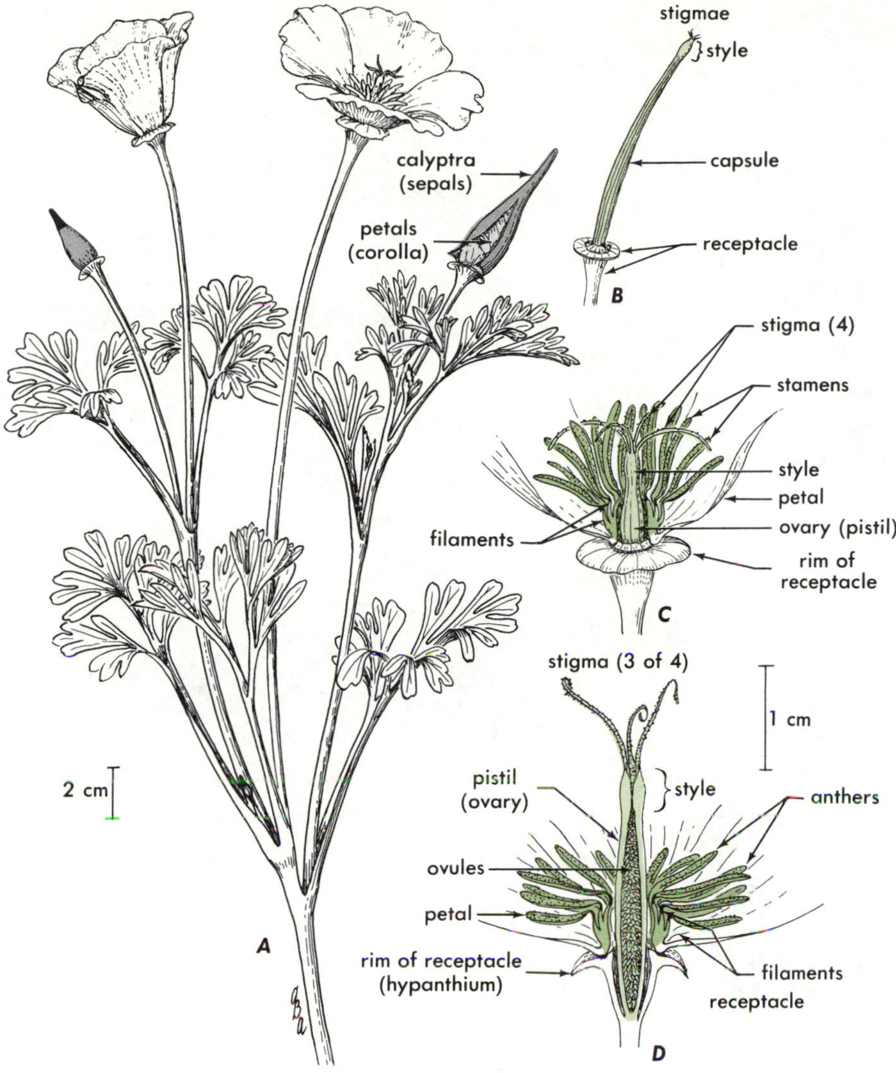

Figure 30.5
A, Eschscholzia californica, habit of plant. *B,* fruit. *C,* flower with exposed pistil. *D,* sectioned flower showing internal view of ovary.

produced by certain of its members. Peppermint, spearmint, thyme, sage (*Salvia,* Fig. 30.9) and lavender are examples. There are about 180 genera with 3500 species well-distributed over the surface of the earth. There are herbs and shrubs in the family, generally with characteristic square stems and opposite leaves.

Magnoliaceae to Asteraceae

The second line of ascent involves the Fabaceae, Rosaceae, Apiaceae, and Asteraceae, with offshoot families, Juglandaceae, Violaceae, and Cucurbitaceae. This large group of families is characterized by an early change from hypogyny to epigyny. Following this there is a division into two sublines, one being marked by a lack of connation and a retention of regular flowers. In the other subline, connate and irregular flowers both occur in advanced forms.

The Rosaceae (rose family) is one of the more primitive families in this line of evolutionary development. There exists a considerable variety of flower structure within the family. The more primitive members are regular, perfect, with numerous parts and with such a slight degree of perigyny as to be recognizable only with careful observation. This is exemplified by the boysenberry (Fig. 30.10). Changes within the family involve a reduction in the number of floral parts, a shift from slight perigyny to true perigyny and to epigyny. Connation also occurs. Members of the family generally have regular flowers.

Fabaceae (pea family) are considered to be a more advanced family than Rosaceae, even though their flowers are hypogynous. There is a reduction in the number of parts, the gynoecium having only one carpel; connation of stamens and of some petals occurs, and the corolla parts are irregular. Apiaceae represent a further advance over Rosaceae in that the flowers are epigynous

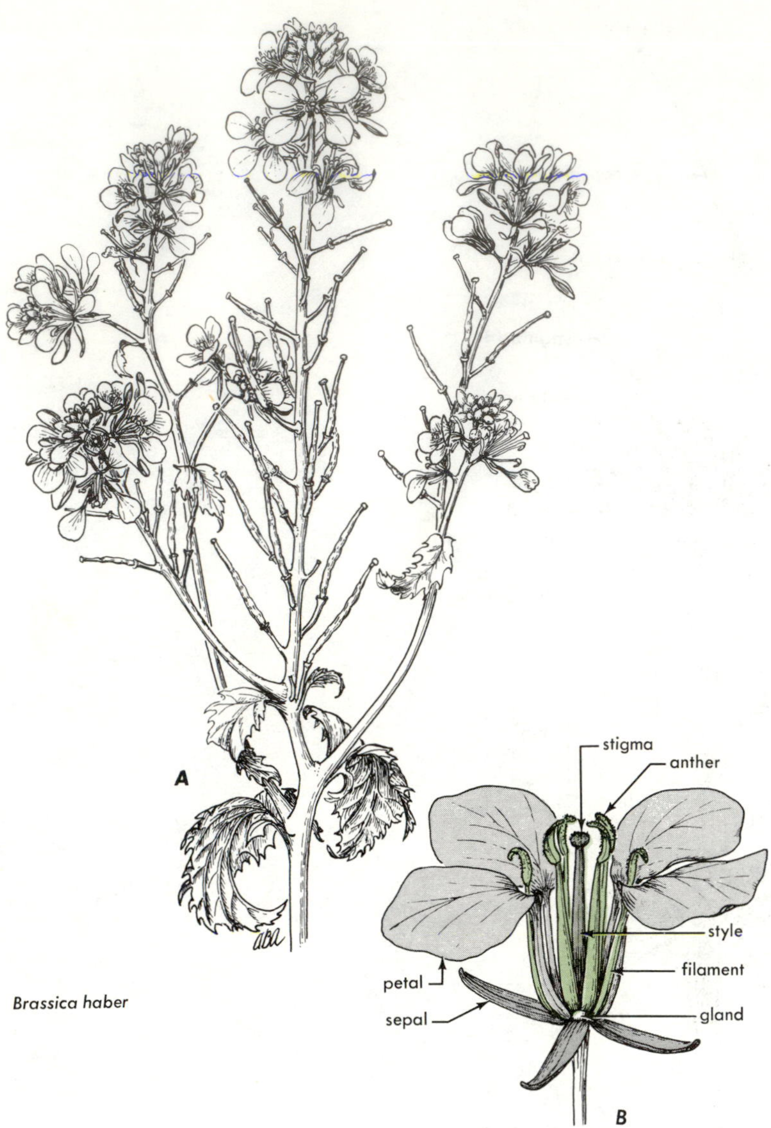

Figure 30.6
Brassica haber. A, branch. *B,* flower.

and exhibit syncarpy. This line of ascent finds its climax in the Asteraceae, whose flowers all possess advanced characters, such as reduction in number of parts, modification of parts (pappus), irregular flowers, and some imperfect flowers, connation in each whorl, and adnation.

The Cucurbitaceae culminate a branch line from the rose family. Flowers of Cucurbitaceae have such advanced characters as loss of floral parts, imperfect flowers, epigyny, and connation.

Rosaceae

Whereas grasses and legumes supply bread and vegetables of basic importance, the rose family supplies fruit for dessert and roses for decoration. There are more than 2000 species in this family and over 100 genera, not counting the almost numberless cultivated forms of roses, peaches, apples, cherries, boysenberries (Fig. 30.10), almonds, and so on. The family is of worldwide distribution and somewhat heterogeneous. Many of the principles of angiosperm evolution can be demonstrated with its members. There are trees, shrubs, and herbs in the family. Leaves are usually alternate and bear stipules.

Cucurbitaceae

The gourd or melon family is of worldwide distribution in at least warmer regions of the world, and various peoples have selected different forms for domestication (Fig. 30.11). It is also probably the only group in which fruits are highly prized for ornamental purposes and for use as containers of various types. Pumpkins and squashes are of Western Hemisphere origin. Cucumbers and melons have probably come from Africa and central Asia. Other species appear to have been first cultivated in the tropics of Asia, Polynesia, and India. They frequently use tendrils to climb and are annual or perennial vines. Most are rapid-growing and frost-tender. There are about 110 genera with 640 species. Stems are usually soft and hairy or prickly. The generally simple leaves are large and sometimes deeply cut. Flowers are usually unisexual.

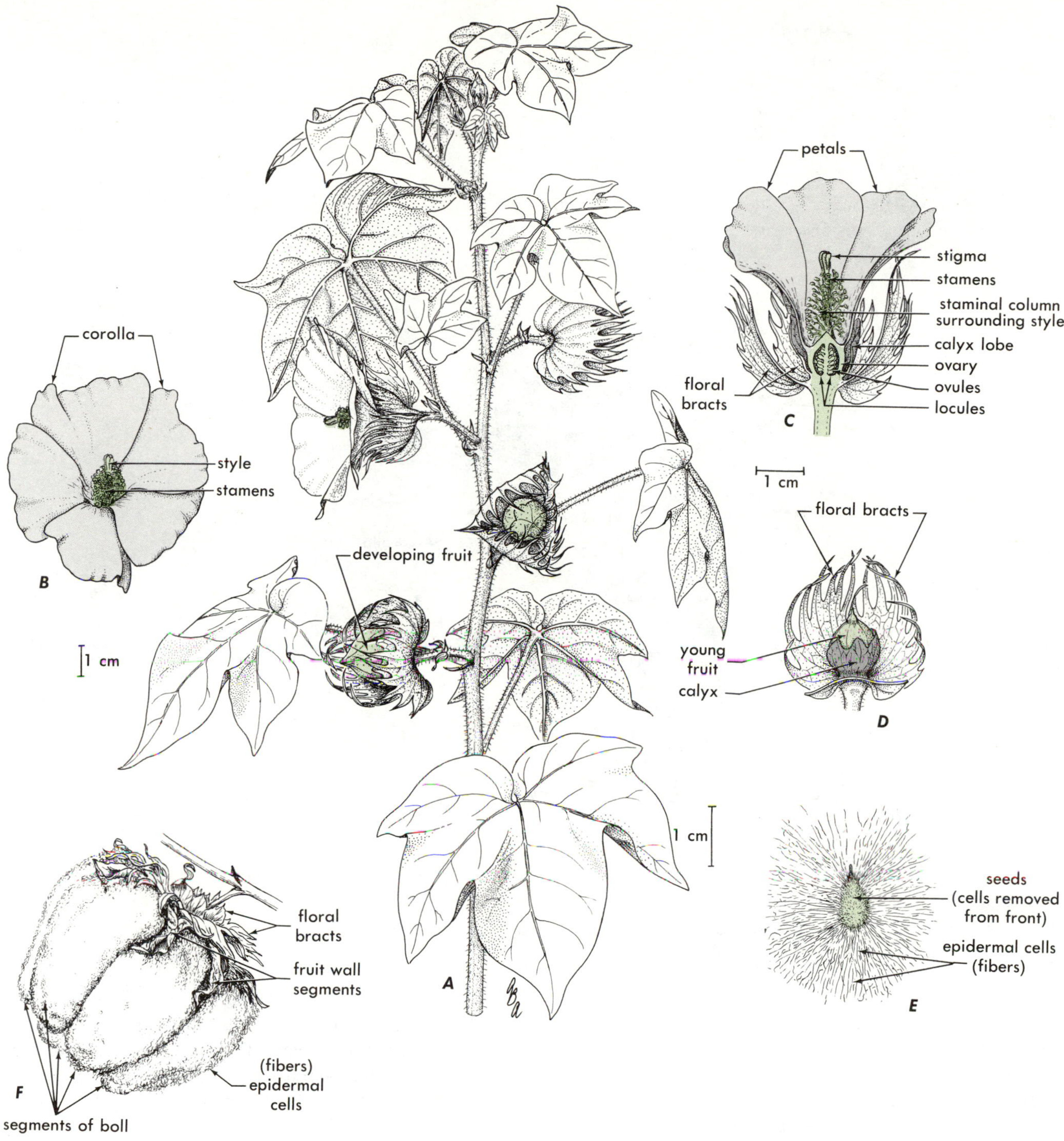

Figure 30.7
Gossypium hirsutum, cotton. *A*, plant habit. *B*, flower. *C*, flower sectioned. *D*, young fruit. *E*, single seed showing seed coat fibers. *F*, cotton bole.

Fabaceae (Leguminosae)

The legume family has such distinctive characteristics that its members can frequently be recognized with only little experience (Fig. 30.12). It is one of the three largest families, with 600 genera and 12,000 species. All major growth forms are represented: herbs, both annuals and perennials, shrubs, vines, and trees. It is of worldwide distribution. While less heterogeneous than the rose family, considerable variation is found among its various members. It is a family of considerable importance in supplying food for humans and their animals. Many legumes are used for ornaments, from shade trees to cut flowers. Some lumber is obtained from the black locust, and some of the tropical species furnish wood for fine

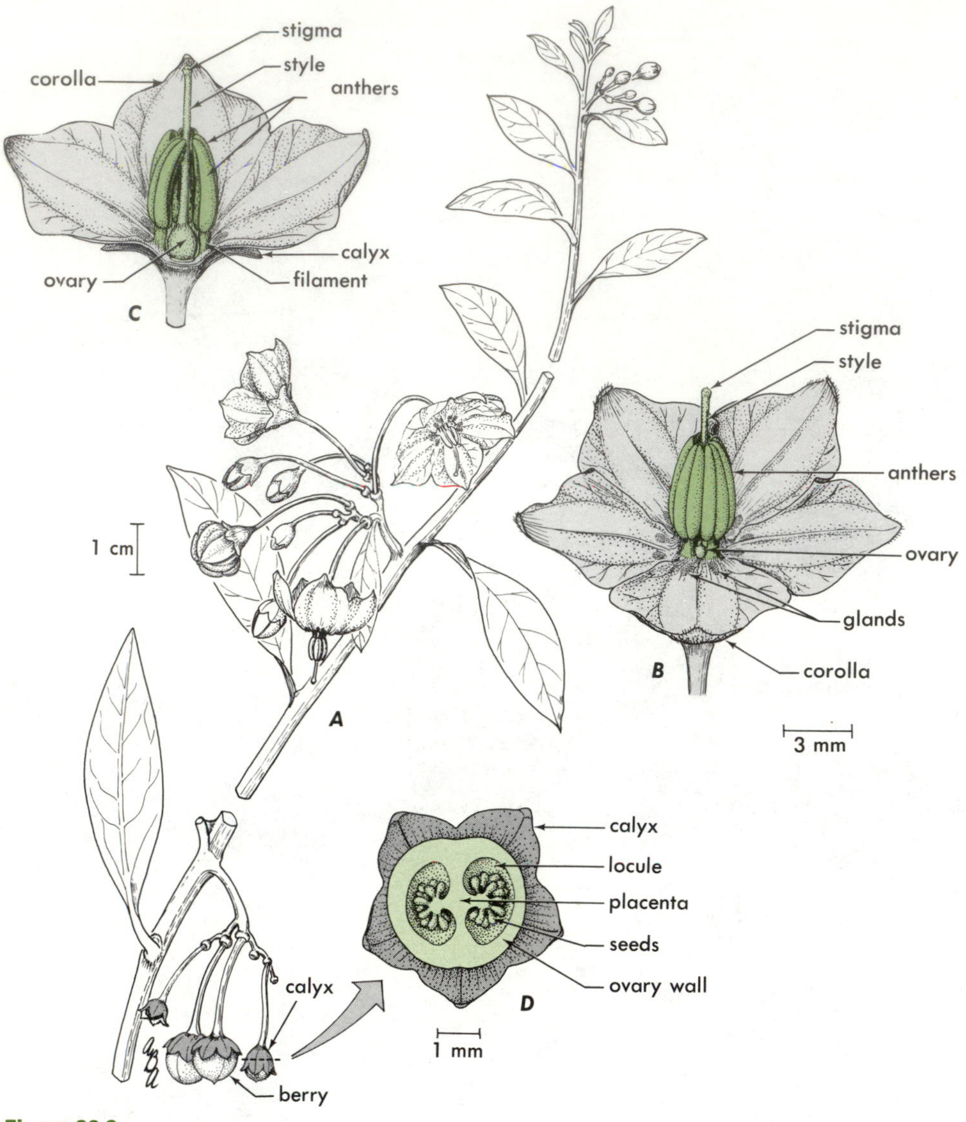

Figure 30.8
Solanum parishii. A, plant habit. *B,* flower. *C,* section of flower. *D,* ovary cross section.

cabinet work. Association of the nitrogen-fixing bacteria with roots of legumes places this family in a unique position relative to maintenance of soil fertility. Peas, beans, peanuts, clovers, and lupines are common herbaceous legumes; wisteria is a vine, brooms and redbuds are shrubs or low trees, and locusts and acacias are trees. The *Mimosa* genus alone has some 450 species, including the sensitive plant of the florist shop; there are others of varying habit from tall trees to low herbs. Leaves of legumes are prevailingly pinnately compound and quite generally with stipules. Sometimes a leaf will be modified to a tendril.

Violaceae

This is a widespread family of nearly 1000 species, found on all continents. The five petals are free but irregular, and the lowermost are recurved to form a spur in which nectar may accumulate (Fig. 30.13). The five stamens tightly surround the pistil. The fruit is a capsule that sometimes opens explosively. Several species of the genus *Viola* have ornamental value.

Juglandaceae

This is a small family of about 60 species of great ecological importance to deciduous forests of eastern Asia and eastern North America. Most members are trees and include walnut (*Juglans,* see Fig. 15.11), hickory (*Carya*), and pecan (*Carya*). Many species also have economic importance as food or lumber. Although the flowers are unisexual and lack a colorful perianth, most taxonomists view them as being advanced, not primitive, having become simplified in the course of evolution. Leaves are pinnately compound.

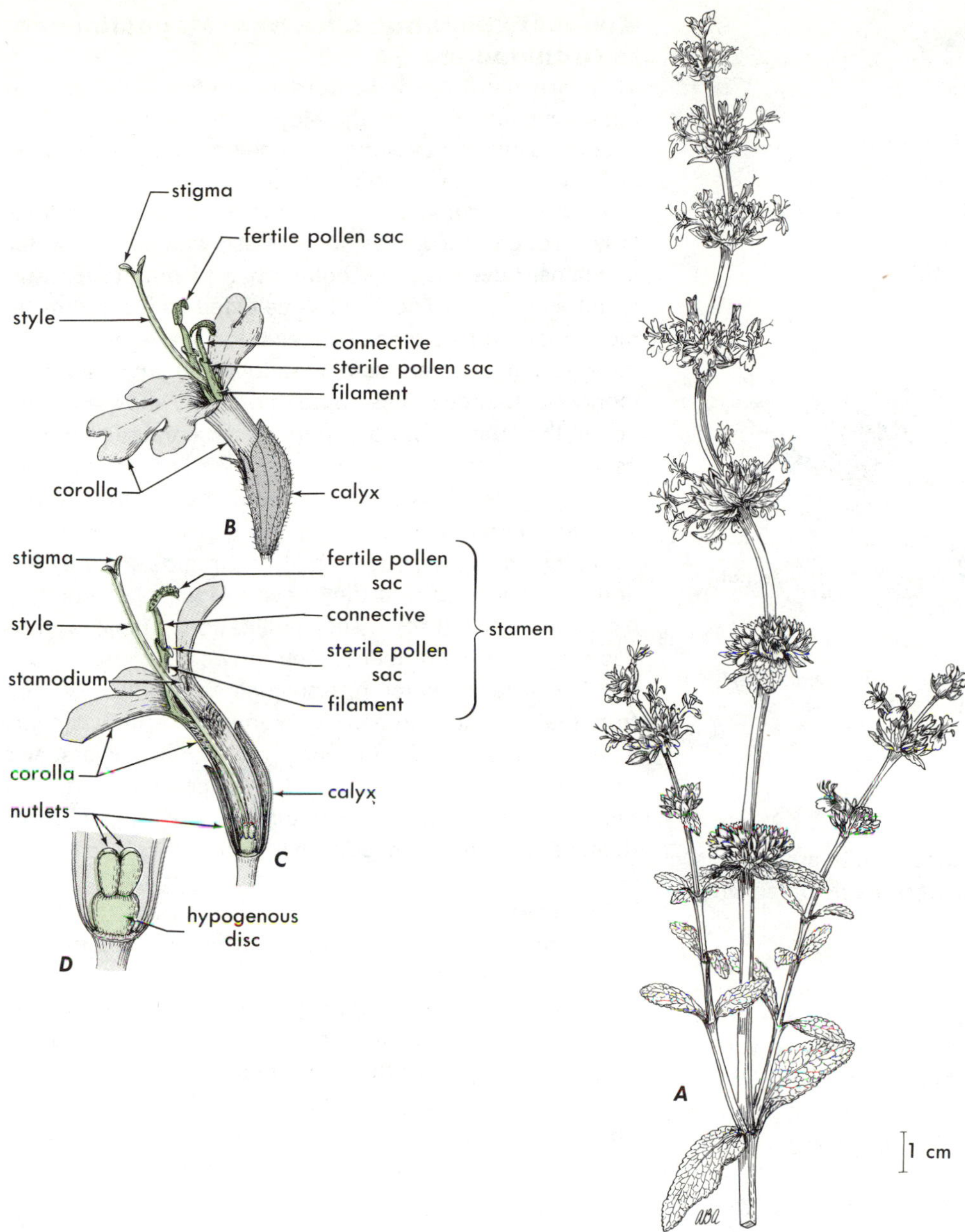

Figure 30.9
A, Salvia mellifera, plant habit. *B,* flower. *C,* longitudinal section of flower. *D,* developing fruit.

Apiaceae (Umbelliferae)

The parsley family forms a very distinctive group of plants, many of which can be recognized with little experience. There are over 200 genera and 3000 species growing in all regions of the world, though confined to mountains in the tropic zone. The name is derived from the typical umbel type of inflorescence (Fig. 30.14). The family contains many crop plants, as well as some that produce drugs and a number of kitchen herbs. Poison hemlock, which is famous because of its use by the Greeks as a means of carrying out the death sentence and which was given as such to Socrates, is a member of this family. The genus is widespread, constituting a hazard to livestock not only in pastures of Greece, but on open ranges of California. Carrots, celery, and parsnips as well as parsley are members of this family. They are mostly herbs, rarely small shrubs with generally hollow stems. Leaves are alternate, mostly compound, and frequently much dissected. The petioles expand at the base and may somewhat sheath the stem.

Flowers are small, being borne in umbels, which are in turn grouped in umbels, so a large inflorescence is formed. Frequently, umbels at all levels are subtended by bracts, forming a characteristic involucre. Flowers are regular or the outer flowers irregular, always epigynous and perfect, although they may not always be complete. There are five stamens, alternating with the same number of petals. The inferior gynoecium is composed of two

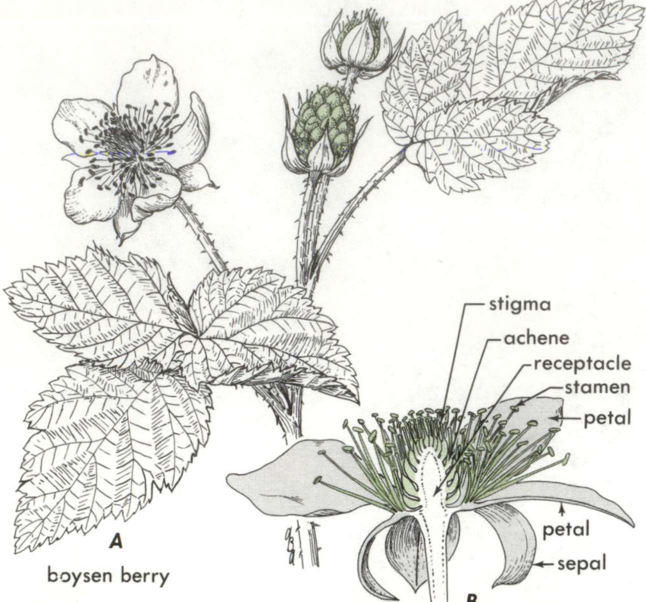

Figure 30.10
Rubus ursinus var. *loganobaccus* (boysenberry). *A*, flowering branch with *B*, one flower split open to show the fleshy receptacle.

carpels, each with but a single seed. Fruits are indehiscent, although carpels separate from each other when mature.

Asteraceae (Compositae)

The Asteraceae is the largest family of angiosperms (Fig. 30.15, 15.21). Only a few of these plants are woody. In this family of herbaceous plants there are around 900 genera and about 13,000 known species. The family is not only of worldwide distribution but in most places is relatively abundant. The family is not noted for its food plants; endive (*Cichorium*), artichokes (*Cynara*), chicory, lettuce, and sunflower form most of its contribution in this respect. Neither is it famous as a producer of drugs or other commercial products. One species, safflower (*Carthamus*), is now being grown for oil and members of the genera *Taraxacum* and *Parthenium* are being considered as possible sources of latex for rubber. Paging through a seed catalog or a visit to a florist shop will show members of this family in their full grandeur. *Dahlia, Chrysanthemum, Aster, Zinnia, Tagetes, Gaillardia, Ageratum* and many others are representatives of this family. Dandelions color lawns, and various wild species add to fall infest of meadows and fields. There are 400 species of the genus *Artemisia,* that cover arid portions of the world and supply, among other things, tarragon for fancy vinegar and nectar for desert honey. Leaves are of various shapes.

Monocotyledonous Line from Magnoliaceae to Orchidaceae

There are many obvious differences between the dicotyledonous plants such as the Magnolias and the monocotyledonous line of evolution. These have been discussed and are briefly reviewed in Table 30.3.

However, some primitive monocotyledons, particularly certain water weeds such as arrowheads and water plantains, have much in common with buttercups and marsh marigolds. The families selected to represent the monocots are Liliaceae, Poaceae, Iridaceae, and Orchidaceae. Lilies differ from magnolias in all distinctive monocot characteristics. In addition, they show a reduction in the number of floral parts and a connation of carpels. Irises show a single advance over lilies in that they are epigynous. Many irises are regular, perfect, and with separate perianth parts. Stamens show no connation. Syncarpy occurs as it does in lilies. Orchids show evolutionary advance in that they have irregular flowers very highly specialized for insect pollination. Stamens have been reduced in number to one or two.

The order to which grasses belong may have arisen from the order to which lilies belong. The primitive characters of grasses include hypogyny, regular flowers, and only connation of carpels. The loss of perianth parts and the reduction in number of stamens and carpels mark them as a more advanced family than the Liliaceae.

Liliaceae

Unlike grasses, lilies are grown largely for ornamental purposes, although two genera, *Allium* (e.g., onions, garlic, and leeks) and *Asparagus,* are grown extensively for food. Some species yield drugs, and others have poisonous properties that may cause trouble in pastures or on ranges of western states (Fig. 30.16). There are about 4,100 species in some 280 genera. Most lilies grow from a bulb or bulblike organ and flower in a single grow-

Table 30.3
TABULATION OF DIFFERENCES BETWEEN MONOCOTYLEDONS AND DICOTYLEDONS

Monocotyledons
1. One cotyledon or seed leaf.
2. Generally marked parallel leaf venation.
3. Flower parts typically in groups of three or multiples of three.
4. Vascular bundles of stems scattered throughout a cylindrical mass of ground tissue.
5. Vascular cambium lacking in most forms.
Dicotyledons
1. Two cotyledons or seed leaves.
2. Generally marked netted venation of leaves.
3. Flower parts typically in groups of four or five.
4. Vascular bundles of stems usually arranged in the form of a cylinder.
5. Vascular cambium present in forms having secondary growth.

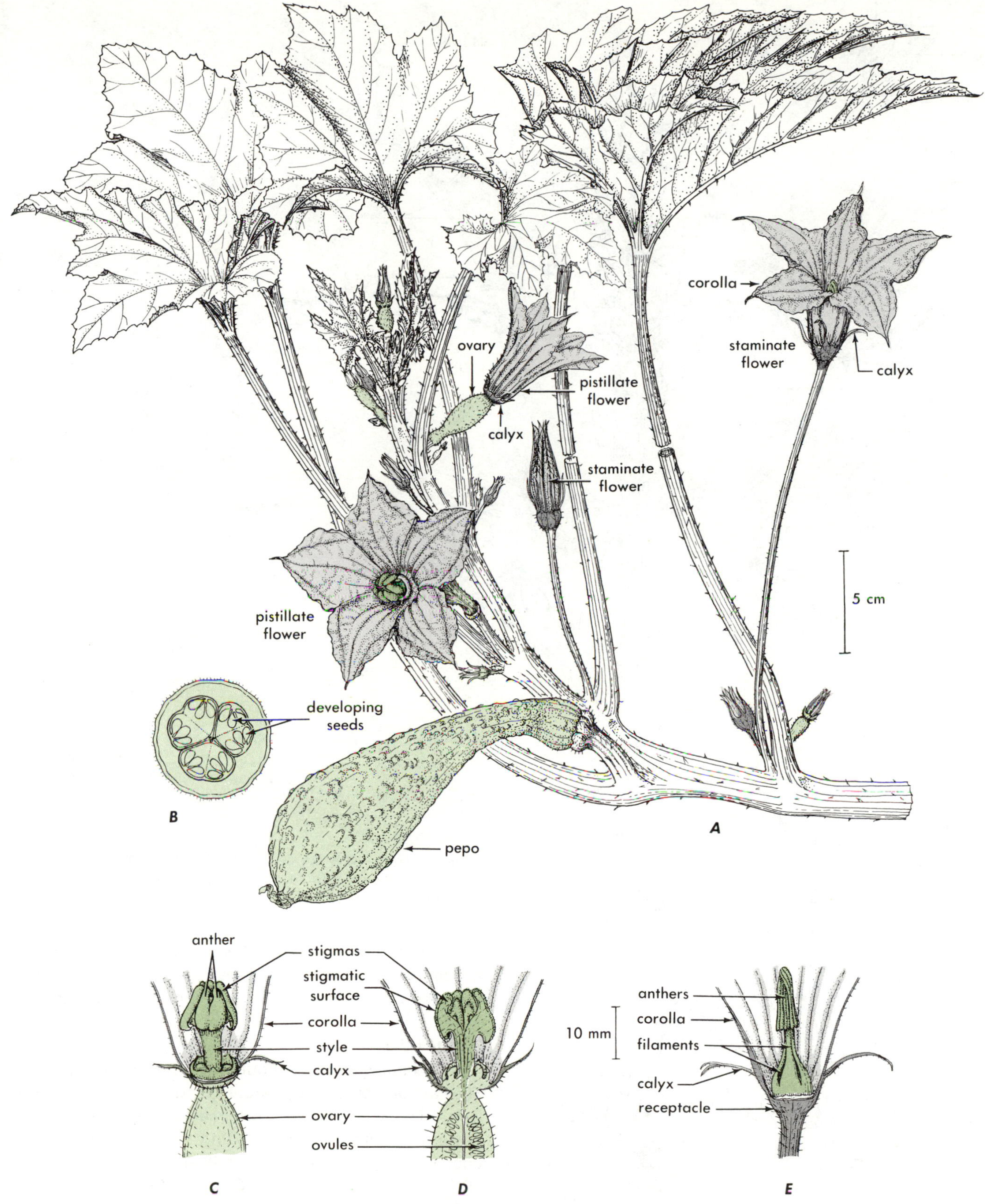

Figure 30.11
A, Cucurbita pepo, branch with leaves, flowers, and fruit. *B,* male flower. *C,* female flower. *D,* female flower sectioned. *E,* cross section of ovary.

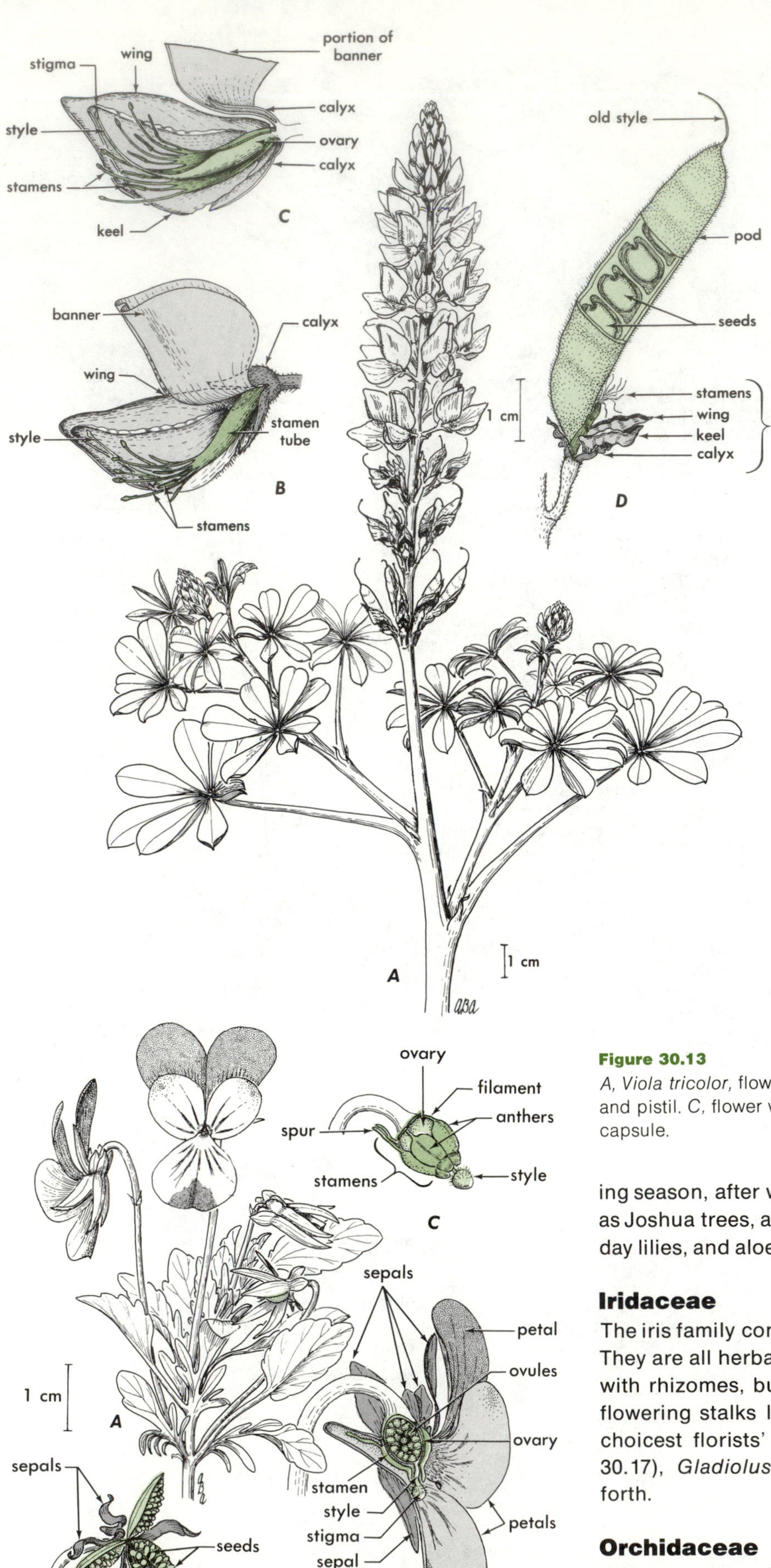

Figure 30.12
A, Lupinus succulentus, plant habit. *B,* flower. *C,* flower sectioned. *D,* fruit pod.

Figure 30.13
A, Viola tricolor, flowering plant habit. *B,* detail view of stamens and pistil. *C,* flower with sectioned view of ovary. *D,* opened capsule.

ing season, after which the shoot dies down. A few, such as Joshua trees, are woody perennials. Tulips, hyacinths, day lilies, and aloes are other examples of the lily genera.

Iridaceae

The iris family contains about 60 genera and 800 species. They are all herbaceous, largely perennial forms, usually with rhizomes, bulbs, or corms, as in lilies. Leaves and flowering stalks last for only one season. Some of the choicest florists' plants occur in this family—*Iris* (Fig. 30.17), *Gladiolus, Freesia, Crocus, Watsonia,* and so forth.

Orchidaceae

It is agreed by all that orchids represent the most advanced family in the Monocotyledonae (Fig. 30.18).

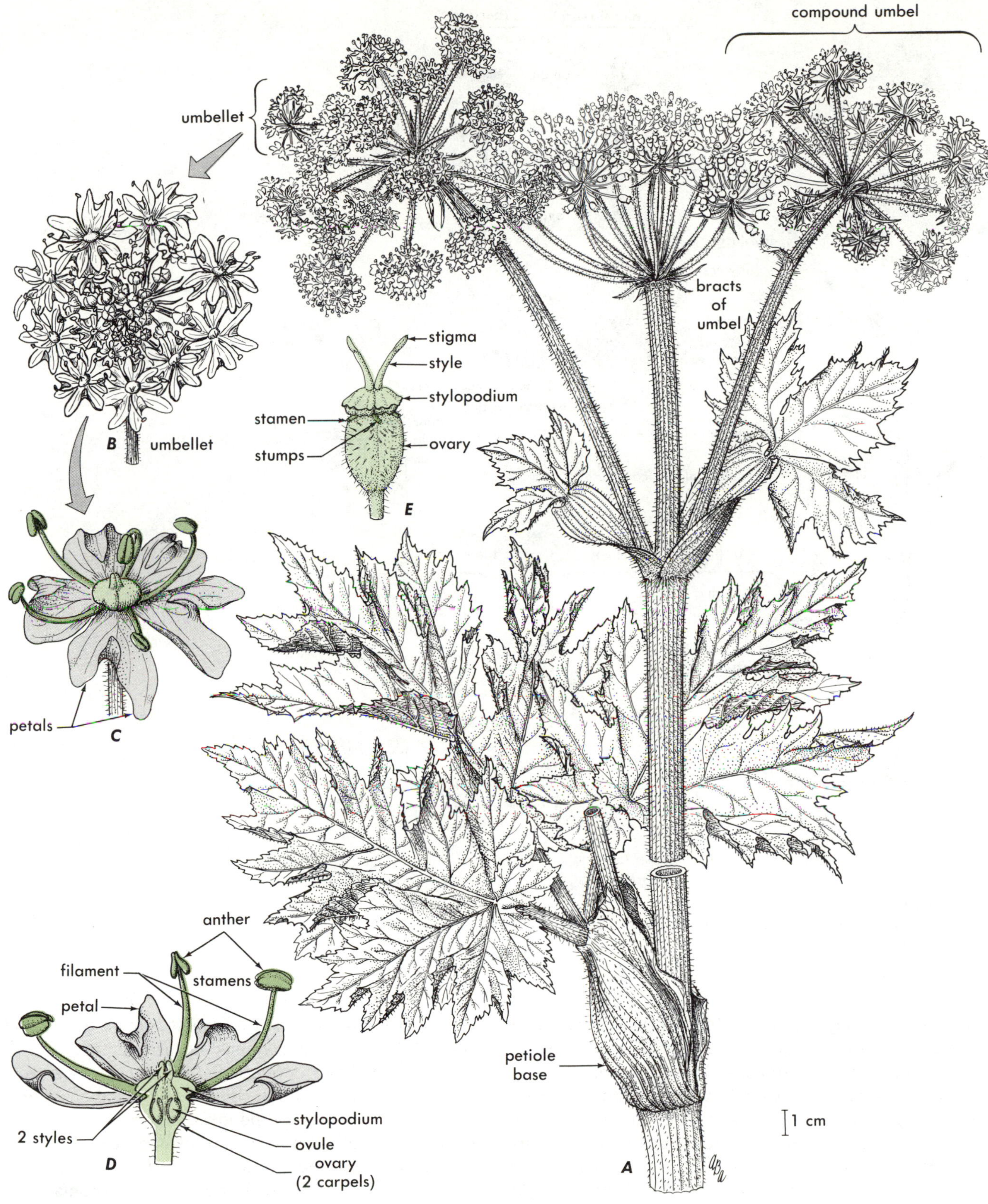

Figure 30.14
A, *Heraculeum lanatum*, plant habit. *B*, sectioned view of flower. *C*, flower. *D*, umbell. *E*, young fruit.

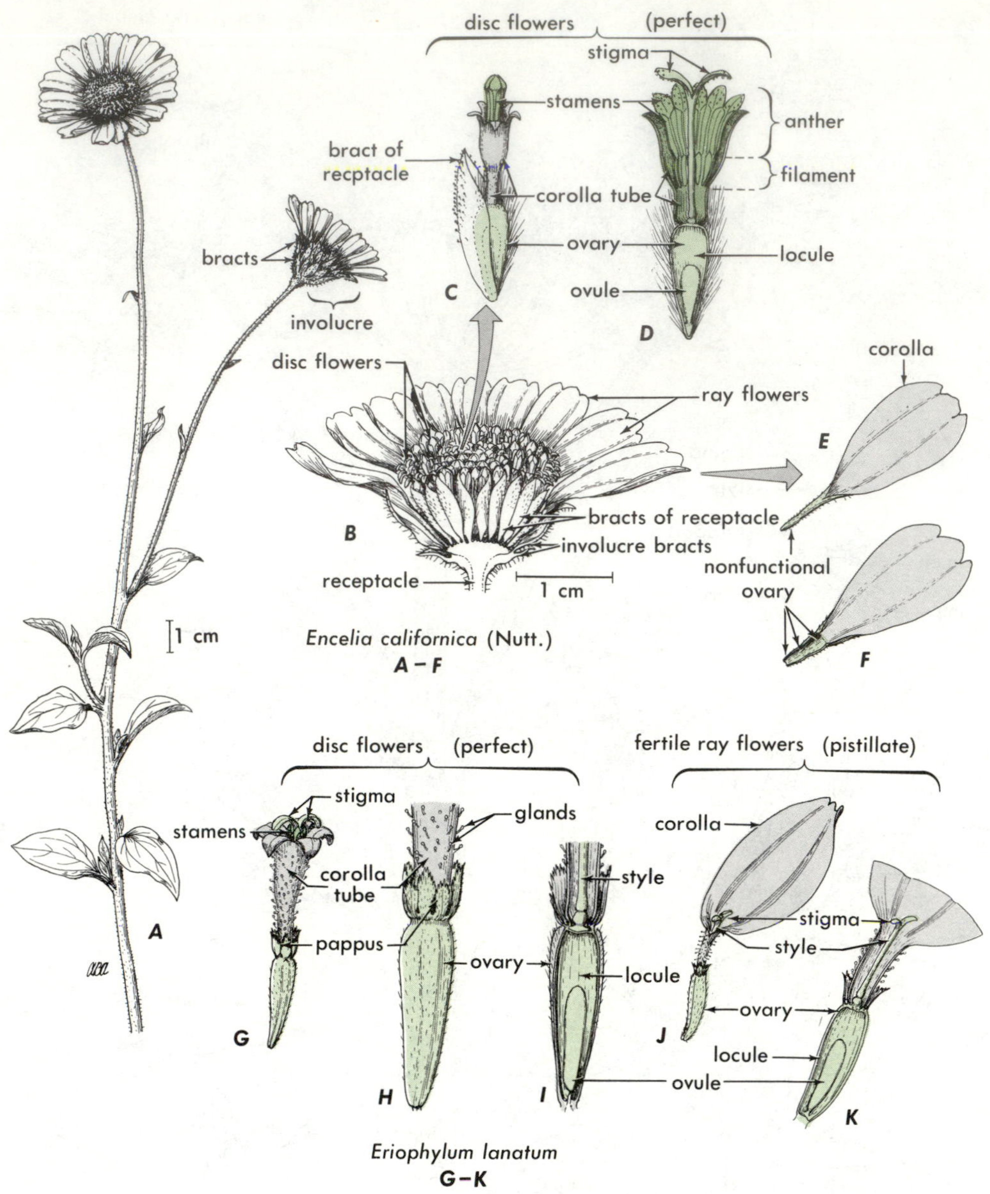

Figure 30.15

A, Encelia californica, plant habit. *B,* flower sectioned view. *C* and *D,* perfect disk flowers. *E* and *F,* sterile ray flowers. *G* through *K, Eriophylum lanatum* var. *achillaeoides. G* to *I,* perfect disk flowers. *J* to *K,* female ray flowers.

They are all herbaceous, occurring throughout the world largely in the tropics, but with a few genera extending into colder temperate regions. There are some 17,000 recognized species in several hundred genera, making this one of the three largest families of angiosperms. All forms are perennial, with tuberous, bulbous, or otherwise thickened roots. They may be erect, prostrate, or climbing. A few are saprophytic and lack chlorophyll; others are epiphytes, growing on trees without benefit of soil.

Poaceae (Gramineae)

Humans and other mammals have been associated with the grass family for far more years than those of recorded history. It is probably not an overstatement to say that without the grass family human civilization, as we know it, could not have developed. Different grasses were domesticated by the three centers of civilization; wheat, rye, barley, and oats by peoples of the Mediterranean region and Southwest Asia; corn by natives of the Western Hemisphere; and rice and millet (*Eleusine*) by peoples in the Orient. Grasses are grown for human and animal consumption, for ornament, and for uses in arts and industry. There are over 620 genera and about 10,000 species. Most grasses are herbaceous, either annuals or perennials; a few bamboos are climbers and some are woody and up to 20 m tall. Grasses of one kind or another grow in all kinds of soil and situations. Grasses may be

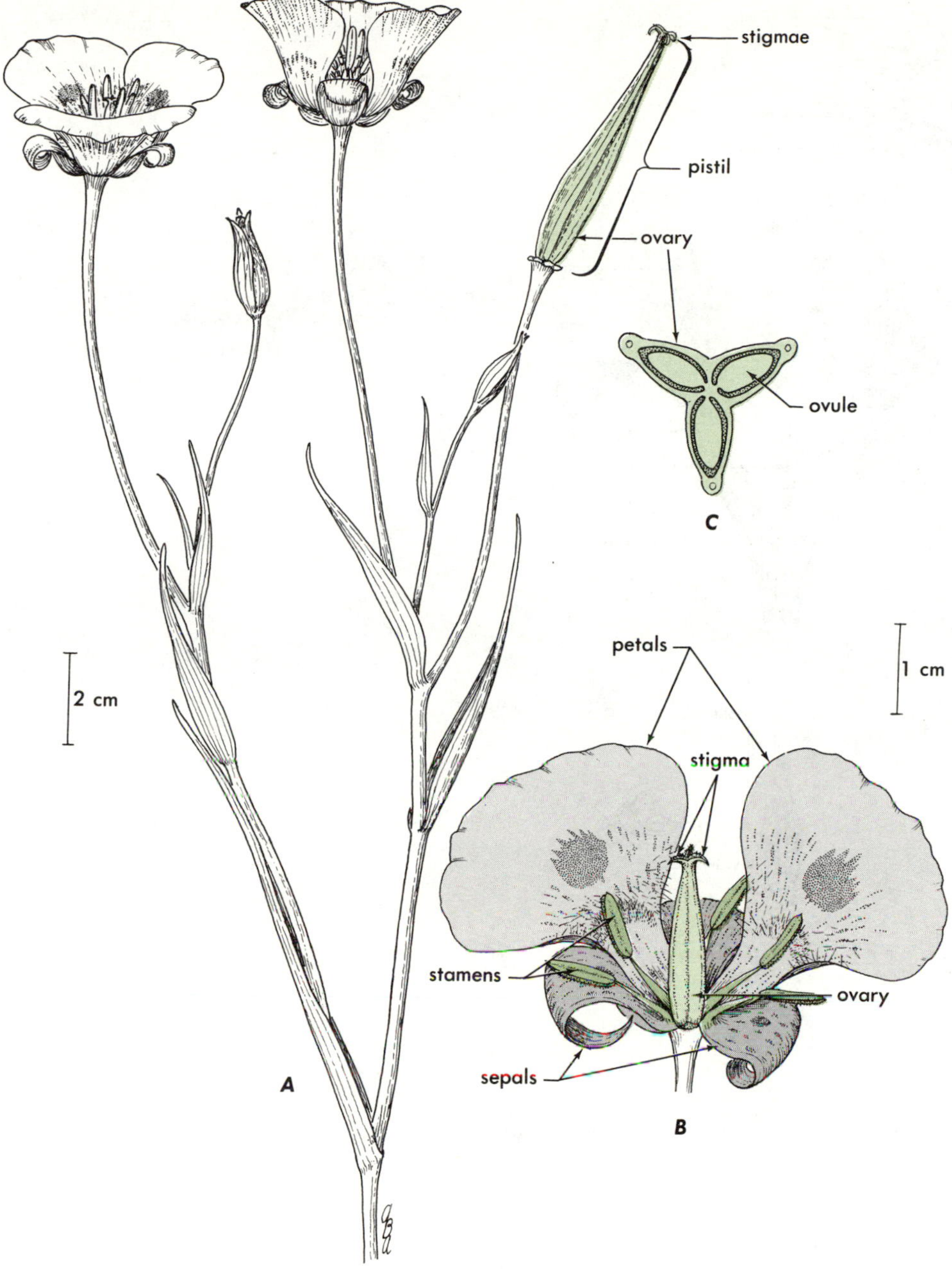

Figure 30.16
Calachortus luteus. A, flowering branch. *B,* flower showing interior ovary. *C,* section through ovary.

conveniently divided into six groups according to their use. (*a*) Bamboos, mostly evergreen, stout perennials, are used for construction in many parts of the Orient, and some have been introduced into warmer parts of the United States as ornamentals. (*b*) Cereals and some other annual grasses supply grain and forage for both human beings and animals. (*c*) Sugar-producing species, such as sugar cane, are strong, upright perennials growing only in the tropics. (*d*) Sod-forming grasses, perennials that cover many square miles of the earth's surface, are used in lawns and meadows and for forage. (*e*) Bunch grasses, common in semi-arid regions, are perennials that do not form rhizomes and do not produce a turf as sod grasses do. (*f*) There is a large group of grasses grown for ornamental purposes, such as pampas grass. Flowers and vegetative characteristics of grasses have been described in some detail in Chapter 15 and shown in Fig. 15.19.

SUMMARY OF ANGIOSPERM EVOLUTION

1. Angiosperm families, according to the Besseyan system of classification, are all derived from the primitive order Ranales, to which buttercups and magnolias belong.

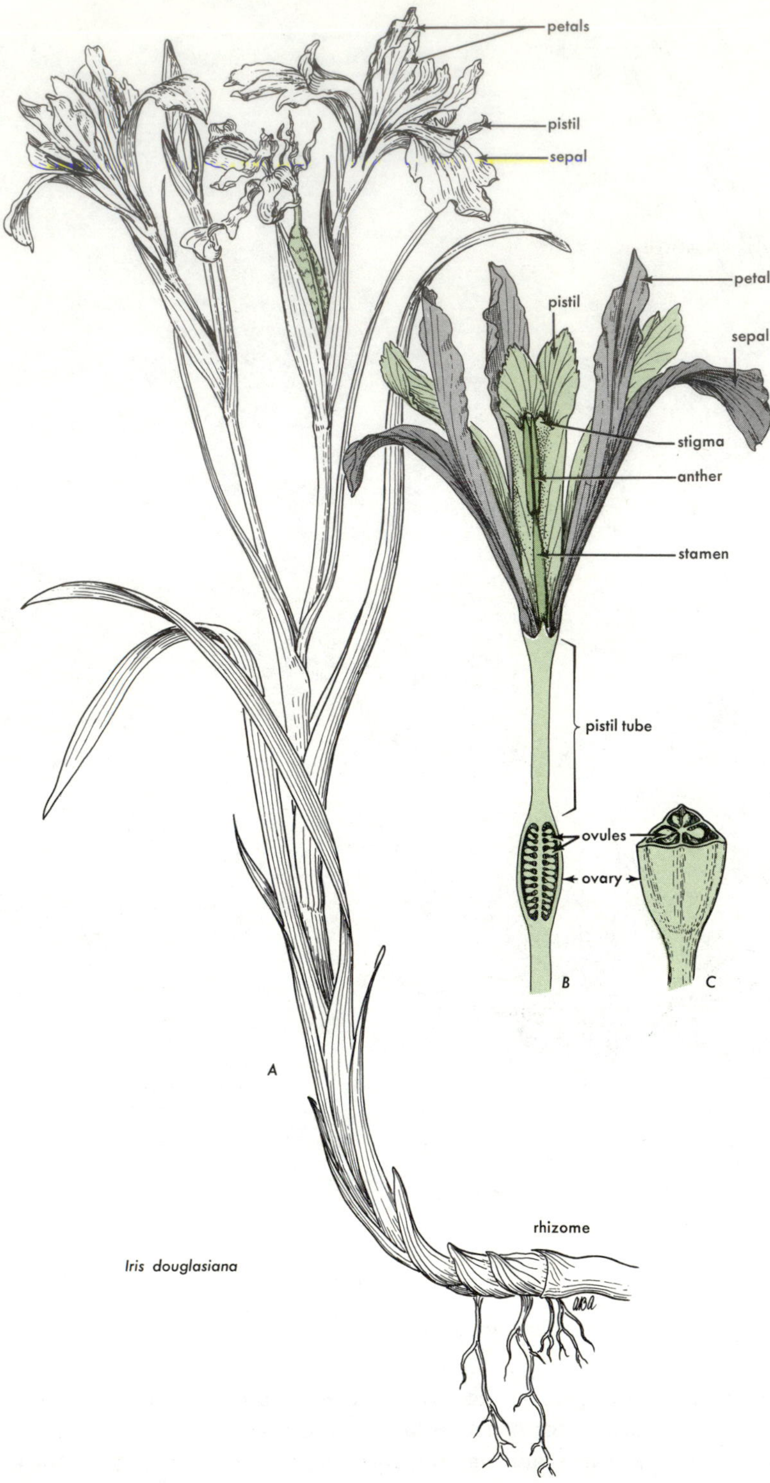

Figure 30.17
Iris douglasiana. A, entire flowering plant. *B,* flower showing interior ovary. *C,* section through ovary.

2. Important changes in floral evolution, according to the Besseyan system, are as follows: (*a*) from a spiral to a whorled arrangement of floral parts; (*b*) from many parts to few or even to a loss of parts; (*c*) from separate floral parts to connate parts; (*d*) from a regular flower to an irregular flower; (*e*) from hypogyny to epigyny.

3. According to the Besseyan system, there are three branched lines of ascent from the Ranales.

4. The first line of ascent includes: Malvaceae (mallow family). Papaveraceae, Brassicaceae (mustard family), Solanaceae (potato family), Salicaceae (willow family), and Lamiaceae (mint family).

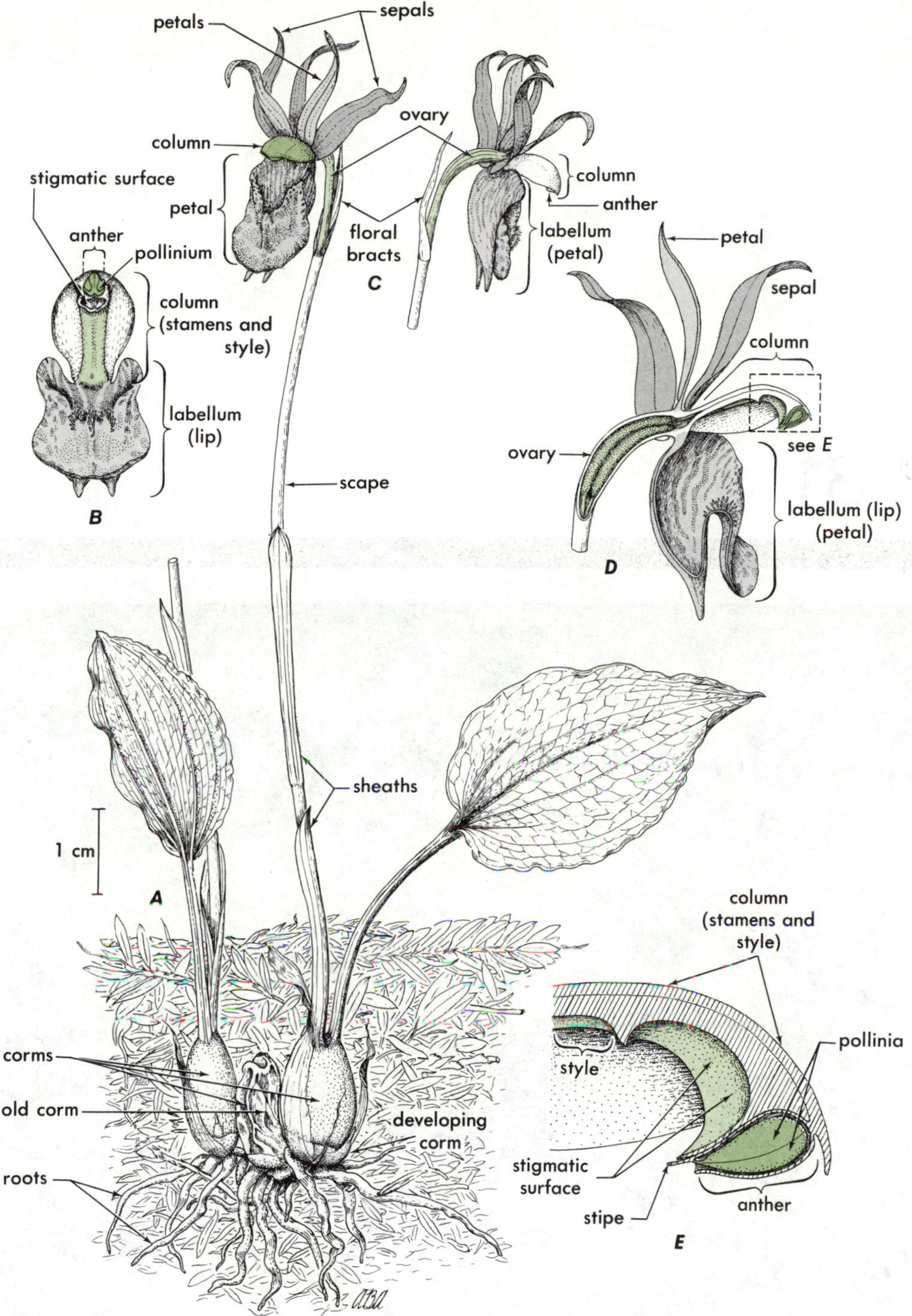

Figure 30.18
Calypso bulbosa orchid. *A*, plant habit showing leaves, stem, corms, and flower. *B*, face view of flower. *C*, flower. *D*, sectioned flower. *E*, detail of column.

5. The second line comprises: Rosaceae (rose family), Juglandaceae, Violaceae, Fabaceae (pea family), Apiaceae (carrot family), and Asteraceae, with Cucurbitaceae (melon family) as an offshoot from Rosaceae.

6. The third line comprises the Monocotyledonae in the following order: Lilaceae (lily family), Iridiaceae (iris family), and Orchidaceae (orchid family). The Poaceae (grass family) is an offshoot from the Liliaceae.

CHAPTER 31

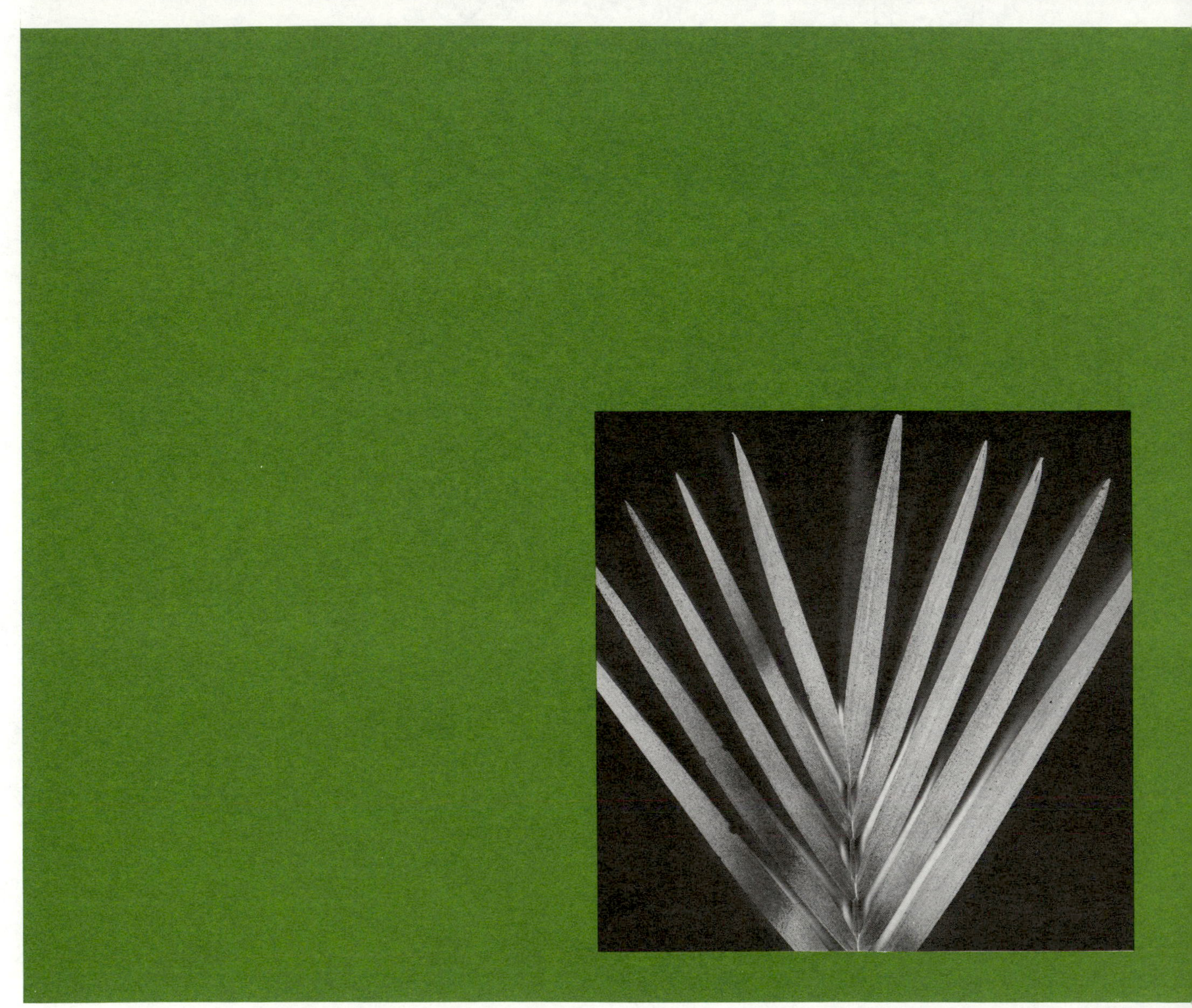

EVOLUTION

In a broad sense the term **evolution** refers to a process involving gradual changes. It is well-known that neither animals, nor plants, nor cities, nor states remain the same. We may speak of the evolution of means of transportation, the evolution of human clothing, the evolution of mountains and valleys, and the evolution of many other nonorganisms. **Organic evolution** pertains to the gradual changes that have taken place in living organisms. Coal is formed from the remains of plants that were different in appearance from those growing today. The student of organic evolution may be interested in, among other things, accounting for the disppearance from the earth's flora of the Coal Age plants and the appearance of the modern flowering plants.

Organic evolution is closely allied to genetics and plant breeding. We have seen that the factors which determine the characteristics and activities of plants are associated with the cell nuclei. Genes are transmitted from generation to generation by the chromosomes during the processes of fertilization and meiosis. If the origin of new plants is to be understood, experiments on genetics will furnish important information. Many such experiments have been carried out, and they have achieved two ends: (*a*) they have made possible clearer insight into some of the mechanisms of evolution, and (*b*) they have greatly increased the yield and improved the quality of agricultural crops.

THE GEOLOGIC HISTORY OF PLANTS

Since plants cannot be dissociated from their environment, some knowledge of the principal changes in the earth's crust is essential to an understanding of plant evolution. Broadly speaking, two different types of rocks are found in the earth's crust. One of them, like lava, was formed from molten material. The oldest known rocks are of this nature; granite is an example. Because of their molten origin, no fossils are ever found in these or in other **igneous** rocks.

The second type of rock (**sedimentary** rocks) is formed by the deposition of large amounts of sediment in bodies of water. As layer upon layer of sediment accumulates, the pressure upon the lower layers increases and they are turned to rock. Shale, limestone, and sandstone are examples of sedimentary rock. If plants or animals are buried in the sediment, and if decay does not occur, the buried individuals are preserved as fossils.

However, because great changes have taken place in the earth's crust over time, sedimentary and igneous rock are not always found where you would expect to find them.

We Live on a Restless Earth

The billions of years of earth's history have been characterized by great changes in the atmosphere and the crust. During this time, entire floras and faunas have evolved, been almost completely eliminated, and new forms arisen in response to these geophysical changes. Dramatic evidence for change is displayed in the desert canyon country of southern Utah and northern Arizona. Monument Valley, in northeastern Arizona lies between 1600 and 2300 m above sea level, yet is consists entirely of sedimentary rocks that were deposited on the bottom of some primeval sea! The sedimentary sandstone must have been thrust up thousands of feet from sea level, and now the rock lies exposed to the erosive action of rain and wind. The buttes are mute evidence that already several hundred feet of sandstone have been eroded from the surface of Monument Valley.

About 240 km to the west, the Colorado River has cut a gorge 1.5 km deep, the Grand Canyon, and all the rock exposed is sedimentary. We can surmise that a similar mile-deep deposit of sandstone and shale underlies Monument Valley, all deposited by an open ocean. About 130 km north of the Grand Canyon, the Towers of the Virgin River in Zion National Park, Utah, rise 1 km into the air—that is, 1 km higher than the buttes in Monument Valley. These towers are also made of sedimentary rock. The inference is that a layer of sedimentary rocks, 3 km thick, covers the entire region; the bottom of a primeval sea has been lifted hundreds of meters and its eroded remains form much of what we today call Utah and Arizona.

But that is not all. Less than 1000 years ago, Sunset Crater, 160 km south of Monument Valley, was formed by a burst of lava boiling through the marine sediments. This was the latest of many volcanic eruptions in the region. The San Francisco Mountains just north of Flagstaff were formed from cubic kilometers of lava. In Monument Valley itself, Mt. Agaltha and some smaller adjacent peaks are lava plugs, once the throats of volcanoes. The volcanoes themselves and the lava they spread out in the Monument Valley have since eroded away. All this sedimentation, eruption, deposition, uplift, and erosion has taken place in the past 350 million years. And trapped within the sediments are fossil remains of many plants and animals that existed during that period of time.

How Fossils Are Formed

A **fossil** may be a shell or bone, little changed from its original aspect; it may be the microscopic cell wall of a spore or pollen grain, which are also very resistant. But most plant parts are not hard, and plant fossils seldom consist of unchanged structures. They often are simply the impressions of leaf or stem fragments that were trapped in mud—mud that later was compressed into sedimentary rock. To yield a good fossil, plant parts must settle in quiet waters, then be buried under silt or volcanic ash to lie trapped in conditions unfavorable to decay. They are seldom entombed where shed, but are more typically carried by water or wind for some distance first. In some cases, the protoplast or lumen of the cell may be filled by silica quartz crystals. Anatomical details may be faithfully preserved (Fig. 31.1). In addition, degradation products from tissue decay may remain trapped as "chemical fossils." The porphyrins from chlorophyll degradation, for example, are very inert and long-lasting. Their presence in sediments indicate that photosynthetic plants were present at the time the sediments were laid down.

It is important to note that plant fossils consist almost entirely of plant remains that grew in or near water or were carried by streams into lakes, bays, or other sites of deposition. The remains of plants growing in arid or mountainous regions are rarely deposited where conditions are favorable for their preservation as fossils. Either it is too dry or the site is undergoing erosion and the deposits are of very short duration. This means that the geological record of plants must probably remain incomplete and that the records we do have usually represent only the floras of lowlands and moist habitats. Considerable evidence indicates that the great environmental extremes prevalent in arid or semiarid mountainous areas are especially favorable to rapid evolution. This factor, we shall see, may be very important in connection with the problem of determining the origin of the angiosperms and may serve to explain peculiarities that surround the first records of the angiosperms.

When Did the World Begin?

When did the world begin? Earth scientists have no definite clues, but the calculations of astronomers and physicists, coupled with known age of meteorite fragments and of rocks from the moon, lead us to estimate that the surface of the earth cooled to a crust about 4.5 to 5.0

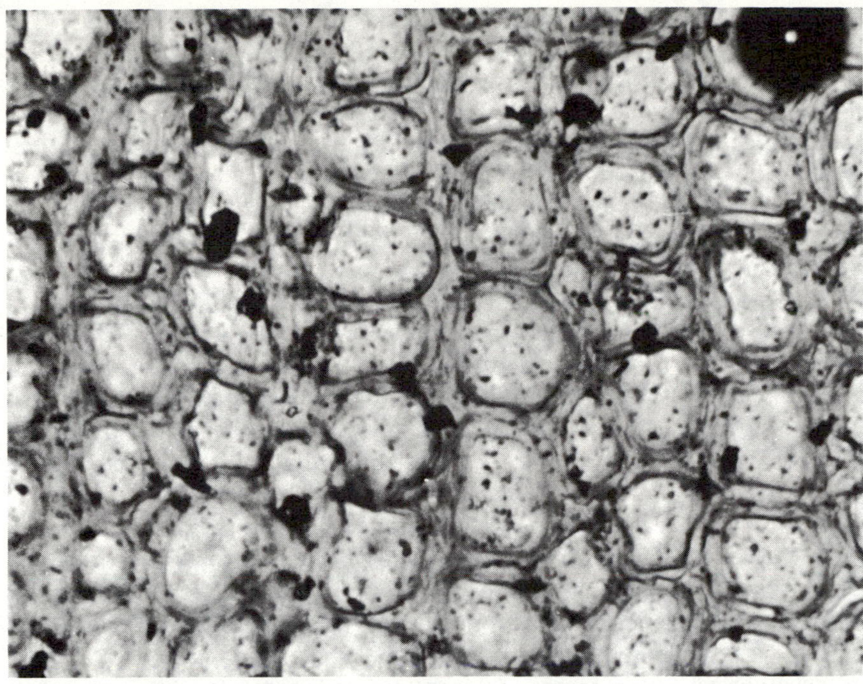

Figure 31.1
A thin section of petrified wood, as seen through a light microscope. Tracheid lumens are filled with silica stone.

billion years ago. This primordial crust probably contained uplands, ocean basins, plateaus, and mountainous volcanoes. Volcanic eruptions filled the atmosphere with a variety of gases, possibly including carbon dioxide, methane, toxic carbon monoxide, nitrogen, hydrogen, and strong-smelling ammonia and hydrogen sulfide. Free oxygen was absent. The first rains washed, dissolved, and eroded the uplands carrying many nutrients to the young oceans.

Life, or some sort of halfway form of life, probably originated very early. It may have consisted of aquatic, small, self-replicating, heterotrophic cells that lived on the rich, anaerobic ocean soup around them. However, our present methods of detecting remains of ancient life require that the fossils be preserved in rocks. So far, the oldest rocks found on earth are 3.4 billion years old. This means that we have no record of the first 1 to 1.5 billion years of earth's history (Fig. 31.2).

Aquatic life did exist 3 billion years ago, for we have fossil evidence of it. Some of the oldest rocks known lie exposed near the gold-mining town of Barberton, on the border between the Republic of South Africa and Swaziland. About 3.2 billion years ago, this area was possibly a shallow, warm sea or embayment. Living things may have existed in thin sheets at the bottom of a silica-rich sea. Apparently, conditions were perfect for preservation of the organisms. They were preserved in the sediment in a siliceous solution that later crystallized into rock called chert, much as a modern biological specimen is pre-

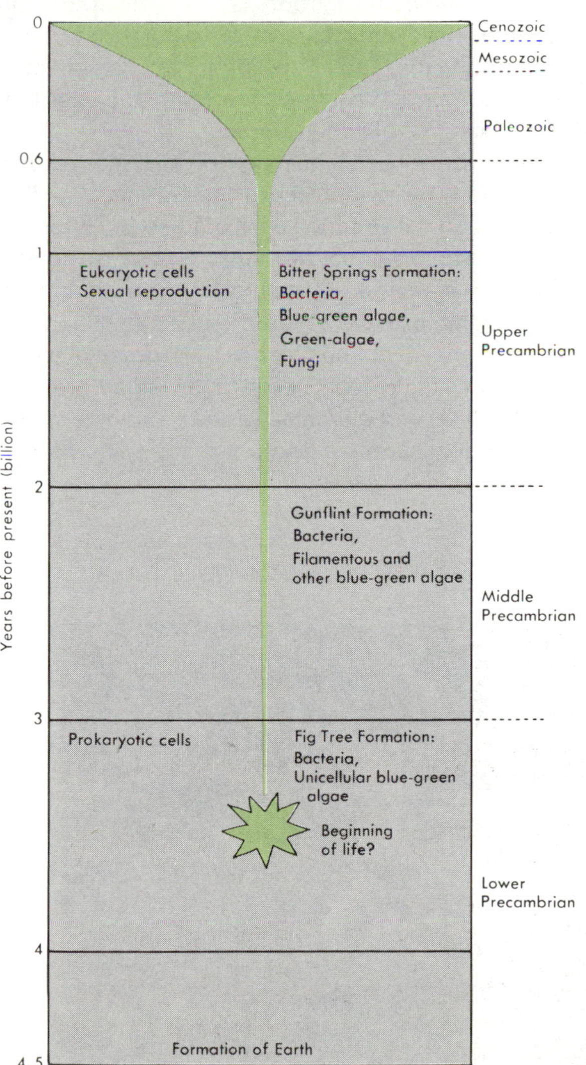

Figure 31.2
The diversity of life through geologic time. Increasing width of the green band indicates increasing diversity.

served by being embedded in plastic. There was no distortion in the soft bodies of these early organisms because the silica matrix of chert is incompressible. Today, these beds of chert are exposed. When thin sections are placed on a microscope slide, the perfectly preserved organisms **(microfossils)** can be seen. The chert itself has not been dated, but rock layers above and below it have. Its age is estimated to be 3.2 billion years.

Many unusually shaped structures, which may or may not be organisms, have been seen in this chert (called the Fig Tree Formation), and some are seen frequently enough and in great enough detail to leave no doubt of their once-living nature. One is a rod-shaped bacteriumlike cell, called *Eobacterium isolatum* (Fig. 31.3), the other is possible a blue-green alga, *Archaeosphaeroides barbertonensis.* Organic residues were also found in the chert. Analysis of them showed the presence of certain hydrocarbons that can most reasonably be regarded as breakdown products of chlorophyll, and that the ratio of carbon-13 to carbon-12, two natural isotopes of carbon, was lower than the ratio in today's atmosphere. When plants photosynthesize, they have preference for carbon-12, so their tissue and fossil residues have a low carbon-13 to carbon-12 ratio, corresponding to the ratio in the chert. This combination of direct and indirect evidence strongly suggests that plant life existed 3.2 billion years ago.

A similar, but richer, collection of microfossils has been preserved in another chert deposit, the Gunflint Formation, along the Ontario shore of Lake Superior. By dating layers above and below it, this deposit is thought to be 2 billion years old. Filaments are most abundant, and some resemble modern blue-green algae (Fig. 31.4). Small spheres that might be colonial blue-green algae or their spores or single-celled blue-green algae are also common. Analysis of the organic residue and determination of the carbon isotope ratio show that photosynthetic organisms were present: however, they were still prokaryotic organisms.

Figure 31.3
Oldest known bacterium. *Eobacterium isolatum* from the Fig Tree Formation, about 3.2 billion years old. (Courtesy of E. S. Barghoorn.)

The first evidence for eukaryotic plants did not appear until 1 billion years ago; their fossils appear in the Bitter Springs Formation, about 70 km northeast of Alice Springs, in the heart of Australia. These fossils are rich and varied, and indicate that at least four groups of organisms existed at that time: (*a*) prokaryotic filamentous blue-green algae, akin to modern *Oscillatoria* and *Nostoc;* (*b*) bacteria; (*c*) eukaryotic fungi (Fig. 31.5); and (*d*) eukaryotic green algae. Cytological details are preserved so well, that stages of cell division can even be detected (Fig. 31.6), but you should be aware that the interpretation of cellular details is not uniform. Some biologists claim that the "nuclei" are artifacts and that the cells are those of prokaryotes.

This early appearance of eukaryotes is impressive, because the differences between prokaryotes and eukaryotes are relatively enormous. These changes, as you may recall from earlier chapters, involve structure of DNA, organization of a nucleus, endoplasmic reticulum, ribosomes, chloroplasts, mitochondria, development of new chlorophylls, the elaboration of sexual reproduction (with meiosis, chromosome recombination, and mitosis), and the formation of many new enzynes in new pathways of metabolism. The evolution of eukaryotes from prokaryotes, even in the long time span of 2 billion years, was a considerable accomplishment.

The evolution of more modern plants during the past 1 billion years may seem rapid by comparison, but it is based on the permutations and combinations of life processes that were already set during the first 2 billion years of life.

Sexual reproduction brings with it variation in offspring and a more rapid rate of evolution. It made possible a greater diversity of forms to develop and proliferate. A spurt of evolution resulted. By the end of the Proterozoic era (Table 31.1), the seas had become crowded with life, including many forms of primitive animals and plants. According to one theory, the rate of oxygen formation had increased, and by the end of the Proterozoic era the level of oxygen in the air may have been 1% of its present level. Oxygen not only supports life, it also screens out ultraviolet radiation from the sun. The radiation can be damaging to cell activities. This amount of oxygen—1% of present level—produced a sufficient filter of radiation to permit life in all but the top inch of water. Life on land, except in sheltered places, would still have been impossible.

Algal Diversity

The "sudden" evolutionary spurt 600 million years ago marked the end of the Proterozoic era and the beginning of the long Paleozoic era, a time when the land was lowlying, and inland seas moderated the climate so that seasonal and latitudinal effects were minor. Apart from

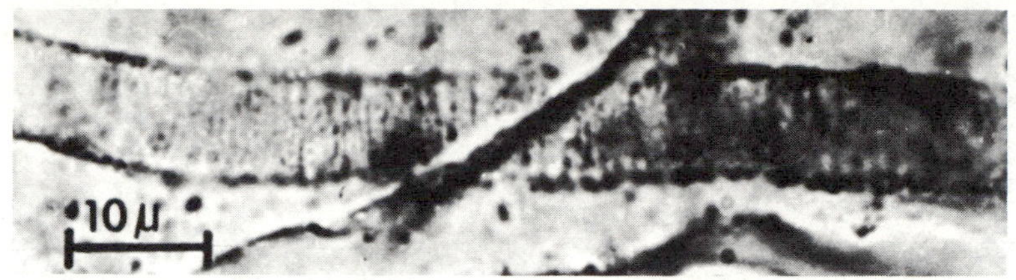

Figure 31.4
The filamentous blue-green alga, *Animikiea septata,* from the Gunflint Formation, 2 billion years old (Courtesy of J. W. Schopf.)

periods of glaciation in the Silurian, Devonian, and Permian, it was a time when a climate much like that in the tropics today dominated the entire world. It was a time of enormous evolutionary change in the plant kingdom. However, the rate of evolution did not proceed at a constant pace during the entire Paleozoic era. Most of the new and successful experiments in plant form appeared in only 25 to 50 million years, during the late Silurian to mid-Devonian periods. It was then that plants came to dominate the land; it was then that the oxygen level of the air may have reached 10% of the present level, according to one hypothesis.

Figure 31.5
Hyphae of a fungus, *Eomycetopsis robusta,* from the Bitter Springs Formation, 1 billion years old. (Courtesy of J. W. Schopf.)

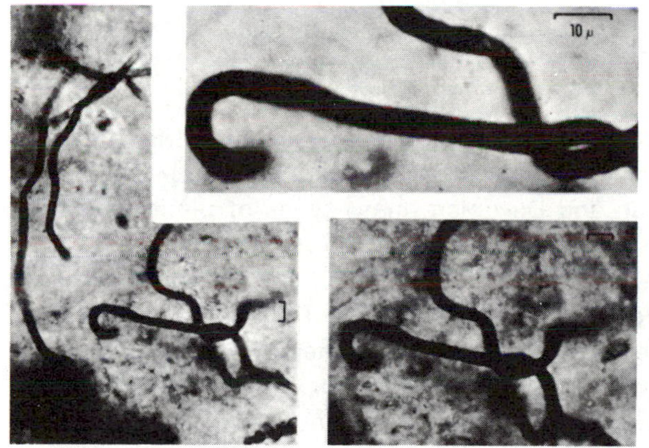

Through the Cambrian, Ordovician, and most of the Silurian periods (570 to 400 million years ago) algal forms continued to dominate the plant kingdom. Blue-green and green algae were abundant, but desmids and lime-encrusted reds and some browns resembling *Laminaria* and *Fucus* were also present. The families Codiaceae and Dasycladaceae in the Chlorophyta were especially prominent; and the early forms closely resemble living members of those families. Many of these showed branching, coenocytic filaments; larger forms had a central axis with radiating branches whose ends united to form an enveloping net. The lime-encrusted forms secreted calcium carbonate just as modern forms do. Fresh water forms as well as salt water forms evolved.

Of particular interest are fossils from this time of *Chara*-like algae (Fig. 31.7), for *Chara* is today only

Figure 31.6
Sequence of fossil impressions from the Bitter Springs Formation arranged to show cell division. (From J. W. Schopf, *J, Paleontol.* **42,** 651. 1968. Reprinted with permission of the publisher.)

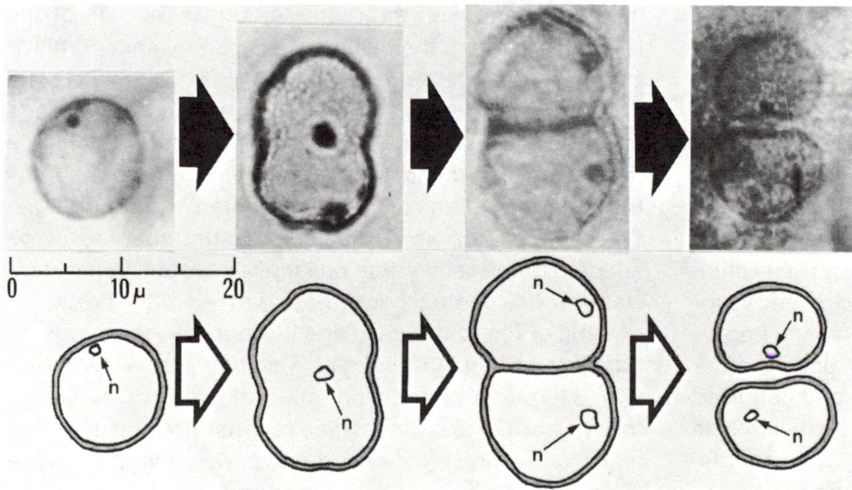

Table 31.1
GEOLOGIC TIME AND THE DOMINANCE OF DIFFERENT PLANT GROUPS THROUGH TIME[a]

Era	Period	Epoch or Part	Began (Millions of Years Ago)	Dominants
Cenozoic	Quaternary	Recent	Last 12,000 years	Flowering plants
		Pleistocene	2.5	
	Tertiary	Pliocene	7	
		Miocene	26	
		Oligocene	38	
		Eocene	54	
		Paleocene	65	
Mesozoic	Cretaceous	Late	90	
		Early	136	Gymnosperms
	Jurassic	Late	166	
		Early	190	
	Triassic	Late	200	
		Early	225	
Paleozoic	Permian	Late	260	
		Early	280	Lower vascular plants
	Carboniferous	Pennsylvanian	325	
		Mississippian	345	
	Devonian	Late	360	
		Middle	370	
		Early	395	Algae
	Silurian		430	
	Ordovician		500	
	Cambrian		570	
Proterozoic	Precambrian		4500–5000	

[a] Shaded sections were times of great evolutionary changes in the plant kingdom. The time scale is not drawn to scale. All named epochs do not appear on this chart.

known from freshwater habitats. The fossil forms had a very thick-walled resting zygospore that would have been resistant to drought. Such spores are rare in marine algae. In addition, there is evidence that some other algae may have been terrestrial, forming thick, mosslike mats on wet ground. So, in the early Paleozoic era, the invasion of the land began. It should be recalled from earlier chapters that the green algae are thought to be a logical starting point for the evolution of a terrestrial, vascular flora.

Invasion of the Land: Silurian and Devonian Periods (430 to 345 Million Years Ago)

What are some of the adaptations required for plant life on dry land? In water, plants need no complex structures for support nor for the uptake of nutrients. The surrounding water buoys them up and bathes them with soluble nutrients. In contrast, land plants must not only develop roots for the absorption of nutrients and the elaboration of stiff tissue for support, they must provide pathways for water and nutrient transport—xylem and phloem. Reproductive cells on land must be carried by agents other than water, and they must be thick-walled and cutinized to avoid desiccation. All aerial surfaces of land plants, in fact, must be adapted to reduce water loss; these adaptations can take the form of stomata and cuticle.

The first undisputed fossils of terrestrial, vascular plants do not appear until the late Silurian period. In the Silurian and Devonian, many types of plants began to occur in the fossil record: representatives of the Psilophyta, Lycophyta, Sphenophyta, and pre-ferns. Did they all evolve from something like *Fritschiella,* or did they evolve from several separate lines? At this time there is not enough evidence to rule out either possibility.

The early vascular plants consisted of slender dichotomously branching stems, with or without leaflife appendages, and with sporangia either at the tips of the branches or along their sides. Perhaps the most primitive example from the end of Silurian period (Fig. 31.8) is *Cooksonia,* discovered in 395 million-year-old fossil beds in Czechoslovakia and Wales. It was a plant probably less than 10 cm tall, made up of dichotomizing branches less than 3 mm in diameter that terminated with sporangia. The spores had a waxy cuticle, indicating that they were adapted for dissemination on land. The xylem in the stem was a solid, central core; there was no pith. We know nothing of the root system and it is possible there was no true root system. Another early vascular plant was *Rhynia* (Fig. 31.9), and its excellent state of preservation in the fossil record allows for detailed reconstructions that have made it the best-known primitive vascular plant. The plant consisted of a slender rhizome upon which were

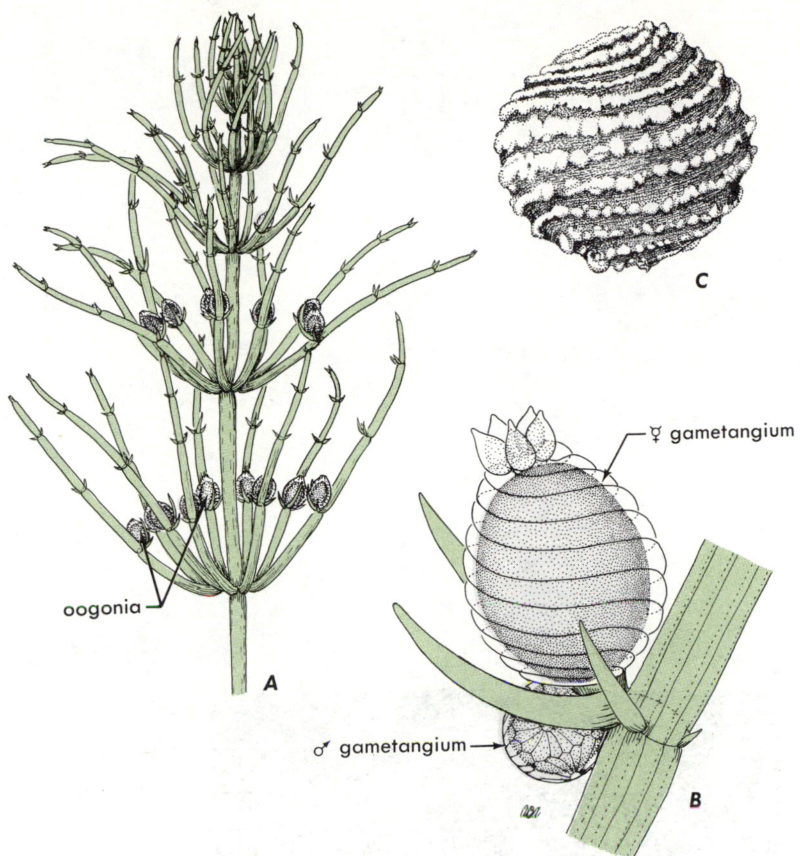

Figure 31.7
Living and fossil *Chara,* a green alga. *A,* living *Chara,* showing whorled branches and location of oogonia, ×70. *B,* one oogonium, ×700. *C,* a fossil oogonium or zygospore from the lower Devonian, ×400. (*C,* redrawn from H. P. Banks, *Evolution and Plants of the Past.* © 1968 by Wadsworth Publishing Company, Inc. Reprinted with permission of the publisher.)

borne erect, cylindrical stems that branch dichotomously and sparingly. Leaves and roots were absent; the stem cortex (Fig. 31.9*B*) was probably photosynthetic, and rhizoids growing from the rhizome absorbed moisture and nutrients from the soil. Modern *Psilotum* shares many of these characteristics with *Rhynia.*

From this primitive, herbaceous start, evolution progressed rapidly to the first trees only 25 million years later. *Pseudosporochnus nodosus* was one of the first forest trees of the world, though comparison with trees of today would make us call it a sapling at best (Fig. 31.10). It had a main trunk up to 8 cm in diameter and 1 m tall, topped with a crown of finely branched limbs. There were no leaves. Xylem in the branches no longer was in a central bundle (as with *Rhynia* and *Cooksonia*); instead it exhibited a more complicated, dissected pattern.

There were several other tree species in the mid- to late-Devonian period, but they were much larger. *Aneurophyton,* for example, was up to 13 m tall. *Archeopteris* was more than 30 m tall and up to 1.5 m in diameter at the base (Fig. 31.11). Both had evolved cambium, and the capacity to produce considerable secondary xylem. Annual rings are absent from petrified wood

Figure 31.8
A reconstruction of *Cooksonia,* one of the first vascular plants. The above-ground portion, shown here, was 1 dm tall. (Redrawn from H. P. Banks, *Evolution and Plants of the Past.* © 1968 by Wadsworth Publishing Company. Reprinted with permission of the publisher.)

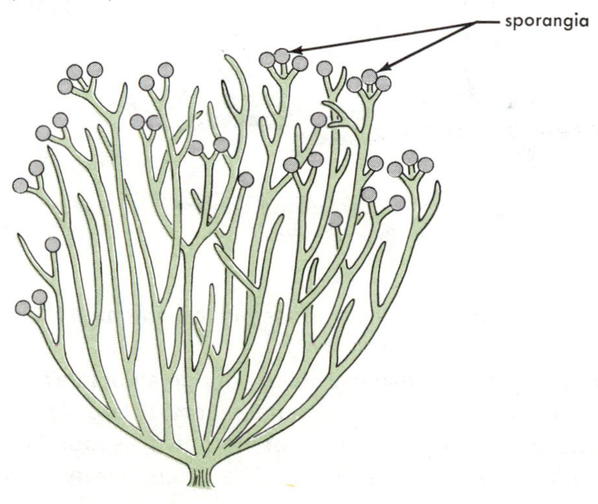

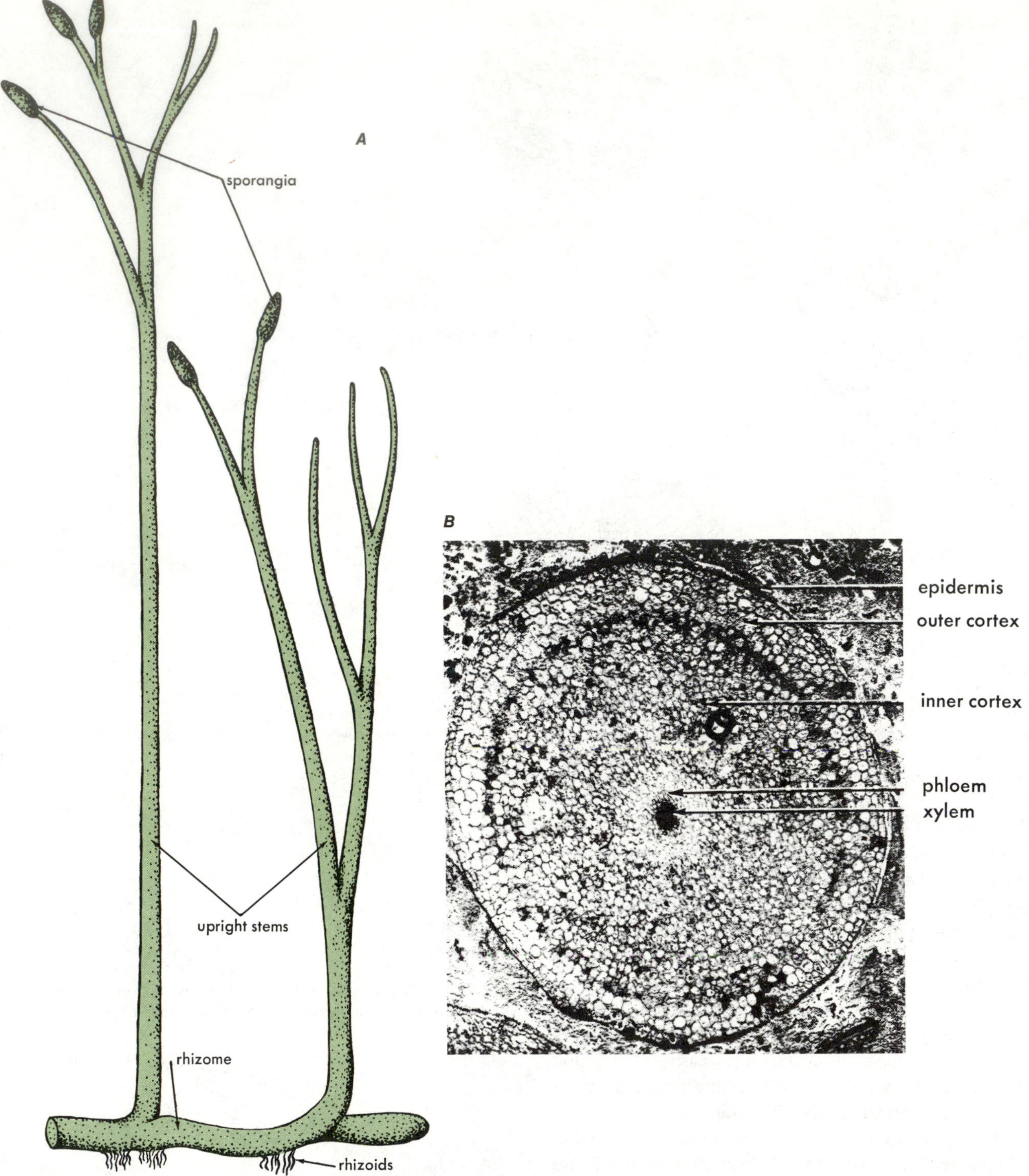

Figure 31.9
A, presumed habit of *Rhynia*, a plant that lived over 300 million years ago. *B*, cross section of stem. (Redrawn from R. Kidston and W. H. Lang, *Trans. Roy, Soc. Edinburgh* **52,** 603, 1919 and 831, 1921.)

samples, however, so it is not possible to determine their age.

Leaves had also evolved from very small epidermal outgrowths of the primitive vascular plants, to larger true leaves having a vascular system of their own, as found in *Aneurophyton* and *Archeopteris*. The "true" leaves are called **megaphylls;** the epidermal outgrowths are called **microphylls.** Some botanists theorize that megaphylls evolved as modifications of branch systems: starting with a system of bifurcating branches, one section of branches first becomes reduced in size (it is overtopped by the other portions); then the branches come to lie all

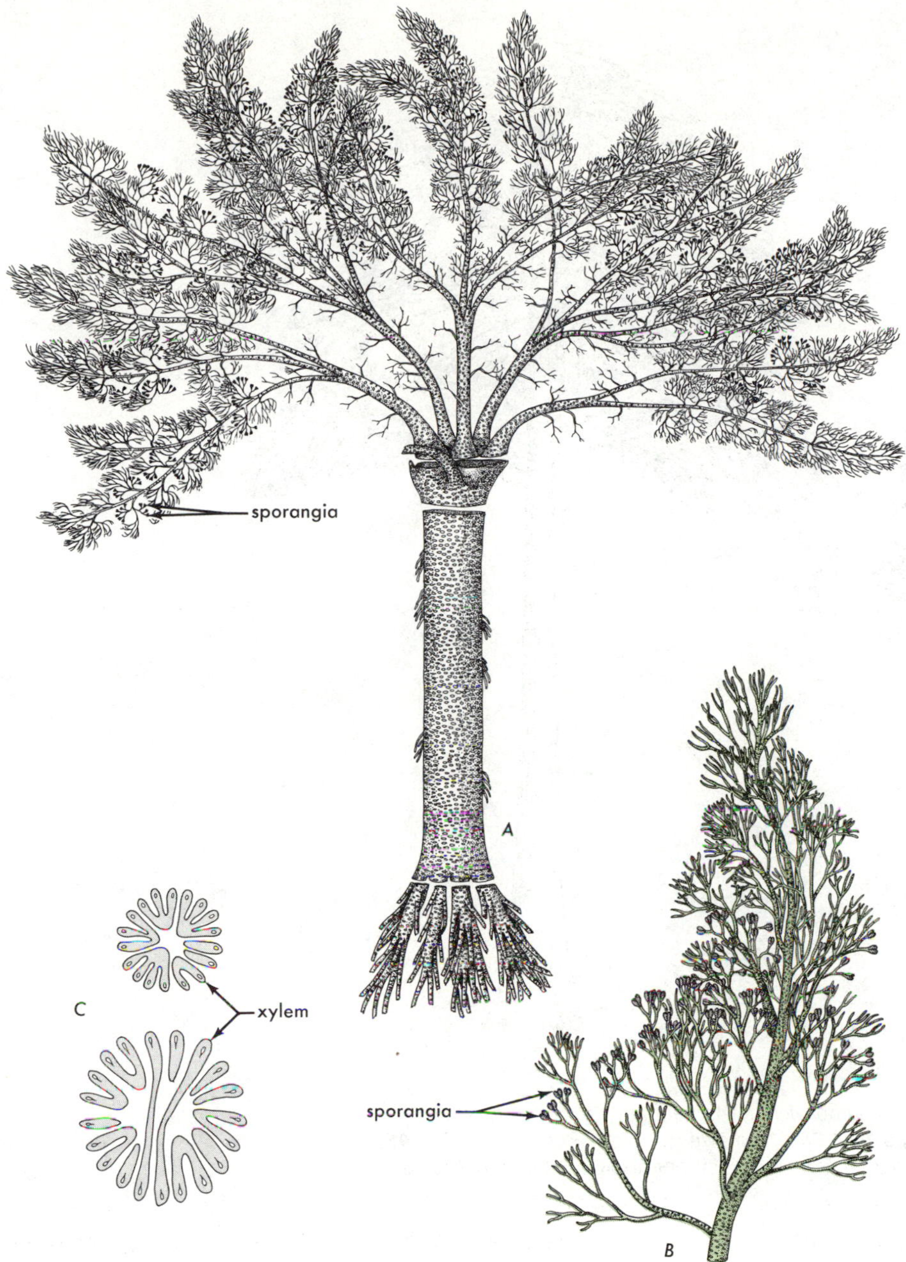

Figure 31.10
Pseudosporochnus nodosus, one of the first trees. *A*, reconstruction of the entire plant, about 1 m tall. *B*, detail of a terminal branch with sporangia. *C*, xylem pattern in cross sections of small and large branches. (Redrawn from S. Leclerq and H. B. Banks, *Palaeontographica* **110**(B), 1, 1962. Reprinted with permission of the publisher.)

in one plane; then epidermal outgrowths from marginal meristems connect the branches, much like webbing on a duck's foot. The result is a leaf blade with many veins (Fig. 31.12). Megaphylls, then, are considerably different in form and evolution from the superficial microphylls that were on the first vascular plants and that still exist on some modern lower vascular plants. Megaphylls today are only found in ferns and seed plants.

The evolution of large plants created another problem for paleobotanists. They rarely find a whole plant preserved as a unit. Leaf impressions may be found in one location in 1910, petrified trunk fragments somewhere else in 1932, and spores or roots still elsewhere at other times. Until the fragments can be connected by collections that show the parts attached to each other, each piece gets its own taxonomic name: a **form genus.** For example, *Archeopteris* at first was only known by fossil branch systems, and this name was the form genus for that part of the unknown whole plant. *Callixylon* was the form genus for petrified wood fossils that all had the same anatomical details. Only in 1960 was a fossil frond found attached to a fossil stem and then it was possible

Figure 31.11
Reconstructions of large Devonian trees. *A, Aneurophyton. B, Archeopteris,* about 30 m tall. (*A,* redrawn from W. Goldring, *Sci. Monthly* **24,** 515, 1927. *B,* redrawn from C. B. Beck, *Am. J. Bot.* **49,** 373. © 1962 by the Botanical Society of America. Reprinted with permission of the publishers.)

to put the two together. *Archeopteris* was the name adopted for the entire plant.

Coal Age Forests and Seed Plants (345 to 280 Million Years Ago)

Plant evolution during the Devonian period not only resulted in a modification in the vegetative size of plants, it affected their reproduction as well. Early vascular plants were homosporous, but *Archeopteris,* in the late Devonian period, was heterosporous. Two sizes of meiospores were formed: many small ones that produced the male gametes, and a few large ones that produced the female gametes. With four additional steps, heterospory can lead to the development of seeds: (*a*) the number of functional female meiospores (megaspores) produced inside the megasporangium is reduced to one; (*b*) the single megaspore is not shed from the megasporangium, but is retained within it; (*c*) the fertilized egg, or some later stage in embryo development becomes dormant; *(d)* protective structures (integuments) grow around the old megasporangium (with its megaspore or later, the embryo).

The first seeds developed on plants during the upper Devonian period. A reconstruction of *Archeosperma* (a form genus for the seed) appears in Fig. 31.13. Outer branches were beginning to fuse and form a protective layer around the several seeds. Unfortunately, we do not know what kind of plant produced this seed. It is not until the Coal Age (Mississippian and Pennsylvanian epochs, also known as the Carboniferous period) that seed plants become better known.

Extremely lush, swamp forests dominated the **Coal Age** landscape (Fig. 31.14). Because the land was low, minor changes in sea level successively inundated, buried, then supported one forest after another. The

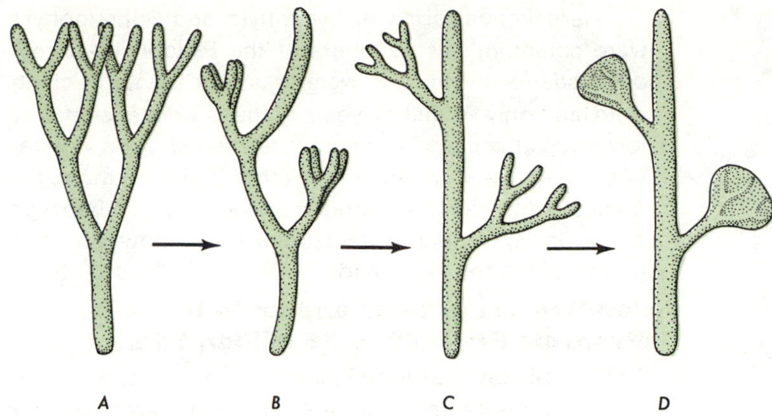

Figure 31.12
Hypothetical origin of true leaves (megaphylls). *A*, bifurcating branches. *B*, overtopping. *C*, the small branches lie in a plane. *D*, webbing occurs. (Redrawn from G. M. Smith, *Cryptogamic Botany*, Vol. **II**, *Bryophytes and Pteridophytes*. © 1955 by McGraw-Hill Book Company. Reprinted with permission of the publisher.)

buried organic remains have become compressed and changed through time. Today, they form the coal reserves of Europe and the eastern United States. Most oil and gas and other coal deposits, however, date to the Mesozoic and Cenozoic. These sources of energy are the chemically changed, fossil remains of Coal Age forests, and for that reason they are referred to as **fossil fuels**. Burial of plants and transformation into fossil fuel has taken place all through time, but not to the extent it did during the Coal Age.

These forests were dominated by Lycophyta. In the Devonian period, this division evolved rapidly into both herbaceous and woody types; in the Coal Age, the woody types reached tree stature. One example is *Lepidodendron* (Fig. 31.15), up to 35 m tall and with a trunk 1 m across. Straplike leaves and sporangia occurred near the ends of the branches. A cross section of the trunk revealed pith, primary and secondary xylem, cambium, and an enormous amount of cortex and cork. Very little of the stem area served for conduction or strengthening. The roots were dichotomously branched.

Second in abundance were giant horsetails, such as

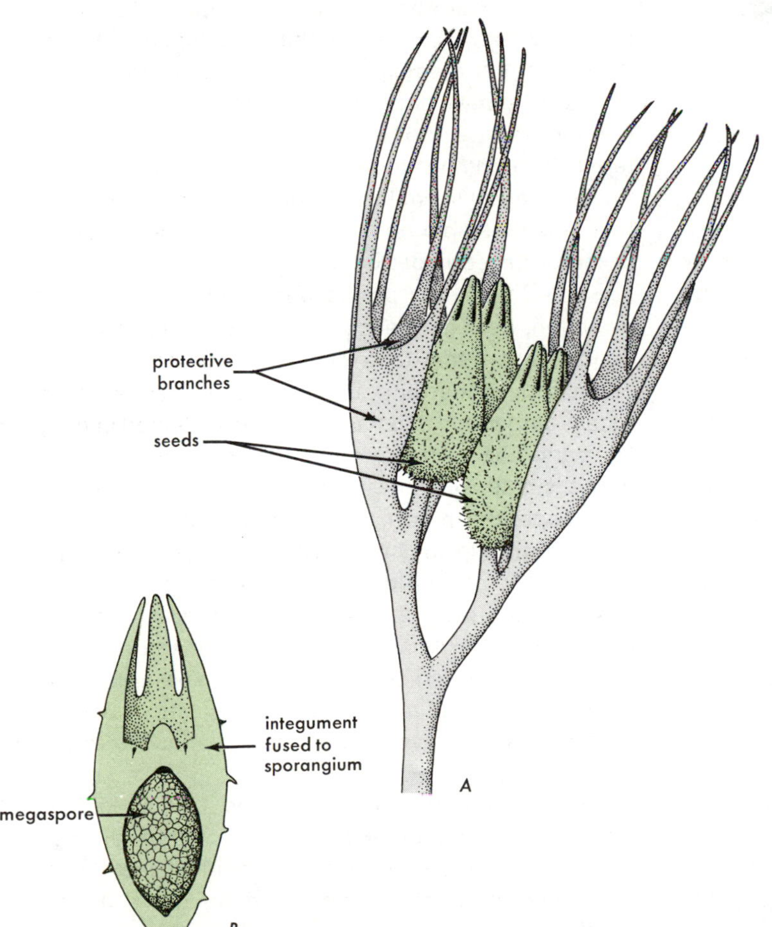

Figure 31.13
Reconstruction of the first seed, *Archeosperma*. *A*, cluster of seeds within protective branches. *B*, section of one seed. (Redrawn from H. P. Banks, *Evolution and Plants of the Past*. © 1968 by Wadsworth Publishing Company, Belmont, California. Reprinted with permission of the publisher.)

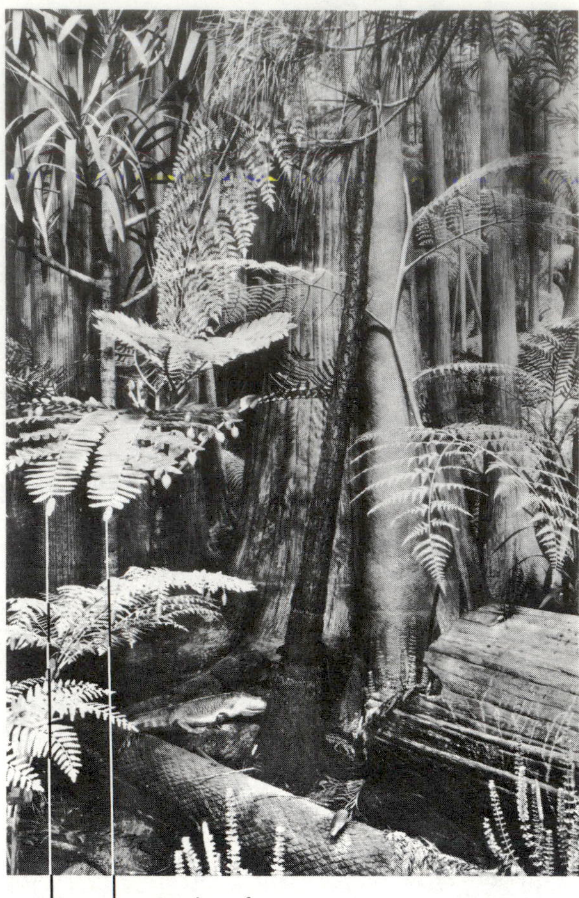

Figure 31.14
Reconstruction of a Coal Age forest. Note the seeds attached to fernlike leaf in left center. (Courtesy of Field Museum of Natural History.)

Calamities (Fig. 31.16). *Calamites* was smaller than *Lepidodendron,* but still was 9 m tall and had a trunk about 30 cm in diameter. Whorls of branches, and smaller branches from those, and leaves from those, arose at nodes. The upright stems developed from a horizontal rhizome system.

Third in abundance—not tall, but dominating the forest floor—were ferns and seed ferns. Seed ferns were fernlike but, in addition, possessed seeds on their fronds instead of spores (Fig. 31.14). Exactly what the ferns evolved from is not clear, and their fossil record began in late Devonian. Some botanists think *Archeopteris* of the Devonian period is a likely ancestor of the ferns because of its fernlike branch systems, but other evidence tends to show that *Archeopteris* is more closely related to gymnosperms. Perhaps ferns and gymnosperms both came from the same stock?

Gymnosperms were fourth in abundance. But many groups make up the gymnosperm category (including the seed ferns), according to some taxonomists. Two groups of gymnosperms evolved in the Coal Age, precursors of the Cycadophyta and Coniferophyta (Fig. 31.17). Gymnosperms did not become abundant and modern in appearance until 50 million years later.

Herbaceous forms of Lycophyta and Sphenophyta were common, but members of the Psilophyta—direct descendants of the first, weak, vascular plants to climb onto land only 70 million years earlier—were absent. The lower vascular plants dominated terrestrial world vegetation during the Coal Age just as thoroughly as the algae dominated the aquatic world in previous ages. The reign of the lower vascular plants was short, however, and great extinctions lay ahead.

Permian Extinctions and on to the Mesozoic Era (260 to 65 Million Years Ago)

The 375 million-year-long Paleozoic era was marked with two concentrated periods of evolution:a 25-million-year period of creativity in the lower Devonian period during which weak vascular herbs became forest trees; and a 45-million-year period of major extinctions in the Permian period. The creativity could have been associated with adaptation to land and to increasing oxygen in the air; the extinction may have been caused by a cooling and drying climate and uplifting of the land (Fig. 31.18). In many ways, the rise and fall of the Coal Age forests is just as striking and mysterious as the rise and fall of the dinosaurs many years later. The Permian extinctions marked the start of the **Mesozoic** era.

The plants that replaced the lower vascular plants in dominance were gymnosperms: the true conifers (Coniferophyta), Cycadophyta, and Ginkgophyta. Most of the Mesozoic era was the age of the gymnosperms. However, the group that dominates the world today, the flowering plants, must have been evolving all through the Mesozoic era. Pieces of wood, leaf impressions, and pollen scattered through the geologic record as far back as the Triassic period seem to be angiosperm in nature. But the first uncontestable appearance of fossil angiosperms is in the Cretaceous period.

The Cretaceous period was a third interval of rapid evolution, perhaps because of major climatic and geologic changes that took place and adaptation to insect pollinators. According to the best evidence we have today—and much of it has only been collected in the past 20 years—the world's continents were once connected into a single land mass near the equator. During the Triassic period, this primeval mass, called Pangaea, began to separate (Fig. 31.19). North and South America split from Europe and Africa, and the Atlantic Ocean formed; Australia and India and Antarctica separated, and India rammed into the Asian continent; Saudi Arabia split from Africa, and the Red Sea was formed. These movements did not happen all at once, but at the rate of only a few centimeters each year, and they continue today. The climate cooled.

Two major biological changes that accompanied these great continental and climatic shifts during the Cretaceous period were (*a*) a spread of the flowering plants and (*b*) an extinction of the dinosaurs. A number of other changes in the plant world accompanied all this (for example, the extinction or near-extinction of several

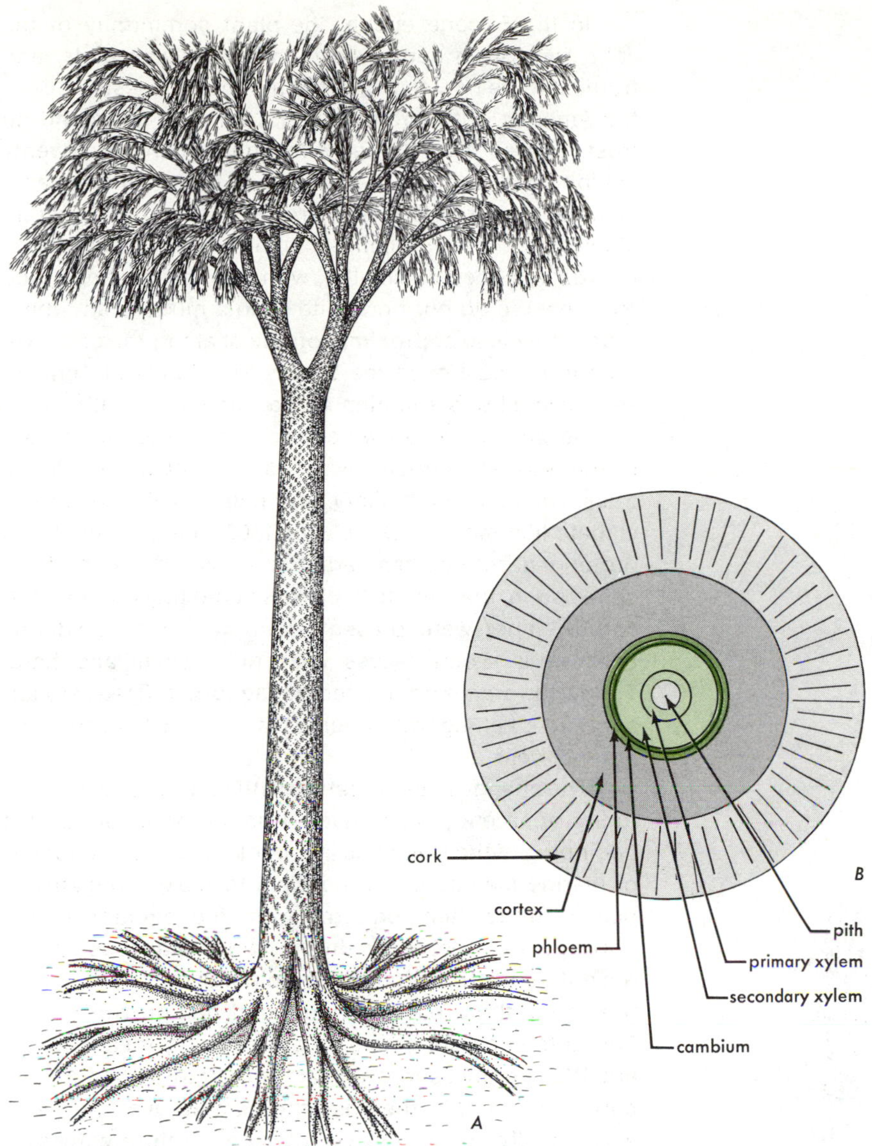

Figure 31.15
Lepidodendron, a dominant of the Coal Age forest. *A,* reconstruction of the entire plant, about 30 m tall. *B,* a cross section of the trunk. (Redrawn from M. Hirmer, *Handbuch der Palöbotanik* 1, 182. Munich, Germany, 1927. By permission of R. Oldenbourg Verlang GmbH, Munich.)

gymnosperm groups), but the rise of the angiosperms was the greatest change. Where had they come from, and why did they evolve so quickly? We may never know the answers. One theory is that flowering plants evolved over a long period of time. This took place in tropical uplands where fossil preservation in sediments is rare. These early angiosperms may have occupied warm, seasonally wet, rocky slopes with great variations in microhabitats because of differences in exposure, elevation, drainage, and soil type. This environmental variability could have been a stimulus to evolution. By the time the Cenozoic era began, 65 million years ago, the conifers dominated only the temperate and polar regions, and to the angiosperms went everything else.

Climatic Change in the Tertiary Period (65 to 2.5 Million Years Ago)

We have already seen in Chapter 18 how fossil assemblages of plants can be used to estimate past climate. Quite simply, the theory is that the present is the key to the past. Where one finds modern relatives of fossil plants today, one finds a climate that must have been similar to the one at the time and place where the fossils were deposited. A good example of the fossil plant record for western North America during the Cenozoic is in the John Day Basin of northeaster Oregon. If the present is indeed the key to the past, this fossil record shows a cooling and drying climatic trend that has continued to the present.

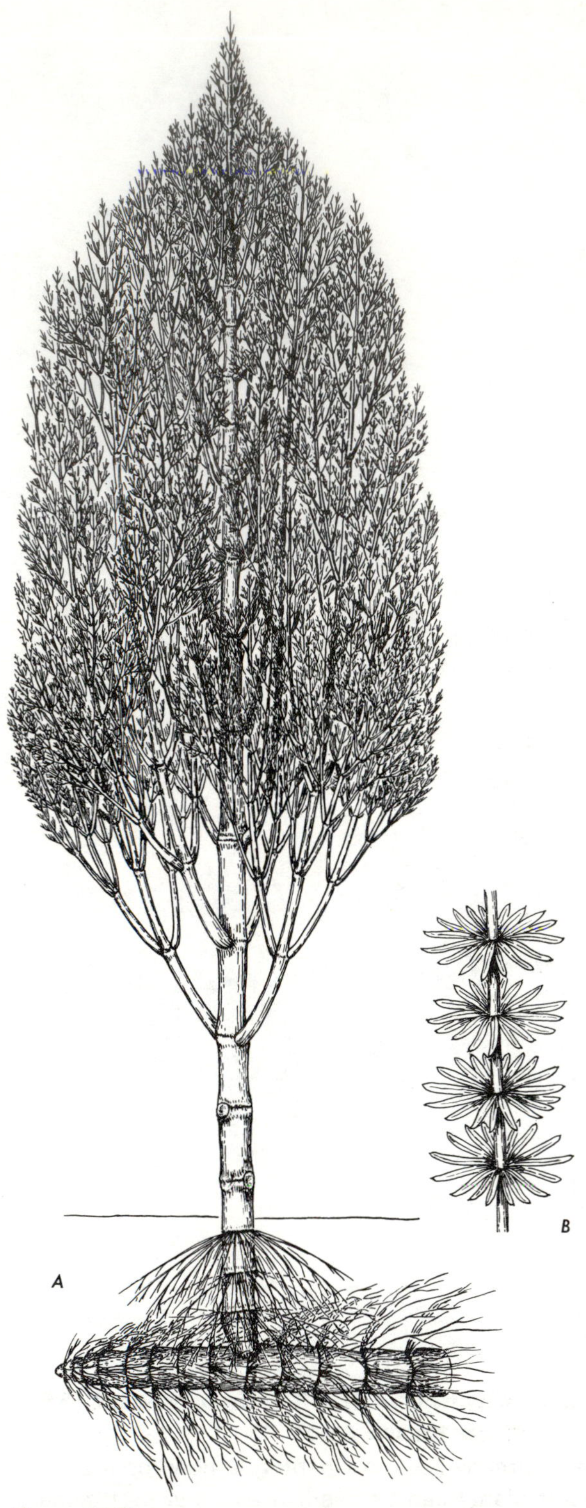

Figure 31.16
Calamites, another Coal Age forest tree. *A,* reconstruction of an entire plant, including rhizome and roots, of *C. carinatus*. *B,* detail of whorled leaves of the form species *Annularia radianta,* which are very similar to the leaves of this species of *Calamites*. The leaves are about 1 cm long. (Redrawn from M. Hirmer, *Handbuch der Paläobotanik* 1, 409, Munich, Germany, 1927. By permission of R. Oldenbourg Verlang GmbH, Munich.)

In the Eocene epoch, the plant community of the John Day Basin contained cinnamon, fig, cycads, and tropical ferns now found in the mountain forests of Central America, with a rainfall of 1500 mm or more and no frost. Leaves were large, with entire margins. Twenty million years later, in the Oligocene epoch, there was a change to mixed conifer-hardwood forest with birch, alder, oak, redwood, elm, sycamore, beech, and others. Leaves had become smaller, with dentate or convoluted margins. We do not find quite such a mixture anywhere today, but close approximations exist along the cool, wet California coast or in the Smoky Mountains of Tennessee. Rainfall was still high at that time, about 1250 mm, but the climate had grown cooler. Ten million years later, in the Miocene epoch, there was a shift to deciduous trees, such as oak, hickory, and maple. This indicates a climate like modern Ohio, with 1000 mm of rainfall and prolonged freezing temperatures in winter. By the Pliocene epoch, the climate became so cold that conifers (fir, spruce, pine) were present along with the deciduous hardwoods. Some leaves were quite small and hard. Today, the area is dominated by sagebrush. Trees are absent except along watercourses, and rainfall is about 250 mm per year.

The change in temperature with latitude that we have today apparently was much less pronounced at the beginning of the Cenozoic era. Looking at fossil deposits of marine plankton, it is possible to make estimates of ocean surface temperatures from the equator to the north pole at past times. In the mid-Cretaceous period, the difference from equator to pole was only 11°C, but by the end of the Cretaceous period, it had grown to 15°C. The difference continued to grow during the Cenozoic era. We can guess that similar temperature changes occurred on land. In the past, the climate over the globe was broadly zoned and without temperature extremes, but today many climatic zones exist between the pole and the equator. Fig. 31.20 shows how these zones may have formed and shifted during the last 40 million years. During Pleistocene glaciation, the temperature gradient was at a peak.

Climatic Changes in the Quaternary Period (2.5 to 0 Million Years Ago)

The Pleistocene **Ice Age,** which ended only a moment in geologic time ago—12,000 years or so—produced another period of extinctions. Glaciated areas were scraped clean of plants, and they are still today being slowly revegetated. The pattern of change in the north is not uniform, however. There is evidence from central Canada, for example, that timberline there is actually being pushed south, and the tundra is advancing at a rate of 400 km in the past 900 years; but in other areas of the world, glaciers are still retreating from a "minor ice age" in the fifteenth century. Pollen diagrams, such as the one shown in Chapter 18, are good illustrations of climatic and vegetation changes that have continued through the

Figure 31.17
Reconstructions of some now-extinct gymnosperms. *A,* member of the Cordaitales, 9 to 30 m tall from the carboniferous. *B* and *C,* members of the Bennetitales. (*C* was about 2 m tall and is from the Jurassic and *B* is from the Cretaceous and drawn to the same scale as *C.*) (*A, Dorycardaites.* Redrawn from D. H. Scott, *Studies in Fossil Botany.* © 1920 by Adam and Charles Black, London. *C, Williamsonia.* Redrawn from C. A. Arnold, *An Introduction to Paleobotany.* © 1947 by McGraw-Hill Book Company. Reprinted with permission of the publisher.)

past 12,000 years.

Many large mammals disappeared from North America during and after glaciation. They include the mastodon, mammoth, giant ground sloth, llama, peccary, giant armadillo, sabertooth tiger, ox, yak, and horse. Some of these survived in other parts of the world, but many became extinct. A debate has raged over the cause of the extinctions, one side attributing them to climatic change and stress, another side attributing them to the hunting prowess of emerging man.

In the quaternary, conifers have become more and more restricted in the land area they dominate. The flowering plants, however, continue to evolve, especially in harshly cold or dry environments. Today, only two groups of vascular plants appear to be expanding in terms of diversity and abundance: ferns and angiosperms. Both have survived great physical changes of the earth, yet their evolution seems to have been stimulated by these stresses, while many other forms have become extinct or survive as tenuous remnants.

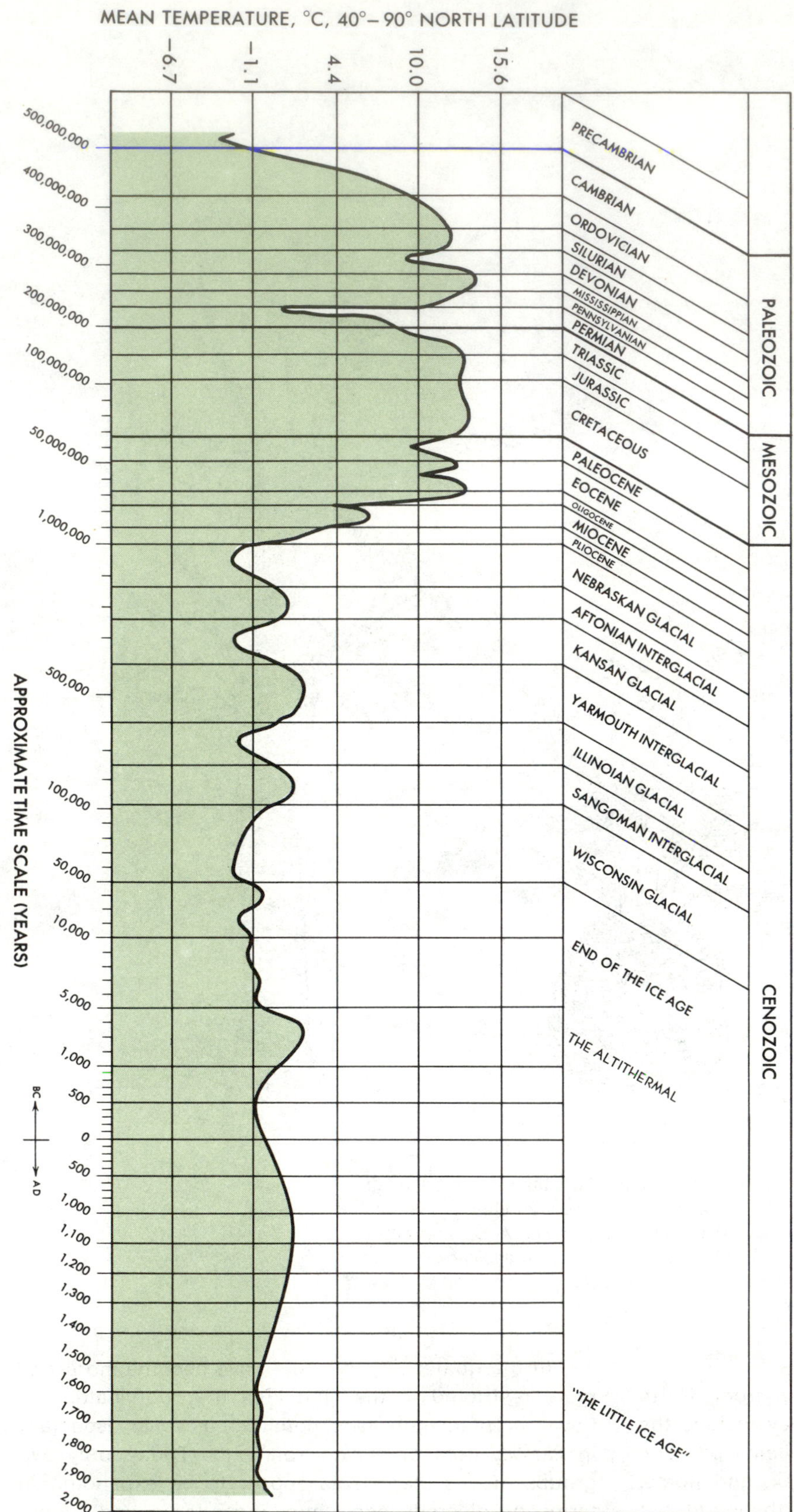

Figure 31.18
Average temperature during geologic time for the earth at 40-90° north latitude. The time scale is distorted to show more detail for the last 1 million years. (Redrawn from E. Dorf, *Am. Sci.* **48,** 341. © 1960 by Sigma Xi. Reprinted with permission of the publisher.)

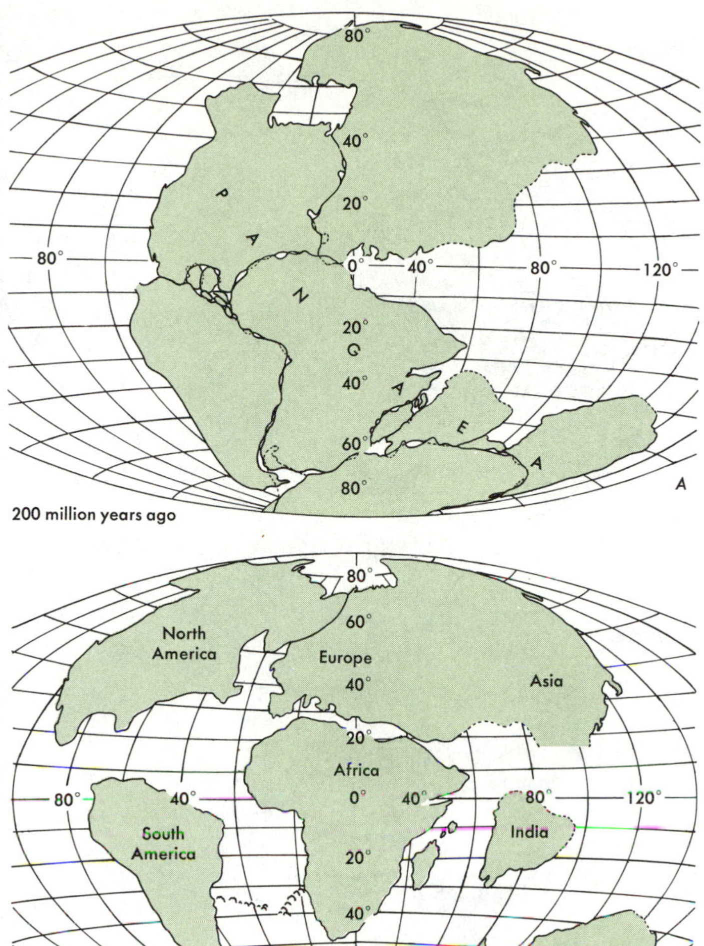

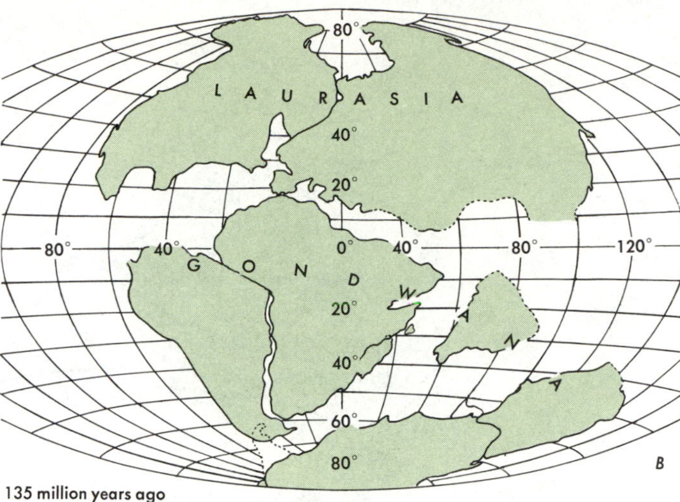

Figure 31.19
Continental drift. *A*, position of the continents in the lower Triassic period. *B*, lower Cretaceous period. *C*, Paleocene epoch.

Man, Cultivation, and Weeds

Man himself began to play a role in the evolution and distribution of plants by the end of the Ice Age. Although the earliest archeological evidence for seed agriculture (cultivation of annuals such as grain, beans, squash) takes us back to 9000 BC, humans may have been cultivating root crops (perennials propagated by cuttings and harvested for starch in the "roots" of cassava, sweet potato, and taro) in southern Asia as long ago as 13,000 BC. Some of these root crops have been asexually propagated for so many years that they have lost or nearly lost the capacity for sexual reproduction, and it is doubtful that they could survive in nature. Annuals have also been selected for productivity, rather than natural survival, and the result is they have become so changed from their wild relatives that it is difficult to determine where, and from what stock, they were first domesticated. Man has also accidentally domesticated and favored the evolution of certain other plants, the weeds. These are plants that grow well in disturbed or trampled soil, in waste areas rich in nitrogen, or in cropland. Some have evolved seeds that are the same size as crop seeds, so they may not be separated during the threshing or sieving of crop seeds. When the next season's crop is sown, the weeds are sown inadvertently right along with it.

In our travels, we have taken these domesticated plants with us. Some strains of crop plants are so distinctive and peculiar to a given group of people, that their past migrations can be traced by determining the present location of the crop plants. This sort of evidence for ***ti*** plant, taro, and candlenut has been used to trace the movement of Polynesians from island to island in the South Pacific.

Weeds, of course, also are taken along inadvertently by moving people. In modern times, Europeans coming to the New World brought along both crops and weeds. As much as a third of the flora of some states is composed of weeds from various parts of the world, mainly Europe. The accelerated pace of land disturbance, which opens the way for weeds, combined with the planting of the same crops and ornamentals everywhere, are making the floras of the world increasingly similar.

CHARLES DARWIN AND EVOLUTION

Many biologists since the time of the Greek philosophers have attempted to explain the diversity of organisms. Great progress has been made during the nineteenth and twentieth centuries. Our better understanding was made possible chiefly because the work of Charles Darwin (see later passages in this chapter) and Gregor Mendel provided a basis that enabled biologists to approach the problem of diversity in an entirely new light.

Darwin, as a result of many years of careful study and observation of a large number of plants and animals, laid emphasis upon the following points:

Figure 31.20
Climatic zones in North America during the past 40 million years. *A*, Oligocene. *B*, Miocene. *C*, Pliocene. *D*, at the peak of the Pleistocene Ice Age. *E*, present. (Redrawn from E. Dorf, *Am. Sci.* **48,** 341. © 1960 by Sigma Xi. Reprinted with permission of the publisher.)

1. Numbers of plants or animals may increase in a geometric ratio. For example, a given plant may produce 1000 seeds. If each seed grows into a new plant and each new plant produces 1000 seeds, 1 million new plants could result. A third generation from the one original plant would result in 1 billion plants. It may be seen from this example

that plants (and animals also) potentially could increase in numbers at a tremendous rate.

2. Actually, the number of individuals of a given plant or animal remains fairly constant. There is no such tremendous increase in the number of individuals as seed production seemingly makes possible.

3. No two individual plants or animals are identical; there is variation.

Reasoning from these observations, Darwin arrived at these conclusions:

1. Any given population is usually able to reproduce many more young individuals than can adequately be raised in the region it occupies. Therefore, a struggle for existence occurs among the individuals.

2. In the struggle for existence, only those individuals survive that, because of their particular variation, are best adapted to their immediate environment. Thus, a natural selection takes place; the unfit do not survive.

3. These selected variations may be inherited, that is passed from one generation to another, and thus may gradually give rise to new species.

The Sources of Variation

Since variation plays such an important part in plant evolution, it will be well to examine the sources of variation a little more closely.

One source of variation is **mutation,** the spontaneous transformation of a gene (Fig. 31.21). Mutation does not create new genetic material (new DNA), it simply changes the arrangement of the existing material so that enzymes and other proteins coded for by the DNA are no longer quite the same. A mutated gene can back-mutate to the original condition, as well. A number of environmental factors, such as radiation, can cause a mutation; but most mutations seem to result from cellular "mistakes" in copying the DNA molecule during cell division.

Mutations are a universal fact of life. They are known to occur in every plant and animal that has been studied. This does not mean that a given gene is relatively unstable and is likely to mutate very often. Rates of mutation for an individual gene, or locus, probably average one for every 100,000 cells. The mutation rate does vary from gene to gene, however, as shown for corn (Table 31.2), where one gene may have a mutation rate 500 times that of another.

However, one mutation in every 100,000 copies is still a small input of variation. Is it significant enough to account for major evolutionary change through time? Is it significant enough to create a new, quite different species, from a preexisting one? If we do a little multiplying, the answer is, yes. Consider that, although a single gene may mutate only once every 100,000 copies, there are probably 10,000 genes in one plant. Each one can

Table 31.2
MUTATION RATES OF DIFFERENT GENES IN CORN *(zea)*

Gene	Number of Mutations per 1,000,000 Gametes
Seed color, not purple	492
Seed color inhibitor	106
Purple seed color	11
Sugary seed	2.4
Shrunken seed	1.2
Waxy seed	less than 0.1

mutate independently of the others, so the chance of one gamete, or sex cell, containing any mutant gene is reduced to one in ten (1/100,000 genes mutated × 10,000 genes). Most mutations are not beneficial and might be so detrimental that they would not be passed on to future generations. Let us assume that only one mutation out of a thousand is beneficial. This means that one gamete in 10,000 has a beneficial mutation (1/10 gametes with any mutation × 1/1000 beneficial gene mutations).

How many gametes, or plants in a species? Let us give one gamete to each plant, and estimate an average species size of 1,000,000 plants. The gametes fuse and produce a new generation. This new generation will then have 100 new, useful mutations scattered among its members (1/10,000 beneficial mutants per gamete × 1,000,000 gametes, or parents). A few mutations will be identical and a few will appear in a single individual, but most will be unique and scattered through the population. The rate of mutation will be repeated each generation. In 100 generations, the total number of new mutations could be as high as 10,000—the same number of mutations as original genes. Of course, some mutants will back-mutate, and some mutations will be identical, so there will be somewhat fewer than 10,000 separate mutations. How long a time is 100 generations? For an annual plant, like sunflower, it is only 100 years; for a large tree, it might be 30,000 years. Even 30,000 years is well within the time limits of species evolution as revealed in the fossil record.

But can mutation provide enough variation to account for the great differences between larger taxonomic groups? Great discontinuities seem to exist between families, classes, and divisions of plants, but these discontinuities are more apparent than real, and result from incomplete information. The more study that is done, the more intermediate forms (fossil and living) are found and the fuzzier the distinctions become between the major groups.

Other sources of variation have been discussed in Chapter 17. These include chromosome aberrations such as deletions, duplications, inversions, translocations, and crossing-over. A fine powerful source of variation is **recombination,** the reshuffling of chromosomes during sexual reproduction.

Figure 31.21
Examples of mutations. *A*, common wild sunflower. *B*, mutant sunflower known as sun gold. *C*, variation in the seed coats of beans. (*A* and *B*, courtesy of C. Heiser; *C*, courtesy of F. Smith.)

The Role of Natural Selection

Mutation and recombination are the raw materials for evolution. They provide the variation among *individuals.* However, for an entire population of individuals in a species to progressively change into something new, this raw material is not enough. Some force, some pressure, must be exerted on the *population* so that the abundance of certain mutations becomes higher and higher, until all or most members of the species possess it. How does this happen? The presently accepted answer had its own evolution, over the course of 150 years of debate and experimentation.

At the beginning of the nineteenth century, the French naturalist Jean Baptiste Lamarck (1744-1829) was a liberal—for his day—in his thoughts on evolution. Lamarck did not believe that all species had the same age, that all were created together at one time. Neither did he believe that species were constant; new ones were forming all the time as a result of changing environments. This stand was quite heretical, considering that Linnaeus 50 years earlier had practically founded "modern" taxonomy on the principles that there was only one time of creation, and that species from that time on were fixed and constant. Lamarck thought that new traits and new species could evolve from old ones if the species were placed under stress (Fig. 31.22). For example, if a tall plant at sea level were transplanted to a severe, timberline habitat, the climate would stunt it. This stunted plant would shed seeds and the new seedlings would also be stunted, even if grown back at sea level. Characters acquired during the lifetime of one plant would be passed on to succeeding generations.

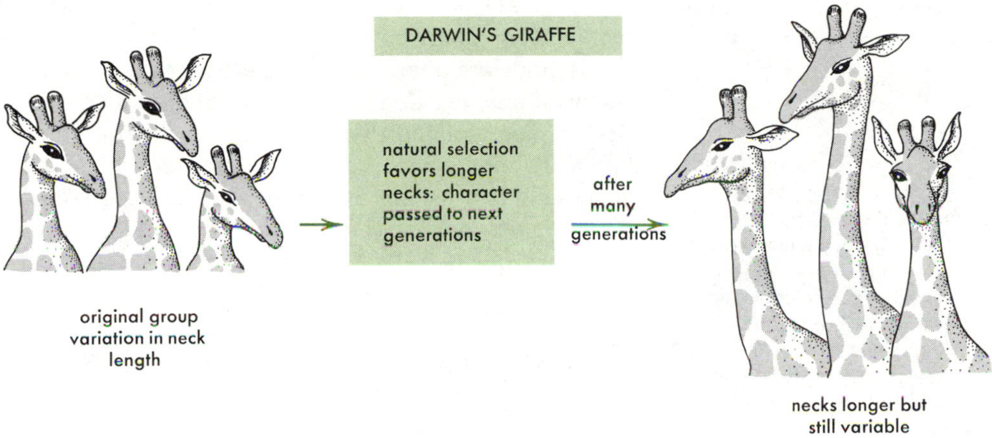

Figure 31.22
Comparison of Lamarck's and Darwin's concepts of evolution. (Redrawn from J. M. Savage, *Evolution.* © 1963 by Holt, Rinehart and Winston. Reprinted with permission of the publisher.)

Lamarck himself did no experimentation to bolster his theory. Other botanists did, however. Bonnier in France made reciprocal transplants of alpine and sea-level species. He grew a number of plants of each species to a convenient size at Paris, then cloned them (split them into pieces) and planted them in plots in the Alps, the Pyrenees, and Paris. They were not planted in gardens, but were put in among natural vegetation. The plots were sometimes fences, sometimes not; watering and fertilization were not practiced. The plots were periodically visited for a number of years (1884-1920) and any changes in the plants noted. He concluded that a number of lowland species *were* transformed into related alpine or subalpine species during the years of the study. Also around 1920, the American ecologist Frederic Clements established a similar series of "gardens" from Pike's Peak to the California shore. He too concluded that some lowland species were transformed to related high-elevation species, and vice versa.

However, the most carefully documented and controlled transplant experiments were conducted last, between 1920 and 1940, by Clausen, Keck, and Hiesey in California (Chapter 19). Their detailed study of some 60 taxonomically diverse species showed not one case of a lowland species becoming an alpine species, or the reverse. A lowland species at timberline might become dwarf or prostrate or stunted, but if its seeds were collected and sown back at sea level, normal tall plants would result. There had been no genetic change—only a temporary, plastic response to a harsh environment. Such temporary changes are called **phenotypic** changes, in contrast to **genotypic** changes that are genetically fixed and are passed on to offspring.

Darwin's concept of evolution, expressed over a hundred years ago, is the one accepted today; not Lamarck's. Its differences with Lamarck's are illustrated in Fig. 31.22. Basically, Darwin's concept is this: Variation exists in the initial population; an environmental stress places certain individuals at an advantage; because those individuals survive or reproduce more successfully, they leave more offspring that carry the same genetic traits; the abundance of the advantageous traits in this way increases in every generation, but variation still persists. This process of directed change is called **natural selection.**

How Species Remain Distinct

Evolution proceeds most rapidly in a population of organisms if it is "isolated" from other populations. But we mean a very special kind of isolation: **reproductive isolation.** If plants in one isolated population cannot breed with similar plants in other populations, then natural selection will produce a distinctive species in the shortest amount of time. This species will remain distinct from all others, once it has formed (Fig. 31.23). Without isolation, crossbreeding would dilute the abundance of new genes that are of most value in one particular environment; all members of a species would continue to be more or less alike and not as well-adapted to extremes within their range, as they could be if isolated. Mediocrity—a population of generalists—would result. One large population of generalists is apparently not as successful a strategy for survival as several small populations of specialists.

How can reproduction isolation be established? Table 31.3 lists the most important isolating mechanisms. The prezygotic mechanisms have to do with preventing pollen of one population from reaching the stigmas of another. This can be accomplished by separating the populations in space (isolated valleys or separating continents), time (different seasons of flowering), or biology (different pollinating insects, which do not visit both flowers). On a world scale, continental drift may have been the most important isolating mechanism. Postzygotic mechanisms prevent normal offspring from developing, even though pollination occurs. The hybrids may be weak or sterile, for example, or the developing embryos can abort.

Hybrids are often weak in nature, or at least they are less well-adapted to a given habitat than either parent. So the most common type of hybrid in nature is not a "pure" hybrid, halfway between each parent. Instead, the successful hybrids are often the result of back-crossing between the original pure hybrid and either or both parental types (Fig. 31.24). The few pure hybrids in this

Figure 31.23
The sequence of events that leads to the formation of new races, subspecies, and species. One species (green, at top) extends into new habitats, and populations at the extremes become modified into races of subspecies. If the races or subspecies become isolated, they may evolve further from the original species and become incompatible with it; at this point, they are recognized as distinct species. The reproductive barrier will keep the species distinct, even if later migration brings them back into contact. (Redrawn from G. L. Stebbins, *Processes of Organic Evolution*. © 1979 by Prentice-Hall, Englewood Cliffs, New Jersey. Reprinted by permission of the publisher.)

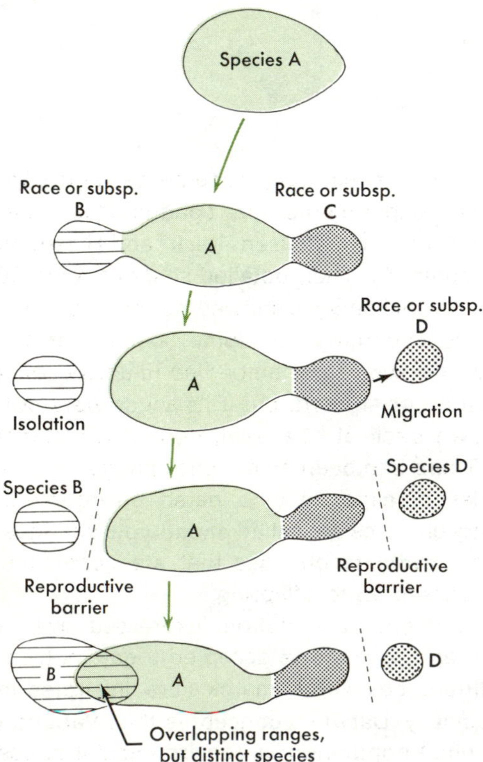

Table 31.3
SUMMARY OF SOME IMPORTANT REPRODUCTIVE ISOLATING MECHANISMS

A. Prezygotic mechanisms: prevention of pollination or fertilization

1. Separation in space: two populations live far from each other so that pollen cannot be transferred, or they live close but are separated by barriers (mountains, bodies of water), or they occupy different habitats (lowlands versus uplands), or drifting continents move them apart
2. Separation in time: two populations flower at different times of the year, or if they flower at the same time, the pollen is shed before the stigmas are receptive
3. Biological separation: the pollinating insect or animal is different for each population, so that it is unlikely cross pollination will occur, even if the plants are close to each other; or the flowers may open at different times of the day (morning versus evening)
4. Physiological differences: the pollen of one is unable to grow through the style of the other, or it grows more slowly than pollen of the other

B. Postzygotic mechanisms: prevention of normal offspring development

1. Incompatability of zygotic or embryonic tissue with that of the mother plant produces seed abortion.
2. Hybrids (F_1) are completely inviable, or considerably weakened.
3. Hybrids are vigorous but sterile.
4. F_1 is vigorous, but successive generations (F_2, etc.) are weak or sterile.

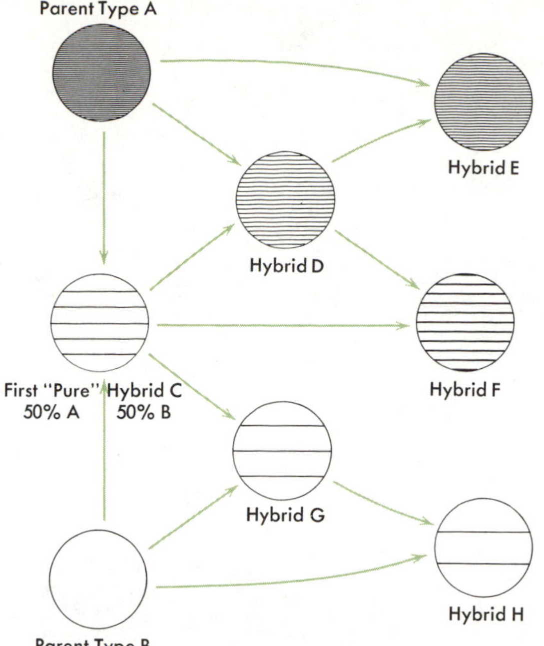

Figure 31.24
Introgressive hybridization. The degree of shading (horizontal lines in the circles) shows the degree of similarity of hybrids to parental types A and B. Introgressive hybrids D, E, F, G, and H would be more numerous in nature than "pure" hybrid C.

way produce an entire series of partial hybrids between themselves and original parental types. This is called **introgressive hybridization,** and it creates great diversity (hybrid swarms) that may spell success for the group as a whole.

Hybridization may be especially difficult to achieve if two populations of a species have come to differ in chromosome number. Polyploidy, as explained in Chapter 17, is a multiplication of the chromosome number. It appears to be a common phenomenon among plants, and it seems to increase their tolerance of climatic extremes. Perhaps it is no mere coincidence that the two vascular groups showing most rapid evolution today, flowering plants and ferns, also show the highest incidence of polyploidy. Exactly how or why polyploidy increases tolerance of environmental extremes is not known, but the high incidence of polyploids in arctic and desert habitats has been well-documented (Table 31.4). It has been estimated that 25 to 33% of the angiosperms are polyploid. Polyploids include some important commercial crops (cotton, wheat, oats, potato, tobacco, alfalfa, many pasture grasses, apples, olives) and ornamentals (poplars, violets, asters, and many ferns).

Table 31.4
INCREASING INCIDENCE OF POLYPLOIDY WITH INCREASING HARSHNESS OF THE ENVIRONMENT

Region	Polyploidy Species (% of All Vascular Plants)
West Africa (tropical)	26
Mediterranean shore	38
Great Britain	53
Iceland	66
Southern Greenland	71
Northern Greenland (edge of the tundra)	86

SUMMARY

A modern theory of the mechanism of evolution can be outlined as follows:

1. Genes, contained in the chromosomes, are largely responsible for the development, structure, and metabolism of plants and animals.
2. The complement of genes does not remain absolutely constant. Mutations (changes in chromosomes and genes) occur, which modify the structure and metabolism of the individuals containing them.
3. Mutations causing considerable phenotypic change are likely to kill or to greatly weaken the plant because they upset the delicate equilibrium existing between the plant and its environment.
4. Hybrids, because of the resulting new combinations of genes, differ from their parents. Hybrid swarms, resulting from introgressive hybridization, are common.
5. If the variants produced by mutations or hybridization are better adapted to the environment than the parent plants, the parent plants may be replaced by the new forms. Many complicated factors are involved in this replacement.
6. As mountains are elevated or eroded, as glaciers advance or recede, as the climate of the earth becomes more or less arid, or as the habitats of plants are changed in other ways, the plants best adapted to the new environment replace the forms less well-adapted.
7. Evolution thus results from slow changes in the earth's surface, variation in plants and animals, and an adjustment between the changes in the earth and changes in the living organisms.
8. In summarizing evolutionary mechanisms, Dobzhansky, writing in *Scientific American,* has said, "Evolution is a creative response of the living matter to the challenges of the environment. Evolution is due neither to chance nor to design; it is due to a natural creative process."

GLOSSARY

Technical Terms Including Their Origin

Abbreviation	Meaning
A.S.	Anglo-Saxon
D.	Dutch
dim.	diminutive
F.	French
Gr.	Greek
It.	Italian
L.	Latin
Lapp.	Lappland
M.E.	Medieval English
M.L.	Medieval Latin
N.L.	New Latin
O.F.	Old French
O.E.	Old English
R.	Russian
Sp.	Spanish

The glossary serves two purposes, (*a*) to define terms used in the text and (*b*) to give their derivation. It should be noted that in carrying out the first purpose, with few exceptions, only the meanings actually used in the text are given. Reference to dictionaries will give additional meanings, while other textbooks may define the same words in slightly different ways.

Scientific language and slang have much in common. They are both vital, growing, and changing phases of modern English. The same words will have different meanings in different parts of the country.

Scientific language grows in a number of ways. Some botanical terms, such as "seed," are used by scientist and layman alike and can be traced back to Anglo-Saxon days, when the word *sed* was used to designate anything sown in the ground, or, in our terms, a seed. In other instances, words from Greek or Latin have been taken into the scientific language with their identical original meaning; for instance, the Greeks called wood *xylon,* and we use this term, changed to xylem, to designate the woody conducting tissue of plants. Other words have a more complex history. *Metabolos* is the Greek word "to change." When the scientists of Europe wrote in Latin they took this Greek word, made a Medieval Latin word from it, and used it to describe the changes undergone by some insects. It was used in 1639 to mean changes in health, and in 1845 the German scientist who elaborated the cell doctrine made a German word of it and first used it as we use it here, designating the changes taking place within a cell.

In using the glossary, pay as much attention to derivation as you do to the definition. To memorize the definition alone is to learn, parrot-fashion, only one word. To learn the derivation is to understand the word and possibly to introduce yourself to a whole family of new words. Pay particular attention to such combining forms as *hetero, auto, micro, phyll, angio, plast* or *plasm, spore,* and the like.

Å abbreviation for an **angstrom,** 0.0001 of a micron; there are 10,000,000 angstroms in a millimeter, 10 in a millimicron, and 10 in a nanometer

ABA abscisic acid

Abscisic acid a hormone variously inducing abscission, dormancy, stomatal closure, growth inhibition, and other responses in plants

Abscission zone (L. *abscissus,* cut off) zone of delicate thin-walled cells extending across the base of a petiole, the breakdown of which disjoins the leaf or fruit from the stem

Absorb (L. *ab,* away + *sorbere,* to suck in) to suck up, to drink up, or to take in; in plant cells materials are taken in (absorbed) in solution

Absorption spectrum a graph relating the ability of a substance to absorb light of various wavelengths

Accessory (M.L. *accessorius,* an additional appendage) something aiding or contributing in a secondary way

Accessory bud a bud located above or on either side of the main axillary bud

Accessory pigment a pigment that absorbs light energy and transfers energy to chlorophyll *a*

Achene simple, dry, one-seeded indehiscent fruit, with seed attached to ovary wall at one point only

Acid (F. *acide,* from L. *acidus,* sharp) a substance that can donate a hydrogen ion; most typical acids are sour and are compounds of hydrogen with another element or elements

Actinomorphic (Gr. *aktis,* ray + *morphe,* form) said of flowers of a regular or star pattern, capable of bisection in two or more planes into similar halves

Action spectrum (F. *acte,* a thing done) a graph relating the degree of physiological response (e.g., phototropism, photosynthesis) caused by different wavelengths of light

Active solute absorption the intake of dissolved materials by cells against a concentration gradient and requiring an expenditure of metabolic energy

Adaptation (L. *ad,* to + *aptare,* to fit) adjustment of an organism to the environment

Adenine a purine base present in nucleic acids and nucleotides.

Adenosine triphosphate a substance formed in metabolism from ADP and inorganic phosphate. The most prominent and universal molecule that acts as a carrier of energy in metabolism.

Adhesion (L. *adhaerere,* to stick to) a sticking together of unlike things or materials

Adnation (L. *adnasci,* to grow to) in flowers, the growing together of two or more whorls to a greater or less extent; compare adhesion

ADP adenosine diphosphate

Adsorption (L. *ad,* to + *sorbere,* to suck in) the concentration of molecules or ions of a substance at a surface or an interface (boundary) between two substances

Advanced (M.E. *advaunce,* to forward) said of a taxonomic trait thought to have evolved late in time from some more primitive trait

Adventitious (L. *adventicius,* not properly belonging to) referring to a structure arising from an unusual place; buds at other places that leaf axils, roots growing from stems or leaves

Aeciospore (Gr. *aikia,* injury + spore) one of the binucleate asexual spores of rust fungi formed as a result of the sexual fusion of cells but not of nuclei

Aecium (plural, **aecia**) (gr. *aikia,* injury) in rust a sorus, on the upper surface of leaves, that produces aeciospores

Aerate to supply or impregnate with common air, such as by bubbling air through a culture solution

Aerobe (Gr. *aer,* air + *bios,* life) an organism living in the presence of molecular oxygen and using it in its respiratory process

Aerobic respiration respiration involving molecular oxygen

Agar (Malay *agaragar*) a gelatinous substance obtained mainly from certain species of red algae

Aggregate fruit (L. *ad,* to + *gregare,* to collect; to bring together) a fruit developing from the several separate carpels of a single flower; for example, a strawberry.

Akinette (Gr. *a,* not + *kinein,* to move) enlarged, thick-walled nonmotile reproductive cell produced by some blue-green algae

Alcohol (M.L. from Arabic *al-kuhl,* a powder for painting eyelids; later applied, in Europe, to distilled spirits that were unknown in Arabia) a product of the distillation of wine or malt; any one of a class of compounds analogous to common alcohol; the ending **ol** designates a member of this class of compounds

Aleurone layer (Gr. *aleurone,* flour) the outer most cell layer of the endosperm of wheat and other grains

Alfisol modified podsol soil, typical of the northern part of the deciduous forest

Alga (plural, **algae**) (L. *alga,* seaweed) a member of the large group of thallus plants containing chlorophyll and thus able to synthesize carbohydrates

Algin a long-chain polymer of mannuronic acid found in the cell walls of the brown algae

Alkali (Arabic *alqili,* the ashes of the plant saltwort), a substance with marked basic properties

Allele (Gr. *allelon,* of one another, mutually, each other) variant form of a gene; **multiple-,** several alleles at a single locus; **pseudo-,** factors that recombine within a single locus

Allopolyploidy (Gr. *allos,* other + polyploidy) ploidy in which sets of chromosomes are derived from different parents, generally not closely related

Alpine (L. *Alpes,* the Alps Mountains) meadowlike vegetation at high elevation, above tree line

Alternate referring to bud or leaf arrangement in which there is one bud or one leaf at a node

Alternation of generations the alternation of haploid (gametophytic) and diploid (sporophytic) phases in the life cycle of many organisms; the phases (generations) may be morphologically quite similar or very distinct, depending on the organism

Amensalism (L. *a,* not + *mensa,* table) a form of biological interaction in which one organism is inhibited by another, but the other is neither inhibited nor stimulated

Amino acid (Gr. *Ammon*, from the Egyptian sun god, in N.L. used in connection with ammonium salts) an acid containing the group $-NH_2$; one of the building blocks of a protein.

Ammonification (*Ammon*, Egyptian sun god, near whose temple ammonium salts were first prepared from camel dung + L. *facere*, to make) decomposition of amino acids, resulting in the production of ammonia

Amoeboid (Gr. *amoibe*, change) eating or moving by means of temporary cytoplasmic extensions from the cell body

Amyloplast (L. *amylum*, starch + *plastos*, formed) cytoplasmic organelle specialized to store starch. Abundant in roots and in storage organs such as tubers.

Anabolism (Gr. *ana*, up + metabolism) the constructive phase of metabolism, in which more complex molecules are built from simpler substances.

Anaerobe (Gr. *a*, without + *aer*, air + *bios*, life) an organism able to respire in the absence of free oxygen, or in greatly reduced concentrations of free oxygen

Anaphase (Gr. *ana*, up + *phais*, appearance) that stage in mitosis in which half chromosomes or sister chromatids move to opposite poles of the cell

Androecium (Gr. *andros*, man + *oikos*, house) the aggregate of stamens in the flower of a seed plant

Aneuploid (Gr. *aneu*, without + *ploid*) the condition in which the number of chromosomes differs from the normal by less than a full set (n), for example, $3n + 1$, $2n - 1$, $2n + 4$; compare Polyploid

Angiosperm (Gr. *angion*, a vessel + *sperma* from *speirein*, to sow, hence a seed or germ) literally a seed borne in a vessel, thus a group of plants whose seeds are borne within a matured ovary

Angstrom (after A. J. Angstrom, a Swiss physicist, 1814-1874) a unit of length equal to 0.0001 of a micron (0.1 nanometer), 10,000 angstroms = 1 micron = 0.001 millimeter

Anisogamy (Gr. *an*, prefix meaning not + *isos*, equal + *gamete*, spouse) the condition in which the gametes, though similar in appearance, are not identical

Annual (L. *annualis*, within a year) a plant that completes its life cycle in one year and then dies.

Annual ring in wood, a layer of growth formed during one year and consisting of spring and summer wood

Annular vessels (L. *annularis*, a ring) vessels with lignified rings of secondary wall material

Annulus (L. *anulus* or *annulus*, a ring) in ferns, a row of specialized cells in a sporangium, of importance in opening of the sporangium; in mosses, thick-walled cells along the rim of the sporangium to which the peristome is attached

Anther (M.L. *anthera*—from the Gr. *anthros*, meaning flower—a medicine extracted from the internal whorls of flowers by medieval pharmacists; confined to pollen-producing parts by herbalists in 1700) pollen-bearing portion of stamen

Antheridium (anther + *-idion*, a Gr. dim, ending, thus a little anther) male gametangium or sperm-bearing organ of plants other than seed plants

Anthocyanin (Gr. *anthros*, a flower + *kyanos*, dark blue) a blue, purple, or red vacuolar pigment

Antibiotic (Gr. *anti*, against or opposite + *biotikos*, pertaining to life) a natural organic substance that retards or prevents the growth of organisms; generally used to designate substances formed by microorganisms that prevent growth of other microorganisms

Antibody (Gr. *anti*, against + body) a protein produced in an organism, in response by the organism to a contact with a foreign substance, and having the ability of specifically reacting with the foreign substance

Anticlinal cell division (Gr. *anti*, against + klinein, incline) cell division where the newly formed cell wall is perpendicular to the axis of the organ surface

Antipodal (Gr. *anti*, opposite + *pous*, foot) generally, referring to one object situated opposite another object. Specifically, referring to cells or nuclei at the end of the embryo sac opposite that of the egg apparatus

Apex (L. *apex*, a tip, point, or extremity) the tip, point, or angular summit of anything: the tip of a leaf; that portion of a root or shoot containing apical and primary meristems

Apical dominance the inhibition of lateral buds or meristems by the apical meristem

Apical meristem a mass of meristematic cells at the very tip of a shoot or root

Apomixis (Gr. *apo*, away from + *mixis*, a mingling) the production of offspring in the usual sexual structures without the mingling and segregation of chromosomes

Apothecium (Gr. *apotheke*, a storehouse) a cup-shaped or saucer-shaped open ascocarp

Archegonium (L. dim. of Gr. *archegonos*, literally a little founder of a race) female gametangium or egg-bearing organ, in which the egg is protected by a jacket of sterile cells

Aril (M.L. *arillus*, a wrapper for a seed) an accessory seed covering formed by an outgrowth at the base of the ovule in *Taxus*

Ascocarp (Gr. *askos*, a bag + *karpos*, fruit) a fruiting body of the Ascomycetes, generally either an open cup, a vessel, or closed sphere lined with special cells called asci (see Ascus)

Ascogenous hyphae hyphae arising from the ascogonium, after the formation of $n+n$ paired nuclei; the hymenial layer of the ascocarp develops from the ascogenous hyphae

Ascogonium the oogonium or female gametangium of the Ascomycetes

Ascomycetes (Gr. *askos,* a bag + *mykes,* fungus) a large group of true fungi with septate hyphae producing large numbers of asexual conidiospores and meiospores called ascospores, the latter in asci

Ascospore (Gr. *askos,* a bag + spore) meiospore produced within an ascus

Ascus (plural, **asci**) (Gr. *askos,* a bag) a specialized cell, characteristic of the Ascomycetes, in which two haploid nuclei fuse, immediately after which three (generally) divisions occur, two of which constitute meiosis, resulting in eight ascospores still contained within the ascus

Asepsis (Gr. *a,* not + *septos,* putrid) the condition of being germ-free

Asexual (Gr. *a,* without + L. *sexualis,* sexual) any type of reproduction not involving the union of gametes or meiosis

Aspect (L. *aspectus,* appearance) the direction of slope of a surface, as a hillside with a south-facing aspect

Assimilation (L. *assimilare,* to make like) the transformation of food into protoplasm

Atoms (F. *atome,* from the Gr. *atomos,* indivisible) the smallest particles in which the elements combine either with themselves or with other elements, and thus the smallest quantity of matter known to possess the properties of a particular element; a unit of matter consisting of a dense, central nucleus surrounded by a number of negatively charged electrons that are in constant motion. The nucleus consists of several positively charged protons and uncharged neutrons.

ATP adenosine triphosphate, a high-energy organic phosphate of great importance in energy transfer in cellular reactions

Auricles (L. *auricula,* dim. of *auris,* ear) earlike structures; in grasses, small projections that grow out from the opposite side of the sheath at its upper end where it joins the blade.

Autoecious (Gr. *auto,* self + *oikia,* dwelling) having a complete life cycle on the same host

Autoradiograph (Gr. *auto,* self + L. *radiolus,* a ray + Gr. *graphe,* a painting) a photographic print made by a radioactive substance acting upon a sensitive photographic film

Autotetraploidy the condition in which the doubling of the chromosome number occurs in one cell or between cells on the same plant

Autotrophic (Gr. *auto,* self + *trophein,* to nourish with food) pertaining to a plant that is able to manufacture its own food

Auxin (Gr. *auxein,* to increase) a plant growth-regulating substance regulating cell elongation

Axil (Gr. *axilla,* armpit) the upper angle between a petiole of a leaf and the stem from which it grows

Axillary bud a bud formed in the axil of a leaf

Bacillus (L. *baculum,* a stick) a rod-shaped bacterium

Backcross the crossing of a hybrid with one of its parents; when the genes of an individual are to be tested, the crossing of that individual with an individual that is homozygous recessive for all genes being tested

Bacteria (Gr. *bakterion,* a stick) common name for the class Schizomycetes

Bacteriochlorophyll a light-harvesting pigment found in certain bacteria; the molecular structure is similar to that of chlorophyll a, but certain side groups and the absorption spectrum are unique

Bacteriology (bacteria + Gr. *logos,* discourse) the science of bacteria

Bacteriophage (bacteria + Gr. *phagein,* to eat) literally, an eater of bacteria; a virus that infects specific bacteria, multiples therein, and usually destroys the bacterial cells

Banner (M.L. *bandum,* a standard) large, broad, and conspicuous petal of legume type of flower

Bark (Swedish *bark,* rind) the external group of tissues, from the cambium outward, of a woody stem or root

Base a substance that can accept a proton (H^+). Also the purine and pyrimidine groups in nucleic acids and nucleotides are collectivey called bases

Base pair the nitrogen bases that pair in the DNA molecule, adenine with thymine, and guanine with cytosine

Basidiomycetes (M.L. *basidium,* a little pedestal + Gr. *mykes,* fungus) group of fungi characterized by the production of meiospores on special cells, the basidia

Basidiospore (M.L. *basidium,* a little pedestal + spore) type of meiospore borne by basidia in the Basidiomycetes

Basidium (plural, **basidia**) (M.L. *basidium,* a little pedestal) a specialized reproductive cell of the Basidiomycetes in which nuclei fuse and meiosis occurs. It may be a special club-shaped cell, a short filamentous cell, or a short four-celled filament

Benthon (Gr. *benthos,* the depths of the sea) attached aquatic plants and animals, collectively

Berry a simple fleshy fruit, the ovary wall fleshly and including one or more carpels and seeds

Biennial (L. *biennium,* a period of two years) a plant that requires two years to complete its life cycle. Flowering is normally delayed until the second year.

Bifacial leaf (L. *bis,* twice + *facies,* face) a leaf having upper and lower surfaces distinctly different

Binomial (L. *binominis,* two names) two-named; in biology each species is generally indicated by two names, the genus to which it belongs and its own species name

Bioassay (Gr. *bios,* life + L. *exagere,* to weigh or test) to test for the presence or quantity of a substance by using an organism's response as an indicator.

Biological barrier a barrier to crossing (hybridization) of plants caused by differences in pollination vector or timing in flower opening, in contrast to physiological barriers (incompatibility of pollen with stigma or style) or ecological barriers (habitats too far apart)

Biology (Gr. *bios,* life + *logos,* word, speech, discourse) the science that deals with living things

Biosystematics (Gr. *bios,* life + *synistanai,* to place together) a field of taxonomy that emphasizes breeding behavior and chromosome characteristics

Biotic (Gr. *biokitos,* relating to life) relating to life

Biotin a vitamin of the B complex

Blade typically the thin expanded portion of a leaf; in some algae, the leaflike frond

Bladder (O.E. *bladre,* a blister) a gas-filled sac whose buoyancy keeps some aquatic plants upright

Bloom (Gr. *blume,* flower; also Indo-European *bhlo,* to spring up) an increase in phytoplankton density sufficient to color bodies of fresh water

Bordered pit a pit in a tracheid or vessel member having a distinct rim of the cell wall overarching the pit membrane

Botany (Gr. *botane,* plant, herb) the science dealing with plant life

Bract (L. *bractea,* a thin plate of precious metal) a modified leaf, from the axil of which arises a flower or an inflorescence

Bud (M.E. *budde,* bud) an undeveloped shoot, largely meristematic tissue, generally protected by modified scale-leaves; also a swelling on a yeast cell that will become a new yeast cell when released.

Bud scale a modified protective leaf of a bud

Bud scar a scar left on a twig when the bud or bud scales fall away

Bulb (L. *bulbus,* a modified bud, usually underground) a short, flattened, or disk-shaped underground stem, with many fleshy scale-leaves filled with stored food

Bundle scar scar left where conducting strands passing out of the stem into the leaf stalk were broken off when the leaf fell

Bundle sheath sheath of parenchyma cells that surround the vascular bundles of leaves, sometimes called border parenchyma

C_3 cycle the Calvin Benson cycle of photosynthesis, in which the first products after CO_2 fixation are three-carbon molecules

C_4 cycle the Hatch-Slack cycle of photosynthesis, in which the first products after CO_2 fixation are four-carbon molecules

Callose (L. *callum,* thick skin + *ose,* a suffix indicating a carbohydrate) an amorphous carbohydrate deposited around pores in sieve-tube members and in other areas in cell walls

Callus (L. *callum,* thick skin) mass of large, thin-walled cells, usually developed as the result of wounding.

Calorie (L. *calor,* heat) the amount of heat needed to raise the temperature of 1 g of water 1°C (usually from 14.5 to 15.5°C), also called gram-calorie; 1000 calories = 1 kilocalorie

Calyptra (Gr. *kalyptra,* a veil, covering) in bryophytes, an envelope covering the developing sporophyte, formed by growth of the venter of the archegonium

Calyx (Gr. *kalyx,* a husk, cup) sepals collectively; outermost flower whorl

Cambium (L. *cambium,* one of the alimentary body fluids supposed to nourish the body organs) a layer, usually regarded as one or two cells thick, of persistently meristematic tissues, giving rise to secondary tissues, resulting in growth in diameter

Canopy (Gr. *kanopeion,* a cover over a bed to keep off gnats) the leafy portion of a tree or shrub

Capillaries (L. *capillus,* hair) very small spaces, or very fine bores in a tube

Caprification an artificial method of pollinating certain cultivated varieties of figs

Capsule (L. *capsula,* dim. of *capsa,* a case) a simple, dry, dehiscent fruit, with two or more carpels

Carbohydrate (chemical combining forms, *carbo,* carbon + *hydrate,* containing water) a food composed of carbon, hydrogen, and oxygen, with the general formula $C_nH_{2n}O_n$

Carbon fixation the enzymatic reaction in which CO_2 is attached to a receiver compound such as ribulose bisphosphate, thereby adding to the supply of organic carbon; occurs chiefly in photosynthesis

Carboxydismutase (F. *carbone,* carbon + Gr. *oxys,* acidic + dismutation, reduction of one metabolite by another) the enzyme responsible for the fixation of inorganic CO_2 into organic compounds in the dark reactions of the C_3 photosynthesis cycle; also called ribulose bisphosphate carboxylase

Carboxysome (F. *carbone,* carbon + Gr. *oxys,* acidic + Gr. *soma,* body) polyhedral bodies rich in RuBP carboxylase, found in some photosynthetic bacteria and blue-green algae

Carcinogenic (Gr. *karkinoma,* cancer + L. *genitalis,* to beget) that which induces cancer

Carotene (L. *carota,* carrot) a reddish-orange plastid pigment

Carpel (Gr. *karpos,* fruit) a floral leaf bearing ovules along the margins

Carpogonium (Gr. *karpos,* fruit + *gonos,* producing) female gametangium (in red algae)

Carpospore (Gr. *karpos,* fruit + spore) one of the spores produced in a carpogonium

Carrageenan a polysaccharide found in the walls of red algae that reacts with milk proteins to make a stable, creamy, thick solution or gel

Caruncle (L. *caruncula,* dim. of *caro,* flesh, wart) a spongy outgrowth of the seed coat, especially prominent in the castor bean seed

Caryopsis (Gr. *karyon,* a nut + *opsis,* appearance) a simple, dry, one-seeded, indehiscent fruit, with pericarp firmly united all around to the seed coat

Casparian suberized strip that impregnates the radial and transverse wall of endodermal cells

Catalyst (Gr. *katelyein,* to dissolve) a substance that accelerates a chemical reaction but that is not used up in the reaction

Cation exchange (Gr. *kata,* downward) the replacement of one positive ion (cation) by another, as on a negatively charged clay particle

Catkin (literally a kitten, apparently first used in 1578 to describe the inflorescence of the pussy willow) a type of inflorescence, really a spike, generally bearing only pistillate flowers or only staminate flowers, which eventually fall from the plant entire

Caulescent (Gr. *kaulos,* a plant stem) a plant whose stem bears leaves separated by visibly elongated internodes, as opposed to a rosette plant

Cell (L. *cella,* small room) a structural and physiological unit composing living organisms, in which take place the majority of complicated reactions characteristic of life; it is surrounded by a plasma membrane, contains a metabolic system, and has a store of DNA

Cell plate a structure that forms at the equatorial plane of the cell at right angles to the spindle fibers during cytokinesis; the precursor of the middle lamella

Cellulose (cell + *ose,* a suffix indicating a carbohydrate) a complex carbohydrate occurring in the cell walls of the majority of plants; cotton is largely cellulose; it is composed of hundreds of simple sugar molecules, glucose, linked together in a characteristic manner

Cenozoic (Gr. *kainos,* recent + *zoe,* life) the geologic era extending from 65 million years ago to the present

Chalaza (Gr. *chalaza,* small tubercle) the region on a seed at the upper end of the raphe where the funiculus spreads out and unites with the base of the ovule

Chaparral (Sp. *chaparro,* an evergreen oak) a vegetation type characterized by small leaved, evergreen shrubs growing together into a nearly impenetrable scrub

Chemoautotroph (Gr. *chymeia,* to pour, later alchemy, chemistry + *autos,* self + *trophein,* to feed) bacteria that oxidize reduced inorganic compounds such as H_2S to obtain energy and that use CO_2 as a carbon source

Chemoheterotroph (Gr. *chymeia,* to pour + *heteros,* other + *trophein,* to feed) bacteria that oxidize or reduce a variety of organic compounds to obtain both energy and carbon

Chemotropism (chemo + Gr. *tropos,* a turning) influence of a chemical substance on the direction of growth

Chernozem (R. *cherny,* black + *zem,* earth) a soil characteristic of some grassland vegetation in warm areas with moderate rainfall; dark in color because of a high content of organic matter

Chiasma (Gr. *chiasma,* two lines placed crosswise) the cross formed by breaking, during prophase 1 of meiosis, of two nonsister chromatids of homologous chromosomes and the rejoining of the broken ends of different chromatids

Chitin (Gr. *chiton,* a coat of mail) a polymer in which the monomer unit is the modified sugar *N*-acetyl glucosamine; it is the principal stiffening material in the cell walls of most fungi and in the exoskeleton of insects and crustaceans

Chlamydospore (Gr. *chlamys,* a horseman's or young man's coat + spore) a heavy-walled resting asexual spore

Chlorenchyma (Gr. *chloros,* green + *-enchyma,* a suffix meaning tissue) parenchyma tissue possessing chloroplasts

Chlorophyll (Gr. *chloros,* green + *phyllon,* leaf) the green pigment found in the chloroplast, important in the absorption of light energy in photosynthesis

Chloroplast (Gr. *chloros,* green + *plastos,* formed) specialized cytoplasmic body, containing chlorophyll, in which occur important reactions of starch or sugar synthesis

Chlorosis (Gr. *chloros,* green + *osis,* diseased state) failure of chlorophyll development, because of a nutritional disturbance or because of an infection of virus, bacteria, or fungus

Chromatid (chromosome + *-id,* L. suffix meaning daughters of) the half-chromosome during prophase and metaphase of mitosis, and between prophase 1 and anaphase 11 of meiosis.

Chromatin (Gr. *chroma,* color) substance in the nucleus which readily takes artificial staining; also, that portion which bears the determiners of hereditary characters made up of DNA

Chromatin bodies bodies in bacteria that give some of the histochemical reactions that are associated with the chromosomes of higher organisms

Chromatophores (Gr. *chromo,* color + *phorus,* a bearer) in algae, bodies bearing chlorophyll; in bacteria, small bodies, about 100 nanometers in diameter, containing chlorophyll, protein, and a carbohydrate

Chromoplast (Gr. *chroma,* color + *plastos,* formed) specialized plastid containing yellow or orange pigments

Chromosome (Gr. *chroma,* color + *soma,* body) a nuclear body containing genes in a linear order and undergoing characteristic division stages; one of the threads or rods of condensed chromatin visible during cell division

Cilia (singular, **cilium**) (Fr. *cil,* an eyelash) protoplasmic hairs that, by a whiplike motion, propel certain types of unicellular organisms, gametes, and zoospores through water

Cisterna (plural, **cisternae**) (L. *cistern,* a reservoir) generally referring to sections of the endoplasmic reticulum that appear as parallel membranes, each about 5 nanometers thick bounding a space about 40 nanometers wide

Cistron (L. preposition *cis,* on this side of, relating to a genetical test of gene location + *tron* or *ton,* from the last syllable of proton, electron, and other such words indicating an elementary particle) a complex unit of the gene, probably consisting of from a hundred to thousands of nucleotide pairs

Cladode (Gr. *kladodes,* having many shoots) a cladophyll

Cladophyll (Gr. *klados,* a shoot + *phyllon,* leaf) a branch resembling a foliage leaf

Class (L. *classis,* one of the six divisions of Roman people) a taxonomic group below the division level but above the order level

Clay soil particles less than 2 microns in diameter, composed mainly of aluminum (Al), oxygen (O), and silicon (S)

Cleistothecium (plural, **cleistothecia**) (Gr. *kleistos,* closed + *thekion,* a small receptacle) the closed, spherical ascocarp of the powdery mildews

Climax community the last stage of a natural succession; a community capable of maintaining itself as long as the climate does not change

Clone (Gr. *klon,* a twig or slip) the aggregate of individual organisms produced asexually from one sexually produced individual

Closed bundle a vascular bundle lacking cambium

Coal Age the Carboniferous period, beginning 345 million years ago and ending 280 million years ago

Coalescence (L. *coalescere,* to grow together) a condition in which there is union of separate parts of any one whorl of flower parts; synonyms are **connation** and **cohesion**

Coccus (plural, **cocci**) (Gr. *kokkos,* a berry) a spherical bacterium

Codon a sequence of three bases (nucleotides) along the RNA molecule that code for a single amino acid

Coenobium (Gr. *koinois,* common + Gr. *bios,* life) a colony of unicellular organisms surrounded by a common membrane

Coenocyte (Gr. *koinos,* shared in common + *kytos,* a vessel) a plant or filament whose protoplasm is continuous and multinucleate and without any division by walls into separate protoplasts

Coenogamete (Gr. *koinos,* shared in common + gamete) a multinucleate gamete, lacking cross walls

Coenzyme a substance, usually nonprotein and of low molecular weight, necessary for the action of some enzymes

Citric acid cycle a system of reactions that contributes to the catabolic breakdown of foods in respiration and that provides building materials for a number of anabolic pathways; also called the **Krebs cycle** and the **tricarboxylic acid (TCA) cycle.**

Cohesion (L. *cohaerere,* to stick together) union or holding together of parts of the same materials; the union of floral parts of the same whorl, as petals to petals

Coleoptile (Gr. *koleos,* sheath + *ptilon,* down, feather) the first leaf in germination of grasses which sheaths the succeeding leaves

Coleorhiza (Gr. *koleos,* sheath + *rhiza,* root) sheath that surrounds the radicle of the grass embryo and through which the young root bursts

Collenchyma (Gr. *kolla,* glue + *-enchyma,* a suffix, derived from parenchyma and denoting a type of cell tissue) a tissue composed of cells that fit rather closely together and with walls thickened at the angles of the cells, found in young stems and petioles

Colloid (Gr. *kolla,* glue + *eidos,* form) referring to matter composed of particles, ranging in size from 0.0001 to 0.000001 millimeter, dispersed in some medium. Milk and mayonnaise are examples

Colony (L. *colonia,* a settlement) a growth form characterized by a group of closely associated, but poorly differentiated, cells; sometimes filaments can be associated together in a colony (as in *Nostoc*), but more typically unicells are associated in a colony

Community (L. *communitas,* a fellowship) all the populations within a given habitat; usually the populations are thought of as being interdependent

Companion cell cell associated with sieve-tube members

Compensation point (L. *compensare,* to counterbalance) the light intensity (light compensation point) or the carbon dioxide concentration (CO_2 compensation point) at which photosynthesis just equals respiration

Compensation depth (L. *compensare,* to counterbalance) that depth, in a body of water, at which light intensity is so low that photosynthesis of floating or submerged plants just equal respiration

Competition (L. *competere,* to strive together) a form of biological interaction in which both organisms (at least initially) decline in growth or success because of the insufficient supply of some necessary factor(s)

Complete flower a flower having four whorls of floral leaves: sepals, petals, stamens, and carpels

Compound leaf a leaf whose blade is divided into several distinct leaflets

Conceptacle (L. *conceptaculum,* a receptacle) a cavity or chamber of a frond (of *Fucus,* for example) in which gametangia are borne

Conduction (L. *conducere,* to bring together) act of moving or conveying a substance through the plant; generally the movement of water through the xylem or food through the phloem

Cone (Gr. *konos,* a pine cone) a fruiting structure composed of modified leaves or branches, which bear sporangia (microsporangia, megasporangia, pollen sacs, or ovules), and frequently arranged in a spiral or four-ranked order; for example, a pine cone

Cone scale the flat, woody parts of pine cones that spiral out from the central axis and bear the ovules (and later seeds) on their upper surfaces; each is subtended by a sterile bract

Conidia (singular, **conidium**) (Gr. *konis,* dust) asexual reproductive cells of fungi, arising by fragmentation of hyphae, by the cutting off of terminal or lateral cells of special hyphae, or by being pushed out from a flask-shaped cell

Conidiophore (conidia + Gr. *phoros,* bearing) conidium-bearing branch of hypha

Conidiosporangium (Gr. *konis,* dust + sporangium) sporangium formed by being cut off from the end of a terminal or lateral hypha

Conidiospore (conidia + spore) spore formed as described for conidia

Conifer (cone + L. *ferre,* to carry) a cone-bearing tree

Conk the fruiting body of a bracket fungus

Connation (L. *connatus,* to be born together) condition in a flower where there is a union of similar parts of any one whorl of appendages; synonym of **coalescence**

Conservative (L. *conservare,* to keep) said of a taxonomic trait whose expression is not modified to any great extent by the external environment; a trait that is constant unless its genetic base is changed

Convergent evolution process of successive progeny, originally of quite distinct parents, coming to appear more and more alike through time because of selection pressure

Conjugation (L. *conjugatus,* united) process of sexual reproduction involving the fusion of isogametes

Cork (L. *quercus,* oak) an external, secondary tissue impermeable to water and gases

Cork cambium the cambium from which cork develops

Corm (Gr. *kormos,* a trunk) a short, solid, vertical, enlarged underground stem in which food is stored

Corolla (L. *corolla,* dim. of *corona,* a wreath, crown) petals, collectively; usually the conspicuous colored flower whorl

Cortex (L. *cortex,* bark) region of primary tissue in a stem or root bounded externally by the epidermis and internally in the stem by the phloem and in the root by the pericycle

Cotyledon (Gr. *kotyledon,* a cup-shaped hollow) seed leaf; two, generally storing food in dicotyledons; one, generally a digestive organ in the monocotyledons

Cristae (L. *crista,* a crest) crests or ridges, used here to designate the infoldings of the inner mitochondrial membrane

Cross pollination the transfer of pollen from a stamen to the stigma of a flower on another plant, except in clones

Crossing-over the exchange of corresponding segments between chromatids of homologous chromosomes

Crustose lichens that are firmly attached, along their entire lower surface, to a substrate

Cuticle (L. *cuticula,* dim. of *cutis,* the skin) waxy layer on outer wall of epidermal cells

Cutin (L. *cutis,* the skin) waxy substance that is but slightly permeable to water, water vapor, and gases

Cutinization impregnation of cell wall with a substance called cutin

Cyclosis (Gr. *kyklosis,* circulation) cytoplasmic streaming in a cell

Cyme (Gr. *kyma,* a wave, a swelling) a type of inflorescence in which the apex of the main stalk or the axis of the inflorescence ceases to grow quite early, relative to the laterals

Cystocarp (Gr. *kystos,* bladder + *karpos,* fruit) a peculiar diploid spore-bearing structure formed after fertilization in certain red algae

Cytochrome (Gr. *kytos,* a receptacle or cell + *chroma,* color) a class of several electron-transport proteins serving as carriers in mitochondrial oxidations and in photosynthetic electron transport

Cytokinesis (Gr. *kytos,* a hollow vessel + *kinesis,* motion) division of cytoplasmic constituents at cell division

Cytokinin (Gr. *kytos,* a receptacle or cell + *kinetos,* to move) a class of growth hormones important in the regulation of nucleic acid and protein metabolism, in cell division, delaying senescence, and organ initiation

Cytology (Gr. *kytos,* a hollow vessel + *logos,* word, speech, discourse) the science dealing with the cell

Cytoplasm (Gr. *kytos,* a hollow vessel + *plasma,* form) all the protoplasm of a protoplast outside the nucleus

Cytosine a pyrimidine base found in DNA and RNA

Deciduous (L. *deciduus,* falling) referring to trees and shrubs that lose their leaves in the fall

Decomposer (L. *de,* from + *componere,* to put together) an organism that obtains food by breaking down dead organic matter into simpler molecules

Decomposition (L. *de,* to denote an act undone + *componere,* to put together) a separation or dissolving into simpler compounds; rotting or decaying

Dehiscent (L. *dehiscere,* to split open) opening spontaneously when ripe, splitting into definite parts

Deletion (L. *deletus,* to destroy, to wipe out) used here to designate an area, or region, lacking from a chromosome

Dendrogram (Gr. *dendron,* tree + *gramme,* what is written or drawn) a graph showing relationship between things at different levels of similarity; the graph resembles the limbs of a tree

Denitrification (L. *de,* to denote an act undone + *nitrum,* nitro, a combining form indicating the presence of nitrogen + *facare,* to make) conversion of nitrates into nitrites, or into gaseous oxides of nitrogen, or even into free nitrogen

Deoxyribose nucleic acid (DNA) hereditary material; the DNA molecule carries hereditary information

Desert shrub (M.E. *schrubbe,* shrub) a vegetation type characterized by evergreen or drought-deciduous shrubs growing together rather openly, generally in an area with annual precipitation below 25 cm

Detritus (L. *detritus,* worn away) particulate organic matter released in the processes of decomposition of dead organisms or parts of organisms (such as plant liter)

Deuterium or heavy hydrogen; a hydrogen atom, the nucleus of which contains one proton and one neutron; it is written as ^2H; the common nucleus of hydrogen consists only of one proton

Development (F. *developper,* to unfold) developmental changes of a cell, tissue or organ leading to the presence of features that equip that cell, tissue or organ for performing specialized functions

Diastase (Gr. *diastasis,* a separation) a complex of enzymes that brings about the hydrolysis of starch with the formation of sugar

Diatom (Gr. *diatomos,* cut in two) member of a group of golden brown algae with silicious cell walls fitting together much as do the halves of a pill box

Diatomite (Gr. *diatomos,* cut in two) fossil deposits of diatom cell walls; currently mined for such commercial purposes as filters, extenders, stabilizers

Dichotomy (Gr. *dicha,* in two) the forking of an axis into two branches

Dicotyledon (Gr. *dis,* twice + *kotyledon,* a cup-shaped hollow) a plant whose embryo has two cotyledons

Dictyosome (Gr. *diktyon,* a net + *soma,* body) one of the component parts of the Golgi apparatus; in plant cells a complex of flattened double lamellae

Differentially permeable referring to a membrane through which different substances diffuse at different rates; some substances may be unable to diffuse through such a membrane

Differentiation (L. *differre,* to carry different ways) development from one cell to many cells, accompanied by a modification of the new cells for the performance of particular functions

Diffuse porous wood with an equal and random distribution of large xylem vessels members throughout the growth season

Diffusion (L. *diffusus,* spread out) the movement of molecules, and thus a substance, from a region of higher concentration of those molecules to a region of lower concentration

Digestion (L. *digestio,* dividing, or tearing to pieces, an orderly distribution) the processes of rendering food available for metabolism by breaking it down into simpler compounds, chiefly through actions of enzymes

Dihybrid cross (Gr. *dis,* twice + *hybrida,* the offspring of a tame sow and a wild boar, a mongrel) a cross between organisms differing in two characters

Dikaryon (Gr. *di,* two + *karyon,* nut) the $n + n$ paired nuclei, each usually derived from a different parent one male, one female

Dioecious (Gr. *dis,* twice + *oikos,* house) unisexual; having the male and female elements in different individuals

Diploid (Gr. *diploos,* double + *oides,* like) having a double set of chromosomes, or referring to an individual containing a double set of chromosomes per cell; usually a sporophyte generation

Disease (L. *dis,* a prefix signifying the opposite + M.E. *aise,* comfort, literally the opposite of ease) any alteration from state of metabolism necessary for the normal development and functioning of an organism

Distromatic (Gr. *di* a prefix meaning two + *stroma*, a bed, currently meaning a supporting framework), referring to a thallus, two cells in thickness

Divergent evolution process of successive progeny, originally of quite similar parents, coming to appear more and more different through time because of isolation and selection pressure

Division a major portion of the plant kingdom; equivalent to phylum

DNA deoxyribose nucleic acid

Dominant (L. *dominari,* to rule) referring, in ecology, to species of a community that receive the full force of the macroenvironment; usually the most abundant of such species, and not all of them; referring, in heredity, to that gene (or the expression of the character it influences) that, when present in a hybrid with a contrasting gene, completely dominates in the development of the character; in peas, tall is dominant over dwarf

Donor one who gives

Dormant (L. *dormire,* to sleep) being in a state of reduced physiological activity such as occurs in seeds, buds, etc.

Dorsiventral (L. *dorsum,* the back + *venter,* the belly) having upper and lower surfaces distinctly different, as a leaf does

Double bond a covalent bond that involves four electrons

Double fertilization in the embryo sac, the fusion of the egg and sperm and the simultaneous fusion of the second male gamete with polar nuclei

DPN diphosphopyridine nucleotide, or more recently, NAD—nicotinamide adenine dinucleotide

DPNH reduced diphosphopyridine nucleotide or, more recently, NADH—reduced nicotinamide adenine dinucleotide

Drupe (L. *drupa,* an overripe olive) a simple, fleshy fruit, derived from a single carpel, usually one-seeded, in which the exocarp is thin, the mesocarp fleshy, and the endocarp stony

Ecology (Gr. *oikos,* home + *logos,* discourse) the study of plant life in relation to environment

Ecosystem (Gr. *oikos,* house + *synistanai,* to place together) an inclusive term for a living community and all the factors of its nonliving environment

Ecotype (Gr. *oikos,* house + *typos,* the mark of a blow) genetic variant within a species that is adapted to a particular environment yet remains interfertile with all other members of the species

Ectomycorrhiza (Gr. *ektos,* outside + *mykos,* fungus + *riza,* root) a type of mycorrhiza that produces a felty mantle of hyphae around the outside of a young root; hyphae do not penetrate cortex cells

Ectoplast (Gr. *ektos,* outside + *plastos,* formed) cytoplasmic membrane surrounding the outside of the protoplast

Edaphic (Gr. *edaphos,* soil) pertaining to soil conditions that influence plant growth

Egg (A. S. *aeg,* egg) a female gamete

Elater (Gr. *elater,* driver) an elongated, spindle-shaped, sterile, hygroscopic cell in the sporangium of a liverwort sporophyte

Electron (Gr. *elektron,* gleaming in the sun, by way of L. *electum,* a bright alloy of gold and silver, and finally amber, from which the first electricity was produced by friction) an elementary particle of matter bearing a unit of negative electrical charge. Low in mass and in constant motion electrons surround the atom's positively charged nucleus and define the size and chemical properties of the atom or molecule

Electron transport chain a membrane-bound series of electron carriers that controls the flow of electrons from reduced to oxidized compounds, so that some of the energy carried by the electrons is used to form ATP. The chain consists of several compounds (carriers) that alternately accept and donate electrons. Found in mitochondria and chloroplasts

Electron microscope a microscope that uses a beam of electrons rather than light to produce a magnified image

Electrophoresis (Gr. *electron,* amber + *phora,* motion + *esis,* drive) the process of causing charged molecules (e.g., proteins) to move between positively and negatively charged poles

Element (L. *elementa,* the first principles; according to one system of medieval chemistry as recent as 1700, there were four elements composing all material bodies: earth, water, air, and fire) in modern chemistry, a substance that cannot be divided or reduced by any known chemical means to a simpler substance; 92 natural elements are known, of which gold, carbon, oxygen, and iron are examples; several, including plutonium, have been formed in atomic piles

Embryo (Gr. *en,* in + *bryein,* to swell) a young sporophytic plant, while still retained in the gametophyte or in the seed

Embryo sac the female gametophyte of the angiosperms; generally a seven-celled structure; the seven cells are two synergids, one egg cell, three antipodal cells (each with a single haploid nucleus), and one endosperm mother cell with two haploid nuclei

Emulsion (L. *emulgere,* to milk out) a suspension of fine particles of a liquid in a liquid

Endocarp (Gr. *endon,* within + *karpos,* fruit) inner layer of fruit wall (pericarp)

Endodermis (Gr. *endon,* within + *derma,* skin) the layer of living cells, with various characteristically thickened walls and no intercellular spaces, which surrounds the vascular tissue of certain plants and occurs in nearly all roots and certain stems and leaves

Endogenous (Gr. *endon,* within + *genos,* race, kind) developed or added from outside the cell

Endomycorrhiza (Gr. *endon,* within + *mykos,* fungus + *riza,* root) a type of mycorrhiza in which fungal hyphae penetrate cortex cells; an external fungal mantle is absent, though hyphae do extend out the root into the soil

Endoplasmic reticulum (Gr. *endon,* within + *plasma,* anything formed or molded; L. *reticulum,* a small net) originally, a cytoplasmic net adjacent to the nucleus, made visible by the electron microscope; now, any system of paired membranes within the cytoplasm; frequently abbreviated to ER

Endosperm (Gr. *endon,* within + *sperma,* seed) the nutritive tissue formed within the embryo sac of seed plants; it is often consumed as the seed matures, but remains in the seeds of corn and other cereals

Endosperm mother cell one of the seven cells of the mature embryo sac, containing the two polar nuclei and, after reception of a sperm cell, giving rise to the primary endosperm cell from which the endosperm develops

Endospore (Gr. *endon,* within + *spora,* seed) a thick-walled resting spore, formed within the cells of certain bacteria after a complex process of nuclear division and cytoplasmic reorganization

Enzyme (Gr. *en,* in + *zyme,* yeast) a protein of complex chemical constitution produced in living cells, which, even in very low concentration, speeds up certain chemical reactions but is not used up in the reaction

Epicotyl (Gr. *epi,* upon + *kotyledon,* a cup-shaped hollow) the upper portion of the axis of embryo or seedling, above the cotyledons

Epiblast (Gr. *epi,* upon + *blastos,* a sprout or shoot) a small lateral appendage in the embryo of grasses

Epidermis (Gr. *epi,* upon + *derma,* skin) a superficial layer of cells occurring on all parts of the primary plant body: stems, leaves, roots, flowers, fruits, and seeds; it is absent from the root cap and not differentiated on the apical meristems

Epigeal (Gr. *epi,* upon + *ge,* the earth) type of germination where cotyledons rise above the ground

Epigyny (Gr. *epi,* upon + *gyne,* woman) the arrangement of floral parts in which the ovary is embedded in the receptacle so that the other parts appear to arise from the top of the ovary

Epilimnion (Gr. *epi,* upon + *limne,* marsh) an upper, aerated, warm zone of water that lies above a lower, less aerated, cold zone (the hypolimnion); common in large bodies of fresh water such as deep lakes

Epiphyte (Gr. *epi,* upon + *phyton,* a plant) a plant that grows upon another plant, yet is not parasitic

Equatorial rain forest vegetation with several tree strata; characteristic of warm, wet regions; synonymous with tropical rain forest of other texts

ER endoplasmic reticulum

Ergot (F. *argot,* a spur) a fungus disease of cereals and wild grasses in which the grain is replaced by dense masses of purplish hyphae, the ergot

Erosion (L. *e,* out + *rodere,* to gnaw) the wearing away of land, generally by the action of water

Ethylene C_2H_4, a growth hormone regulating fruit ripening, various aspects of vegetative growth, and also important in the abscission process

Etiolation (F. *etioler,* to blanch) a condition involving increased stem elongation, poor leaf development, and lack of chlorophyll, found in plants growing in the absence, or in a greatly reduced amount, of light

Eukaryote (L. *eu,* true + *karyon,* a nut referring in modern biology to the nucleus) any organism characterized by having the cellular organelles, including the nucleus, bounded by membranes

Eutrophication (Gr. *eu,* good, well + *trephein,* to nourish) pollution of bodies of water resulting from slow, natural, geological, or biological processes such as siltation or encroachment of vegetation or accumulation of detritus, also called natural eutrophication

Evapotranspiration (L. *evaporare, e,* out of + *vapor,* vapor + F. *transpirer,* to perspire) the process of water loss in vapor form from a unit surface of land both directly and from leaf surfaces

Evolution (L. *evolutio,* an unrolling) the development of a race, species, genus, or other larger group of plants or animals

Exine (L. *exterus,* outside) outer layer of pollen grain wall

Exocarp (Gr. *exo,* without, outside + *karpos,* fruit) outermost layer of fruit wall (pericarp)

Exogenous (Gr. *exe,* out, beyond + *benos,* race, kind) produced outside of, originating from, or due to external causes

F_1 first filial generation in a cross between any two parents

F_2 second filial generation, obtained by crossing two members of the F_1, or by self-pollinating the F_1

Facultative (L. *facultas,* capability) referring to an organism having the power to live under a number of certain specific conditions, for example, a facultative parasite may be either parasitic or saprophytic

Family (L. *familia,* family) in plant taxonomy, a group of genera; families are grouped in orders

Fascicle (L. *fasciculus,* a small bundle) a bundle of pine or other needle-leaves of gymnosperms

Fascicular cambium cambium within vascular bundles

Fat (A.S. *faett,* fatted) one of the three major types of foods (the other two are carbohydrates and proteins); nonpolar, containing carbon, hydrogen, and small amounts of oxygen; rich in energy

Fermentation (L. *fermentum,* a drink made from fermented barley, beer) an oxidative process in foods in which molecular oxygen is not involved, such as the production of alcohol from sugar by yeasts

Ferredoxin an electron-transferring protein containing iron, involved in photosynthesis and in the biological production and consumption of hydrogen gas

Fertilization (L. *fertilis,* capable of producing fruit) that state of a sexual life cycle involving the union of egg and sperm and hence the doubling of chromosome numbers

Fiber (L. *fibra,* a fiber or filament) an elongated, tapering, thick-walled strengthening cell occurring in various parts of plant bodies

Fiber-tracheid xylem elements found in pine that are structurally intermediate between tracheids and fibers

Fibril (dim. of L. *fibra,* fiber) submicroscopic threadlike units of cellulose found in cell walls

Field capacity the amount of water retained in a soil (generally expressed as percent by weight) after large capillary spaces have been drained by gravity

Filament (L. *filum,* a thread) stalk of stamen bearing the anther at its tip; also, a slender row of cells (certain algae)

Fission (L. *fissilis,* easily split) asexual reproduction involving the division of a single-celled individual into two new single-celled individuals of equal size

Flagellum (plural, **flagella**) (L. *flagellum,* a whip) a long, slender whip of protoplasm

Flora (L. *floris,* a flower) an enumeration of all the species that grow in a region; also, the collective term for all the species that grow in a region

Floret (F. *fleurette,* a dim. of *fleur,* flower) one of the small flowers that make up the composite flower or the spike of the grasses

Flower (F. *fleur,* L. *flos,* a flower) floral leaves grouped together on a stem and adapted for sexual reproduction in the angiosperms

Foliose (L. *folium,* a leaf) lichens that are leaflike, attached to the substrate only along part of their surface

Follicle (L. *folliculus,* dim. of *follis,* bag) a simple, dry, dehiscent fruit, with one carpel, splitting along one suture

Food (A. S. *foda*) any organic substance that furnishes energy and building materials directly for vital processes

Food chain the path along which caloric energy is transferred within a community (from producers to consumers to decomposers)

Foot (O.E. *fot,* foot) that portion of the sporophyte of bryophytes and lower vascular plants that is sunk in gametophyte tissue and absorbs food parasitically from the gametophyte

Form genus a scientific name given to an organism from the fossil record, when only a portion of the entire plant has been recovered and is known

Fossil (L. *fossio,* a digging) any impression, natural or impregnated remains, or other trace of an animal or plant of past geological ages that has been preserved in the earth's crust

Fret (O. F. *frette,* latticework) flattened membrane sacs that connect grana in chloroplasts

Frond (L. *frons,* branch, leaf) a synonym for a large divided leaf, especially a fern leaf; also the leaflike blades of some algae

Fruit (L. *fructus,* that which is enjoyed, hence product of the soil, trees, cattle, etc.) a matured ovary; in some seed plants other parts of the flower may be included; also applied, as **fruiting body,** to reproductive structures of other groups of plants

Frustule (L. *frustulum,* little piece) a diatom cell, composed of two overlapping halves (valves)

Fruticose (L. *frutex,* a shrub) lichens that are erect and shrublike or pendant from branches; may have radial symmetry

Fucoxanthin (Gr. *phykos,* seaweed + *xanthos,* yellowish brown) a brown pigment found in brown algae

Fungus (plural, **fungi**) (L. *fungus,* a mushroom) a thallus plant unable to make its own food, exclusive of the bacteria

Funiculus (L. *funiculus,* dim. of *funis,* rope or small cord) a stalk of the ovule, containing vascular tissue

Fusiform initials (L. *fusus,* spindle + form) meristematic cells in the vascular cambium that develop into xylem and phloem cells comprising an axial system

Gametangium (Gr. *gametes,* a husband, *gamete,* a wife + *angeion,* a vessel) organ bearing gametes

Gamete (Gr. *gametes,* a husband, *gamete,* a wife) a protoplast that fuses with another protoplast to form the zygote in the process of sexual reproduction

G_1 period of cell cycle preceding DNA synthesis

G_2 period of cell cycle preceding mitosis

Gametophyte (gamete + Gr. *phyton,* a plant) the gamete-producing plant

Gel (L. *gelare,* to freeze) jellylike, colliodal mass

Gemma (plural, **gemmae**) (L. *gemma,* a bud) a small mass of vegetative tissue; an outgrowth of the thallus

Gene (Gr. *genos,* race, offspring) a group of base pairs in the DNA molecule in the chromosome that determines or conditions one or more hereditary characters

Gene recombination the appearance of gene combinations in the progeny different from the combinations present in the parents

Generation (L. *genus,* birth, race, kind) any phase of a life cycle characterized by a particular chromosome number, as the gametophyte generation and the sporophyte generation

Genetics (Gr. *genesis,* origin) the science of heredity

Genotype (gene + type) the assemblage of genes in an organism

Genus (plural, **genera**) (Gr. *genos,* race, stock) a group of structurally or phylogenetically related species

Geotropism (Gr. *ge,* earth + *tropos,* turning) a growth curvature induced by gravity

Germination (L. *germinare,* to sprout) the beginning of growth of a seed, spore, bud, or other structure

Gibberellins a group of growth hormones (not identical with auxins), the most characteristic effect of which is to increase the elongation of stems in a number of kinds of higher plants

Gill (M.E. *gile,* a lip, probably to its resemblance in shape and arrangement to gills of fishes) in mushrooms, thin spore-bearing plates on the underside of the cap

Girdle (O.E. *gyrdel,* enclosure, girdle) that region of a stem from which a ring of bark extending to the cambium has been removed

Glucose (Gr. *glykys,* sweet + *-ose,* a suffix indicating a carbohydrate) a simple sugar, grape sugar, $C_6H_{12}O_6$

Glume (L. *gluma,* husk) an outer and lowermost bract of a grass spikelet

Glycogen (Gr. *glykys,* sweet + *gen,* of a kind) a carbohydrate related to starch but found generally in the liver of animals

Glycolysis (Gr. *glykys,* sweet + *lysis,* a loosening) decomposition of sugar compounds without involving free oxygen; early steps of respiration

Glyoxisome (Gr. *glyks,* sweet + *soma,* body) a type of microbody that contains enzymes involved in the conversion of fats to carbohydrates during germination of fat-storing seeds

Golgi body or **zone** (Italian cytologist Camillo Golgi 1844-1926, who first described the organelle) in animal cells, a complex perinuclear region thought to be associated with secretion; in plant cells a series of flattened plates, more properly called dictyosomes

Grana (singular, **granum**) (L. *granum* a seed) structures within chloroplasts, seen as green granules with the light microscope and as a series of parallel lamellae with the electron microscope

Ground cover the area of ground covered by a plant when its canopy edge is projected perpendicularly down

Ground meristem (Gr. *meristos,* divisible) a primary meristem that gives rise to cortex, pit rays, and pith

Grow (A. S. *growan,* probably from Old Teutonic *gro,* from which grass is also derived) of living bodies generally: to increase gradually in size by natural development

Growth retardant a chemical (such as cycocel, CCC) that selectively interferes with normal hormonal promotion of growth—but without appreciable toxic effects

Guanine a purine base found in DNA and RNA

Guard cells specialized epidermal cells found on young stems and leaves; between each pair of guard cells is a small pore through which gases enter or leave; a pair of guard cells plus the pore constitute a stoma

Guttation (L. *gutta,* drop, exudation of drops) exudation of water from plants, in liquid form

Gynoecium (Gr. *gyne,* woman + *oikos,* house) the aggregate of carpels in the flower of a seed plant

Habitat (L. *habitare,* inhabit, dwell) the place or natural environment where an organism naturally grows

Hallucinogenic (L. *hallucinari,* to mentally wander + *genitalis,* to beget) that which induces hallucinations

Haploid (Gr. *haploos,* single + *oides,* like) having a single complete set of chromosomes, or referring to an individual or generation containing such a single set of chromosomes per cell; usually a gametophyte generation

Haptera fingerlike, tubular projections (sing., hapteron) that make up the holdfast of certain kelps

Haustorium (plural, **haustoria**) (M. L. *haustrum,* a pump) a projection of hyphae that acts as a penetrating and absorbing organ

Head in inflorescence, typical of the composite family, in which flowers are grouped closely on a receptacle

Heartwood wood in the center of old secondary stems that is plugged with resins and tyloses and is not active

Helix (Gr. *helix,* anything twisted) anything having a spiral form; here, quite generally refers to the double spiral of the DNA molecule

Hemicellulose (Gr. *hemi,* half + cellulose) a class of polysaccharides of the cell wall, built of several different kinds of simple sugars linked in various combinations

Herb (L. *herba,* grass, green blades) a seed plant that does not develop woody tissues

Herbaceous (L. *herbaceus,* grassy) referring to plants having the characteristics of herbs

Herbal (L. *herba,* grass) a book that contains the names and descriptions of plants, especially those that are thought to have medicinal uses

Herbarium (L. *herba,* grass) a collection of dried and pressed plant specimens

Herbicide (L. *herba,* grass or herb + *cidere,* to kill) a chemical used to kill plants, frequently chemically related to a hormone (as the herbicide 2,4-D is related to the hormone IAA); an herbicide may have narrow or wide selectivity (range of target organisms)

Heredity (L. *herditas,* being a heir) the transmission of morphological and physiological characters of parents to their offspring

Hermaphrodite flower (Gr. *hermaphroditos,* a person having the attributes of both sexes, represented by Hermes and Aphrodite) a flower having both stamens and pistils

Heterobasidiomycotinia (Gr. *heteros,* other + Basidiomycete) a subclass of Basidiomycetes with variable basidia, never club-shaped cells

Heterocyst (Gr. *heteros,* different + *cytis,* a bag) an enlarged colorless cell that may occur in the filaments of certain blue-green algae

Heteroecious (Gr. *heteros,* different + *oikos,* house) referring to fungi that cannot carry through their complete life cycle unless two different host species are present

Heterogametes (Gr. *heteros,* different + gamete) gametes dissimilar from each other in size and behavior, like egg and sperm

Heterogamy (Gr. *heteros,* different + *gamos,* union or reproduction) reproduction involving two types of gametes

Heterophyte (Gr. *heteros,* different + *phyton,* a plant) a plant that must secure its food ready-made

Heterosis (Gr. *heteros,* different + *-osis,* suffix indicating a state of) the state of a genotype having a large degree of heterozygosity

Heterospory (Gr. *heteros,* different + spore) the condition of producing microspores and megaspores

Heterothallic (Gr. *heteros,* different + thallus) referring to species in which male gametangia and female gamentangia are produced in different filaments or by different individual plant bodies

Heterotrichy (Gr. *heteros,* different + *trichos,* a hair) in the algae, the occurrence of two types of filaments, erect and prostrate

Heterotrophic (Gr. *heteros,* different + *trophein,* to nourish with food) referring to a plant obtaining nourishment from outside sources

Heterozygous (Gr. *heteros,* different + *zygon,* yoke) having different genes of a Mendelian pair present in the same cell or organism; for instance, a tall pea plant with genes for tallness (*T*) and dwarfness (*t*)

Hexose (Gr. *hexa,* six + *-ose,* suffix indicating, in this usage, carbohydrate) a carbohydrate with six carbon atoms

Hfr gene gene in *Escherichia coli* including high frequency of gene recombination

Hilum (L. *hilum,* a trifle) scar on seed, which marks the place where the seed broke from the stalk

Histology (Gr. *histos,* cloth, tissue + *logos,* discourse) science that deals with the microscopic structure of animal and vegetable tissues

Holdfast an anchoring organ in certain seaweeds; not a true root because it lacks vascular tissue; furthermore, most absorption occurs elsewhere on the thallus

Homobasidiomycetidae (Gr. *homo,* the same + Basidiomycete) a subclass of Basidiomycotinia with a typical club-shaped cell as a basidium

Homologous chromosomes (Gr. *homologos,* the same) members of a chromosome pair; they may be heterozygous or homozygous

Homospory (Gr. *homos,* one and the same + spore) the condition of producing one sort of spore only

Homothallic (Gr. *homos,* one and the same + thallus) referring to species in which male gametangia and female gamentangia are produced in the same filament or by the same individual plant body

Homozygous (Gr. *homos,* one and the same + *zygon,* yoke) having similar genes of a Mendelian pair present in the same cell or organism; for instance, a tall pea plant with genes for tallness (*TT*) only

Hormogonia (singular, **homogonium**) (Gr. *hormos,* necklace + *gonos,* offspring) short filaments, the result of a breaking apart of filaments of certain blue-green algae at the heterocysts

Hormone (Gr. *hormaein,* to excite) a specific organic product, produced in one part of a plant or animal body, and transported to another part where, effective in small amounts, it controls or stimulates another and different process

Host (L. *hospes,* host, guest) an organism on or in which another organism lives

Humidity, relative (L. *humidus,* moist) the ratio of the weight of water vapor in a given quantity of air, to the total weight of water vapor that quantity of air is capable of holding at the temperature in question, expressed as percent

Humus (L. *humus,* the ground) decomposing organic matter in the soil

Hyaloplasm (Gr. *hyalinos,* glossy + *plasma,* anything formed) the clear background portion of the cytoplasm, forming the continuous substance in which other protoplasmic bodies are embedded

Hybrid (L. *hybrida,* offspring of a tame sow and a wild boar, a mongrel) the offspring of two plants or animals differing in at least one Mendelian character; or the offspring of plants or animals differing in many characters

Hydathode (Gr. *hydro,* water + O.E. *thoden,* stem or *thyddan,* to thrust) a structure, usually on leaves, that releases liquid water during guttation

Hydrogen acceptor a substance capable of accepting

hydrogen atoms or electrons in the oxidation-reduction reactions of metabolism

Hydrogen bond a weak bond between a hydrogen atom attached to a strongly electronegative atom and another strongly electronegative atom (e.g. oxygen, nitrogen)

Hydroid (Gr. *hydor,* water + *eidos,* a shape) a water-conducting cell found in some mosses

Hydrolysis (Gr. *hydro,* water + *lysis,* loosening) union of a compound with water, attended by decomposition into less complex compounds; usually controlled by enzymes

Hydrophyte (Gr. *hydro,* water + *phyton,* a plant) a plant that grows wholly or partly submerged in water

Hymenium (Gr. *hymen,* a membrane) spore-bearing tissue in various fungi

Hypantheum (L. *hypo,* under + Gr. *anthos,* flower) fusion of calyx and corolla part way up their length to form a cup, as in many members of the rose family

Hypertonic (Gr. *hyper,* above, over + *tonos,* to stretch) having a concentration high enough so that water will move into it across a membrane from another solution

Hypertrophy (Gr. *hyper,* over + *trophein,* to nourish with food) a condition of overgrowth or excessive development of an organ or part

Hypha (plural, **hyphae**) (Gr. *hyphe,* a web) a fungal thread or filament

Hypocotyl (Gr. *hypo,* under + *kotyledon,* a cup-shaped hollow) that portion of an embryo or seedling between the cotyledons and the radicle or young root

Hypogeal (Gr. *hypo,* under + *ge,* the earth) type of germination where cotyledons remain below ground

Hypogyny (Gr. *hypo,* under + *gyne,* female) a condition in which the receptacle is convex or conical, and the flower parts are situated one above another in the following order, beginning with the lowest: sepals, petals, stamens, carpels

Hypolimnion (Gr. *hypo,* under + *limne,* marsh) a lower, cold, relatively nonaerated zone of water that lies below a warmer zone (the epilimnion); common in large bodies of fresh water, such as deep lakes

Hypothesis (Gr. *hypothesis,* foundation) a tentative theory or supposition provisionally adopted to explain certain facts and to guide in the investigation of other facts

Hypotrophy (Gr. *hypo,* under + *trophein,* to nourish with food) an underdevelopment of an organ or part

IAA indoleacetic acid

Imbibition (L. *imbibere,* to drink) the absorption of liquids or vapors into the ultramicroscopic spaces or pores found in such materials as cellulose or a block of gelatine; an adsorption phenomenon

Imperfect flower a flower lacking either stamens or pistils

Imperfect fungi fungi reproducing only by asexual means

Inclusion body a body found in the cells of organisms with a virus infection

Incomplete flower a flower lacking one or more of the four kinds of flower parts

Indehiscent (L. *in,* not + *dehiscere,* to divide) not opening by valves or along regular lines

Indicator species a species that has a narrow range of tolerance for one or more environmental factors so that, from its occurrence at a site, one can predict these factors at that site (e.g., nutrient availability or summer temperatures)

Indoleacetic acid a naturally occurring growth regulator, an auxin

Indusium (plural, **indusia**) (L. *indusium,* a woman's undergarment) membranous growth of the epidermis of a fern leaf that covers a sorus

Infect (L. *infectus,* to put into, to taint with morbid matter) specifically to produce disease by such agents as bacteria or viruses

Inferior ovary an ovary more or less, or even completely, attached to or united with the calyx

Inflorescence (L. *inflorescere,* to begin to bloom) a flower cluster

Inheritance (O. F. *enheritance,* inheritance) the reception or acquisition of characters or qualities by transmission of parent to offspring

Inorganic referring in chemistry to compounds that do not contain carbon

Integument (L. *integumentum,* covering) external layer of ovule that later develops into the seed coat

Inter a prefix, from the Latin preposition *inter,* meaning between, in between, in the midst of

Intercalary (L. *intercalare,* to insert) descriptive of meristematic tissue or growth not restricted to the apex of an organ, that is, growth at nodes

Intercellular (L. *inter,* between + cells) lying between cells

Interfascicular cambium (L. *inter,* between + *fasciculus,* small bundle) cambium that develops between vascular bundles

Internode (L. *inter,* between + *nodus,* a knot) the region of a stem between two successive nodes

Interphase (L. *inter,* between + Gr. *phasis,* appearance) the period of preparation for cell division; state between two mitotic or meiotic cycles

Intertidal zone the strip of coastal land that is alternately inundated and exposed as tides rise and fall

Intine (L. *intus*, within) the innermost coat of a pollen grain

Intra a prefix from the Latin preposition *intra* meaning on the inside, within

Intracellular (L. *intra*, within + cell) lying within cells

Introgressive hybridization (L. *intro*, to the inside + *gress*, walk + *hybrida*, halfbreed) backcrossing between complete or partial hybrids and the original parental stock

Involucre (L. *involucrum*, a wrapper) a whorl or rosette of bracts surrounding an inflorescence

Ion (Gr. *ienai*, to go) an electrified particle formed by the breakdown of substances able to conduct an electric current

Irregular flower a flower in which one or more members of at least one whorl are of different form from other members of the same whorl; zygomorphic flower

Isidia rigid protuberances on some lichens that readily detach and vegetatively spread the lichen

Isobilateral leaf (Gr. *isos*, equal + L. *bis*, twice, two-fold + *lateralis*, pertaining to the side) a leaf having the upper and lower surfaces essentially similar

Isodiametric (Gr. *isos*, equal + diameter) having diameters equal in all directions, as a ball

Isogametes (Gr. *isos*, equal + gametes) gametes similar in size and behavior

Isogamy (Gr. *isos*, equal + *gamete*, spouse) the condition in which the gametes are identical

Isomers (Gr. *isos*, equal + *meros*, part) two or more compounds having the same molecular formula, for example, glucose and fructose are both $C_6H_{12}O_6$

Isotonic (Gr. *isos*, equal + *tonos*, to stretch) having equal osmotic concentration

Isotope (Gr. *isos*, equal + *topos*, place) any of two or more forms of an element having the same or very closely related chemical properties

K selection natural selection that favors long-lived, late-maturing individuals that devote a small fraction of their resources into reproduction; tree species are K strategists

Karyogamy (Gr. *karyon*, nut + *gamos*, marriage) the fusion of two nuclei

Karyolymph (Gr. *karyon*, nut + *lympha*, water) the ground substance of a nucleus

Karyon (Gr. *karyon*, nut) term used in conjunction with the nuclei in cells of Ascomycetes and Basidiomycetes; **dikaryon,** two nuclei per cell, each derived from a different parent, $n + n$; **heterokaryon,** the situation in which members of a dikaryon pair carry different alleles

Keel (A.S. *ceol*, ship) a structure of the legume type of flower, made up of two petals loosely united along their edges

Kelp (M.E. *culp*, seaweed) a collective name for any of the large brown algae

Kinetochore (Gr. *kinein*, to move + *chorein*, to move apart) specialized body of a chromosome, which seems to direct its movement

Kinetin (Gr. *kinetikos*, causing motion) a purine that acts as a cytokinin in plants but probably does not occur in nature

Krebs cycle see **Citric acid cycle**

Lamella (plural, **lamellae**) (Gr. *lamin*, a thin blade) cellular membranes, frequently those seen in chloroplasts

Lamina (L. *lamina*, a thin plate) blade or expanded part of a leaf

Lateral bud a bud that grows out of the side of a stem

Laterite (L. *later*, a brick) a soil characteristic of rain forest vegetation; color is red from oxidized iron in the A horizon

Latex (L. *latex*, juice) a milky secretion

Leaf (O. E. *leaf*) lateral outgrowth of stem axis, which is the usual primary photosynthetic organ, and in the axil of which may be a bud

Leaf axil angle formed by the leaf stalk and the stem

Leaf primordium (L. *primordium*, a beginning) a lateral outgrowth from the apical meristem, which will become a leaf

Leaf gap the region composed of parenchyma that is located in the primary vascular cylinder above the point of departure of the leaf vascular tissue

Leaflet separate part of the blade of a compound leaf

Leaf scar characteristic scar on stem axis made after leaf abscission

Leaf trace the vascular bundle extending from the stem to the base of a leaf

Legume (L. *legumen*, any leguminous plant, particularly bean) a simple, dry dehiscent fruit with one carpel, splitting along two sutures

Lemma (Gr. *lemma*, a husk) lower bract that subtends a grass flower

Lenticel (M.L. *lenticella*, a small lens) a structure of the bark that permits the passage of gas inward and outward

Leptoid (Gr. *leptos*, small, thin + *eidos*, a shape) a sugarconducting cell found in some mosses

Leucoplast (Gr. *leukos*, white + *plastos*, formed) a colorless plastid

Liana (F. *liane* from *lier*, to bind) a plant that climbs upon other plants, depending upon them for mechanical support; a plant with climbing shoots

Lichen (Gr. *leichen*, thallus plants growing on rocks and trees) a composite plant consisting of a fungus living symbiotically with an alga

Lignification (L. *lignum*, wood + *facere*, to make) impregnation of a cell wall with lignin

Lignin (L. *lignum*, wood) an organic substance or group of substances impregnating the cellulose framework of certain plant cell walls

Ligule (L. *ligula*, dim. of *lingua*, tongue) in grass leaves, an outgrowth from the upper and inner side of the leaf blade where it joins the sheath

Line transect a method of sampling vegetation by stretching a tape along a straight line and measuring the canopy cover of plants beneath that line or which cut through a vertical plane described by that line

Linkage the grouping of genes on the same chromosome

Linked characters characters of a plant or animal controlled by genes grouped together on the same chromosome

Lipid (Gr. *lipos*, fat + L. *ides*, suffix meaning son of; now used in sense of having the quality of) any of a group of fats or fatlike compounds insoluble in water and soluble in fat solvents

Lipid body lipid storage organelle found in seeds

Liverwort (liver + M.E. *wort*, a plant; literally, a liver plant, so named in medieval times because of its fancied resemblance to the lobes of the liver) common name for the Class Hepaticae of the Bryophyta

Loam (O.E. *lam* or Old Teutonic *lai*, to be sticky, clayey) a particular soil texture class, referring to a soil having 30 to 50 % sand, 30 to 40% silt, and 10 to 25% clay

Lobed leaf (Gr. *lobos*, lower part of the ear) a leaf divided by clefts or sinuses

Locule (L. *loculus*, dim. of *locus*, a place) a cavity of the ovary in which ovules occur

Lodicules (L. *lodicula*, a small coverlet) two scalelike structures that lie at the base of the ovary of a grass flower

Longevity (L. *longaevus*, long-lived) length of life

Lumen (L. *lumen*, light, an opening for light) the cavity of the cell within the cell walls

Lysis (Gr. *lysis*, a loosening) a process of disintegration and cell destruction

Lysogen (from *lysin*, a substance that brings about lysis, + N.L. *genic*, to give rise to) a substance, virus, or bacteria that stimulates the formation of lysins

Lysogenic bacteria (lysis + N.L. *genic*, combining form meaning to give rise to) here used to indicate bacteria carrying phage that eventually disrupt the bacteria

Lysogenic conversion the influence of prophage DNA on the host bacterium, including colony morphology, and on the presence of particular enzymes and other compounds

Macroenvironment (Gr. *makros*, large + O.F. *environ*, about) the environment due to the general, regional climate; traditionally measured some 4 ft above the ground and away from large obstructions

Macronutrient (Gr. *makros*, large + L. *nutrire*, to nourish) an essential element required by plants in relatively large quantities

Map distance on a chromosome, the distance in crossover units between designated genes

Medulla (L. *medulla*, marrow) the filamentous center of certain lichens and kelp blades and stipes

Megaphyll (Gr. *megas*, great + *phyllon*, leaf) a leaf whose trace is marked with a gap in the stem's vascular system; megaphylls are thought to represent modified branch systems

Megasporangium (Gr. *megas*, large + sporangium) sporangium that bears megaspores

Megaspore (Gr. *megas*, large + spore) the meiospore of vascular plants, which gives rise to a female gametophyte

Megasporocyte (Gr. *megas*, large + *spora*, seed + L. *cyta*, vessel) a diploid cell in which meiosis will occur, resulting in four megaspores; synonymous with **megaspore mother cell**

Megasporophyll (Gr. *megas*, large + spore + Gr. *phyllon*, leaf) a leaf bearing one or more megasporangium.

Meiocyte (meiosis + Gr. *kytos*, currently meaning a cell) any cell in which meiosis occurs

Meiosis (Gr. *meioun*, to make smaller) two special cell divisions occurring once in the life cycle of every sexually reproducing plant and animal, halving the chromosome number and effecting a segregation of genetic determiners

Meiospore (meiosis + spore) any spore resulting from the meiotic divisions

Meristem (Gr. *meristos*, divisible) undifferentiated tissue, the cells of which are capable of active cell division and differentiation into specialized tissues

Meristoderm (meristem + epidermis) the outer meristematic cell layer (epidermis) of some Phaeophyta

Mesocarp (Gr. *mesos*, middle + *karpos*, fruit) middle layer of fruit wall (pericarp)

Mesophyll (Gr. *mesos*, middle + *phyllon*, leaf) parenchyma tissue of leaf between epidermal layers

Mesophyte (Gr. *mesos*, middle + *phyton*, a plant) a plant avoiding both extremes of moisture and drought

Mesosome (Gr. *mesos*, middle + *soma*, body) one of a series of paired membranes occurring in many bacteria

Mesozoic (Gr. *mesos*, middle + *zoe*, life) a geologic era beginning 225 million years ago and ending 65 million years ago

Metabolism (M.L. from the Gr. *metabolos,* to change) the process, in an organism or a single cell, by which nutritive material is built up into living matter, or aids in building living matter, or by which protoplasm is broken down into simple substances to perform special functions

Metabolite (Gr. *metabolos,* changeable + *ites,* one of a group) a chemical that is a normal cell constituent capable of entering into the biochemical transformations within living cells

Metamorphic rock (Gr. *meta,* change + *morphe,* shape or form) one of three major categories of rock; rocks whose original structure or mineral composition has been changed by pressures or temperatures in the earth's crust

Metaphase (Gr. *meta,* after + *phasis,* appearance) stage of mitosis during which the chromosomes, or at least the kinetochores, lie in the central plane of the spindle

Metaxylem (Gr. *meta,* after + xylon, *wood*) last formed primary xylem

Microbody (Gr. *mikros,* small + body) a cellular organelle, always bound by a single membrane, frequently spherical, from 20 to 60 nanometers in diameter, containing a variety of enzymes

Microcapillary space exceedingly small spaces, such as those found between microfibrils of cellulose

Microenvironment (Gr. *mikros,* small + O.F. *environ,* about) the environment close enough to the surface of a living or nonliving object to be influenced by it

Microfibrils (Gr. *mikros,* small + fibrils, dim. of fiber; literally, small little fibers) the translation of the name expresses the concept very well; microfibrils are exceedingly small fibers visible only with the high magnifications of the electron microscope

Microfossil (Gr. *mikros,* small + L. *fossilis,* dug up) fossils of microscopic organisms, only visible when thin sections of rock are examined

Micrometer (Gr. *mikros,* small + *metron,* measure) one millionth (10^{-6}) of a meter, or 0.001 millimeter; also called a micron, and abbreviated μm

Micron (Gr. *mikros,* small) a unit of distance, 0.001 millimeter or 0.000039 inch; symbol μ (Greek letter mu)

Micronutrient (Gr. *mikros,* small + L. *nutrire,* to nourish) an essential element required by plants in relatively small quantities

Microphyll (Gr. *mikros,* small + *phyllon,* leaf) a leaf whose trace is not marked with a gap in the stem's vascular system; microphylls are thought to represent epidermal outgrowths

Micropyle (Gr. *mikros,* small + *pyle,* orifice, gate) a pore leading from the outer surface of the ovule between the edges of the two integuments down to the surface of the nucellus

Microsporangium (plural, **microsporangia** (Gr. *mikros,* little + sporangium) a sporangium that bears microspores

Microspore (Gr. *mikros,* small + spore) a spore that, in vascular plants, gives rise to a male gametophyte

Microsporocyte (Gr. *mikros,* small + *spora,* seed + L. *cyta,* vessel) a diploid cell in which meiosis will occur, resulting in four microspores; synonymous with **microspore mother cell**

Microsporophyll (Gr. *mikros,* little + spore + Gr. *phyllon,* leaf) a leaf bearing microsporangia

Microtubule (Gr. *mikros,* small + *tubule,* dim. of tube) a tubule 25 nm in diameter and of indefinite length, occurring in the cytoplasm of many types of cells

Micurgical dissections (Gr. *mikros,* small + surgical) surgical experiments done on individual cells or groups of cells, such as a shoot apex

Middle lamella (L. *lamella,* a thin plate or scale) original thin membrane separating two adjacent protoplasts and remaining as a distinct cementing layer between adjacent cell walls

Millimeter the 0.001 part of a meter, equal to 0.0394 inch

Mitochondrion (plural, **mitochondria**) (Gr. *mitos,* thread + *chondrion,* a grain) a small cytoplasmic particle associated with intracellular respiration

Mitosis (plural, **mitoses**) (Gr. *mitos,* a thread) nuclear division, involving appearance of chromosomes, their longitudinal duplication, and equal distribution of newly formed parts to daughter nuclei

Mitospore (mitosis + spore) a spore forming after mitosis

Mixed bud a bud containing both rudimentary leaves and flowers

Molecular biology a field of biology which emphasizes the interaction of biochemistry and genetics in the life of an organism

Molecule (F. *môle,* mass + *cule,* dim,; literally, a little mass) a unit of matter, the smallest portion of an element or a compound that retains chemical identity with the substance in mass; the molecule usually consists of a union of two or more atoms, some organic molecules containing a very large number of atoms

Mollisol (L. *mollis,* soft + *solum,* soil, solid) one of the 10 world soil orders, characterized by containing more than 1% organic matter in the top 17.5 cm and associated with grassland vegetation; synonymous with chernozem

Monocotyledon (Gr. *monos,* solitary + *kotyledon,* a cup-shaped hollow) a plant whose embryo has one cotyledon

Monoecious (Gr. *monos,* solitary + *oikos,* house) having the reproductive organs in separate structures, but borne on the same individual

Monohybrid (Gr. *monos,* solitary + L. *hybrida,* a mongrel) a cross involving one pair of contrasting characters

Monophyletic (Gr. *mono,* single + *phyle,* tribe) said of organisms having a common (but sometimes quite ancient) ancestor

Monostromatic (Gr. *monos,* single, solitary + *stroma,* a bed, currently meaning a supporting framework) referring to a thallus, one cell in thickness

Morphogenesis (Gr. *morphe,* form + L. *genitus,* to produce) the structural and physiological events involved in the development of an entire organism of part of an organism

Morphology (Gr. *morphe,* form + *logos,* discourse) the study of form and its development

Moss (L. *muscus,* moss) a bryophytic plant

Mu μ, Greek symbol used to indicate a micron; there are 1000 microns in 1 millimeter

Multiciliate (L. *multus,* many + F. *cil,* an eyelash) having many cilia present on a sperm or spore or other type of ciliated cell

Multiple fruit a cluster of matured ovaries produced by separate flowers; for example, a pineapple

Mutation (L. *mutare,* to change) a sudden, heritable change appearing in an individual as the result of a change in genes or chromosomes

Muton (mutation + *tron,* or *on,* from the last syllable of proton, electron, and other such words indicating an elementary particle) the smallest element, the alteration of which can cause a mutation; about 0.08 map unit, possibly a single nucleotide pair

Mutualism (L. *mutuus,* reciprocal) a form of biological interaction in which both organisms must associate together for the continued success of both

Mycelium (Gr. *mykes,* mushroom) the mass of hyphae forming the body of the fungus

Mycobiont (Gr. *mykes,* mushroom + *bios,* life) the fungal partner in a lichen

Mycology (Gr. *mykes,* mushroom + *logos,* discourse) the branch of botany dealing with fungi

Mycorrhiza (Gr. *mykos,* fungus + *riza,* root) a symbiotic association between a fungus and usually the root of a higher plant

Myxomycophyta (Gr. *myxa,* mucus + *mykes,* mushroom + *phyton,* plant) a division comprising the "slime fungi"

NAD nicotineamide adenine dinucleotide, a coenzyme capable of being reduced; also called DPN, diphosphopyridine nucleotide

NADH reduced NAD

NADP nicotineamide adenine dinucleotide phosphate, a coenzyme capable of being reduced; also called TPN, triphosphopyridine nucleotide

NADPH reduced NADP

Naked bud a bud not protected by bud scales

Nanometer (Gr. *nanos,* small) one millionth (10^{-6}) of a millimeter, similar to a millimicron; equals 10 angstroms; abbreviated nm

Natural classification a classification scheme that is based on the phylogenetic nature of the organisms classified; contrasts with an artificial classification, which separates organisms on the basis of convenient traits, but fails to show the evolutionary relationships among the organisms

Natural selection the effect of the environment in channeling the genetic variation of organisms down particular pathways

Nectar (Gr. *nektar,* drink of the gods) a fluid rich in sugars secreted by nectaries, which are often located near or in flowers

Nectar guide a mark of contrasting color or texture that may serve to guide pollinators to nectaries within the flower

Nectary (Gr. *nektar,* the drink of the gods) a nectar-secreting gland, found in flowers

Neritic zone a subtidal but relatively shallow offshore zone, often dominated by large kelps

Net productivity the arithmetic difference between calories produced in photosynthesis and calories lost in respiration

Net radiation the arithmetic difference between incoming solar radiation and outgoing terrestrial radiation

Net venation veins of leaf blade visible to the unaided eye, branching frequently and joining again, forming a network

Neutron (L. *neuter,* neither) an uncharged particle found in the atomic nucleus of all elements except hydrogen; the helium nucleus has two protons and two neutrons; mass of a neutron is equal to 1.67×10^{-24} gram

Niche (It. *nicchia,* a recess in a wall) the functional position of an organism in its ecosystem

Nitrification (L. *nitrum,* nitro, a combining form indicating the presence of nitrogen + *facere,* to make) change of ammonium salts into nitrates through the activities of certain bacteria

Nitrogen base the nitrogen-containing, positively charged, components of nucleic acids and phospholipids

Nitrogen fixation the process of reducing N_2 gas into ammonia and incorporating it into the protoplast; accomplished only by certain prokaryotes

Node (L. *nodus,* a knot) slightly enlarged portion of the stem where leaves and buds arise, and where branches originate

Nodule (L. *nodulus* dim. of *nodus,* a knot) knot or swelling on a root, especially one containing nitrogen-fixing bacteria

Nonseptate descriptive of hyphae lacking cross walls

Nucellus (L. *nucella,* a small nut) tissue composing the chief part of the young ovule, in which the embryo sac develops; megasporangium

Nucleic acid an acid found in all nuclei, first isolated as part of a protein complex in 1871 and separated from the protein moiety in 1889; all known nucleic acids fall into two classes, DNA and RNA; they differ from each other in the sugar, in one of the nitrogen bases, in many physical properties, and in function

Nucleolus (L. *nucleolus,* a small nucleus) dense protoplasmic body in the nucleus

Nucleosides components of nucleic acids consisting of a nitrogen base and a sugar; in DNA, the sugar is deoxyribose, and in RNA, ribose; adenine, guanine, and cytosine occur in both DNA and RNA, thymine occurs in DNA, and uracil occurs in RNA

Nucleotides components of nucleic acid: nucleoside (nitrogen base + sugar) + phosphoric acid

Nucleus (L. *nucleus,* kernel of a nut) a dense protoplasmic body essential in cellular synthetic and developmental activities; present in all eukaryotic cells except mature sieve-tube members

Numerical taxonomy a field of taxonomy that does not place subjective weight on any particular type of evidence that shows relationships between taxa

Nut (L. *nux,* nut) a dry, indehiscent, hard, one-seeded fruit, generally produced from a compound ovary

Obligate anaerobe an organism obliged to live in the absence of oxygen

Obligate parasite an organism obliged to live strictly as a parasite

Obligate saprophyte an organism obliged to live strictly as a saprophyte

Ontogeny (Gr. *on,* being + *genesis,* origin) the development of an individual organism or of a part of it

Oogamy (Gr. *oion,* egg + *gamete,* spouse) the condition in which the gametes are different in form and activity, that is, sperms and eggs

Oogonium (L. dim. of Gr. *oogonos,* literally, a little egg layer) female gametangium of egg-bearing organ not protected by a jacket of sterile cells, characteristic of the thallophytes

Oospore (Gr. *oion,* an egg + spore) a resistant spore developing from a zygote resulting from the fusion of heterogametes

Open bundle a vascular bundle with cambium

Operculum (L. *operculum,* a lid) in mosses, cap of sporangium

Opposite referring to bud or leaf arrangement in which there are two buds or two leaves at a node

Order (L. *ordo,* a row of threads in a loom) a taxonomic category below class and above family

Organ (L. *organum,* an instrument or engine of any kind, musical, military, etc.) a part or member of an animal or plant body or cell adapted by its structure for a particular function

Organelle a membrane-bound specialized region within a cell such as the mitochondrion or dictyosome

Organic referring in chemistry to the carbon compounds, many of which have been in some manner associated with living organisms

Osmosis (Gr. *osmos,* a pushing) diffusion of a solvent through a differentially permeable membrane

Osmometer (Gr. *osmos,* pushing + *meter,* measure) a devise for measuring the magnitude of osmotic force

Osmotic pressure the maximum theoretical pressure that can be developed in a solution as a result of osmosis when the solution is placed in an osmometer surrounded by pure water; it is a measure of the concentration of a solution

Ovary (L. *ovum,* an egg) enlarged basal portion of the pistil, which becomes the fruit

Ovulate referring to a cone, scale, or other structure bearing ovules

Ovule (F. *ovule,* from L. *ovulum,* dim. of *ovum,* egg) a rudimentary seed, containing, before fertilization, the female gametophyte, with egg cell, all being surrounded by the nucellus and one or two integuments

Ovuliferous (ovule + L. *ferre,* to bear) referring to a scale or sporophyll-bearing ovules

Oxidation (F. *oxide,* oxygen + *-tions,* suffix denoting action) to increase the positive valence or decrease the negative valence of an element or ion; loss of an electron by an atom

P_{fr} and P_r abbreviations for the far-red (FR) or red (R) absorbing form of phytochrome (P)

Palea (or **palet**) (L. *palea,* chaff) upper bract that subtends a grass flower

Paleobotany (Gr. *palaios,* ancient, prehistoric + *botane,* a plant) the study of fossil plants

Paleoecology (Gr. *palaios,* ancient) a field of ecology that reconstructs past vegetation and climate from fossil evidence

Paleozoic (Gr. *palaios,* ancient + *zoe,* life) a geologic era beginning 570 million years ago and ending 225 million years ago

Palisade parenchyma elongated cells, containing many chloroplasts, found just beneath the upper epidermis of leaves

Palmately veined (L. *palma,* palm of the hand) descriptive of a leaf blade with several principal veins spreading out from the upper end of the petiole

Panicle (L. *panicula,* a tuft) an inflorescence, the main axis of which is branched, and whose branches bear loose racemose flower clusters

Pappus (L. *pappus,* woolly, hairy seed or fruit of certain plants) scales or bristles representing a reduced calyx in composite flowers

Paradermal section (Gr. *para,* beside + *derma,* skin) a section cut parallel to a flat surface, such as a leaf section cut parallel to the surface of the blade

Parasexual cycle (Gr. *para,* beside) a sexual cycle involving changes in chromosome number differing in time and place from the usual sexual cycle; occurring in fungi in which the normal cycle is suppressed or apparently absent

Parallel venation type of venation in which veins of a leaf blade that are clearly visible to the unaided eye are parallel to each other

Paraphysis (plural, **paraphyses**) (Gr. *para,* beside + *physis,* growth) a slender, multicellular hair (*Fucus,* etc.); one of the sterile branches or hyphae growing beside fertile cells in the fruiting body of certain fungi

Parasite (Gr. *parasitos,* one who eats at the table of another) an organism deriving its food from the living body of another plant or an animal

Parenchyma (Gr. *parenchein,* an ancient Greek medical term meaning to pour beside and expressing the ancient concept that the liver and other internal organs were formed by blood diffusing through the blood vessels and coagulating, thus designating ground tissue) a tissue composed of cells that usually have thin walls of cellulose, and that often fit rather loosely together, leaving intercellular spaces

Parent material the original rock or depositional matter from which the soil of a region has been formed

Parietal (F. *pari*étal, attached to the wall, from L. *paries,* wall) belonging to, connected with, or attached to the wall of a hollow organ or structure, especially of the ovary or cell

Parietal placentation a type of placentation in which placentae are on the ovary wall

Parthenocarpy (Gr. *parthenos,* virgin + *karpos,* fruit) the development of fruit without fertilization

Parthenogenesis (Gr. *parthenos,* virgin + *genesis,* origin) the development of a gamete into a new individual without fertilization

Passive absorption of water the absorption of water by roots due to forces of transpiration in leaves

Passive solute absorption absorption due only to forces of simple diffusion

Pathogen (Gr. *pathos,* suffering + *genesis,* beginning) an organism that causes a disease

Pathology (Gr. *pathos,* suffering + *logos,* account) the study of diseases, their effects on plants or animals, and their treatment

Peat (M.E. *pete,* of Celtic origin, a piece of turf used as fuel) any mass of semicarbonized vegetable tissue formed by partial decomposition in water

Pectin (Gr. *pektos,* congealed) a white amorphous substance which, when combined with acid and sugar, yields a jelly; the substance cementing cells together; the middle lamella

Pedicel (L. *pediculus,* a little foot) stalk or stem of the individual flowers of an inflorescence

Peduncle (L. *pedunculus,* a late form of *pediculus,* a little foot) stalk or stem of a flower that is borne singly; or the main stem of an inflorescence

Penicillin an antibiotic derived from the mold *Penicillium*

Pentose (Gr. *pente,* five, + *-ose,* a suffix indicating a carbohydrate) a five-carbon-atom sugar

Peptidoglycan a particular macromolecule that makes up the walls of bacteria and blue-green algae

Peptide (Gr. *pepton,* cooked or digested, a substance remaining after the digestion of proteins) two or more amino acids linked end to end

Perennial (L. *perennis,* lasting the whole year through) a plant that lives from year to year

Perfect flower a flower having both stamens and pistils; hermaphroditic flower

Perianth (Gr. *peri,* around + *anthos,* flower) the petals and sepals taken together

Pericarp (Gr. *peri,* around + *karpos,* fruit) fruit wall, developed from ovary wall

Pericycle (Gr. *peri,* around + *kyklos,* circle) tissue, generally of root, bound externally by the endodermis and internally by the phloem

Periderm (Gr. *peri,* around + *derma,* skin) protective tissue that replaces the epidermis after secondary growth is initiated; consists of cork, cork cambium, and phelloderm

Periclinal cell division cell division where the newly formed cell wall is parallel to the axis of the organ

Peridium (plural, **peridia**) (Gr. *peridion,* a little pouch) external covering of the hymenium of certain fungi; in Myxomycota, the hardened envelope that covers the sporangium

Perigyny (Gr. *peri,* about + *gyne,* a female) a condition in which the receptacle is more or less concave, at the margin of which the sepals, petals, and stamens have their origin, so that these parts seem to be attached around the ovary; also called half-inferior

Peristome (Gr. *peri,* about + *stoma,* a mouth) in mosses, a fringe of teeth about the opening of the sporangium

Perithecium (Gr. *peri,* around + *theke,* a box) a spherical or flask-shaped ascocarp having a small opening

Permafrost (L. *permanere,* to remain + A.S. *freosan,* to freeze) soil that is permanently frozen; usually found some distance below a surface layer that thaws during warm weather

Permanent wilting the condition of wilting of a plant when it can no longer obtain moisture from the soil

Permanent wilting percentage the maximum amount of water (expressed as % of the dry weight of the soil) that a soil can hold that is unavailable to a plant

Permeable (L. *permeabilis,* that which can be penetrated) said of a membrane, cell, or cell system through which substances may diffuse

Peroxysome an organelle of the microbody class that contains enzymes capable of making and destroying hydrogen peroxide, glycolic oxidase, and catalase

Petal (Gr. *petalon,* a flower leaf) one of the flower parts, usually conspicuously colored

Petiole (L. *petiolus,* a little foot or leg) stalk of leaf

PGA phosphoglyceric acid

pH (Fr. *p(ouvoir),* to be able + *h(ydrogené),* hydrogen) a symbol for the degree of acidity (values from 0 to 7) or alkalinity (values from 7 to 14); values representing the relative concentration of the hydrogen ion in solution

Phage (Gr. *phago* to eat) a virus infecting bacteria; originally bacteriophage

Phelloderm (Gr. *phellos,* cork + *derma,* skin) a layer of cells formed in the stems and roots of some plants from the inner cells of the cork cambium

Phellogen (Gr. *phellos,* cork + *genesis,* birth) cork cambium, a cambium giving rise externally to cork and in some plants internally to phelloderm

Phenotype (Gr. *phaneros,* showing + type) the external visible appearance of an organism

Phloem (Gr. *phloos,* bark) food-conducting tissue, consisting of sieve tubes with companion cells or sieve cells, phloem parenchyma, and fibers

Phosphoenolpyruvate carboxylase the enzyme responsible for the fixation of inorganic CO_2 into oxaloacetic acid in a dark reaction of the C_4 photosynthesis cycle

Phosphoglyceric acid (PGA) a three-carbon compound formed by the interaction of carbon dioxide (CO_2) and a five-carbon compound, ribulose bisphosphate; the reaction yields two molecules of PGA for each molecule of the ribulose bisphosphate; the first step in the C_3 carbon cycle of photosynthesis

Phosphorylation a reaction in which phosphate is added to a compound, for example, the formation of ATP from ADP and inorganic phosphate

Photoautotroph (Gr. *photos,* light + *autos,* self + *trophein,* to feed) bacteria that use light as an energy source and CO_2 as a carbon source

Photoheterotroph (Gr. *photos,* light + *heteros,* other + *trophein,* to feed) bacteria that use light as an energy source and various organic compounds as a carbon source

Photon a quantum of light; the energy of a photon is proportional to its frequency; $E = hv$, where E is energy; h, Planck's constant, 6.62×10^{-27} erg-second; and v is the frequency

Photoperiod (Gr. *photos,* light + period) the optimum length of day or period of daily illumination required for the normal growth and maturity of a plant

Photophosphorylation a reaction in which light energy is converted into chemical energy in the form of ATP produced from ADP and inorganic phosphate

Photoreceptor (Gr. *photos,* light + L. *receptor,* a receiver) a light-absorbing molecule involved in converting light into some metabolic (chemical energy) form, for example, chlorophyll and phytochrome

Photosynthesis (Gr. *photos,* light + *syn,* together + *tithenai,* to place) a process in which carbon dioxide and water are brought together chemically to form a carbohydrate, the energy for the process being radiant energy

Phototropism (Gr. *photos,* light + *tropos,* turning) a growth curvature in which light is the stimulus

Phycobiliproteins pigments found in the red and blue-green algae, similar to bile pigments and always associated with proteins

Phycobilisomes (Gr. *phykos,* seaweed + L. *bilis,* relating to greenish bile + Gr. *soma,* body) rods or discs of accessory pigments that are attached to photosynthetic membranes in blue-green algae; they absorb light in the green to orange part of the spectrum

Phycobiont (Gr. *phykos,* seaweed + *bios,* life) the algal partner in a lichen

Phycocyanin (Gr. *phykos,* seaweed + *kyanos,* blue) a blue phycobilin pigment occurring in blue-green algae

Phycoerythrin (Gr. *phykos,* seaweed + *erythros,* red) a red phycobilin pigment occurring in red algae

Phycomycetes (Gr. *phykos,* seaweed + *mykes,* mushroom or fungus) a class of fungi that approaches the algae in some characters

Phylogenetic classification see **natural classification**

Phycology (Gr. *phykos,* seaweed + *logos,* word, thought) the science of the study of algae

Phylogeny (Gr. *phylon,* race or tribe + *genesis,* beginning) the evolution of a group of related individuals

Phylum (Gr. *phylon,* race or tribe) a primary division of the animal or plant kingdom

Physiology (Gr. *physis,* nature + *logos,* discourse) the science of the functions and activities of living organisms

Phytobenthon (Gr. *phyton,* a plant + *benthos,* depths of the sea) attached aquatic plants, collectively

Phytochrome a reversible pigment system of protein nature found in the cytoplasm of green plants; it is associated with the absorption of light that affects growth, development, and differentiation of a plant, independent of photosynthesis, for example, in the photoperiodic response

Phytoplankton (Gr. *phyton,* a plant + *planktos,* wandering) free-floating plants, collectively, usually algae

Pi an abbreviation for inorganic orthophosphate (a mixture of H_2OP_{-4} and HPO_4^{2-}, depending on the pH)

Pigment (L. *pingere,* to paint) a substance that absorbs visible light, hence, appears colored

Pileus (L. *pileus,* a cap) umbrella-shaped cap of fleshy fungi

Pinna (plural, **pinnae**) (L. *pinna,* a feather) leaflet or division of a compound leaf (frond)

Pinnately veined (L. *pinna,* a feather + *vena,* a vein) descriptive of a leaf blade with single midrib from which smaller veins branch off, somewhat like the divisions of a feather

Pinocytosis (Gr. *pinein,* to drink + *kytos,* hollow vessel) the process by which a cell may take in food or other material by forming an invagination of the plasma membrane that is pinched off into the cytoplasm

Pioneer community the first stage of a succession

Pistil (L. *pistillum,* a pestle) central organ of the flower, typically consisting of ovary, style, and stigma

Pistillate flower a flower having pistils but no stamens

Pit a minute thin area of a secondary cell wall

Pith the parenchymatous tissue occupying the central portion of a stem

Placenta (plural, **placentae**) (L. *placenta,* a cake) the tissue within the ovary to which the ovules are attached

Placentation (L. *placenta,* a cake + *-tion,* state of) manner in which the placentae are distributed in the ovary

Plankton (Gr. *planktos,* wandering) free-floating aquatic plants and animals, collectively

Plaque (D. *plak,* flat piece of wood) here, the clear area in a plate culture of bacteria caused by the lysis of the bacteria

Plasmalemma (Gr. *plasma,* anything formed + *lemma,* a husk or shell of a fruit) a delicate cytoplasmic membrane found on the outside of the protoplast adjacent to the cell wall

Plasma membrane, see **Plasmalemma**

Plasmid (Gr. *plasma,* to form) a piece of extrachromosomal DNA, found in some bacteria

Plasmodesma (plural, **plasmodesmata**) (Gr. *plasma,* something formed + *desmos,* a bond, a band) fine protoplasmic thread passing through the wall that separates two protoplasts

Plasmodium (Gr. *plasma,* something formed + mod. L. *odium,* something of the nature of) in Myxomycetes, a slimy mass of protoplasm, with no surrounding wall and with numerous free nuclei distributed throughout

Plasmogamy (Gr. *plasma,* anything molded or formed + *gamos,* marriage) the fusion of protoplasts, not accompanied by nuclear fusion

Plasmolysis (Gr. *plasma,* something formed + *lysis,* a loosening) the separation of the cytoplasm from the cell wall due to removal of water from the protoplast

Plastid (Gr. *plastis,* a builder) the cellular organelle in which carbohydrate metabolism is localized

Plastoquinone a quinone, one of a group of compounds involved in the transport of electrons during photosynthesis in chloroplasts

Plumule (L. *plumula,* a small feather) the first bud of an embryo or that portion of the young shoot above the cotyledons

Pneumatophores (Gr. *pneuma,* breath + *phore,* Fr. *pherein,* to carry) extensions of the root systems of some plants growing in swampy areas; in contrast to most roots, they are negatively geotropic and grow out of the water and thus assure adequate aeration

Podzol (R. *pod,* under + *zola,* ashes) a soil characteristic of taiga vegetation; color of the A horizon is gray because of excessive leaching

Polar transport the directed movement within plants of compounds (usually hormones) predominantly in one direction; polar transport overcomes the tendency for diffusion in all directions

Polarity (Gr. *pol,* an axis) the observed differentiation of an organism, tissue, or cell into parts having opposed or contrasted properties or form

Pollen (L. *pollen,* fine flour) the germinated microspores or partially developed male gametophytes of seed plants

Pollen mother cell see microsporocyte

Pollen profile a diagrammatic summary of the sequence and abundance of pollen types that have been chronologically trapped in sediments

Pollination the transfer of pollen from a stamen or staminate cone to a stigma or ovulate cone

Pollinium (L. *pollentis*, powerful or *pollinis*, fine flour + *ium*, group) a mass of pollen that sticks together and is transported by pollinators as a mass; present in orchids and milkweeds

Polygenes many genes influencing the development of a single trait; results in continuous variability; compare allele

Polymerization the chemical union of monomers such as glucose, or nucleotides to form starch or nucleic acid

Polynomial (Gr. *polys*, many + L. *nomen*, name) scientific name for an organism composed of more than two words (as in binomial)

Polynucleotides (Gr. *polys*, much, many) long-chain molecules composed of units (monomers) called nucleotides, nucleic acid is a polynucleotide

Polyphyletic (Gr. *polys*, many + *phyle*, tribe) referring to organisms having more than one common ancestor

Polyploid (Gr. *polys*, many + *ploos*, fold) referring to a plant, tissue, or cell with more than two complete sets of chromosomes, for example, $4n$, $6n$

Polyribosome (Gr. *polys*, many + ribosomes) an aggregation of ribosomes; frequently simply *polysome*

Polysaccharides (Gr. *polys*, much, many + *sakcharon*, sugar) long-chain molecules composed of units (monomers) of a sugar; starch and cellulose are polysaccharides

Pome (L. pomum, apple) a simple fleshy fruit, the outer portion of which is formed by the floral parts that surround the ovary

Population (L. *populus*, people) a group of closely related, interbreeding organisms

p-protein a mass of protein material formerly called slime found in sieve-tube members

Prairie (L. *pratum*, meadow) grassland vegetation, with trees essentially absent; often considered to have more rainfall than does the steppe

Predation (L. *predatio*, plundering) a form of biological interaction in which one organism is destroyed (by ingestion); parasitism is a form of predation

Primary (L. *primus*, first) first in order of time or development

Primary endosperm cell a cell of the embryo sac after fertilization, generally containing a nucleus resulting from fusion of the two polar nuclei with a sperm nucleus; the endosperm develops from this cell

Primary meristems meristems of the shoot or root tip giving rise to the primary plant body

Primary pit field thin areas of primary cell walls

Primary tissues those tissues, epidermis, xylem, phloem, and ground tissues, that form from primary meristems

Primary wall the first formed cell wall layer formed during cell expansion

Primitive (L. *primus*, first) referring to a taxonomic trait thought to have evolved early in time

Primordium (L. *primus*, first + *ordiri*, to begin to weave; literally beginning to weave, or to put things in order) the beginning or origin of any part of an organ

Procambium (L. *pro*, before + cambium) a primary meristem that gives rise to primary vascular tissues and, in most woody plants, to the vascular cambium

Producer (L. *producere*, to draw forward) an organism that produces organic matter for itself and other organisms (consumers and decomposers) by photosynthesis

Proembryo (L. *pro*, before + *embryon*, embryo) a group of cells arising from the division of the fertilized egg cell before those cells that are to become the embryo are recognizable

Prokaryotes (L. *pro*, before + Gr. *karyon*, a nut, referring in modern biology to the nucleus) primitive organisms, bacteria, and blue-green algae, that do not have the DNA separated from the cytoplasm by an envelope

Prophage (Gr. *pro*. prefix meaning before) a noninfectious phage unit that multiplies with the growing and dividing bacteria but does not bring about lysis of the bacteria; prophage is a stage in the life cycle of a temperate phage

Prophase (Gr. *pro*, before + *phasis*, appearance) an early stage in nuclear division, characterized by the shortening and thickening of the chromosomes and their movement to the metaphase plate

Proplastid (Gr. *pro*, before + plastid) a type of plastid, occurring generally in meristematic cells, that will develop into a chloroplast

Protease (protein + *-ase*, a suffix indicating an enzyme) an enzyme breaking down a protein

Protein (Gr. *proteios*, holding first place) naturally occurring complex organic substances (egg albumen, meat) composed of amino acids, which are associated to form submicroscopic chains, spirals, or plates

Protein body protein storage organelle in seeds, sometimes called aleurone grain

Proterozoic (Gr. *protero*, before in time + *zoe* life) the earliest geologic era, beginning about 4.5 to 5 billion years ago and ending 570 million years ago; also called Precambrian era

Prothallium (Gr. *pro*, before + *thallos*, a sprout) in ferns, the haploid gametophyte generation

Protochlorophyll (Gr. *protos*, first + *chloros*, green + *phyllos*, leaf) one of the precursors of chlorophyll; it accumulates in dark-grown and potentially green tissue

Protochlorophyllide holochrome a light-sensitive compound or complex composed of photochlorophyll and a protein; absorption of light converts the protochlorophyll part to chlorophyll

Protoderm (Gr. *protos,* first + *derma,* skin) a primary meristem that gives rise to epidermis

Proton (Gr. *proton,* first) the nucleus of a hydrogen atom is a single positively charged particle, the proton; the nucleus of all other elements consists of protons and neutrons; the mass of a proton is 1.67×10^{-24} gram

Protonema (plural, **protonemata**) (Gr. *protos,* first + *nema,* a thread) an algal-like filamentous growth; an early stage in development of the gametophyte of mosses

Protoplasm (Gr. *protos,* first + *plasma,* something formed) living substance

Protoplast (Gr. *protoplastos,* formed first) the organized living unit of a single cell

Protoxylem (Gr. *protos,* first + *xylon,* wood) first formed primary xylem

Pseudopodium (Gr. *pseudes,* false + *podion,* a foot) in Myxomycetes, an armlike projection from the body by which the plant creeps over the surface

Purine a group of nitrogen bases having a double-ring structure, one five-carbon, the other six-carbon

Pyrenoid (Gr. *pyren,* the stone of a fruit + L. *oïdes,* like) a denser body occurring within the chloroplasts of certain algae and liverworts and apparently associated with starch deposition

Pyrimidine a nitrogen base having a single-ring structural formula

Quadrat (L. *quadrus,* a square) a frame of any shape which, when placed over vegetation, defines a unit sample area within which the plants may be counted or measured

Quantum (L. *quantum,* how much) an elemental unit of energy; its energy value is $\hbar v$, where $\hbar$, Planck's constant, is 6.62×10^{-27} erg-second and v is the frequency of the vibrations or waves with which the energy is associated

Quiescent center (L. *quiescere,* to rest) disk-shaped region of root apex containing slowly dividing cells

r selection natural selection that favors short-lived, early-maturing individuals that devote a large fraction of their resources into reproduction; annual herbs are r strategists

Raceme (L. *racemus,* a bunch of grapes) an inflorescence in which the main axis is elongated but the flowers are born on pedicels that are about equal in length

Rachilla (Gr. *rhachis,* a backbone + L. dim. ending *-illa*) shortened axis of spikelet

Rachis (Gr. *rhachis,* a backbone) main axis of spike; axis of fern leaf (frond) from which pinnae arise; in compound leaves, the extension of the petiole corresponding to the midrib of an entire leaf

Radicle (L. *radix,* root) portion of the plant embryo that develops into the primary or seed root

Random plant distribution a distribution of a plant species within an area such that the probability of finding an individual at one point is the same for all points

Raphe (Gr. *rhaphe,* seam) ridge on seeds, formed by the stalk of the ovule, in those seeds in which the funiculus is sharply bent at the base of the ovule

Raphides (Gr. *rhaphis,* a needle) fine, sharp, needlelike crystals

Ray initials (L. *radius,* a beam or ray) meristematic cells in the vascular cambium that develop into xylem and phloem cells comprising the ray system

Ray system system of cells in secondary tissues that are oriented perpendicular to the long axis of the stem, formed from ray initials of the vascular cambium

Receptacle (L. *receptaculum,* a reservoir) enlarged end of the pedicel or peduncle to which other flower parts are attached

Recessive character that member of a pair of Mendelian characters which, when both members of the pair are present, is subordinated or suppressed by the other, dominant character

Recombination (L. *re,* repeatedly + *combinatus,* joined) the mixing of genotypes that results from sexual reproduction

Red tide a coloring of offshore marine water caused by dense phytoplankton populations; often accompanied by toxic byproducts

Reduction (F. *reduction,* L. *reductio,* a bringing back) originally "bringing back" a metal from its oxide, that is, iron from iron rust or ore; any chemical reaction involving the removal of oxygen from or the addition of hydrogen or an electron to a substance; energy is required and may be stored in the process as in photosynthesis

Regular flower a flower in which the corolla is made up of similarly shaped petals equally spaced and radiating from the center of the flower; star-shaped flower; actinomorphic flower

Replication the production of a facsimile or a very close copy; here used to indicate the production of a second molecule of DNA exactly like the first molecule

Reproduction (L. *re,* repeatedly + *producere,* to give birth to) the process by which plants and animals give rise to offspring

Reproductive isolation the separation of populations in time or space so that genetic flow between them is cut off

Residual meristem meristematic region near shoot tip that remains after differentiation of the pith and cortex

Resin duct (L. *resina*, rosin + *ductus*, led) resin canal; in conifers, continuous tubes lined with secretory cells that run through the sap wood; they function as repositories for metabolic by-products

Respiration (L. *re*, repeatedly + *spirare*, to breathe) a chemical oxidation controlled and catalyzed by enzymes that in living protoplasm break down carbohydrate and fats, thus releasing energy to be used by the organism in doing work

Reticulum (L. *reticulum*, a small net) a small net

Rhizoid (Gr. *rhiza*, root + L. *oïdes*, like) one of the cellular filaments that perform the functions of roots

Rhizome (Gr. *rhiza*, root) an elongated, underground, horizontal stem

Rhizophores (Gr. *rhiza*, root + *phoros*, bearing) leafless branches that grow downward from the leafy stems of certain Lycophyta and give rise to roots when they come into contact with the soil

Ribose a five-carbon sugar, a component of RNA

Ribonucleic acid a nucleic acid containing the sugar ribose, phosphorus, and the bases adenine, guanine, cytosine, and uracil; present in all cells and concerned with protein synthesis in the cell

Ribosomes (*ribo*, from RNA + Gr. *somatos*, body) small particles 10 to 20 nanometers in diameter, containing RNA

Ring porous wood wood with large xylem vessel members mostly in early wood; compare with **diffuse porous wood**

Ripening (A.S. *rifi*, perhaps related to reap) changes in a fruit that follow seed maturation and that prepare the fruit for its function of seed dispersal

RNA ribonucleic acid

Root (A.S. *rot*) the descending axis of a plant, normally below ground, serving to anchor the plant and absorb and conduct water and mineral nutrients

Root cap A thimblelike mass of cells covering and protecting the apical meristems of a root, also site of gravity perception

Root hairs epidermal projections of root cells in region of maturation that increase the absorptive surface of the root

Root pressure pressure developed in the root as the result of osmosis and inducing bleeding in stem wounds

Rootstock an elongated, underground, horizontal stem

Rosette (dim. of L. *rose*, rose) a shoot with a very short stem, composed of several unelongated internodes but with fully expanded leaves

Runner a stem that grows horizontally along the ground surface

Samara (L. *samara*, the fruit of the elm) simple, dry, one- or two-seeded indehiscent fruit with pericarp bearing a winglike outgrowth

Sand soil particles between 50 and 2000 microns in diameter

Saprophyte (Gr. *sapros*, rotten + *phyton*, a plant) an organism deriving its food from the dead body or the nonliving products of another plant or animal

Sapwood peripheral wood that actively transports

Savannah (Sp. *sabana*, a large plain) vegetation of scattered trees in a grassland matrix

Scalariform vessel (L. *scala*, ladder + form) a vessel with secondary thickening resembling a ladder

Schizocarp (Gr. *schizein*, to split + *karpos*, fruit) dry fruit with two or more united carpels that split apart at maturity

Sclereids (Gr. *skleros*, hard) more or less isodiammetric cells having heavily lignified cell walls

Sclerenchyma (Gr. *skleros*, hard + *-echyma*, a suffix denoting tissue) a strengthening tissue composed of cells with heavily lignified cell walls

Scrub (A.S. *scrob*, a shrub) vegetation dominated by shrubs; described as thorn forest in areas with moderate rainfall, or as chaparral or desert in areas with low rainfall

Scutellum (L. *scutella*, a dim. of *scutum*, shield) single cotyledon of grass embryo

Seaweed large marine algae, generally reds and browns

Secondary tissue those tissues, xylem, phloem, and periderm, that form from secondary meristems

Secondary wall wall material deposited on the primary wall in some cells after elongation has ceased

Sedimentary rock (L. *sedere*, to sit) rock formed from material deposited as sediment, then physically or chemically changed by compaction and hardening while buried in the earth crust

Seed (A.S. *sed*, anything that may be sown) popularly as originally used, anything that may be sown; that is, "seed" potatoes, "seeds" of corn, sunflower, etc.; botanically, a seed is the matured ovule without accessory parts

Self-pollination transfer of pollen from the stamens to the stigma of either the same flower or flowers on the same plant

Seminal root the root or roots forming from primordia present in the seed

Sepals (M. L. *sepalum*, a covering) outermost flower structures that usually enclose the other flower parts in the bud

Septate (L. *septum,* fence) divided by cross walls into cells or compartments

Septicidal dehiscence (L. *septum,* fence + *caedere,* to cut; *dehiscere,* to split open) the splitting open of a capsule along the line of union of carpels

Septum (L. *septum,* fence) any dividing wall or partition; frequently a cross wall in a fungal or algal filament

Serpentine (L. *serpens,* a serpent) referring to soil derived from metamorphic parent material characterized (among other things) by low calcium (Ca), high magnesium (Mg), and a greenish-gray color

Sessile (L. *sessilis,* low, dwarf, from *sedere,* to sit) sitting, referring to a leaf lacking a petiole or a flower or fruit lacking a pedicel

Seta (plural, **setae**) (L. *seta,* a bristle) in bryophytes, a short stalk of the sporophyte, which connects the foot and the capsule

Sexual reproduction reproduction that requires meiosis and fertilization for a complete life cycle

Sheath part of leaf that wraps around the stem, as in grasses

Shoot (derivation uncertain, but early referring to new plant growth) a young branch that shoots out from the main stock of a tree, or the young main portion of a plant growing above ground

Shoot tip portion of the shoot containing apical and primary meristems and early stages of differentiation

Sibling species species morphologically nearly identical but incapable of producing fertile hybrids

Sieve cells a long and slender sieve element without a companion cell, with relatively unspecialized sieve areas, and with tapering end walls that lack sieve plates

Sieve plate wall area in a sieve-tube member containing a region of pors through which pass strands connecting sieve-tube protoplasts

Sieve tube a series of sieve-tube members forming a long cellular tube specialized for the conduction of food materials

Sieve-tube members portion of a sieve-tube comprised of a single protoplast and separated from other sieve-tube members by sieve plates

Silique (L. *siliqua,* pod) the fruit characteristic of Brassicaceae (mustards); two-celled, the valves splitting from the bottom and leaving the placentae with the false partition stretched between

Silt soil particles between 2 and 50 microns in diameter

Simple pit pit not surrounded by an overarching border; on contrast to bordered pit

Siphonous line a line of evolutionary development in the algae in which mitosis is not followed by cytokinesis; this results in an elongated multinucleate, coenocytic filament

Soil (L. *solum,* soil, solid) the uppermost stratum of the earth's crust, which has been modified by weathering and organic activity into (typically) three horizons: an upper A horizon which is leached, a middle B horizon in which the leached material accumulates, and a lower C horizon, which is unweathered parent material

Soil texture refers to the amounts of sand, silt, and clay in a soil, as a sandy loam, loam, or clay texture

Solute (L. *solutus,* from *solvere,* to loosen) a dissolved substance

Solution (M.E. *solucion,* from O.F. *solucion,* to loosen) a homogeneous mixture, the molecules of the dissolved substance (e.g., sugar), the solute, being dispersed between the molecules of the solvent (e.g. water)

Solvent (L. *solvere,* to loosen) a substance, usually a liquid, having the properties of dissolving other substances

Soralium (Gr. *soros,* a heap) an eruption of the surface of a lichen where soredia are released

Soredium (plural, **soredia**) (Gr. *soros,* a heap) asexual reproductive body of lichens, consisting of a few algal cells surrounded by fungous hyphae

Sorus (plural, **sori**) (Gr. *soros,* a heap) a cluster of sporangia

Species (L. *species,* appearance, form, kind) a class of individuals usually interbreeding freely and having many characteristics in common

Sperm (Gr. *sperma,* the generative substance or seed of a male animal) a male gamete

Spermagonium (plural, **spermagonia**) (Gr. *sperma,* sperm + *gonos,* offspring) flask-shaped structure characteristic of the sexual phase of the rust fungi; bearing receptive hyphae and spermatia

Spermatium (plural, **spermatia**) (Gr. *sperma,* sperm) in rust fungi, a cell borne at the tip of the hyphae that line the interior of spermagonia (on barberry leaves)

Spermatophyte (Gr. *sperma,* seed + *phyton,* plant) a seed plant

Spike (L. *spica,* an ear of grain) an inflorescence in which the main axis is elongated and the flowers are sessile

Spikelet (L. *spica,* an ear of grain + dim. ending *-let*), the unit of inflorescence in grasses; a small group of grass flowers

Spindle (A.S. *spinel,* an instrument employed in spinning thread by hand) referring in mitosis and meiosis to the spindle-shaped intracellular structure in which the chromosomes move

Spirillum (L. *spira,* a coil) bacterial cell that has a spiral shape

Spodosol (Gr. *spodos,* wood ashes; R, *pod,* under + *zola,* ashes) one of the 10 world soil orders, characterized by an ashy, sandy, bleached, acidic A_2 horizon and associated mainly with coniferous forest vegetation; synonomous with podzol

Sporangiophore (sporangium + Gr. *-phore,* a root of *phorein,* to bear) a branch bearing one or more sporangia

Sporangium (spore + Gr. *angeion,* a vessel) spore case

Spore (Gr. *spora,* seed) a reproductive cell that develops into a plant without union with other cells; some spores such as meiospores occur at a critical stage in the sexual cycle, but others are asexual in nature

Sporidium (dim. meaning a little spore) the basidiospore of smut fungi

Sporophore (spore + Gr. *phorein,* to bear) the fruiting body of fleshy and woody fungi, which produces spores

Sporophyll (spore + Gr. *phyllon,* leaf) a spore-bearing leaf

Sporophyte (spore + Gr. *phyton,* a plant) in alternation of generations, the plant in which meiosis occurs and which thus produces meiospores

Stamen (L. *stamen,* the standing-up things or a tuft of thready things) flower structure made up of an anther (pollen-bearing portion) and a stalk or filament

Staminate flower a flower having stamens but no pistils

Starch (M.E. *sterchen,* to stiffen) a complex insoluble carbohydrate, the chief food storage substance of plants, that is composed of several hundred hexose sugar units and that easily breaks down on hydrolysis into these separate units.

Statolith (Gr. *statos,* standing + *lithos,* stone) a starch grain that moves to its position in a cell as a result of gravity, thus providing an initial sensing of, or orientation to, gravity by a cell

Stele (Gr. *stele,* a post) the central vascular cylinder, inside the cortex, of roots and stems of vascular plants

Stem (O.E. *stemn*) the main body of the portion above ground of tree, shrub, herb, or other plant; the ascending axis, whether above or below ground, of a plant, in contradistinction to the descending axis or root

Steppe (R. *step,* a lowland) an arid grassland vegetation

Stereid (Gr. *stereos,* solid) thick-walled cells at or near the epidermis of certain mosses

Sterigma (plural, **sterigmata**) (Gr. *sterigma,* a prop) a slender, pointed protuberance at the end of a basidium, which bears a basidiospore

Sterilization (L. *sterilis,* barren) the process of making something germ-free

Stigma (L. *stigma,* a prick, a spot, a mark) receptive portion of the style to which pollen adheres

Stipe (L. *stipes,* post, tree trunk) the stem portion of a kelp, to which are attached bladders and blades

Stipule (L. *stipula,* dim. of *stipes,* a stock or trunk) a leaflike structure from either side of the leaf base

Stolon (L. *stolo,* a shoot) a stem that grows horizontally along the ground surface

Stoma (plural, **stomata**) (Gr. *stoma,* mouth) a minute opening plus two guard cells in the epidermis of leaves and stems, through which gases pass

Strobilus (Gr. *strobilos,* a cone) a number of modified leaves (sporophylls) or ovule-bearing scales grouped together on an axis

Stroma (Gr. *stroma,* a bed or covering) a mass of protecting vegetative filaments; the background substance of chloroplasts, probably the location of the carbon cycle of photosynthesis

Style (Gr. *stylos,* a column) slender column of tissue that arises from the top of the ovary and through which the pollen tube grows

Suberin (L. *suber,* the cork oak) a waxy material found in the cell walls of cork tissue

Succession (L. *successio,* a coming into the place of another) a sequence of changes in time of the species that inhabit an area, from an initial pioneer community to a final climax community

Succulent (L. *sucus,* juice) a plant having juicy or watery tissues

Sucrose (F. *sucre,* sugar + *-ose,* termination designating a sugar), cane sugar ($C_{12}H_{22}O_{11}$)

Superior ovary an ovary completely separate and free from the calyx

Suspensor (L. *suspendere,* to hang) a cell or chain of cells developed from a zygote whose function is to place the embryo cells in an advantageous position to receive food

Suture (L. *sutura,* a sewing together; originally the sewing together of flesh or bone wounds) the junction, or line of junction, of contiguous parts

Symbiosis (Gr. *syn,* with + *bios,* life) an association of two different kinds of living organisms involving benefit to both

Sympetaly (Gr. *syn,* with + *petalon,* leaf) a condition in which petals are united

Synandry (Gr. *syn,* with + *andros,* a man) a condition in which stamens are united

Syncarpy (Gr. *syn,* with + *karpos,* fruit) a condition in which carpels are united

Synergids (Gr. *synergos,* toiling together) the two nuclei at the upper end of the embryo sac, which, with the third (the egg), constitute the egg apparatus

Synsepaly (Gr. *syn,* with + sepals) a condition in which sepals are united

Taiga (Teleut, *taiga,* rocky mountainous terrain) a broad northern belt of vegetation dominated by conifers; also, a smiliar belt in mountains just below alpine vegetation

Tannin a substance that has an astringent, bitter taste

Tapetum (Gr. *tapes,* a carpet) nutritive tissue in the sporangium, particularly an anther

Taxon (plural, **taxa**) (Gr. *taxis,* order) a general term for any taxonomic rank, from subspecific to divisional

Taxonomy (Gr. *taxis,* arrangement + *nomos,* law) systematic botany; the science dealing with the describing, naming, and classifying of plants

Teliospore (Gr. *telos,* completion + spore) resistant spore characteristic of the Heterobasidiomycetidae, in which karyogamy and meiosis occur and from which a basidium develops

Telium (plural, **telia**) (Gr. *telos,* completion) a sorus of teliospores

Telophase (Gr. *telos,* completion + phase) the last stage of mitosis, in which daughter nuclei are reorganized

Temperate phage a phage that does not necessarily cause lysis of the bacterial cells in which it reproduces

Tendril (L. *tendere,* to stretch out, to extend) a slender coiling organ that aids in the support of stems

Terminal bud a bud at the end of a stem

Testa (L. *testa,* brick, shell) the outer coat of the seed

Tetrad (Gr. *tetradeion,* a set of four) a group of four, usually referring to the meiospores immediately after meiosis

Tetraploid (Gr. *tetra,* four + *ploos,* fold) having four sets of chromosomes per nucleus

Tetraspores (Gr. *tetra,* four + spores) four spores formed by division of the spore mother cell, used particularly for meiospores in certain red algae

Tetrasporine line (tetraspore + L. suffix *-ine,* like) a line of evolutionary development in the algae in which mitosis is directly followed by cytokinesis, resulting in a filament, thallus, or complex plant body of varied form

Thallophytes (Gr. *thallos,* a sprout + *phyton,* plant) a division of plants whose body is a thallus, that is, lacking roots, stems, and leaves

Thallus (Gr. *thallos,* a sprout) plant body without true roots, stems, or leaves

Thermocline (Gr. *therme,* heat + *klino,* slope) a zone of rapidly changing temperature that separates the epilimnion from the hypolimnion in large bodies of fresh water

Thylakoid (Gr. *thylakos,* sack + N.L. *oid,* a thing that is like) one of membranes in the chloroplasts

Thymidine a nucleoside incorporated in DNA, but not in RNA

Thymidine ^{3}H tritiated or radioactive thymidine

Thymine a pyrimidine occurring in DNA, but not in RNA

Tiller (O.E. *telga,* a branch) a grass stem arising from a lateral bud at a basal node; tillering is the process of tiller formation

Tissue a group of cells of similar structure that performs a special function

Tonoplast (Gr. *tonos,* stretching tension + *plastos,* molded, formed) the cytoplasmic membrane bordering the vacuole; so-called by de Vries, as he thought it regulated the pressure exerted by the cell sap

Toxin (L. *toxicum,* poison) a poisonous secretion of a plant or animal

TPN triphosphopyridine nucleotide, or, more recently, NADP, nicotinamide adenine dinucleotide phosphate

TPNH reduced triphosphopyridine nucleotide or, more recently, NADPH, reduced nicotinamide adenine dinucleotide phosphate

Tracheid (Gr. *tracheia,* windpipe) an elongated, tapering xylem cell with lignified pitted walls, adapted for conduction and support, does not have open end walls

Tracheophytes (Gr. *tracheia,* windpipe + *phyton,* plant) vascular plants

Trait a distinctive definable characteristic; a mark of individuality

Transcription (L. *trans,* across + *scribere,* to write) the process of RNA formation from a DNA code

Transduction gene transfer in bacteria, in which bacterial genes are carried from one bacteria to another by way of a temperate phage

Transfer cells specialized cells modified by their cell wall projections that may facilitate short distance transport

Transformation mutation or change in genes of bacteria by the direct intervention of extracellular DNA

Translocation (L. *trans,* across + *locare,* to place) the transfer of food materials or products of metabolism; in genetics, the exchange of chromosome segments between nonhomologous chromosomes

Transmit to pass or convey something from one person, organism, or place to another person, organism, or place

Transpiration (F. *transpirer,* to perspire), the giving off of water vapor from the surface of leaves

Trichogyne (Gr. *trichos,* a hair + *gyne,* female) receptive hairlike extension of the female gametangium in the Rhodophyta and Ascomycetes

Trichome (Gr. *trichoma,* a growth of hair) a short filament of one or more cells extending from the epidermis

Triose (Gr. *treis,* three + *-ose,* suffix indicating a carbohydrate) any three-carbon sugar

Trisomic (Gr. *treis*, three + *soma*, body) a plant containing one additional chromosome; $2n + 1$

Tritium a hydrogen atom, the nucleus of which contains one proton and two neutrons; it is written as ^3H; the more common hydrogen nucleus consists only of a proton

Tropism (Gr. *trope*, a turning) movement of curvature due to an external stimulus that determines the direction of movement

Tuber (L. *tuber*, a bump, swelling) a much-enlarged, short, fleshy underground stem

Tundra (Lapp. *tundar*, hill) meadowlike vegetation at low elevation in cold regions that do not experience a single month with average daily maximum temperatures above 50°F

Turgid (L. *turgidus*, swollen, inflated) swollen, distended; referring to a cell that is firm due to water uptake

Turgor pressure (L. *turgor*, a swelling) the pressure within the cell resulting from the absorption of water into the vacuole and the imbibition of water by the protoplasm

Tylose (plural, **tyloses**) (Gr. *tylos*, a lump or knot) a growth of one cell into the cavity of another

Type specimen the herbarium specimen selected by a taxonomist to serve as a basis for the naming and descriptions of a new species

Umbel (L. *umbella*, a sunshade) an inflorescence, the individual pedicels of which all arise from the apex of the peduncle

Unavailable water water held by the soil so strongly that root hairs cannot readily absorb it

Unicell (L. *unus*, one + cell) an organism consisting of a single cell; generally used in describing algae

Uniseriate (L. *unus*, one + M.L. *seriatus*, to arrange in a series) said of a filament having a single row of cells

Uracil a pyrimidine found in RNA but not in DNA

Uredium (plural, **uredinia**) (L. *uredo*, a blight) a sorus of uredospores

Uredospore (L. *uredo*, a blight + spore) a red, one-celled summer spore in the life cycle of the rust fungi

Vaccine (L. *vacca*, cow) a suspension of weakened or dead bacteria or other pathogens injected into the body to immunize against the same species of pathogen or their toxins

Vacuole (L. dim. of *vacuus*, empty) a watery solution of various substances forming a portion of the protoplast distinct from the protoplasm

Vascular (L. *vasculum*, a small vessel) referring to any plant tissue or region consisting of or giving rise to conducting tissue, for example, bundle, cambium, ray

Vascular bundle a strand of tissue containing primary xylem and primary phloem (and procambium if present) and frequently enclosed by a bundle sheath of parenchyma or fibers

Vascular cambium cambium giving rise to secondary phloem and secondary xylem

Vector (L. *vehere*, to carry) an organism, usually an insect, that carries and transmits disease-causing organisms

Vegetation (L. *vegetare*, to quicken) the plant cover that clothes a region; it is formed of the species that make up the flora, but is characterized by the abundance and life form (tree, shrub, herb, evergreen, deciduous plant, etc.) of certain of them

Venation (L. *vena*, a vein) arrangement of veins in leaf blade

Venter (L. *venter*, the belly) enlarged basal portion of an archegonium in which the egg cell is borne

Ventral canal cell the cell just above the egg cell in the archegonium

Ventral suture (L. *ventralis*, pertaining to the belly) the line of union of the two edges of a carpel

Vernalization (L. *vernalis*, belonging to spring + *izare*, to make) the promotion of flowering by naturally or artificially applied periods of extended low temperature; seeds, bulbs, or entire plants may be so treated

Vessel (L. *vasculum*, a small vessel) elongate or short xylem conducting cell with lignified secondary walls and open ends called perforation plates

Vessel element a portion of a vessel derived from a single cell of the vascular cambium or procambium

Virion (L. *virulentus*, full of poison) infectious virus particle as it exists outside a host

Virulence (L. *virulentia*, a stench) the relative infectiousness of a bacteria or virus, or its ability to overcome the resistance of the host metabolism

Virus (L. *virus*, a poisonous or slimy liquid) a disease principle that can be cultivated only in living tissues, or in freshly prepared tissue brei

Vitamins (L. *vita*, life + amine) naturally occurring organic substances, akin to enzymes, necessary in small amounts for the normal metabolism of plants and animals

Volva (L. *volva*, a wrapper) cup at base of stipe or stalk of fleshy fungi

Volvocine line (*Volvox* + L. suffix *-ine*, like) a line of evolutionary development in the algae in which the cells remain separate, as in *Volvox*, and are never connected to form a filament or flattened thallus

Water potential refers to the difference between the activity of water molecules in pure distilled water at atmospheric pressure and 30°C (standard conditions) and the activity of water molecules in any other system; the activity of these water molecules may be greater (positive) or less (negative) than the activity of the water molecules under standard conditions

Weed (A.S. *wēod,* used at least since 888 in its present meaning) generally a herbaceous plant or shrub not valued for use or beauty, growing where unwanted, and regarded as using ground or hindering the growth of more desirable plants

Whorl a circle of flower parts, or of leaves

Whorled referring to bud or leaf arrangement in which there are three or more buds or three or more leaves at a node

Wild-type in genetics, the gene normally occurring in the wild population, usually dominant

Wings lateral petals of legume type of flower

Wood (M.E., *wode, wude,* a tree) a dense growth of trees, or a piece of a tree, generally the xylem

Xanthophyll (Gr. *xanthos,* yellowish brown + *phyllon,* leaf) a yellow chloroplast pigment

Xerophyte (Gr. *xeros,* dry + *phyton,* a plant) a plant very resistant to drought, or that lives in very dry places

Xylem (Gr. *xylon,* wood) a plant tissue consisting of tracheids, vessels, parenchyma cells, and fibers; wood

Zoology (Gr. *zoon,* an animal + *logos,* speech) the science having to do with animal life

Zoosporangium (Gr. *zoon,* an animal + sporangium) a sporangium bearing zoospores

Zoospore (Gr. *zoon,* an animal + spore) a motile spore

Zygomorphic (Gr. *zygo,* yoke, pair + *morphe,* form) referring to bilateral symmetry; said of organisms, or a flower, capable of being divided into two symmetrical halves only by a single longitudinal plane passing through the axis

Zygospore (Gr. *zygon,* a yoke + spore) a thick-walled resistant spore developing from a zygote resulting from the fusion of isogametes

Zygote (Gr. *zygon,* a yoke) a protoplast resulting from the fusion of gametes (either isogametes or heterogametes)

TABLE OF METRIC EQUIVALENTS

Exponents

Very large or very small numbers are often written in exponential form to save space. Thus, 0.1 can be written exponentially as 1×10^{-1}, or simply as 10^{-1}; $0.0000035 = 3.5 \times 10^{-6}$; $1000 = 1 \times 10^{3}$ or 10^{3}.

Common Prefixes

nano = 10^{-9} (one billionth)
micro = 10^{-6} (one millionth)
milli = 10^{-3} (one thousandth)
kilo = 10^{3} (one thousand times)
mega = 10^{6} (one million times)

I. Units of Length

kilometer (km) = 10^{3} m = 0.62 mile
meter (m) = 1.09 yd = 3.28 ft
decimeter (dm) = 10^{-1} m = 3.90 in.
centimeter (cm) = 10^{-2} m = 0.39 in.
millimeter (mm) = 10^{-3} m = 0.04 in.
micron, micrometer (μ) = 10^{-6} m
millimicron, nanometer (mμ or nm) = 10^{-9} m
angstrom (Å) = 10^{-10} m

II. Units of Weight

metric ton = 10^{3} kg = 10^{6} g = 1.10 short tons = 0.98 long ton = 2200 lb
kilogram (kg) = 10^{3} g = 2.20 lb
gram (g) = 0.035 oz
milligram (mg) = 10^{-3} g = 0.015 grain

III. Units of Area

are(a) = 100 m^2 = 1075.84 ft^2
hectare (ha) = 100 are = 2.47 acres

IV. Units of Volume

liter (l) = 10^{3} ml = 1.06 U.S. liquid qt = 0.91 U.S. dry qt = 0.88 British qt
milliliter, cubic centimeter (ml or cm^3) = 0.03 fluid oz = 0.06 $in.^3$

V. Miscellany: Units of Pressure, Light Intensity, Energy, Work, Force

bar = 0.99 atm of pressure = 14.54 lb/$in.^2$
lux (L) = 0.09 ft-c of light intensity
joule (j) = 0.735 ft-lb of work = 9.5×10^{-4} British Thermal Units (BTU)
erg = 10^{-7} joule
watt (w) = 1.3×10^{-3} hp
calorie, gram-calorie (c) = amount of heat needed to raise the temperature of 1 g of water 1°C from 14.5 to 15.5°C
kilocalorie, kilogram-calorie (kc) = 1000 c = 3.97 BTU (1 BTU is the amount of heat needed to raise 1 lb of water 1°F
solar constant (radiation at Earth's outer atmosphere edge) = 2 cal/cm^2/min = 13,400 ft-c = 4.06×10^{7} BTU/m^2/yr
dyne = force required to accelerate 1 g at the rate of 1 cm/sec/sec = 1 kg-m/sec^2
newton = 10^{5} dyne = 1 kg-m/sec^2

VI. Units of Temperature

one degree Celsius, Centigrade (1°C) = 1°K = 1.8°F
absolute zero = −273°C = 0°K = −460°F
boiling point of water = 100°C = 373°K = 212°F
freezing point of water = 0°C = 273°K = 32°F
to convert from °F to °C: (°F − 32) × 0.56 = °C
to convert from °C to °F: (°C x 9/5) − 32

SUBJECT INDEX

Illustrations are indicated by boldface type.

INDEX TO GENERA

Illustrations are indicated by **boldface type.** Text discussion often appears on adjacent pages. Families are given for the Anthophyta, classes for the Mycota, and divisions for all other genera. Fossil plants are indicated by (f).

Common names are given for the vascular plants and for some of the bryophytes and thallophytes. The common names selected are those preferred by Bailey (*Manual of Cultivated Plants*) or Jepson (*Manual of the Flowering Plants of California*) for either the genus or a common species of the genus. It is probable that different common names will be used for the same species in different regions.